The Physical Geography of North America

The PHYSICAL GEOGRAPHY of NORTH AMERICA

Edited by

Antony R. Orme

2002

OXFORD
UNIVERSITY PRESS

Oxford New York
Athens Auckland Bangkok Bogotá Buenos Aires Cape Town
Chennai Dar es Salaam Delhi Florence Hong Kong Istanbul Karachi
Kolkata Kuala Lumpur Madrid Melbourne Mexico City Mumbai Nairobi
Paris São Paulo Shanghai Singapore Taipei Tokyo Toronto Warsaw

and associated companies in
Berlin Ibadan

Published by Oxford University Press, Inc.,
198 Madison Avenue, New York, New York 10016

Library of Congress Cataloging-in-Publication Data
The physical geography of North America / edited by Anthony R. Orme.
p. cm.
ISBN 0-19-511107-9
1. Physical geography—North America. 2. Natural history—North America. I. Orme, A. R.

GB115 .P58 2000

917'02—dc21 00-028557

9 8 7 6 5 4 3 2

Printed in Hong Kong
on acid-free paper

To Our Students

Preface

This book, *The Physical Geography of North America,* is a contribution to the Oxford Regional Environments series being published by Oxford University Press. The series involves a finite number of volumes devoted to major regions of the world. Each volume presents a detailed and current statement of knowledge written by specialists in the many research fields of physical geography. With this book, we aspire to fill a void in recent scientific literature, namely, the lack of high-quality interpretive and correlative work that seeks to integrate across the environmental spectrum.

By convention, the land mass of the Western Hemisphere is often separated at the isthmus of Panama into two continents, North America and South America. For the purposes of this series, however, three principal divisions are recognized, with Mesoamerica interposed between the two larger masses. North America is defined here in terms of the extratropical landscapes that reach from the Tropic of Cancer northward to the Arctic Ocean. North America thus defined reflects both the continent's tectonic nucleus and its present, mostly temperate, climatic and biogeographic character. In contrast, Mesoamerica, though much smaller in area, contains distinctively humid tropical landscapes superposed onto a mostly recent tectonic framework. Human impacts on these environments also differ.

This book contains 25 chapters that are broadly divided into three groups: systematic, regional, and human impacts. The first twelve chapters focus systematically on the broad physical and biogeographic character of North America. Of these, chapters 1–4 examine landscape evolution from earliest times to the rich legacy of Pleistocene and Holocene events, whereas chapters 5–12 explain the contemporary landscape in terms of climate, water, soils, plants, and animals, culminating in a perspective on North America's ecoregions.

Chapters 13–21 examine some of the continent's more distinctive regions. Rather than use one criterion, these regions are distinguished in terms of a special attribute that produces a distinctive landscape. The presence of permafrost, boreal forest, two prominent cordilleras, midcontinent plains and lakes, persistent or seasonal aridity, and coasts are such attributes. Despite some overlap, for example, among boreal forest, cordilleras, and permafrost, these are undeniably distinctive regions that any student of North American environments must surely recognize.

Last, chapters 22–25 examine how the North American environment has been affected by human impacts, producing landscapes which in some areas are quite different from those inherited from nature. Although modest human impacts have occurred throughout the Holocene, massive immigration and settlement over the past 500 years have generated major changes in the primeval landscape, most notably through the impacts of vegetation change, agriculture, and urbanization. The book concludes with a perspective on environmental management and conservation, the focus of so much continuing debate regarding people and nature.

A book of this nature inevitably involves much synthesis and some subjectivity. Whereas the chapters are framed within an agreed context, individual authors have been encouraged to be original, flexible, and selective in their approach. To aid the reader in pursuing various themes, each chapter offers a substantial bibliography, which includes both selected classic works and much that has been published within the past two decades.

Los Angeles, 2000 Antony R. Orme

Acknowledgments

The authors thank the following who have kindly given permission for the use of copyrighted material:

Bellwether Publishing, V. H. Winston and Son, Inc., and Antony R. Orme, Editor-in-Chief, *Physical Geography,* for figs. 1.12, 5.4, 21.13, 21.17, 21.19, 21.20, and 22.15.

Blackwell Publishers and The Association of American Geographers for fig. 7.10.

Canadian Meteorological and Oceanographic Society for figs. 7.4 and 7.7.

Elsevier Science for fig. 7.9.

The Geological Society of America for figs. 10.4 and 10.10.

UNESCO for figs. 7.1 and 7.5

University of California, Los Angeles (UCLA), Department of Geography, for photographs from the Spence Collection, namely: figs. 1.5, 1.8, 1.9, 1.10, 1.11, 1.15, 1.16, 3.2, 3.3, 3.4, 3.7, 3.9, 19.8, 20.7, 20.10, 21.14, 21.15, 22.16, 22.18; and from the Fairchild Collection, Fig. 22.4.

We are also grateful to individual authors and friends, and to various government agencies for freely providing many other figures used in the book. Although every effort has been made to trace and contact copyright holders, we apologize for any apparent negligence. Finally, we thank the many reviewers and cartographers, including Chase Langford of UCLA, who provided generous advice and assistance in the preparation of this book.

Contents

III. Nature in the Human Context

Contributors

Robert Bailey (Ph.D., 1971, University of California, Los Angeles) is Director of the Ecosystem Management Analysis Center, Forest Service, United States Department of Agriculture, Fort Collins, Colorado. His research concerns geomorphology and ecosystem management in the United States, and global ecosystem classification.

Roger Barry (Ph.D., 1965, University of Southampton, England) is Professor of Geography and Director of the National Snow and Ice Data Center at the University of Colorado, Boulder. His research interests involve polar and mountain climates, synoptic and dynamic climate processes, cryosphere-climate interactions, and global snow and ice information systems.

Mark Blumler (Ph.D., 1992, University of California, Berkeley) is Assistant Professor of Geography in Binghamton University, State University of New York. His research interests involve biogeography, early agriculture, and environmental history, particularly with respect to California grasslands.

David Butler (Ph.D., University of Kansas, 1982) is Professor of Geography and Planning at Southwest Texas State University, San Marcos. His research interests are in geomorphology, biogeography, and natural resources, and he has worked extensively in the Rocky Mountains.

John Dixon (Ph.D., 1983, University of Colorado, Boulder) is Professor of Geography at the University of Arkansas, Fayetteville, His research interests are geomorphology, soils, and Quaternary studies. He has worked extensively in Arctic and alpine environments in western North America and Scandinavia.

Reid Ferring (Ph.D., 1980, Southern Methodist University; 1993, University of Texas, Dallas) is Professor of Geography at the University of North Texas, Denton. His research interests lie in geoarchaeology, geology, and physical geography.

Jonathan Harbor (Ph.D., 1990, University of Washington) is Professor of Earth and Atmospheric Sciences at Purdue University, West Lafayette, Indiana. His research includes the hydrologic and geomorphic impacts of urbanization. Budhendra Bhaduri, Matt Grove, Martha Herzog, Shankar Jaganapathy, Marie Minner, and John Teufert are graduate students at Purdue University.

Vance Holliday (Ph.D., 1982, University of Colorado, Boulder) is Professor of Geography at the University of Wisconsin, Madison. His research interests are in geomorphology, geoarchaeology, and soils, and he has worked extensively across the Great Plains.

Kenneth Hinkel (Ph.D., 1986, University of Michigan) is Professor of Geography at the University of Cincinnati, Ohio. His research interests are permafrost, periglacial geomorphology, climate, and field instrumentation. He has worked extensively in Canada and Alaska.

Scott Isard (Ph.D., Indiana University, 1984) is Professor of Geography in the University of Illinois, Urbana. His research interests concern physical climatology and aerobiology.

Allan James (Ph.D., 1988, University of Wisconsin, Madison) is Associate Professor of Geography in the University of South Carolina, Columbia. His research interests are geomorphology, hydrology, water resources, and Quaternary studies, and include studies of mining impacts on river systems.

James Knox (Ph.D., 1970, University of Iowa) is Professor of Geography at the University of Wisconsin, Madison. His research interests are in geomorphology, paleohydrology, and water resources, and he has worked extensively on Holocene rivers of the upper Midwest and Mississippi basin.

Julie Laity (Ph.D., 1982, University of California, Los Angeles) is Professor of Geography at California State University, Northridge. Her research interests are in geomorphology and climatology, with a focus on the effects of aeolian systems in the deserts of the Southwest.

Glen MacDonald (Ph.D., 1984, University of Toronto) is Professor of Geography in the University of California, Los Angeles. His research interests are biogeography and Holocene environments, with emphasis on forest-tundra ecotones, palynology, and dendroclimatology across the Northern Hemisphere.

George Malanson (Ph.D., 1983, University of California, Los Angeles) is Professor of Geography at the University of Iowa, Iowa City. His research interests are in biogeography, landscape ecology, riparian vegetation, and ecological modeling. He has worked extensively in the Rocky Mountains and California.

Rolfe Mandel (Ph.D., 1991, University of Kansas) is a private consultant in soils and geomorphology in Topeka, Kansas, with special interests in the landscapes and geoarchaeology of the Great Plains.

John Menzies (Ph.D., 1976, University of Edinburgh, Scotland) is Professor of Geography at Brock University, St. Catherines, Ontario. His research interests involve geomorphology, glaciology, and soil science, and include studies of past and present glacial environments and drumlin fields.

Frederick Nelson (Ph.D., 1982, University of Michigan) is Professor of Geography at the University of Delaware, Newark. His research interests include permafrost, periglacial geomorphology, and spatial analysis, and he has worked extensively in Canada and Alaska.

John Oliver (Ph.D., 1969, Columbia University, New York) is Professor of Geography at Indiana State University, Terre Haute. His research interests involve climatology and physical geography, and he has worked on dynamic and synoptic climatology, as well as on climate classification and regionalism.

Amalie Jo Orme (Ph.D., 1983, University of California, Los Angeles) is Professor of Geography at California State University, Northridge. Her research interests involve geomorphology, Quaternary studies, and biogeography, and she works on the coastal, lake, and chaparral environments of California.

Antony Orme (Ph.D., 1961, University of Birmingham, England) is Professor of Geography at the University of California, Los Angeles, and Editor-in-Chief of the journal *Physical Geography*. His main research interests are geomorphology, Quaternary studies, and environmental management, and he has worked in western North America, Africa, and the British Isles.

Albert Parker (Ph.D., 1980, University of Wisconsin, Madison) is Professor of Geography at the University of Georgia, Athens. His research interests are North American forest composition, structure, and dynamics.

Kathleen Parker (Ph.D., 1982, University of Wisconsin, Madison) is Professor of Geography at the University of Georgia, Athens. Her research interests are the arid lands of western North America, vegetation dynamics, and biogeomorphology.

John Pitlick (Ph.D., 1988, Colorado State University, Fort Collins) is Associate Professor of Geography at the University of Colorado, Boulder. His research concerns hydrology and fluvial geomorphology, and he has worked on relations between precipitation, runoff, and flooding in the Mississippi River basin and western North America.

Garry Running (Ph.D., 1997, University of Wisconsin, Madison) is Associate Professor of Geography at the University of Wisconsin, Eau Claire. His research interests are in geomorphology, soils, and geoarchaeology, with a focus on the northern Great Plains.

Louis Scuderi (Ph.D., 1983, University of California, Los Angeles) is Associate Professor of Earth and Planetary Sciences at the University of New Mexico, Albuquerque. His research interests are Holocene paleoclimatology, dendroclimatology, lichenometry, geographic information systems, and remote sensing.

Randall Schaetzl (Ph.D., 1987, University of Illinois) is Professor of Geography at Michigan State University, East Lansing. His research interests focus on soil geomorphology, plant geography, and Quaternary studies, especially of forest soils in the Great Lakes region.

David Shankman (Ph.D., 1986, University of Colorado, Boulder) is Professor of Geography at the University of Alabama, Tuscaloosa. His research interests are in biogeography, bioclimatology, and environmental management, with a focus on the southeastern United States.

Thomas Vale (Ph.D., 1971, University of California, Berkeley) is Professor of Geography at the University of Wisconsin, Madison. His research interests are North American biogeography, resource conservation, and images of nature.

Ellen Wohl (Ph.D., 1990, Colorado State University, Fort Collins) is Professor of Earth Resources at Colorado State University. Her research focuses on river systems and paleohydrology, with emphasis on channel hydraulics, morphology, sediment transport, and human impacts in western North America.

Ming-ko Woo (Ph.D., 1972, University of British Columbia) is Professor of Geography at McMaster University, Hamilton, Ontario. His research interests involve hydrology and wetlands, and he has worked extensively on Arctic, subarctic, and prairie wetlands in Canada.

I

SYSTEMATIC FRAMEWORK

1

Tectonism, Climate, and Landscape

Antony R. Orme

Earth's physical landscape is primarily an expression of tectonism and climate functioning within a gravitational context. Tectonism, namely, earth movements and the rocks and structures involved therein, forms the physical framework of the continents and ocean basins, and the environment for subsequent erosion and sedimentation. Climate, the synthesis of weather, generates the surface processes that reshape this framework by sculpting the landscape and providing habitat for plants and animals, mainly through the agency of water in its various states. In geodynamic terms, this is also a distinction between endogenic forces within Earth's crust and exogenic forces at the surface. Tectonism and climate have played these roles from early in Earth history when the nascent crust began to reorganize itself beneath a soupy atmosphere and primordial ocean. Much later, when 90% of Earth time had passed, organic activity became sufficiently organized beneath the atmospheric umbrella to begin clothing continental landscapes with vascular plants, which in turn encouraged soil formation, further climate change, and expansion of land animals. Recently, during but a small fraction of Earth time, human beings have come to refashion the landscape, like ants scurrying industriously across the surface, manipulating its resources, and generating fresh suites of environmental consequences to augment those caused by natural processes.

Tectonism and climate are not wholly independent forces. Through its impact on the distribution and shape of continents and ocean basins, tectonism influences climate, for example, by enhancing precipitation against windward mountains while reducing it in their lee and by influencing ocean circulations so important to atmospheric processes. Over time, tectonism also influences climate change, by promoting uplift favorable to prolonged cooling and eventual glaciation, by opening or closing seaways to ocean circulation and by affecting the composition of the atmosphere through the generation and consumption of crustal rocks. Though more subtle, climate in turn may affect tectonism by redistributing continental mass through erosion and deposition, thereby generating isostatic adjustments to crustal loading and unloading. Apart from tectonism's influence on plant and animal distributions, climate and vegetation are also interlinked, through biogeochemical cycles and the effect of vegetation cover on such variables as albedo and greenhouse gases, and thus on the exchange of energy between the atmosphere and the ground.

This chapter examines the interactive roles of tectonism and climate in shaping the North American landscape, without dwelling on the mechanics of tectonism or on tectonic events of the distant past. Nor does this chapter focus specifically on weather dynamics and regional expressions of climate—these are discussed in later chapters. Instead, the chapter outlines the continent's evolving tectonic framework and then focuses on the relationship between tectonism and climate that has emerged over the past 200 million years, as the Pangea supercontinent ruptured and North America shifted toward its present location. The implications of these changes for the continent's geomorphology and biogeography are then examined, providing the context for the landscape processes and expressions discussed subsequently.

1.1 Tectonism and Climate: Paradigms and Linkages

Tectonism involves a suite of crustal and subcrustal processes that find primary expression at Earth's surface in the location and shape of continents and ocean basins and, on continents such as North America, in the size and distribution of secondary relief features such as mountain cordilleras and ancient shields. Nestled within these are tertiary components such as individual mountain ranges and structural basins that reflect three-dimensional adjustments within larger contexts. Climate then functions to modify these tectonic units through the agencies of erosion and deposition, producing smaller relief features such as escarpments and river valleys. This work is accomplished mainly through the impact of precipitation in a thermal context, as percolating water, streamflows, and glaciers, and through the work of wind, waves, and currents.

Tectonic paradigms have changed often during the growth of the earth sciences, but it is the concept of plate tectonics, once espoused by few and rejected by many, that emerged during the later twentieth century as the basis for explaining much of Earth's crustal behavior. In its modern form, this concept recognizes that Earth's lithosphere comprises rocks of varying density that mobilize as relatively rigid plates that shift in response to both deep-seated forces, such as convection in the upper mantle, and crustal forces, which involve push and pull mechanics between plates. Thinner, denser, more basaltic plates (ρ 2.9–3.4 g cm^{-3}) lie lower in the crust and are found mostly beneath oceans. Thicker, lighter, more granitic plates (ρ 2.6–2.8 g cm^{-3}) maintain a higher freeboard to form continents that rise above sea level but also include submerged continental margins. Although plate architecture has changed over geologic time, Earth's lithosphere is presently organized into seven major plates, including North America, and numerous smaller plates and slivers.

Because the concept of plate tectonics implies crustal mobility and because mobility is expressed most dramatically between plates, scientific interest often focuses more on plate boundaries than on plate interiors. In this context, it is usual to distinguish between passive margins, where plates are diverging, and active margins, where plates are either converging head-on or obliquely, or shearing laterally alongside one another. At passive or divergent margins, severe surface deformation is rare but crustal flexuring (epeirogeny), faulting, and volcanism occur as plates shift away from spreading centers where new crust is forming. Epeirogenic uplift may also occur where plates move across plumes of upwelling mantle rocks. In contrast, at active or convergent margins, mountain building (orogeny) commonly results from subduction of oceanic plates, accretion of displaced terranes, or collision between continental plates. Convergence also implies that much of the crust is progressively consumed and replaced as older plates descend into the mantle, often accompanied by intense magmatic and seismic activity, even as newer crust is generated at distant spreading centers. In its continuing evolution, the North American plate is presently flanked to the north and east by passive margins and trailing oceanic crust, and to the south and west by active margins in which oceanic plates are subducting beneath or shearing alongside the mostly continental plate. Compared with these margins, the nucleus of the North American plate, the continental craton, has experienced much less change.

Climatic paradigms also changed during the growth of the atmospheric sciences. As the twentieth century progressed, linkages between the various components of the climate system became better understood, not only in space but over time, relative to the changing distribution of land and sea and varying atmospheric composition induced by plate tectonics. Paleoclimatology, the study of past climates, was revolutionized by the integration of plate tectonic concepts in the quest for explanations of climate change. Recognition of the relationships between tectonism and climate change led in turn to a reappraisal of geomorphic and biogeographic paradigms, notably those based on simplistic models of change through time.

Earth's climate is driven by solar forcing, by the interception of solar radiation by the planetary atmosphere and surface, and by the continuing exchange of energy between surface, atmosphere, and space. Energy transfers by radiation, convection, and conduction, notably the transfer of excess warmth from the tropics via the atmosphere and oceans toward the poles, eventually impart specific climatic character to a region. Though dependent on solar radiation, Earth's climate system also reflects complex linkages or couplings between its various components (fig. 1.1). The atmosphere is linked with the oceans, the vast reservoir of water under present planetary conditions; with the frozen water of ice sheets and sea ice; with the biomass of plants and animals with which it exchanges water and carbon; and with the physical framework of the continents that generates dust and interferes with climate near the ground. There are also linkages between surface components: between ice sheets and ocean volume as reciprocal controls of sea level, between organic activity and parent material in soil formation, and between continents and oceans in the water cycle. These linkages among atmosphere, hydrosphere, cryosphere,

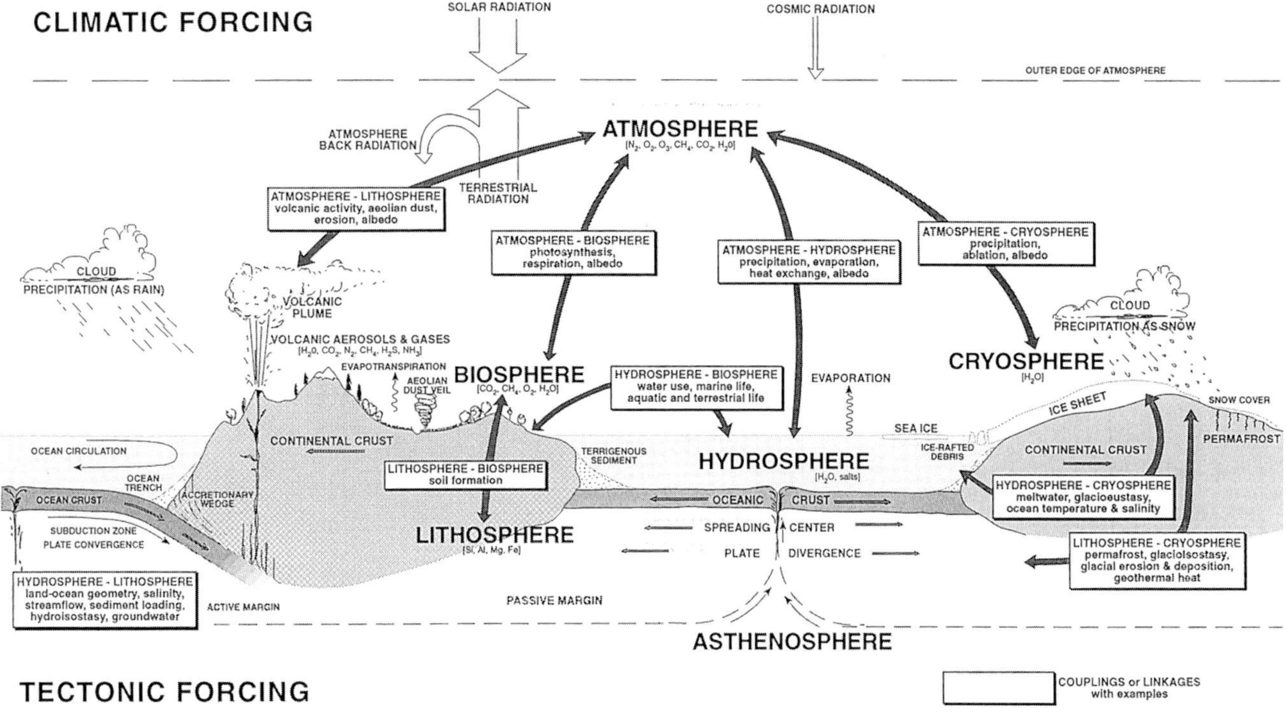

Figure 1.1 Linkages within Earth's systems.

biosphere, and lithosphere, so critical to understanding the climate of a place, also change through time. Some such changes may reflect variations in solar energy output and in Earth's orbital relations with the Sun, but changes at the surface are also important. The latter include complex feedbacks triggered by changes in surface albedo induced by variations in ice cover and biomass, or by changes in ocean volume and temperature as ice sheets grow and decay. The basic linkage, however, remains that between the atmosphere and climate system as a whole and the nature and distribution of land and sea attributable to tectonism.

1.2 The Tectonic Framework of North America

The North American continent covers 24×10^6 km^2 or 16% of Earth's land area. Relative to other continents, it is of middling size, but inclusion of its continental shelf and slope expand this area to over 37×10^6 km^2, making the continental block second in size only to Asia (table 1.1). The land mass is roughly triangular in shape, broadest in the north, where it extends over 6000 km from Alaska to Newfoundland, and tapering southward, rapidly so south of 30°N (fig. 1.2). Again, if continental shelf and slope are

Table 1.1 Comparison of the dimensions of North America to those of other continents

Continent	Land Area (10^6 km^2)	Continental Shelf (10^6 km^2)	Continental Slope (10^6 km^2)	Continental Block[a] (10^6 km^2)	Freeboard[b] (%)	Mean Altitude (m)	Coastline (10^3 km)
North America	24.06	6.74	6.68	37.48	64.19	720	76
South America	17.79	2.43	2.14	22.36	79.56	590	29
Africa	29.83	1.28	2.25	33.36	89.42	750	30
Antarctica	14.17	0.36	5.41	19.94	71.06	2200	
Asia	44.19	9.38	7.01	60.58	72.94	960	71
Australia	8.90	2.70	1.64	13.24	67.22	340	20
Europe	10.01	3.11	3.13	16.25	61.60	340	37

Modified from Kossinna, 1933.

[a] The Continental Block is defined as Land Area plus Continental Shelf plus Continental Slope. It is not coterminous with the continental plate (see text).
[b] Freeboard refers to the land area above sea level as a percentage of the entire Continental Block. Antarctica is a special case because most of its land area is covered by ice sheets that extend onto the continental shelf. All values are approximations.

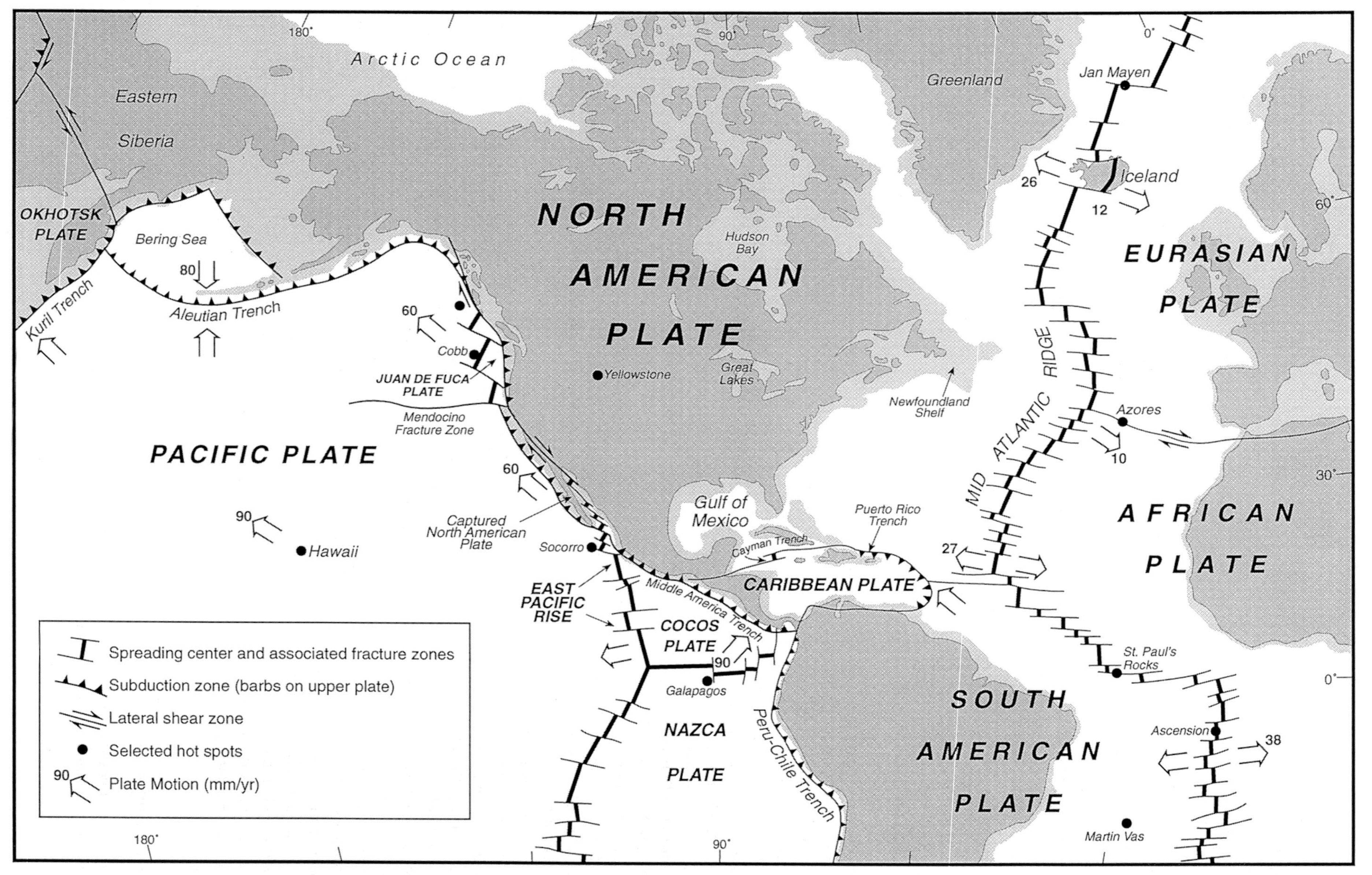

Figure 1.2 North America's present plate relations.

included, the North American block extends over 80° of latitude, from northern Greenland to the isthmus of Panama, and over 170° of longitude from 20°W off eastern Greenland to 170°E in the Aleutian Islands. Because most land lies between 30°N and 70°N, the continent has a predominantly temperate character, merging northward with vast subpolar terrains and southward with restricted tropical areas. This character, together with the tectonic attributes of the associated plates and the very different landscapes of Mesoamerica, recommends that this treatment of North America, together with other chapters in the book, should focus on the extratropical landscapes of the continent.

Whereas the previous dimensions are readily measurable, the continental block thus outlined is not coterminous with the North American plate defined by tectonism. Thus, although the submerged eastern edge of the North American block more or less coincides with the passive margin of the original North American plate as it parted from the Mid-Atlantic Ridge, this margin has been augmented since separation by widening oceanic crust. To the south, this passive plate margin, augmented by carbonate platforms and oceanic crust, now terminates against the eastward-thrusting Caribbean plate along a shear zone flanking the Cayman Trough. To the southwest, from Cabo San Lucas to Cape Mendocino, a sliver of the Californias has been captured by the subducting Pacific plate, which elsewhere still lies seaward of the west coast. In the far northwest, the North American plate margin is less certain but is thought to extend into eastern Siberia.

The imperfect fit between the continent's physical presence and the North American plate is, like its major relief features, the product of complex tectonic evolution. For present purposes, the record is divided into three major intervals: a lengthy period before the assembly of Pangea during which the Precambrian and Paleozoic framework of the continent was constructed; the assembly of the Pangea supercontinent between 330 and 230 Ma (million years before present); and, after the rupture of Pangea, the shaping of the continent's present structural features. This discussion involves the concept of terrane, a lithotectonic entity of regional extent characterized by geologic development distinct from nearby terranes. Accreted terranes are those that have been incorporated into continental crust; exotic terranes are of far-traveled origin; and suspect terranes are of uncertain origin. The term *terrane* implies geologic origins distinct from the term *terrain*, which refers to surface character, but the two may coincide by virtue of the former's distinctive rocks and structures that influence their relative resistance to mass wasting and erosion.

1.2.1 Pre-Pangea Terranes

Laurentia, the ancestral nucleus of the North American continent, was formed by collisional tectonic processes around 1 Ga (billion years before present) (Hoffman, 1988, 1989; Moores, 1991). It was an amalgam of earlier terranes that were assembled into Rodinia, a supercontinent of uncertain dimensions and components. Laurentia broke away from Rodinia during Neoproterozoic time (≈800–700 Ma), and in the earlier Paleozoic (≈570–460 Ma) became isolated from its former neighbors by the Iapetus and other oceans (Powell et al., 1994). At this time, the continent's Greenland margin was also joined to Rockall (northern Britain) and Barentia (Svalbard) (Dalziel, 1997). Then, in the mid-Paleozoic (≈460–400 Ma), prolonged collision with exotic terranes brought Laurentia into contact with the plate slivers of west Avalonia (Maritime Provinces), east Avalonia (southern Britain), and Baltica (northern Europe). The larger continent thus formed, Laurussia, was sutured by collisional orogens joining east Greenland to Scandinavia, and Rockall to east Avalonia (uniting Britain!). With Grenvillian terrane in eastern Canada now welded to west Avalonia, mountains flanked eastern Laurentia from Greenland to the Appalachians. During the later Paleozoic (≈400–245 Ma), the oceans separating Laurussia from Gondwana and Siberia closed, more fold mountains rose along the broad collisional front from Mexico to Newfoundland, and the Florida basement terrane from Gondwana welded onto the southern margins of Laurentia. Pangea was now assembled. However, its subsequent rupture in early Mesozoic time (≈245–180 Ma) allowed the nucleus of Laurentia, with certain gains and losses, to reassert its control over the shaping of North America.

Laurentia's place in the pre-Pangea world is much debated. The traditional view, building on the belief that a proto-Atlantic Ocean (Iapetus) closed with the assembly of Pangea and then reopened, places Laurentia off northwest Africa throughout the Paleozoic (Wilson, 1966). Alternatively, as shown in figure 1.3, Laurentia may have lain alongside east Gondwana in the Neoproterozoic, later migrated clockwise relative to South America's proto-Andean margin, forming collisional or transpressional orogens whenever the continents came together, and eventually docking against Africa in late Paleozoic time (Moores, 1991; Dalziel et al., 1994; Dalziel, 1997). Paleomagnetic and paleobiogeographic data, indicative of the latitude but not the longitude of plate arrangements, are invoked to support both concepts and others. Whatever the precise geometry, Laurentia probably lay around the South Pole in the Neoproterozoic and migrated northward, at rates of up to 20 cm yr^{-1}, into equatorial latitudes during the Paleozoic (Dalziel, 1997; Torsvik et al., 1997).

The amalgam of Precambrian terranes represented in Laurentia still finds expression in the modern landscape, directly in the subdued relief of the Canadian Shield, indirectly as basement supporting or protruding through later cover rocks (fig. 1.4). In general, the Precambrian terranes and their buried extensions comprise an Archean protocraton (≈4.0 and 2.5 Ga) welded together by Paleo-

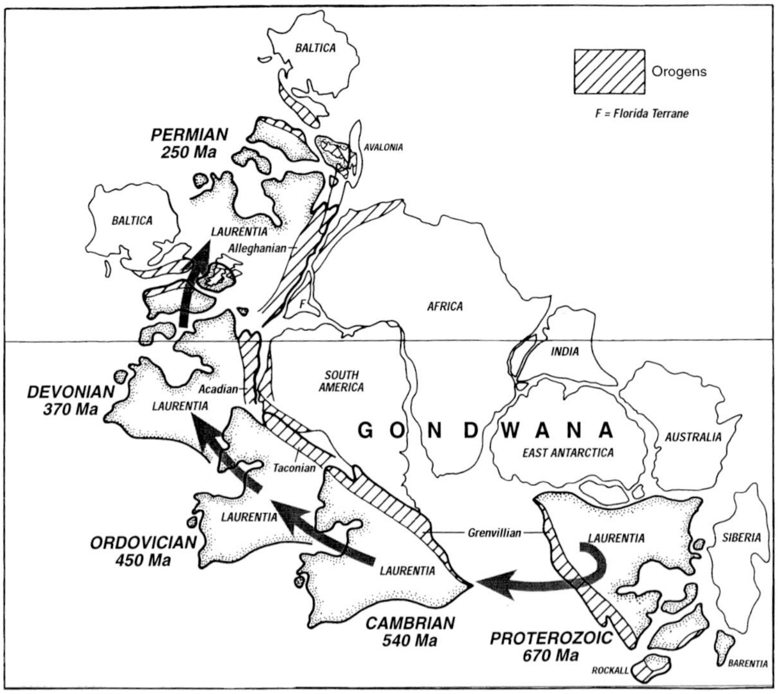

Figure 1.3 Hypothetical relative motion of Laurentia during Neoproterozoic and Paleozoic time. Africa is in its present position, with Gondwana reassembled for the terminal Paleozoic. For clarity, continents are defined by present coastlines where possible, and Laurentia's northern outposts are not shown for early Paleozoic time (developed from Dalziel et al., 1994; Dalziel, 1997; Ziegler et al., 1997).

proterozoic collisional orogens (2.0–1.8 Ga), flanked by later terranes that accreted to the protocraton (≈1.9 and 1.6 Ga), overlapped by Mesoproterozoic sedimentary and igneous assemblages (≈1.8–0.9 Ga), reworked in part by the Grenvillian (≈1.3–1.0 Ga) and later orogenies, and finally broken apart in the Neoproterozoic (≈0.8–0.6 Ga). Because most Precambrian rocks have been subjected to repeated orogenic and magmatic activity, they occur today mainly as complex metamorphic and igneous suites of variable structure, lithology, and erosional resistance. They include some of Earth's oldest known intact rocks, notably a 3.96-Ga gneiss in the western Canadian Shield and a 3.8-Ga metamorphic sequence in west Greenland (Bowring et al., 1989).

The Archean protocraton is an assemblage of seven former microcontinents, of which the Superior, Hearne, Rae, and Slave provinces make up most of the Canadian Shield whereas the smaller Nain and Burwell provinces are now split by post-Pangean rifting between Labrador and Greenland (Hoffman, 1989). The Wyoming province, mostly buried beneath later rocks, is exposed in the Beartooth Mountains and Wind River Range (fig. 1.5). Similarly, various Proterozoic accreted terranes and overlap assemblages are exposed around the margins of the Archean protocraton and also underlie much of the continent to the south and west. Here, Proterozoic rocks have been exposed locally by subsequent tectonism and erosion, notably in the >1.8-Ga Monashee complex of British Columbia, the <1.6-Ga clastic sequence of Utah's Uinta Mountains, the 1.1–0.8-Ga shelf strata in the Grand Canyon of the Colorado River, the 1.2-Ga San Gabriel batholith of California, and the 1-Ga Pikes Peak batholith of Colorado. Of the later orogenic events, the Grenvillian orogen (≈1.3–1.0 Ga) forms a 300–500-km-wide metamorphic belt that is exposed for 2000 km along the southeast margin of the Canadian Shield from Labrador to the Adirondack Mountains, and continues beneath later rocks to Mexico.

The Neoproterozoic rifting of Rodinia isolated the Laurentian platform on which Phanerozoic cover rocks later accumulated and the continental margins against which Phanerozoic orogenies occurred and exotic terranes accreted. Paleozoic and later sediments cover Precambrian basement across the midcontinent from the Appalachians to beyond the Rockies and extend discontinuously northeast along the St. Lawrence Valley and northwest to the Beaufort Sea (Quinlan, 1987; Bally, 1989) (fig. 1.4). They also occur in and around Hudson Bay. These cover rocks, mostly shallow-water carbonates and evaporites and deeper water muds, are of variable thickness—thinning where arches occur in the basement, thickening where basinal subsidence has occurred. The Michigan Basin contains 5000 m of Paleozoic sedimentary rocks, the Illinois Basin nearly 7600 m, the Williston Basin of North Dakota and Saskatchewan nearly 3000 m plus 2000 m of post-Paleozoic sediment, and the Hudson Bay Basin some 2000 m (Buschbach and Kolata, 1991). Near Laurentia's margins, sedimentary covers are thicker, exceeding 8000 m against the central Appalachians. These and similar deposits elsewhere occupy foredeeps at the continental margin that were to be incorporated, at least in part, into later orogens.

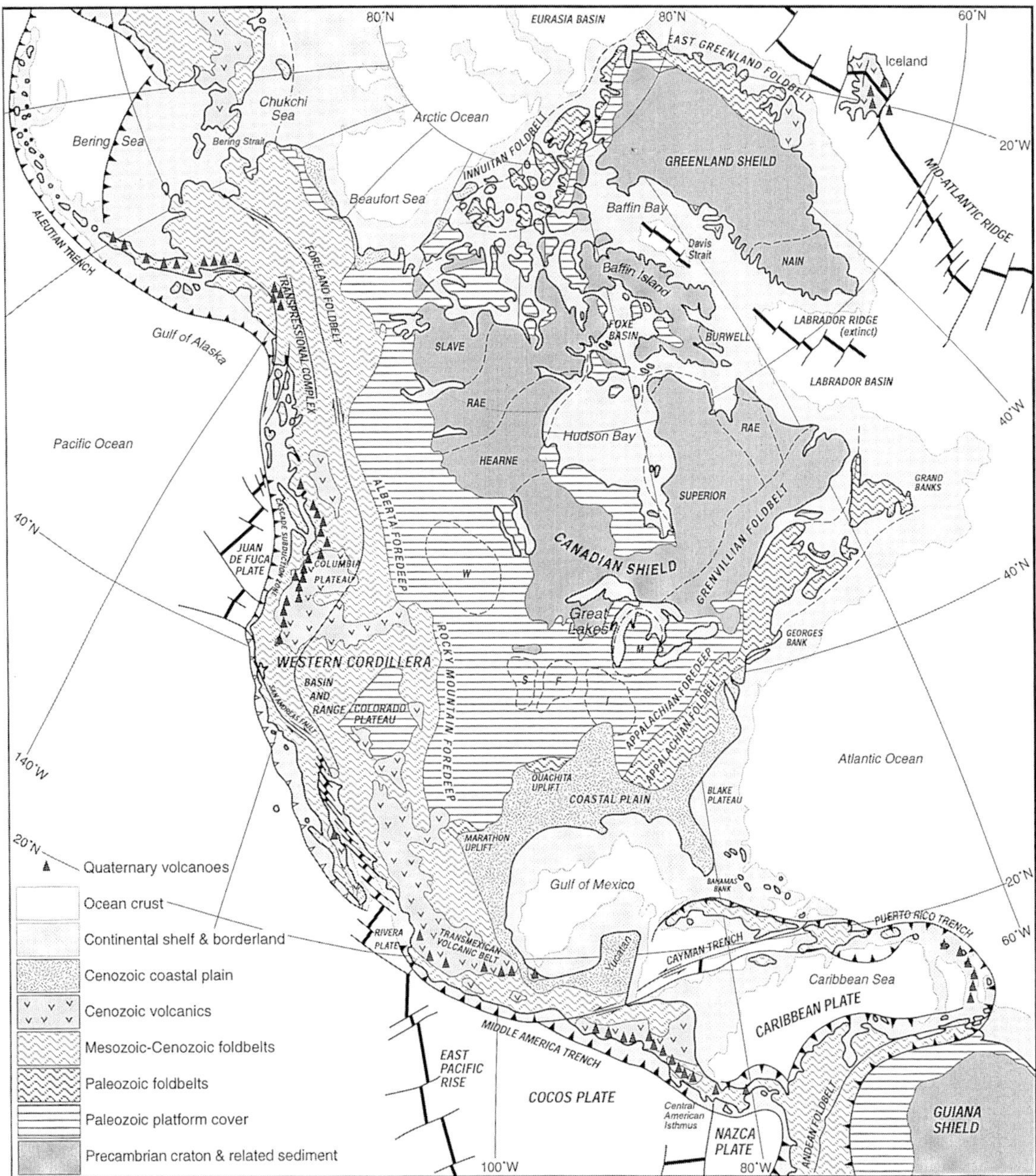

Figure 1.4 Tectonic provinces of North America. Selected Phanerozoic basins: F, Forest City; I, Illinois; M, Michigan; S, Salina; W, Williston.

The sequence of Paleozoic orogenic events that shaped the eastern margin of Laurentia is strongly expressed in the present landscape (table 1.2). The collisional tectonics of mid-Paleozoic time that generated the Taconian orogeny (≈480–440 Ma) are reflected in the "older" Appalachians. Farther north, the continent–continent collision of Laurentia and Baltica that generated the prolonged Scandian-Caledonian orogeny (≈440–400 Ma) is seen in east Greenland. Still later, the accretion of Avalonia composite terrain and the Acadian transpressional orogeny (≈400–360 Ma) further augmented the Appalachians. Along Laurentia's northern margin, episodic orogenic activity began in late Silurian time (≈420 Ma) with continental convergence and the accretion of Pearya exotic terrane to Ellesmere Island, and culminated in the Ellesmerian orogeny (370–350 Ma) that generated a 400-km-wide fold belt across northern Greenland and the Queen Elizabeth Islands.

Figure 1.5 The heart of a continent, looking southwest up Sunlight Creek, Wyoming. In the foreground, the North American basement is seen in 3.4-Ga Archean gneiss exposed in Clark's Fork Canyon of the Yellowstone River. This basement was later covered by Paleozoic marine carbonates whose resistant members form the more distant cliffs. In Eocene time, oceanic plate subduction farther west led to massive terrestrial volcanism, the Absaroka Volcanic Group (50–44 Ma), which buried rising Laramide relief toward the skyline. Windy Mountain, just north of Sunlight Creek, reaches 3128 m above sea level (Spence Collection, UCLA).

During the early Paleozoic (≈570–380 Ma), the western margin of Laurentia remained passive, accumulating up to 8000 m of shallow-water carbonates and siliciclastic deposits. However, volcanic island-arc systems began forming farther west and these later collided with the continent to generate the mid-Paleozoic Antler orogeny (≈380–330 Ma) and subsequent clastic sedimentation (Oldow et al., 1989). The effects of the Antler orogeny, though much disturbed by later events, are seen today in the Klamath terrane of California, the Kootenay terrane of British Columbia, and the Yukon-Tanana terrane farther north.

1.2.2 Assembly and Rupture of Pangea

Mid-Paleozoic orogenesis had expanded the framework of North America by surrounding the Precambrian basement and its early Paleozoic cover with a horseshoe arc of mountainous collisional terrain open only to the south. Foredeeps between the covered basement and these peripheral mountains came to resemble moats, which, depending on the relative roles of subsidence and uplift, gradually filled with clastic and carbonate sediment. For example, the downwarped central Appalachian foredeep filled with Ordovician deep-water clastics, Siluro-Devonian shallow-water carbonates, and Devonian-Pennsylvanian alluvial-deltaic clastics and coal-forming swamp peats.

The assembly of Pangea was completed during Carboniferous and Permian time when the approach of Gondwana triggered prolonged orogenic activity culminating in the Alleghanian orogeny (≈330–260 Ma) and epeirogenic warping of North America's cratonic interior (Hatcher et al., 1989; Howell and van der Pluijm, 1990). Simultaneously, the southern gulf was closed along the 2000-km Ouachita front by approaching Gondwana, either by obduction of allochthonous deep-water sediment (Cebull et al., 1976) or subduction of an accretionary thrust system (Lowe, 1985). However, excepting the Ouachita Mountains, the Marathon uplift, and small inliers in Mexico, the Ouachita fold and thrust belt is buried beneath later Coastal Plain sediment. Related intracontinental transpression farther west formed the ancestral Rocky Mountains at this time, but these were later eroded and covered by Cretaceous seas. The reconstruction of Pangea in figure 1.6 shows North America's location relative to other continents at the Paleozoic-Mesozoic transition (≈245 Ma) and the remarkably long belt of late Paleozoic fold mountains from Mexico to the Urals, flanked to the north by the older Scandian-Caledonian fold belts (Ziegler et al., 1997).

The Appalachian fold belt, the Eastern Cordillera of North America, is thus the product of repeated orogenic events during Paleozoic time, of which the Taconian, Acadian, and Alleghanian were most important. Although often overprinted by later structures, the Taconian event is expressed along the western margins of the northern Appalachians. The Acadian event is also seen in the northern Appalachians but farther south is strongly overprinted by the Alleghanian orogeny. These events involved widespread deformation, often intense metamorphism, and frequent plutonic and volcanic activity. They also incorporated older orogens, such as the Grenvillian, and a succession of accreted terranes, such as Avalonia, in addition to later clastic and carbonate deposits. Thus Grenvillian gneiss occurs in the Long Range of Newfoundland, the Green Mountains of Vermont, and the Blue Ridge of Virginia and North Carolina. Neoproterozoic metasediment from Avalonia is exposed in southeast Newfoundland and the Maritime Provinces.

The precise fit of Pangea's component parts is the

Table 1.2 Principal tectonic and climatic events affecting North America over geologic time, together with selected global biotic events. The logarithmic timescale condenses the distant past, thereby enhancing Mesozoic and Cenozoic events of significance to the present landscape.

Age (Ma)	Eon	Era	Period *[Epoch]*	Plate Motion	Orogenic Events	Climate-Forcing Events	Biotic Events
1						Glaciation	
			Quaternary		Pasadenan		
						Arctic sea ice Sierra Nevada uplift	
					Panamanian	Central American isthmus forms	proboscideans enter the Americas
			[Pliocene]	Gulf of California opens			
					Cascadian	Cascade volcanism	bears enter the Americas
		Cenozoic					Beringia pathways fully open
10						Columbia Plateau basalts	
			[Miocene]				
	PHANEROZOIC			Basin & Range extension peaks		Tethys closing Himalayan orogeny peaks	hominids appear in Africa
			Tertiary	Caribbean plate moves east		Colorado Plateau uplift begins	
				East Pacific Rise impacts west		Global cooling	
			[Oligocene]	Labrador rift stops		Antarctic ice	rodents appear
			[Eocene]	Greenland rifts from NW Europe	Laramide ends	India hits Asia North Atlantic volcanism	early horses appear grasses appear mammal diffusion
			[Paleocene]				primates appear
				Labrador Sea opens	Laramide begins		
100		Mesozoic	Cretaceous			Midcontinental seaway	angiosperms spread placental mammals
				Grand Banks part from Iberia	Sevier	South Atlantic opens	birds appear
			Jurassic		Nevadan	Central North Atlantic opens	
				Pangea ruptures			early mammals appear
			Triassic		Sonoma	Newark lavas	
			Permian			Gondwana glaciation	
			Carboniferous	Pangea forms	Alleghanian		conifers appear
		Paleozoic	Devonian	Iapetus closes	Acadian		gymnosperms appear
			Silurian				land plants appear
			Ordovician		Taconian		fish appear
			Cambrian	Iapetus opens	Avalonian		
							multicelled organisms appear
1000 (1 Ga)	PROTEROZOIC				Grenvillian		
				Laurentia forms			
		Precambrian			several Proterozoic orogenies		
				North American protocraton forms			bacteria and algae oldest life forms
4.6 Ga	ARCHEAN			Origin of Earth			

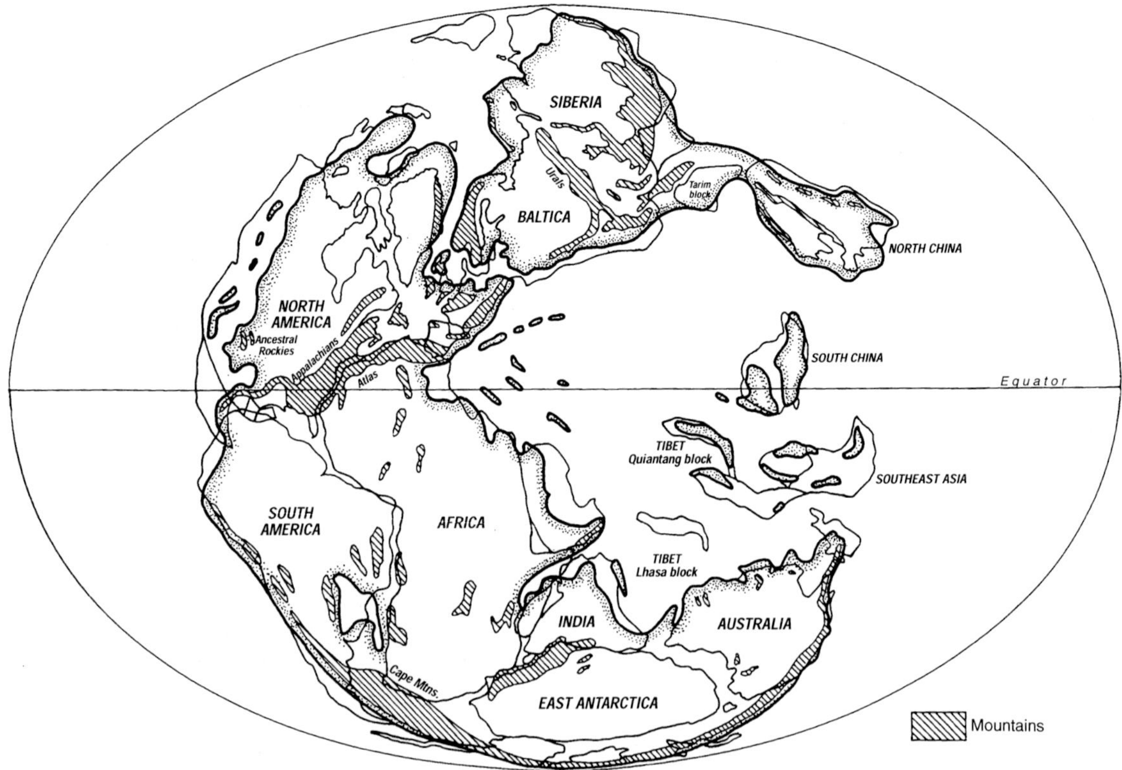

Figure 1.6 Pangea at the close of the Paleozoic, 245 Ma (after Ziegler et al., 1997).

subject of continuing debate. However, the docking of Gondwana introduced Laurentia to exotic terranes, parts of which remained after Pangea broke apart, notably the sutured Gondwana basement beneath modern Florida (Opdyke et al., 1987; figs. 1.3, 1.6). But Laurentia also lost terranes in the subsequent rupture, notably parts of the Caledonian orogen in Scotland and Ireland. Among the sediments that accumulated on the assembled supercontinent are the spectacular reef limestones of the Permian Basin of west Texas and the widespread aeolian sandstones of Jurassic age in Utah and Nevada.

The rupture of Pangea along the eastern margin of North America began with crustal extension and rifting in mid-Triassic time (≈240–230 Ma). Rifting more or less paralleled the NE–SW strike of the Appalachian orogen and, lacking structural support to the east, produced a series of elongate half-grabens (Sheridan, 1989). The earlier, more westerly rift basins, inland from the present coast, contain up to 10 km of terrigenous "red beds," mostly fanglomerates fining upward and outward into alluvial and lacustrine deposits, interbedded with volcanics. A 6770-m core spanning 30 Ma of the late Triassic from the Newark Basin reveals an exceptional record of "red bed" sedimentation indicating cyclical tropical climate change (Olsen et al., 1996). As later basins developed farther east, terrigenous "red beds" came to be overlain with shallow-water carbonates and evaporites, and these dominate the most easterly basins such as those beneath the Blake Plateau and the Grand Banks. Rifting of the continent's northern margin probably began much earlier, in mid-Carboniferous time (≈340 Ma), judging from deposits in the Sverdrup Basin of the northern Arctic archipelago, but the implications of this are far from certain.

1.2.3 Post-Pangea Evolution

Events in North America following the rupture of Pangea may be grouped into five categories: progressive separation of the continent from its Pangean neighbors; structural adjustments to existing terranes over the eastern two-thirds of the continent; accumulation of additional platform covers; development of the Western Cordillera; and the shaping of the present landscape through denudation and drainage evolution. Compared with earlier developments, the post-Pangean interval has been relatively brief—barely 180 Ma, or 4% of Earth time—but its events, because they are comparatively recent and ongoing, are more boldly defined in the present landscape than the subtler expressions of earlier times.

The progressive separation of North America from its Pangean neighbors is well illustrated along its eastern passive margin (fig. 1.7). Here, dated magnetic lineations show

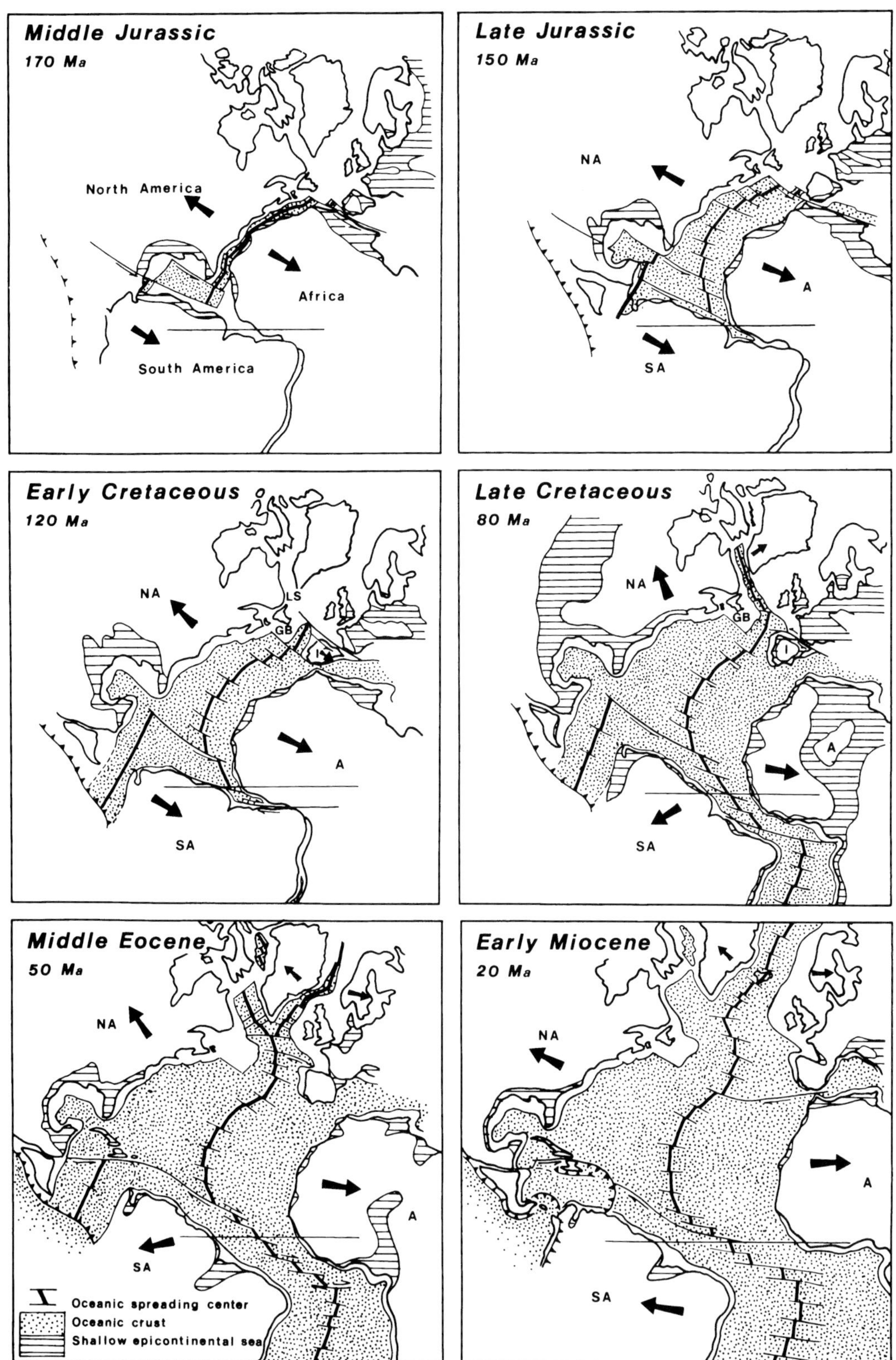

Figure 1.7 The rupture of Pangea and opening of the North Atlantic Ocean during later Mesozoic and Cenozoic time. Modern coastlines are used, where possible, for reference, but the best fit between continents occurs around the 2000-m isobath on the continental slope. Compare with figures 1.6 (before rupture) and 1.2 (present).

how basaltic oceanic crust spewing from the Mid-Atlantic Ridge has been added to the trailing edge of the North American plate since at least mid-Jurassic time (≈180–160 Ma). Flood basalts and related dike rocks exposed from Nova Scotia to Virginia are probably remnants of linear fissure eruptions related to the early production of new oceanic crust (McHone, 1996). The continental portion of the plate has since been rafted, pushed, and pulled north and west as the central North Atlantic Ocean basin has widened at rates varying from 1 to 4 cm yr^{-1}. From a narrow mid-Jurassic rift comparable to the present Red Sea, this ocean basin has expanded to ≈5000 km between North America and Africa. The ocean-floor basalts have in turn been covered by postrift sediment, thickest near continental sources and thin or absent close to the active spreading center.

From late Jurassic time onward, the North Atlantic basin opened progressively northward as the Grand Banks separated from Iberia (≈150–120 Ma), the Labrador Sea opened in late Cretaceous time (≈90–70 Ma), and Greenland was detached from Scandinavia in the early Cenozoic (≈60–40 Ma) but, when sea-floor spreading in the Labrador Sea ceased around 35 Ma, remained attached to North America. The "unzipping" of the northern North Atlantic was accompanied by massive magmatic venting, notably at the Iceland hot spot. The links thus established with the Arctic Ocean were to have a profound effect on Cenozoic ocean circulation and therefore on climate. Meanwhile, with the "unzipping" of the South Atlantic basin that began between Argentina and South Africa around 130 Ma and culminated with the opening of the Gulf of Guinea around 80 Ma, the Atlantic as a whole began taking shape. The Gulf of Mexico formed initially in the shear zone between North and South America and later linked southward with an Atlantic–Pacific seaway as early Cretaceous rifting (≈130 Ma) stranded Yucatan alongside North America.

Whereas major orogenic activity over the eastern two-thirds of North America was curtailed by the assembly of Pangea, tectonic activity did not cease. The rifting of Pangea generated thermal changes in the underlying mantle and released crustal stresses that led locally to uplift along the continent's rim. The Cenozoic uplift of Precambrian terranes on Labrador and Baffin Island was perhaps a delayed response to the opening of the Labrador Sea. The Atlantic Coastal Plain probably owes its onshore presence in the southeastern United States but its submergence off eastern Canada to post-Pangean flexural warping related to a hinge zone oblique to the coast (Steckler et al., 1988). The northern rim of the Gulf of Mexico experienced localized warping, faulting, and salt mobility throughout Cretaceous and Cenozoic times, generating structures favorable to hydrocarbon accumulation amid an accumulating mass of carbonate and clastic deposits. The evolving post-Pangean landscapes of the eastern two-thirds of North America have likely seen repeated episodes of gentle but prolonged uplift and erosion, or of subsidence and deposition, on which the shorter eustatic and isostatic events of the late Cenozoic have been superimposed.

The platform covers added to North America during post-Pangean time may be grouped broadly into carbonate and clastic lithofacies, the former common to the continent's passive margins and shallow seas, the latter dominated by debris shed from a rising Western Cordillera. In mid-Jurassic time, while the incipient central North Atlantic Ocean and Gulf of Mexico remained warm and shallow, carbonate and evaporite deposits accumulated in basins straddling the continental margin. Later, as these marginal basins subsided from lithospheric cooling and sediment loading, these sequences thickened until by mid-Cretaceous time (≈100 Ma) vast carbonate platforms stretched from Yucatan across the Gulf of Mexico to Florida and the Bahamas. Meanwhile, depending on the flux between tectonic and eustatic forces along the ocean front, the foundations of the present Gulf and Atlantic coastal plains were forming as shallow marine sediments lapped onto the continent and terrigenous clastic debris prograded seaward. Indeed, for a while during the Cretaceous, high eustatic sea levels and low structural freeboard combined to provide a shallow continuous seaway, 100–300 m deep and 1500–2000 km wide, from the Gulf of Mexico northward to the Arctic Ocean. Related carbonate deposition provided the limestone for the karst landscapes of the southern Great Plains, notably on the Edwards Plateau of Texas (see chapter 17). Farther west, however, the rising Western Cordillera began shedding fluvial and deltaic waste into a foredeep along the developing Rocky Mountains. Aided by falling eustatic sea levels, this clastic debris led to progressive closure of the seaway in later Cretaceous time. With Cretaceous seas expelled and uplift farther west, terrigenous clastic sedimentation came to dominate most of the continent during the Cenozoic Era.

The prolonged uplift of the Western Cordillera that peaked with the Laramide orogeny has been the most dramatic expression of post-Pangean tectonism, forming mountain ranges and intermontane plateaus and basins in a zone 800–1800 km wide stretching over 8000 km from Alaska to Mexico (fig. 1.8). The rupture of Pangea had transformed western North America from a passive to an active margin along which plate convergence generated repeated orogenic activity and related magmatism. Relative to the mantle, this active margin now moved north and west over eastward-dipping subduction zones, causing compression in the form of a bow wave along the continent's leading edge (Russo and Silver, 1996; Dalziel, 1997). Earlier events in the region were now overprinted by the Nevadan (≈160–140 Ma), Sevier (≈120–100 Ma), and Laramide (≈80–40 Ma) orogenies. Initially, from Jurassic to Paleocene time (≈200–50 Ma), oblique plate convergence favored thrusting and

Figure 1.8 The Rocky Mountains in Montana showing Proterozoic sedimentary and volcanic rocks, massively uplifted by the Laramide orogeny, subsequently dissected, and glaciated (Spence Collection, UCLA).

transpression as successive terranes were displaced mostly northward along transcurrent faults (Oldow et al., 1989). Later, as this active western margin impinged on the East Pacific Rise spreading center, transtension became more important, notably stretching the Basin and Range Province after ≈35 Ma.

Vast quantities of rock became involved in shaping the Western Cordillera: Precambrian basement, Neoproterozoic and Paleozoic sedimentary and volcanic rocks from Laurentia's formerly passive western margin, accretionary wedges from displaced terranes, magmatic rocks generated along the now active margin, and clastic sediment shed from the rising cordillera were all affected by ongoing deformation. Indeed, apatite fission-track data suggest that the Permian rocks forming the modern surface of the Colorado Plateau were originally overlain by up to 4500 m of Mesozoic strata that were to be eroded during and after the Laramide orogeny (Dumitru et al., 1994). Of the 200 or so terranes that accreted to the continent, most had North American affinities, such as the Kootenay and Yukon-Tanana terranes of the Canadian Rockies; but some like the Wrangellia and Alexander terranes around the Gulf of Alaska were of more suspect origin, whereas the subduction-related Franciscan terrane of California involved oceanic basalts and pelagic cherts of equatorial provenance (Isozaki and Blake, 1994). The Western Cordillera thus came to reflect a tectonic mix of oceanic and terrestrial rocks of differing age and provenance whose disparate terranes became involved in eastward subduction or westward overthrusting, even as they were displaced to a greater or lesser extent northward and subject to often intense metamorphism.

In general, the Western Cordillera comprises two major tectonic belts: a Foreland Fold-and-Thrust Belt of unquestioned North American rocks to the east, and a Transpressional Complex, whose displaced terranes become increasingly suspect to the west (Oldow et al., 1989). The Foreland Fold-and-Thrust Belt extends from Alaska's Brooks Range, through the Rocky Mountains, Great Basin, and Colorado Plateau, to Mexico's Sierra Madre Oriental. It also includes outliers like the Black Hills of South Dakota. Compression of this belt caused crustal shortening of 100–200 km in the Canadian Rockies and perhaps 500 km in the Brooks Range, accompanied by great thickening of continental crust to 35–50 km in the southern Rockies. The Transpressional Complex embraces three belts that have experienced varying displacement. The Central or Intermontane Belt exhibits strong North American affinities, severe crustal shortening, and intense metamorphism, for example, in the Yukon-Tanana terrane of central Alaska, the Blue Mountains of eastern Oregon, and the Sierra Nevada of California. The Columbia-Coast Belt includes the metamorphic core of the Cascade Range and batholiths in British Columbia and Washington whose magnetic lineations suggest a Cretaceous origin in the latitude of Baja California (Ague and Brandon, 1996). This implies ≈3000 km of cumulative northward displacement along the Yalakom and other faults (Wynne et al., 1995), although Archean zircons derived from the Canadian Shield suggest less offset (Mahoney et al., 1999). The Insular Belt of western British Columbia and southern Alaska includes Wrangellia and other terranes that may have originated even farther south and only docked against North America during the Eocene.

Magmatism broadly linked with plate convergence and Laramide events led to emplacement of composite granite plutons and extensive volcanism. Notable plutons since exposed by erosion include batholiths in peninsular Alaska (≈80–45 Ma), coastal British Columbia (≈130–60 Ma), Idaho (≈90–70 Ma), the Sierra Nevada (≈140–80 Ma), peninsular California, and Mexico's Sierra Madre Occidental (≈60–40 Ma). Major volcanic fields include the Absaroka volcanics (≈50–45 Ma) within the Challis magmatic arc that stretches from western Canada to Wyoming. If 100–Ma plutons in the Sierra Nevada crystallized 20,000 m below the surface, uplift of at least that magnitude has occurred since (Sams and Saleeby, 1988).

Later Cenozoic tectonism and volcanism in the Western Cordillera are linked to the oblique subduction of fragmented oceanic plates beneath the westward-moving North American plate—events that have produced some spectacular landscapes across the region. In the far north, subduction of the Cretaceous Kula plate beneath eastern Siberia and peninsular Alaska paved the way for the later subduction of the Pacific plate in the Aleutian Trench (>7000 m deep) and formation of the Aleutian volcanic arc, which became particularly active after 15 Ma. This subduction zone links southward, through the Queen Charlotte dextral transform system, with the East Pacific Rise. East of this spreading center, the Juan de Fuca plate and other Farallon plate fragments are sliding at a mean rate of 4 cm yr^{-1} into the Cascadia subduction zone, dipping 5–12° ENE to depths of >70 km beneath the southwest-moving North American plate (DeMets et al., 1990). Melting of the subducting plate has caused magma to rise through the overlying continental plate to form the Cascade volcanic arc from British Columbia to northern California in which over 4000 volcanic vents have developed over the past 15 Ma (Guaffanti and Weaver, 1988). Among the 19 recently active composite volcanoes are Mount Rainier (4392 m) and Mount Shasta (4317 m), both high enough to carry numerous glaciers. Mount St. Helens, formerly 2950 m high, lost 400 m in elevation and 2.73 km^3 in mass during its 1980 eruption, the most recent of nine eruptive episodes over the past 40 ka (thousand years before present) (Lipman and Crandell, 1981).

Farther south, between Cape Mendocino and the tip of Baja California, the much dislocated Farallon plate and the East Pacific Rise have long since passed beneath the North American plate. In so doing, a lengthy sliver of the latter has been captured by the Pacific plate, the present plate boundary being marked by the East Pacific Rise beneath the Gulf of California, where it appeared around 6 Ma, and onshore to the northwest by the 1200–km long San Andreas fault zone (fig. 1.9). This dextral strike-slip zone and related faults, the focus of many historic earthquakes in California, are active extensions of transform structures that offset the East Pacific Rise to the north and south. To the west, the captured sliver has moved northwest relative to the North American plate at a mean rate of 5–6 cm yr^{-1} over the past 6 Ma, with movement of 3–4 cm yr^{-1} on the San Andreas fault and a further 2–3 cm yr^{-1} distributed among similar structures within the broad system of lateral shear between the offshore continental borderland and the Basin and Range Province (Wallace, 1990; Orme, 1992; DeMets, 1995). Displacement of volcanic rocks at the Pinnacles and Neenach, which originated from a common eruptive center around 23 Ma, suggests 310 km of total offset along the San Andreas fault during later Cenozoic time (Matthews, 1976). Oligo-Miocene conglomerates near Los Angeles are thought to represent paleodeltas that have been offset from an ancestral Colorado River by ≈300 km of dextral shear over the past 15 Ma (Howard, 1996). Continuing transpression within the captured part of the Californias is reflected in uplifted Quaternary shorelines (see chapter 21; Orme, 1980, 1998). The western Transverse Ranges are unique in their east–west orientation nearly orthogonal to other Pacific coastal ranges. This is attributed to ≈100° of clockwise rotation during the capture of Farallon plate fragments by the Pacific plate between ≈24 Ma and 12 Ma (Nicholson et al., 1994). South from Baja California, the East Pacific Rise still lies west of the mainland coast, such that the Rivera (Farallon) and Cocos plates continue to subduct beneath Mesoamerica, generating the Middle America Trench and extensive volcanic activity in the Transmexican Volcanic Belt.

Later Cenozoic tectonism is also expressed farther east in the extension of the Basin and Range Province and Rio Grande Rift, and in the uplift of the Colorado Plateau. These events, possibly related to mantle plume upwelling beneath the continent (Parsons and McCarthy, 1995), involved up to 3000 m of regional uplift, since reduced by erosion, basinal subsidence, and isostatic adjustments. The Basin and Range Province stretches from Idaho to the Gulf of California and is flanked to the east by the Colorado Plateau and to the west by the Sierra Nevada. Here, some 50–250 km of crustal extension are reflected in the north–south alignment of horst-and-graben relief bounded by normal high-angle and low-angle detachment faults descending to depths of 8000–15,000 m. Cenozoic basin fills suggest that extension began in Oligocene time (≈35–30 Ma), peaked in the Miocene (≈20–12 Ma), and, except in the quiescent Sonoran region, still continues at a reduced rate. Opening of the Rio Grande Rift began ≈32–26 Ma and, judging from active fault scarps in Quaternary sediment, continues today. In contrast, the Colorado Plateau retained its structural integrity during uplift that began around 24 Ma but in the process was separated from nearby regions by marginal faults, such as the Hurricane fault in Utah, and deeply scored by the Colorado River and its tributaries to form spectacular canyons, now incised up to 1000 m below the plateau surface (fig. 1.10).

Figure 1.9 The San Andreas fault, the Pacific-North America plate boundary, slicing through the southern California landscape. To the right or south, the San Gabriel Mountains (3068 m) are now part of the Pacific plate, whereas to the left or north the Mojave Desert (≈800 m) remains, for a while at least, part of the North American plate (Spence Collection, UCLA).

Extensional tectonics and changes in relative plate motions throughout these regions also favored widespread volcanism, for example, in the San Juan Mountains of Colorado (≈31–23 Ma), the Jemez field of New Mexico (≈13–0.1 Ma) with its notable Valles caldera, central British Columbia (≈20 Ma to recent), and in the extrusion of 170,000 km^3 of flood basalts across the Columbia Plateau (≈17–11 Ma) (fig. 1.11) (Hooper, 1988). As the North American plate has passed westward across eruptive centers thought to be related to the mantle plume or "hot spot," so volcanism propagated eastward from the Columbia Plateau up the Snake River Plain into the Yellowstone region of northwest Wyoming, still the scene of much geothermal activity. Extensional tectonics and volcanism in the Sierra Madre Occidental (≈24–12 Ma) and later in the Gulf of California correlate well with pulses of rapid spreading at the East Pacfic Rise (Nieto-Samaniego et al., 1999).

1.3 Tectonism and Climatology

Earth's climates are driven by solar radiation. Assuming that a significant amount of this energy penetrates the atmosphere, how it is treated depends largely on the nature of Earth's surface. In general, this implies differences between land and sea, mountain and plain, bare soil and vegetation cover, and between surfaces that are covered by snow and ice and those that are not. These differences are expressed through such variables as albedo and soil conductivity and are reflected in complex feedback mechanisms between the surface and the atmosphere.

However, climates also change through time in complex response to Earth's orbital relations with the Sun and to the impact on the atmosphere of surface changes related to tectonism, ocean circulation, and biogeochemical cycling. Orbital variations include the eccentricity of

Figure 1.10 The Grand Canyon of the Colorado River, Arizona, looking northeast. Precambrian basement rocks in the canyon floor and their cover of Paleozoic marine sedimentary rocks have been exposed by late Cenozoic uplift and dissection. Stepped canyon walls reflect alternations of resistant Paleozoic limestone and sandstone with less durable shale, capped here by plateau-forming Permian limestone (Spence Collection, UCLA).

Earth's orbit, the tilt of Earth's axis of rotation, and the wobble on that rotational axis, changes that occur, respectively, on cycles of ≈95 ka, ≈41 ka, and ≈22 ka. Permutations of these variations are reflected in climate changes repeated on timescales ranging from 10^4 to 10^5 years (see chapter 3). This chapter focuses, however, on climate changes attributable directly or indirectly to tectonism and most readily measurable on timescales of 10^5 to 10^7 years. Climate changes caused by dust generation resulting from volcanism and meteorite impact function at much shorter timescales (≈10^0–10^2 years) but may trigger far-reaching surface responses. This relationship between tectonism and climate is examined at two levels. First, major events in the evolving climate scene of North America over the past 200 Ma are discussed, events that are reflected to varying extent in the present landscape. Secondly, the sig-

Figure 1.11 Ship Rock, a late Cenozoic volcanic neck with radiating dikes, on the Colorado Plateau, New Mexico (Spence Collection, UCLA).

nificance of past tectonic events to the present climate is outlined as a prelude for discussions of climate in subsequent chapters.

1.3.1 Evolving Climate Patterns

Tectonism and climate interact at three scales: (1) the global scale, which is related to the generation, location, and shape of continents and ocean basins; (2) continental scales, which are related to secondary relief features like cordilleras; and (3) regional scales, which involve tertiary relief features such as mountain ranges. The effect of tectonism through time is to change the playing field on which climate performs and to influence the composition and mobility of the atmosphere overlying the playing field.

At the global scale, the assembly of Pangea caused North America to become embedded in a supercontinent, much of it far removed from deep oceans. The climate was profoundly continental, as shown by the Triassic "red beds" and evaporites of the Newark Basin. It was also mainly tropical. Most reconstructions of Pangea place the bulk of North America in the northern tropics, from the Equator to around 40°N (fig. 1.6). Mid-Jurassic rupture of Pangea introduced warm shallow epicontinental seas to the continent's southern and eastern margins, reflected in the growth of carbonate reefs. Further opening of the North Atlantic basin, initiation of deep-water linkages with the South Atlantic after 100 Ma, and the persistence of warm oceans in the west allowed marine influences to permeate the continent. The climatic impact of these warm tropical waters was enhanced by poleward ocean-heat transport in the absence of a major link to the Arctic Ocean (Rind and Chandler 1991) and by eustatic sea levels higher than today, ≈300 m higher in the Cretaceous and ≈200 m higher in the Paleocene (Haq et al., 1987). Such high sea levels may have been induced by delayed crustal response to the rupture of Pangea or by increased ocean-ridge volume produced by rapid sea-floor spreading. Also, the absence of an isthmus between North America and Gondwana ensured the persistence around the continent's southern margin of a tropical east–west seaway, the Tethys, whose circulation was driven by warm winds, high sea-surface temperatures, and saline deep water. These conditions favored strong zonal circulation, whereas warmer land temperatures encouraged swift onshore winds, intensified Hadley cell circulations, higher precipitation, winters warmer than today in Canada (O'Connell et al., 1996), and intense chemical weathering as seen in increased kaolinite production in New Jersey (Gibson et al., 1993). Deep weathering residues in the Canadian Shield may have survived until the arrival of Pleistocene ice sheets. Indeed, late Mesozoic and early Cenozoic climates were warmer globally than today, probably due to increased atmospheric CO_2 (Crowley and North, 1991), which, for the Eocene at least, may have been favored by crustal degassing during accelerated ocean spreading and Himalayan and Pacific-rim orogenesis and metamorphism (Berner, 1990). In a largely ice-free world, paleobotanical and paleoceanographic evidence suggests that mean annual temperatures peaked around 50 Ma (Prentice and Matthews, 1988; Wing et al., 1991).

These warm equable climate conditions over North America were to change as the Cenozoic progressed, in large measure due directly to tectonism. First, the northward "unzipping" of the North Atlantic as Greenland separated from Scandinavia (≈60–40 Ma) allowed cool bottom waters and surface currents from the Arctic Ocean to penetrate southward. By early Oligocene time (34–30 Ma), this influx was establishing the massive transfers of energy in an enlarged Atlantic circulation that were the necessary precursors to later Cenozoic cooling (Berggren and Prothero, 1992). Second, the gradual closure of the Tethys, through Caribbean and Cocos plate motions and emergence of the Central American isthmus around 3 Ma, separated the Atlantic and Pacific Oceans into discrete bodies, discouraged zonal atmospheric and oceanic flows in favor of more meridional transfers, strengthened the Gulf Stream, and favored the formation of North American Deep Water (Mikolajewicz et al., 1993). The revised thermohaline circulation of these oceans pumped more warm surface water into higher latitudes, which in turn raised the saturation vapor content of the overlying air, increased snowfall, and thus favored glaciation. This process was enhanced by the opening of the Bering Strait around 5 Ma (Marincovich and Gladenkov, 1999). Third, the distant uplift of the Tibetan Plateau, which began around 50–45 Ma as India collided with Asia and propagated northward from the Himalayan foreland over the next 10–15 Ma (Yin and Harrison, 1996), effectively excluded tropical marine influences from central Asia (Prell and Kutzbach, 1992). This affected atmospheric circulation and planetary wave motions across the entire northern hemisphere, including those across the North Pole into North America (Ruddiman and Kutzbach, 1989). Fourth, in Antarctica, where significant glaciers first appeared in the early Oligocene, the growth of the East Antarctic Ice Sheet toward its present form by 14 Ma had a profound global impact on ocean temperatures and climate cooling. Last, while these events were occurring, North America continued to drift generally north and west into higher latitudes.

These changes were reflected in the southward shift of the boreal forest treeline across Canada during Cenozoic time. For example, the northern limit of forest soils (podzols, luvisols) was 80°N on Axel Heiberg Island in the Eocene, but it retreated to ≈70°N in the Yukon in the Pliocene and to ≈60°N during Pleistocene interglacials (Tarnocai, 1997). Inasmuch as fossil animals may be used as climate proxies and by analogy with living American alligators (*Alligator mississippiensis*), which prefer air temperatures of 25°–35°C and use water to buffer extremes, this cooling is also reflected in the southward retreat of crocodilians from the Eocene of Ellesmere Island, at 78°N, to the Holocene Gulf of Mexico (Markwick, 1994).

These events were augmented by changes in atmospheric composition, notably in reduced concentrations of greenhouse gases, also related in part to tectonism. For example, the Himalayan and Laramide orogenies, having perhaps triggered earlier degassing, may now have provided major sinks for atmospheric CO_2 by enhancing, through uplift and erosion, the weatherability of silicate minerals and sequestering CO_2 in sediment and by subducting the deep-water carbonates produced by pelagic marine organisms as the Cenozoic progressed (Caldeira, 1992; Raymo and Ruddiman, 1992). In another feedback mechanism, the expansion of deciduous angiosperms (flowering plants) at the expense of coniferous gymnosperms during Cenozoic time probably led to increased rates of weathering and soil formation which also consumed large amounts of CO_2 (Volk, 1989). In short, by early Miocene time (≈23 Ma), global conditions had become favorable to the significant cooling that was to be reflected in the late Cenozoic landscapes of North America and the formation of continental ice sheets.

At the continental scale, three post-Pangean tectonic events made a notable impact on climate, namely, erosion of the Appalachians, uplift of the Western Cordillera and related volcanism, and uplift of the continent's northeast margin. When North America separated from Gondwana, its formerly active eastern margin was relatively high, whereas its passive western margin was quite low. As the Appalachians, now at a passive margin, were denuded, their impact on climate lessened. In contrast, as the Western Cordillera rose along the now-active western margin, its impact on climate increased, and Eocene mountains as far south as Wyoming became high enough (>3000 m) to sustain snowpacks (Norris et al., 1996). This role reversal between eastern and western North America had a major influence on transfers of warmth and moisture across the continent, rendering much of the west increasingly dry, whereas the center became more vulnerable to extremes of heat and cold caused by outbreaks of tropical and polar air, respectively. Tectonic inducement toward aridity in the west is reflected in the sedimentary record—from the marine sediment of warm Cretaceous seas (e.g., Mancos Shale of Colorado), to extensive early Cenozoic lake deposits (e.g., Eocene Green River Formation of Wyoming), to more restricted late Cenozoic basin fills (e.g., Miocene Salt Lake Formation of Utah), and finally to the alternating lacustrine, evaporite, alluvial, and aeolian deposits of confined Quaternary lake basins (e.g., Lake Bonneville). Indeed, the variable uplift of 500–2500 m that has occurred across the west over the past 5 Ma alone (Unruh, 1991) has had a profound impact on climate and biota throughout the region.

Uplift along North America's passive northeast margin may be invoked to explain, at least in part, initiation of late Cenozoic glaciation in the region (e.g., Eyles, 1996). Cenozoic uplift of 1500–2000 m has occurred around the margins of the Labrador Basin, Baffin Bay, and eastern Greenland, perhaps as a delayed response to oceanic crustal motions, deep-seated magmatism, or the Eurekan orogeny (≈80–60 Ma). This uplift may have raised mountains and plateaus above a climatic threshold to altitudes where, in the cooling conditions of the late Cenozoic, perennial snowfields might develop into glaciers.

At the regional scale, the uplift of the Sierra Nevada provides a lucid example of how tectonism interacts with climate. This tilted block has risen rapidly during late Cenozoic time, adding as much as 2000 m to its height over the past 10 Ma and 1000 m during the last 3 Ma alone (Huber, 1981; Unruh, 1991). In doing so, it has significantly increased rain-shadow effects to the east, transforming late Pliocene savanna into temperate desert. Whereas the Sierra Nevada was probably raised by asthenosphere welling up beneath a thin crust (Ducea and Saleeby, 1996), uplift may have been accentuated by isostatic response to massive erosion induced by late Cenozoic cooling and glaciation (Small and Anderson, 1995). Feedback relationships between tectonic uplift and climate change thus pose many questions, aptly referred to as the "chicken or egg" scenario (Molnar and England, 1990).

1.3.2 Present Climatic Patterns

North America now lies in middle to high latitudes, mostly between 30° and 70°N. This location inevitably imposes a temperate to subpolar character on its climate. North America also lies between the broad Pacific and narrower Atlantic Oceans, which should impart significant marine influences to the continent. This is generally true in the south and east where the relatively warm waters of the Atlantic Ocean and Gulf of Mexico introduce warm, moist air to the coastlands and, during summer, to the lowland interior of the continent. It is less true in the west where, despite the generally eastward movement of the atmospheric circulation, moist Pacific air masses are abruptly modified by the Western Cordillera. Further, the presence of cold terrain and Arctic sea ice to the north means that, in the absence of topographic barriers, outbreaks of cold, dry air have ready access to the continental interior and, in winter, to the Gulf of Mexico. Thus the morphotectonic configuration of North America, with two cordilleras and an intervening lowland aligned north–south, exposes the continent to massive transfers of warm, moist air from the tropics and cold, dry air from high latitudes, the conflict between which creates a zone of atmospheric disturbance, including severe thunderstorms and tornadoes, that fluctuates seasonally across the midcontinent. In this respect,

North America is unlike any other continent and quite the opposite of Eurasia, whose morphotectonic configuration discourages north–south transfers.

Within the continent, the climatic impacts of individual ranges in the Western Cordillera are impressive. In the far west, the modest Coast Ranges of California are sufficiently high in a seasonal Mediterranean regime to squeeze enough moisture from Pacific winter storms to render the western margins of the Central Valley semiarid, though it is only 100 km inland. Farther east, the high Sierra Nevada exhibits humid temperate conditions on its forested western slopes but semiarid conditions in the shrubland immediately to the east. Similar contrasts are repeated across the cordillera such that, despite its location between 35° and 45°N, the Great Basin is essentially a temperate desert. The presence of the Rocky Mountains in turn strongly influences cyclogenesis over the Great Plains farther east (see chapter 5). The impact of the Appalachians is now much muted, but they are still high enough to influence low-level pressure gradients and atmospheric disturbances. For example, their eastern flanks often dam cold surface flows emanating from winter anticyclones farther north. When moist air from the south advects across this dam, rain falling through the cold air becomes supercooled and freezes on contact with the ground (fig. 1.12). Such ice storms play havoc with eastern forests, leading to long-term changes in forest composition and dynamics (Nicholas and Zedaker, 1989; Lafon et al., 1999).

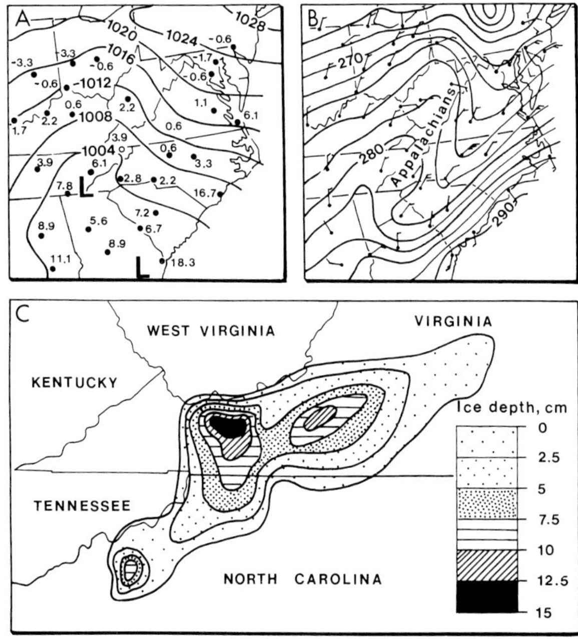

Figure 1.12 The Appalachians as a cold-air dam that resulted in the ice storm of 2 March 1994.

A, surface pressure (mb) and temperature (°C), 1200 hr Coordinated Universal Time (UTC); B, potential temperature (°K) and surface winds, 1200 hr UTC; C. total ice deposition (cm) during storm (from Lafon et al., 1999).

1.4 Tectonism and Geomorphology

The relationship between past tectonism and the present physique of North America is exemplified at two scales. At the continental scale, spatial links exist among tectonic events, continental shape, and rock distributions, and temporal links exist between orogenic events and modern relief (figs. 1.4, 1.13). At regional scales, temporal and spatial links exist among tectonism, denudation, and sedimentation. Now, as in the past, climate is the critical connector. Continental location and relief dictate regional climates, which in turn, within the context of gravity and rock type, influence the efficacy of mass wasting, erosion, and sediment transfers. Because tectonism and climate also vary over time, however, modern landscapes often reveal imprints of earlier formative episodes, palimpsests of past events that may or may not have modern analogs (Bloom, 1998).

1.4.1 Megageomorphology

The spatial link between tectonism and the shape of North America is best expressed along the eastern margin. Here, the plate boundary at the Mid-Atlantic Ridge is, despite asymmetric spreading and ridge jumps, reflected in the shape of the continent–ocean crustal boundary at the continental slope and in the shape of recent coastlines. These in turn reflect the structural lineations of the pre-Pangean Alleghanian, Acadian, Taconian, and Grenvillian fronts. Departure of the crustal boundary from Appalachian structural control occurs most notably where Gondwana's Florida terrane accreted to the continent and has since become the locus of widespread carbonate sedimentation. Similarly, owing to its later separation from Europe, the east coast of Greenland closely mimics the alignment of the Mid-Atlantic Ridge. The northern and southern margins of North America are more complex because of post-Pangean plate adjustments and orogenic activity. However, the western margin does reflect the repeated subduction and terrane accretion involved in forming the Western Cordillera, although much plate motion was angled toward the coast and plate linkages are now further confused by the oblique lateral shear of the Californias relative to the East Pacific Rise.

The spatial distribution of rock types, a direct or indirect reflection of tectonism, is important in terms of resistance to weathering and erosion. Of the major rock types exposed on land, North America comprises 31% metamorphic, 6% intrusive, 11% extrusive, and 52% sedimentary (Blatt and Jones, 1975). The high proportion of metamor-

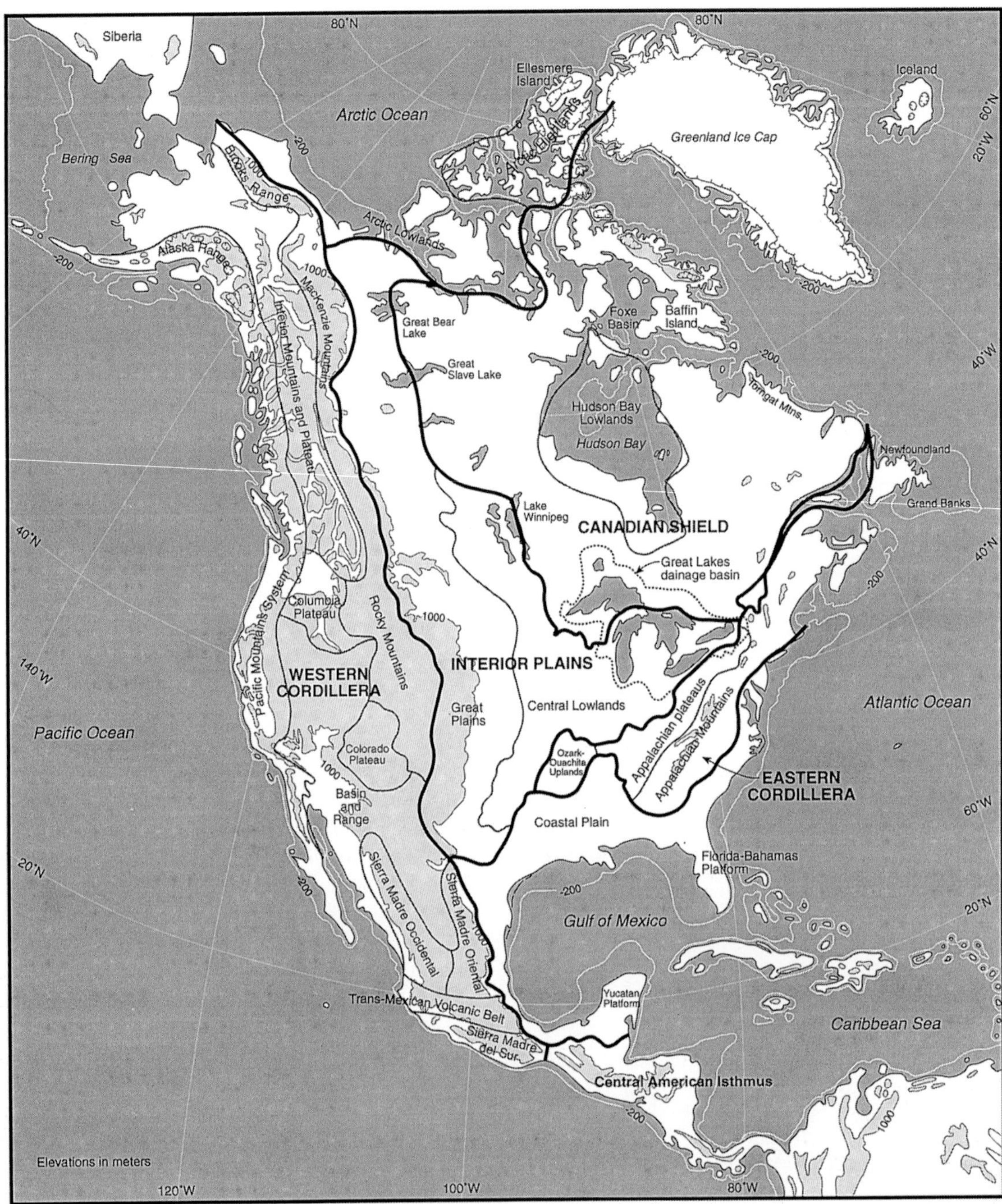

Figure 1.13 Geomorphic provinces of North America.

phic rocks compared to other continents (global average 17%) is due mainly to the exposure of Precambrian basement in the Canadian Shield. The high value for extrusive rocks (global average 8%) reflects widespread volcanism caused by successive plate subductions beneath the Western Cordillera. Intrusive and sedimentary outcrops are less that the global average (9% and 66%, respectively), but the latter are well represented by the Phanerozoic platform covers of the midcontinent.

Temporal links between orogenic events and present relief are often simple. The oldest exposed orogens, those of Precambrian age in the Canadian Shield, are mostly subdued. Paleozoic orogens are more prominent, notably the Appalachians, which reach 2037 m above sea level on Mt. Mitchell on the Blue Ridge; indeed the central Appalachians are divided into geomorphic provinces that reflect the nature and intensity of tectonic deformation and the rocks involved. The youngest orogens of late Mesozoic and

Cenozoic age in the Western Cordillera are spectacular, reaching 6194 m on Mt. McKinley in the Alaska Range, 6050 m on Mt. Logan in the St. Elias Mountains, and 4418 m on Mt. Whitney in the Sierra Nevada. Exceptions to this age–height relationship occur where older orogens and ancient rocks have been raised by later crustal movements. Thus the northeast margin of the Canadian Shield rises to 2491 m on Baffin Island and to 1652 m in the Torngat Mountains of Labrador, whereas Precambrian rocks are exposed in several western mountains, such as the Front Range of the Rockies, where Pikes Peak reaches 4301 m. Conversely, the little deformed Paleozoic cover rocks that have been raised into the Colorado Plateau are a notable exception to the intimate involvement of Phanerozoic sedimentary sequences in the intense orogenies of the Western Cordillera.

1.4.2 Tectonism, Denudation, and Sedimentation

Earlier in the twentieth century, attempts to explain landscape evolution were resolved into a number of models. Based on many antecedents, the Davisian model defined landforms in terms of geologic structure, geomorphic process, and time (Davis, 1899). It assumed rapid tectonic uplift followed by prolonged structural quiescence during which denudation reduced the landscape to a peneplain near sea level, the base level of subaerial erosion. Such peneplains could be later elevated by falling sea level or renewed uplift to form upland plains that were progressively incised by a new cycle of erosion. The Penckian model defined landforms in terms of the relationship between tectonic uplift and erosion (Penck, 1924). When uplift exceeded erosion, "waxing" landforms typified by convex slopes developed; when erosion exceeded uplift, "waning" landforms with concave slopes developed; when uplift and erosion were balanced, straight slopes emerged. Both models had their adherents, and the Davisian model was particularly popular in North America where its disciples believed ancient peneplains could be seen, notably in the Appalachians. It was later thought that widespread denudation could be achieved by Penckian slope retreat, the pediplain model, or by prolonged weathering and subsequent regolith removal, the etchplain model, concepts that found particular favor in the ancient landscapes of Gondwana (e.g., Wayland, 1933; King, 1962). Though much debated, all these models remained untested until the advent of quantitative data on rates of uplift and erosion.

The advent of plate tectonic models during the later twentieth century, aided by improved dating techniques and better understanding of geomorphic processes, led to abandonment of the Davisian model. Lengthy structural quiescence was incompatible with continuing, if variable, plate motion. Prolonged orogenic events and crustal instability between orogenies simply could not favor peneplain development. At best, the model could be considered an end member in a spectrum of possible scenarios. The Penckian model groped toward tectonic reality, but its details were difficult to vindicate in terms of observable geomorphic processes. Pediplains and etchplains, though sometimes invoked for the American west and the Canadian Shield, respectively, gained little credence in North America.

The continuing search for explanation in geomorphology can invoke a model in which changes in surface relief (ΔH), especially at the regional scale, are seen as a net response to processes induced by tectonism and climate over time (fig. 1.14). Thus;

$$\Delta H = \frac{[T_u + S] - [T_d + D] +/- [M + I + E + G]}{Z}$$

where T_u = tectonic uplift, S = sedimentation, T_d = tectonic subsidence, D = denudation, M = magmatism, including volcanism, I = isostatic adjustment, E = eustatic change, G = geoidal effect, and Z = time. This model implies that, over time, landforms are raised by orogenic or epeirogenic uplift and augmented by sedimentary covers, reduced by tectonic subsidence, mass wasting and erosion, and variably affected by deep-seated magmatism, surface volcanism, isostatic adjustments related to crustal loading and unloading, eustasy, which determines the relative importance of subaerial and marine processes, and geoidal effects related to the uneven distribution of mass beneath the surface. The linkages are complex and the timing uncertain. Orogenesis creates mountain ranges that load the lithosphere, which in turn adjusts isostatically. Meanwhile, mass wasting and erosion are favored by the potential and kinetic energy generated by high relief and steep slopes, respectively, modulated by vegetation type and cover, and lead to the erosional unloading of orogens. The transfer of small amounts of eroded waste to nearby basins may occur without isostatic compensation, but the massive transfer of waste from an orogen to a large sedimentary basin commonly triggers major crustal adjustments. Isostatic uplift in response to erosional unloading thus prolongs the time needed to lower landscapes. Over the millions of years needed for denudation, isostatic adjustments probably occur more or less continuously, dependent largely on the effective elastic thickness of the lithosphere (Gilchrist and Summerfield, 1991). Because the latter ranges from as little as ≈4 km in the Basin and Range Province to ≈128 km in the cratonic core (Bechtel et al., 1990), such effects are highly variable. At local—as distinct from regional—scales, landforms over the short term are of course shaped by mass wasting, erosion, and sedimentation without the need to invoke other forces.

The time factor in geomorphology has long been a focus of contention. Early attempts to resolve the role of time, by invoking catastrophes within a brief "biblical" timescale,

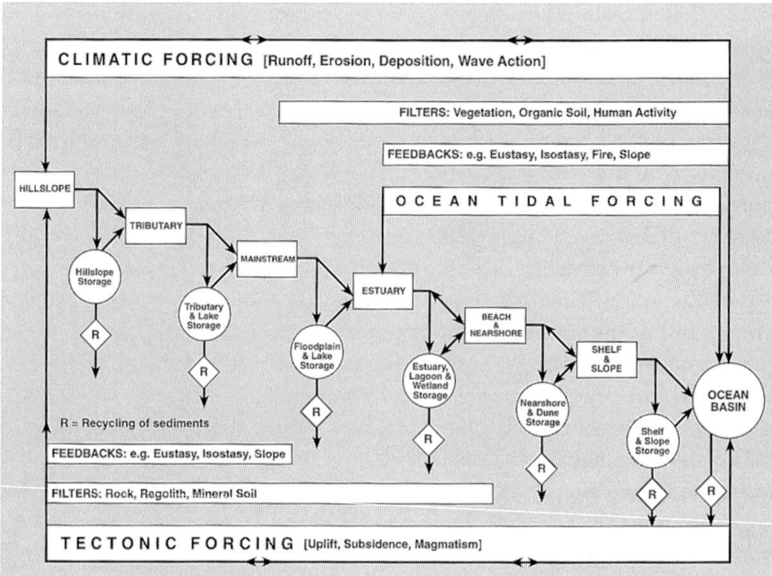

Figure 1.14 An erosional-depositional cascading system reflective of tectonic and climatic forcing under nonglacial conditions.

were later rejected in favor of more uniform development over a longer time interval. Later, rigid applications of uniformitarianism were modified by recognizing that geomorphic processes rarely operated gradually, except over brief intervals, and that rare events could indeed have significant impacts on landforms. Borrowing the concept of dynamic equilibrium from classical thermodynamics, many scientists sought to view landforms as seeking some type of balance between tectonic forces and geomorphic processes (e.g., Hack, 1960), allowing for rapid accelerations in geomorphic response when some threshold was crossed (e.g., Schumm, 1977). However, because tectonic activity, climate change, geomorphic processes, and isostatic adjustments to crustal loading and unloading function at different rates, it is unlikely that the landscape ever attains equilibrium between these forces. Further, because these rates also vary over time, it is difficult to infer long-term landscape changes from short-term measurements, such as those available from geodetic surveys and satellite altimetry. Nevertheless, over the longer term, tectonic activity appears to raise mountains far more quickly than denudation lowers them, ensuring, in association with isostatic adjustments, the persistence of quite old orogens, such as the Appalachians, in the present landscape.

To expand on this model, the landforms of western North America have been shaped predominantly during Cenozoic time by tectonic uplift, volcanism, and basin sedimentation; tectonic subsidence is restricted, and, although local erosion is often impressive, denudation is regionally variable and isostatic adjustment very complex. The net effect, therefore, has been to increase absolute relief, with some notable exceptions such as the Salton Trough and Death Valley. It may be further argued that, because valley erosion rates are often an order of magnitude faster than summit erosion rates for many ranges, relative relief within those ranges is presently increasing at about 100 m Ma^{-1} (Small and Anderson, 1998).

In contrast, over much of northern and eastern North America, the reverse is true. Prolonged Cenozoic denudation has mostly exceeded tectonic uplift to lower ancient orogens, the net effect of which is decreased relief only partly offset by isostatic adjustments and climatically induced valley incisions during Quaternary time. In between, in the Mississippi drainage basin, the Cenozoic has witnessed widespread sedimentation as the waste from adjacent cordilleras has been deposited, at least temporarily, en route to the Gulf of Mexico, but, with epeirogenic subsidence offsetting this vast basin fill, the net effect has been little change in relief. Give or take 100–200 m, the low relief of central North America relative to sea level has changed little during the Cenozoic. Sea level has, of course, itself changed in response to global tectonic and climatic forcing, the effect of which is well expressed in transgressive and regressive sedimentary sequences straddling the Gulf coast.

The Colorado Plateau is a dramatic example of relief formed where late Cenozoic uplift rates of ≈100 m Ma^{-1}, prompted by mantle plume activity, have exceeded all other variables, resulting in a structural tableland, which, despite later erosion, still rises 1500–3000 m above sea level. Responding to uplift, the reorganized Colorado River has gutted much of the plateau over the past 6 Ma to form the spectacular Grand Canyon (figs. 1.10, 1.15). In contrast, the Gulf Coastal Plain has maintained its low elevation over the past 60 Ma because, despite the accumulation of ≈6000 m of sediment, tectonic subsidence rates induced by thermal contraction and sediment loading have more or less kept pace with sedimentation rates and indeed favored continued deposition. The Atlantic Coastal Plain is more complex. In the south, the plain has risen epeirogenically and the continental shelf has subsided from sediment loading at rates of 10–30 m ka^{-1} during late Cenozoic time (Cronin, 1981). The hinge line between uplift and subsidence parallels the coast, and Cenozoic marine sediments have been raised and tilted seaward inland from the coast. In the north, however, this hinge line lies farther west, against the northern Appalachian orogen, and the thick post-Pangean sedimentary record lies submerged offshore beneath the Grand Banks (Steckler et al., 1988). Furthermore, although the Appalachians have been massively unroofed during the ≈300 Ma since the Alleghanian orogeny, the process of denudation is incomplete and recent surveys indicate that the region is rising at 6 m ka^{-1} relative to the Atlantic coast, probably related in part to isostatic adjustment. Finally, the surface of the Great Plains from Nebraska to Texas was formed mainly by Neogene sands and gravels, the Ogallala Formation, that buried earlier river valleys to depths of 150 m (Gustafson and Winkler, 1988). These deposits were generated by a combination of tectonic uplift farther west and climatic change, perhaps increasing aridity, which caused ancestral rivers to convey fluvial waste onto but not necessarily beyond the Great Plains. This sedimentary cover was later dissected in response to changing stream regimes for which continuing epeirogenic uplift and climate change can again be invoked.

Geomorphic relationships between tectonism and climate are also revealed by short-term measurements of sediment discharged to the ocean by the continent's rivers. Table 1.3, based on Milliman and Syvitski (1992), shows average annual sediment loads for the larger North American rivers, a selection of smaller ones and, for comparison, some great rivers elsewhere. Several inferences emerge from these data. First, much sediment is, or was before dam construction, discharged from a few major rivers such as the Mississippi and Colorado. Second, the largest sediment flux derives from rivers draining the Western Cordillera, the continent's active margin throughout Cenozoic time. Third, in terms of sediment yield (load divided by basin area), the wet mountains of the most active part of this margin discharge vast amounts of debris, remarkably so for such small basins as California's Eel River, British Columbia's Stekine River, and Alaska's Copper River. Fourth, in contrast, rivers draining the continent's passive margins yield much less sediment, despite a humid climate and abundant runoff. Some, such as the Apalachicola, drain modest uplands and run out over broad coastal plains. More striking, rivers such as the St. Lawrence, Saguenay, Moose, and Nottaway carry little sediment because they drain the resistant ice-scoured cratonic rocks of the Canadian Shield. In short, sediment yields correlate well with tectonism, through the medium of high relief and incompetent rocks, but less well with climate and runoff. In addition to major rivers, small basins along the active Pacific margin are also major conveyors of waste (Milliman and Syvitski, 1992).

Attempts to derive long-term denudation rates from short-term sediment and solute loads are more problematic. However, such studies have merit if they allow for the inherent variability of such records, the role of sediment

Figure 1.15 Monument Valley, Utah, showing residual mesas and buttes of Permian and Triassic sandstone on the Colorado Plateau (Spence Collection, UCLA).

Table 1.3 Sediment loads, calculated sediment yields, and runoff for selected rivers in North America and elsewhere

River	Basin Area 10^6 km^2	Sediment Load 10^6 t yr^{-1}	Sediment Yield t km^{-2} yr^{-1}	Runoff mm yr^{-1}
Mississippi	3.27	400 (210)	122	150
Mackenzie	1.81	42	23	170
St. Lawrence	1.03	4	4	435
Yukon	0.84	60	71	230
Rio Grande	0.67	20 (0.8)	30	5
Columbia	0.67	15 (10)	22	375
Colorado	0.63	120 (0.1)	190	32
Fraser	0.22	20	91	510
Brazos	0.11	16	145	65
Saguenay	0.078	0.4	5	
Nottaway	0.066	1	15	270
Moose	0.06	0.4	7	410
Copper	0.06	70	1167	650
Apalachicola	0.044	0.17	4	470
Skeena	0.042	11	262	690
Potomac	0.025	0.72	29	310
Stekine	0.018	20	1111	690
Eel	0.008	14	1750	915
Amazon	6.15	1200	195	1024
Congo	3.82	43	11	340
Nile	2.96	120 (0)	41	30
Paraná	2.60	79	30	165
Yenesei	2.58	13	5	220
Ganges	0.98	520	531	373
Huang He	0.77	1100	1429	77

Source: Selected from Milliman and Syvitski, 1992.

[a]Sediment loads used for compiling sediment yields are pre-dam values. Post-dam sediment loads are shown in parentheses where available.

storage in floodplains, the inflationary impacts of human land use, and the effect of dams on recent data. As early as 1909, Dole and Stabler concluded that North America was being denuded at a mean rate of ≈33 mm ka^{-1} and stressed the contrast between the dissolved and solid load carried by various rivers. Table 1.4 is a synthesis of more recent attempts to address this complex issue. Tectonic uplift, rock erodibility, climatic forcing, and vegetation cover are all reflected in these data. Total denudation aside, the high ratio of chemical denudation seen in dissolved loads coming off the Canadian Shield contrasts markedly with the high ratio of mechanical denudation and low dissolved loads from the Western Cordillera. The denudation rate for the semiarid Colorado basin, little protected by vegetation, is particularly noteworthy (fig. 1.16). Denudation rates for the Eel and Mad basins in northern California suggest that their landscapes are unraveling from erosion as swiftly as they are being raised by tectonic forces.

1.5 Tectonism and Biogeography

Because biomes are closely linked to climate, much that has been said regarding tectonism and climate also applies to biogeography. The tectonic forces that brought North America to its present location, determined its shape, and influenced its climate also affected the nature and distribution of its plants and animals. There is thus a broad correlation among terrain location, climate, and biome. But there are also significant differences. Plants and animals exhibit preferences for particular habitats that may become inaccessible because of tectonic or climatic change. Further, through time, direct correlation of biota with climate may be clouded by evolution, opportunism, competition, extinction, lagging response to climate and sea-level fluctuations, or simply by chance. And although land plants tend to be more responsive to environmental change, land mammals, being mobile, are more adaptable. Thus it can-

Table 1.4 Estimated denudation rates for selected drainage basins in North America and elsewhere

River	Basin Area 10^6 km^2	Total Denudation mm ka^{-1}	Mechanical Denudation mm ka^{-1}	Chemical Denudation mm ka^{-1}	Mechanical Denudation % of Total
Mississippi	3.27	44	35	9	80
Mackenzie	1.81	30	20	10	67
St. Lawrence	1.03	13	1	12	8
Yukon	0.84	37	27	10	73
Rio Grande	0.67	9	6	3	67
Columbia	0.67	29	16	13	55
Colorado	0.63	84	78	6	93
Amazon	6.15	70	57	13	81
Congo	3.82	7	4	3	57
Nile	2.96	15	13	2	87
Paraná	2.60	19	14	5	74
Yenisei	2.58	9	2	7	22
Ganges	0.98	271	249	22	92
Huang He	0.77	529	518	11	98

Sources: Meybeck, 1976; Summerfield, 1991; Milliman and Syvitski, 1992.

Figure 1.16 The Goosenecks of the San Juan River, Utah, are incised into Carboniferous (Pennsylvanian) marine limestone, sandstone, and shale beneath the surface of the Colorado Plateau (Spence Collection, UCLA).

not be assumed that tectonism will generate a predictable temporal sequence and spatial distribution of plants and animals in quite the same way that climate responds to tectonic change. This discussion now examines the relationship between tectonism and biogeography over the past 180 Ma, with a brief preface to the modern plant and animal patterns evaluated in chapters 11 and 12.

1.5.1 Evolving Biogeographies

The breakup of Pangea initiated a prolonged episode of biogeographic fragmentation set against a backcloth of continuing biological evolution, diversification, and extinction. Several tectonic events conspired to influence the changing biogeography of North America and set the stage for present patterns. First, the rupture of Pangea around 180 Ma favored evolutionary divergence between Laurussia and Gondwana. Second, within Laurussia, the much later separation of North America from Europe ensured some continuity across the northern landmass until the early Cenozoic, continuity that was later reestablished as North America linked with eastern Asia across Beringia. Third, while these separations progressed, North America continued to drift away from the tropics, setting the scene for the mostly temperate biomes of today. Fourth, the emergence of the Central America isthmus in the Pliocene re-established links with part of Gondwana, favoring renewed exchanges of terrestrial biota between North and South America, even as Atlantic and Pacific marine biota were separated. Fifth, the episodic liberation and sequestration of massive amounts of CO_2, related in part to tectonism, together with many chance events, created challenges for all forms of life.

Although the Mesozoic flora and fauna of North America have little relevance to the modern landscape, two events will serve to illustrate the impact of tectonism during that time. First, the initial rupture of Pangea may have triggered the faunal extinctions that closed the Paleozoic era. One hypothesis for these extinctions postulates that the ocean surrounding Pangea was meromictic—it did not mix seasonally but remained stratified for long periods because of solute density and was oxygenated only at the surface (Knoll et al., 1996). When Pangea broke apart, this world ocean may have turned over, liberating massive amounts of CO_2 and H_2S, and resulting in widespread extinction of marine species. Second, onshore, Laurussia's separation from Gondwana resulted in the evolution of distinct dinosaur faunas on each landmass during the later Mesozoic, with tyrannosaurs, hadrosaurs, and ceratopsids dominating Laurussia while titanosaurs and ceratosaurs developed in Gondwana (Sereno, 1991). Many causes, including tectonic and volcanic forcing, have been invoked to explain the rapid demise of dinosaurs and other organisms at the Cretaceous-Tertiary boundary (Officer and Page, 1996). One hypothesis that ignores tectonism invokes global climate change caused by a massive asteroid impact around 65 Ma that formed the Chicxulub crater in Yucatan. The widespread diffusion of impact ejecta (dust, soot, CO_2, SO_2, H_2O) may have shut down photosynthesis for months and reduced insolation for decades (Alvarez et al., 1980).

Whatever the cause, the demise of the dinosaurs undoubtedly paved the way for the expansion and evolution of mammals during the Cenozoic. This expansion was favored by early Cenozoic warmth, which promoted climatic equability, opened high-latitude corridors to mammal migrations across the Northern Hemisphere, and favored the evolution of new orders, such as artiodactyls, perissodactyls, and primates that, from the Eocene to the present, came to dominate North American landscapes (Clyde and Gingerich, 1998).

The late Mesozoic flora of North America was dominated by gymnosperms, mostly conifers and cycads, which flourished under mostly warm subtropical conditions. However, even as gymnosperms gave way to angiosperms during the Cretaceous, orogenic events were raising the Western Cordillera, and the continent continued to shift poleward. Thus Cenozoic plant life evolved against a climatic background that became cooler, locally drier, and certainly more variable as the Cenozoic progressed. A few representative floras from the geologic record reveal these changes (Tidwell, 1975). For late Paleocene time, the coal-bearing Fort Union Formation of the Dakotas, Montana, and Wyoming mixes gymnosperms (conifers, cycads, and ginkos) with angiosperms (palms, willow, maple, ash, and *Arctocarpus* or breadfruit) indicative of swampy subtropical lowlands and drier hillsides, which also supported a rich mammalian fauna (Wing et al., 1991). Tree ferns, cycads, and palms, plants incapable of withstanding prolonged freezes, occur in western assemblages around this time. Eocene floras reveal conditions ranging from warm, humid subtropical (e.g., the lacustrine Green River Formation, which covered 130,000 km^2 of Wyoming; Wilf, 2000) to cool temperate at higher elevations (e.g., Copper Basin, Nevada). Conditions were, however, changing. Paleosols from South Dakota badlands indicate a change from dense rainforest to more open forest and scrub across the Eocene-Oligocene transition (Retallack, 1983). The Oligocene Florissant beds of Colorado contain over 100 plant species indicative of a warm subtropical lowland to temperate upland climate, including a dominant oak (*Fagopsis longifolia*), now extinct but reflective of expanding broadleaf deciduous forest, and genera such as *Zelkova* (keaki tree) and *Ailanthus* (tree of heaven) no longer native to North America (MacGinitie, 1953).

The onset of global cooling in the Miocene and, in the west, of drier conditions induced by Cordilleran uplift, is reflected in many biota. Temperate forest dominated by *Sequoia* now occupied much of the Great Basin; elsewhere, local conditions became more important, for example, in the baldcypress (*Taxodium*) swamp represented by Oregon's Mascall flora. Herbaceous angiosperms expanded under the drier conditions induced by the rising Sierra Nevada and Cascade Range, and sagebrush, *Artemesia tridentata*, appeared in the late Miocene, eventually to dominate the temperate western deserts. Similarly, uplift was instrumental in transforming the late Pliocene savanna of California's Tecopa Basin into the desert of today (Hillhouse, 1987). Cenozoic uplift also led to increased provincialism and divergent evolution between the Western Cordillera and the plains farther east. The late Miocene fauna of the Dove Spring Formation (≈14–7 Ma) in the Mojave Desert thus reveals the first appearance of some restricted species, notably shoveltusk gomphotheres and horses quite distinct from their cousins on the Great Plains, and the last appearance of older, more widespread assemblages (Whistler and Burbank, 1992). Miocene cooling and Cordilleran uplift have also been invoked to explain the late Cenozoic retreat of crocodilians from the increasingly dry northern High Plains to the Atlantic and Gulf coastal plains (Markwick, 1994). Meanwhile, with the rapid evolution and spread of grasses during later Cenozoic time, especially after 7 Ma, much of the lowland vegetation west of the Mississippi had changed to grassland, with poplar, sycamore, and willow growing alongside streams, deciduous savannas at intermediate elevations, and drought-resistant conifers on higher mountains. In time, herbivorous mammals such as the horse (*Equus*) and camel (*Camelops*) developed high-crowned teeth in response to their expanding grass diet.

Once widely invoked to explain similar but disjunct biotic distributions between continents, the concepts of land bridges and island-hopping were given new meaning by the plate tectonic revolution. Land bridges are far more important for the migration of land animals than for plants, for whom seed dispersal over narrow seas is entirely feasible. Conversely, their absence between continents facilitates exchanges of marine biota through ocean gateways. For the animals that came to dominate the late Cenozoic landscape, the links with South America and Eurasia were especially important.

By late Miocene time, Central America existed as a complex island-arc archipelago. Three principal corridors linked Pacific and Caribbean waters: the San Carlos Strait across Nicaragua, the Panama Strait, and the Atrato Strait across northwest Colombia (Coates and Obando, 1996). These straits varied in size from ≈25 Ma to ≈3 Ma, after which the Central America isthmus was fully established. These fluctuations are seen in marine sediment in Panama where benthic foraminifera in the late Miocene Gatun Formation (≈12–8 Ma) show strong Caribbean affinity, whereas those in the overlying Chagres Formation (≈8–5 Ma) have a strong Pacific affinity (Collins et al., 1996). After ≈3 Ma, the isthmus allowed continuing biotic interchanges between the Americas but prevented exchange of marine animals between the Atlantic and Pacific Oceans (Stehli and Webb, 1985). For example, the isthmus now limited the migration of pinniped seals (Phocidae) that had evolved in the Atlantic realm and had only begun reaching the north Pacific realm in late Pliocene time (Barnes and Goedert, 1996). In contrast, the land bridge afforded the opportunity for North American land mammals to explode southward into South America where they began displacing the long resident fauna. The northward migration of South American land mammals, accustomed to mostly tropical habitats, was less emphatic, restricted probably by cooler conditions beyond southern Mexico. Of those that migrated and survived, an opossum (*Didelphis marsupialis*) and a porcupine (*Erethizon dorsatum*) became widely dispersed in North America, and an armadillo (*Dasypus novemcinctus*) inflated its intestines to cross rivers to reach Missouri and Florida, but various ground sloths (Edentata), despite reaching northwest Canada, did not survive the Pleistocene (see table 3.3).

Cenozoic links between North America and Eurasia have been quite different in character and complexity. Here the initial rupture of Pangea left the two continents more or less attached until the final "unzipping" of the North Atlantic in early Cenozoic time. Even then links across the Arctic may have persisted while Beringia was forming from the lodgment of the North American plate against eastern Siberia, at least until the initial opening of the Bering Strait around 5 Ma by a combination of tectonic events and eustatic sea-level rise (Marincovich and Glodenkov, 1999). Many Eurasian animal immigrants thus figure strongly alongside indigenous evolution in defining the 20 Cenozoic land mammal ages recognized in North America. Indeed, over the past 30 Ma, widespread overland intercontinental dispersals between Eurasia and North America have been a prominent, if episodic, feature of the faunal record (Woodburne, 1987). Faunal exchanges were particularly favored by low sea levels, such as those during Pleistocene glacial maxima when a continental shelf at least 1000 km wide was exposed across the Bering and Chukchi Seas. Even with high interglacial seas, such as today, the two continents were separated by a shallow Bering Strait less than 50 km wide. Thus it is thought that the westward migration of the three-toed hipparion horse from North America to Asia, one of few pre-Pliocene overland migrations in that direction, was favored in late Miocene time (≈11–9 Ma) by low glacioeustatic seas associated with global cooling and Antarctic ice expansion (Garces et al., 1997). More significant to the present landscape were the late Cenozoic migrations eastward from Asia. Noteworthy among the larger mammal immigrants were cat (*Felis*) around 6 Ma, bear (*Ursus*) around 4 Ma, mammoth (*Mammuthus*), and saber-toothed cat (*Smilodon*) around 2 Ma, and bison (*Bison*) sometime after 0.3 Ma (Woodburne, 1987). Some of these were to evolve into modern forms; some became extinct (see table 3.3). Ironically, some genera that evolved indigenously in North America, such as horse and camel, which appeared around 4 Ma, later became extinct within the continent, but not before their relatives had migrated to Eurasia. Finally, the Bering land bridge afforded a favored route for the immigration of that most predatory of mammals, *Homo*!

1.5.2 Present biogeographic patterns

Under the changing conditions of later Cenozoic time, evolution, migration, and extinction thus combined to give the North American landscape the primeval flora and fauna that people encountered when they reached the continent. Influenced by Cenozoic tectonic and climatic events, including sea-level changes, the flora and fauna of North America became distinctive but rarely unique. At the continental scale, the geographical affinities and evolutionary origins of these plants and animals define the Nearctic realm, as distinct from the Neotropical realm of Central and South America whose biota have made only limited incursions farther north. Because of prolonged biotic exchange across the Northern Hemisphere, the Nearctic is often linked in turn with the Palearctic of Eurasia into a Holarctic realm spanning the temperate and polar environments of the Northern Hemisphere. At lesser scales, the present distribution of plants and animals continues to reflect the legacies of tectonic and climate forcing, for example, in regional contrasts among tundra, forest, prairie, and desert, and in the more localized zonation of vegetation with elevation and water availability.

1.6 Conclusion

The fundamental character of the North American landscape is driven by tectonic and climatic forcing, the earth beneath and the sky above. Tectonic activity has provided the continent with its location and general architecture. Climate has generated the contexts whereby this crude tectonic framework has been refashioned by mass wasting, running water, ice, and wind. Abetted by evolution, migration, and extinction, biological responses to these physical scenarios have in turn clothed this landscape with vegetation and colonized it with animals.

Today, as in the past, the North American landscape is subject to continuing change. Some of these changes are so slow as to be imperceptible to the human eye. Thus plate motions and denudation rates of a few millimeters to a few centimeters a year mean little to the general population. On the other hand, embedded within these imperceptible changes are a variety of accelerations, of thresholds reached and strains released, which serve as reminders of more profound physical processes.

Earthquakes along the continent's Pacific Rim, such as the Magnitude (M) 8.3 San Francisco earthquake of 1906, which caused surface ruptures along 435 km of the San Andreas fault (Lawson, 1908), and the M9.2 Anchorage earthquake of 1964, which deformed 400,000 km^2 of southern Alaska (Plafker, 1965), are instances of the sudden release of tectonic stresses that accumulate along the active margin. Dislocated coastal wetlands and tsunami deposits from British Columbia to California show that great earthquakes (M > 8) recur every few hundred years in the Cascadia subduction zone (Clague and Bobrowsky, 1994; Atwater, 1996). Although such events have not occurred recently, there is evidence for a great earthquake (M9?) in Puget Sound around A.D. 1700, for 7 m of offset on the Seattle fault 1000 years ago (Bucknam et al., 1992), and for 150 m of uplift on Whidbey Island over the past 250 ka (Johnson et al., 1996). The M6.7 Northridge earthquake that occurred on a buried thrust fault in California in 1994 emphasized the uncertainty implicit in predicting tectonic activity. In a related vein, the 40 or so active volcanoes in the Alaska Peninsula, where Mt. Katmai disgorged 28 km^3 of tephra in 1912 (Hildreth and Fierstein, 2000), and the eruption of Mount St. Helens in 1980 (Lipman and Crandell, 1981) (fig. 1.17) are but reminders of the many late Cenozoic volcanic events that have not only shaped many western landscapes but which, through their dust veils, may have affected regional climates.

Even the continent's relatively passive interior and eastern margins are not immune from tectonic activity. The M8.7? New Madrid earthquakes of 1811–12 in the midcontinent disrupted drainage in the central Mississippi valley, deformed Reelfoot Lake, and stimulated the growth of baldcypress (Johnson and Schweig, 1996). Crustal adjustments to shifting stresses also generate occasional earthquakes along the St. Lawrence Valley where, in 1663, perhaps the largest earthquake yet witnessed in eastern North America caused 3 km^3 of Holocene sediment to collapse into Saguenay Fjord, Québec (Syvitski and Schafer, 1997). Earthquakes also occur along the east coast from the Carolinas, where a M6.6 earthquake shook Charleston in 1886, through the Grand Banks, to Baffin Bay where a M7.3 event in 1933 was the largest event yet recorded in Arctic Canada (*National Atlas of Canada*, 1996).

In climatic terms, North America is virtually unique among continents in the high frequency and magnitude of tornado activity. Why? Because the continent's tectonic geometry places a high Western Cordillera to windward of a lowland interior exposed to extreme inputs of arctic and tropical air. Thus, in spring months in particular, cold, dry air, depleted of moisture in crossing the Cordillera, meets warm, moist air from the Gulf of Mexico along a broad front, "tornado alley," extending diagonally from the southern Great Plains to the St. Lawrence Valley, often with devastating consequences. Further, although not unique, the placement of the continent's southeast quadrant west of a warm subtropical ocean generates optimal conditions for late summer and autumn hurricanes to move onshore, bringing coastal storm surges and inland floods from

Figure 1.17 Eruption of Mount St. Helens, Washington, 18 May 1980. This view toward the northeast shows the vertical ash column, spreading dust veil, pyroclastic flows, and melting snow and ice (U.S. Geological Survey).

Mexico to Canada. Other flood events typically reflect hydrologic responses to climatic forcing in a tectonic setting. Thus spring snowmelt in the Rocky Mountains generates flood peaks that move progressively downstream through the Mississippi system, augmenting flood hazards across the lowland interior. Summer thunderstorms over high tectonic relief farther west cause flash floods to descend with great force into nearby desert basins.

Thus, tectonism and climate conspire to create in North America a continent of vast grandeur and remarkable extremes. From high mountains to vast plains, from arctic tundra to coral reefs, from majestic forests to prairie grasslands and desert scrub, the continent is a collage of landscapes that owes much to past tectonic events and past climates, even as these landscapes are being reshaped today by contemporary processes and human impacts.

References

Ague, J.J., and M.T. Brandon, 1996. Regional tilt of the Mount Stuart batholith, Washington, determined using aluminum-in-hornblende barometry: Implications for northward translation of Baja British Columbia. *Geological Society of America Bulletin*, 108, 471–488.

Alvarez, L.W., W. Alvarez, F. Asaro, and H.V. Michel, 1980. Extraterrestrial cause for the Cretaceous-Tertiary extinction. *Science*, 208, 1095–1108.

Atwater, B.F., 1996. Coastal evidence for great earthquakes in western Washington. *U.S. Geological Survey Professional Paper*, 1560, 77–90.

Bally, A.W., 1989. Phanerozoic basins of North America. In: A.W. Bally and A.R. Palmer (Editors), *The geology of North America—An overview*. Geological Society of America, Boulder, 397–446.

Barnes, L.G., and J.L. Goedert, 1996. Marine vertebrate

paleontology on the Olympic Peninsula, Washington. *Geology*, 24, 17–25.
Bechtel, T.D., D.W. Forsyth, V.L. Sharpton, and R.A.F. Grieve, 1990. Variations in effective elastic thickness of the North American lithosphere. *Nature*, 343, 636–638.
Berggren, W.A., and D.R. Prothero (Editors), 1992. *Eocene-Oligocene climatic and biotic evolution.* Princeton University Press, Princeton, New Jersey.
Berner, R.A., 1990. A model for atmospheric CO_2 over Phanerozoic time. *American Journal of Science*, 291, 339–376.
Blatt, H., and R.L. Jones, 1975. Proportions of exposed igneous, metamorphic, and sedimentary rocks, *Geological Society of America Bulletin*, 86, 1085–1088.
Bloom, A.L., 1998. *Geomorphology: A systematic analysis of late Cenozoic landforms*, 3rd edition. Prentice Hall, Upper Saddle River, New Jersey.
Bowring, S.A., I.S. Williams, and W. Compston, 1989. 3.96 Ga gneisses from the Slave Province, Northwest Territories, Canada. *Geology*, 17, 971–975.
Bucknam, R.C., E. Hemphill-Haley, and E.B. Leopold, 1992. Abrupt uplift within the past 1,700 years at southern Puget Sound, Washington. *Science*, 258, 1611–1614.
Buschbach, T.C., and D.R. Kolata, 1991. Regional setting of Illinois Basin. In: M.W. Leighton, D.R. Kolata, D.F. Oldtz, and J.J. Eidel (Editors), Interior cratonic basins. *American Association of Petroleum Geologists Memoir 51*, 29–55.
Caldeira, K., 1992. Enhanced Cenozoic chemical weathering and the subduction of pelagic carbonate. *Nature*, 357, 578–581.
Cebull, S.E., D.H. Shurbet, G.R. Keller, and L.R. Russell, 1976. Possible role of transform faults in the development of apparent offsets in the Ouachita-southern Appa-lachian tectonic belt. *Journal of Geology*, 84, 107–114.
Clague, J.J., and P.J. Bobrowsky, 1994. Tsunami deposits beneath tidal marshes on Vancouver Island, British Columbia. *Geological Society of America Bulletin*, 106, 1293–1303.
Clyde, W.C., and P.D. Gingerich, 1998. Mammalian community response to the latest Paleocene thermal maximum: An isotaphonomic study in the northern Bighorn Basin, Wyoming. *Geology*, 26, 1011–1014.
Coates, A.G., and J.A. Obando, 1996. The geologic evolution of the Central American isthmus. In: J.B.C. Jackson et al. (Editors), *Evolution and environment in tropical America.* University of Chicago Press, Chicago.
Collins, L.S., A.G. Coates, W.A. Berggren, M.P. Aubry, and J. Zhang, 1996. The late Miocene Panama isthmian strait. *Geology*, 24, 687–690.
Cronin, T.M., 1981. Rates and possible causes of neotectonic vertical crustal movements of the emergent southeastern United States Atlantic Coastal Plain. *Geological Society of America Bulletin*, 92, 812–833.
Crowley, T.J., and G.R. North, 1991. *Paleoclimatology*, Oxford University Press, New York.
Dalziel, I.W.D., 1997. Neoproterozoic-Paleozoic geography and tectonics: Review, hypothesis, environmental speculation. *Geological Society of America Bulletin*, 109, 16–42.
Dalziel, I.W.D., L.H. Dalla Salda, and L.M Gahagan, 1994. Paleozoic Laurentia-Gondwana interaction and the origin of the Appalachian-Andean mountain system. *Geological Society of America Bulletin*, 106, 243–252.
Davis, W.M., 1899. The Geographical Cycle. *Geographical Journal*, 14, 481–504.
DeMets, C., 1995. A reappraisal of seafloor spreading lineations in the Gulf of California: Implications for the transfer of Baja California to the Pacific plate and estimates of Pacific-North America motion. *Geophysical Research Letters*, 22, 3545–3548.
DeMets, C., R.G. Gordon, D.F. Argus, and S. Stein, 1990. Current plate motions. *Geophysical Journal International*, 101, 425–478.
Dole, R.B., and H. Stabler, 1909. Denudation. *U.S. Geological Survey Water Supply Paper*, 234, 78–93.
Ducea, M.N., and J.B. Saleeby, 1996. Buoyancy sources for a large, unrooted mountain range, the Sierra Nevada, California: Evidence from xenolith thermobarometry. *Journal of Geophysical Research*, 101, 8229–8244.
Dumitru, T.A., I.R. Duddy, and P.F. Green, 1994. Mesozoic-Cenozoic burial, uplift, and erosion history of the west-central Colorado Plateau. *Geology*, 2, 499–502.
Eyles, N., 1996. Passive margin uplift around the North Atlantic region and its role in Northern Hemisphere late Cenozoic glaciation. *Geology*, 24, 103–106.
Garces, M., L. Cabrera, J. Agusti, and J.M. Parés, 1997. Old World first appearance datum of "Hipparion" horses: Late Miocene large mammal dispersal and global events. *Geology*, 25, 19–22.
Gibson, T.G., L.M. Bybell, and J.P. Owens, 1993. Late Paleocene lithologic and biotic events in neritic deposits of southern New Jersey. *Paleo-Oceanography*, 8, 495–514.
Gilchrist, A.R., and M.A. Summerfield, 1991. Denudation, isostasy and landscape evolution. *Earth Surface Processes and Landforms*, 16, 555–562.
Guaffanti, M., and C.S. Weaver, 1988. Distribution of late Cenozoic volcanic vents in the Cascade Range: Volcanic arc segmentation and regional tectonic considerations. *Journal of Geophysical Research*, 93, 6513–6529.
Gustafson, T.C., and D.A. Winkler, 1988. Depositional facies of the Miocene-Pliocene Ogallala Formation, northwestern Texas and eastern New Mexico. *Geology*, 16, 203–206.
Hack, J.T., 1960. Interpretation of erosional topography in humid temperate regions. *American Journal of Science*, 258–A, 80–97.
Haq, B.U., J. Handenbol, and P.R. Vail, 1987. The chronology of fluctuating sea levels since the Triassic (250 million years ago to present). *Science*, 235, 1156–1167.
Hatcher, R.D., T.A. Thomas, and G.W. Viele (Editors), 1989. *The Appalachian-Ouachita orogen in the United States.* The Geology of North America, F-2, Geological Society of America, Boulder.
Hildreth, W., and J. Fierstein, 2000. Katmai volcanic cluster and the great eruption of 1912. *Geological Society of American Bulletin*, 112, 1594–1620.
Hillhouse, J.W., 1987. *Late Tertiary and Quaternary geology of the Tecopa Basin, southeastern California.* U.S. Geological Survey, Miscellaneous Investigations, Map I-1728.
Hoffman, P.F., 1988. United plates of America, the birth of a craton: Early Proterozoic assembly and growth of Laurentia. *Annual Review of Earth and Planetary Sciences*, 16, 543–603.
Hoffman, P.F., 1989. Precambrian geology and tectonic

history of North America. In: A.W. Bally and A.R. Palmer (Editors), *The geology of North America—An overview*. Geological Society of America, Boulder, 447–512.

Hooper, P.R., 1988. The Columbia River basalt. In: J.D. McDougall (Editor), *Continental flood basalts*. Kluwer Academic Publishers, Dordrecht, 331–341.

Howard, J.L., 1996. Paleocene to Holocene paleodeltas of ancestral Colorado River offset by the San Andreas fault system, southern California. *Geology*, 24, 783–786.

Howell, P.D., and B.A. van der Pluijm, 1990. Early history of the Michigan Basin: Subsidence and Appalachian tectonics. *Geology*, 18, 1195–1198.

Huber, N.K., 1981. Amount and timing of late Cenozoic uplift of the Sierra Nevada—Evidence from the upper San Joaquin River basin. *U.S. Geological Survey Professional Paper*, 1197, 28 p.

Isozaki, Y., and M.C. Blake, 1994. Biostratigraphic constraints on formation and timing of accretion in a subduction complex: An example from the Franciscan complex of northern California. *Journal of Geology*, 102, 283–296.

Johnson, A.C., and E.S. Schweig, 1996. The enigma of the New Madrid earthquakes of 1811–1812. *Annual Review of Earth and Planetary Sciences*, 24, 339–384.

Johnson, S.Y., C.J. Potter, J.M. Armentrout, J.J. Miller, C. Finn, and C.S. Weaver, 1996. The southern Whidbey Island fault: An active structure in the Puget Lowland, Washington. *Geological Society of America Bulletin*, 108, 334–354.

King, L.C., 1962. *The morphology of the Earth*. Oliver & Boyd, Edinburgh.

Knoll, A.H., R.K. Bambach, D.E. Canfield, and J.P. Grotzinger, 1996. Comparative earth history and late Permian mass extinction. *Science*, 273, 452–457.

Kossinna, E., 1933. Die Erdoberflache. In: B. Gutenberg (Editor), *Handbuch der Geophysik*, 2, Berlin, Borntraeger, 869–954.

Lafon, C.W., D.Y. Graybeal, and K.H. Orvis, 1999. Patterns of ice accumulation and forest disturbance during two ice storms in southwestern Virginia. *Physical Geography*, 20, 97–115.

Lawson, A.C., 1908. *The California earthquake of April 18, 1906*. Carnegie Institution of Washington, Report of the State Earthquake Investigation Commission, 451 p.

Lipman, P., and D.R. Crandell, 1981. The eruptive history of Mount St. Helens. In: P. Lipman and D.R. Mullineaux (Editors), The 1980 eruption of Mount St. Helens, Washington, *U.S. Geological Survey Professional Paper* 1250, 3–15.

Lowe, D.R., 1985. Ouachita trough: Part of a Cambrian failed rift system. *Geology*, 13, 790–793.

MacGinitie, H.D., 1953. The fossil plants of the Florissant beds. Carnegie Institution, Washington, 599, 1–198.

Mahoney, J.B., P.S. Mustard, J.W. Haggart, R.M. Friedman, C.M. Fanning, and V.J. McNicholl, 1999. Archean zircons in Cretaceous strata of the western Canadian Cordillera: The "Baja-B.C." hypothesis fails a crucial test. *Geology*, 27, 195–198.

Marincovich, L., and A.Y. Gladenkov, 1999. An early opening of the Bering Strait. *Nature*, 397, 149–151.

Markwick, P.J., 1994. Equability, continentality, and Tertiary "climate": The crocodilian perspective. *Geology*, 22, 613–616.

Matthews, V., 1976. Correlation of Pinnacles and Neenach volcanic formations and their bearing on San Andreas fault problem. *American Association of Petroleum Geologists Bulletin*, 60, 2128–2141.

McHone, J.G., 1996. Broad-terrane Jurassic flood basalts across northeastern North America. *Geology*, 24, 319–322.

Meybeck, M., 1976. Total annual dissolved transport by world major rivers. *Hydrological Sciences Bulletin*, 21, 265–289.

Mikolajewicz, U., E. Maier-Reimer, T.J. Crowley, and K-Y. Kim, 1993. Effect of Drake and Panamanian gateways on the circulation of an ocean model. *Palaeoceanography*, 8, 409–426.

Milliman, J.D., and J.P.M. Syvitski, 1992. Geomorphic/tectonic control of sediment discharge to the ocean: The importance of small mountainous rivers. *Journal of Geology*, 100, 525–544.

Molnar, P., and P. England, 1990. Late Cenozoic uplift of mountain ranges and global climate change: Chicken or egg? *Nature*, 346, 29–34.

Moores, E.M., 1991. Southwest U.S.–East Antarctica (SWEAT) connection: A hypothesis. *Geology*, 19, 425–428.

National Atlas of Canada, Ottawa, 1996.

Nicholas, N.S., and S.M. Zedaker, 1989. Ice damage in spruce-fir forests of the Black Mountains, North Carolina. *Canadian Journal of Forest Research*, 19, 1487–1491.

Nicholson, C., C.C. Sorlien, T. Atwater, J.C.Crowell, and B.P.Luyendyk, 1994. Microplate capture, rotation of the western Transverse Ranges, and initiation of the San Andreas transform as a low-angle fault system. *Geology*, 22, 491–495.

Nieto-Samaniego, A.F., L. Ferrari, S.A. Alaniz-Alvarez, G. Labarthe-Hernandez, and J. Rosas-Elguera, 1999. Variation of Cenozoic extension and volcanism across the southern Sierra Madre Occidental volcanic provice, Mexico. *Geological Society of America Bulletin*, 111, 347–363.

Norris, R.D., L.S. Jones, K.M. Corfield, and J.E. Cartlidge, 1996. Skiing in the Eocene Uinta Mountains? Isotope evidence in the Green River Formation for snow melt and large mountains. *Geology*, 24, 403–406.

O'Connell, S., M.A. Chandler, and R. Ruedy, 1996. Implications for the creation of warm saline deep water: Late Paleocene reconstructions and global climate model simulations. *Geological Society of America Bulletin*, 108, 270–284.

Officer, C.B., and J. Page, 1996. *The great dinosaur extinction controversy*. Addison-Wesley-Longman, Reading, 209 pp.

Oldow, J.S., A.W.Bally, H.G.A.Lallemant, and W.P. Leeman, 1989. Phanerozoic evolution of the North American Cordillera; United States and Canada. In: A.W. Bally and A.R. Palmer (Editors), *The Geology of North America—An Overview*, Geological Society of America, Boulder, 139–232.

Olsen, P.E., D.V. Kent, B. Cornet, W.K. Witte, and R.W. Schlische, 1996. High resolution stratigraphy of the Newark rift basin (early Mesozoic, eastern North America), *Geological Society of America Bulletin*, 108, 40–77.

Opdyke, N.D., D.S. Jones, B.J. MacFadden, D.L. Smith, P.A. Mueller, and R.D. Shuster, 1987. Florida as exotic terrane: Paleomagnetic and geochronologic investiga-

tion of lower Paleozoic rocks from the subsurface of Florida. *Geology*, 15, 900–903.
Orme, A.R., 1980. Marine terraces and Quaternary tectonism, northwest Baja California, Mexico. *Physical Geography*, 1, 138–161.
Orme, A.R., 1992. The San Andreas Fault. In: D.G. Janelle (Editor), *Geographical Snapshots of North America*, 27th International Geographical Congress, Guilford Press, New York, 143–149.
Orme, A.R., 1998. Late Quaternary tectonism along the Pacific coast of the Californias: A contrast in style. In: I.S. Stewart and C. Vita-Finzi (Editors), *Coastal Tectonics*. Geological Society, London, Special Publications, 146, 179–197.
Parsons, T., and J. McCarthy, 1995. The active southwest margin of the Colorado Plateau: Uplift of mantle origin. *Geological Society of America Bulletin*, 107, 139–147.
Penck, W., 1924. *Die Morphologische Analyse*. J. Engelhorn's Nachfolger, Stuttgart.
Plafker, G., 1965. Tectonic deformation associated with the 1964 Alaskan earthquake. *Science*, 148, 1675–1687.
Powell, C.M., Z.X. Li, M.W. McElhinny, J.G. Meert, and J.K. Park, 1994. Paleomagnetic constraints on timing of the Neoproterozoic break-up of Rodinia and the Cambrian formation of Gondwana. *Geology*, 22, 889–892.
Prell, W.L., and J.E. Kutzbach, 1992. Sensitivity of the Indian monsoon to forcing parameters and implications for its evolution. *Nature*, 360, 647–652.
Prentice M.L., and R.K. Matthews, 1988. Cenozoic ice-volume history: Development of a composite oxygen isotope record. *Geology*, 16, 963–966.
Quinlan, G.M., 1987. Models of subsidence mechanisms in intercratonic basins and their applicability to North American examples. In: C. Beaumont and A.J. Tankard (Editors), *Sedimentary basins and basin-forming mechanisms*. Canadian Society of Petroleum Geologists Memoir 12, 463–481.
Raymo, M.E., and W.F. Ruddiman, 1992. Tectonic forcing of the late Cenozoic climate. *Nature*, 359, 117–122.
Retallack, G.J., 1983. *Late Eocene and Oligocene fossil paleosols from Badlands National Park, South Dakota*. Geological Society of America Special Paper 193.
Rind, D., and Chandler, M., 1991. Increased ocean heat transports and warmer climate. *Journal of Geophysical Research*, 96, 7437–7461.
Ruddiman, W.F. (Editor), 1997. *Tectonic uplift and climate change*. Plenum Press, New York.
Ruddiman, W.F., and J.E. Kutzbach, 1989. Forcing of the late Cenozoic northern hemisphere climate by plateau uplift in southern Asia and the American west. *Journal of Geophysical Research*, 94, 18,409–18,427.
Russo, R.M., and P.G. Silver, 1996. Cordillera formation, mantle dynamics, and the Wilson cycle. *Geology*, 24, 511–514.
Sams, D.B., and J.B. Saleeby, 1988. Geology and petrotectonic significance of crystalline rocks of the southernmost Sierra Nevada, California. In: W.G. Ernst (Editor), *Metamorphism and crustal evolution of the western United States*. Prentice Hall, Englewood Clffs, New Jersey, 866–893.
Schumm, S.A., 1977. *The fluvial system*. John Wiley and Sons, New York.
Sereno, P.C., 1991. Ruling reptiles and wandering continents: A global look at dinosaur evolution. *GSA Today*, 1 (7), 141–145.
Sheridan, R.E., 1989. The Atlantic passive margin. In: A.W. Bally and A.R. Palmer (Editors), *The Geology of North America—An overview*. Geological Society of America, Boulder, 81–96.
Small, E.E., and R.S. Anderson, 1995. Geomorphically driven late Cenozoic rock uplift in the Sierra Nevada, California. *Science*, 270, 277–280.
Small, E.E., and R.S. Anderson, 1998. Pleistocene relief production in Laramide mountain ranges, western United States. *Geology*, 26, 123–126.
Steckler, M., A.B. Watts, and J.A. Thorne, 1988. Subsidence and basin modeling at the U.S. passive margin. In: R.E. Sheridan and J.A. Grow (Editors), *The Atlantic continental margin*, Geological Society of America, Boulder, 399–416.
Stehli, F.G., and S.D. Webb (Editors), 1985. *The Great American biotic interchange*. Plenum Press, New York.
Summerfield, M.A., 1991. *Global geomorphology*. Longman, London.
Syvitski, J.P.M., and C.T. Schafer, 1997. Evidence of an earthquake-triggered basin collapse in Saguenay Fjord, Canada. *Sedimentary Geology*, 104, 127–153.
Tarnocai, C., 1997. Paleosols of the northern part of North America. In: I.P. Martini (Editor), *Late glacial and postglacial environmental changes: Quaternary, Carboniferous-Permian, and Proterozoic*. Oxford University Press, New York, 276–293.
Tidwell, W.D., 1975. *Common fossil plants of western North America*. Brigham Young University Press, Provo.
Torsvik, T.H., M.A. Smethurst, J.G. Meert, R. Vander Voo, W.S. McKerrow, M.D. Brasier, B.A. Sturt, and H.J. Walderhaug, 1997. Continental break-up and collision in the Neoproterozoic and Paleozoic—A tale of Baltica and Laurentia. *Earth Science Reviews*, 40, 229–258.
Unruh, J.R., 1991. The uplift of the Sierra Nevada and implications for late Cenozoic epeirogeny in the Western Cordillera. *Geological Society of America Bulletin*, 103, 1395–1404.
Volk, T., 1989. Rise of angiosperms as a factor in long-term climatic cooling. *Geology*, 17, 107–110.
Wallace, R.E. (Editor), 1990. *The San Andreas Fault System, California*. U.S. Geological Survey Professional Paper 1515, 283 pp.
Wayland, E.J., 1933. Peneplains and some other erosional platforms. *Bulletin of the Geological Society of Uganda*, Annual Report 1933, Notes 1, 74, 366.
Whistler, D.P. and D.W. Burbank, 1992. Miocene biostratigraphy and biochronology of the Dove Spring Formation, Mojave Desert, California, and characterization of the Clarendonian mammal age (late Miocene) in California. *Geological Society of America Bulletin*, 104, 644–658.
Wilf, P., 2000. Late Paleocene-early Eocene climate changes in southwestern Wyoming: Paleobotanical analysis. *Geological Society of America Bulletin*, 112, 292–307.
Wilson, J.T., 1966. Did the Atlantic close and then re-open? *Nature*, 211, 676–681.
Wing, S.L., T.M. Bown, and J.D. Obradovich, 1991. Early Eocene biotic and climatic change in interior western North America. *Geology*, 19, 1189–1192.

Woodburne, M.O. (Editor), 1987. *Cenozoic mammals of North America: Geochronology and biostratigraphy.* University of California Press, Berkeley.

Wynne, P.J., E. Irving, J.A. Maxson, and K.L. Kleinspehn, 1995. Paleomagnetism of the Upper Cretaceous strata of Mount Tatlow: Evidence for 3000 km of northward displacement of the eastern Coast Belt, British Columbia. *Journal of Geophysical Research*, 100, 6073–6091.

Yin, A. and T.M. Harrison (Editors), 1996. *Tectonic evolution of Asia.* Cambridge University Press, Cambridge.

Ziegler, A.M., M.L. Hulver, and D.B. Rowley, 1997. Permian world topography and climate. In: I.P. Martini (Editor), *Late glacial and postglacial environmental changes: Quaternary, Carboniferous-Permian, and Proterozoic.* Oxford University Press, New York, 111–146.

2

The Pleistocene Legacy: Glaciation

John Menzies

It has been less than 10,000 years, and in some places less than 6000 years, since the two major ice sheets that once covered North America finally disappeared leaving behind a legacy of sediments and landforms that cover slightly more than 60% of the continental land surface (fig. 2.1). The consequence of this last glaciation, and the several previous glaciations, that affected North America is immense in terms of the impact on soils, groundwater, river systems, topography, vegetation, fauna, and ultimately human society.

Debate as to when the Pleistocene Epoch began is fraught with considerable controversy because the criteria for identifying the start of the Pleistocene are debatable. In general, it is thought to be marked by (1) the onset of colder conditions than prevailed during the Pliocene, as evidenced by the increased presence of cold-water species in the waters of the Atlantic and Pacific Oceans or by reductions in ring growth in trees in the southwestern United States, and (2) significant changes in oxygen isotope ratios of waters that form stalactites and stalagmites in caves or in deep sea sediments. It is now generally agreed that the boundary between the Pliocene and the Pleistocene can be fixed at about 1.8 Ma at a time when cold-water mollusks are detected in sediments at Vrica, Calabria, on the shores of the Mediterranean Sea (Pasini and Colalongo, 1997; Van Couvering, 1997). In North America, this date may be accepted with the proviso that conditions prevalent in Europe were somewhat out of phase with those in continental North America. Lindsay (1997) notes that in North America several changes in the mammalian fauna around this time reveal slight climatic shifts symptomatic of changing climatic and oceanographic conditions with the onset of the Pleistocene. It is fundamental, however, to realize that these changes occurred nonpervasively throughout the continent. In some places, no changes are noted, whereas elsewhere dramatic changes occurred (Van Couvering, 1997).

The impact of continental glaciation is broadly twofold. First, those areas directly overrun by glacier ice have suffered a consequent devastating set of effects that is the substance of this chapter; second, those areas that remained unglaciated were affected indirectly through major environmental alterations whose imprint can also be detected today (see chapter 3). The effect of glaciation on any landscape is devastating as is the aftermath of glacial retreat when soils develop, plants and animals begin colonization, river systems are established, and groundwater systems evolve. Glaciation results in large volumes of sediment being deposited, much of it from distant sources, in topographic changes that disrupt older river systems, in lakes being formed, and in new landscapes that either replace or are imposed on earlier landscapes. Within the areas covered by glacier ice, zones of terrain can be differentiated where distinctive processes occurred and dominated

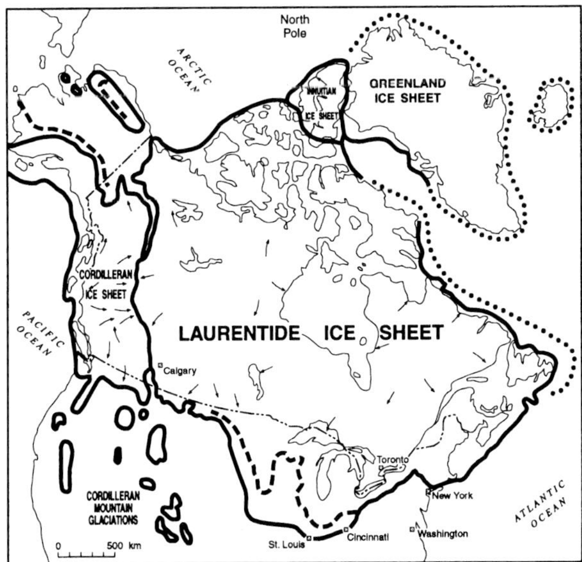

Figure 2.1 Extent of the Laurentide, Cordilleran, and Greenland Ice Sheets. Dashed lines indicate the extent of glaciation at the Last Glacial Maximum (LGM) approximately 18,000 years ago. Dotted line indicates uncertain extent of marine ice margin.

as in the following examples: along ice margins in the midwestern United States and the Canadian prairie provinces; in areas where fast-moving ice streams developed as in the Puget lowlands of Washington; where coalescing ice lobes formed as in the Oak Ridges moraine north of Toronto; where ice streams surged into the Lake Michigan basin; where vast downwasting of the ice sheet occurred as in the prairie provinces; and where extensive, proglacial lakes such as Lake Agassiz formerly existed (Clark et al., 1993).

It is fundamental to view the impact of glaciation at this continental scale as part of a dynamic and ever-changing environmental system that was influenced and "forced," in part, by global climatic and oceanic pattern changes that have their origins in variations in Earth's relationship to the Sun and the surrounding Solar System in terms of orbital fluctuations, spin velocities, and sunspot cycles. These variations must also be coupled to fluctuations in ocean level (eustatic variations) and areal extent, global carbon dioxide levels, continental plate positioning, and tectonic histories of the plates as related to mantle rheology and thus to the total areal expanses of highland and lowland terrains (isostatic variations) (COHMAP, 1988; Broecker et al., 1989; Imbrie et al., 1992; Clark et al., 1993).

2.1 Glacial Chronology

At one time or another during the Pleistocene, major ice sheets and smaller ice caps could be found from the southern Rocky Mountains to the high Arctic and from the continental ice shelf off eastern Canada and New England to the Pacific Ocean and the Bering Sea (fig. 2.1). The largest ice mass, the Laurentide Ice Sheet, effectively stretched from 42.5°N to 75°N, that is, from the central plains of the United States to the Canadian Arctic Archipelago, an enormous climatic gradient if measured today in mean annual temperature (≈30°C). It also extended from the continental shelf off the coast of Canada and the northeastern United States to the foothills and high plains of Alberta, Montana, and Wyoming. This ice sheet at its maximum extent during the late Wisconsinan accounted for 35% of the world's ice volume (Calkin, 1995) (table 2.1).

At least 17 major glacial stages have affected North America over the past 1.8 million years (Richmond and Fullerton, 1986a; Clark et al., 1993; Calkin, 1995) (see table 2.2). As each major glacial stage developed, so the margins of ice sheets expanded and retreated, usually through series of recurring shorter glacial phases (stadials) and warmer intervals (interstadials); these glacial stages then were followed by much warmer and longer stages—the major interglacials (tabl 2.2) (Bowen et al., 1986; Calkin, 1995). The North American ice sheets appear to have been closely linked to a dynamic glacial–ocean–atmosphere system that responded remarkably quickly in terms of Earth's orbital variations (20-, 40-, and 100-ka [thousand years] periodicities) (Denton and Hughes, 1981; Imbrie et al., 1992) and shorter fluctuations (≤10 ka) (Bond et al., 1992; Andrews, 1997). These responses in turn led to large oscillations in regional ice margins that were at times synchronous but also often asynchronous (Richmond and Fullerton, 1986b). Such oscillations, when viewed from a glacigenic sediment and landform viewpoint, caused overprinting and reworking of existing features that led to complex glaciated landscapes. Presently the cause of these dynamic fluctuations is thought to be a function of climatic and oceanic responses linked to glacier-bed behavior that results in variations in ice-sheet volume during very short time periods (Thomas, 1979;

Table 2.1 Estimated values of Pleistocene ice sheet volumes and areas (note values for present day Greenland and Antarctica)

Ice Sheet	Volume ($km^3 \times 10^6$)	Area ($km^2 \times 10^6$)
Laurentide	29.5	16.0
Cordilleran	3.6	1.4
Greenland	3.5 *2.6*#	2.3 *1.7*#
Scandinavia	13.3	6.7
Antarctica	26.0 *23.5*#	13.8 *12.5*#

#Present Day values

Data derived from Flint, 1971; Nilsson, 1983

Table 2.2 Correlation of Quaternary Glaciations in Canada and the United States of America (modified from Bowen et al., 1986)

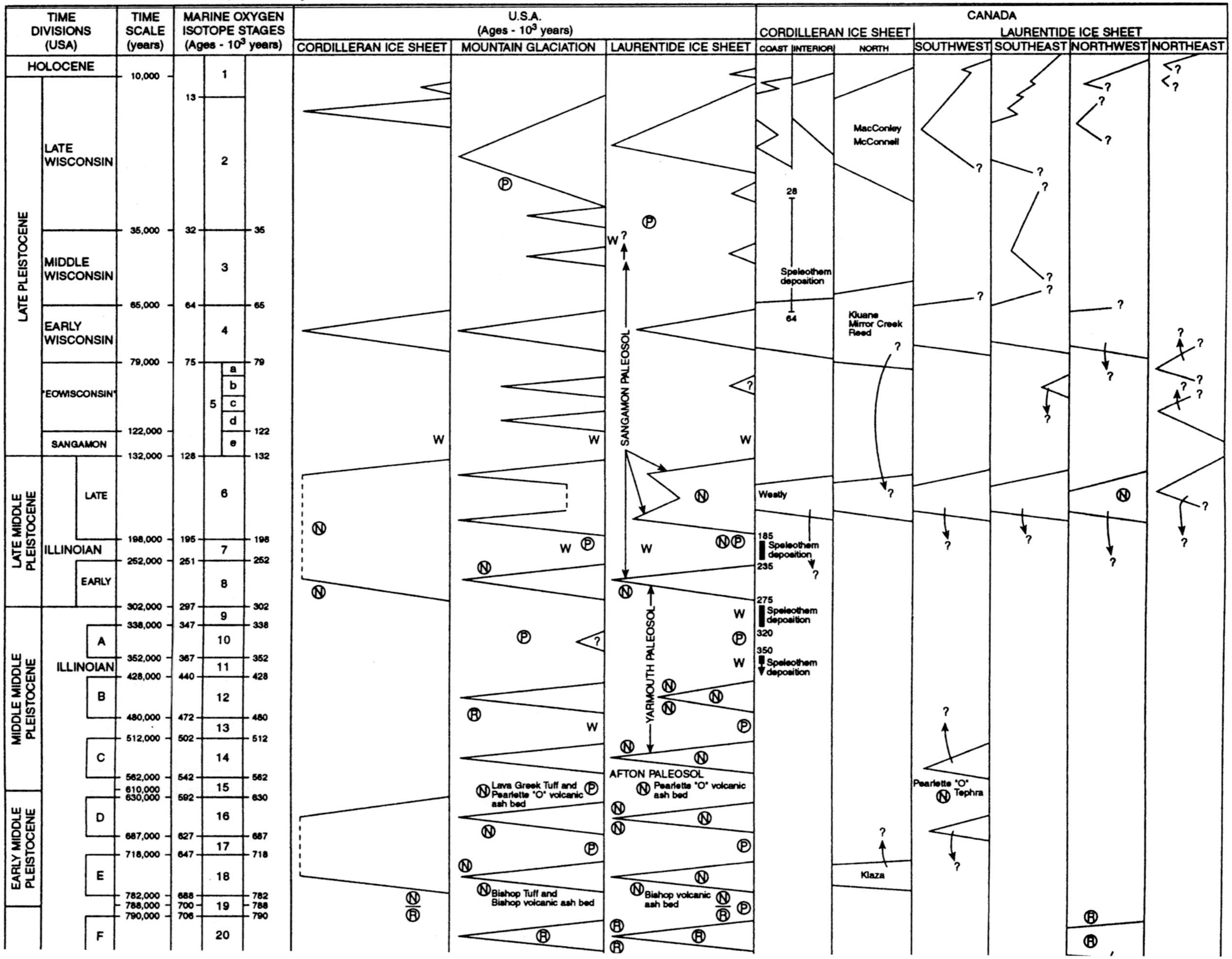

Time Divisions (USA)		Time Scale (years)	Marine Oxygen Isotope Stages (Ages - 10^3 years)			U.S.A. (Ages - 10^3 years) Cordilleran Ice Sheet	U.S.A. Mountain Glaciation	U.S.A. Laurentide Ice Sheet	Canada Cordilleran Ice Sheet Coast / Interior	Canada Cordilleran Ice Sheet North	Canada Laurentide Ice Sheet Southwest	Canada Laurentide Ice Sheet Southeast	Canada Laurentide Ice Sheet Northwest	Canada Laurentide Ice Sheet Northeast
HOLOCENE		10,000	13	1										? ?
LATE PLEISTOCENE	LATE WISCONSIN	35,000	32	2	35		P		28	MacConley McConnell	?	? ?	? ?	
LATE PLEISTOCENE	MIDDLE WISCONSIN	65,000	64	3	65			W ? P	Speleothem deposition		?	? ?		
LATE PLEISTOCENE	EARLY WISCONSIN	79,000	75	4	79			SANGAMON PALEOSOL	64	Kluane Mirror Creek Reed ?			?	? ?
LATE PLEISTOCENE	"EOWISCONSIN"	122,000		5 a b c d	122			?				?	?	? ? ?
LATE PLEISTOCENE	SANGAMON	132,000	128	5 e	132	W	W	W						
LATE MIDDLE PLEISTOCENE	ILLINOIAN LATE	198,000	195	6	198	N		N	Westly	?			N	?
LATE MIDDLE PLEISTOCENE	ILLINOIAN	252,000	251	7	252		W P	W N P	185 Speleothem deposition 235 ?		?	?	?	?
LATE MIDDLE PLEISTOCENE	ILLINOIAN EARLY	302,000	297	8	302	N	N	N						
MIDDLE MIDDLE PLEISTOCENE	A	338,000	347	9	338			W YARMOUTH PALEOSOL	275 Speleothem deposition 320					
MIDDLE MIDDLE PLEISTOCENE	A	352,000	367	10	352		P ?	P						
MIDDLE MIDDLE PLEISTOCENE	ILLINOIAN	428,000	440	11	428			W	350 Speleothem deposition					
MIDDLE MIDDLE PLEISTOCENE	B	480,000	472	12	480		R	N N N						
MIDDLE MIDDLE PLEISTOCENE		512,000	502	13	512		W	P			?			
MIDDLE MIDDLE PLEISTOCENE	C	562,000	542	14	562			N N AFTON PALEOSOL						
EARLY MIDDLE PLEISTOCENE		610,000 630,000	592	15	630		N Lava Creek Tuff and Pearlette "O" volcanic ash bed P	N Pearlette "O" volcanic ash bed			Pearlette "O" N Tephra			
EARLY MIDDLE PLEISTOCENE	D	687,000	627	16	687		N	N/N N						
EARLY MIDDLE PLEISTOCENE		718,000	647	17	718		P N	P		? Klaza	?			
EARLY MIDDLE PLEISTOCENE	E	782,000	688	18	782		N Bishop Tuff and Bishop volcanic ash bed	N N Bishop volcanic ash bed						
EARLY MIDDLE PLEISTOCENE		788,000 790,000	700 706	19	788 790	N/R		N/R P					R	
EARLY MIDDLE PLEISTOCENE	F			20			R	R/R R					R	

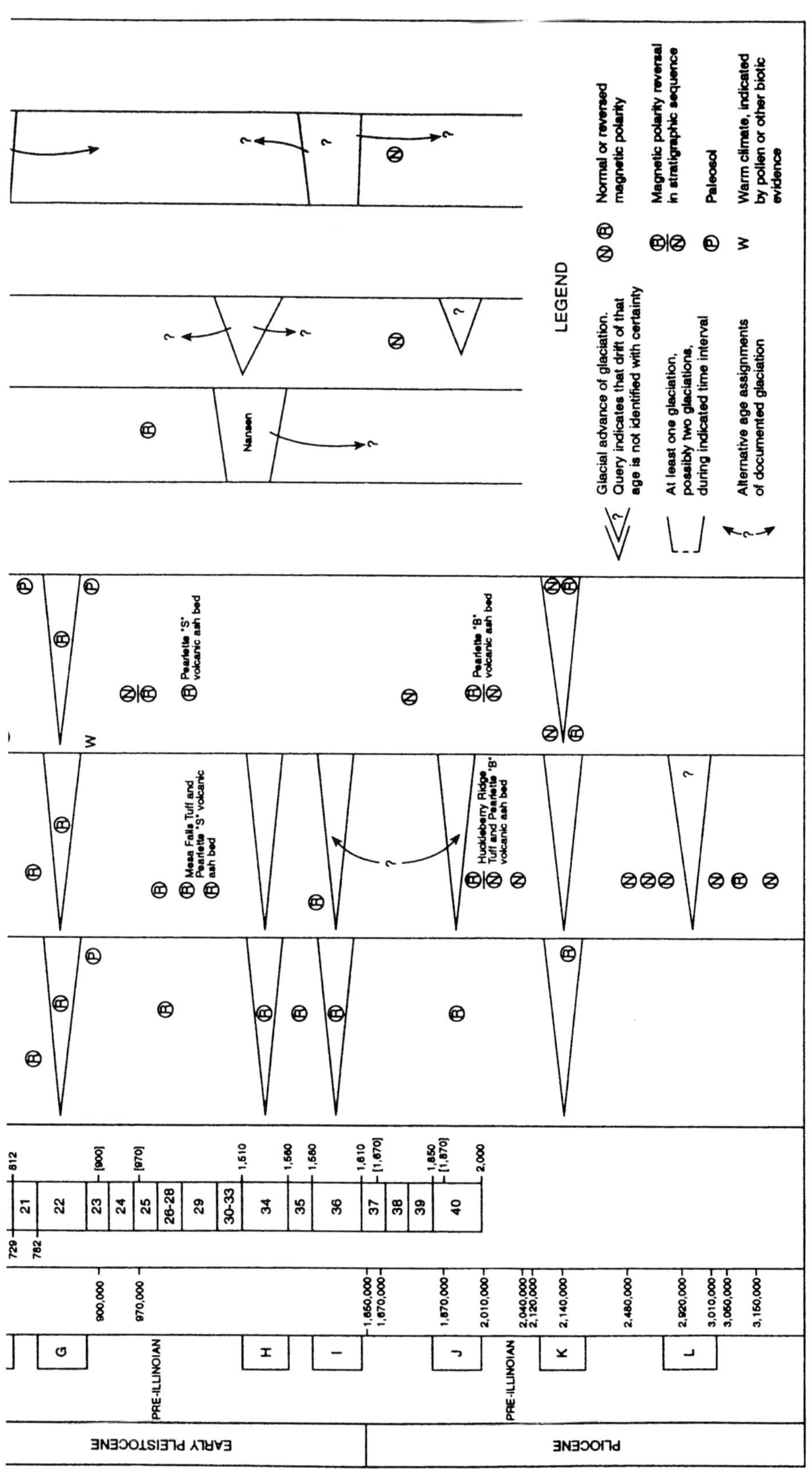
EARLY PLEISTOCENE
PLIOCENE
PRE-ILLINOIAN
PRE-ILLINOIAN
Nansen
Mesa Falls Tuff and Pearlette 'S' volcanic ash bed
Pearlette 'S' volcanic ash bed
Huckleberry Ridge Tuff and Pearlette 'B' volcanic ash bed
Pearlette 'B' volcanic ash bed
LEGEND
Glacial advance of glaciation. Query indicates that drift of that age is not identified with certainty
At least one glaciation, possibly two glaciations, during indicated time interval
Alternative age assignments of documented glaciation
Normal or reversed magnetic polarity
Magnetic polarity reversal in stratigraphic sequence
Paleosol
Warm climate, indicated by pollen or other biotic evidence

Hughes, 1992a,b; MacAyeal, 1992, 1993; Clark et al., 1993; Clark, 1994, 1997).

By the beginning of the late Pliocene and possibly earlier in continental North America, extensive glaciation was established within most mountainous areas of the Western Cordillera and probably on the mountains east of Hudson Bay in Labrador and Ungava, on Baffin Island, and on other islands of the Canadian Arctic Archipelago. Glaciation in Alaska seems to have been developed by 6 Ma, possibly in response to the convergence of the North American and Pacific plates (Eyles and Eyles, 1989; Eyles, 1993; Hamilton, 1994). Extensive glacial advances had occurred within the Rocky Mountains of Wyoming and the Cascade Range of Washington by 3 Ma (glaciations L, K, and/or J [J can be equated with isotope stage 40]) (Richmond and Fullerton, 1986b; Wright, 1989; Lindsay, 1997). It seems likely that mountain glaciation was established in the eastern and western mountain ranges of Greenland by 2.7 Ma and an ice sheet as large as today formed by 2.4 Ma (Robin, 1988; Funder, 1989).

By the beginning of the Pleistocene around 1.8 Ma, both the Laurentide and the Cordilleran ice sheets appear to be in existence. Limited evidence of the Pliocene-Pleistocene boundary exists within Canada. In a few places, extensive erratic trains have been regarded as pre-Pleistocene; vertebrate fossil remains and palaeomagnetic data from southern Saskatchewan are dated around the Pliocene-Pleistocene boundary. Identifiable glacial deposits are known to have been laid down by the Cordilleran Ice Sheet in the Puget Lobe by late Pliocene or early Pleistocene time (Easterbrook, 1986); similarly, deposits in Iowa and Wisconsin may be linked to Glaciation K (Matsch and Schneider, 1986; Richmond and Fullerton, 1986b; Clark et al., 1993). However, other evidence for the expansion of the Laurentide Ice Sheet is unknown until 0.9 Ma (Kukla, 1989; Calkin, 1995). In the Canadian Arctic Archipelago, the presence of a separate Innuitian Ice Sheet engenders some controversy (Fulton, 1989). It seems likely that a separate ice sheet could have come into existence before the maximum development of the Laurentide Ice Sheet because, once the Arctic Ocean became frozen over, a dramatic diminution of moisture to feed the Innuitian Ice Sheet would have led to a marked negative mass balance and the subsequent overrunning of the latter ice sheet by the larger Laurentide Ice Sheet with its positive mass balance. The Innuitian Ice Sheet possibly developed asynchronously as the Laurentide Ice Sheet was still growing, and therefore the advance and retreat of the Laurentide Ice Sheet was out of phase with the Innuitian Ice Sheet's growth and decay (Clark et al., 1993; Andrews, 1997).

During the period 1.5 to 0.9 Ma, there appears to have been no major ice sheet expansion into the coterminous United States, although large parts of Canada must have been ice covered (van Donk, 1976; Fullerton and Richmond, 1986). Around 0.9 Ma, Cordilleran and Laurentide ice expanded from Canada and from the inner mountains of the Western Cordillera. Thereafter, repeated glacial stages can be chronicled in phase intervals of 40 to 140 thousand years, corresponding to the marine oxygen isotope stages (OIS) (table 2.2) (Bowen et al., 1986).

For much of the twentieth century, the repeated glaciations of North America were defined in terms of the Nebraskan, Kansan, Illinoian, and Wisconsinan stages. However, when evidence garnered from the Atlantic and Pacific Oceans began establishing a much larger number of cold and warm (glacial and interglacial) intervals, this classical subdivision became untenable. The basis for the earlier stratigraphic framework hinged on multiple till sheets and the spatial distribution of end moraines to identify glacial stages, whereas interglacial stages were identified by the presence, especially in the Midwest, of extensive paleosols (buried soils) (Frye et al., 1968). As more data were gathered, the old classical subdivisions no longer held. For example, whereas previously a Yarmouthian-Kansan age had once been assigned to the Pearlette Ash, a tephra derived from the Yellowstone volcanic region, such a division became untenable when three separate but similar ash beds of different age were distinguished (Hallberg, 1986). In fact, across the type areas of the Kansan and Nebraskan glaciations and the Aftonian and Yarmouthian interglacials, the degree of miscorrelation became such that all four terms have had to be abandoned. For instance, beneath the "type" Nebraskan till, at least three other separate tills have been distinguished (Richmond and Fullerton, 1986b). As shown in table 2.2, before OIS 8 we now use the term Pre-Illinoian. Between OIS 36 and 9, at least nine major glacial stages are now recognized before the base of the Illinoian is reached (Clark et al., 1993; Lindsay, 1997).

From the close of OIS 9 (≈300 ka) to the end of OIS 2 (≈13 ka), the accepted glaciations are more clearly differentiated since extensive sedimentological and fossil evidence remains extant. As table 2.2 illustrates, after OIS 9, two major glacial stages occurred, the Illinoian and Wisconsinan, that can be conclusively subdivided into repeated glacial advances and retreats with nonglacial intervals of markedly cool climates (interstadials) and others of much warmer conditions (interglacials). It should be remembered that the margins of both the Cordilleran Ice Sheet and Laurentide Ice Sheet stretched for thousands of kilometers, through a multiplicity of regional topographic and extraglacial climatic settings such that, while the ice margin retreated in one area, it may well have been advancing elsewhere. It is possible, for example, that the Laurentide Ice Sheet retreating from central New York State might well have been advancing simultaneously across the plains of North Dakota. These vagaries of the ice margin are a function of many disparate factors associated with ice sheet dynamics and mass balance that generally come under the heading of "response time," at least for periods of margin changes less than 1000 years in length (Menzies, 1995). For periods of marginal fluctuations greater than approximately 1000 years, a more general synchroneity of response tended to occur such that the margin along vast stretches of the ice sheets would begin to respond similarly. How ice sheet margins responded along the Arctic, Pacific, and Atlantic coastal margins where floating ice tongues and shelves existed is more difficult to establish. It seems likely that rather rapid variations in ice mass balance may have translated into fast changes of the positions of ice sheet margins (Andrews et al., 1996; Andrews, 1997).

2.2 Causes of Glaciation

Perennial discussion surrounds the question of what caused the massive ice sheets to develop in North America and elsewhere on Earth. Evidence now indicates that not only were there many more glaciations that spread across much of the North American continent during the Pleistocene than was thought in the past but that global glaciations have occurred at fairly regular intervals as far back in geologic time as the Precambrian (Young, 1996). This regularity suggests one or more mechanisms that can be "turned on or off" at some global level. Details of the many hypotheses of causes of global glaciation are discussed in the literature (Calkin, 1995; Ehlers, 1996). In broad terms, agreement appears to suggest that a complex relationship exists between the Earth's orbital relationship with the Sun, expressed in terms of orbital eccentricity, axial tilt, and precession of the equinoxes, and oceanic conditions, especially ocean water temperatures, the elevation and geographic positioning of the continental land masses, and the level of carbon dioxide in the atmosphere. Different combinations of these factors appear in some instances to predispose Earth toward global glaciation, whereas under other circumstances glaciations may be brief, rapidly concluded, or simply localized mountain glaciations (Clark et al., 1993). No single set of parameters appears to explain global glaciation in all cases. For example, increasing evidence of low-latitude near-maritime glaciations during the Precambrian is difficult to explain based on the past idea that glaciation originated at high polar latitude's and/or high land elevations (Young, 1996). It is possible that shifting continental land mass positions, tectonics, and miscalculation of polar positions may individually or together combine to permit the initiation of glaciation at these "apparent" low latitudes (Eyles, 1993).

2.3 Ice Dynamics and the North American Ice Sheets

The major emphasis on the ice dynamics of the North American ice sheets has focussed on the Laurentide Ice Sheet. As the larger of the two ice sheets, with both terrestrial and marine margins and at its maximum probably only a single ice center, this ice sheet is much more readily studied than the complex ice masses that made up the Cordilleran Ice Sheet with its many ice centers and enormous topographic variations. The Cordilleran Ice Sheet also had marine and terrestrial margins, but these remain to be researched in detail (Clague, 1989; Clark et al., 1993).

In the past, considerable debate has focused on whether the Laurentide Ice Sheet had a single, dual, or multidomed ice center (fig. 2.2) (Boulton et al., 1985; Fisher et al., 1985; Hughes, 1992b; Clark et al., 1993). It seems likely, based on erratic trains and dispersal fans, that the ice sheet, as it evolved initially had several subcenters over Ungava, Keewatin, and Newfoundland; based on isostatic depression data, as it approached maximum development, it had a single main dome center over Hudson Bay (Shilts et al., 1979; Dyke et al., 1982; Hughes, 1992b; Clark et al., 1993, 1996; Andrews and Peltier, 1989). During subsequent deglaciation, this pattern of domes and saddles reappeared and shifted location as changes in ice stream drawdown, subglacial bed dynamics, patterns of snow accumulation, and peripheral ocean level all affected the overall dynamics of the Laurentide Ice Sheet.

To understand the many fluctuations of the North American ice sheets within the context of their total size and dimensions and resultant marginal fluctuations during ice-sheet growth and decay, ice-sheet glaciology must be considered (Clark, 1992; Clark et al., 1993; Murray, 1997). The key to understanding any ice sheet's development and subsequent behavior is knowledge of its basal thermal and topographic condition (Clark, 1992; Hughes, 1992b; Clark et al., 1996). Beneath the Laurentide Ice Sheet, basal conditions were in general probably polythermal, implying that thermal states were in a constant flux both spatially and temporally across the ice-sheet bed; central areas were essentially warm based but surrounded by a cold-based zone that was succeeded toward the margin of the ice sheet by a warm-based zone (Sugden, 1977; Denton and Hughes, 1981; Fisher et al., 1985; Hughes, 1992b; Payne, 1995; Clark et al., 1993, 1996; Greve, 1997). Figure 2.3 suggests such a pattern, based on a maximum stage development of the late Wisconsinan ice sheet around 18,000 BP (radiocarbon years before AD 1950) (Sugden, 1977). It seems likely that a patchwork pattern of fluctuating basal thermal regimens existed below the Laurentide Ice Sheet wherein certain dominant thermal states would have persisted at specific times and locations, depending on ice dynamics and bed topographic control. Significantly, marginal areas of the ice sheet at any time during its growth and decay were characterized by a warm-based temperate bed regime under which meltwater production and deformable bed conditions existed (Denton and Hughes, 1981; Boulton et al., 1985; Fisher et al., 1985; Hughes, 1992b; Menzies and Shilts, 1996; Marshall and Clarke, 1997a,b). The Laurentide Ice Sheet was probably

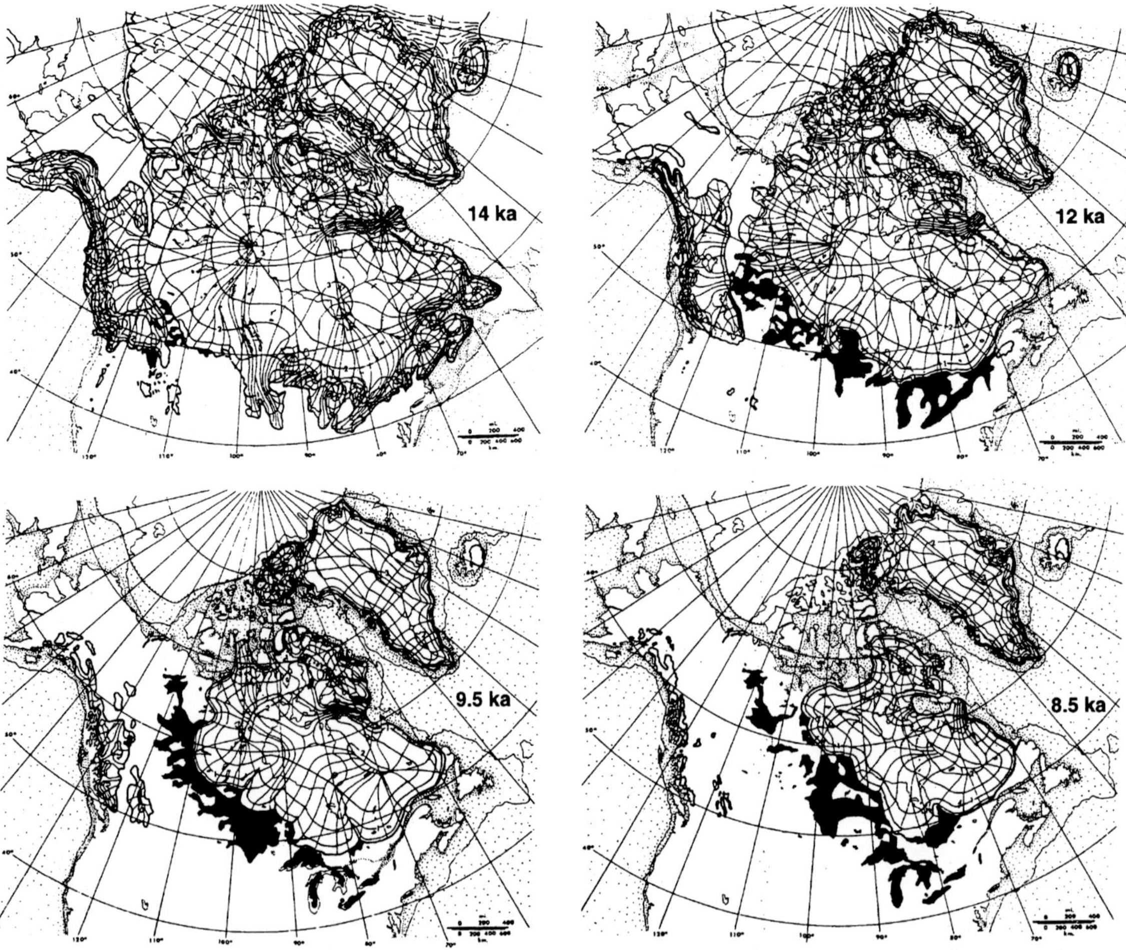

Figure 2.2 Potential ice surface topography of the Laurentide Ice Sheet, showing domes and saddles at various stages of deglaciation at 14, 12, 9.5, and 8.5 ka (modified from Hughes, 1992b, Fig. 18). Isolines, at a contour interval of 500 m, reveal changing topographic patterns on the ice sheet during deglaciation. Note the development of large proglacial lakes at the ice-sheet margins. Ice flowlines, transverse to the contours, indicate fast ice flow where they are close together.

never in complete equilibrium with the controlling parameters that influenced its existence such that instability was built into the system (Budd and Smith, 1987; Greve and MacAyeal, 1996).

In general, the influence of basal topography on ice-sheet motion is, on a large scale, limited (Paterson, 1994); however, on a smaller scale (10^2–10^4 m), bed topography begins to assert an increasing control on ice motion, basal ice temperatures, subglacial meltwater regimens, and basal sediment transport pathways (Kamb, 1970; MacAyeal et al., 1995; Jenson et al., 1996; Marshall et al., 1996). Several authors have explored the possible relationships between ice sheet movement and the topographic controls exerted by the bed on the Laurentide Ice Sheet (Clark et al., 1993, 1996; Jenson et al., 1995, 1996). Two distinct bed types are perceived, namely, (1) those areas of the bed that act as pinning points (sticky points) creating frictional retardation, which causes basal ice-sheet drag, and (2) those areas of bed decoupling where enhanced flow and vastly reduced basal drag occur. On the basis of modeling, Marshall et al. (1996) have suggested that the Laurentide Ice Sheet can be subdivided into these categories (fig. 2.4). Where enhanced flow occurs, the possibility of fast sliding (ice streams) and/or deformable conditions is likely to have been present (MacAyeal, 1993). This model, when compared to that produced by Fisher et al. (1985), suggests that the Hudson Bay Lowlands, the Great Lakes region, and large parts of the Canadian prairies fall within the realm of enhanced fast flow (Clark et al., 1993; Lowell and Brockman, 1994; Maher and Mickelson, 1997). The implications of such spatial differentiation of basal ice flow are likely to be revealed in terms of sediment, landforms, and land system types (Eyles

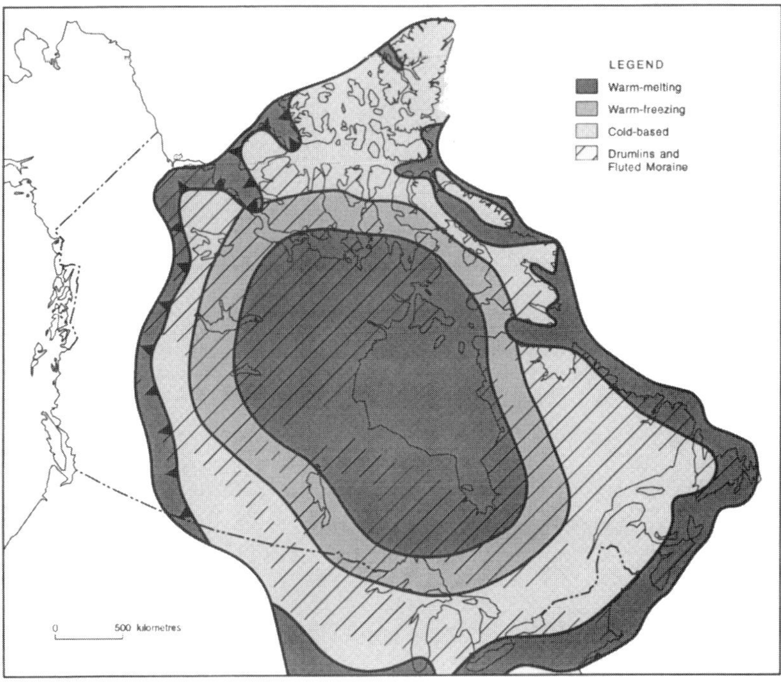

Figure 2.3 Possible basal ice temperatures beneath the Laurentide Ice Sheet at the Last Glacial Maximum (modified from Sugden, 1977).

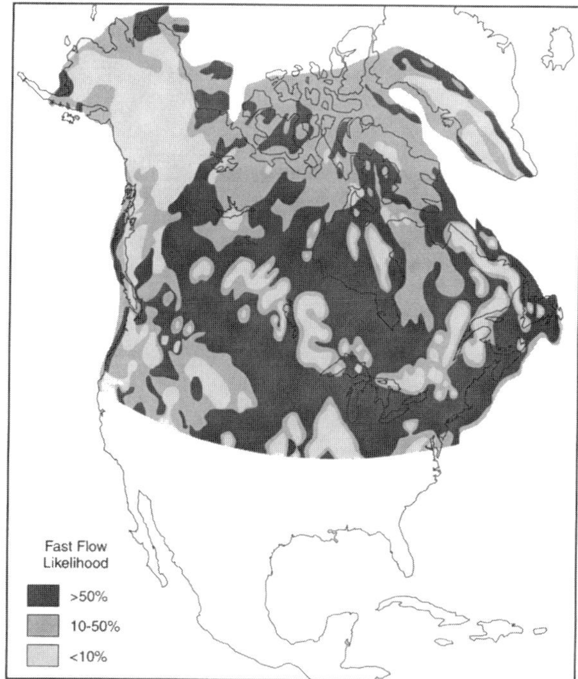

Figure 2.4 Differential basal ice velocities below the Laurentide Ice Sheet as predicted from modeling by Marshall et al. (1996). Fast basal ice flow is likely, for example, in the Canadian prairies, the Great Lakes, Hudson Bay, and Puget lowlands (modified from Marshall et al., 1996, Fig. 10).

and Miall, 1984; Clark, 1997; Kleman et al., 1997; Colgan and Principato, 1998).

2.4 Regional Descriptions

Over the years, many regional descriptions of the Pleistocene glaciations of North America have appeared; as information and dating methods have improved so descriptions have become increasingly refined (e.g., Porter, 1983; Šibrava et al., 1986; Ruddiman and Wright, 1987; Fulton, 1989; Hamilton, 1994). This chapter does not provide a catalogue of these descriptions; the reader is referred to the previous references. Unfortunately, very few attempts have been made to coordinate these descriptions beyond adjacent areas. These regional descriptions indicate that, over thousands of kilometers along the margins of ice sheet, conditions varied immensely in reaction time to changes in ice-sheet dynamics, topographic constraints, and regional climatic conditions. Since the terrestrial evidence for earlier Pleistocene glaciations is patchy because it was often erased or overprinted by later glaciations, the following discussion focuses on the nature and extent of the last major continental glaciation of North America, namely, the late Wisconsinan that reached its maximum limits, often asynchronously, between 22 and 14 ka—the so-called Last Glacial Maximum (LGM).

In the far west, along the coasts of British Columbia and Washington, evidence suggests that during the LGM the Cordilleran Ice Sheet extended beyond the present coast to the edge of the continental shelf (Booth, 1987; Clague, 1989; Clark et al., 1993). In many places, outlet lobes extending from fjords and estuaries coalesced. Ice from the southern Coast Mountains and Vancouver Island, for example, joined and flowed as a huge piedmont lobe into the Puget Lowland of Washington (Waitt and Thorson, 1983; Brown et al., 1987). Farther north, along the western seaboard of British Columbia, ice appears to have advanced to the outer edge of the continental shelf in many places. For example, the Cordilleran Ice Sheet extended across the Queen Charlotte Islands to their western edge, overwhelming the Alexander Archipelago in the Alaskan Panhandle (Hamilton, 1994).

In Alaska, a major ice center developed in the Alaska Range and nearby mountains and stretched to edge of the continental shelf in the Gulf of Alaska, while ice also streamed north into the lowlands occupied by the Yukon River (Hamilton and Thorson, 1983; Hamilton, 1994). Because of diminished precipitation north and east of the central mountains, ice did not extend into interior Alaska nor into central and northern Yukon Territory, these regions remaining ice-free throughout the Quaternary (Hughes et al., 1969; Hamilton, 1994). The part played by this ice-free area and its connection to the land bridge of Beringia to the west were crucial in the movement of nomadic peoples and animal herds as a major route into North America (Jopling et al, 1981; Ritchie, 1984; Schweger, 1989). The Brooks Range in northern Alaska supported an independent ice cap at various times during the Pleistocene. Along the eastern margin of the Cordilleran Ice Sheet, ice lobes descended into the eastern foothills of the Rocky Mountains. It appears that the Cordilleran and the Laurentide ice sheets, at their maximum, were in contact for about 700 km, stretching from the southeast corner of the Yukon across northeast British Columbia into Alberta as far south as near Hinton, where another ice-free zone developed (fig. 2.5) across the Great Plains of Alberta and into Montana.

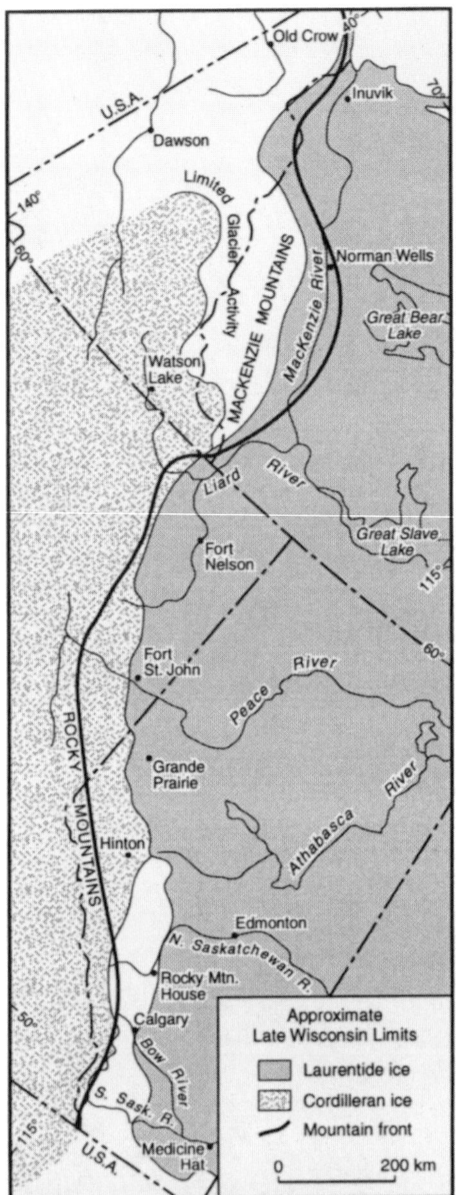

Figure 2.5 Contact zone and ice-free corridor between the Laurentide and Cordilleran ice sheets (modified from Schweger, 1989).

Farther east, across the northern islands of the Canadian Arctic Archipelago (the Queen Elizabeth Islands), a complex set of conditions developed, with local ice masses probably extending to coalesce into a large ice complex referred to as the Innuitian Ice Sheet (Prest et al., 1969, Prest, 1984; Clark et al., 1993) or a slightly different build-up of ice from separate islands formed the Franklin Ice Complex or the Queen Elizabeth Islands Glacier Complex (England, 1976) (fig. 2.6). At approximately the same time, the Laurentide Ice Sheet from mainland Canada extended northward across the Mackenzie Lowlands and the Arctic Plain of Keewatin to the present shoreline and beyond (fig. 2.7). The relationship between the Laurentide Ice Sheet and the more northerly Innuitian ice remains controversial, but it is likely that during the LGM both ice sheets joined along a west-to-east line stretching from McClure Strait to Lancaster Sound and that, at approximately the same time, the Innuitian and Greenland ice sheets merged just to the east of Ellesmere and Devon Islands (Hodgson, 1989). As the Innuitian Ice Sheet waned, various smaller ice complexes remained on individual islands, for example, Ellesmere (fig. 2.8), Bathhurst, Devon, and Axel Heiberg Islands (Hodgson, 1989).

In the Baffin Island area, including Bylot and Southampton Islands, separate ice caps appear to have developed, coalescing at the LGM into the northeastern part of

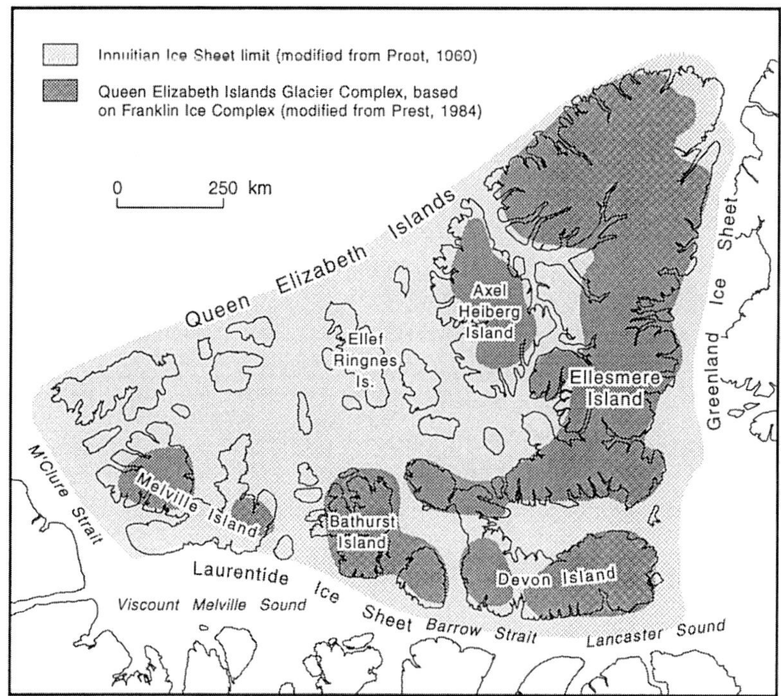

Figure 2.6 (*at left*) Maximum Innuitian Ice Sheet and minimum Franklin Ice Complex; map of the extent of the Last Glacial Maximum in the Canadian Arctic Archipelago (modified from Hodgson, 1989).

Figure 2.7 (*below*) Drumlins near Coronation Gulf, Northwest Territories (modified from Prest, 1983).

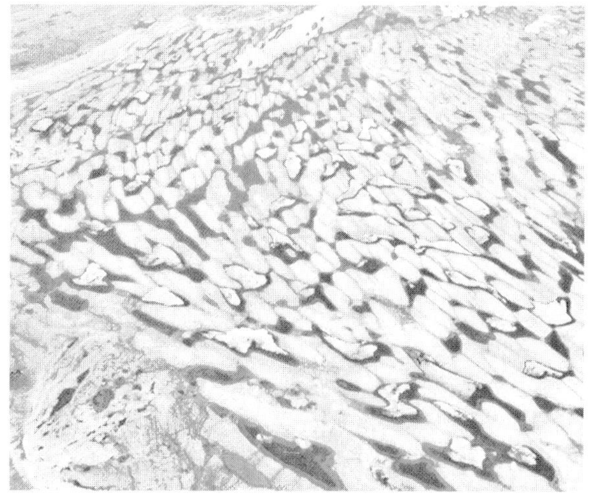

the Laurentide Ice Sheet (Andrews, 1989). During the LGM, a large subsidiary ice dome within the Laurentide Ice Sheet appears to have formed over the Foxe Basin, in the area of Prince Charles Island (Andrews, 1989, 1997; Andrews et al., 1996).

Along the eastern seaboard of Canada, from the mouth of Hudson Bay across the Ungava Peninsula and Labrador to the north edge of the Gulf of St. Lawrence, the seaward extent of the Laurentide Ice Sheet is still unclear (Vincent, 1989) (Fig. 2.9). Increasing evidence points to Hudson Bay as a major corridor during the drawdown of this ice sheet, causing ice-rafted debris to be discharged into the North Atlantic where Heinrich events can be tied to this major outsource of ice and sediment (Andrews et al., 1996). In general, the Laurentide Ice Sheet appears to have reached the edge of the continental shelf off Labrador, but its extent beyond this topographic break remains questionable. A radial flow of ice spreading from a Labrador Dome is readily detected from striae, drumlin, and esker lineations (Vincent, 1989) (Figs. 2.10, 2.11). Other dispersal centers have been recognized, for example, in southern Labrador (Klassen and Bolduc, 1984). In the Torngat Mountains and Newfoundland, separate ice caps flourished and were perhaps never totally overwhelmed by Laurentide ice (Clark, 1984; Clark et al., 1993).

Evidence suggests that a separate ice dispersal center developed over central Newfoundland (Grant, 1989). Some controversy exists as to whether ice extended beyond the present shoreline or was restricted to the island (Figs. 2.9, 2.12) (Grant, 1989). A small ice dispersal center may have developed over the Magdalen Shelf, forming the Escuminac Ice Center (Vincent, 1989; Stea et al., 1998). This ice center, located over a shallow area of the Gulf of St. Lawrence, affected ice dispersal from Québec and New Brunswick, and contributed to ice moving south across Prince Edward Island and Nova Scotia. The development of the Escuminac Center further affected regional ice flow by diverting the Appalachian ice from the uplands of New Hampshire and Maine to the south across the Bay of Fundy (fig. 2.12). By late Wisconsinan times, ice from the combined Laurentide and Appalachian ice sheets extended onto the continen-

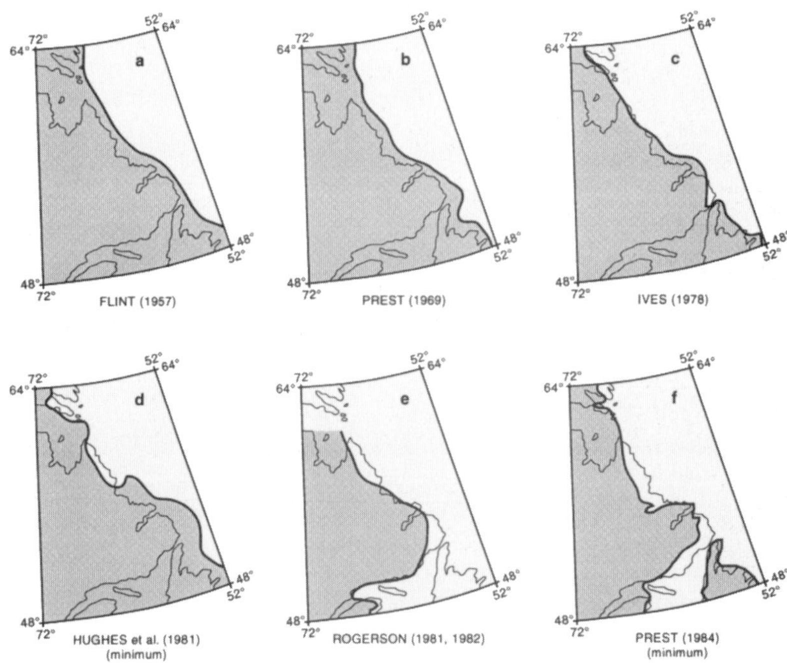

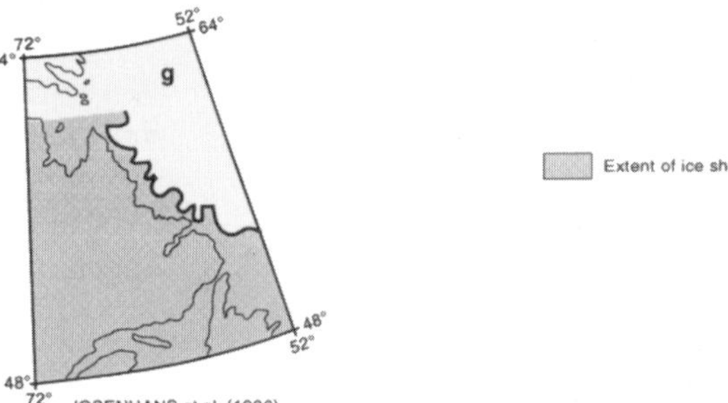

Figure 2.8 (*top*) An outlet glacier of the Agassiz Ice Cap descending into Cañon Fiord, Ellesmere Island, NWT, Canada. (Courtesy of the Geological Survey of Canada's World Wide Web site: Canadian Landscapes).

Figure 2.9 (*at right*) Possible ice limits of the Laurentide Ice Sheet along the east coast of Canada at the Last Glacial Maximum (modified from Vincent, 1989).

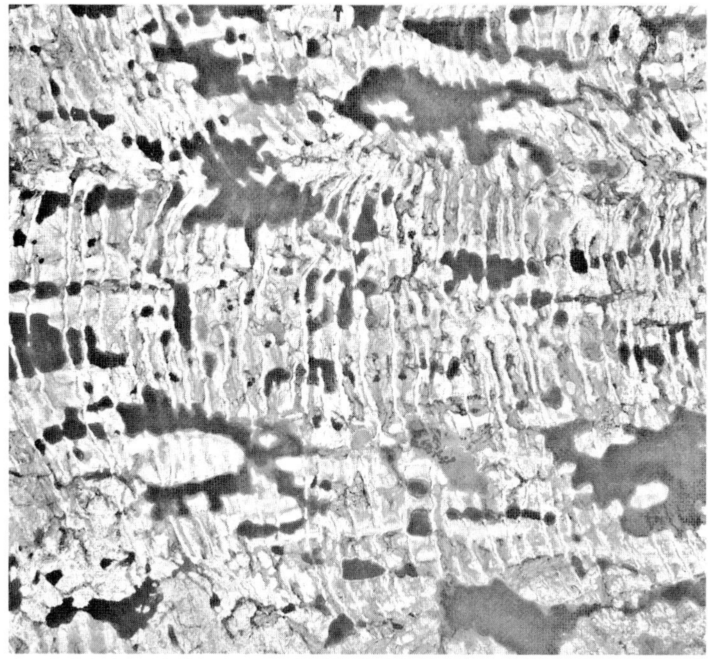

Figure 2.10 (*top*) Esker ridge built into Lake Agassiz, Manitoba (modified from Prest, 1983).

Figure 2.11 (*at left*) De Geer moraines east of Hudson Bay, Quebec (modified from Prest, 1983).

tal shelf south of present day Nova Scotia. By around 18 ka, this ice had retreated to the midshelf area of the Scotian Shelf (Grant, 1994; Stea et al., 1998).

Appalachian ice from the uplands of Maine and New Hampshire joined with the Laurentide Ice Sheet to form a vast ice barrier along the eastern seaboard of the United States from Maine to Long Island, New York. The well-known erratics of Central Park, New York City, are testaments to the efficiency of long-distance ice transport. The precise location of the late Wisconsinan ice margin is the subject of considerable debate. Off the coast of Boston, for example, ice appears to have extended to the edge of the continental shelf where subaqueous proximal deltas formed, often associated with massive submarine slumps and erosive scars.

On land, the late Wisconsinan ice margin stretched from Long Island across southern New York State into northern Pennsylvania, then westward across central Ohio, Indiana, Illinois, Iowa, Wisconsin, and Minnesota to the Dakotas, into the foothills of the Rockies in Montana, and finally

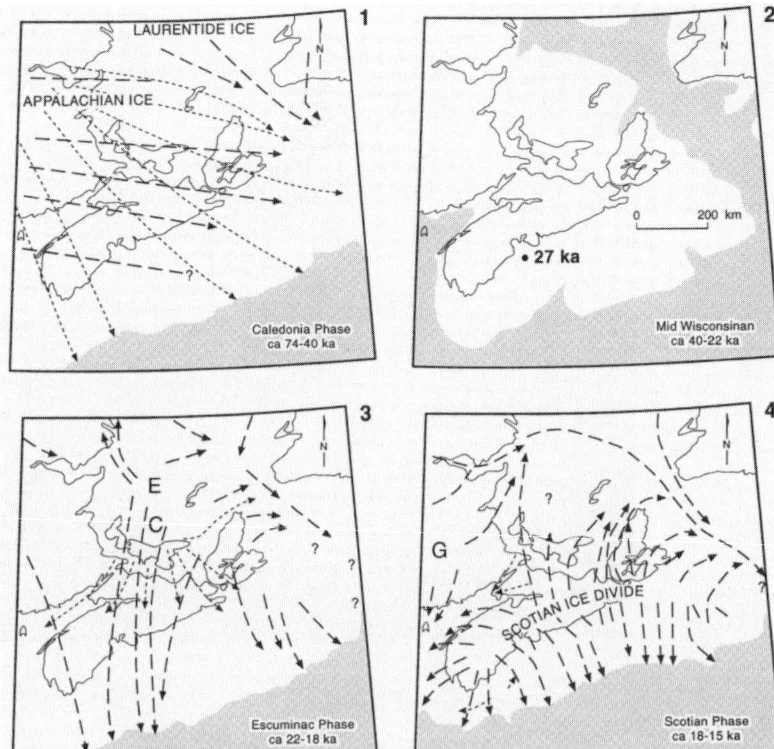

Figure 2.12 Development of ice divides and domes in Maritime Canada during the Wisconsinan (75–10 ka) (modified from Stea et al., 1998). White represents ice sheet cover; dashed and dotted arrows represent ice flow lines; E: Escuminac Ice Center on the Magdalen Shelf; C: Chignecto glacier; G: Gaspereau Ice Center in New Brunswick.

across the continental divide to Washington (see also chapters 16 and 17). For much of its distance across the midwestern United States, this late Wisconsinan ice margin thus lay to the north of the earlier Pleistocene margins, which, at their maximum extent, looped across the midcontinent from Cincinnati through St. Louis to Kansas City before turning northwest along the line of the present Missouri valley to Montana (fig. 2.1, see also chapter 17). Most notably, the Mississippi valley from St. Louis northward to near Minneapolis remained ice-free during the LGM, and part of southwest Wisconsin, the so-called "Driftless Area," may never have been glaciated. Much literature has been published on these margins and their frequent large and often very localized vacillations (Matsch and Schneider, 1986; Richmond and Fullerton, 1986b; Šibrava et al., 1986; Wright, 1987; Clark et al, 1993)). In some places, the late Wisconsinan ice margin must have been well over 200 m in height, whereas elsewhere a thin wedge of ice no more than a few meters high marked the edge of the ice. In other areas, vast expanses of the margins would have been buried below supraglacial debris, making the location of the margin much less definite (notably in parts of Illinois and Iowa). Finally, where large proglacial lakes existed, ice margins may, depending on water depths, have been floating. It seems unlikely that, at the southern edge of the Laurentide Ice Sheet, any ice shelves would have existed for more than a few weeks at a time. Unlike maritime locations along the coasts of North America, the southern margins of both the Cordilleran and Laurentide ice sheets during the late Wisconsinan are clearly delineated and dated. Debate now focuses on the many fluctuations of these ice margins over this vast stretch of territory, their timing, their correlation to local and regional climatic indicators, and their correspondence with periods of faunal and vegetation recolonization, rather than on precise geomorphological arguments over ice-margin location (Clark et al., 1993).

2.5 Aftermath of Glaciation—The Legacy

The impact of the glaciation of North America has been immense. Figure 2.13 uses an overlay of the maximum extent of glacial deposits when compared with the proportional distribution of the population of both Canada and the United States (Berg et al., in prep). On a map of the normal distribution of the population for the continent, the impact of glaciation seems rather meager, restricted to perhaps 50% or less of the land area; but when the map is redrawn in terms of the proportional population distribution, the impact of glaciation on the largest areas of population becomes readily apparent. The com-

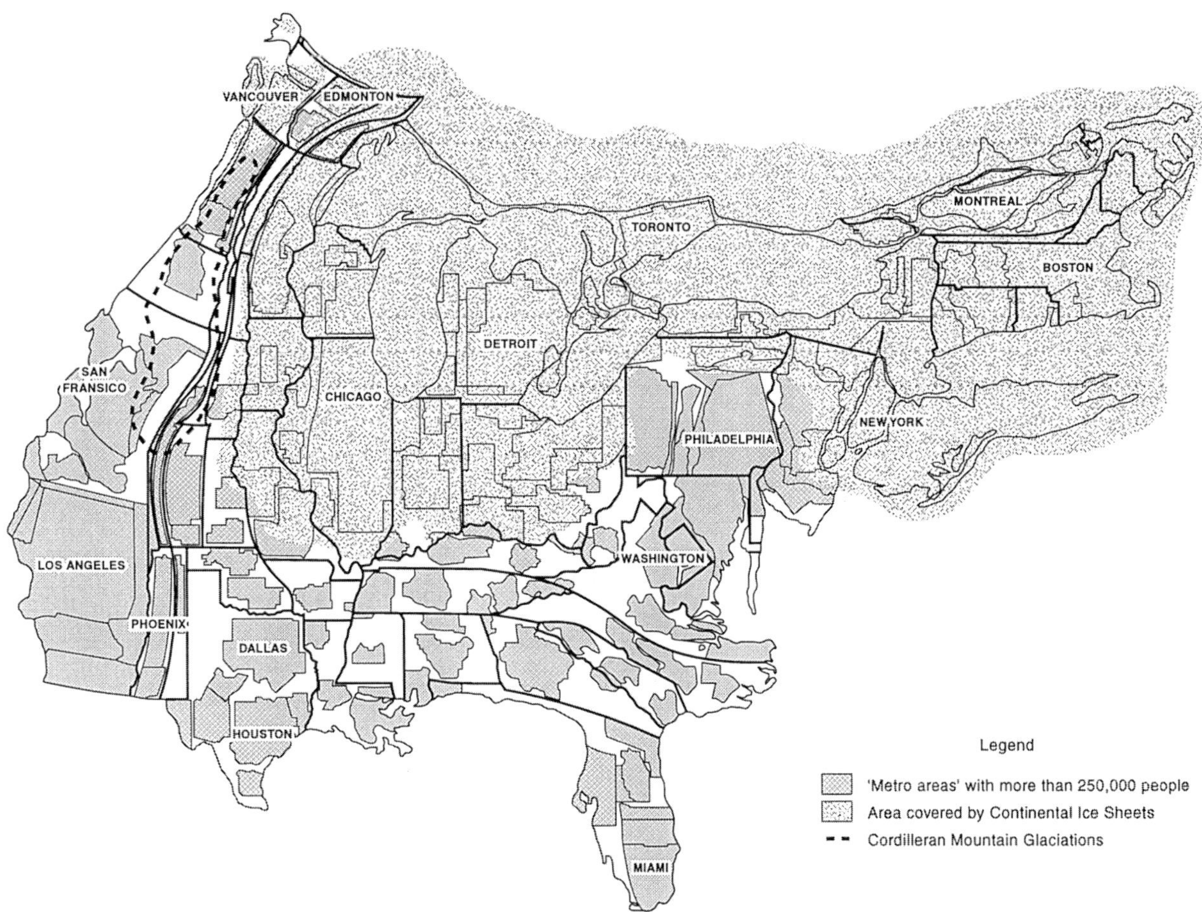

Figure 2.13 A composite map showing the isodemographic map of North America (1975–1976) overlain by the areal extent of the Laurentide and Cordilleran Ice Sheets at the Last Glacial Maximum limits. The composite map illustrates the impact of glacial geology on the spatially proportional population distribution of North America. [The population map was drawn by the Cartographic Lab., Queen's University, Kingston, Ontario: this compilation map is modified from Berg et al.]

posite map illustrates the extraordinary impact that glacial deposits have on the human population of North America in terms of agriculture, groundwater, and construction. This map does not include those populated areas covered by glacioaeolian loess deposits or riverine sediments that are derivatives of the effects of glaciation. In one sense, the glaciations that have affected the continent have been of such a large geographical and temporal scale as to be overwhelming. All aspects of the environment of the continent, both those areas covered by the ice and those that remained ice-free, have been affected in some measure. If we begin with the landscape itself, all terrain within the boundaries of the glaciations has been largely altered, covered in glacial sediments, or suffered varying degrees of glacial erosion, while terrain beyond the margins of the ice has been covered by vast proglacial lakes, suffered glacial lake outbursts, such as those from Lake Missoula (see chapter 3), or covered by glacioaeolian sediments. River systems far removed from the active ice margins, marine basins such as the Gulf of Mexico, estuaries such as the Chesapeake Bay, and coral reefs off the present coast of Florida have all been influenced to a lesser or greater extent by sediment plumes, meltwater discharges, sea-level changes, and associated geochemical exchanges. At the maximum extent of glaciation, climatic, vegetational, and faunal ecozones were "compressed" into the southern United States such that tundra-like conditions and associated plant and animal life prevailed as far south as Oklahoma (Wright, 1987). With the onset of deglaciation, rapid climatic changes and much slower vegetational and faunal recolonization occurred in the formerly glaciated regions of the continent. In some measure, these changes are still occurring today in the far north of Canada, just as isostatic rebound continues over more northerly parts of the continent in response to the loss of ice mass from the crust (fig. 2.14).

Figure 2.14 Raised beaches in an ascending flight sequence on the southwest shore of Hudson Bay, Ontario (modified from Prest, 1983).

Perhaps the greatest impact of glaciation has been in the production and eventual deposition of quantities of glacial sediments from a wide set of lithological sources, involving all particle sizes in various stratigraphical positions. Sediments range from tills and dense subaqueous diamicts, to coarse sands and gravels from glaciofluvial sources, to finely laminated clays and muds, to blankets of massive unstratified glacioaeolian dusts of yellow loess (Brodzikowski and Van Loon, 1991). These sediments provide a vast array of parent materials on which soils can develop and evolve (Birkeland, 1984). These same soil types in association with regional climates tolerate a wide range of vegetation communities that are a vast natural resource, a byproduct of continental glaciation. The agricultural and forestry industries of much of North America are the beneficiaries of this complex pattern of glacial sediments, climates, and plant recolonization dynamics.

In those areas once glaciated, no building or structure of any form can be erected, nor can a road, pavement, dam, sewer pipe, or harbor jetty be constructed, without some heed being given to the influence of glaciation and glacial sediments. The range of sediment types, their particle sizes, their geotechnical variety and variability all require intense examination before a structure of any weight or complexity can be built. Changes in clay mineral type and volume is just one example of the engineering soil patterns that must be studied closely (Scott, 1976). Glacial sediments are often deemed to be overconsolidated, but this assumption is more often than not erroneous. Likewise, where glacial clays occur, a wider range in void ratios and internal angles of friction occurs than might be typically expected. Because of the complexity of glacial sediment stratigraphy, foundation engineering requires extensive borehole and geophysical investigations before beginning construction.

With respect to groundwater sources, pathways, and fluxes, glacial sediments typically are extremely complicated in terms of permeability, porosity, and recharge capacity (Freeze and Cherry, 1979). Perched aquifers, localized aquitards, sediment dykes, and distinctive fracture patterns are common within these sediments yet are often unpredictable in their location and extent (Fredericia, 1993). The result is that ground investigations within glacial sediments must be prudent and extensive (Lloyd, 1983; De Mulder and Hageman, 1989). The movement of toxic substances within groundwater as the result of accidental pollution or deliberate emplacement of toxic wastes in landfill sites is one of the most crucial concerns in studying the hydrology of glacial sediments (Cooper et al., 1989). The ability to trace and predict the path of such toxic substances remains a very difficult yet fundamental task. In heavily populated areas dependent on groundwater for domestic use, concerns for pollutant pathways and contamination within glacial sediments remain acute. During the 1980s, it was recognized that a centralized facility for the dumping of toxic wastes, especially PCBs, was much needed in the metropolitan Toronto area. Investigations of potential locations for such a huge dump, close to the major cities in the lower Great Lakes region as well as to good transport routes, resulted in several initial sites being rejected, often as a consequence of specific concerns regarding glacial stratigraphy and sediment characteristics (Cooper et al., 1989).

Last, glaciation has had an impact on the mining industry. Because of the ability of glacial ice to transport sediments over considerable distances, both precious metals and productive mineral deposits can be detected. These metals and mineral resources can often be discovered from examination of sediment-transport pathways that act as indicator trails leading back up the ice flow line to ore bodies and deposits of economic value. Such methods have been successfully used, for example, in Canada and in the mountains of Alaska and Idaho (Repsher et al., 1980; DiLabio and Coker, 1989; Stephens et al., 1990; Shilts, 1996; Levson and Morison, 1996;). Ore or mineral-bearing rocks can often be found as placer deposits in ancient glacial river valleys typically within glaciofluvial sands and gravels. Excellent examples of such deposits abound in British Columbia, Alaska, Montana, Idaho, and the Yukon Territory (Levson, 1992; Hamilton, 1994; Bobrowsky et al., 1995; Levson and Morison, 1996; Nokleberg, 1997).

In summary, the impact in geologic, geomorphic, biotic, and human terms of the Pleistocene glaciations of North America has been and continues to be far reaching, intrusive, and unremitting. In some instances, the impact and the legacy of these continental glaciations remain to be fully understood. Much knowledge concerning these dramatic and wide-ranging events remains to be learned.

Acknowledgments In the compilation of this chapter, I especially thank several individuals who kindly assisted with advice and information: John Andrews, Richard Berg, Julie Brigham-Grette, Norm Catto, John Clague, John England, Shawn Marshall, David Piper, Bill Shilts, and Ralph Stea. Finally, my thanks to Loris Gasparotto for his superb cartographic skills.

References

Andrews, J.T., 1989. Quaternary Geology of the Northeastern Canadian Shield. In Fulton, R.J. (ed.) *Quaternary geology of Canada and Greenland.* Geological Survey of Canada No. 1 (also Decade of North American Geology Project, Geological Society of America, The Geology of North America, Vol. K-1), Chapter 3, 276–302.

Andrews, J.T., 1997. Northern Hemisphere (Laurentide) deglaciation: Processes and responses of the Ice Sheet/ Ocean Interaction. In Martini, I.P. (ed.) *Late Glacial and Postglacial Environmental Changes*, Oxford University Press, New York Chapter 2, 9–27.

Andrews, J.T. and W.R. Peltier, 1989. Quaternary Geodynamics in Canada. In Fulton R.J. (ed.) *Quaternary geology of Canada and Greenland.* Geological Survey of Canada No. 1 (also Decade of North American Geology Project, Geological Society of America, The Geology of North America, Vol. K-1) Chapter 8, 543–572.

Andrews, J.T., L.E. Osterman, A.E. Jennings, J.P.M., Syvitski, G.H. Miller, and N. Wiener, 1996. Abrupt changes in maritime conditions, Sunneshine Fiord, eastern Baffin Island, NWT during the last deglacial transition: Younger Dryas and H-0 events. In Andrews, J.T., W.E.N., Austin, H. Bergsten, and A.E. Jennings, (eds.) *Late Quaternary Palaeoceanography of the North Atlantic Margins.* Geological Society Special Publication No. 111, The Geological Society, 1–6.

Berg, R.C., N.K, Bleuer, B.E. Jones, K.A. Kincare, R.R. Pavey, and B.D. Stone. Mapping the glacial geology of the central Great Lakes region in three dimensions: A model for state-federal cooperation: *U.S. Geological Survey Open-file Report.*

Birkeland, P.W., 1984. *Soils and Geomorphology.* Oxford University Press, New York 372 p.

Bobrowsky, P.T., S.J. Sibbick, J.M. Newell, and P.F. Matysek, 1995. Drift exploration in the Canadian Cordillera. *British Columbia Ministry of Energy, Mines and Petroleum Resources*, Paper 1995–2, 304p.

Bond, G., H. Heinrich, W. Broecker, L. Labeyrie, J. McManus, J.T. Andrews, S. Huon, R. Jantschik, S. Clasen, C. Simet, K. Tedesco, M. Klas, G. Bonani, and S. Ivy, 1992. Evidence of massive discharges of icebergs into the North Atlantic Ocean during the last glacial period. *Nature*, 360, 245–249.

Booth, D.B., 1987. Timing and processes of deglaciation along the southern margin of the Cordilleran ice sheet. In Ruddiman, W.F. and H.E. Wright Jr., (eds.), *North America and Adjacent Oceans During the Last Deglaciation.* Geological Society of America, Boulder, Colorado. The Geology of North America, K-3, 71–90.

Boulton, G.S., G.S. Smith, A.S. Jones, and J. Newsome, 1985. Glacial geology and glaciology of the last midlatitude ice sheets. *Journal of Geological Society of London*, 142, 447–474.

Bowen, D.Q., G.M. Richmond, D.S. Fullerton, V. Šibrava, R.J. Fulton, and A.A. Velichko, 1986. Correlation of Quaternary Glaciations in the Northern Hemisphere. In: Šibrava, V., D.Q. Bowen, and G.M. Richmond, (eds.), Quaternary Glaciations in the Northern Hemisphere. *Quaternary Science Reviews*, 5, 509–510 and Chart 1.

Brodzikowski, K., and A.J. Van Loon, 1991. *Glacigenic Sediments.* Developments in Sedimentology, 49. Elsevier Science Publications, Amsterdam. 674 p.

Broecker, W.S., J.P. Kennett, J. Teller, S. Trumbore, G. Bonani, and W. Wolfli, 1989. Routing of meltwater from the Laurentide Ice Sheet during the Younger Dryas cold episode. *Nature*, 341, 318–321.

Brown, N.E., B. Hallet, and D.B. Booth, 1987. Rapid soft bed sliding of the Puget Glacial Lobe. *Journal of Geophysical Research*, 92: 8985–8997.

Budd, W.F., and I.N. Smith, 1987. Conditions for growth and retreat of the Laurentide ice sheet. *Géographie Physique et Quaternaire*, 41, 279–290.

Calkin, P.E., 1995. Global glacial chronologies and causes of glaciation. In Menzies, J. (ed.) *Modern Glacial Environ-*

ments, Glacial Environments: Volume I, Butterworth-Heineman, Chapter 2, 9–76.

Clague, J.J., 1989. Quaternary geology of the Canadian Cordillera. In Fulton R.J. (ed.) *Quaternary geology of Canada and Greenland.* Geological Survey of Canada No. 1 (also Decade of North American Geology Project, Geological Society of America, The Geology of North America, Vol. K-1) Chapter 1, 1–95.

Clark, P.U., 1984. *Glacial geology of the Kangalaksiorvik-Abloviak region, northern Labrador, Canada.* Ph.D. Thesis, University of Colorado, Boulder, 240p.

Clark, P.U., 1992. Surface form of the southern Laurentide ice sheet and its implications to ice-sheet dynamics. *Geological Society of America Bulletin*, 104: 595–605.

Clark, P. U., 1994. Unstable behavior of the Laurentide Ice Sheet over deforming sediment and its implications for climate change. *Quaternary Research*, 41, 19–25.

Clark, P.U., 1997. Sediment deformation beneath the Laurentide Ice Sheet. In Martini, I.P. (ed.) *Late Glacial and Postglacial Environmental Changes*, Oxford University Press, New York, Chapter 6, 81–97.

Clark, P.U., C.M. Licciardi, D.R. MacAyeal, and J.W. Jenson, 1996. Numerical reconstruction of a soft-bedded Laurentide Ice Sheet during the last glacial maximum. *Geology*, 23, 679–682.

Clark, P.U., J.J. Clague, B.B. Curry, et al., 1993. Initiation and development of the Laurentide and Cordilleran Ice Sheets following the last Interglaciation. *Quaternary Science Reviews*, 12, 79–114.

COHMAP Members 1988. Climatic changes of the last 18,000 years: Observations and model simulations: *Science*, 241, 1043–1052.

Colgan, P.M., and S. Principato, 1998. Distribution of glacial landforms and sediments in Wisconsin and the Upper Peninsula of Michigan: An application of GIS to glacial geology. *Geological Society of America, Abstracts with Programs*, 30, 11.

Cooper, A.J., G.H. Funk, and E.G. Anderson, 1989. Using Quaternary stratigraphy to help locate a hazardous waste treatment site. In De Mulder, E.F.J., and B.P. Hageman (eds.), *Applied Quaternary Research*, A.A. Balkema Publishers, Rotterdam. pp. 1–13.

De Mulder, F.J., and B.P. Hageman, 1989. *Applied Quaternary Research.* A.A. Balkema Publishers, Rotterdam, 185 p.

Denton, G.H., and T.J. Hughes, (eds.), 1981. *The Last Great Ice Sheets.* Wiley-Interscience, New York, 484 p.

DiLabio, R.N.W., and W.B. Coker, 1989. Drift Prospecting. *Geological Survey of Canada*, Paper 89–20, 169 p.

Dyke, A.S., L.A. Dredge, and J.-S. Vincent, 1982. Configuration and dynamics of the Laurentide Ice Sheet during the Late Wisconsin maximum. *Géographie Physique et Quaternaire*, 36, 5–14.

Easterbrook, D.J., 1986. Stratigraphy and chronology of Quaternary deposits of the Puget Lowland and Olympic Mountains of Washington and the Cascade Mountains of Washington and Oregon. In Šibrava, V., D.Q. Bowen, and G.M. Richmond (eds.), Quaternary Glaciations in the Northern Hemisphere. *Quaternary Science Reviews*, 5: 145–159.

Ehlers, J., 1996. *Quaternary and glacial geology*; translated from *Allgemeine und historische Quartärgeologie*; English version by Gibbard, P.L., J. Wiley & Sons, New York, 578 p.

England, J.H., 1976. Late Quaternary glaciation of the Eastern Queen Elizabeth Islands, NWT, Canada: Alternative models. *Quaternary Research*, 6, 185–203.

Eyles, C.H., and N. Eyles, 1989. The upper Cenozoic White River "tillites" of southern Alaska: Subaerial slope and fan-delta deposits in a strike-slip setting. *Geological Society of America, Bulletin*, 101, 1091–1102.

Eyles, N., 1993. Earth's glacial record and its tectonic setting. *Earth Science Reviews*, 35 (½), 1–248.

Eyles, N., and A.D. Miall, 1984. Glacial Facies. In Walker, R.G. (ed), *Facies Models.* Geoscience Canada Reprint Series 1, 2nd ed., 15–38.

Fisher, D.A., N. Reeh, and K. Langley, 1985. Objective reconstructions of the Late Wisconsinan Laurentide Ice Sheet and the significance of deformable beds. *Géographie Physique et Quaternaire*, 39, 229–238.

Flint, R.F., 1957 Glacial and Pleistocene Geology. John Wiley & Sons, New York and London, 553 p.

Fredericia, J., 1993. Fractures in Clayey Till in Denmark: Occurrence, Genesis and Hydrogeological Significance. *Geological Society of America, 1993 Annual Meeting, Boston, Massachusetts*, October 25–28, p. A-426.

Freeze, R.A., and J.A., Cherry, 1979. *Groundwater.* Prentice Hall Englewood Cliffs, New Jersey, 604 p.

Frye, J.C., H.D. Glass, and H.B. Willman, 1968. Mineral zonation of Woodfordian loesses of Illinois. *Illinois Geological Survey* Circular 427, 44 p.

Fullerton, D.S., and G.M. Richmond, 1986. Comparison of the marine oxygen isotope record, the eustatic sea level record, and the chronology of glaciation in the United States of America. In Šibrava, V., D.Q. Bowen, and G.M. Richmond, (eds.) Quaternary Glaciations in the Northern Hemisphere, Report of the IGCP Project No. 24. *Quaternary Science Reviews*, 5, 197–205.

Fulton R.J., (ed.) 1989. *Quaternary geology of Canada and Greenland.* Geological Survey of Canada No. 1 (also Decade of North American Geology Project, Geological Society of America, The Geology of North America, Vol. K-1), 839 p.

Funder, S., 1989. Quaternary geology of the ice-free areas and adjacent shelves of Greenland. In Fulton R.J., (ed.) *Quaternary geology of Canada and Greenland.* Geological Survey of Canada No. 1 (also Decade of North American Geology Project, Geological Society of America, The Geology of North America, Vol. K-1) Chapter 13, 743–792.

Grant, D.R., 1989. Quaternary Geology of the Atlantic Appalachian Region of Canada. In Fulton, R.J., (ed.) *Quaternary geology of Canada and Greenland.* Geological Survey of Canada No. 1 (also Decade of North American Geology Project, Geological Society of America, The Geology of North America, Vol. K-1) Chapter 5, 393–440.

Grant, D.R., 1994. Quaternary geology, Cape Breton Island. *Geological Survey of Canada*, Bulletin 482, 159 p.

Greve, R., 1997. A continuum mechanical formulation for shallow polythermal ice sheets. *Philosophical Transactions of the Royal Society London*, A355, 921–974.

Greve, R., and MacAyeal, D.R., 1996. Dynamic/thermodynamic simulations of the Laurentide ice-sheet instability. *Annals of Glaciology*, 23, 328–335.

Hallberg, G.R., 1986. Pre-Wisconsin glacial stratigraphy of the central plains region in Iowa, Nebraska, Kansas, and Missouri. In Šibrava,V., D.Q. Bowen, and G.M. Richmond, (eds), Quaternary Glaciation in the Northern Hemisphere. Report of the IGCP Project No. 24. *Quaternary Science Reviews*, 5: 11–15.

Hamilton, T.D., 1994. Late Cenozoic glaciation of Alaska. In Plafker, G., and H.C. Berg, (eds.) *Geology of Alaska*. Decade of North American Geology Project, Geological Society of America, Volume G-1, Chapter 27, 813–844.

Hamilton, T.D., and R.M. Thorson, 1983. The Cordilleran ice sheet in Alaska. In Porter, S.C., (ed.), *Late-Quaternary Environments of the United States*, Volume 1—The Late Pleistocene. University of Minnesota Press, Minneapolis, 38–52.

Hodgson, D.A., 1989. Quaternary Geology of the Queen Elizabeth Islands. In Fulton, R.J., (ed.) *Quaternary geology of Canada and Greenland*. Geological Survey of Canada No. 1 (also Decade of North American Geology Project, Geological Society of America, The Geology of North America, Vol. K-1) Chapter 6, 443–459.

Hughes, O.L., R.B. Campbell, J.E. Muller, and J.O. Wheeler, 1969. Glacial limits and flow patterns, Yukon Territory, south of 65 degrees north latitude. *Geological Survey of Canada*, Paper 68–34, 9 p.

Hughes, T.J., 1992a. On the pulling power of ice streams. *Journal of Glaciology*, 38, 125–151.

Hughes, T.J., 1992b. Abrupt climatic changes related to unstable ice-sheet dynamics: Toward a new paradigm. *Palaeogeography, Palaeoclimatology, Palaeoecology*, 97, 203–234.

Imbrie, J., E.A. Boyle, S.C. Clemens, et al. 1992. On the structure and origin of major glaciation cycles, 1. Linear responses to Milankovitch forcing. *Paleoceanography*, 7, 701–738.

Jenson, J.W., P.U. Clark, D.R. MacAyeal, C.L. Ho, and J.C. Vela, 1995. Numerical modeling of advective transport of saturated deforming sediment beneath the Lake Michigan Lobe, Laurentide Ice Sheet. *Geomorphology*, 14, 157–166.

Jenson, J.W., D.R. MacAyeal, P.U. Clark, C.L. Ho, and J.C. Vela, 1996. Numerical modeling of subglacial sediment deformation: Implications for the behaviour of the Lake Michigan Lobe, Laurentide Ice Sheet. *Journal of Geophysical Research*, 101(B4), 8717–8728.

Jopling, A.V., W.N. Irving, and B.F. Beebe, 1981. Stratigraphic, sedimentological and faunal evidence for the occurrence of pre-Sangamonian artefacts in northern Yukon. *Arctic*, 34, 3–33.

Kamb, B., 1970. Sliding motion of glaciers: Theory and observation. *Reviews of Geophysics and Space Physics*, 8, 673–728.

Klassen, R.A., and A. Bolduc, 1984. Ice flow directions and drift composition, Churchill Falls, Labrador. In Current Research, Part A, *Geological Survey Canada*, Paper 84–1A, 255–258.

Kleman, J., C. Hättestrand, I. Bergström, and A. Stroeven, 1997. Fennoscandian paleoglaciology reconstructed using a glacial geological inversion model. *Journal of Glaciology*, 43, 283–299.

Kukla, G.J., 1989. Long continental records of climate—An introduction. *Palaeogeography, Palaeoclimatology, Palaeoecology*, 72, 1–9.

Levson, V.M., 1992. The sedimentology of Pleistocene deposits associated with placer gold bearing gravels in the Livingstone Creek area, Yukon Territory. In Bremner, T.J., (ed.) *Yukon Geology*, 3, 99–132.

Levson, V.M., and S.R. Morison, 1996. Geology of placer deposits in glaciated environments, In Menzies, J., (ed.) *Past Glacial Environments—Sediments, Forms and Techniques*. Volume II, Butterworth-Heineman, Oxford. Chapter 16, 441–478.

Lindsay, E.H., 1997. The Pliocene-Pleistocene boundary in continental sequences of North America. In Van Couvering, J.A., (ed.), *The Pleistocene Boundary and the beginning of the Quaternary*. Cambridge University Press, Cambridge, Chapter 30, 278–289.

Lloyd, J.W., 1983. Hydrogeological investigations in glaciated terrains. In Eyles, N., (ed.) *Glacial Geology*. Pergamon Press, Elmsford, New York, Chapter 15, 349–368.

Lowell, T.V., and C.S. Brockman, 1994. Quaternary sediment sequences in the Miami lobe and environs. *Midwest Friends of the Pleistocene Annual Meeting: Cincinnati*, University of Cincinnati, Cincinnati, 68 p.

MacAyeal, D.R., 1992. Irregular oscillations of the West Antarctic ice sheet. *Nature*, 359: 29–32.

MacAyeal, D.R., 1993 Binge/Purge oscillations of the Laurentide ice sheet as a cause of the North Atlantic's Heinrich events. *Paleoceanography*, 8, 775–784.

MacAyeal, D.R., R.A. Bindschadler, and T.A. Scambos, 1995. Basal friction of ice stream E, West Antarctica. *Journal of Glaciology*, 41, 247–262.

Maher, L.J., and D.M. Mickelson, 1997. Palynological and radiocarbon evidence for deglaciation event in the Green Bay Lobe, Wisconsin. *Quaternary Research*, 46, 251–259.

Marshall, S., and G.K.C. Clarke, 1997a. A continuum mixture model of ice stream thermomechanics in the Laurentide Ice Sheet 1. Theory. *Journal of Geophysics Research*, 102(B9), 20,599–20,613.

Marshall, S., and G.K.C. Clarke, 1997b. A continuum mixture model of ice stream thermomechanics in the Laurentide Ice Sheet 2. Application to the Hudson Strait Ice Stream. *Journal of Geophysics Research*, 102(B9), 20,615–20,637.

Marshall, S., G.K.C., Clarke, A.S. Dyke, and D.A. Fisher, 1996. Geological and topographic controls on fast flow in the Laurentide and Cordilleran Ice Sheets. *Journal of Geophysics Research*, 101(B8), 17,827–17,839.

Matsch, C.L., and A.F. Schneider, 1986. Stratigraphy and correlation of the glacial deposits of the glacial lobe complex in Minnesota and northwestern Wisconsin. In Šibrava, V., D.Q. Bowen, and G.M. Richmond, (eds.) Quaternary Glaciations in the Northern Hemisphere, Report of the IGCP Project No. 24. *Quaternary Science Reviews*, 5, 59–64.

Menzies, J., 1995. Glaciers and Ice Sheets. In Menzies, J., (ed.) *Modern Glacial Environments*, Glacial Environments, Volume I, Butterworth-Heineman, Chapter 4, 101–138.

Menzies, J., and W.W. Shilts, 1996. The Subglacial Environment. In Menzies, J., (ed.) *Past Glacial Environments*, Volume II, Butterworth-Heineman, Chapter 2, 15–136.

Murray, T., 1997. Assessing the paradigm shifts: Deformable glacier beds. *Quaternary Sciences Reviews*, 16, 995–1016.

Nokleberg, W.J., 1997. Significant metalliferous and selected non-metalliferous lode deposits and placer deposits from the Russian Far East, Alaska, and the Canadian Cordillera. *United States Geological Survey*, Open-File Report, OF-96–0513 (disc).

Pasini, G., and M.L. Colalongo, 1997. The Pliocene-Pleistocene boundary-stratotype at Vrica, Italy. In Van Couvering, J.A., (ed.), *The Pleistocene Boundary and the beginning of the Quaternary*. Cambridge University Press, Cambridge, Chapter 2, 15–45.

Paterson, W.S.B., 1994 *The Physics of Glaciers*. 3rd Edition, Pergamon Press, Oxford, 480 p.

Payne, T., 1995. Limit cycles in the basal thermal regime of ice sheets. *Journal of Geophysical Research*, 100(B3), 4249–4263.

Porter, S.C., (ed.), 1983. *Late-Quaternary Environments of the United States*, Volume 1—The Late Pleistocene. University of Minnesota Press, Minneapolis.

Prest, V.K., 1983. Canada's Heritage of Glacial Features. Geological Survey of Canada, Misc. Rep. 28, 119 p.

Prest, V.K., 1984. The late Wisconsinan glacier complex. In Fulton, R.J., (ed.), *Quaternary Stratigraphy—A Canadian Contribution to IGCP Project 24. Geological Survey of Canada*, Paper 84–80, pp. 21–36.

Prest, V.K., D.R. Grant, and V.N. Rampton, 1969. Glacial map of Canada. *Geological Survey of Canada* Map 1253.

Repsher, A.A., J.F.P. Cotter, and E.B. Evenson, 1980. Glacial history, provenance and mineral exploration in an area of alpine glaciation, Custer and Blaine counties, Idaho. *Geological Society of America, Abstracts with Programs*, Rocky Mountain section, 33rd annual meeting Ogden, Utah, 301–302.

Richmond, G.M., and D.S. Fullerton, 1986a. Introduction to Quaternary Glaciations in the United States. In Šibrava, V., D.Q. Bowen, and G.M. Richmond, (eds.) Quaternary Glaciations in the Northern Hemisphere, Report of the IGCP Project No. 24. *Quaternary Science Reviews*,5, 3–10.

Richmond, G.M., and D.S. Fullerton, 1986b. Summation of Quaternary Glaciations in the United States of America. In Šibrava, V., D.Q. Bowen, and G.M. Richmond, (eds.) Quaternary Glaciations in the Northern Hemisphere, Report of the IGCP Project No. 24. *Quaternary Science Reviews*, 183–196.

Ritchie, J.C., 1984. Past and present vegetation of the far northwest of Canada. University of Toronto Press, Toronto, 251 p.

Robin, G. deQ., 1988. The Antarctic ice sheet, its history and response to sea level and climatic changes over the past 100 million years. *Palaeogeography, Palaeoclimatology, Palaeoecology*, 67, 31–50.

Ruddiman, W.F., and H.E. Wright Jr., (eds.) 1987. *North America and Adjacent Oceans during the Last Deglaciation*. The Geology of North America; an overview. Decade of North American Geology Project, Geological Society of America, Volume K-3, 501 p.

Schweger, C.E., 1989. Paleoecology of the western Canadian ice-free corridor. In Fulton, R.J., (ed.) *Quaternary geology of Canada and Greenland*. Geological Survey of Canada No. 1 (also Decade of North American Geology Project, Geological Society of America, The Geology of North America, Vol. K-1) Chapter 7, 491–498.

Scott, J.S., 1976. Geology of Canadian tills. In Legget, R.F. (ed.) *Glacial Till*, The Royal Society of Canada Special Publication No. 12, Ottawa, 50–66.

Shilts, W.W., 1996. Drift Exploration, In Menzies, J., (ed.) *Past Glacial Environments—Sediments, Forms and Techniques*. Volume II, Butterworth-Heineman, Oxford. Chapter 15, 411–439.

Shilts, W.W., C.M. Cunningham, and C.A. Kaszycki, 1979. The Keewatin ice sheet: Reevaluation of the traditional concept of the Laurentide ice sheet. *Geology*, 7, 537–541.

Šibrava, V., D.Q. Bowen, and G.M. Richmond, (eds.) 1986. Quaternary Glaciations in the Northern Hemisphere, Report of the IGCP Project No. 24. *Quaternary Science Reviews*, 5, 510 p.

Stea, R.R., D.J.W. Piper, G.B.J. Fader, and R. Boyd, 1998 Wisconsinan glacial and sea-level history of Maritime Canada and the adjacent continental shelf: A correlation of land and sea events. *Geological Society of America, Bulletin*, 110, 821–845.

Stephens, G.C., E.B. Evenson, and D.E. Detra, 1990. A geochemical sampling technique for use in areas of active alpine glaciation; an application from central Alaska Range. *Journal of Geochemical Exploration*, 37, 301–321.

Sugden, D.E., 1977. Reconstruction of the morphology, dynamics, and thermal characteristics of the Laurentide Ice Sheet at its maximum. *Arctic and Alpine Research*, 9, 21–47.

Thomas, R.H., 1979. The dynamics of marine ice sheets. *Journal of Glaciology*, 24, 167–177.

Van Couvering, J.A., (ed.) 1997. *The Pleistocene Boundary and the Beginning of the Quaternary*. Cambridge University Press, Cambridge, 269 p.

van Donk, J., 1976. O^{18} record of the Atlantic Ocean for the entire Pleistocene Epoch. In Cline, R.M., and J.D. Hays, (eds.) Investigation of Late Quaternary Paleoceanography and Paleoclimatology. *Geological Society of America, Memoir*, 145, 147–163.

Vincent, J.-S., 1989. Quaternary Geology of the Southeastern Canadian Shield. In Fulton, R.J., (ed.) *Quaternary Geology of Canada and Greenland*. Geological Survey of Canada No. 1 (also Decade of North American Geology Project, Geological Society of America, The Geology of North America, Vol. K-1) Chapter 3, 249–275.

Waitt Jr., R.B., and R.M. Thorson, 1983 The Cordilleran ice sheet in Washington, Idaho, and Montana. In Porter, S.C., (ed.), *Late-Quaternary Environments of the United States*, Volume 1—The Late Pleistocene. University of Minnesota Press, Minneapolis, 53–70.

Wright Jr., H.E., 1987. Synthesis; The Land south of the ice sheets. In Ruddiman, W.F., and H.E. Wright Jr., (eds.) *North America and adjacent Oceans during the Last Deglaciation*. The Geology of North America; an overview. Decade of North American Geology Project, Geological Society of America, Volume K-3, Chapter 22, 479–488.

Wright Jr., H.E, 1989. The Quaternary. In Bally, A.W., and A.R. Palmer, (eds.) *The Geology of North America; an overview*. Decade of North American Geology Project, Geological Society of America, Volume A, 513–536.

Young, G.M., 1996. Glacial environments of Pre-Pleistocene age. In Menzies, J., (ed.) *Past Glacial Environments*, Volume II, Butterworth-Heineman, Oxford, Chapter 7, 239–252.

3

The Pleistocene Legacy: Beyond the Ice Front

Antony R. Orme

Among the varied landscapes of North America, a basic distinction exists between those regions that were subject to repeated Pleistocene glaciation and those that were not. The regions beyond the southern and northwestern margins of the vast ice sheets that once covered much of the continent are important because only here was sufficient land exposed above sea level to house the many terrestrial physical and biological processes at work beyond the ice front. Elsewhere, the ice sheets either calved into the deeper waters of the Atlantic and Pacific Oceans, or grounded on now-submerged continental shelves. The southern margin between glaciated and unglaciated terrain forms a broad irregular arc, convex southward, from the Atlantic continental shelf off New England to the Pacific shelf beyond Juan de Fuca Strait (fig. 3.1). The northwest margin separates glaciated terrain in southern Alaska and the Brooks Range from unglaciated Beringia, which, during eustatic lowstands, formed a broad land connection with Asia.

The ice front as defined here is not a synchronous margin but a composite of different glacial maxima that collectively represents the farthest advance of continental ice and separates glaciated terrain from those landscapes shaped by nonglacial processes. Glaciers did of course exist at higher elevations farther south. Some, such as the ice caps over the Yellowstone Plateau, the Colorado Rockies, and the Sierra Nevada, waxed and waned repeatedly during the Pleistocene, sculpting terrain and influencing regional hydrology (fig. 3.2). But beyond the main ice front, landscapes, even during glacial stages, were dominated by nonglacial processes; such glaciers as did exist were but small parts of a vast and changing landscape.

This chapter examines those components of the modern landscape that have been inherited from Pleistocene events that took place beyond the main ice front. Evidence and dating of the climate changes that generated these landscape responses are reviewed. A perspective on Pleistocene climates and vegetation is then given, emphasizing the last interglacial-glacial cycle, ≈130–10 ka (thousand years before present). Two themes are elaborated, namely, the paleohydrological implications of climate change for selected rivers and lakes, and the aeolian landscapes formed from blowing loess and sand. Other impacts of glaciation beyond the ice front, namely, eustatic and isostatic effects and some biotic anomalies, are briefly discussed. The chapter concludes with a perspective on nonglaciated North America within the broader context of late Cenozoic climate change. These events were of course played out against a backdrop of continuing tectonism, notably in the west, and of isostatic crustal adjustments, upland denudation, and downstream sediment transfers.

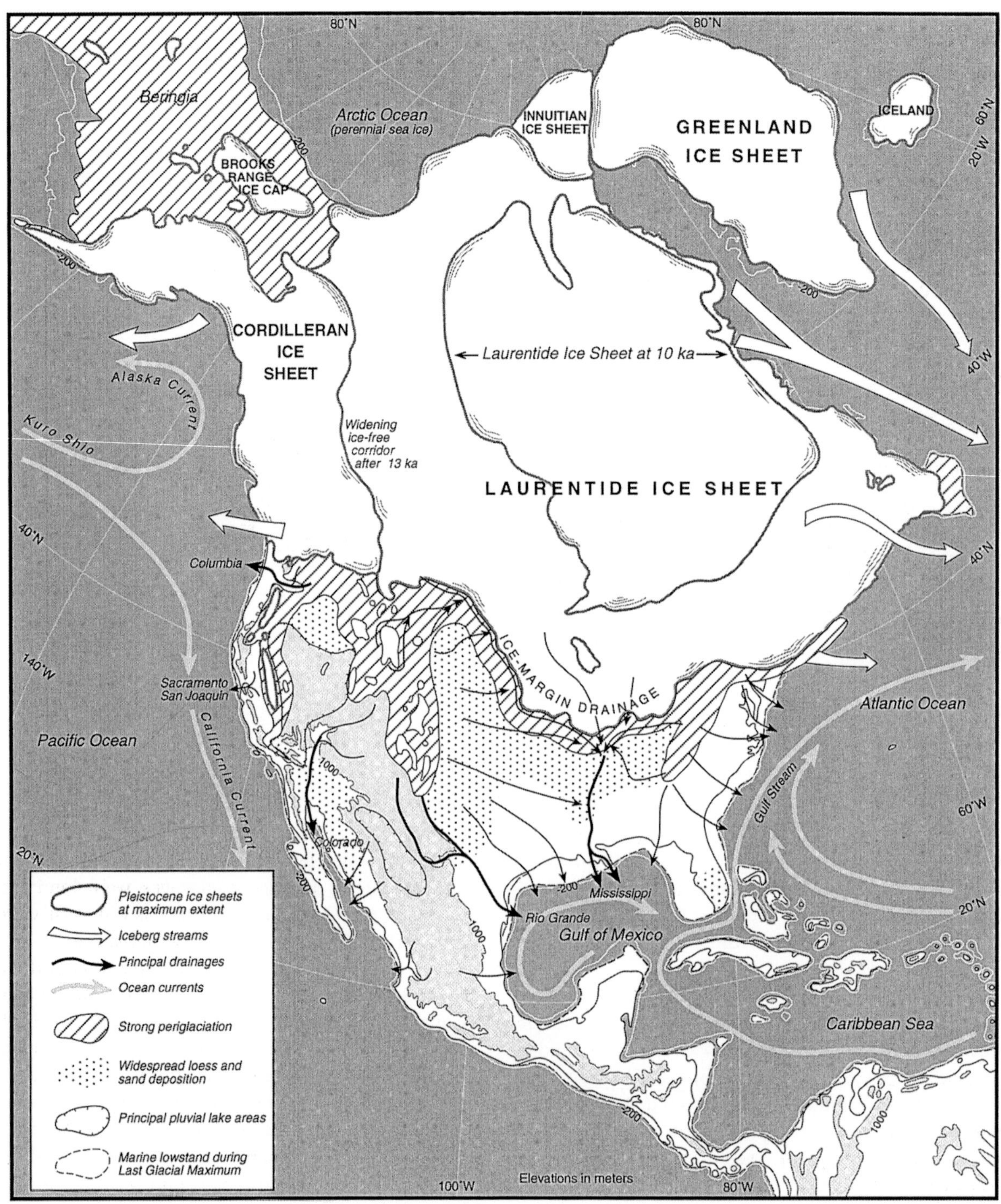

Figure 3.1 North America during the Last Glacial Maximum showing extent of continental ice sheets, larger ice caps, periglacial and aeolian activity, and main drainages beyond the ice front.

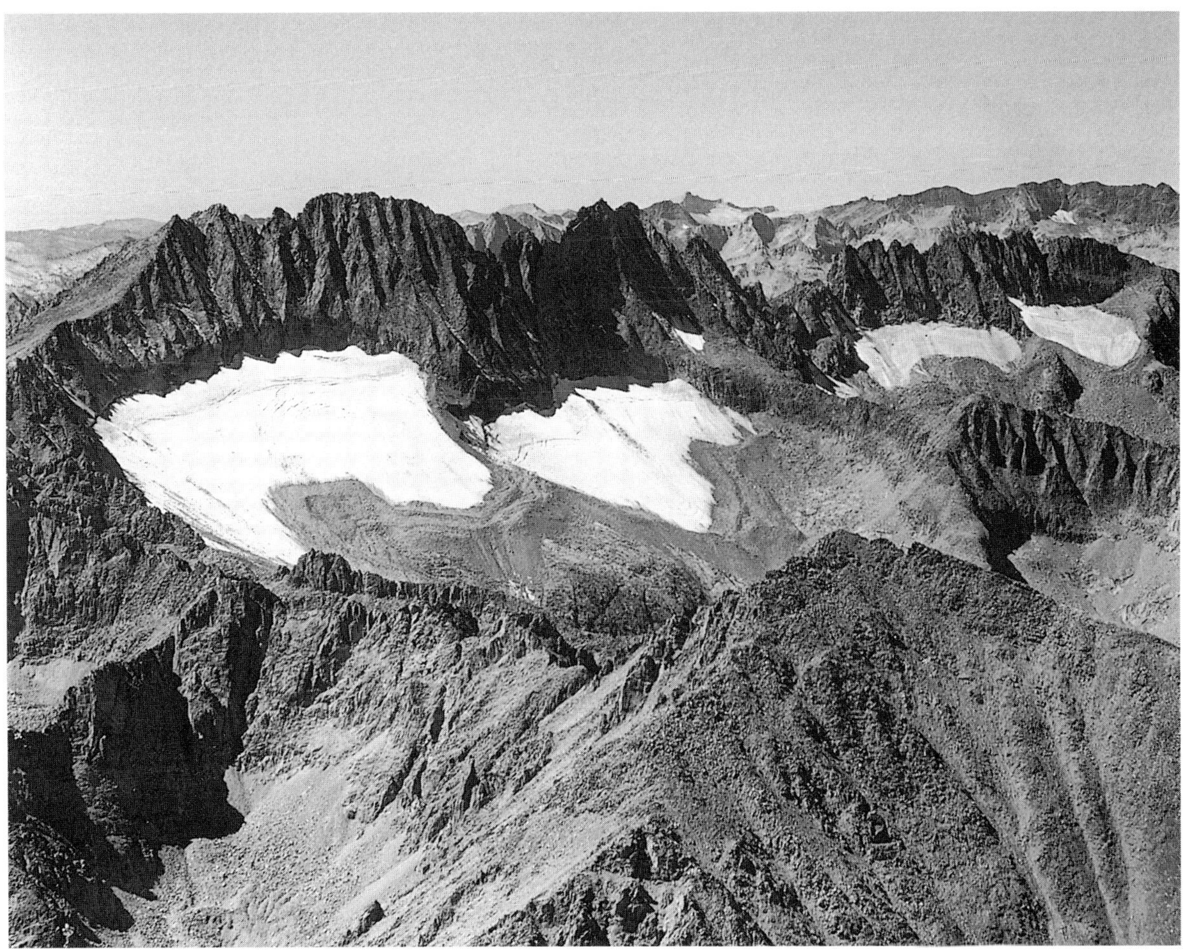

Figure 3.2 The Palisade glacier complex in the Sierra Nevada, California. Although far south of the main ice front, the Sierra Nevada, rising above 4000 m, supported ice caps during several Pleistocene cold stages. Today's small glaciers are reminders of this legacy, their recent retreat from late Holocene advances shown by fresh morainic ridges (photo: Spence Collection, UCLA).

3.1 Pleistocene Environments: Evidence and Dating

The Pleistocene Epoch constitutes most of the Quaternary Period, all but the last 10 ka designated as the Holocene Epoch. Once viewed as synonymous with the Ice Age, the Pleistocene is now known to have embraced numerous warm-cold cycles of varying duration and intensity. If the Pleistocene began around 1.8 Ma (million years before present), then some 18–20 warm-cold cycles have occurred since then (table 3.1). If the base of the Pleistocene is linked with magnetic geochronology and placed at the 2.5-Ma Gauss-Matuyama reversal, a recognizable marker in many nonglacial sedimentary sequences, or relocated to a major shift in the oxygen isotope record around 2.6 Ma, then many more warm-cold cycles have occurred since Pliocene cooling trends were first noted from Greenland and elsewhere. Certainly, some nine cycles have occurred since the Matuyama-Brunhes magnetic reversal around 0.79 Ma, the boundary between the early and middle Pleistocene. This sequence of climatic oscillations is a far cry from the notion of a simple monoglacial ice age that ushered in the glacial theory in the early nineteenth century, as later popularized by Agassiz (1840). Indeed, continental glaciation is but one expression of the many environmental changes that occurred during the Pleistocene. Glaciation was certainly important, but, beyond the ice front, climatic oscillations generated a broad spectrum of changes whose effects are essential to understanding the present landscape.

Evidence for Pleistocene landscapes beyond the ice front comes from a variety of terrestrial, glacial, and marine records. Terrestrial evidence is least satisfactory because, owing to nondeposition, burial, or subsequent erosion, it commonly lacks spatial and temporal continuity.

Table 3.1 A simplified Quaternary timescale. The logarithmic scale emphasizes the significance of later Quaternary events to present landscapes. Normal polarity chrons and subchrons, and relatively warm oxygen isotope stages are shown in bold type.

Age (yr B.P.)[a]	Magnetic Polarity Chron	Epoch	Sub-Epoch	Midcontinent Stage	Oxygen-Isotope Stage	Lower Age (ka)
1000		HOLOCENE	Late Holocene		**1**	3
			Middle Holocene			7
			Early Holocene			
10,000				terminal oscillations		10
		PLEISTOCENE	Late Pleistocene			14
					2	30
	BRUNHES			Wisconsinan	**3**	50
					4	75
					5a	85
					5b	95
100,000					**5c**	105
					5d	115
				Last Interglacial	**5e**	130
			Middle Pleistocene	Illinoian	6–8	300
				Pre-Illinoian	9–39	
						790
1,000,000	**Jaramillo**		Early Pleistocene			
	MATUYAMA					
	Olduvai	?	?	?		1.80 Ma
3,000,000	**GAUSS**	PLIOCENE			40	2.48 Ma

[a]Years before present.

Ice cores from Greenland and Antarctica offer more continuous records for the later Pleistocene, but basal deformation and melting have largely destroyed evidence from before 250 ka. Thus ocean cores, especially those obtained from intermediate depths beyond the continental shelf, offer the best archives for long-term landscape reconstruction, essentially because oceans are sumps for much that happens on Earth, for such proxy climatic evidence as atmospheric fallout, terrigenous debris, and dead marine organisms.

Dating of this evidence was long based on relative-age techniques such as lithostratigraphy, biostratigraphy, geomorphology, and soil development. Indeed, as long ago as 1839, Lyell used mollusk frequency to distinguish the Pleistocene, with 70% of living forms, from earlier and later epochs (Lyell, 1839). Sidereal methods such as tree rings and historical records may provide precise dates for Holocene events but not for the Pleistocene, although varve chronologies may extend back into the Pleistocene. Dating of Pleistocene events has, however, been stimulated in recent decades by development of sophisticated numerical-age techniques involving isotopic and radiogenic methods, initially with the application of ^{14}C and U-series in the 1950s, later with other cosmogenic isotopes such as ^{36}Cl and ^{10}Be. These in turn have been aided by improved correlative methods, notably the use of stable isotopes, tephrochronology, and magnetostratigraphy, and by developments in mass spectrometry (Dorn and Phillips, 1991; Rosholt et al., 1991). Nevertheless, despite continuing advances in dating techniques, the late Pleistocene, the last interglacial-glacial cycle (130–10 ka) remains much better known and its timescale more robust than the middle Pleistocene (790–130 ka). The early Pleistocene (before 790 ka) remains enigmatic, its record blurred by later events, but even here the temporal resolution of terrestrial and marine events is improving. All age estimates for the Quaternary continue to be revised and fine-tuned as fresh field data emerge, new methods are refined, and older data are reevaluated.

3.1.1 The Terrestrial Record

Terrestrial evidence in the form of glacial deposits and glaciated landforms provided the first affirmation of glacial theory in North America. Similarly, glacial features and intervening soils and peats in the midcontinent later provided evidence for the several glacial and interglacial stages that were defined toward the close of the nineteenth century. However, although once much used, the terms for the four glacial stages then recognized (from oldest to youngest, Nebraskan, Kansan, Illinoian, Wisconsinan), and for the three intervening interglacials (Aftonian, Yarmouthian, Sangamonian) are now nearly obsolete, discouraged in favor of marine oxygen-isotope stages, although Illinoian, Sangamonian, and Wisconsinan retain regional significance (Richmond and Fullerton, 1986).

Even as the glacial record was being refined, terrestrial evidence from beyond the ice front was being evaluated. The term *pluvial* was introduced for wet intervals in deserts that were thought to equate with glacial stages farther poleward. Scientific exploration of the American West identified many pluvial lakes, culminating in the classic works of Gilbert (1890) on Lake Bonneville and of Russell (1885) on Lake Lahontan. Today, sediments preserved within deeper lake basins are among North America's best terrestrial archives of Pleistocene climate change. As sumps for internal drainage basins, terminal lakes are particularly suited to record regional hydrologic responses to climate forcing, ranging from individual storms to full interglacial-glacial cycles, and many geomorphological, geochemical, geophysical, and chronostratigraphic tools are now available for deciphering these archives.

Among other terrestrial evidence of Pleistocene environments, relict periglacial features beyond the ice front and on distant mountains clearly indicate former frost climates (fig. 3.2). Extensive loess sheets in the Mississippi valley and sand dunes in the High Plains and the Southwest, now mostly stabilized, illustrate former aeolian processes favored by high wind velocities, abundant loose sediment, and limited vegetation cover. River terraces, valley fills, and alluvial fans likewise reflect changing hydrologies that may relate to climate, although other factors such as tectonism must be evaluated. Buried soils, weathering profiles, and rock varnish, again not wholly dependent on climate, often signify environmental change, as do interbedded fluvial, paludal, and aeolian deposits (Orme, 1992). Speleothems, mostly calcite or aragonite that precipitate in oxygen-isotopic equilibrium with parent seepage waters, provide paleotemperature and paleomagnetic information datable by ^{14}C and U-series methods and, with related clastic sediments, may enclose pollen and macrofossil evidence of past environments. Vein calcite lining the walls of Devil's Hole, Nevada, has yielded a $\partial^{18}O$ paleotemperature record for the past 500 ka (Winograd et al., 1992).

With respect to biological evidence, terrestrial and lake sediments may entomb environmental indicators in the form of plant and animal macrofossils, insects, ostracods, diatoms, pollen, and other organic remains, revealing the changing conditions around pluvial lakes and climate-induced fluctuations of alpine treeline. In the desert Southwest, middens built by *Neotoma* woodrats have yielded useful data. Woodrats forage locally and, for food, curiosity, or protection, bring edible plants and insects to their dens, where, cemented by urine, they accumulate as tarry masses over long periods, a record datable by ^{14}C methods.

For many reasons, however, biological evidence should be treated with caution. For example, because of biological inertia, once plants colonize an area, they may survive long after favorable climatic conditions change, as relict stands supported by self-generated microclimates and seedling dynamics. Conversely, an area may be climatically receptive to new organisms, but colonization may be retarded by lagging migrants. Such discordancy between climate change and environmental response may range from several hundred to a few thousand years. Nor can it be assumed that modern analogs can be found for all fossil pollen assemblages, some of which may have been responding to climatic factors very different from today (Webb, 1987). Further, even as climate change favors plant migration, that very migration may generate climatic feedback. Thus the northward expansion of boreal forest during interglacials may have enhanced climate warming by reducing surface albedo. Additionally, changing ecological perspectives, especially regarding vegetation dynamics, favor continuing revision of paleoecological interpretations.

3.1.2 The Ocean Record

Challenged by ambiguous terrestrial evidence and by pioneer oceanographic studies from the *Challenger* (1872–76), *Meteor* (1925–27), and *Albatross* (1947–48) expeditions, scientists began to understand the environmental significance of ocean sediments around the mid-twentieth century. Research was aided by advances in ocean technology and drilling methods, involving specialized vessels such as *Glomar Challenger* (retired 1983) and *JOIDES Resolution*, and the introduction in 1979 of hydraulic piston coring, which permitted soft sediment to be recovered virtually undisturbed from the ocean floor. Analyses of several thousand cores recovered by the Deep Sea Drilling Project (1968–83) and its successor since 1985, the Ocean Drilling Program, have begun deciphering the mysteries of the deep. Whereas such cores may not paint a direct picture of continental environments, they do provide a better temporal framework for explaining onshore climatic oscillations.

The chemistry and isotope signature of marine organisms are important surrogates for ocean circulation, water temperature, and nutrient supply, and thus for the ocean-atmosphere interaction in a changing climate. In particular, the oxygen-isotope ratios ($\partial^{18}O/^{16}O$) within the calcareous tests of planktonic and benthic foraminifera and other microfossils recovered from ocean sediments reflect the composition and temperature of the water in which the organisms lived. Variations in oxygen isotope ratios are then used to infer changes in the volume of ice stored onshore and thus the magnitude of glaciation or interglaciation. Enrichment of ^{18}O in marine organisms implies glacial conditions as lighter ^{16}O, evaporated from the oceans and precipitated on land, is locked up in ice sheets. Conversely, depletion of heavier ^{18}O in such organisms implies dilution as ^{16}O is returned by melting ice sheets to the oceans during interglacials. Complications in this record may be induced by an organism's disequilibrium with its environment, by lateral and vertical variations in ocean temperature and salinity, and later by selective dissolution of organisms and disturbance of sediment by submarine landslides, turbidity currents, and bioturbation.

The concept of an isotopic thermometer beneath the ocean floor, first noted by Urey (1947), was later developed by Emiliani (1955, 1966) into a relative chronology based on oxygen isotopes, with cold stages assigned even numbers and warm stages odd numbers (table 3.1). Paleotemperature implications were later modified as ice-storage effects became better understood. With the development of suitable dating methods, this record and its inferred changes in ice volume were placed on an absolute timescale, essential to understanding the frequency and magnitude of climate change (Williams et al., 1988). For example, cores from Deep Sea Drilling Projects 552 and 607 in the north Atlantic Ocean reveal abrupt changes in both percentage $CaCO_3$ and $\partial^{18}O$ from benthic foraminifera around 2.5 Ma, the Gauss-Matuyama magnetic boundary, that indicate the onset of ice rafting into the ocean induced by the growth of Northern Hemisphere ice sheets after the relatively warm equable climate of Pliocene time (Raymo, 1992). For the period since 0.79 Ma, the celebrated V28-238 core from the Solomon Plateau in the equatorial Pacific Ocean reveals nine full interglacial-glacial cycles for the middle and late Pleistocene at intervals of 80–120 ka (Shackleton and Opdyke, 1973). Each full interglacial, characterized by conditions similar to the present, lasted 10–20 ka and was succeeded by a gradual descent over 60–80 ka to a full glacial condition lasting 10–20 ka, terminated in turn by rapid deglaciation, often within 10 ka (Broecker and Denton, 1990). Closer to North America, cores from the northwest Atlantic, northeast Pacific, Gulf of Mexico, and Caribbean, and from nearshore locations, such as the Santa Barbara Basin, have tended to confirm the approximate global synchroneity of these climatic events. The similarity of these climatic cycles has in turn invited correlation with Earth's orbital variations, suggested long ago by Croll (1864), revived by Milankovitch (1930), and later rejected, but resurrected as a major research focus since the 1960s.

In addition to their oxygen-isotope signatures, faunal assemblages may also indicate past ocean temperature and salinity. Further, the geochemistry and texture of ocean sediment may indicate sources from ice rafting, rivers, and deserts, and thus the nature and extent of sea ice, continental erosion, prevailing winds, and ocean currents. Accumulation rates may indicate the persistence and intensity of such processes. The ratio of terrigenous debris to biogenic carbonate in ocean sediment is another climate proxy. Thus large amounts of ice-rafted debris and low mass accumulation rates of biogenic carbonate in the so-called Heinrich layers of bottom sediment reflect the high iceberg production and low primary productivity (due to ice-blocked insolation) found in surface waters during cold stages (Heinrich, 1988; Van Kreveld et al., 1996). At ocean margins, marine terraces and coral reefs also indicate environmental change, though tectonic signatures must be distinguished from climatic factors. Where such features can be dated, sea-level data can be reconciled with oxygen-isotope records.

3.1.3 The Ice-Core Record

In terms of oxygen-isotope records, ice cores are the reciprocal of ocean sediment. During glaciation, ^{16}O is preferentially evaporated from the oceans, precipitated within snow, and preserved as glacier ice, which is thus depleted in ^{18}O. During interglacials, ^{16}O returns to the oceans, leading to ^{18}O enrichment in the ice. Thus, although ice sheets have neither existed so long as oceans nor offered such ultimate sumps, cores from the larger masses may provide a valuable record for the later Quaternary. To minimize interpretive problems, such cores are commonly sought from interior areas of minimal ablation, but the record at depth is always limited by basal ice compression and deformation. Corrections must also be made for snow source and ice-sheet altitude, but, unlike ocean cores, bioturbation is not a problem. Some excellent cores have been obtained from the Greenland Ice Cap (e.g., Camp Century, 1964–66; Dye 3, 1981–82; Summit, 1992–93), the East Antarctica Ice Sheet (e.g., Byrd, 1966–68; Vostok, since 1974), and Canada's Arctic islands. As with ocean sediment, if ice cores can be dated, for example, by counting annual layers or identifying volcanic tephra for the recent past, by radiometric methods, deuterium analyses, correlative marine and lake oxygen isotope chronologies, and ice-flow modeling for earlier times, then links with Earth's orbital variations can be tested.

Ice cores also contain trace elements and microparticles from precipitation or dry fallout onto the ice surface. The relative abundance of sea salts (Na, Cl), continental dust (Al, Fe, Mg, Mn, Ca), and volcanic aerosols reveals much about atmospheric composition, turbidity, and prevailing winds (fig. 3.3). Greenland's Dye 3 core has yielded much more dust and alkaline material from the Last Glacial Maximum (LGM) than from the Holocene, probably because of less precipitation, expanded upwind source areas including exposed continental shelves, and more vigorous atmospheric circulation during the LGM (Dansgaard et al., 1984). Atmospheric CO_2 from air bubbles trapped within glacier ice in the Vostok core from East Antarctica also correlates with the oxygen-isotope record over the past 130 ka.

Figure 3.3 Crater Lake, Oregon, a caldera formed by the Plinian eruption of Mount Mazama around 7627 +/–150 cal yr B.P., ejecting 50 km^3 of tephra and depositing ash over 1.7 $\times 10^6$ km^2, including the Greenland Ice Sheet (Zdanowicz et al., 1999). Such eruptions in western North America often affected regional atmospheric processes and temporarily reduced mean temperatures during Quaternary time (photo: Spence Collection, UCLA).

CO_2 is higher during the Last Interglacial (greenhouse warming) and lower during the LGM (Lorius et al., 1985). CO_2 records from Greenland are more problematic, probably because of melting and refreezing, but two cores from near the ice cap's summit, the 3029-m-long GRIP (1992) and the 3053-m-long GISP2 (1993) cores, correlate well in terms of $\partial^{18}O$ ratios, dust layers, and other markers over the last interglacial-glacial cycle (GRIP, 1993; Grootes et al., 1993; Hammer et al., 1997). They also reveal frequent brief $\partial^{18}O$ fluctuations of 6–9°C at intervals of ≈2–3 ka (Dansgaard-Oeschger cycles) during the last glaciation, thought to reflect complex feedback mechanisms between the ice cap and the ocean-atmosphere system. The cold minima seem to correlate with ice-rafted debris in North Atlantic sediment and with increased terrestrial dust and sea salt, which indicate storminess.

3.2 Pleistocene Climates and Vegetation Patterns

Several problems conspire to confuse climate reconstructions based on the previous evidence. Some are problems inherent to the evidence, such as fragmentary terrestrial records or deformed marine sediments and ice cores. Some are related to the resolution of dating methods, which for the Holocene can be refined into annual and decadal records but for the Pleistocene become less certain with age. Even where high temporal resolution exists for unimpeachable data, the links between marine and ice-core records on the one hand and terrestrial events on the other are problematic. The latter were often out of phase with the former because of the variable time lags of terrestrial, oceanic, and glacial systems, and the time-transgressive nature of terrestrial boundaries. Such lags may explain the uncertain start of the Last Interglacial, the marine oxygen isotope stage (OIS) 6/5 transition, which may have begun as early as 140 ka or as late as 128 ka. Further, the Last Glacial Maximum, OIS 2, peaked in ocean cores around 18–20 ka, but the southern fronts of the Laurentide and Cordilleran ice sheets reached their maximum limits at various times between 22 ka and 14 ka. Similarly, as a result of regional and local conditions, the highstands of pluvial lakes were often out of phase, not only with glacial maxima, but with one another. Also, for reasons of biological inertia or migrational lag noted previously, biomes may not correspond to predicted climates.

For these reasons, recent decades have seen a concerted attempt to reconstruct and explain Pleistocene climates by combining analyses of field data, which by their very nature tend to have local or regional significance, with climate models developed at continental and global scales. Some models apply variations in solar radiation calculated from Milankovitch-type orbital relations to the prediction of global ice volumes and other variables, giving a continuous image of climate variation over time but no spatial resolution (e.g., Bryson and Goodman, 1986). Other models, the General Circulation Models, apply inferences regarding ocean-atmosphere dynamics and boundary conditions, including ice-sheet dimensions, to simulate climate as a three-dimensional system for discrete time periods (e.g., CLIMAP, 1981; COHMAP, 1988; Wright et al., 1993). Such work is aided by many methods, for example, by the application of transfer-function techniques to pollen analysis whereby a taxon's modern frequency over its range is compared with climatic parameters using mutliple linear regression, thereby permitting climate reconstruction based on comparable fossil pollen assemblages.

Clearly, such models simplify the real world, based as they are on estimates of boundary conditions for specific periods derived from proxy data of varying quality. Assumptions regarding ice-sheet dimensions, sea-ice cover, ocean bathymetry, and surface temperature and albedo are among the many variables likely to affect the final product. Furthermore, such models are usually global or continental in scale and, constrained by coarse grids, are less useful at regional and local scales, where topography may generate different responses. For example, prevailing winds may be predicted from modeled pressure systems, but field data, such as paleodune axes, will reflect local surface winds as influenced by topography. Further, biological checks assume a knowledge of an organism's ecology and physiology even though plants may respond to increased rainfall as though climate had cooled, or to higher temperatures as though rainfall had diminished (Bryson, 1991). Despite their limitations, however, such models have been refined progressively during recent decades, offering a valuable and complementary approach to climate reconstruction. Improved understanding of the past must combine careful evaluation of field data with climate modeling; the higher the quality and resolution of the field data, the better the model, and ultimately the more acceptable the resulting hypotheses.

3.2.1 The Early and Middle Pleistocene

The growing body of information about this time indicates that the lengthy Miocene and shorter Pliocene Epochs of relatively stable climate and ecosystem development changed around 2.7–2.5 Ma with the onset of more variable climates, which repeatedly displaced existing ecosytems and established new habitats for migrating plants and animals. These changes were linked with lower atmospheric CO_2 levels, which lowered the Pliocene "greenhouse" warmth, and reduced ocean heat transport, which lowered sea-surface temperatures in higher latitudes (Crowley, 1996). Climate change was more marked in higher than in lower latitudes (Thompson and Fleming, 1996). Thus, whereas palynological evidence for mixed coniferous-hardwood vegetation near the subarctic Labrador Sea suggests Pliocene January temperatures of 4–10°C warmer than today, vegetation in Pliocene Florida's Pinecrest beds is similar to the longleaf and slash pine forests (*Pinus palustris, P. elliottii*) of modern Florida (Willard, 1995). The Gauss-Matuyama magnetic reversal around 2.5 Ma is a critical stratigraphic marker for distinguishing between relatively warm Pliocene conditions and approaching Pleistocene variability.

Among the best indicators of climatic cooling was the retreat of the boreal forest treeline, which still reached above 80°N during the late Pliocene but today lies south of 70°N in western Canada and below 60°N in Labrador, 2500 km south of its former location in northern Greenland (see chapter 14). In Arctic Alaska around 2.5 Ma, open spruce-birch forest was replaced by lowland shrub tundra, whereas ice-rafted debris increased in north Pacific cores and Alaska's Yakataga Formation in response to the onset of continental glaciation (Lagoe and Zellers, 1996). The abrupt increase in terrigenous clastic sediment and ice-rafted debris in the North Pacific around 2.67 Ma also coincided with a tenfold increase in volcanic tephra, suggesting that incipient glaciation may have been intensified by explosive volcanism in nearby island arcs (Prueher and Rea, 1998). The opening of the Bering Strait to Arctic waters around 3 Ma and the continuing uplift of the Alaska Range and St. Elias Mountains accentuated this cooling trend. Around 2 Ma, the forest tundra of northern Greenland, dominated by larch (*Larix occidentalis*) and black spruce (*Picea mariana*), began retreating as the treeless polar tundra expanded, probably in response to increasing perennial sea ice (Funder et al., 1985; Reppening and Browers, 1992). Much farther south, increasing continentality was reflected in the retreat of midcontinental Pliocene woodlands before the advancing steppe and grassland and their herbivorous mammals (Stebbins, 1981). By the close of the early Pleistocene, around 0.8 Ma, biota in the midcontinent were indicating temperate conditions similar to the present (Miller et al., 1994).

When it occurred, glaciation was characterized by a southward or downslope retreat of climate zones. Based on coniferous and herbaceous pollen, much of nonglaciated North America experienced essentially cool temperate, continental conditions. In the High Plains of Kansas and Oklahoma, for example, late Illinoian cold (OIS 6) saw invasion from the west by spruce (*Picea*), pine (*Pinus*) and Douglas-fir (*Pseudotsuga*) more typical of the modern pine savannas of the lower Rocky Mountains (Kapp, 1965). On the Yellowstone Plateau, this cold stage was dominated by *Artemisia*, indicating a cold open tundra, followed as warming began by a cool subalpine forest of spruce, pine, and fir (*Abies*) (Baker, 1981).

Interglacials saw climate zones migrate northward and upslope, resulting in more diverse and regionally distinctive ecosystems similar to the present. In the Ohio valley, for example, deposits from middle Pleistocene interglacials (e.g., OIS 7) contain pollen of pine, oak (*Quercus*), hornbeam (*Ostrya* and *Carpinus*), and hazel (*Corylus*), overlain by sediment containing beech (*Fagus*), hickory (*Carya*), elm (*Ulmus*), herbs, and grasses, indicating a shift from closed to open deciduous forest in a climate warmer than today (Kapp and Gooding, 1964). Deposits of similar age from the High Plains of Nebraska are dominated by grass and herb pollen with little oak or pine, suggesting open grassland in a moist, temperate environment not unlike the present (King, 1991).

3.2.2 The Last Interglacial

The Last Interglacial was once equated stratigraphically with the Sangamon geosol, which formed across the midcontinent on Illinoian and earlier deposits and was later buried beneath Wisconsinan material. This soil was later correlated with OIS 5, ≈130–75 ka, an age range supported by ^{10}Be data (Curry and Pavich, 1996). However, the isotope record also showed that only during the earliest part of this interval did global temperatures compare with the Holocene and that the period after 115 ka saw erratic climate deterioration toward the next glaciation (Chapman and Shackleton, 1999). Thus the Last Interglacial is now restricted to OIS 5e, with later oscillations assigned to OIS 5d, 5c, 5b, and 5a. The Sangamon geosol, formed by weathering and pedogenesis, was initiated late in OIS 6, continued to develop throughout OIS 5, and persisted later until buried by advancing ice or other deposits (Fullerton, 1997).

Numerous sites exist throughout North America and adjacent oceans to reveal the character of the Last Interglacial and its aftermath. Pollen and ocean-core data indicate fairly rapid warming after Illinoian cold, including an increase in ice-rafted debris in ocean cores analogous to terminal Pleistocene events, then an interval of pronounced warmth (OIS 5e), followed by a slow irregular deterioration (OIS 5d-5a) toward the next major cold stage (OIS 4). In Last Interglacial sites straddling the Mississippi valley near the Wisconsinan ice front, the pollen record begins with boreal coniferous forest dominated by spruce and pine augmented by cool-temperate species such as alder (*Alnus*), birch (*Betula*), and willow (*Salix*), followed by a warm-temperate deciduous forest of oak, elm, hickory, ash (*Fraxinus*), hackberry (*Celtis*), sweetgum (*Liquidambar*), sycamore (*Platanus*), and various herbs (Grüger, 1972; King and Saunders, 1986). Then the pollen record is dominated by mostly nonarboreal species indicative of savanna conditions under peak interglacial warmth and decreased soil moisture, followed by a partial return to deciduous forest before conifers reappear as a prelude to the Wisconsinan glaciation. In the Don beds beneath Toronto, interglacial taxa include not only deciduous trees common to the area today but warm-temperate genera such as swamp cypress (*Chamaecyparis*), water locust (*Gleditsia*), locust (*Robinia*), and bodark (*Maclura*) that now grow much farther south, indicating mean temperatures at least 3°C warmer than today (Terasmae, 1960). Lying on these, but separated by an uncertain interval, are the Scarborough beds dominated by spruce, pine, and various deciduous trees, indicating a mean annual temperature 6°C lower than today prior to glaciation. In Atlantic Canada, OIS 5e saw annual temperatures at least 4°C warmer than today (Mott, 1990), whereas data from the GRIP ice core suggest that the temperature in central Greenland rose 2°C above present during warm spells (GRIP, 1993).

The Last Interglacial saw spruce and pine vacate the High Plains for the Rocky Mountains and beyond, where they reoccupied former tundras, for example, on the Yellowstone Plateau. Their place was taken by sagebrush, bur-sage (*Ambrosia*), and grasses as warmer drier conditions favored the expansion of semiarid grassland and deciduous trees were confined to watercourses. Reduced aquatic taxa and the development of calcrete confirm this drying trend. In central Alaska, the Last Interglacial saw deep thawing of permafrost, major erosion of loess, and boreal forest expansion as seen in the Eva Forest Bed.

General circulation and related biome models confirm that conditions during the Last Interglacial were warmer, drier, and more continental than today, and that the ensuing OIS 5d saw warm but wetter winters conducive at height to increased snowfall and ice-cap growth, for example, over the Yellowstone Plateau (Richmond, 1986; Harrison et al., 1995). The transition from full interglacial to full glacial conditions was erratic. Again, isotopic evidence from Greenland's GRIP ice core suggests that a 9°C fall in air temperature from OIS 5e to OIS 5d was largely reversed in OIS 5c (Johnsen et al., 1992) and that as many as six warm interludes characterized the 115–70 ka interval (Dansgaard et al., 1993).

The last interglacial-glacial cycle is well revealed at two western sites: Clear Lake in northern California and the Santa Barbara Basin off southern California. Clear Lake, in an active tectonic basin subject to almost continuous lacustrine deposition over the past 500 ka, now lies near the ecotone between oak woodland at lower elevations and mixed coniferous forest on intermediate slopes, which in turn passes upward into montane pine forest (Sims, 1988). Pollen and aquatic organisms trapped in dated lake sediment show that these ecotones shifted vertically, often rapidly, in response to climate change. After late Illinoian cold (OIS 6), rapid warming saw oak woodland colonize the basin from 130 ka to 120 ka (OIS 5e), after which oscillating changes occurred over 50 ka (OIS 5d-5a) until more persistent cold returned around 70 ka (OIS 4) (Adam, 1988). Under warm conditions, oak predominated; with cooler conditions, coast redwood (*Sequoia sempervirens*), incense cedar (*Calocedrus decurrens*), cypress (*Cupressus*), and juniper (*Juniperus*) of the mixed conifer assemblage flourished alongside pine, with oak returning during mid-Wisconsinan (50–30 ka, OIS 3) and Holocene times (OIS 1). The slow climatic deterioration after warm intervals, a swifter return of warm temperatures after cool intervals, and modest temperature shifts of up to 3°C within colder intervals are noteworthy features of this record.

Beneath the Santa Barbara Channel, terrigenous and marine sediments have been trapped in a semienclosed tectonic basin 600 m below the present sea level. Because a sill only 400 m deep separates this basin from open Pacific Ocean to the west, sea-level changes influence the penetration of oxygen-deficient intermediate waters. With high sea levels, anoxic conditions favor laminated sedimentation (varves) within the basin by discouraging bottom-dwelling organisms and thus bioturbation. With low sea levels, more oxic conditions favor bioturbation and nonlaminated sedimentation. Thus sediment cycles and biotic changes correlate with interglacial-glacial cycles. One core from only 20 km offshore has yielded a high-resolution record for the past 160 ka (Kennett et al., 1994). Oxygen isotopes clearly reflect late Illinoian (OIS 6) and late Wisconsinan (OIS 2) cold, Last Interglacial (OIS 5e) and Holocene (OIS 1) warmth, and instability between 115 ka and 30 ka. The dextral-coiling planktonic foraminifer *Neogloboquadrina pachyderma* dominates the warm stages, when subtropical taxa increase species diversity, whereas its sinistral form correlates with cold stages, when the cool California Current was intensified by subarctic waters and sea-surface temperature was 8°C lower than today. Whereas oak and pine pollen dominate warm and cool intervals, respectively, the paucity of redwood and western hemlock pollen from the latter suggests that cool episodes did not see a simple southward displacement of the coniferous forests.

3.2.3 The Last Glacial Maximum

The Last Glacial Maximum was the culmination of climatic cooling that began after 115 ka and led to early Wisconsinan cold (OIS 4, ≈75–50 ka) and the Altonian Glaciation in midcontinent, then to mid-Wisconsinan warmth (OIS 3, ≈50–30 ka), the Farmdalian interstadial, when boreal forest invaded northern Illinois, and finally to late Wisconsinan cold (OIS 2, 30–10 ka), the Woodfordian Glaciation. Unlike the Last Interglacial, for which comparisons may be found in Holocene landscapes, the Last Glacial Maximum has no modern analogs and is thus a major challenge for modeling and analysis.

For the Last Glacial Maximum, reconstructions for 18 ka assume a combined Laurentide-Cordilleran Ice Sheet at least 3000 m high, 16 million km^2 in area and 32–38 million km^3 in volume (COHMAP, 1988; Kutzbach and Webb, 1993; Kutzbach et al., 1993). The effect of this vast ice mass and nearby sea ice on the overlying atmosphere was to generate a permanent glacial anticyclone characterized by clockwise outflows of cold air, which brought great cold and strong aeolian activity to peripheral areas. The ice also formed a major barrier to the eastward flow of tropospheric westerlies, causing the polar jet stream to bifurcate. The more northerly jet swept northeast across Alaska and Arctic Canada and then southeast between the Laurentide and Greenland ice masses to reach the North Atlantic. The more southerly jet was diverted into the Southwest, then east along the Gulf coast before trending northeast to converge with the more northerly jet over the northern Atlantic.

Off both the western and eastern shores of North America, sea ice may have extended as far south as 40–45°N in winter and, although it retreated to around 60°N in summer, the proximity of sea ice to warmer subtropical waters created much steeper temperature gradients and enhanced cyclonic activity. In the North Pacific Ocean, the Aleutian subpolar low-pressure cell was more pronounced and farther south in winter; with a weaker Hawaiian subtropical high-pressure cell, conditions favored the onshore flow of moisture-laden cyclonic systems into the Southwest, where wetter conditions caused lakes to expand. Nearer shore, subarctic foraminifers from the Santa Barbara Basin indicate an intensification of the cool California Current (Kennett et al., 1994). In the North Atlantic, steep temperature gradients where the northern and southern jets converged favored cyclonic activity, which fed snow onto the eastern Laurentide and Greenland ice masses. In the Arctic, extensive sea ice between these grounded ice masses combined with restricted oceanic circulation caused by lowered sea levels and closure of the Bering Strait to promote atmospheric subsidence and stability. In the resulting cold arid conditions, precipitation starvation probably limited the growth of ice caps in the Canadian Arctic and northern Greenland at this time.

Below the Brooks Range ice cap in Alaska, similar cold and aridity existed in areas cut off by the Cordillera to the south from precipitation sources over the Pacific Ocean. Permafrost became widespread, and Last Interglacial deposits, including the Eva Forest Bed, were frozen. Pollen and faunal remains from the Yukon to the exposed Chukchi Sea shelf show that unglaciated Beringia was mantled by a vegetation mosaic grading from sparse alpine tundra to moist meadow-like lowland tundra dominated by grasses and sedges with *Salix* and *Betula* shrubs (Anderson, 1985) and inhabited by such large mammals as mammoth (*Mammuthus primigenius, M. columbi*), bison (*Bison crassicornis*), horse (*Equus*), camel (*Camelops hesternus*), and American lion (*Panthera leo atrox*) (Harrington, 1980).

Far to the south, beyond the main ice front, permafrost indicators such as pingos and ice-wedge polygons point to a 200-km-wide periglacial zone from Appalachia to the Rockies during the Last Glacial Maximum (Péwé, 1983; Clark and Ciolkosz, 1988). Macrofossil and pollen evidence suggest a dry, nearly treeless tundra, swept by easterly surface winds and inhabited by collared lemming (*Dicro-*

stonyx) and other creatures adapted to severe cold. This periglacial zone was narrow by modern standards, which may explain why lemming occurred alongside prairie vole, ground squirrel, and other species with modern prairie and deciduous woodland affinities. To the south lay the tundra-boreal forest ecotone, which today in northern Canada reflects the 10–12°C July isotherm. From here, open spruce-pine forest stretched southward to pass into oak savanna and open grassland from the southern Appalachians to the southern High Plains (Holliday, 1987; Hall and Valastro, 1995). Pollen and speleothem data show that during the mid-Wisconsinan interstadial around 40 ka, the southern Appalachians had been characterized by an open forest of oak, hickory, and southern pines, but that the pines retreated during the Last Glacial Maximum when mean annual temperatures were around 0°C (Delcourt, 1979; Delcourt and Delcourt, 1981; Brook and Nickmann, 1996). Even as warmth returned, winter temperatures remained low. Pollen data from the central Appalachians suggest that, as late as 14 ka, although mean July temperatures rose to 19–20°C, mean January temperatures remained around −14 to −17°C, or 4–5°C and 15–18°C, respectively, below present temperatures (Shane and Anderson, 1993).

In the Southwest, *Neotoma* woodrat middens and insect evidence for the Last Glacial Maximum indicate a downslope displacement of desert shrubs by pinyon-juniper woodland and cool temperate conifers. On the Colorado Plateau, coniferous forests were dominated by Engelmann spruce (*Picea engelmannii*) by 20 ka, suggesting summers 7°C cooler than today (Anderson, 1993). In the Chihuahuan Desert, between 43 and 12 ka, temperate insects lived in pinyon woodlands (*Pinus monophylla*) that reached as low as 600 m, whereas Douglas-fir descended to 1200 m. Desert tortoise (*Xerobates agassizii*) and a mix of northern and southern mammals also indicate equable conditions (Elias and Van Devender, 1992). In the Mojave Desert, pinyon-juniper woodland descended to 1000 m and subalpine woodland, containing limber pine (*Pinus flexilis*) and bristlecone pine (*P. longaeva*), to 1800 m (Spaulding and Graumlich, 1986). California's coastal vegetation during the Last Glacial Maximum ranged from cypress-pine woodland (*Cupressus macrocarpa, Pinus muricata, P. radiata*) with sage scrub in the south, to cool coniferous forest containing mountain hemlock (*Tsuga mertensiana*) in the north. The paucity of broadleaved trees, such as oak, and of temperate conifers, such as coast redwood and western hemlock (*Tsuga heterophylla*), suggests that precipitation had increased by about 20 cm and evapotranspiration had decreased. The Oregon coast was dominated by a cool temperate parkland of spruce, fir, pine, and hemlock.

3.2.4 The Close of the Pleistocene

As North American ice masses decreased and the ice front retreated into Canada in response to increasing summer insolation, the size and intensity of the glacial anticyclone diminished, the more northerly jet waning in favor of the more southerly jet, which in turn shifted farther north (COHMAP, 1988). Westerly winds were reestablished, and by 12 ka moist temperate conditions had invaded the Pacific margin of the retreating Cordilleran ice sheet while the widening corridor between Cordilleran and Laurentide ice was swept by strong surface winds (see chapter 4).

Farther east, the still formidable presence of the Laurentide Ice Sheet at 12 ka favored periglacial tundra at its margins, notably along the St. Lawrence Valley, but farther south, with the jet stream encouraging cyclonic activity and weak onshore flows, conditions were moister and less cold. Spruce forest moved north along a broad front from New England through southern Ontario to the High Plains and the Yellowstone Plateau (Ritchie, 1987; Whitlock and Bartlein, 1993). Behind spruce, northern pine forests migrated northward to New England, oak woodland expanded from sanctuaries in the Southeast, and southern pines reappeared in Florida (Webb et al., 1987). From New England to the Great Lakes, summer temperatures rose rapidly from 12°C to 17°C between 13 ka and 11 ka, nearing present summer values even as winters remained cold (Peteet et al., 1993).

While deciduous forests expanded northward in the east, the area west of the Mississippi valley saw similar forest from the Southeast battle with steppe from the Southwest, although in the Great Plains the forest episode was soon ended by increasing aridity and fire, and prairie grasslands expanded. In the Southwest, pinyon-juniper woodland and coniferous forest were retreating upslope by 12 ka as summer warmth and aridity increased and desert thermophiles began returning from Mexican refugia. Several millennia were to pass, however, before desert communities acquired their present composition. Creosote bush (*Larrea tridentata*), so typical of the present Mojave Desert, did not appear until after 8 ka. In Oregon, spruce parkland retreated, replaced by 13 ka by a mixed woodland of Douglas fir, alder, and hazel, whose dominance by 10.5 ka confirmed the return of warmth. By 10 ka, although Laurentide ice maintained a diminished presence, the close of the Pleistocene saw most of North America beginning to assume a reasonably modern appearance, with fluvial processes dominant in landscapes characterized by the north-south sequence from tundra, through boreal forest, to mixed forest and deciduous woodland, with prairie, steppe, and desert farther west.

Ideally, climate models should seek to explain not only optimal situations, such as the Last Glacial Maximum, but also transitional conditions. One intriguing challenge in this respect is the evidence for cold snaps between 14 ka and 10 ka during the warming trend that followed the Last Glacial Maximum. Glacial readvances during this interlude occurred around the Great Lakes, although whether these were due to climatic forcing or to the internal dynamics of a waning ice sheet is unclear. Farther afield, however, distinct climatic oscillations have long been known from northern Europe, where warm spells are reflected in peat beds at Bølling and Allerød in Denmark, and where cold snaps are linked to the reappearance in the pollen record of mountain avens (*Dryas octopetala*), a small flowering plant of Northern Hemisphere tundras. Similar oscillations are now known from peats and lake deposits in North America, most notably related to the Younger Dryas interval between 11 and 10 ka at the very end of the Pleistocene. Across Atlantic Canada, for example, a warming trend between 13 ka and 11 ka was followed by a distinct cold snap and retreat of the mixed forest during the Killarney Oscillation around 11.2–10.9 ka before warming resumed (Levesque et al., 1993; Peteet et al., 1993). Farther south, from the central Appalachians to Florida, warming was interrupted by a cold snap around 12 ka but a Younger Dryas event is less certain, perhaps because warm oceanic influences were now increasing in these latitudes (Kneller and Peteet, 1999).

There has been much speculation regarding these oscillations. The Younger Dryas cold snap may have been triggered by the cooling effect of glacial meltwaters from southern Canada's proglacial lakes, notably Lake Agassiz, suddenly reaching the North Atlantic directly down the St. Lawrence Valley rather than via the Mississippi and the warmer Gulf of Mexico (Teller, 1990). This may have led to a more stable stratification of sensitive surface waters in the North Atlantic and a southward expansion of sea ice and polar front, causing a brief return of cold conditions over much of the Northern Hemisphere and possibly beyond (Wright, 1989; Peteet, 1993). Alternatively, ocean cores from the North Atlantic reveal maximum delivery of ice-rafted debris every 10 ka over the last interglacial-glacial cycle, a frequency that correlates with maxima and minima of the cyclic precession of the equinoxes (Heinrich, 1988). Cyclic ice rafting would be favored as minimum winter warmth favored ice accumulation while maximum summer warmth delivered more icebergs. Many Heinrich layers have been recognized for the past 200 ka from ice-rafted debris and low biogenic carbonate accumulations in North Atlantic sediments (Hillaire-Marcel et al., 1994; Van Kreveld et al., 1996; Chapman and Shackleton, 1999). Such layers are prominent in late Pleistocene cores and their mineralogy suggests that most icebergs came from the Laurentide ice sheet, floating south from the Hudson Strait with the Labrador and Canary currents (Baas et al., 1997). A dense iceberg cover would have increased ocean albedo and, allied to a massive meltwater influx, would have cooled the sea surface, reduced the production of North Atlantic Deep Water, and modified the thermohaline circulation of the Atlantic Ocean. Later, as iceberg production decreased, these conditions would have been reversed, perhaps quickly, and the scene set for rapid Holocene warming.

Off the Pacific coast, similar decreases in biogenic carbonate and silica and changes in biotic assemblages may be related to changes in the California Current synchronous with late Pleistocene climate oscillations. The foraminiferal record from the Santa Barbara Basin reveals numerous warm-cold episodes, the so-called Dansgaard-Oeschger cycles, during late Pleistocene times that correlate with Greenland ice-core records, suggesting ocean-atmosphere coupling across North America (Hendy and Kennett, 1999). The Younger Dryas cold snap is thus but one of many oscillations of varying duration that continue to challenge scientific enquiry.

3.3 Paleohydrology and Pleistocene Drainage Legacies

Major hydrological distinctions exist between those regions of North America that were glaciated and those that were not. In glaciated regions, networks of streams, lakes, and wetlands originated anew as ice sheets withdrew, sometimes following earlier paths, often not, but usually poorly integrated. Near former ice fronts, the proximity of glaciers and related deposits severely disrupted and sometimes reversed preglacial drainage patterns. Farther afield, however, drainage networks had evolved over long intervals of time in response to tectonism and climate. Whereas the tectonic framework, with some exceptions, was in place by the onset of the Pleistocene, climate changes throughout the epoch were to cause substantial shifts in effective precipitation and water storage, which translated into altered streamflows, lake budgets, water tables, and water for plants and animals.

3.3.1 Water Budgets

Field data suggest that hydroclimates during warm stages were similar to the present. In contrast, cold stages had very different hydroclimates whose legacies are quite unrelated to modern conditions. Warm or cold, the situation is further complicated in the West where continuing uplift and volcanism changed the playing field for precipitation and runoff during the Pleistocene.

During cold stages, the strong glacial anticyclone, which split the jet stream over northern North America, diverted cool, moist Pacific air masses into the Southwest, into areas where prolonged warmth and aridity now prevail. Ephemeral and intermittent streamflows typical of interglacial seasonality became intermittent and perennial, respectively. Summer drought lessened, more winter precipitation fell as snow, water tables rose, and mass movement increased. Dated landslides indicate that mass movement peaked around the Last Glacial Maximum but diminished as warm, dry conditions returned in the Holocene (Orme, 1991). In the Great Basin, lake fluctuations were closely related to the location of the polar jet stream while monsoonal flows from the south were curtailed (Benson, 1981). Farther east, northward incursions of warm, moist tropical air masses from the Gulf of Mexico were also restricted, such that the Southeast was cooler and drier than today. However, despite reduced precipitation during cold stages, midlatitude runoff during the Last Glacial Maximum may well have intensified as diminished evaporation and high glacier-meltwater production combined to generate runoff two to three times the values of modern river fluxes (Marshall and Clarke, 1999).

3.3.2 Drainage Near the Ice Front

Cold-stage hydrologies have left many legacies near former ice fronts: disrupted drainages, outwash debris terraced by later streams, and proglacial and ice-dammed lakes. Ice-sheet proximity usually disrupted preglacial rivers, some of which never reoccupied earlier paths. In the northern Great Plains, Illinoian and Wisconsinan ice blocked the earlier courses of the upper Missouri, Yellowstone, and Cheyenne rivers toward Hudson Bay, so that, when the ice retreated, the revised drainage used ice-marginal channels to drain ultimately southward, although broad northeast-trending segments survived south of the ice front (Alden, 1932; Wayne et al., 1991). Farther east, the net effect of glaciation was to increase the volume of water and sediment discharged to the Mississippi system, whose main stem between Illinois and Iowa only assumed its present course after Wisconsinan ice had blocked earlier routes. Preglacial relief around the Great Lakes was so profoundly modified by repeated glaciation that the present drainage only emerged after withdrawal of the last ice sheet. Preglacial drainage from the Appalachians north to the St. Lawrence and west via the Teays-Mahomet river system to the Mississippi was so disrupted by glacial incursions that the modern Ohio River system only began to develop in mid-Pleistocene time, surviving south of later ice fronts (Teller, 1973; Swadley, 1980).

Elsewhere, drainage changes were less dramatic. Thus, the Allegheny, Delaware, and Susquehanna rivers in the Appalachians, though supplied with discharge and debris from Wisconsinan ice lodged against their headwaters, maintained south-flowing courses during and after glaciation. In Pennsylvania, seven terraces ranging from 3 m to 51 m above the modern Susquehanna River reveal episodic fluvial degradation over the past 2.4 Ma (Engel et al., 1996). In West Virginia and Kentucky, an unglaciated tributary of the preglacial Teays River survives in the abandoned meanders, underfit streams, terraces, and lake deposits of the Big Sandy Valley. In the West, where the Puget lobe of Cordilleran ice had diverted streams south into the Chehalis River, drainage north to Puget Sound was reset after deglaciation, albeit much deranged. In Alaska, outwash fans flank glaciated mountain ranges whereas fluvial deposition was enhanced in downstream basins like the Yukon Flats.

Among the most spectacular effects of drainage disruption by Pleistocene ice were the superfloods (jökulhlaupe) released across the Columbia Plateau by the breaching of ice dams created by pulsating Cordilleran ice lobes. These occurred many times, but those linked to the last deglaciation have left the most dramatic imprints. Glacial Lake Missoula was the largest of these ice-dammed lakes, forming repeatedly as the Clark Fork drainage was dammed by the Lake Pend Oreille ice lobe in western Montana. Just before it last spilled around 14 ka, this 7500-km^2 lake contained 2000 km^3 of water up to 700 m deep against its ice dam (Pardee, 1910; Baker and Bunker, 1985). On breaching, floodwaters spilled westward into the Columbia River basin, depositing Cordilleran granite debris, entraining local basalt, streamlining the loessic Palouse Formation, shaping deltas, boulder bars, and other giant bedforms, and sculpting the Channeled Scabland, a remarkable network of anastomosing channels in eastern Washington (fig. 3.4; Bretz, 1923, 1969). Spilling through the Wallula Gap, floodwaters gouged the Columbia River bed to depths of >200 m, deposited a massive delta near Portland, and transported finer debris far seaward across the Astoria submarine fan (Waitt, 1985; Brunner et al., 1999.). At its peak, this superflood may have had a phenomenal discharge of 21 million $m^3\ s^{-1}$.

Meltwaters from successive Laurentide ice sheets also did much to reshape landscapes across the midcontinent. During glacial maxima, discharge from the ice front was probably more seasonal, with summer ablation yielding sediment into braided streams that flowed south across tundra and boreal forest. Deglaciation initially favored massive fluvial aggradation in the upper Mississippi valley, with the alluvial wedge feathering out southward to intersect the present floodplain south of the Ohio confluence. Later, drainage from proglacial lakes dissected and terraced this wedge en route to the Gulf of Mexico (Knox, 1996). Spikes around 14–11 ka in ^{18}O foraminiferal records from Gulf cores suggest that some $18 \times 10^6\ km^3$ of meltwater came down the Mississippi to the Gulf, with peak accumulations of terrigenous sediment in offshore basins around 12 ka (Teller, 1990; Brown and Kennett, 1998).

Lake Agassiz, the largest proglacial lake, began forming around 13 ka as the Red River lobe of Laurentide ice retreated into Canada. Initially, like its forebears along this ice front, Lake Agassiz discharged into the Mississippi, probably at a rate of around 100,000 $m^3\ s^{-1}$. Later, the lake's drainage switched several times among the Mississippi, St. Lawrence, and Mackenzie river systems in response to pulsating regional ice lobes and isostatic adjustments (Teller, 1990; Leverington et al., 2000). By 11 ka, perhaps earlier, all meltwater delivery to the south had ceased (Brown and Kennett, 1998), and the Mississippi converted from a late-glacial braided pattern to a dissecting meander-

Figure 3.4 Grand Coulée at Dry Falls, Washington. The 80-km long, 100–300-m deep main channel and its bars, falls, cataracts, and feeders were sculpted across Columbia Plateau basalts by glacial meltwaters and Lake Missoula superfloods when the Columbia River was diverted southward by the Okanogan lobe of the Cordilleran Ice Sheet, most recently during the terminal Pleistocene (photo: Spence Collection, UCLA).

ing system. Further contraction of Lake Agassiz into the Winnipeg Basin stabilized around 8 ka, by which time much of the former lake bed was exposed in the Red River Plain (Dyke and Prest, 1987).

3.3.3 Drainage Far Beyond the Ice Front

Beyond the direct influence of nearby ice sheets, most drainage systems were very old and were able to maintain their preglacial courses. Even so, streams everywhere had to adapt to the changing Pleistocene climates and vegetation covers, yielding a legacy of terraces, entrenched meanders, and valley fills along their courses. Pleistocene rivers of the southern High Plains and Osage Plains, for example, dissected and entrenched the great blanket of Neogene floodplain and channel deposits, the Ogallala Formation, that had been deposited by earlier streams draining from the Rocky Mountains. In the Colorado River of Texas, the smaller but more frequent floods of the LGM created narrower, more sinuous channels than exist under today's drier conditions (Baker and Penteado-Orellana, 1977). Farther west, wetness during Pleistocene cold stages probably explains many of the relict mass wasting and fluvial dissection features of the Colorado Plateau and Basin and Range Province. At higher elevations where outwash debris had choked valleys below local glaciers, later fluvial incision produced some fine terrace sequences, notably along the Animas River in Colorado and the Snake River in Wyoming (fig. 3.5). Terrace evidence also suggests that the upper Grand Canyon of the Colorado River has deepened some 135 m over the past 0.7 Ma (Patton et al., 1991).

3.3.4 Lake Systems

The several hundred lakes of late Pleistocene age that once existed across western North America, testimony to a more effective water presence, have left a visible legacy of shorelines and sediments in their now largely dry basins. Such features should not be viewed in temporal isolation, however, because they were often just the most recent expressions of lakes that had existed episodically throughout the Cenozoic. Initially, the Laramide orogeny favored Paleogene lake formation by expelling Cretaceous seas, raising the Rocky Mountains, and forming intermontane basins, such as the Uinta Basin, which harbored the lacustrine Eocene Green River Formation. Later, Neogene block faulting created many closed basins for lakes in the Basin and Range Province. In Utah's Sevier Desert basin, 3000 m of Neogene alluvial, lacustrine, and volcanogenic sediment accumulated, including 1500 m of salt and gypsum (Mitchell and McDonald, 1987). Such basins were to see lakes often during the Pleistocene, many of them larger earlier in the epoch than later, because of either changing climate, including rain-shadow effects, or tectonism (Reheis, 1999).

As our understanding of Pleistocene events improved during the nineteenth century, we began to assume that the conditions generating glaciation in higher latitudes also favored more rainfall in lower latitudes. Glacial stages were equated with pluvial stages and lake expansion beyond the ice front; interglacial stages with aridity and lake desiccation. Qualified support for this view soon emerged when Russell (1885) and Gilbert (1890) suggested a close, if imperfect, relationship between glacial moraines and high lake levels. Later work on salt deposits (Gale, 1913) and

Figure 3.5 The Snake River in Jackson Hole, Wyoming, flowing southward at an elevation around 2000 m above sea level east of the glaciated Teton Range (4197 m). The river here is incising extensive late Pleistocene fluvioglacial outwash deposits derived from the Jackson Lake lobe of the former Yellowstone Ice Cap (photo: A.R. Orme).

mammaliferous lake beds (Buwalda, 1914) renewed interest in these lakes, and maps were refined (Meinzer, 1922; Snyder et al., 1964). More recently, however, improved dating of lake deposits has served to caution against the easy equation of glacial and pluvial stages. Certainly, many basins that have dried out during the Holocene were occupied by lakes during the Pleistocene, but high lake levels often preceded or followed glacial maxima. Nor can synchronous water-level fluctuations between neighboring lake systems be assumed, essentially because forcing factors varied from region to region. Such asynchronous behavior is due mainly to the shifting tracks of rain-bearing storms and desiccating winds, to water storage in local ice caps and ground ice during glacial maxima, and to the linkages between lake systems. Pluvial conditions may also result from decreased evapotranspiration, caused by increased cloudiness or lower temperatures, as much as from increased precipitation.

Lake Bonneville was the largest late Pleistocene lake in the Basin and Range Province, successor to earlier lakes and predecessor to the now modest Great Salt Lake of northwest Utah (fig. 3.6, table 3.2; Gilbert, 1890; Currey et al., 1983; Benson et al., 1990). A 307-m core from beneath the south shore of Great Salt Lake has revealed many lacustrine and alluvial events over the past 3 Ma, including, above the 99-m deep Matuyama-Brunhes boundary (≈790 ka), several deep-lake cycles and discrete soils indicating lake desiccation (Eardley et al., 1973; Oviatt et al., 1999). Meanwhile, tectonic and volcanic activity continued reshaping the Bonneville basin, notably through eruptions in southeast Idaho, which diverted the Bear River away from the Snake River and into Utah after 130 ka. Today, Bear River contributes half the streamflow reaching Great Salt Lake, flow which in turn accounts for 66% of total input to the lake, with direct precipitation yielding an additional 31% and groundwater 3%.

Lake Bonneville has been much studied for its shoreline features and complex clastic and evaporite stratigraphy (fig. 3.7). Though controversial, most authorities agree that, after a prolonged interval of low lake conditions and soil development after 130 ka, the lake began rising around 30 ka. It may have paused to form one or more shorelines around 23–20 ka, when a brackish lake covered 24,000 km^2 at 1372 m above sea level, or these may be a later recessional feature (Currey and Oviatt, 1985; Morrison, 1991; Sack, 1999). Renewed rise after 20 ka saw the lake reach 1551 m by 16 ka, covering 51,700 km^2 in a 140,000-km^2 drainage basin (Currey et al., 1983). At this level, the freshwater lake began spilling at Zenda, beyond Red Rock Pass, to the Snake River and thus to the ocean. Maintaining this level for a millennium or more, it cut the intricate Bonneville shoreline. Isostatic downwarping beneath 9500 km^2 of water up to 372 m deep was such that later crustal unloading has raised the Bonneville shoreline to 1626 m in the basin's center (Currey, 1988). Then, around 14.5 ka, sudden failure of weak deposits in the outlet channel released a superflood of 1 million m^3 s^{-1} into the Snake River, lowering the lake rapidly by more than 100 m onto a firmer, bedrock-controlled outlet at 1448 m, at which level the conspicuous Provo shoreline formed over the next millennium, enclosing a lake of 37,300 km^2. By 13 ka, Lake Bonneville was falling rapidly, probably to below historic levels in Great Salt Lake by 12 ka. Water loss was now entirely by evaporation and, in the drying climate, salinity increased. The lake rose again briefly to the Gilbert shoreline at 1295 m during the Younger Dryas interval between 11 and 10 ka

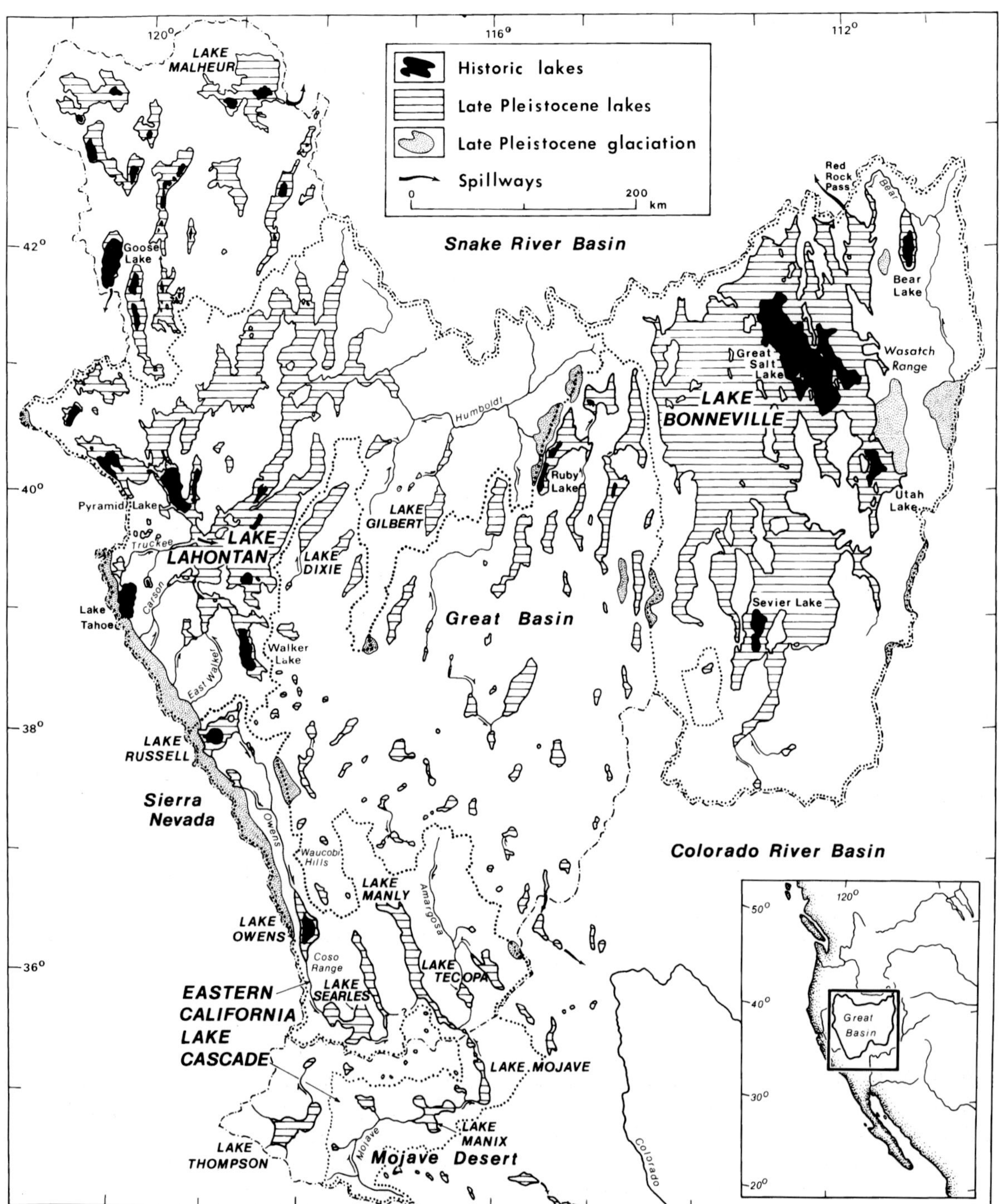

Figure 3.6 Late Pleistocene lakes of the Great Basin and adjacent areas (developed from Snyder et al., 1964; Currey et al., 1983; Benson et al., 1990; and author).

Table 3.2 Selected late Pleistocene lakes of western North America

Lake	Basin Area Maximum (km²)	Lake Area Maximum (km²)	Lake Depth Maximum (m)	Lake Volume Maximum (km³)	Spillway Elevation[1] (m)	Historic Lake/ Playa Surface (m, asl)
Bonneville	140,000	51,700	372	9,500	1,551	1,280
Lahontan	114,700	22,300	276 (Pyramid)	2,020	none[2]	1,177 (Pyramid)
Other Contributing Lakes						
Diamond	8,192	761	40			
Gilbert	1,406	541	76			
Madeline	2,163	777	38			
Tahoe	1,427	546				
Eastern California						
Russell	2,056	790	229			
Adobe	2,901	52	24			
Owens	10,969	531	67		1,145	1,081[3]
Searles	16,408	914	195		689	493
Panamint	19,930	765	283		355	317
Manix	9,363	407	116		571	dissected
Mojave	10,502	199	12		287	276
Tecopa	10,567	254	drained by Late Pleistocene		eroded	dissected
Manly	45,000	1,601	183		none	−86
Other Lakes in Arizona, New Mexico, and Nevada						
Cochise		190				
Dixie	6,314	1,088	72		none	
Waring	9,197	1,334	53		none	
Newark	3,590	925	87		none	
Railroad	9,233	1,360	96		none	
Franklin	4,913	1,220	64		?	
Spring	4,250	860	81		none	
Steptoe	4,623	1,189	107		?	

After Snyder et al., 1964; Currey et al., 1983; Benson et al., 1990; and author.

Notes

[1]Present elevation above sea level (i.e., the level to which the upstream lake must fill today before spilling downstream). Because of subsequent isostatic and tectonic deformation, erosion, and alluviation, this does not necessarily represent the Late Pleistocene spillway elevation.

[2]Lake Lahontan contained the following major internal sills that determined linkages within the larger lake: Mud Lake Slough, 1177 m; Emerson Pass, 1207 m; Astor Pass, 1222 m; Chocolate Pass, 1262 m; Darwin Pass, 1265 m; Pronto Pass, 1292 m; Adrian Valley, 1308 m.

[3]In 1913, before diversion of Owens River water to the Los Angeles Aqueduct, the surface of Owens Lake stood about 1088 m above sea level.

and then fluctuated at lower levels during the Holocene. Since first measured in 1843, the surface of Great Salt Lake has ranged from a high elevation of 1284 m in 1873 to a low of 1278 m in 1963, needing only to reach 1285 m to spill westward into Great Salt Lake Desert and increase in area by one-third at that level (Currey et al., 1983).

Lake Lahontan in northwest Nevada embraces seven basins covering 114,700 km^2, linked by sills and fed by several rivers including the Truckee, Carson, and Walker from the Sierra Nevada, and the Humboldt from the northeast. The sills allow discrete lakes to form in each subbasin before spilling to lower levels, and to persist even as the larger lake system desiccates. Lake Lahontan has been the focus of much research, initially by Russell (1885) and later by others (e.g., Morrison, 1964, 1991; Benson et al., 1990). As at Bonneville, the Lahontan basin has seen alternating lacustral and interlacustral events throughout the Pleistocene, with some lake beds in the Walker basin predating the Matuyama-Brunhes boundary. Unlike Bonneville, Lake Lahontan did not spill to the outside during the later Pleistocene. Because the lake ranges from 38° to 42°N, its deposits thus provide useful archives for climate change, supported by datable tephra, tufa, and mollusks. These suggest that the size of individual lakes depended on effective precipitation determined by proximity to the polar jet stream as it was forced north or south by the pulsating continental ice sheets farther north (Benson and Thompson, 1987; Benson et al., 1995).

As at Bonneville, the interval after 130 ka saw low lake levels and soil formation, but dated tufas show an expanded lake in Lahontan's northwest basins as early as 40 ka (Lao and Benson, 1988). Thereafter, the lake rose to 1265 m above sea level, spilling over Darwin Pass into the Carson Sink to form an expanded lake from 22 to 16.5 ka (Benson

Figure 3.7 Shorelines of Lake Bonneville, Utah, looking east toward the Wasatch Range (Mt. Timpanogos, 3581 m). The highest shoreline lies at ≈1600 against the slab-sided Traverse Mountains (≈2000 m) beyond the railway line; the lower shoreline, at ≈1500 m. Great Salt Lake to the north and Utah Lake to the south, successors to Lake Bonneville, are now linked by the Jordan River, which has dissected old lake deposits and provided a route for road, railway, and irrigation canals in recent times. This shoreline series, seen here in 1940, is now much altered by freeway construction and gravel pits (photo: Spence Collection, UCLA).

et al., 1990). After a brief recession to 1240 m, the lake then rose rapidly from 15.5 to 14 ka, integrating southward with Walker Lake across the 1308 m Adrian Pass. At its highest stand of 1330 m around 14 ka, Lake Lahontan had a surface area of 22,300 km², a volume of 2020 km³, and a maximum depth in Pyramid Lake of 276 m (Benson and Thompson, 1987). Around 13.5 ka, the lake fell rapidly to below 1208 m, rose again to above 1220 m around 10.8 ka during the Younger Dryas, then renewed its fall. By 10 ka, the lake was again segregated into seven shrinking water bodies, ceasing to exist as an integrated system. Although Pyramid Lake persisted throughout the Holocene, Walker Lake dried out around 16 ka, 5 ka, and 2 ka.

The Eastern California Lake Cascade comprises a series of lakes along three rivers, the Owens, Mojave, and Amargosa, which with sufficient discharge may link to an ultimate sump below sea level in Death Valley. Unlike Lake Bonneville and intricate Lake Lahontan, however, this cascade never formed a single continuous lake. Instead, each lake had to fill to a threshold before spilling downstream toward the next lower lake in the system. Lesser discharges linked only the higher lakes, leaving the lower lakes to respond to local runoff. From its complex salines, Lake Searles must have served more often than not as the sump for the Pleistocene Owens River, whereas Death Valley depended mostly on local precipitation and Amargosa River runoff (Yang et al., 1999). During the arid Holocene, the cascade has atrophied, although unusually wet winters may still regenerate higher lakes, whereas summer thunderstorms often yield flash floods.

There is abundant evidence in eastern California for earlier lake systems whose ages are constrained by dated tephras and paleomagnetic reversals. The ≈2 Ma lake beds in the Waucobi Hills, for example, indicate a westward shift in deposition caused by uplift of the White-Inyo Mountains relative to a subsiding Owens Valley (Bachman, 1978). Farther south, interbedded volcanic, lacustrine, and alluvial deposits of similar Plio-Pleistocene age have been deformed by uplift of the Coso Range. A 323-m core from beneath Owens Lake, which penetrated the 790-ka Matuyama-Brunhes magnetic reversal and the 760-ka Bishop ash near its base, revealed a shallow lake until 450 ka and a deeper lake thereafter, thought to reflect tectonic subsidence (Smith and Bischoff, 1997). Clay mineralogy and carbonate stratigraphy suggest climate cycles of roughly 100-ka duration. The top 693 m of a 930-m core from Searles Lake reveal climate oscillations over the past 3.2 Ma, with lacustrine marl and clay indicating deep water in a wet climate conducive to Owens River inflow, whereas evaporites and playa sediment reflect dry conditions (Smith, 1984). Along the Amargosa

River, from 3 Ma to 160 ka, a climate much wetter than today sustained a deep 250-km^2 lake in the Tecopa Basin, which attracted many animals, including elephant, horse, camel, llama, and flamingo (Hillhouse, 1987; Morrison, 1991). Plio-Pleistocene spring-fed carbonates and Mg-rich clays in the Amargosa Desert also indicate a wetter climate (Hay et al., 1986), favored by a Sierra Nevada some 1000 m lower 3 Ma ago (Huber, 1981). Sierra Nevada uplift later reduced moisture inputs, the resulting depletion of deuterium from eastbound Pacific storms being reflected in travertines. In Death Valley, evaporite, lacustrine, and alluvial facies in the Confidence Hills indicate a fluctuating Plio-Pleistocene lake, which strontium-isotope ratios suggest was fed by local runoff rather than by the Owens River (Stewart et al., 1994). Lake Manix and its wetland along the Mojave River attracted a rich middle to late Pleistocene fauna similar to that at Tecopa, plus now extinct ground sloth and dire wolf (Jefferson, 1991).

The late Pleistocene cascade is well documented, though linkages have been disrupted by recent tectonism, notably around Owens Lake where late Pleistocene shorelines are markedly deformed. At its maximum, the Owens River system directed overflow from lakes Russell (Mono) and Adobe into a cascade involving lakes Long Valley, Owens, Searles, and Panamint and perhaps Manly (Death Valley) (fig. 3.8). The Mojave system linked drainage from the San Bernardino Mountains through lakes Manix and Mojave, with the Amargosa drainage from the Spring Mountains through the Pahrump Valley and Tecopa Basin, again to Death Valley. Because higher lakes had to fill before spilling to lower levels, lakes in the cascade did not fill or desiccate simultaneously. Lower lakes experienced shorter freshwater episodes and more prolonged evaporite deposition than higher lakes. Also, lakes Tecopa and Manix were both breached and drained before the Pleistocene closed. In general, however, lake levels were relatively low between 120 ka and 30 ka, rose to a maximum around 22–18 ka, fell slightly around 18–14 ka partly because water was locked in local glaciers, and then oscillated from 14–10 ka before falling in the Holocene. Thus a 186-m core from Badwater reveals that Death Valley sustained perennial lakes up to 90 m deep during cool moist intervals from 186 ka to 120 ka, and again from 35 ka to 10 ka, whereas saline pans and mudflats characterized intervening hot, dry conditions from 120 ka to 35 ka and again after 10 ka (Li et al., 1996; Lowenstein et al., 1999). The salines reveal interbedded halite and mud, and repeated dissolution and reprecipitation of halite. Lake beds comprise initial mud-halite cycles, followed by bedded thenardite (Na_2SO_4) and mud, and capped by massive halite as the lake desiccated, all of which reflect variable salinities and lake levels in a perennial lake. Today, were it not for the loss of water to the Los Angeles Aqueduct, Owens River would maintain a ≈10-m deep Owens Lake but not spill farther south; shallow water would pond in lower lake basins only from local flash floods or winter rainstorms. Reconstruction of late Pleistocene water budgets is complicated by changing basin geometry, caused by tectonism and sedimentation.

In the now mostly arid Southwest, these and other Pleistocene lakes are relics of a past age, archives of profound environmental changes. Their strandlines tell much about basin responses to climate forcing and isostatic and tectonic deformation (Orme and Orme, 1991). Preserved deposits reveal much about repeated wet and dry phases, rain-shadow effects, and regional ecology, as reflected in pollen and faunal remains.

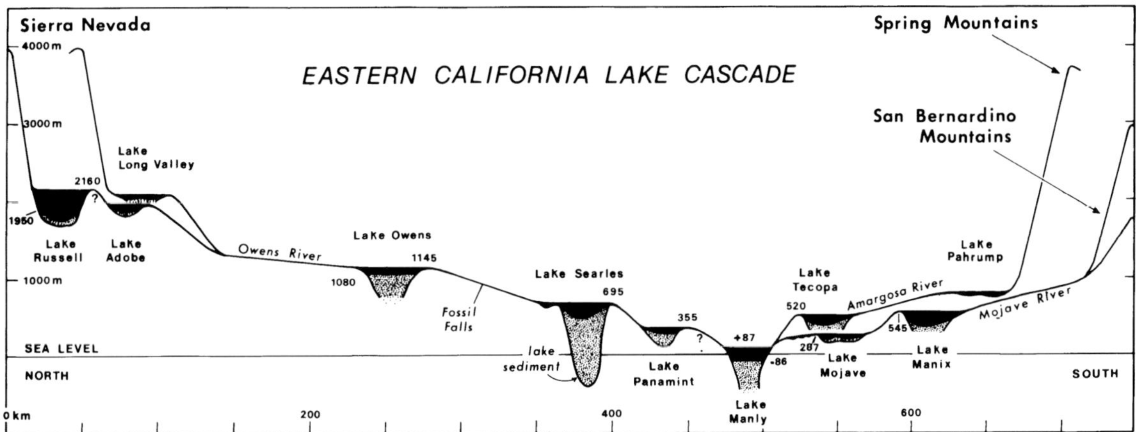

Figure 3.8 The Eastern California Lake Cascade. The lakes did not necessarily coexist simultaneously, nor did the river systems always reach Lake Manly.

3.4 Aeolian Landscapes

When ice sheets covered northern North America, the steeper temperature and pressure gradients of each cold stage generated strong winds beyond the ice front. Where vegetation cover was negligible, these winds deflated the surface, entrained sediment, shaped ventifacts and yardangs, and fashioned extensive loess sheets and sand dunes. When interglacial warmth and humidity returned and pressure gradients declined, many aeolian landscapes became clothed in vegetation and dissected by streams, but reduced wind activity often continued in less vegetated or disturbed areas. Beyond the ice front, such landscapes were created many times during the Pleistocene; however, as with other features, it is those forms shaped during and shortly after the last cold stage that survive most prominently today. Dunes also developed within former glaciated margins as ice fronts retreated, but, because they were usually destroyed during later glaciation, surviving dunes are mostly postglacial. Although aeolian landscapes are not easily dated, rock-varnish geochemistry of erosional features and cosmogenic, luminescence, and correlative dating of depositional sequences offer useful age constraints.

3.4.1 Loess Sheets

Loess is a wind-transported silt that is deposited as a poorly consolidated blanket whose thickness and grain size diminish downwind from source areas. The main sources of loess during Pleistocene cold stages were bare outwash deposits, till sheets, and floodplains along the windswept margins of advancing or retreating ice fronts, especially during the summer ablation season when outwash maximized and the landscape was neither frozen nor snow-covered. As ice fronts retreated and vegetation invaded these source areas, loess deposition diminished and soil began forming. Loess normally accumulates through dry deposition of tropospheric dust but it may also be washed out in precipitation. Though readily observed on land, where it is trapped by plants and acquires cohesion from soil moisture and carbonate cementation, loess also settles through the water column of lakes and ocean basins.

South of the ice front, loess 20–60 m deep blankets much of the midcontinent from the Appalachians to the Rockies, thinning southward across the Great Plains and down the Mississippi valley. As many as seven loess and loess-derived units of middle and late Pleistocene age have been found but few deposits older than the Brunhes magnetic chron have as yet been recognized, probably because of later erosion and reworking (Leigh and Knox, 1994). In the High Plains of Nebraska and Colorado, radiocarbon and luminescence ages indicate the deposition of five loesses: (1) pre-Illinoian loess; (2) Loveland Loess around 163 ka (Illinoian) on which Sangamon soil later developed; (3) Gilman Canyon Loess around 40–30 ka (mid-Wisconsinan); (4) Peoria Loess around 24–12 ka (late Wisconsinan) on which the Brady Soil developed; and (5) Bignell Loess from 9–3 ka, below the modern soil (Maat and Johnson, 1996; Muhs et al., 1999). The Loveland and Peoria loesses are widely distributed across the midcontinent, the latter continuing to accumulate as late as 10.5 ka (Pye et al., 1996). Lesser loess units and paleosols are not easily correlated with climate changes.

Loess also occurs west of the Rockies and in Alaska. Loess up to 75 m thick covers 60,000 km^2 of the Columbia Plateau, where it is interstratified with paleosols, tephra, superflood deposits, and flood-cut unconformities. Some loess units with reversed magnetic polarity reflect deposition before 790 ka and the total loess record may exceed 2 Ma (Busacca, 1989). For the later Pleistocene, texture and mineralogy point to slackwater sediment of episodic superfloods as the likely source of most loess, whereas four distinct paleosols and datable tephra, mostly from Mount St. Helens, show that loess deposition always followed these superfloods (McDonald and Busacca, 1992). In central Alaska, fission-track and paleomagnetic studies of interbedded tephra reveal a 3-Ma record of intermittent loess deposition during cold dry stages when adjacent areas were extensively glacierized, followed by erosion during warm interglacial stages (Preece et al., 1999).

3.4.2 Sand Dunes and Sand Sheets

Because sand is larger than silt, its entrainment and transport are more constrained by effective winds. Aeolian sands thus accumulate closer to sediment sources than loess, clustering as dunes or blanketing the landscape as sheets. During Pleistocene cold stages, aeolian sands typically accumulated downwind from ice-marginal outwash deposits, unstable floodplains, pluvial lakes, and river mouths. Dune occurrence and form varied with sand supply and effective wind velocity and direction. These supply-limited or transport-limited dunes ranged in form from transverse through barchanoid to barchan as sand supplies decreased, or from parabolic to longitudinal under more varied conditions involving vegetation change and fluctuating transport efficacy. Optimal conditions were favored by strong winds, low humidity, and bare source areas. However, dune formation did not necessarily cease with a return to interglacial warmth and humidity because, provided sand was available, migrating vegetation might stabilize one area while thinning elsewhere. In short, any changes that reduced the binding effect of vegetation encouraged dune formation, notably during transitions from glacial to interglacial climates, or vice versa, or as a result of fire, animal grazing, and later human activity.

Bare glaciogenic deposits at windy ice margins were particularly favorable to dune formation, more so as the ice retreated and before outwash and proglacial lake beds were clothed by vegetation. Just as ice-front oscillations were diachronous, so were various episodes of dune building. Dunes both formed and stabilized earlier along the farther margins of the ice front than in areas subject to late deglaciation. Thus, as Wisconsinan ice withdrew from the Appalachians, strong winds blowing across deltaic and lacustrine sands of proglacial Lake Albany created parabolic dunes in the Mohawk valley. In Connecticut, dunes were initially formed by anticyclonic northeast winds blowing across the floor of former glacial Lake Hitchcock and later reshaped by northwest winds between 12 and 11 ka (Thorson and Shile, 1995). Later, as first the Laurentide Ice Sheet and then the Champlain Sea withdrew from the St. Lawrence Valley, parabolic dunes were shaped by anticyclonic northeast winds blowing across southern Québec (Filion, 1987). Parabolic dunes also formed on Lake Saginaw's exposed floor in central Michigan following deglaciation around 12 ka (Arbogast et al., 1997). Farther west, wind-abraded bedrock surfaces and extensive parabolic dune fields were shaped initially by anticyclonic southeast winds in the ice-free corridor between the retreating Laurentide and Cordilleran ice sheets in Saskatchewan and Alberta (David, 1981). Most such features formed between 11 and 9 ka and later stabilized, but aeolian activity has persisted locally throughout Holocene times, notably south of Lake Athabaska.

On the Great Plains, where downslope winds have long provided high sand-transport potentials, extensive dune sheets and parabolic dunes formed downwind of braiding floodplains, such as the South Platte and Arkansas rivers, and poorly vegetated sandy outcrops, such as the Cretaceous Laramie Formation. At least three late Quaternary dune-building episodes have been recognized: extensive sand sheets between 27 and 11 ka, and less-extensive parabolic dunes between 11 and 4 ka and also within the past 1.5 ka (Muhs et al., 1996). Dune form indicates paleowinds from the northwest, not much different from modern winds. Recurrent aeolian activity in this region has been linked to regional drought and perhaps bison grazing, both of which caused a vegetation shift from grassland to scattered shrubs, notably disturbance indicators like goosefoot (Chenopodiaceae) (Forman et al., 1996). Even so, the size of the Nebraska Sand Hills north of the Platte River and the vast sand sheets of the southern Great Plains indicate the preeminence of aeolian processes far back in the Pleistocene and probably earlier (e.g., Holliday, 1991, 1997).

Although high tectonic relief dictates a more gravelly than sandy landscape, sand dunes are locally prominent in the desert Southwest. The larger dunes commonly occur downwind of former lakes and alluvial basins, for example, the 250 km^2 Samalyuca dunes in Mexico's Chihuahan Desert. Many formed during or shortly after pluvial episodes when abundant lacustrine and fluvial sediment was available and when southward displacement of the jet stream generated more frequent runoff and strong surface winds. As the jet stream retreated northward with returning warmth, many dune fields became supply limited or transport limited, often both, and either stabilized or dispersed thinly downwind. Many sand ramps of the Mojave Desert are thus relicts of the last pluvial maximum around 25–12 ka (Tchakerian, 1989). Relict dunes around Lake Thompson also stabilized as sand supplies diminished, but Holocene winds later stripped these to their structural roots (Orme, 1998). Under present arid conditions, the region's larger dune clusters remain active, but most have adopted stable locations in equilibrium with dominant wind regimes (fig. 3.9). Thus the Death Valley dunes, north of former Lake Manly, reflect a balance between northerly winter winds and southerly (monsoonal) summer winds. The small, 175-km^2 Kelso dunefield in the Mojave River basin east of former Lake Manix, partly active under present westerly winds, reveals evidence of earlier, very different wind regimes. In the Salton Trough, transverse and barchanoid forms of the active Algodones Dunes and the more stable East Mesa dune field are superimposed across a weathered mass of Pleistocene sand, in place by 31 ka and largely derived from deflation of beach and nearshore deposits of former Lake Coahuilla (Stokes et al., 1997). In the 5700-km^2 Gran Desierto sand sea of northwest Mexico, active dunes mantle several generations of Pleistocene dunes. These dunes are formed from sand discharged to the Salton Trough by the Colorado River, inputs that varied with delta switching, eustatic changes, and tectonism (Lancaster, 1995). Sand sheets, longitudinal dunes, and deflation "hogwallows" also occur in California's Central Valley, shaped by strong northwest winds sweeping across lake beds and alluvial fans fed by Sierra Nevada glacial meltwaters. Preserved on undissected mesas, these features reflect several aeolian episodes. Along the Pacific coast, changing late Quaternary sea levels, sand budgets, and aeolian regimes are reflected in differing dune forms (Orme and Tchakerian, 1986).

In Beringia, glaciation of the Brooks Range led to extensive fluvial aggradation in the Yukon, Koyukuk, and Kobuk lowlands, which, under the semiarid conditions of the late Pleistocene, provided abundant sand for dunes and sand sheets. Most dunes were stabilized by calcrete formation by 10 ka (Dijkmans et al., 1986), but others were reworked well into the Holocene. On northern Alaska's coastal plain, a widespread Pleistocene sand sea has seen repeated Holocene reactivation as parabolic and longitudinal dunes (Carter, 1981; Galloway and Carter, 1992).

3.5 Other Pleistocene Legacies

3.5.1 Glacioeustatic Sea-level Change

The recurrent eustatic changes that accompanied the growth and decay of Pleistocene ice sheets are but one variable in the complex equation that defines the relative elevation of

Figure 3.9 Active sand dunes on the western flanks of the Saddle Peak Hills, eastern California. Typical of many dune fields that have formed downwind from desiccating rivers and lakes, the sand here is derived from reworking of late Quaternary alluvium being transported by the Amargosa-Mojave river system toward nearby Death Valley (photo: Spence Collection, UCLA).

land and sea. Ignoring the separate effects of geoidal variation, tectonism, and isostasy on regional sea levels, the maximum glacioeustatic range of sea level between full glacial and full interglacial conditions probably approached 180 m. During the Last Interglacial, substage 5e, global sea level rose to 6 m above the present. During the Last Glacial Maximum, stage 2, sea level fell to about 130 m below present, somewhat less than in earlier larger glaciations. Apart from coastal impacts (see chapter 21), glacioeustatic fluctuations had three broad effects beyond the ice front. First, they impacted the lower reaches of coastal rivers, causing incision during glacial stages and aggradation during interglacials, as shown in the lower Mississippi valley. Second, fluctuating land-sea ratios influenced regional climates, affecting continentality as sea level rose and fell, most markedly in the Southeast where the continental shelf is at its widest. Third, the fluctuating land area affected the biota, offering more space and refuge for migrating plants and animals during glacial stages but less during interglacials.

3.5.2 Glacioisostatic and Hydroisostatic Crustal Adjustments

Earth's lithosphere adjusts to repeated glacier loading and unloading by subsiding beneath ice sheets during glaciation and rising during deglaciation. Because isostasy involves compensating lateral transfers of mass in the upper mantle, these adjustments are slow and, for larger loads, completion may lag several thousand years behind the triggering force. Postglacial isostatic uplift within the melting Laurentide Ice Sheet reached a maximum rate of 10 cm yr^{-1} during early deglaciation and later slowed, but uplift around Hudson Bay has approached 120 m over the past 6000 years and is projected to continue until the bay becomes dry land. In the midcontinent, crustal rebound from Laurentide ice loading has tilted shorelines from the Great Lakes to the Winnipeg Basin. Lake gauges show that the region continues to tilt up to the northeast, at rates of up to 0.5 cm yr^{-1}, and aided by forebulge collapse, down to the southwest (Clark and Persoage, 1970; Tackman et al.,

1999). Such tilting poses problems for lake management and, in about 3200 years, may redirect the Great Lakes into the Mississippi system, unless humans intervene. In Cascadia and the St. Lawrence Valley, ice loading may have caused tectonic stresses to accumulate beyond the strength of faults in their unglaciated condition, leading to increased seismic activity on deglaciation (Thorson, 1996).

Isostatic adjustments are not confined to glaciated areas. Because of lithospheric flexuring, ice loading extended peripheral subsidence 100–200 km beyond the ice front. Beyond that, displacement of mass during glaciation caused the emergence of a forebulge up to 300 km from the ice front. South of the Laurentide ice front, this forebulge was about 18 m high, but, following deglaciation and the return of ocean water, it gradually collapsed, prolonging the rise of postglacial seas along the mid-Atlantic coast (see chapter 21). Further, as ocean water moved into voids left by the collapsing forebulge, so coasts far distant from the former ice front saw modest emergence (e.g., Mitrovica and Peltier, 1991).

Large proglacial and pluvial lakes also generated hydroisostatic adjustments. Gilbert (1890) early recognized the nature of crustal warping induced by Lake Bonneville. Where it was deepest, the Bonneville (≈16 ka) and Provo (≈14 ka) shorelines have rebounded isostatically as much as 75 m and 59 m, respectively, above their present elevations along the lake's undeformed outer margins (Currey, 1988). Lesser lakes generated less isostatic response. Central Lake Lahontan has risen 22 m since its last highstand at 13 ka, mostly as a result of hydroisostatic rebound but aided by tectonism (Adams et al., 1999). Small Pleistocene lakes were usually supported without crustal deformation.

3.5.3 Biotic Anomalies and Enigmas

Pleistocene climate fluctuations caused major changes in the composition and range of the flora and fauna, in community structure, ecological succession, and habitat variety. As the Pleistocene closed, increasing warmth saw most surviving taxa return to habitats favored in previous interglacials (Pielou, 1991). Thus the boreal forest returned northward to Canada, whereas desert shrubs vacated Mexican refugia for the southwestern United States. However, it cannot be assumed that alternating warm and cold stages saw an orderly northward or southward migration of species. There were also less orderly east–west migrations, reflecting available moisture, and significant stochastic elements in the changing distributions as individual taxa responded to chance events. Inevitably there were some casualties. Many Neogene taxa that saw the onset of the Pleistocene, such as the warmth-loving *Magnolia* that lingered in southern California, did not survive to witness its end. Other plants, driven south during cold stages, became stranded and isolated by returning warmth. Thus, along the Pacific coast, Pleistocene cold saw the southward retreat of closed-cone pine forest to southern California and Mexico, to habitats they should have vacated with returning Holocene warmth and aridity. Yet today, relict stands of Bishop pine (*Pinus muricata*) and Monterey pine (*P. radiata*) occur on Guadalupe and Cedros islands off Baja California, although their preferred habitat lies 1000–1500 km farther north. Associated plants such as Torrey pine (*P. torreyana*) and Gowen cypress (*Cupressus goveniana*) never reestablished former ranges and are now restricted. Late Pleistocene pairings, such as sugar pine (*P. lambertiana*) with mountain hemlock (*Tsuga mertensiana*) or bristlecone pine (*P. longaeva*) with shadscale (*Atriplex canescens*), are now well separated by elevation.

With respect to animals, the Pleistocene legacy is even more dramatic. Although many Pleistocene animals, such as the northern bog lemming (*Mictomys borealis*) of the Great Basin and Appalachians, retreated north to Canada with returning warmth, others became isolated in mountain refuges, and many large animals simply died out. The megafaunal extinctions that occurred across the Pleistocene-Holocene transition are one of the most intriguing enigmas in the Quaternary record, particularly because animals tend to migrate when facing environmental stress. Indeed, over 40 of the 135 mammal genera living in late Pleistocene North America became extinct between 14 ka and 8 ka. Massive faunal extinctions have of course occurred at intervals throughout the Phanerozoic, notably among fish and marine invertebrates, but the extinctions that befell larger animals, such as Cretaceous dinosaurs and Pleistocene mammals, have evoked much speculation. Because the chronology of extinctions rests on the discovery of datable deposits containing the last member of a genus, care must be taken to avoid exaggeration; deposits may be misinterpreted or new evidence may contradict the belief that previously known deposits really contain the last of a genus. Indeed, improved temporal resolution suggests that lingering demise may be more common than sudden extinction.

Table 3.3 illustrates these extinctions from the North American type site for the late Pleistocene (Rancholabrean) land mammal age. Of 58 mammal species that were trapped in asphalt deposits at Rancho La Brea, Los Angeles, California, sometime after 40 ka, 22 are now extinct; of 138 bird species, 20 are now extinct (Stock and Harris, 1992). This fauna is noteworthy for the high proportion of large predators (e.g., saber-toothed cat, dire wolf, teratorn) and scavengers (e.g., condor, vulture), presumably because they were lured to victims already mired in the tar seeps. Of the 22 mammals listed, all were extinct locally before 11 ka, but some lingered regionally into the Holocene (e.g., camel in the Mojave Desert); of the birds, the passenger pigeon only disappeared in 1914.

Explanation for the megafaunal extinctions has long polarized on two very different, but not mutually exclusive causes: climate change and human predation (e.g.,

Table 3.3 Late Pleistocene mammals and birds, now extinct, from Rancho La Brea, California

Class Mammalia (mammals)		Class Aves (birds)	
Order Proboscidea (proboscideans)		Order Ciconiformes (storks, teratorns, vultures)	
Mammuthus columbi	Columbian mammoth	*Ciconia maltha*	Asphalt stork
Mammut americanum	American mastodont	*Mycteria wetmorei*	
		Teratornis merriami	Giant teratorn
Order Carnivora (placental carnivores)		*Cathartornis gracilis*	Small teratorn
Smilodon fatalis	Saber-toothed cat	*Breagyps clarki*	Long-beak condor
Homotherium serum	Scimitar cat	*Coragyps occidentalis*	Western black vulture
Panthera leo atrox	American lion	*Gymnogyps amplus*	Condor
Panthera onca agusta	Jaguar		
Arctodus simus	Short-faced bear	Order Accipitriformes (diurnal birds of prey)	
Canis dirus	Dire wolf	*Buteogallus fragilis*	Slender-limbed eagle
		Amplibuteo woodwardi	Morphnine eagle
Order Perissodactyla (odd-toed ungulates)		*Wetmoregyps daggetti*	Long-legged eagle
Equus occidentalis	Western horse	*Spizaetus grinnelli*	Crested eagle
Equus conversidens	Small horse	*Neogyps errans*	
Tapirus californicus	Californian tapir	*Neophrontops americanus*	
Order Artiodactyla (even-toed ungulates)		Order Anseriformes (waterfowl)	
Platygonus compressus	Peccary	*Anabernicula gracilenta*	Brea pigmy goose
Camelops hesternus	Large camel		
Hemiauchenia macrocephala	Stilt-legged llama	Order Galliformes (quails, turkeys)	
Capromeryx minor	Small antelope	*Meleagris californica*	California turkey
Euceratherium collinum	Shrub ox		
Bison latifrons	Long-horned bison	Order Strigiformes (owls)	
Bison antiquus	Ancient bison	*Strix brea*	Brea owl
Order Edentata (edentates)		Order Columbiformes (pigeons, doves)	
Glossotherium harlani	Grazing ground sloth	*Ectopistes migratorius*	Passenger pigeon
Megalonyx jeffersonii	Flat-footed ground sloth		
Nothrotheriops shastensis	Browsing ground sloth	Order Passeriformes (songbirds)	
		Pipilo angelensis	Towhee
Order Rodentia (rodents)		*Euphagus magnirostris*	Blackbird
Peromyscus imperfectus	Imperfect mouse	*Pandanaris convexa*	Icterid

After Stock and Harris, 1992.

Martin and Klein, 1984; Agenbroad et al., 1990). Variations of the climatic hypothesis invoke rapidly changing temperature, precipitation, seasonality, water availability, and vegetation patterns across the Pleistocene-Holocene transition (Guthrie, 1984). Such changes would have impacted habitat diversity, growing-season duration, and forage, which in turn would have affected animal nutrition, breeding potential, and gestation time. Monogastrics such as mammoth, sloth, and horse would have been more vulnerable than ruminants such as deer and bison (Owen-Smith, 1988). Animals either adjusted or migrated, or, when stranded in unfavorable habitats, they spiraled toward extinction. At Rancho La Brea, increased seasonality across the Pleistocene-Holocene transition, especially longer summer droughts, probably decreased the food available to large mammalian herbivores, especially the monogastrics adapted to high-fiber, low-protein diets. This in turn reduced the food source for large carnivores. Rising sea level also restricted coastal habitats and sometimes isolated land mammals, such as the mammoths that died out on southern California's offshore islands (Cushing et al., 1984).

Human predation hypotheses point to the intriguing stratigraphic coincidence of the first serious hunters to reach North America with the last Pleistocene megafauna. Some scientists have emphasized overkill, even blitzkrieg, as large mammals, naive and unprepared, were suddenly exposed to big-game hunters from Beringia penetrating the widening corridor between the Laurentide and Cordilleran ice sheets sometime after 14 ka. Beyond the retreating ice, a death front advanced southward across North America as large mammals were driven to extinction by hunting groups armed with Clovis-point technology. Mammals long resident in the Americas, such as sloths, may have been more vulnerable to hunters than later animal migrants from Asia, long accustomed to and more wary of human ways.

Both hypotheses pose problems. If climate change alone was to blame, why did earlier glacial-interglacial transitions not generate comparable extinctions? Or did they, but the evidence has yet to be found? If human predation was solely to blame, why did late Pleistocene megafaunal extinctions elsewhere, notably in Australia with its naive marsupials, not coincide with human arrival? In face of such questions, compromise is often suggested. For example, climate change may have softened many species, making them more vulnerable to human predation—the proverbial straw breaking the camel's back. Hunters may have simply applied the coup de grâce to animals already weakened by rapid late Pleistocene ecological changes. Extinction of large herbivores may have so reduced habitat diversity that the demise of other animals was inevitable. Whatever their cause, these extinctions greatly changed the composition and variety of the Holocene fauna. Extinction for some was an opportunity for others. For example, the loss of many large predators favored smaller predators, such as wolf, agile ruminants, such as pronghorn and deer, and descendants of extinct species, such as bison. This Holocene fauna was, however, to suffer great indignity at human hands in historic time (see chapter 22).

3.6 Conclusion

From skeptical beginnings, the glacial theory of the early nineteenth century has since blossomed into a complex and sophisticated body of knowledge concerning the Quaternary Period. The Quaternary, itself a relic term from geology's heroic age, is no longer equated simply with the Great Ice Age or with glaciation but with multiple climatic oscillations with implications far beyond the ice front. Again, from fledgling beginnings in European mountains and academies, the quest for knowledge about the Quaternary has become global in scope, fundamental to understanding present landscapes and predicting future ones. The North American contribution to this work has been significant—from early studies of glaciation and pluvial lakes to recent research into ocean sediments and ice cores.

Inevitably, explanations of field data and modeling of climate scenarios have generated controversy. Initially, many scholars were reluctant to espouse glacial theory; later they often disavowed the concept of multiple glaciations. Still later, explanations of climate change refocused, often heatedly, on the forcing role of Earth's orbital variations, proposed so long ago by Croll and Milankovitch, relative to essentially Earth-bound forces, such as tectonism. Present consensus suggests that Earth's orbital variations do indeed correlate, more or less, depending on prevailing rhythms, with major Quaternary climate changes and successive interglacial-glacial cycles (e.g., Ruddiman et al., 1986; Imbrie et al., 1993; Lowe and Walker, 1997). The eccentricity of Earth's orbit around the Sun, varying from nearly circular to strongly elliptical on a ≈95-ka cycle, correlates quite well, presumably through its impact on solar radiation received, with the periodic expansion and contraction of ice sheets over the past 790 ka. The precession of the equinoxes, which determines the perihelion and aphelion and thus seasonality on a variable 19–23-ka cycle, is reflected in substages within glacial and interglacial stages, notably over the past 20 ka. The ≈41-ka obliquity cycle wherein Earth's axial tilt varies from 21.4° to 24.5° from vertical, which controls the intensity of seasonality, most effectively in higher latitudes, is seen in Pleistocene glacial fluctuations and in oxygen-isotope ratios, carbonate contents, and sea-surface temperatures derived from ocean sediment. The effect of these cycles is expressed directly in solar radiation received by Earth's atmosphere and surface, and indirectly through such derived climatic variables as temperature, precipitation, and wind. Complexity is added by the phase relations of these cycles and by less predictable feedbacks in the coupled ocean-atmosphere system. These cycles may combine to change insolation by as much as 14% in any month, leading to more persistent snow cover and sea ice during summer, which, by positive feedback, may initiate prolonged cold. But controversy continues: the 500-ka record from Devil's Hole, Nevada, has been used both to support (Johnson and Wright, 1989) and refute (Winograd et al., 1992) the concept of orbital cycles.

But why did North America experience such dramatic climate changes in the Quaternary after the relative warmth and stability of the Pliocene? The answer is complex. Certainly, orbital cycles may conspire to generate glacial conditions, notably by lowering summer temperatures, but if this was so, comparable cyclic events would have been more common in the distant past. Nor does orbital forcing readily explain the many rapid but brief oscillations in climate that occurred within the longer cycles. Thus much of the explanation must be sought on Earth, through complex feedback mechanisms in which orbital variations interact with other surface changes, notably tectonism and surface-atmosphere couplings.

Several tectonic events that had begun earlier continued during the Quaternary (see chapter 1). The Atlantic Ocean continued to widen and its meridional circulation probably intensified under the forcing influence of polar ice; the Western Cordillera continued to rise and expand, increasing continentality and meridional circulation over central North America; and the Tibetan Plateau remained high. Further, increased explosive volcanism in the Aleutian and Kurile arcs after 3 Ma may have intensified trends toward Northern Hemisphere glaciation through the dust scattering of incoming solar radiation and resultant atmospheric cooling (Stewart, 1975; Prueher and Rea, 1998). Inevitably, there were complex feedbacks to these changes, including fluctuations in atmospheric carbon and sulfate aerosols.

With respect to surface-atmosphere interaction, North Atlantic cores have revealed a close relationship between influxes of glacial meltwaters and the position of the polar front, notably toward the close of the Pleistocene (Ruddiman and McIntire, 1981). Further, although ice sheets may have been initiated and maintained by orbital forcing, their later behavior may have been related more to internal dynamics than to external factors, resulting in fluxes of meltwater and icebergs to the oceans that changed climate through their impact on sea-surface temperatures (Heinrich, 1988; MacAyeal, 1993). It is also likely that air temperatures were influenced by changes in ocean circulation triggered by salinity and temperature gradients in response to fluctuating freshwater inputs from rivers, sea ice, and distant fluctuations of the West Antarctic Ice Sheet and its more robust East Antarctic neighbor. These changes are reflected in marine sediment geochemistry and the isotope signatures of benthic organisms (Broecker and Denton, 1990). For example, changes in the production of North Atlantic Deep Water appear to correlate with climatic variability between 130 ka and 70 ka, as indicated by ice cores from Greenland. Also, orbitally induced climate changes may have been amplified, through intricate feedback loops, by fluctuations in atmospheric greenhouse gases between interglacial and glacial stages (Dansgaard and Oeschger, 1989). During interglacial stages, the relative extent of frozen ground and wetland over northern North America would have significantly influenced the carbon flux involving the release or sequestering of CO_2 and CH_4 at the surface.

In short, for a variety of reasons, the Pleistocene was an epoch of frequent and complex change, which ensured that the landscapes bequeathed to the Holocene were very different from those inherited from the Pliocene. Legacies of the Pleistocene are to be found throughout the continent, essentially because the brevity of the Holocene to date has allowed relatively little time for their removal.

References

Adam, D.P., 1988. Pollen zonation and proposed informal climatic units for Clear Lake, California, cores CL-73-4 and CL-73-7. In: J.D. Sims (Editor), *Late Quaternary Climate, Tectonism, and Sedimentation in Clear Lake, northern California Coast Ranges*. Geological Society of America, Special Paper 214, Boulder, 63-80.

Adams, K.D., S.G. Wesnousky, and B.G. Bills, 1999. Isostatic rebound, active faulting, and potential geomorphic effects in the Lake Lahontan basin, Nevada and California. *Geological Society of America Bulletin*, 111, 1739–1756.

Agassiz, L., 1840. *Etudes sur les glaciers*. Neuchâtel.

Agenbroad, L., et al., (Editors), 1990. Megafauna and man. Northern Arizona University, Flagstaff.

Alden, W.C., 1932. Physiography and glacial geology of eastern Montana and adjacent areas. *U.S. Geological Survey Professional Paper* 174.

Anderson, P.M., 1985. Late Quaternary vegetational change in the Kotzebue Sound area, northwestern Alaska. *Quaternary Research*, 24, 307–320.

Anderson, R.S., 1993. A 35,000 year vegetation and climate history from Potato Lake, Mogollon Rim, Arizona. *Quaternary Research*, 40, 351–359.

Arbogast, A.F., P. Scull, R.J. Schaetzl, J. Harrison, T.P. Jameson, and S. Crozier, 1997. Concurrent dune stabilization of some interior dune fields in Michigan. *Physical Geography*, 18, 63–79.

Baas, J.H., J. Meinert, F. Abrantes, and A. Prins, 1997. Late Quaternary sedimentation on the Portuguese continental margin: Climate related processes and products. *Palaeogeography, Palaeoclimatology, and Palaeoecology*, 130, 1–23.

Bachman, S., 1978. Pliocene-Pleistocene break-up of the Sierra Nevada—White-Inyo Mountains block and formation of Owens Valley. *Geology*, 6, 461–463.

Baker, R.G., 1981. Interglacial and interstadial environments in Yellowstone National Park. In: W.C. Mahaney (Editor), *Quaternary paleoclimate*. GeoBooks, Norwich, 361–375.

Baker, V.R., and R.C. Bunker, 1985. Cataclysmic late Pleistocene flooding from glacial Lake Missoula: A review. *Quaternary Science Reviews*, 4, 1–41.

Baker, V.R., and M.M. Penteado-Orellana, 1977. Adjustment to Quaternary climate change by the Colorado River in central Texas. *Journal of Geology*, 85, 395–422.

Benson, L.V., 1981. Paleoclimatic significance of lake-level fluctuations in the Lahontan Basin. *Quaternary Research*, 16, 390–403.

Benson, L.V., and R.S. Thompson, 1987. Lake-level variation in the Lahontan Basin for the past 50,000 years. *Quaternary Research*, 28, 69–85.

Benson, L.V., M. Kashgarian, and M. Rubin, 1995. Carbonate deposition, Pyramid Lake subbasin, Nevada. 2. Lake levels and polar jet stream positions reconstructed from radiocarbon ages and elevations of carbonates (tufas) deposited in the Lahontan basin. *Palaeogeography, Palaeoclimatology, and Palaeoecology*, 117, 1–30.

Benson, L.V., D.R. Currey, R.I. Dorn, K.R. Lajoie, C.G. Oviatt, S.W. Robinson, G.I. Smith, and S. Stine, 1990. Chronology of expansion and contraction of four Great Basin lake systems during the past 35,000 years. *Palaeogeography, Palaeoclimatology, and Palaeoecology*, 78, 241–286.

Bretz, J.H., 1923. The Channeled Scabland of the Columbia Plateau. *Journal of Geology*, 31, 617–649.

Bretz, J.H., 1969. The Lake Missoula floods and the Channeled Scabland. *Journal of Geology*, 77, 505–543.

Broecker, W.S., and G.H. Denton, 1990. The role of ocean-atmosphere reorganizations in glacial cycles. *Quaternary Science Reviews*, 9, 305–341.

Brook, G.A., and R.J. Nickmann, 1996. Evidence of late Quaternary environments in northeast Georgia from sediments preserved in Red Spider Cave. *Physical Geography*, 17, 465–484.

Brown, P.A., and J.P. Kennett, 1998. Megaflood erosion and meltwater plumbing changes during the last North American deglaciation recorded in Gulf of Mexico sediments. *Geology*, 26, 599–602.

Brunner, C.A., W.R. Normark, G.G. Zuffa, and F. Serra, 1999. Deep-sea sedimentary record of the late Wisconsin cataclysmic floods from the Columbia River. *Geology*, 27, 463–466.

Bryson, R.A., 1991. Modeling past climates. In: R.B. Morrison (Editor), *Quaternary nonglacial geology: Conterminous U.S.* The Geology of North America, K-2, Geological Society of America, Boulder, 15–19.

Bryson, R.A., and B.M. Goodman, 1986. Milankovitch and global ice-volume simulation. *Theoretical and Applied Climatology*, Series B, 37, 22–28.

Busacca, A.J., 1989. Long Quaternary record in eastern Washington, U.S.A., interpreted from multiple buried soils in loess. *Geoderma*, 45, 105–122.

Buwalda, J.P., 1914. Pleistocene lake beds at Manix in the eastern Mohave Desert region. University of California Publications, *Bulletin of the Department of Geological Sciences*, 7, 443–464.

Carter, L.D., 1981. A Pleistocene sand sea on the Alaskan Arctic coastal plain. *Science*, 211, 381–383.

Chapman, M.R., and N.J. Shackleton, 1999. Global ice-volume fluctuation, North Atlantic ice-rafting events, and deep-ocean circulation changes between 130 and 70 ka. *Geology*, 27, 795–798.

Clark, G.M., and E.J. Ciolkosz, 1988. Periglacial geomorphology of the Appalachian highlands and interior highlands south of the glacial border: A review. *Geomorphology*, 1, 191–220.

Clark, R.H., and N.P. Persoage, 1970. Some implications of crustal movement in engineering planning. *Canadian Journal of Earth Sciences*, 7, 628–633.

CLIMAP Project Members, 1981. Seasonal reconstructions of the Earth's surface at the Last Glacial Maximum. *Geological Society of America Map and Chart Series*, MC-36.

COHMAP Members, 1988. Climatic changes of the last 18,000 years: Observations and model simulations. *Science*, 241, 1043–1052.

Croll, J., 1864. On the physical cause of the change of climate during geological epochs. *Philosophical Magazine*, 28, 121–137.

Crowley, T.J., 1996. Pliocene climates: The nature of the problem. *Marine Micropaleontology*, 27, 3–12.

Currey, D.R., 1988. Seismotectonic kinematics inferred from Quaternary paleolake datums, Salt Lake City seismopolitan region, Utah. *U.S Geological Survey Open-File Report* 88-673, 457–461.

Currey, D. R., and C.G. Oviatt, 1985. Durations, average rates, and probable causes of Lake Bonneville expansions, stillstands, and contractions during the last deep-lake cycle, 32,000 to 10,000 years ago. In: P.A. Kay and H.F. Diaz (Editors), *Problems and prospects for predicting Great Salt Lake levels.* Center for Public Affairs and Administration, University of Utah, Salt Lake City, 1–9.

Curry, B.B., and M.J. Pavich, 1996. Absence of glaciation in Illinois during marine oxygen isotope stages 3 through 5. *Quaternary Research*, 46, 19–26.

Currey, D.R., G. Atwood, and D.R. Mabey, 1983. Major levels of Great Salt Lake and Lake Bonneville. *Utah Geological and Mineral Survey*, Map 73.

Cushing, J., M. Daily, E. Noble, V.L. Roth, and A. Wenner, 1984. Fossil mammoths from Santa Cruz Island, California. *Quaternary Research*, 21, 376–384.

Dansgaard, W., and H. Oeschger, 1989. Past environmental long-term records from the Arctic. In: H. Oeschger and C.C. Langway (Editors), *The environmental record in glaciers and ice sheets.* John Wiley, Chichester, 287–317.

Dansgaard, W., S.J. Johnsen, H.B. Clausen, D. Dahl-Jensen, N. Gundestrup, C.U. Hammer, and H. Oeschger, 1984. North Atlantic oscillations revealed by deep Greenland ice cores. *Geophyiscal Monographs*, 29, 288–298.

Dansgaard, W., S.J. Johnsen, H.B. Clausen, D. Dahl-Jensen, N.S. Gundestrup, C.U. Hammer, C.S. Hvidberg, J.P. Steffensen, A. E. Sweinbjörnsdottir, J. Jouzel, and G. Bond, 1993. Evidence of general instability of past climate from a 250–kyr ice-core record. *Nature*, 364, 218–220.

David, P.P., 1981. Stabilized dune ridges in northern Saskatchewan. *Canadian Journal of Earth Sciences*, 18, 286–310.

Delcourt, H.R., 1979. Late Quaternary vegetation history of the eastern Highland Rim and adjacent Cumberland Plateau of Tennessee. *Ecological Monographs*, 49, 255–280.

Delcourt, P.A., and H.R. Delcourt, 1981. Vegetation maps for eastern North America: 40,000 yr. B.P. to the present. In: R.C. Romans (Editor), *Geobotany II*, Plenum Press, New York, 123–165.

Dijkmans, J.W.A., E.A. Koster, J.P. Galloway, and W.G. Mook, 1986. Characteristics and origin of calcretes in a subarctic environment, Great Kobuk sand dunes, northwestern Alaska, U.S.A. *Arctic and Alpine Research*, 18, 377–387.

Dorn, R.I., and F.M. Phillips, 1991. Surface exposure dating: Review and critical evaluation. *Physical Geography*, 12, 303–333.

Dyke, A.S., and V.K. Prest, 1987. Late Wisconsin and Holocene history of the Laurentide ice sheet. *Géographie Physique et Quaternaire*, 41, 237–263.

Eardley, A.J., R.T. Shuey, V. Gvosdetsky, W.P. Nash, M. D. Picard, D.C. Grey, and G.J. Kukla, 1973. Lake cycles in the Bonneville Basin, Utah. *Geological Society of America Bulletin*, 84, 211–216.

Elias, S.A., and T.R. Van Devender, 1992. Insect fossil evidence of late Quaternary environments in the northern Chihuahuan Desert of Texas and New Mexico: Comparisons with the paleobotanical record. *Southwestern Naturalist*, 37, 101–116.

Emiliani, C., 1955. Pleistocene temperatures. *Journal of Geology*, 63, 538–578.

Emiliani, C., 1966. Paleotemperature analysis of Caribbean cores P6304-8 and P6304-9 and a generalized temperature curve for the past 425,000 years. *Journal of Geology*, 74, 109–126.

Engel, S.A., T.W. Gardner, and E.J. Ciolkosz, 1996. Quaternary soil chronosequences on terraces of the Susquehanna River, Pennsylvania. *Geomorphology*, 17, 273–294.

Filion, L., 1987. Holocene development of parabolic dunes in the central St. Lawrence Lowland, Quebec. *Quaternary Research*, 28, 196–209.

Forman, S.L., R. Oglesby, V. Markgraf, and T. Stafford, 1996. Paleoclimatic significance of late Quaternary eolian deposition on the Piedmont and High Plains, central United States. *Global and Planetary Change*, 11, 35–55.

Fullerton, D.S., 1997. Sangamon and Sangamonian time divisions, chronostratigraphic unit, and Geosol: Clarification of terminology. *Quaternary Research*, 48, 247–248.

Funder, S., N. Abrahamsen, O. Bennike, and R.W. Feyling-Hanssen, 1985. Forested Arctic: Evidence from North Greenland. *Geology*, 13, 542–546.

Gale, H.S., 1913. Salines in the Owens, Searles, and Panamint basins, southeastern California. *U.S. Geological Survey Bulletin*, 580-L (1914).

Galloway, J.P., and L.D. Carter, 1992. Late Holocene longitudinal and parabolic dunes in northern Alaska: Preliminary interpretations of age and paleoclimatic significance. In: C. Dusel-Bacon and A.B. Till (Editors), Geologic studies in Alaska by the U.S. Geological Survey, *U.S. Geological Survey Bulletin* 2068, 3–11.

Gilbert, G.K., 1890. Lake Bonneville. *U.S. Geological Survey Monograph 1*, Washington, 1–438.

GRIP (Greenland Ice-core Project), 1993. Climate instability during the last interglacial period recorded in the GRIP ice core. *Nature*, 364, 203–207.

Grootes, P.M., et al., 1993. Comparison of oxygen isotope records from the GISP2 and GRIP Greenland ice cores. *Nature*, 366, 552–554.

Grüger, E., 1972. Late Quaternary vegetation development in south-central Illinois. *Quaternary Research*, 2, 217–231.

Guthrie, R.D., 1984. Mosaics, allelochemics and nutrients: An ecological theory of late Pleistocene megafaunal extinctions. In: P.S. Martin and R.G. Klein (eds.), *Quaternary extinctions: A prehistoric revolution*. University of Arizona Press, Tucson, 259–298.

Hall, S.A., and S. Valastro, 1995. Grassland vegetation in the southern Great Plains during the Last Glacial Maximum. *Quaternary Research*, 44, 237–245.

Hammer, C., P.A. Mayewski, D. Peel, and M. Stuiver (eds.), 1997. Greenland Summit ice cores. *Journal of Geophysical Research*, 102, 26, 315–26, 886.

Harrington, C.R., 1980. Faunal exchanges between Siberia and North America: Evidence from Quaternary land mammal remains in Siberia, Alaska, and the Yukon Territory. *Canadian Journal of Anthropology*, 1, 45–49.

Harrison, S.P., J.E. Kutzbach, I.C. Prentice, P.J. Behling, and M.T. Sykes, 1995. The response of the northern hemisphere extratropical climate and vegetation to orbitally induced changes in insolation during the last interglacial. *Quaternary Research*, 43, 174–184.

Hay, R.L., R.E. Pexton, T.T. Teague, and T.K. Kyser, 1986. Spring-related carbonate rocks, Mg-clays, and associated minerals in Pliocene deposits of the Amargosa Desert, Nevada and California. *Geological Society of America Bulletin*, 97, 1488–1508.

Heinrich, H., 1988. Origin and consequences of cyclic ice-rafting in northeast Atlantic Ocean during the past 130,000 years. *Quaternary Research*, 29, 142–152.

Hendy, I.L., and J.P. Kennett, 1999. Latest Quaternary North Pacific surface-water responses imply atmosphere-driven climatic instability. *Geology*, 27, 291–294.

Hillaire-Marcel, C., A. de Vernal, G. Bilodeau, and G. Wu, 1994. Isotope stratigraphy, sedimentation rates, deep circulation, and carbonate events in the Labrador Sea during the last ~200 ka. *Canadian Journal of Earth Sciences*, 31, 63–89.

Hillhouse, J.W., 1987. Late Tertiary and Quaternary geology of the Tecopa Basin, southeastern California. *U.S. Geological Survey Miscellaneous Investigations*, Map I-1728.

Holliday, V.T., 1987. A re-examination of late Pleistocene boreal forest reconstructions for the southern High Plains. *Quaternary Research*, 28, 238–244.

Holliday, V.T., 1991. The geological record of wind erosion, eolian deposition, and aridity on the southern High Plains. *Great Plains Research*, 1, 1–25.

Holliday, V.T., 1997. Origin and evolution of lunettes on the High Plains of Texas and New Mexico. *Quaternary Research*, 47, 54–69.

Huber, N.K., 1981. Amount and time of late Cenozoic uplift and tilt of central Sierra Nevada, California—Evidence from the upper San Joaquin River. *U.S. Geological Survey Professional Paper* 1197, 1–28.

Imbrie, J., A. Berger, and N.J. Shackleton, 1993. Role of orbital forcing: A two-million-year perspective. In: J.A. Eddy and H. Oeschger (Editors), *Global changes in the perspective of the past*. John Wiley, Chichester, 263–277.

Jefferson, G.T., 1991. The Camp Cady local fauna: stratigraphy and paleontology of the Lake Manix basin. *San Bernardino County Museum Association Quarterly*, 38, 93–99.

Johnsen, S.J., H.B. Clausen, W. Dansgaard, K. Fuhrer, N. Gundestrup, C.U. Hammer, P. Iverson, J. Jouzel, B. Stauffer, and J.P. Steffensen, 1992. Irregular glacial interstadials recorded in a new Greenland ice core. *Nature*, 359, 311–313.

Johnson, R.G. and H.E. Wright, 1989. Great Basin calcite vein and the Pleistocene timescale. *Science*, 246, 262.

Kapp, R.O., 1965. Illinoian and Sangamon vegetation of southwestern Kansas and adjacent Oklahoma. *University of Michigan, Contributions from the Museum of Paleontology*, 19, 167–255.

Kapp, R.O., and A.M. Gooding, 1964. Pleistocene vegetational studies in the Whitewater Basin, southeastern Indiana. *Journal of Geology*, 72, 307–326.

Kennett, J.P., and 19 others, 1994. Initial Reports: Santa Barbara Basin. *Proceedings of the Ocean Drilling Program*, 146, Part 2, College Station, Texas.

King, J.E., 1991. Early and Middle Quaternary vegetation. In R.B. Morrison (Editor), *Quaternary nonglacial geology: Conterminous U.S.* The Geology of North America, K-2, Geological Society of America, Boulder, 19–26.

King, J.E., and J.J. Saunders, 1986. *Geochelone* in Illinois and the Illinoian-Sangamonian vegetation in the type region. *Quaternary Research*, 25, 89–99.

Kneller, M., and D. Peteet, 1999. Late-glacial to early Holocene climate changes from a central Appalachian pollen and macrofossil record. *Quaternary Research*, 51, 133–147.

Knox, J.C., 1996. Late Quaternary upper Mississippi River alluvial episodes and their significance to the lower Mississippi River system. *Engineering Geology*, 45, 263–285.

Kutzbach, J.E., and T. Webb, 1993. Conceptual basis for understanding late Quaternary climates. In: H.E. Wright, J.E. Kutzbach, T. Webb, W.F. Ruddiman, F.A. Street-Perrott, and P.J. Bartlein (Editors), *Global climates since the Last Glacial Maximum*. University of Minnesota Press, Minneapolis, 5–11.

Kutzbach, J.E., P.J. Guetter, P.J. Behling, and R. Selin, 1993. Simulated climatic changes: Results of the COHMAP climate-model experiments. In: H.E. Wright, J.E. Kutzbach, T. Webb, W.F. Ruddiman, F.A. Street-Perrott, and P.J. Bartlein (Editors), *Global climates since the Last Glacial Maximum*, Minneapolis, University of Minnesota Press, 24–93.

Lagoe, M.B., and S.D. Zellers, 1996. Depositional and microfaunal response to Pliocene climate change and tectonics in the eastern Gulf of Alaska. *Marine Micropaleontology*, 27, 121–140.

Lancaster, N., 1995. Origin of the Gran Desierto sand sea, Sonora, Mexico: Evidence from dune morphology and sedimentology. In: V.P. Tchakerian (Editor), *Desert aeolian processes*, Chapman and Hall, London, 11–35.

Lao, Y., and L.V. Benson, 1988. Uranium-series age estimates and paleoclimatic significance of Pleistocene tufas from the Lahontan Basin, California and Nevada. *Quaternary Research*, 30, 165–176.

Leigh, D.S., and J.C. Knox, 1994. Loess of the upper Mississippi Valley Driftless Area. *Quaternary Research*, 42, 30–40.

Leverington, D.W., J.D. Mann, and J.T. Teller, 2000. Changes in the bathymetry and volume of glacial Lake Agassiz between 11,000 and 9300 ^{14}C yr B.P. *Quaternary Research*, 54, 174–181.

Levesque, A., F.E. Mayle, and L.C. Cwynar, 1993. The amphi-Atlantic oscillation: A proposed late-glacial climatic event. *Quaternary Science Reviews*, 12, 629–644.

Li. J., T.K. Lowenstein, C.B. Brown, T-L. Ku, and S. Luo, 1996. A 100 ka record of water tables and paleoclimates from salt cores, Death Valley, California. *Palaeogeography, Palaeoclimatology, Palaeoecology*, 123, 179–203.

Lorius, C., J. Jouzel, C. Ritz, L. Merlivat, N.I. Barkov, Y.S. Korotkevich, and V.M. Kotlyakov, 1985. A 150,000 year climate record from Antarctic ice. *Nature*, 316, 591–596.

Lowe, J.J., and M.J.C. Walker, 1997. *Reconstructing Quaternary environments*. 2nd edition, Addison Wesley Longman, Harlow.

Lowenstein, T.K., J. Li, C.B. Brown, S.M. Roberts, T-L. Ku, S. Luo, and W. Yang, 1999. 200 k.y. paleoclimate record from Death Valley salt core. *Geology*, 27, 3–6.

Lyell, C., 1939. *Nouveaux éléments de géologie*. Pitois-Levrault, Paris.

Maat, P.B., and W.C. Johnson, 1996. Thermoluminescence and new ^{14}C age estimates for late Quaternary loesses in southwestern Nebraska. *Geomorphology*, 17, 115–128.

MacAyeal, D.R., 1993. Binge/purge oscillations of the Laurentide ice sheet as a cause of the North Atlantic's Heinrich events. *Palaeoceanogaphy*, 8, 775–784.

Marshall, S.J., and G.K.C. Clarke, 1999. Modeling North American freshwater runoff through the last glacial cycle. *Quaternary Research*, 52, 300–315.

Martin, P.S. and R.G. Klein, 1984. *Quaternary extinctions: A prehistoric revolution*. University of Arizona Press, Tucson.

McDonald, E.V., and A.J. Busacca, 1992. Late Quaternary stratigraphy of loess in the Channeled Scabland and Palouse regions of Washington state. *Quaternary Research*, 38, 141–156.

Meinzer, O.E., 1922. Map of the Pleistocene lakes of the Basin and Range Province and its significance. *Geological Society of America Bulletin*, 33, 541–552.

Milankovitch, M., 1930. Mathamatische Klimalehre und astronomische Theorie der Klimaschwankungen. In W. Köppen and R. Geiger (Editors), *Handbuch der Klimatologie*, Gerbrüder Borntraeger, Berlin, 1–176.

Miller, B.B., R.W. Graham, A.V. Morgan, N.G. Miller, W.D. McCoy, D.F. Palmer, A.J. Smith, and J.J. Pilny, 1994. A biota associated with Matuyama-age sediments in west-central Illinois. *Quaternary Research*, 41, 350–365.

Mitchell, G.C., and R.E. McDonald, 1987. *Subsurface Tertiary strata, depositional model, and hydrocarbon potential of the Sevier Desert basin, west-central Utah*. Utah Geological Association, Publication 16.

Mitrovica, J.X., and W.R. Peltier, 1991. On postglacial geoid subsidence over equatorial oceans. *Journal of Geophysical Research*, 96, 20,053–20,071.

Morrison, R.B., 1964. Lake Lahontan: Geology of southern Carson Desert, Nevada. *U.S. Geological Survey Professional Paper* 401.

Morrison, R.B., 1991. Quaternary stratigraphic, hydrologic, and climatic history of the Great Basin, with emphasis on Lakes Lahontan, Bonneville, and Tecopa. In: R.B. Morrison (Editor), *Quaternary nonglacial geology: Conterminous U.S.* The Geology of North America, Volume K-2, Geological Society of America, Boulder, 283–320.

Mott, R.J., 1990. Sangamonian forest history and climate in Atlantic Canada. *Géographie Physique et Quaternaire*, 44, 257–270.

Muhs, D.R., J.N. Aleinikoff, T.W. Stafford, R. Kihl, J. Been, S.A. Mahan, and S. Cowherd, 1999. Late Quaternary loess in northeastern Colorado: Part I—Age and paleoclimatic significance. *Geological Society of America Bulletin*, 111, 1861–1875.

Muhs, D.R., T.W. Stafford, S.D. Cowherd, S.A. Mahan, R. Kihl, P.B. Maat, C.A. Bush, and J. Nehring, 1996. Origin of the late Quaternary dune fields of northeastern Colorado. *Geomorphology*, 17, 129–149.

Orme, A.R., 1991. Mass movement and seacliff retreat along the southern California coast. *Bulletin, Southern California Academy of Sciences*, 90, 58–79.

Orme, A.R., 1992. Late Quaternary deposits near Point Sal, south-central California. In: C.H. Fletcher and J.F. Wehmiller (Editors), *Quaternary coasts of the United States: Marine and lacustrine systems*, Society for Sedimentary Geology, 48, 309–315.

Orme, A.R., 1998. *Geomorphic Map of Rosamond Dry Lake, Mojave Desert, California*. U.S. Army Corps of Engineers, Cold Regions Research and Engineering Laboratory, Hanover, New Hampshire.

Orme, A.J., and A.R. Orme, 1991. Relict barrier beaches as paleoenvironmental indicators in the California Desert. *Physical Geography*, 12, 334–346.

Orme, A.R., and V.P. Tchakerian, 1986. Quaternary dunes of the Pacific coast of the Californias. In: W.G. Nickling (Editor), *Aeolian geomorphology*, Allen and Unwin, London, 149–175.

Oviatt, C.G., R.S. Thompson, D.S. Kaufman, J. Bright, and R.M. Forester, 1999. Reinterpretation of the Burmester core, Bonneville Basin, Utah. *Quaternary Research*, 52, 180–184.

Owen-Smith, R.N., 1988. *Mega-herbivores*. Cambridge University Press, New York.

Pardee, J.T., 1910. The glacial lake Missoula. *Journal of Geology*, 18, 376–386.

Patton, P.C., N. Biggar, C.D. Condit, M.L. Gillam, D.W. Love, M.N. Machette, L. Mayer, R.B. Morrison, and J.N. Rosholt, 1991. Quaternary geology of the Colorado Plateau. In: R.B. Morrison (Editor), *Quaternary nonglacial geology: Conterminous U.S.* The Geology of North America, K-2, Geological Society of America, Boulder, 373–406.

Peteet, D.M., 1993. Global Younger Dryas? *Quaternary International*, 28, 93–104.

Peteet, D.M., R.A. Daniels, L.E. Heusser, J.S. Vogel, J.R. Southon, and D.E. Nelson, 1993. Late-glacial pollen, macrofossis, and fish remains in northeastern USA—The Younger Dryas oscillation. *Quaternary Science Reviews*, 12, 597–612.

Péwé, T.L., 1983. The periglacial environment in North America during Wisconsin time. In S.C. Porter (Editor), *Late Quaternary environments of the United States, Volume 1: The Late Pleistocene*. University of Minnesota Press, Minneapolis, 157–189.

Pielou, E.C., 1991. *After the ice age: The return of life to glaciated North America*. University of Chicago Press, Chicago.

Preece, S.J., J.A. Westgate, B.A. Stemper, and T.L. Péwé, 1999. Tephrochronology of late Cenozoic loess at Fairbanks, central Alaska. Geological Society of America Bulletin, 111, 71–90.

Prueher, L.M. and D.K. Rea, 1998. Rapid onset of glacial conditions in the subarctic North Pacific region at 2.67 Ma: Clues to causality. *Geology*, 26, 1027–1030.

Pye, K., N.R. Winspear, and L.P. Zhou, 1996. Thermolumniscence ages of loess and associated sediments in central Nebraska, USA. *Palaeogeography, Palaeoclimatology, and Palaeoecology*, 118, 73–87.

Raymo, M.E., 1992. Global climate change: A three-million year perspective. In: G.J. Kukla and E. Went (Editors), *Start of a glacial*, Springer-Verlag, Berlin.

Reheis, M., 1999. Highest pluvial-lake shorelines and Pleistocene climate of the western Great Basin. *Quaternary Research*, 52, 196–205.

Reppening, C.A., and E.M. Browers, 1992. Late Pliocene-early Pleistocene ecologic changes in the Arctic Ocean borderland. *U.S. Geological Survey Bulletin*, 2036, 37.

Richmond, G.M., 1986. Stratigraphy and chronology of glaciation in Yellowstone National Park, *Quaternary Science Reviews*, 5, 83–98.

Richmond, G.M., and D.S. Fullerton, 1986. Introduction to Quaternary glaciations in the United States of America. In: V. Sibrava, D.Q. Bowen, and G.M. Richmond (Editors), Quaternary glaciations in the Northern Hemisphere, *Quaternary Science Reviews*, 5, 3–10.

Ritchie, J.C., 1987. *Postglacial vegetation of Canada*. Cambridge University Press, Cambridge.

Rosholt, J.N., and 11 others, 1991. Dating methods applicable to the Quaternary. In R.B. Morrison (Editor), *Quaternary nonglacial geology; Conterminous U.S.* The Geology of North America, K-2, Geological Society of America, Boulder, 45–74.

Ruddiman, W.F., and A.F. McIntyre, 1981. Ice-age thermal response and climatic role of the surface Atlantic Ocean, 40° to 63°N. *Geological Society of America Bulletin*, 95, 381–396.

Ruddiman, W.F., A.F. McIntyre, and M.E. Raymo, 1986. Matuyama 41,000-year cycle: North Atlantic Ocean and northern hemisphere ice sheets. *Earth and Planetary Science Letters*, 80, 117–129.

Russell, I.C., 1885. Geological history of Lake Lahontan, a Quaternary lake of northwestern Nevada. *U.S. Geological Survey Monograph 11*, 288 p.

Sack, D., 1999. The composite nature of the Provo level of Lake Bonneville, Great Basin, western North America. *Quaternary Research*, 52, 316–327.

Shackleton, N.J., and N.D. Opdyke, 1973. Oxygen isotope and paleomagnetic stratigraphy of equatorial Pacific core V28–238: Oxygen isotope temperatures and ice volumes on a 10^5 and 10^6 year scale. *Quaternary Research*, 3, 39–55.

Shane, L.K.C., and K.H. Anderson, 1993. Intensity gradients and reversals in late-glacial environmental change in east-central North America. *Quaternary Science Reviews*, 12, 307–320.

Sims, J.D. (Editor), 1988. *Late Quaternary Climate, Tectonism, and Sedimentation in Clear Lake, northern California Coast Ranges*. Geological Society of America, Special Paper 214, Boulder.

Smith, G.I., 1984. Paleohydrologic regimes in the southeast Great Basin, 0.32 my ago, compared with other long records of "global" climate. *Quaternary Research* 22, 1–17.

Smith, G.I., and J.L. Bischoff, 1997. *An 800,000–year paleoclimatic record from core OL-92, Owens Lake, southeast California*. Geological Society of America Special Paper 317, Boulder.

Snyder, C.T., G. Hardman, and F.F. Zdenek, 1964. Pleistocene lakes of the Great Basin. *U.S. Geological Survey Miscellaneous Geological Investigations*, Map I-416.

Spaulding, W.G., and L.J. Graumlich, 1986. The last pluvial climatic episodes in the deserts of southwestern North America. *Nature*, 3320, 441–444.

Stebbins, G.L., 1981. Coevolution of grasses and herbivores. *Missouri Botanical Gardens Annals*, 68, 75–86.

Stewart, B., J.C.C. Hsieh, and B.C. Murray, 1994. Paleohydrology of the late Pliocene Basin and Range Province using strontium isotopes. Abstracts, 8th *International Conference on Geochronology, Cosmochronology, and Isotope Geology*, 805.

Stewart, R.J., 1975. Late Cainozoic explosive eruptions in the Aleutian and Kurile Island arcs. *Nature*, 258, 505–507.

Stock, C., and J.M. Harris, 1992. *Rancho La Brea: A record of Pleistocene life in California*, 7th edition, Natural History Museum of Los Angeles County.

Stokes, S., G. Kocurek, K. Pye, and R. Winspear, 1997. New evidence for the timing of aeolian sand supply to the Algodones dunefield and East Mesa area, southeastern California, USA. *Palaeogeography, Palaeoclimatology, and Palaeoecology*, 128, 63–75.

Swadley, W.C., 1980. New evidence supporting Nebraskan age for origin of Ohio River in north-central Kentucky. *U.S. Geological Survey Professional Paper* 1126-H.

Tackman, G.E., B.G. Bills, T.S. James, and D.R. Currey, 1999. Lake-gauge evidence for regional postglacial tilting in southern Manitoba. *Geological Society of America Bulletin*, 111, 1684–1699.

Tchakerian, V.P., 1989. *Later Quaternary aeolian geomorphology of the central Mojave Desert, California*. Ph.D. dissertation, University of California, Los Angeles.

Teller, J.T., 1973. Preglacial (Teays) and early glacial drainage in the Cincinnati area, Ohio, Kentucky, and Indiana. *Geological Society of America Bulletin*, 84, 3677–3688.

Teller, J.T., 1990. Meltwater and precipitation runoff to the North Atlantic, Arctic, and Gulf of Mexico from the Laurentide ice sheet and adjacent regions during the Younger Dryas. *Paleoceanography*, 5, 897–905.

Terasmae, J., 1960. A palynological study of Pleistocene interglacial beds at Toronto, Ontario. *Geological Survey of Canada Bulletin*, 56, 23–41.

Thompson, R.S., and R.F. Fleming, 1996. Middle Pliocene vegetation: Reconstructions, paleoclimatic inferences, and boundary conditions for climate modeling. *Marine Micropaleontology*, 27, 27–49.

Thorson, R.M., 1996. Earthquake recurrence and glacial loading in western Washington. *Geological Society of America Bulletin*, 108, 1182–1191.

Thorson, R.M., and C.A. Shile, 1995. Deglacial eolian regimes in New England. *Geological Society of America Bulletin*, 107, 751–761.

Urey, H.C., 1947. The thermodynamic properties of isotopic substances. *Journal of the Chemical Society*, 562–581.

Van Kreveld, S.A., M. Knappertsbusch, J. Ottens, G.M. Ganssen, and J.E. van Hinte, 1996. Biogenic carbonate and ice-rafted debris (Heinrich layers) accumulation in deep-sea sediments from a northeast Atlantic piston core. *Marine Geology*, 131, 21–46.

Waitt, R.B., 1985. Case for periodic, colossal jökulhlaups from glacial Lake Missoula. *Geological Society of America Bulletin*, 95, 1271–1286.

Wayne, W.J., J.S. Aber, S.S. Agard, R.N. Bergantino, J.P. Bluemle, D.A. Coates, M.E. Cooley, R.F. Madole, J.E.

Martin, B. Meras, R.B. Morrison, and W. M. Sutherland, 1991. Quaternary geology of tho northern Great Plains. In: R.B. Morrison (Editor), *Quaternary nonglacial geology: Conterminous U.S.* The geology of North America, K-2, Geological Society of America, Boulder, 441–476.

Webb, T., 1987. The appearance and disappearance of major vegetational assemblages: Long-term vegetational dynamics in eastern North America. *Vegetation*, 69, 17–187.

Webb, T., P.J. Bartlein, and J.E. Kutzbach, 1987. Climatic change in eastern North America during the last 18,000 years: Comparisons of pollen data with model results. In: W.F. Ruddiman and H.E. Wright (Editors), *North America and adjacent oceans during the last deglaciation.* The geology of North America, K-3, Geological Society of America, Boulder, 447–463.

Whitlock, C., and P.J. Bartlein, 1993. Vegetation and climate change in northwest North America during the past 125 kyr. *Nature*, 388, 57–61.

Willard, D.A., 1995. Palynological record from the North Atlantic region at 3 Ma: Vegetational distribution during a period of global warmth. *Review of Palaeobotany and Palynology*, 83, 275–297.

Williams, D.F., R.C. Thunell, E. Tappa, D. Rio, and I. Raffi, 1988. Chronology of oxygen isotope record, 0–1.88 million years before present. *Palaeogeography, Palaeoclimatology, Palaeoecology*, 64, 221–240.

Winograd, I.J., T.B. Coplen, J.M. Landwehr, A.C. Riggs, K.R. Ludwig, B.J. Szabo, P.T. Kolesan, and K.M. Revesz, 1992. Continuous 500,000 year climate record from vein calcite in Devil's Hole, Nevada. *Science*, 258, 255–260.

Wright, H.E., 1989. The amphi-Atlantic distribution of the Younger Dryas palaeoclimatic oscillation. *Quaternary Science Reviews*, 8, 295–306.

Wright, H.E., J.E. Kutzbach, T. Webb, W.F. Ruddiman, A.F. Street-Perrott, and P.J. Bartlein (Editors), 1993. *Global climates since the Last Glacial Maximum.* University of Minneapolis Press, Minnesota.

Yang, W., H.R. Krouse, R.J. Spencer, T.K. Lowenstein, I.E. Hutcheon, T-L. Ku, S.M. Roberts, and C.B. Brown, 1999. A 200,000-year record of change in oxygen isotope composition of sulfate in a saline sediment core, Death Valley, California. *Quaternary Research*, 51, 148–157.

Zdanowicz, C.M., G.A. Zielinski, and M.S. Germani, 1999. Mount Mazama eruption: Calendrical age verified and atmospheric impact assessed. *Geology*, 27, 621–624.

4

The Holocene Environment

Louis A. Scuderi

Holocene environmental conditions in North America are closely linked to the prevailing atmospheric circulation pattern. Climatically, North America can be divided into several distinct regions on the basis of each region's relationship to large-scale features of the general circulation such as the Arctic Front and the average polar jet location. Although the meteorology of these regions varies on an annual basis, they may be viewed as relatively stable when averaged over decadal and centennial timescales. However, over longer intervals, the position and extent of these regions varies in response to large-scale changes in the general circulation pattern.

Long-term variations in the location and extent of climatic regions in North America are, in turn, responses to variations in energy balance conditions (Berger, 1978; Kutzbach, 1987; Kutzbach and Webb, 1993; Felzer et al., 1998). For example, changes related to Milankovitch cycles alter solar insolation receipt, modify the flow of energy over Earth's surface, and in the North American sector initiate changes in ice-cap size with resultant variation in the position of the Arctic Front and jet stream. An understanding of these interactions can help explain many of the disparate findings that result from time-transgressive changes in circulation patterns over North America (Davis, 1984).

4.1 Environmental Determinants of North American Climate

Solar radiation changes relative to the present, as derived from equations that describe Earth's orbital eccentricity, precession, and obliquity (Berger, 1978), indicate that summer insolation (June, July, and August) in the Northern Hemisphere has varied in a predictable way from the end of the Pleistocene to the present (fig. 4.1). Summer insolation was approximately 8% greater than present at 9000 B.P. (radiocarbon years before A.D. 1950), and overall was at least 4% greater than present between 12,000 and 6000 B.P. This enhanced summer energy receipt was due to the combination of a maximal tilt of Earth's axis (24.5°) and a change in the timing of perihelion to July at 9000 B.P. Winter insolation (December, January, and February) had the opposite sign at the same times (fig. 4.1). The results of these combined orbital factors and the contrast between summer and winter insolation produced an enhanced seasonality at 9000 B.P.

As summer insolation increased to its Northern Hemisphere maximum at 9000 B.P., the Laurentide Ice Sheet melted and retreated. This had a profound effect on atmospheric circulation over North America. Earlier, during the height of the Wisconsinan Glaciation at 18,000 B.P., the

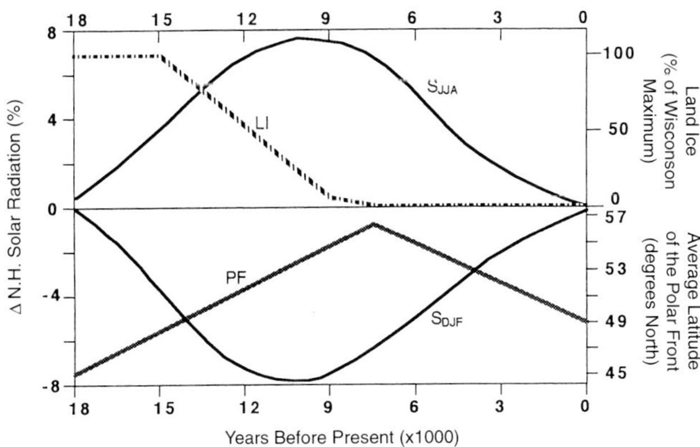

Figure 4.1 Climate-forcing factors, 18,000 B.P. to present. Solid lines indicate summer (JJA) and winter (DJF) solar radiation receipt as a percentage of current receipt. Dashed line represents North American land ice (LI) as a percentage of the Wisconsinan maximum. Stippled line represents the average latitude of the polar front (PF) over the North American continent.

Laurentide Ice Sheet had produced a significant cold air mass and resultant high-pressure area centered over the Hudson Bay region (see chapters 2 and 3). The circulation around the high produced a clockwise rotation opposite to that of the prevailing westerly winds along the Arctic Front. Cold, dry air dominated both the ice sheet and adjacent areas, and the equatorward extension of the ice sheet shifted the Arctic Front southward to a position around 45°N. As summer insolation increased toward the beginning of the Holocene, the Laurentide Ice Sheet waned and the Arctic Front shifted northward (fig. 4.1).

This circulation change led to a time-transgressive start of the Holocene in different areas, as well as to variable timing of the warm interlude around the middle of the Holocene. Other factors may have also played a significant role in the timing of the specific climatic events during the Holocene. For example, as the melting of the Laurentide Ice Sheet uncovered the darker surface of the continent and as vegetation migrated into these now ice-free regions, the albedo of the North American continent changed from the maximum value at 18,000 B.P. to near present values by 9000 B.P. (fig. 4.1). This resulted in a significant change in both local and regional energy balances and in circulation changes, which produced additional melting of sea ice and warming of the oceans (Kutzbach and Ruddiman, 1993; Bartlein et al., 1998).

In addition to specific changes in the atmospheric and oceanic circulation, seasonal changes in insolation may have also had a significant impact on the timing, magnitude, and pattern of Holocene environmental change in North America. As Davis (1984) has shown, thermal maxima occurred at multiple times during the Holocene because each season had its own insolation peak. Distributions of plant species with different climatic tolerances can result in an apparent peak growth, depending on the climatic variable controlling growth. This can lead to apparent timing differences in the onset of environmental change and may be a prime factor in some of the uncertainty of environmental reconstructions discussed subsequently.

4.2 Environmental Conditions of the Terminal Pleistocene: 12,000–10,000 B.P.

The Laurentide Ice Sheet in North America persisted near its late Pleistocene maximum until approximately 15,000 B.P.; however, by 12,000 B.P. it was 30% less extensive than during its Wisconsinan maximum and sea ice in the North Atlantic was less extensive (Kutzbach and Ruddiman, 1993; COHMAP, 1988) (see also chapter 2). This ice and ocean surface arrangement produced a weaker, single-strand jet stream located along the southern margin of the Laurentide Ice Sheet (fig. 4.2). By 12,000 B.P., the southern margin of the ice sheet was dominated by cool and wet conditions and cyclonic storms (Bartlein et al., 1998). Increased summer insolation established warmer conditions

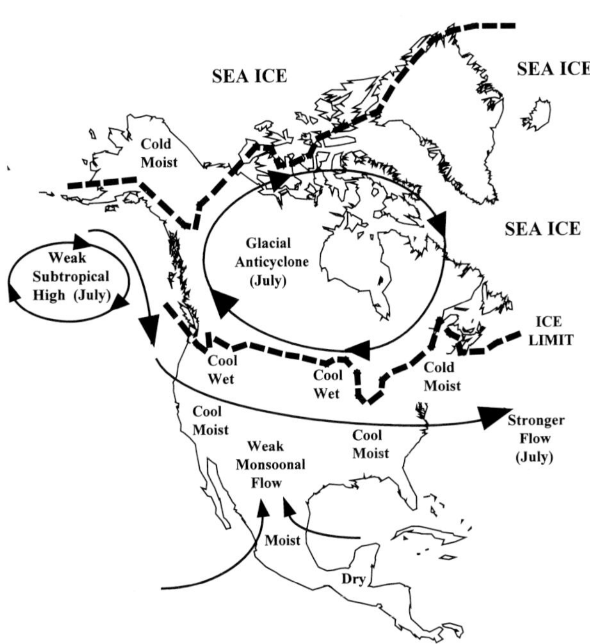

Figure 4.2 Climatic conditions at 12,000 B.P. Arrow size reflects relative strength of atmospheric circulation.

farther west. As a result, a summer thermal low formed in the southwestern United States, and a weak monsoonal flow began to bring enhanced moisture into this portion of the continent.

Insolation-induced warming west of the Laurentide Ice Sheet and the rise in sea level that opened the seaway in the Bering Strait allowed cool, moist summer air masses to enter western Alaska and penetrate inland to the Brooks Range. This increased moisture availability may have led to a glacial advance in the western Brooks Range dating to ≈12,800 B.P. (Hamilton, 1986). By 12,000 B.P. pollen and macrofossil data indicate that both temperature and precipitation had increased throughout Alaska (Anderson and Brubaker, 1993). Before 11,000 B.P., the vegetation of the central Alaskan lowlands was tundra, but with the change to warmer conditions scrub forest began invading the tundra, with spruce (*Picea*) forests appearing by ≈9500 B.P. (Billings, 1987).

Although most of eastern Canada from Hudson Bay eastward remained ice-covered at the end of the Pleistocene, the earliest deglaciation at the southern edge of the ice sheet occurred about 13,000 B.P. The Laurentide ice margin at this time was in Ontario, but by 12,000 B.P. southern Ontario was ice-free (Ritchie, 1987). By 11,300 B.P., spruce (*Picea*) reached its peak in the area, but farther east treeless vegetation communities prevailed north of the Champlain Sea and east of Quebec City. By the time the ice had retreated to north of Lake Superior, all areas to the west were registering warmer conditions and a *Picea*-dominated vegetation assemblage (Ritchie, 1987).

In the western United States, early stages of forestation of the Yellowstone region are indicated as early as 12,000 B.P. Prior to this, pollen spectra indicate that the area was rapidly filled with meadow and tundra vegetation after deglaciation (Whitlock and Bartlein, 1993). Warmer and wetter conditions led to the establishment of *Picea* by 12,000 B.P. and by 11,000 B.P. to the migration of subalpine fir (*Abies lasiocarpa*) and whitebark pine (*Pinus albicaulis*) into the region. Between 11,000 and 9500 B.P., *Pinus* (specifically lodgepole pine, *Pinus contorta*) was dominant throughout the region. Farther south, in the San Juan Mountains of Colorado, deglaciation was complete by 10,000 B.P. (Elias et al., 1991).

An insect assemblage in the Marias Pass area of northern Montana dating to 12,200 B.P. suggests that July temperatures approached modern values, and by 10,000 B.P. several regional insect assemblages suggest temperature values exceeding mean modern summer and winter temperatures (Elias, 1996). The thermal maximum in the Rocky Mountains appears to have occurred between 11,000 and 9000 B.P. Summer temperatures appear to have peaked between 3.7° and 6.7°C above present temperatures at about 9900 B.P. By 9000 B.P., insect assemblages indicate that summer temperatures were declining from earlier highs but were still warmer than present (Elias, 1996).

Packrat midden evidence, based on residual accumulations by *Neotoma* species, in the White Mountains of California and Nevada suggests that the transition from glacial climate to a cool modern climate began about 10,000 B.P. (Jennings and Elliott-Fisk, 1993). The vegetation in midden samples at that time is similar to current pinyon-juniper woodland vegetation found 1000 m higher. This suggests temperatures approximately 4°C lower and precipitation 60% higher than present at the close of the Pleistocene. In the North Cascade Range of Washington, glacial deposits are mantled with pumice from Mount Mazama (Crater Lake) deposited around 6700 to 6900 B.P. but overlie tephra deposits from Glacier Peak eruptions dating from 11,250 to 12,500 B.P. This suggests that cirques in this region were ice-free by the time of Glacier Peak eruptions (Beget, 1983).

Although much of the North American continent south of the ice sheet had lower precipitation totals than today between 18,000 and 12,000 B.P., colder temperatures made conditions moister than at present (Webb et al., 1993). The northward movement of many taxa suggest a warming that was gradual before 12,000 B.P. and rapid between 12,000 and 9000 B.P.. By 12,000 B.P., *Picea* was the dominant tree in the northeastern United States (Webb et al., 1993). The modern north-south sequence of taxa in this region (*Picea*, *Pinus*, *Quercus*, forbs) was beginning to appear during this period, although pollen records from eastern North America show that modern patterns were not completely established until after 9000 B.P. (Miller and Morgan, 1982; Webb III et al., 1993b). Between 12,000 and 8000 B.P., possibly in response to atmospheric circulation changes, populations of *Picea* experienced a major collapse in the Great Lakes region but advanced across New England and eastern Canada (Delcourt and Delcourt, 1987).

Lake-level data suggest that conditions were moister than at present at 12,000 B.P. for eastern North America, then rapidly became drier than current conditions after 10,000 B.P. (Harrison and Metcalfe, 1985). In the southeast portion of the continent, north of Florida, high lake levels between 12,000 and 10,000 B.P. coupled with lower temperatures inferred from pollen (up to 6°C lower in January and 1–4°C in July) produced a level of effective moisture higher than the present. In the north-central region, wetter than present conditions existed at 12,000 B.P., followed by a decline to near present values by 9000 B.P. when the prairie-forest border moved eastward into Minnesota and Iowa (Webb III et al., 1993b).

Much farther south, freshwater marshes around Lake Chapultepec in central Mexico indicate increased moisture levels between 14,000 and 10,000 B.P., probably due to increased spring discharge. However, most of Mesoamerica appears to have experienced late Pleistocene aridity (Markgraf, 1989). Cores from Chalco Lake, dating between 12,000 and 10,000 B.P., in the Basin of Mexico reveal the presence of freshwater diatoms and suggest that the water level in this lake increased to a few meters depth and the forests surrounding the lake reached their maximum development with mixed stands of *Pinus* and *Quercus* (Lozano-Garcia et al., 1993). About 10,000 B.P., saline-water deposits suddenly dominated the Chalco Lake sediments, sugggesting the onset of drier conditions at the beginning of the Holocene (Watts and Bradbury, 1982). Throughout the higher elevations of Mesoamerica, modern climatic conditions and vegetation assemblages became established between 10,000 and 9000 B.P. (Markgraf, 1989).

4.3 The Early to Middle Holocene: 10,000–6000 B.P.

The Holocene Epoch formally began at 10,000 B.P., at which time the Laurentide Ice Sheet was much diminished. By 9000 B.P., the ice sheet had been reduced in size to approximately 35% of its maximum Wisconsinan size, and North Atlantic sea ice had almost completely melted (COHMAP, 1988; Kutzbach and Ruddiman, 1993). The actual configuration of the ice sheet by 9000 B.P. probably included small ice domes over Labrador, Keewatin, and Foxe Basin, with a thinner cover over Hudson Bay; by 7800 B.P., an ice-free embayment was present in southern Hudson Bay (Kutzbach and Ruddiman, 1993) (fig. 4.3). Orbitally produced enhancement of summer conditions and overall enhanced seasonality produced a stronger thermal low and a greatly enhanced monsoonal flow pattern over southwestern North America.

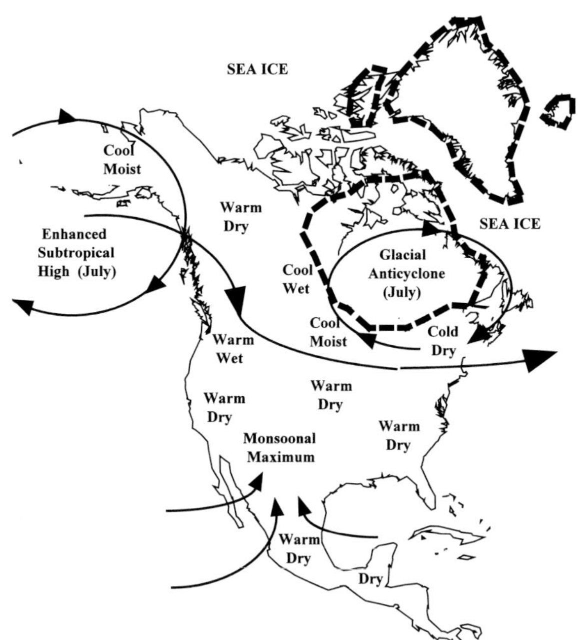

Figure 4.3 Climatic conditions at 9000 B.P. See figure 4.2 for key.

The early Holocene in North America was characterized by increasing aridity, which led to widespread low lake levels by 9000 B.P. in the midlatitudes. Because lake behavior reflects a balance between precipitation and evaporation, regional patterns of lake levels are related to airmass distribution and associated frontal systems. Generally speaking, high lake levels are characteristic of the frontal zone between Pacific and Arctic air (Arctic Front) and also the region influenced by tropical air (Harrison and Metcalfe, 1985).

By 9000 B.P., increasing aridity had led to widespread low lake levels in both the eastern and western portions of the continent. However, higher lake levels indicate relatively moist conditions in the Midwest and North American tropics. This pattern appears to indicate a northward displacement of the equatorial trough and less zonal flow, resulting from increased monsoonal flow into the region. This interpretation is consistent with modeling results, which show an increase in precipitation associated with an enhanced seasonal land-ocean temperature contrast and a stronger monsoon circulation (Kutzbach and Guetter, 1984).

Although the early Holocene is often thought to be a time of gradual uninterrupted warming and relatively dry conditions in North America, evidence from glacial deposits suggests otherwise (Beget, 1983). For example, this warm interval may have been interrupted between 8500 to 7500 B.P. by a global episode of cooler climate similar in scope to more recent intervals characterized by glacial advances (Rothlisberger, 1986). Even though the underlying reasons for this period of glacial advance are obscure, both Beget (1983) and Scuderi (1994) have suggested that a decrease in solar output might be responsible.

Alaska and western Canada experienced a period of relatively warm summers between 11,000 and 8000 B.P. as a result of the pronounced maximum in summer insolation in the northern high latitudes (Ritchie and Harrison, 1993; Hu et al., 1996). This is evidenced in range extensions of *Picea* and *Populus*, increased aeolian activity, and possibly lowered lake levels (Billings, 1987; Anderson and Brubaker, 1994). In Alaska, white spruce (*Pinus glauca*) experienced a significant population decline between 8500 and 7500 B.P., possibly in association with a climatic cooling at this time and did not reach its modern limits until 4000 B.P. (Anderson and Brubaker, 1993, 1994; Hu et al., 1996, 1998).

In northwest Canada and eastern Yukon from 8600 to 7700 B.P., the region was dominated by *Betula* tundra (MacDonald, 1983). The tundra became increasingly productive toward 8000 B.P., but shortly thereafter climatic conditions changed and both *Picea glauca* and *Picea mariana* (black spruce) expanded into the area and reached their present extent with highest treelines between 7700 and 5000 B.P. (MacDonald et al., 1993) (see chapter 14).

Pollen in lake sediments from central Canada indicate that after deglaciation at 9000 B.P., the boreal treeline advanced and then retreated in response to shifts in the position of the Arctic frontal zone (Edwards et al., 1996). Vegetation shifted from dwarf shrub tundra to *Picea mariana* forest as the Arctic Front migrated northward across the region at 5000 B.P., allowing Pacific air masses to move into central Canada. The forest cover remained until 3000 B.P. when the Arctic Front again shifted south and the tundra cover returned.

In eastern Canada, the Laurentide Ice Sheet retreated for more than 500 km from localities near the Great Lakes to central Laurentide Canada between 11,000 and 8500 B.P. (Beget, 1983). Between 8500 and 7500 B.P., moraines were deposited during stillstands or readvances of the ice sheet. In the Keewatin area of Arctic Canada, a large moraine on the northern flank of the Keewatin ice dome has been radiocarbon dated at 8160 ± 140 B.P. (Beget, 1983). On Baffin Island, multiple moraines that have been radiocarbon dated between 7800 to 8400 B.P. are probably indicative of a readvance in several areas. The $\partial^{18}O$ values from lake sediments and fossil wood cellulose for southern Ontario indicate that the postglacial cold, dry period in eastern Canada lasted until ≈7400 B.P. (Edwards et al., 1996). This was followed by a relatively short period from 7400 to 6000 B.P. of warm, dry conditions.

In western North America, insect and plant macrofossils suggest that tree line was higher than today during the early Holocene with a warmer than present climate between 9000 and 6000 B.P. (Elias et al., 1991). Radiocarbon ages on wood fragments indicate that, between 9600 and 5400 B.P., tree line in the northern San Juan Mountains was at least 80 m above present levels with a maximum reached at ≈8000 B.P. up to 140 m above present levels (Carrara et al., 1991). Deposits from Lake Emma in Colorado indicate that, from 9600 to 7800 B.P., treeline was at least 70 m higher than at present (Carrara et al., 1984).

The distribution of pika (*Ochotona princeps*) in the western United States in the early Holocene suggests that the range of this small rabbit-like, short-legged mammal may have extended between 100 and 300 m above present levels between 9500 and 6000 B.P. (Hafner, 1993). The distribution of this species is a function of temperature and snowline elevation and thus indicates warmer temperatures and higher tree lines during this period. Insect assemblages from the Rocky Mountains suggest that summer temperatures appear to have peaked at about 9850 B.P. (Elias, 1996). These assemblages indicate that, by 9000 B.P., although summer temperatures were declining from earlier highs, they were still warmer than at present. Between 7800 and 3000 B.P., these assemblages indicate a gradual summer cooling trend.

Tree-line altitudes inferred from fossil insect species indicate that the climatic optimum in the western United States, inferred from maximum forest-tundra insect ratios, occurred between 9000 and 7000 B.P. (Elias, 1985). Deuterium concentrations determined from the cellulose of conifer fragments indicate a decrease in deuterium from 9600 B.P. to the present, possibly indicating that the increase in July radiation at 9000 B.P. may have resulted in an intensification of the Arizona monsoon (Carrara et al., 1991).

Early Holocene moraines near the head of the White Chuck River in the North Cascade Range of Washington incorporated charcoal during an advance about 8300 to 8400 B.P. (Beget, 1983). Evidence of a moraine in the Enchantment Lakes Basin suggests a climatic episode of increased precipitation and cooling sometime around 8000 B.P. (Waitt et al., 1982). This is consistent with evidence from the San Juan Mountains of Colorado and suggests a glacier advance between 7000 and 8300 B.P. (Andrews et al., 1975).

In the Yellowstone region after 9500 B.P., the vegetational history diverges, with the southern region experiencing warmer and drier summer conditions than present, whereas the northern region was warmer and wetter than today (Whitlock and Bartlein, 1993). Model simulations for 9000 B.P. suggest that the eastern Pacific subtropical high was intensified in summer, bringing warmer summers to the Pacific Northwest. In the Southwest, the summer heating enhanced the thermal low, which resulted in increased onshore flow from the Gulf of Mexico and Gulf of California and an enhanced monsoon. At 5000 B.P., pollen evidence suggests a cooler, moister climate than before to the south of Yellowstone, whereas to the north at 7000 B.P. climate became drier than before with the drying continuing to the present.

Packrat (*Neotoma* spp.) midden evidence from the White Mountains of California and Nevada suggests that, as the Arctic Front shifted northward between 9000 and 7500 B.P., climate became similar to the present (Jennings and Elliott-Fisk, 1993). By the early Holocene, the Arctic Front may have become less important in generating precipitation in the western United States and conditions became warmer and drier. In the southern Sierra Nevada, at Balsam Meadow, pollen data suggest dry conditions between 11000 and 7000 B.P. (Davis et al., 1985). Farther to the south in the Mojave Desert, conditions were favorable for the formation of lakes until ≈8000 B.P. (Enzel et al., 1989).

Pollen and macrofossil evidence from Potato Lake in Arizona (Anderson, 1993) suggests that the early Holocene was marked by a cessation of lake sedimentation at ≈10,000 B.P. Increases in effective precipitation occurred only after 3000 B.P. This record is similar to that from Montezuma Well in central Arizona, where Davis and Shafer (1992) have suggested that there was an early Holocene monsoonal maximum for northern Arizona and areas to the north and west, with the driest conditions and lower lake levels occurring between 7000 and 4000 B.P. This is supported by evidence from Walker Lake in Arizona, which dried entirely at 6000 B.P. (Hevly, 1985).

The general trend toward increasing aridity during the early and mid-Holocene is consistent with independent evidence for expansion of the prairie in the Midwest of the United States between 10,000 and 7000 B.P. During early Holocene warming, the western Midwest dried out, and the prairie formed and moved eastward (Webb et al., 1983; Dean et al., 1984). The movement was gradual before 12,000 B.P., rapid between 12,000 and 9000 B.P., and then gradual again. Pollen types show that modern vegetation patterns in the Midwest and Northeast did not develop until after 9000 B.P. when moisture conditions began to approach present levels (Webb et al. 1993). By 8000 B.P., the northward and eastward advance of the prairie displaced forests throughout the Great Lakes region (Delcourt and Delcourt, 1987). Pollen of southern pines became more abundant in the Southeast as winter temperatures rose after 9000 B.P.

Reconstructions of mean annual precipitation and moisture balance from fossil pollen data in the northeastern United States indicate that low annual precipitation coupled with warm temperatures and high solar insolation resulted in increased evaporation and decreased soil moisture levels at 9000 B.P. (Harrison, 1989; Qin et al., 1998; Webb et al., 1998). Pollen data indicate that pine reached its maximum elevation and highest latitude during this interval. Lake levels were somewhat lower at 9000 B.P., a time when northern pine populations were most abundant in the East and inferred precipitation was lower than at present. In the Northeast, *Pinus* pollen was replaced by birch (*Betula*) and beech (*Fagus*) pollen after 9000 B.P. as this region became moister and summer temperatures fell. By 8000 B.P., *Pinus* had advanced southward into central and southern Florida (Delcourt and Delcourt, 1987). These changes in the regional moisture balance reflect changes in the seasonal distribution of insolation in the Northern Hemisphere and resultant changes in sea surface temperatures.

In subtropical Mesoamerica, diatom analysis and stratigraphic sections of lake and marsh sediments from the Cuenca de Mexico suggest that dry climates with reduced infiltration characterized the early Holocene (Bradbury, 1989). About 10,000 B.P., saline-water deposits suddenly dominate the Chalco Lake sediments in the Basin of Mexico (Watts and Bradbury, 1982) and highly saline conditions persisted until about 6000 B.P. Cores show that between 9000 and 6500 B.P., this lake began to develop into a shallow saline and alkaline marsh, suggestive of general drying conditions (Lozano-Garcia et al., 1993).

4.4 The Middle to Late Holocene: 6000–3000 B.P.

By 7000 B.P., the Laurentide Ice Sheet had all but disappeared from the North American continent (COHMAP, 1988). In response, the Arctic Front shifted farther north and produced significant changes in the atmospheric circulation. In a study of lake levels in North America, Harrison and Metcalfe (1985) found that the early to middle Holocene was characterized by increasing aridity, which led to widespread low lake levels by 9000 B.P. in the mid-latitudes. By 6000 B.P., this zone of low lake levels extended from 32° to 57°N (Harrison and Metcalfe, 1985).

The well-developed zone of aridity that formed over North America possibly reflects the further poleward displacement of the frontal zone between the Arctic and Pacific air masses, possibly to north of 56°N (figs. 4.1 and 4.4). Low lake levels may indicate poleward migration of this front, culminating at approximately 6000 B.P. when the Arctic Front may have reached its most northerly position. The general trend toward increasing aridity in this region during the early and mid-Holocene is consistent with pollen and other evidence for eastward expansion of the prairie across the Midwest between 9000 and 7000 B.P. (Webb et al., 1983). This early Holocene drying phase may have peaked by 7000 B.P. in the northern and eastern portions of the continent, and between 6000 and 5000 B.P. in the western portion.

After 7700 B.P., vegetation in northwest Canada took on an appearance similar to the current vegetation assemblage. At 5000 B.P., a decline in *Picea* indicates a downward movement of the tree line, whereas a decline in total tree pollen may indicate a climatic deterioration after 5000 B.P. (MacDonald, 1983). Since 6000 B.P., taiga, a swampy coniferous subarctic forest, has dominated central Alaskan lowland areas (Elias, 1982; Billings, 1987).

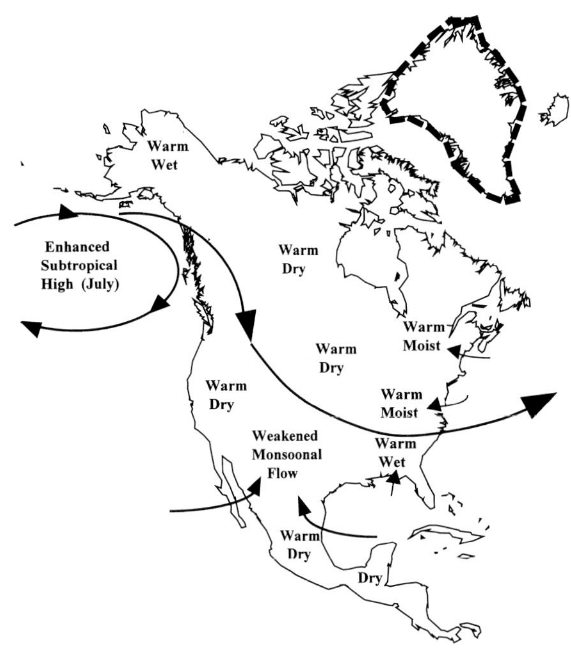

Figure 4.4 Climatic conditions at 6000 B.P. See figure 4.2 for key.

Pollen from lake sediments from west-central Canada indicate that vegetation shifted from dwarf shrub tundra to black spruce (*Picea mariana*) forest as the Arctic Front passed the region moving northward at 5000 B.P., allowing Pacific air masses to move into the region (Edwards et al., 1996). The forest cover remained until 3000 B.P. when the front moved southward and the tundra cover returned. To the south, in the Rocky Mountains, pollen and fossil logs show that timberlines were higher than at present between 8700 and 5200 B.P. (Kearney and Luckman, 1983). A major interval of timberline recession occurred between 6700 and 5900 B.P. Since 5200 B.P., timberlines have receded to their current levels.

The $\partial^{18}O$ values for southern Ontario (Edwards et al., 1996) indicate that conditions in eastern Canada between 6000 and 2000 B.P. were both warmer and moister than today, probably because of a shift from drier Arctic air masses to moister Atlantic air masses (Dean et al., 1996). The decline to modern conditions, which are slightly cooler and drier than earlier, had begun by 4000 B.P.

Post-glacial vegetation was established across the lowlands around the margins of Hudson Bay in northwestern Québec around 6000 B.P. after deglaciation and the isostatic uplift of this initially submerged terrain (Payette and Morneau, 1993). In southern Quebec, white pine (*Pinus strobus*) and hemlock (*Tsuga*) arrived in the St. Lawrence Valley between 6000 and 5700 B.P. and continued to form mixed stands until 3200 B.P. (Filion and Quinty, 1993). Hemlock reached a large size and formed dense stands by 4800 B.P. but declined significantly afterward, possible as a result of the onset of wetter, cooler conditions (Bhiry and Filion, 1996). This is unlike the American Midwest, where drier conditions may have persisted until 3500 B.P. (Winkler et al., 1986; Baker et al., 1992).

In the western United States, insect and plant macrofossils suggest that tree line was higher than the present during the middle Holocene, with a warmer than present climate between 6000 and 3100 B.P. (Elias et al., 1991; Scuderi, 1987b). In central Colorado, deposits in Lake Emma indicate that from 6700 to 5600 B.P. and from 3100 to 3000 B.P., tree line was at least 70 m higher than at present (Carrara et al., 1984). Pollen evidence from the Front Range in Colorado suggests that tree line was above its modern limit from 6500 to 3500 B.P. (Short, 1985; Elias, 1985, 1996). In the southern Yellowstone region, pollen evidence indicates that by 5000 B.P. the earlier warm and dry conditions had been subplanted by a cooler and moister climate. However, in northern Yellowstone at 7000 B.P., the climate was drier than before, conditions that have continued to the present (Whitlock and Bartlein, 1993).

Packrat midden evidence from the White Mountains suggests that during the middle Holocene pinyon and juniper migrated upward into the subalpine zone (Jennings and Elliott-Fisk, 1993). The total movement upward indicates temperatures as much as 2°C higher between 7350 and 3450 B.P. At the same time, in the Sierra Nevada, tree line was between 65 and 70 m above present levels between 6300 and 3500 B.P. (Scuderi, 1987b). Wide rings preserved in relict wood samples suggest a significantly warmer climate and excellent growing conditions centered around 6000 B.P. A rapid decline in tree line elevations by 30 m occurred between 3400 and 3200 B.P., with tree line stabilizing 35 m above current levels by 3200 B.P.

A strong east to west precipitation gradient developed around 6000 B.P. with maximum aridity in the Midwest and Southwest, and moister conditions to the east south of the Great Lakes. The prairie-forest boundary moved farther eastward in the northern Midwest between 10,000 and 6000 B.P. and thereafter retreated westward. Forest taxa reached their most northerly extent, probably representing the warmest conditions at 6000 B.P. Later, several taxa including *Picea* and *Abies* expand in abundance to the south, suggesting a significant climatic reversal.

In the northeastern United States, pollen reconstructions indicate high annual precipitation between 6000 and 3000 B.P. (Webb et al., 1993). This increase in the regional moisture supply reflects changes in the seasonal distribution of insolation in the Northern Hemisphere and resultant changes in sea surface temperatures. Maximum temperatures in the eastern United States were reached around 6000 B.P., with levels some 5–10°C higher than at the height of the Wisconsinan Glaciation (Webb et al., 1993). Mean July temperatures at 6000 B.P. were approximately 1–2°C higher than present. After 6000 B.P., mean July temperatures fell and the 20°C isotherm shifted south of the Canadian border. At the same time, mean January temperatures increased from 9000 B.P. to the present, especially in the southeastern United States (Webb III et al., 1993a).

In eastern North America, lake level data indicate that peak Holocene aridity was centered around 6000 B.P. when almost all lakes in the Midwest and Southeast were at low levels (Harrison and Metcalfe, 1985). After 4000 B.P., these areas began to return to more humid conditions, but lake levels at 4000 B.P. were still not as high as at the beginning of the Holocene. In Florida and the southeastern United States, the low effective moisture at 6000 B.P. occurred when July temperatures were similar to the present, but annual precipitation was as much as 20% lower than the present.

In the north-central United States by 6000 B.P., when lake levels were at their lowest in the Midwest and the prairie-forest border was farthest east, precipitation was at its lowest in the northwestern Midwest and July temperatures were slightly higher than present (Webb III et al., 1993b). After 6000 B.P., lake levels began to return to near present levels with increases in precipitation and a decrease in temperatures. Ostracodes, diatoms, and varved sediments from Elk Lake in Minnesota suggest that, between 6700 and 4000 B.P., mean annual temperatures approached present levels and precipitation was between 85 and 90% of current values (Dean et al., 1984; Forester et al., 1987).

In Mesoamerica, upland forest pollen shows increases in *Quercus* and decreases in *Pinus*, suggesting drier conditions at this time. By 5000 B.P., slight increases in precipitation established the modern climatic regime (Bradbury, 1989). After 5000 B.P., climatic reconstructions are severly compromised by the presence of humans and the cultivation of corn (Watts and Bradbury, 1982; Curtis et al., 1996).

4.5 The Late Holocene: 3000 B.P. to the Present

During the past 3000 years, the present-day patterns of both air-mass movements and the position of the Arctic Front have become established (figs. 4.1 and 4.5). By 3000 B.P., high to intermediate lake levels are found between 46° and 56°N. This probably reflects a southward displacement of the Arctic Front to near its present-day mean position (Harrison and Metcalfe, 1985). Although the Great Basin was relatively dry during this period, the eastern United States experienced wetter conditions, indicative of increased penetration by moist tropical air as far north as 45°N. This resulted in the high lake levels along the eastern margins of North America, which have persisted to the present.

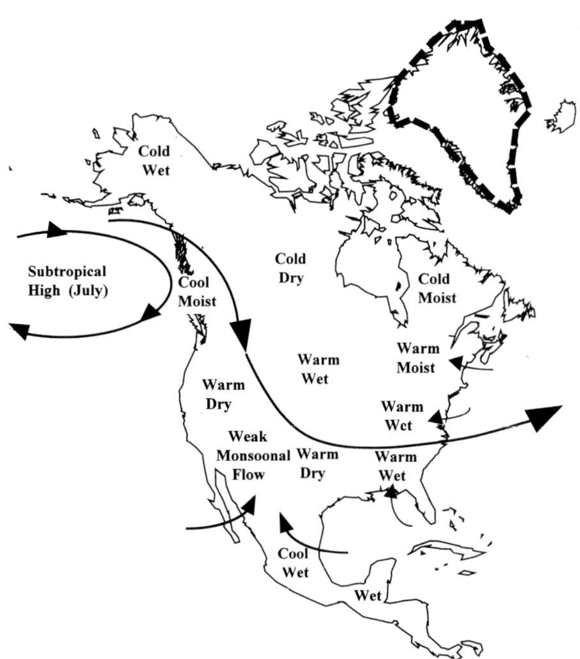

Figure 4.5 Climatic conditions at 3000 B.P. See figure 4.2 for key.

High lake levels in Florida may indicate stronger trade wind circulation (Harrison and Metcalfe, 1985). The more meridional flow at this time would have been encouraged by the establishment of a semipermanent ridge in the westerlies over the western portion of the continent and a trough east of the Rockies (Bartlein et al., 1984).

In Alaska and western Canada, modern synoptic conditions are linked to shifts in North Pacific storm tracks, with a more northerly position linked to beach erosion and a more southerly position linked to beach progradation. Beach ridges along the Chukchi Sea shoreline indicate less frequent storms prior to 3300 B.P. (Mason and Jordan, 1993). Between 3300 and 1700 B.P., coarse sediment facies in storm horizons indicate an increase in storminess possibly linked to a northerly shift in Pacific Ocean storm tracks, but later, between 1700 and 1200 B.P., these beach complexes prograded as a result of less stormy conditions. Between 1200 and 900 B.P., stormier conditions returned and have persisted to the present. These events have also been shown to be contemporaneous with glacial advances throughout Alaska (Wiles and Calkin, 1994; Heusser, 1995; Hu et al., 1998). Lichenometric dating of moraines in the central Brooks Range indicates that moraines stabilized at 4400, 3500, and 2900 B.P. In the last 2000 years, advances are indicated at 1150, 800, and 390 B.P. (Ellis and Calkin, 1984).

In the central Canadian Rockies, evidence from Watchtower Basin in the Jasper area suggests that the tree line dropped below present levels after 4500 B.P. (Luckman and Kearney, 1986; Luckman, 1990). Forests were overridden by glaciers between 3100 and 2500 B.P. with the tree line higher than present just before this advance (Luckman et al., 1993). Additional glacial advances are indicated at 1550 B.P. and between 800 and 600 B.P. Glacial advances in this area over the last 1000 years are the most extensive of the Holocene, leading to the conclusion that the major trigger of these glacial advances was a decrease in summer insolation during the Holocene (Luckman, 1993).

Pollen from lake sediments from north-central Canada indicate that the *Picea mariana* forest cover remained until 3000 B.P. when the Arctic Front moved southward (Edwards et al., 1996). This reduced the influence of moist Pacific air masses in the region and resulted in a return to tundra cover. In southern Ontario, that portion of the late Holocene between 3000 and 2000 B.P. was characterized by conditions that were both warmer and moister than current conditions, probably as a result of a shift from Arctic air masses to moister Atlantic air masses (Dean et al., 1996). The decline to modern conditions, which are slightly cooler and drier than the previous period, had begun by 4000 B.P. and had reached modern levels by 2000 B.P.

In southern Québec, the rise of both hemlock and white pine at 3800 B.P. is attributable to the return of a drier climate. Two peaks in the *Pinus* record at 3500 and 3000 B.P.

followed by an increase in eastern larch at 2600 B.P. indicate a change from cold dry to cold moist conditions (Filion and Quinty, 1993). In northern Québec, the existence of relict lichen-spruce woodlands between 2000 and 900 B.P. (Payette and Morneau, 1993) suggests that temperatures in this area have declined by at least 1°C since 900 B.P.

In western North America from 3200 to 1100 B.P., the species composition of mammals in the Yellowstone region indicates wetter conditions (Hadly, 1996). After 1200 B.P., the fauna becomes more representative of xeric conditions with a maximum of xeric-indicator taxa occurring between 1000 to 650 B.P., which coincides approximately with the medieval warm interlude recognized in Europe. Cooler and wetter conditions are indicated by the fauna for most of the subsequent Little Ice Age between 650 and 100 B.P. (Hadly, 1996).

Deposits at Lake Emma in Colorado indicate that from 3100 to 3000 B.P., the tree line was at least 70 m higher than at present. After 3000 B.P., cooling took place in the San Juan Mountains as indicated by decreases in the percentage of *Picea* pollen and the *Picea:Pinus* pollen ratio and by the absence of large wood (megafossil) fragments (Carrara et al., 1984). Tree line altitudes inferred from fossil insects (Elias et al., 1991) and pollen (Short, 1985) in central Colorado show that significant tree-line decline occurred after 4500 B.P. with some amelioration between 3000 and 2000 B.P. Forest-tundra ratios declined between 2000 to 1000 B.P. when current levels were reached (Elias, 1985). Insect records from the Rocky Mountains suggest a progression from warmer than modern to cooler than modern conditions and back again over the past 3000 years (Elias, 1996).

In Lake Mojave, eastern California, a significant lake episode occurred around 3600 B.P. and a second highstand occurred during the Little Ice Age around 390 B.P., both periods indicating wetter conditions (Enzel et al., 1989, 1992). Lake Mojave's highstands correspond to similar highstands at Mono Lake around 3770 and 290 B.P. and Stine (1990) has suggested that many recent oscillations of the lake may be linked to solar variations.

Packrat (*Neotoma* spp.) midden evidence in the White Mountains of California and Nevada suggests that, during the late Holocene, the climatic variability was not large enough to alter the vegetation community significantly even though cooler and drier conditions prevailed (Jennings and Elliott-Fisk, 1993). A 150-m lowering of the woodland boundary between 4000 and 3000 B.P. in southwestern Nevada (Spaulding, 1985) corresponds to the only significant groundwater recharge in the southwest region in the late Holocene (Benson and Klieforth, 1989).

In the southern Sierra Nevada, following the descent of tree line to 35 m above present levels by 3200 B.P., further reductions in tree-line elevation occurred at 2400 and 1400 B.P. These changes appear to be linked to glacial advances dating between 2800 and 2000 B.P. (Scuderi, 1987a; Graumlich, 1991) and may be related to variations in solar activity (Scuderi, 1993).

Analysis of pollen and snail shells from northeastern Oklahoma indicates that the period from 2000 to 1000 B.P. was moister than today on the Great Plains (Hall, 1982). A drying trend began about 1000 B.P. and has persisted to the present. Existing colonies of small mammals became disjunct about 1000 B.P., leading to the establishment of the current fauna. A similar moisture decline in the Great Basin also resulted in a decrease in small mammal richness (Grayson, 1998).

Pollen evidence indicates that by 4000 B.P. *Pinus* had become the dominant tree species in the northern boreal forest region of the central United States (Delcourt and Delcourt, 1987; Webb III et al., 1993b). After 4000 B.P., lake level data from eastern North America (Harrison and Metcalfe, 1985) suggest that this area began to return to more humid conditions, but lake levels did not rise as high as they did at the beginning of the Holocene. By 3000 B.P., most eastern lakes reached levels similar to the present, indicating effective moisture patterns similar to current conditions.

In Mexico, cores from Chalco Lake show that between 3000 B.P. and the present, water levels increased significantly with fluctuations between freshwater and alkaline marsh (Lozano-Garcia et al., 1993). This evidence suggests a return to cooler, wetter conditions, even as human influences began to be reflected in the vegetation (Watts and Bradbury, 1982).

Based on $\partial^{18}O$ measurements in ostracods and gastropods from a lake core from the Yucatan Peninsula of Mexico, Curtis et al. (1996) found that the late Holocene could be divided into three distinct periods. From 3300 to 1800 B.P., low $\partial^{18}O$ values suggest relatively wet conditions (low evaporation to precipitation ratio). Between 1800 and 900 B.P., mean isotopic values increased, suggesting a distinctly drier period with some periods of consistent dryness concentrated between 1200 and 1100 B.P. Evidence from other areas of Mexico also records this dry interval (Metcalfe et al., 1994; Metcalfe, 1995). After 900 B.P., isotopic values decreased abruptly, indicating the return of wetter conditions that have persisted to the present, with the possible exception of a dry interval about 550 B.P. (Curtis et al., 1996).

4.6 Conclusion

Environmental conditions in North America during the Holocene are thought to be strongly influenced by variations in solar energy receipt, which are determined in turn by Milankovitch-type orbital variations. The peak in summer insolation at 9000 B.P. coincides with generally warm and dry conditions throughout North America. Variations in these orbital parameters from 9000 B.P. to the present have resulted in significant environmental change.

Variations in climatic conditions during the Holocene are also linked to variations in the size and location of the Laurentide Ice Sheet, which control the position of the Arctic Front and jet stream. The ice sheet represents a strong surface influence reflecting and supplementing the orbital variations. The waning of the Laurentide Ice Sheet at the end of the Pleistocene resulted in a northward migration of the Arctic Front and a shift in air-mass source regions for most of North America. Around 9000 B.P., this resulted in a significant drying of most of the central portion of the continent. After ≈6000 B.P., in response to decreased summer solar insolation and the absence of ice sheet–type glaciers in the northern portion of the continent, the Arctic Front migrated southward again, causing significant shifts in air masses over different regions of North America. By 3000 B.P., most areas of the continent began to approach modern climatic conditions. Superimposed on the recent climate have been several interludes of renewed mountain glaciation centered near 3000–2500 B.P. and 500 B.P. During these interludes, cirque glaciers coalesced into valley glaciers and moved downslope in the Rocky Mountains, the Sierra Nevada, the Cascade Range, and in British Columbia and Alaska.

The most recent interlude, the Little Ice Age, so well expressed around the globe, ended only some 150 years ago with the onset of strictly modern climatic patterns (see chapter 5).

References

Anderson, E.S., 1993. A 35,000 year vegetation and climate history from Potato Lake, Mogollon Rim, Arizona. *Quaternary Research* 40, 351–359.

Anderson, P.M., and L.B. Brubaker, 1993. Holocene vegetation and climatic histories of Alaska. In, H.E. Wright Jr., J.E. Kutzbach, T. Webb III, W.F., Ruddiman, F.A. Street-Perrott, and P.J. Bartlein, eds., *Global Climates Since the Last Glacial Maximum.* Minneapolis: University of Minnesota Press, 386–400.

Anderson, P.M., and L.B. Brubaker, 1994. Vegetation history of northcentral Alaska: A mapped summary of late-Quaternary pollen data. *Quaternary Science Reviews* 13, 71–92.

Andrews, J.T., P.E. Carrara, F.B. King, and R. Stuckenrath, 1975. Holocene environmental changes in the alpine zone, northern San Juan Mountains, Colorado: Evidence for long stratigraphy and palynology. *Quaternary Research* 8, 173–197.

Baker, R.G., L.J. Maher, C.A. Chumbley, and K.L. Van Zant, 1992. Patterns of Holocene environmental change in the Midwestern United States. *Quaternary Research* 37, 379–389.

Bartlein, P.J., T. Webb III, and E. Fleri, 1984. Holocene climatic change in the northern Midwest: Pollen derived estimates. *Quaternary Research* 22, 361–374.

Bartlein, P.J., K.H. Anderson, P.M. Anderson, M.E. Edwards, C.J. Mock, R.S. Thompson, R.S. Webb, T. Webb, and C. Whitlock, 1998. Paleoclimate simulations for North America over the past 21,000 years: Features of the simulated climate and comparisons with paleoenvironmental data. *Quaternary Science Reviews* 17, 549–585.

Beget, J.E., 1983. Radiocarbon-dated evidence of worldwide early Holocene climate change. *Geology* 11, 389–393.

Benson, L.V., and H. Klieforth, 1989. Stable isotopes in precipitation and ground water in Yucca Mountain region, southern Nevada: Paleoclimatic implications. In, D.H. Peterson, ed., *Aspects of Climate Variability in the Pacific and Western Americas.* Boulder: Geologic Society of America, 241–260.

Berger, A.L., 1978. Long-term variations in caloric insolation resulting from the earth's orbital elements. *Quaternary Research* 9, 139–167.

Bhiry, N., and L. Filion, 1996. Characterization of the soil hydromorphic conditions in a paludified dunefield during the mid-Holocene Hemlock decline near Quebec City, Quebec. *Quaternary Research* 46, 281–297.

Billings, W.D., 1987. Carbon balance of Alaskan tundra and taiga ecosystems: Past, present and future. *Quaternary Science Reviews* 6, 165–177.

Bradbury, J.P., 1989. Late Quaternary lacustrine paleoenvironments in the Cuenca de Mexico. *Quaternary Science Reviews* 8, 75–100.

Carrara, P.E., D.A. Trimble, and M. Rubin, 1991. Holocene treeline fluctuations in the northern San Juan Mountains, Colorado, U.S.A., as indicated by radiocarbon-dated conifer wood. *Arctic and Alpine Research* 23, 233–246.

Carrara, P.E., W.N. Mode, M. Rubin, and S.W. Robinson, 1984. Deglaciation and postglacial timberline in the San Juan Mountains, Colorado.*Quaternary Research* 21, 42–55.

COHMAP, 1988. Climatic changes of the last 18,000 years: Observations and model simulations. *Science* 241, 1043–1052.

Curtis, J.H., D.A. Hodell, and M. Brenner, 1996. Climate variability on the Yucatan peninsula (Mexico) during the past 3500 years, and implications for Maya cultural evolution. *Quaternary Research* 46, 37–47.

Davis, O.K., 1984. Multiple thermal maxima during the Holocene. *Science* 225, 617–619.

Davis, O.K., and D.S. Shafer, 1992. A Holocene climatic record for the Sonoran Desert from pollen analysis of Montezuma Well, Arizona, USA. *Palaeogeography, Palaeoclimatology, Palaeoecology* 92, 107–119.

Davis, O.K., R.S. Anderson, P.L. Fall, M.K. O'Rourke, and R.S. Thompson, 1985. Palynological evidence for early Holocene aridity in the southern Sierra Nevada, California. *Quaternary Research* 24, 322–332.

Dean, W.E., T.S. Ahlbrandt, R.Y. Anderson, and J.P. Bradbury, 1996. Regional aridity in North America during the middle Holocene. *The Holocene* 6, 145–155.

Dean, W.E., J.P. Bradbury, R.Y. Anderson, and C.W. Barnosky, 1984. The variability of Holocene climate change: Evidence from varved lake sediments. *Science* 226, 1191–1194.

Delcourt, P.A., and H.R. Delcourt, 1987. Late-Quaternary dynamics of temperate forests: Applications of paleoecology to issues of global environmental change. *Quaternary Science Reviews* 6, 129–146.

Edwards, T.W.D., B.B. Wolfe, and G.M. MacDonald, 1996. Influence of changing atmospheric circulation on precipitation ∂^{18}O-temperature relations in Canada during the Holocene. *Quaternary Research* 46, 211–218.

Elias, S.A., 1982. Holocene insect fossils from two sites at Ennadai Lake, Keewatin, Northwest Territories, Canada. *Quaternary Research* 17, 371–390.

Elias, S.A., 1985. Paleoenvironmental interpretations of Holocene insect fossil assemblages from four high-altitude sites in the Front Range, Colorado, U.S.A. *Arctic and Alpine Research* 17, 31–48.

Elias, S.A., 1996. Late-Pleistocene and Holocene seasonal temperatures reconstructed from fossil beetle assemblages in the Rocky Mountains. *Quaternary Research* 46, 311–318.

Elias, S.A., P.E. Carrara, L.J. Toolin, and A.J.T. Jul, 1991. Revised age of deglaciation of Lake Emma based on new radiocarbon and macrofossil analyses. *Quaternary Research* 36, 307–321.

Ellis, J.M., and P.E. Calkin, 1984. Chronology of Holocene glaciation, central Brooks Range, Alaska. *Geological Society of America Bulletin* 95, 897–912.

Enzel, Y., D.R. Cayan, R.Y. Anderson, and S.G. Wells, 1989. Atmospheric circulation during Holocene lake stands in the Mojave Desert: Evidence of regional climate change. *Nature* 341, 44–47.

Enzel, Y., W.J. Brown, R.Y. Anderson, L.D. McFadden, and S.G. Wells, 1992. Short-duration Holocene lakes in the Mojave River drainage basin, Southern California. *Quaternary Research* 38, 60–73.

Felzer, B., T. Webb, and R.J. Oglesby, 1998. The impact of ice sheets, CO_2, and orbital insolation on late Quaternary climates: Sensitivity experiments with a general circulation model. *Quaternary Science Reviews* 17, 507–534.

Filion, L., and F. Quinty, 1993. Macrofossil and tree-ring evidence for a long-term forest succession and mid-Holocene hemlock decline. *Quaternary Research* 40, 89–97.

Forester, R.M., L.D. Delorme, and J.P. Bradbury, 1987. Mid-Holocene climate of Northern Minnesota. *Quaternary Research* 28, 263–273.

Graumlich, L.J., 1991. A 1000-year record of temperature and precipitation in the Sierra Nevada. *Quaternary Research* 39, 249–255.

Grayson, D.K., 1998. Moisture history and small mammal community richness during the late Pleistocene and Holocene, northern Bonneville Basin, Utah. *Quaternary Research* 49, 330–334.

Hadly, E.A., 1996. Influence of late-Holocene climate on northern Rocky Mountain mammals. *Quaternary Research* 46, 298–310.

Hafner, D.J., 1993. North American Pika (*Ochotona princeps*) as a late Quaternary biogeographic indicator species. *Quaternary Research* 39, 373–380.

Hall, S.A., 1982. Late Holocene paleoecology of the Southern Plains. *Quaternary Research* 17, 391–407.

Hamilton, T.D., 1986. Late Cenozoic glaciation of the central Brooks Range. In, T.D. Hamiliton, K.M. Reed, and R.M. Thorson, eds., *Glaciation in Alaska: The Geologic Record*. Anchorage: Alaska Geological Society, 9–50.

Harrison, S.P., 1989. Lake level and climatic change in eastern North America. *Climate Dynamics* 3, 157–167.

Harrison, S.P., and S.E. Metcalfe, 1985. Variations in lake levels during the Holocene in North America: An indicator of changes in atmospheric circulation patterns. *Geographie physique et Quaternaire* 39 (2), 141–150.

Heusser, C.J., 1995. Late-Quaternary vegetation response to climatic-glacial forcing in north Pacific America. *Physical Geography* 16, 118–149.

Hevly, R.H., 1985. A 50,000 year record of Quaternary environments, Walker Lake, Coconino Co., Arizona. In, B.F. Jacobs, P.L. Fall, and O.K. Davis, eds., *Late Quaternary Vegetation and Climates of the American Southwest*. Dallas: American Association of Stratigraphic Palynologists Contribution 16, AASP, 141–154.

Hu, F.S., L.B. Brubaker, and P.M. Anderson, 1996. Boreal ecosystem development in the northwestern Alaska Range since 11,000 B.P. *Quaternary Research* 45, 188–201.

Hu, F.S., E. Ito, L.B. Brubaker, and P.M. Anderson, 1998. Ostracode geochemical record of Holocene climate change and implications for vegetational response in the northwestern Alaska Range. *Quaternary Research* 49, 86–95.

Jennings, S.A., and D.L. Elliott-Fisk, 1993. Packrat midden evidence of late-Quaternary vegetation change in the White Mountains, California-Nevada. *Quaternary Research* 39, 214–221.

Kearney, M.S., and B.H. Luckman, 1983. Holocene timberline fluctuations in Jasper National Park, Alberta. *Science* 221, 261–263.

Kutzbach, J.E., 1987. Model simulations of the climatic patterns during the deglaciation of North America. In, W.F. Ruddiman and H.E. Wright Jr., eds., *North America and the Adjacent Oceans During the Last Deglaciation. The Geology of North America*, Vol. K-3. Boulder: Geological Society of America, 425–446.

Kutzbach, J.E., and P.J. Guetter, 1984. The sensitivity of monsoon climates to orbital parameter changes for 9000 years B.P.. Experiments with the NCAR General Circulation model. In, A. Berger and J. Imbrie, eds., *Milankovitch and Climate*. Dordrecht: D. Reidel, 801–820.

Kutzbach, J.E., and W.F. Ruddiman, 1993. Model description, external forcing, and surface boundary conditions. In, H.E. Wright Jr., J.E. Kutzbach, T. Webb III, W.F. Ruddiman, F.A. Street-Perrott, and P.J. Bartlein, eds., *Global Climates Since the Last Glacial Maximum*. Minneapolis: University of Minnesota Press, 12–23.

Kutzbach, J.E., and T. Webb III, 1993. Conceptual basis for understanding late-Quaternary climates. In, H.E. Wright Jr., J.E. Kutzbach, T. Webb III, W.F. Ruddiman, F.A. Street-Perrott, and P.J. Bartlein, eds., *Global Climates Since the Last Glacial Maximum*. Minneapolis: University of Minnesota Press, 5–11.

Lozano-Garcia, M., B. Ortega-Guerrero, M. Caballero-Miranda, and J. Urrutia-Fucugauchi, 1993. Late Pleistocene and Holocene paleoenvironments of Chalco Lake, Central Mexico. *Quaternary Research* 40, 332–342.

Luckman, B.H., 1990. Mountain areas and global change—A view from the Canadian Rockies. *Mountain Research and Development* 10, 183–195.

Luckman, B.H., 1993. Glacier fluctuations and tree-ring records for the last millennium in the Canadian Rockies. *Quaternary Science Reviews* 12, 441–450.

Luckman, B.H., and M.S. Kearney, 1986. Reconstruction of Holocene changes in alpine vegetation and climate in the Maligne Range, Jasper National Park, Alberta. *Quaternary Research* 26, 244–261.

Luckman, B.H., G. Holdsworth, and G.D. Osborn, 1993. Neoglacial glacier fluctuations in the Canadian Rockies. *Quaternary Research* 39, 144–153.

MacDonald, G.M., 1983. Holocene vegetation history of the Upper Natla River area, Northwest Territories, Canada. *Arctic and Alpine Research* 15, 169–180.

MacDonald, G.M., T.W.D. Edwards, K.A. Moser, R. Pienitz, and J.P. Smol, 1993. Rapid response of treeline vegetation and lakes to past climatic warming. *Nature* 361, 243–246.

Markgraf, V., 1989. Paleoclimates in Central and South America since 18,000 B.P. based on pollen and lake-level records. *Quaternary Science Reviews* 8, 1–24.

Mason, O.K., and J.W. Jordan, 1993. Heightened North pacific storminess during synchronous late Holocene erosion of Northwest Alaska beach ridges. *Quaternary Research* 40, 55–69.

Metcalfe, S.E., 1995. Holocene environmental change in the Zacapu Basin, Mexico: A diatom based record. *The Holocene* 5, 196–208.

Metcalfe, S.E., F.A. Street-Perrott, S.L. O'Hara, P.E. Hales, and R.A. Perrott, 1994. The palaeolimnological record of environmental change: Examples from the arid frontier of Mesoamerica. In, A.C. Millington and K. Pye, eds., *Environmental Change in Drylands: Biogeographical and Geomorphological Perspectives.* Chichester: Wiley, 131–145.

Miller, R.F., and A.V. Morgan, 1982. A postglacial Coleopterous assemblage from Lockport Gulf, New York. *Quaternary Research* 17, 258–274.

Payette, S., and C. Morneau, 1993. Holocene relict woodlands at the eastern Canadian treeline. *Quaternary Research* 39, 84–89.

Qin, B., S.P. Harrison, and J.E. Kutzbach, 1998. Evaluation of modelled regional water balance using lake status data: A comparison of 6-ka simulations with the NCAR CCM. *Quaternary Science Reviews* 17, 535–548.

Ritchie, J.C., 1987. *Postglacial Vegetation of Canada.* Cambridge: Cambridge University Press.

Ritchie, J.C., and S.P. Harrison, 1993. Vegetation, lake levels, and climate in western Canada during the Holocene. In, H.E. Wright Jr., J.E. Kutzbach, T. Webb III, W.F. Ruddiman, F.A. Street-Perrott, and P.J. Bartlein, eds., *Global Climates Since the Last Glacial Maximum.* Minneapolis: University of Minnesota Press, 401–414.

Rothlisberger, F., 1986. *10000 Jahre Gletschergeschichte der Erde.* Aarau: Verlag Sauerlander.

Scuderi, L.A., 1987a. Glacier variations in the Sierra Nevada California, as related to a 1200-year tree-ring chronology. *Quaternary Research*, 27, 220–231.

Scuderi, L.A., 1987b. Late-Holocene upper timberline variation in the southern Sierra Nevada. *Nature* 325, 242–244.

Scuderi, L.A., 1993. A 2,000-year tree ring record of annual temperatures in the Sierra Nevada Mountains. *Science* 259, 1433–1436.

Scuderi, L.A., 1994. Solar influences on Holocene treeline altitude variability in the Sierra Nevada. *Physical Geography* 15, 146–165.

Short, S.K., 1985. Palynology of Holocene sediments Colorado Front Range: Vegetation and treeline changes in the subalpine forest. *American Association of Stratigraphic Palynologists Contribution Series* 16, 7–30.

Spaulding, W.G., 1985. *Vegetation and Climates of the Last 45,000 Years in the Vicinity of the Nevada Test Site, South-Central Nevada.* United States Geological Survey Professional Paper 1329.

Stine, S., 1990. Late Holocene fluctuations of Mono Lake, eastern California. *Palaeogeography, Palaeoclimatology, Palaeoecology* 78, 333–381.

Waitt, R.B., J.C. Yount, and P.T. Davis, 1982. Regional significance of an early Holocene moraine in Enchantment Lakes Basin, North Cascade Range, Washington. *Quaternary Research* 17, 191–210.

Watts, W.A., and J.P. Bradbury, 1982. Paleoecological studies of Lake Patzcuaro on the west-central Mexican Plateau and at Chalco in the Basin of Mexico. *Quaternary Research* 17, 56–70.

Webb, R.S., K.H. Anderson, and T. Webb III, 1993. Pollen response-surface estimates of late-Quaternary changes in the moisture balance of the Northeastern United States. *Quaternary Research* 40, 213–227.

Webb, R.S., K.H. Anderson, P.J. Bartlein, and R.S. Webb, 1998. Late Quaternary climate change in eastern North America: A comparison of pollen-derived estimates with climate model results. *Quaternary Science Reviews* 17, 587–606.

Webb III, T., E.J. Cushing, and H.E. Wright, Jr., 1983. Holocene changes in the vegetation of the Midwest. In, Wright Jr., H.E., ed., *Late Quaternary Environments of the United States, Volume 2: The Holocene.* Minneapolis: University of Minnesota Press, 142–165.

Webb III, T., P.J. Bartlein, S.P. Harrison, and K.H. Anderson, 1993a. Vegetation, lake levels, and climate in eastern North America for the past 18,000 years. In, H.E. Wright Jr., J.E. Kutzbach, T. Webb III, W.F. Ruddiman, F.A. Street-Perrott, and P.J. Bartlein, eds., *Global Climates Since the Last Glacial Maximum.* Minneapolis: University of Minnesota Press, 415–467.

Webb III, T., W.F. Ruddiman, F.A. Street-Perrott V. Markgraf, J.E. Kutzbach, P.J. Bartlein, H.E. Wright Jr., and W.L. Prell, 1993b. Climatic changes during the past 18,000 years: Regional syntheses, mechanisms, and causes. In, H.E. Wright Jr., J.E. Kutzbach, T. Webb III, W.F. Ruddiman, F.A. Street-Perrott, and P.J. Bartlein, eds., *Global Climates Since the Last Glacial Maximum.* Minneapolis: University of Minnesota Press, 5–11.

Whitlock, C., and P.J. Bartlein, 1993. Spatial variations in Holocene climatic change in the Yellowstone region. *Quaternary Research* 39, 231–238.

Wiles, G.C., and P.E. Calkin, 1994. Late Holocene high-resolution glacial chronologies and climate, Kenai Mountains, Alaska. *Geological Society of America Bulletin* 106, 281–303.

Winkler, M.G., A.M. Swain, and J.E. Kutzbach, 1986. Middle Holocene dry period in the northern Midwestern United States. *Quaternary Research* 25, 235–250.

5

Dynamic and Synoptic Climatology

Roger G. Barry

Tectonic processes and subsequent geologic history have established the continental framework of North America, but it is the interaction of the coupled atmosphere–ocean system with this framework that generates its climate. In turn, climatic conditions greatly influence the hydrologic cycle, vegetation cover, and soils discussed in subsequent chapters. The climate has also changed over time, most dramatically after the disappearance of the last Laurentide Ice Sheet and associated Cordilleran ice caps and glaciers between about 12,000 and 8,000 years ago (see chapters 2–4). Subsequent centennial- and multidecadal-scale climatic fluctuations, evidenced by neoglacial episodes in the Western Cordillera and recurrent episodes of severe drought in the western United States, have been no less important, however, in human affairs.

This chapter surveys the dynamic and synoptic aspects of North America's present climate, namely, the planetary-scale controls of atmospheric circulation and the embedded synoptic weather systems, recurrent mesoscale phenomena, and the resulting climatic conditions that affect the continent. The discussion is organized according to four topics: the planetary-scale setting, continental-scale influences and their climatic effects, synoptic regimes, and regional climatic features and anomalies. We begin by considering the overall climatic setting of the continent.

The climate of North America is shaped by geographical, oceanographic, and atmospheric factors and their interactions. The continent and adjacent Canadian Arctic Archipelago extend approximately from 15° to 80°N, and from 170°W in Alaska to 50°–60°W in eastern Canada, narrowing to between 120°W in southern California and 80°W in Florida before tapering southward through Mexico. High mountains along the continent's west coast join with intermontane plateaus and the Rocky Mountains farther east to form the Western Cordillera that extends southward from Alaska to Mexico. Mountains also dominate the eastern Canadian Arctic Archipelago and Labrador. The Arctic Ocean and year-round sea ice are key factors in the climate of the Canadian Arctic Archipelago and northern Alaska. The west coast waters are cold as a result of the southward-flowing California Current, whereas the warm Gulf Stream follows the Atlantic coast of the United States. However, the western Baffin Bay–Labrador Current transports subpolar water and, in late winter and spring, sea ice and icebergs southward to 48°W off Newfoundland. Maritime influences are also present in the heart of the continent in summer and autumn around Hudson Bay and the Great Lakes, as well as more generally as a result of the large number of smaller lakes and bogs in the boreal forest of Canada and the mixed forest of the upper Midwest. Until now, the effects of these smaller water bodies have not been represented in general atmospheric circulation models. The Gulf of Mexico is a prominent factor in the climate of the southern and southwestern United States and much of

Mexico by supplying heat and moisture for subtropical low pressure systems and warm-season precipitation.

5.1 Planetary-scale Setting

5.1.1 The Circumpolar Vortex and Planetary Wave Structure

The dominant feature of the troposphere in the Northern Hemisphere is a vast circumpolar vortex and its associated zonal (westerly) air circulation over the midlatitudes. The vortex is quite symmetrical about the pole in summer when the circulation is weak, but in winter the vortex intensifies, expands, and becomes more asymmetric, with three major troughs and ridges producing northerly and southerly (meridional) components in the airflow. The winter circulation is not only stronger but also more variable interannually and intraseasonally. This wave structure provides the dynamic setting for North America's climate, both its mean features and its characteristic anomalies (Wallace, 1983). Empirical studies show that the 30-day mean 500-mb (millibar) patterns are closely similar to those obtained using band-pass filters to distinguish periods longer than 10 days. In contrast, 2.5–6-day filters identify a different spectral structure representing synoptic anomalies (Wallace and Blackmon, 1983). This result demonstrates the existence of two sets of circulation features and anomalies, each with a distinctive time and space scale, and it establishes a sound rationale for their separate treatment in the following discussion.

During winter months, the mean wave structure of the middle troposphere features a pronounced trough over eastern North America near 70°W and a ridge over the Western Cordillera (fig. 5.1.). This pattern reflects both orographic forcing and the effects of diabatic heat sources, particularly the contrasting cold land surface and warm western North Atlantic surface water (Kasahara, 1980). Nevertheless, the orographic effect is not solely dynamic, but includes the thermal effects of latent heat released especially over the windward slopes through cloud formation and precipitation. The eastern North America trough, located over Baffin Island, is the upper-level counterpart of the sea-level Icelandic low, which tilts westward with height where there are lower values of 1000–500-mb thickness in response to the cold air to the west of the surface low. The upper tropospheric jet stream is farthest south in January and February when mean monthly geostrophic wind speeds at 500 mb exceed 28 ms^{-1} over the Atlantic coast near 38°N (Harman, 1991). The upper-level westerlies extend over Mexico, sometimes as far south as Mexico City, with associated upper-level troughs, and surface westerlies prevail over the Mexican Plateau (Mosiño Alemán and García, 1974). Over the west coast of Mexico,

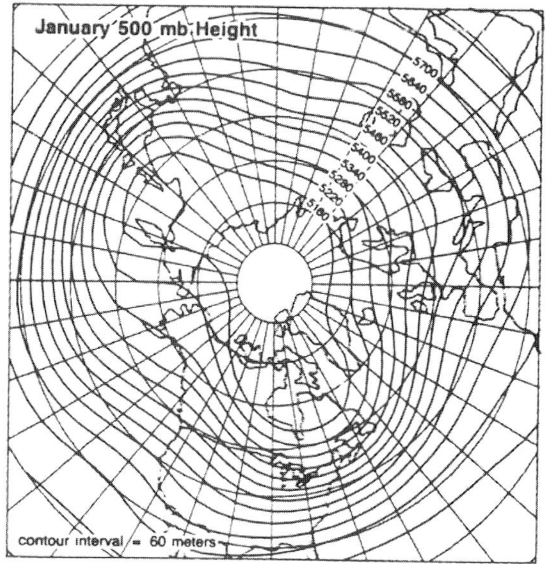

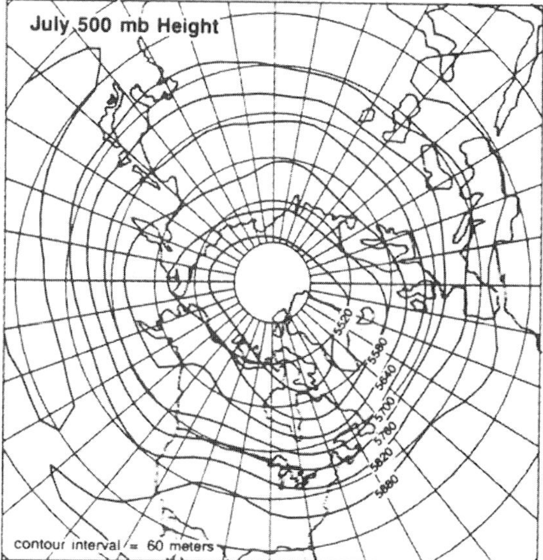

Figure 5.1 The structure of the Northern Hemipshere's middle troposphere revealed by mean 500-mb contours for January and July, 1946–87 (from Harman, 1991).

subsiding air from the North Pacific high gives low-level northwesterly flow.

The 500-mb pattern is much more zonal in summer although a weak eastern North American trough is still evident, and in midlatitudes there are some seven long-wave troughs around the hemisphere instead of the three or four in winter (Harman, 1991). During the summer, the pattern of low-level diabatic heating is now very different as the land mass acts as a heat source. The subtropical ridge of high pressure, which is located over the Gulf of Mexico in winter, extends in summer across the southern United States, and there is a 500-mb ridge at 45°N, 90°–100°W. The 500-mb geostrophic wind averages only half of its winter value with a broad maximum of 12–14 ms^{-1} over the Great Lakes and New England.

In summer and fall, deep, moist easterly trade winds, associated with the subtropical high to the north and east, prevail over Mexico (Davis et al., 1997). These trade winds account for the May to November rainy season (Mosiño Alemán and García, 1974), which has been described as a summer monsoon by Douglas et al., (1993). Moisture inflows from the Pacific and the Gulf of Mexico are directed toward the thermal low formed over the northern plateau of Mexico in May and June (Dilley, 1996). Midsummer dry spells that affect the coastal plains of eastern Mexico appear to be a result of southwestward extensions of an upper trough from the eastern seaboard of the United States, which causes disturbances in the tropical easterlies to recurve toward the Gulf coast. A stationary upper trough is commonly located over the west coast, related to the thermal low of the desert southwestern United States.

The southwestern United States also experiences a monsoon-like regime beginning in late June. Moisture enters the region from the Gulf of California in association with a lower tropospheric Sonoran jet and also with southeasterly flows from the Gulf of Mexico (Carleton et al., 1990). Summer rainfall is related to anticyclonic circulations at 500 mb; Carleton (1987) shows the variety of synoptic patterns involved and their specific characteristics in terms of wet and dry episodes and "burst and break" types of monsoonal activity. Precipitation falls from convective cells, often in mesoscale clusters, locally enhanced by orography. Interannual variability of summer precipitation in the Southwest arises from northward and southward displacements of the subtropical ridge and Pacific sea-surface temperatures. Northward (southward) ridge positions are associated with wetter (drier) summers. Wetter summers also tend to show an enhanced longitudinal gradient of sea-surface temperature (SST) along the west coast of the United States, Baja California, and the Gulf of California.

5.1.2 Circulation Modes

The quasi-stationary waves discussed previously account for many of the broad elements of North American climate through their roles in fostering cyclogenesis and cyclolysis, and in influencing the locations of storm tracks. Thus, a "normal" winter pattern favors generally mild conditions in the northwestern United States; these conditions are associated with the upper ridge and onshore southwesterly flow, and cold, snowy conditions in the eastern and central United States as a result of northwesterly flows of arctic air from Canada. One type of anomaly is an intensification of this pattern when the waves are amplified. This circulation pattern is characteristic of the positive mode of the Pacific North American (PNA) pattern identified by Wallace and Gutzler (1981). This comprises a strong ridge over western North America with amplified troughs over the central North Pacific and eastern North America. It is present during all months except June and July.

This positive mode is a major component of cold-season midtropospheric circulation, giving rise to warm anomalies in the Pacific Northwest and Great Plains. The winter of 1991–92 illustrates such a regime (fig. 5.2.); mean temperature departures from normal in January 1992 exceeded 7°C in Montana and North Dakota, for example. Negative departures of precipitation occur over much of the central and eastern United States, but increased precipitation occurs in Alaska (Yarnal and Diaz, 1986; Keables, 1992). In the Mississippi Valley and Ohio Valley, precipitation is below normal. Cold outbreaks in the southeastern United States are also favored under positive PNA patterns (Leathers et al., 1991). Over western Canada, snow cover is much reduced during positive PNA winters (Brown and Goodison, 1996). In the southeastern United States, Serreze et al. (1998) find that snowfall increases (decreases) with a positive (negative) PNA mode. The corresponding circulation over the eastern United States features a strong 500-mb trough (zonal flow and weak ridge), respectively, which results in lower (higher) maximum temperatures on precipitation days in the Southeast. There is an opposing snowfall-PNA signal in the upper Midwest, where changes in precipitation amount rather than temperature are decisive.

Another important influence on North American climate is associated with El Niño-Southern Oscillation (ENSO) events over the tropical Pacific Ocean. A composite of ENSO events indicates they are commonly triggered from January to June when sea-surface temperatures rise in the equatorial Pacific from the South American coast to the International Date Line (Harrison and Larkin, 1998). From March to June, a pressure rise over northern Australia-Indonesia gives a reversal of the normal Tahiti-Darwin sea level pressure gradient (the negative or low phase of the Southern Oscillation). During the July to December peak of the warm El Niño phase, convective activity normally located over Indonesia shifts into the central equatorial Pacific where there is strong wind convergence. Concurrently, westerly wind anomalies develop over the west-central subtropical North Pacific, continuing through the following April. ENSO events tend to favor frequent and stronger westerly flow across western North America and into Canada during the subsequent winter. Positive temperature anomalies affect the west coast from California to Alaska and all across southern Canada (Kiladis and Diaz, 1989). Conversely, during the corresponding winters of the La Niña (cold) phase with strong Pacific easterly trade winds, weaker zonal circulation in the midlatitudes allows arctic outbreaks and negative anomalies over the Canadian

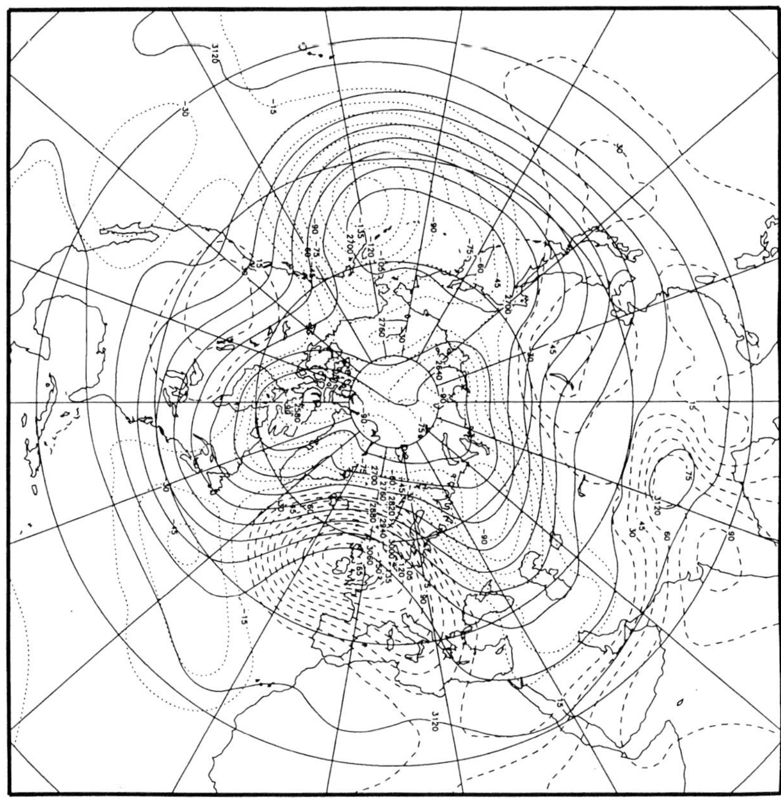

Figure 5.2 Northern Hemisphere 700-mb mean height and height anomalies (g.p.m.) for January 1992. Mean heights are solid contours (6-dam interval); anomalies are departures from the 1961–90 means. Positive (negative) anomalies are indicated by dashed (dotted) contours (15-m interval). The zero anomaly contour is omitted (from Kousky, 1993).

Prairies. The strongest precipitation signal of ENSO is for wetter conditions during the winter following the El Niño peak, over the Gulf Coast of the United States and northern Mexico (see fig. 5.3) (Kiladis and Diaz, 1989). This anomaly is related to a stronger subtropical jetstream over the Gulf of Mexico and activity along a southward-displaced extratropical cyclone track to the north of the jet.

The significance of the ENSO-PNA pattern linkage for climatic anomalies in the United States is strongly dependent on the PNA mode. Keables (1992) shows that ENSO warm events may be associated with positive or negative PNA or there may be no PNA signature. With negative (reversed) PNA, 500-mb height anomalies are positive (negative) over Hudson Bay (the northern North Pacific and Gulf of Mexico). Anomalous easterly flow occurs across southern Canada and much of the United States, with anomalous southerly flow over the west coast. Winter temperatures are above average from the Pacific Northwest to the upper Midwest and below average in the southeastern United States (see Fig. 5.3). Precipitation is above average in the Southeast, along the Atlantic seaboard, and in New England, and below average over the northern Great Plains and the Midwest. For ENSO warm events with no PNA signature, Keables (1992) shows that the 500-mb height anomalies resemble those of the reversed PNA except that the negative height anomaly is displaced to the east coast. This trough displacement causes northwesterly flow over the eastern United States, giving large negative departures of temperature in the southeast and of precipitation south of the Great Lakes.

Temporal trends of the PNA index (Leathers and Palecki, 1992) show that it was negative from 1947 through 1957, indicating zonal regimes, but generally positive for 1958–87, and this tendency has generally continued. The PNA pattern together with ENSO and regional low-pressure systems over the southwestern United States are also shown to account for much of the variance in maximum winter temperatures and rain-day events in the Sonoran Desert (Woodhouse, 1997). Regional and planetary-scale circulation features evidently interact to determine winter climate in this region. Burnett (1994) finds that the reverse PNA sets up a 500-mb trough over the Southwest. This pattern is common in spring and the cold season generally. Although it occurs on only about 25% of days, this trough pattern accounts for 50–60% of the precipitation over much of the southern United States and High Plains (fig. 5.4.)

The North Pacific Oscillation (NPO), involving the Aleutian Low and the high pressure over the North Pacific, is a regional phenomenon and a feature of cold-season months (Rogers, 1981). The oscillation has a related temperature

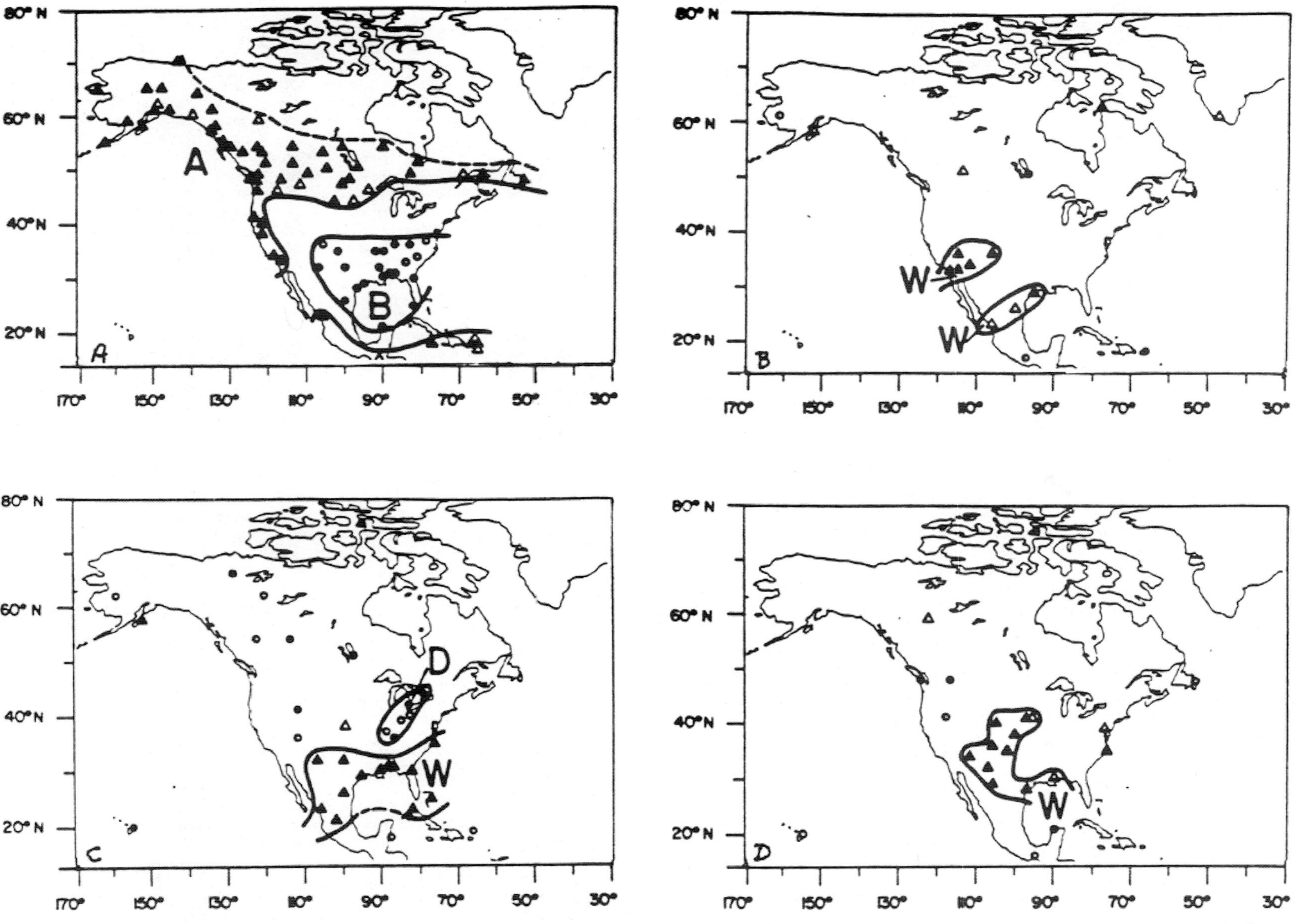

Figure 5.3 Composite differences of (a) temperature anomalies and (b, c, d) precipitation anomalies for warm minus cold ENSO. The differences are statistically significant at the 1% (solid symbols) or 5% (open symbols) level. W = wetter, D = dryer, A = above normal, B = below normal during warm events. (a) December–February of year +1 for temperature; (b) September–November of year 0 when the warm or cold event first occurs; (c) December–February; and (d) March–May of year +1 for precipitation (after Kiladis and Diaz, 1989).

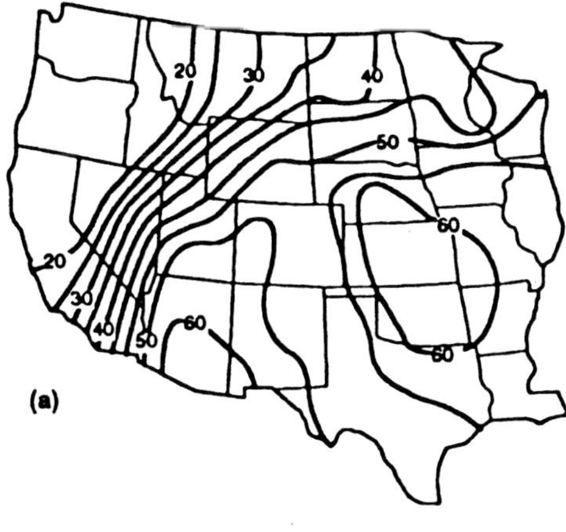

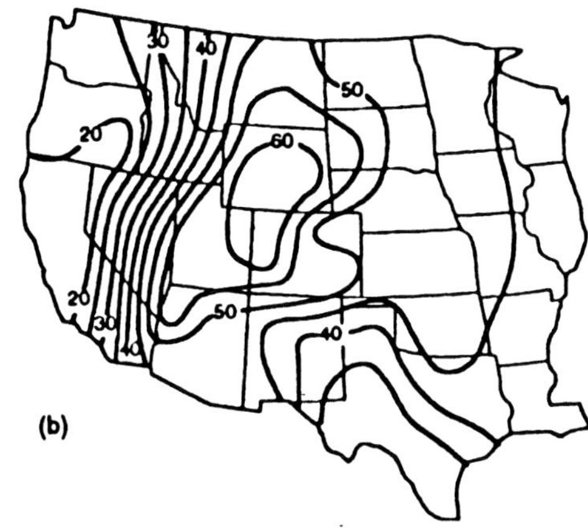

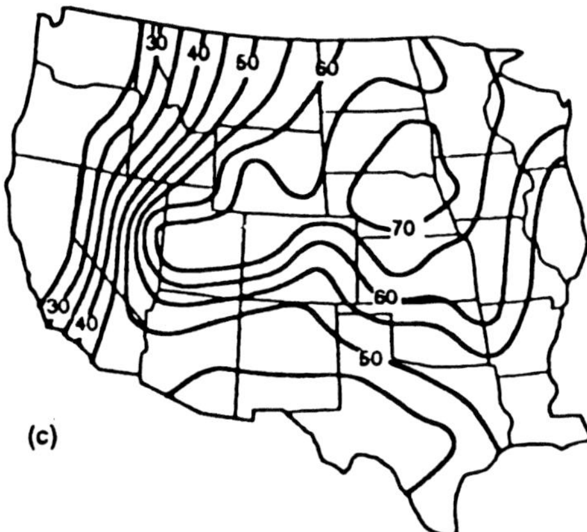

Figure 5.4 Percentage of total precipitation associated with 500-mb southwestern troughs during 1948–63, 1969–76 for (a) January, (b) April, and (c) October (from Burnett, 1994).

seesaw between Alberta and western Alaska; in the Aleutians below-normal (AB) mode, mild winter conditions in western Canada result from enhanced zonal westerlies. The spring of 1993 was a typical positive NPO mode corresponding to the AB temperature pattern (Halpert and Smith, 1994).

The North Atlantic Oscillation (NAO) involves a cold-season negative correlation between the Icelandic low and the subtropical Atlantic high. Although its primary effects are felt from Greenland to Europe, enhanced northerly flow over Davis Strait and the Labrador Sea, when the Icelandic low is deep, transports sea ice and icebergs far south of normal. This pattern gives below normal temperatures in eastern Canada and New England. The NAO also correlates with snow cover variability in southern Ontario and Québec (Brown and Goodison, 1996). Since 1988, it has been in a predominantly positive mode, where a strong zonal gradient gives mild winters in northwest Europe and cold anomalies in west Greenland (Hurrell, 1995).

5.2 Continental-Scale Influences

5.2.1 Continentality and Oceanicity

The dynamic processes of the atmosphere operate within the context of the continental geography and are in turn modified by thermodynamic processes activated by the physical setting. The principal factors, as outlined in the introduction, are continentality, orography, and sea-surface temperatures. Driscoll and Yee Fong (1992) emphasize that continentality is determined primarily by distance from the ocean. Thus, maximum continentality for North America occurs in southern Saskatchewan and North Dakota.

Continentality may also be enhanced by extensive and persistent snow cover. Studies indicate that winter air temperatures are lowered by 5–7°C during intervals when a snow cover is present compared with snow-free ground. Conversely, the recent observed reduction in duration of springtime snow cover in the Northern Hemisphere is shown to be a potential contributor to the higher spring temperatures observed over the continents (Groisman et al., 1994), as a result of the albedo-temperature feedback effect. Less extensive snow cover reduces the surface reflection of short-wave solar radiation, increasing the amount of solar radiation absorbed by the surface and thereby raising the surface temperature. The presence of a persistent extensive snow cover over North America modifies the 1000–500-mb thickness pattern and leads to the southward displacement of cyclone tracks (Dewey, 1987), whereas significant reductions in snow extent are expected to permit a northward displacement.

The climates of northern and eastern Canada are shaped significantly by the snow cover, which lasts 8 months in the subarctic and 10 months in the High Arctic, and by the ice-covered waters of the Beaufort Sea, Arctic Archipelago, Baffin Bay, Hudson Bay, and the western Labrador Sea (Barry, 1993; Bailey et al., 1997). Persistent temperature inversions under anticyclonic regimes in winter and spring favor severe ice conditions, whereas in summer low cloud and fog inhibit ice melt. The Baffin Bay–Davis Strait–Labrador Sea coastal margins have year-round cyclonic activity and arctic-subarctic maritime climates with cool, cloudy summers. The land loses its snow cover in a rapid 10–15-day transition during early June in central Québec and Labrador and in Keewatin, and only in early July in the Queen Elizabeth Islands. However, landfast sea ice persists in the narrow channels of the Arctic Archipelago and off the coasts of Baffin Island and Labrador for much of the summer, depressing coastal air temperatures.

The effects of sea-surface temperature on the atmosphere operate on a hierarchy of scales, from the near-global influence of ENSO teleconnections to local-scale effects such as those of the cold California Current and the Labrador Current on coastal fogs in summer. In California, coastal mountains limit the inland penetration of maritime air flows, and, more generally, diurnal heating rapidly modifies the air as it moves inland. Sea-surface temperature anomalies in the tropical Pacific Ocean appear to be linked, through atmospheric teleconnections, with precipitation in North America. Montory (1997), for example, shows strong associations between positive SST anomalies in the central and eastern tropical Pacific and November–March rainfall in the southeastern United States and also in Texas. In the southern Canadian Prairies, November–January precipitation is related to negative SST anomalies in the same ocean areas, as is January–March precipitation over the Great Lakes–Ohio region. However, such results remain to be interpreted in terms of the associated atmospheric circulation modes.

5.2.2 *Orography*

Orographic effects on the regional and local climates of the North American continent are clearly evident in the most fundamental climatic elements—mean temperature, annual precipitation, snowfall amount, and snow-cover duration—as shown in standard climatographies and atlases (Bryson and Hare, 1974). The combined effects of altitude, local relief, orientation, and aspect result in terrain-related topoclimates and recurring mesoscale wind and weather systems that are detailed in numerous site-specific reports and case studies (Barry, 1992). The dynamic influence of orography is also apparent in the distribution of cyclogenesis, with important centers located east of the Rocky Mountains in Alberta and Colorado (Whittaker and Horn, 1981). On a regional scale, the eastward slope from the Rocky Mountains across the High Plains sets up the warm-season, nocturnal, low-level southerly jet. This is located about 500 m above the terrain and occurs north of the Mexican border at about 100°W, extending into the upper Midwest. The associated transport of moisture from the Gulf of Mexico fosters the nocturnal summer thunderstorms and precipitation excess over daytime totals observed over the central Great Plains (Higgins et al., 1997). Strong (> 20 m s^{-1}) jets played a major role in the transport of moisture into the upper Midwest of the central United States during the 1993 record summer floods (Arritt et al., 1997).

In the cold season, the Rocky Mountains often cause lee troughs to form during upper-level northwesterly flow. Surges of cold air propagate southward east of the mountains with strong northerly winds and sharp temperature decreases (Colle and Mass, 1995). These outbreaks of polar air affect the southern United States (northers) and western Gulf of Mexico (el norte) during winter. The shallow continental polar airflow, with overlying northeasterly winds, is unable to climb the slopes of the Sierra Madre Oriental of Mexico and is deflected into the southern Gulf of Mexico until it can cross the isthmus (Mosiño Alemán and García, 1974). Cold-air damming also occurs east of the Appalachian Mountains in winter and early spring, with high pressure over New England and a low over the southeastern United States generating an along-barrier pressure gradient and advection of cold air southward.

5.2.3 *Air Masses and Biomes*

Air-mass analyses have been widely used to identify continent-wide and regional climatic features. One of the most comprehensive studies is that of Bryson (1966), who used a combination of mean streamlines and partial collectives of daily maximum air temperature frequency distributions (see Barry and Perry, 1973) to define air mass types and frequencies. Figure 5.5 illustrates the areas dominated by the major air-mass types in summer. The median line of individual air-mass frequencies denotes the average climatological frontal boundaries. This approach was extended to the high terrain of the western United States by Mitchell (1976), using potential temperature analyses. The most significant aspect of these investigations is the demonstration of a close spatial association between air-mass boundaries (the median location) and the limits of several major biomes. The southern limit of tundra in North America is closely approximated by the summer location of Bryson's Arctic Front, for example, a pattern confirmed for Eurasia by Krebs and Barry (1970). Hare and Ritchie (1972) showed that the annual surface net radiation decreases from about 65–80 Wm^{-2} (Watts per square meter) over the boreal forest to 13–20 Wm^{-2} over the tundra, with a concomitant drop in thaw season duration from 180 days to 100–120 days. However, the surface-atmosphere interactions that this spatial association represents may involve not only temperature and net radiation influences on the vegetation, but also the effect of surface cover type on turbulent transfers

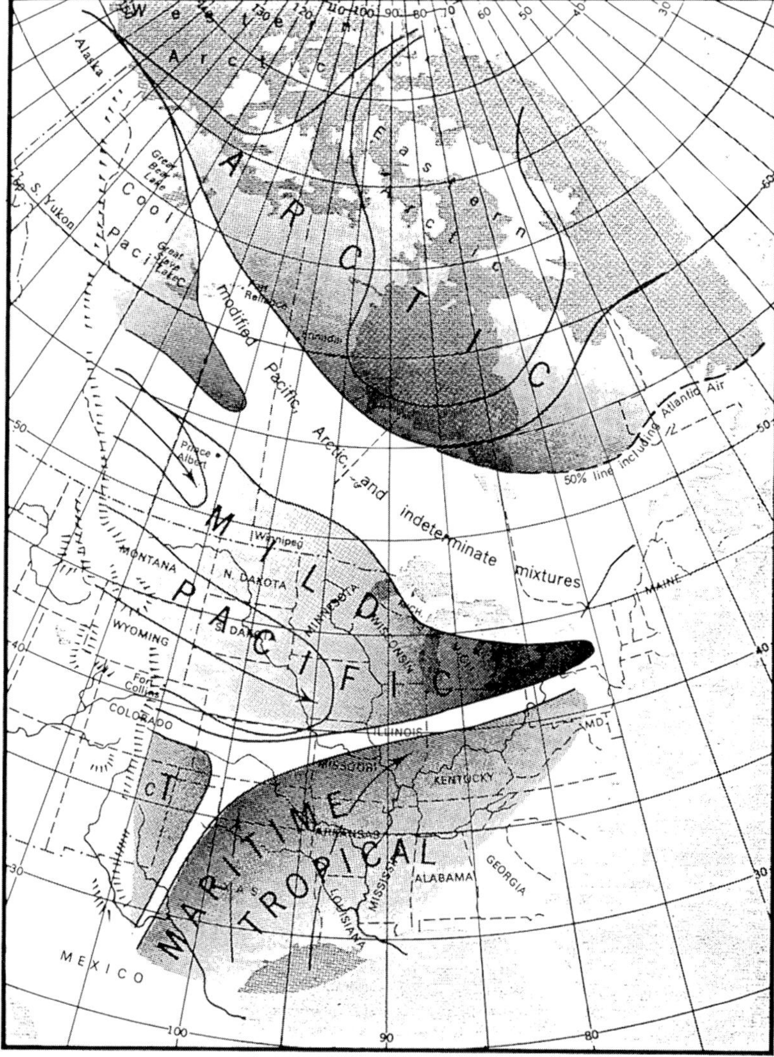

Figure 5.5 The frequency of air mass types over North America in summer as identified by R.A. Bryson (from Bryson, 1966)

to the atmosphere, thereby helping to determine the air-mass characteristics (Pielke and Vidale, 1995). They show that a paradigm shift has occurred concerning atmosphere-biosphere interactions. Rather than the atmosphere (air-mass frequencies and surface energy budgets) driving the location of the boreal forest and tundra transition, the pattern of vegetation forces the atmospheric regime, or at least exerts profoundly significant feedback effects on it.

5.3 Synoptic Regimes

5.3.1 Cyclogenesis and Storm Tracks

Much of the year-round weather is determined by transient synoptic and mesoscale disturbances. Midlatitude cyclones forming or deepening over the North Pacific Ocean commonly move toward the west coast of North America along tracks just south of the Aleutian Islands and then into the Gulf of Alaska. Another path is directed across the Pacific Ocean toward either the Gulf of Alaska or Vancouver Island (Bryson and Hare, 1974; Zishka and Smith, 1980). Precipitation totals, especially in the cold season, are large all along the coastal mountains and trailing cold fronts may bring precipitation far to the south in California. Some of these systems draw additional moisture from the subtropical Pacific, particularly during ENSO warm events, when a rapid succession of storms brings frequent, often intense, precipitation to California and northern Baja California. The noteworthy storms of winters 1982–83 and 1997–98, for example, brought widespread flooding and mass movement throughout California.

If Pacific storm systems cross the Western Cordillera, they do so as upper-level disturbances. In winter, there are two principal tracks for cyclones crossing the continent. One locus originates over Alberta and the other over eastern Colorado, both in the lee of the Rocky Mountains.

The continental systems form or regenerate mostly through lee cyclogenesis (Whittaker and Horn, 1981). Generally, the Alberta and Colorado systems move eastward toward the Great Lakes and Newfoundland. Systems from the Gulf of Mexico on another track toward the east coast also move northeastward toward Newfoundland. Some of these may develop explosively into intense systems over the Atlantic. However, the 12–14 March 1993 superstorm (Kocin et al., 1995) deepened explosively over the Gulf of Mexico and the southeastern United States. It brought widespread heavy snowfall, coastal flooding, and severe thunderstorms and tornadoes (Halpert and Smith, 1994). The east coast storms continue toward Iceland or recurve northward into Baffin Bay. Departures from these mean patterns are associated with shifts in the wave structure noted previously.

Anticyclones in winter are most common over the Mackenzie–Yukon area and the Great Basin of the southwestern United States. The cold anticyclones of the northwest may spawn traveling high-pressure cells, which move into the northern plains causing outbreaks of cold air. The Great Basin high pressure is a more static feature.

In summer, cyclone activity is displaced well north. Pacific cyclones affect primarily northern British Columbia and Alaska. Systems forming over Alberta move eastward across Canada and recurve northeastward toward Davis Strait. Cyclogenesis continues to occur along the eastern United States coast, but the systems are much less intense than in other seasons of the year.

Between June and November, the Gulf of Mexico and southeastern United States are affected by Atlantic tropical disturbances (Pielke, 1990). There are, on average, nine hurricanes a year with maximum sustained winds exceeding 25 m s^{-1}. September is the peak month, averaging about three disturbances of tropical storm intensity. Systems approaching Florida or the Gulf Coast typically recurve northward and eastward, steered by the upper-level circulation on the western end of the subtropical anticyclone (see chapter 21). Although most systems decay rapidly over land, a few may acquire extratropical baroclinicity and cause heavy rainstorms across the eastern United States and even southern Ontario. Tropical storms are equally numerous in the eastern tropical Pacific where they move northwestward along the coast of Mexico. Primarily, these represent storms that cross the narrow isthmus of southern Mexico and Central America and reintensify. They affect the west coast of Mexico, notably in August and September, and may bring warm, humid conditions and occasional thunderstorms to the otherwise dry southern California and the desert Southwest.

5.3.2 Mesoscale Weather Systems

The spring and summer climate of the United States east of the Rocky Mountains is also strongly affected by mesoscale convective systems (MCSs) that involve either a cluster of convective thunderstorm cells or a squall line of such cells extending several hundred kilometers. These bring severe weather with rain, hail, lightning, and tornadoes. During their roughly 12-hour life cycle, beginning in the early evening, such systems may travel from eastern Colorado to the Mississippi River or the Great Lakes, or from the Missouri–Mississippi River valley to the east coast (Easterling and Robinson, 1985). MCSs account for between 30 and 70% of April–September rainfall over most of the area from the Rocky Mountains to the Missouri River (Fritsch et al ., 1986). The frequency of these systems and their prominent contribution to summer rainfall appear to be comparable in both "normal" and drought years. Nevertheless, the average precipitation area and the volumetric totals are substantially less in a drought year, such as 1983, than in a "normal" year like 1982.

The role of mesoscale convective systems in the Midwest climate is also reflected in the spatial distribution of thunderstorm activity. For example, in 1993 the maximum annual cloud-to-ground flash density (about 11–13 flashes per square kilometer) within the contiguous United States was observed over the Midwest. Lightning is most frequent during June through August, and in the summer of 1993 it occurred in association with abnormal summer moisture advection (discussed subsequently). The flash density over this area is high most years, although data for 1989–95 show maximum flash densities located over Florida in 5 of the 7 years (Orville and Silver, 1997). In Florida, lightning is associated mainly with individual convective cells or with convective bands set up by the convergence of sea breezes over the peninsula.

5.3.3 Synoptic Classifications

Numerous regional studies of North America have used synoptic pressure pattern and air-mass classification methods. Overviews are provided by Barry and Perry (1973) and Yarnal (1993). Figure 5.6 shows a categorization of the various approaches (Kalkstein et al., 1996). They include early manual studies for Labrador-Ungava (Barry, 1960), Baffin Island (Barry et al., 1975), Mexico (Mosiño Alemán, 1964), and the Gulf Coast (Muller, 1977); manually derived types for the United States based on the classical Norwegian cyclone model (Yarnal, 1993); "objective" methods using the Kirchhofer sums-of-squares approach for Alaska (Moritz, 1979), the Canadian Arctic Archipelago (Bradley and England, 1979), western North America (Barry *et al.*, 1981) and the Pacific Northwest (Yarnal, 1985); and principal components analysis combined with cluster analy-

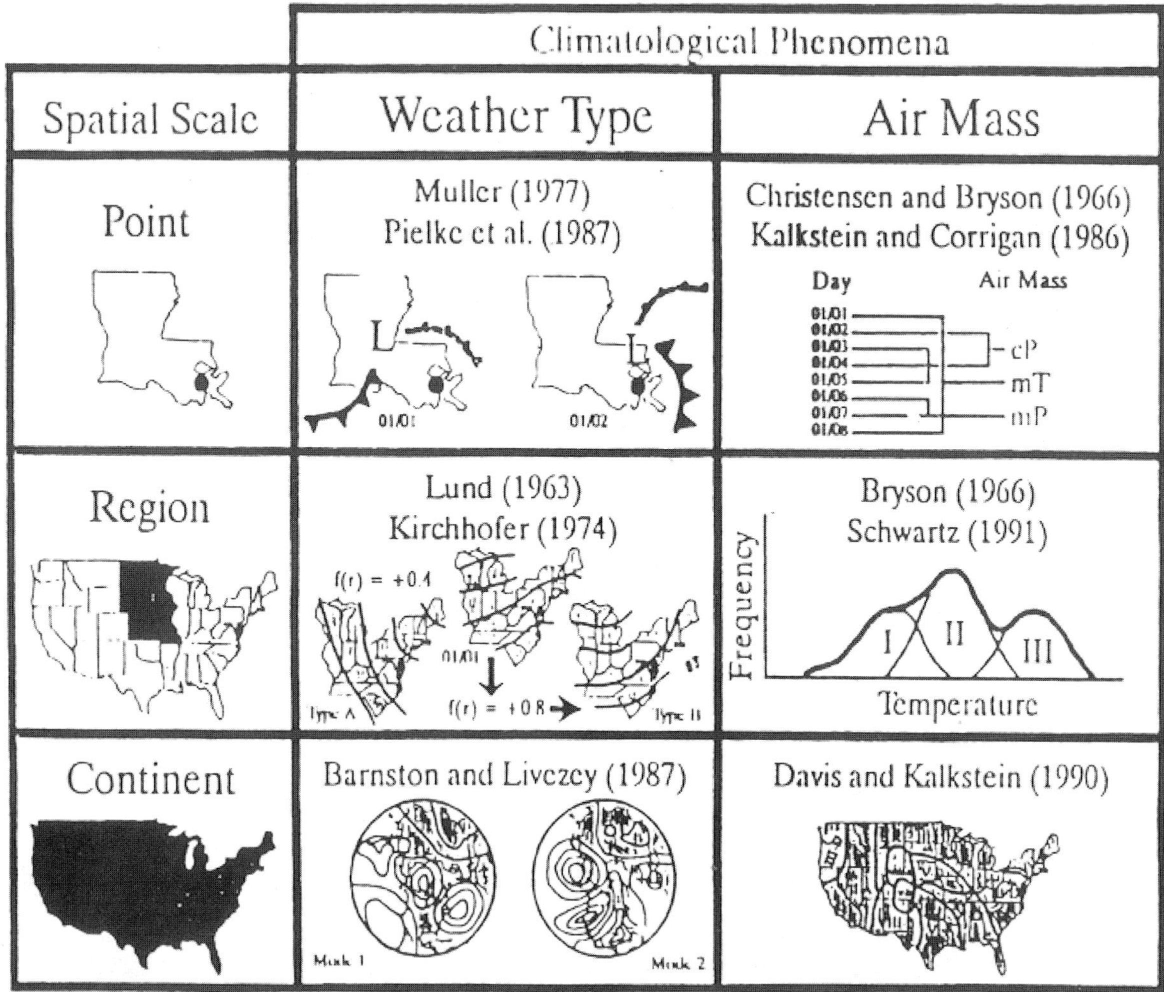

Climatological Phenomena
Spatial Scale
Weather Type
Air Mass
Point
Muller (1977)
Pielke et al. (1987)
Christensen and Bryson (1966)
Kalkstein and Corrigan (1986)
Day
Air Mass
cP
mT
mP
Region
Lund (1963)
Kirchhofer (1974)
Bryson (1966)
Schwartz (1991)
Frequency
Temperature
I
II
III
Continent
Barnston and Livezey (1987)
Davis and Kalkstein (1990)

Figure 5.6 A categorization of synoptic approaches to climate analysis for the United States (Kalkstein et al., 1996).

sis for a zonation of the conterminous United States (Fovell and Fovell, 1993). Synoptic catalogs have found particular use in studies of regional climate, atmospheric transport of aerosols, and other environmental factors that are strongly affected by air-flow conditions and sources. However, many studies have focused on methodological issues or rather limited local or regional applications (Yarnal, 1993). Moreover, regional air-flow classifications are sometimes rendered inadequate for distinguishing distinctive climatic conditions as a result of within-type variability, which can arise when planetary-scale circulation regimes cause the mixing of data within one regional air-flow type.

For some purposes, synoptic categories that are thermodynamically homogeneous are desirable. Recent investigations by Schwartz (1991) and Kalkstein et al. (1996) focus on this problem. Kalkstein et al. (1996) make an initial identification of six major air masses (dry or moist variants of polar, temperate, and tropical) and their typical characteristics at a number of sites. A sample of "seed days" exhibiting these characteristics is used as input to a linear discriminant analysis to obtain a daily categorization of air masses. Kalkstein et al. (1996) illustrate this approach for the eastern United States. Schwartz (1991) uses the 850-mb temperature and dew point as the air mass identifiers and separates partial collectives from the curves of temperature frequency for the north-central United States. Schwartz and Skeeter (1994) develop a procedure to relate these air-mass categories to mean sea-level pressure and 500-mb zonal and meridional circulation patterns over North America and the western Pacific by determining a cluster of days for each air mass and then averaging the circulation fields. This work suggests a useful blending of procedures to account for the local thermodynamic and advective components of regional climate.

5.4 Regional Climatic Anomalies

5.4.1 *Droughts and Floods—El Niño and La Niña*

Droughts typically occur with persistent anticyclonic patterns, which result in subsiding, warm, dry air and above-average solar radiation. Displacements of the North Pacific and Bermuda high pressure cells over the continent are the usual proximate factors involved. The monthly frequency of severe or extreme drought conditions exceeds 15% in southern California, western Utah–northern Nevada, eastern Oregon, the High Plains from Wyoming to west Texas, and the Ohio Valley; maximum values of 25% occur in western Kansas and western Wyoming (Court, 1974).

With respect to excessive rainfall, atmospheric moisture budget analyses for the sector 30°–50°N, 80°–105°W shows that a typical Midwest summer cyclone provides a net moisture convergence of 3.6 mm per day and that, if there is no soil storage of precipitation, only 15 to 16 days with Midwest cyclones are sufficient to provide around 0.5 mm per day of runoff for the Mississippi River (Chen et al., 1996). For the five winter months, the net precipitation minus evaporation requires 30 typical cyclone days. The Midwest of the United States experienced unprecedented flooding in late spring and summer 1993. Up to twice the normal January–July precipitation fell, with point values having return periods of 100–500 years or more. The flood levels may represent a 1000-year event (see chapter 7). Strong, moist southwesterly airflows affected the area throughout the summer, with a quasi-stationary cold front extending from southwest to northeast over the region (Trenberth and Guillemot, 1996). The 1993 events were associated with a mature ENSO. In contrast, the April–July 1988 drought over the northern plains, upper Midwest, and Ohio Valley occurred during strong La Niña conditions and the displacement of the storm track northward into Canada.

Across the northern plains, April–October precipitation has been significantly above average during 20 of the 23 ENSO warm events since the late nineteenth century, whereas during the La Niña cold phase there was a strong decrease in May–August precipitation (Bunkers and Miller, 1996). Drought conditions in California, which are principally a factor during the winter, occur mainly with La Niña cold phases of ENSO. In southern California, six of the eleven ENSO events during the period 1950–82 were very wet, whereas the remaining four were near normal. Schonher and Nicholson (1989) show that during the wet cases there were large positive sea-surface temperature anomalies east of the International Date Line, whereas the near-normal cases had a more extensive broad warming. One extreme drought occurred with weak positive anomalies in the eastern tropical Pacific.

5.4.2 *Unusual Heat and Cold*

Cold-air outbreaks over the southeastern United States are clearly related to both planetary- and synoptic-scale antecedent conditions. Konrad (1996) identifies persistent positive mean sea-level pressure anomalies over western Canada 6 to 12 days before cold outbreaks, with negative 500-mb-height anomalies over the Great Lakes and cold air over the Midwest a few days before. Temperatures in the Southeast during such outbreaks are on the order of 10–15^0 C below the mean for the time of year; cold-air outbreaks are most common during December through February. These events occur less frequently than the cold surges over eastern Asia, where the Siberian winter anticyclone is a stronger and more persistent feature than the high-pressure cells over northwest Canada. Nevertheless, the cold outbreaks in the southeastern United States can have serious consequences for citrus cultivation. Rogers and Rohli (1991) show that the absence of citrus freezes during 1948–57, compared with the high frequencies from 1880 to 1909 and since 1977, can be related to changes in the PNA pattern and the mean 500-mb standing wave over North America.

Heat waves have only recently begun to receive attention following high mortality rates in Chicago during July 1995. The high temperatures were a result of subsidence in a warm core anticyclone, accompanied by high moisture, apparently caused by local evaporation, and exacerbated by urban heating (Kunkel et al., 1996). Until now, however, little attention has been given to either the synoptic climatology of such events on a regional basis or to the specific local and temporal factors that lead to excessive levels of heat stress. Other recent summers with heat waves and drought conditions in the Midwest were 1980 and 1988; similar conditions prevailed in Texas in July–August 1998. Lyon and Dole (1995) show that dynamic forcing of anomalous wave trains was important in the early stages of each Midwest event, reinforced later by low evapotranspiration rates resulting from anomalous heating.

5.4.3 *Climatic Hazards and Society*

To put meteorological hazards in their societal context, we note that collectively they caused about 6,000 fatalities and 50,000 injuries in the United States during the period 1975–94 (Forrest and Nishenko, 1996). Tornadoes, floods, and excessive heat accounted for 23%, 14%, and 11%, respectively, of the fatalities; tornadoes, severe winds, and hurricanes for 51%, 11%, and 10%, respectively, of the injuries. In terms of property damage (totaling between \$52 billion and \$520 billion for 1975–94), hurricanes and floods accounted for 42% and 33%, respectively, whereas crop damage (totaling between \$29 billion and \$292 billion) was attributed to floods (27%), droughts (26%) and hurricanes (20%). The tenfold range in the cost estimates is an artifact of the scaling used in these assessments. The costs resulting from Hurricane Andrew's impact on southeastern Florida and Louisiana in 1992 were a staggering $30 billion, with the next costliest storm, Hurricane Hugo in 1989, accounting for $8.5 billion (adjusted to 1996 dollars).

5.5 Conclusion

North America exhibits a wide variety of climatic conditions from polar desert and Arctic tundra in the far north to tropical forests in southern Mexico. In between are the subarctic boreal forest, the continental climate of the High Plains, and the southwestern subtropical deserts. There is a corresponding variety in the west–east direction, related to topography and distance from the Pacific Ocean and Gulf of Mexico (see chapter 6). These climatic differences and the different seasonal regimes are variously reflected in the patterns of landscape, vegetation cover, and land use across the continent. For example, the dry climate in much of the western half of the United States has led to a heavy dependence on water transfers over long distances (see Riebsame et al., 1998) and to the use of groundwater for local irrigation and consumption. Recurring droughts heavily impact farming and also encourage wildfires in the montane forests, where summer lightning storms, with or without precipitation, are frequent. In the central and northern Rocky Mountains, as well as in the Pacific Northwest, 60% or more of the annual precipitation falls as snow, giving rise to a runoff hydrograph that has a pronounced early summer peak. Across most of Canada and Alaska, the upper Midwest, and the northern states of the eastern United States, snow blankets the ground for 3–4 months per year, affecting all aspects of human activity. In contrast, in the Sonoran desert and northwestern Mexico, temperatures may top 38° C 10–20% of the time. The consequences of this variety of climatic and weather regimes for other aspects of the physical geography of North America are addressed in subsequent chapters.

References

Arritt, R.W., T.D. Rink, M. Segal, D.P. Todey, C.A. Clark, M.J. Mitchell, and K.M. Labas, 1997. The Great Plains low level jet during the warm season of 1993. *Monthly Weather Review*. 125(9): 2176–2192.

Bailey, W.G., T.R. Oke, and W.R. Rouse, editors. 1997. *The Surface Climates of Canada*, Montreal: McGill–Queen's University Press, 369 p.

Barry, R.G. 1960. A note on the synoptic climatology of Labrador-Ungava. *Quarterly Journal of the Royal Meteorological Society*, 86: 557–565.

Barry, R.G. 1992. *Mountain Weather and Climate*. London and New York: Routledge.

Barry, R.G. 1993. Canada's Cold Seas. In *Canada's Cold Environments*, ed. H.M. French and O. Slaymaker, 29–61. Montreal: McGill-Queen's University Press.

Barry, R.G. and A.H. Perry. 1973. *Synoptic Climatology: Methods and Applications*. London: Methuen.

Barry, R.G., J.D. Jacobs, and R.S. Bradley. 1975. Synoptic climatological studies of the Baffin Island area. In *Climate of the Arctic*, ed. G. Weller and S.A. Bowling, 82–90. Geophysical Institute, Fairbanks: University of Alaska.

Barry, R.G. G. Kiladis, and R.S. Bradley. 1981. Synoptic climatology of the western United States in relation to climatic fluctuations during the twentieth century. *Journal of Climatology* 1(2): 97–113.

Bradley, R.S., and J. England. 1979. Synoptic climatology of the Canadian High Arctic. *Geografiska Annaler* 61A: 187–201.

Brown, R.D., and B.E. Goodison. 1996. Interannual variability in reconstructed Canadian snow cover, 1915–1992. *Journal of Climate* 9(5): 1299–1318.

Bryson, R.A. 1966. Air masses, streamlines and the boreal forest. *Geographical Bulletin*, 8: 228–269.

Bryson, R.A., and F.K. Hare. 1974. The climates of North America. In *Climates of North America*, ed. R.A. Bryson and F.K. Hare, Vol. 11, World Survey of Climatology, 1–67. Amsterdam: Elsevier.

Bunkers, M.J., and J.R. Miller Jr. 1996. An examination of El Niño-La Niña–related precipitation and temperature anomalies across the Northern Plains. *Journal of Climate* 9(1): 147–160.

Burnett, A.W. 1994. Regional-scale troughing over the southwestern United States: Temporal climatology, teleconnections and climate impact. *Physical Geography*, 15: 80–98.

Carleton, A.M. 1987. Summer circulation climate of the American southwest, 1945–1984. *Annals of the Association of American Geographers* 77: 619–34.

Carleton, A.M. Carpenter, and P.J. Weser. 1990. Mechanisms of interannual variability of the southwest United States summer rainfall maximum. *Journal of Climate* 3(9): 999–1015.

Chen, T.-C., M.-C. Yen, and S. Schubert. 1996. Hydrological processes associated with cyclone systems over the United States. *Bulletin of the American Meteorological Society*, 77(7): 1557–1567.

Colle, B.A., and C.F. Mass. 1995. The structure and evolution of cold surges east of the Rocky Mountains. *Monthly Weather Review*, 123(9): 2577–2610.

Court, A. 1974. Climate of the conterminous United States. In *Climates of North America*. ed. R.A. Bryson and F.K. Hare, Vol. 11, World Survey of Climatology, 193–343. Amsterdam: Elsevier.

Davis, R.E., B.P. Hayden, D.A. Gay, W.L. Phillips, and G.V. Jones. 1997 The North Atlantic subtropical anticyclone. *Journal of Climate*, 10(4): 728–744.

Dewey, K.F. 1987. Snow cover–atmospheric interactions. In: *Large-Scale Effects of Seasonal Snow Cover, IAHS (International Association of Hydrological Sciences) Publication* No. 166, ed. B.E. Goodison, R.G. Barry, and J. Dozier, 27–42. International Association of Hydrological Sciences, Wallingford, U.K.

Dilley, M. 1996. Synoptic controls on precipitation in the valley of Oaxaca, Mexico. *International Journal of Climatology*, 16(9): 1019–1031.

Douglas, M.W., R.W. Madden, K.W. Howard, and S. Reyes.

1993. The Mexican monsoon. *Journal of Climate* 6(8): 1665–1677.

Driscoll, D.M., and J.M. Yee Fong. 1992. Continentality: A basic climate parameter reexamined. *International Journal of Climatology*, 12(2): 185–192.

Easterling, D.R., and P.J. Robinson. 1985. The diurnal variation of thunderstorm activity in the United States. *Journal of Climatology and Applied Meteorology*, 24: 1048–1058.

Forrest, B., and S. Nishenko. 1996. Losses due to natural hazards. *Natural Hazards Observer* (University of Colorado, Boulder) 21(1): 16–17.

Fovell, R.G., and M.-Y.C. Fovell. 1993. Climate zones of the conterminous United States defined using cluster analysis. *Journal of Climate* 6(11): 2103–2135.

Fritsch, J. M., R.J. Kan, and C.R. Chelius, 1986. The contribution of mesoscale convective weather systems to the warm-season precipitation in the United States. *Journal of Climate and Applied Meteorology*, 25: 1333–1345.

Groisman, P.Ya, T.R. Karl, and R.W. Knight. 1994. Observed impact of snow cover on the heat balance and the rise of continental spring temperature. *Science* 263: 198–300.

Halpert, M.S., and T.M. Smith. 1994. The global climate for March–May 1993: Mature ENSO conditions persist and a blizzard blankets the eastern United States. *Journal of Climate* 7(11): 1772–1793.

Hare, F.K., and J.C. Ritchie. 1972. The boreal bioclimates. *Geographical Review* 62: 333–365.

Harman, J.R. 1991. *Synoptic climatology of the westerlies: Process and patterns.* Washington, D.C. Association of American Geographers, 80 p.

Harrison, D.E., and N.K. Larkin. 1998. El Niño–Southern Oscillation sea surface temperature and wind anomalies, 1946–1993. *Reviews of Geophysics*, 36: 353–399.

Higgins, R.W., Y. Yao, E.S. Yarosh, J.E. Janowiak, and K.C. Mo. 1997. Influence of the Great Plains low-level jet on summertime precipitation and moisture transport over the central United States, *Journal of Climate*, 10(3): 481–507.

Hurrell, J.W. 1995. Decadal trends in the North Atlantic Oscillation: Regional temperatures and precipitation. *Science*, 269: 676–679.

Kalkstein, L.S., M.C. Nichols, C.D. Barthel, and J.S. Greene. 1996. A new spatial synoptic classification: Application to air-mass analysis. *International Journal of Climatology*, 16(9): 983–1004.

Kasahara, A. 1980. Influence of orography on the atmospheric general circulation. In *Orographic Effects in Planetary Flow*, ed. R. Hide and P.W. White, GARP Publ. Ser. 23, 1–49, World Meteorological Organization, Geneva.

Keables, M.J. 1992. Spatial variability of midtropospheric circulation patterns and associated surface climate in the United States during the ENSO winters. *Physical Geography*, 13: 331–348.

Kiladas, G.N., and H.F. Diaz. 1989. Global climatic anomalies associated with extremes of the Southern Oscillation. *Journal of Climate* 2: 1069–1090.

Kocin, P.J., P.N. Schumacher, F.F. Morales Jr., and L.W. Uccellini. 1995. Overview of the 12–14 March 1993 superstorm. *Bulletin of the American Meteorological Society*, 76(2): 165–199.

Konrad, C.E., II. 1996. Relationships between the intensity of cold-air outbreaks and the evolution of synoptic and planetary-scale features over North America. *Monthly Weather Review*, 124(6): 1067–1083.

Kousky, V.E. 1993. The global climate of December 1991–February 1992: Mature-phase warm (ENSO) episode conditions develop. *Journal of Climate* 6(8): 1639–1655.

Krebs, S.J., and R.G. Barry. 1970. The Arctic front and the tundra-taiga boundary in Eurasia. *Geographical Review*, 60: 548–54.

Kunkel, K.E., S.A. Chagnon, B.C. Reimke, and R.W. Arritt. 1996. The July 1995 heat wave in the Midwest: A climatic perspective and critical weather factors. *Bulletin of the American Meteorological Society*, 77(7): 1507–1518.

Leathers, D.J., B. Yarnal, and M.A. Palecki. 1991. The Pacific/North American teleconnection pattern and United States Climate: Part I: Regional correlations. *Journal of Climate* 4 (5): 517–528.

Leathers, D.J. and M.A. Palecki. 1992. The Pacific/North American teleconnection pattern and United States Climate: Part II: Temporal characteristics and index specification. *Journal of Climate* 5 (7): 707–716.

Lyon, B., and R.M. Dole. 1995. A diagnostic comparison of the 1980 and 1988 U.S. summer heat wave-droughts. *Journal of Climate* 8(6): 1658–1675.

Mitchell, V. 1976. The regionalization of climate in the western United States. *Journal of Applied Meteorology*, 15(9): 920–927.

Montory, D.L. 1997. Linear relations of central and eastern North American precipitation to tropical Pacific surface temperature anomalies. *Journal of Climate*, 10(6): 541–558.

Mosiño Alemán, P.A. 1964. Surface weather and upper-airflow patterns in Mexico. *Geofisica International*, 4(3): 117–168.

Mosiño Alemán, P.A. and E. García. 1974. The climate of Mexico. In *Climates of North America.* ed. R.A. Bryson and F.K. Hare, Vol. 11, World Survey of Climatology, 345–404. Amsterdam: Elsevier.

Moritz, R.E. 1979. Synoptic climatology of the Beaufort Sea coast. *Occasional Paper* no. 30, Institute for Arctic and Alpine Research, University of Colorado, Boulder.

Muller, R.A. 1977. A synoptic climatology for environmental baseline analysis: New Orleans. *Journal of Applied Meteorology*, 16: 20–33.

Orville, R.E., and Silver, A.C. 1997. Lightning ground flash density in the contiguous United States, 1992–95. *Monthly Weather Review*, 125: 631–638.

Pielke, R.A. 1990. *The Hurricane.* London: Routledge.

Pielke, R.A. and P.L. Vidale. 1995. The boreal forest and the polar front. *Journal of Geophysical Research*, 100(D12): 25,755–25,758.

Riebsame, W.E. (ed.) with H. Gosnell and D. Theobald, 1998. *Atlas of the New West*, University of Colorado, Center of the American West, Boulder.

Rogers, J.C. 1981. The North Pacific Oscillation. *Journal of Climatology* 1(1): 39–85.

Rogers, J.C., and Rohli, R.V. 1991. Florida citrus freezes and polar anticyclones in the Great Plains. *Journal of Climate* 4(11): 1103–1113.

Schonher, T., and S.E. Nicholson. 1989. The relationship between California rainfall and ENSO events. *Journal of Climate* 2(11): 1258–1269.

Schwartz, M.D., 1991. An integrated approach to air mass classification in the north central United States. *Professional Geographer*, 43: 77–91.

Schwartz, M.D., and B.R. Skeeter. 1994. Linking air-mass analysis to daily and monthly midtropospheric flow patterns. *International Journal of Climatology*, 14(4): 439–464.

Serreze, M.C., M.P. Clark, D.L. McGinnis, and D.A. Robinson, 1998. Characteristics of snowfall over the eastern half of the United States and relationships with principal modes of low-frequency atmospheric circulation. *Journal of Climate*, 11(2): 234–250.

Trenberth, K.E., and C.J. Guillemot. 1996. Physical processes involved in the 1988 drought and 1993 floods in North American. *Journal of Climate* 9(6): 1288–1298.

Wallace, J.M. 1983. The climatological mean stationary waves: Observational evidence. In *Large-Scale Dynamical Processes in the Atmosphere*, ed. B. Hoskins and R. Pearce, 27–53. New York: Academic Press.

Wallace, J.M., and M.P. Blackmon. 1983. Observations of low-frequency atmospheric variability. In *Large-Scale Dynamical Processes in the Atmosphere*, ed. B. Hoskins and R. Pearce, 55–94. New York: Academic Press.

Wallace, J.M., and D.S. Gutzler. 1981. Teleconnections in the geopotential height field during the Northern Hemisphere winter. *Monthly Weather Review* 109(4): 784–812.

Whittaker, L.M., and L.H. Horn. 1981. Geographical and seasonal distribution of North American cyclogenesis, 1958–1977. *Monthly Weather Review*, 109(11): 2312–2322.

Woodhouse, C.A. 1997. Winter climate and atmospheric circulation patterns in the Sonoran Desert region, USA *International Journal of Climatology*, 17(8): 859–873.

Yarnal, B. 1985. A 500 mb synoptic climatology of Pacific north-west coast winters in relation to climatic variability, 1948–49 to 1977–78. *Journal of Climatology*, 5(3): 237–352.

Yarnal, B. 1993. *Synoptic Climatology in Environmental Analysis*. London: Bellhaven Press.

Yarnal, B. and H.F. Diaz. 1986. Relationships between extremes of the Southern Oscillation and the winter climate of the Anglo-American Pacific coast. *Journal of Climatology*, 6(2): 197–219.

Zishka, K.M., and P.J. Smith. 1980. The climatology of cyclones and anticyclones over North America and surrounding ocean environs for January and July 1950–77. *Monthly Weather Review*, 108(4): 387–401.

6

Climatic Regionalization

John E. Oliver

Captain John Smith proved that he was much more than a colonial adventurer when he wrote the following account of Virginia in 1607:

> The sommer here is hot as in Spaine; the winter colde as in Fraunce or England. The heat of sommer is in June, Julie, and August, but commonly the coole Breeses asswage the vehemencie of the heat. The chiefe of winter is halfe December, January, February, and halfe March. The colde is extreame sharp, but here the proverb is true that no extreame long continueth (Tyler, 1907).

In this succinct paragraph, this early observer outlined the essence of climatic regionalism. First, he identified a baseline: the previously experienced conditions encountered in Spain, France, and England; then, by comparing the nature of the seasons of the new area with those already known, Smith characterized the climatic identity of early Virginia.

This climatic identity differentiates one climatic region from another. But to recognize exactly what separates one regional identity from another is not a simple task. The climatic elements of any region distinguish that region not by their presence or absence, but by a difference in their character. Climatic elements do vary systematically from place to place, sometimes rather abruptly but mostly over considerable distance so that boundary definition is often arbitrary. Adding to the problem is that climate is an abstract concept that represents the summation of all interacting atmospheric processes at a location over a stated, usually lengthy, period of time. As such, it does not exist at any given moment. Thus, although it is necessary to systematize the long-term effect of interacting atmospheric processes, the manner in which they may be grouped is variable. As such, the objective of this contribution is to provide a rational description of the climatic regions of North America and to examine their climatic identities.

6.1 Changing Perceptions

Climatologists today are in the fortunate position of having available an enormous amount of climatic data. The development of the World Wide Web and CD-ROMs has opened new vistas for those who use climatic data. Of course, this has not always been the case, with the result that perceptions of climates of North America have changed over time. The charitable view of the early Norse explorers, who found Greenland and Vinland nominally attractive, differed from the perceptions of nineteenth century Russians, whose image of Alaska as a chaos of snow-covered mountains and frigid climates induced them to sell off rather cheaply. The descriptions of Ponce de Leon's discovery of Florida and Coronado's quest for gold in the desert Southwest gave images of lands and climates of fascinating diversity. In Canada, the searches for Sir John Franklin and the Northwest Passage gave rise to the portrait of the "cold and tragic shores." In fact, the story of the exploration of North America is often one of untold hardships in harrowing climatic conditions and ever-changing perceptions (Goetzmann and Williams, 1992).

A fine example of such changing perceptions of climate is the emergence of the "Great American Desert." Until the Civil War (1861–65), the public was convinced that a Great American Desert lay between the Missouri River and the Rocky Mountains. Between 1820 and 1850, this desert was widely reproduced in atlases, texts, and histories. The idea was implanted by such explorers as Zebulon Pike and Stephen Long, the latter being the first to use the term "Great Desert." The concept was not put to rest until the railroad surveys of the 1850s and the settlement of the Great Plains. However, even before this time, the U.S. Army set up many forts across the United States and collected climatic data. One of the first maps using the acquired data is shown in figure 6.1. Published in Forry's 1842 text, it provides average summer (isothural), winter (isocheimal), and average (isothermal) temperatures. The text itself is testimony to the perceived relationships between climate and health, for its major focus was to provide the medical conditions as related to the climatic laws of the period. Forry classified the climate of the United States using the data he had available. Although simple (based on northern, middle, and southern divisions), it was a major step toward understanding the climate of the United States.

Another example of changing ideas for regionalization can be seen in the history of climate divisions across the United States as identified by the Weather Bureau (under a variety of names). Such regions are used to enable data collection and compilation of applied features such as drought indices. Currently, the National Climate Data Center (NCDC), which maintains the data set, uses 344 divisions in the contiguous United States. The boundaries are mostly structured to correspond to county boundaries so that each state is covered fully. Unlike the 344 units of today, those of 1908 identified 12 climatological sections, conforming to the 12 principal drainage basins of the United States. Nonpolitical boundaries were abandoned in 1914 and replaced by 106 climatology divisions designed to overcome the difficulty of dissemination of data. Further divisions over time led to the current form (Guttman and Quayle, 1995), the divisions being used to distinguish regional climate characteristics.

A recent addition to the climatic/political division of the United States is the location of six regional climate centers that are administratively associated with the Climate Analysis Center of the National Weather Service.

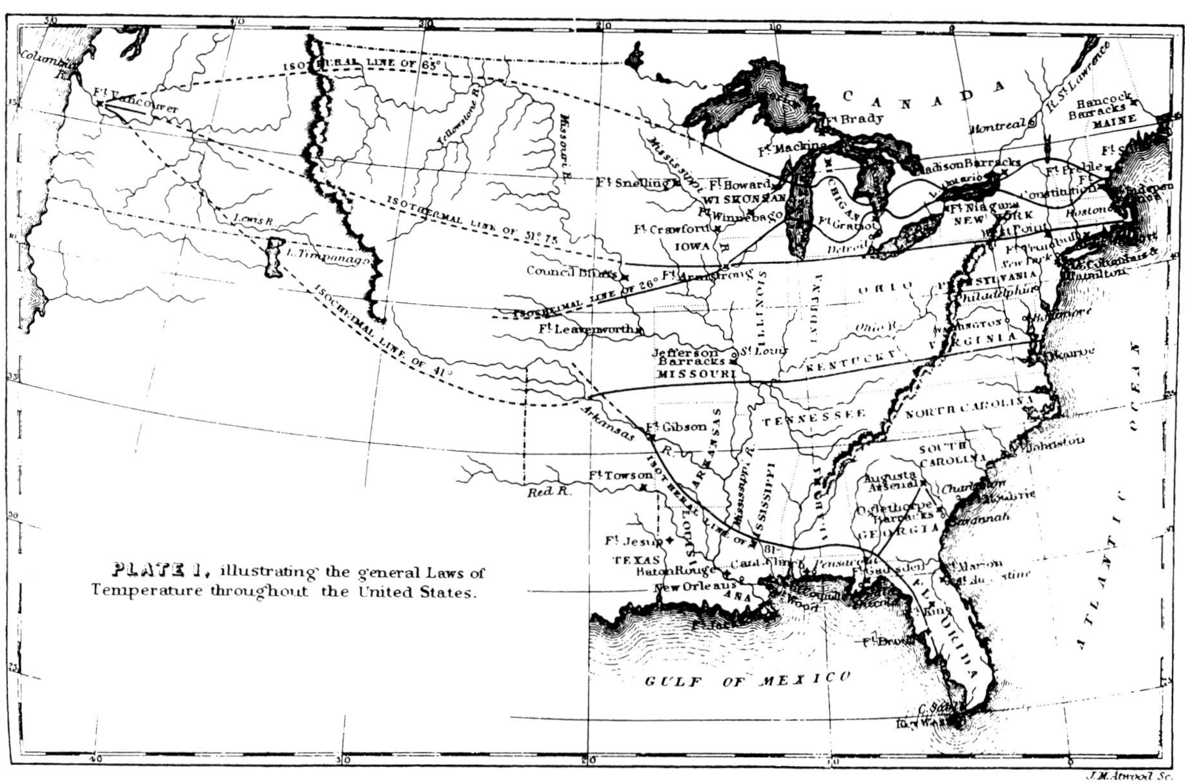

Figure 6.1 Temperature regions of the United States as identified in 1842 (from Forry, 1842).

These centers are the Western Center located at the Desert Research Institute in Nevada, the High Plains Center at the University of Nebraska, Lincoln, the Midwestern Center at the Illinois State Water Survey, the Southern Regional Center at Louisiana State University, Baton Rouge, the Northeast Center at Cornell University, and the Southeastern Center at the South Carolina Water Commission. Each of these centers is expected to monitor climatic variability, interpret climatic impacts, and provide services for the states that they represent (e.g., Muller et al., 1990). The division of the country into six climatic regions using state boundaries is in itself an interesting approach to regionalization (Changnon et al., 1990). Climates of Canada have also been described on a political basis using province boundaries (Phillips, 1990).

The role of modern communications in providing regional perceptions of climate is well seen in the coverage of the 1997–98 El Niño event. Any adverse climatic or hydrologic event that happened on the west coast was immediately attributed to this, whereas, right or wrong, other parts of the nation were informed that El Niño might ease, for example, the rigors of winter. A regional consciousness of climate appears to have been revived by the many reports that made El Niño one of the few climatic terms that are widely known (Glantz, 1996).

6.2 Identifying the Regions

To describe successfully the variety of climates in North America, it is necessary to identify and group similar types. This may be done in many ways. Climate may, for example, be divided politically as in the identification of the regional climate centers in the United States. Perhaps the most fundamental regional division is to separate climates east of the Rocky Mountains from those to the west. Such a division, as simple as it seems, allows some climatic features to be discussed in a meaningful way. Similarly, when considering the fate of runoff from precipitation falling over the continent, climates east and west of the continental divide may be differentiated. Consider too, for example, the hazard of tornadoes. Although tornadoes have been observed in every state, the Great Plains region of the United States is the foremost region for tornado formation in the world. Cold air advancing from the Canadian source, warm, moist air from the Gulf of Mexico, a source of dry continental air, and a stream of cold, dry air above this mix provide the perfect physical setting (Marshall, 1992). But when we consider the hazard of tornadoes as a whole, even though the Great Plains states are foremost in the number formed, we see in table 6.1 that they do not necessarily have the most fatalities or damage. Texas, largely because of its size, leads the list in all three categories, but, in other states, no exact correlation exists among number of tornadoes, fatalities, and injuries. The states shown in the list of "top ten" indicate the east–west division. A regional map of high tornado hazard would reflect this distribution (Morgan and Moran, 1997).

Another simple way to regionalize climate is to identify a single important climatic element and use it as the basis for the grouping. A well-chosen approach should certainly provide clear images of what makes up at least one part of the climate. Given that temperature and precipitation are used by most people to characterize the climate of the place in which they live, these two variables serve as apt examples.

6.2.1 *Single Variable Groups*

A highly generalized portrait of the continent is shown in temperature distribution maps, which are widely available in texts and atlases. The isothermal patterns exhibit dif-

Table 6.1 Tornado Data for the United States: 1950–94

Number of Tornadoes			Fatalities			Injuries		
RANK	STATE	TOTAL	RANK	STATE	TOTAL	RANK	STATE	TOTAL
1	TX	5490	1	TX	475	1	TX	7452
2	OK	2300	2	MS	386	2	MS	5344
3	KS	2110	3	AR	279	3	AL	4483
4	FL	2009	4	AL	275	4	OH	4156
5	NE	1673	5	MI	237	5	AR	3697
6	IA	1374	6	IN	218	6	IN	3641
7	MO	1166	7	OK	217	7	IL	3599
8	SD	1139	8	KS	199	8	MT	3214
9	IL	1137	9	IL	182	9	OK	3184
10	CO	1113	10	TN	181	10	GA	2662

Source: Storm Prediction Center/National Climate Data Center (www.ncdc.noaa.gov).

ferences between coastal regions and those in the center of the continent, where a large summer-to-winter-temperature range occurs. Study of the thermal gradients indicates that winter is a time when a north-south traverse of the continent would take a traveler from temperatures less than −30°C to perhaps 20°C, a range of 50C°. In summer, a range of perhaps 30°C may be experienced.

A large number of climate groupings based on temperature have been devised, often expressed in a form or an index derived from temperature data for applied use. Often, a significant temperature is selected, and by mapping its distribution a wealth of information may be obtained about temperature-related conditions. For example, regions could be identified based on the distribution of the length of the growing season, and these could be related to agricultural patterns (Santibáñez, 1994). Such specific-use maps can identify all types of regions, particularly those that can occur over large areas of the continent.

Figure 6.2a provides the pattern of annual average precipitation totals for North America. It is an interesting pattern; areas with less than an average total of 50 cm extend from the northern fringes of the Arctic to the arid southwest. In contrast, some regions of the Pacific rim receive more than 200 cm. The annual total alone, however, does not prove a complete key to the nature of the precipitation region identified. In some respects, the time of the year that precipitation occurs is almost as important as the amount. For example, rainfall during the high sun season is less effective because of high evaporation rates; further, winter precipitation may occur as snow and not become run off until the spring thaw. As a result, many schemes for distinguishing the precipitation regimes of North America have been devised. Some authors use statistical methods to define regions, others identify regions based on some sort of precipitation index. In figure 6.2b, a regionalization based on precipitation regime is shown. This is a simple but effective classification that is based on a system suggested a number of years ago by Kendrew (1961). North America is divided into 11 major regions, depending on both average annual totals of precipitation and, significantly, the season at which maxima and minima occur. The regions are briefly described in table 6.2.

6.2.2 Global Systems

Temperature and precipitation can be combined to identify regions, a combination often used in world climate classification schemes, which, in turn, may then be applied to continental areas. Of the global climate classification systems currently in use, that developed in 1918 by climatologist Wladimir Köppen probably is the best known. Based essentially on early vegetation classifications, the Köppen system provides temperature values to differentiate tropical, midlatitude, and polar climates and uses precipitation to identify deserts and semideserts. The complete Köppen system is described in many texts (e.g.,

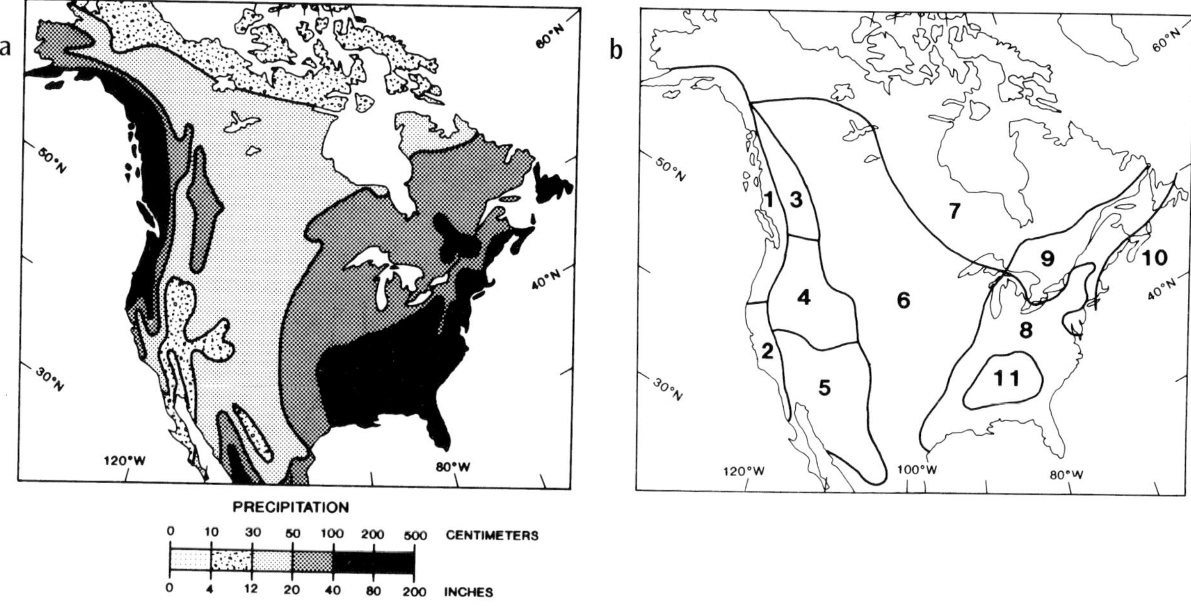

Figure 6.2 (a) Regions of precipitation described by total annual precipitation distribution; (b) Precipitation regions based on seasonality. Table 6.2 provides the explanation of identified regions (after Kendrew, 1961).

Table 6.2 Legend for Precipitation Regions shown in figure 6.2b

1. Pacific Coast Type	A strong winter precipitation maximum resulting from cyclonic rainfall and enhanced by topography.
2. California Type	Like the preceding type, this too has a strong winter maximum, but a very dry summer and a much lower annual total. A Mediterranean-type regime.
3. British Columbia Transition	Winter precipitation in the west is replaced by a summer maximum in the east. Whereas spring is relatively dry, precipitation is fairly evenly distributed throughout the year.
4. Snake River Type	South of the British Columbia transition, the coastal influence is stronger, so this region receives more precipitation in the winter. Amounts are variable, both from east to west and from north to south.
5. Arizona Type	The driest tract in the continent, this desertic area receives its limited rain during two periods, late winter and late summer, the latter from a regional monsoon effect.
6. Plains Type	Early summer precipitation occurs over most parts of this enormous area extending from subarctic Canada to central Mexico and from the Rockies to the Great Lakes. Amounts are quite variable, from place to place and from year to year.
7. Hudson Bay Type	In the north and northeast of Canada, frozen ground delays the rise in spring temperature with the result that maximum precipitation occurs in early autumn.
8. Gulf Type	With rainfall abundant during all seasons, the maximum occurs in late summer. This type dominates the Atlantic and Gulf coasts and extends inland to the Great Lakes.
9. St. Lawrence Type	Characterized by remarkable precipitation uniformity throughout the year. This distribution, together with that of the next region, is somewhat anomalous for east coast climates at this latitude.
10. Nova Scotia Type	The pronounced winter precipitation maximum here is a result of vigorous cyclonic activity influenced, in part, by the Great Lakes, the Gulf of St. Lawrence, and the warm North Atlantic Drift.
11. South Appalachian Type	This area has more rain in winter than summer, which differentiates it from the Gulf Type.

After Kendrew (1961).

Hidore and Oliver, 1993) and is not described in detail herein.

Many classifications, like that of Köppen, use observed variables, such as temperature and moisture, to identify regions. Others, often referred to as genetic classifications, are based on the cause of the observed climate. Such divisions distinguish regions by air-mass frequency. These classifications mostly do not provide numerically defined boundaries, but rather present an image of the different causes of the North American regional climates. Strahler and Strahler (1996) provide an excellent example of this type of classification, as does the world grouping suggested by Lauer et al. (1996). Another widely used approach to comprehending the climate of a region is through water budget analysis. In this, the various components of the hydrologic cycle are evaluated to derive indices. The best known of such classifications is that formulated by Thornthwaite and used to provide the distribution of various components of the water balance (Mather, 1985).

A very useful set of groupings is given by what may be called "special purpose classifications." These are constructed to show distributions of applied climatic variables and indices. Thus it is possible to identify regions based on a host of groupings that range from include air-pollution potential (Davis and Kalkstein, 1990) to human physiological stress (Kalkstein and Valimont, 1986). The list is extensive, and each classification in a special way is an important informational set relating to the climate of a region. In some instances, the grouping of climates based on the selected variables is completed through statistical analysis using clustering techniques (Kalkstein et al., 1987). The approaches available for such analyses are well described by Balling (1984).

6.3 The Regions

In his thorough description of the climates of the United States, Court (1974) stated:

> Three of the world's five major climatic types occur over extensive portions of the conterminous United States, and a fourth, the cold or polar, is found in higher mountains; only the equatorial rainy type is absent. The western deserts are dry, the Pacific coast and the southeastern quarter of the country are warm, and the northern third is cool. Further subdivisions have been the subject of many fruitless discussions of various proposed classifications.

Although this is perhaps an extreme approach to climatic regionalism, the point that emerges is that an overall description of the climates of North America can, and perhaps should, be given in relatively simple form. Such a course is followed herein.

Almost any of the classification systems mentioned thus far could be used to display a meaningful grouping of the regions of North America. For discussion purposes, the distribution shown in figure 6.3 is selected. The regions identified are based essentially on the Köppen grouping, but do not attempt to use any of the keys or codes of that system. Instead, a set of descriptive terms is applied, selected to characterize the climate and the location. A quick look at the map immediately indicates some of the problems of using any classification scheme to describe a continent. If the Köppen classification were used with no changes, the area designated as High Plains would be grouped in the same type as the semidesert area of the intermontane region. The two areas do have a similar climatic regime, but they are differentiated here. Consider, too, the eastern boundary of the High Plains as shown here. This boundary separates the humid climates of North America (the C and D climates of Köppen) from the dry (B climates), and the somewhat arbitrary nature of its location may be demonstrated. If, instead of using the 30-year mean temperature and precipitation values to define the boundary, each individual year is classified (either as arid or humid), then a zone rather than a single line will be identified. In figure 6.4, the extreme limits of the B/H boundary for two 4-year periods are shown. In both periods, the boundary changes over a fairly broad area, and, even for these short intervals, the limits vary. This simple analysis indicates that at any location within the broad boundaries, some years with be classed as arid, others will be labeled humid.

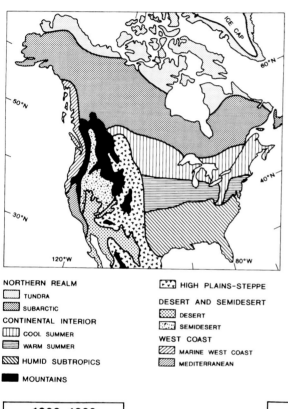

Given that it is not feasible to describe in detail each area identified in figure 6.3, the following account provides an overview of the core conditions of each region and some of their vulnerabilities. More information is available in a number of publications concerning the regions emphasized. Some sources, for example, Lydolph (1985), provide an overview of the entire continent, whereas others de-

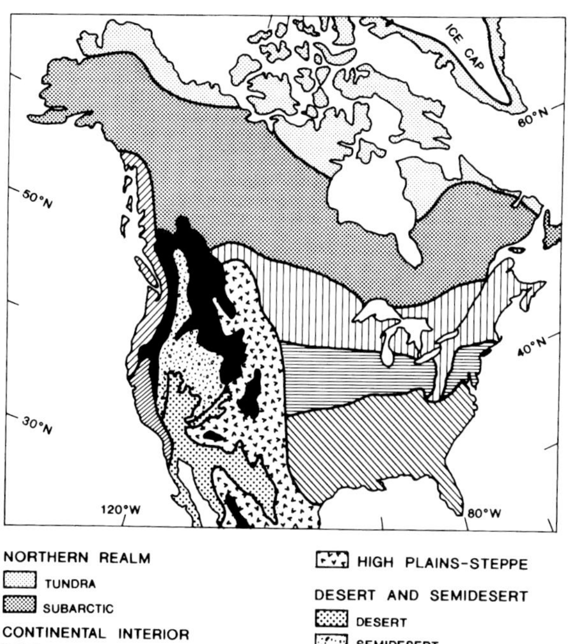

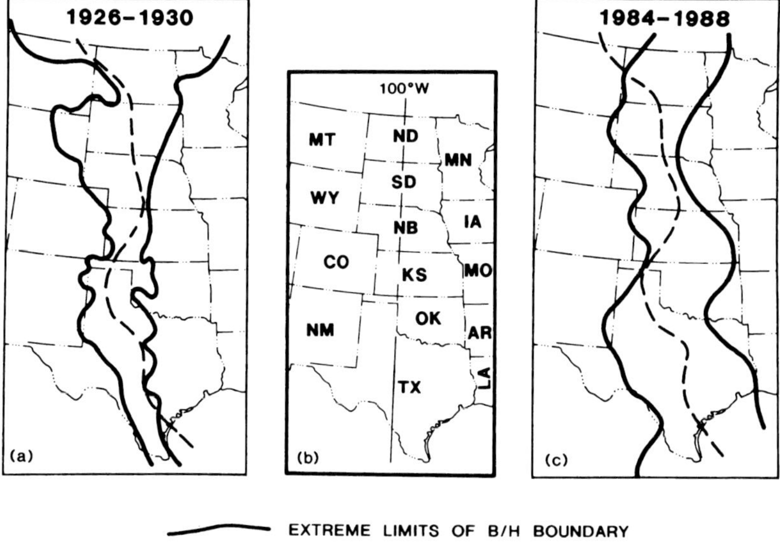

Figure 6.3 (*top*) Location of regional climates described in text. The distribution is based on the Köppen system of climate classification.

Figure 6.4 (*at left*) An illustration of the variability of the humid (H) and desert (B) boundary of the Köppen system. Extremes of the boundary for two 4-year periods are shown. The center map provides a locational guide (after Oliver, 1992)

scribe regional areas (e.g., Eichenlaub, 1979) or climates by states (e.g., Felton, 1965; Bomar, 1990; Henry, 1994). Additionally, more specialized topics such as severe weather (Schmidlin and Schmidlin, 1996) and regional climatic change (North et al., 1996) provide interesting insights into the regionalization of climate.

6.3.1 The Northern Realm

Stretching across a large part of Canada and into Alaska are two climatic regions that are a function of high latitude and continental location: the frigid tundra and subarctic climates.

The lowland plain that is the tundra reflects both past and present influences of climate (Burn and Lewkowicz, 1990). Gone is the preglacial rolling topography. Instead, the landscape is dominated by the legacies of Pleistocene glaciation and Holocene deglaciation: thousands of lakes and swamps, frozen during the long winter, and features that result from frost action. Pingoes, permafrost, and patterned ground occur over much of the region (Woo and Gregor, 1992; also chapter 13, this volume).

This bleak image is essentially the result of the high-latitude climatic regime. Consider a location at 60°N, a latitude that passes through the region. The length of daylight ranges from 5.7 hours at the winter solstice to 18.8 hours at the summer solstice. Clearly, the very short winter days provide minimum inputs of energy, but what about the long summer days? Unfortunately, the angle of the sun in the sky, the factor that basically determines the intensity of solar radiation, is only 53° at the highest elevation, whereas at the winter solstice it is only 6°. Low energy inputs and consequent low temperatures are the result (Rouse and Bello, 1983). At Barrow, Alaska, located at about 71°N, for example, the case is even more extreme: the sun remains continuously below the horizon from late November to mid-January.

The prevailing cold air has a low moisture content. Given that the area is largely dominated by high pressure, with minimal invasion by cyclonic storms, precipitation is low, with most falling during the summer. Because of the small amount of winter precipitation and the high winds in this area, there is surprisingly little snow cover.

Vegetation in the subarctic climatic zone comprises the vast coniferous forests that extend across Canadian North America, a sparsely populated zone also known as the taiga. The area is bounded to the north by the 10°C isotherm of the warmest month, the approximate poleward limit of tree growth, and stretches from the Atlantic Ocean to the Rocky Mountains. It gives way in the south to the warmer areas of more humid continental climates. Although there is an east–west climatic contrast, the region as a whole is characterized by short summers and long, cold winters. The transitions between the two seasons are abrupt. With high pressure dominant during the long, cold winter, most precipitation occurs during the summer; even so, total amounts are characteristically low.

A major change in temperature from summer to winter is characteristic of climates that experience continental regimes. With annual temperature ranges among the largest in the world, this is a region of high continentality. To express the concept of continental versus maritime location, indices of continentality have been derived. The essential ingredient of most formulas is the annual amplitude of temperature, with some indication of latitude. Conrad's Index, for example, is derived from $k = (1.7A) / (\sin(\phi + 10)) - 14$, where k is continentality, A is average annual temperature range in °C, and ϕ is the latitude angle (Barry and Chorley, 1992). The highest values of continentality, as measured by Conrad's Index, are in this northern realm, with high values extending southward to the center of the continent. The traditional approach to determining continentality, as exemplified by Conrad's contribution, has been challenged (Driscoll and Fong, 1992), and a new map based on residuals from a regression line of annual temperature range on latitude produced. This also shows higher values in the northern realms.

6.3.2 The Continental Interior

South of the subarctic and east of the Rockies, two large, continental climatic regimes are found. In the north, the continental regime consists of cool summers, whereas, to the south, the mean summer temperatures exceed 22°C. In such vast regions, there are major differences between both the northernmost and southernmost areas and between those in the east and those in the west. A traverse across the northern tier of the region provides a view of both this region's climate and the variations within it, as seen in table 6.3. The data clearly indicate that, although grouped within the same region, the coastal and continental locations experience somewhat different conditions. Differences are heightened in many ways by the special types of conditions experienced. The infamous "nor'easters" of the Atlantic coast result from intense low-pressure systems moving up the coast bringing high winds, rain, and snow onshore (Davis et al., 1993). Also of note is the occurrence of fog caused by the offshore advection of warm air over the cold Labrador Current. At about 50°N, water associated with this current has sea surface temperatures about 1°C in January and between 5° and 7°C in July; by comparison, the mid-Atlantic Ocean sea surface temperature is about 12°C in January and 16°C in July. The infamous fogs of the Grand Banks off Newfoundland have been the topic of films and novels, and, unfortunately, of many, real maritime disasters. In Nova Scotia, coastal fogs occur mostly in summer and early spring, when the adjacent sea temperatures are cold because of the influx of Labrador Current water. Hare and Thomas (1979) report that, in the summer of 1967, the town of Yarmouth, at the southwestern tip of Nova Scotia,

Table 6.3 Sample Data for the Continental Interior Climate with cool summers

	Winnipeg, Manitoba	North Bay, Ontario	Halifax, Nova Scotia
Latitude[a]	50°N	46°N	44°N
Longitude[a]	97°W	79°W	64°W
January Mean Temperature, °C	−17.7	−12.2	−3.3
July Mean Temperature, °C	20.2	18.7	18.5
Maximum Monthly Precipitation, mm	70 (August)	114 (September)	141 (January)
Minimum Monthly Precipitation, mm	20 (February)	68 (April)	94 (July)
Annual Average Precipitation, mm	517	1034	1384

Source: National Climate Data Center (www.ncdc.noaa.gov).

[a]Given to nearest degree.

experienced fog for at least 1 hour on 85 of the 92 days of the summer months of June, July, and August.

Within this realm are the Great Lakes. The size of these lakes together with the prevailing winds and air masses give rise to the phenomenon known as the "lake effect," which ameliorates summer temperatures and increases winter snowfall (Eichenlaub, 1979; Eichenlaub et al., 1990; also chapter 16, this volume). The lakes remain unfrozen far into the winter; in some winters, none of them freeze over. Whenever the air flowing over the lakes is colder than the water beneath, large amounts of water evaporate into the air. As the air streams over the land on the downwind side of the lakes, it cools and the moisture precipitates as snow. The greater the difference in the temperature of the air and the water, the greater the evaporation and the more likely a lake-effect snowfall will result. Regions that receive unusually large amounts of snowfall thus occur downwind of the Great Lakes, for example, in distinctive snowbelts immediately east of Lake Michigan and Lake Erie (Schmidlin, 1993). When a lake freezes, it shuts off the supply of water, and lake-effect snow declines.

Areas in the more moderate southern continental interior region experience rain all year, with a slight summer maximum, cool winters, and warm summers. Such a climate proved optimal for the growth of corn, and hence the identification of the "corn belt." Although the agriculture of the region is now more mixed, it remains a climatic region where, favored by warm summers and precipitation at all seasons, agricultural productivity is characteristically high. Mountain systems, particularly the Appalachians, interrupt this pattern to provide a regional climate modified by altitude (Weisman, 1990). Like most mountain climates, there is appreciably more variation in the conditions, with climate regimes changing from mountain to valley in an endless array.

6.3.3 The Humid Subtropics

The climate of the South is the feature of many classic movies and stories. The long, hot, moist summers give way to short winters, which, at least in the southern part of the region, seldom experience deep freezes and snows. Precipitation patterns and events in the South are the topic of many recent studies (e.g., Easterling, 1991; Gamble and Meetemeyer, 1997). One way to describe the climate of this region is to relate it to human comfort. In a classification based on what he terms the physiological climate, Terjung (1966) classifies July in the coastal area of this region as S_3a. The "S_3" means that the day–night combination is oppressive-hot, whereas the "a" suggests that any wind effect may be discomforting rather than cooling. Thus, although the high prevailing temperatures provide benefits, such as a long growing season, the climatic conditions in summer can be uncomfortable for inhabitants.

Table 6.4 provides sample data for the region. Along an west–east traverse, the longitudinal extent encompassed by this region, from 80° to 97°W in this sample, is quite large. Accordingly, areas to the west, such as Dallas, receive appreciably lower precipitation amounts than those in the east. The rainfall maximum occurs mostly during late summer. Notice too how the annual range of temperatures varies, with maritime stations, exemplified by Charleston, experiencing cooler summers but warmer winters than, for example, Dallas.

One major vulnerability of this southern region is that it is in the zone of maximum hurricane incidence. Hurricanes form over tropical waters during the summer. The central Atlantic Ocean, between Africa and the Americas, and the adjacent Caribbean Sea and Gulf of Mexico, comprise one of the world's major areas for tropical cyclone formation. Because hurricanes derive their energy from the oceans, once they pass over land they lose much of their ferocity. It follows that the areas most devastated by hurricanes occur along the coastal strip that borders the formation areas. In the United States, most such locations occur in the humid subtropical region, although hurricanes on a more northerly track impinge on coasts farther north. As shown in table 6.5, which provides a listing of the most intense, costliest, and deadliest of the storms, coastal Texas, Louisiana, and Florida figure significantly in the listings.

Table 6.4 Sample Data for Humid Subtropical Climates

	Dallas, Texas	Vicksburg, Mississippi	Charleston, South Carolina
Longitude[a]	97°W	91°W	80°W
Latitude[a]	33°N	32°N	33°N
Elevation, m	146	71	12
January Mean Temperature, °C	7.7	9.4	10.8
July Mean Temperature, °C	29.4	27.7	26.7
Maximum Monthly Precipitation, mm	123 (May)	146 (March)	196 (June)
Minimum Monthly Precipitation, mm	59 (January)	52 (September)	53 (Nobember)
Annual Average Precipitation, mm	879	1258	1250

Source: National Climate Data Center (www.ncdc.noaa.gov).

[a]Given to nearest degree.

Table 6.5 Hurricanes that affected the United States between 1900 and 1994, ranked as Most Intense, Costliest, and Deadliest

A. The Five Most Intense Hurricanes

Hurricane and Rank	Year	Category	Pressure[a] (mb)
1. Florida Keys	1935	5	892
2. Louisiana/Mississippi, "Camille"	1969	5	909
3. Southeast Florida, "Andrew"	1992	4	922
4. Florida Keys/S. Texas	1919	4	927
5. Florida	1928	4	929

B. The Five Costliest Hurricanes

Hurricane and Rank	Year	Category	Damage[b] (millions)
1. Southeast Florida, "Andrew"	1992	4	$30,475
2. South Carolina, "Hugo"	1989	4	$8,491
3. Northeast USA, "Agnes"	1972	3	$7,500
4. Florida/Louisiana, "Betsy"	1965	3	$7,425
5. Louisiana/Florida, "Camille"	1969	5	$6,096

C. The Five Deadliest Hurricanes

Hurricane and Rank	Year	Category	Deaths
1. Texas (Galveston)	1900	4	6000+
2. Florida (Lake Okeechobee)	1928	4	1836
3. Florida Keys/S. Texas	1919	4	600+
4. New England	1938	3	600
5. Florida Keys	1935	5	408

Source: National Climate Data Center (www.ncdc.noaa.gov).

[a]At time of landfall.
[b]Adjusted to 1996 dollars.

The hurricanes are ranked from 1 to 5 on the Saffir-Simpson Scale, with a level-5 storm being the most disastrous (table 6.6) (see chapter 21).

Despite the climatic problems, oppressive summer temperatures with high humidities and the hazard of hurricanes, this region is part of the rapidly developing Sun Belt. Although modern technology provides efficient cooling devices that provide a respite from the heat, the vulnerability to hurricanes, as seen in the devastation of Hurricane Andrew in Florida in 1992 and Hurricane Floyd along the southeast coast in 1999, remains a threat to this developing area (Williams et al., 1992).

6.3.4 *The High Plains, The Steppe*

In contrast to the regions previously described, this region has a north–south rather than an east–west orientation. The identifying feature is not temperature, but rather moisture, for this is a semiarid region that experiences low inputs of precipitation. This is well illustrated in the data of table 6.7 which provides a south–north transect using the stations selected. The elevation at each station is about 1000 m. Spring and summer rainfall maxima, large annual ranges of temperature, and low precipitation totals are the rule in this region. Not only is the total rainfall low, it is also extremely variable from year to year. This year-to-year variability makes identification of rigid climatic boundaries somewhat difficult. It was previously noted that the boundary is better shown as a zone and that the climate is steppe or semidesert (figure 6.4). Here, the region is called the High Plains to identify it as a region associated with the Great Plains. The latter is both a vegetative and a geomorphic region, but its relationship to the steppe identified in some climatic classifications is not a perfect one.

The region and its adjacent areas suffer recurring droughts. Documentary and dendroclimatic evidence shows that between 1825 and 1865, rainfall was low in many parts of North America. During this period, the plains were being explored but were not yet settled. Pioneer wagon trains ground tracks into the sediment on the dry beds of some western lakes in the 1840s. The lakes refilled and covered the route until around 1900. Wet and dry periods continued until, after World War I, another wave of farmers moved into the Great Plains on the heels of some wetter than normal years. Inevitably, the dry years of the 1930s spelled real disaster for the land and for the socioeconomic structure of the Great Plains (Hurt, 1981). The drought of the 1930s was not of equal intensity over the entire plains or through the decade. The first period, from mid-1933 to early 1935, was the worst. Dust began to blow during this time, with topsoil blowing away by the billions of tonnes. The dust storms of 1934 and 1935 demonstrated to people as far east as the Atlantic coast that severe problems existed farther west. An estimated 150,000 people moved out of the plains states. John Steinbeck's novel, *The Grapes of Wrath*, dramatized the impact of the drought. Two more periods of intense drought occurred in 1936 and 1939–40. The 1936 period was very intense, but short; the later spell was long, but less intense. Since that time, droughts of varying degrees of intensity have occurred (Fryread, 1981; McGregor, 1985). The relative intensity and length of these droughts is well illustrated by the derived Palmer Index (Palmer, 1965), which is used by a number of U.S. government agencies to depict moisture conditions throughout the country. For example, a long-term depiction of drought for each climate division of the United States is available from the Climate Visualization (CLIMVIS) depiction produced by the National Climate Data Center (NCDC, 1997) and available at their web site.

Table 6.6 Explanation of Categories Using the Saffir-Simpson Scale for Hurricanes

Scale Number	Central Pressure mb	Winds mph	Storm Surge ft	Storm Surge m	Damage
1	≥980	74–95	4–5	1.5	Minimal
2	965–979	96–110	6–8	2.0–2.5	Moderate
3	945–964	111–130	9–12	2.5–4.0	Extensive
4	920–944	131–155	13–18	4.0–5.5	Extreme
5	<920	>155	>18	>5.5	Catastrophic

Heat waves and heat stress (Cooter, 1990) are summer hazards, whereas winter brings blizzards. Although blizzards do occur across much of the interior of North America, they are a particular feature of this region. As Hare and Thomas (1979) state in their account of the Prairie region of Canada, "No description of the winter climate of the Prairies would be complete without mention of the blizzards—a storm of winds of 11.2 ms^{-1}, temperatures of −12°C or lower, and visibilities of 1 km or less in snow and/or blowing snow." During and immediately after these storms, which must last for 6 hours or more to qualify as blizzard, transportation and communications are disrupted, and there may be reports of suffering and even death of travelers unable to obtain shelter. Winter conditions may be tempered in some locations by warm chinook winds. In the Calgary area of Alberta, there are some 50 chinook days between November and February; these offer socioeconomic benefits ranging from warmer winters to reduced snow-removal costs (Nkemdirim, 1997).

6.3.5 *The Western Mountains and Basins*

In the mountain regions, maximum climatic diversity is encountered over short distances (Barry, 1992). Differences in altitude alone cause pronounced changes in atmospheric properties. With increasing altitude, moisture and dust

Table 6.7 Sample Data for the Plains

	Amarillo, Texas	Rapid City, South Dakota	Regina, Alberta
Latitude[a]	35°N	44°N	50°N
Longitude[a]	102°W	103°W	112°W
Elevation, m	1099	965	280
January Mean Temperature, °C	2.3	−5.6	−8.2
July Mean Temperature, °C	26.2	22.2	18.9
Maximum Monthly Precipitation, mm	86 (May)	78 (June)	81 (June)
Minimum Monthly Precipitation, mm	16 (February)	8 (December)	20 (December)
Annual Average Precipitation, mm	502	373	439

Source: National Climate Data Center (www.ncdc.noaa.gov).

[a]Given to nearest degree.

content diminish, as well as atmospheric density and barometric pressure. The atmosphere's heat capacity also decreases, resulting in lower temperatures that vary depending on existing lapse rates. Given these basic facts, the diversity of climates makes a formal description redundant because each area has its own character. Cross sections of the western mountain chains provide an interesting variety of conditions that illustrate this variability. Often, precipitation increases eastward from the Pacific coast across the lower foothills almost to the crests of the major ranges. Thereafter, in the rainshadow farther east, a rapid drop in precipitation amount is seen. The example shown in figure 6.5 provides a visual relationship between elevation and annual precipitation for an 800-km transect near the United States–Canadian border. Similar relationships are seen farther south across most of the more westerly ranges, although precipitation amounts fall significantly in the Great Basin before rising again across the Rocky Mountains.

Precipitation in the form of snow varies from one winter to the next, particularly in the more southerly mountains. Throughout this region, snow is a major source of water for irrigation and power. In North America, the areas with the most snowfall are the mountain areas of the far west. The mean permanent snow line in mountains in the humid tropics is about 4700 m. The snow line increases slightly poleward to 5200 m near 30°N. From there it drops to near 3000 m at 45°N latitude and to about 1400 m at 60°N. The actual height of the snow line on a given mountain depends on its orientation to the sun and to the wind direction. The snowiest location in North America is Paradise Ranger Station on Mount Rainier in Washington. At an elevation of 1692 m and facing the Pacific winds, snowfall there averages more than 15 m a year. In the 1971–72 season, a phenomenal 28.5 m of snow fell.

But just as mountains influence their own local climates, so they influence regions in their lee. Mountains also form a significant barrier to airstreams and, as a result, play a role in the climates of far distant locations. This is evident in the discussion of synoptic climatology in chapter 5.

6.3.6 Deserts and Semideserts

As figure 6.3 exhibits, this region actually comprises desert and semidesert or steppe. Most North American semideserts are in the elevated uplands between the Cascade Range–Sierra Nevada and the Rockies. Extending from the Great Sandy Desert in central Oregon through the Great Basin of Nevada and Utah, and from the dry Colorado Plateau to the plateaus of northern Mexico, these semiarid areas are at elevations above 1500 m. Deserts, such as the Mojave Desert of southern California and the Sonoran of northwest Mexico, are at lower elevations, the latter reaching down to sea level around the northern shores of the Gulf of California and the Pacific coast of central Baja California. During the summer, most of the western deserts have cloudless skies, high temperatures, and low humidities. During the winter, temperatures are generally above freezing except at higher elevations. Such an environment presents particular challenges to plant and animal life and several hazards to humans, notably from sunstroke, heat exhaustion, blowing sand, and dust storms (Lee et al., 1993).

The hottest areas within this desert realm are found in the down-faulted valleys of the Basin and Range Province east of the Sierra Nevada. The best known is Death Valley, at 86 m below sea level, where the world's second highest recorded temperature, 54°C was measured. Despite locations such as Death Valley, some observers do not consider North American deserts to be very dry. The plant geographer Nicholas Polunin (1960) noted that, to many who have experienced the deserts of, for example, northern Africa and southwestern Asia, it seems unreasonable to refer to the more productive of the so-called deserts of North America as properly desertic. Although mean precipitation is generally less than 25 cm a year, the main characteristic of these dry regions is precipitation variability. Apart from the climatic factors, such variability is also explained by major differences in elevation, often as much as 3000 m, between mountain crests and valley floors. The higher mountains, such as the Panamint Range, which rises 3454

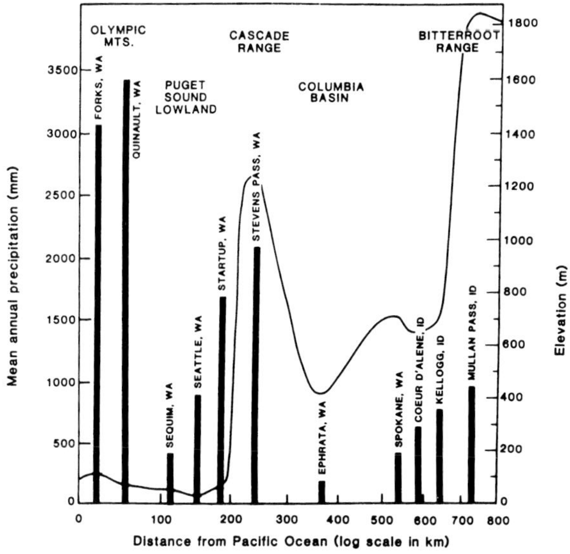

Figure 6.5 A cross section showing the relationship between mean annual precipitation and elevation across the western United States along the 48th parallel (after Granger, 1987).

m above the floor of Death Valley, readily intercept a portion of the residual moisture in winter rainstorms, releasing it downstream to support modest vegetation covers at relatively low altitudes. In the Great Basin and in deserts to the north, winter precipitation falls as snow about 30% of the time, whereas in the south snow is infrequent.

An interesting aspect of the climate of the deserts is a singularity known as the Southwest Monsoon (Andrade and Sellers, 1988). This summertime event generally occurs between early June and mid-September when a "monsoon-type" circulation exists over western North America. High surface temperatures coupled with an influx of moisture from both the Pacific and the Gulf of Mexico give rise to frequent thunderstorms in Arizona and New Mexico. These thunderstorms provide a significant portion of the total rainfall of the area (Harrington et al., 1992).

6.3.7 The West Coast Climates, The Pacific Rim

From southern Alaska to northwestern Mexico, the climates of western North America form a narrow, elongated region confined to a belt some 80 to 160 km from the ocean. For much of this distance, from Cook Inlet at latitude 60°N to just north of San Francisco at 38°N, the marine west coast climate type, with its extensive cover of coastal evergreen forest, is dominant. This vegetation results from the primarily maritime climate of the region with abundant precipitation, high humidity, and cool to mild temperatures. This region has the lowest continentality index in North America. South of about 38°N, the midlatitude dry summer regime occurs. Often referred to as a Mediterranean-type climate, its dominant feature is that 95% of the annual precipitation occurs in the period between October and April, with the length of the rainy season shortening southward (see chapter 20).

Air masses that dominate in western North America vary seasonally. Relatively dry subsiding air from the subtropical Hawaiian high-pressure cell and maritime polar air from the North Pacific Ocean dominate the weather in the West. During summer, the westerlies are farther north and relatively weak. The coastal states of California and Oregon are under the influence of the subtropical high and experience clear, warm, dry weather (Lydolph, 1985). During some years, no measurable precipitation has fallen from California to Washington in June, July, or August. Even Vancouver, British Columbia, has experienced a July with no measurable precipitation. Along the immediate coast, advection fog is common, generated as moist oceanic air passes onshore across the cool California Current.

A winter concentration of precipitation distinguishes the west coast of the continent, from northwest Baja California into Canada. California typically receives an average of 85% of its annual precipitation in the winter, and more than half of the annual total falls during December, January, and February. Farther north along the coast, the precipitation increases. Most of this increase occurs in the spring and fall and is due to the longer season in which the westerlies flow over the coast. Oregon averages only 73% of its annual total during the 6 winter months. Even Prince Rupert, British Columbia, at 54°N, receives 62% of its annual total of 2400 mm during the 6 winter months. This characteristic winter maximum extends inland into the Great Basin of the United States and into the interior valleys of western Canada. Although the annual total of precipitation tends to increase poleward, the actual amount received at any given location depends on site characteristics.

Table 6.8 provides sample data for the west coast from 34° to 48°N. The precipitation distribution, as previously described, is evident. Temperatures are distinctively maritime. Winter means are above freezing, and, at least north of Los Angeles, cool summers are prevalent along the coast. Farther inland, notably in California's Central Valley, much higher summer temperatures prevail; even in the greater Los Angeles area, there is considerable thermal variation among coastal lowlands, interior valleys, and high mountains. Temporal variations from the norm also occur. In California, for example, heat waves may occur either when the subtropical high is displaced northwestward, producing tropical desert conditions for long periods, or when the hot, dry Santa Ana winds flow down from the interior plateaus of the western United States in summer and fall. Santa Ana winds result from the development of a high-pressure system over the relatively high Nevada desert. The air, warmed by subsidence and the hot desert surface, flows

Table 6.8 Sample Data for the West Coast

	Los Angeles, California	Eureka, California	Tatoosh Island, Washington
Latitude[a]	34°N	41°N	48°N
Longitude[a]	118°W	124°W	125°W
Elevation, m	95	13	31
January Mean Temperature, °C	13.2	8.6	5.6
July Mean Temperature, °C	22.8	13.5	13.1
Maximum Monthly Precipitation, mm	85 (February)	170 (January)	309 (December)
Minimum Monthly Precipitation, mm	Trace (June)	3 (July)	50 (August)
Annual Average Precipitation, mm	373	975	1973

Source: National Climate Data Center (www.ncdc.noaa.gov).

[a]Given to nearest degree.

outward and downward from the desert. As air currents descend into the Los Angeles basin, they are heated even more by compression. Most frequent in September and October, the Santa Ana winds raise maximum daytime temperatures to the 38°–43°C range. These winds are commonly dust-laden, as in November 1969, when winds reached velocities of 122 kph through the passes. This desiccating wind can raise the fire hazard in the vegetation to a dangerous level. The sudden change in weather and the extreme dryness of the air often make these weather events—and the people who experience them—highly disagreeable. The influence of weather and climate on human activities and health is a topic about which much has been written (e.g., Munn, 1987).

Weather in this region often sees a cyclic alternation of summer drought, fall fire, and winter rain. Natural vegetation is chaparral woodland and scrub, which is explosively dry after the long, hot summer. Dry thunderstorms in the fall, augmented by human carelessness and arson, trigger wildfires that often destroy large areas of vegetal cover. Where fires have raged, the arrival of torrential winter rains often turns the unprotected soil into mud. Landslides and mudflows destroy much valuable property and leave behind denuded hillsides, which are recolonized by weeds and, more slowly, by chaparral plants that appear to be adjusted to recurrent flood and fire cycles. Nearly every year in California, residential areas are incinerated, buried in sediment, or simply carried down the hillsides in a mass of mud, kindling, and plasterboard. Raphael et al. (1994) provide a graphic example of the impacts of an unusual storm series during February 1992.

6.4 Climatic Variations

In recent years, there has been extensive discussion, in both the professional literature and in the popular press, of global climate change as a result of a modified greenhouse effect. The culprit in the scenario is the addition to the atmosphere of greenhouse gases (such as carbon dioxide and methane that contribute to the warming effect) generated through human activities. An interesting sidelight of this discussion concerns the changes that have occurred in regional climate in North America. To date, almost all discussion has focused on past climates, because the forecasting of future regional climates is not yet accurate enough to warrant extensive discussion.

One very useful source for regional climate data and trends is available through the publications of Oak Ridge National Laboratory in the United States. In one report (CDIAC, 1994), Canada and the United States are divided into climatic regions, and the climate for each region is assessed through the years of record. In the United States, 23 regions are identified, whereas Canada has 11 regions (see figure 6.6). For each region, annual average temperature and precipitation data have been derived; the mean temperature for the years 1961–90 (1951–80 for the Canadian data) was generated and compared with every other year, with the result expressed as a temperature anomaly.

Examples of temperature patterns over time are shown in figure 6.7. The examples were specifically selected to indicate the great variety of trends that occur. In the Mackenzie District (region 9 in figure 6.6a), there appears to be a distinct upward trend. As can be seen, the Canadian contributors provide a best-fit line to the data. A very similar trend is seen in the northern and continental regions of Canada. In the South Coastal Plain of the United States (region 14 of figure 6.6b), a quite different picture emerges. Here, there appears to be a downward trend in temperature. In fact, a variety of trends are found throughout North America. Some areas have become slightly warmer, some slightly cooler. It remains to be seen how temperature trends will vary in the future.

An interesting approach to analyzing how precipitation has varied over time can be seen in the area-averaged latitude bands of figure 6.8. For each designated latitudinal region, a graph shows annual precipitation from 1950 to 1990. An upward trend, indicating a wetter climate, is

a

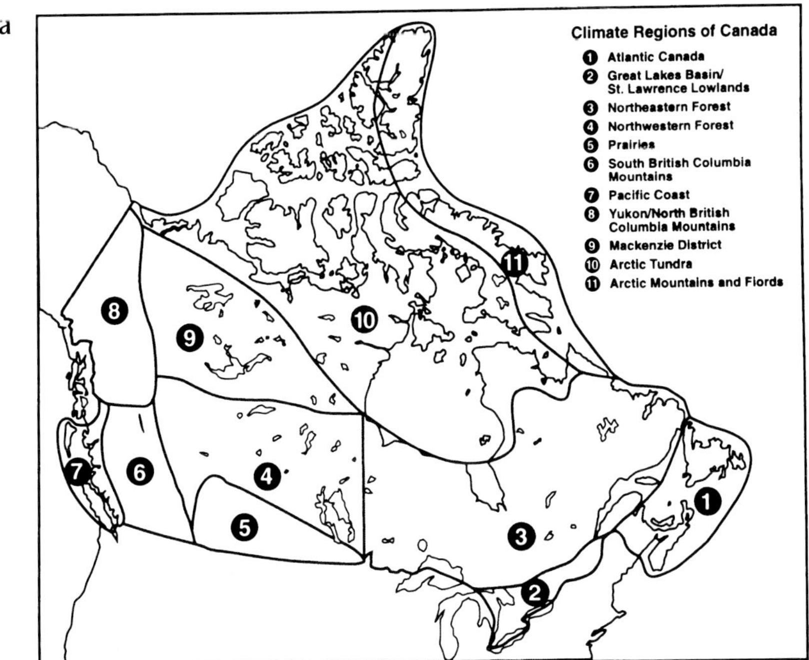

b

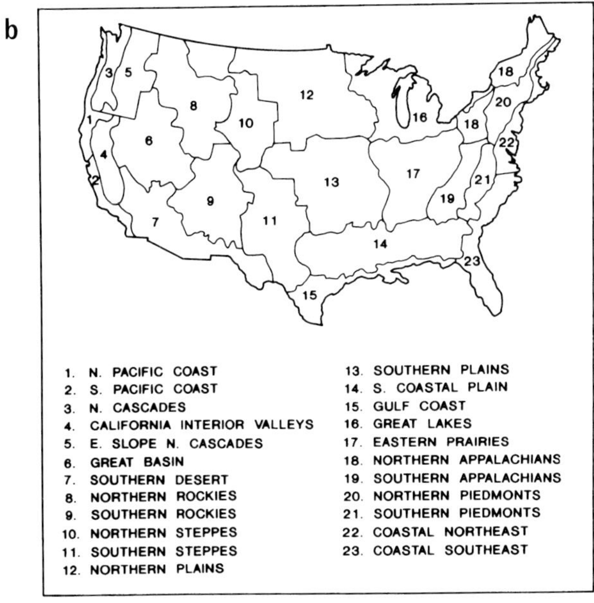

Figure 6.6 Climatic regions of (a) Canada and (b) the conterminous United States identified to describe regional climatic variations (after CDIAC, 1994).

particularly evident in data representing northern Canada (the 55°–70°N band).

Although the indicated precipitation trends are interesting, any results drawn from them may be misleading because of the relatively short period of record examined. In fact, the analysis of any climatic data to determine trends and potential patterns of change is fraught with potential problems. One obvious problem relates to the changes of climate associated with human constructions such as cities. To examine climatic change, researchers must take into account such facts as the warming influence of the urban climate. Since Landsberg's (1981) classic summary of urban-influenced climates, many research studies have been completed, and urban climatology is now a distinctive component of climatology. The type of analysis varies from modeling urban canyons (Arnfield, 1990) to assessing the role of CO_2 in climate change (Idso et al., 1998).

Within the identified climatic regions, many studies of the potential impact of global warming have been completed. Research sites range from Alaska (Esch and Osterkamp, 1989) and the Arctic (Maxwell, 1992) to the southeastern United States (Stahle and Cleaveland, 1992).

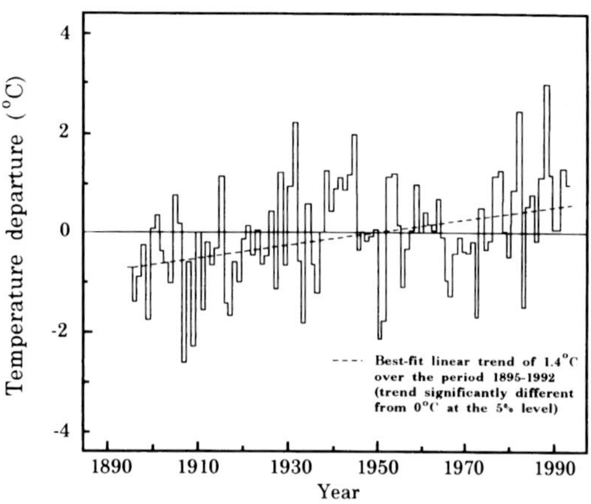

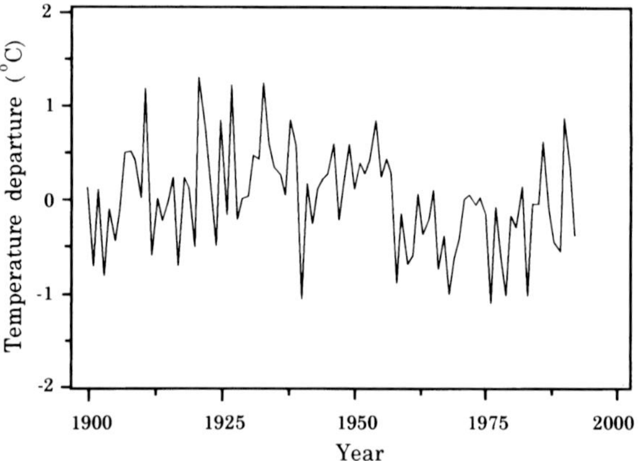

Figure 6.7 Temperature departures for selected regions identified in figure 6.6. (a) the Mackenzie Region of northwest Canada (region 9 in fig. 6.6a) and (b) the South Coastal Plain (region 14 in fig. 6.6b) (after CDIAC, 1994).

Variables examined range from snow-cover changes (Karl et al., 1993) to widespread impacts (Smith and Tirpak, 1989; CCCPB, 1991).

6.5 Climate and Environment

A discussion of the climatic regions of North America would be incomplete without mention of the relationships between climate and other features of the environment. That early climatic classifications were based on the nature of the natural vegetation is testimony to this (Oliver, 1991). Such a viewpoint is also clearly expressed by Mannion (1997), who describes the mutually reinforcing relationship between climate and vegetation. In examining the distribution of natural vegetation in North America, the climatic influence is so clear that some authors have used vegetation as the basis for discussion of the climates of the continents (Rumney, 1987).

Vegetation is also an important proxy for describing climates of the past. The analysis of tree rings has been used to reconstruct climates over wide areas. Of particular interest are the reconstructed climates of the Southwest completed by the Tree Ring Laboratory at the University of Arizona (Fritts, 1987). Similarly, the development of pollen analysis has also proved to be a useful tool in deciphering past climate, provided that interpretation of the pollen record is approached with caution (see chapter 3).

The role of climate in geomorphic activity is a significant field of study. Climatic geomorphology examines the relationship between landforms and climate at a variety of scales using a variety of methods. An appropriate example of the climate–geomorphology interface is the interpretation of past climates from surviving geomorphic evidence, most notably the inferences regarding major cooling derived from glacial and periglacial landscapes. Conversely, the role of past tectonic activity in generating climate change and contemporary climatic regionalism has already been discussed in chapter 1. Soil formation may be related to climate by studying pedogenic regimes, the regimes under which given soil-forming processes can occur (Bridges, 1997; see also chapter 9). Climate related to other disciplines provides a fertile area for study of the environmental regions in which we live.

6.6 Conclusion

North America contains a fascinating variety of climatic types. In attempting to find order in the diversity, it is essential that similar climate types be grouped together. However, there is a wide range of ways in which different climates can be grouped. Single variables, such as rainfall or temperature, certainly provide a basic way of identifying similar types, but single-variable classification clearly cannot provide more than a glimpse of the climatic complexities. Perhaps the easiest way to provide a systematic outline of the regions is to select widely used variables, usually temperature and precipitation, as a starting point. A classification based on these two variables provides a framework for examination of other climatic features. Such a course has been followed in this chapter. It is noteworthy that the climatic division map used in this description does not provide exact boundaries for the regions despite the fact that most formal climate classification systems must provide a guide as to where to locate a boundary.

Climate is a significant variable in the physical geographic mix that gives a location its identity. The relationships between climate and, for example, vegetation, geomorphic processes, agriculture, and human comfort provide the basis for continuing studies of the regions of North America.

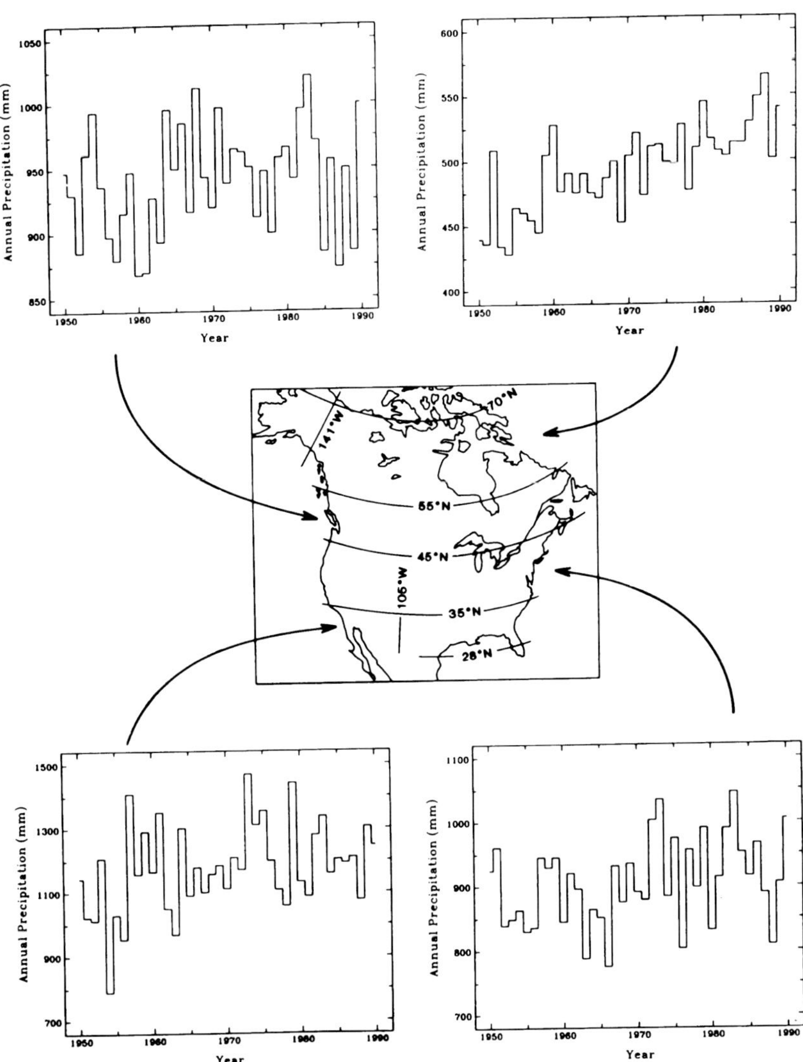

Figure 6.8 Annual precipitation variations, 1950–90, for four latitudinal bands.

Acknowledgments I acknowledge the helpful comments provided by John J. Hidore and William A. Dando in the preparation of this chapter.

References

Andrade, E.R., Jr., and W.D. Sellers, 1988. El Niño and its effect on precipitation in Arizona and western New Mexico. *Journ. Climatol.*, Vol. 8, 403–410.

Arnfield, A.J., 1990. Canyon geometry, the urban fabric and nocturnal cooling: A simulation approach. *Physical Geography*, 11, 220–239.

Balling, R.C., 1984. Classification in Climatology, in G.L. Gaile and C.J. Willmott, eds., *Spatial Statistics and Models*, Dordrecht: Reidel.

Barry, R.G., 1992. *Mountain Weather and Climate*, 2nd ed., New York: Routledge.

Barry, R.G., and R.J. Chorley, 1992. *Atmosphere, Weather and Climate*, 6th ed., New York: Routledge.

Bomar, G.W., 1990. *Texas Weather*, 3rd Printing, Austin: Univ. of Texas Press.

Bridges, M., 1997. Soils, in R.D. Thompson and A. Perry (eds.), *Applied Climatology*: New York: Routledge.

Burn, C.R. and A.G. Lewkowicz, 1990. Canadian Landform Examples, *Canadian Geographer*, Vol. 34, 273–276.

CCCPB (Canada Climate Control Program Board),1991. Climate change and Canadian impacts: The scientific perspective. *Climate Change Digest, CCD 91–01.* Ottawa: Canada Climate Centre, Environment Canada.

CDIAC (Carbon Dioxide Information Analysis Center,) 1994. *Trends '93: A Compendium of Data on Global Change*, U.S. Department of Energy, Oak Ridge National Laboratory, Publication ORNL/CDIAC-65, ESD Publication No. 4195, Oak Ridge.

Chang, F-C and J.M. Wallace, 1987. Meterological conditions during heat waves and drought in the United States Great Plains. *Monthly Weather Review*, Vol. 115, 1253–1269.

Changnon, S.A., P.L. Lamb, and K.C. Hubbar, 1990. Regional Climate Centers: New institution for climate service and climate-impact research, *Bull. Amer. Meteor. Soc.*, 71, 527–537.

Cooter, E., 1990. A heat stress climatology for Oklahoma. *Physical Geography*, Vol. 11, 17–35.

Court, A., 1974. Climate of the conterminous United States, in R.A. Bryson and F.K. Hare, eds., *Climates of North America*, World Survey of Climates, vol. 11, New York: Elsevier.

Davis, R.E. and L.S. Kalkstein, 1990. Using a spatial climatological classification to assess changes in atmospheric pollution concentrations, *Physical Geography*, 11, 320–342.

Davis, R.E., R., Dolan, and G. Demme, 1993. Synoptic climatology of Atlantic coast North-Easters. *International Journal of Climatology*, Vol. 13, 171–189.

Driscoll, D.M., and J.M. Yee Fong, 1992. Continentality: A basic climatic parameter re-examined. *International Journal of Climatology*, 12, 185–192.

Easterling, D.R. 1991. Climatological patterns of thunderstorm activity in the South-East USA. *International Journal of Climatology*, Vol. 11, 213–221.

Eichenlaub, V.L., 1979. *Weather and Climate of the Great Lakes Region*, Notre Dame: University of Notre Dame Press.

Eichenlaub, V.L., J.R. Harmon, F.V. Nurnberger, and H.J. Stolle, 1990. *The Climatic Atlas of Michigan*, Notre Dame: University of Notre Dame Press.

Esch, D., and Osterkamp, T. 1989. Arctic and cold region engineering: Climatic warming concerns for Alaska. *Journal of Cold Regions Engineering*, Vol. 4, 6–14.

Felton, E.L., 1965. *California's Many Climates*, Palo Alto: Pacific Books.

Forry, S., 1842. *The Climate of the United States and Its Endemic Influences*, Boston: Little and Brown. (Reprinted AMS edition, 1978).

Fritts, H.C., 1987. Tree Ring Analysis, in J.E. Oliver and R.W. Fairbridge, eds, *The Encyclopedia of Climatology*, New York: Van Nostrand Reinhold.

Fryread, D.W., 1981. Dust Storms in the Southern Great Plains, *Trans. American Soc. of Agric. Engineers*, 24, 991–994.

Gamble, D.W., and V.G. Meentemeyer, 1997. A synoptic climatology of extreme unseasonable floods in the southeastern United States. *Physical Geography*, 18, 496–524.

Glantz, M.H., 1996. *Currents of Change*, New York: Cambridge University Press.

Goetzmann, W.H., and G. Williams, 1992. *The Atlas of North American Exploration*, New York: Prentice Hall.

Granger, O., 1987. Precipitation Distribution, in J.E. Oliver and R.W. Fairbridge, eds., *The Encyclopedia of Climatology*, New York: Van Nostrand Reinhold.

Guttman, N.B., and R.G. Quayle, 1995. A Historical Perspective of U.S. Climate Divisions, *Bull. American Met. Soc.*, 77, 293–303.

Hansen-Bristow, K.J., 1986. Influence of increasing elevation on growth characteristics at timberline. *Canadian Journal of Botany*, Vol. 64, 2517–2523.

Hare, F.K. and M.K. Thomas, 1979. *Climate Canada*, 2nd ed., Toronto: Wiley.

Harrington, J.A., R.S. Cerveny, and R.C. Balling, 1992. Impact of the Southern Oscillation on the North American Southwest Monsoon, *Physical Geography*, 13, 318–330.

Henry, J.A., 1994. *The Climate and Weather of Florida*, Coyne: Pineapple Press.

Hidore, J.J., and J.E. Oliver, 1993. *Climatology: An Atmospheric Science*, New York: Macmillan.

Hurt, R.D., 1981. *The Dust Bowl: An Agricultural and Social History*, Chicago: Nelson-Hall.

Idso, C.D., S.B. Idso, and R.C. Balling, 1998. The urban CO_2 dome of Phoenix, Arizona, *Physical Geography*, 19, 95–108.

Kalkstein, L., and K.M. Valimont, 1986. An evaluation of summer discomfort in the United States using a relative climatological index, *Bulletin of the American Meteorological Society*, Vol. 67, 842–848.

Kalkstein, L.S., G. Tan, and J.A. Skindlov, 1987. An evaluation of three clustering procedures for use in synoptic climatological classification, *Journ. of Climate and Applied Meteorology*, 26, 717–730.

Karl, T.R., P.Y. Groisman, R.W. Knight, and R.R. Heim, 1993. Recent variations of snow cover and snowfall in North America and their relations to precipitation and temperature variations. *Journal of Climate*, Vol. 6, 1327–1343.

Kendrew, W.G., 1961. *The Climates of the Continents*, 5th ed. London: Oxford University Press.

Landsberg, H.E., 1981. *The Urban Climate*, New York: Academic Press.

Lauer, W., M. Daud Rafiqpoor, and P. Frankenberg, 1996. Die Klimate Der Erde, *Erdkunde*, 50, 275–300.

Lee J.A., K.A. Wigner, and J.M. Gregory, 1993. Drought, Wind and Blowing Dust on the Southern High Plains of the United States, *Physical Geography*, 14, 56–67.

Lydolph, P.E., 1985. *The Climate of the Earth*, Totowa, N.J.: Rowman & Allanheld.

Mannion, A.M., 1997, Vegetation, in R.D. Thompson and A. Perry, eds., *Applied Climatology: Principles and Practice*, London: Routledge.

Marshall, T., 1992. Dryline Magic. *Weatherwise*, 45, 25–28.

Mather, J.R., 1985. The Water Budget and the Distribution of Climates, Vegetation and Soils, *Publications in Climatology*, 38(2), University of Delaware, Center for Climatic Research.

Maxwell, J.B., 1992. Arctic climate: Potential for change under global warming. In F.S. Chapin et al., eds., *Arctic Ecosystems in a Changing Climate: An Ecophysiological Perspective*. San Diego: Academic Press, pp. 11–34.

McGregor, K.M., 1985. Drought during the 1930's and 1950's in the central United States. *Physical Geography*, Vol. 6, 288–301.

Morgan M.D., and J.M. Moran, 1997. *Weather and People*, Upper Saddle River, N.J.: Prentice Hall.

Muller, R.A., B.D. Keim, and J.L. Hoff, 1990. Application of Climatic Divisional Data to Flood Interpretations: An Example from Louisiana, *Physical Geography*, 11, 353–362.

Munn, R.E., 1987. Biometeorology, in Oliver, J.E., and R.W. Fairbridge, eds., *Encyclopedia of Climatology*, New York: Van Nostrand Reinhold.

NCDC (National Climate Data Center), 1997. Web site: www.ncdc.noaa.gov

Nkemdirim, L.C., 1997. On the frequency and sequencing of Chinook events, *Physical Geography*, 18, 101–113.

North, G., J. Schmandt, and J. Clarkson, 1996. *The Impact of Global Warming on Texas*, Austin: University of Texas Press.

Oliver, J.E., 1991. The History, Status and Future of Climatic Classification, *Physical Geography*,12, 231–251.

Oliver, J.E., 1992. The Great American Desert and the Arid/Humid Boundary, in D.G. Janelle, ed., *Geographical Snapshots of North America*, New York: Guilford Press.

Palmer, W.C. 1965. *Meteorological Drought.* Washington, D.C.: U.S. Government Printing Office, Research Paper No. 45.

Phillips, D., 1990. *The Climates of Canada*, Minister of Supply and Services, Ottawa, Canada.

Polunin, N. 1960. *Introduction to Plant Geography and Some Related Sciences*, New York: McGraw-Hill.

Raphael, M., J., Feddema, A.J. Orme, and A.R. Orme, 1994. The Unusual Storms of February 1992 in Southern California, *Physical Geography*, 15, 442–464.

Rouse, W.R., and R.L. Bello, 1983. The radiation balance of typical terrain units in the low Arctic, *Annals of the Association of American Geographers*, 73, 538–549.

Rumney, G.R., 1987. Climate of North America, in J.E. Oliver and R.W. Fairbridge, eds., *The Encyclopedia of Climatology*, New York: Van Nostrand Reinhold.

Santibáñez, F., 1994. Crop requirements—Temperate Crops, in J.F. Griffiths, ed., *Handbook of Agricultural Meteorology*, New York: Oxford University Press.

Schmidlin, T.W., 1993. Impacts of Severe Winter Weather during December 1989 in the Lake Erie Snowbelt, *Journal of Climate*, 6, 761–767.

Schmidlin T.W., and J.A. Schmidlin, 1996. *Thunder in the Heartland: A Chroncle of Outstanding Weather Events in Ohio*, Kent: Kent State University Press.

Smith, J.B., and D.A. Tirpak, 1989. *The Potential Effects of Global Climatic Change on the United States: Appendix G—Health.* Washington, D.C.: United States Environmental Protection Agency.

Stahle, D.W., and M.K. Cleaveland, 1992. Reconstruction and analysis of spring rainfall over the southeastern U.S. for the past 1000 years. *Bull. Amer. Meteor. Soc.* Vol. 73, 1947–1961.

Strahler A., and A.N. Strahler, 1996. *Physical Geography: Science and Systems of the Human Environment*, New York: Wiley.

Terjung, W., 1966. Physiological Climates of the United States, *Annals of the Association of American Geographers*, 56, 141–179.

Tyler, L.G., 1907. Narratives of Early Virginia, in M. Ridge and R.A. Billington (eds.), *Americas's Frontier Story*, New York: Holt, Rinehart and Winston.

Weisman, R.A., 1990. An observational study of warm season southern Appalachian lee troughs. Part II: Thunderstorm genesis zones. *Monthly Weather Review*, Vol. 118, 2020–2041.

Williams, J., H. Brandli, and W. Frindley, 1992. Hurricane Andrew in Florida, *Weatherwise*, 45, 7–17.

Woo, M.K., and D.J. Gregor, eds., 1992. *Arctic Environment: Past, Present and Future.* Hamilton, Ontario McMaster University.

7

Surface-Water Hydrology

John Pitlick

North America is a place of hydrologic extremes: it includes some of the wettest, driest, warmest, and coldest places on Earth. This diversity arises because the continent is relatively large, spanning almost 60° of latitude between polar and subtropical regions. As such, virtually every hydrologic phenomenon—from extreme rainfall to extreme aridity—occurs within North America. The challenge thus becomes one of distilling the diverse aspects of North America's hydrology into these few pages. That challenge is made simpler because of the sheer number of published scientific studies and the amount of data available for this continent. In the United States, for example, about 8000 surface-weather stations and 7000 surface-water gauging stations are currently in operation; thousands more are no longer in operation but provide historical data. Several high-quality subsets of these data (numbering more than 1000 stations) have been compiled for use in evaluating long-term trends in temperature, precipitation, and runoff, and for validating results from general circulation models (Karl et al., 1990; Wallis et al., 1991; Slack and Landwehr, 1992). These data are available in digital format, and often can be obtained from government agencies free of charge or at minimal cost (the chapter appendix includes sources of hydrologic data, including Internet addresses). It is safe to say that the job of every hydrologist in North America is made simpler by the abundance and accessibility of information, whether in the form of raw data or published reports and studies.

This chapter summarizes major geographic patterns and long-term temporal trends in the surface-water hydrology of North America. It begins with a brief discussion of the water balance and summarizes previous estimates of the amount of water moved through the North American continent as precipitation, evaporation, and runoff. Subsequent sections examine each of these components, emphasizing important physical processes, regional patterns, and long-term trends; some sections include discussion of extreme events or events of local interest. The chapter attempts a comprehensive treatment of these topics, but it does not include much discussion of the related topics of groundwater hydrology and water quality. Readers interested in groundwater processes and resources should consult volume O-2 of the Decade of North American Geology, edited by Back et al. (1988). Similarly, the article by Hem et al. (1990) and the report compiled by Paulson et al. (1993) summarize important water quality issues in the United States and Canada. Useful companions to this chapter include volume O-1 of the Decade of North American Geology, edited by Wolman and Riggs (1990), and the book *Water in Crisis*, edited by Gleick (1993).

7.1 Water Balance

This chapter is organized around the concept of water balance, which may be expressed in simplified form as

$$P = ET + R + \Delta S$$

where P is precipitation, ET is evapotranspiration, R is runoff, and ΔS is the change in water stored in lakes, snowfields, glaciers, and aquifers. Each of these terms can be expanded to account for separate components. Precipitation, for example, includes both rain and snow; evapotranspiration includes not just evaporation from lakes and transpiration from plants, but also changes in soil moisture; and runoff includes both surface-water and groundwater flow. The water balance can be applied to spatial scales ranging from small plots to large regions. In many parts of North America, one component clearly dominates over the others, making it possible to emphasize certain processes within the context of specific regions.

The water balance can also be applied to any time period of interest, from hours to months and years. Over certain time periods, some terms are considered negligible, and other terms are considered to be invariant. For example, it is often assumed that over periods of decades, $\Delta S \approx 0$. It might also be reasonable to assume that, on an annual basis, the change in soil moisture is zero, but this is obviously not a good assumption when considering short-term (seasonal) trends.

The water balance of North America has been estimated in a handful of studies (table 7.1). The numbers derived in these studies differ more than one might expect, most likely because of differences in the data sets and methods used to estimate individual components. On average, the continent as a whole receives about 680 mm of precipitation annually. Of this, approximately 416 mm (61%) are returned to the atmosphere by evapotranspiration, and 266 mm (39%) reach the oceans as runoff. The following sections discuss each of these components in more detail, examining both temporal and regional patterns.

7.2 Precipitation

The spatial distribution of precipitation in North America and the mechanisms that generate rainfall or snowfall are closely linked to weather systems that deliver moisture from the oceans and the interaction of these weather systems with major topographic barriers. The connection between atmospheric processes and precipitation is described in chapter 5. Other short summaries of the hydroclimatology of North America can be found in Hirschboeck (1991) and Lins et al. (1990). Key points of these articles are summarized herein.

Most of the moisture for precipitation in North America originates in the Pacific Ocean, Atlantic Ocean, or Gulf of Mexico. Large inland bodies of water such as Hudson Bay and the Great Lakes can act as local sources of moisture (Barry and Chorley, 1987). There is some evidence that evaporation can enhance rainfall locally (Anthes, 1984; Barnston and Schickedanz, 1984; Brubaker et al., 1993), but otherwise little of the precipitation falling in North America originates on the land itself. Moist air masses from the Pacific Ocean enter the continent along several pathways, which shift in position from about 35°N in winter to about 60°N in summer. Soon after they enter the continent, these air masses are forced upward by a series of mountain ranges, resulting in steep gradients in precipitation along much of the west coast (fig. 7.1; see also fig. 6.5). The rain-shadow effect of these mountain ranges produces precipitation minima in the Basin and Range Province immediately to the east. Not all the moisture is lost here, however, and farther east orographic effects again come into play to produce secondary maxima throughout the Rocky Mountain region. Moisture from the Atlantic Ocean and Gulf of Mexico is delivered from the east and south by clockwise circulation around a quasi-stationary, high-pressure system in the Atlantic Ocean. These air masses contain abundant precipitable water vapor and can penetrate far into the interior of the continent, particularly in spring and summer. The wide band of relatively high annual precipitation in the eastern United States and Canada

Table 7.1 Estimates of average annual water balance (mm/yr) for North America

P	*E*	*E/P*	*R*	*R/P*	Source
645	403	0.62	242	0.38	Baumgartner and Reichel, 1975
660	396	0.60	264	0.40	AWRC, 1976
756	418	0.55	339	0.45	UNESCO, 1978
670	383	0.57	287	0.43	L'vovich, 1979
563	427	0.76	136	0.24	Willmot et al., 1985
800	470	0.59	330	0.41	Budyko, 1986
Averages:					
682	416	0.61	266	0.39	

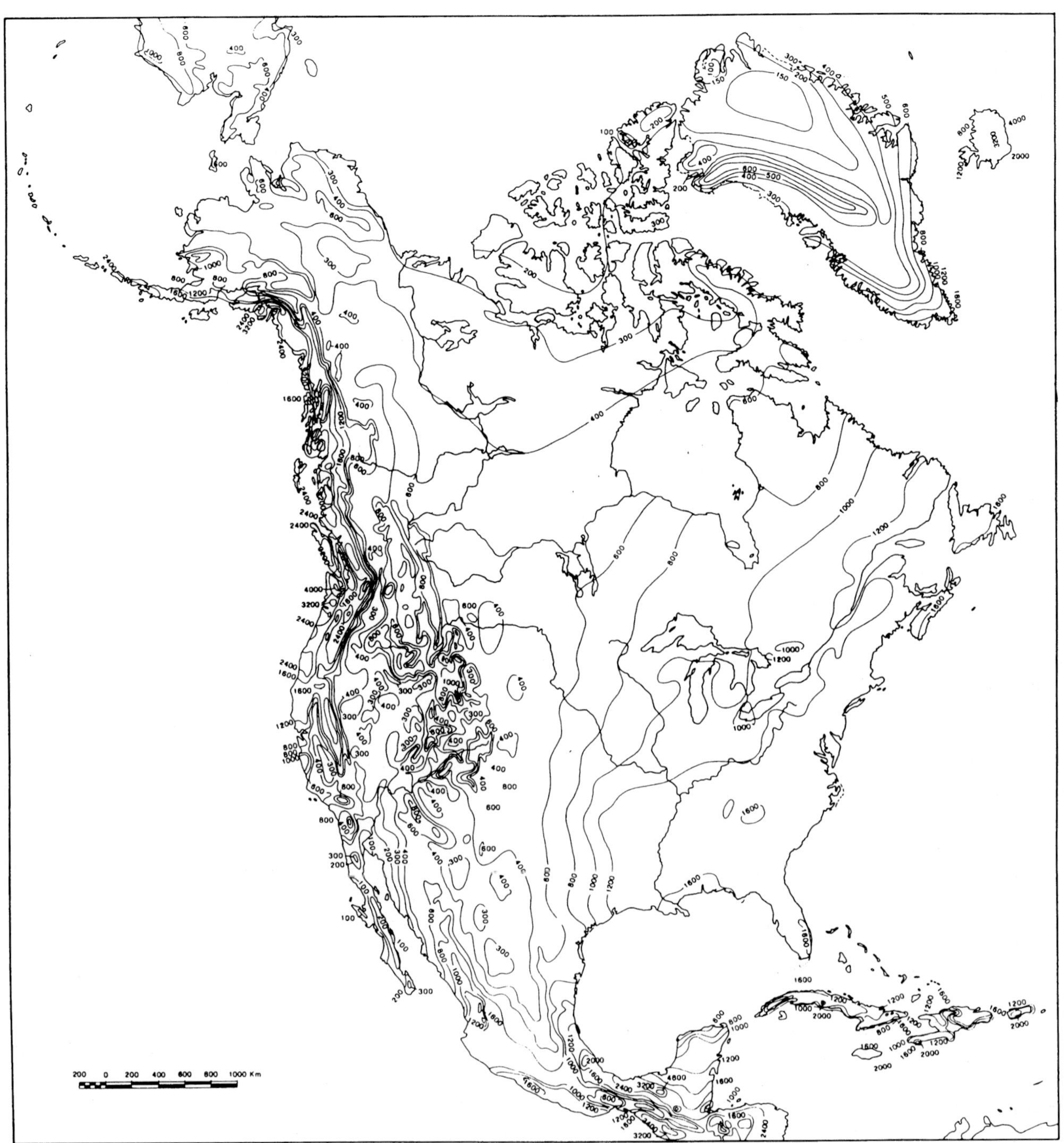

Figure 7.1 Mean annual precipitation in North America (modified from the Atlas of World Water Balance, UNESCO, 1978).

(fig. 7.1) is the result of this circulation pattern. Air masses from the Arctic generally do not contain much precipitable water vapor, because cold air holds less water vapor, but these Arctic air masses contrast sharply with warm, moist air moving up from the south and can therefore trigger widespread precipitation along zones of atmospheric convergence (fronts).

Precipitation types are classified as *cyclonic*, *convective*, and *orographic*, according to the principal mechanism producing vertical uplift (see chapter 5). Some regions of North America, such as the Rocky Mountains and Appalachian Mountains, experience all three types of precipitation, whereas other regions have more restricted regimes. Along the west coast, for example, convective storms are uncommon, and almost all of the precipitation is produced by cyclonic storms originating in the Pacific Ocean. These storms are often widespread, generating low-intensity rain at lower elevations, and snow or high-intensity rain at higher elevations. In most other regions of North America, precipitation is produced by a combination of cyclonic and convective storms. In the Midwest, most of the annual precipitation results from low-intensity storms associated with synoptic-scale systems, as opposed to thermal convection (Keables, 1989). As you move farther to the southeast and closer to the Gulf of Mexico, convective storms (including thunderstorms, mesoscale convective complexes, and tropical cyclones) become more common (Miller et al., 1983). Thunderstorms often occur as isolated events, but particularly severe weather, including lightning, tornadoes, hail, and intense rainfall, results when these storms occur in conjunction with cyclonic storms or orographic lifting.

Although it is common to associate severe weather and high-intensity rainfalls with arid areas, some of the most intense rainfalls in North America have occurred in the humid east (Maddox et al., 1979). A recent rain storm in the Blue Ridge Mountains of Virginia, for example, produced more than 600 mm in 6 hr on 27 June 1995 (Smith et al., 1996), which is near the maximum intensity for storms of that duration worldwide (Dingman, 1994). In this case, the movement of the storm relative to major topographic features played a key role in the production of rainfall. Other severe storms and high-intensity rainfalls are produced in this part of North America by the remnants of tropical cyclones that move inland and northward along the eastern seaboard (see fig. 21.8). These tropical cyclones can produce copious rainfall over widespread areas, independent of orographic effects. Storms caused by Hurricane Camille in 1969 dropped more than 700 mm of rainfall in 8 hr over parts of central Virginia (Williams and Guy, 1973). Along the Gulf Coast and in the Caribbean, tropical storms and hurricanes can be prolific rain-producers, and most of the records for short-term rainfall in North America have been set in this region. A hurricane passing over Florida on 5 September 1950 produced 980 mm of rain in 24 hr at Yankeetown. That record stood as the highest 24-hr rainfall in North America until 25 July 1979, when another tropical storm dropped 1092 mm in 24 hr near Alvin, Texas (Ludlum, 1982). Some tropical storms originate in the Pacific Ocean and move inland across Mexico and the southwestern United States. These storms, or *chubascos*, are an important source of moisture for agricultural regions along the west coast of Mexico (Anthes, 1982), and are the source of late-summer monsoon rainfall in the southwestern United States.

The seasonal distribution of precipitation varies considerably across North America, depending on how moisture sources and precipitation-generating mechanisms change through the year. Using monthly averages, we can distinguish four general patterns of annual precipitation: *winter maximum*, *summer maximum*, *evenly distributed*, and *seasonal augmentation*. These four patterns are remarkably consistent within broad regions of the continent (fig. 7.2; see also fig. 6.2b). A winter maximum in precipitation characterizes much of the west coast of North America (fig. 7.2a). As described previously, this pattern reflects more intense cyclogenesis in the North Pacific during winter. The summer maximum pattern is characteristic of a broad region in the interior part of the continent, extending from northern Mexico into the Great Plains (fig. 7.2b). This pattern results from the movement of warm, moist air masses from the Gulf of Mexico, which are able to penetrate into the interior in summer and displace the cool, dry Arctic air masses that dominate in winter. Next, there is a broad swath across eastern North America where precipitation is distributed more or less evenly throughout the year (fig. 7.2c). This pattern reflects a combination of climatic factors, but especially the influence of extratropical cyclones, which can be spawned in several different parts of the continent, the Gulf of Mexico, or the Atlantic Ocean. Finally, there is an area encompassing the southeastern United States, the Caribbean, and the region surrounding the Gulf of Mexico where the summer maximum in precipitation is augmented by rainfall from tropical storms and hurricanes, which reach their peak activity in September (fig. 7.2d).

7.2.1 Snow

Large parts of North America are seasonally snow covered, and all but the southernmost regions may experience snowfall at some time during the winter months. Not surprisingly, the greatest individual snowfalls, as well as the highest ratios of snowfall to annual precipitation, occur in the mountain ranges of western North America. Snowfall accounts for 60–70% of the total annual precipitation in the Rocky Mountains and Sierra Nevada (Serreze et al., 1999), and more than 90% of the annual precipitation in some high mountain basins (Minnich, 1986; Barry and Chorley, 1987; Kattelmann and Elder, 1991). High-latitude areas are likewise characterized by a high percentage of snowfall and extended seasonal snow cover. In Alaska and northern Canada above 55°N latitude, snow accounts for 50% of the annual precipitation (Karl et al., 1993) and covers the ground for about half of the year (Walsh, 1984). In southern Canada below 55°N latitude, about 30% of the annual precipitation falls as snow, and the duration of snow cover varies from 3 to 6 months. Much less of the United States experiences seasonal snow cover, and overall, snowfall accounts for only about 16% of the annual precipitation in the United States (Karl et al., 1993). However, snowfall and snowmelt are very important for water resources in the arid parts of the West (Diaz and Anderson, 1995), and snow is a potentially disruptive weather phenomenon in the major metropolitan areas of the East. The statistics of individual snowstorms and annual snowfalls are also most

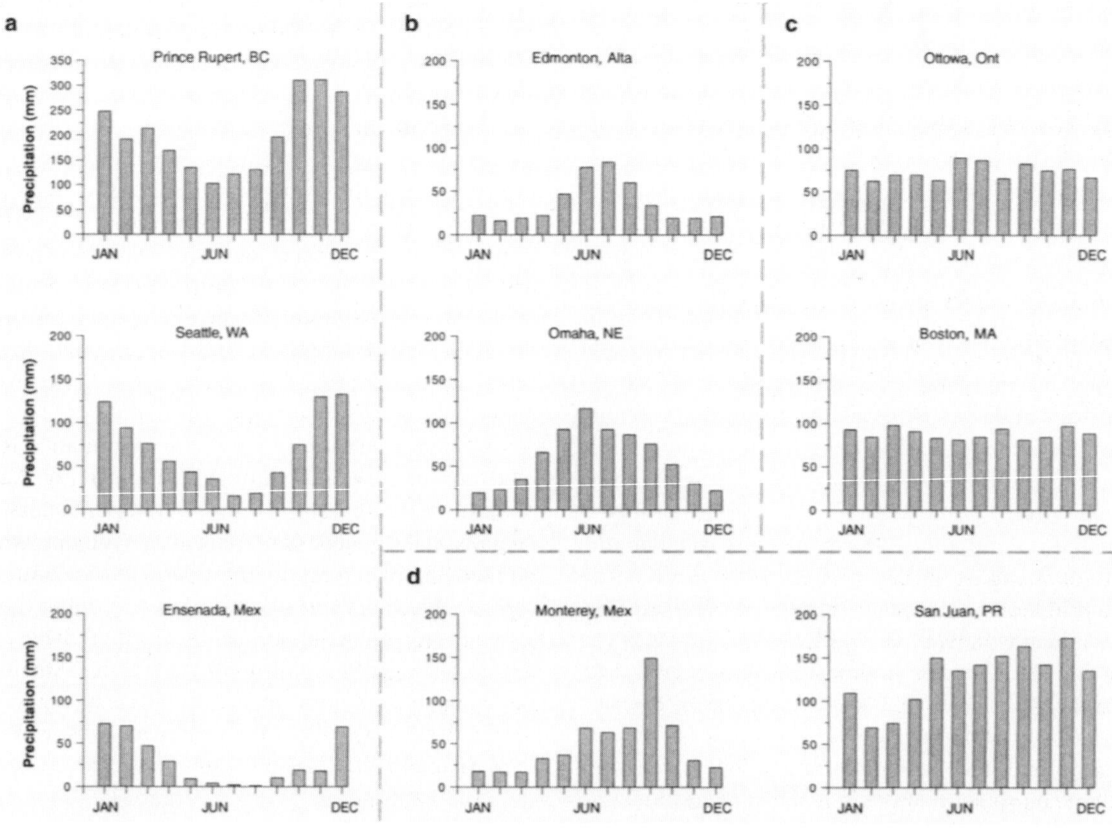

Figure 7.2 Seasonal distribution of precipitation in different regions of North America, including (a) the far West, (b) the Great Plains, (c) the East, and (d) the Southeast and Caribbean.

impressive in the mountains of the West. The record for 24-hr snowfall in North America is 193 cm, measured 14–15 April 1921 at Silver Lake, Colorado (Doesken and Judson, 1997). The record for annual snowfall is 2850 cm, measured at the Paradise Ranger Station on Mt. Rainier, Washington, in 1971–72 (Doesken and Judson, 1997). The western mountain ranges have always posed a challenge to travelers, and it is not uncommon for the major east–west transportation corridors through these ranges, such as Snoqualmie Pass (921 m) through the Cascade Range and Donner Pass (2203 m) through the Sierra Nevada, to be shut down for several days each winter because of heavy snow and avalanches.

7.2.2 Long-Term Trends in Precipitation

The prospect that elevated CO_2 levels will lead to global warming has important hydrologic consequences, particularly in moisture-limited agricultural regions. For this reason, there has been much interest and discussion over whether we are indeed seeing the signals of climate change in the long-term records of temperature and precipitation in North America. The strongest evidence for climate change is demonstrated by the apparent increase in mean annual temperatures over most continental areas, including North America (NRC, 1991; NAS, 1992; IPCC, 1990, 1996). Although there are uncertainties and biases underlying these data (see Karl et al., 1989), the general consensus appears to be that global temperatures have risen about 0.5°C during the twentieth century (NRC, 1991). It is not at all clear, however, that precipitation is following suit. Figure 7.3 shows a 102-year record of annual precipitation for the United States taken from the Climate Variations Bulletin, a monthly circular published by the NCDC. This time series was constructed by aggregating data from the 8000-plus weather stations that are part of the Cooperative Observer network of the National Weather Service (NWS). This record shows the decadal-scale variations in annual precipitation that have been discussed in several studies (Karl and Riebsame, 1984; Diaz and Anderson, 1995), but overall it does not appear that precipitation is increasing. If we consider only the last 40 years, there is some indication that precipitation is increasing, but most scientists have stopped short of saying this trend offers positive proof that

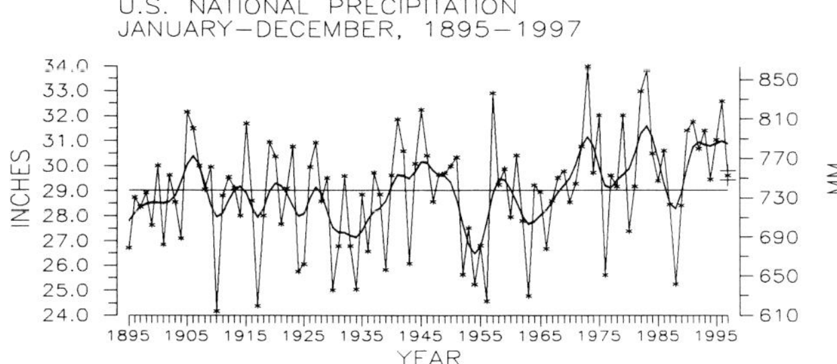

Figure 7.3 Trend in average annual precipitation for the United States, 1895–1997. Thick smooth curve is 9-point binomial filter; horizontal line is long-term average (Source: National Climatic Data Center, NOAA, 1997).

the climate is changing (Bradley et al., 1987; Karl et al. 1996). One problem with aggregated data sets such as this is that they may not show changes that are taking place at smaller scales or over shorter (seasonal) time intervals. Thus, certain areas, such as California and Texas, appear to be getting drier, whereas other areas, such as the upper Midwest, appear to be getting wetter (Karl et al., 1996). Another distinct possibility, not really brought out by these aggregate data sets, is that temperature and precipitation are becoming increasingly more variable, and thus we may see more extreme events—floods and droughts—in the future (see, for example, Katz and Brown, 1992; Karl et al., 1996).

Regardless of whether precipitation is increasing, increases in surface temperature could have an important effect on the proportion of snow versus rain, the duration of snow cover, and the timing and amount of snowmelt runoff. Results from models of global warming scenarios certainly suggest this may happen (see Gleick, 1989). However, comparatively few studies have looked at the historic record to see whether there have been significant changes in snowfall and snow cover over large areas. Karl et al. (1993) used ground and satellite observations to examine recent variations in snow cover and snowfall in North America, and found that, despite continent-wide increases in precipitation (including snow), there have been reductions in the extent of snow cover. Using 19 years of satellite observations, these authors showed that the area of snow cover in North America decreases by about 0.6×10^6 km^2 (or ≈7% of the total mean annual cover) for every 1°C increase in average temperature. Their analysis of snow and rain gauge records indicates that, in the last 40-plus years, precipitation has been increasing over all of Canada, but south of 55°N, the percentage of annual precipitation falling as snow has been decreasing. The Great Plains of the United States and Canada were identified as a temperature-sensitive region because the mean maximum temperature in fall and winter is not far above 0°C, so relatively small increases in temperature result in more rain and less snow.

7.3 Evapotranspiration

Most of the precipitation that falls on the continents is cycled through the soil and plant cover, with very little going directly into rivers and lakes. The characteristics of the land surface, such as soil type, vegetation density, and topography, govern the rates of interception and infiltration, which in turn affect both the quantity and quality of water generated as runoff. Water that is not intercepted by plants or does not run off immediately enters the soil and goes into temporary storage. Some of this stored water may percolate lower, eventually reaching a lake or stream as groundwater, and some of it may remain in the soil where it is slowly returned to the atmosphere via evaporation and transpiration. Hydrologists have known of the importance of evaporation and transpiration for a long time, but they have tended to treat them quantitatively as simply the residual of precipitation minus runoff. That situation has changed significantly in the last 20 years because of advances in modeling capabilities and improvements in remote sensing and other techniques used to characterize land surfaces (see Giorgi and Avissar, 1997).

7.3.1 Energy Balance Considerations

The importance of evaporation and the physics behind this process are often described in terms of the energy balance. On a global basis, about half the solar energy reaching Earth is reflected or absorbed by the atmosphere, and the other half is absorbed by the surface where it warms the soil and water. Of the amount absorbed by the surface, roughly one-third is returned to the atmosphere as outgoing long-wave radiation Q_l, one-sixth is transferred as sensible heat Q_h, and one-half is transferred as latent heat Q_e released in the process of evaporation (Kiehl and Trenberth, 1997). Evaporation is thus a key process for redistributing energy received from the Sun. The redistribution of energy via radiation and heat transfer is not uniform across Earth, however. In general, Q_e is high wherever there is a source of moisture, that is, over the oceans or in hot humid regions with lush vegetation, and Q_h is high over arid, sparsely vegetated regions of the continent. The relative contribution of the two can be expressed by the Bowen ratio, $B = Q_h/Q_e$. A plot of the Bowen ratio for North America (fig. 7.4) shows that sensible heat transfer dominates ($B > 1$) over much of the arid West, and latent heat transfer dominates ($B < 1$) in the northwestern, central, and southeastern parts of the continent. The prominent band of low values of B in the Great Plains region reflects the in-phase relation between precipitation and net radiation, both of which are high in summer and low in winter.

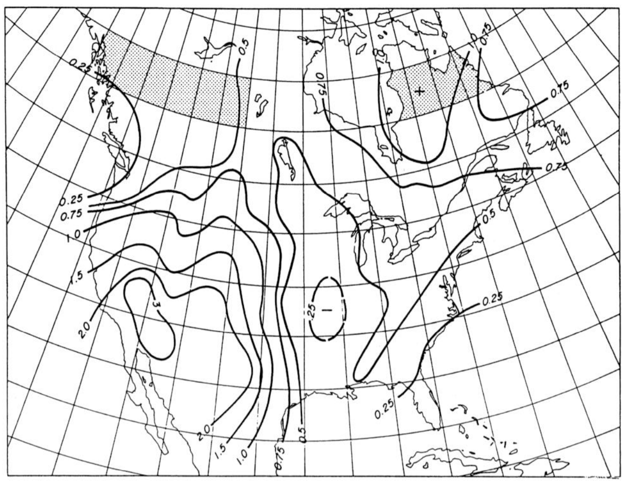

Figure 7.4 Plot of the ratio of sensible heat to latent heat (Bowen ratio, *B*) for North America. Shading indicates areas with insufficient data (from Hare, 1980).

7.3.2 *Actual Evapotranspiration*

The amount of water that actually evaporates from soils and vegetated surfaces cannot be measured directly, except at very small scales (e.g., by using lysimeters). Estimates of actual evapotranspiration for small- to moderate-size areas (10–1,000 km^2) can be modeled using physically based equations, such as the Penman–Montieth or Priestly–Taylor equations, driven by real-time weather observations. For larger and more heterogeneous areas, the best estimates of actual evapotranspiration are determined from the water balance equation by taking the difference between precipitation and runoff. Figure 7.5 shows a map of actual evapotranspiration for North America estimated in this way. The areas of high annual evapotranspiration in the southeastern portions of the United States and Central America coincide with high annual precipitation, lush vegetation, and high net radiation. The areas of low annual evapotranspiration in the western interior and in northern Canada reflect contrasting conditions of limited moisture, sparse vegetation, and/or low net radiation.

7.3.3 *Lake Evaporation*

Evaporation from open water bodies (lakes and wetlands) contributes relatively little to the overall water balance of North America because most of the lakes that could potentially contribute moisture are in high latitudes where the input from solar radiation is lower. At regional scales, however, evaporation from unfrozen lakes may exert significant influence on downwind precipitation, as demonstrated by enhanced lake-effect snowfalls south and east of the Great Lakes (see chapters 6 and 16). Furthermore, evaporative loss from individual lakes and reservoirs can have a significant impact on water resources, particularly in areas such as the southwestern United States where a few large reservoirs store much of the region's water supply. Studies conducted by the U.S. Geological Survey (USGS) in the early 1950s indicated that about one billion cubic meters of water evaporate annually from Lake Mead, Arizona; this represents an equivalent loss of more than 2 m of water from the entire surface of the lake each year (Harbeck et al., 1958).

7.3.4 *Soil Moisture*

Seasonal changes in soil moisture affect the water balance of large regions and play a key role in governing the short-term hydrologic response to individual precipitation events. Certain parts of North America, especially the continental interior, are particularly vulnerable to persistent surpluses and deficits in soil moisture, as evidenced by the catastrophic floods in 1993 and 1997, and the extended periods of drought in the late 1980s (Bonan and Stillwell-Soller, 1998).

The volumetric water content of soils fluctuates with time in response to precipitation and evapotranspiration. After a precipitation event, most soils (except those that are permanently saturated) go through a progression of soil-moisture states defined by specific water-holding mechanisms and moisture contents. Starting from an initially moist or perhaps saturated state, water will drain from the soil under the influence of gravity. However, a certain amount of water that was initially present will be retained in the root zone by capillary tension. At this point, the soil is said to be at *field capacity.* As time progresses, plants extract this water from the soil, reducing its moisture content until, eventually, all of the capillary water is exhausted. The water remaining in the soil at this point is held so tightly to the soil particles that it is not available to plants and they begin to wilt, thus bringing the soil to the *wilting point.* The difference in moisture content between field capacity and the wilting point is termed the *available water capacity.* Coarse-textured, sandy soils typically drain more rapidly than fine-textured soils because of the large difference in pore size. As a result, sandy soils reach their field capacity or wilting point at much lower water contents than silty or clayey soils.

Estimates of soil properties such as depth to bedrock, texture, permeability, and specific moisture contents, such as available water capacity (AWC), are normally included in published soil survey reports and maps. Recently, much of this map-based information has been converted from hard-copy to digital format, making it easier to incorporate soil data into climatological and hydrological models. In the United States, the U.S. Department of Agriculture Natural Resources Conservation Service has developed the State Soil Geographic (STATSGO) database for use in regional

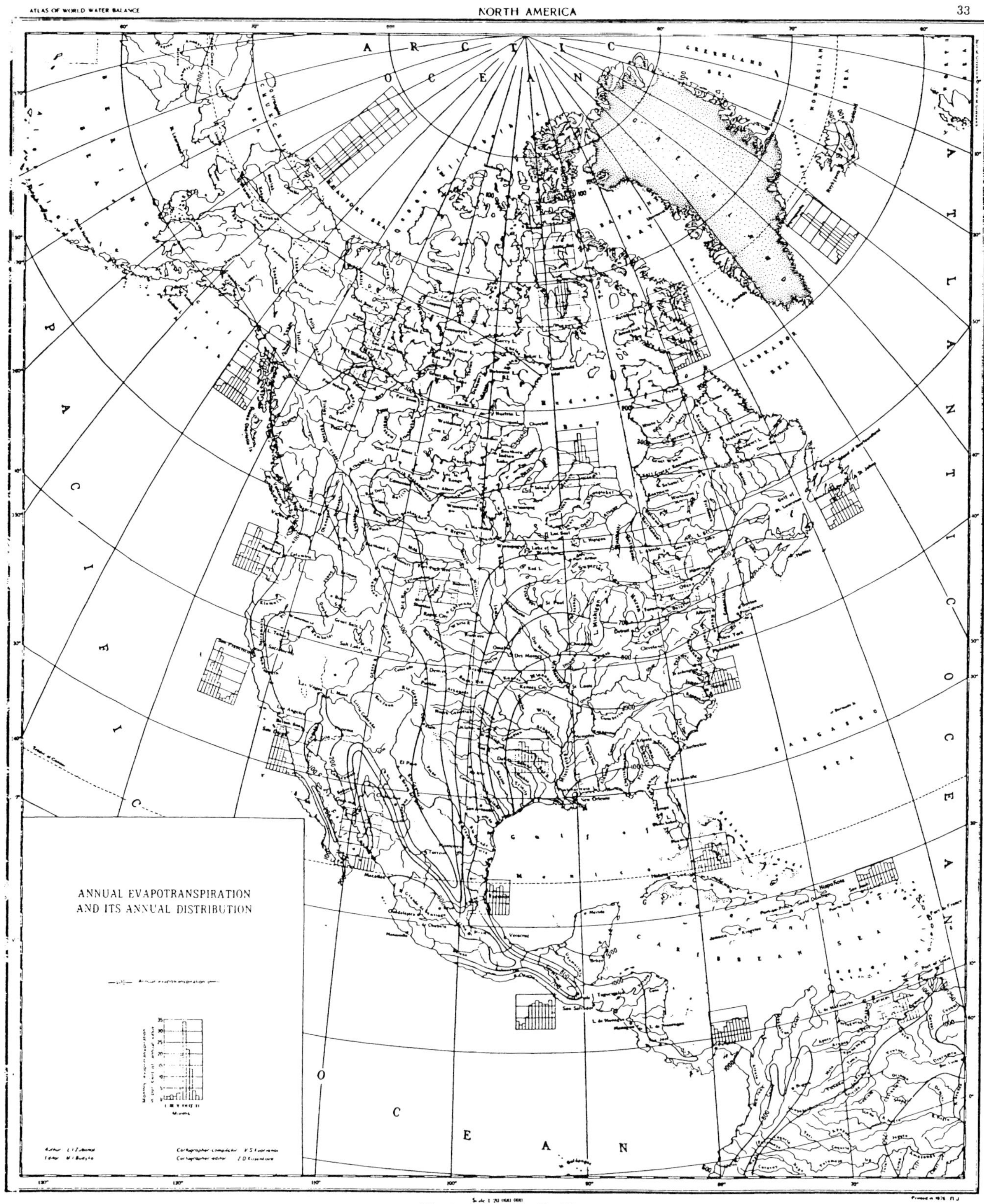

Figure 7.5 Plot of actual evapotranspiration over the North American continent (modified from the Atlas of World Water Balance, UNESCO, 1978).

planning and land management. In Canada, similar soils and land resource data are maintained in the National Soil Database of the Canadian Soil Information System (CanSIS). Figure 7.6, which shows a map of AWC for soils in the contiguous United States, is an example of the type of information that can be extracted from these databases. The general patterns in AWC observable in this map reflect regional variations in soil texture and thickness, which are in turn related to parent material and land-surface topography. Values of high AWC in the region southwest of the Great Lakes and down the Mississippi River valley coincide with areas of relatively low relief and thick/silty soils formed from transported parent materials; values of low AWC in the Rocky Mountains and far West coincide with thin rocky soils developed on bedrock residuum; and values of intermediate AWC in the Southeast and East reflect a mixture of soils developed on both residual and transported parent materials. Similar digital maps of soil texture, porosity, and depth to bedrock are available for much of North America (Miller and White, 1998). In addition to being useful for agricultural and land-management purposes, these and related data on land-surface characteristics (Loveland et al., 1991) are now being used in soil-vegetation-atmosphere transfer schemes to model changes in soil moisture in real time (Capehart and Carleson, 1994; Giorgi and Avissar, 1997).

7.4 Runoff

7.4.1 Annual Streamflow

The geographical patterns of annual runoff in North America reflect the combined influence of the weather systems that deliver precipitation and the sources of energy that drive evapotranspiration. A map of the ratio of mean annual runoff R to mean annual precipitation P (fig. 7.7) shows areas of high R/P in western and eastern Canada and in the northwestern United States. These regions are characterized by high precipitation and low net radiation, leading to annual surpluses in soil moisture and abundant runoff ($R/P > 0.6$). The wide swath of low R/P values in the interior part of the continent reflects the opposite case of moisture-limited conditions. Throughout much of the Great Plains and intermountain regions, less than 20% of the annual precipitation ends up as runoff. This pattern is particularly significant from the standpoint of agricultural resources because some of the most productive farmland

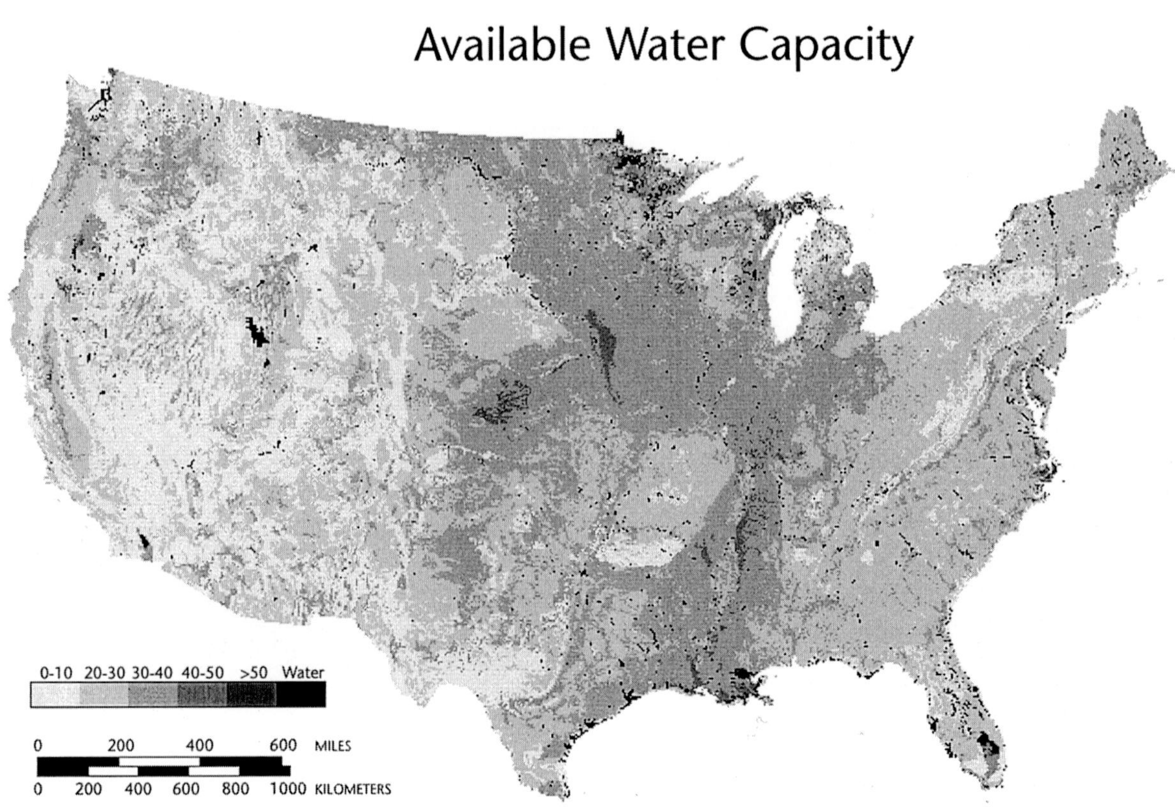

Figure 7.6 Digital map of available water capacity (AWC) for soils in the contiguous United States. This map was developed using information from the State Soil Geographic (STATSGO) database (from Miller and White, 1998).

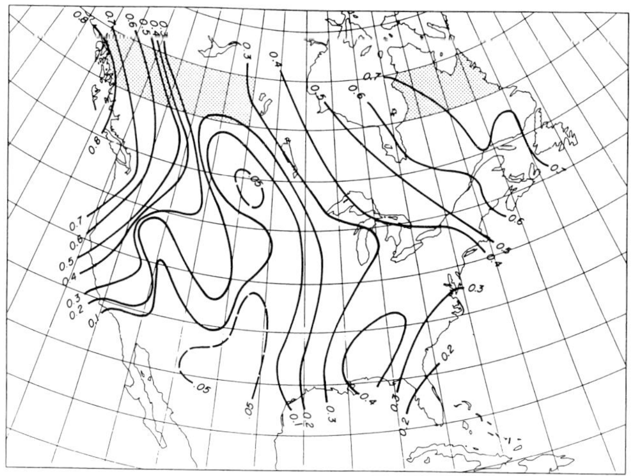

Figure 7.7 Ratio of mean annul runoff R to mean annual precipitation P for North America. Shading indicates areas with insufficient data (from Hare, 1980).

in North America is contained within this swath. Values of R/P are moderate to low in the southeastern United States, not because of a lack of precipitation, but because of high net radiation and high evapotranspiration.

The year-to-year variability of runoff in North America generally mirrors the variability in precipitation, with annual streamflow being more variable in semiarid regions and less so in humid regions (Riggs and Harvey, 1990). To illustrate, figure 7.8 shows time series trends in average annual discharge, Q_{ave}, for the Smith River near Bristol, New Hampshire, a stream in the northeastern United States, and Bear Creek at Morrison, Colorado, a stream in the midcontinent. The time series from the Smith River (fig. 7.8a) indicates that annual discharges rarely deviate from the mean by more than 50%, whereas the time series from Bear Creek (fig. 7.8b) indicates that mean deviations of 50% are common, and deviations of more than 100% have occurred several times since 1920. Both streamflow records also exhibit a certain amount of persistence, with some wet or dry periods lasting for about 10 years. These decadal-scale variations in streamflow have been noted in other studies, for example, by Meko and Stockton (1984), Diaz et al. (1985), and Probst and Tardy (1987). In related studies of spatial patterns in streamflow variability, Bartlein (1982) and Lins (1985, 1997) used principal components analysis to identify distinct geographic regions of the United States and Canada characterized by recurrent streamflow anomalies. The location and size of these regional flow anomalies appear to be linked to seasonally varying tracks of primary and secondary storm systems. Another key point highlighted in these studies is the tendency for anomalies of opposite sign to occur in adjoining or nearby regions. Perhaps the best example of this is the so-called "western opposition pattern," where streamflow anomalies of one sign in the Pacific Northwest coincide with anomalies of weaker but opposite sign in the Southwest (Lins, 1997). Meko and Stockton (1984) note, for example, that during the period 1932–80, five of the ten driest years in the lower Colorado River region were among the ten wettest in the Pacific Northwest, and of the ten driest years in the Pacific Northwest, five were among the ten wettest in the lower Colorado region. Another striking example of opposing precipitation and streamflow anomalies occurred in the midwestern and southeastern United States in 1993. At the same time that large parts of the upper Midwest were experiencing extreme rainfall and flooding in June and July 1993, most of the southeastern United States was under a drought (Lott, 1993). These two events were clearly related, and both owed their history to the positioning of a high-pressure system (the Bermuda High) over the southeast that repeatedly steered warm, humid air masses from the Caribbean northward into the Mississippi River valley (NOAA, 1994).

7.4.2 Droughts

Persistent dry spells, or droughts, have played a more important role in the history of North America than perhaps

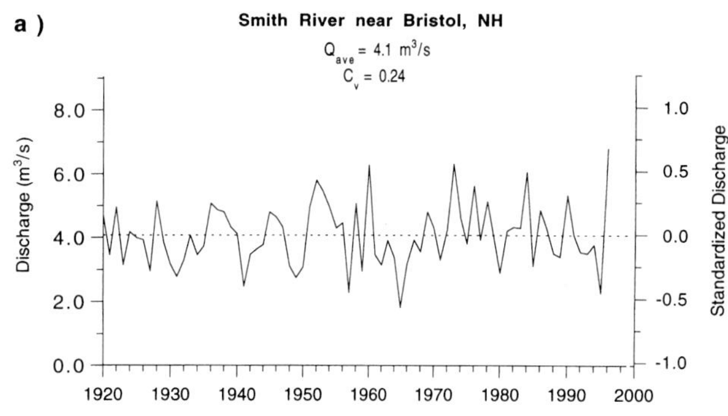

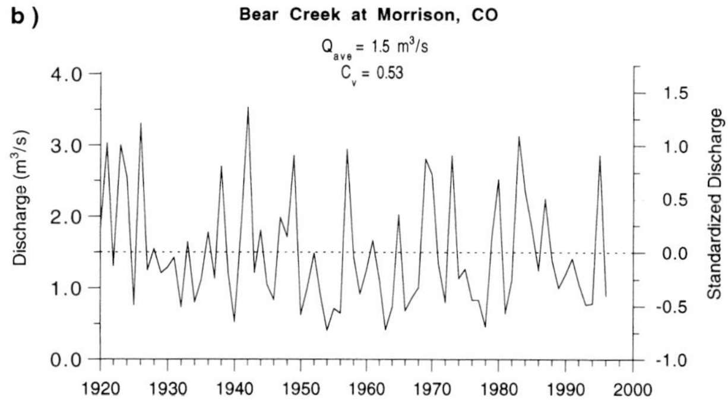

Figure 7.8 Long-term trends in average annual discharge, Q_{ave}, for (a) the Smith River near Bristol, New Hampshire, and (b) Bear Creek at Morrison, Colorado (source: USGS, Water Resources of the United States, *http://water.usgs.gov*).

any other climatological or hydrological phenomena. The severe drought of the 1930s caused tremendous hardship and financial losses in many parts of North America, but it also led to important changes in land-use practices and the establishment of soil conservation programs. A more recent drought in the Midwest in the late 1980s resulted in over $40 billion in losses, and perhaps as many as 10,000 fatalities (http://www.ncdc.noaa.gov/ol/reports/billionz.html). The latter event, coupled with intense heat and forest fires in the western United States, raised public awareness of the potential impact of climate change and transformed the debate about global warming into a more broad-based discussion involving scientists, politicians, and the public at large (although many scientists wondered at the time whether news media coverage of these events was more sensational rather than informative; see Schneider, 1988).

Although the droughts of the 1930s and late 1980s stand out as important hydrological events, analyses of historical and paleoclimatic data indicate that droughts are a recurring phenomenon in North America. Diaz (1983) used monthly precipitation and temperature data for each state in the contiguous United States to calculate values of the Palmer Drought Index for the period 1895–1981. His analysis shows that, besides the 1930s event, moderate to severe droughts occurred in the far West and Rocky Mountain regions from 1900 to 1905, in the southeastern United States in the mid-1920s, across much of the country in the mid-1950s, and in the northeastern United States in the mid-1960s. In addition to studying the instrumental record, a number of researchers have used dendrochronologic techniques to reconstruct the paleoclimate of many different parts of North America for the last several centuries (Bradley and Jones, 1992). These studies suggest that severe multiyear droughts equal to or worse than the 1930s event have occurred several times in the last 300–600 years (Bark, 1978; Duvick and Blasing, 1981; Meko et al., 1991). Almost no part of North America is immune from drought, with the Great Plains being undoubtedly the most drought-prone region of the continent. The atmospheric conditions favoring droughts, as well as the floods discussed subsequently, are described in chapter 5.

7.4.3 Floods

During the decade of the 1990s, people in many different parts of North America witnessed unusually large and destructive floods. Examples include the 1997 floods in the northern Great Plains, the 1996 floods in California, Oregon, and Washington, and the 1993 floods in the upper Mississippi River basin. Because of improved flood-warning and prediction systems, relatively few fatalities occurred as a direct result of these floods. However, property damages and monetary losses resulting directly and indirectly from these events were unprecedented. The NWS reports that since 1980 there have been nine floods in the United States with losses exceeding one billion dollars (http://www.ncdc.noaa.gov/ol/reports/billionz.html). In terms of monetary loss and impact on society, the 1993 Mississippi River basin floods were by far the worst of these events. This flood, and some of the controversy surrounding it, are discussed in more detail herein, but first a series of flood frequency curves for specific hydroclimatic regions of the western United States are presented to illustrate how the magnitudes of peak discharges vary systematically with precipitation and runoff-generating mechanisms.

Floods in North America can be produced by rainfall from frontal storms, convective storms, or hurricanes; by snowmelt; or by combinations of "rain-on-snow" events. In some regions, flooding occurs at roughly the same time of year in response to seasonal shifts in storm tracks or changes in temperature. Spring snowmelt floods in alpine and high-latitude areas are an example of this cyclical pattern. In other regions, floods can occur at almost any time of year and can be produced by more than one of the mechanisms listed previously. These "mixed population" flood regimes are characteristic of areas such as the upper Mississippi Valley (Knox, 1988), the northeastern United States (Magilligan and Graber, 1996), and the coastal and interior mountain ranges of the far West (Church, 1988).

Using these distinctions in flood-producing mechanisms, it is possible to define hydroclimatic regions within which flood frequency curves for individual rivers tend to have similar slopes and shapes. (Statistically, this means that the flood frequency distributions of individual streams in a region have similar coefficients of variation and skewness.) Figure 7.9 shows a set of flood frequency curves developed by Pitlick (1994) for five hydroclimatic regions in the western United States. These curves represent groups of rivers draining areas of 10–2300 km^2 within regions that are similar with respect to climatology and physiography. The data used in this study were derived from alpine areas above 2300 m elevation in the Colorado Front Range (Colorado Alpine); montane basins below 2300 m elevation in the Colorado Front Range (Colorado Foothills); montane basins in central and northern California (Sierra Nevada and Klamath Mountains); and moderate relief mountain ranges near the Pacific Ocean in northwestern California (Coast Ranges).

Although these flood frequency curves were developed for relatively small regions of the western United States, they characterize conditions within many other parts the continent. The steep flood frequency curve of the Colorado Foothills region is typical of semiarid areas where the largest floods are produced by intense thunderstorms (which in certain areas are sometimes associated with tropical storms; see Baker, 1977). In a region such as this, the 100-year flood may be more than ten times the mean annual flood. The very flat flood frequency curve of the Colorado Alpine region is typical of alpine or high-latitude areas

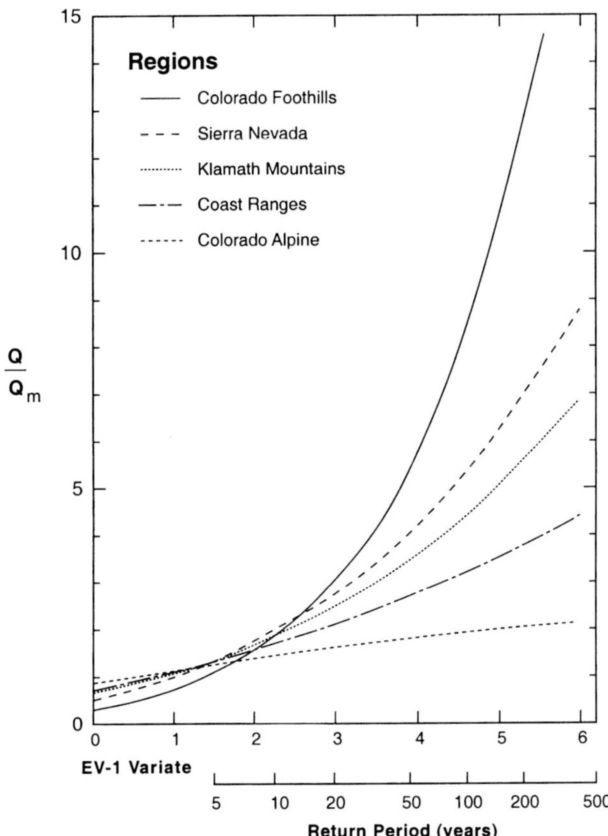

Figure 7.9 Normalized flood frequency curves for five regions in the western United States. These curves were obtained by normalizing annual flood data from a number of sites within each region by the mean annual flood, Q_m, and then fitting the regional curves with weighted moments of their respective flood frequency distributions (from Pitlick, 1994).

where floods are produced almost entirely by snowmelt. In this region, the 100-year flood is not even twice the mean annual flood (compare Church, 1988). The flood frequency curves for the three regions in California with intermediate slopes typify areas where large-scale frontal storms or rain-on-snow events produce the largest floods. In areas such as these, the 100-year flood ranges from three to six times the mean annual flood (compare Waylen and Woo, 1982). Of course, individual streams within regions may not follow these generalizations because of local differences in topography, soil types, and vegetation, but at regional scales, physiographic factors do not appear to be nearly as important in controlling the size of floods from moderate to large drainage basins as precipitation characteristics are (Pitlick, 1994).

The most extraordinary and costly individual flood event in the recent history of North America was the Mississippi River basin flood of late spring and summer 1993. This flood, or series of floods, ranks along with Hurricane Andrew and the Northridge, California, earthquake as one of the three largest historic natural disasters in the United States. The 1993 flood caused $15 to $20 billion in damage and resulted in 48 fatalities (NOAA, 1994). The report of the Interagency Floodplain Management Review Committee (IFMRC)—also known as the Galloway Report—indicated that more than 500 counties in nine states were declared disaster areas. Approximately $1.4 billion in crop disaster payments were issued to farmers who could not plant or harvest their crops because of excessive soil moisture. The Mississippi and Missouri rivers remained above flood stage for many weeks, causing extensive damage to communities and individual homes built in low-lying areas along the floodplain (IFMRC, 1994). News media coverage of these events dramatized the plight of people in the Midwest and heightened public awareness of the hazards posed by floods, but at the same time, encouraged hasty assessments of the causes of the floods, especially the role played by levees and agricultural practices.

The 1993 Mississippi River floods were caused by excessive rainfall, which occurred over an unusually large area characterized by high antecedent soil moisture (Kunkel et al., 1994; Changnon, 1996). The reoccurring and widespread nature of the rainfall were key factors in this event. Atmospheric convergence along a quasi-stationary ridge of high pressure to the southeast and a trough to the northwest spawned very large thunderstorms that tracked repeatedly across the Upper Midwest. Some individual storms covered areas of 250,000 km^2 or more (NOAA, 1994). Precipitation totals in the states of Iowa, Minnesota, and Illinois for 1-, 3-, and 12-month periods in 1993 were the highest in 100 years of record (Kunkel et al., 1994). Scores of cities and towns throughout the region received the equivalent of a year's worth of rainfall (or about twice the normal amount) from May to August 1993 (Pitlick, 1997).

Peak discharges, return periods, and durations of flooding in 1993 were greatest along the larger rivers in the region, namely, the Des Moines, Missouri, and Mississippi Rivers (Parrett et al., 1993). At St. Louis, Missouri, the Mississippi River remained above flood stage for more than 2 months, and it remained above the previous record flood level for almost 1 month (NOAA, 1994). Flooding along the Des Moines River at Des Moines, Iowa, caused major damage to the city's water treatment plant, leaving over 250,000 people without potable water for almost 2 weeks (Lott, 1993). Peak discharges exceeded the 100-year flood at six locations along the Missouri River and at two locations along the Mississippi River (Parrett et al., 1993). Major flooding also occurred on some tributaries of the Missouri and Mississippi rivers, but, in general, the return periods of peak discharges on the smaller rivers in the region were not nearly as high as they were on the larger rivers. Pitlick (1997) examined peak discharges at 100 USGS gauging stations distributed evenly across the flood-affected region and found that, in approximately two-thirds of the cases, peak discharges in 1993 were less than two times the mean annual flood, and fewer than 10% of the stations experienced floods greater than three times the mean annual flood. To emphasize this point, figure 7.10 shows a plot of peak discharge versus drainage area for the largest floods that have occurred in the United States (from Costa, 1987), and the 154 largest floods that occurred in the Midwest in 1993 (from Parrett, et al., 1993). None of the 1993 peak discharges exceeded the envelope curve for the largest floods in the United States, and only a few exceeded the envelope curve for the state of Iowa (dashed line, fig. 7.10). Most of the points lying above the Iowa curve correspond to main-stem Mississippi and Missouri river sites. The vast majority of points lie well below the envelope curve for Iowa, indicating again that floods on the smaller rivers were not as extreme in 1993 as they have been in the past. The difference in 1993, of course, was the fact that moderate flooding was occurring on streams and rivers all across the region, and at roughly the same time. In normal years, peak discharges equivalent to, or even greater than, the 1993 floods might occur on any individual tributary stream, but these events would be distributed in time and space such that they would have little effect on main-stem river discharges. Thus, although many people have argued that land-use practices in the upper Mississippi River basin and levees along the main-stem Mississippi and Missouri rivers exacerbated flooding, the data appear to indicate that it was the scale and persistence of the rainfall and excess soil moisture, more than any anthropogenic factor, that made the Great Flood of 1993 such a unique hydrologic event (see Pitlick, 1997, for further discussion of these issues).

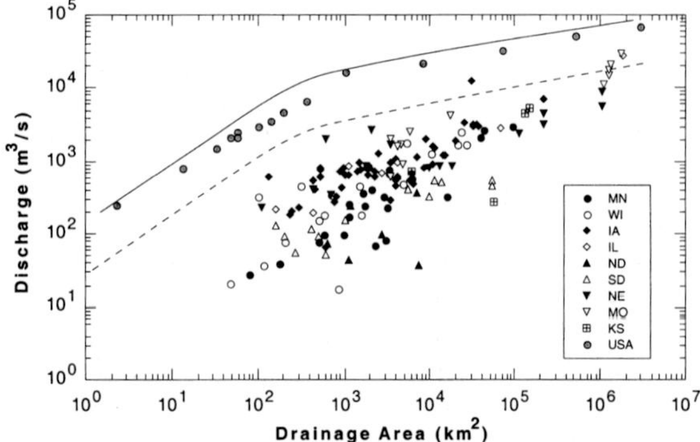

Figure 7.10 Plot of peak discharge versus drainage area for the largest floods in the United States, and the 154 largest floods in the Midwest in 1993. The upper solid line indicates an upper limit to peak discharges in the United States (modified from Costa, 1987); the lower dashed line indicates an upper limit to peak discharges that occurred in Iowa before 1993 (after Pitlick, 1997).

7.5 Conclusion

North America's streams, rivers, and lakes are influenced by a wide array of hydrologic processes, which are governed at regional scales by the weather patterns that deliver precipitation, the energy sources that drive evapotranspiration, and the mechanisms that generate runoff. The seasonal patterns of precipitation, evaporation, and runoff thus tend to be very consistent within broad geographic regions of the continent. For example, virtually all of the west coast of North America experiences a winter maximum in precipitation, and since plants are not utilizing much of the available water at this time, runoff volumes are moderate to abundant (> 70% of the annual precipitation above latitude 45°N) and the potential for flooding in winter is high. In contrast, most of the interior part of the continent experiences a summer maximum in precipitation. Much of this precipitation goes into satisfying soil-moisture deficits that have developed over the winter and into meeting the demands of plants as they grow and transpire water through the summer. The volumes of runoff carried by streams in the Midwest thus tend to be low (< 30%) in comparison to the annual rainfall. It is no coincidence that the worst droughts in North America have occurred in the Midwest because this area, with its vast agricultural resources, is very vulnerable to imbalances between moisture supply and demand. The southeastern United States–Caribbean region represents yet another interesting case where precipitation is abundant (1200–2000 mm across most of the region), but high rates of evapotranspiration driven by high net solar radiation result in moderate to low volumes of runoff (< 50% of annual precipitation).

In the last two decades, the emergence of new technologies and the availability of new information, or new ways to portray and utilize old data, have dramatically changed the way hydrologists in North America go about the business of predicting precipitation-runoff patterns. Key among these advances is the ability to provide spatially explicit (gridded) data on important hydrological variables such as rainfall intensity, snow cover, soil depth, available water capacity, land-surface slope, drainage patterns, and so on. New hydrologic models that utilize these data and incorporate additional real-time data from remote-sensing or ground-based instruments have recently been developed, and, when coupled with atmospheric models, these new models appear to offer significant potential for predicting seasonal trends in precipitation, evaporation, and runoff. Undoubtedly, this new generation of hydrologic models will not do so well in predicting the very extreme events (floods or droughts) or in predicting events very far into the future. But given the advances of just the last 10 years, there are many indications that the science of hydrology, as practiced in North America and elsewhere, is moving away from stochastically based approaches and more toward truly deterministic, physically based approaches.

Acknowledgments I thank Doug Miller of the Earth Systems Science Center, Pennsylvania State University, for providing the STATSGO image of available water capacity, and Nel Caine and Antony Orme for reviewing and commenting on previous versions of this manuscript.

References

Anthes, R.A., 1982, Tropical cyclones—Their evolution, structure, and effects, Meteorological Monographs, 19, 208 p.

Anthes, R.A., 1984, Enhancement of convective precipitation by mesoscale variations in vegetative covering in semiarid regions, Journal of Climate and Applied Meteorology, 23, 541–554.

AWRC (Australian Water Resources Council), 1976, Review of Australia's Water Resources, 1975, Australian Government Publishing Service, Canberra.

Back, W., J.S. Rosenshein, P.R. Seaber, 1988, *Hydrogeology*, v. O-2, Geology of North America, Geological Society of America, Boulder.

Baker, V.R., 1977, Stream-channel response to floods with examples from central Texas, Geological Society of America Bulletin, 88, 1057–1071.

Bark, L.D., 1978, History of American Droughts, *in* N.J. Rosenberg (ed.), *North American Droughts*, Westview Press, Boulder.

Barnston, A.G., and P.T. Schickedanz, 1984, The effect of irrigation on warm season precipitation in the southern Great Plains, Journal of Climate and Applied Meteorology, 23, 865–888.

Barry, R.G., and R.J. Chorley, 1987, *Atmosphere, Weather and Climate*, Methuen, New York.

Bartlein, P.J., 1982, Streamflow anomaly patterns in the U.S.A. and southern Canada, Journal of Hydrology, 57, 49–63.

Baumgartner, A., and E. Reichel, 1975, *The World Water Balance: Mean Annual Global, Continental and Maritime Precipitation, Evaporation and Runoff*, Elsevier, Amsterdam.

Bonan, G.B., and L.M. Stillwell-Soller, 1998, Soil water and the persistence of floods and droughts in the Mississippi River Basin, Water Resources Research, 34, 2693–2702.

Bradley, R.S., H.F. Diaz, J.K. Eischeid, P.D. Jones, P.M. Kelley, and C.M. Goodess, 1987, Precipitation fluctuations over Northern Hemisphere land areas since the mid-19th century, Science, 237, 171–175.

Bradley, R.S., and P.D. Jones, 1992, *Climate Since A.D. 1500*, Routledge, New York, 679 pp.

Brubaker, K.L., D. Entekhabi, and P.S. Eagleson, 1993, Estimation of continental precipitation recycling, Journal of Climate, 6, 1077–1089

Budyko, M.I., 1986, *The Evolution of the Earth's Biosphere*, Reidel, Dordrecht.

Capehart, W.J., and T.N. Carlson, 1994, Estimating near-surface soil moisture availability using a meteorologically driven soil-water profile model, Journal of Hydrology, 160, 1–20.

Changnon, S.A. (ed.), 1996, *The Great Flood of 1993*, Causes, Impacts and Responses, Westview Press, Boulder.

Church, M., 1988, Floods in cold climates, *in* V.R. Baker, R.C. Kochel, and P.C. Patton (eds.), *Flood Geomorphology*, Wiley, New York.

Costa, J.E., 1987, A Comparison of the Largest Rainfall-Runoff Floods in the United States with Those of the People's Republic of China and the World, Journal of Hydrology, 93, 313–338.

Diaz, H.F., 1983, Some aspects of major dry and wet periods in the contiguous United States, 1895–1981, Journal of Climate and Applied Meteorology, 22, 3–16.

Diaz, H.F., R.L. Holle, and J.W. Thorn Jr., 1985, Precipitation trends and water consumption related to population in the southwestern United States, 1930–83, Journal of Climate and Applied Meteorology, 24, 145–153.

Diaz, H.F., and C.A. Anderson, 1995, Precipitation trends and water consumption related to population in the southwestern United States: A reassessment, Water Resources Research, 31, 713–720.

Dingman, S.L., 1994, *Physical Hydrology*, Prentice Hall, New Jersey.

Doesken, N., and A. Judson, 1997, The Snow Booklet: A Guide to the Science, Climatology, and Measurement of Snow in the United States, Department of Atmospheric Science, Colorado State University, Fort Collins.

Duvick, D., and T.J. Blasing, 1981, A dendroclimatic reconstruction of annual precipitation amounts in Iowa since 1680, Water Resources Research, 17, 1183–1189.

Giorgi, F., and R. Avissar, 1997, Representation of heterogeneity effects in Earth system modeling: Experience from land surface modeling, Reviews of Geophysics, 35, 413–437.

Gleick, P.H., 1989, Climate change, hydrology, and water resources, Reviews of Geophysics, 27, 329–344.

Gleick, P.H. (ed.), 1993, *Water in Crisis*, Oxford University Press, Oxford.

Harbeck, G.E., 1958, Water-loss investigations: Lake Mead studies, U.S. Geological Survey Professional Paper 298.

Hare, F.K., 1980, Long-term annual surface heat and water balances over Canada and the United States south of 60°N: Reconciliation of precipitation, runoff and temperature fields, Atmospheres and Oceans, 18, 127–153.

Hem, J.D., A. Demayo, and R.A. Smith, 1990, Hydrogeochemistry of rivers and lakes, *in* M.G. Wolman and H.C. Riggs (eds.), *Surface Water Hydrology*, v. O-1, Geology of North America, Geological Society of America, Boulder, 189–231.

Hirschboeck, K., 1991, Climate and Floods, *in* R.W. Paulson, E.B. Chase, R.S. Roberts, and D.W. Moody (eds.), *National Water Summary 1988–1989, Hydrologic Events and Floods and Droughts*, U.S. Geological Survey Water Supply Paper 2375, Washington, D.C, pp. 67–88.

IFMRC (Interagency Floodplain Management Review Committee), 1994, *Sharing the Challenge: Floodplain Management Into the 21st Century*, Report of the Interagency Floodplain Management Review Committee, U.S. Government Printing Office, Washington, D.C.

IPCC (Intergovernmental Panel on Climate Change), 1990, Climate Change: The IPCC Scientific Assessment, J.T. Houghton, G.J. Jenkins, and J.J. Ephraums (eds.), Cambridge University Press, Cambridge.

IPCC (Intergovernmental Panel on Climate Change), 1996, Climate Change 1995: The Science of Climate Change, J.T. Houghton, L.G. Meira Filho, B.A. Callender, N. Harris, A. Katternberg, and K. Maskell (eds.), Cambridge University Press, Cambridge.

Karl, T.R., and W.E. Riebsame, 1984, The identification of 10– to 2–year temperature and precipitation fluctuations in the contiguous United States, Journal of Climate and Applied Meteorology, 23, 950–966.

Karl, T.R., D. Tarpley, R.G. Quayle, H.F. Diaz, D.A. Robinson, and R.S. Bradley, 1989, The recent climate record: What it can and cannot tell us, Reviews of Geophysics, 27, 405–430.

Karl, T.R., C.N. Williams, F.T. Quinlan, and T.A. Borden, 1990, United States Historical Climatology Network (HCN) Serial Temperature and Precipitation, Department of Energy, Oak Ridge National Laboratory ORNL/CDIAC-30, NDP-019/R1, 83 pp.

Karl, T.R., P.Y. Groisman, R.W. Knight, and R.R. Heim, 1993, Recent variations of snow cover and snowfall in North America and their relation to precipitation and temperature variations, Journal of Climate, 6, 1327–1344.

Karl, T.R., R.W. Knight, D.R. Easterling, and R.G. Quayle, 1996, Indices of climate change for the United States, Bulletin of the American Meteorological Society, 77, 279–292.

Kattelmann, R., and K. Elder, 1991, Hydrologic characteristics and water balance of an alpine basin in the Sierra Nevada, Water Resources Research, 27, 1553–1562.

Katz, R.W., and B.G. Brown, 1992, Extreme events in a changing climate: Variability is more important than averages, Climate Change, 21, 289–302.

Keables, M.J., 1989, A synoptic climatology of the bimodal precipitation distribution of the upper Midwest, Journal of Climate, 2, 1289–1294.

Kiehl, J.T., and K.E. Trenberth, 1997, Earth's annual global mean energy budget, Bulletin of the American Meteorological Society, 78, 197–208.

Knox, J.C., 1988, Climatic influence on upper Mississippi Valley floods, *in* V.R. Baker, R.C. Kochel, and P.C. Patton (eds.), *Flood Geomorphology*, Wiley, New York.

Kunkel, K.E., S.A. Changnon, and J.R. Angel, 1994, Climatic aspects of the 1993 upper Mississippi River basin flood, Bulletin of the American Meteorological Society, 75, 811–822.

Lins, H.F., 1985, Streamflow variability in the United States: 1931–78, Journal of Climate and Applied Meteorology, 24, 463–471.

Lins, H.F., 1997, Regional streamflow regimes and hydroclimatology of the United States, Water Resources Research, 33, 1655–1667.

Lins, H.F., F.K. Hare, and K.P. Singh, 1990, Influence of the atmosphere, *in* M.G. Wolman and H.C. Riggs (eds.), *Surface Water Hydrology*, v. O-1, Geology of North America, Geological Society of America, Boulder, 11–53.

Lott, N., 1993, The summer of 1993: Flooding in the Midwest and drought in the Southeast, Technical Report 93–04, National Climatic Data Center, Asheville, NC.

Loveland, T.R., J.W. Merchant, D.O. Ohlen, and J.F. Brown, 1991, Development of a land-cover characteristics database for the conterminous U.S., Photogrammetry and Remote Sensing, 57, 1453–1463.

Ludlum, D.M., 1982, Updating weather records: The American weather book, Weatherwise, 35, 123–126.

L'vovich, M.I., 1979, *World Water Resources and Their Future*, American Geophysical Union, Washington, D.C.

Maddox, R.A., C.F. Chappell, and L.R. Hoxit, 1979, Synoptic and mesoscale aspects of flash flood events, Bulletin of the American Meteorological Society, 60, 115–123.

Magilligan, F.J., and B.E. Graber, 1996, Hydroclimatological and geomorphic controls on the timing and spatial variability of floods in New England, USA, Journal of Hydrology, 178, 159–180.

Meko, D.M., and C.W. Stockton, 1984, Secular variations in streamflow in the western United States, Journal of Climate and Applied Meteorology, 23, 889–897.

Meko, D., M. Hughes, and C. Stockton, 1991, Climate change and climate variability: The paleo record, *in* Managing Water Resources in the West Under Conditions of Climate Uncertainty, National Academy Press, Washington, D.C.

Miller, A., J.C. Thompson, R.E. Peterson, and D.R. Haragan, 1983, *Elements of Meteorology*, Macmillan, Columbus, OH.

Miller, D.A., and R.A. White, 1998, A conterminous United States multi-layer soil characteristics data set for regional climate and hydrology modeling, Earth Interactions, v. 2 (available on-line at http://Earthinteractions.org).

Minnich, R.A., 1986, Snow levels and amounts in the mountains of Southern California, Journal of Hydrology, 89, 37–58.

NAS (National Academy of Sciences), 1992, *Policy Implications of Greenhouse Warming*, National Academy Press, Washington, D.C.

NOAA (National Oceanic and Atmospheric Administration) 1994, The Great Flood of 1993, Natural Disaster Survey Report, U.S. Department of Commerce, Washington, D.C., U.S. Government Printing Office.

NRC (National Research Council), 1991, *Managing Water Resources in the West Under Conditions of Climate Uncertainty*, National Academy Press, Washington, D.C.

Parrett, C., N.B. Melcher, and R.W. James, 1993, Flood discharges in the upper Mississippi River basin, 1993, U.S. Geological Survey Circular 1120–A, U.S. Government Printing Office, Washington, D.C.

Paulson, R.W., E.B. Chase, J.S. Williams, and D.W. Moody, 1993, National water summary 1990–91: Hydrologic events and stream water quality, U.S. Geological Survey Water Supply Paper 2400, Washington, D.C.

Pitlick, J., 1994, Relation between peak flows, precipitation and physiography for five mountainous regions in the western USA, Journal of Hydrology, 158, 219–240.

Pitlick, J., 1997, A regional perspective of the hydrology of the 1993 Mississippi River basin floods, Annals, Association of American Geographers, 87, 135–151.

Probst, J.L., and Y. Tardy, 1987, Long range streamflow and world continental runoff fluctuations since the beginning of this century, Journal of Hydrology, 94, 289–311.

Riggs, H.C., and K.D. Harvey, 1990, Temporal and spatial variability of streamflow, *in* M.G. Wolman and H.C. Riggs (eds.), *Surface Water Hydrology*, v. O-1, Geology of North America, Geological Society of America, Boulder, 81–96.

Schneider, S.H., 1988, The greenhouse effect and the U.S. summer of 1988: Cause and effect of media event? Climatic Change, 13, 113–115.

Serreze, M.C., M.P. Clark, R.L. Armstrong, D.A. McGinnis, and R.S. Pulwarty, 1999, Characteristics of the western United States snowpack from Snowpack telemetry (SNOTEL) data, Water Resources Research, 35, 2145–2160.

Slack, J.R., and J.M. Landwehr, 1992, Hydro-Climatic Data Network (HCDN): A U.S. Geological Survey Streamflow Data Set for the United States for the Study of Climate Variations, 1874–1988, U.S. Geological Survey Open File Report 92–129, Reston, Virginia.

Smith, J.A., M.L. Baeck, M. Steiner, and A.J. Miller, 1996, Catastrophic rainfall from an upslope thunderstorm in the central Appalachians: The Rapidian storm of June 27, 1995, Water Resources Research, 32, 3099–3113.

UNESCO, 1978, World Water Balance and Water Resources of the Earth, including Atlas of World Water Balance, UNESCO, Paris.

Wallis, J.R., D.P. Lettenmaier, and E.F. Wood, 1991, A daily hydroclimatological data set for the continental United States, Water Resources Research, 27, 1657–1663.

Walsh, J.E., 1984, Snow cover and atmospheric variability, American Scientist, 72, 50–57.

Waylan, P. and M.-K. Woo, 1982, Prediction of annual floods generated by mixed processes, Water Resources Research, 18, 1283–1286.

Williams, G.P., and H.P. Guy, 1973, Erosional and depositional aspects of Hurricane Camille in Virginia, 1969, U.S. Geological Survey Professional Paper 804, Washington, D.C.

Willmot, C.J., C.M. Rowe, and Y. Mintz, 1985, Climatology of the terrestrial seasonal water cycle, Journal of Climatology, 5, 589–606.

Wolman, M.G., and H.C. Riggs, 1990, *Surface Water Hydrology*, v. O-1, Geology of North America, Geological Society of America, Boulder.

Appendix: Website information

Data, reports, and other useful information pertaining to the hydrology of North America can be accessed through the following World Wide Web sites:

Weather and climate

http://www.ncdc.noaa.gov
http://www.cmc.ec.gc.ca/climate

Soils

http://www.essc.psu.edu/soil_info/
http://www.nrcs.usda.gov/
http://res.agr.ca/PUB/CANSIS/_overview.html

Hydrology and surface water

http://water.usgs.gov
http://www.cmc.ec.gc.ca/climate

8

Wetlands: A Hydrological Perspective

Ming-ko Woo

In Canada and the United States, it is generally agreed that wetlands are lands saturated for most of the growing season which allow the development of hydric soil, or the support of hydrophytes, or prolonged flooding to a depth of 2 m [Cowardin et al., 1979; National Wetland Working Group (NWWG) 1987]. The areas of wetlands are estimated to be 1.27×10^6 km^2 in Canada (Zoltai, 1988) and 0.28×10^6 km^2 in the United States (Hofstetter, 1983). These estimates should be qualified by the shrinking trend of wetland areas. The loss of wetlands in the United States, for example, may have amounted to about 2.2×10^3 km^{2}a^{-1} during the nineteenth and the twentieth centuries (Mitsch and Gosselink, 1986).

Mapping the distribution of wetlands in North America is not an easy exercise. Canadian wetland maps show the percentage of land occupied by wetlands (NWWG, 1988), whereas United States maps often indicate the distribution of various wetland types (e.g., marshes, swamps, bogs) or areas designated as peatlands, without indicating the percentage of wetland cover (see maps in Hofstetter, 1983). It is extremely difficult to compile different maps into a single wetland distribution sheet, particularly across the border between the two countries. Satellite mapping on a global scale cannot be harmonized easily with ground-based maps unless extensive ground-truthing is performed. In view of these difficulties, the wetland distribution map given in figure 8.1 is a subjective reconciliation from several sources, a sketch of wetland distribution on a continental scale subject to considerable error, including all inaccuracies inherent in the source maps. The overall pattern reveals a concentration of wetlands in the subarctic and boreal zones of Canada and Alaska, in glacial depositional terrain from Minnesota to southern Ontario, and on the coastal plain that lies to the south and southeast of the Fall Line in the United States (fig. 8.1).

Wetland literature has expanded greatly over the past few decades. Wetlands in North America have been studied from different viewpoints. Traditionally, much work has been done on the engineering aspects, for trafficability, peat production, or drainage and, nowadays, restoration of wetlands (e.g., MacFarlane, 1969; Galatowitsch and van der Valk, 1994). Wetlands have received considerable attention by biologists who primarily maintain an ecological focus (e.g., Moore and Bellamy, 1974). More recently, wetlands have been examined from multiple perspectives, concerning different wetland functions (e.g., Greeson et al., 1979) or the management of wetland resources (Williams, 1990). The hydrological factor has always been given consideration because the distribution, storage, and movement of water into, out of, and within the wetlands are of fundamental importance to their initiation, preservation, or degradation. In addition, the role of wetlands in climatic change, through their emission of "greenhouse gases," has interested soil scientists and climatologists (e.g., Matthews

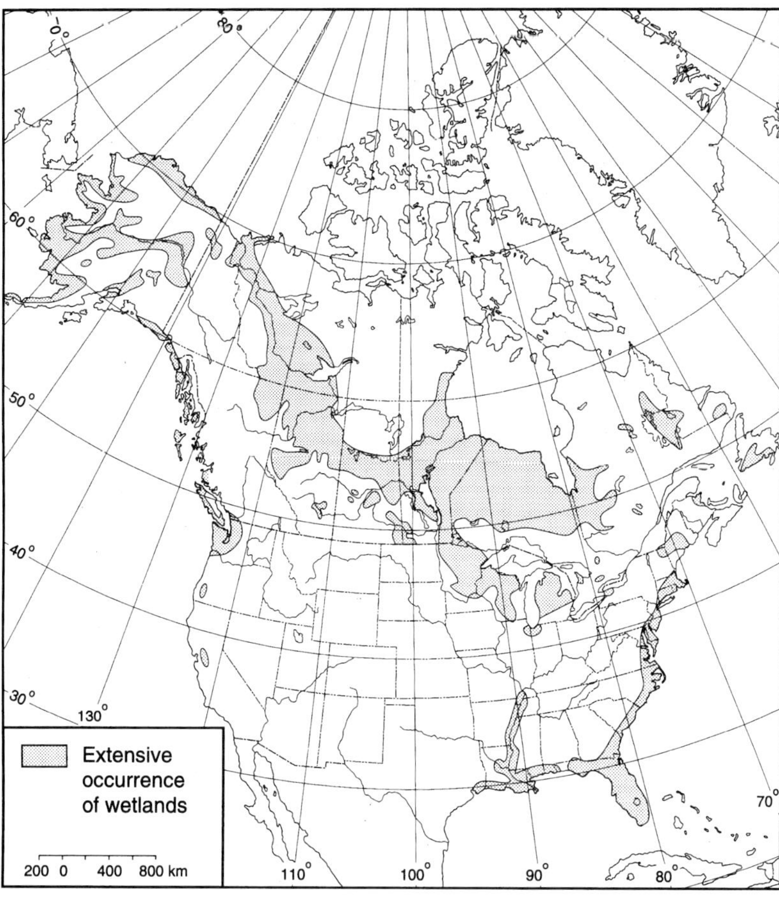

Figure 8.1 Generalized distribution of wetlands, based on various sources including maps that show wetland ecosystem distribution (Matthews and Fung, 1987), areas with >25% of wetlands in Canada (after Zoltai and Pollett, 1983), peat lands in continental United States (Hofstetter, 1983), and various soil maps in Alaska Regional Profiles (ARP, 1975).

and Fung, 1987; IPCC, 1996). Clearly, a review of all the environmental aspects of wetlands is beyond the scope of this chapter. Emphasis will be placed on the geographical setting and the hydrologic roles of wetlands. The hydrologic aspect is emphasized because water is the primary driving force for wetlands (Carter, 1986), and hydrology underlies all aspects of wetland investigations.

8.1 Wetland Classification

A remarkable number of classification systems have been devised for wetlands, as well as an inordinate collection of terms for different types of wetlands. The purposes of classification have been to describe wetland units with similar physical attributes as an aid to inventory, mapping, and management of resources or ecology. A mixture of climatic, hydrologic, biologic, and pedologic criteria are employed, using the gradients of various factors to delineate subgroups. These criteria are often interrelated, sometimes causing overlaps among groups (Bridgham et al., 1996). To cope with the profusion of wetland terminology, some terms are now disfavoured (e.g., "muskeg," a loose term referring to the organic terrain), whereas others are retained and better defined (Bridgham et al., 1996).

Two broad categories of classification can be distinguished: those developed at continental or national levels and those formulated for local regions or environments. National classification schemes cater to a broad range of users, hence are based on multi-disciplinary criteria and are hierarchical in structure, from the general to the specific level. In Canada, the widely adopted system was provided by the National Wetland Working Group (NWWG, 1987) of the Canada Committee on Ecological Land Classification. At the most generalised level, the Canadian scheme distinguishes five classes based on physiognomy. They include bogs, fens, marshes, swamps, and shallow open waters. At the next level, surface morphology of wetlands provides the modifiers for wetland forms. At a lower level, plant cover provides the designator for wetland types (Zoltai and Pollett, 1983). In the United States, the most commonly adopted scheme was designed by the Office of Biological Services of the U.S. Fish and Wildlife Service (Cowardin et al., 1979). Habitat is the criterion used to group wetlands into five major systems: marine, estuarine, riverine, lacustrine, and palustrine. These systems are then divided into subsystems according to location with respect to tides, permanence of flooding, or position from the lakeshore. Further subdivision into classes is based on bed materials or vegetation. Modifiers are added according to the water regime, water chemistry, and soil. Two features underline the difference between the Canadian and the United States schemes: (1) many wetlands in Canada have peat deposits, but in the United States, the presence of absence of organic soils does not play a major role in classification except as a minor modifier; and (2) four of the five systems in the latter scheme apply to wetlands as well as deep-water (water depth >2 m) habitats; only the palustrine system has a wetland component without its deep-water counterpart.

Different classification schemes have been devised for particular regions or to meet the needs of specific environments. One example is the various classifications proposed for the wetlands of the prairie region, most of which occupy depressions in glacial terrain, known as potholes in the United States or sloughs in Canada. These classification schemes cater to the needs of wetland inventory and mapping and the management of waterfowl habitats in a predominantly agricultural region with a history of wetland conversion to cultivation. Stewart and Kantrud (1971) distinguished eight classes according to vegetation and water-cover conditions at the center of the potholes: potholes that are ephemeral, temporary, seasonal, semipermanent, or tilled, alkaline wetland, and lakes that are permanent or intermittent and saline. Sloan (1972) lent some hydrologic support to the four pothole categories, noting that the average water depths in early June were dry, 0.23, 0.5, and 0.79 m, respectively, for the ephemeral, temporary, seasonal, and semipermanent potholes in parts of North Dakota. Millar (1976) noted the frequent occurrence of different vegetation zones that change from the edge to the center of a pothole and offered a classification scheme according to the predominant vegetation at the pothole center. His scheme was modified by Adams (1988) to place it in the context of the national classification system of Canadian wetlands (NWWG, 1987) to include wet meadow, shallow or deep marsh, intermittent or permanent open water, and intermittent saline lake. Woo et al. (1993) gave a hydrological representation of these wetlands by recognizing that the various pothole categories have different probabilities of being inundated for varying durations each year. The studies quoted demonstrate parallelism of the Canadian and United States classifications to suit the local environment and to address the regional needs. They also indicate the convergence of vegetation, salinity, and hydrologic criteria as the basis for wetland classification.

Many classification schemes are available but they are artificial or intended for convenience and should be open to change to accommodate new knowledge (Clymo, 1983). Some schemes may be considered too simple and do not recognize certain critical differences among wetland groups. For example, early schemes often ignored ecological differences between fresh and mixosaline inland wetlands (Hofstetter, 1983). Otherwise, a system formulated for wide geographical areas may be quite complex in its entirety (Cowardin et al., 1979) and yet offer little indication of the dynamics involved, such as wetland history or interactions with the atmosphere, hydrosphere, or geosphere. Despite these difficulties, some form of classification must be adopted for this chapter. Keeping the format simple and general, I herein distinguish wetlands primarily according to their hydrological attributes. Three major types are recognized, namely, bogs, fens, and inundated land. Bogs and fens are wetlands where the water table is at or near the wetland surface for a large part of the growing season, and inundated land is subject to frequent flooding by lateral spreading of surface water. Inundated land includes shallow-water wetlands (ponds, frequently flooded riverine or shore zones), marshes, and swamps. Their distinction is due to vegetation. Ponds have less than 2 m of water and may have aquatic plants or floating mats. Marshes may have aquatic vegetation, grasses, sedges, rushes, reeds, or shrubs, whereas swamps support trees and shrub. The presence of peat is not a prerequisite for this group of wetlands.

Hydrologically, a bog receives its water from rainfall and snowmelt, with an insignificant contribution from lateral input, though it has been observed that some bogs may receive groundwater discharge during parts of the year (Siegel and Glaser, 1987). A fen receives water from vertical as well as lateral sources, including groundwater inflow. Because their water sources differ, bogs (considered to be ombrogenous or rain-fed) tend to be acidic and low in nutrients, whereas fens (considered to be geogenous or groundwater-fed) are richer in minerals. The vegetation response to such differences in hydrological and chemical conditions is noticeable. Gorham and Janssens (1992), for example, observed a distinct bimodal distribution of the pH of surface water in peatlands across North America, corresponding with the dominance of Sphagnaceae at low pH and the prevalence of Amblystegiaceae above a pH of 5.7 (fig. 8.2), thereby allowing a bipartite division of peat lands into bogs and fens. This example illustrates the interrelationships among the hydrology, chemistry, and vegetation, confirming that a distinction between bogs and fens has implications beyond their hydrological control.

8.2 Factors Favoring Wetland Occurrence

The occurrence of wetlands depends on a surplus of water supply over water loss for a protracted period during the growing season. Water balance is therefore a primary consideration governing the formation and preservation of wetlands. Water sources include precipitation and lateral inflows from surface runoff and groundwater. Water losses are due to evaporation and lateral outflow, both as surface and subsurface drainage. The relative magnitudes of

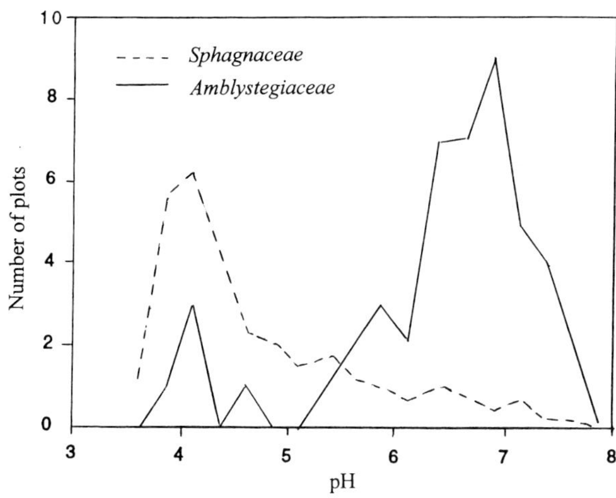

Figure 8.2 Gorham and Janssens' (1992) sample plots from five northern North American peatlands, which suggest a bipartite separation of moss communities along a pH gradient (bogs with high percentages of Sphagnaceae and fens with high percentages of Amblystegiaceae).

water sources and sinks vary for different types of wetlands under different climates (see fig. 2.11 in Orme, 1990). The presence of an impervious substrate and low gradient retard drainage losses, thus prolonging the duration of excess water storage to sustain inundation or persistent saturation of the land. Several factors contribute to such hydrological conditions.

8.2.1 Climate

Climate dictates the precipitation and evaporation regimes, thus determining the hydrological wetness of a region and its relative abundance of wetlands. One measure of wetness is the difference between annual precipitation and evaporation, using shallow lake evaporation as an approximate surrogate for wetland evaporation (fig. 8.3). Note that the pattern of lake evaporation is highly generalized for open-water surfaces, whereas the evaporation rates for wetlands, especially on a local scale, may deviate substantially from the regional trend. Figure 8.4, based on Winter and Woo's (1990) map, shows the difference between precipitation and open-water evaporation for North America. The zones with excess precipitation correspond to large parts of Alaska, the Hudson Bay Lowlands, and the eastern part of the continent with abundant wetlands (fig. 8.1). Under a very wet climate, as in southeast Newfoundland, blanket bogs can form on sloping terrain with gradients of 20°. On the other hand, despite a large surplus of precipitation over evaporation, the Western Cordillera in Canada does not support extensive wetlands because much of this water is shed quickly as runoff on steep slopes. Where the local relief is reduced, wetland formation is favored on flat sites or gentle slopes (e.g., parts of the Queen Charlotte Islands in British Columbia). Where there is a large water deficit, namely, where evaporation greatly exceeds precipitation, wetland formation is limited, as is evidenced in the Basin and Range Province of the western United States.

Besides controlling the rates of evaporation, the heat balance indirectly affects wetland development through peat production and decay, and the formation and thawing of ground frost. Under cold climates, peat production is limited by the rate of vegetation growth, but the decomposition process also slows; the opposite occurs in subtropical or warm temperature latitudes. The occurrence of ground frost, particularly permafrost, provides an impervious substrate that curtails vertical seepage, facilitating saturation of the materials that lie above the frost table. Many wetlands in the Arctic owe their preservation to the presence of such an impervious frozen zone not far below ground (see chapter 13).

8.2.2 Topography

Topography and location sometimes present hydrological conditions favorable to the occurrence of wetlands. Generalized interactions between topography and water supply may be modified from Novitzki's (1979) hydrologic classes of standing water and the interpretation broadened to encompass most wetland hydrologic environments. Topographic depressions collect precipitation, including drifting snow and surface runoff, thus favoring wetland formation at these surface-water depressions. For example, the presence of wetlands in some kettle holes in glaciated terrain in North Dakota allows water collected in depressions to contribute to groundwater recharge (LaBaugh et al., 1987) (fig. 8.5a). Littoral or riparian zones are surface-water slope margins prone to overflow of sea, lake, or stream waters, and their frequent flooding may produce wetlands. For example, along the Platte River south of Grand Island, Nebraska, groundwater in the floodplain wetland is generally lower than the level of water in the nearby river (Hurr, 1983). Groundwater flow tends to be perpendicular to the river banks as water flows from the river to the wetland aquifer (fig. 8.5b). Where topographic depressions receive groundwater discharge, fens or water bodies may be found (fig. 8.5c). Some prairie potholes are thus fed by groundwater of local or regional sources (Swanson et al., 1988). The emergence of groundwater in the depressions enriches the wetlands with minerals, sometimes creating saline ponds, including those found in the playas. Where slope concavities intercept groundwater flow (Novitzki's groundwater slopes), conditions favor wetland development. Many headwater wetlands owe their existence to the emergence of groundwater at a change of slope (Roulet, 1990) (fig. 8.5d).

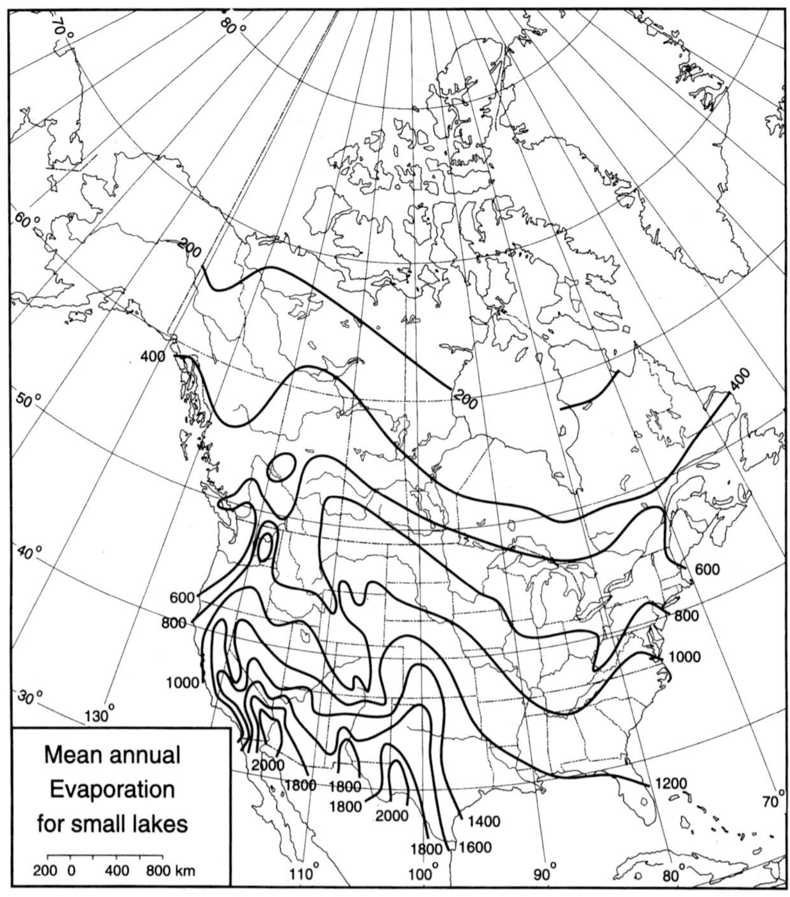

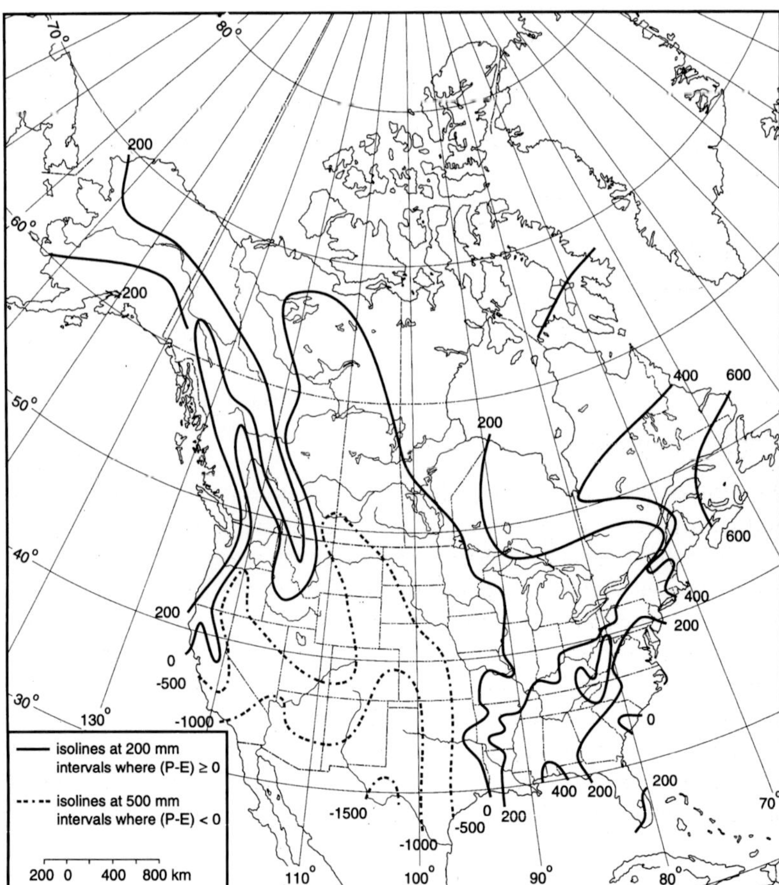

Figure 8.3 (*top*) Annual evaporation from lakes, in mm (after Lins et al., 1990).

Figure 8.4 (*at right*) Difference between precipitation and open-water evaporation (or P – E), in mm (after Winter and Woo, 1990). Isolines are at 200-mm intervals when the values are positive and at 500-mm intervals when precipitation is less than evaporation.

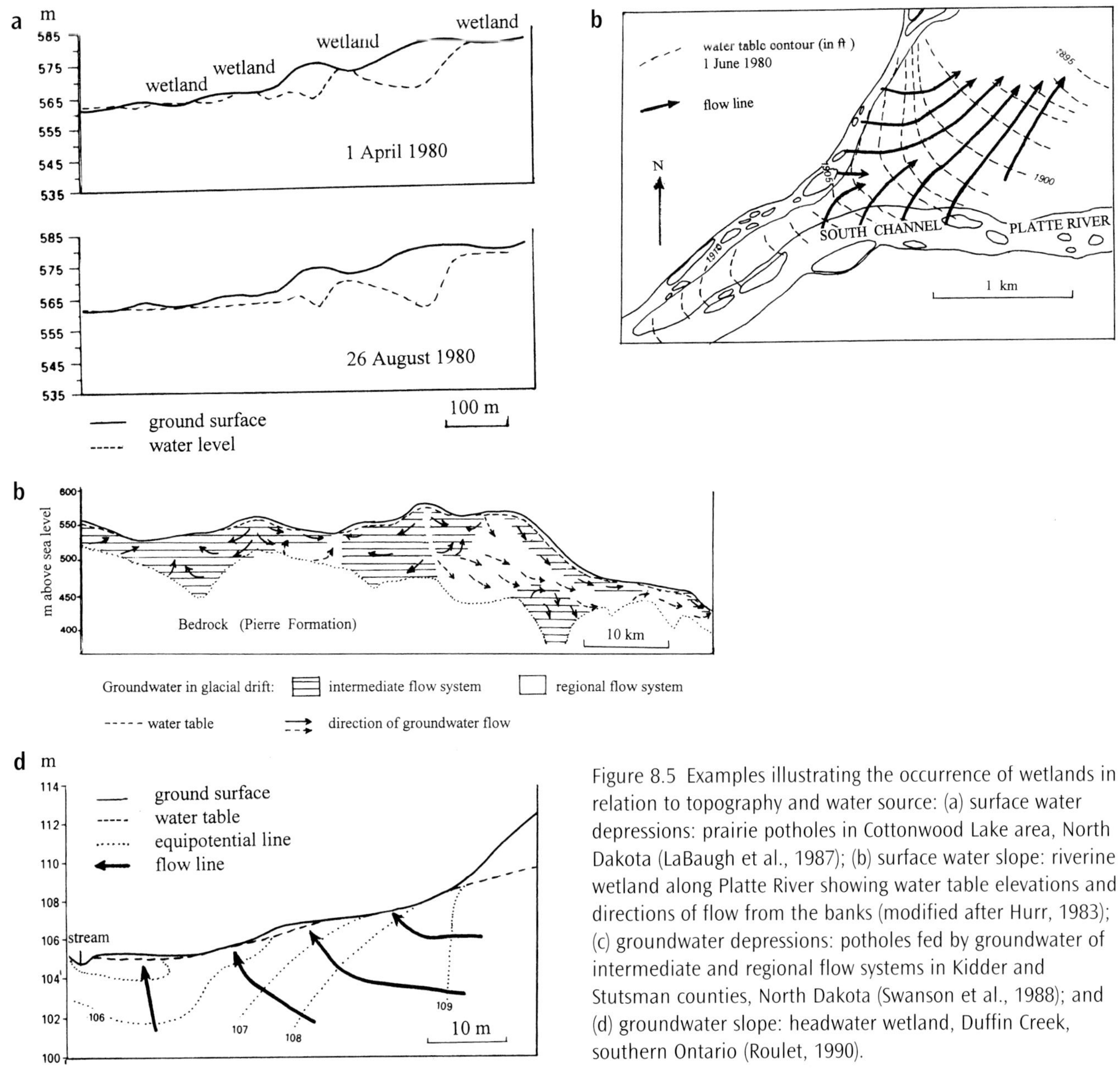

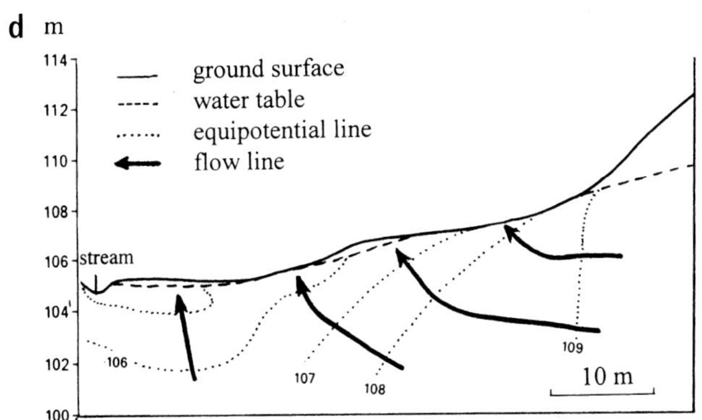

Figure 8.5 Examples illustrating the occurrence of wetlands in relation to topography and water source: (a) surface water depressions: prairie potholes in Cottonwood Lake area, North Dakota (LaBaugh et al., 1987); (b) surface water slope: riverine wetland along Platte River showing water table elevations and directions of flow from the banks (modified after Hurr, 1983); (c) groundwater depressions: potholes fed by groundwater of intermediate and regional flow systems in Kidder and Stutsman counties, North Dakota (Swanson et al., 1988); and (d) groundwater slope: headwater wetland, Duffin Creek, southern Ontario (Roulet, 1990).

8.2.3 Stratigraphy

Both the stratigraphy of the peat and its underlying mineral substrate strongly influence the occurrence of wetlands. An impermeable substrate composed of fine-grained materials or impervious bedrock inhibits vertical percolation, allowing water to stay close to the ground surface for long time periods. The geologic setting and the geomorphic history also play a role. Many parts of the Canadian Shield where the crystalline bedrock is little fractured provide an impervious base on which wetlands have developed. Formerly glaciated areas with their legacy of glacio-lacustrine or other clayey deposits also offer impermeable foundations for the establishment of wetlands. Figure 8.6 shows the stratigraphy of Beverly Swamp in southern Ontario where a lacustrine marl inhibits downward drainage below the peat. In permafrost areas, the frozen materials also act as an impervious substrate to favor the development of wetlands wherever the water supply is adequate. The abundance of wetlands in Alaska is favored by low relief in the coastal plains and valleys, ample water supply over loss, and the presence of permafrost (Ford and Bedford, 1987). The stratigraphy of bedrock and unconsolidated sediment also controls groundwater movement, sometimes retarding the flow, sometimes directing it toward the surface to sustain wetlands. Winter (1976) simulated the inter-

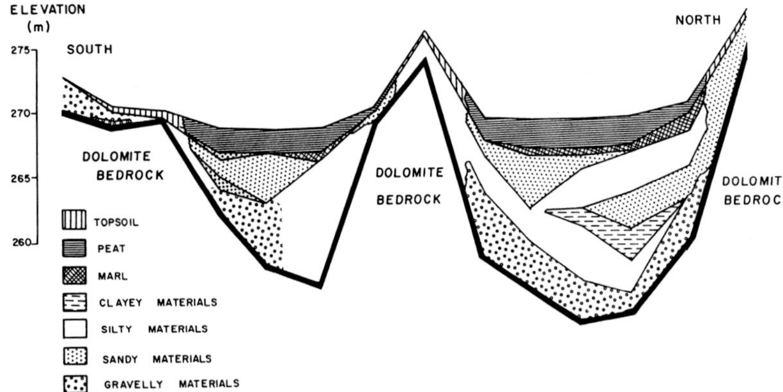

Figure 8.6 Stratigraphy of Beverly Swamp showing two bedrock depressions first infilled by clastic sediments, topped by an impermeable lacustrine marl layer with shell fragments, and then capped by peat.

action of lakes and groundwater discharge associated with variations in subsurface stratigraphy. Such quantitative exercises can be applied equally to wetlands.

In peatlands, the stratigraphy of the peat is important. In a typical profile, the peat is increasingly humified downward since the organic materials laid down earlier experience longer periods of humification. The bulk density (dry weight expressed as a ratio of sample volume) increases but the permeability decreases with humification (Baden and Eggelsmann, 1963). A downward decrease in the permeability of peat greatly restricts vertical water movement, but the pore space may still retain substantial amounts of water. This creates an environment conducive to the growth of mosses (e.g., *Sphagnum*), but it inhibits microbial activities, resulting in incomplete decomposition of plant remains (Foster, 1984). Based on plant records preserved in Labrador and Minnesota peatlands, Foster et al. (1983) proposed the evolution of wetlands with strings and flarks to involve peat accumulation where drainage is impeded, smoothing surface irregularities as adjoining depressions merge, followed later by peat degradation when pools of standing water reduce plant growth and pools deepen and widen as the peat decomposes. Although this evolutionary trajectory may not necessarily apply to wetlands of other locations, the accumulation and degradation of peat can alter the characteristics of the wetlands.

Autogenic (internally driven) successional development of peat leads to increasing control over its hydrology (Bridgham, 1996). Eventually, the peat layer as an aquifer may have its own perched water table. The chemical characteristics and circulation of water in the perched aquifer then become distinct from those of the regional groundwater below, exemplified by the evolution of pocosin wetlands in the eastern United States. During the last deglaciation, streams on the coastal plain were drowned by the rising sea level. Aquatic vegetation grew and accumulated in the estuaries and depressions where the flow was sluggish, and, as peat deposits extended beyond the confines of the depressions to cover interstream areas, a broad elevated topography and perched water tables developed (Daniel, 1981). Some wetlands comprise vegetation mats floating on water (fig. 8.7). Such wetlands have formed on small lakes occupying formerly glaciated terrain of the north-central United States, the southern Canadian Shield, the prairie potholes, and the waterways of the subarctic. Racine and Walters (1994) have also described floating mat wetlands in the Tanana lowlands of interior Alaska, where artesian groundwater discharge in the nonpermafrost bottomland provides mineral-rich water for the vegetation of these floating "fens."

8.2.4 Other Factors

On a regional scale, coastal areas undergoing uplift favor wetland development, notably in the Hudson Bay Lowlands where extensive wetlands are found alongside raised beaches as the land rebounds from deglaciation (fig. 8.8). Elsewhere, geomorphic processes may lead to impeded

Figure 8.7 A floating mat forming in a northern Ontario wetland.

Figure 8.8 Formation of new wetlands on the west coast of James Bay, northern Ontario, as the land rebounds after deglaciation. Raised beaches offer drier sites that favor tree growth in this wetland environment.

drainage or the blocking of water flow, causing prolonged water stagnation or flooding and eventually the formation of wetlands. Wetlands may develop on oxbow lakes or abandoned drainage channels when streamflow becomes sluggish or ceases and vegetation growth follows. Active sedimentation in deltaic or estuarine areas gives rise to shoaling, creating an environment suitable for wetland formation (Orme, 1990). Myriads of open-water wetlands on the Mackenzie River delta (fig. 8.9) have formed as levees grow around the perimeters of shallow lakes, some of which are still connected to the channels whereas others are cut off from river flow except during spring floods (Marsh and Hey, 1989). Sedimentation still occurs in the shallow lakes, but the aggradation of permafrost in the levees prevents or limits groundwater exchange between these wetlands and the river. Along the coast, barrier beaches frequently evolve to protect shoreward lagoons from wave action. Vegetation grows in the stagnant water or sluggish flow conditions of these lagoons and wetlands form, not only along the seacoast, but also around the Great Lakes (Geis, 1985).

Biological activities can lead to wetland formation. By damming the waterways, beavers alter the drainage (Naiman et al., 1986) and create ponds or shallow, open-water wetlands. These may develop into fens when organic debris and sediments accumulate to support aquatic and nonaquatic plant growth. The flow pattern and water balance of the resulting wetlands change as the beaver dams pass through various stages of decay (Woo and Waddington, 1990). Human activities, such as highway construction or damming of headwater streams, may also inadvertently alter the drainage to cause ground saturation, leading to wetland formation.

8.3 Hydrologic Behavior

Hydrology plays a central role in wetlands for these reasons: (1) it is fundamental to the preservation or degradation of the wetland environment; (2) it strongly influences the physical, chemical, and biological processes operating in the wetland ecosystem; and (3) the impact of land use on wetlands is mostly transmitted through an alteration of the hydrologic regime. An examination of the hydrologic characteristics of wetlands will elucidate the interaction between water and wetlands.

8.3.1 Peat Properties

Although peatland and wetland are not synonymous, most wetlands have a peat layer that accumulates above the mineral soil. Peat can be classified according to its botanical composition and state of decomposition (Clymo, 1983). One common botanical approach distinguishes mossy herbaceous (mainly grasses and sedges) peat and woody peat (fig. 8.10). The slow decomposition of organic matter is facilitated by the frequently saturated and anaerobic conditions of the wetland; the degree of paludification or humification is often represented by the von Post scale (see glossary at the end of this chapter). The fiber content of peat, defined in terms of particles greater than 0.15 mm, decreases with increasing humification. The peat is considered to be fibric, hemic, or sapric when the fiber content is above 67%, between 33 and 67%, and below 33%, respectively (Boelter, 1969).

The hydrologic behavior of peat depends on its structure, which is a function of the botanical composition, the

Figure 8.9 Myriads of shallow lakes, representing open-water wetlands, formed on the Mackenzie River delta; some lakes are connected to the river via channels, others are completely cut off from river flow except during spring floods.

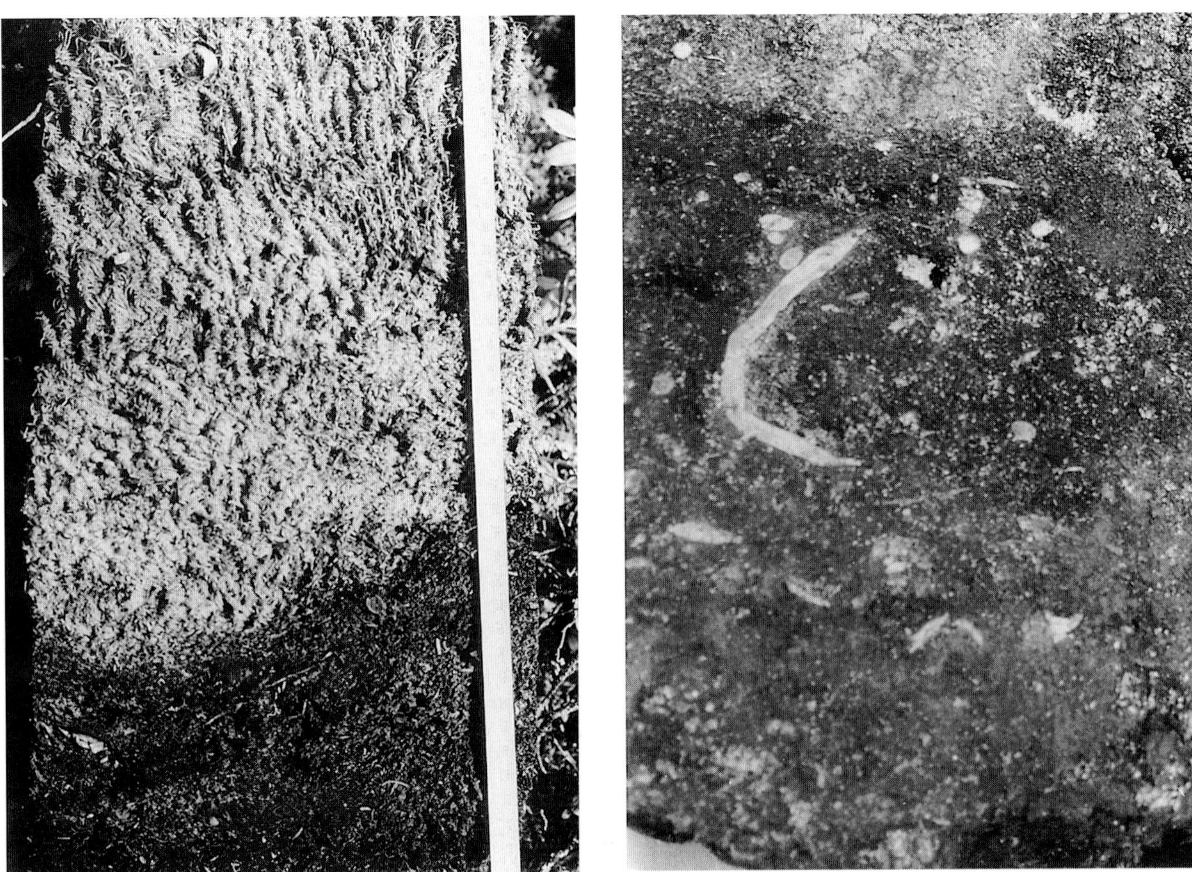

Figure 8.10 (left) *Sphagnum* peat overlying herbaceous peat; (right) woody peat with many plant roots.

manner of deposition and compaction by overburden, degree of humification and the amount of dilation by roots, desiccation and frost cracks. Peat can store large quantities of water in the pore space, often reaching over 80% by volume at saturation (Boelter, 1964). However, water release by gravitational drainage (measured by specific yield, expressed as a fraction of the peat sample volume) varies greatly, depending on the state of peat decomposition. As a result of smaller pore size, more humified peat has a lower specific yield because it can withhold more water against gravity drainage. Thus, for a Wyoming mountain bog, the well-decomposed peat at a depth of 0.36–0.48 m has a specific yield of 0.076 compared with 0.223 for the surface peat (Sturges, 1968), whereas for the bogs in north-central Minnesota, Boelter (1965) obtained a range of 0.79 for live moss to 0.10 for well-decomposed peat at a depth of 0.5–0.6 m.

Vertical change in peat structure is accompanied by a downward decrease in permeability and hydraulic conductivity expressed as a rate of flow. Boelter (1965), for instance, found that the value dropped from 3.8 mm s^{-1} for undecomposed mosses at the surface, to 7.5×10^{-5} mm s^{-1} for moderately decomposed herbaceous peat at a depth of 0.7–0.8 m in Minnesota bogs. Rycroft et al. (1975) provided additional field data from various parts of the world that generally support this trend. Large variations exist, however, partly due to the differential rates of sedimentation or decomposition at various depths, and to the local presence of fissures or plant roots as preferred flow conduits. In addition to vertical variability, horizontal hydraulic conductivity may be notably different from the vertical values. Dai and Sparling's (1973) study of an Ontario wetland suggests that the ratio of horizontal to vertical component is larger for herbaceous than for mossy peat. An additional complication was reported by Ingram et al. (1974), who observed that hydraulic conductivity increases as the hydraulic gradient steepens, notably for the humified peat, therefore violating an assumption of Darcy's Law, which requires the two variables to be independent. On the other hand, Hemond and Goldman (1985) suggested that such observed behaviour is an artifact of the conductivity measurement; for practical purposes, Darcy's Law remains appropriate for wetland hydrologic modeling.

For most peat profiles, the near-surface layer is hydrologically much more dynamic, being able to store and transmit more water than the lower zone. Based on the idea proposed by Ivanov and elaborated by Romanov (1968), Ingram (1978) proposed a distinction of the peat profile into an upper layer, which includes the living moss or lichen, called the acrotelm ("topmost marsh"), overlying a saturated layer of low hydraulic conductivity, called the catotelm ("down marsh"). Hydrologic features of the acrotelm include an oscillating water table, variable moisture content, and high hydraulic conductivities.

Under a cold climate, water held in the peat freezes in the winter to yield abundant ground ice. The large thermal conductivity of ice further enhances heat loss, but, during the thawing period, the ice occupying the pore space will be replaced by water or air, both of which have lower thermal conductivity to inhibit ground thaw. Some permafrost patches in the subarctic region are created this way. The presence of permafrost will impede vertical drainage and perpetuate wetlands such as those found in the bottomland of subarctic valleys.

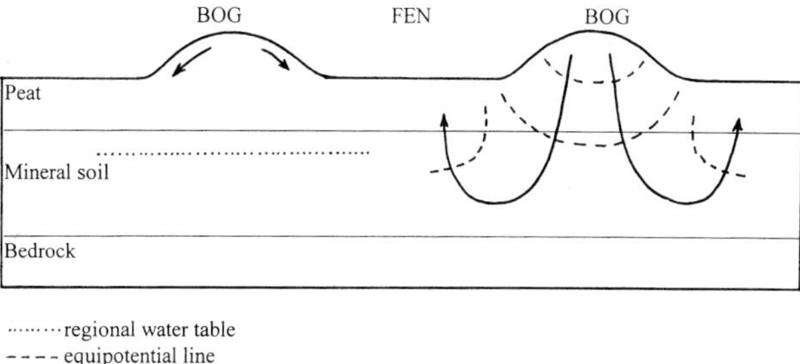

Figure 8.11 Conceptual flow directions for bogs and fens (modified after Bridgham et al., 1996) showing (left) how a perched water table in the bog may shed water laterally or (right) how a groundwater mound in the bog may create sufficient hydraulic head to drive vertical flow cells.

8.3.2 Water Sources

Wetlands receive water from precipitation, with rainfall being a major source, surpassed by snowmelt farther north and supplemented by fog under humid maritime climates such as those along the temperate Pacific Coast, southeastern Newfoundland (Price, 1992a), and along the Arctic coast of Alaska (Dingman et al., 1980). Precipitation is the primary, if not the only, source of water for bogs. For fens and inundated land, lateral water supply is important, often overwhelming the direct contribution from atmospheric sources. Lateral inflow to lacustrine, riverine, and coastal wetlands may be via groundwater discharge, but surface flow that results from flooding of the land is usually more important. Flooding may be caused by a large seasonal input of meltwater or rainwater from the catchment, increased lake levels, tidal effects, high river flows, or backwater effects whereby streamflow is blocked by high water levels downstream.

In terms of the pattern of groundwater flow, alternative conceptualizations have been proposed for bogs and fens (fig. 8.11). In one view, a perched water table occurs in the bog to shed water to the adjacent fen (Ingram, 1982). In the other, a groundwater mound creates a strong enough hydraulic head to drive the water downward, allowing the flow to pass through the peat column and into the underlying mineral soil (Siegel, 1983). This sets a convective flow in motion, and water resurfaces from beneath the adjacent fens. Both situations may occur (Bridgham et al., 1996), probably depending on the stratigraphy and local hydrologic conditions.

The connection between prairie wetlands and groundwater flow systems has received much attention. Working in the Canadian prairies, Toth (1963) and Meyboom (1967) postulated that there are several groundwater flow systems overlying one another: a shallow, local flow system is recharged at high water-table locations and discharged at adjacent depressions; an intermediate system is located at greater depth and discharges at major lowlands to feed the rivers, lakes, or wetlands; and a regional flow system underlies these two systems, discharging at some distance beyond the recharge zone of the region. Lissey (1971) modified the concept to suggest that both recharge and discharge occur mostly in the depressions where lakes and wetlands are found (fig. 8.12). The fact that the uplands contribute little to groundwater recharge was verified in Manitoba (Mills and Zwarich, 1986) and in the Cottonwood Lake area of North Dakota (LaBaugh et al. 1987). Both studies found that water beneath the uplands does not move very far downward before it is taken up by evapotranspiration. Also, some wetlands showed seasonal reversal of groundwater flow, seeping from the wetland to the groundwater system when the water level is high in the spring, then receiving groundwater discharge in the summer when evapotranspiration from the wetland depresses the wetland water level. With regard to wetlands in the depressions, those related to the local flow system tend to have a less reliable water supply. Those fed by the regional system remain wet for prolonged periods, even during the very dry years, but their water tends to be highly mineralized (Winter, 1989; Winter and Rosenberry, 1995).

8.3.3 Evaporation

Where the wetland surface is entirely or largely covered by open water, evaporation proceeds at a rate similar to that of small lakes. The map given by Lins et al. (1990, reproduced herein as fig. 8.3) indicates the regional pattern for Canada and the conterminous United States. However, several wetland conditions modify the local evaporation regime. Under ice-covered conditions, evaporation from open water is inhibited, therefore the duration of ice on wetlands affects the annual evaporation total. The advec-

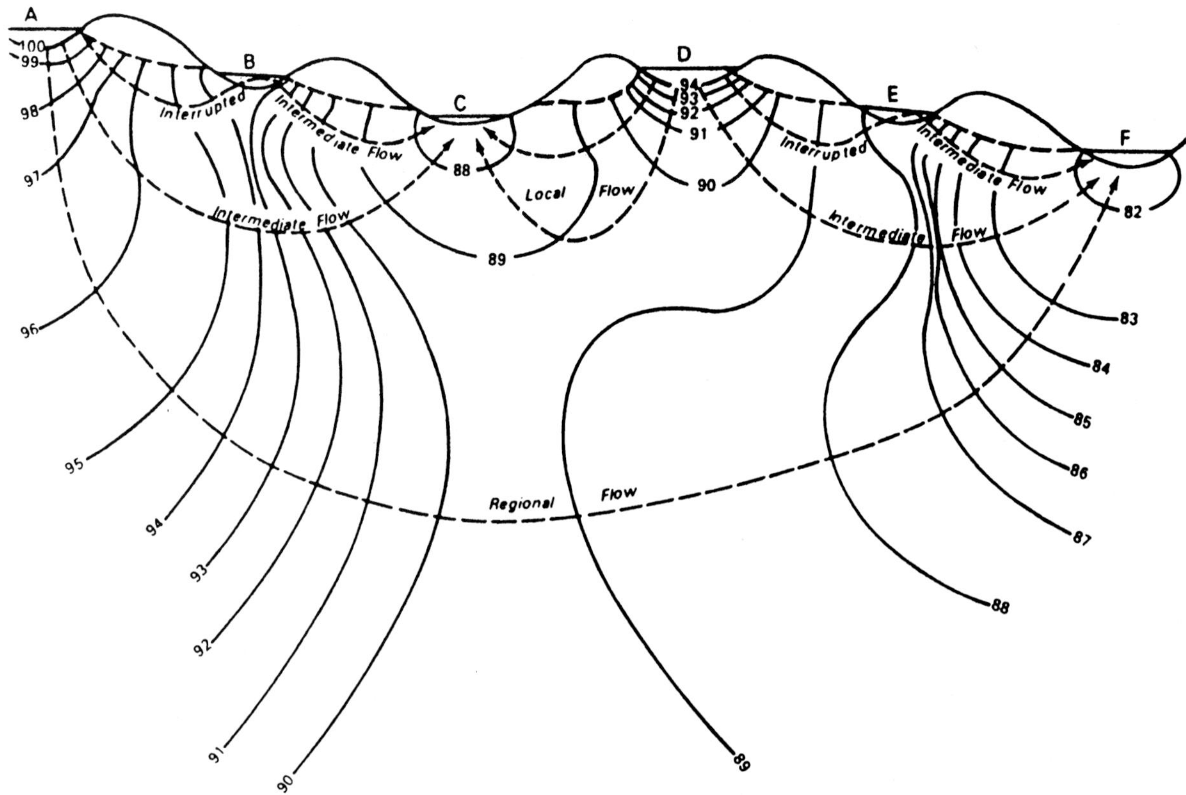

Figure 8.12 Groundwater flow systems in the prairie showing lines of equal head, flow lines, and depression-focused groundwater recharge and discharge (after Lissey, 1971).

tion of heat from nonwetland areas can enhance the evaporation from small wetlands or along the edge of large wetlands downwind of the air flow. Another consideration is the changing surface area of the open water in the wetlands as the water level rises or falls. For example, in the course of a summer, the declining water level in a prairie slough or pothole causes shrinkage of the flooded area, so evaporation of the wetland has to be determined as the areally weighted mean of the changing inundated and noninundated zones (Woo and Rowsell, 1993).

The presence of vegetation complicates evaporation processes. Vascular plants transpire and withdraw water from the soils through their roots, and atmospheric humidity governs stomatal resistance to vapor loss. Thus, transpiration rates are related to plant physiology (Lafleur, 1988). For wetlands dominated by nonvascular, nontranspiring plants, evaporation proceeds freely when the water table is at the surface and vapor loss is limited by energy considerations alone. As the water level falls, the acrotelm offers physical resistance to vapor diffusion (Lafleur, 1990) because the *Sphagnum* moss layer has limited capillary suction and the nontranspiring plants lack water-conducting tissues to draw moisture to the surface (Ingram, 1983). When cells in the *Sphagnum* dry out, the plant takes on a lighter shade, thus increasing the albedo and reducing the radiation receipt at the wetland surface. This is accompanied by a large reduction in evaporation and by a warming of the surface layer as some of the energy not consumed by evaporation is used to heat the ground.

8.3.4 Water Storage

Wetlands have traditionally been considered as areas with considerable capacity for water storage, either as subsurface storage in the peat or as surface storage in topographic depressions. With respect to subsurface storage, although the large amount of pore space in peat accommodates considerable water, the frequent saturated status of wetlands, particularly the catotelm, prevents them from absorbing additional influxes of water. Only after a prolonged dry spell when much water in the pores is lost to evaporation will subsurface storage be available to allow infiltration of rainwater. In cold regions, the presence of ground frost complicates meltwater infiltration during the snowmelt period. Where the peat is not filled with ice, infiltration is possible (e.g., Roulet and Woo, 1986, estimated that 4 mm d^{-1} entered the frozen peat for 2 days at an Arctic site west of Baker Lake, Canada) until the water freezes in the pore space to block further entry. Lichens and mosses may be desiccated in the fall or become dehydrated in the winter through upward loss of vapor into the overlying snow cover (Woo, 1982). However, they can absorb some of the snowmelt that infiltrates in the spring. Kane et al. (1981) demonstrated experimentally that about half of the meltwater generated in a subarctic environment can be taken up by lichens and mosses without generating runoff. This finding is relevant to wetland acrotelms with a thick moss and lichen cover.

Wetland surfaces are marked by micro- to macrotopographic expressions that range in scale from centimeters to hundreds of meters. These negative topographic elements include irregularities on the wetland surface (fig. 8.13), strings and flarks commonly found in subarctic and boreal wetlands (fig. 8.14), troughs between hummocks or tussocks, tundra ponds, and depressions in the patterned ground of permafrost terrain, or shallow and interconnected open-water ponds (fig. 8.15). In forested wetlands, the presence of living and dead trees accentuates the wetland floor topography (fig. 8.16). Animals can also modify the depression storage of wetlands. Examples include beaver dams and alligator holes. All these features provide depression storage, which serves the functions of water detention and retention. Detention storage represents the top layer of depression storage that eventually drains laterally, but it lengthens the travel time of water across the wetland. Retention storage is the water that remains in the hollows without generating outflow, and this water may ultimately be lost to evaporation or infiltration.

For palustrine wetlands, the linkage between surface and subsurface storage follows a general sequence. During a long dry spell, water in the depressions disappears or stagnates in isolated pools while the water table drops below ground. Subsequent rain events must first satisfy the subsurface storage deficit before raising the water table to the surface. Then, the depressions are filled and the various water-filled pockets are joined to initiate surface flow. For riverine or coastal wetlands subject to inundation by horizontal inflow, the flood may travel overland, but it will lose water to depression storage and to infiltration if the wetland is not already saturated.

From the vertical profile of water storage and water yield, it is clear that the catotelm retains water for the longest period but has a low yield to runoff. The acrotelm has large storage capacity and can yield much water to runoff once it is saturated. Surface storage can deliver water in larger quantities and more rapidly than subsurface flow. Thus, wetland discharge increases rapidly when the water level rises (e.g., Bay, 1970; Goode et al., 1977). Figure 8.17a provides such an example from Beverly Swamp in southern Ontario. Riverine wetlands also produce an accelerated increase in discharge when the river level exceeds bankfull stage (fig. 8.17b, after Shankman and Kortright, 1994). This is related to an expanding flow-contributing area on the floodplain.

Figure 8.13 (*top*) Surface of a small arctic fen in central Keewatin, Nunavut, showing microscale irregularities that provide depression storage.

Figure 8.14 (*middle*) Strings and flarks on a subarctic bog, which drains into several shallow lakes, northern Ontario. The spruce-covered strips that flank the bog are former raised beaches, now covered by about 1 m of sphagnum peat (see Woo and Heron, 1987b, for the peat profile).

Figure 8.15 (*bottom*) Ponds and shallow lakes offer ample surface storage capacities for a wetland in northern Ontario.

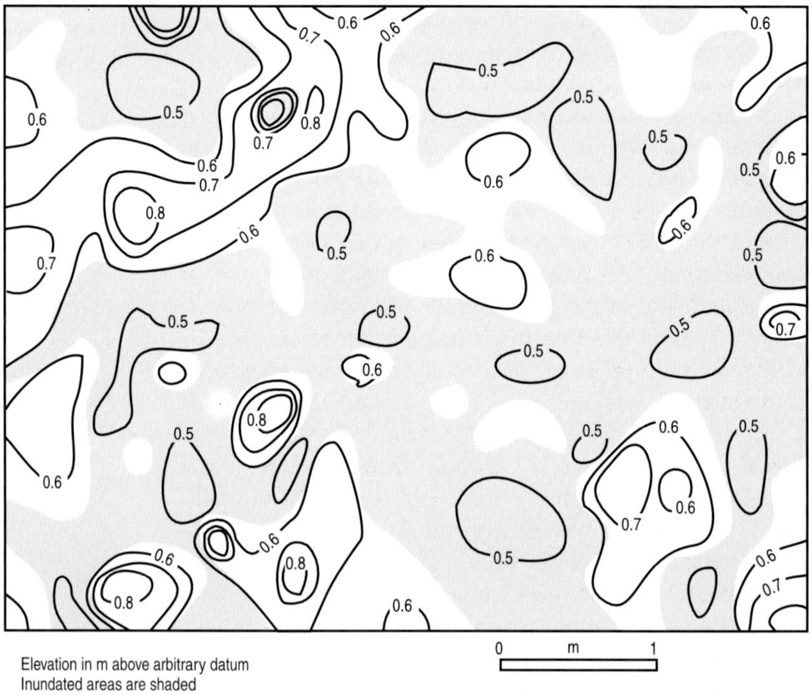

Elevation in m above arbitrary datum
Inundated areas are shaded

Figure 8.16 Depressions on the floor of Beverly Swamp, southern Ontario, with living and dead trees accentuating the topographic roughness. Water level was mapped on 15 November 1990 when the depression storage, much depleted by summer evapotranspiration, was being replenished by rainfall and water released from upstream.

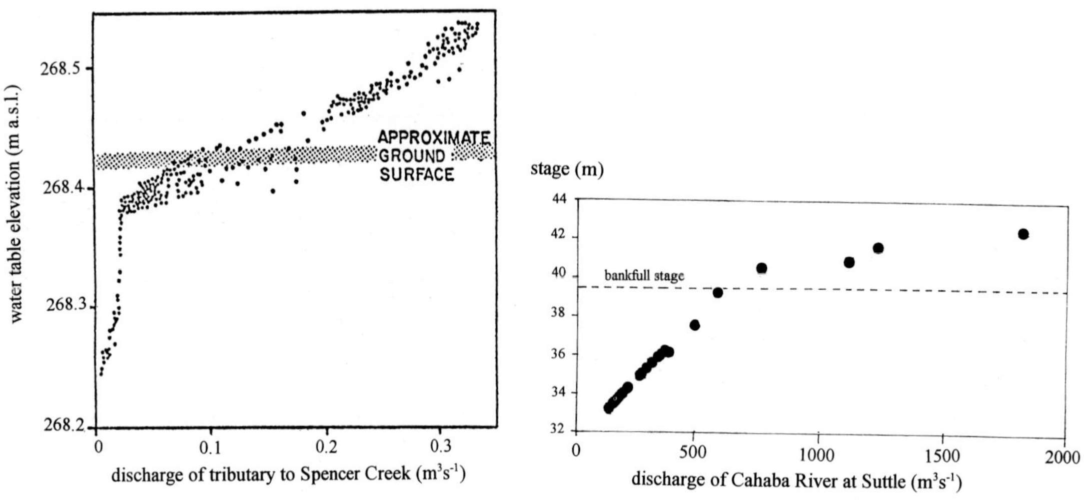

Figure 8.17 (a) A large increase in wetland discharge is associated with a rise in the water table in the acrotelm and above the wetland surface of Beverly Swamp, southern Ontario (a.s.l. = above sea level); (b) a rapid increase in river discharge occurs as the bankfull stage is exceeded on a riverine wetland (after Shankman and Kortright, 1994).

8.3.5 *Water Table*

The water table marks the top of the saturated zone. It serves as an easily observable proxy of wetland storage, though the storage status of the nonsaturated zone cannot be obtained easily without measurements of the soil-moisture content. The mean position and the magnitude of fluctuation of the water table or the flood level are significant considerations in wetland soil formation, in water storage and runoff, and in engineering and ecological concerns. According to Bay (1970), the position of the water table in a wetland is affected by (1) the amount of precipitation, (2) vegetation, which affects interception and evapotranspiration losses, (3) hydrogeology, which controls whether the groundwater is perched or linked to the regional source (ground frost as an impervious substrate is a special case), and (4) peat properties, which affect the response to rainfall because larger rises of the water table are expected when the saturated zone is in the lower peat horizon (which has low storage capacities). To these must be added another factor: (5) lateral water input from streams, lakes, and tidewater, which causes water-level fluctuations in riparian, estuarine, or shore zones.

For palustrine wetlands, distinctions between the behavior of the water table of a perched bog and a fen can be illustrated from two small wetlands in northern Minnesota (Bay, 1970). Figure 8.18 reproduces the records for a normal (1964) and a wet (1966) year. The perched bog exhibits large fluctuations in its water table, rising to greater heights in response to rainfall but falling lower during the dry period than the fen. The water table position in the fen is notably higher during a wet year when increased groundwater discharge into the wetland smooths the water-level fluctuations also. In a normal year, the water table fluctuates more widely in the fen but still does not match the bog in amplitude.

Water levels of nonpalustrine wetlands are subject to strong lateral water exchanges in addition to the vertical fluxes produced by snowmelt, rainfall, and evaporation.

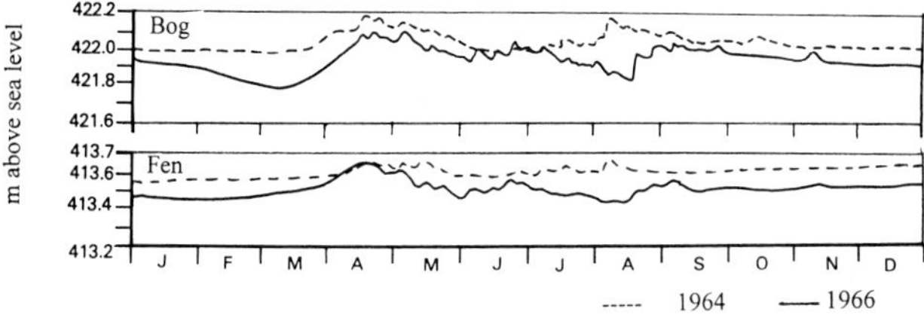

Figure 8.18 Water-table records for a perched bog and a groundwater fen, northern Minnesota, showing that the fen has fewer fluctuations during a wet year (1964) than a normal year (1966), but the bog shows larger variations than the fen in both years (data from Bay, 1969).

Figure 8.19 (after Moore and Martin, 1993) shows the response of the deltaic marsh water table to tidal fluctuations in the Musqueam Marsh of southwestern Vancouver, British Columbia. Rising tides cause water flow to the marsh but the water table rise lags behind the tidal rise. Ebb tides steepen the hydraulic gradient along the shoreline, quickening drainage from the marsh, but the decline in the water table farther inland lags behind the immediate shore. Although these tide-induced events fluctuate on a short and reasonably predictable time scale, the response to lateral water inputs from rivers or lakes may be less regular or may occur at longer time intervals. Wetlands adjacent to small streams are more responsive to individual rainstorms than the riverine wetlands of large rivers; consequently, water tables in the former areas are expected to vary more frequently. Regular flooding of riverine and

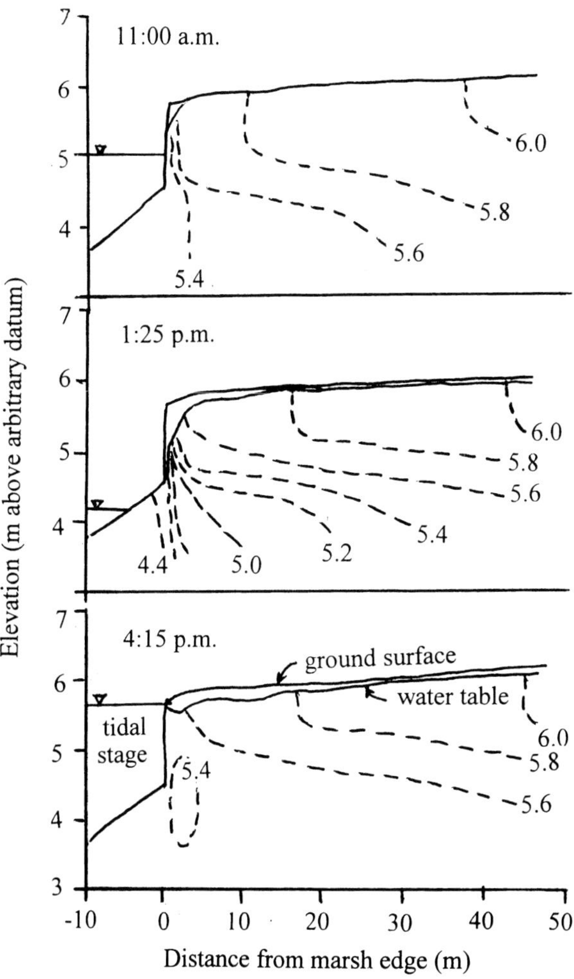

Figure 8.19 Water-table fluctuations in a coastal marsh, southwest Vancouver, British Columbia, during a tidal cycle (after Moore and Martin, 1993). Values of the equipotential lines are in meters.

lacustrine zones during the rainy or snowmelt season will result in recurrent seasonal rises of the water table in the zones of inundation.

8.3.6 Groundwater Movement in Wetlands

Diffuse or Darcian flow is often assumed for groundwater movement in wetlands. The flow is determined by the hydraulic conductivity and the hydraulic gradient, the former being affected by the structure of the peat and its underlying substrate. The direction of flow may not be deduced easily from the topographical configuration of the wetlands for several reasons. First, except for wetlands on steep slopes, many wetlands have low topographic gradients too subtle to be discerned from maps. Second, local groundwater mounds and depressions do not necessarily correspond to the surface contours (fig. 8.5a and 8.5c). Third, reversal of flows can occur when water infiltration during a wet period is followed by evapotranspiration withdrawal, which produces a hydraulic gradient that causes upward groundwater flow during the dry period (Meyboom, 1966; Devito et al., 1997). Fourth, artesian flow may occur in a wetland; for example, in swampland along the Passaic River, New Jersey, Vecchioli et al. (1962), observed artesian heads in the wells that penetrate an impermeable clay and silt layer to reach the confined aquifers of sand and gravel or bedrock.

Steepening of the hydraulic gradient, such as along the stream banks, will increase local discharge, sometimes accompanied by erosion, leading to the formation of first-order channels along the main stream (Woo and diCenzo, 1989). Pipes may also be encountered in peat (fig. 8.20), often forming near the boundary between the peat and the mineral soil (O'Brien, 1977). Pipe flow offers a mechanism that can quickly deliver water in the subsurface zone (Woo and diCenzo, 1988; Price, 1992b).

Figure 8.20 Stream bank exposing a pipe (to the right of the gun) formed at the base of the acrotelm, near Ekwan Point, northern Ontario.

During the winter, groundwater discharge in the wetlands of the boreal and subarctic regions can maintain low flows in the streams under an ice cover. Where the discharged water reaches the ground or river ice surface, it freezes into a sheetlike mass of layered ice, called icing, which is often tinted brownish-yellow by the organic matter present. Price and FitzGibbon (1987) described the occurence of icing on the surface of a wetland in central Saskatchewan when groundwater is forced aboveground, under pressure, as the frost deepens in a fen. Kane and Slaughter (1972) reported icing on Caribou-Poker Creek, which drains a central Alaskan basin with wetland occupying the bottomland. The abundance of icing and river ice, together with snow in the channels, often blocks meltwater runoff, contributing to the flooding of wetlands along streams (Woo and Heron, 1987a).

8.3.7 Surface-Water Movement in Wetlands

Within a wetland, water moves as surface or subsurface flows, with the likelihood that the water switches between these two modes of flow. Surface runoff across wetlands follows overland flow, flow between hummocks or tussocks, and channeled flow along rills or streams. For some wetlands, such as the pocosins, nonchanneled runoff is the predominant mode of surface water flow (Daniel, 1981). For most wetlands, overland flow is particularly widespread during the flooding season, be it caused by heavy rainfall, snowmelt, or large water releases from upstream. The sequence of events during a typical spring thaw is illustrated by Thom's (1972) description for a subarctic wetland. Snow slushing is followed by discontinuous sheet flow from one standing pool to another until the pools coalesce. Extensive overland flow ensues until the ice and snow blockages are overcome and the flow becomes more channelized. Overland flow then subsides as drainage along channels improves and the snow is cleared. As the water level drops, the flow zone disintegrates into patches (Woo and diCenzo, 1989) that become stagnant or linked by flow along shallow rills between hummocks and tussocks. Rainfall events often cause an expansion of the ponded area or the overland flow zones, integrating some of the surface flow routes to speed water delivery within and out of the wetland. Sheet flow in wetlands, however, is affected by the vegetation standing in the path of flow. Dingman (1996) noted that the mosses in a valley bottom wetland near Fairbanks, Alaska, retard surface flow. Christensen (1976) analyzed the flow hydraulics and concluded that vegetation increases the flow resistance, the coefficient of roughness being a function of vegetation type and density and the depth of sheet flow.

For northern wetlands with flarks and strings (fig. 8.14),

the general flow direction is perpendicular to the orientation of these shallow pools and vegetated ridges. The pointed ends of some tree islands in wetlands, such as those of the boreal zone or the Everglades, also point to the direction of flow (Hofstetter, 1983). In detail, the flow paths are much more complicated, as illustrated by several examples taken from sections of a small arctic fen and a subarctic marsh (fig. 8.21). The flow on intertidal wetlands follows the shallow tidal channels or spreads across the sand and mud flats (fig. 8.22). For these wetlands, there is a reversal of surface flow during floods and ebbs (Orme, 1990).

8.3.8 Wetland and Streamflow Interaction

Four types of streams are associated with wetlands: those originating from the wetland, those terminating in the wetland, those flowing through wetland along well-defined channels, and those that disappear and then reemerge in the wetland, undergoing considerable mixing with the groundwater. Figure 8.23 gives examples of the hydrographs for three streams in the coastal wetland near Ekwan Point, northern Ontario, which comprises a series of raised beaches separating marshy depressions. Site 4 is along a stream that originates in the wetland, with overland flow as the main water source augmented by groundwater discharge. This site had low discharge in 1987, but there was no flow the following summer when the water table was below the wetland surface for a long period. Between site 1 and site 2 is a well-defined channel along which groundwater and a tributary channel supply additional input to the stream. The hydrographs between these sites show only minor differences, both sites being highly responsive to rainfall.

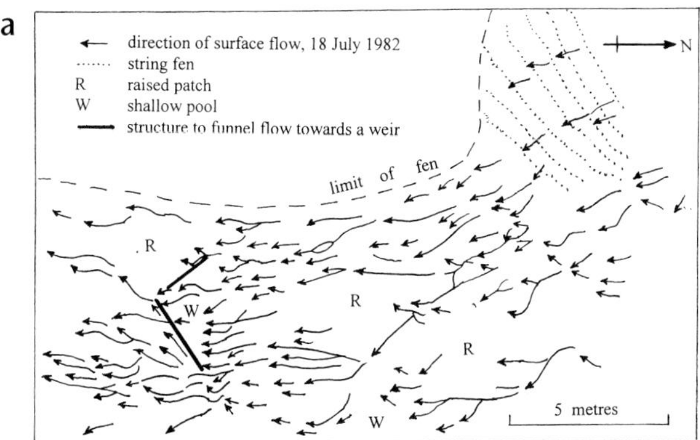

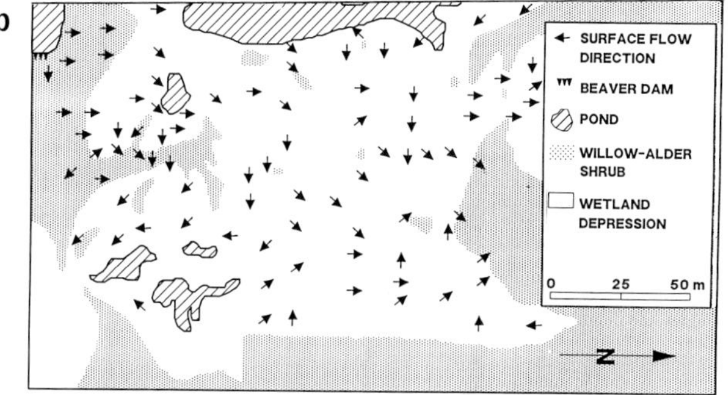

Figure 8.21 Paths of overland flow: (a) arctic fen west of Baker Lake, Nunavut; (b) subarctic wetland north of Ekwan Point, Ontario.

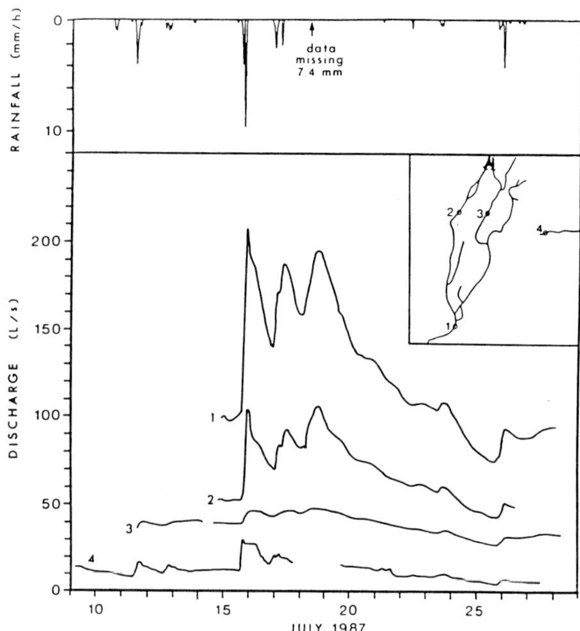

Figure 8.22 (*top*) Channels on a tidal flat near Anchorage, Alaska, indicating the occurrence of sheet flow and channeled flow during ebb and flood tides.

Figure 8.23 (*bottom*) Hydrographs of three streams in a wetland near Ekwan Point, northern Ontario, for July 1987. Site 2 is where a stream flows along a well-defined channel through the wetland; site 3 is where a stream reemerges after it has branched upstream; site 4 is where a stream originates in the wetland; site 1 combines the flows of the channels for site 2 and site 3.

In contrast, the channels in the vicinity of site 3 have several beaver dams, which divert the stream water into the surrounding wetland, but there is also water returning from the wetland to the channel at site 3. The hydrograph shows little fluctuation and is not so sensitive to rainstorms as streams at sites 1 and 2. For streams that disappear in the wetland, their channels may branch into distributaries or simply shrink and merge with wetland depressions.

The extent of interaction between streamflow and wetland strongly influences the effectiveness of the wetland as a regulator of flow, as shown along two tributaries of Spencer Creek that pass through Beverly Swamp, southern Ontario (Woo and Valverde, 1981). One tributary branches and disappears as it enters the swamp, then reemerges downstream, its flow pattern changed considerably after mixing with the surface and subsurface storage water of the wetland. The flow of the other tributary that cuts a definite channel across the swamp remains little modified through the wetland; in this case, the wetland exerts less control on its streamflow regime.

The function of streamflow regulation in wetlands usually refers to their ability to attenuate floods and enhance low flows. The ability of wetlands to attenuate high flows is attributed to the low gradients of the wetlands, the high porosity of peat to absorb and retain water, the microtopography that offers depression storages, and the presence of vegetation to retard surface flow. The effectiveness of peakflow detention depends on how saturated the wetland was before the flood event and the interaction between streamflow and wetland storage. When the water table is high, there is little excessive capacity to retain additional water, and the wetland is ineffective in flood attenuation. Indeed, given similar amounts of rainfall, a higher flow is generated by a wetland catchment when its water table is high than when it is low (Bay, 1969; Daniel, 1981). Where the streams carves distinct and deep channels in the wetland, there will be little interaction between streamflow and wetland storage other than inundation adjacent to the channels. Then, the wetland cannot exercise its flood-control function.

Wetlands are thought to be able to sustain low flows through a gradual release of storage water to outflow. Bay's (1969) study of four forested bog watersheds in northern Minnesota demonstrated that their flow recession limbs are long, indicating their effectiveness as storage areas for short-term runoff. However, zero flow events occur in the summer, and there are large variations in streamflow, as indicated by a steepening of the flow duration curve (which plots the probability that streamflow may equal or exceed a certain magnitude). These phenomena led to the conclusion that the perched bogs do not offer effective long-term storage. On the other hand, for fens that are largely fed by groundwater, the wetlands are more effective as regulators of streamflow. Boelter and Verry (1977) presented the flow duration curves for the streamflow from two small north-central Minnesota basins, one with about 30% bog, the other with a similar percentage of fen. The bog basin shows larger fluctuations of flow, including periods with zero runoff. The fen basin has a more reliable water supply and maintains high and relatively uniform outflow (fig. 8.24). Generally speaking, baseflow competes with summer evapotranspiration for the water stored in the wetland. Prolonged evapotranspiration losses during the growing season may lead to a reduction of low flow compared with nonwetland catchments (Carter et al., 1979). Many small arctic fens, with storage limited by shallow ground thaw, may be able to moderate the streamflow response to rain when the water table is low. However, their limited storage capacity as a result of shallow ground thaw often fails to sustain low flow (Roulet and Woo, 1986) unless there are continued upstream water supplies from melting snowbanks, glaciers, or lakes.

8.3.9 *Wetland Drainage Effects*

Wetlands have been altered or destroyed to provide land for agriculture, tree plantations, or urbanization (e.g.,

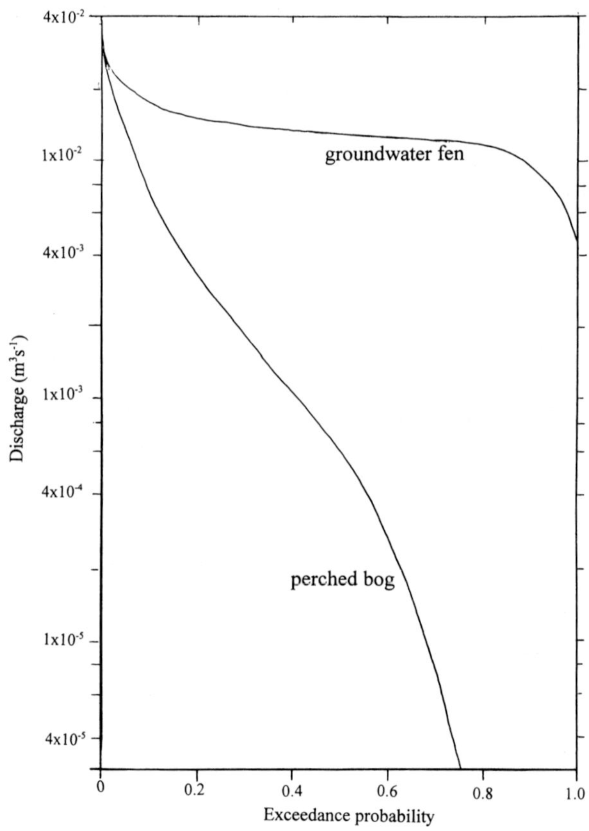

Figure 8.24 Flow duration curves from basins with about 30% cover of bog or fen, north-central Minnesota. The fen basin has more uniform flow than the bog basin (after Boelter and Verry, 1977).

Richardson and McCarthy, 1994; Holland et al., 1995). These disturbed wetlands are frequently drained to alleviate the waterlogged condition of the soil. This lowers the water table and is followed inevitably by a change in the vegetation, be it artificially introduced, as in the case of agriculture or silviculture, or from natural replacement of the hydrophytes by other species. Evapotranspiration rates change accordingly. Drainage also frees up the storage capacity below the former water table to be available to absorb a new influx of rainfall or meltwater. However, drainage often involves the straightening of channels and the excavation of ditches, reducing the opportunity for streamflow to interact with the groundwater storage in the wetland. The presence of an open ditch or the pumping of a well in the wetland causes a local steepening of the water table slope (Boelter, 1972; Winter, 1988), which alters the configuration of the groundwater flow field (fig. 8.25). Ditches also create an efficient conveyance system for the flow, reducing the opportunity for streamflow to interact with wetland storage. Thus, the new drainage network will shorten the travel time of floods through the wetland and steepen the flood peaks (compare with the effects of channelizing the bottomlands of the Obion River, Tennessee, reported by Shankman and Pugh, 1992). Baseflow may increase as more groundwater is discharged into deeper channels created by a steepening of the hydraulic gradients along the channel banks. Thus, after the creation of deep drainage ditches, both the peak flows and the base flows increase at the expense of the middle-range flows, as was observed in the pocosins of North Carolina (Daniel, 1981). Should land-use change, such as vegetation clearing, the amount of annual runoff would also be altered as a consequence of changes in evapotranspiration.

Lowering the water table leads to drying of the peat, resulting in the formation of desiccation cracks. This is accompanied by deepening of the aerated zone and accelerated decomposition of the surface peat layer, the decomposition rate being greater in woody and herbaceous peat than in *Sphagnum* peat. Destruction of the original peat structure follows, accompanied by a decrease in permeability and an increase in bulk density. Compaction of the peat is followed by subsidence of the original wetland surface, particularly when the peat was less decomposed before drainage. Subsidence increases with a lower water table. Harris et al. (1962), for example, found that in the peatland of Walkerton, Indiana, the subsidence rate linearly increased with a drop in the water table. The incremental subsidence rate was 0.03 m a^{-1} for every 1-m drop in the water table. They also found that the subsidence rate of 20 mm a^{-1} for Indiana was less than the rates of 32 mm a^{-1} for Florida (Stephens, 1956) or 76 mm a^{-1} for central California (Weir, 1950). Under warmer climates, oxidation is an important consideration in addition to peat compaction and desiccation. Thus, altering the hydrologic behavior affects the entire wetland environment.

8.4 Wetland Regions

Various schemes to divide wetlands into regions have been proposed. Some are based on geographical locations. For example, Hofstetter (1983) divided the wetland regions of the United States into Alaska, Pacific Coast, Western Interior, Prairie, North-Central and Northeast, and Atlantic and Gulf Coastal Plains. Climatic criteria are also used, including Bailey's (1978) ecoregions, which follow the Köppen climatic zones. The Canadian land classification scheme uses a mixture of climate, vegetation, and permafrost variables to distinguish between Arctic, subarctic, boreal, temperate, prairie, mountain, and oceanic regions (Zoltai, 1988). In the discussion to follow, wetland regions are modified after Winter and Woo (1990), with boundaries based primarily on physiography, vegetation, or subsurface considerations. This demarcation of regions makes no pretense of being rigorous; it merely offers a spatial framework for examining the varying hydrologic attributes of wetlands in different parts of the continent. Of interest is the seasonal variation of hydrological processes in different regions. Wetland water sources include precipitation, surface inflow, and groundwater inflow; water losses include

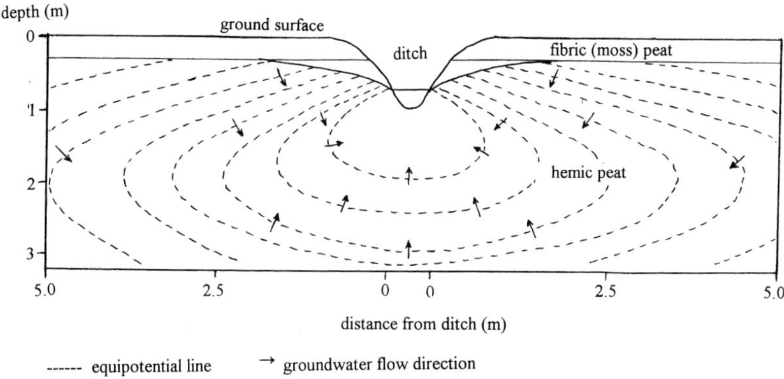

Figure 8.25 Groundwater flow direction and equipotential lines (at 0.06-m intervals) around an open ditch in the Marcell Bog, Minnesota, 10 August 1967 (modified after Boelter, 1972).

evapotranspiration, streamflow, and groundwater outflow. Water is stored as snow and ice on and in the wetlands, as liquid water retained on the surface, and as moisture in wetland soils. These processes of water input, loss, and storage change all operate at different intensities during different times of the year, producing an annual rhythm of hydrologic events known as the hydrologic regime, which varies according to different wetland environments.

8.4.1 Arctic Region

The Arctic is the area that lies north of the treeline (fig. 8.26). Intense and persistent cold gives rise to three unifying features of arctic wetlands, namely, the presence of permafrost underlying all land areas, the absence of trees, and a period of 9 months or more when the wetlands are frozen (see chapter 13). The most extensive arctic wetland stretches along the coastal plain bordering the Beaufort Sea. Elsewhere, local strips or patches of wetlands are supported by late-lying snowbanks, overflowing lakes or streams, discharge of suprapermafrost groundwater, and summer thawing of ground ice.

The active layer above the permafrost is usually less than 1 m thick, and most hydrologic activities are confined within this seasonal freeze–thaw zone. The frozen substrate invariably provides an impervious barrier to deep percolation, facilitating waterlogged conditions when ample moisture is available. Evapotranspiration is limited by the energy supply, the presence of snow and ice cover, and the energy demand for ground thaw during the growing season.

Arctic wetlands are frozen for up to 10 months each year. Even in the summer, shallow ground thaw in the permafrost terrain strongly affects wetland storage and the position of the water table. A small valley-bottom wetland on Ellesmere Island (fig. 8.27) (Glenn and Woo, 1997) shows that water level is aboveground during snowmelt when the large influx of meltwater cannot easily infiltrate the frozen soil (Kane and Stein, 1983; Woo and Marsh, 1990). The water table declines as the ground thaws (see the relative positions of the water table, the frost table, or the top of the frozen zone, with respect to the ground surface in fig. 8.27) and as evaporation continues. The water table rises quickly in response to rainfall, but the water level may or may not reach the surface, depending on the rainfall amount and the antecedent moisture condition of the wetland.

Streamflow pattern closely follows variations in the water table. A typical example of streamflow is from a small

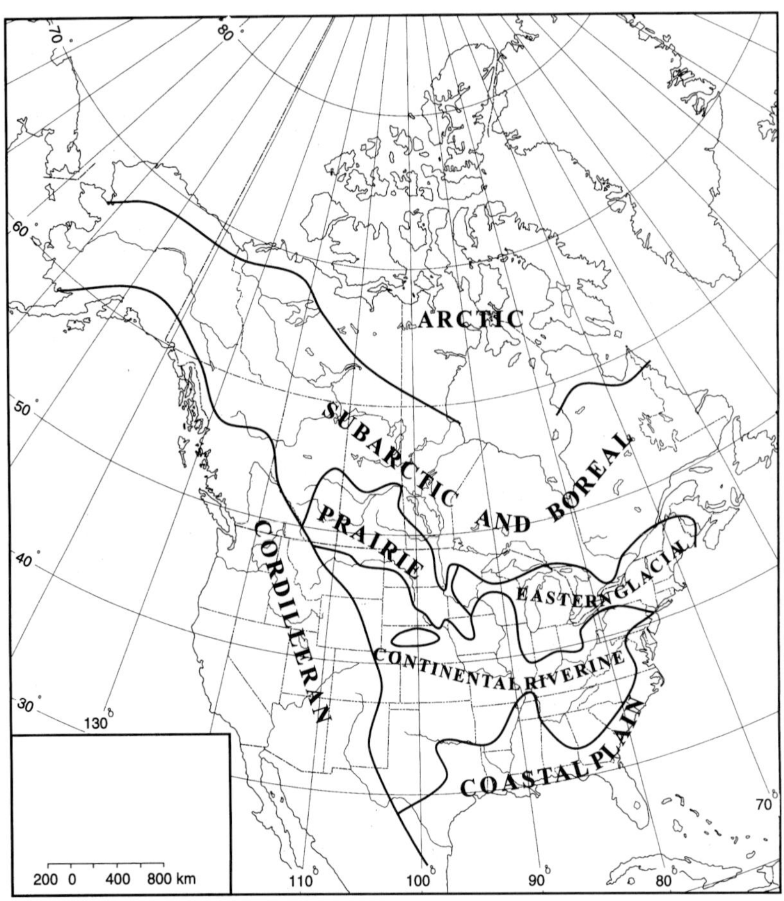

Figure 8.26 Wetland regions of North America.

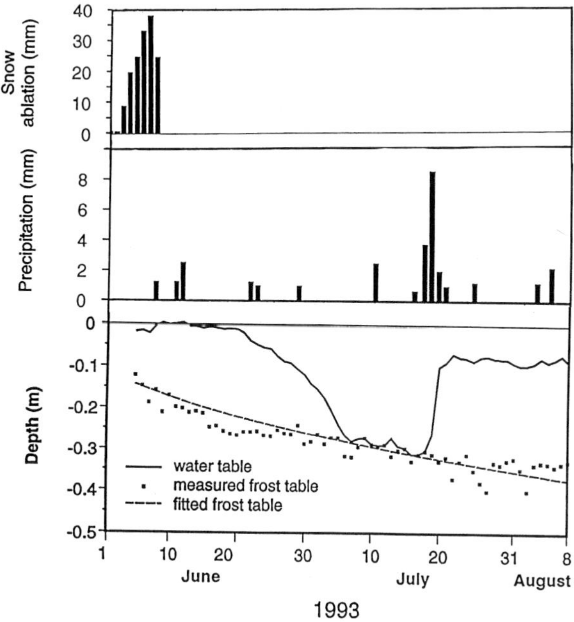

Figure 8.27 Daily snowmelt, rainfall, water-table variations, and thawing of the active layer (indicated by position of the frost table) in a small arctic valley-bottom wetland, central Fosheim Peninsula, Ellesmere Island, Northwest Territories.

wetland at Barrow, Alaska (Brown et al., 1968). The wetland produces no flow at all during the winter. During the snowmelt period, which is short but intense, frozen soil conditions inhibit meltwater infiltration, and the bulk of this water is released to outflow (fig. 8.28). After the spring melt, the ground thaws, allowing water to be stored in the subsurface. Vegetation growth increases the resistance to surface flow. Consequently, runoff response to rainfall is retarded, giving rise to lower peaks and longer recession flows than in the spring. Flow recession after rainfall is about 50 hours but can be extended to as much as 160 hours. Peak flows, though attenuated by the wetland, remain flashy. A similar streamflow regime has been reported for wetlands elsewhere in the region (Rydén, 1977; Roulet and Woo, 1986).

8.4.2 Subarctic and Boreal Region

South of the Arctic, the permafrost is discontinuous in the subarctic belt. Farther south, permafrost occurs sporadically, then in, isolated patches or not at all in the boreal zone. This region (fig. 8.26) encompasses the largest extent of wetlands in North America and is the second largest in the world, next to northern Russia. Wetlands are found in vast areas from the deltas of the Yukon and Kuskokwim Rivers and interior Alaska, across the Mackenzie valley, the Canadian Shield and its overlying sedimentary beds beneath the Hudson Bay lowlands, to Newfoundland (fig. 8.29).

Several factors lead to distinctions in the hydrologic behavior of the Arctic versus the subarctic and boreal regions: (1) compared with the Arctic, this region has longer and warmer periods of thaw that moderate the frost conditions and extend the growing season; (2) where permafrost occurs, the active layer is thicker, but where permafrost is absent, seasonal frost can still penetrate to considerable depth, often exceeding 0.5 m (Woo and Winter, 1993); (3) groundwater discharge in the winter favors icing on the wetlands and in stream channels (Kane and Slaughter, 1972); (4) the subarctic has an open-canopied forest with stands of black or white spruce (*Picea mariana, P. glauca*) and a lichen and moss ground cover, whereas the boreal forests are denser and have closer canopies than the subarctic and an undergrowth of shrubs (see chapter 14); (5) with larger contributions of vegetation, peat thickness increases south of the Arctic; and (6) sustained discharge during cold winters produces a thick (reaching 1.5 m) ice cover on wetland rivers, accentuating the spring floods when the ice breaks up (Woo and Heron, 1987a).

The seasonal rhythm of hydrological activities is often as follows. Winter is a period of snow accumulation with few intervening melt episodes. The presence of forest complicates the snow distribution through snow interception and interference of snow-drift processes. Frost descends into the wetlands in winter, but, except for permafrost terrain, groundwater movement continues below the seasonal frost. The water table may be well below the frost line in the bogs, or, in the case of fens, the hydrostatic level may lie within the frost or higher than the ground elevation, causing icing when the frozen cap is breached (Price and FitzGibbon, 1987). Snowmelt in the spring releases months of winter snow accumulation over a period of several days or weeks. The forests affect meltwater delivery through differential snowmelt and modification of flow directions (Woo and Heron, 1987b). With only a small amount of

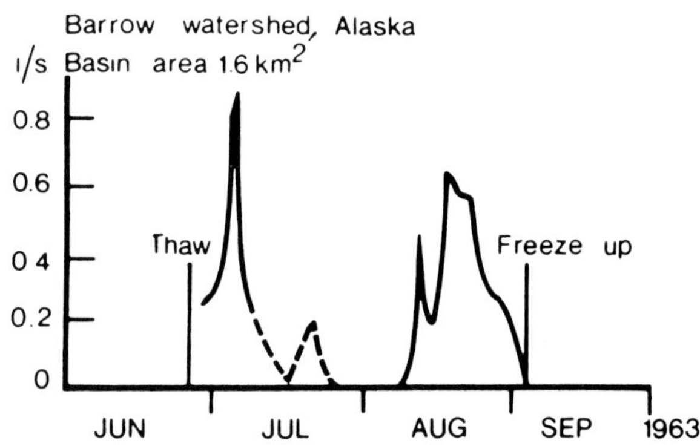

Figure 8.28 Hydrograph of a wetland catchment, Barrow, Alaska (after Brown et al., 1968).

Figure 8.29 Wetlands in the subarctic and boreal regions: (a) Kuskokwim delta with an airstrip providing the scale; (b) rivers meandering through frozen wetlands west of Lynn Lake, Manitoba; (c) Hudson Bay lowland south of Churchill, Manitoba, with the railway providing the scale; and (d) west coast of Newfoundland north of Cornerbrook showing the occurrence of fens on the coastal plain.

meltwater infiltrating the frozen ground, flooding of the wetlands prevails, particularly where the flow channels are blocked by snow and ice. Peak flows from wetland catchments also occur. In summer, evapotranspiration increases as the growing season advances. The water table falls below the surface unless sufficient rainfall or lateral inflow replenishes the losses to evapotranspiration and outflow. Some sites, such as local depressions, continue to be flooded throughout the summer (Price and Woo, 1988).

Typical streamflow hydrographs for this region are presented in figure 8.30, which includes three rivers draining parts of the extensive northern Ontario wetlands: Washkugaw River (basin area 175 km^2), Kwataboahegan River (area 4250 km^2), and Winisk River (area 50, 000 km^2).

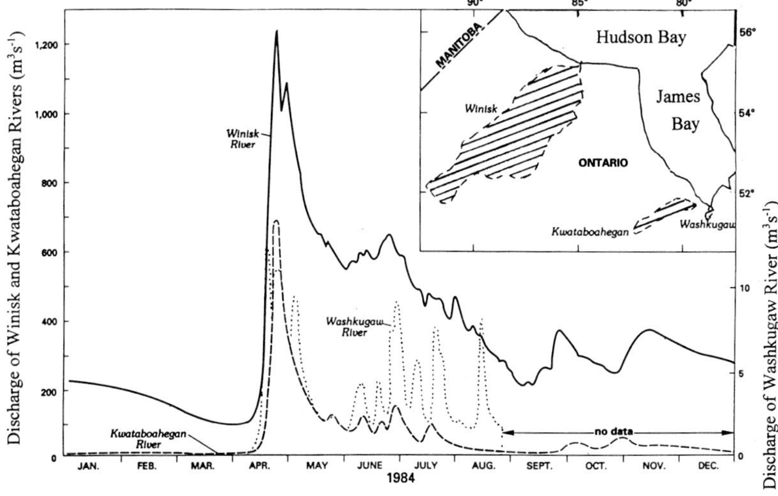

Figure 8.30 Hydrographs of three subarctic wetland rivers with different catchment areas, northern Ontario. All rivers exhibit winter low flow, prominent snowmelt peaks, and summer responses to rainfall. Inset shows location of the rivers.

Winter is a period with low flows for the Winisk and minimal flows for the other rivers. All rivers have spring peak flows in response to snowmelt, and these flows are modified by wetland surface storage as well as channel snow and ice breakup. Streamflow responses to summer rainfall are noticeable but are dampened by larger wetland storage in the bigger catchments.

Wetlands along large river valleys are inundated in the spring. These floods are accentuated by the break up of river ice, which involves an impoundment of flow upstream from an ice jam followed by the sudden release of water to the downstream areas when the jam breaks (Prowse, 1986). Riverine wetlands thus affected are also subject to considerable erosion by ice scour. On the other hand, many riverine and deltaic wetlands, such as those on the deltas of the Mackenzie River or the Peace River, owe their existence to the recurrence of spring floods (Marsh and Hey, 1989).

8.4.3 Prairie Region

The Prairie region is considered here to comprise those parts of the temperate grassland characterized by deposition beneath or just beyond the Pleistocene ice sheets. Thus, it also includes the Nebraska Sandhills. Most uplands have been converted to farming. Wetlands in the entrenched valleys, such as those along the Saskatchewan River, are subject to snowmelt flood, with the meltwater derived largely from the Rocky Mountains. Of particular interest are the prairie potholes or sloughs, wetlands that occupy the kettle holes formed during deglaciation or depressions in the sand dune terrain. These wetlands are subject to large annual and interannual water-level variations, and frequent aerobic conditions prevent thick accumulations of peat.

The drainage network in the region is not well integrated, and many potholes are not connected to main drainage systems. However, the depressions occupied by the wetlands are important sites of groundwater recharge or discharge (figure 8.12). Individual potholes may show seasonal reversals of flow interaction with the groundwater in a manner postulated by Meyboom (1966). Figure 8.31 shows schematically two Saskatchewan sloughs in a hummocky moraine area with groundwater following the regional gradient. When the sloughs are inundated, notably by meltwater in the depression or runoff from the slopes (Woo and Rowsell, 1993), a groundwater mound is formed. Water infiltrates from the sloughs to recharge the groundwater. As many sloughs have a coppice of willow trees (*Populus tremuloides*), the willow ring (fig. 8.32) depletes the moisture through evapotranspiration, creating cones of depression around the phreatophytic willows. Groundwater flow then reverses, and water is discharged to the sloughs.

Most prairie wetlands therefore experience seasonal cycles of high and low water levels. Winter and Rosenberry (1995) presented 12 years of water-level measurements for a wetland around Cottonwood Lake, North Dakota. The prairies have cold winters that freeze all the potholes unless the water in the pond is deeper than the frost line. Snowmelt gives rise to high water levels in the spring, followed by a drop in water level in the summer, with occasional augmentation from rain. The annual rise and fall of the water level can vary greatly from year to year; one year the water level may reach a record high, whereas the next year is completely dry. Besides annual variability, the pattern of water-level fluctuation also varies spatially. Summarizing the 1990 water-level measurements for 46 sloughs near Saskatoon, Saskatchewan, Woo et al. (1993) noted that they fall into groups in accordance with the vegetation-based classification scheme of Millar (1976). As shown in

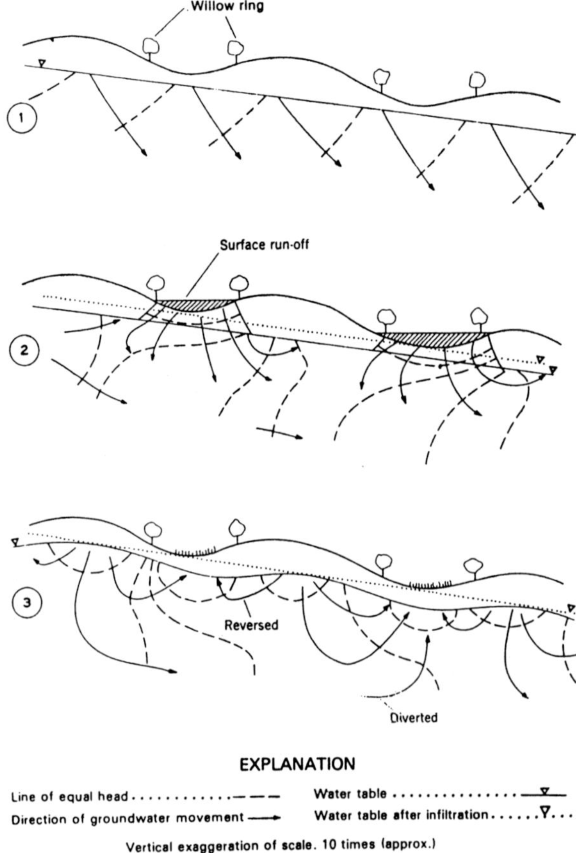

Figure 8.31 (*top*) Seasonal reversal of shallow groundwater flow, with recharge from the prairie sloughs in the snowmelt season and groundwater discharge to the sloughs after a long spell of evapotranspiration (after Meyboom, 1966).

Figure 8.32 (*bottom*) A prairie slough (pothole) near Saskatoon, Saskatchewan, showing a willow ring around the depression occupied by the shallow, open-water wetland.

figure 8.33, these include (1) shallow open-water wetlands, largely fed by groundwater discharge, which remain wet throughout the summer, (2) emergent deep marsh and shallow marsh that are flooded for about 3 or 4 months of summer, and (3) ephemeral wet meadows flooded only in spring.

8.4.4 Eastern Glacial Region

South of the Canadian Shield and east of the prairies is a continuation of the depositional terrain associated with Wisconsinan continental glaciation. Wetlands have formed on glacial and glacio-lacustrine deposits, in depressions in morainic terrain, along postglacial drainage channels, and around lakes (see chapters 15 and 16). Winter temperatures are low enough to develop seasonal frost, and both snow and rain events can occur between November and March. In summer, vegetation growth, together with high temperature and radiation receipt, favor high evapotranspiration from the wetlands (Munro, 1979). The region has mixed stands of hardwood and coniferous forests. Peat accumulation often reaches 5 m, and it is sometimes harvested to supply horticultural, industrial, or energy demands.

The water-level fluctuation pattern for Beverly Swamp, southern Ontario, illustrates the seasonal variation of wetland hydrologic conditions (fig. 8.34). Like other wetlands in the region, this wetland has a snow and ice cover during the winter. Groundwater flow continues throughout the year, and midwinter snowmelt or rain can flood the ice, generating rising water levels and surface runoff (Smith and Woo, 1986). Spring snowmelt produces general flooding of the swamp and yields high flow to the streams. As the growing season arrives, evapotranspiration increases, resulting in a total water loss of up to 500 mm (Woo and Valverde, 1981). The water table falls below the surface, occasionally revived by summer rainfall. Flooding occurs again in October as a result of low evaporation and the release of water from upstream. Over the year, the water level fluctuates between 0.2 m above to 0.65 m below ground level.

Similar hydrologic regimes were reported by Vecchioli et al. (1962) for small wetlands west of Boston, and these can be compared with the water-table variations reported by Bay (1969) in Minnesota (fig. 8.18), although streamflows from catchments with bogs or fens can be quite different (fig. 8.24). Long periods of human occupancy in this region have led to the conversion of many small wetlands to farmland, pasture, or urban areas, and wetland hydrology has been modified by drainage or upstream land-use changes.

Many freshwater marshes and swamps are found along the shorelines of the Great Lakes. Flooding by the lake water causes the wetland water table to rise above ground. Although nontidal, these wetlands experience water level fluctuations attributed to (1) short-term seiche activities

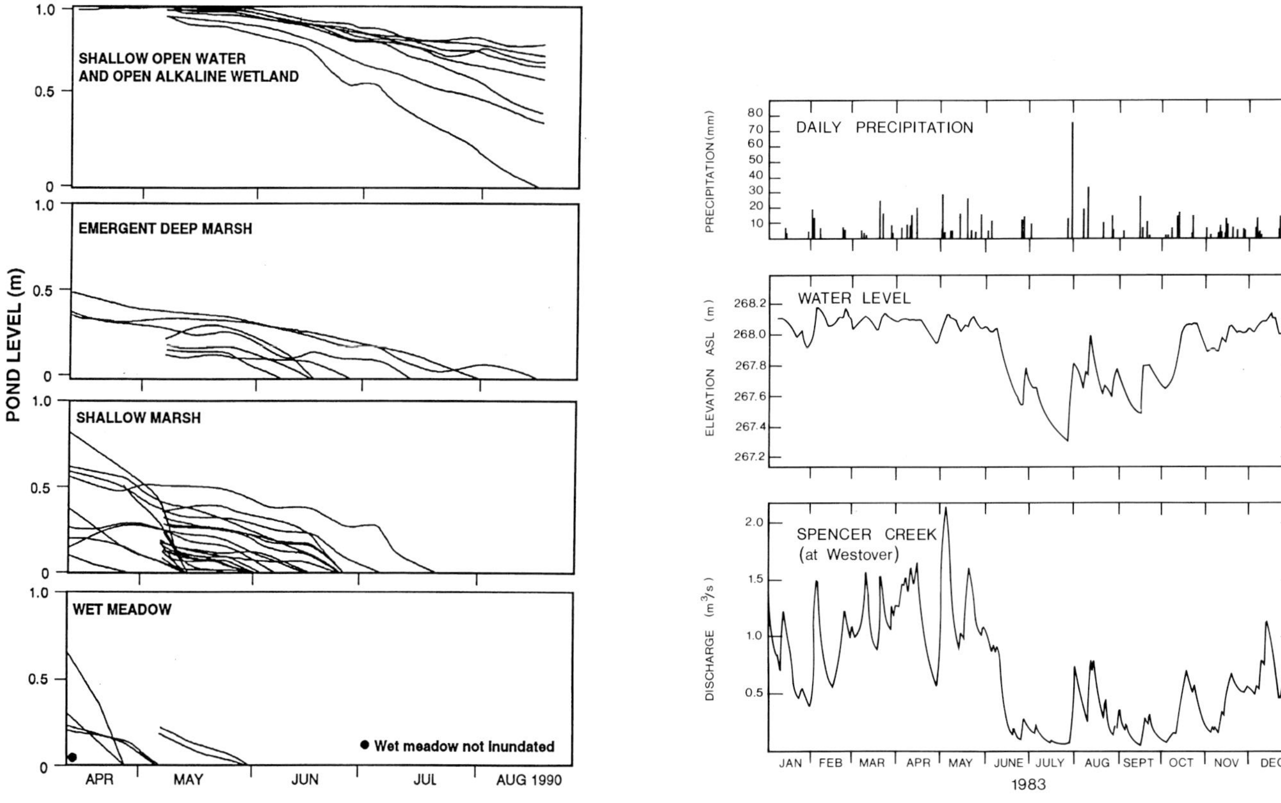

Figure 8.33 (*left*) Water levels of 46 prairie sloughs in St. Denis National Wildlife Area east of Saskatoon, Saskatchewan.

Figure 8.34 (*right*) Daily rainfall, wetland water-level fluctuation in Beverly Swamp, southern Ontario, and daily streamflow of Spencer Creek, which is fed by the swamp (ASL = above sea level).

with hourly periodicity and ranges from centimeters to over a meter, (2) seasonal water balances that give rise to annual cycles, and (3) long-term variability, which produces year-to-year differences in lake level. Figure 8.35 provides segments of the lake level record for Lake Erie; these segments illustrate the annual and interannual variability that has an impact on the water level of the lakeshore wetlands.

8.4.5 Cordilleran Region

The Western Cordillera stretches from the Great Plains to the Pacific coast but, for the purpose of wetland analysis, only the area south of the discontinuous permafrost zone is included in this region. For this region, the only unifying feature is its ample relief. This, together with the latitudinal extent, gives rise to greatly diverse environments. Topographically, this region encompasses some of the most prominent mountain ranges and plateaus of the continent, but it also has many coastal plains, estuaries, and fjords. The temperature ranges from the coldness of the alpine climates to the considerable summer heat of the Basin and Range Province. The latter includes deserts with extremely low precipitation, in direct contrast with the Alaska panhandle and British Columbia coast where annual precipitation exceeds 3500 mm. As a consequence of the diversity in climate, topography, and soils, the vegetation shows remarkable altitudinal, latitudinal, and maritime versus continental zonation (see chapter 18).

The cool, oceanic climate favors wetland formation along the northern section of the Pacific coast. Swamps and marshes occur in embayments or estuarine areas, such as around Puget Sound, Washington, and the Fraser delta, British Columbia. Many bogs and fens are found on slopes, valleys, and flat grounds on uplands and lowlands (Banner et al., 1988) because precipitation greatly exceeds evapotranspiration. From Oregon to the British Columbian coast, the highest precipitation occurs during winter. For the Alaska panhandle, the greatest amount comes in the fall. Siegel (1988) noted that the water level in a fen and a forested wetland near Juneau, Alaska, closely followed the precipitation pattern during the period between spring thaw and autumn freeze-up. Despite high precipitation in August, the water level in one bog did not increase due to the large soil-moisture deficit, which needed to be replenished in the upper, dry peat layer.

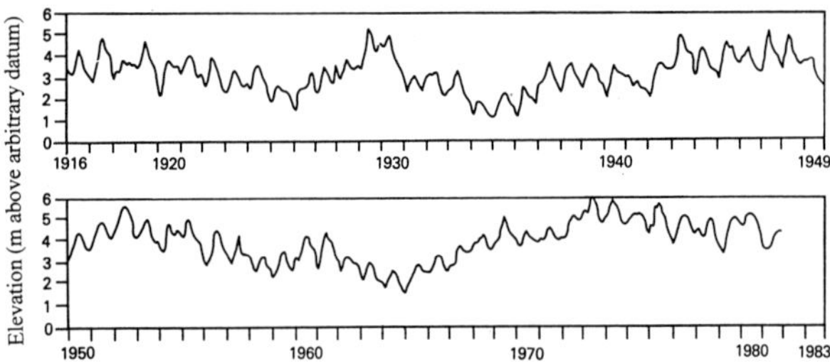

Figure 8.35 Seasonal and yearly variations of the level of Lake Erie, indicating that the shoreline wetlands are subject to these intra- and interannual water-level fluctuations.

In the interior zone, wetlands are found along streams or lakes located on the plateaus and in the valleys. Large parts of the northern sector have been modified by mountain glaciation, and they are underlain by glacier-related deposits. An example of the streamflow regime for a wetland-dominated catchment is the Muskeg River in north-central British Columbia (fig. 8.36). Winter is a period of low flow and ice conditions on the river. The magnitude of snowmelt flood depends on the winter snow accumulation and the rapidity of snowmelt. Summer hydrographs rise in response to rainfall, but these peaks are much lower than the snowmelt floods. The recessions from high flows are more gradual than those from nonwetland mountain streams because of flow attenuation by the wetland storage.

For obvious climatic reasons, there are few significant wetlands farther south, in the more arid parts of the Western Cordillera. Restricted marshes do, however, occur along many rivers, around groundwater seeps, and in internal drainage basins such as Death Valley, and stratigraphy records larger wetlands during Pleistocene pluvial episodes (see chapter 3). Of the once extensive interior wetlands along California's Central Valley, only 4% still exist, the remainder having been converted to irrigation agriculture (Orme and Orme, 1998). On the California coast, though never extensive south of San Francisco Bay, some 80% of the coastal wetlands that existed in 1850 have been reclaimed for urban and industrial use.

8.4.6 Continental Riverine Region

This region lies to the south of the continental glaciation limit and east of the Western Cordillera, with the Fall Line marking its southeastern boundary. Most of the wetlands, ranging from the small headwater wetlands in the Appalachians to those on the Mississippi floodplain, are found along rivers. A large part of the region is within the Mississippi basin, which regularly receives snowmelt during the spring from its headwater areas and from rainfall that can be intense or persistent enough to cause widespread flooding, for example, in 1973 (Chin et al., 1975) and 1993 (see chapter 7). Many wetlands are flooded not only once, but several times a year. Invariably, the number of tree species increases from the most frequently flooded to the less frequently flooded (higher) parts of the floodplain (Brinson, 1990). Transpiration is efficient because water at or near the surface is readily accessible to the plants (Winter and Woo, 1990). Large sections of the Mississippi and its tributaries flow along alluvial valleys, which have rapid interaction between surface flow and groundwater. The hydrologic attributes of an alluvial aquifer include (1) thick, extensive sand and gravel deposits, (2) limited width relative to length, laterally constrained by valley width, and (3) hydraulic connection with a stream (Heath, 1984).

The hydrologic regimes of riverine wetlands are strongly influenced by the discharge rhythms of the rivers but are modified by local evapotranspiration and water withdrawal. Hurr (1983) provided an example from the riverine wetland on the Platte River floodplain near Grand Island, Nebraska. River flow is usually highest in the spring because of snowmelt in the headwater areas and spring rain, whereas streamflow decreases in the summer as a result of upstream diversions for irrigation. The riverine wetlands receive water from rainfall, snowmelt, and recharge from the Platte River. The water table beneath wetlands in the riparian zone is generally lower than the river level (fig. 8.5b), mainly because of evapotranspiration by riparian vegetation. Groundwater response to the river stage is rapid, usually within 24 hours. In addition, diurnal water-table fluctuations are caused by evapotranspiration, and periodic sharp rises are produced by rainfall events (fig. 8.37).

Groundwater exchanges between a riverine wetland and its aquifers may be illustrated by Gonthier's (1996) study of the Black Swamp adjacent to Cache Creek in eastern Arkansas. At different locations and at different times of the year, the wetland exchanges water through discharge

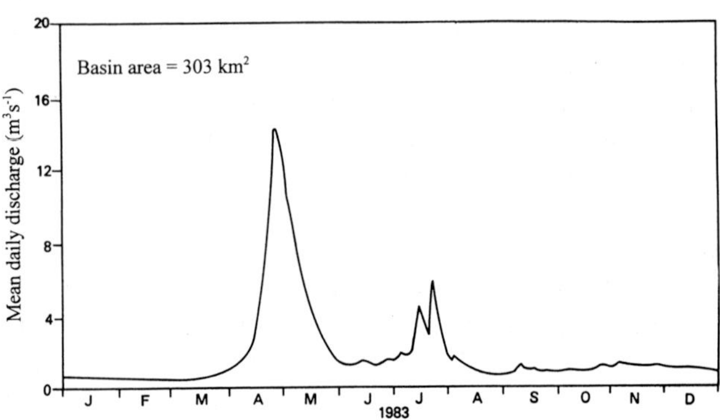

Figure 8.36 Monthly streamflow of Muskeg River, which drains a wetland-dominated catchment in north-central British Columbia.

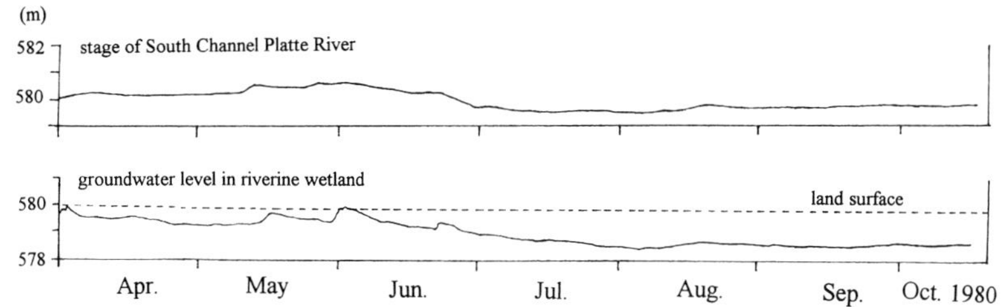

Figure 8.37 Fluctuations in the water table in several wells across a riverine wetland, Platte River south of Grand Island, Nebraska (modified after Hurr, 1983). Daily fluctuations are not shown.

and recharge with the "intermediate aquifer," which is the aquifer larger than the extent of the wetland but not as broad as the "regional aquifer" that encompasses the entire alluvial system. Black Swamp also discharges to the "local aquifer" (within the confines of the wetland), mostly at locations close to Cache Creek, which presumably is a ready water source.

Many rivers of the Mississippi system have been modified by engineering works and water management so that the natural flows are altered. Water diversion and interbasin transfer are practiced, groundwater is pumped along the valleys, channels are dredged or dammed to enable navigation, banks are protected or levées are built for flood and erosion control, and water is returned to the river from irrigation or hydropower releases. The resulting effects on streamflow may be a reduction of high flows, an increase in the minimum level of low flows, and a flattening of the flow-duration curve (Kircher and Karlinger, 1983). Although there is usually little reduction in high flows, significant enhancements of low flows are experienced downstream from the reservoirs. Such changes in the flow regime inevitably affect the hydrological rhythms of wetlands connected to the rivers.

8.4.7 Coastal Plain

The nonglaciated part of the Coastal Plain rises from the Atlantic and Gulf coasts to the Fall Line, which has an elevation of 250 m in Georgia, 110 m near Little Rock, Arkansas, and 210 m at Austin, Texas (Walker and Coleman, 1987). The second largest concentration of wetlands in the continent occurs on this area of low topography. Marshes and swamps typify most coastal and deltaic areas, areas that are also prone to the attacks of hurricanes and their associated storm surges. Riverine wetlands form along parts of valleys and are subject to channel shifting and the attendant consequences of sedimentation and scouring, drainage alteration, and transformation of wetland vegetation (Shankman, 1993). For large rivers, wetlands are found on floodplains or trapped as "back swamps" on low ground between levées and upland areas. Extensive wetlands also develop on peatlands, such as the pocosins (Daniel, 1981), the Everglades (Gleason and Stone, 1994), or on marl lands, such as the Big Cypress Swamp in Florida.

This region traverses a range of climates, from warm temperate in the north to tropical conditions in the south. Evaporation increases southward, and rainfall is higher between May and October than during the winter months. Water-level variations in palustrine wetlands are influenced by the balance between rainfall and evapotranspiration. In winter, when rainfall exceeds evapotranspiration, wetland water tables stay close to the surface, as shown by the pocosin wetlands of the Albemarle-Pamlico peninsula, North Carolina (Daniel, 1981) and the Great Dismal Swamp that extends farther north into Virginia (Kirk, 1979; Carter, 1990). Increasing evapotranspiration in summer leads to water-level decline but it rises significantly in response to rainfall (fig. 8.38). Water depth on the Everglades is influenced by the vertical water balance between rainfall and evapotranspiration, as well as lateral flow driven by the hydraulic gradient, such as it is. Localized rainfall creates transient water mounds that shed water to all sides, especially downstream (Kushlan, 1991). Lying on carbonate bedrock, the Everglades also receive discharge from a number of karst groundwater springs.

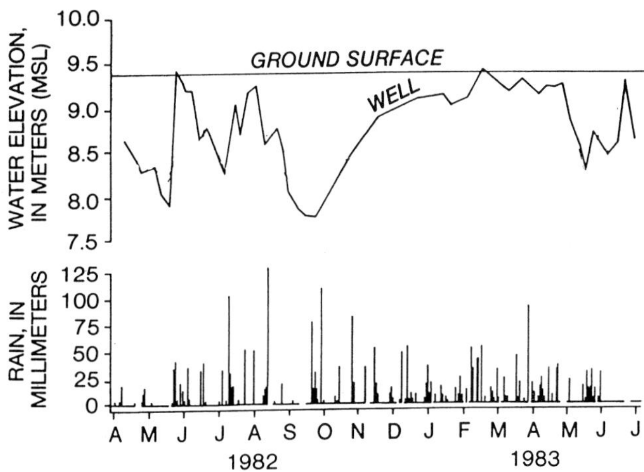

Figure 8.38 Variation in the water table in Great Dismal Swamp, showing winter high level and summer decline due to evapotranspiration losses but with occasional rises in response to rainfall (after Carter, 1990) (MSL = mean sea level).

Additionally, wetlands in this coastal zone are affected by tides. Furthermore, the appearance of salt-water mangroves in peninsular Florida and southwest Texas heralds the transition from temperate salt marshes to tropical mangrove swamps and wet forests farther south in the Caribbean and along the Mexican Gulf coast (see chapter 21). Figure 8.39 shows the monthly variations of rainfall, evapotranspiration, tidal ranges, and inflow into the Barataria area on the Mississippi delta, as well as the water levels of several wetland types, including fresh open water, salt marsh, freshwater marsh, and cypress-tupelo swamp (Costanza et al., 1983). The distinct summer maximum in evapotranspiration is of similar magnitude to the summer rainfall. Inflow is low in the summer. The level of fresh open water reflects the bimodal tidal pattern, and the level of the salt marsh is low in the winter, corresponding to the seasonal low tides. For other types of wetlands, variations in the water level are complex and suggest the influence of multiple factors.

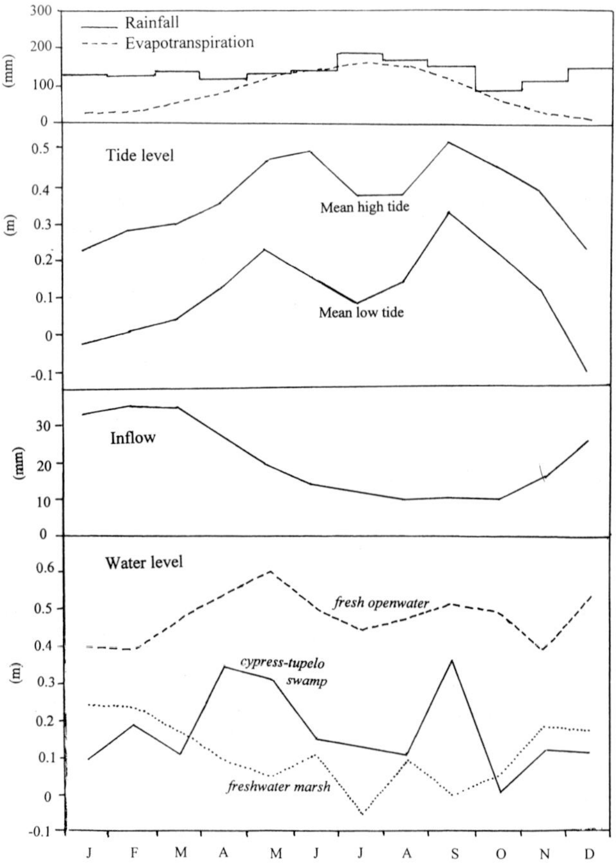

Figure 8.39 Monthly variations of rainfall, evapotranspiration, tidal ranges, and inflow into Barataria area, Mississippi Delta; water levels of fresh open water, freshwater marsh, and cypress-tupelo swamp (data from Costanza et al., 1983).

One major hydrologic consideration of the Coastal Plain wetlands is the influence of human activities over the years. Large tracts of wetland have been lost to agriculture, including many shallow, elliptical open-water wetlands known as the Carolina Bays (Lide et al., 1995), and large portions of the pocosins (Richardson et al., 1981). Over the past century, flood control, water management, and reclamation have changed the Everglades (Light and Dineen, 1994) to accommodate agriculture and human settlement. About 65% of the original marsh was drained, and water movement is now regulated by engineering structures such as canals, levées, and gates (Kushlan, 1991). Thus the natural flows in the Everglades are altered, the peaty soil subsides, some marshes are lost, wetland habitats are ruined, and the amount and timing of freshwater inflow to the Florida Bay are modified (McIvor et al., 1994). The Mississippi delta has numerous lakes, bayous (complex assemblages of water courses), and forested swamps (Devall, 1990), many of them now lost or disappearing. About half of the original forested wetland in Louisiana has been logged, burnt, drained, and converted to other land use. Streamflow regulation of the Mississippi and its tributaries also changes the inflow to the delta. The hydrologic behavior, indeed the preservation, of its deltaic wetlands, both the swamps and the marshes, are subject to external and internal impacts, be they natural or human-induced.

8.5 Conclusion

Much of society has long believed wetlands to be insect- and disease-infested, barriers to movement, and deterrents to development. The redeeming features of wetlands are perceived to be their role as refuges from persecution or attack and their regulatory effects on runoff, acting as sponges to soak up floodwaters and then to release this water over the dry season. Wetlands have been viewed as wastelands and, generally speaking, to many a farmer or developer, a good wetland is a drained wetland. Wetland researchers, however, have recognized the multiple functions served by wetlands. They provide habitats for fish, wildlife, and migratory birds. Wetlands near population centers offer recreational opportunities. Their peat is a source of fuel and industrial raw materials. Wetlands are a major source of methane, which, as a greenhouse gas, has potential effects on global warming.

Since the last century, many wetlands in the United States and southern Canada have undergone degradation or conversion to other forms of land use. The conversion for agricultural, silvicultural, pastoral, or urban usage is preceded by a transformation in wetland hydrology. Wetland restoration, therefore, calls for a reversal of such artificial hydrologic behavior. Restoration does not end with hydrology alone—other wetland attributes must be considered. The engineering work required to change the water storage and flow in a wetland may take up to 10 years, as does the alteration of topography that accompanies these hydrologic changes. Sedimentation and vegetation change may take up to 100 years to attain a new equilibrium. Chemical and physical changes of the sediments and accumulation of peat will take on the order of 100 to 1000 years, depending on climate and the biogeochemical environment. In addition, the hydrology of many wetlands is influenced by conditions upstream of, or surrounding, the wetlands. Even after the internal morphology, hydrology, and ecology are restored to a wetland, it may still be affected by stream inflow and/or lateral runoff from adjacent slopes, both of which could have been irreversibly changed by human activities.

All wetlands, be they in the natural or altered state, are subject to short-term fluctuations and long-term changes in the physical, chemical, and biological environments. Hydrologic processes play a crucial role in determining the preservation and the nature of a wetland. Hydrology responds to complex external forcing, including the following factors: precipitation; evaporation; tidal regimes; frequency, duration, and intensity of floods and droughts; subsidence or uplift of the coastal zone; persistent or catastrophic geomorphic processes; activities of animals, including human beings; and gradual or sudden shifts of vegetation types and patterns. Many of these factors are interrelated, and there is strong feedback between hydrology and these external conditions. To understand properly the hydrologic dynamics of wetlands, it is essential to analyze related aspects of hydroclimatology, hydrogeology, hydrochemistry, and hydroecology.

Glossary

(Most of the definitions are abridged from the list provided by Stanek, 1977.)

Acrotelm (topmost marsh layer) is the top layer of a peat profile, including the living plants, that is the principal zone of matter and energy exchange in the peatland ecosystem (Ingram, 1978). This term replaces the term *active layer*, which is easily confused with the annual freeze-thaw zone in the permafrost literature.

Anaerobic refers to the condition of no molecular oxygen in the environment.

Catotelm (lower marsh layer) is the lower layer of peat, which is more inert in terms of matter and energy exchanges (Ingram, 1978).

Eutrophic (well-nourished) refers to soils with high nutrient content and high biological activity.

Flark is a depression in string peatlands, usually elongated with the long axis perpendicular to the direction of flow, and is bounded by strings.

Humification is the process by which organic matter decomposes to form humus.

Humus is the organic matter formed by the decomposition of plant or animal residues.

Hydroperiod denotes the seasonal pattern of the water-level variation in a wetland (Nuttle, 1997). In this regard, *period* is a misnomer (the term *hydroperiod* is seldom used by nonwetland hydrologists).

Mesotrophic describes soils with nutrient content intermediate between eutrophic and oligotrophic.

Minerotrophic (nourished by mineral water) refers to peatlands nourished by mineral-rich groundwater.

Muskeg (word of Algonquin Indian origin) is peatland or organic terrain; a term commonly applied to northern wetlands.

Oligotrophic refers to soils with low nutrient content and low biological activity.

Ombrogenous (produced by rain) refers to wetlands that rely on precipitation for nutrients.

Ombrotrophic (nourished by rain) refers to peatlands that rely upon precipitation for nutrient supply.

Organic soils are naturally occurring soil bodies produced by an accumulation of plant remains. Organic soils in wetlands are usually represented by peat. Three organic soil materials are included: (1) *fibric* materials contain much fiber and are the least decomposed; (2) *sapric* materials have few fiber and are the most decomposed; and (3) *hemic* materials are intermediate.

Paludification is the process of peat formation, generally under anaerobic conditions caused by waterlogging.

Palustrine system "includes all non-tidal wetlands dominated by trees, shrubs, persistent emergents, emergent mosses or lichens, and all such wetlands that occur in tidal areas where salinity due to ocean-derived salts is below 0.5%" (Cowardin et al., 1979).

Peat is an organic soil formed under waterlogged conditions. The dead plant materials are incompletely decomposed due to the prevalent anaerobic conditions. The organic matter content should be no less than 20% of the dry weight.

Peatland is land with at least 0.4 m of peat accumulation on which organic soils develop.

Permafrost is ground that has a temperature at or below 0°C for at least two consecutive summers.

Pocosin (Algonquin Indian term meaning swamp on a hill) is an evergreen shrub bog, occurring on the southeast coastal plain from Virginia to north Florida, usually found on the youngest marine terraces and removed from large streams.

Pothole (or *slough* in Canada) is a depression in the glaciated terrain of the prairies occupied by a wetland.

Soligenous (produced by soil) refers to being nourished by flowing water from higher ground.

String is a narrow strip, elevated and better drained than the adjacent depressions called *flarks*.

Topogenous (produced by relief) refers to being nourished by water collected in a depression.

von Post humification scale provides a field guide to describe the stages of peat decomposition, ranging from completely unhumified (H_1) to completely humified (H_{10}) state. The stages are described by what happens when the peat sample is squeezed, as follows:

H_1: yields only colorless, clear water
H_2: yields almost clear but yellowish brown water
H_3: yields turbid water but the residue is not mushy
H_4: yields strongly turbid water and leaves a somewhat mushy residue
H_5: some substance passes between the fingers together with mucky water, leaving mushy residues in the hand (plant remains are recognizable in the peat, but not distinct)
H_6: at most, one-third of the peat passes between the fingers (plant remains are not distinct)
H_7: about half of the peat passes between the fingers (plant remains can still be seen)
H_8: about two-thirds of the peat passes between the fingers, leaving behind mainly resistant root fibers (plant remains not recognizable in the peat)
H_9: nearly all of the peat passes between the fingers like a mush (hardly any plant remains are apparent in the peat)
H_{10}: all the peat passes between the fingers (no plant remains are apparent)

References

Adams, G.D., 1988. "Wetlands of the prairies of Canada." In: *Wetlands of Canada.* Ecological Land Classification Series No. 24, Environment Canada and Polysciences Publications Inc., Montreal, 155–198.

ARP (Alaska Regional Profiles), 1975. Arctic Environmental Information and Data Center, vol. I–IV, University of Alaska, Anchorage, Alaska.

Baden, W., and R. Eggelsmann, 1963. "Zur Durchlässigkeit der Moorböden." *Zeitschrift für Kulturtechnik und Flurbereiningung* 4, 226–254.

Bailey, R.G., 1978. *Description of the Ecoregions of the United States.* U.S. Department of Agriculture, Forest Service, Ogden, Utah, 77 p.

Banner, A., R.J. Hebda, E.T. Oswald, J. Pojar, and R. Trowbridge, 1988. "Wetlands of Pacific Canada." In: *Wetlands of Canada.* Ecological Land Classification Series No. 24, Environment Canada and Polysciences Publications Inc., Montreal, 306–346.

Bay, R.R., 1969. "Runoff from small peatland watersheds." *Journal of Hydrology* 9, 90–102.

Bay, R.R., 1970. "The hydrology of several peat deposits in northern Minnesota, U.S.A." *Third International Peat Congress Proceedings*, Quebec, Canada, 212–218.

Boelter, D.H., 1964. "Water storage characteristics of several peats *in situ.*" *Proceedings Soil Science Society of America* 28, 433–435.

Boelter, D.H., 1965. "Hydraulic conductivity of peat." *Soil Science* 100, 227–231.

Boelter, D.H., 1969. "Physical properties of peat as related to degree of decomposition." *Proceedings Soil Science Society of America* 33, 606–609.

Boelter, D.H., 1972. "Water table drawdown around an open ditch in organic soils." *Journal of Hydrology* 15, 329–340.

Boelter, D.H., and E.S. Verry, 1977. "Peatland and water in the Northern Lake States." *U.S. Department of Agriculture, Forest Service General Technical Report* NC-31, 22 p.

Bridgham, S.C., J. Paster, J.A. Janssens, C. Chapin, and T.J. Malterer, 1996. "Multiple limiting gradients in peatlands: A call for a new paradigm." *Wetlands* 16, 45–65.

Brinson, M.M., 1990. "Riverine forests." In: E. Ariel, M. Brinson, and S. Brown (eds.), *Forested Wetlands: Ecosystems of the World* 15, Elsevier, Amsterdam, 87–141.

Brown, J., S.L. Dingman, and R.I. Lewellen, 1968. "Hydrology of a drainage basin on the Alaska coastal plain." *U.S. Army CRREL Research Report* 240, 18 p.

Carter, V., 1986. "An overview of the hydrologic concerns related to wetlands in the United States." *Canadian Journal of Botany* 64, 364–374.

Carter, V., 1990. "The Great Dismal Swamp: An illustrated case study." In: E. Ariel, M. Brinson, and S. Brown (eds.): *Forested Wetlands: Ecosystems of the World* 15, Elsevier, Amsterdam, 201–211.

Carter, V., M.S. Bedinger, R.P. Novitzki, and W.O. Wilen, 1979. "Water resources and wetlands." In: P.E. Greeson, J.R. Clark, and J.E. Clark (eds.), *Wetland Functions and Values: The State of our Understanding.* Proceedings, National Symposium on Wetlands, American Water Resources Association, Technical Publication Series S79–2, 344–376.

Chin, E.H., J. Skelton, and H.P. Guy, 1975. "The 1973 Mississippi River Basin flood—Compilation and analysis of meteorologic, streamflow, and sediment data." *U.S. Geological Survey Professional Paper* 937, 137 p.

Christensen, B.A., 1976. "Hydraulics of sheet flow in wetlands." *Proceedings of ASCE Symposium on Inland Waters for Navigation, Flood Control, and Water Diversions.* Colorado State University, August 10–12, 1976, 746–759.

Clymo, R.S., 1983. "Peat." In: A.J.P. Gore (ed.), *Ecosystems of the World 4A: Swamp, Bog, Fen and Moor.* Elsevier, Amsterdam, 159–224.

Costanza, R., C.Neill, S.C. Leibowitz, J.R. Fruci, L.M. Bahr Jr., J.W. Day Jr., and M.W. Young, 1983. *Ecological models of the Mississippi deltaic plain region: data collection and presentation.* U.S. Fish and Wildlife Service, Division of Biological Services, Washington, D.C., 342 p.

Cowardin, L.M., V. Carter, F.C. Golet, and E.T. LaRoe, 1979. *Classification of wetlands and deepwater habitats of the United States.* Fish and Wildlife Service, U.S. Department of the Interior, Washington, D.C., 103 p.

Dai, T.S., and J.H. Sparling, 1973. "Measurement of hydraulic conductivity of peat." *Canadian Journal of Soil Science* 53, 21–26.

Daniel, C.C. III., 1981. "Hydrology, geology, and soils of pocosins: A comparison of natural and altered systems." In: C.J. Richardson (ed.), *Pocosin Wetlands—An Integrated Analysis of Coastal Plain Freshwater Bogs in North Carolina.* Hutchinson Ross, Stroudsburg, Penn., 69–108.

Devall, M.S., 1990. "Cat Island Swamp: Window to a fading Louisiana ecology." *Forest Ecology and Management* 33/34, 303–314.

Devito, K.J., J.M. Waddington, and B.A. Branfireun, 1997. "Flow reversals in peatlands influenced by local groundwater systems." *Hydrological Processes* 11, 103–110.

Dingman, S.L., 1996. "Hydrologic studies of the Glenn Creek drainage basin." *U.S. Army Cold Regions Research and Engineering Laboratory Special Report* 86, 30 p.

Dingman, S.L., R.G. Barry, G. Weller, C. Benson, E.F. LeDrew, and C.W. Goodwin, 1980. "Climate, snow cover, microclimate, and hydrology." In: J. Brown et al. (eds.), *An Arctic Ecosystem: The Coastal Tundra at Barrow, Alaska.* Dowden, Hutchinson and Ross, Stroudsburg, Penn., 30–65.

Ford, J., and B.L. Bedford, 1987. "The hydrology of Alaskan wetlands, U.S.A.: A review." *Arctic and Alpine Research* 19, 209–229.

Foster, D.R., 1984. "The dynamics of *Sphagnum* in forest and peatland communities in southeastern Labrador, Canada." *Arctic* 37, 133–140.

Foster, D.R., G.A. King, P.H. Glaser, and J.E. Wright Jr., 1983. "Origin of string patterns in boreal peatlands." *Nature* 306, 256–258.

Galatowitsch, S.M., and A.G. van der Valk, 1994. *Restoring Prairie Wetlands: An Ecological Approach.* Iowa State University Press, Ames, Iowa, 246 p.

Geis, J.W., 1985. "Environmental influences on the distribution and composition of wetlands in the Great Lakes basin." In: H.H. Price and F.M. D'Itri (eds.), *Coastal Wetlands.* Lewis Publishers Inc., Chelsea, Mich., 15–31.

Gleason, P.J., and P. Stone, 1994. "Age, origin, and landscape evolution of the Everglades peatland." In: S.M. Davis and J.C. Ogden (eds.), *Everglades: The Ecosystem and its Restoration.* St. Lucie Press, Delray Beach, Florida, 149–197.

Glenn, M.S., and M.K. Woo, 1997. ASpring and summer hydrology of a valley-bottom wetland, Ellesmere Island, Northwest Territories, Canada." *Wetlands* 17, 321–329.

Gonthier, G.J., 1996. "Ground-water-flow conditions within a bottomland hardwood wetland, eastern Arkansas." *Wetlands* 16, 334–346.

Goode, D.A., A.A. Marsan, and J.-R. Michaud, 1977. "Water resources." In: N.W. Radforth and C.W. Brawner (eds.), *Muskeg and the Northern Environment in Canada.* University of Toronto Press, Toronto, 299–331.

Gorham, E., and J.A. Janssens, 1992. "Concepts of fen and bog re-examined in relation to bryophyte cover and the acidity of surface waters." *Acta Societatis Botanicorum Poloniae* 61, 7–20.

Greeson, P.E., J.R. Clark, and J.E. Clark, 1979. *Wetland Functions and Values: The State of our Understanding.* Proceedings, National Symposium on Wetlands, American Water Resources Association, Technical Publication Series S79-2, 674 p.

Harris, C.J., H.T. Erickson, N.K. Ellis, and J.E. Larson, 1962. "Water-level control in organic soil, as related to subsidence rate, crop yield and response to nitrogen." *Soil Science* 94, 158–161.

Heath, R.C., 1984. "Ground-water regions of the United States." *U.S. Geological Survey Water-Supply Paper* 2242, 78 p.

Hemond, H.F., and J.C. Goldman, 1985. "On non-Darcian water flow in peat." *Journal of Ecology* 73, 579–584.

Hofstetter, R.H., 1983. "Wetlands in the United States." In: A.J.P. Gore (ed.): *Ecosystems of the World 4B: Swamp, Bog, Fen and Moor.* Elsevier, Amsterdam, 201–244.

Holland, C.C., J. Honea, S.E. Gwin, and M.E. Kentula, 1995. "Wetland degradation and loss in the rapidly urbanizing area of Portland, Oregon." *Wetlands* 15, 336–345.

Hurr, R.T., 1983. "Ground-water hydrology of the Mormon Island Crane Meadows Wildlife Area near Grand Island, Hall County, Nebraska." *U.S. Geological Survey Professional Paper* 1277–H, 12 p.

Ingram, H.A.P., 1978. "Soil layers in mires: Function and terminology." *Journal of Soil Science* 29, 224–227.

Ingram, H.A.P., 1982. "Size and shape in raised mire ecosystems: A geophysical model." *Nature* 297, 300–303.

Ingram, H.A.P., 1983. "Hydrology." In: A.J.P. Gore (ed.), *Ecosystems of the World 4A: Swamp, Bog, Fen and Moor.* Elsevier, Amsterdam, 67–158.

Ingram, H.A.P., D.W. Rycroft, and D.J.A. Williams, 1974. "Anomalous transmission of water through certain peat." *Journal of Hydrology* 22, 213–218.

IPCC (Intergovernmental Panel on Climate Change), 1996. "Non-tidal wetlands." In: *Climate Change 1995—Impacts, Adaptations and Mitigation of Climate Change.* Cambridge University Press, Cambridge, 215–239.

Kane, D.L., and C.W. Slaughter, 1972. "Seasonal regime and hydrological significance of stream icings in central Alaska." In: *The Role of Snow and Ice in Hydrology.* Proceedings of the Banff Symposium, IAHS Publication 107, 528–540.

Kane, D.L., and J. Stein, 1983. "Water movement into seasonally frozen soils." *Water Resources Research* 19, 1547–1557.

Kane, D.L., S.R. Bredthauer, and J. Stein, 1981. "Subarctic snowmelt runoff generation." *Proceedings of the Specialty Conference on the Northern Community: A Search for a Quality Environment, ASCE,* Seattle, Wash., 591–601.

Kircher, J.E., and M.R. Karlinger, 1983. "Effects of water development on surface-water hydrology, Platte River basin in Colorado, Wyoming, and Nebraska upstream from Duncan, Nebaraska." *U.S. Geological Survey Professional Paper* 1271–B, 99 p.

Kirk, P.W., 1979. *The Great Dismal Swamp.* University Press of Virginia, Charlottesville, 427 p.

Kushlan, J.A., 1991. "The Everglades." In: R.J. Livingston (ed.), *The Rivers of Florida.* Springer-Verlag, New York, 121–142.

LaBaugh, J.W., T.C. Winter, V. Adomaitis, and G.A. Swanson, 1987. "Geohydrology and chemistry of prairie wetlands, Stutsman County, North Dakota." *U.S. Geological Survey Professional Paper* 1431, 26 p.

Lafleur, P.M., 1988. "Leaf conductance of four species growing in a subarctic marsh." *Canadian Journal of Botany* 66, 1367–1375.

Lafleur, P.M., 1990. "Evaporation from wetlands." *Canadian Geographer* 34, 79–82.

Lide, R.F., V.G. Meentemeyer, J.E. Pinder III, and L.M. Beatty, 1995. "Hydrology of a Carolina bay located on the upper coastal plain of western South Carolina." *Wetlands* 15, 47–57.

Light, S.S., and J.W. Dineen, 1994. "Water control in the Everglades: A historical perspective." In: S.M. Davis and J.C. Ogden (eds.), *Everglades: The Ecosystem and its Restoration.* St. Lucie Press, Delray Beach, Florida, 47–84.

Lins, H.F., K.F. Hare, and K.P. Singh, 1990. "Influence of the atmosphere." In: M.G. Wolman and H.C. Riggs (eds.), *Surface Water Hydrology*. The Geology of North America, Vol. 0-1. Geological Society of America, Boulder, Colo., 11–53.

Lissey, A., 1971. "Depression-focused transient groundwater flow patterns in Manitoba." *Geological Association of Canada Special Paper* 9, 333–341.

MacFarlane, I.C., 1969. *Muskeg Engineering Handbook*. University of Toronto Press, Toronto, 297 p.

Marsh, P., and M. Hey, 1989. "The flooding hydrology of Mackenzie Delta lakes near Inuvik, N.W.T., Canada." *Arctic* 42, 41–49.

Matthews, E., and I. Fung, 1987. "Methane emission from natural wetlands: Global distribution, area, and environmental characteristics of sources." *Global Biogeochemical Cycles* 1, 61–86.

McIvor, C.C., J.A. Ley, and R.D. Bjork, 1994. "Changes in freshwater inflow from the Everglades to Florida Bay including effects on biota and biotic processes: a review." In: S.M. Davis and J.C. Ogden (eds.), *Everglades: The Ecosystem and its Restoration*. St. Lucie Press, Delray Beach, Florida, 117–146.

Meyboom, P., 1966. "Unsteady groundwater flow near a willow ring in hummocky moraine." *Journal of Hydrology* 4, 38–62.

Meyboom, P., 1967. "Mass-transfer studies to determine the ground-water regime of permanent lakes in hummocky moraine of western Canada." *Journal of Hydrology* 5, 117–142.

Millar, J.B., 1976. "Wetland classification in western Canada: A guide to marshes and shallow open water wetland in the grasslands and parklands of the prairie provinces." Canadian Wildlife Service Report Series No. 37, Ottawa, 38 p.

Mills, J.G., and M.A. Zwarich, 1986. "Transient groundwater flow surrounding a recharge slough in a till plain." *Canadian Journal of Soil Science* 66, 121–134.

Mitsch, W.J., and J.G. Gosselink, 1986. *Wetlands*. Reinhold, N.Y., 722 p.

Moore, P.D., and D.J. Bellamy, 1974. *Peatlands*. Elek Science, London, 221 p.

Moore, R.D., and J. Martin, 1993. AGroundwater flow in estuarine sediments." In: A.P. Farrell (ed.), *Towards Environmental Risk Assessment and Management of the Fraser River Basin*, Technical Report for the B.C. Ministry of the Environment, Centre for Excellence in Environmental Research, Simon Fraser University, Burnaby, B.C., 273–285.

Munro, D.S., 1979. "Daytime energy exchange and evaporation from a wooded swamp." *Water Resources Research* 15, 1259–1265.

Naiman, R.J., J.M. Melillo, and J.E. Hobbie, 1986. "Ecosystem alteration of boreal forest streams by beaver (*Castor canadensis*)." *Ecology* 67, 1254–1269.

NWWG (National Wetland Working Group), 1987. The Canadian wetland classification system. Lands Conservation Branch, Canadian Wildlife Service, Environment Canada, Ecological Land Classification Series No. 21, 18 p.

NWWG (National Wetland Working Group), 1988. "*Wetlands of Canada*." Ecological Land Classification Series No. 24, Environment Canada and Polysciences Publications Inc., Montreal, 452 p.

Novitzki, R.P., 1979. "Hydrological characteristics of Wisconsin's wetlands and their influence on floods, streamflow and sediment." In: P.E. Greeson, J.R. Clark, and J.E. Clark (eds.), *Wetland Functions and Values: The State of our Understanding*. Proceedings, National Symposium on Wetlands, American Water Resources Association, Technical Publication Series S79–2, 377–388.

Nuttle, W.K., 1997. "Measurement of wetland hydroperiod using harmonic analysis." *Wetlands* 17, 82–89.

O'Brien, A.L., 1977. "Hydrology of two small wetland basins in eastern Massachusetts." *Water Resources Bulletin* 13, 325–340.

Orme, A.R., 1990. "Wetland morphology, hydrodynamics and sedimentation." In: M. Williams (ed.), *Wetlands: A Threatened Landscape*. The Institute of British Geographers Special Publication Series 25, Basil Blackwell Inc., Cambridge, Mass., 42–94.

Orme, A.R., and A.J. Orme, 1998. "Greater California." In A.J. Conacher, and M. Sala (eds.): *Land Degradation in Mediterranean Environments of the World: Nature and Extent, Causes and Solutions*. Wiley, Chichester, 109–122 and 214 et seq.

Price, J.S., 1992a. "Blanket bog in Newfoundland. Part 1. The occurrence and accumulation of fog-water deposits." *Journal of Hydrology* 135, 87–101.

Price, J.S., 1992b. "Blanket bog in Newfoundland. Part 2. Hydrological processes." *Journal of Hydrology* 135, 103–119.

Price, J.S., and J.E. FitzGibbon, 1987. "Groundwater storage–streamflow relations during winter in a subarctic wetland, Saskatchewan." *Canadian Journal of Earth Sciences* 24, 2074–2081.

Price, J.S., and M.K. Woo, 1988. "Studies of a subarctic coastal marsh, I. Hydrology." *Journal of Hydrology* 103, 275–292.

Prowse, T.D., 1986. "Ice jam characteristics, Liard-Mackenzie Rivers confluence." *Canadian Journal of Civil Engineering* 13, 653–665.

Racine, C.H., and J.C. Walters, 1994. "Groundwater-discharge fens in the Tanana Lowlands, Interior Alaska, U.S.A." *Arctic and Alpine Research* 26, 418–426.

Richardson, C.J., and E.J. McCarthy, 1994. "Effect of land development and forest management on hydrologic response in southeastern coastal wetlands: A review." *Wetlands* 14, 56–71.

Richardson, C.J., R. Evans, and D. Carr, 1981. "Pocosins: An ecosystem in transition." In: C.J. Richardson (ed.), *Pocosin Wetlands*. Hutchinson Ross Publishing Co., Stroudsburg, Penn., 3–19.

Romanov, V.V., 1968. *Hydrophysics of Bogs*. Israel Programme for Scientific Translations, Jerusalem, 299 p.

Roulet, N.T., 1990. "Hydrology of a headwater basin wetland: Groundwater discharge and wetland maintenance." *Hydrological Processes* 4, 387–400.

Roulet, N.T., and M.K. Woo, 1986. "Hydrology of a wetland in the continuous permafrost region." *Journal of Hydrology* 89, 73–91.

Rycroft, D.W., D.J.A. Williams, and H.A.P. Ingram, 1975. "The transmission of water through peat. I. Review." *Journal of Ecology* 63, 535–556.

Rydén, B.E., 1977. "Hydrology of Truelove Lowland." In: L.C. Bliss (ed.), *Truelove Lowland, Devon Island, Canada: A High Arctic Ecosystem*. University of Alberta Press, Edmonton, Alberta, 107–136.

Shankman, D., 1993. "Channel migration and vegetation patterns in the southeastern coastal plain." *Conservation Biology* 7, 176–183.

Shankman, D., and R.M. Kortright, 1994. "Hydrogeomorphic conditions limiting the distribution of baldcypress in the southeastern United States." *Physical Geography* 15, 282–295.

Shankman, D., and T.B. Pugh, 1992. "Discharge response to channelization of a coastal plain stream." *Wetlands* 12, 157–162.

Siegel, D.I., 1983. "Ground water and the evolution of patterned mires, Glacial Lake Agassiz peatlands, northern Minnesota." *Journal of Ecology* 71, 913–921.

Siegel, D.I., 1988. "The recharge-discharge function of wetlands near Juneau, Alaska: Part I. Hydrogeological investigations." *Ground Water* 26, 427–434.

Siegel, D.I., and P.H. Glaser, 1987. "Groundwater flow in a bog-fen complex, Lost River Peatland, northern Minnesota." *Journal of Ecology* 75, 743–754.

Sloan, C.E., 1972. "Ground-water hydrology of prairie potholes in North Dakota." *U.S. Geological Survey Professional Paper* 585–C, 28 p.

Smith, S.L., and M.K. Woo, 1986. "Ground and water temperature in a mid-latitude swamp." *Canadian Water Resources Journal* 11, 76–88.

Stanek, W., 1977. "A list of terms and definitions." In: N.W. Radforth and C.O. Brawner (eds.), *Muskeg and the Northern Environment in Canada.* University of Toronto Press, Toronto and Buffalo, 367–387.

Stephens, J.C., 1956. "Subsidence of organic soils in the Florida Everglades." *Soil Science Society of America Proceedings* 20, 77–80.

Stewart, R.E., and H.A. Kantrud, 1971. "Classification of natural ponds and lakes in the glaciated prairie region." Bureau of Sport Fishing and Wildlife Research Publication No. 92, 57 p.

Sturges, D.L., 1968. "Hydrologic properties of peat from a Wyoming mountain bog." *Soil Science* 106, 262–264.

Swanson, G.A., T.C. Winter, V.A. Adomaitis, and J.W. LaBaugh, 1988. "Chemical characteristics of prairie lakes in south-central North Dakota; their potential for impacting fish and wildlife." *U.S. Fish and Wildlife Service Technical Report* 18, 44 p.

Thom, B.G., 1972. "The role of spring thaw in string bog genesis." *Arctic* 25, 236–239.

Toth, J., 1963. "A theoretical analysis of groundwater flow in small drainage basins." *Proceedings Hydrology Symposium No. 3: Groundwater.* Queen's Printer, Ottawa, 75–96.

Vecchioli, J., H.E. Gill, and M.L. Solomon, 1962. "Hydrologic role of the Great Swamp and other marshland in Upper Passaic River basin." *Journal of American Water Works Association* 54, 695–701.

Walker, H.J., and J.M. Coleman, 1987. "Atlantic and Gulf Coastal Provinces." In: W.L. Graf (ed.), *Geomorphic System of North America.* Geological Survey of America, Centennial Special Vol. 2, Boulder, Colorado, 51–110.

Weir, W.W., 1950. "Subsidence of peat lands of the Sacramento–San Joaquin Delta, California." *Hilgardia* 20, 37–56.

Williams, M., 1990. "Understanding wetlands." In: M. Williams (ed.), *Wetlands: A Threatened Landscape.* The Institute of British Geographers Special Publication Series 25, Basil Blackwell Inc., Cambridge, Mass., 1–41.

Winter, T.C., 1976. "Numerical simulation analysis of the interaction of lakes and ground water." *U.S. Geological Survey Professional Paper* 1001, 45 p.

Winter, T.C., 1988. "A conceptual framework for assessing cumulative impacts on the hydrology of nontidal wetlands." *Environment Management* 12, 605–620.

Winter, T.C., 1989. "Hydrologic studies of wetlands in the northern prairies." In: A. van der Valk (ed.), *Northern Prairie Wetlands.* Iowa State University Press, Ames, Iowa 16–54.

Winter, T.C., and D.O. Rosenberry, 1995. "The interaction of ground water with prairie pothole wetlands in the Cottonwood Lake area, east-central North Dakota, 1979–1990." *Wetlands* 14, 193–211.

Winter, T.C., and M.K. Woo, 1990. "Hydrology of lakes and wetlands." In: M.G. Wolman and H.C. Riggs (eds.), *Surface Water Hydrology.* The Geology of North America, Vol. 0-1. Geological Society of America, Boulder, Colo., 159–187.

Woo, M.K., 1982. "Upward flux of vapor from frozen materials in the High Arctic." *Cold Regions Science and Technology* 5, 269–274.

Woo, M.K., and P.D. diCenzo, 1988. "Pipe flow in James Bay coastal wetlands." *Canadian Journal of Earth Sciences* 26, 625–629.

Woo, M.K., and P.D. diCenzo, 1989. "Hydrology of small tributary streams in a subarctic wetland." *Canadian Journal of Earth Sciences* 26, 1557–1566.

Woo, M.K., and R. Heron, 1987a. "Breakup of small rivers in the subarctic." *Canadian Journal of Earth Sciences* 24, 784–795.

Woo, M.K., and R. Heron, 1987b. "Effects of forests on wetland runoff during spring." In: *Forest Hydrology and Watershed Management.* Proceedings of the Vancouver Symposium, IAHS Publication 167, 297–307.

Woo, M.K., and P. Marsh, 1990. "Response of soil moisture change to hydrological processes in a continuous permafrost environment." *Nordic Hydrology* 21, 235–252.

Woo, M.K., and R.D. Rowsell, 1993. "Hydrology of a prairie slough." *Journal of Hydrology* 146, 175–207.

Woo, M.K., and J. Valverde, 1981. "Summer streamflow and water level in a mid-latitude swamp." *Forest Science* 27, 177–189.

Woo, M.K., and Waddington, M.J., 1990. "Effects of beaver dams on subarctic wetland hydrology." *Arctic* 43, 223–230.

Woo, M.K., and T.C. Winter, 1993. "The role of permafrost and seasonal frost in the hydrology of northern wetlands in North America." *Journal of Hydrology* 141, 5–31.

Woo, M.K., R.D. Rowsell, and R.G. Clark, 1993. "Hydrological classification of Canadian prairie wetlands and prediction of wetland inundation in response to climatic variability." *Canadian Wildlife Service Occasional Paper* No. 79, 24 p.

Zoltai, S.C., 1988. "Wetland environments and classification." *Wetlands of Canada.* Ecological Land Classification Series No. 24, Environment Canada and Polysciences Publications Inc., Montreal, 2–26.

Zoltai, S.C., and F.C. Pollett, 1983. "Wetlands in Canada: Their classification, distribution and use." In: A.J.P. Gore (ed.), *Ecosystems of the World 4B: Swamp, Bog, Fen and Moor.* Elsevier, Amsterdam, 245–268.

9

Weathering and Soils

John C. Dixon

The landscape of North America is mantled with varying types and thicknesses of unconsolidated materials collectively referred to as regolith. Although much of the regolith has been transported, some of it has been derived in situ. This residual material is the product of the combined effects of chemical, physical, and biological weathering processes. These processes operating synergistically have transformed unstable primary minerals in rocks at and near Earth's surface to secondary products. The weathering residua accumulate under favorable geomorphic conditions to give rise to a wide variety of weathering-dominated landscapes.

Regolith is variably modified by a complex set of processes operating at and near the boundary layer. These processes encompass the addition, translocation, transformation, and removal of materials from the regolith, resulting in soil horizons with diagnostic physical, chemical, and biological characteristics. Spatial variability in the degree of soil development is determined by the variability of the soil-forming factors of climate, organisms, relief, parent material, and time. These factors operate at different spatial and temporal scales, but at the continental scale reflect their influence in the mosaic of diverse soils of North America.

This chapter examines the weathering mantles of the North American landscape and discusses the major soil types, their origins, and their distributions across the continent.

9.1 Weathered Landscapes of North America

Weathering-dominated landscapes, reflected in the occurrence of thick weathering mantles, occur extensively across large areas of North America (Hunt, 1986). The first section of this chapter focuses on the evolution of some of the distinctive landscapes in which weathering processes have dominated their development.

When the rate of weathering exceeds the rate of removal and transportation of weathered waste, then transport-limited landscapes develop with characteristic accumulations of weathered debris. These landscapes are not limited to the stable core of the continent but can also be found associated with mobile mountain belts and ancient collision zones.

9.1.1 Weathering Mantles of Arctic and Alpine North America

Arctic North America Expansive areas of far northern North America are characterized by the accumulation of vast amounts of both unconsolidated, weathered bedrock debris known as felsenmeer and grus associated with deeply weathered continental basement bedrock. Throughout the

Arctic lowlands, bedrock outcrops display a great diversity of weathering forms. Undoubtedly, glacial processes, as well as permafrost and seasonal ice, have modified these landscapes, but they are nonetheless fundamentally of a weathering origin.

Traditionally, these felsenmeer-covered surfaces have been attributed to frost action (Bird, 1967; Watts, 1986), but in recent years the applicability of traditional volumetric expansion models associated with freeze-thaw cycles has been questioned (Thorn, 1979, 1982; French, 1981, 1993). Recent experimental work by Hallet et al. (1991) has suggested that an alternative explanation for frost weathering might be segregation ice growth in microcracks at sustained subzero temperatures.

Extensive areas of Arctic Canada also display accumulations of grus derived from the weathering of igneous and metamorphic rocks. These accumulations, which may be as much as a meter thick, have been largely attributed to the development of microfractures resulting from the interaction of frost action, hydration, and salt crystallization (Watts, 1981, 1983, 1986). The intensity of grussification is strongly linked to such inherent rock properties as grain size, mineralogy, porosity, and structure. Accompanying the formation of grus is the development of granitic and metasedimentary tors and microweathering forms, such as weathering pits, spalled surfaces, and tafoni. Although Watts's research has mostly focused on the origin of these forms by physical weathering, there is a growing body of evidence that chemical processes contribute significantly to landscape evolution in these environments as exemplified by extensive solution of limestone terrains (Ford, 1993) and the accumulating evidence of chemical weathering processes associated with soil formation in the Arctic (Ugolini, 1986; Campbell and Claridge, 1992).

Figure 9.1 Coarse debris mantling alpine valley slopes, upper Green Lakes Valley, Colorado Front Range.

Alpine North America Coarsely fractured bedrock debris accumulated along valley side walls is a hallmark of high alpine environments (fig. 9.1). Traditionally, this debris has been interpreted as the result of frost weathering in fractured bedrock. Slaymaker (1974) stressed the role of frost action in both barren and covered alpine landscape evolution in Canada. Gardner (1992), working in the Canadian Rockies, has suggested that indeed freeze-thaw weathering is exceedingly important in the recession of cirque headwalls, but may be more spatially restricted than generally believed.

In contrast, Thorn (1979, 1980) questioned the efficacy of freeze-thaw weathering of bedrock surfaces. His measurements of rock surface temperature and moisture availability led him to conclude that, when bedrock temperatures were low enough to induce the freezing of water, water was not available, and when moisture was available in the form of snow accumulation, temperatures failed to drop to critical values. White (1976), also working in the Front Range of the Rocky Mountains of Colorado, has likewise questioned the efficacy of freeze-thaw processes in bedrock weathering and has instead proposed hydration shattering, an idea recently elaborated by Hallet et al. (1991). Our understanding of the processes responsible for the production of the large quantities of coarse angular debris found in alpine environments remains limited. Recent studies suggest that chemical weathering processes are far more widespread and quantitatively more significant to the production and modification of weathering mantles in alpine environments than long believed.

Early studies of weathering in alpine and subalpine environments recognized the interaction of both physical and chemical processes in the breakdown of a wide variety of lithologies. For the Cascade Mountains of Washington, Reynolds (1971) and Reynolds and Johnson (1972) documented both extensive and intensive chemical weathering of metamorphic bedrock with accompanying high denudation rates and the formation of a diverse assemblage of secondary clay minerals. Thorn (1979) documented the chemical enrichment of waters draining a snow patch hollow in the Colorado Front Range. Chemical and mineralogical transformations associated with soil formation in the Rocky Mountains of Colorado have also been documented by Dixon (1983, 1986) and Birkeland et al. (1987). Caine (1992, 1995a,b), and Caine and Thurman (1990) have demonstrated the significance of chemical denudation in alpine landscape evolution. Gallie and Slaymaker (1984, 1985) report extensive chemical weathering in the high mountains of British Columbia. Chemical denudation of limestone terranes in alpine environments has been documented by Ford (1993). Dixon et al. (1984), working in the Coast Ranges of Alaska, documented extensive chemical losses and the formation of secondary clay minerals associated with the weathering of granodiorite. Even though absolute rates of chemical weathering are often lower than those of more temperate environments, these studies have demonstrated that landscape evolution and soil formation in alpine North America are strongly affected by geochemical processes.

The role of biological weathering processes operating in alpine environments has received little attention. However, Hall and Otte (1990), working in the Juneau Icefield in the Coast Ranges of Alaska, have suggested that widespread rock-surface flaking may be the result of the expansion and contraction of algal mucilage, associated with episodes of wetting and drying.

9.1.2 *Weathering Mantles of the Western Cordillera*

Landscapes developed on deeply weathered grus occur extensively in the montane regions of the Western Cordillera (Hunt, 1986; Graham et al., 1994). In the Rocky Mountains, for example, Varnes and Scott (1967) report grussified granite, which extends to depths exceeding 15 m, from the Rampart Range. In the southern Laramie Range, the Sherman erosion surface is cut across the deeply weathered Trail Creek Granite, which is weathered to depths of 70 m. The deep weathering is the result of the exploitation of hydrothermally induced microcracks associated with the oxidation of magnetite, ilmenite, and biotite (Eggler et al., 1969) (fig. 9.2).

Deeply weathered landscapes also occur in the Sierra Nevada where micaceous residuum dominated by kaolinitic clays occur (Hunt, 1986), particularly on the western slopes of the range. The Sierra Nevada typically displays a pronounced stepped topography, with bold granitic escarpments separated from each other by deeply weathered steps, which are underlain by grus to depths in excess of 30 m (Wahrhaftig, 1965). Farther to the east, in the White Mountains, Marchand (1974) reports the occurrence of grus associated with adamellite and a residual lag accumulation associated with dolomite, accumulations that he ascribes to extensive ground ice development. Secondary mineral alteration with accompanying grain-size reduction is attributed to chemical processes. In the case of the dolomite, solution is the dominant chemical process. Secondary alteration of the adamellite involves mineralogical and chemical transformation of biotite, microcline, plagioclase, allenite, and magnetite. In the Cascade Mountains, deep, clay-rich regolith occurs on volcanic lithologies (Hunt, 1986). The weathering of basalt and andesite involves the alteration of most of the primary minerals to a mixture of allophane, iron oxyhydroxides, and poorly crystallized clay minerals. Accompanying these mineralogical transformations are chemical changes involving depletion of alkalis, alkali earths, silicon, and sesquioxides (Colman, 1982).

Figure 9.2 Deeply grussified landscape, Sherman Erosion Surface, Laramie Range, Wyoming.

Along the Pacific coast, south of the glacial limit, mountain landscapes are strongly influenced by a variety of weathering processes, leading to the formation of deep weathering mantles (Hunt, 1986). Dethier (1986) estimates that in the Pacific Northwest saprolite production is taking place at a rate of 30 mm ka^{-1} (ka = thousand years), which is similar to that in the Appalachian Piedmont (Pavich, 1986). He further estimates that approximately 60% of the dissolved load of rivers in the Pacific Northwest is derived from the transformation of fresh bedrock to saprolite, with only 40% being derived from soil biogeochemical processes. Ugolini et al. (1977) suggest that as much as 70% of the stream solutes may be derived from saprolite formation. However, the rate of saprolite formation varies considerably, depending on lithology. Consequently, the relative contributions of chemical and physical denudation to landscape evolution also vary throughout the region, being about equal in western Washington and Oregon, whereas physical denudation (mass wasting) is some 20 times higher in the Franciscan terranes (Dethier, 1986).

In the Pacific mountain systems of California, saprolite-dominated landscapes have also been reported (Nettleton et al., 1970; Hunt, 1986; Graham et al., 1988, 1994; Jones and Graham 1993). In their study of tonalite weathering, Nettleton et al. (1970) report the transformation of biotite to vermiculite and kaolinite and occasionally montmorillonite, and the transformation of plagioclase and hornblende to montmorillonite. For both hornblende and biotite, considerable physical weathering accompanies their transformation to secondary clays as a result of crystal lattice expansion.

9.1.3 *Weathered Landscapes of the Appalachian Piedmont*

The Appalachian Piedmont extends from New York to Alabama and varies in width from 20 to 200 km. The province is a geologically complex part of the Appalachian Highlands, which are extensively underlain by metasedimentary and other metamorphic rocks intruded by mafic dikes and granitic plutons (Markewich et al., 1990). The landscape is essentially an incised sloping plain with accordant interfluves broken by granitic inselbergs. Beneath the interfluves, the bedrock is deeply weathered to 10–30 m (Pavich 1986, 1989; Stolt et al., 1992). For the most part, the saprolite is isovolumetrically weathered and maintains the basic fabric and structure of the unaltered bedrock. It has experienced much disaggregation and softening as a result of weathering (Stolt and Baker, 1994). Hunt (1986) recognizes two distinctive types of saprolite dominating the Piedmont. One type, a micaceous residuum with little quartz, is associated principally with metamorphic rocks. The second is a quartz-rich residuum with substantially less mica associated with granitic bedrock. However, recent studies have recognized greater variability in the grus (Stolt et al., 1993). The development of great thicknesses of saprolite exerts a profound influence on the geomorphology of the province, which turn influences the thickness and extent of the saprolite (Costa and Cleaves, 1984; Pavich et al., 1989; Graham et al., 1990).

Stolt and Baker (1994) characterize saprolite formation as occurring in two stages, involving the formation of secondary minerals from readily weatherable primary minerals, followed by iron oxidation, desilication, and removal of bases. The formation of the saprolite is accompanied by losses of calcium, sodium, and silicon, which are directly related to the weathering of plagioclase feldspar. In addition, magnesium is lost from the weathering of mafic constituents. There are accompanying relative increases in iron, aluminum, and zirconium, but losses of aluminum and iron have also been reported from some saprolites (Pavich, 1985, 1986; Pavich et al., 1989; Stolt et al., 1992). Mineralogical transformations primarily involve the alteration of plagioclase feldspar to kaolinite. In the soil developed in the saprolite, however, the dominant clay mineral is vermiculite (Pavich, 1986). Beneath the saprolite, the weathering profile consists of some 5 m of weathered bedrock characterized by the oxidation of iron, which Pavich (1986) has argued is evidence for the important role played by oxidation in the initiation of weathering. This oxidation appears to primarily affect biotite and iron sulfides. Other factors, in addition to landscape position, have been shown to influence the rate at which a saprolite forms and the depths that it obtains. These include bedrock texture, fabric, mineralogy, and structure (e.g., Velbel, 1985, 1986). In general, saprolite thicknesses are greater on the metamorphic rocks of the piedmont than on the granitic rocks.

9.1.4 Residual Mantles of the Ozark Plateaus and Interior Low Plateaus

Extensive areas of the Ozark Plateaus of northwest Arkansas and southwest Missouri, and the Interior Low Plateaus of west Tennessee and central Kentucky are underlain by calcareous bedrock. These plateau surfaces carry a residual mantle of weathered limestones and dolomites; this mantle is dominated by kaolinitic red clays and, in places, by a considerable amount of chert (Hunt, 1986). This clayey regolith is quite variable in thickness. Where the residuum lacks chert, it is generally less than 3 m thick. Chert-rich regolith is generally less than 30 m thick in the western part of the region, but as much as 100 m thick near the Mississippi Valley. Parse (1995) demonstrated that the thickness of regolith developed on the Springfield Plateau in northern Arkansas displayed considerable variability in thickness, but commonly reached thicknesses of 30–50 m. Regolith thickness is also strongly related to landscape position. In the Missouri Ozarks, the regolith is thickest on older, stable landscape surfaces and thinner on younger, unstable surfaces (Madole et al., 1991). On the Springfield Plateau of the Arkansas Ozarks, regolith thicknesses are greater on low-angle slopes and where overall relief is low, whereas the thinnest regolith occurs on valley sidewalls and where overall relief is great. Locally, Parse (1995) ascribed the great variability in regolith thickness to strong cutter and pinnacle development.

The clay mineralogy of the regolith is dominated by kaolinite, halloysite, and illite (Madole et al., 1991), with clay abundances in the <2-mm size fraction of the regolith accounting for as much as 95% by weight. Clay mineral assemblages of the soils developed on the regolith differ from those of the regolith, with surface horizons being dominated by vermiculite and subsurface horizons by smectite (Gamble and Mausbach, 1982). The regolith displays considerable evidence for clay illuviation: clay films are widely developed on fractures and along pore linings, and clay skins on ped faces are widespread (Parse, 1995).

Regolith formation appears to be largely the result of the accumulation of residual clay, insoluble chert, and other primary minerals consequent upon the dissolution of the limestones and other carbonate bedrocks. Weathering is not isovolumetric, as there is considerable evidence for regolith subsidence and deformation (fig. 9.3). However, extensive areas of the regolith display essentially continuous and relatively undeformed beds of chert and clay. Progressive thickening of the regolith appears to follow a series of predictable stages in which the original bedrock is progressively replaced by clay and chert fragments as dissolution progresses along bedding planes and down joints with accompanying translocation of clay and silt (fig. 9.4).

9.1.5 Weathered Mantles of Arid North America

In the arid and semiarid western United States and northern Mexico, deep weathering has produced mantles of saprolite or grus often accompanied by accumulations of calcium carbonate, silica, and gypsum (Hunt, 1986). In the Mojave Desert, granite pediments commonly carry mantles

Figure 9.3 (*left*) Mantled karst landscape, Springfield Plateau, northwest Arkansas.

Figure 9.4 (*right*) Chert-rich regolith, Springfield Plateau, northwest Arkansas.

of grus that are over 3 m thick. Atop these weathered materials are aridisols, 20–100 cm thick, which have been indurated by calcium and silica derived from the weathering of the underlying granitic rocks (Boettinger and Southard, 1991). The weathering profiles display marked losses of calcium and silicon from the grus, with accompanying accumulations in the overlying soils.

Pope et al. (1995) report weathering profiles in excess of 40 m in granodiorite underlying pediments in the Phoenix area. These mantles are also accompanied by calcareous crusts, or calcretes (Dixon, 1994a), which are particularly well developed along joints within the grus. The preservation of these mantles in the landscape is due to the combined influence of strongly indurated calcrete deposits and the relative tectonic stability of the local landscape. Graham and Franco-Vizcaino (1992) report deeply weathered landscapes in the Sonoran desert of Mexico, where a diversity of igneous and metamorphic rocks have been grussified.

Landscape surfaces in the arid regions of North America characteristically carry a mantle of coarse stones referred to as desert pavement. The origin of desert pavement has received considerable attention (Dixon, 1994b). Basically, theories of their evolution fall into five groups: (1) stone concentration as a result of fine-particle deflation, (2) stone concentration as a result of the removal of fines by running water, (3) stone concentration as a result of upward migration due to fine-grain, swell-shrink activity, (4) birth of stone pavements at the surface by predominantly weathering processes, and (5) differential subsurface weathering. Recently, the role of rain splash detachment of fine particles on desert surfaces has been proposed as an important process of desert pavement formation (Parsons et al., 1992; Wainwright et al., 1995). One essential component of all of these models is the determination of the origin of the coarse debris.

Although some stone surfaces may be primary fluvial plains (Cooke, 1970), it is more generally accepted that they are fundamentally the result of weathering (Dixon, 1994b). A variety of weathering processes have been invoked to explain their origin. Cooke (1970) ascribed pavement gravels in California to the combined efects of insolation, frost riving, and salt crystallization. More recent detailed work on the origin of pavements in the Cima volcanic field of California has demonstrated that volumentric expansion of aerosolic salts and clays deposited in rock and gravel fractures is exceedingly important in both producing coarse gravel and reducing the sizes of gravel clasts over time (McFadden et al., 1987; Dohrenwend, 1987). The role of salt in clast-size reduction has also been stressed by Goudie and Day (1980) from their studies of Death Valley pediment surfaces. Clast shape is undoubtedly related to chemical weathering processes and is most likely manifest as clast rounding (Dixon, 1994b).

9.2 Soil Classification in North America

9.2.1 Soil Classification in the United States of America

The soil classification system used in the United States is the Comprehensive Soil Classification System or Soil Taxonomy (Soil Survey Staff, 1975, 1999). This hierarchical classification system consists of six principal categories (from highest to lowest levels of generalization): order, suborder, great group, subgroup, family, and series. At its highest level of generalization, the classification system consists of 12 *soil orders* based on diagnostic soil horizons or combinations of soil horizons present within a soil profile. The soil orders are subdivided into 55 *suborders*, primarily on the basis of their temperature and moisture regimes, the nature of their parent materials, and processes of soil formation. Suborders are subdivided into approximately 250 *great groups* on the basis of the degree of expression of soil horizons and soil horizon sequences, base status, soil temperature and moisture regimes, and the presence or absence of diagnostic soil horizons. The great groups are further subdivided into approximately one thousand *subgroups*. The subgroups are the central taxa of the great groups and indicate integrades to other great groups, suborders, and orders. Subgroups are subdivided into *families*, which are broad classes based on texture, mineralogy, and temperature of the solum. Finally, the families are broken into *series*, which number approximately 12,000 in the United States and are based on the type and arrangement of horizons, soil color, texture, structure, consistence, and reaction of horizons, as well as chemical and mineralogical properties of the horizons. Most of the discussion of this section will focus on soil orders and suborders because these two categories most completely characterize soils at the regional scale adopted in this chapter.

9.2.2 Canadian Soil Classification System

In Canada, a different soil classification system is used, yet it probably bears closer resemblance to the U.S. Comprehensive Soil Classification System than to any other system used in the world. Like the U.S. system, the Canadian System is hierarchical in that classes or taxa are based on observable and measurable soil properties that reflect soil-forming processes and environments. The taxa of the Canadian System of Soil Classifications are based on the dominant soils of Canada: unlike the U.S. classification, the Canadian System is not designed to be taxonomically comprehensive. Nor is the Canadian System as strongly based on subsurface soil horizons as the U.S. system because in the United States the surface horizons have been so strongly modified by tillage. The Canadian Soil Classification System does not include a category of suborder, but it does raise to order the status of some soils that in the United States are assigned suborder or great group status. In the following discussion, soil orders of the Comprehensive Soil Classification System are arranged alphabetically, with equivalent Canadian soil orders in brackets.

9.3 Soil Orders of North America (Canadian Equivalents)

9.3.1 Alfisols (Luvisols and Solonetz)

Classification Alfisols possess an argillic (clay-rich), natric (salt-rich), or kandic (low activity clay-dominated) horizon and either a frigid temperature regime or a base saturation below the argillic horizon of 35% or more (Buol et al., 1989). These soils are subdivided into five suborders: (1) Aqualfs, which display pronounced wetness; (2) Boralfs, which are freely drained alfisols in cool environments; (3) Udalfs, which are red to brown freely drained alfisols; (4) Ustalfs, which possess a ustic (dry) moisture regime, a massive and hard epipedon when dry, and a calcic horizon within 150 cm of the surface; and (5) Xeralfs, which have a xeric moisture regime or an epipedon that is massive and hard when dry (Steila and Pond, 1989). Luvisols may be either Gray Brown Luvisols or Gray Luvisols, the essential difference being mean annual soil temperature (MAST). Gray Brown Luvisols have MAST of >8°C and Gray Luvisols have a MAST of <8°C. These two great groups correspond to udalfs and boralfs, respectively. Alfisols with natric horizons correspond to solonetzic soils in the Canadian Soil Classification System.

Morphology and Genesis The dominant morphological element of Alfisols and Luvisols is the presence of strongly developed eluvial and illuvial horizons (Canada Soil Survey Committee, 1978; Rust, 1983). These strongly texture-contrast soils have been widely attributed to the strong eluviation of clay from the A and E horizons to the B horizon (Canada Soil Survey Committee, 1978; Buol et al., 1989). It is also probable that this strong texture contrast is the result of lateral contrasts in weathering intensity (Buol et al., 1989), or of the addition of fine-grained materials to the A horizon by eolian addition especially in Alfisols of the lower Mississippi valley (Rutledge et al., 1985; West and Rutledge, 1987). This strong textural contrast between the upper and lower solum may also represent parent material contrasts (Schaetzl, 1996). Alfisols are characterized by the occurrence of an ochric or umbric epipedon, development of an argillic or kandic horizon, possess moderate to high base saturation, possess contrasting soil horizons including the presence of natric, petrocalcic, duripan and fragipan horizons, and plinthite under appropriate climatic regimes, they display little organic accumulation, and have water available to mesophytic plants for most of the year (Soil Survey Staff, 1975, 1999; Foth and Schafer, 1980; Birkeland, 1984; Buol et al., 1989). Although the accumulation of soluble chemical constituents is generally associated with alfisols in drier environments (Holliday, 1988), these chemical constituents have also been reported in alfisols from the Midwest (Schaetzl et al., 1996). The soils also display considerable effects of bioturbation, including that associated with both tree throw and faunal burrowing (Buol et al., 1989).

Distribution In the conterminous United States, about 40% of the Alfisols occur in the Midwest, especially in the valleys of the Mississippi and Ohio Rivers and their major tributaries (Foth and Schafer, 1980; Schaetzl, 1996; Schaetzl et al., 1996). Additionally, 25% occur in the western states, in particular in the Central Valley of California and the

Rocky Mountains, and 20% lie in the south-central United States, especially in the lower Mississippi valley and in eastern Texas and Oklahoma. These soils have also been reported from several of the Appalachian provinces, including the Appalachian Plateau, the Triassic Lowland and the Ridge and Valley Province, (Ciolkosz et al., 1989) and from southeastern states (Foth and Schafer, 1980; Rust, 1983) (fig. 9.5). In Canada, Luvisolic soils cover a relatively small area, occurring primarily in the Western Cordillera, with secondary occurrences in the northern Interior Plains in association with Chernozems. In southern Ontario, they represent the northern extension of the Alfisols of Michigan, New York (fig. 9.6, 9.7), and the St. Lawrence Valley (Foth and Schafer, 1980).

9.3.2 Andisols

Andisols are one of the most recently recognized soil orders of the U.S. Soil Taxonomy. Andisols are soils developed on tephra recently erupted from active volcanoes, and they

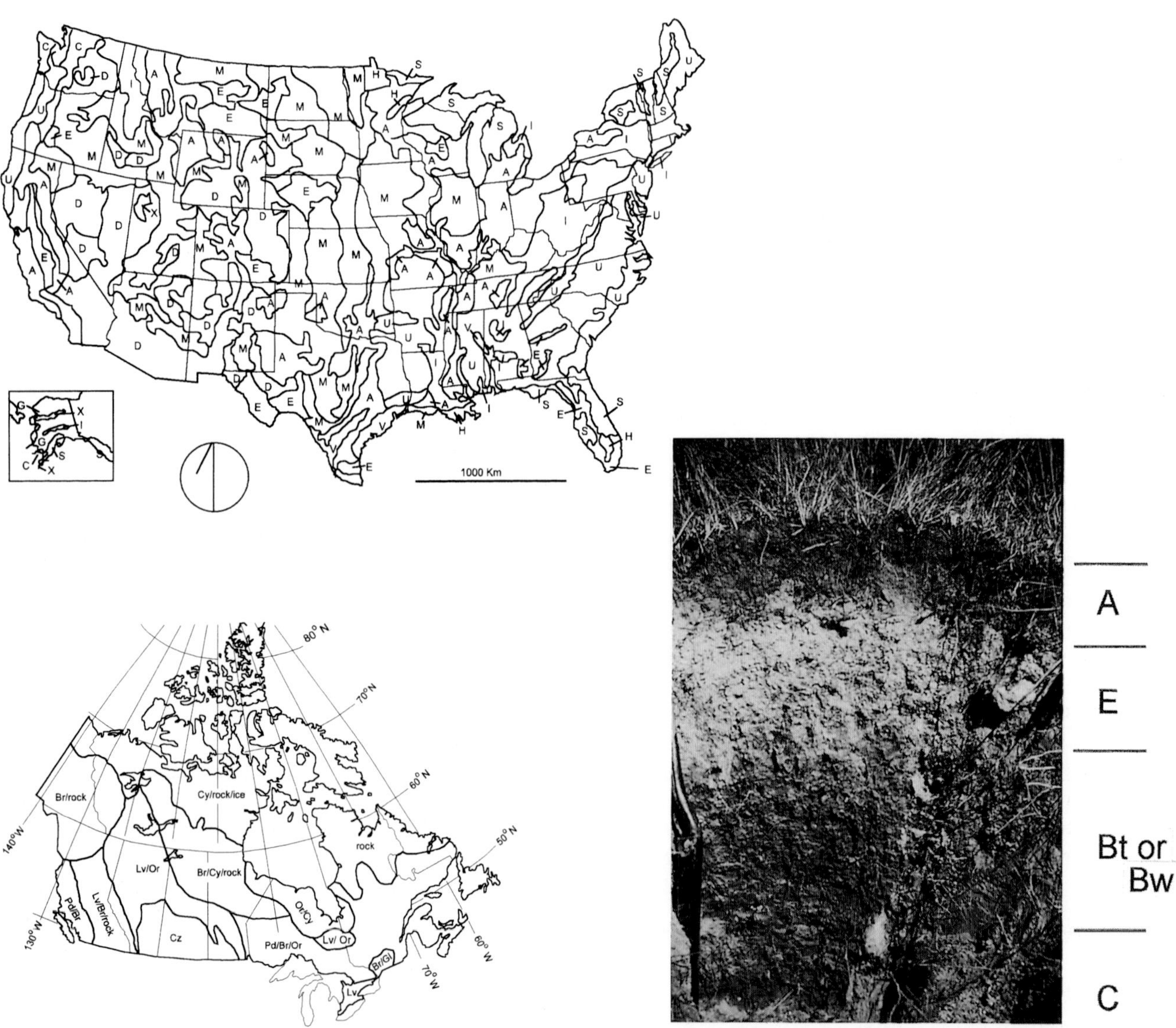

Figure 9.5 (*top*) Generalized soil map of the United States showing distribution of soil orders. A–Alfisols; C–Andisols; D–Aridisols; E–Entisols; G-Gelisols; H–Histosols; I–Inceptisols; M–Mollisols; S–Spodosols; U–Ultisols; V–Vertisols; (USDA Natural Resources Conservation Service, National Soil Survey Center, 1998).

Figure 9.6 (*bottom left*) Generalized soil map of Canada showing distribution of soil orders. Br–Brunisols; Cy–Cryosols; Cz–Chernozems; Gl–Gleysols; Lv–Luvisols; Or–Organics; Pd–Podsols (Canada Soil Survey Committee, 1998).

Figure 9.7 (*right*) Alfisol, northern Michigan.

are characterized by properties intimately derived from these unique parent materials. Most of the soils that are now included in this order are the members of the previous suborder of Inceptisols referred to as Andepts and great groups of Andaquepts. In this section, discussion will relate to the soils previously classified in these two groups.

Classification Andisols are subdivided into seven suborders. Aquands possess aquic moisture regimes; Torrands, Xerands, Ustands, and Udands possess aridic, xeric, ustic, and udic moisture regimes, respectively. Cryands possess either cryic or pergelic temperature regimes (Buol et al., 1989).

Morphology and Genesis These soils are associated with parent materials possessing unique physical, chemical, and mineralogical properties, which may change from one eruptive event to the next. Their surface horizons frequently display little evidence of weathering. These soils are mineralogically and chemically distinctive because volcanic ash is dominated by volcanic glass, which, upon weathering, produces solubility products that are predominantly noncrystalline aluminosilicates known as allophane. In the presence of humus, these products produce allophane-humus complexes. Another secondary product associated with the alteration of volcanic glass is imogolite. Both allophane and imogolite occur predominantly in the subsoil (Buol et al., 1989). Some of these subsoil horizons may develop into spodic horizons under favorable conditions. These soils typically have low bulk densities, are very freely drained, and display weak profile development (Foss et al., 1983).

Distribution In North America, Andisols are associated with the volcanic landscapes of the Cascade Range and adjacent mountain systems, where they reach the highest elevations near recently active volcanoes (Foth and Schafer, 1980) (fig. 9.5). They also occur in association with volcanoes of the Aleutian Islands and Alaska Peninsula (Simonson and Rieger, 1967). Andisols that occur at particularly high elevation or latitude are usually members of the Cryand suborder.

9.3.3 *Aridisols*

Classification Aridisols are characterized by a lack of moisture for extended periods of time but nonetheless display considerable pedogenic development. They are typified by the occurrence of an ochric epipedon, a surface crust, and accompanying development of one or more subsurface soil horizons including cambic, argillic, natric and salic, calcic and petrocalcic, and duripans (fig. 9.8). They are further characterized by the lack of organic matter and the occurrence of a stone pavement at the surface typically one or two stones thick. Their thermal regimes vary from cryic to isohyperthermic (Soil Survey Staff, 1975, 1999; Foth and Schafer, 1980; Nettleton and Peterson, 1983; Birkeland, 1984; Buol et al., 1989).

Figure 9.8 Aridisol, central Arizona.

Aridisols are subdivided into two suborders, Argids, which have an argillic or natric horizon, and Orthids, which lack either of these horizons. Argids are subdivided into five great groups on the basis of the depth to the argillic or natric horizon and the abundance of clay in the argillic horizon. Orthids are subdivided into four great groups on the basis of the depth to diagnostic subsurface horizons (Soil Survey Staff, 1975, 1999; Nettleton and Peterson, 1983; Steila and Pond, 1989).

Morphology and Genesis The surface horizon is typically light colored, thin, and vesicular in structure. The vesicular structure has been attributed to seasonal wetting and entrapment of air (Nettleton and Peterson, 1983) and more recently to entrapment of air in aeolian infall with subsequent thermal expansion (Wells et al., 1985; McFadden et al., 1986, 1987). This structure is repeatedly formed and destroyed as a result of the combined effects of alternating wetting and drying, and trampling by animals (Nettleton and Peterson, 1983). The vesicular horizon is also frequently accompanied by the development of a distinctive columnar structure as a result of the accumulation of clays and fine silts from both aeolian infall and weathering. The ochric horizon typically lacks organic matter because of the common association of these soils with xeric vegetation communities.

Aridic soils are commonly dominated by the accumulation of calcite and/or gypsum, which are predominantly aeolian in origin. With progressive accumulation of these minerals, petrocalcic and petrogypsic horizons develop. Calcic and gypsic horizons may be indurated or nonindurated and exhibit a variety of morphological forms, including powder and nodules as well as massive crusts with laminar surfaces. Descriptions of the great diversity of form and origin of calcic horizons in aridisols of North America are to be found in Gile et al. (1981), Harden et al. (1991), and Dixon, (1994a). Calcic horizons are commonly stacked on top of one another, leading to great thicknesses of indurated carbonate such as those found in west Texas (fig. 9.9). Gypsic horizons are less common, but they have been reported from several areas by Nettleton et al. (1982), Harden et al. (1991), and Reheis (1987).

More soluble constituents also accumulate in some Aridisols that form both natric and salic horizons. Natric horizons are special argillic horizons that display the development of prismatic, columnar, and occasionally blocky structure; in addition, they possess 15% or greater saturation with exchangeable sodium (Nettleton et al., 1975). Salic horizons are at least 15 cm thick and possess 2–3% soluble salts (Soil Survey Staff, 1975). The soluble salts are derived from airfall (Alexander and Nettleton, 1977; Peterson, 1980; and McFadden et al., 1991) and/or shallow groundwater movement (Nettleton and Peterson, 1983).

Figure 9.9 Stacked calcretes, eastern New Mexico.

Cambic horizons commonly occur beneath ochric horizons and are characterized by relatively fine textures, loss of original structure, presence and alteration of weatherable minerals, accumulation of clays and iron oxides, and accumulation and translocation of carbonates (Nettleton and Peterson, 1983; Steila and Pond, 1989). The development of this horizon is intimately related to the destruction of the vesicular structure of ocric epipedons and the associated decrease in water percolation and thus decreased accumulation of translocated materials and authigenic mineral transformation (McFadden et al., 1986).

Duripans are subsurface horizons cemented predominantly by silica, but they may also grade into petrocalcic horizons. This horizon appears to be most strongly developed in profiles associated with volcanic ash (Nettleton and Peterson, 1983). However, duripans may also result from the concentration of silica released by the weathering of silicate minerals (Boettinger and Southard, 1991).

Distribution Aridisols dominate the arid and semiarid regions of North America where evaporation exceeds precipitation, mostly in the western United States but extending north to the Canadian border and south into northern Mexico (fig. 9.5). They extend eastward into western Nebraska and into far south-central Texas. In the southwestern United States, these soils are adjacent to, and interfinger with, xeric moisture regimes of Alfisols, Entisols, and Inceptisols. Aridisols occur on a great variety of landscapes, but more commonly are associated with older Pleistocene surfaces (Nettleton and Peterson, 1983).

9.3.4 Entisols (*Regosols*)

This soil order is characterized by the absence of a distinctive imprint of soil-forming processes, expressed by the predominance of soil mineral matter and the lack of distinctive horizons (fig. 9.10).

Classification The distinctive criterion in the classification of these soils is the absence of diagnostic soil horizons within stipulated depths for inclusion in other soil orders (Grossman, 1983). To differentiate Entisols from other soil orders, an ochric epipedon must be present. In addition, soils of this order may display the presence of anthropic, albic, and agric horizons and may also accumulate soluble salts and iron oxides (Buol et al., 1989).

Entisols are subdivided into five suborders: Aquents, Arents, Psamments, Fluvents, and Orthents. Aquents are perennially or seasonally wet soils that display gleyed or strongly mottled horizons. Arents are better drained and lack gleyed or mottled horizons. They frequently display

Figure 9.10 Entisol developed on floodplain alluvium.

highly disturbed fragments of deeper diagnostic horizons. Psamments are better drained than Aquents and are characterized by textures that are loamy fine sands or coarser. Fluvents are loamy or clayey alluvial soils with very simple profiles, commonly displaying primary fluvial stratification with accompanying irregular distributions of organic matter. Orthents are better drained than aquents, are loamy and clayey in texture, and display regular decreases in organic matter with depth (Buol et al., 1989). Regosols are subdivided into two great groups on the basis of the degree of development of the Ah (humic A horizon). Regosols with Ah horizons that are <10 cm thick are called Regosols, and those with Ah horizons that are thicker than 10 cm are called humic Regosols (Canada Soil Survey Committee, 1978).

Morphology and Genesis Entisols lack horizon development because they are simply too young. They are commonly associated with parent materials compacted by either natural or human processes, especially when the parent materials have been deposited under water or by aeolian processes. Frequently, these same processes lead to soils that are cumulic in nature from the continuous or episodic addition of new parent material, for example, during recurrent floods. These soils also are typically rocky where they form on coarse parent materials, such as glacial tills, hillslope deposits, or bedrock in the early stages of weathering. Entisols also typically display the accumulation of iron sulfide and associated oxidation, especially where they are developed on lagoonal parent materials and coal-mine spoils (Canada Soil Survey Committee, 1978; Foth and Schafer, 1980; Grossman, 1983).

Distribution These soils develop on a wide variety of parent materials under a broad range of climatic environments (fig. 9.5). They are comonly associated with very young parent materials, waterlogged environments, steep topographic settings, sandy parent materials, and actively eroding landscapes. These soils occur throughtout the United States, but they are particularly dominant in the arid and semiarid west, in the sandhills of Nebraska, Colorado, and South Dakota, in the glaciated landscapes of Minnesota, Wisconsin, and Michigan, and along the Atlantic coast, Gulf Coastal Plain, Florida peninsula, and Alaska (Gersmehl, 1977; Tedrow, 1977; Foth and Schafer, 1980). In addition, they also occur extensively in the Rocky Mountains and Sierra Nevada–Cascade Range systems (Burns, 1980). In Canada, this soil order covers less than 1% of the land area, mostly in the Interior Plains, but also occurs as a subdominant soil across much of the country. Entisols particularly associated with boreal and cryoboreal climatic regions (Foth and Schafer, 1980).

9.3.5 Gelisols (Cryosols)

Gelisols, the most recently designated soil order in the Soil Taxonomy (Soil Survey Staff, 1999), are soils that developed on permafrost and/or display the effects of cryoturbation in the upper parts of their profiles.

Classification These soils have either permafrost within 100 cm of the soil surface or gelic materials within 100 cm of the soil surface and permafrost within 200 cm of the surface. They are subdivided into three suborders: Histels, Turbels, and Orthels. Histels are saturated with water for 30 or more days each year and have 80% or more organic material in the upper 50 cm of the profile. Turbels are characterized by the presence of one or more cryoturbated horizons and display an accumulation of organic matter above permafrost, the presence of ice or sand wedges, and oriented rock fragments. Orthels possess between 40% and 80% organics in the upper 50 cm of the profile and display a variety of diagnostic mineral horizons, including mollic and umbric epipedons and argillic horizons at depth (U.S. Department of Agriculture, 1998). This order primarily incorporates soils previously classified as Pergelic Cryaquepts.

The Canadian System of soil classification recognizes the cryosolic order for soils underlain by permafrost. These soils have permafrost within 1 m of the surface or 2 m if the soil has been strongly disturbed by cryoturbation. They have a MAST of less than 0°C and may be either mineral or organic soils. They are subdivided into three great groups on the basis of the degree of cryoturbation. Turbic Cryosols are mineral soils with marked cryoturbation and accompanying patterned ground and permafrost within 2 m of the ground surface. Static Cryosols are mineral soils with no marked cryoturbation and permafrost within 1 m of the surface. Organic Cryosols are organic soils with permafrost within 1 m of the surface.

Morphology and Genesis Gelisols are characterized by the presence of permafrost within 2 meters of the soil surface, with a MAST of 0°C for 2 years in succession. They display evidence of cryoturbation. Cryoturbation takes many forms, including irregular and fractured horizons, involutions, organic matter accumulation on and within permafrost, oriented rock particles, and silt-enriched layers. These soils also display distinctive soil structures, including, platy, blocky, and granular macrostructures and orbiculic, conglomeritic, banded, and vesicular microfabrics. The processes primarily responsible for the development of the gelic horizons are volumetric changes associated with state changes in soil water, moisture migration along thermal gradients in frozen soil, and thermal contraction within frozen materials (U.S. Department of Agriculture, 1998).

Distribution Gelisols occur extensively in Alaska and their distribution corresponds strongly with the major permafrost zones of the state (fig. 9.5). Gelisols dominate the landscape from the Brooks Range north to the Arctic coast. Between the Brooks and Alaska Ranges, where permafrost is discontinuous, Gelisols occur in association with other soils, particularly Inceptisols. South of the Alaska Range, where permafrost is sporadic to nonexistent, Gelisols occur only in isolated pockets and at high elevations. In all, some 72 million hectares of Gelisols occur in Alaska.

Cryosols are the most extensive soils in Canada, covering approximately 40% of the land area (fig. 9.6). They occur predominantly in the Northwest Territories on the Arctic lowlands, but also in association with Organic soils on the Hudson Bay lowlands and with Brunisols in the southwestern Northwest Territories, northern Saskachewan, and Manitoba (Tedrow, 1977; Foth and Schafer, 1980).

9.3.6 *Histosols* (*Organics*)

Histosols and Organics are dominated by organic matter (fig. 9.11). They are predominantly formed from organic material deposited in water, principally lakes, but they may also form from forest litter (Soil Survey Staff, 1975, 1999; Canada Soil Survey Committee, 1978).

Classification Histosols are subdivided into four suborders, namely, Fibrists, Hemists, Saprists, and Folists, depending on the degree of decomposition of the organic matter. Fibrists are composed of fibrous plant remains and are the least decomposed and wettest of the suborders. Hemists contain organic matter that is decomposed to such a degree that its biologic origin cannot be recognized in two-thirds of the soil. The fibrous material can be largely destroyed by simply rubbing between the fingers. The moisture content of soils in this suborder is between 45% and 85% by weight. Saprists consist of highly decomposed organic material. Few plant remains can be readily identified and the fiber content is less than 33% of the total organic volume. Folists are composed of a diversity of organic materials in varying stages of decomposition, but they contain at least 20% organic carbon. These soils are never saturated with water for more than a few days and maintain moisture at or near field capacity (Everett, 1983). In the Canadian System of Soil Classification, Folists are subdivided into four great groups: Fibrisols, Mesisols, Humisols, and Folisols. The first three of these represent progressively increasing degrees of decomposition of organic matter. Folisols are thin, organic soils developed under forests, are dominated by leaf litter, and are rarely saturated (Canada Soil Survey Committee, 1978).

Figure 9.11 Histosol, southern Michigan.

Morphology and Genesis These soils are characteristically wet, at least intermittently, and commonly saturated. Their characteristics vary markedly, depending on the nature of the organic material accumulating in the wet environment. Histosols exhibit an extremely low bulk density, possess cation exchange capacities typically higher than those of mineral soils, and have a variable pH, although they are commonly strongly acid (Foth and Schafer, 1980).

Distribution Histosols occur in virtually all climatic regimes in North America. Extensive areas occur in the glaciated terrains of the Upper Midwest, northern New England, and the Adirondack Mountains, and also along the Atlantic and Gulf coasts (fig. 9.5) (Gersmehl, 1977; Ciolkosz et al., 1989). Extensive areas of organic-rich soils also occur in Alaska (Everett, 1983), but many of these soils are now incorporated into the Histel suborder of the Gelisols as they are underlain by relatively shallow permafrost. Approximately 30 million hectares of Alaska are underlain by Histosols. In Canada, organic-rich soils are found in all provinces, but only about 4% fall into the Organic soil order of the Canadian soil classification system. The remainder have permafrost within 1 meter of the ground surface and so are classified as organic Cryosols, occurring extensively across northeast Canada (fig. 9.6).

9.3.7 *Inceptisols* (*Brunisols, Cryosols*)

Classification These soils display profiles that are more strongly developed than Entisols and Regosols. They exhibit one or more horizons of concentration or alteration, with little accumulation of translocated material except for carbonates and silica (fig. 9.12). Inceptisols contain some weatherable minerals and possess a moderate to high cation exchange capacity. In these soils, water is available to plants more than half the year or more than 3 months of the summer. Inceptisols are characterized by textures finer than sandy loam. Where they have developed on residual rather than transported materials, they commonly contain rock at shallow depth (Soil Survey Staff, 1975, 1999; Canada Soil Survey Committee, 1978). They also develop on gravelly parent materials such as alpine tills and hillslope deposits (Burns, 1980).

Figure 9.12 Inceptisol, West Virginia.

Inceptisols are subdivided into six suborders: Aquepts formed in wet environments; Ochrepts, which are freely drained and light in color; Umbrepts, which are rich in organic matter and possess low base saturation; Plaggepts, which are characterized by a plaggen or anthropic epipedon; and Tropepts of the intertropical regions of the world.

Brunisols are subdivided into four great groups, including Melanic, Eutric, Sombric, and Dystric Brunisols, based on soil reaction and the thickness of the Ah horizon. Melanic Brunisols have an Ah horizon greater than 10 cm thick with a pH greater than 5.5. Eutric Brunisols have a thin to nonexsistent Ah horizon with a pH greater than 5.5. Sombric Brunisols have a thick Ah horizon with a pH less than 5.5. Dystric Brunisols have a thin or nonexsistent Ah horizon and a pH less than 5.5.

Morphology and Genesis Typically, Inceptisols display the development of an ochric or umbric epipedon over a cambic horizon. The ochric epipedon is generally a light colored, organic-poor surface horizon. Under some circumstances, this horizon may be quite dark in color because of abundant organic matter, but it fails to meet thickness requirements to be classified as mollic or umbric. These soils may develop umbric horizons if they have dark horizons, but they have base saturations too low to be classified as mollic epipedons. Cambic horizons are B horizons that have textures of very fine sand or loamy very fine sand or finer and have weatherable minerals present. Originally, this horizon was defined on the basis of color, but because of the difficulties this often presented, it has been redefined on the basis of the degree of weathering. Inceptisols may also contain duripans, fragipans, and mollic epipedons. However they may not possess any of the other major diagnostic horizons. Aquepts in temperate environments typically display gray, strongly mottled subsoils, whereas in cold climates they reveal various frost and associated patterned ground features (Foss et al., 1989; Campbell and Claridge, 1992).

Distribution These soils are concentrated in the Pacific Northwest, Colorado Rockies, southern Great Plains of Texas and Oklahoma, lower Mississippi River valley, and Appalachian Mountains (Gersmehl, 1977; Ciolkosz et al., 1989; Foss et al., 1983) (fig 9.5). There are strong geographical patterns of distribution of these soils at the suborder level. Aquepts occur extensively on young landscapes with perpetually high water tables and are therefore widely distributed in the lower Mississippi River valley and its tributaries, the lacustrine areas of the Midwest, and the lower coastal plains of the Atlantic and Gulf coasts (Foss et al., 1983). Aquepts also occur extensively in Alaska, where they constitute the dominant soil suborder (Foth and Shafer, 1980; Foss et al., 1983), and in the Rocky Mountains under tundra vegetation on the floors of cirques and along river valleys farther downstream (Burns and Tonkin, 1982).

Ochrepts are freely drained, light brown soils developed on acidic glacial till and various sedimentary parent materials. They occur extensively throughout the northeastern United States and are the dominant soils of southern New England (excluding the Connecticut River valley), the glaciated Appalachian plateau, and the unglaciated Ridge and Valley Province (Ciolkoscz et al., 1989). Ochrepts also occur extensively in cold regions of the United States (Tedrow, 1977; Burns and Tonkin, 1982; Foss et al., 1989).

Umbrepts are the least extensive of the Inceptisols (Soil Survey Staff, 1975, 1999). They are freely drained dark reddish brown acidic soils high in organic matter (Soil Survey Staff, 1975), and they occur extensively under forest vegetation. Umbrepts are most widely distributed in western Washington and Oregon on a variety of parent materials (Foss et al., 1983), and above tree line in the alpine zone of the Rocky Mountains (Burns and Tonkin, 1982).

Brunisols are the dominant soils in northern British Columbia and the southern Yukon Territory, where they occur in association with exposed bedrock. In the western Northwest Territories and northern Saskatchewan and Manitoba, they are associated with Cryosols and bedrock. There is also a small occurrence in the vicinity of Ottawa in Ontario (Foth and Schafer, 1980). Brunisols also occur in association with Podsols, Luvisols, and Organics in western and east-central Canada (fig 9.6).

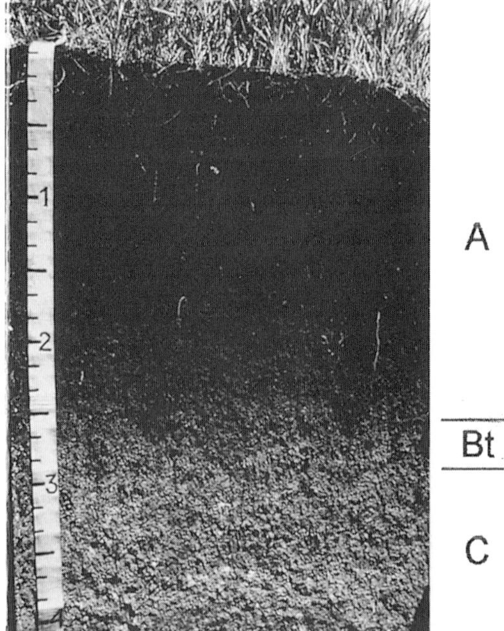

Figure 9.13 Mollisol, central Iowa.

9.3.8 *Mollisols (Chernozems and Solonetz)*

Classification Mollisols are characterized by very dark brown to black mollic epipedons that constitute more than one-third of the combined thickness of the A and B horizons or that are greater than 25 cm thick (fig. 9.13). The molllic epipedon typically possesses a crumb structure and a soft consistency when dry. These soils commonly display accumulations of calcium carbonate along with other exchangeable cations at depth in the profile. Their base saturation is commonly greater than 50% throughout the profile (Birkeland, 1984). Mollisols possess an abundance of clay minerals of moderate to high cation exchange capacity, but with clay abundances of less than 30% in horizons above 50cm (Soil Survey Staff, 1975, 1999).

Mollisols are subdivided into seven suborders. Albolls possess an albic (light-colored horizon) and occur in environments with fluctuating water tables. They are commonly saturated at or near the surface during winter or spring. Aquolls are naturally wet with low chroma colors and occur over a wide temperature range. Rendolls possess a udic moisture regime or a MAST greater than 0°C but less than 8°C. They occur in association with calcareous parent materials and broadly correspond to the old rendzinas of the Great Soil Group classification. Xerolls possess xeric moisture regimes but some have aridic moisture regimes, with a MAST of less than 22°C. Borolls are freely drained and occur under cool to cold temperature regimes. Moisture regimes are udic to ustic and the MAST is less than 8°C. Ustolls possess ustic or aridic moisture regimes that border on ustic. They have a MAST greater than 8°C. They develop in subhumid to semiarid climates and typically accumulate lime in their profiles. Udolls are freely drained soils possessing a udic moisture regime and a MAST of greater than 8°C. They occur in midlatitude humid-continental climates (Soil Survey Staff, 1975).

In Canada, Chernozems are well to imperfectly drained with dark surface horizons resulting from the accumulation of organic matter. Most Chernozems are frozen for part of the winter and dry for part of the summer, with MASTs of greater than 0°C but less than 5.5°C. Like Mollisols, they accumulate some soluble salts low in their profiles during the summer (Canada Soil Survey Committee, 1978). Chernozems are subdivided into four great groups on the basis of differences in climate and vegetation. Brown Chernozems occur under subarid to semiarid climates, Dark Brown Chernozems in semiarid environments, Black Chernozems in subhumid environments, and Dark Gray Chernozems in subhumid environments.

Closely associated with the Chernozems are the solonetzic soils, which develop on saline parent materials as a result of progressive desalinization. Three great groups of solonetzic soils are recognized. Solonetz possess an eluviated A horizon that is less than 2 cm thick and display the development of a Solonetzic B horizon. Solodized Solonetz have an eluvial A horizon that is at least 2 cm thick and a well-developed solonetzic B horizon. Solods possess an eluvial A horizon that is at least 2 cm thick, with AB or BA horizons with disintegrating solonetzic horizons (Canada Soil Survey Committee, 1978). These three Great Soil Groups represent an evolutionary sequence of progressively more strongly leached and less-saline soils.

Morphology and Genesis The dominant process operating in the formation of Mollisols is melanization, the darkening of the soil as a result of the addition of organic matter. This is especially the case in the surface soil and accompanying formation of the diagnostic mollic epipedon. Melanization is a combination of processes, including extension of roots of grassland vegetation into the soil, partial decay of organic matter, mixing of the soil by a variety of soil organisms, eluviation and illuviation of organic and mineral matter with accompanying formation of dark-colored cutans, and the formation of resistant lignoproteins, which also impart dark colors to the soil (Buol et al., 1989).

Because these soils typically occur under semiarid climates, they display seasonal accumulation of calcium carbonate, and under extreme conditions they accumulate gypsum. These calcareous horizons are the result of the translocation of calcium present in much of their parent material. In western Canada, the accumulation of salts leads to the formation of solonetzic soils, which characteristically possess saline parent materials. These soils have salts distributed throughout their entire thickness in the early stages of their formation and essentially lack distinct horizons. As salts are progressively leached, soil-horizon differentiation begins because fine particle translocation can take place once the soils deflocculate.

Another important process affecting the formation of Mollisols is cumulization, which is the result of the accumulation of material from upslope locations to form overthickened mollic epipedons. This accumulation is frequently associated with the redistribution of soil matter by water erosion, but may also be accompanied by the addition of aeolian materials (Fenton, 1983; Buol et al., 1989).

Distribution Mollisols are intimately associated with the grasslands of North America. They occur in climates that are semiarid to subhumid and that possess a distinct seasonality with respect to the distribution of precipitation. Their distribution, while being extensive across the western half of the United States (fig. 9.5) and Canada (fig. 9.6), is particularly concentrated in the Great Plains and the Prairie Peninsula (Gersmehl, 1977; Foth and Schafer, 1980). In the United States, suborder distribution shows a marked climatic influence. Borolls cover a large area of cool to cold climates with relatively low precipitation. They occur in Minnesota, North and South Dakota, and Montana. Udolls occur in areas of slightly more abundant precipitation and under mesic thermal regimes, from Illinois to Nebraska, and from Minnesota to Oklahoma. Ustolls occur in subhumid and semiarid climates with a concentration of precipitation in spring and summer from South Dakota to Texas. In the Rolling Red Plains of western Oklahoma, the dominant soils of flood plains and terraces are Ustolls with diagnostic clay-rich and carbonate-rich horizon. To the south on the southern High Plains of Texas, Ustolls similarly account for a substantial area of surface soils, which again display strongly developed clay-rich and carbonate-rich subsurface horizons (Holliday, 1988). Xerolls occur where most moisture falls in the winter and where summers are virtually rainless. In the United States, these soils occur in eastern Washington and Oregon, northeastern Nevada, and southeastern Idaho (Fenton, 1983). Aquolls and Rendolls are more restricted in their distribution, but Aquolls occur widely in the Red River valley between Minnesota and North Dakota as well as along the Mississippi, Missouri, Ohio, and Wabash river valleys, and in the coastal marshes of Louisiana and Texas (Foth and Schafer, 1980).

Chernozemic soils are the dominant soils of the Interior Plains of Alberta, Manitoba, and Saskatchewan. Minor areas of Black and Dark Grey Chernozemic soils occur in valleys and on mountain slopes in the Cordilleran region of southern British Columbia, sometimes extending above tree line (Canada Soil Survey Committee, 1978). Solonetzic soils occur throughout western Canada, especially in central Alberta, southern Saskatchewan, and southern Manitoba (Foth and Schafer, 1980).

9.3.9 Spodosols (Podzolics)

Spodosols typically develop on sandy to loamy parent materials in cool, temperate environments. They are characterized by the occurrence of four diagnostic horizons, including a dark colored, organic-rich surface horizon, a bleached eluvial (E) horizon, a reddish, brownish or black illuvial horizon enriched in amorphous material, and a sandy C horizon (fig. 9.14) (McKeague et al., 1983).

The diagnostic horizon of Spodosols is the development of a spodic horizon. This horizon is characterized by the accumulation of iron, aluminum, and organic matter, and it possesses a high pH-dependent cation exchange capacity. The spodic horizon typically has a loamy or sandy texture and low base status (Soil Survey Staff, 1975, 1999). Spodosols typically display the presence of a pronounced bleached, eluvial E horizon from which the formative components of the spodic horizon have been derived.

Classification Spodosols are subdivided into four suborders, namely, Aquads, Ferrods, Humods, and Orthods. Aquads are Spodosols that have an aquic moisture regime or which, if artificially drained, display evidence of wetness (McKeague et al., 1983). Such features might include the presence of a histic epipedon, mottling in albic and

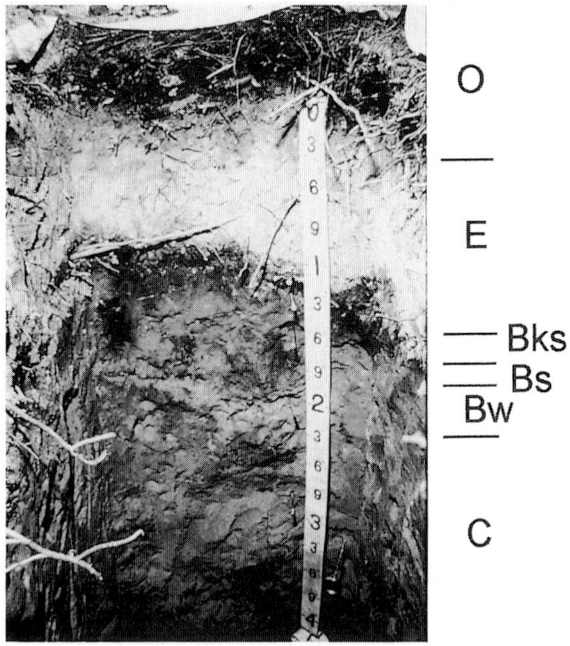

Figure 9.14 Spodosol, northern New York.

upper spodic horizons, and the development of a duripan in the albic horizon (Buol et al., 1989). Ferrods are not as wet as Aquads and contain abundant iron in all horizons. Humods are freely drained Spodosols; they are not as wet as Aquods and contain large amounts of organic carbon in the spodic horizon together with aluminum but little iron (McKeague et al., 1983; Buol et al., 1989). Orthods are freely drained Spodosols with a spodic horizon in which organic matter, iron, and aluminium have accumulated.

Podsolics are subdivided into three Great Groups on the basis of the nature of their B horizon. Humic Podsols have humic B horizons at least 10 cm thick, with abundant carbon relative to iron. Ferro-humic Podsols have B horizons that are abundant in both organic carbon and iron and aluminium. Humo-ferric Podsols have less organic carbon than ferro-humic Podsols, but about the same amount of iron and aluminium.

Morphology and Genesis Spodosols display a wide variety of morphological features and are formed by a broad range of interacting pedogenic processes. These soils have traditionally been viewed as the result of the formation of sesquioxide-organic matter complexes, translocation of these complexes, and precipitation and subsequent degradation of the complexed materials (McKeague et al., 1978). Formation of the sesquioxide-organic matter complexes takes place in both the O horizon and the A horizon, as water soluble organic material reacts with iron, aluminium, and silica. Reaction of the sesquioxides and organic matter continues until the ratio of iron and aluminium to organic carbon becomes sufficiently great that the complex precipitates and the spodic B horizon is formed (McKeague et al., 1983). Should the ratio decrease due to the addition of soluble organic material to the B horizon, the spodic horizon will be destroyed and transported deeper in the soil. Spodic horizons may also form as a direct result of the eluviation and illuviation of iron and aluminium under favorable conditions, usually alternating reduction and oxidation.

Spodosols occasionally display unique morphological features, including placic horizons, which are thin horizons of strongly cemented accumulations of iron oxides, iron-manganese oxides, or iron-organic matter complexes that are black or dark red in color (McKeague et al., 1968; Brewer et al., 1973). Orstein may also develop in some Spodosols. This is a horizon of continuously or discontinuously cemented nodules and masses of iron or aluminum-complex material (McKeague and Wang, 1980; Barrett and Schaetzl, 1992). Some Spodosols display subsoil horizons cemented by silica or by amorphous oxides of iron and/or aluminum. These cementing materials may be locally derived or translocated laterally or vertically (McKeague and Sprout, 1975). Some Spodosols display particularly thick E horizons, especially those with porous parent materials, low base and sesquioxide concentrations, and high precipitation climates. In such porous materials, argillic horizons also develop by the accumulation of multiple clay lamelli (Schaetzl, 1992).

Distribution In the United States, these soils occur in southern Alaska, southwestern Oregon (Nettleton et al., 1982), the Upper Midwest (Schaetzl and Isard, 1991), and New England. Small areas also occur along the Atlantic Coastal Plain (fig. 9.5). The most extensive Spodosols are the freely drained Orthods, which display a well-developed spodic horizon but lack placic horizons or fragipans. Some Fragiothods occur in the northeastern United States, mostly on coarse-textured, acidic Quaternary deposits under coniferous and occasionally hardwood forests. Most Orthods have formed under cryic or frigid temperature regimes. Orthods of the Upper Midwest have formed on gently sloping glacial outwash plains and tend to be thicker in Canada than in the United States (Foth and Schafer, 1980). In New England, Spodosols are primarily found in the more northerly glaciated terrains, as well as in the Adirondack Mountains of New York, though they also occur south of the glacial limit. They are associated primarily with sandy parent materials and frigid soil temperature regimes (Ciolkosz et al., 1990). Spodosols found beyond frigid and mesic soil temperature regimes include the Aquods and Humods of the Atlantic Coastal Plain and the glacial outwash plains of New York (Ciolkosz et al., 1990). Aquods also occur in limited areas of the Carolinas and Georgia. Most of the Aquods and Humods of the Atlantic and Gulf Coastal Plain are found in Florida, where they are predominantly associated with thermic and hyperthermic temperature regimes (Rourke et al., 1988; Goldin and Edmonds, 1996) and sandy and humus-rich parent materials.

Podzolics occur throughout Canada but are most extensive on the southern Canadian Shield in eastern Canada. They also occur in a narrow belt along the west coast where they are closely associated with volcanic ash (Foth and Schafer, 1980; Goldin, 1983) (fig. 9.6).

9.3.10 *Ultisols*

These soils display pronounced clay translocation with evidence of strong mineral weathering. The distinguishing characteristics of Ultisols include an argillic horizon and low base status, especially in deeper horizons. The clay content of these soils increases and then decreases with increasing depth, but the argillic horizon may be of variable thickness. Cation exchange capacity is moderate to low (Soil Survey Staff, 1975, 1999). Ultisols are confined to areas where mean annual soil temperature is 8°C or warmer and precipitation is abundant (Buol et al., 1989).

Figure 9.15 Ultisol, northwest Arkansas.

Classification Five suborders of Ultisols are recognized, based on profile wetness and organic matter, namely, Aquults, Ustults, Humults, Udults, and Xerults (Soil Survey Staff, 1975). Aquults are either saturated with water year round or have been drained. This suborder is characterized by the development of mottles with low chroma colors, the presence of iron-manganese concretions >2 mm in diameter, and an overall soil color with a chroma of 2 or less immediately below an A horizon (Buol et al., 1989). Ustults possess ustic moisture regimes in which moisture is limited. Humults display relatively abundant accumulations of organic matter that extend into argillic horizons and are freely drained. Udults develop in humid environments where moisture is not limiting and where organic-matter content and water tables are low. Xerults occur under xeric moisture regimes where summers are dry and winters moist (Buol et al., 1989).

Morphology and Genesis Ultisols typically display ochric epipedons over a red, yellowish brown, or reddish brown argillic horizon. The argillic horizon is primarily the result of the translocation of clay from upper A and E horizons to form Bt horizons at depth (fig. 9.15). Evidence of this translocation is the development of argillans (clay skins) on mineral grains (Buoll et al., 1989). Some clay may result from mineral weathering in situ. Accompanying the development of argillic horizons is the weak development of E horizons, which in addition to experiencing loss of fines also experience the mobilization and translocation of iron and organics by the combined processes of pozolization. Two other commonly developed morphological characteristics of Ultisols are the presence of plinthite horizons, accumulations of sesquioxides that lead to the development of iron hardpans, and fragipans, brittle horizons of high bulk density.

Distribution Ultisols are concentrated in the southeastern United States, particularly on the Piedmont Plateau and the Atlantic and Gulf Coastal Plain (fig. 9.5). However, they also occur extensively in the lower Mississippi River valley and its tributaries, the Ozarks, the Appalachian Mountains, and the mountains of the Pacific Northwest (Gersmehl, 1977; Foth and Schafer, 1980; Ciolkosz et al., 1989; Markewich et al., 1990). Vast areas of the Piedmont Plateau support the Ultisols, which are predominantly Hapludults developed primarily as residual soils associated with deep weathering of the underlying bedrock (Rebertus and Buol, 1985; Pavich, 1989; Markewich et al., 1990). Piedmont soils tend to thicken with increasing distance from the mountains, primarily as a result of the thickening of the argillic horizon to become paleudults (Markewich et al., 1990). Ultisols on the Atlantic Coastal Plain typically develop on coarser and more mineralogically mature parent materials than those on the piedmont (Markewich et al., 1990). Ultisols on the Coastal Plain are more typically Paleudults, with thick sandy epipedons, than Hapludults as a result of the coarser parent materials. These soils also typically display plinthic horizons. Ultisols near the coast are more commonly waterlogged and display accumulations of organic matter. Soils with high chroma argillic horizons are restricted to higher topographic positions (Markewich et al., 1990). Hapludults and Paleudults are also widely distributed on the dissected plateaus on either side of the lower Mississippi River valley (Foth and Schafer, 1980). Humults develop under moderately well-drained conditions in middle to low latitudes in rainy mountain environments. In the United States, Humults occur in the mountains of the Pacific Northwest and typically exhibit dark epipedons. Xerults also occur in similar areas to Humults, but differ from them by occurring under more strongly seasonal climatic environments and possessing a xeric moisture regime. Xerults occur under coniferous vegetation in the Sierra Nevada and Cascade Range of California and Oregon (Foth and Schafer, 1980).

9.3.11 Vertisols

Vertisols are characterized by regular mixing, which prevents the strong development of diagnostic horizons (fig. 9.16). A unique property of vertisols is the high clay content, which leads to high bulk density when dry and very slow conductivity when wet. Vertisols exhibit pronounced changes in volume with changes in moisture content with accompanying swelling, shrinking, and cracking. They display considerable evidence of movement within the soil profile in the form of slickensides, gilgai microrelief, and the presence of wedge-shaped structural aggregates tilted at an angle from the horizontal (Soil Survey Staff, 1975).

Figure 9.16 Vertisol, Puerto Rico.

Classification Six suborders of vertisols are defined on the basis of soil behavior and climate. They are Xererts, Torrerts, Uderts, Usterts, Aquerts, and Cryerts (Soil Survey Staff, 1999). Xererts develop cracks that open for 60 consecutive days of the 3 months after the summer solstice. Their temperature regimes are thermic or mesic. Torrerts possess cracks that remain open for most of the year and close for no more than 60 days when affected by precipitation. Because of the permanance of the open cracks, they tend to fill substantially with debris. Uderts are usually moist, but cracks open sometime during the year and remain open for less than 90 days. Usterts have cracks that open and close more than once during the year and remain open for between 90 and 305 days. Aquerts are moist with moist chroma of 2 or less if redox concentrations are present, or 1 or less if there is sufficient active ferrous iron present. Cryerts are vertisols with a cryic temperature regime.

Morphology and Genesis The formation of Vertisols is dominated by a set of processes that collectively act to inhibit the development and manifestation of soil horizons (haplidization). Dominant among these processes is the mixing of clay-dominated horizons. These horizon-inhibiting processes are very strongly linked to the periodic swelling and shrinking of 2:1 expandable clay minerals and the accompanying development of desiccation cracks, which permit the addition of fine-grained material to the soil matrix. The addition of material to the deeper soil matrix means that when the 2:1 clay-dominated matrix swells, a greater volume is required for the expansion process; as a result, distinctive soil properties emerge. The first of these is the development of a lentil angular blocky structure and accompanying slickensides in the B horizon. The second is the development of distinctive ridges and swales, known as gilgai, on the ground surface. Several models have been proposed to explain the origin of the abundant fine clay dominated by 2:1 expandable clay mineral species. One model proposes that these soils are relatively young and have undergone rapid destruction of expandable clays, followed by accumulation of illuviated clay. A second model proposes that these soils are very old and represent the long operation of swell-shrink cycles that eventually consume the A horizon. Some support for this model is afforded by vertisols older than 200 ka on marine terraces on San Clemente Island off the coast of California (Muhs, 1982). Younger terraces were dominated by Mollisols and Alfisols. The third model relates to soil formation that is in equilibrium with basic parent materials in strongly seasonal wet-dry climates.

Distribution The primary environmental explanation for the distribution of vertisols is a seasonally wet-dry climate, aided secondarily by the occurrence of basic parent materials. The distribution of this soil order is mainly limited to central Texas, Arizona, and coastal California (fig. 9.5) (Gersmehl, 1977). The largest area of Vertisols in North

America is in the Blackland Prairie of north-central Texas and southern Oklahoma. Other important areas include the Gulf Coast Prairie in Texas, parts of Alabama and Mississippi, and Arizona, where these soils cover some 160,000 hectares (Buol et al., 1989). In all, some 10.5% of Texas is covered by vertisols.

9.3.12 Gleysolic Soils

Gleysolic soils of the Canadian Soil Classification System have no direct equivalent in the Soil Taxonomy of the United States. They do, however, correspond to various aquic suborders of the taxonomy. Gleysols are mineral soils that are characterized by periods of prolonged or episodic waterlogging with accompanying reducing environments. They are associated with high water tables at certain times of the year or with temporary saturation above impermeable substrate.

Classification Three great groups of Gleysols are recognized on the basis of the degree of development of diagnostic soil horizons. Humic Gleysols posess humic A horizons that are at least 10 cm thick and lack the development of a textural B horizon. Gleysols display no humic A horizon, or if an A horizon is present it is less than 10 cm thick, and they lack the development of a textural B horizon. Luvic Gleysols display the development of gleyed textural B horizons and possess grey eluvial A horizons or eluviated humic A horizons.

Morphology and Genesis As their name suggests, these soils commonly display gleying in their B and C horizons as a result of waterlogging. They also display humic A horizons, which may or may not be sufficiently leached to meet the criteria to be classified as Ahe horizons. Eluviation of the A horizon may in some settings be strong enough to develop Ae horizons such as those typically associated with Podsolics (Spodosols).

Distribution Gleysols cover just a little over 1% of the land area of Canada, and in no part of the country do they represent a dominant mapping unit. Because these soils possess a mesic soil temperature regime, they have frequently been drained and used for agriculture. The two principal areas of occurrence are the St. Lawrence Lowlands and the lower Fraser Valley (Foth and Schafer, 1980).

9.4 Conclusion

Weathering and soil formation are essential links between the physical landscape represented by rocks, sediment, rain, and runoff, and the biological environment represented by plants and animals. Weathering of in situ rock and transported sediment provides the regolith, which is much more readily removed by mass wasting and erosion than hard rock. The regolith, is in turn modified by complex biogeochemical processes to form soil, the prerequisite for the sustenance of plants and other organisms, including humans through the medium of agriculture. As this chapter has shown, soils are complex phenomena, subject to detailed classification, and distributed in great variety across the landscape. Ultimately, however, soil is the quintessential link between the physical and biological world and its occupants.

References

Alexander, E.B., and W.D. Nettleton (1977), "Post-Mazama Natriargids in Dixie Valley, Nevada," *Soil Science Society of America Journal*, 41:1210–1212.

Barrett, L.R., and R.J. Schaetzl (1992), "An Examination of Podsolization near Lake Michigan using Chronofunctions," *Canadian Journal of Soil Science*, 72:527–541.

Bird, J.B. (1967), *The Physiography of Arctic Canada*, Baltimore, Johns Hopkins.

Birkeland, P.W. (1984), *Soils and Geomorphology*, New York, Oxford University Press.

Birkeland, P.W., R.M. Burke, and R.R. Shroba (1987), "Holocene Alpine Soils in Gneissic Cirque Deposits, Colorado Front Range," *United States Geological Survey Bulletin*, 1590E:21.

Bockheim, J.G., C.L. Ping, P.J. Moore, and J.M. Kimble (1994), "Gelisols: A New Proposed Order for Permafrost-Affected Soils," in: Kimble, J. and R.J. Ahrens (eds) *Proceedings of the Meeting on the Classification, Correlation, and Management of Permafrost Affected Soils*, U.S. Department of Agriculture, Soil Conservation Service, Lincoln, Nebraska, 25–44.

Boettinger, J.L., and R.J. Southard (1991), "Silica and Carbonate Sources for Aridisols on a Granitic Pediment, Western Mojave Desert," *Soil Science Society of America Journal*, 55:1057–1067.

Brewer, R., R. Protz, and J.A. McKeague (1973), "Microscopy and Electron Microprobe Analysis of Some Iron-Manganese Pans from Newfoundland," *Canadian Journal of Soil Science*, 53:349–361.

Buol, S.W., F.D. Hole, and R .J. McCracken (1989), *Soil Genesis and Classification*, Ames, Iowa State University Press.

Burns, S.F. (1980), "Alpine Soil Distribution and Development, Indian Peaks, Colorado," Ph.D. Dissertation, University of Colorado, Boulder.

Burns, S.F., and P.J. Tonkin (1982), "Soil-Geomorphic Models and the Spatial Distribution and Development of Alpine Soils," in: C.E. Thorn (ed.), Space and Time in Geomorphology, Allen and Unwin, London, 25–43.

Caine, N. (1992), "Spatial Patterns of Geochemical Denudation in a Colorado Alpine Environment," in: J.C. Dixon and A.D. Abrahams (eds), *Periglacial Geomorphology*, John Wiley and Sons, Chichester, 63–88.

Caine, N. (1995a), "Temporal Trends in the Quality of Streamwater in an Alpine Environment: Green Lakes Valley, Colorado Front Range, U.S.A.," *Geografiska Annaler*, 77:207–220.

Caine, N. (1995b), "Snowpack Influences on Geomorphic Processes in Green Lakes Valley, Colorado Front Range," *Geographical Journal*, 161:55–68.

Caine, N., and E.M. Thurman (1990), "Temporal and Spatial Variations in the Solute Content of an Alpine Stream, Colorado Front Range," *Geomorphology*, 4:44–72.

Campbell, I.B., and G.C. Claridge (1992), "Soils of Cold Climate Regions," in: I.P. Martini and W. Chesworth (eds.), *Weathering, Soils and Paleosols*, Elsevier Amsterdam, 183–201.

Canada Soil Survey Committee (1978), *The Canadian System of Soil Classification*, Ottawa, Canada, Department of Agriculture.

Ciolkosz, E.J., W.J. Waltman, T.W. Simpson, and R.R. Dobos (1989), "Distribution and Genesis of Soils of the Northeastern United States," *Geomorphology*, 2:285–302.

Ciolkosz, E.J., B.J. Carter, M.T. Hoover, R.C. Cronce, W.J. Waltman, and R.R. Dobos (1990), "Genesis of Soils and Landscapes in the Ridge and Valley Province of Central Pennsylvania," *Geomorphology*, 3:245–261.

Colman, S.M. (1982), "Chemical Weathering of Basalts and Andesites: Evidence from Weathering Rinds," *United States Geological Survey Professional Paper* 1246.

Cooke, R.U. (1970), "Stone Pavements in Deserts," *Annals of the Association of American Geographers*, 60:560–577.

Costa, J.E., and E.T. Cleaves, 1984. "The Piedmont Landscape of Maryland: A New Look at an Old Problem," *Earth Surface Process and Landforms*, 9:59–74.

Dethier, D.P. (1986), "Weathering Rates and the Chemical Flux from Catchments in the Pacific Northwest, U.S.A.," in: S.M. Colman and D.P. Dethier (eds.), *Rates of Chemical Weathering of Rocks and Minerals*, Academic Press, Orlando, 503–530.

Dixon, J.C. (1983), "Chemical Weathering of Late Quaternary Cirque Deposits in the Colorado Front Range," Ph.D. Dissertation, University of Colorado, Boulder.

Dixon, J.C. (1986), "Solute Movement on Hillslopes in the Alpine Environment of the Colorado Front Range," in: A.D. Abrahams (ed.), Hillslope Processes, Allen and Unwin, Boston, 139–159.

Dixon, J.C. (1994a), "Aridic Soils, Patterned Ground, and Desert Pavements," in: A.D. Abrahams, and A.J. Parsons (eds), *Geomorphology of Desert Environments*, Chapman and Hall, London, 64–81.

Dixon, J.C. (1994b), "Duricrusts," in: A.D. Abrahams and A.J. Parsons (eds.) *Geomorphology of Desert Environments*, Chapman and Hall, London, 82–105.

Dixon, J.C., C.E. Thorn, and R.G. Darmody (1984), "Chemical Weathering Processes on the Vantage Peak Nuntak, Juneau Icefield, Southeastern Alaska," *Physical Geography*, 5, 111–131.

Dohrenwend, J.C. (1987), "Basin and Range," in W.L. Graf (ed.), *Geomorphic Systems of North America*, Geological Society of America, Boulder, 303–342.

Eggler, D.H., E.E. Larson, and W.C. Bradley (1969), "Granites, Grusses, and the Sherman Erosion Surface, Southern Laramie Range, Colorado-Wyoming," *American Journal of Science*, 267, 510–522.

Everett, K. (1983), "Histosols," in: L.P. Wilding, N.E. Smeck, and G.F. Hall (eds.), *Pedogenesis and Soil Taxonomy: The Soil Orders*, Amsterdam, Elsevier, 1–53.

Fenton, T.E. (1983), "Mollisols," in: L.P. Wilding, N.E. Smeck, and G.F. Hall (eds.), *Pedogenesis and Soil Taxonomy: The Soil Orders*, Amsterdam, Elsevier, 125–163.

Ford, D.C. (1993), "Karst in Cold Environments," in: H.M. French and O. Slaymaker (eds.), *Canada's Cold Environment*, Montreal, McGill-Queens University Press, 199–222.

Foss, J.E., F.R. Moormann, and S. Riege, (1983), "Inceptisols," in: L.P. Wilding, N.E. Smeck, and G.F. Hall (eds.), *Pedogenesis and Soil Taxonomy: The Soil Orders*, Amsterdam, Elsevier, 355–381.

Foth, H.D., and J.W. Schafer (1980), *Soil Geography and Land Use*, New York, John Wiley and Sons.

French, H.M. (1981), "Periglacial Geomorphology and Permafrost," *Progress in Physical Geography*, 5:267–273.

French, H.M. (1993), "Cold Climate Processes and Landforms," in: H.M. French and O. Slaymaker (eds.), *Canada's Cold Environments*, Montreal, McGill-Queens University Press, 143–167.

Gallie, T.M., and O. Slaymaker (1984), "Variable Solute Sources and Hydrologic Pathways in a Coastal Subalpine Environment," in: T.P. Burt and D.E. Walling (eds.), *Catchment Experiments in Fluvial Geomorphology*, Norwich, Geobooks, 347–357.

Gallie, T.M., and O. Slaymaker (1985), "Hydrological Controls in Alpine Stream Chemistry," *Proceedings of the 14th N.R.C. Hydrology Symposium*, Quebec City, 287–306.

Gamble, E.E., and M.J. Mausbach (1982), "Summary report, Missouri Ozarks Soil-Geomorphology Study, Laclede County NSSL Project, CP 81-Mac28." National Soil Survey Laboratory, USDA, Lincoln, 21p.

Gardner, J.S. (1992), "The Zonation of Freeze-Thaw Temperatures at a Glacier Headwall, Dome Glacier, Canadian Rockies," in: J.C. Dixon and A.D. Abrahams (eds.), *Periglacial Geomorphology*, Chichester, John Wiley and Sons, 89–102.

Gerrard, A.J. (1990), *Mountain Environments*, Cambridge, MIT Press.

Gersmehl, P.J. (1977), "Soil Taxonomy and Mapping," *Annals of the Association of American Geographers*, 67:419–429.

Gile, L.H., J.W. Hawley, and R.B. Grossman (1981) "Soils and Geomorphology in the Basin and Range Area of Southern New Mexico: *Guidebook to the Desert Project*," New Mexico Bureau of Mines and Minoir Resources Memoir 39.

Goldin, A. (1983), "Comparison of Some Podzols and Spodosols in the Fraser Lowland, British Columbia and Adjacent Washington State," *Canadian Journal of Soil Science*, 63:579–591.

Goldin, A., and J. Edmonds (1996), "A Numerical Evaluation of Some Florida Spodosols," *Physical Geography*, 17:242–252.

Goudie, A.S., and M.J. Day (1980), "Disintegration of Fan Sediments in Death Valley California, by Salt Weathering," *Physical Geography*, 1:126–137.

Graham, R.C., and E. Franco-Vizcaino (1992), "Soils on Igneous and Metavolcanic Rocks in the Sonoran Desert of Baja California, Mexico," *Geoderma*, 54:1–21.

Graham, R.C., M.M. Diallo, and L.J. Lund (1990), "Soils and Mineral Weathering on Phyllite Colluvium and Serpentinite in Northwestern California," *Soil Science Society of America Journal*, 54:1682–1690.

Graham, R.C., B.E. Herbert, and J.O. Ervin, (1988), "Mineralology and Incipient Pedogensis in Anorthosite Terrane of the San Gabriel Mountains, California," *Soil Science Society of America Journal*, 52:738–746.

Graham, R.C., K.R. Tice, and W.R. Guetal (1994), "The Pedogenic Nature of Weathered Rock," in: D.L. Cremeens, R.B. Brown, and J. H. Huddleston (eds.),

Whole Regolith Pedology, *Soil Science, Society of America Special Publication 34*, Madison, 21–40.

Grossman, R.B. (1983), "Entisols," in: L.P. Wilding, N.E. Smeck, and G.F. Hall (eds.), *Pedogenesis and Soil Taxonomy, The Soil Orders*, Amsterdam, Elsevier.

Hall, K., and W. Otte (1990), "A Note on Biological Weathering on Nunataks of the Juneau Icefield, Alaska," *Permafrost and Periglacial Processes*, 1:189–196.

Hallet, B., J.S. Walder, and C.W. Stubbs (1991), "Weathering by Segregated Ice Growth in Microcracks at Sustained Sub-zero Temperatures: Verification from an Experimental Study Using Acoustic Emissions," *Permafrost and Periglacial Processes*, 2:283–300.

Holliday, V.T. (1988), "Genesis of a Late Holocene Soil Chronosequence at the Lubbock Lake Archeological Site, Texas," *Annals of the Association of American Geographers*, 78:594–610.

Harden, J.W., E.M. Taylor, L.D. McFadden, and M.C. Reheis (1991), "Calcic, Gypsic and Siliceous Soil Chronosequences in Arid and Semi-Arid Environments," in W.D. Nettleton (ed.), *Occurrence, Characteristics and Genesis of Carbonate, Gypsum Accumulations in Soils*, Soil Science Society of America Special Publication 26:1–16.

Hunt, C.B. (1986), "Surficial Deposits of the United States," New York, Van Nostrand/Reinholt.

Jones, D.P., and R.C. Graham (1993), "Water-Holding characteristics of Weathered Granite Rock in Chaparral and Forest Ecosystems," *Soil Science Society of America Journal*, 57:256–261.

Madole, R.F., C.R. Ferring, M.J. Guccione, S.A. Hall, W.C Johnson, and C.J. Sorenson (1991), "Quaternary Geology of the Osage Plains and Interior Highlands," in: R.B. Morrison (ed.), *Quaternary Non-Glacial Geology; Conterminous U.S.,* Boulder, Colorado, Geological Society of America, The Geology of North America J. K-2 pp. 503–546.

Marchand, D.E. (1974), "Chemical Weathering, Soil Development and Geochemical Fractionation in a Part of the White Mountains, Mono and Inyo Counties, California," *United States Geological Survey Professional Paper*. 352–J.

Markewich, H.W., M.J. Pavich, and G.R. Buell (1990), "Contrasting Soils and Landscapes of the Piedmont and Coastal Plain, Eastern United States," *Geomorphology*, 3:417–447.

McFadden, L.D., R.G. Amundson, and O.A. Chadwick (1991), "Numerical Modelling, Chemical and Isotopic Studies of Carbonate Accumulation in Soils of Arid Regions," in W.D. Nettleton (ed.), Occurrence, Characteristics, and Genesis of Carbonate, Gypsum and Silica Accumulations in Soils. *Soil Science Society of America, Special Publication* 26, 17–35.

McFadden, L.D., S.G. Wells, and J.C. Dohrenwend (1986), "Influences of Quaternary Climate Changes on Processes of Soil Development on Desert Loess Deposits of the Cima Volcanic Field, California," *Catena*, 13:361–389.

McFadden, L.D., S.G. Wells, and M.J. Jercinovich (1987), "Influences of Eolian and Pedogenic Processes on the Origin and Evaluation of Desert Pavements," *Geology*, 15:504–508.

McKeague, J.A., and P.N. Sprout (1975), "Cemented Subsoils (Duric Horizons) in some Soils of British Columbia," *Canadian Journal of Soil Science*, 55:189–203.

McKeague, J.A., A.W.H. Dumman, and P.K. Heringa 1968. "Iron manganese and other pans in some soils of Newfoundland." *Canadian Journal of Soil Science*, 48, 243–253.

McKeague, J.A., G.J. Ross, and D.S. Gamble, 1978. "Properties, criteria of classification and genesis of Podzolic soils in Canada." In: W.C. Mahaney (ed.) *Quaternary Soils*. Geo Books, Norwich. 26–70.

McKeague, J.A., F. DeConinck, and D.P. Franzmeier 1983. "Spodosols." In: Wilding, L.P., Smeck, N.E., and Hall G.F. (eds.) *Pedogenesis and Taxonomy*. II The Soil Orders. Amsterdam. Elsevier. 410p.

Muhs, D.R., 1982. "A Soil chronosequence on Quaternary marine terraces, San Clemente Island, California." *Geoderma*, 28, 257–283.

Nettleton, W.D., and F.F. Peterson (1983), "Aridisols," in: L.P. Wilding, N.E. Smeck, and G.G. Hall (eds.), *Pedogenesis and Soil Taxonomy: The Soil Orders*, Elsevier, Amsterdam, 165–215.

Nettleton, W.D., K.W. Flach, and R.E. Nelson (1970), "Pedogenic Weathering of Tonalite in Southern California," *Geoderma*, 4:387–403.

Nettleton, W.D., R.E. Nelson, B.R. Brasher, and P.S. Deer (1982), "Gypsiferous Soils in the Western United States," in: J.A. Kittrick, D.S. Fanning, and L.S. Hossner (eds.), *Acid Sulphate Weathering*, Soil Science of America Special Publication 10:147–168.

Nettleton, W.D., R.B. Parsons, A.O. Ness, and F.W. Gelderman, (1982), "Spodosols Along the Southwest Oregon Coast," *Soil Science Society of America Journal*, 46: 593–598.

Nettleton, W.D., J.G. Witty, R.E. Nelson, and J.W. Hanley (1975), "Genesis of Argillic Horizons in Soils of Desert Areas of the Southwestern United States," *Soil Science Society of America Journal*, 39:919–926.

Parse, M.W. (1995), "Geomorphic Analysis of the Role of Regolith in Karst Landscape Development," M.A. Thesis, University of Arkansas, Fayetteville.

Parsons, A.J., A.D. Abrahams, and J.R. Simonton (1992), "Microtopography and Soil Surface Materials on Semi-Arid Piedmont Hillslopes, Southern Arizona," *Journal of Arid Environments*, 12:107–115.

Pavich, M.J. (1985), "Appalachian Piedmont Morphogenesis: Weathering, Erosion, and Cenozoic Uplift," in M. Morisawa and J.T. Hack (eds.), *Tectonic Geomorphology*, London, Allen and Unwin, 299–319.

Pavich, M.J. (1986), "Processes and Rates of Saprolite Production and Erosion on a Foliated Granitic Rock in the Virginia Piedmont," in: S.M. Coleman and D.P. Diether (eds.), *Rates of Chemical Weathering of Rocks and Minerals*, Orlando, Academic Press, 551–590.

Pavich, M.J. (1989), "Regolith Residence Time and the Concept of Surface Age of the Piedmont "Peneplain," *Geomorphology*, 2:181–196.

Pavich, M.J., G.W. Leo, S.F. Obermeier, and J.R. Estabrook (1989), "Investigations of the Characteristics, Origin, and Residence Time of the Upland Residual Mantle of the Piedmont of Fairfax County, Virginia," *United States Geological Survey Professional Paper*, 1352.

Peterson, F.F. (1980), "Holocene Desert Soil Formation Under Sodium Salt Influence in a Playa-Margin Environment," *Quaternary Research* 13:172–186.

Pope, G.A., R.I. Dorn, and J.C. Dixon (1995), "A New Conception Model for Understanding Geographical Variations in Weathering," *Annals of the Association of American Geographers*, 85:38–64.

Rebertus, R.A., and S.W. Buol (1985a), "Iron Distribution in a Development Sequence of Soils from Mica Gneiss and Schist," *Soil Science Society of America Journal*, 49:713–720.

Reheis, M.C. (1987), "Gypsic Soils on the Kane Alluvial Fans, Big Horn County, Wyoming," *U.S. Geological Survey Bulletin*, 1590C.

Reynolds, R.C. (1971), "Clay Mineral Formation in an Alpine Environment," *Clays and Clay Minerals*, 19: 361–374.

Reynolds, R.C., and N.M. Johnson (1972), "Chemical Weathering in the Temperate Glacial Environment of the Northern Cascade Mountains," *Geochimica et Cosmochimica Acta*, 36:537–556.

Rourke, R.V., B.R. Brasher, R.D. Yeck, and F.T. Miller, (1988), "Characteristic Morphology of United States Spodosols," *Soil Science Society of America Journal*, 52:445–449.

Rust, R.H. (1983), "Alfisols," in: L.P. Wilding, N.E. Smeck, and G.F. Hall (eds.), *Pedogenesis and Soil Taxonomy: The Soil Orders*, Amsterdam, Elsevier.

Rutledge, E.M., L.T. West, and M. Onakupt, (1985), "Loess Deposits on a Pleistocene Age Terrace in Eastern Arkansas," *Soil Science Society of America Journal*, 49, 1231–1238.

Schaetzl, R.J. (1992), "Texture, Mineralogy, and Lamellae Development in Sandy Soils in Michigan," *Soil Science Society of America Journal*, 56:1538–1545.

Schaetzl, R.J. (1996), "Spodosol–Alfisol Intergrades: Bisequal Soils in NE Michigan, U.S.A.," *Geoderma*, 74:23–47.

Schaetzl, R.J., and S.A. Isard (1991), "The Distribution of Spodosol Soils in Southern Michigan: A Climatic Interpretation," *Annals of the Association of American Geographers*, 81:425–442.

Schaetzl, R.J., W.E. Frederick, and L. Tornes (1996), "Secondary Carbonates in Three Fine and Fine Loamy Alfisols in Michigan," *Soil Science Society of America Journal*, 60:1862–1870.

Simonson, R.W., and S. Rieger (1967), "Soils of the Andept Suborder in Alaska," *Soil Science Society of America Proceedings*, 31:692–699.

Slaymaker, O. (1974), "Alpine Hydrology," in: J.D. Ives and R.G. Barry (eds.), *Arctic and Alpine Environments*, London, Methuen.

Soil Survey Staff (1975), "Soil Taxonomy," *U.S. Department of Agriculture Handbook 436*, U.S. Government Printing Office, Washington, D.C.

Soil Survey Staff (1999), "Soil Taxonomy," 2nd edition. *U.S. Department of Agriculture Handbook 436*, U.S. Government Printing Office, Washington, D.C.

Steila, D., and T.E. Pond (1989), "*The Geography of Soils: Formation, Distribution, and Management*, Maryland, Rowman and Littlefield Publishers.

Stolt, M.H., and J.C. Baker (1994), "Strategies for Studying Saprolite and Saprolite Genesis," in D.L. Cremeens, R.B. Brown, and J.H. Huddleston (eds.), Whole Regolith Pedology, *Soil Science of America Special Publication,* 34, Madison, Wisconsin.

Stolt, M.H., J.C. Baker, and T.W. Simpson (1992), "Characterization and Genesis of Saprolite Derived from Gneissic Rocks of Virginia," *Soil Science Society of America Journal*, 56:531–539.

Stolt, M.H., J.C. Baker, and T.W. Simpson (1993), "Soil-Landscape Relationships in Virginia: I. Soil Variability and Parent Material Uniformity," *Soil Science Society of America Journal*, 57:414–421.

Tedrow, J.C.F. (1977), *Soils of the Polar Landscapes*, New Brunswick, Rutgers University Press.

Thorn, C.E. (1979), "Bedrock Freeze-Thaw Weathering Regime in an Alpine Environment," *Earth Surface Processes* 4:211–228.

Thorn, C.E. (1980), "Alpine Bedrock Temperatures; An Empirical Study," *Arctic and Alpine Research*, 12:73–86.

Thorn, C.E. (1982), "Nivation: A Geomorphic Chimera," in: M.J. Clark (ed.), *Advances in Periglacial Geomorphology*, Chichester, U.K., John Wiley and Sons, 3–31.

Ugolini, F.C. (1986), "Processes and Rates of Weathering in Cold and Polar Desert Environments" in: S.M. Colman and D.P. Dethier (eds), *Rates of Chemical Weathering of Rocks and Minerals*, Orlando, Academic Press, 193–235.

Ugolini, F.C., H. Dawson, and J. Zachera, 1977. Direct evidence of particle migration in the soil solution of a podsol. *Science*, 603–605.

USDA (U.S. Department of Agriculture). 1998. National Soil Survey Center, National Resources Conservation Service, Washington, D.C.

Varnes, D.L., and G.R. Scott, 1967. General and engineering geology of the United States Air Force Academy site, Colorado. *U.S. Geological Survey Professional Paper,* 551.

Velbel, M.A. (1985), "Geochemical Mass Balances and Weathering Rates in Forested Watersheds of the Southern Blue Ridge," *American Journal of Science*, 285: 904–930.

Velbel, M.A. (1986), "A Mathematical Basis for Determining Rates of Geochemical and Geomorphic Processes in Small Forested Watersheds by Mass Balance: Examples and Implictions" in: S.M. Colman and D.P. Dethier (eds.), *Rates of Chemical Weathering of Rocks and Minerals*, Orland, Academic Press, 439–451.

Wahrhaftig, C. (1965), "Stepped Topography of the Southern Sierra Nevada, California," *Geological Society of America Bulletin*, 76:1165–1190.

Wainwright, J., A.J. Parsons, and A.D. Abrahams (1995), "A Simulation Study of the Role of Raindrop Erosion in the Formation of Desert Pavements," *Earth Surface Processes and Landforms*, 20:277–291.

Watts, S.H. (1981), "Bedrock Weathering Features in a Part of Eastern High Arctic Canada: Their Nature and Significance," *Annals of Glaciology*, 2:170–175.

Watts, S.H. (1983), "Weathering Processes and Products Under Arid Arctic Conditions," *Geografiska Annaler*, 65:85–98.

Watts, S.H. (1986), "Intensity Versus Duration of Bedrock Weathering Under Periglacial Conditions in High Arctic Canada," *Biuletyn Peryglacjalny*, 30:141–152.

Wells, S.G., J.C. Dohrenwend, L.D. McFadden, B.D. Turrin, and K. D. Mahrer (1985), "Late Cenozoic Landscape Evolution on Lava Flow Surfaces of the Cima Volcanic Field, Mojave Desert, California," *Geological Society of America Bulletin*, 96:1518–1529.

West, L.T., and E.M. Rutledge (1987), "Silty Deposits of a Low, Pleistocene-Age Terrace in Eastern Arkansas," *Soil Science Society of America Journal*, 511:709–715.

White, S.E., 1976. "Is Frost Action Really Only Hydration Shattering?" *Artic and Alpine R*esearch, 8:1-6.

10

Rivers

Ellen E. Wohl

Rivers are important components of the water cycle, essential links between the atmospheric and hydrologic processes discussed previously and the massive reservoir of water represented by the world's oceans. North America receives variable quantities of precipitation, mostly from nearby oceans. With the exception of water lost to evapotranspiration and long-term groundwater and glacier storage, North American rivers sooner or later transfer this precipitation back to the oceans. In doing so, rivers perform important additional work in eroding the landscape, including their own channels, and conveying the resultant sediments downstream. Some of this sediment is stored for longer or shorter periods in channel and floodplain deposits, and some sediment reaches the coast and ocean basins fairly quickly. In North America, a small number of major river systems accomplish most of this work, most notably the Mississippi-Missouri-Ohio system, which drains much of the central area to the Gulf of Mexico (table 10.1). Other important rivers, such as the Mackenzie flowing to the Arctic Ocean, and the Colorado draining to the Gulf of California, also contribute significantly to the movement of water and sediment. Even relatively small streams, such as the Eel River in northern California, may transport sediment out of proportion to their basin area. This chapter examines (1) the environmental factors that affect rivers and their channels, and (2) regional differences in river behavior and related issues across the North American continent.

10.1 The River Environment

10.1.1 Basin-Scale Controls on Channel Characteristics

North America is characterized by a highly diverse array of natural river channels and channel networks. This diversity of channel forms results from diversity in the variables that control channel characteristics. Channel networks respond to the movement of water and sediment from hill slopes and along channels, as controlled by numerous variables. At the broadest level are those variables that influence channel processes over the scale of entire drainage basins and that are largely independent of influences from other variables. These independent variables are (1) the geologic framework of the basin (lithology, structure, tectonic regime), (2) climate, and (3) human land use (fig. 10.1).

The geologic framework determines the substrate on which processes of weathering and erosion act, and consequently the characteristics of water and sediment supplied to the river channel network. North America may be broadly described as having belts of folded and faulted rocks along its eastern and western margins, with a Precambrian Shield and tectonically stable craton in its interior, and recent sedimentary deposits along its southeastern margin (Bally et al., 1989; see chapter 1). Variations in

Table 10.1 Some major rivers and their characteristics

River	Length (km)	Drainage basin area ($km^2 \times 10^3$)	Mean discharge (m^3s^{-1})	Coefficient of variation of annual mean flow (%)
Mississippi-Missouri	6000	2929	15040	39
Mackenzie	4200	2000	9100	15
Yukon	3200	832	6390	7
Rio Grande	3100	457	50	72
St. Lawrence	3100	1025	10700	13
Nelson	2600	1132	2270	16
Arkansas	2300	410	1120	49
Colorado	2300	279	480	100
Columbia-Snake	2200	665	5490	25
Churchill	2100	298	1000	16
Ohio	2100	528	7710	—
Platte	1600	222	190	43
Fraser	1400	230	2730	13
Other major rivers of the world for comparison				
Nile	6700	3000	2800	—
Amazon	6300	6300	175000	—
Ob	5600	2929	12400	—
Congo	4700	4000	39200	—
Murray	3800	1057	330	—
Yangtze-kiang	3600	1958	32190	—
Volga	3500	1360	8060	—

Source: Herschy, 1998.

lithology and in tectonic regime from the active western margin and interior to the quiescent eastern margin in turn influence topography.

Climate also controls the processes of weathering and erosion, and it determines the magnitude and frequency of water supplied to the channel networks. Mean annual precipitation decreases from 1400–1600 mm along the eastern margins to 400–600 mm in the continental interior (see fig. 7.1 and chapter 5). Precipitation displays extreme spatial variability across the Western Cordillera, ranging from 300 mm to 1600 mm, and then rises to peaks of 2400–3200 mm along the northwestern coast. Annual runoff is correspondingly variable, ranging from a few mm in the west-central United States and central Mexico, to 4000 mm along the Pacific coast from southeastern Alaska to northern Washington (Riggs and Wolman, 1990). Annual

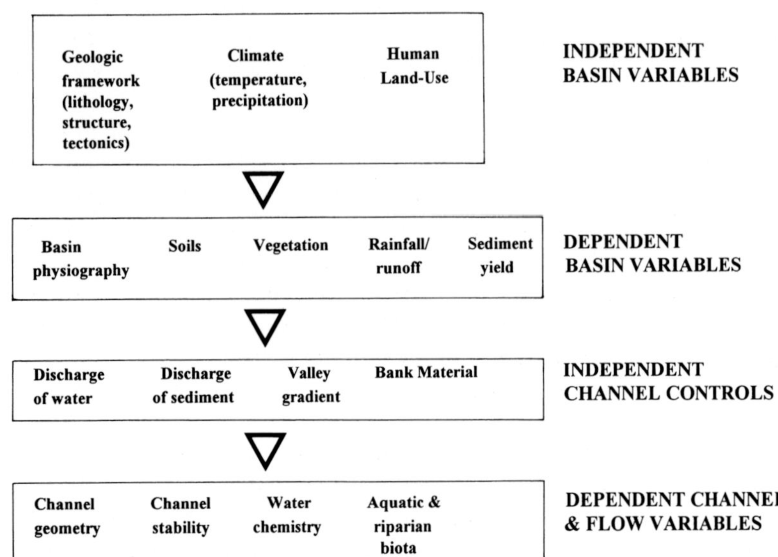

Figure 10.1 Schematic diagram of relations among variables affecting river morphology.

mean streamflow in the humid regions also tends to be less variable than that in the arid regions (Riggs and Harvey, 1990).

Human land use may be influenced by changes in the drainage network, or it may be largely independent of network characteristics. By potentially altering the movement of water and sediment into and along channels irrespective of drainage-basin characteristics, human land use plays the role of an independent basin-scale variable. Very few channels in North America may be considered completely free from the influence of human land use. Agricultural clearing of native vegetation and planting of crops began around A.D. 1000 in the Atlantic and Gulf coastal region, causing soil erosion and increased sediment yield to channels. The ancient Native American civilizations of the Southwest had diverted the flow of channels through extensive irrigation canals by A.D. 500 (Comeaux, 1981). The arrival and expansion of Euroamericans across the continent accelerated the rate of alteration as fur trappers removed beavers from channel networks, hill-slope forests were cleared for lumber, crops were planted, and channels were altered for navigation or irrigation, or disrupted to mine placer metals (see chapters 22 and 23). Springs dried up and channels incised as regional water tables dropped, and denser human settlement patterns produced contaminated runoff that polluted streams and altered aquatic communities.

Also acting at the scale of the drainage basin, but partly dependent on geology, climate, and land use, are the dependent basin variables, namely, basin physiography, soils, vegetation, rainfall-runoff relations, and sediment yield (fig. 10.1). These variables may be thought of as filters that mediate the movement of water and sediment from hill slopes to and through channels. Drainage networks may be thought of as vast conveyor belts for water and sediment, but different conveyor belts operate at different speeds, and an individual conveyor belt does not operate uniformly through time.

Sediment yield is a variable that is particularly indicative of the processes operating within a drainage basin. Sediment yield generally does not equal rates of weathering and hill-slope erosion within a drainage basin because of the potential for sediment storage at various sites (fig. 10.2). The storage of sediment in a basin may be controlled by seasonal fluctuations in discharge; the presence of natural or artificial dams; changes in sediment supply associated with glaciation, land use, climate change, mass movement, or tectonics; or by internal thresholds in the channel network (Meade et al., 1990). Langbein and Schumm (1958) used data from across North America to develop a curve of sediment yield versus precipitation that showed the highest sediment yields in semiarid regions (approximately 400-mm mean annual precipitation). These regions optimize sediment yield because of their low vegetation densities, moderate precipitation levels, and low frequency,

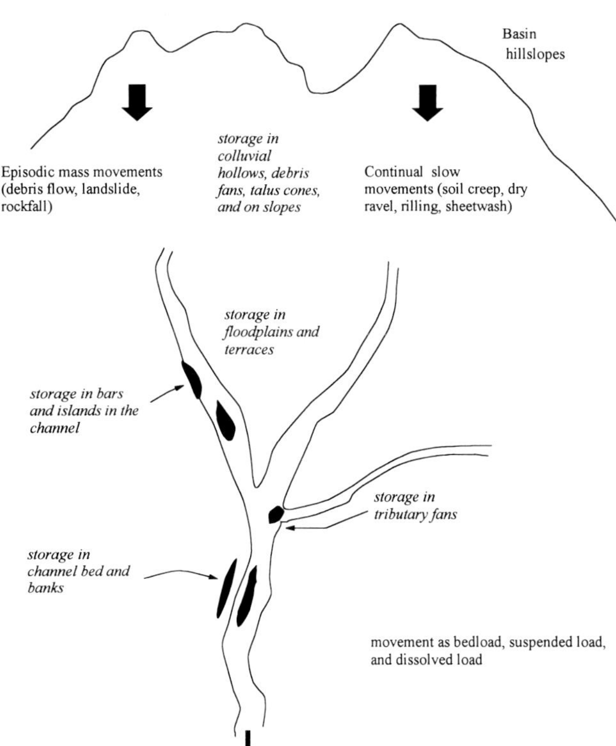

Figure 10.2 Schematic illustration of sediment movement from hill slopes into and along channels.

high magnitude floods. Each of these factors facilitates erosion of sediment from hill slopes and transport of sediment through a drainage basin. Drainage basins with high relief, intensive land use, or very erodible rocks may also have high sediment yields. Suspended sediment (fig. 10.3) is often used as a surrogate for sediment yield because the majority of sediment moving along channels is transported in suspension. In North America, the highest levels of suspended sediment discharge come from the semiarid savannas and plateaus, and from the Arctic plains (fig. 10.4).

10.1.2 Reach-Scale Variables of Channels

The following characteristics of individual channels or channel segments occur in response to basin-scale variables: (1) valley gradient; (2) discharges of sediment and water; (3) bank material; (4) channel geometry; (5) channel stability through time in both horizontal and vertical dimensions; (6) water chemistry; and (7) aquatic and riparian biota adapted to the physical and chemical characteristics of channels.

Valley gradient influences reach-scale channel morphology by constraining the channel gradient. Individual channel reaches may have longitudinal gradients very different from the valley gradient, as, for example, when a

Figure 10.3 View of the junction of a tributary channel (flowing from bottom toward top of view) with the Colorado River (flowing left to right) in the Grand Canyon of Arizona. The Colorado, which drains a basin with several highly erodible sedimentary rock units, has a much larger suspended sediment load and appears turbid. The tributary channel drains a small basin formed on resistant crystalline rocks and has a low suspended sediment load. Three boats 5 m in length are tied to the right bank of the tributary near the junction.

beaver dam traps sediment and creates a low gradient reach upstream or when a knickpoint along the channel creates a reach with a gradient much higher than the valley gradient. A channel may also meander back and forth along a stream valley, creating a channel gradient lower than the valley gradient. In most rivers, however, the reach-scale channel gradient reflects controls imposed by the valley gradient.

Figure 10.4 Average discharges of suspended sediment by major rivers of North America, estimated as of circa A.D. 1700, before the advent of any significant human impact. Width of river represents suspended-sediment discharge (Meade et al., 1990, reproduced by permission of the Geological Society of America).

The discharge of water along a channel reach will partly control the movement of sediment along the channel and thus the geometry and stability of the channel. For completely adjustable alluvial channels, water discharge is the dominant variable controlling velocity, width, depth, and channel gradient (Leopold and Maddock, 1953). Water discharge may be characterized in terms of magnitude, frequency, and duration. These discharge characteristics will be a function of precipitation supplied to the drainage basin and of the surface and subsurface processes by which water moves from hill slopes into and along channels. The perennial rivers of eastern North America tend to have lower spatial and temporal discharge variability than the ephemeral, intermittent, or perennial rivers of the arid regions of western North America.

Sediment discharge may be characterized by the volume of sediment transported, by the grain-size distribution of the sediment, and by whether the sediment is transported in solution as dissolved load, in suspension as wash load, or in fairly continuous contact with the channel bed as bed-material load. The majority of sediment transported by North American rivers moves as suspended sediment, but rivers such as the St. Lawrence may carry 91% of their total sediment load in solution (Knighton, 1984).

Bank material composition and strength exert an important control on channel geometry and stability by determining the resistance to erosion of channel boundaries. Channel characteristics such as sinuosity, meander wavelength, and width-depth ratio directly reflect the ability of water and sediment moving along the channel to erode the channel boundaries. Channels with highly erodible banks tend to have high width-depth ratios, whereas channels with resistant boundaries are more likely to be sinuous and to have lower width/depth ratios.

Channel geometry refers to three-dimensional channel morphology. Channel geometry may be characterized by describing the bedforms or regularly repeated variations along the channel bed. Alternating pools and riffles, steps and pools, dunes, or ripples are types of bedforms that, by creating irregularities along the channel boundaries, affect channel boundary resistance and energy expenditure by flowing water. Riparian vegetation, individual large clasts, or variations in the channel banks also affect boundary roughness. Channel geometry may also be described in terms of cross-sectional dimensions, including width, depth, width-depth ratio, and cross-sectional asymmetry. Channel geometry at the reach-scale may be described in two dimensions, for example, straight, meandering, or braided channels, or in three dimensions if channel gradient and longitudinal profile are quantified. These increasing scales of channel geometry, from bedforms and boundary roughness to channel planform and gradient, represent various levels at which a channel can adjust to variations in water and sediment discharge. Initial, relatively rapid adjustments will occur at the scale of bedforms, with progressively larger and longer-lasting changes being absorbed by adjustments in cross-sectional geometry or gradient. Rivers within specific regions of North America are often characterized by fairly consistent geometries, such as straight, steep, step-pool channels in the mountains of western North America, or low-gradient, sinuous, silt-clay channels in the lower Mississippi River basin.

Most natural channels have an inherent level of horizontal and vertical instability. Braided channels (fig. 10.5) tend to aggrade and to shift laterally in abrupt movements at timescales of hours to years, depending on water and sediment discharges. Meandering channels (fig. 10.6) migrate laterally at more constant rates, with the exception of meander cutoffs or avulsions. Decadal averages of meander migration vary widely among rivers and along a

Figure 10.5 (*top*) Upstream view of a braided channel, the Muddy River, near Mount St. Helens, Washington. The active channel occupies only a small portion of the valley bottom at any given time, but shifts repeatedly back and forth across the valley, which is approximately 100 m wide here.

Figure 10.6 (*at left*) A meandering channel in the Wind River Range of Wyoming. Higher velocity, deeper flow along the outside of the meander bend is causing bank erosion and progressive channel migration to the right, whereas deposition is occurring on the point bar at left.

river; Schumm (1977) reports averages from 0.6 m yr^{-1} to 300 m yr^{-1} along different portions of the Mississippi River. Ephemeral alluvial channels in arid and semiarid regions have cycles of aggradation and incision that alternate over periods of decades to centuries. In general, bedrock channels or alluvial channels carrying predominantly suspended sediment and flowing through cohesive, fine-grained sediments tend to be more stable than alluvial channels carrying coarser bed load (Schumm, 1981). Alluvial channels in the lower Mississippi River drainage basin and in the arid and semiarid regions of the western United States have been particularly unstable during the past century, changing dramatically in response to human land use.

Water chemistry will be a function not only of geology, climate, and land use within a drainage basin, but also of the flow paths by which water reaches a channel. Channels that have a large component of base flow will tend to have more solutes because the base flow has a longer period of contact with geologic materials, including weathered regoliths, than does surface runoff. Channels flowing through warmer climatic regimes will also have higher solute concentrations because of the enhanced chemical weathering. In North America, rivers south of approximately 40°N and west of approximately 90°W tend to have the highest solute concentrations.

As a physical system, a river functions as a conveyor moving water and sediment from higher toward lower elevations. Superimposed on this physical system, and affecting the system through complex feedback mechanisms, are the aquatic biota living within the channel and the riparian biota along the channel margins. The aquatic biota of North American river channels vary systematically with water temperature, chemistry, and nutrients, as well as channel hydraulics, substrate, and morphology (Allan, 1995). At the level of primary production, mosses, lichens, bacteria, phytoplankton, algae, and organic matter from the terrestrial environment provide the food base for the consumers. Consumers begin with aquatic macroinvertebrates, such as larval insects, that are adapted to specific microenvironments within the channel. Shredders such as caddisflies (*Trichoptera* spp.) break down coarse particulate organic matter (> 1 mm in size), entering the channel as twigs or leaves. Grazers such as mayflies (*Ephemeroptera* spp.) feed on algae growing on the channel substrate, whereas collectors such as true flies (*Diptera* spp.) filter bits of organic detritus being carried by the current or gather the detritus from sediments. At the next level of predation, various fish are adapted to different physical environments (coldwater versus warmwater, fast versus slow currents, gravel versus silty substrate) and to different food sources (invertivores, planktivores, piscivores).

Systematic variations in aquatic biota may be described in terms of the river continuum concept (Vannote et al., 1980). The aquatic biota of upland or headwater channels are adapted to abundant terrestrial inputs of coarse, particulate organic matter from riparian vegetation and to cooler water temperatures produced by shading from this vegetation. In the central portion of a drainage basin, aquatic biota depend more on primary production by algae and rooted vascular plants, and on organic transport from upstream. In the center of a drainage basin, fluctuations in water temperature are greater, and species diversity is higher than in headwater regions. Channels lower in a drainage basin have warmer water and maximum autochthonous primary production, and species diversity decreases relative to reaches upstream. The specifics of these downstream trends vary with climatic regime, but the general patterns appear to be similar among North American rivers.

Extending beyond the stream channel is the riparian corridor—the bottomland, floodplain, and stream-bank communities within the 100–year floodplain (Swift, 1984). The riparian corridor and the stream channel are intricately connected (Hupp and Osterkamp, 1996). Riparian vegetation (fig. 10.7) attenuates incoming sunlight and influences ground surface and water temperatures (Barton et al., 1978); contributes organic material that provides both nutrients and roughness to the channel (Keller and Swanson, 1979; Thompson, 1995; Wohl et al., 1997); provides a low-velocity, shallow-water nursery habitat for young fish (Petts, 1990); increases out-of-bank roughness, attenuating flood peaks and promoting accumulation of sediments and nutrients (McKenney et al., 1995); and increases bank stability via the root network (Fetherston et al., 1995).

The specific species and patterns of riparian vegetation vary among North American channels. A river flowing through a coastal valley of central California may have grasses and herbaceous annuals growing on its banks, low thickets of willow, alder, elder, and cottonwood on its gravel bars, and a riparian forest with a 25-m overstory of cottonwood, elder, and alder on its floodplains (Brinson et al., 1981). A channel in the subarctic region may have willow and cottonwood along its banks and the proximal floodplain, with spruce and fir growing on the distal floodplain. However, the general pattern of younger vegetation that is resistant to disturbance and inundation growing close to the channel, with older successional communities farther from the channel, applies to rivers across North America.

10.1.3 Spatial and Temporal Variations

Both basin-scale and channel-scale variables may vary across time and space. All channels in North America have been affected directly or indirectly by changes associated with Pleistocene climate changes, including glacial advances and retreats (see chapter 3). Channels within the margins of the Laurentide or Cordilleran ice sheets, or the numerous smaller alpine glaciers, may have ceased to flow

Figure 10.7 A drainage network outlined by darker riparian vegetation in the arid Little Harquhalla Mountains, Arizona. The riparian corridor along each of these ephemeral channels is very narrow; along perennial rivers with broad floodplains in wetter climatic regimes, the riparian corridor may extend for kilometers on either side of the channel.

during episodes of glacial advance or may have been completely reformed by the meltwater and sediment pulses associated with glacial retreat. Channels beyond the ice margins were indirectly affected by meltwater, by changing sea level associated with glacial fluctuations, or by changes in climatic circulation patterns. During the Holocene Epoch, lower magnitude climatic variability has continued to alter the supply of water and sediment to channels, with consequent changes in channel characteristics.

Channels may also change through time irrespective of climatic variability, as a result of instabilities inherent to the channel's functioning. Schumm and Parker (1973) demonstrated that alluvial channels in arid and semiarid regions alternately incise and aggrade on a timescale of several hundred years independent of climatic controls (fig. 10.8).

Finally, changes in the geologic framework through time may alter basin- and channel-scale processes. Active tectonic uplift or subsidence, volcanic eruptions that alter surface topography and introduce large quantities of sediment to the channel, or channel incision that exposes different lithologies may each alter the substrate on which processes of weathering and erosion operate.

For drainage basins larger than a few square kilometers, both basin- and channel-scale variables are likely to vary spatially. Even in the absence of variation among controlling variables, most drainage basins have predictable downstream changes (fig. 10.9). In the case of North America's major drainage basins, these predictable downstream trends are also overprinted with changes caused by varying geology, climate, or land use.

Figure 10.8 Actively incising channels in the western plains of Colorado. These channels average 5 m in depth and 15–20 m in width at the top. Infilled paleochannels exposed along the modern channel walls indicate that this drainage area has had multiple episodes of channel incision and filling during the Quaternary.

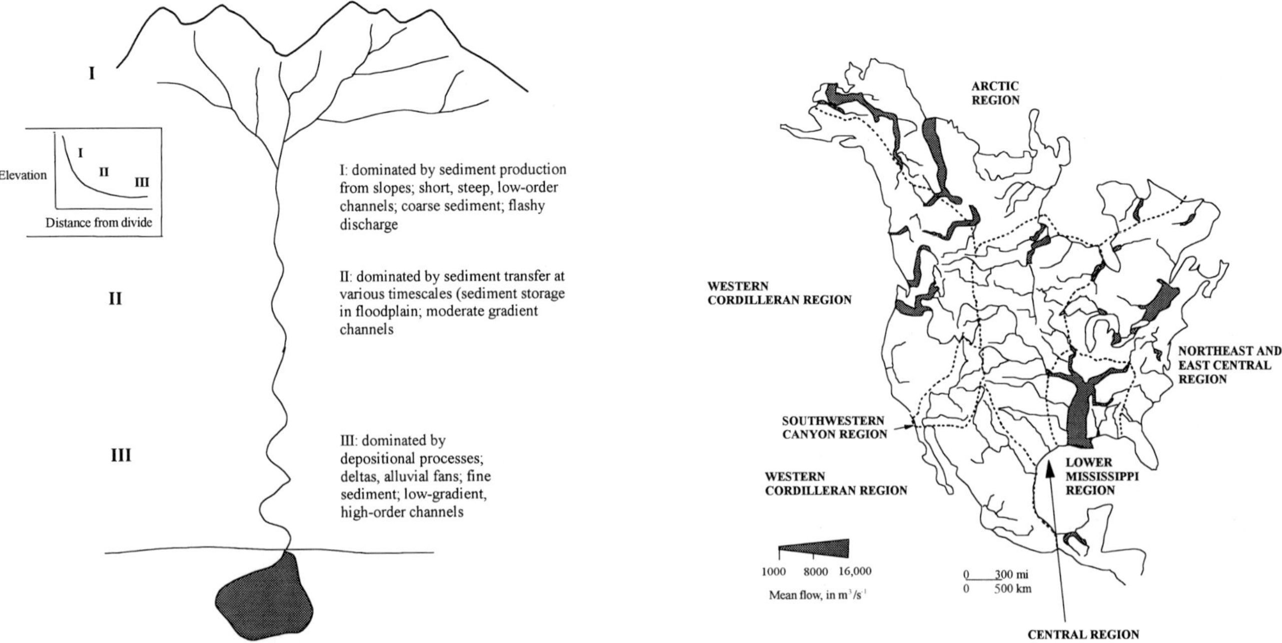

Figure 10.9 (*left*) Schematic illustration of generalized downstream trends in drainage basins.

Figure 10.10 (*right*) Principal river systems and those channels carrying mean flows of more than 1000 m^3s^{-1} (Riggs and Wolman, 1990, reproduced by permission of the Geological Society of America). Regional subdivisions discussed in the text added.

Bearing in mind this conceptual framework of basin-scale variables that influence the movement of water and sediment into and along channels, I have devoted the remainder of this chapter to a systematic description of the general characteristics of river channels in each of six major subdivisions of North America (fig. 10.10, table 10.2). Each region has been designated on the basis of relative internal consistency of geologic framework and regional climate, which in turn produce certain consistencies among the river channels of that region.

10.2 Regional Subdivisions of North American Rivers

Although six major regions may be defined in terms of river regime, there is a broader fundamental distinction in the North American landscape, namely, the distinction between rivers that have developed in areas of former Pleistocene glaciation and rivers that evolved beyond the ice front. In the former case, the present drainage network is essentially a Holocene response to a Pleistocene glacial legacy of bare rock, bedrock hollows, and morainal topography. In the latter case, the drainage network has evolved over a much longer period and, although often fed by glacial meltwater, preserves extensive legacies of much earlier river systems.

10.2.1 Rivers of the Arctic Region

The channels of the low (< 300 m) relief Arctic lowlands drain to Hudson Bay or to the Arctic Ocean. These channels flow across regions of extensive permafrost where the frozen ground supports numerous shallow lakes during summer. The channels are predominantly alluvial, and individual valleys often contain multiple, laterally mobile channels. The Mackenzie River, the largest drainage in this region, exemplifies regional channel characteristics. The Mackenzie drainage basin is underlain by continuous and discontinuous permafrost. The channel has created an enormous delta plain (13,000 km^2) at the edge of the Arctic Ocean, and thermokarst lakes are scattered across the plain between meandering and anastomosing channels (Wolman et al., 1990). Similarly, the Yukon River meanders across a low-relief terrain underlain by glaciogenic sediments, loess, and volcanic ash.

Flooding in the dry Arctic lowlands is usually associated with spring snow melting, or in some basins with jökulhlaups, outburst floods that occur when glacial meltwater from an ice-dammed lake is released catastrophically. Jökulhlaups such as those in 1982 and 1984 from Strandline Lake, Alaska, have peaked hydrographs with very abrupt recessional limbs (Sturm et al., 1987). These repetitive drainings of a lake create and maintain extensive proglacial braid plains unless the outflow is confined, in which case the jökulhlaups may cause substantial channel erosion (Church, 1988). The rivers of the arctic region are presently relatively pristine, although major interbasin water transfers along the region's southern border are altering channel and riparian characteristics

10.2.2 Rivers of the Northeast and East-Central Region

This region includes two major physiographic provinces, the Appalachian Mountains and Plateaus, and the Atlantic coastal province, and it impinges on the Canadian Shield farther north. Channels throughout the region drain east to the Atlantic Ocean. Channels crossing the folded sedimentary rocks of the Appalachians have long stretches formed in bedrock and are strongly influenced by lithology; more resistant rock types produce steeper channels with shorter meander lengths. Channels in the glaciated uplands toward the north may have steep profiles and bouldery beds where they flow across glacial till. Channels in the unglaciated upland regions farther south periodically undergo intervals of enhanced bedrock-valley erosion (Mills et al., 1987). The entire Appalachian region is characterized by wind and water gaps. In contrast, alluvial channels of the low-relief coastal plains tend to meander broadly, forming a complex ridge and swale topography of natural levees, back swamps, and oxbow lakes.

The northern two-thirds of this region were covered by glacial ice during the Wisconsinan glacial maximum circa 18,000 years ago. As the ice retreated, the forest vegetation became established across the southern half of the region by 10,000 years ago, but it reached the northern portion only 6,000 years ago (Davis, 1983). Climatic fluctuations during the past 10,000 years have affected forest composition and relative abundance of different species, but the persistent forest cover maintained relatively constant water and sediment yields (Knox, 1983). The dominant Quaternary paleohydrologic events on channels of the northeast and east-central region before European settlement were the glacial meltwater floods.

As the Laurentide Ice Sheet began to retreat, glacial meltwater initially flowed primarily down the channels of the Mississippi River basin to the Gulf of Mexico. With further retreat of the ice front, some of the meltwater was diverted through the Great Lakes and the St. Lawrence River to the North Atlantic Ocean (Teller, 1987, 1990). Evidence of these floods persists in the form of large potholes, high terraces, proglacial lake sediments, and deeply incised river channels in such areas as the Susquehanna River valley, the Grand Valley of Michigan (Kehew, 1993), and the basins of Lakes Ontario and Erie (Shaw and Gilbert, 1990).

The channels of the humid temperate northeast and east-central region tend to be perennial. Flow variability is relatively low, and floods may occur at any time during the year, with dissipating hurricanes most commonly producing floods during the summer and autumn. Flow storage for human use is generally relatively low; the ratio of reservoir storage ranges from 0.04 to 0.28, except for one portion of northeast Canada. However, channels have been dramatically affected by changes in land-use patterns during the past two centuries. Wolman (1967) conducted one of the first studies of land-use impacts on channels. Working on channels in the Maryland Piedmont, he documented threefold to ninefold increases in sediment yield with the replacement of forest by agricultural lands and associated channel aggradation. During the construction phase of urbanization, sediment yield increased by a thousand times relative to sediment yield from forested lands, and then dropped precipitously once construction was completed, triggering channel scour and bank erosion. Similar alterations in channel morphology have occurred along many other channels in the region.

10.2.3 Rivers of the Lower Mississippi River Basin and Gulf Coastal Province

The low relief coastal plains from Georgia to east Texas are characterized by elongate drainage basins composed of meandering river systems that have prograded seaward over the past few million years. During major episodes of Pleistocene glaciation, much of the continental shelf was exposed and the rivers became braided and incised. These same channels later aggraded and formed broad alluvial terraces during interglacial episodes (Walker and Coleman, 1987). As the Wisconsinan glaciation waned, the Mississippi River was the principal drainage for much of the glacial meltwater from the Laurentide Ice Sheet and the Rocky Mountain glaciers. This influx of meltwater and sediment caused the Mississippi River to aggrade the incised valley that had been shaped during successive intervals of low sea level, forming a braided river. As the ice margin continued to retreat, discharge and aggradation decreased, and the present meandering channel pattern developed in finer grained sediments (Saucier, 1974; Baker, 1983; Porter and Guccione, 1994). Today, despite relatively high sea level, the coast continues to prograde seaward over longer time intervals as the region's many rivers transport sediment to their mouths and numerous deltas. These deltas include that of the Mississippi River, at 28,600 km^2, one of the world's largest deltas (see chapter 21).

Water consumption takes a fairly low proportion of streamflow in this region, but channels have been dramatically affected by land use. The lower Mississippi River drainage basin has been extensively channelized to decrease overbank flooding and improve navigation (Harvey et al., 1983; Tobin, 1995), and some of the continent's highest historic losses of riparian zones have occurred in this region (table 10.2). Channelization and wetland loss have affected even relatively small basins, as exemplified by the

Table 10.2 Summary descriptions of rivers in six regions of North America

Region	Geology, Soils, & Topography	Climate	Runoff & Discharge Characteristics
Arctic regions	Canadian Shield & Interior Platform; resistant crystalline bedrock: Cambisols, Histosols, Gleysols: Arctic coastal plains of northern Canada & Alaska—low relief (<300m) plains & hills; glacial topography—moraines, eskers, drumlins, & post-glacial fluvial valleys	300–800 mm mean annual precip; P > E by 0–20 cm; cold & dry	Summer peak Q; 10–300 mm mean annual runoff; runoff ratio 0.4–0.7; coefficient of variation of mean annual flow 15%-31%; Mackenzie River has high suspended sediment Q; runoff is nival (snowmelt), wetland (peaks in early & late summer), or spring-fed (peaks in mid-summer—carbonate terrains); largest channels are Mackenzie, Liard, Peace, & Yukon; region drains north to Arctic Ocean
Northeast & east-central	Canadian Shield in north; Interior Platform in central; Paleozoic folded belts in south; Cretaceous and Tertiary age marine sedimentary rocks dip gently seaward along coastal margins: Lithosols, Cambisols, Histosols, Podzols, Luvisols, Acrisols; mountains & plateaus in north, low relief (300m) plain to south & east	800–1600 mm mean annual precip (north to south); P > E by 0–60 cm; hurricane incursions bring infrequent torrential rain	Summer Q peaks in north, autumn & winter Q peaks in south & central; 100–1000 mm mean annual runoff; runoff ratio 0.3–0.7; coefficient of variation of mean annual flow 13%-50%; low suspended sediment Q; most Q records < 50 yr long; largest channels are St. Lawrence & upper Mississippi; most of region drains east to Atlantic
Lower Mississippi drainage	Passive margin & coastal plain; Cretaceous and Tertiary age marine sedimentary rocks dip gently seaward; alternating periods of emergence & submergence until sea level stabilized 5–6 ka: Acrisols, Gleysols: very low relief plains	1400–1600 mm mean annual precip; P > E by 0–40 cm	Autumn Q peaks; 100–600 mm mean annual runoff; runoff ratio 0.2–0.4; coefficient of variation of mean annual flow 14%-59%; largest suspended sediment Q in continent in Mississippi River (but mostly originates from Missouri River); Mississippi River is largest channel; region drains south to Gulf of Mexico
Central	Interior Platform; Cretaceous-Tertiary sedimentary rocks with eolian, fluvial, & glacial overlying sediments: Chernozems, Kastanozems, Phaeozems: plains & gently rolling hills	400–1200 mm mean annual precip; P ≤ E by 0 to -60 cm	Spring-summer Q peaks; 0–300 mm mean annual runoff; runoff ratio 0.05–0.3; coefficient of variation of mean annual flow 5%-112%; very large suspended sediment Q; sediment yield ranges from 16–610 $m^3/km^2/yr$; largest channels are the Missouri and Platte; region drains northeast to Hudson Bay, or east to the Mississippi & thence to the Gulf of Mexico
Western Cordillera	Mesozoic-Tertiary fold belts; Tertiary & Quaternary volcanics; Laramide basement uplifts; subduction complexes of plutons, metamorphics, volcanics, faulted & folded sediments: Yermosols, Kastanozems, Luvisols, Lithosols: high relief (> 300m), alpine glaciation, alluvial &	300–4000 mm mean annual precip; P > E by 0–30 cm in north, P < E by 0 to -150 cm in south; subtropical desert scrub through temperate montane & alpine, to boreal forest & tundra	Spring-summer peak Q; 0–1000 mm mean annual runoff (south to north); runoff ratio 0.1–0.8; coefficient of variation of mean annual flow 12%-126%; low suspended sediment Q; flashy flood peaks in southwest; jokulhlaups in north; largest channels are

Channel Morphology	Aquatic & Riparian Biota	Water Chemistry	Water Use	Historical Channel Change
Channels may be ephemeral, or cut in snow; bed scour & lateral migration important; multiple-channel valleys	Tundra & boreal forest dominated by *Populus* & *Salix*, with succession to spruce or other conifers; cold-water streams & wetlands; suckers, minnows, pike, salmon, trout	Low solute concentrations	0–0.35 ratio of reservoir storage; 96–100% current/natural outflow	River diversions for hydro-electric in Canadian Intermontane Belt
Appalachians—meander freely, ingrown meanders, or straight channels with little or no net valley-floor alluviation; Atlantic coast—meandering channels with natural levees, backswamps, ridge & swale topography	Tundra & boreal forest in far north, mixed forest elsewhere includes humid broadleaf forest; warmwater streams in US, coldwater in Canada; pike, suckers, minnows, sturgeon, garfish	Low-moderate solute concentrations—dominated by HCO_3 & Ca	0.04–0.71 ratio of reservoir storage; 86–105% current/natural outflow	Appalachians: Euroamerican settlement from late 1700s to early 1800s caused fluvial sedimentation due to forest clearing, burning, grazing, & cropping—trend reversed in 1900s; urbanization beginning in 1930s
Meandering channels with natural levees, backswamps, ridge & swale topography; mostly perennial except in south Texas or Florida karst regions; elongate basins; extensive delta development	Southeastern mixed forest, river bottom forest (bottomland hardwood), & cypress swamps; warmwater streams; suckers, pike, minnows, garfish	Low-moderate solute concentrations	0.15–0.32 ratio of reservoir storage; 93–99% current/natural outflow	Soil erosion and sedimentation associated with agriculture began as early as AD 1000; up to 96% loss of riparian ecosystems since c. 1750
Perennial channels from the Rockies, ephemeral or perennial channels originating on the plains; mostly alluvial—braided & meandering; sapping networks on southern plains; catastrophic flood features along Cordilleran margins	Hardwood forest, tall & shortgrass prairie, *Salix* & *Populus* dominant; intermittent & warm-water streams; suckers, minnows, chub, goldeye, garfish	Moderate-high solute concentrations—dominantly SO_4, Mg, Na+K, Ca	0.14–1.19 ratio of reservoir storage; 73–96% current/natural outflow	Channel metamorphosis since late 1800s due to irrigation diversions, reservoirs, & flow regulation; more than 50% loss of riparian ecosystems along some channels since 1880
Braided outwash channels in glacial regions; step-bed channels in mountains; internal drainage present in arid regions; arroyos (incised ephemeral channels in southwest); both bedrock & alluvial	Sagebrush steppe, montane coniferous forest, desert shrubs, subalpine forest; coldwater streams; suckers, minnows, salmon, trout	Low solute concentrations	0–1.45 ratio of reservoir storage; 68–101% current/natural outflow	Placer mining (eg. Sierra Nevada & Colorado Rockies); deforestation & sedimentation; beaver trapping; tie drives for railroad ties; dams; in Basin & Range, drying & incision of channels; up to 98% riparian loss

Table 10.2 Continued

Region	Geology, Soils, & Topography	Climate	Runoff & Discharge Characteristics
	outwash fans, pluvial lakes & playas, fluvial terraces		Columbia, Fraser; east of the Continental Divide the region drains to the Atlantic, to the west it drains to the Pacific Ocean
Southwest canyons	Colorado Plateau—gently dipping sedimentary rocks & volcanics: Kastanozems, Yermosols: dissected uplands & canyons	100–400 mm mean annual precip; P < E by 40–120 cm; semiarid & arid	Summer peak Q; 0–10 mm mean annual runoff; runoff ratio 0.05–0.1; coefficient of variation of mean annual flow 35%-44%; high suspended sediment Q; Colorado is largest channel; region drains southwest to Gulf of California

Notes
P > E: precipitation exceeds evaporation; P < E: evaporation exceeds precipitation; Q: discharge
Runoff ratio: the ratio of mean annual runoff to mean annual precipitation
Coefficient of variation of annual mean flow taken from selected rivers in region
Water chemistry is a function of [composition of rain & snow; climate & weathering—eg. snowmelt is low in solutes because the surface soil is frozen & biological activity is minimal; soil; vegetation; land use; discharge (particulate material increases with discharge, while dissolved load decreases)].
Ratio of reservoir storage to annual supply
Ratio of current outflow to natural outflow in m/s

(Sources: Thomas, 1978; Brinson et al., 1981; Bovis, 1987; Carter et al., 1987; Graf et al., 1987; Madole et al., 1987; Mills et al., 1987; Muhs et al., 1987; Osterkamp et al., 1987; Shilts et al., 1987; Walker and Coleman, 1987; Bally et al., 1989; Hem et al., 1990; Meade et al., 1990; Riggs and Harvey, 1990; Riggs and Wolman, 1990; Malanson, 1993; Mitsch and Gosselink, 1993)

response of the 100-km² Oaklimiter Creek watershed in Mississippi after channelization in the 1960s (Harvey et al., 1983). Straightening of the naturally sinuous (1.3–2.5 sinuosity) channel resulted in steeper gradients. Retention of larger flows within the channel increased the sediment transport capacity by a factor of 50, causing severe channel incision and widening. Fifteen years after channelization, channel capacity had increased by a factor of ten.

10.2.4 Rivers of the Central Region

Channels of the gently rolling hills and plains of central North America flow predominantly eastward to join the Mississippi River or, in Canada, northeast via the Churchill or Nelson Rivers to Hudson Bay. The region has been characterized throughout the Quaternary by alternating cycles of erosion and deposition, which have been controlled primarily by fluctuations in water and sediment discharge associated with climatic changes (Knox, 1983, 1985; Schumm and Brakenridge, 1987). Drainage networks in the northern portion of the region were substantially rearranged by repeated advances and retreats of the Laurentide ice sheet, and glacial-lake outburst floods along the southern margins of the ice sheet left huge trench-shape channels (Kehew and Lord, 1987). Channels in the central and southern portions were more directly affected by fluctuations in the alpine glaciers of the Rocky Mountains. The contemporary topography of this region has been shaped predominantly by fluvial and aeolian processes working across mostly prairie grassland but with boreal forest to the north and mesquite savanna to the south (Osterkamp et al., 1987).

Precipitation decreases steadily from east to west across the region. Channels are more likely to be ephemeral or intermittent at the western margins, although the larger channels originating in the Rocky Mountains are perennial because of high snowmelt-driven discharges in later spring to early summer. Channels originating on the plains are more likely to be ephemeral and to have flashy flood peaks generated by summer thunderstorms. The alluvial channels of the prairies have some of the continent's highest suspended sediment discharges because of the region's

Channel Morphology	Aquatic & Riparian Biota	Water Chemistry	Water Use	Historical Channel Change
channels; high rates of sediment input & coarse sediment from steep terrain; glacio-fluvial channels in north; short, steep, braided to meandering in lower reaches				(eg. Sacramento River, CA) since 1850
Many ephemeral channels	Desert shrubs & grasslands; warmwater & coldwater streams	High solute concentrations—dominated by HCO_3,SO_4, Mg, Na+K	2.3–3.3 ratio of reservoir storage; 25–68% current/natural outflow	Quaternary cycles of incision & aggradation—channels wide & entrenched 1880–1940, aggrading to 1980, incision presently; late 1800s to present—invasion by tamarisk; dams on Colorado River system; up to 50% riparian loss since the 1600s

semiarid climate. A diversity of weather systems, including dissipating hurricanes, frontal cyclones, summer thunderstorms, and rapid snowmelt, may give rise to flooding in this region at any time (Hayden, 1988).

The ratio of reservoir storage is quite high (up to 1.19) in the Missouri River drainage basin, and all of the major channels on the western Great Plains have undergone channel metamorphosis during the past century. This generally involves a change from broad, braided channels with little riparian forest to narrow, meandering, heavily forested channels as a result of reduction in peak flows, increase in base flows, and local rise in water tables. These changes have been described in detail for the North and South Platte Rivers (Williams, 1978; Nadler and Schumm, 1981; Eschner et al., 1983). Many smaller, ephemeral channels in the western Great Plains incised to form deep arroyos during the late nineteenth to early twentieth centuries (Bull, 1997). There has been extensive debate over whether this incision was triggered by land use, climate change, or multiple factors. Along the eastern margin of the central region, increases in sediment yield after Euroamerican settlement have caused some channels to develop broad meander belts where floodplain accretion confined the channel and increased its ability to erode laterally, as described by Lecce (1997) for the Blue River in the Driftless Area of Wisconsin.

10.2.5 Rivers of the Western Cordillera

The Western Cordillera includes the broad band of mountain ranges and intermontane basins and plateaus, from the eastern margin of the Rocky Mountains to the Pacific coast, that stretches from Alaska south through central Mexico. The region thus includes several different climatic zones: the Sonoran Desert and semiarid Sierra Madre and plateau lands of northern Mexico, the Basin and Range Province of the southwestern United States, the Mediterranean climate of California, and the temperate rainforest of the Canadian and Alaskan coasts. In general, precipitation decreases with distance from the Pacific coast, although annual totals are spatially variable and usually increase with elevation. Lithology is equally variable and includes accretionary terranes, volcanics, metamorphics, plutons, and faulted and folded sedimentary rocks. The region is unified by high (> 3000 m) relief, ongoing tectonic activity, a legacy of Quaternary alpine glaciation, and frequent mass movement as landslides, debris flows, and rockfalls move sediment from hill slopes into channels.

Channels draining these mountains are often relatively straight and incised into bedrock canyons. The canyons of the Columbia River and the Fraser River are particularly impressive. Valley and channel morphology along smaller channels may change abruptly in a downstream direction as a result of glacial erosion and deposition (Day, 1972), lithologic changes, or beaver activity (Naiman et al., 1988; Butler and Malanson, 1995). The relatively low discharges and coarse sediment loads of the eastern and interior ranges prevent substantial coarse sediment movement except during large floods. In contrast, the coastal ranges, where higher precipitation enhances both weathering and erosion, experience highly variable sediment transport both temporally and spatially, with processes and responses occurring more rapidly and frequently than in the drier eastern ranges. The Eel River of northern California, for example, has the highest recorded average suspended sediment yield per drainage area of any river in the United States (Wolman et al., 1990). The combination of active tectonic uplift, high precipitation (1500 mm mean annual precipitation basinwide, with 2800 mm at high elevations), fractured and weathered bedrock, grazing, and timber harvesting create a very large sediment supply within the 81-km^2 drainage basin. Numerous studies have demonstrated that large woody debris plays a vital role in regulating boundary roughness and sediment transport and in providing habitat diversity in mountain channels in forested basins within the western Cordillera (Muhs et al., 1987; Abbe and Montgomery, 1996) (fig. 10.11).

The mountain channels of the northern and central Cordillera have been subject to outburst floods at intervals throughout the Quaternary. Some of the largest outburst floods ever described occurred toward the end of the Pleistocene when glacial Lake Missoula (Baker, 1973, 1983; O'Connor and Baker, 1992) and pluvial Lake Bonneville (Jarrett and Malde, 1987; O'Connor, 1993) drained, creating vast scablands and deeply eroded canyons (see chapter 3). For many bedrock rivers with Pleistocene outburst floods, subsequent flows have not been competent to substantially alter channel morphology, as in the Big Lost River of Idaho (Rathburn, 1993). Outburst floods still commonly dominate channel morphology where they occur under present conditions, as described for channels on Mount Rainier, Washington (Driedger and Fountain, 1989) and in the Homathko River basin, British Columbia (Blown and Church, 1985).

Reservoir storage varies widely throughout the Cordilleran region, from essentially no reservoirs in the Alaskan Cordillera to high storage rates (up to 1.45) in the southern Canadian Rockies and southern Rockies in the United States. The Columbia River basin is heavily dammed, with 65 large dams along its channel network; the number of Pacific salmon (*Onchorhyncus* spp.) present in the drainage basin during the twentieth century has severely declined. In contrast, the undammed Fraser River to the north has the largest remaining populations of Pacific salmon in the region (Wolman et al., 1990). Similarly, mountain channels range from being fairly pristine, to those that are heavily impacted by placer mining, flow diversions, or increased sedimentation associated with timber harvesting (Wohl et al., 1998; Wohl, 2000, 2001). One of the classic examples of land-use impacts on channels comes from the Sierra Nevada of California. Hydraulic placer mining for gold took place from the 1850s to the 1880s along several west-flowing streams, including the Bear, Yuba, Sacramento, and American Rivers. The huge sediment influxes associated with mining caused up to 5 m of channel aggradation and an increase in channel braiding and avulsion (James, 1991). Channel responses to past mining activities continue today, more than a hundred years after hydraulic mining officially ceased (James, 1989), and channel responses continue to impact aquatic biota (Wagener and LaPerriere, 1985).

10.2.6 Canyon Rivers of the Southwestern United States

The Colorado Plateau and many of the nearby uplands have been characterized by epeirogenic uplift for the past 35 Ma (million years before present) (Graf et al., 1987). The channels of this region, which drain to the Colorado River, flow through deeply incised canyons that are geologically quite young, the Colorado River having incised the Grand Canyon within the past 5 Ma. The major channels are perennial, but many of the tributaries are ephemeral and these contribute large amounts of suspended and bed-load sediment. At present, the 12 major dams along the Colorado River trap much of this sediment, creating such problems as bank and beach erosion within the Grand Canyon (Collier et al., 1996).

Many of the canyon rivers had highly variable flow regimes before the reservoir construction of the twentieth century. The larger tributaries of the Colorado River are dominated by late spring–early summer snowmelt floods, whereas moderate and small channels may have flash floods caused by convective storms, dissipating tropical cyclones, or North Pacific frontal storms (Webb and Betancourt, 1992). The frequency of flooding has fluctuated during the late Holocene. Intervals of cool, moist climate and frequent El Niño events coincided with numerous large floods 4800–3600 years ago, around 1000 years ago, and during the last 500 years (Ely et al., 1993; Ely, 1997).

The southwestern canyon region has the most anthropogenically disrupted flow patterns of any region of North America, as evidenced by very high ratios of reservoir storage and low ratios of present to natural outflow (table 10.2). Few major channels have escaped alterations associated

Figure 10.11 Large, woody debris in a channel of the Cascade Mountains, Washington. The wood, which has lodged transverse to flow, stores sediment upstream and causes plunging flow and scour of a pool downstream. Channel is approximately 8 m wide.

with flow regulation or introduced vegetation (principally tamarisk, *Tamarix chinensis*). The Green River of Utah, for example, has responded variably to the completion of Flaming Gorge Dam in 1962 (Andrews, 1986). Mean annual sediment discharge has decreased by 45% since dam completion. The capacity of the river to transport sediment is greater than the sediment supply immediately downstream from the dam, and the channel bed has thus incised along a 35-km reach. Approximately 170 km downstream, a 100-km reach is in quasiequilibrium with respect to sediment, although channel width has narrowed by approximately 10% because of reduced peak flows. Some 460 km downstream from the dam, transport capacity has decreased below the sediment supply from small tributaries, and the reach is aggrading. Portions of the Green River in Canyonlands National Park have narrowed by an average of 27% because of the combination of reduced peak flows and colonization by tamarisk, which effectively traps sediment (Graf, 1978). Reduction of peak flows, overbank flooding, and fine-sediment flushing from spawning bars have also contributed to the decline of the endangered Colorado squawfish (*Ptychocheilus lucius*) in the Green River (Harvey et al., 1993).

10.3 Conclusion

Rivers have played a vital role in shaping the topography of North America. They have incised canyons and transported sediment across landforms newly created by glaciation, orogeny, or volcanism, in the process reshaping the landscape into integrated drainage networks. In many regions of the continent, the forms and processes of river channels have changed dramatically during the Holocene Epoch, as glacial ice melted, regional climates warmed, and sea level began stabilizing approximately 6000 years ago. Even drainage basins far distant from the glaciated regions were affected by fluctuating sea level and climate, so that in a sense the rivers of North America may be said to have been recreated during the past few thousand years.

When early human migrants reached North America across the Bering Land Bridge, they found a continent rich in surface waters supporting a diverse and abundant riparian and aquatic biota. Those early settlers, and subsequent waves of human migrants from Europe and the other continents, began to alter the landscape to suit their needs. As these alterations accelerated during the twentieth century, the rivers of North America once more entered a phase of re-creation. As of 1987, 27% of surface flow in North America was regulated by dams and reservoirs (Walling, 1987). More than 80%of the original riparian vegetation along channels in the arid west, Midwest, and lower Mississippi alluvial valley has been destroyed (Swift, 1984). During the past 150 years, channelization in the United States has reduced the length of major rivers by at least 320,000 km, or about 5% (Swift, 1984), and more than 40,000 km of levees, floodwalls, embankments, and dikes constrain overbank flooding (Tobin, 1995). As contemporary North American societies consciously and unconsciously reshape the continent's rivers, it becomes increasingly difficult, and imperative, to protect fairly natural portions of the river basins as natural laboratories, biological refugia, functioning relics of once-widespread drainage basin conditions, and sources of inspiration for humans (Ligon et al., 1995; Graf, 1996).

Acknowledgments Douglas Thompson and Antony Orme provided helpful reviews of this manuscript. The Geological Society of America kindly provided permission to reproduce figures 10.4 and 10.10.

References

Abbe, T.B., and D.R. Montgomery. 1996. Large woody debris jams, channel hydraulics and habitat formation in large rivers. Regulated Rivers: Research and Management 12: 201–221.

Allan, J.D. 1995. Stream ecology: Structure and function of running waters. Chapman and Hall, London.

Andrews, E.D. 1986. Downstream effects of Flaming Gorge Reservoir on the Green River, Colorado and Utah. Geological Society of America Bulletin 97: 1012–1023.

Baker, V.R. 1973. Paleohydrology and sedimentology of Lake Missoula flooding in eastern Washington. Geological Society of America Special Paper 144.

Baker, V.R. 1983. Late-Pleistocene fluvial systems. In, S.C. Porter, ed., Late-Quaternary environments of the United States, vol. 1, The Late Pleistocene. University of Minnesota Press, Minneapolis, p. 115–129.

Bally, A.W., C.R. Scotese, and M.I. Ross. 1989. North America; plate-tectonic setting and tectonic elements. In, A.W. Bally and A.R. Palmer, eds., The geology of North America—An overview. Geological Society of America, Boulder, Colorado, p. 1–15.

Barton, D.R., W.D. Taylor, and R.M. Biette. 1978. Dimensions of riparian buffer strips required to maintain trout habitat in southern Ontario streams. North American Journal of Fisheries Management 5: 364–378.

Blown, I., and M. Church. 1985. Catastrophic lake drainage within the Homathko River basin, British Columbia. Canadian Geotechnical Journal 22: 551–563.

Bovis, M.J. 1987. The interior mountains and plateaus. In, W.L. Graf, ed., Geomorphic systems of North America. Geological Society of America, Boulder, Colorado, p. 469–515.

Brinson, M.M., B.L. Swift, R.C. Plantico, and J.S. Barclay. 1981. Riparian ecosystems: Their ecology and status. U.S. Fish and Wildlife Service, FWS/OBS-81/17, 154 p.

Bull, W.B. 1997. Discontinuous ephemeral streams. Geomorphology 19: 227–276.

Butler, D.R., and G.P. Malanson. 1995. Sedimentation rates and patterns in beaver ponds in a mountain environment. Geomorphology 13: 255–269.

Carter, L.D., J.A. Heginbottom, and M. Wook. 1987. Arctic Lowlands. In, W.L. Graf, ed., Geomorphic systems of North America. Geological Society of America, Boulder, Colorado, p. 583–628.

Church, M. 1988. Floods in cold climates. In, V.R. Baker, R.C. Kochel, and P.C. Patton, eds., Flood geomorphology. John Wiley and Sons, New York, p. 205–229.

Collier, M., R.H. Webb, and J.C. Schmidt. 1996. Dams and rivers: Primer on the downstream effects of dams. U.S. Geological Survey Circular 1126, 94 p.

Comeaux, M.L. 1981. Arizona: A geography. Westview Press, Boulder, Colorado, 336 p.

Davis, M.B. 1983. Holocene vegetational history of the eastern United States. In, H.E. Wright, Jr., ed., Late-Quaternary environments of the United States, vol. 2, The Holocene. University of Minnesota Press, Minneapolis, p. 166–181.

Day, T.J. 1972. The channel geometry of mountain streams. In, H.O. Slaymaker and H.J. McPherson, eds., Mountain geomorphology: Geomorphological processes in the Canadian Cordillera. Tantalus Research, Vancouver, p. 141–149.

Driedger, C.L., and A.G. Fountain. 1989. Glacier outburst floods at Mount Rainier, Washington State, USA. Annals of Glaciology 13: 51–55.

Ely, L.L. 1997. Response of extreme floods in the southwestern United States to climatic variations in the late Holocene. Geomorphology 19: 175–201.

Ely, L.L., Y. Enzel, V.R. Baker, and D.R. Cayan, 1993. A 5000–year record of extreme floods and climate change in the southwestern US. Science 262: 410–412.

Eschner, T.R., R.F. Hadley, and K.D. Crowley. 1983. Hydrologic and morphologic changes in channels of the Platte River basin in Colorado, Wyoming, and Nebraska: A historical perspective. U.S. Geological Survey Professional Paper 1277–A, 39 p.

Fetherston, K.L., R.J. Naiman, and R.E. Bilby. 1995. Large woody debris, physical process, and riparian forest development in montane river networks of the Pacific Northwest. Geomorphology 13: 133–144.

Graf, W.L. 1978. Fluvial adjustments to the spread of tamarisk in the Colorado Plateau region. Geological Society of America Bulletin 89: 1491–1501.

Graf, W.L. 1996. Geomorphology and policy for restoration of impounded American rivers: What is 'natural'? In, B.L. Rhoads and C.E. Thorn, eds., The scientific nature of geomorphology. John Wiley and Sons, New York, p. 443–473.

Graf, W.L., R. Hereford, J. Laity, and R.A. Young. 1987. Colorado Plateau. In, W.L. Graf, ed., Geomorphic systems of North America. Geological Society of America, Boulder, Colorado, p. 259–302.

Harvey, M.D., R.A. Mussetter, and E.J. Wick. 1993. A physical process–biological response model for spawning habitat formation for the endangered Colorado squawfish. Rivers 4: 114–131.

Harvey, M.D., C.C. Watson, and S.A. Schumm. 1983. Channelized streams: An analog for the effects of urbanization. 1983 International Symposium on Urban Hydrology, Hydraulics and Sediment Control. University of Kentucky, p. 401–409.

Hayden, B.P. 1988. Flood climates. In, V.R. Baker, R.C. Kochel, and P.C. Patton, ed., Flood geomorphology. Wiley and Sons, New York, p. 13–26.

Hem, J.D., A. Demayo, and R.A. Smith. 1990. Hydrogeochemistry of rivers and lakes. In, M.G. Wolman and H.C. Riggs, eds., Surface water hydrology. Geological Society of America, Boulder, Colorado, p. 189–231.

Herschy, R.W. 1998. Rivers. In, R.W. Herschy and R.W. Fairbridge, eds., Encyclopedia of Hydrology and Water Resources. Kluwer Academic Publishers, Dordrecht, p. 571–583.

Hupp, C.R., and W.R. Osterkamp. 1996. Riparian vegetation and fluvial geomorphic processes. Geomorphology 14: 277–295.

James, L.A. 1989. Sustained storage and transport of hydraulic gold mining sediment in the Bear River, California. Annals of the Association of American Geographers 79: 570–592.

James, L.A. 1991. Incision and morphologic evolution of an alluvial channel recovering from hydraulic mining sediment. Geological Society of America Bulletin 103: 723–736.

Jarrett, R.D., and H.E. Malde. 1987. Paleodischarge of the late Pleistocene Bonneville Flood, Snake River, Idaho, computed from new evidence. Geological Society of America Bulletin 99: 127–134.

Kehew, A.E. 1993. Glacial-lake outburst erosion of the Grand Valley, Michigan, and impacts on glacial lakes in the Michigan Basin. Quaternary Research 39: 36–44.

Kehew, A.E., and M.L. Lord. 1987. Glacial-lake outbursts along the mid-continent margins of the Laurentide ice-sheet. In, L. Mayer and D. Nash, eds., Catastrophic flooding. Allen and Unwin, Boston, p. 95–120.

Keller, E.A., and F.G. Swanson. 1979. Effects of large organic debris on channel form and fluvial processes. Earth Surface Processes 4: 361–380.

Knighton, D. 1984. Fluvial forms and processes. Edward Arnold, London, 218 p.

Knox, J.C. 1983. Responses of river systems to Holocene climates. In, H.E. Wright, Jr., ed., Late-Quaternary environments of the United States, vol. 2, The Holocene. University of Minnesota Press, Minneapolis, p. 26–41.

Knox, J.C. 1985. Responses of floods to Holocene climatic change in the Upper Mississipi Valley. Quaternary Research 23: 287–300.

Langbein, W.B., and S.A. Schumm. 1958. Yield of sediment in relation to mean annual precipitation. Transactions, American Geophysical Union 39: 1076–1084.

Lecce, S.A. 1997. Spatial patterns of historical overbank sedimentation and floodplain evolution, Blue River, Wisconsin. Geomorphology 18: 265–277.

Leopold, L.B., and Maddock,T., 1953. The hydraulic geometry of stream channels and some physiographic implications. U.S. Geological Survey Professional Paper 252, 56p.

Ligon, F.K., W.E. Dietrich, and W.J. Trush. 1995. Downstream ecological effects of dams. BioScience 45: 183–192.

Madole, R.F., W.C. Bradley, D.S. Loewenherz, D.F. Ritter, N.W. Rutter, and C.E. Thorn. 1987. Rocky Mountains. In, W.L. Graf, ed., Geomorphic systems of North America. Geological Society of America, Boulder, Colorado, p. 211–257.

Malanson, G.P. 1993. Riparian landscapes. Cambridge University Press, Cambridge, 296 p.

McKenney, R., R.B. Jacobson, and R.C. Wertheimer. 1995. Woody vegetation and channel morphogenesis in low-gradient, gravel-bed streams in the Ozark Plateau, Missouri and Arkansas. Geomorphology 13: 175–198.

Meade, R.H., T.R. Yuzyk, and T.J. Day. 1990. Movement and storage of sediment in rivers of the United States and Canada. In, M.G. Wolman and H.C. Riggs, eds., Surface water hydrology. Geological Society of America, Boulder, Colorado, p. 255–280.

Mills, H.H., G.R. Brakenridge, R.B. Jacobson, W.L. Newell, M.J. Pavich, and J.S. Pomeroy. 1987. Appalachian mountains and plateaus. In, W.L. Graf, ed., Geomorphic systems of North America. Geological Society of America, Boulder, Colorado, p. 5–50.

Mitsch, W.J., and J.G. Gosselink. 1993. Wetlands. 2nd ed. Van Nostrand Reinhold, New York, 722 p.

Muhs, D.R., R.M. Thorson, J.J. Clague, W.H. Mathews, P.F. McDowell, and H.M. Kelsey. 1987. Pacific coast and mountain system. In, W.L. Graf, ed., Geomorphic systems of North America. Geological Society of America, Boulder, Colorado, p. 517–581.

Nadler, C.T., and S.A. Schumm. 1981. Metamorphosis of South Platte and Arkansas Rivers, eastern Colorado. Physical Geography 2: 95–115.

Naiman, R.J., C.A. Johnston, and J.C. Kelley. 1988. Alteration of North American streams by beaver. BioScience 38: 753–762.

O'Connor, J.E. 1993. Hydrology, hydraulics and geomorphology of the Bonneville Flood. Geological Society of America Special Paper 274, 83 p.

O'Connor, J.E., and V.R. Baker. 1992. Magnitudes and implications of peak discharges from glacial Lake Missoula. Geological Society of America Bulletin 104: 267–279.

Osterkamp, W.R., M.M. Fenton, T.C. Gustavson, R.F. Hadley, V.T. Holliday, R.B. Morrison, and T.J. Toy. 1987. Great Plains. In, W.L. Graf, ed., Geomorphic systems of North America. Geological Society of America Bulletin, Boulder, Colorado, p. 163–210.

Petts, G. 1990. Forested river corridors: A lost resource. In, D. Cosgrove and G. Petts, eds., Water, engineering, and landscape. Belhaven Press, London, p. 12–34.

Porter, D.A., and M.J. Guccione. 1994. Deglacial flood origin of the Charleston alluvial fan, lower Mississippi alluvial valley. Quaternary Research 41: 278–284.

Rathburn, S.L. 1993. Pleistocene cataclysmic flooding along the Big Lost River, east central Idaho. Geomorphology 8: 305–319.

Riggs, H.C., and K.D. Harvey. 1990. Temporal and spatial variability of streamflow. In, M.G. Wolman and H.C. Riggs, ed., Surface water hydrology. Geological Society of America, Boulder, Colorado, p. 81–96.

Riggs, H.C., and M.G. Wolman. 1990. Introduction. In, M.G. Wolman and H.C. Riggs, eds., Surface water hydrology. Geological Society of America, Boulder, Colorado, p. 1–9.

Saucier, R.T. 1974. Quaternary geology of the lower Mississippi Valley. Arkansas Archeological Survey Research Series 6.

Schumm, S.A. 1977. The fluvial system. John Wiley and Sons, New York, 338 p.

Schumm, S.A. 1981. Evolution and response of the fluvial system: Sedimentologic implications. Society of Economic Paleontologists and Mineralogists Special Publication 31, p. 19–29.

Schumm, S.A., and G.R. Brakenridge. 1987. River responses. In, W.F. Ruddiman and H.E. Wright, Jr., eds., North America and adjacent oceans during the last deglaciation. Geological Society of America, Boulder, Colorado, p. 221–240.

Schumm, S.A., and R.S. Parker. 1973. Implications of complex response of drainage systems for Quaternary alluvial stratigraphy. Nature 243: 99–100.

Shaw, J., and R. Gilbert. 1990. Evidence for large-scale subglacial meltwater flood events in southern Ontario and northern New York State. Geology 18: 1169–1172.

Shilts, W.W., J.M. Aylsworth, C.A. Kaszycki, and R.A. Klassen. 1987. Canadian Shield. In, W.L. Graf, ed., Geomorphic systems of North America. Geological Society of America, Boulder, Colorado, p. 119–161.

Sturm, M., J. Beget, and C. Benson. 1987. Observations of jökulhlaups from ice-dammed Strandline Lake, Alaska: Implications for paleohydrology. In, L. Mayer and D. Nash, eds., Catastrophic flooding. Allen and Unwin, Boston, p. 79–94.

Swift, B.L. 1984. Status of riparian ecosystems in the United States. Water Resources Bulletin 20: 223–228.

Teller, J.T. 1987. Proglacial lakes and the southern margin of the Laurentide Ice Sheet. In, W.F. Ruddiman and H.E. Wright, Jr., eds., North America and adjacent

oceans during the last deglaciation. Geological Society of America, Boulder, Colorado, p. 39–69.
Teller, J.T. 1990. Volume and routing of late-glacial runoff from the southern Laurentide Ice Sheet. Quaternary Research 34: 12–23.
Thomas, B. 1978. American rivers: A natural history. W.W. Norton and Company, New York.
Thompson, D.M. 1995. The effects of large organic debris on sediment processes and stream morphology in Vermont. Geomorphology 11: 235–244.
Tobin, G.A. 1995. The levee love affair: A stormy relationship. Water Resources Bulletin 31: 359–367.
Vannote, R.L., G.W. Minshall, K.W. Cummins, J.R. Sedell, and C.E. Cushing. 1980. The river continuum concept. Canadian Journal of Fisheries and Aquatic Sciences 37: 130–137.
Wagener, S.M., and J.D. LaPerriere. 1985. Effects of placer mining on the invertebrate communities of interior Alaska streams. Freshwater Invertebrate Biology 4: 208–214.
Walker, H.J., and J.M. Coleman. 1987. Atlantic and Gulf coastal province. In, W.L. Graf, ed., Geomorphic systems of North America. Geological Society of America, Boulder, Colorado, p. 51–110.
Walling, D.E. 1987. Hydrological processes. In, K.J. Gregory and D.E. Walling, eds., Human activity and environmental processes. Wiley and Sons, Chichester, p. 53–85.
Webb, R.H., and J.L. Betancourt. 1992. Climatic variability and flood frequency of the Santa Cruz River, Pima County, Arizona. U.S. Geological Survey Water-Supply Paper 2379, 40 p.
Williams, G.P. 1978. The case of the shrinking channels—The North Platte and Platte Rivers in Nebraska. U.S. Geological Survey Circular 781, 48 p.
Wohl, E. E. 2000. Mountain rivers. American Geophysical Union Press, Washington, D.C., 320 p.
Wohl, E. E. 2001. Virtual rivers: Lessons from the mountain rivers of the Colorado Front Range. Yale University Press, New Haven, 210 p.
Wohl, E.E., S. Madsen, and L. MacDonald. 1997. Characteristics of log and clast bed-steps in step-pool streams of northwestern Montana, USA. Geomorphology 20: 1–10.
Wohl, E.E., R. McConnell, J. Skinner, and R. Stenzel. 1998. Inheriting our past: River sediment sources and sediment hazards in Colorado. Colorado Water Resources Research Institute, Water in the Balance, no. 7, 26 p.
Wolman, M.G. 1967. A cycle of sedimentation and erosion in urban river channels. Geografiska Annaler 49A: 385–395.
Wolman, M.G., M. Church, R. Newbury, M. Lapointe, M. Frenette, E.D. Andrews, T.E. Lisle, J.P. Buchanan, S.A. Schumm, and B.R. Winkley. 1990. The riverscape. In, M.G. Wolman and H.C. Riggs, eds., Surface water hydrology. Geological Society of America, Boulder, Colorado, p. 281–328.

11

Plant and Animal Ecology

Albert J. Parker
Kathleen C. Parker
Thomas R. Vale

Ecology is the study of interactions of organisms with the physical and biological elements of their environments. The breadth of the North American continent and the diversity of the biota that inhabits it frustrate efforts to summarize these patterns of interaction in simple or tidy ways. In this chapter we attempt to synthesize the ecological factors that structure the biota of North America while acknowledging at the outset that an even and comprehensive coverage of the vast array of interesting and important interactions that shape this structure must remain elusive. Our purpose here is to address the principal processes that influence the geographic heterogeneity of plant cover and animal life in North America. The scope is generally restricted to patterns evident from the landscape scale to the continental scale. Overall, then, the intent is to explore recent ecological themes and trends that are deemed important in understanding spatial patterns of biotic response to the physical environment as it varies across the face of the continent.

Four broad categories of influence act over time to shape geographic variation in the North American biota: genetic constraints and evolutionary history, physical environmental gradients, biotic interactions, and environmental history (which includes disturbance regimes, dispersal dynamics, unique events, and chance). The role of human action in shaping the North American biota is well documented (Marsh, [1864] 1965, Goudie, 1981). Whether human influence should be treated as separate from these four—or subsumed within one or more of them—is a debate best left to environmental philosophers; some elements of human influence will be considered in the discussion of environmental history.

11.1 Genetic Traits and Evolutionary Heritage

Genes provide an inherited set of encoded instructions (collectively, a genotype) that impose limits on the range of expression of form and function (or phenotype) of individuals. To the extent that genetic material dictates the manner in which organisms acquire the energy and material necessary to survive, it imposes limits on the range of physical environments in which these organisms might be found.

For a population, the gene pool will vary over time as natural selection, genetic drift, gene flow, and other evolutionary forces modify the frequency of different genes. Such changes in gene frequency in a population generally accrue through the operation of demographic processes:

birth, death, and migration or disperal. Although there remains considerable debate among geneticists as to the relative importance of the various evolutionary forces that sponsor genetic change in populations, natural selection (differential inheritance of genes that encode phenotypic traits that are favored in the context of the ambient environment of individual organisms) traditionally has been viewed as crucial to understanding and interpreting organic evolution. Natural selection, acting in concert with sexual recombination of genetic material and sporadic mutations, permits ongoing phenotypic adjustment (adaptation) to environmental conditions that change over space and time.

Genetic diversity reflects the amount of variation preserved within the gene pool of a population. At the landscape scale, conservation biologists have identified genetic diversity as a crucial resource, insofar as it represents the raw material on which natural selection works to winnow and sift the traits of a population over time. Those populations that possess high genetic diversity are more likely to have some members persist in the face of inhospitable change, whereas populations of narrower genetic stock (typically, although not restricted to, populations of few individuals) are less likely to persist in similar circumstances. This has critical import for the maintenance of domesticated crops and livestock, as well as for undomesticated species in the advent of increasing landscape fragmentation and global climatic change. Hence, prominent thrusts of conservation biology include documenting the effective population size necessary to avoid extinction (Soulé, 1987) and examining patterns of genetic structuring and gene flow among members of both sessile and mobile populations (Hedrick, 1995, Godt et al., 1996).

At the continental scale for North America, evolutionary opportunities during the Cenozoic have varied geographically, with differing degrees of historical isolation prompting regional differentiation in the flora and fauna (Ricklefs, 1987). The floristic evolutionary roots of the continent are traditionally linked with three centers of action: Holarctic taxa, with affinities that span northern temperate and polar environments of North America (Nearctic) and Eurasia (Palearctic); Neotropical taxa, with affinities to humid environments of Central and South America; and Madrean taxa, with affinities to arid and semiarid settings of Mexico and the American Southwest. Faunistic patterns across North America vary substantially among taxonomic groups, but they generally emphasize varying degrees of contribution from Nearctic and Holarctic taxa (species with North American and circumboreal distributional affinities) and Neotropical taxa (species with Central and South American roots). The interaction between Nearctic and Holarctic mammals and Neotropical mammals poignantly illustrates the role of historical isolation and evolutionary opportunity in structuring modern animal assemblages. With the erratic emergence of the isthmus of Panama as an avenue of terrestrial migration between the Americas over the past 5–10 million years (see chapter 1), the more evolutionarily primitive fauna of South America has minimally penetrated North America (opossums, *Didelphis virginiana*, and armadillos, *Dasypus novemcinctus*, are exceptions), whereas more advanced placental mammals of North American stock have had great success in diffusing across South America (Marshall, 1988). Despite the prevalence of seasonal migration among many temperate bird species, the avifaunas of the Americas also show contrasts that reflect divergent evolutionary histories in the Cenozoic.

Lest the limitations imposed by genetic stock and evolutionary history on the biota of North America be overstated, this section closes by briefly considering convergent evolution, the physiognomic convergence of vegetation in similar environments, such as Mediterranean-type climates, on different continents with divergent evolutionary stock. Convergent evolution provides powerful support for the role of the environment in shaping patterns of regional differentiation of plant and animal forms (Cody and Mooney, 1978), especially as it underscores the importance of selective constraints on the range of possible forms and functions that permit sufficient energy conversion and material capture for survival in stressful settings (Mooney and Dunn, 1970).

11.2 Physical Gradients of the Environment

Ecologists often conceptualize variation in vegetation as a response to physical gradients dictated by site or habitat factors. Prominent among these factors are climate, soil, and landform. In turn, animal assemblages can be viewed as varying in response to local or regional contrasts in habitat factors and associated vegetation structure. In the past several decades, most ecologists have treated the physical gradients of energy, water, and mineral nutrients as varying continuously in nature; furthermore, they have embraced the notion that each species possesses a unique set of tolerances to these physical gradients. The result has been acceptance of two related concepts: individualistic species response and continuous variation in composition of species assemblages along prominent physical gradients (Gleason, 1926, Whittaker, 1956; Curtis, 1959; Chapin and Shaver, 1985, but see Westman, 1983). This attention to physical gradients is especially attractive to biogeography because such environmental variation is firmly rooted in the traditions of physical geography, including continental-scale variation in air mass dominance patterns (Bryson and Hare, 1974) or water balance (Mather and Yoshioka, 1968), daily or weekly variation in latitudinal temperature gradients and the associated configuration of the polar front jet stream (Harman, 1991), regional patterns of soil acidity and base saturation along effective moisture gradients (Sowell, 1985), local variation in water-table depth and soil-drainage class along gradients of landscape position (Buol et al., 1989), or microclimatic gradients of energy and water balance associated with topographic exposure, aspect, and prevailing wind field (Hack and Goodlett, 1960).

Whether physical gradients and associated vegetation and faunal patterns are envisioned as continuously varying across diffuse boundaries or abruptly discontinuous along discrete boundaries is largely a matter of scale. What appear to be sharp boundaries when viewed from a broader context, for example, the prairie-forest ecotone of the Midwest (Weaver, 1980) and the tundra-forest ecotone of the Canadian Arctic (Elliott-Fisk, 1983), often transform into a fuzzy mosaic when scrutinized at more local scales. Ecologists and biogeographers working at regional and landscape scales have provided numerous empirical studies of ecotonal vegetation composition and structure. The sharpness of these boundaries varies, but is often more abrupt for ecotones imposed by edaphic constraints. Examples of ecotonal studies include the flora of the temperate and boreal tension zone in Wisconsin (Griffin, 1997); patterns of elevation zonation in mountains of western North America (Peet, 1981; A. Parker, 1994); substrate-related discontinuities in vegetation on glacial drift (lacustrine sediments vs. till vs. outwash) in the Midwest (Pregitzer et al., 1983) and on serpentine soils of the Pacific Rim (Whittaker, 1954); and bajada vegetation in relation to depositional history and pedogenic development on alluvial fan surfaces in the American Southwest and northern Mexico (McAuliffe, 1994; K. Parker, 1995).

Different vegetation assemblages offer different habitat structures and food supplies for animals. The increase in the diversity of bird species with increasing vertical and horizontal complexity of vegetation along a gradient from open grassland and tundra to closed forest is well documented (MacArthur and MacArthur, 1961; Vale et al., 1982). Similar patterns are also evident for other vertebrates (Vale et al., 1989). The combination of greater architectural complexity (which enhances the variety of foraging substrates for animals) and greater primary productivity (which generally enriches the diversity of dietary options) increases the opportunities for resource partitioning among animal species, with regard to both foraging behaviors and nesting preferences. Patterns of feeding and nesting guild occurrence, which directly relate the energy and material needs of animals to vegetation structure and associated habitat features, provide a valuable tool for linking animals with the landscape (Vale et al., 1982; K. Parker, 1987).

Although climate, soil, landform, and vegetation structure constitute the basic site or habitat factors, individual organisms do not often respond directly to these larger landscape elements. Rather, organisms respond to resource levels or physical conditions of the environment that directly impinge on their energy supply, water balance, or biogeochemistry. For clarity, resources are commodities that may be consumed by an individual in the sense that use of a resource by an individual preempts its use by others. Consequently, resources may become the object of competitive exploitation or interference within an animal or plant assemblage. By contrast, physical conditions of the environment are not discretely consumable. For plants, the most prominent physical resources are light, nutrients, and water. For animals, the range of potentially important physical resources is broader, but generally includes food, water, feeding and nesting substrates, and mates. For all organisms, temperature (and affiliated heat energy) is the most obvious example of a condition that is imposed on organisms but cannot be preemptively consumed by an individual.

Direct links between species occurrence and environmental resources or conditions are often more apparent for plants than animals. Fundamentally, this tighter link among plants is related to differences in energy acquisition and geographic mobility. Autotrophic (green) plants convert sunlight to metabolic energy by photosynthesis; they are sessile. Therefore, they must accommodate the environmental vagaries of a specific setting. As a consequence of their position at a higher trophic level in food webs, animals are less directly tied to sunlight for their energy supply. Instead, animals must harvest a variety of foods to provide metabolic energy; this mandates mobility to facilitate search and capture of prey (except in those rare circumstances where food is predictably delivered to the animal, as occurs among marine invertebrates of the intertidal zone). Mobility allows animals to move (within the limits of size, locomotion skill, and life stage) to more amenable environments during stressful periods, either on a broad scale as animals migrate or on a local scale as they use microclimatically different parts of their habitat. Although mobility among animals buffers them somewhat from the physical exigencies of a particular site, environmental conditions, nevertheless, impose limits on animals by stressing their metabolic demands. Ectotherms, animals that rely primarily on heat sources other than metabolic energy for thermoregulation, often exhibit behavioral traits that modulate internal temperature changes by seeking exposed or sheltered sites depending on ambient temperature (e.g., regal horned lizards, *Phrynosoma solare*) or by remaining inactive in dry or cold seasons (e.g., Great Basin spadefood toad, *Scaphiopus intermontanus*). This strong reliance on environmental heat excludes some ectotherms from extremely cold environments, as evidenced by the paucity of amphibians in the arctic. The use of metabolic heat for thermoregulation by endotherms permits them to occupy some extreme environments; the narrow thermoregulatory control in endotherms often decouples activity patterns from surrounding thermal conditions, except during periods of particularly strong heat or cold stress. This flexibility comes at a cost, however, as endotherms have considerably greater energy demands to maintain thermal homeostasis.

Environmental resources and conditions vary in response to processes operating at multiple scales of influence. To highlight these scaling issues in relation to the interaction of organisms with their environment, we present two simplified examples: the water relations of ponderosa pine (*Pinus ponderosa*) on the west slope of the Sierra Nevada of California (fig. 11.1A) and temperature effects on scarlet tanagers (*Piranga olivacea*) that breed in the southern Appalachians (and throughout the eastern deciduous forest) but overwinter in northern South America (fig. 11.1B).

At the continental scale, the delivery of precipitation to the Sierra Nevada is controlled by the seasonal dynamics of atmospheric circulation, with much snowfall and rainfall occurring with the strengthening of the circumpolar vortex and penetration of Pacific cyclones along the polar front in winter (see chapter 5). Dry summers are imposed by the northward displacement and anchoring of the North Pacific subtropical anticyclone off the California coast, as the polar front recedes to higher latitudes. This causes a pronounced seasonal asynchrony of water supply and demand in the ponderosa pine forest, which is partially offset by the delay in runoff associated with ablation of the snowpack into the spring and early summer months at higher elevations in the Sierra. At the regional scale, the processes that cause precipitation in winter are accentuated by orographic lifting on the windward western slope of the Sierra, but are muted by subsidence on the eastern slopes. This engenders sharp vegetation contrasts across the Sierran crest; open sagebrush (*Artemisia* spp.) flats occur in a midlatitude desert scrub on leeward slopes at elevations comparable to the ponderosa pine forest on the windward slope (fig. 11.2).

At the landscape scale, water balance varies in association with soil water-holding capacity, soil drainage patterns, water-table dynamics, and topographic effects (slope aspect, steepness, and topographic position). Ponderosa pine is broadly distributed in the uplands of western North America, from British Columbia to northern Baja California. It is relatively drought tolerant (when compared with other large conifers) and thrives on sandy, well-drained soils and exposed slopes. Where the water table approaches the surface or soil drainage is impeded, other taxa (such as alder, *Alnus* spp., and willow, *Salix* spp.,) replace ponderosa pine. Similarly, on sheltered, cooler slopes, water demand is reduced and more drought-sensitive species, such as white fir (*Abies concolor*), incense-cedar (*Calocedrus decurrens*), or Douglas-fir (*Pseudotsuga menziesii*) assume canopy dominance in Sierran forests.

Locally, water balance varies with biotically mediated influences associated with the canopy and root zone. Water supply to individuals within the ponderosa pine forest is modified by canopy interception of precipitation, stemflow, variable canopy light penetration (which fosters variation in the rate of snowpack melt), and competition with surrounding plants for capillary water in the soil. At all scales, then, water relations are critical to understanding relationships between ponderosa pine and environmental setting, although the processes involved in that understanding differ at each scale.

Scarlet tanagers are canopy-feeding insectivores whose distribution and behavioral patterns illustrate a range of responses to scale-related variation in thermal regime. These responses are more often a product of the effects of temperature on the insects that make up the diet of tanagers rather than direct physiological effects on the birds themselves. At the continental scale, scarlet tanagers migrate between breeding and wintering grounds, ultimately in response to seasonal variation in solar radiation caused by changing sun angles and daylengths linked to Earth's orbital mechanics. Proximate stimuli for migrational movements are debated, but probably include some combination of sensitivity to photoperiodic change, meteorologic triggers, changes in insect productivity, and hormonal cues that promote or suppress breeding behavior. The long, warm days of the temperate spring and summer boost primary and secondary productivity in the southern Appala-

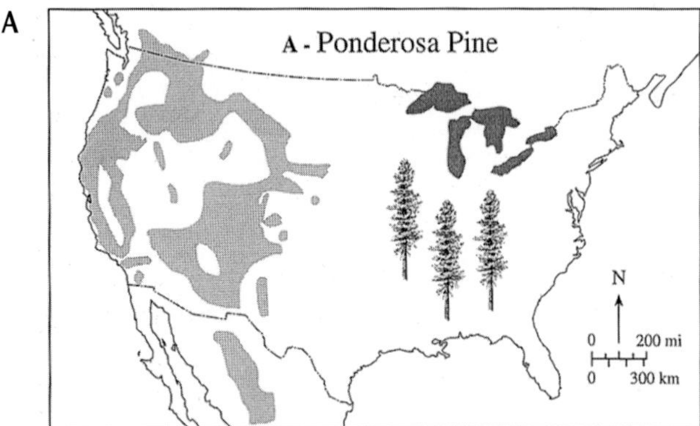

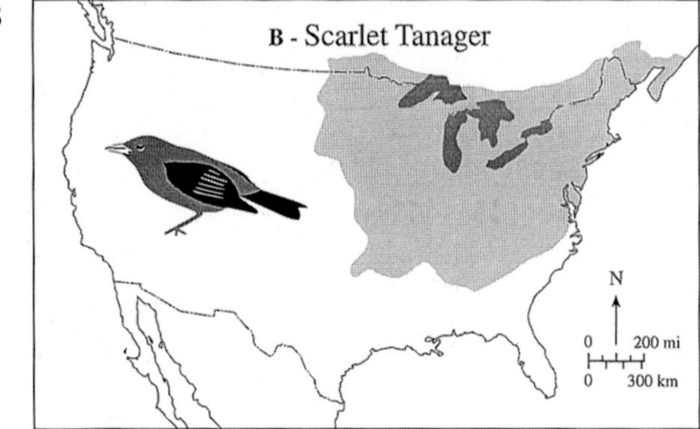

Figure 11.1 Generalized range maps and line drawings of (A) ponderosa pine and (B) scarlet tanager for North America.

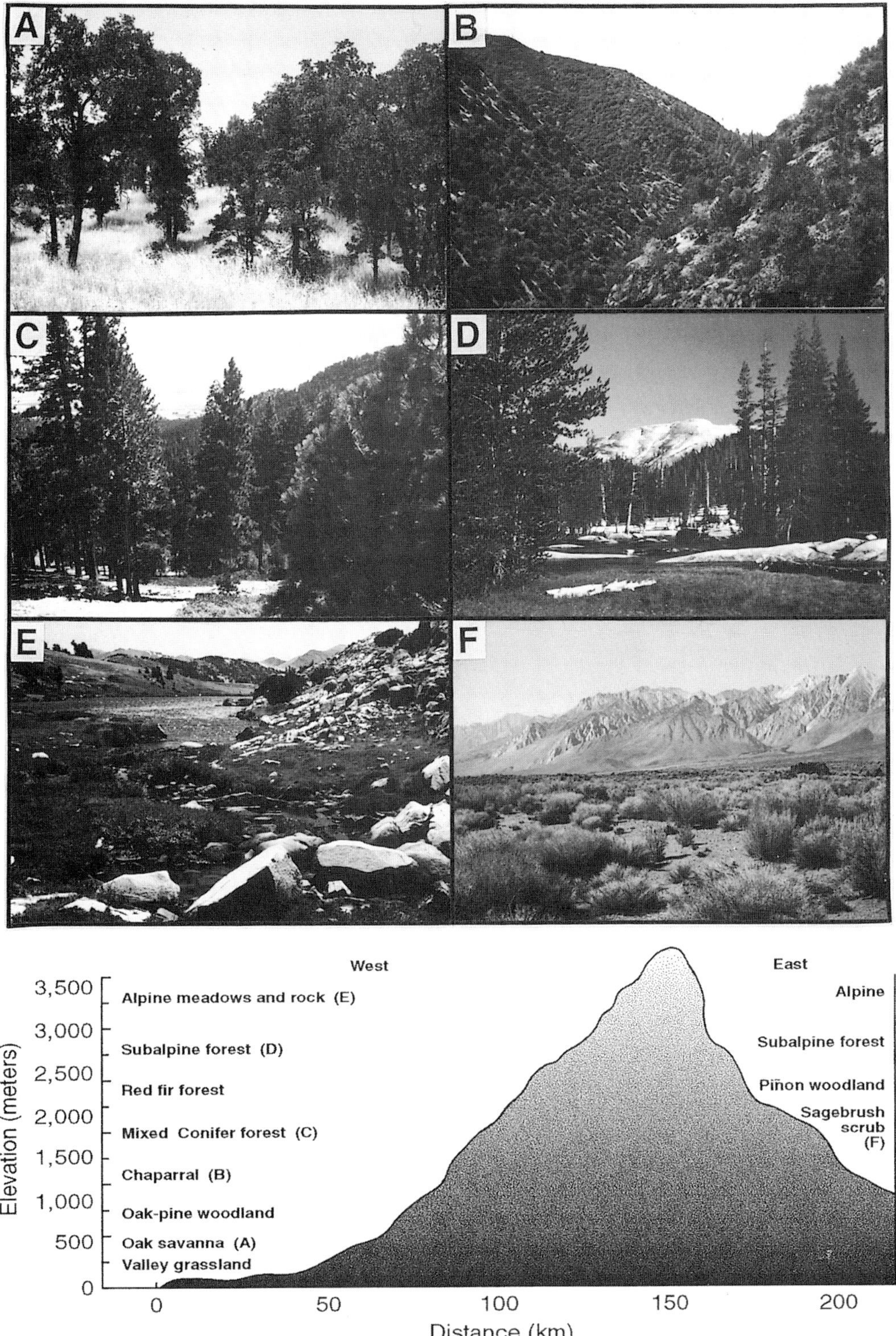

Figure 11.2 Vegetation transect of the Sierra Nevada. All photos taken by Kathleen C. Parker.

chians and provide the energetic reserve necessary to support reproductive activities; the birds, in response, fly in from lower latitudes.

Regional- and landscape-scale variation in temperature in the southern Appalachians reflects the interaction of altitudinal lapse rates and topographic influences on radiation budgets, and this variation influences what parts of the environment the birds occupy. Declining temperature and length of growing season with increased altitude limit scarlet tanagers to lower and midelevation forests of the Appalachians. Aspect (south- vs. north-facing slopes) and topographic position (ridgetop to valley bottom) modify amounts of incoming solar radiation and reinforce preferred advection fields (e.g., valley breezes and nocturnal cold air drainage). These, in turn, alter the amplitude and timing of daily and seasonal temperature cycles. These regional- and landscape-scale temperature influences affect the distribution of the tanager's preferred habitats or dietary options and thus the birds themselves. Appropriately, then, tanagers generally occupy broadleaf deciduous forests with scattered gaps and openings and diverse insect faunas typical of many lower and midslope settings of the southern Appalachians.

Local conditions, too, influence these birds. Horizontal variation in temperature within a forest ensues from heterogeneity in canopy architecture and attendant differences in light extinction. However, vertical contrasts in daily temperature fluctuations commonly overshadow horizontal contrasts, with the greatest diurnal fluctuation at the upper canopy surface in a forest (or at the ground surface in large forest openings) and attenuated ranges beneath the forest canopy or below ground. Such local-scale variation might influence nest location for scarlet tanagers, which prefer sheltered tree branches for nest sites. Tanagers, like ponderosa pine, then, respond to environmental conditions that vary at different scales.

The aggregate effect of these multiple scales of influence on water supply and temperature regime is a cascading sequence of explanation or apparent control of biotic distributions. At a continental or regional scale, macroclimatic and physiographic influences constrain the range of ponderosa pine and the distribution and seasonal migration of scarlet tanagers. At a landscape or local scale, topographic, edaphic, and biotic variability fine-tune the habitats occupied by ponderosa pine and the activity patterns of scarlet tanagers.

11.3 Biotic Interactions

Genetic constraints interact with spatial variation in resources and conditions of the physical environment to impose limits on the potential habitat range of any given plant or animal species. These physiologically defined limits are further modified by biotic interactions, as well as by the ebb and flow of historical events at particular places. Ecologists categorize biotic interactions, interactions between organisms, by specifying the influence of those interactions on the well-being of the two participants. (Well-being might be judged, depending on the context, by net carbon balance, production of offspring, or survival.) Competition, predation–parasitism, and symbiosis are the most prominent forms of biotic interaction. Competition between organisms generates potentially negative effects for both participants, although dominance by trees can cause profound asymmetries in competition for light with members of the forest understory and dominance by individual animals can cause asymmetries in competition for food or mates. Predation–parasitism denotes either consumption of prey by predators or channeling of energy reserves from host to parasite; in either event, this has a positive effect on the consumer and a negative effect on the provider. Symbiosis includes a variety of mutualistic interactions that benefit both participants. Collectively, biotic interactions may play important roles in structuring local-scale, or occasionally landscape-scale, patterns of distribution or activity among organisms, although their effects generally diminish at broader scales. Typically, biotic interactions are more prominent in shaping animal than plant assemblages.

The role of competition in structuring biotic communities is often ambiguous and has been widely debated among ecologists (Connell, 1983; Schoener, 1983). Some have viewed nature as highly organized, with communities of plants and animals largely structured by competition among constituent species. Such views embrace notions of interspecific competition acting through natural selection to fine-tune the niche—the subset of the available resources in a habitat used by a species. This implies that populations are at their carrying capacities and that the community is saturated (all resources are partitioned among the constituent species, each niche is occupied) and in equilibrium (the physical environment is sufficiently stable, or varies with sufficient predictability, that the species composition of the assemblage varies narrowly around a mean). Examples of competitively structured niche differentiation are more common in animal studies; for instance, MacArthur (1958) attributed spatial segregation among wood warblers (Parulidae) during foraging (i.e., different vertical positions and substrates within the canopy of a tree) to interspecific competition and resource partitioning. Others view nature as less well structured, more loosely organized. This view reflects an increasing awareness that environmental fluctuations and progressive change frequently disrupt species assemblages, reducing the probability that they ever approach a stable composition, one characterized by saturation or equilibrium. These nonequilibrium views have gained increasing ascendancy in the past several decades, as the pervasiveness of episodic disturbance and secular climatic change have become more evident (Wiens, 1984).

Competition between plants at the plot level remains an inarguable influence on the growth, spatial pattern, and structure of vegetation assemblages, just as competition among animals is crucial in structuring such behaviors as foraging and territoriality. Indeed, some have argued (Tilman, 1985) that succession, or vegetation change through time following a disturbance, might be viewed as a competitively structured phenomenon where the resources of interest are light (high in open fields, low at ground-level in a forest) and nutrients (presumably low, or inaccessible to plants, in new fields; more available in older plots). This simplistic view has been criticized by Huston and Smith (1987) because it fails to focus on competitive mechanisms at the individual level. In any event, both Tilman (at the population level) and Huston and Smith (at the individual level) argue for competitive interactions in structuring successional processes. This linkage of competition, succession, and life-history traits of plants enjoys considerable precedent in ecology, evident in such seminal concepts as r- versus K-selection (tradeoffs of growth rate and longevity; seed size, mobility, and dormancy potential, see Pianka, 1970, for full explication) or Grime's (1977) triad of plant strategies (stress tolerant, competition tolerant, or ruderal). These organizational frames lay the groundwork for later discussion of vegetation dynamics.

Predation and parasitism structure food webs; many ecologists view energy and material flow through foraging networks (or trophic levels) as the fundamental organizational scheme of ecosystems. Traditional food webs envision carbon flow from green plants, which convert radiant energy to stored chemical energy (carbohydrates) by photosynthesis, through herbivores and various levels of carnivores. This carbon flow from lower to higher trophic levels is continuously recycled by decomposers, which break down dead organic material, releasing carbon dioxide by respiration of carbohydrates and liberating nutrients to the soil that were previously assimilated in the structure of organic tissue. Decomposer food chains have received increasing attention among ecologists because of their influence on processes operating at a wide range of spatial scales, from a possible role in modulating global carbon balance to effects on soil-rhizosphere structure and local ecosystem productivity. One microscale, decomposer-based process that has profound implications for soil fertility over large regions is bioturbation caused by ants, termites, earthworms, and other invertebrates. These soil-forming agents have long been recognized as important in maintaining soil productivity (Buol et al., 1989, see also chapter 9).

Like competition, the role of predation in controlling the structure of a community has been vigorously debated. Some have perceived highly structured food webs whose patterns of energy flow are dominated by one or more keystone predator species, species that are crucial to maintaining the stability of the food web or diversity of the ecosystem's biota (Mills et al., 1993). A commonly cited example is that of the starfish's dominance of marine intertidal zones along parts of the North Pacific coast (Paine, 1966). Others have envisioned more flexible food webs in which dietary switching is common as foraging opportunities change with changes in prey item density and conspicuousness or, over longer time periods, with changes in vegetation structure (Pimm, 1991). In such communities, removal of a predator may cause unpredictable effects that vary over time.

Grazing and browsing of plants by animals is an ecologically important form of predation. The consequences of changing ungulate foraging pressure on vegetation structure and composition are a common research focus in North America. Examples include the expansion of browsing by ungulates, such as that by white-tailed deer (*Odocoileus virginianus*) in the east (Alverson et al., 1989); the introduction of feral exotics such as the wild boar (*Sus scrofa*) in the southern Appalachians (Bratton, 1974); and the introduction of domesticated livestock, such as the effects of cattle on range conditions in the desert Southwest (Hastings and Turner, 1965, Bahre, 1991) or the effects of overstocking followed by removal of sheep, which may have opened turf and triggered invasion of woody species into montane meadows of the West (Vale, 1981; Hansen et al., 1995).

Symbiosis has gained increasing attention as a significant agent in understanding plant and animal assemblages. It would be difficult to overstate the importance of mycorrhizal fungi on plant nutrition (especially phosphorus uptake), productivity, establishment, and survival. Many vascular plant species cannot successfully germinate and grow without innoculating the soil medium with obligate mycorrhizal associates (Richardson and Bond, 1991). Another significant symbiotic interaction operates between nitrogen-fixing bacteria and legumes. These bacteria, clustered on root nodules of many leguminous species, enhance soil fertility by their unusual ability to convert atmospheric molecular nitrogen into nitrates that can be assimilated by plants. This is, of course, the basis for many agricultural systems that advocate crop rotation, with legumes serving as a vital cover crop in facilitating soil replenishment. As a final acknowledgment of the fundamental significance of symbiosis in nature, the ubiquitous linkage of animal pollinators with flowering plants is noteworthy. Flying animals, such as birds, bats, and insects, serve as directed vectors of pollen flow that facilitate out-crossing and sexual reproduction in many angiosperms.

Collectively, biotic interactions add a fascinating layer of complexity to animal and plant assemblages. The relative contribution of physical environmental gradients versus biotic interactions to ecosystem structure varies over space and time. The degree to which importance is attributed to one factor over the other is largely a reflection of the spatial and temporal frames of reference. A local focus in the short term might favor biotic forces in structuring ecosystems. As larger areas and longer time frames are considered, the signature of variability in the physical environment becomes increasingly prominent and partially obscures the finer details imposed by biotic interactions.

11.4 Succession, Disturbance, and Environmental History

Plants and animals respond to environmental change across a broad range of time scales. From a gust of wind that prompts a momentary increase in transpirative vapor flux to the panoply of evolutionary forces that drive genetic alteration of a species gene pool over millions of years, life is continuously responding to a dynamic physical world. For convenience, a distinction is made between those responses that result in gross genetic changes (events that have macroevolutionary consequences) and those that do not. The realm of evolutionary response to change over longer time periods and its genetic and biogeographic consequences for the biota were addressed previously. We now turn our focus to nonevolutionary (or, more correctly, less overtly evolutionary) changes that ensue from ecological interactions that generally occur over shorter time spans. These include the effects of historical actions and events that shape the configuration of the modern landscape and its biota. Elements of environmental history that are of broad ecological significance include successional response of plants and animals to disturbance events (which incorporates the vagaries of seed source and dispersal ecology in plants, migration and colonization in animals), response to secular climatic change (from decadal to multimillennial scales), and the role of unique events and the element of chance in shaping geographic patterns of the biotic landscape.

11.4.1 Succession and Vegetation Dynamics

Modern perspectives on vegetation dynamics have evolved over the past century from single pathway, deterministic successional processes that end in an idealized climax community (Clements, 1916), toward a multiple pathway, state-and-transition model of dynamic response to disturbance and other events that acknowledge the prospect for multiple stable states of a biotic assemblage (Cattelino et al., 1979; Vale, 1988). In part, this shift in perspective parallels developments such as concepts of individualistic species response to environmental variables and continuous variation in composition along environmental gradients (here, time might be perceived as one such gradient). These ideas underscore the role of life-history traits (dispersal and migration capacity, shade tolerance and resource needs, growth rates, age of maturation, and longevity) of plants and animals in mediating participation in vegetation dynamics (Noble and Slatyer, 1980; Sauer, 1988). Another contribution to this shift in viewpoint regarding succession emanates from an awareness of the differential survival of species in different habitats, so that successional possibilities are constrained by ambient environmental resources and conditions (Peet and Loucks, 1977). A third factor that has supported this shift toward less deterministic views of vegetation dynamics is increasing cognizance of disturbance as part of nature—a realization that few places are sheltered from the potentially disruptive effects of disturbance for more than a few centuries and that all plants and animals eventually die and provide colonization opportunities for other organisms (Watt, 1947). This less deterministic view of vegetation dynamics is not one that advocates chaos or unknowing. Rather, it allows for a variety of possible pathways and outcomes, as modulated by the intersection of physical environmental possibilities, the habits of potential plant and animal colonists, and the timing and extent of disturbance events in the site's past (A. Parker, 1993).

11.4.2 Disturbance Agents

Disturbance regimes incorporate one or more disturbance agents in a repetitive, episodic process of mortality followed by recolonization. Disturbance regimes may be characterized by such traits as dominant agent, typical intensity or severity of impact, typical recurrence interval, and typical scale or areal extent (White, 1979). This section focuses on fire, one of the most obvious and widespread disturbance agents in North America, to illustrate spatial variability in disturbance-regime descriptors. The ecological effects of other disturbance agents are then briefly considered.

Wildfires can be classified into three types: crown (or canopy) fires, surface fires, and ground fires. Crown fires burn through the overstory of a forest or shrubland, generating locally high burn temperatures and resulting in relatively high mortality among canopy individuals, at least in areas of high burn intensity (fig. 11.3A). Surface fires consume dry herbaceous and small, woody material immediately above the soil surface. They can move swiftly, but are typically cooler, low-intensity fires. Ground fires consume near-surface peat and litter in organic soils, often in wetlands that have been subject to water-table drawdown by prolonged drought. These fires can smolder for weeks or months, flare up in windy, dry periods, and eliminate the soil seed bank.

Many conifer-dominated forests of the boreal latitudes and western mountains, as well as Mediterranean-type chaparral shrublands, experience episodic crown fires. The burn severity in these events is often highly heterogeneous, even within the perimeter of a single large fire, as shown by the Yellowstone fires of summer 1988 (Anderson and Romme, 1991). Most crown fires are small in areal extent, but with an appropriate fuel mosaic (horizontally continuous fuels to allow spread, vertical connectivity of fuels from understory to canopy to sustain ignition) and weather conditions (prolonged warm, dry, windy periods), individual crown fires, or clusters of simultaneous crown fires, can consume large areas in forest and chaparral shrubland vegetation mosaics.

Recurrence intervals in crown fire ecosystems are influenced by climate, topography, primary productivity, and vegetation structure. In chaparral shrubland, return intervals of 20–30 years typified presettlement conditions, because of chronic summer drought and the formation of a dense shrub cover 15–20 years after the previous fire (Minnich, 1983). In conifer forests affected by crown fire, recurrence intervals range from 50–300 years (Hemstrom and Franklin, 1982; Romme, 1982). Intervals between fires are shorter in more productive climates characterized by periodic drought (some southern pine forests) or in areas subject to disease or pest outbreaks (some boreal forests). Fire recurrence intervals increase to several centuries in areas of low primary productivity (high-elevation forests of the Yellowstone Plateau), in regions or topographic settings that are more mesic (Pacific Northwest, lowlands and north-facing slopes), or in sites protected from the sweep of fires by streams and lakes (Bergeron, 1991).

Surface fires often subdue young, invasive woody stems and favor grasses and other herbs that resprout from belowground perennating organs or reseed on the ash-covered mineral soil that follows fire. Prairie fires may cover extensive areas in the absence of terrain breaks (Wells, 1965) and may recur at short intervals (2–5 years in tallgrass prairie near the prairie forest boundary of the midwestern and south-central states, 10 or more years in dissected landscapes and drier, bunch-grass prairies of the western Plains; Wright and Bailey, 1980). In desert and semidesert landscapes, surface fires can have protracted recurrence intervals; indeed, fires generally occur following wet years that promote dense blooms of desert ephemerals and grasses, which form a continuous surface fuel mosaic that eventually dries out (Baisan and Swetnam, 1990). Where surface fires occur in the understory of savanna, woodland, or forest, larger trees typically survive such fires, especially those with thick, well-insulated barks. In these settings, surface fires retard overstory replacement by favoring an open, herbaceous understory. Surface fires in wooded settings historically occurred at shorter intervals (3–5 year recurrence interval in longleaf pine (*Pinus palustris*) savannas of the Southeast (Christensen, 1981), 5–20–year intervals in many oak (*Quercus* spp.) forests of the mid-Atlantic and Great Lakes region, as well as in lower montane pine forests of western North America. Although fire is now widely regarded as a critical natural disturbance agent in many landscapes, the details of presettlement fire history are still sketchy for many regions, especially as fairly aggressive and successful fire suppression activities by humans in the latter half of the twentieth century have obscured evidence of past wildfire.

Like crown and surface fires, other disturbance agents display geographic patterns of behavior that reflect a strong measure of physical environmental control through the effects of climate and topography. The role of climate in mediating disturbance agents is evident in many ways. Climate governs biomass accumulation (through both primary productivity and decomposition rates; Meentemeyer, 1984) and thereby affects fuel levels; it dictates energy and moisture balances of organisms; it defines storm tracks that subject favored corridors to strong winds, and it generates sporadic anomalies—extreme events—that themselves may be viewed as disturbance events (MacDonald, 1988; K. Parker, 1993). In addition to its effects on spatial variability of fire recurrence intervals, topography often dictates exposure to disturbance agents (K. Parker and Bendix, 1996). Examples include exposure to low-frequency, high-intensity floods in a riparian corridor (fig. 11.3B; Bendix, 1994; Baker and Walford, 1995); to windthrow on exposed, windward slopes and ridges and along woodlot edges (fig. 11.3C; Boose et al., 1994); to disease (Brewer, 1995); to snow avalanches (fig. 11.3D; Johnson, 1987); and to slope-failure events in steep, tectonically active terrain (Miles and Swanson, 1986).

In many settings, disease or pest outbreaks serve as disturbance agents, either in their own right or in conjunction with other agents. Of course, such agents are also relevant to the preceding discussion of predation and parasitism as prominent biotic interactions. Extensive tracts of boreal and montane conifer forest are infected quasi-periodically by insects such as spruce budworm (*Choristoneura* spp.) (see Swetnam and Lynch, 1993), needle miner (*Coleotechnites milleri*), mountain pine beetle (*Dendroctonus ponderosae*), and balsam woolly adelgid (*Adelges piceae*). Eastern hardwood forests are similarly infected by tent caterpillars, *Malacosoma* spp.), and southern pine forests by southern pine beetle (*Dendroctonus frontalis*). To the extent that patches of infested, standing dead trees add dry fuel ladders between canopy and understory, they may impose temporal limits on fire return intervals in many of these forests.

Where more acute disturbance events are uncommon or absent, plants still die, but the scale of events and patterns of response are finer. These processes, historically associated with individual tree fall, are referred to by ecologists as gap-phase replacement, because recruitment opportunities exist in the canopy gaps created by tree fall (Watt, 1947). In a number of old-growth forests of the Atlantic Northeast and Pacific Northwest, pit-and-mound topography provides testimony to the ubiquity of individual tree fall (Beatty, 1984; Schaetz et al., 1989). Some ecologists have observed that gap-phase processes are common in remnants of eastern deciduous forest; they argue that this region is one of less frequent fire, one where canopy-gap

Figure 11.3 A collage of disturbance agents: (A) crown fire, (B) riparian flooding, (C) windthrow, and (D) snow avalanche. All photographs taken by Kathleen C. Parker.

processes can maintain a mature forest over long time periods in a shifting-mosaic, steady-state structure (Bormann and Likens, 1979; Runkle, 1990). Others have insisted on a greater historical role for fire in eastern forests, such as the linkage by Abrams et al. (1995) of oak dominance with frequent fire in forests of the mid-Atlantic states or the comparison by Cowell (1995) of modern and presettlement forests of the Georgia Piedmont, in which he observes that the gap-phase processes that operate in the protected forest remnants of today do not necessarily provide accurate pictures of the historical role of fire in this landscape.

Episodic disturbance and vegetation recovery produce an ever-changing landscape, a spatial mosaic in which there are colonization opportunities for organisms able to reach open patches, as well as opportunities for other species that dominate later in succession to persist on sites that have not been catastrophically disturbed for decades or centuries. In such a dynamic spatial mosaic, the dispersal ecology of plants and animals emerges as a crucial factor in understanding the maintenance of populations over long time periods. As noted in our discussion of competition, ecologists have linked dispersal syndromes to means of population persistence in a patchy landscape by manifestation of r- versus K-selected species traits. For those opportunists that colonize open patches after disturbance events (r-selected species), dispersal success is augmented by the production of large numbers of seeds that are small and easily dispersed (by wind, gravity, water), that exhibit extended periods of dormancy and may remain viable in a seed bank until the site is opened, and that grow quickly when conditions are favorable. For those species that tend to persist on long-undisturbed ground (K-selected species), seed mobility and dormancy potential are less important traits; individual seeds become larger, carrying greater endosperm to improve germination potential in the understory of closed vegetation; growth rates are more conservative, extending reproductive effort over a longer lifespan to improve the likelihood of leaving progeny. Dispersal traits, to the extent that they influence the suite of species that participate in the vegetation recovery process, may exert profound influence on vegetation composition and animal assemblages following a disturbance event (Fastie, 1995).

11.4.3 Environmental History

Recognition of the impact of unique events in shaping the composition and structure of plant and animal assemblages has emerged as a prominent theme in recent years (Christensen, 1989; Botkin, 1990; Sprugel, 1991). The singular effects of history are evident in three categories of biotic responses: (1) climatic change; (2) the nature, timing, and interactions among disturbance agents; and (3) human influence. Although vegetation and animal assemblages respond continuously to atmospheric changes at shorter time scales, the response to climatic change often plays out over centuries and millennia. Grimm (1984) has documented that the Big Woods maple–basswood forests of southern Minnesota, although appearing to be a stable vegetation assemblage to earlier workers (Daubenmire, 1936), have become established relatively recently, in a region formerly occupied by more open oak woodlands, in response to reduced fire frequencies associated with the Little Ice Age (1600–1850). In a similar vein, Clark (1990) in northern Minnesota and Swetnam (1993) in California's Sierra Nevada describe the variation in fire recurrence intervals associated with changes in temperature and effective precipitation throughout the past millennium. Paleoecologists have reconstructed a detailed sequence of Holocene vegetation migration in response to climatic warming and glacial retreat throughout North America (Davis, 1976; Liu, 1990; Prentice et al., 1991). The climatic adjustments from the last full glacial to modern interglacial conditions, perhaps in conjunction with the migration of humans across the Beringian isthmus into North America, have exposed larger mammals, especially ungulates and carnivores, to elevated extinction rates (Grayson, 1984; see chapter 3).

The nature and sequence of disturbance events on a site exert strong influence on postdisturbance patterns of dominance. Eastern white pine (*Pinus strobus*) was a conspicuous, economically important dominant of forests in the Great Lakes region in the eighteenth and nineteenth centuries. Eastern white pine is a long-lived, early successional species that requires a bare mineral seedbed for germination; hence, it was favored after widespread crown fires in this region in the 1600s. As the wave of Euroamerican settlers, aided by expansion of the railroad network, penetrated the region and logged the timber, eastern white pine declined in importance. This decline is variously linked to conversion to oak–aspen (*Populus* spp.) forests on areas exposed to postlogging slash fires that consumed the seedbank over large areas and favored vegetative regeneration of hardwoods; to invasion by eastern hemlock (*Tsuga canadensis*), sugar maple (*Acer saccharum*), American beech (*Fagus grandifolia*) and other more shade-tolerant taxa on more mesic sites where fire frequencies have been depressed by fire suppression activities; and to a change from fire to wind as the dominant disturbance agent, with windthrow favoring hardwoods over eastern white pine in the recovery sequence after disturbance (Oliver, 1981; Whitney, 1987). Similar arguments have linked the importance of Douglas-fir in the modern forests of the Pacific Northwest to unusually widespread fires around 1300 (Dunwiddie, 1986).

In general, there has emerged an increasing cognizance that what ecological historians have traditionally perceived as "natural" or "pristine" is in large measure context-dependent (Cronon, 1995). This recognition of the importance of meaning-related context does not seek to ignore the role of physical gradients or biotic influence in sorting out vegetation recovery processes over time. Rather, it serves as a poignant reminder that these processes operate against a backdrop of environmental change that is mediated by unique events that impose distinctiveness on the landscape at a given place and time. In this view, the biotic configuration of a particular landscape reflects a specific condition within a range of possibilities that is articulated by the sweep of historical events that alter resource levels, physical conditions, or biotic interactions in such a way as to favor one set of species over another in the ambient circumstance.

The imprint of human influences on the biota cuts across all scales from local to global. The ecological impacts may be summarized by observing three obvious and pervasive effects of human action on the ecological integrity of the biota. First, economic activity has remade many landscapes in many places again and again (Cronon, 1983; Whitney, 1994). These structural changes are commonly referred to as habitat alteration (changing the conditions and resources associated with a site). As the network of human action has expanded, it has prompted increasing habitat fragmentation or increasing separation of populations of plants and animals into pockets or enclaves of a formerly more continuous distribution. The ecological and genetic implications of this fragmentation have just recently begun to be thoroughly investigated; they have become a major focus of conservation biology (Harris, 1984).

Second, humans have altered the biota of North America through acceleration of species extinctions and introduction of exotic species. Although only a few North American animal species have become extinct, or all but extinct, in their natural habitats since European contact some 500 years ago, many species have been reduced to threatened or endangered status by human action. The Pacific coast, Texas coast, Florida, and parts of the Southwest and southern Plains host disproportionate numbers of endangered birds, whereas threatened fish species are concentrated in the Southwest and molluscs in the southern Appalachians (Dobson et al., 1997). For all taxa, both plant and animal, the Southwest (including southern California) and Florida are particularly important centers for threatened species (fig. 11.4; Flather et al., 1994). Much of this extinction pressure is associated with habitat alteration and fragmentation activities, as noted previously, although collecting and harvesting pressures, widespread pesticide use, and the introduction of exotic pathogens or competitors have been cited as contributors to this process. Introduction and naturalization of exotic species (either by design or accident) have triggered pronounced changes in vegetation and animal assemblages: witness the effects of chestnut-blight

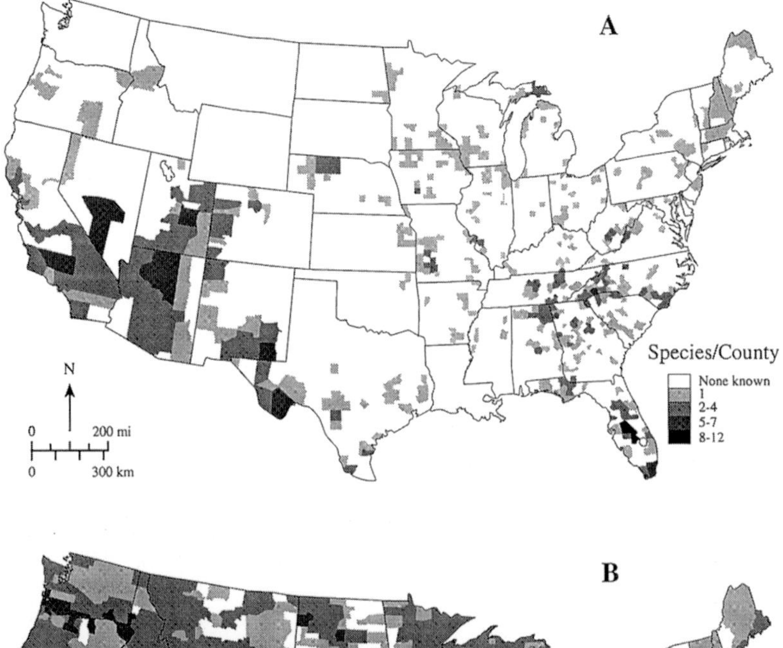

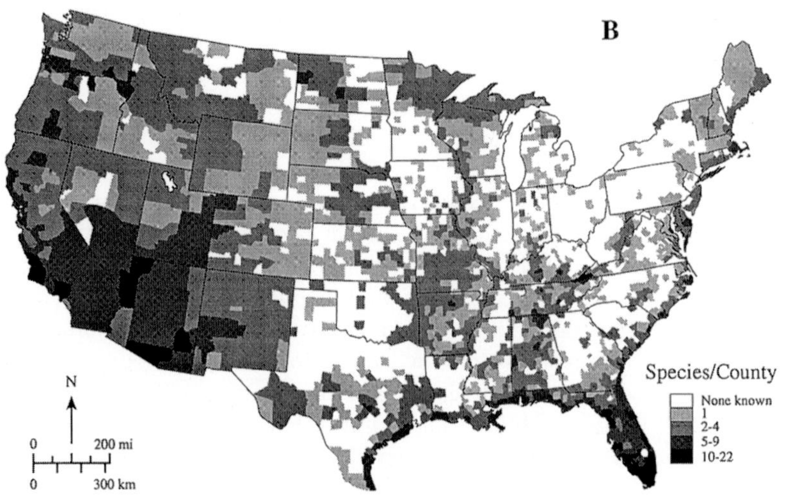

Figure 11.4 County-level distribution of threatened and endangered species of (A) animals and (B) plants within the United States (adapted from Flather et al., 1994).

(*Cryphonectria parasitica*) on American chestnut (*Castanea dentata*), a former dominant of Appalachian forests, the impact of Dutch elm disease (*Ceratocystis ulmi*) on American elm (*Ulmus americana*), a formerly important element of midwestern mesic forests, the influence of water hyacinth (*Eichornia crassipes*) on the structure and hydrologic dynamics of wetlands in Florida, the rapid spread of zebra mussels (*Dreissena polymorpha*) throughout aquatic ecosystems of the Mississippi River basin (fig. 11.5), and the displacement of native birds by starlings (*Sturnus vulgaris*) and English sparrows (*Passer domesticus*) in many suburban and rural settings throughout North America.

Third, humans have both inadvertently and purposefully altered ecological processes that have historically operated at different rates or scales. Examples of pervasive and pernicious alterations of ecological process are legion. Suppression of fire from many vegetation types over much of the past century has altered the composition and structure of many biotas (Loucks, 1970; Vale, 1982). Alteration of streamflow regimes through dams and diversions for flood control, wetland drainage for agricultural or urban development, modification of vegetation cover in the surrounding drainage basin, and channelization activities have dramatically altered riparian dynamics (Shankman, 1993; Miller et al., 1995; see also chapter 22). Alteration of the rates and direction of fluxes in carbon, nitrogen, phosphorus, and other biogeochemical cycles has harnessed the energy and materials required to power human technological advance. These intrusions into material cycles bring often unanticipated consequences, such as elevated atmospheric carbon dioxide as fossil fuel reserves are combusted. To the extent that human effects on ecological processes alter the geography of physical conditions and resources, modify the constituents of the biota, and change the probability of the occurrence of events, they evoke profound and sustained ecological consequences.

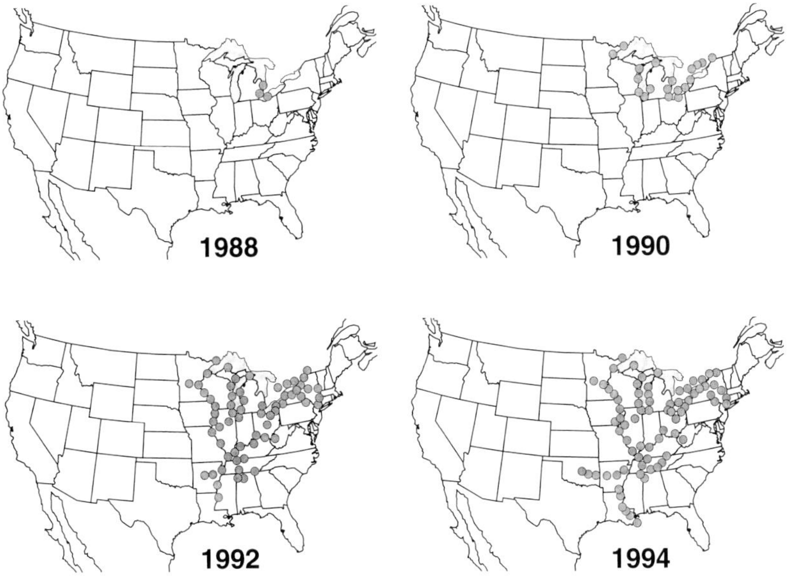

Figure 11.5 Time sequence of rapid range expansion by the introduced zebra mussel.

11.5 Ecological Themes and Trends

Six interrelated themes or trends characterize the recent ecological and biogeographic literature of North America, as follows:

1. Realization of the importance of spatial scale in understanding ecological patterns and processes.
2. Emergence of the field of landscape ecology, supported by geographic information systems and affiliated mapping technologies.
3. Greater awareness of the role of disturbance ecology in mediating vegetation structure and dynamics.
4. Renewed emphasis on ecological linkages of plants and animals with geomorphic processes.
5. Elaboration of the role of historical legacy, contingency, and unique events on biological organization.
6. Reinterpretation of pollen and other evidence of Quaternary vegetation dynamics that embraces a less deterministic view of succession and community structure.

It is fashionable to underscore the effects of scale of observation on the outcome of ecological studies and to emphasize the need to identify more carefully the scale used in individual projects and how it is related to patterns and processes (Meentemeyer, 1989; Wiens, 1989; Allen and Hoekstra, 1990; Levin, 1992). Urban et al. (1987) emphasize the importance of using multiscale analyses to appreciate context (which they suggest is embedded in a broader spatial scale) and to understand mechanism (which they assert is controlled by processes operating more locally). Baker (1989b) and Bendix (1994) highlight multiple scales of variation in physical phenomena that affect vegetation patterns in riparian corridors. Recently developed models of animal population structure also account for processes operating at different scales (Pullium, 1988); these have facilitated assessments of metapopulation stability, especially across highly fragmented landscapes. In these spatial models of population structure, source areas produce more individuals than they can support; migrants from these source areas sustain population levels in sinks, areas of suboptimal conditions inhabited by populations whose death rates exceed birth rates.

Landscape ecology has emerged as an active subdivision of ecology devoted to measuring the spatial structure of landscapes and interpreting how that structure affects the movement and function of organisms within the landscape mosaic (Forman and Godron, 1986). Landscape ecologists have developed geometric indices of map elements (pattern, shape, orientation), summary measures of landscape diversity and heterogeneity (Romme, 1982), and models of landscape dynamics that vary from nonspatial to highly spatially explicit forms (Baker, 1989c). The recent proliferation of landscape ecological studies reinforces attention to scale effects in ecology and is aided by concomitant developments in geographic information technologies (Narumalani et al., 1997). As fuzzy set theory is merged with geographic information technologies, the dependence of measurement outcomes on imposed, a priori classification structures will diminish and the portrayal of compositional and structural variation along environmental gradients will be enhanced.

Physical geographers and ecologists have made substantial strides in the past few decades in understanding disturbance ecology and its role in mediating vegetation structure and dynamics. A small sample of disturbance studies suggests the richness of the work: western conifer forests (Veblen et al., 1991, 1994; Taylor, 1993; Savage, 1994; Huff, 1995), boreal forests (Baker, 1989a, Frelich and Reich, 1995), Mediterranean-type shrublands (Zedler et al., 1983; Westman and O'Leary, 1986), and eastern forests (Foster, 1988; Runkle, 1990). The marriage of disturbance ecology with patch dynamics to form a landscape perspective on the operation of disturbance regimes has forged a strong link between traditional elements of physical geography and the emerging views from landscape ecology.

Geomorphic processes and resulting landforms often dictate the pattern of physical gradients and biotic response from local to landscape scales (Swanson et al., 1988; K. Parker and Bendix, 1996). Biogeomorphic influences have received particular attention in riparian corridors, where the linkage between pattern and process is direct and compelling (Hupp and Osterkamp, 1985; Malanson, 1993). Other kinds of biogeomorphic linkages include the effects of mass wasting and snow avalanches on vegetation patterns (Flaccus, 1959; Suffling, 1993), the effects of geomorphic history on pedogenesis and vegetation patterns in glaciated landscapes (Brubaker, 1975, Graumlich and Davis, 1993), the role of topography in mediating disturbance frequencies (Hadley, 1994), and the geomorphic effects of animals, including burrowing, digging, rooting, and damming behaviors (Butler, 1995).

With increased emphasis on disturbance and landscape dynamics has come the revision of basic successional models. As noted previously, some degree of indeterminacy has been embodied in many models of vegetation dynamics imposed by the unique interaction of events and attendant circumstances on a site over time. These newer models of vegetation dynamics have been termed state-and-transition models, to emphasize that the vegetation on a site might fit one of several alternative stable states and that transitions between states may be driven more by the nature and timing of disturbance events rather than by autogenic changes in composition over time (Cattelino et al., 1979; Laycock, 1991). Choosing one type of vegetation dynamic model over another (state-and-transition vs. successional change) seems counterproductive. Rather, these might be viewed as polar extremes among possible sequences, from the staccato pattern of state-and-transition to the seamless developmental change of succession. In this light, vegetation dynamics models can be recognized for what they are, human attempts to simplify and understand the complexities of biotic response to physical, chemical, and biological forces that act over time to shape the biota (Wu and Loucks, 1995).

As ecologists and physical geographers respond to a dynamic view of plant and animal life that is not strongly deterministic, they have relaxed some of the notions that governed interpretations of paleoenvironmental reconstructions made a few decades ago, when views of discrete communities that migrate in equilibrial lockstep with climate and undergo predictable, progressive successional change were more popular. Reviews by Webb (1986), Schoonmaker and Foster (1991), and Tausch et al. (1993) provide consideration of how modern ecological perspectives can inform paleoecological work. Examples of paleoecological work that embrace these modern ecological precepts include, among many others, the works of MacDonald (1987; see also chapter 14) in western Canada, Liu (1990) in eastern Canada, Whitlock and Bartlein (1993) in the western United States, and Clark et al. (1996) in the eastern United States. As we gain greater understanding of paleoenvironments through pollen, charcoal, diatom, and other stratigraphic reconstructions, we gain a richer sense of the diversity of life and the range of conditions to which it has been subjected during the Quaternary. We also establish a frame of reference for understanding possible biotic responses to an inevitable future of sometimes subtle, sometimes dramatic, environmental change.

References

Abrams, M.D., D.A. Orwig, and T.E. Demeo, 1995. Dendroecological analysis of successional dynamics for a presettlement-origin white-pine–mixed-oak forest in the southern Appalachians, U.S.A. *Journal of Ecology* 83: 123–133.

Allen, T.F.H., and T.W. Hoekstra, 1990. The confusion between scale-defined levels and conventional levels of organization in ecology. *Journal of Vegetation Science* 1: 5–12.

Alverson, W.S., D.M. Waller, and S.L. Solheim, 1989. Forests too deer: Edge effects in northern Wisconsin. *Conservation Biology* 2: 348–358.

Anderson, J.E., and W.H. Romme, 1991. Initial floristics in lodgepole pine *Pinus contorta* forests following the 1988 Yellowstone fires. *International Journal of Wildland Fire* 1: 119–124.

Bahre, C.J., 1991. *A Legacy of Change: Historic Human Impact on Vegetation in the Arizona Borderlands.* University of Arizona Press, Tucson, Arizona. 231 pp.

Baisan, C.H., and T.W. Swetnam, 1990. Fire history on a desert mountain range: Rincon Mountain Wilderness, Arizona, U.S.A. *Canadian Journal of Forest Research* 20: 1559–1569.

Baker, W.L., 1989a. Landscape ecology and nature reserve design in the Boundary Waters Canoe Area, Minnesota. *Ecology* 70: 23–35.

Baker, W.L., 1989b. Macro- and micro-scale influences on riparian vegetation in western Colorado. *Annals of the Association of American Geographers* 79: 65–78.

Baker, W.L., 1989c. A review of models of landscape change. *Landscape Ecology* 2: 111–133.

Baker, W.L., and G.M. Walford, 1995. Multiple stable states and models of riparian vegetation succession on the Animas River, Colorado. *Annals of the Association of American Geographers* 85: 320–338.

Beatty, S.W., 1984. Influence of microtopography and canopy species on spatial patterns of forest understory plants. *Ecology* 65: 1406–1419.

Bendix, J., 1994. Scale, direction, and pattern in riparian vegetation-environment relationships. *Annals of the Association of American Geographers* 84: 652–665.

Bergeron, Y., 1991. The influence of island and mainland lakeshore landscapes on boreal forest fire regimes. *Ecology* 72: 1980–1992.

Boose, E.R., D.R. Foster, and M. Fluet, 1994. Hurricane impacts to tropical and temperate forest landscapes. *Ecological Monographs* 64: 369–400.

Bormann, F.H., and G.E. Likens, 1979. *Pattern and Process in a Forested Ecosystem.* Springer-Verlag, New York. 253 pp.

Botkin, D.B., 1990. *Discordant Harmonies: A New Ecology for the Twenty-first Century.* Oxford University Press, New York. 241 pp.

Bratton, S.P., 1974. The effect of the European wild boar (*Sus scrofa*) on the high-elevation vernal flora in Great Smoky Mountains National Park. *Bulletin of the Torrey Botanical Club* 101: 198–206.

Brewer, L.G., 1995. Ecology of survival and recovery from blight in American chestnut trees (*Castanea dentata* (Marsh.) Borkh.) in Michigan. *Bulletin of the Torrey Botanical Club* 122: 40–57.

Brubaker, L.B., 1975. Postglacial forest patterns associated with till and outwash in north-central Upper Michigan. *Quaternary Research* 5:499–527.

Bryson, R.A., and F.K. Hare., 1974. The climates of North America. Pages 1–47 *in* R. A. Bryson and F. K. Hare, editors. *World Survey of Climatology: Climates of North America.* Volume 11. Elsevier Scientific, New York.

Buol, S.W., F.D. Hole, and R.J. McCracken, 1989. *Soil Genesis and Classification*, 3rd Edition. Iowa State University Press, Ames. 446 pp.

Butler, D.R., 1995. *Zoogeomorphology: Animals as Geomorphic Agents.* Cambridge University Press, Cambridge. 231 pp.

Cattelino, P.J., I.R. Noble, R.O. Slatyer, and S.R. Kessell, 1979. Predicting the multiple pathways of plant succession. *Environmental Management* 3:41–50.

Chapin, F.S., III, and G.R. Shaver, 1985. Individualistic growth responses of tundra plant species to environmental manipulations in the field. *Ecology* 66: 564–576.

Christensen, N.L., 1981. Fire regimes in southeastern ecosystems. Pages 112–136 *in* H.A. Mooney, T.M. Bonnickson, N. L. Christensen, J. E. Lotan, and W.A. Reiners, editors. *Fire Regimes and Ecosystem Properties.* General Technical Report, WO-26. United States Department of Agriculture, Forest Service, Washington, D.C.

Christensen, N.L., 1989. Landscape history and ecological change. *Journal of Forest History* 33: 116–124.

Clark, J.S., 1990. Fire and climate change during the last 750 yr in northwestern Minnesota. *Ecological Monographs* 60: 135–159.

Clark, J.S., P.D. Royall, and C. Chumbley, 1996. The role of fire during climate change in an eastern deciduous forest at Devil's Bathtub, New York. *Ecology* 77: 2148–2166.

Clements, F.E., 1916. *Plant Succession: An Analysis of the Development of Vegetation.* Carnegie Institution of Washington Publ. 242. Carnegie Institution of Washington, Washington, D.C. 512 pp.

Cody, M.L., and H.A. Mooney, 1978. Convergence versus nonconvergence in mediterranean-climate ecosystems. *Annual Review of Ecology and Systematics* 9: 265–321.

Connell, J.H., 1983. On the prevalence and relative importance of interspecific competition: Evidence from field experiments. *American Naturalist* 122: 661–696.

Cowell, C.M., 1995. Presettlement Piedmont forests: Patterns of composition and disturbance in central Georgia. *Annals of the Association of American Geographers* 85: 65–83.

Cronon, W., 1983. *Changes in the Land: Indians, Colonists, and the Ecology of New England.* Hill and Wang, New York. 241 pp.

Cronon, W., 1995. The trouble with wilderness; or, getting back to the wrong nature. Pages 69–90 *in* W. Cronon, editor. *Uncommon Ground: Toward Reinventing Nature.* W.W. Norton, New York.

Curtis, J.T., 1959. *The Vegetation of Wisconsin: An Ordination of Plant Communities.* University of Wisconsin Press, Madison. 657 pp.

Daubenmire, R., 1936. The "Big Woods" of Minnesota: Its structure, and relation to climate, fire and soils. *Ecological Monographs* 6: 233–268.

Davis, M.B., 1976. Pleistocene biogeography of temperate deciduous forests. *Geoscience and Man* 13: 13–26.

Dobson, A.P., J.P. Rodriguez, W.M. Roberts, and D.S. Wilcove, 1997. Geographic distribution of endangered species in the United States. *Science* 275: 550–553.

Dunwiddie, P.W., 1986. A 6000–year record of forest history on Mount Rainier, Washington. *Ecology* 67: 58–68.

Elliott-Fisk, D.L., 1983. The stability of the northern Canadian tree limit. *Annals of the Association of American Geographers* 73: 560–576.

Fastie, C.L., 1995. Causes and ecosystem consequences of multiple pathways of primary succession at Glacier Bay, Alaska. *Ecology* 76: 1899–1916.

Flaccus, E., 1959. Revegetation of landslides in the White Mountains of New Hampshire. *Ecology* 40: 692–703.

Flather, C.H., L.A. Joyce, and C.A. Bloomgarden, 1994. *Species Endangerment Patterns in the United States.* General Technical Report, RM-241. U.S. Department of Agriculture, Forest Service, Rocky Mountain Forest and Range Experiment Station, Fort Collins, Colorado. 42 pp.

Forman, R.T.T., and M. Godron, 1986. *Landscape Ecology.* John Wiley and Sons, New York. 619 pp.

Foster, D.R., 1988. Species and stand response to catastrophic wind in central New England, U.S.A. *Journal of Ecology* 76: 135–151.

Frelich, L.E., and P.B. Reich, 1995. Spatial patterns and succession in a Minnesota southern-boreal forest. *Ecological Monographs* 65: 325–346.

Gleason, H.A., 1926. The individualistic concept of plant association. *Bulletin of the Torrey Botanical Club* 53:7–26.

Godt, M.J.W., B.R. Johnson, and J.L. Hamrick, 1996. Genetic diversity and population size in four rare southern Appalachian plant species. *Conservation Biology* 10: 796–805.

Goudie, A., 1981. *The Human Impact: Man's Role in Environmental Change.* Massachusetts Institute of Technology Press, Cambridge. 316 pp.

Graumlich, L.J., and M.B. Davis, 1993. Holocene variation in spatial scales of vegetation pattern in the upper Great Lakes. *Ecology* 74: 826–839.

Grayson, D.K., 1984. Explaining Pleistocene extinctions: Thoughts on the structure of a debate. Pages 807–823

in P. S. Martin and R. G. Klein, editors. *Quaternary Extinctions: A Prehistoric Revolution.* University of Arizona Press, Tucson.
Griffin, D., 1997. Wisconsin's vegetation history and the balancing of nature. Pages 95–111 *in* R. Ostergren and T.R. Vale, editors. *Wisconsin Land and Life.* University of Wisconsin Press, Madison.
Grime, J.P., 1977. Evidence for the existence of three primary strategies in plants and its relevance to ecological and evolutionary theory. *American Naturalist* 111: 1169–1194.
Grimm, E.C., 1984. Fire and other factors controlling the Big Woods vegetation of Minnesota in the mid-nineteenth century. *Ecological Monographs* 54: 291–311.
Hack, J.T., and J.C. Goodlett, 1960. *Geomorphology and Forest Ecology of a Mountain Region in the Central Appalachians.* United States Geological Survey Professional Paper 347. United States Government Printing Office, Washington, D.C. 66 pp.
Hadley, K.S., 1994. The role of disturbance, topography, and forest structure in the development of a montane forest landscape. *Bulletin of the Torrey Botanical Club* 121:47–61.
Hansen, K., W. Wyckoff, and J. Banfield, 1995. Shifting forests: Historical grazing and forest invasion in southwestern Montana. *Forest and Conservation History* 39: 66–76.
Harman, J.R., 1991. *Synoptic Climatology of the Westerlies: Process and Patterns.* Association of American Geographers, Washington, D.C. 80 pp.
Harris, L.D., 1984. *The Fragmented Forest: Island Biogeography Theory and the Preservation of Biotic Diversity.* University of Chicago Press, Chicago. 211 pp.
Hastings, J.R., and R.M. Turner, 1965. *The Changing Mile: An Ecological Study of Vegetation Change with Time in the Lower Mile of an Arid and Semiarid Region.* University of Arizona Press, Tucson. 317 pp.
Hedrick, P.W., 1995. Gene flow and genetic restoration: The Florida panther as a case study. *Conservation Biology* 9: 996–1007.
Hemstrom, M.A., and J.F. Franklin, 1982. Fire and other disturbances of the forests in Mount Ranier National Park. *Quaternary Research* 18: 32–51.
Huff, M.H., 1995. Forest age structure and development following wildfires in the western Olympic Mountains, Washington. *Ecological Applications* 5:471–483.
Hupp, C.R., and W.R. Osterkamp, 1985. Bottomland vegetation distribution along Passage Creek, Virginia, in relation to fluvial landforms. *Ecology* 66: 670–681.
Huston, M., and T. Smith, 1987. Plant succession: Life history and competition. *American Naturalist* 130: 168–198.
Johnson, E.A., 1987. The relative importance of snow avalanche disturbance and thinning on canopy plant populations. *Ecology* 68:43–53.
Laycock, W.A., 1991. Stable states and thresholds of range conditions on North American rangelands: A viewpoint. *Journal of Range Management* 44:427–433.
Levin, S.A., 1992. The problem of pattern and scale in ecology. *Ecology* 73: 1943–1967.
Liu, K.-B., 1990. Holocene paleoecology of the boreal forest and Great Lakes–St. Lawrence forest in northern Ontario. *Ecological Monographs* 60: 179–212.
Loucks, O.L., 1970. Evolution of diversity, efficiency and community stability. *American Zoologist* 10: 17–25.
MacArthur, R.H., 1958. Population ecology of some warblers of northeastern coniferous forests. *Ecology* 39: 599–619.
MacArthur, R.H., and J.W. MacArthur, 1961. On bird species diversity. *Ecology* 42: 594–598.
MacDonald, G.M., 1987. Postglacial development of the subalpine-boreal transition forest of western Canada. *Journal of Ecology* 75: 303–320.
MacDonald, G.M., 1988. Vegetation disturbance-response: A biogeographical perspective I. *Canadian Geographer* 32: 76–85.
Malanson, G.P., 1993. *Riparian Landscapes.* Cambridge University Press, Cambridge. 296 pp.
Marsh, G.P., [1864] 1965. *Man and Nature; or, Physical Geography as Modified by Human Action.* Edited by D. Lowenthal. Harvard University Press, Cambridge, Masssachusetts. 472 pp.
Marshall, L.G., 1988. Land mammals and the great American exchange. *American Scientist* 76: 380–388.
Mather, J.R., and G.A. Yoshioka, 1968. The role of climate in the distribution of vegetation. *Annals of the Association of American Geographers* 58: 29–41.
McAuliffe, J.R., 1994. Landscape evolution, soil formation, and ecological patterns and processes in Sonoran Desert bajadas. *Ecological Monographs* 64: 111–148.
Meentemeyer, V., 1984. The geography of organic decomposition rates. *Annals of the Association of American Geographers* 74: 551–560.
Meentemeyer, V., 1989. Geographical perspectives of space, time, and scale. *Landscape Ecology* 3: 163–173.
Miles, D.W.R., and F.J. Swanson, 1986. Vegetation composition on recent landslides in the Cascade Mountains of western Oregon. *Canadian Journal of Forest Research* 16: 739–744.
Miller, J.R., T.T. Schulz, N.T. Hobbs, K.R. Wilson, D.L. Schrupp, and W.L. Baker, 1995. Changes in the landscape structure of a southeastern Wyoming riparian zone following shifts in stream dynamics. *Biological Conservation* 72: 371–379.
Mills, L.S., M.E. Soulé, and D.F. Doak, 1993. The keystone-species concept in ecology and conservation. *BioScience* 43: 219–224.
Minnich, R.A., 1983. Fire mosaics in southern California and northern Baja California. *Science* 219: 1287–1294.
Mooney, H.A., and E.L. Dunn, 1970. Convergent evolution of mediterranean-climate evergreen sclerophyll shrubs. *Evolution* 24: 292–303.
Narumalani, S., J.R. Jensen, J.D. Althausen, S. Burkhalter, and H.E. Mackey, 1997. Aquatic macrophyte modeling using GIS and logistic multiple-regression. *Photogrammetric Engineering and Remote Sensing* 63:41–49.
Noble, I.R., and R.O. Slatyer, 1980. The use of vital attributes to predict successional changes in plant communities subject to recurrent disturbance. *Vegetatio* 43: 5–21.
Oliver, C.D., 1981. Forest development in North America following major disturbances. *Forest Ecology and Management* 3: 153–168.
Paine, R.T., 1966. Food web complexity and species diversity. *American Naturalist* 100: 65–75.
Parker, A.J., 1993. Structural variation and dynamics of lodgepole pine forests in Lassen Volcanic National Park, California. *Annals of the Association of American Geographers* 83: 613–629.

Parker, A.J., 1994. Latitudinal gradients of coniferous tree species, vegetation, and climate in the Sierran-Cascade axis of northern California. *Vegetatio* 115: 145–155.

Parker, K.C., 1987. Avian nesting habits and vegetation structure. *Professional Geographer* 39:47–58.

Parker, K.C., 1995. Effects of complex geomorphic history on soil and vegetation patterns on arid alluvial fans. *Journal of Arid Environments* 30: 19–39.

Parker, K.C., and J. Bendix, 1996. Landscape-scale geomorphic influences on vegetation patterns in four environments. *Physical Geography* 17: 113–141.

Peet, R.K., 1981. Forest vegetation of the Colorado Front Range. *Vegetatio* 45: 3–75.

Peet, R.K., and O.L. Loucks, 1977. A gradient analysis of southern Wisconsin forests. *Ecology* 58:485–499.

Pianka, E.R., 1970. On r- and K-selection. *American Naturalist* 104: 592–597.

Pimm, S.L., 1991. *The Balance of Nature: Ecological Issues in the Conservation of Species and Communities.* University of Chicago Press, Chicago. 434 pp.

Pregitzer, K.S., B.V. Barnes, and G.D. Lemme, 1983. Relationship of topography to soils and vegetation in an upper Michigan ecosystem. *Soil Science Society of America Journal* 47: 117–123.

Prentice, I.C., P.J. Bartlein, and T. Webb III, 1991. Vegetation and climate change in eastern North America since the last glacial maximum. *Ecology* 72: 2038–2056.

Pulliam, H.R., 1988. Sources, sinks and population regulation. *American Naturalist* 132: 652–661.

Richardson, D.M., and W.J. Bond, 1991. Determinants of plant distribution: evidence from pine invasions. *American Naturalist* 137: 639–668.

Ricklefs, R.E., 1987. Community diversity: relative roles of local and regional processes. *Science* 235: 167–171.

Romme, W.H., 1982. Fire and landscape diversity in subalpine forests of Yellowstone National Park. *Ecological Monographs* 52: 199–221.

Runkle, J.R., 1990. Gap dynamics in an Ohio *Acer-Fagus* forest and speculations on the geography of disturbance. *Canadian Journal of Forest Research* 20: 632–641.

Sauer, J.D., 1988. *Plant Migration: The Dynamics of Geographic Patterning in Seed Plant Species.* University of California Press, Berkeley. 282 pp.

Savage, M., 1994. Anthropogenic and natural disturbance and patterns of mortality in a mixed conifer forest in California. *Canadian Journal of Forest Research* 24: 1149–1159.

Schaetzl, R.J., S.F. Burns, D.L. Johnson, and T.W. Small, 1989. Tree uprooting: Review of impacts on forest ecology. *Vegetatio* 79: 165–176.

Schoener, T.W., 1983. Field experiments on interspecific competition. *American Naturalist* 122: 240–285.

Schoonmaker, P.K., and D.R. Foster, 1991. Some implications of paleoecology for contemporary ecology. *Botanical Review* 57: 204–245.

Shankman, D., 1993. Channel migration and vegetation patterns in the southeastern coastal plain. *Conservation Biology* 7: 176–183.

Soulé, M.E., editor, 1987. *Viable Populations for Conservation.* Cambridge University Press, Cambridge. 189 pp.

Sowell, J.B., 1985. A predictive model relating North American plant formations and climate. *Vegetatio* 60: 103–111.

Sprugel, D.G., 1991. Disturbance, equilibrium, and environmental variability: What is "natural" vegetation in a changing environment? *Biological Conservation* 58: 1–18.

Suffling, R., 1993. Induction of vertical zones in sub-alpine valley forests by avalanche-formed fuel breaks. *Landscape Ecology* 8: 127–138.

Swanson, F.J., T.K. Kratz, N. Caine, and R.G. Woodmansee, 1988. Landform effects on ecosystem patterns and processes. *BioScience* 38: 92–98.

Swetnam, T.W., 1993. Fire history and climate change in giant sequoia groves. *Science* 262: 885–889.

Swetnam, T.W., and A.M. Lynch, 1993. Multicentury, regional-scale patterns of western spruce budworm outbreaks. *Ecological Monographs* 63: 399–424.

Tausch, R.J., P.E. Wigand, and J.W. Burkhardt, 1993. Viewpoint: Plant community thresholds, multiple steady states, and multiple successional pathways: Legacy of the Quaternary? *Journal of Range Management* 46: 439–447.

Taylor, A.H., 1993. Fire history and structure of red fir (*Abies magnifica*) forests, Swain Mountain Experimental Forest, Cascade Range, northeastern California. *Canadian Journal of Forest Research* 23: 1672–1678.

Tilman, D., 1985. The resource-ratio hypothesis of plant succession. *American Naturalist* 125: 827–852.

Urban, D.L., R.V. O'Neill, and H.H. Shugart Jr., 1987. Landscape ecology. *BioScience* 37: 119–127.

Vale, T.R., 1981. Tree invasion of montane meadows in Oregon. *American Midland Naturalist* 105: 61–69.

Vale, T.R., 1982. *Plants and People: Vegetation Change in North America.* Association of American Geographers, Washington, D.C. 88 pp.

Vale, T.R., 1988. Clearcut loggings, vegetation dynamics, and human wisdom. *Geographical Review* 78: 375–386.

Vale, T.R., A.J. Parker, and K.C. Parker, 1982. Bird communities and vegetation structure in the United States. *Annals of the Association of American Geographers* 72: 120–130.

Vale, T.R., K.C. Parker, and A.J. Parker, 1989. Terrestrial vertebrates and vegetation structure in western North America. *Professional Geographer* 41:450–464.

Veblen, T.T., K.S. Hadley, and M.S. Reid, 1991. Disturbance and stand development of a Colorado subalpine forest. *Journal of Biogeography* 18: 707–716.

Veblen, T.T., K.S. Hadley, E.M. Nel, T. Kitzberger, M. Reid, and R. Villalba, 1994. Disturbance regime and disturbance interactions in a Rocky Mountain subalpine forest. *Journal of Ecology* 82: 125–135.

Watt, A.S., 1947. Pattern and process in the plant community. *Journal of Ecology* 35: 1–22.

Weaver, T., 1980. Climates of vegetation types of the northern Rocky Mountains and adjacent Plains. *American Midland Naturalist* 103: 392–398.

Webb, T., III, 1986. Is vegetation in equilibrium with climate? How to interpret late-Quaternary pollen data. *Vegetatio* 67: 75–91.

Wells, P.V., 1965. Scarp woodlands, transported grassland soils, and concept of grassland climate in the Great Plains region. *Science* 148: 246–249.

Westman, W.E., 1983. Xeric Mediterranean-type shrubland associations of Alta and Baja California and the community/continuum debate. *Vegetatio* 52: 3–19.

Westman, W.E., and J.F. O'Leary, 1986. Measures of resilience: The response of coastal sage scrub to fire. *Vegetatio* 65: 179–189.

White, P.S., 1979. Pattern, process, and natural disturbance in vegetation. *Botanical Review* 45: 229–299.

Whitlock, C., and P.J. Bartlein, 1993. Spatial variations of Holocene climatic change in the Yellowstone region. *Quaternary Research* 39: 231–238.

Whitney, G.G., 1987. An ecological history of the Great Lakes forest of Michigan. *Journal of Ecology* 75: 667–684.

Whitney, G.G., 1994. *From Coastal Wilderness to Fruited Plain: A History of Environmental Change in Temperate North America from 1500 to the Present.* Cambridge University Press, Cambridge. 451 pp.

Whittaker, R.H., 1954. The ecology of serpentine soils. *Ecology* 35: 258–288.

Whittaker, R.H., 1956. Vegetation of the Great Smoky Mountains. *Ecological Monographs* 26: 1–80.

Wiens, J.A., 1984. On understanding a non-equilibrium world: Myth and reality in community patterns and processes. Pages 439–457 *in* D. R. Strong Jr., D. Simberloff, L. G. Abele and A. B. Thistle, editors. *Ecological Communities: Conceptual Issues and the Evidence.* Princeton University Press, Princeton, New Jersey.

Wiens, J.A., 1989. Spatial scaling in ecology. *Functional Ecology* 3: 385–397.

Wright, H.A., and A.W. Bailey, 1980. *Fire Ecology and Prescribed Burning in the Great Plains: A Research Review.* General Technical Report, INT-77. United States Department of Agriculture, Forest Service, Intermountain Forest and Range Experiment Station, Ogden, Utah. 61 pp.

Wu, J., and O.L. Loucks, 1995. From balance of nature to hierarchical patch dynamics: A paradigm shift in ecology. *Quarterly Review of Biology* 70: 439–466.

Zedler, P.H., C.R. Gautier, and G.S. McMaster. 1983. Vegetation change in response to extreme events: The effect of a short interval between fires in a California chaparral and coastal scrub. *Ecology* 64: 809–818.

12

Ecoregions

Robert G. Bailey

North America contains a great variety of ecosystems, including mountains, deserts, tropical savannas, areas of permanently frozen subsoil, forests, and steppes. Ecosystems are primarily products of climate. As a source of energy and moisture, climate acts as the primary control for ecosystem distribution. As the climate changes, the other components of the system change in response, although such responses often lag behind climate change and may, through feedback mechanisms, influence climate (see chapters 3 and 11). Climate influences soil formation, helps shape surface topography, and affects plant growth and animal communities, which in turn influence the suitability of a given system for human habitation. As a result, ecosystems of different climates differ significantly.

This chapter provides a comprehensive summation of much that has been presented in previous chapters. It offers a scheme that embraces climatic factors, vegetation ecology, geomorphology and soils to delineate the ecoregions of North America. These regions may not concide precisely with regions based on other criteria, because ecosystems respond to a complex mix of factors. Nevertheless, the scheme presented herein offers a viable summation of North America's natural landscapes, relevant not only to the subsequent discussion of selected regions, but also to the broader understanding of such issues as environmental management.

12.1 An Ecological Classification Scheme

Based on climatic conditions and on the prevailing plant associations determined primarily by those conditions, we can subdivide the continents into various ecosystems. The climatic classification developed by Köppen (1931) as modified by Trewartha (1968) is adaptable for this purpose. The Köppen-Trewartha classification identifies six main groups of climate, and all but one—the dry group—are thermally defined. These groups are subdivided into 15 types based on seasonality of precipitation or on degree of dryness or cold. They range from the ice caps at high latitudes to the tropical wet climates at low latitudes.

Trewartha's (1968) climate groups are briefly described as follows. The low latitudes contain a winterless, frostless belt with adequate rainfall. This is the tropical-humid climate or *A* group. It is subdivided into two types, tropical wet (*Ar*) and tropical wet and dry (*Aw*). The subtropical belt, or *C* group, occurs on the low-latitude margins of the middle latitudes, where winters are mild and killing frosts are only occasional. Two subdivisions are recognized: subtropical dry summer (*Cs*) and subtropical humid (*Cf*). Poleward from the subtropics is the temperate belt, or *D* group, which contains two types, temperate continental (*Dc*) and temperate oceanic (*Do*). Two subtypes of temperate continental climate are recognized: a more moderate subtype with hot summers and cold winters (*Dca*) and a more severe subtype (*Dcb*), located poleward, which has warm summers and rigorous winters. Still farther poleward is the boreal or subarctic belt, the *E* group, which has not been subdivided. In the very high latitudes are the summerless, polar climates (*F* group), subdivided into the tundra climate (*Ft*) and the ice-cap climate (*Fi*). The dry climates, group *B*, are subdivided into a semiarid or steppe type (*BS*) and an arid or desert type (*BW*). An additional subdivision separates the hot tropical-subtropical deserts and steppes (*BWh*, *BSh*) from the cold temperate boreal deserts and steppes (*BWk*, *BSk*) of middle latitudes. Highland climates, which are low-temperature variants of climates at low altitude in similar latitudes, are designated by the letter *H*.

Although we can define ecosystems climatically, they are most effectively treated by combining and rearranging the 15 climate types to maximize correspondence with major vegetation types or plant formations. Through this process, North America may be subdivided into zones, called *ecoregion provinces*, each of which has characteristic ecosystems. The U.S. Fish and Wildlife Service published a map of these provinces in 1981. The map (Bailey, 1998a), based on Crowley (1967), has been substantially revised using a number of sources, including Milanova and Kushlin (1993), and Bailey (1994). The map shows 59 provinces, but for present purposes they have been grouped together into 15 *divisions* (fig. 12.1). Mountains exhibiting altitudinal zonation are distinguished according to the character of the zonation. This classification of ecosystems is simplified still further by grouping the divisions in four large regions, called *domains* (fig. 12.2). Table 12.1 lists climate types, zonal vegetation, and zonal soil types associated with each division, and a few representative provinces are illustrated in figure 12.3. More details may be found in Bailey (1996, 1998b).

12.2 Polar Domain—100

Polar and arctic air masses control climates of the polar domain, located at high latitudes (fig. 12.2). In general, climates in the polar domain have low temperatures, severe winters, and small amounts of precipitation, most of which falls in summer (see chapters 5 and 6). Polar systems are dominated by a periodic fluctuation of solar energy and temperature, in which the annual range is far greater than the diurnal range. The intensity of the solar radiation is never very high compared to ecosystems of the middle latitudes and tropics. In areas where summers are short and temperatures are generally low throughout the year, thermal efficiency, rather than the effectiveness of

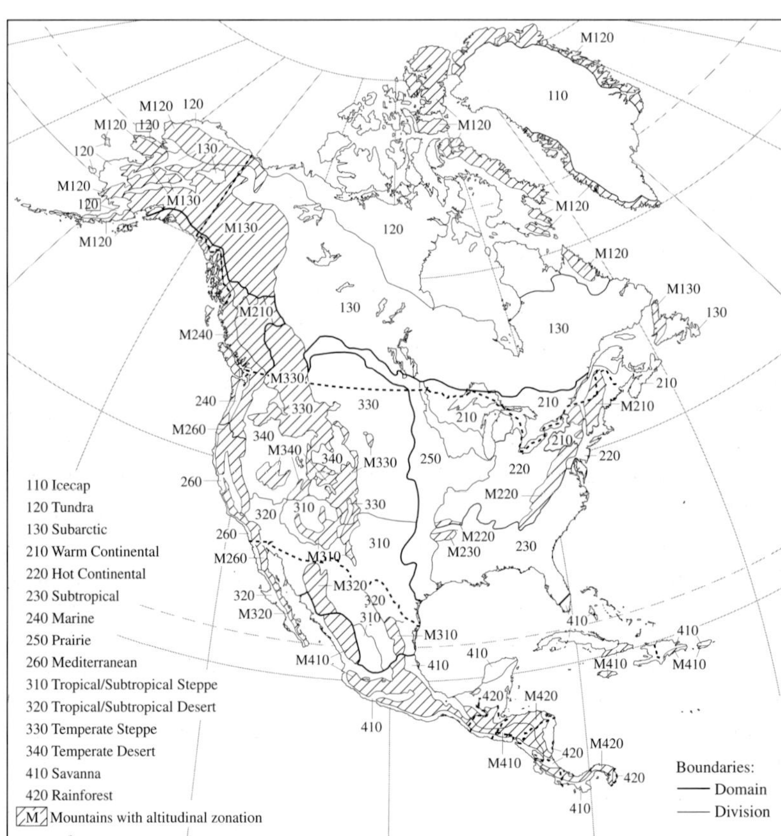

Figure 12.1 Ecoregion divisions of North America.

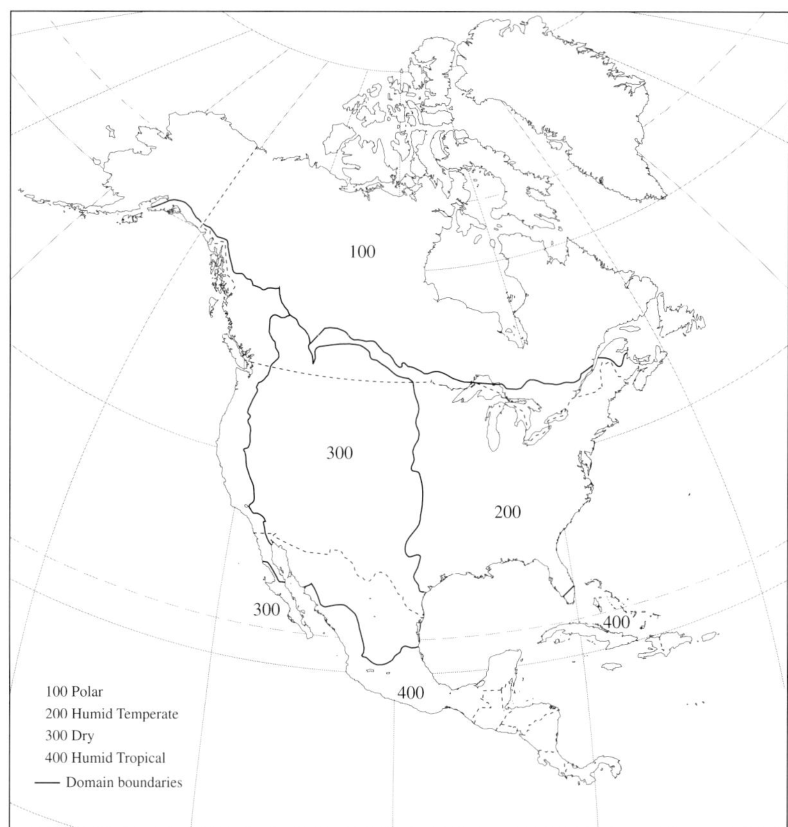

Figure 12.2 Ecoregion domains of North America.

precipitation, is the critical factor in plant distribution and soil development. Three major divisions have been recognized and delimited in terms of thermal efficiency: the ice cap, tundra, and subarctic (taiga).

12.2.1 Ice-cap Division—110

These are the ice sheets of Greenland, Ellesmere Island, and Baffin Island. Mean annual temperature is much lower than that of any other climate, with no month above freezing, defining this climate as *Fi*. Precipitation, almost all occurring as snow, is small but accumulates because of the continuous cold. Driving blizzard winds are frequent. As a result of low monthly mean temperatures throughout the year over the ice sheets, this environment is devoid of vegetation and soils.

12.2.2 Tundra Division—120

The northern continental fringes of North America from the Arctic Circle northward to about the 75th parallel lie within the outer zone of control of arctic air masses. This produces the tundra climate that Trewartha (1968) designated by *Ft*. The average temperature of the warmest month lies between 0°C and 10°C. The tundra climate has very short, cool summers and long, severe winters. No more than 188 days per year, and sometimes as few as 55, have a mean temperature higher than 0°C. Annual precipitation is light, often less than 200 mm, but because potential evaporation is also very low, the climate is humid.

Vegetation on the tundra consists primarily of grasses, sedges, and lichens, with some willow shrubs. Farther south, the vegetation changes into birch-lichen woodland, then into a needle-leaf forest. In some places, a distinct tree line separates the forest from tundra. Köppen (1931) used this line, which coincides approximately with the 10°C isotherm of the warmest month, as a boundary between subarctic and tundra climates.

Soil particles of the tundra derive from mechanical breakup of the parent rock, by continual freezing and thawing, and from chemical weathering, although the relative significance of mechanical and chemical weathering continues to be debated (see chapter 9). Tundra soils (Entisols, Inceptisols and associated Histosols) with weakly differentiated horizons dominate. As in the northern continental interior, the tundra is underlain by permanently frozen ground, permafrost, which is more than 300 m thick throughout the region; seasonal thaw reaches only 10–60 cm below the surface.

Geomorphic processes are distinctive in the tundra, resulting in a variety of curious landforms (see chapter 13). Under a protective layer of sod, ground ice melts in

Table 12.1 General environmental conditions for ecoregional divisions

Name of division	Equivalent Köppen-Trewartha climates	Zonal vegetation	Principal zonal soil type[a]
110 Ice cap	Fi	Not applicable	Not applicable
120 Tundra	Ft	Ice and stony deserts: tundras	Tundra humus soils with solifluction (Entisols, Inceptisols, and associated Histosols)
130 Subarctic	E	Forest-tundras and open woodlands; taiga	Podzolic (Spodosols and Histosols)
210 Warm Continental	Dcb	Mixed deciduous-coniferous forests	Gray-brown podzolic (Alfisols)
220 Hot Continental	Dca	Broadleaf forests	Gray-brown podzolic (Alfisols)
230 Subtropical	Cf	Broadleaf-coniferous evergreen forests; coniferous-broadleaf semi-evergreen forests	Red-yellow podzolic (Ultisols)
240 Marine	Do	Mixed forests	Brown forest and gray-brown podzolic (Alfisols)
250 Prairie[b]	Cf, Dca, Dcb	Forest-steppes and prairies; savannas	Prairie soils, chernozems, and chestnut-brown soils (Mollisols)
260 Mediterranean	Cs	Dry steppe; hard-leaved evergreen forests, open woodland and shrub	Soils typical of semiarid climates associated with grasslands
310 Tropical-subtropical Steppe	BSh	Open woodland and semideserts; steppes	Brown soils and sierozems (Mollisols, Aridisols)
320 Tropical-subtropical Desert	BWh	Semideserts; deserts	Sierozems and desert soils (Aridisols)
330 Temperate steppe	BSk	Steppes; dry steppes	(same as BSh)
340 Temperate desert	BWk	Semideserts and deserts	(same as BWh)
410 Savanna	Aw	Open woodlands, shrubs and savannas; semievergreen forest	Latisols (Oxisols)
420 Rainforest	Ar	Evergreen tropical rain forest (selva)	Latisols (Oxisols)

[a]Great soil group (U.S. Department of Agriculture, 1938). Names in parenthesis are soil taxomony soil orders (Soil Survey Staff 1975); see also chapter 9.
[b]Köppen did not recognize the prairie as a distinct climatic type. The ecoregion classification system represents it at the arid sides of the Cf, Dca, and Dcb types.

summer to produce a thick mud that sometimes flows downslope as solifluction lobes. Freeze and thaw processes also sort the coarse particles from the fine, giving rise to stone polygons, stone stripes, and similar patterns. The coastal plains have numerous lakes of thermokarst origin, formed by melting ground ice.

12.2.3 Subarctic Division—130

The source region for the continental polar air masses is south of the tundra zone between latitudes 50° and 70°N. The climate here exhibits a great seasonal range in temperature. Winters are severe, and the region's small amounts of annual precipitation are concentrated in the three warm months. This cold, snowy, forest climate, referred to in this chapter as the boreal subarctic type, is classified as *E* in the Köppen-Trewartha system. This climate is moist all year, with cool, short summers. Only one month of the year has an average temperature above 10°C.

Winter is the dominant season of the boreal subarctic climate. Because average monthly temperatures are subfreezing for six to seven consecutive months, all moisture in the soil and subsoil freezes solidly to depths of a few meters. Summer warmth is often insufficient to thaw more than a meter or so at the surface, so permafrost prevails under large areas. Seasonal thaw penetrates from 0.5 to 4 m, depending on latitude, aspect, and type of soil.

The subarctic climate zone coincides with a great belt of needle-leaf forest, often referred to as boreal forest, and open lichen woodland known as taiga. Most trees are small, with more value for pulpwood than for lumber. The needle-leaf forest grows on podsols (Spodosols with pockets of wet, organic Histosols). These light gray soils are wet, strongly leached, and acidic. A distinct layer of humus and forest litter lies beneath the top soil layer. Agricultural potential is poor, due to the natural infertility of soils and the bare ice-scoured bedrock, lakes, and wetlands (see chapter 8).

12.3 Humid Temperate Domain—200

Both tropical and polar air masses govern the climate of the humid temperate domain, located in the midlatitudes

a

b

c

d

Figure 12.3 Four ecoregion provinces in North America: (a) broad-leaved forest in Virginia (photograph by U.S. Forest Service); (b) temperate semidesert in southeast Oregon (photograph by Joseph F. Pechanec, U.S. Forest Service); (c) temperate steppe in southern Saskatchewan (photograph by W.G. Pierce, U.S. Geological Survey);
(d) altitudinal zonation beneath Wheeler Peak (3985 m), eastern Nevada: temperate desert mountains with semidesert shrubs in the foreground, coniferous forest on the lower mountain slopes, and alpine tundra toward the crest (photograph by Leland J. Prater, U.S. Forest Service).

between 30° and 60°N, mostly in the southeast quadrant of the continent and along the Pacific Coast (fig. 12.2). The midlatitudes are subject to cyclones; much of the precipitation in this belt comes from rising moist air along fronts within those cyclones. Pronounced seasons are the rule, with strong annual cycles of temperature and precipitation. The seasonal fluctuation of solar energy and temperature is greater than the diurnal fluctuation. Climates of the midlatitudes have a distinctive winter season, which tropical climates do not. The humid temperate domain contains forests of broadleaf deciduous and needle-leaf evergreen trees. The amount of winter frost determines six divisions: warm continental, hot continental, subtropical, marine, prairie, and Mediterranean.

12.3.1 *Warm Continental Division—210*

South of the eastern area of the subarctic climate, between latitudes 40° and 55° N and from the continental interior to the east coast, lies the humid, warm-summer, continental climate. Located between the source regions of polar continental air masses to the north and maritime or continental tropical air masses to the south, it is subject to strong seasonal contrasts in temperature as air masses push back and forth across the continent. The Köppen-Trewartha system designates this area as *Dcb*, a cold, snowy, winter climate with a warm summer. This climate has four to seven months when temperatures exceed 10°C, with no dry season. The average temperature during the coldest month is below 0°C. The warm summer signified by the letter *b* has an average temperature that never exceeds 22°C during its warmest month. Precipitation is ample all year, but substantially greater during the summer.

Needle-leaf and mixed needle-leaf and deciduous forests grow throughout the colder northern parts of this region, extending into the mountains of the Adirondacks and northern New England. Soils are mostly gray-brown podsols (Alfisols). Such soils have a low supply of bases and a horizon in which organic matter, iron, and aluminum have accumulated. They are strongly leached, but they have an upper layer of humus. Cool temperatures inhibit bacterial activity that would destroy this organic matter in tropical regions. Deficient in calcium, potassium, and magnesium, soils are generally acidic. Thus, they are poorly suited to crop production, even though adequate rainfall is generally assured. Conifers thrive in this region.

12.3.2 *Hot Continental Division—220*

South of the warm continental climate lies another division in the humid temperate domain, one with a humid, hot-summer continental climate. It has similar characteristics to the warm continental region except that its summers are much warmer and its winter less cold. The boundary between the two is the isotherm of 22°C for the warmest month. In the warmer sections of the humid temperate domain, the frost-free or growing season continues for five to six months, in the colder sections only three to five months. Snow cover is deeper and lasts longer in the more northerly areas.

In the Köppen-Trewartha system, areas in this division are classified as *Dca* (*a* signifies hot summer). Also included in this division is the northern part of Köppen's *Cf* (subtropical) climate region in the eastern United States. Köppen used the isotherm of -3°C for the coldest month as the boundary between the *C* and *D* climates. Thus, for example, Köppen placed New Haven, Connecticut, and Cleveland, Ohio, in the same climatic region as New Orleans, Louisiana, and Tampa, Florida, despite obvious contrasts in January mean temperatures, soils, and natural vegetation between these northern and southern zones. Trewartha (1968) redefined the boundary between *C* and *D* climates as the isotherm of 0°C for the coldest month, thereby pushing the climate boundary south to a line extending roughly from St. Louis to New York City. Trewartha's boundary is adopted herein to distinguish between humid continental and humid subtropical climates.

Natural vegetation in this region is winter deciduous forest, dominated by tall broadleaf trees that provide a continuous dense canopy in summer, but shed their leaves completely in the winter (fig. 12.3a). Understories of small trees and shrubs are weakly developed. In spring, a luxuriant ground cover of herbs quickly develops, but is greatly reduced after trees reach full foliage and shade the ground. Soils are chiefly red-yellow podzols (Ultisols) and gray-brown podzols (Alfisols), rich in humus and moderately leached with a distinct, light-colored zone under the upper dark layer. The Ultisols have a low supply of bases and a horizon of accumulated clay. Where topography is favorable, diversified farming and dairying are successful.

12.3.3 *Subtropical Division—230*

The humid subtropical climate, marked by high humidity (especially in summer) and the absence of really cold winters, prevails throughout the southern Atlantic and Gulf Coast regions of the United States. In the Köppen-Trewartha system, this area lies within the *Cf* climate, described as temperate and rainy with hot summers. This climate has no dry season; even the driest summer month receives at least 30 mm of rain. The average temperature of the warmest month is greater than 22°C. Rainfall is ample all year, but is markedly greater during summer. Thunderstorms, whether of thermal, squall-line, or cold-front origin, are especially frequent in summer. Tropical cyclones and hurricanes strike the coastal area occasionally, bringing heavy rains. Winter fronts bring precipitation, some in the form of snow. Temperatures are moderately wide in range, comparable to those in tropical deserts, but without the extreme heat of a desert summer.

Soils of the moister, warmer parts of the region are strongly leached red-yellow podzols (Ultisols) related to those of humid tropical climates. Rich in oxides of both iron and aluminum, these soils are poor in many of the plant nutrients essential for successful agricultural production. Forest is the natural vegetation throughout most areas of this division. Much of the sandy coastal region is now covered by a second growth forest of longleaf (*Pinus palustris*), loblolly (*P. taeda*), and slash pines (*P. elliottii*). Inland areas have deciduous forest.

12.3.4 Marine (West Coast) Division—240

Situated on the Pacific coast between latitudes 40° and 60°N, this zone receives abundant rainfall from maritime polar air masses and has a narrow range of temperature because it borders the ocean. Trewartha (1968) classified the marine, west coast climate as *Do*—temperate and rainy, with warm summers. The average temperature of the warmest month is below 22°C, but for at least four months of the year the average temperature is 10°C. The average temperature during the coldest month of the year is above 0°C. Precipitation is abundant throughout the year, but is markedly reduced during the summer. Although total rainfall is not great by tropical standards, the cooler air temperatures reduce evaporation and produce a damp, humid climate with much cloud cover. Mild winters and relatively cool summers are typical. Coastal mountain ranges influence precipitation markedly in these middle latitudes. The mountainous coasts of British Columbia and Alaska annually receive 1500 to 2000 mm of precipitation and more.

Needle-leaf forest is the natural vegetation of the Marine West Coast division. In the coastal ranges of the Pacific Northwest, Douglas-fir (*Pseudotsuga menziesii*), western red cedar (*Thuja plicata*), and Sitka spruce (*Picea sitchensis*) grow to enormous heights, forming some of the densest of all coniferous forests with some of the world's largest trees. Soils are strongly leached, acidic brown forest soils (Alfisols). Because of the region's cool temperatures, bacterial activity is slower than in the warm tropics, so unconsumed vegetative matter forms a heavy surface deposit. Organic acids from decomposing vegetation react with soil compounds, removing bases such as calcium, sodium, and potassium.

12.3.5 Prairie Division—250

Prairies are typically associated with continental, mid-latitude climates designated as *subhumid*. Precipitation in these climates ranges from 500 to 1000 mm per year and is almost entirely offset by evapotranspiration. In the summer, air and soil temperatures are high. Soil moisture in the uplands is inadequate for tree growth, and deeper sources of water are beyond the reach of tree roots. Prairies form a broad belt extending from Texas northward to southern Alberta and Saskatchewan. In a transitional belt on the eastern border of the division, forest and prairie mix.

The prairie climate is not designated as a separate type in the Köppen-Trewartha system, but its recognition by others (e.g., Borchert, 1950) has led to its incorporation into the system presented herein. Prairies lie on the arid western side of the humid continental climate, extending into the subtropical climate at lower latitudes. Temperature characteristics correspond to those of the adjacent humid climates, forming the basis for two types of prairies: temperate and subtropical.

Tall grasses associated with subdominant broad-leaved herbs dominate prairie vegetation. Trees and shrubs are virtually absent, but a few may grow as woodland patches in valleys and other depressions. Deeply rooted grasses form a continuous cover. They flower in spring and early summer, whereas the forbs flower in late summer. In the tall-grass prairie of Iowa, for example, typical grasses are big bluestem (*Andropogon gerardii*) and little bluestem (*Schizachyrium scoparium*), and a typical forb is black-eyed Susan (*Rudbeckia hirta*).

Because rain falls less in the grasslands than in the forest, less leaching of the soil occurs. The pedogenic process associated with prairie vegetation is calcification, because carbonates accumulate in the lower soil layers. Soils include prairie soils, chernozems, and chestnut-brown soils (Mollisols), which have black, friable, organic surface horizons and a high content of bases. Grass roots deeply penetrate these soils. Bases brought to the surface by plant growth are released on the surface and restored to the soil, perpetuating fertility and favoring agricultural productivity.

12.3.6 Mediterranean Division—260

Situated on the Pacific coast between latitudes 30° and 45° N is a zone subject to alternately wet and dry seasons, the transition zone between the dry west coast desert and the wet west coast.

Trewartha (1968) classified the climate of these lands as *Cs*, signifying a temperate, rainy climate with dry, hot summers. The symbol *s* stands for dry summers. The combination of wet winters with dry summers is unique among North America's climate types and produces a distinctive natural vegetation of hard-leaved evergreen trees and shrubs called sclerophyll forest. Various forms of sclerophyll woodland and scrub are also typical. Trees and shrubs must withstand the severe summer drought—two to four rainless months—and severe evaporation, and they are prone to the ravages of fire (see chapter 20). Soils of this region are not susceptible to simple classification in part because of rugged terrain and frequent sediment transfers, but many lowland soils are typical of semiarid grassland climates.

12.4 Dry Domain—300

The essential feature of a dry climate is that annual losses of water through evaporation at the surface exceed annual water gains from precipitation. Thus, no permanent streams originate in dry climate zones. Because evaporation, which depends chiefly on temperature, varies greatly from one area to another, no specific value for precipitation can be used as the boundary for all dry climates.

Two divisions of dry climates are commonly recognized: the arid desert (*BW*), and the semiarid steppe or semidesert (*BS*). Generally, the steppe is a transitional belt surrounding the desert, separating it from the humid climates beyond. The boundary between arid and semiarid climates is arbitrary, but is commonly defined as one-half the amount of precipitation that separates steppe from humid climates.

12.4.1 Tropical-Subtropical Steppe Division—310

Tropical steppes border the tropical deserts in the north from the Gulf coast to the Colorado Plateau, and in the south on the Mexican plateau. Locally, altitude causes a semiarid steppe climate on plateaus and high plains that would otherwise be desert. Trewartha (1968) classified the climate of tropical-subtropical steppes as *BSh*, indicating a hot semiarid climate where potential evaporation exceeds precipitation and where all months have temperatures above 0°C.

Steppes typically are grasslands of short grasses and other herbs, with locally developed shrub and woodland. Pinyon-juniper woodland (*Pinus-Juniperus*) grows on the Colorado Plateau, for example. To the east, in New Mexico and Texas, the grasslands grade into savanna woodland or semideserts composed of xerophytic shrubs and trees, and the climate becomes nearly arid-subtropical. Cactus plants are present in some places. These areas support limited grazing, but they are not generally moist enough for crop cultivation without irrigation. Brown soils and sierozems (Mollisols and Aridisols), containing some humus, are associated with these climates.

12.4.2 Tropical-Subtropical Desert Division—320

The continental desert climates extend south from southern New Mexico and Arizona and eastern California to the high plateaus and mountains of northern Mexico, and also reach the coast around the Gulf of California and the west coast of Baja California. They are not only arid, but have extremely high air and soil temperatures. Direct solar radiation is very high, as is outgoing radiation at night, causing large diurnal temperature variations and rare nocturnal frosts. Annual precipitation is less than 200 mm, sometimes less than 100 mm. These areas have climates that Trewartha (1968) calls *BWh*.

Dry-desert vegetation characterizes the region. Widely dispersed xerophytic plants provide negligible ground cover. During dry periods, visible vegetation is limited to small hard-leaved or spiny shrubs, cacti, or hard grasses. Many species of small annuals may be present, but they appear only after rare, but often heavy, rains have saturated the soil.

In the Mojave and Sonoran Deserts that extend from southeast California across southwest Arizona into Sonora, Mexico, plants are often so large that some places have a near-woodland appearance. They include the treelike saguaro cactus (*Cereus giganteus*), prickly pear cactus (*Opuntia*), ocotillo (*Fouquieria splendens*), creosote bush (*Larrea tridentata*), and smoke tree (*Dalea spinosa*). Elsewhere, however, much of the desert is in fact scrub, thorn scrub, savanna, or steppe grassland. Parts of this region contain shifting dune sands or sterile salt flats with few visible plants. The dominant pedogenic process is salinization, which produces areas of salt crust where only salt-loving plants (halophytic) can survive. Calcification is conspicuous on well-drained uplands, where encrustations and deposits of calcium carbonate (caliche) are common. Humus is lacking and soils are mostly sierozems and desert soils (Aridisols).

12.4.3 Temperate Steppe Division—330

Temperate steppes are areas that have a semiarid continental climatic regime in which, despite maximum summer rainfall, evaporation usually exceeds precipitation. Trewartha (1968) classified the climate as *BSk*. The letter *k* signifies a cool climate with at least one month of average temperature below 0°C. Winters are cold and dry, and summers are warm to hot. Drought periods are common and with the droughts come the dust storms that blow the fertile topsoil from plowed lands being used for dry farming.

The vegetation is steppe, sometimes called shortgrass prairie, and semidesert (fig. 12.3c). Typical steppe vegetation consists of numerous species of short grasses that usually grow in sparsely distributed bunches. Buffalograss (*Buchlow dactyloides*) and blue grama (*Bouteloua gracilis*) are typical grasses; other typical plants include sunflower (*Helianthus annuus*) and locoweed (*Oxytropis*). Scattered shrubs and low trees sometimes grow in the steppe and all gradations of cover are present, from semidesert to woodland. Because ground cover is generally sparse, much soil is exposed. This may be a relatively recent condition due to a large extent, to commercial livestock grazing. The semidesert cover is xerophytic shrub vegetation accompanied by a poorly developed herbaceous layer, most notably the sagebrush vegetation (*Artemisia* spp.) of the middle and southern Rocky Mountain region and the Colorado Plateau. Trees are generally absent.

In this climatic regime, the dominant pedogenic process is calcification, with salinization on poorly drained sites. Soils contain an excess of precipitated calcium carbonate and are rich in bases. Brown soils (Mollisols) are typical in steppe lands. The soils of the semidesert shrub are sierozems (Aridisols) with little organic content, accompanied occasionally by clay horizons and accumulations of various salts. Humus content is low because the vegetation is so sparse.

12.4.4 Temperate Desert Division—340

Temperate deserts of continental regions have low rainfall and strong temperature contrasts between summer and winter. In the intermountain region of the western United States, between the Sierra Nevada–Cascade Range and the Rocky Mountains, this region has the characteristics of a sagebrush semidesert, with a pronounced drought season and a short humid season. Most precipitation falls in the winter, despite a peak in May. Aridity increases markedly in the rain shadow of the western mountains. Even at intermediate elevations, winters are long and cold, with temperatures below 0°C.

Under the Köppen-Trewartha system, this is true desert, *BWk*. The letter *k* signifies that at least one month has an average temperature below 0°C. These deserts differ from those at lower latitudes chiefly in their far greater annual temperature range and much lower winter temperatures. Unlike the dry climates of the tropics, middle-latitude dry climates receive a portion of their precipitation as snow.

Temperate deserts support the sparse xerophytic shrub vegetation typical of semideserts (Fig. 12.3b). One example is the sagebrush (*Artemesia* spp.) vegetation of the Great Basin and northern Colorado Plateau region. Recently, as a result of overgrazing and trampling by livestock, semidesert shrub vegetation has invaded wide areas of the western United States that were formerly steppe grasslands. Soils of the temperate desert are sierozems (Aridisols), low in humus and high in calcium carbonate. Poorly drained areas develop saline soils, and salt deposits cover dry lake beds.

12.5 Humid Tropical Domain—400

Equatorial and tropical air masses largely control the humid tropical group of climates found at low latitudes (fig. 12.2). Every month of the year has an average temperature above 18°C, and no winter season occurs. In these tropical systems, the primary periodic energy flux is diurnal: the temperature variation from day to night is greater than from season to season. Average annual rainfall is heavy and exceeds annual evaporation, but varies in amount and in season and distribution.

Two types of climates are differentiated on the basis of the seasonal distribution of precipitation. The tropical wet, or rainforest, climate has ample rainfall through ten or more months of the year. It extends southward from the Yucatan peninsula into Central America, mainly along the windward Caribbean coast. The tropical wet-and-dry, or savanna, climate has a dry season more than two months long, and extends southward from Tampico on the Gulf of Mexico and Guaymas on the Gulf of California to encompass much of southern Mexico and the Pacific slopes of Central America. Thus, for the purpose of this book, which defines the essence of North America as extratropical, both these divisions lie beyond the area of concern, except the small area of tropical savanna and swampland in southern Florida discussed next (fig. 12.1).

12.5.1 Savanna—410

Globally, the latitudinal belt between 10° and 30° N is intermediate between the equatorial and middle-latitude climates. This produces the tropical wet-dry savanna climate, which has a wet season controlled by moist, warm, maritime tropical air masses at times of high sun, and a dry season controlled by the continental tropical masses at times of low sun. Trewartha (1968) classified the tropical wet-dry climate as *Aw*, the latter *w* signifying a dry winter. Alternating wet and dry seasons result in the growth of a distinctive vegetation known generally as tropical savanna, characterized by open expanses of tall grasses interspersed with hardy drought-resistant shrubs and trees. Some areas have savanna woodland. Soils are mostly latisols (Oxisols), and high rainfall and high temperatures cause heavy leaching.

In North America, the characteristics of this savanna division as found in southern Florida are strongly modified by local conditions. The region, a low-lying, emergent carbonate platform characterized by limestone solution and negligible drainage gradients, is dominated by swamps. Indeed, 80% of the 20,000–km^2 region south of Lake Okeechobee is swampland, whereas the remainder is covered by wet savanna with pines and a few hardwoods. Of these swamps, the Everglades are dominated by so-called sawgrass, actually the sedge *Cadium effusum*.

12.6 Qualifications to the Classification System

12.6.1 Arrangement of Ecological Zones by Altitude

The arrangement of the ecoregions described in this chapter depends largely on latitude. This pattern, however, is overlain by mostly north–south mountain ranges, which cut across latitudinally oriented climatic zones to create their own ecosystems. Altitude creates characteristic ecological zones that are variations of lowland climates, which in turn reflect the location of mountains in the overall pattern of global climatic zones. The Coast Ranges of California, for example, experience strong seasonal energy variations and seasonal moisture regimes similar to neighboring lowlands, but these are modified by altitude so that temperatures are lower and precipitation higher and more effective.

Every mountain within a climatic zone has a typical sequence of altitudinal belts, with different ecosystems at successive levels, generally montane, alpine, and nival, but exhibiting considerable differences according to the zone in which they occur (table 12.2, fig. 12.3d). When a mountain extends over two or more climatic zones, it produces different vertical zonation patterns. This is shown in figure 12.4, which compares locations along the length of the Western Cordillera from Alaska to Colorado.

12.6.2 Patterns Within Zones

The climatic zones give only a broad-brush picture. Variations within a zone break up and differentiate major subcontinental zones. For example, vegetation in the savanna is highly differentiated and is related to variation in the length of the dry season.

Within the same macroclimate, broad-scale landforms break up the zonal pattern and provide a basis for further delineation of ecosystems, known as *landscapes*. In different climates, the same landform results in different landscape patterns.

A landscape may be further subdivided into microscale ecosystems called *sites*. Within a landscape, the sites are arranged in a specific pattern. For example, various ecosystems form a mosaic of riparian, forest, and grassland sites in the Idaho Mountains, a high-mountain landscape of the temperate semiarid zone. Deep dissection of the mountain range has resulted in variously oriented slopes with varying local climates. The result is a striking contrast in vegetation. Local variations in topography will cause small-scale variations in the amount of solar radiation received, create topoclimates, and affect the soil moisture. These variables will subsequently affect the biota, creating ecosystem sites.

Topography, even in areas of uniform macroclimate, leads to differences in local climates and soil conditions. Whereas the climatic climax would occur, theoretically, over an entire region, topography may lead to different local climates.

These deviations occur in various combinations within a region, resulting in three types of sites: (1) *zonal* sites, such as a well-drained sagebrush terrace in a semiarid climate; (2) *azonal* sites, such as a riparian forest; and (3) *intrazonal* sites, that may occur in any zone on extreme types of soil and parent material that override the climatic effect, such as sand dunes.

12.6.3 Boundaries Between Zones

The scheme developed in this chapter provides a general picture of ecological regions in North America. Nevertheless, the boundaries are imprecise. For example, whereas the 10°C isotherm may coincide with the northernmost limits of tree growth, separating boreal forest from treeless tundra, an observer would see not an abrupt line but a transitional zone with trees on favorable sites, muskeg and bog on wetter sites, and tundra on exposed ridges (see chapter 14).

12.7 Conclusion

This chapter has surveyed the ecoregion geography of North America at two levels—the domain and the division.

Table 12.2 Types of altitudinal spectra

Name of division	Altitudinal spectra
110 Ice cap	Polar desert
120 Tundra	Tundra–polar desert
130 Subarctic	Open woodland–tundra; taiga–tundra
210 Warm Continental	Mixed forest–coniferous forest–tundra
220 Hot Continental	Deciduous or mixed forest–coniferous forest–meadow
230 Subtropical	Mixed forest–meadow
240 Marine	Deciduous or mixed forest–coniferous forest–meadow
250 Prairie	Forest–steppe–coniferous forest–meadow
260 Mediterranean	Mediterranean woodland or shrub–mixed or coniferous forest–steppe or meadow
310 Tropical-subtropical steppe	Steppe or semidesert–mixed or steppe coniferous forest–alpine meadow or steppe
320 Tropical-subtropical desert	Semidesert–shrub–open woodland–desert steppe or alpine meadow
330 Temperate steppe	Steppe–coniferous forest–tundra; steppe–mixed forest–meadow
340 Temperate desert	Semidesert woodland–meadow
410 Savanna	Open woodland–deciduous forest–coniferous forest–steppe or meadow
420 Rainforest	Evergreen forest–meadow or paramos

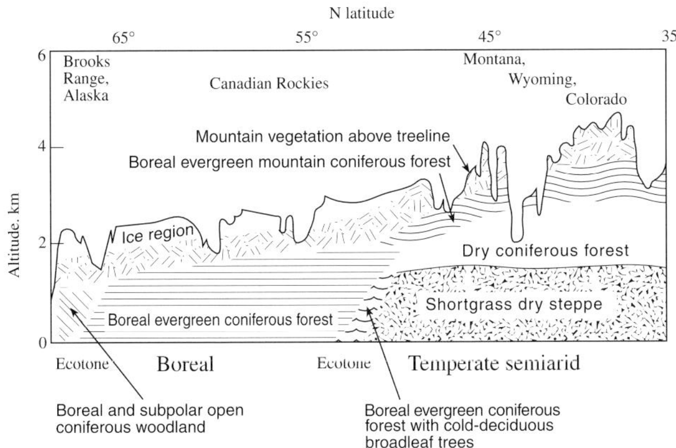

Figure 12.4 Vertical zonation in different ecoclimatic zones along the eastern slopes of the Western Cordillera (from Schmithüsen, 1976).

These levels are based on macrofeatures of the climate. Each type of climate, together with its characteristic vegetation, soils, landforms, and geomorphic processes, comprises a unique ecoregion and supports a distinctive pattern of ecosystems.

The vegetation descriptions are those of "potential" natural vegetation, which may not exist under current land-use practices. Furthermore, the descriptions refer to zonal conditions, namely, the vegetation that would occur if the area were reasonably flat and well drained. Local contrasts, related to elevation, substrate, drainage, and other factors, can exist within a region, forming a complex intraregional mosaic.

As society moves away from concern for single resources toward the management of entire ecosystems, it is essential to know where such ecosystems are and why they are there. The ecoregion perspective presented in this chapter is a contribution to effective environmental management.

Acknowledgments I am indebted to John M. Crowley, who began the task of classifying the natural ecosystems of the continents and their distribution. I extend my thanks to Gordon Warrington for his review of the manuscript.

References

Bailey, R.G., 1994. *Ecoregions of the United States*, revised 1994. Washington, DC: USDA Forest Service. 1:7,500,000.

Bailey, R.G., 1996. *Ecosystem geography*. New York: Springer-Verlag. 216 p.

Bailey, R.G., 1998a. *Ecoregions map of North America: Explanatory note*. Misc. Publ. 1548. Washington, DC: USDA Forest Service. In cooperation with The Nature Conservancy and the U.S. Geological Survey. 10 p. with separate map at 1:15,000,000.

Bailey, R.G., 1998b. *Ecoregions: The ecosystem geography of the oceans and continents*. New York: Springer-Verlag. 186 p.

Borchert, J.F., 1950. The climate of the central North American grassland. *Annals, Association of American Geographers*. 40: 1–39.

Crowley, J.M., 1967. Biogeography [in Canada]. *Canadian Geographer* 11: 312–326.

Köppen, W., 1931. *Grundiss der klimakunde*. Berlin: Walter de Gruyter. 388 p.

Milanova, E.V., and A.V. Kushlin, eds., 1993. *World map of present-day landscapes: An explanatory note*. Moscow: Moscow State University. 33 p. with separate map at 1:15,000,000.

Schmithüsen, J., 1976. *Atlas zur biogeographie*. Mannheim-Wien-Zurich: Bibliographisches Institut. 33 p.

Soil Survey Staff, 1975. *Soil taxonomy: A basic system for making and interpreting soil surveys*. Agric. Handbook 436. Washington, DC: U.S. Department of Agriculture. 754 p.

Trewartha, G.T., 1968. *An introduction to climate*, 4th ed. New York: McGraw-Hill. 408 p.

U.S. Department of Agriculture, 1938. *Soils and men*, 1938 Yearbook of Agriculture. Washington, DC: U.S. Government Printing Office. 1232 p.

II

REGIONAL ENVIRONMENTS

13

The Far North: A Geographic Perspective on Permafrost Environments

Frederick E. Nelson
Kenneth M. Hinkel

The influence of *permafrost*, defined as any subsurface earth material remaining below the freezing point of water for two or more years (Associate Committee on Geotechnical Research, 1988), is interwoven closely with the surface characteristics, biotic environment, and human infrastructure of Alaska and the northern reaches of Canada. Permafrost is a unifying concept between those branches of natural science and engineering concerned with the continental sections of the high latitudes. The frozen condition of the subsurface affects many aspects of the North's physical geography, but the local characteristics of permafrost are also heavily dependent on myriad influences and can be affected profoundly by changes in the natural or built environment.

The literature on permafrost and periglacial geomorphology of North America is vast and extends over a period of more than a century. In this brief review, we illustrate some important permafrost-induced phenomena, stressing results from recent studies. We place special emphasis on those aspects of permafrost science that have facilitated recent progress in understanding the geographic distribution of various phenomena, an important goal of climate-change science that has reinvigorated much of physical geography. The recent emphasis on spatially oriented permafrost research has been facilitated by the intersection of three factors: (1) the development of geographic information systems (GIS) technology; (2) the issues raised in global-change research; and (3) recent recognition of the importance of preservation and access to data (Clark and Barry, 1998), as well as expectations by funding agencies in this regard.

Useful regional accounts of the physiography and Quaternary history of Alaska were provided by Wahrhaftig (1965), Péwé (1975), and the volumes edited by Wright (1983) and Wright and Porter (1983). For northern Canada, comprehensive overviews were provided by Bird (1967), Bostock (1970), Trenhaile (1998), and the edited volume by French and Slaymaker (1993). A landform atlas of Canada (Mollard, 1996) includes a large number of air photographs of typical permafrost features. Carter et al. (1987) provided an excellent review of permafrost geomorphology in lowland areas of the North American Arctic. Recent systematic treatments of permafrost-related phenomena, with many North American examples, include those of Washburn (1980), Andersland and Ladanyi (1994), and French (1996). Articles contained in the *Proceedings* of the seven International Permafrost Conferences held to date contain a wealth of detailed information on specific aspects of frozen-ground conditions in North America.

13.1 Background

13.1.1 *Permafrost*

Although permafrost can contain large volumes of water substance and survive for hundreds of thousands of years, the term is defined solely on the basis of thermal proximity to the ice point over a relatively brief period of time. Despite its etymological construction, modern usage of the term *permafrost* does not connote either the presence of water or permanence; if they remain below 0°C for more than 2 years, bedrock, ice,-rich silt, and buried organic matter all qualify as permafrost. Numerous authors have noted that *perennially frozen ground* is perhaps the most succinct phrase that describes the phenomenon accurately. Many locations have alternated repeatedly between permafrost and nonpermafrost status in response to global climatic changes or more localized environmental factors such as vegetation succession, hydrological changes, and human activities. In recent years, the term *geocryology* has found favor among many writers seeking a concise term to describe the scientific study of permafrost and related landforms (e.g., Washburn, 1980; Williams and Smith, 1989; Yershov, 1998).

Unlike the situation in Russia, where geocryological investigations began several centuries ago (see, for example, Yershov, 1998: 18–26), systematic research on permafrost has a relatively short history in North America. Although officers of the Hudson's Bay Company called for inventories of perennially frozen ground over the area of its domain (Richardson, 1839), observations were both scattered and sparse until the gold rush of 1898 brought development to northwestern North America and highlighted permafrost-related engineering problems (Brown, 1970, 45). Results from subsequent reconnaissance investigations in Alaska and the western Canadian Arctic during the first decades of the twentieth century are documented in many publications of the United States Geological Survey and the Geological Survey of Canada (e.g., Leffingwell, 1919). Interest in permafrost revived during World War II for similar reasons, culminating in the first comprehensive English-language book on the subject (Muller, 1947). With the opening of the Naval Arctic Research Laboratory in Barrow, Alaska, half a century ago (Reed, 1969), permafrost research in North America entered its modern phase. Development of the petroleum reserves at Prudhoe Bay, construction of the Trans-Alaska Pipeline System, iron ore operations at Schefferville, Québec, the Mackenzie pipeline inquiry (Williams, 1986, chapter 5), and many other developments have provided impetus and opportunities for basic and applied research into permafrost in the northern parts of North America.

Permafrost is, in the first instance, a climatically determined phenomenon. Figure 13.1 illustrates a typical thermal profile at a location where permafrost is a consequence of the contemporary surface climate. The thickness of permafrost is a function of mean annual temperature, the geothermal gradient, and the thermal properties of the earth materials between the surface and the bottom of the permafrost. Permafrost thickness in North America ranges from a few centimeters near its southernmost limit to possibly more than 1000 m in northern Baffin Island (Judge, 1973). Permafrost thickness does not map as a smooth field, however (e.g., Osterkamp and Payne, 1981). At Prudhoe Bay in north-central Alaska, permafrost extends more than 600 m below the surface. Near Barrow, about 300 km to the west, permafrost is much thinner (~400 m), despite a similar climatic regime. This discrepancy is attributed to differences in the thermal properties of the sediments at the respective locations (Lachenbruch and Marshall, 1969).

The study of permafrost and periglacial geomorphology is concerned to a large extent with heat transfer in cold environments. Recent geocryological texts reflect this fact (e.g., Lunardini, 1981; Williams and Smith, 1989; Andersland and Ladanyi, 1994; Yershov, 1998). Permafrost is an integrated thermal response to the influence of macro- and microclimate, snow cover, the thermal properties of subsurface materials, vegetation cover, moisture content, and the flow of heat from Earth's interior. The complex interplay between these factors gives rise to heat transfer by both conductive and nonconductive processes, and results in complicated patterns of geocryological phenomena over a wide range of spatial and temporal scales.

13.1.2 *The Active Layer*

An important ancillary concept is the *active layer,* a layer of ground above permafrost that freezes and thaws on an annual basis (fig. 13.1). If the formation of permafrost at a given location is a consequence of the contemporary climate, the bottom of the active layer coincides with the top of permafrost, called the *permafrost table.* Because it achieves a thawed state on an annual basis, most biological and hydrological activity in permafrost regions occurs in the active layer. This relatively thin but spatially and temporally variable layer serves as a boundary across which exchanges of heat, moisture, and gases occur between the atmospheric and terrestrial systems. The active layer provides the rooting zone for plants, and most hydrological and mass-movement processes occur within it. The active layer is the only part of the permafrost system in which major changes in its properties occur during each annual cycle. These changes include moisture content, thermal properties, density, and mechanical strength.

Like permafrost, the thickness of the active layer is an integrated response to a large number of factors, including surface temperature, thermal properties of the ground cover (vegetation and snow), thermal properties and mois-

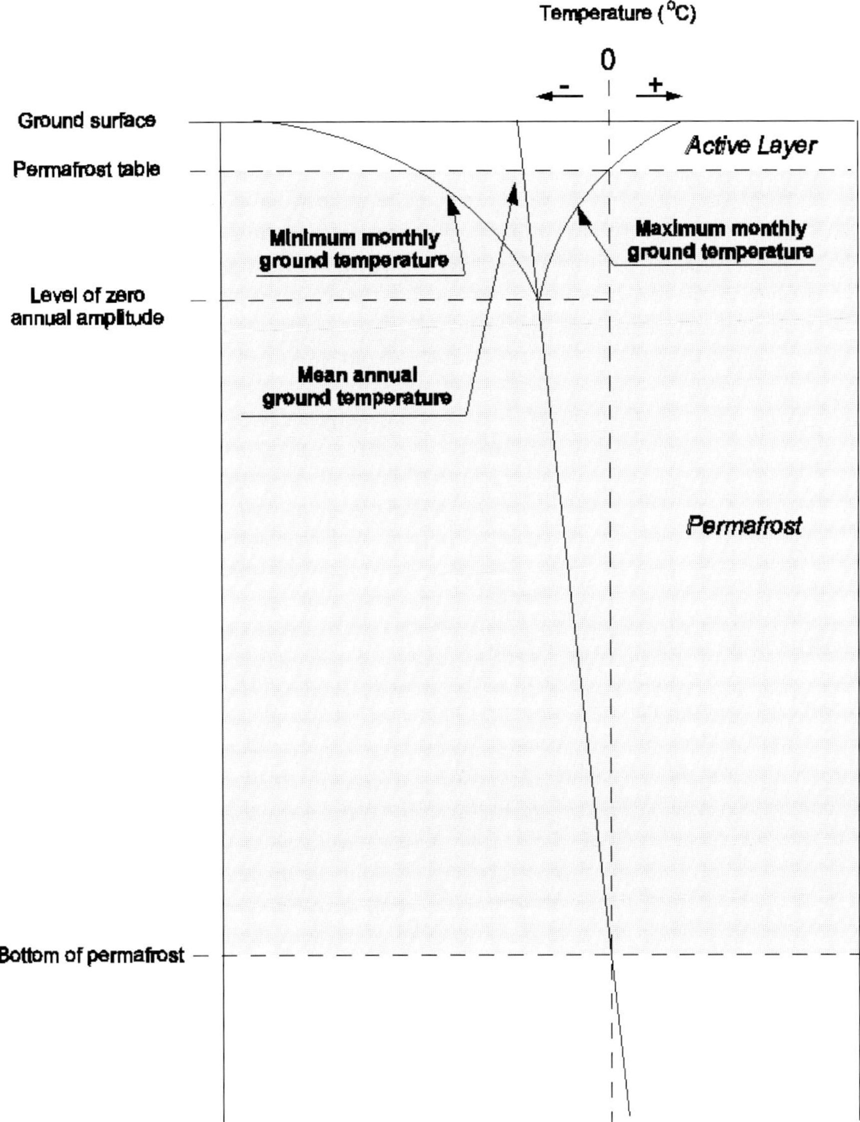

Figure 13.1 Ground thermal regime in permafrost, following Brown (1970).

ture content of the substrate, and topographic position. Moreover, nonconductive heat-transfer processes are known to operate within the active layer (e.g., Outcalt et al., 1990; Hinkel et al., 1997). Strategies for modeling near-surface thermal evolution in permafrost environments are well developed (e.g., Goodrich, 1982; Waelbroeck, 1993), and numerical models are in wide use, particularly in applied work (Nixon, 1998). Despite the complexity of heat-transfer processes in the active layer, it is frequently possible to describe thaw progression using relatively simple solutions, such as the Stefan equation (e.g., Harlan and Nixon, 1978) because the rate of thawing is proportional to the square root of time.

13.1.3 Ground Ice

To the uninitiated of an earlier era, the gently sloping, low-relief landscape of the lowlands of North America's western Arctic may have appeared as an excellent candidate for relatively simple construction techniques. Only a cursory inspection of the scientific literature, however, reveals that subsurface ice is abundant in much of Alaska's North

Slope and the coastal areas of western Canada. This *ground ice* has a variety of origins (Mackay, 1972). Human-induced or natural changes in the energy balance at the surface can lead to thawing of ice-rich permafrost, resulting in differential settlement and development of *thermokarst* terrain. Ground ice is responsible for much of the relief, both positive and negative, in the outermost coastal plain of northwestern North America. In upland areas, ice provides mechanical discontinuities upon which mass movements occur; meltwater generated by the thawing of ice-rich sediments reduces the shear strength of near-surface sediments and may also act as a lubricant.

One of the most widespread forms of ground ice occurs as networks of downward-tapering veins, known as *ice-wedge polygons* (fig. 13.2). Leffingwell (1915) provided the first cogent explanation for these features, which involves repeated cracking of frozen ground in winter, and infiltration and freezing of snow meltwater the following spring. Lachenbruch (1962) and Plug and Werner (1998) provided confirmation of Leffingwell's interpretation from more theoretical perspectives, and many empirical investigations (e.g., Mackay et al., 1978; Mackay, 1993) have added details to our knowledge of this phenomenon.

The stratigraphic profile at the interface between the top of permafrost and the base of the active layer typically shows relatively thick accumulations of segregation ice, a consequence of two-sided freezing (Mackay, 1981) and migration of moisture to freezing fronts near the surface and the base of the active layer. Although the latent heat associated with a significant thickness of ice-rich permafrost is substantial and slows the thawing process considerably (Riseborough, 1990), human-induced changes at the surface, such as removal of the vegetation cover, can result in pronounced subsidence over periods of a decade or less. Similarly, a series of very warm summers can thaw the ice-rich layer considerably and "reset" the depth to which the active layer extends, even in subsequent colder years (Nelson et al., 1998).

Figure 13.2 Polygonal network of ice wedges near Prudhoe Bay, Alaska.

Ground ice can play an extremely important role in reconstructing the Quaternary history of permafrost regions. Thaw unconformities provide information about previous warm intervals (e.g., Burn, 1997), ice crystallography indicates the nature of genetic processes (e.g., Pollard and van Everdingen, 1992), and the geochemistry of ground ice yields information about both genesis and age (e.g., Moorman et al., 1996). The term *cryostratigraphy* has found recent favor among permafrost scientists engaged in interpreting earth history from stratigraphic analysis of ground ice (French, 1998).

13.2 Geographical Distribution of Permafrost Phenomena

13.2.1 Permafrost

The climatic origin of permafrost is easily discernable from a map of its geographic distribution. At small geographical scales (i.e., over large areas), permafrost distribution is usually mapped on the basis of its lateral continuity, using a classification scheme with roots in Dokachaev's notion of physical-geographical zonation (e.g., Tricart and Cailleux, 1972). At the continental scale, permafrost distribution in North America conforms closely to the concept of geographic zonality. The arrangement of permafrost zones can be represented effectively using relatively simple climate-based methods (Nelson, 1986; Anisimov and Nelson, 1997).

Under severely cold climatic regimes in North America, permafrost underlies most locations and is therefore classified as *continuous*. Farther south, in the *discontinuous* zone, permafrost occurs where local ecological, lithological, edaphic, and hydrological conditions are favorable for its formation or maintenance. In the *sporadic* zone and near its southern limit, permafrost occurs in relatively small patches and is usually associated with peat accumulations (e.g., Zoltai, 1971). The continuity scheme has been implemented in a variety of ways, often on the basis of the percentage of the land surface underlain by permafrost. Nelson (1989b) provided a discussion of many of these schemes in the North American context.

Figure 13.3 is based on the most recent and detailed depiction of the distribution of permafrost in North America (Brown et al., 1997). This map is taken from the first coordinated effort to portray the distribution of permafrost in the circumarctic using explicit, standardized criteria. The zonal nature of the permafrost regions is readily apparent, with concentric continuity categories corresponding roughly with the parallels of latitude.

One of the primary factors behind the geographic distribution of permafrost near its southernmost extent in

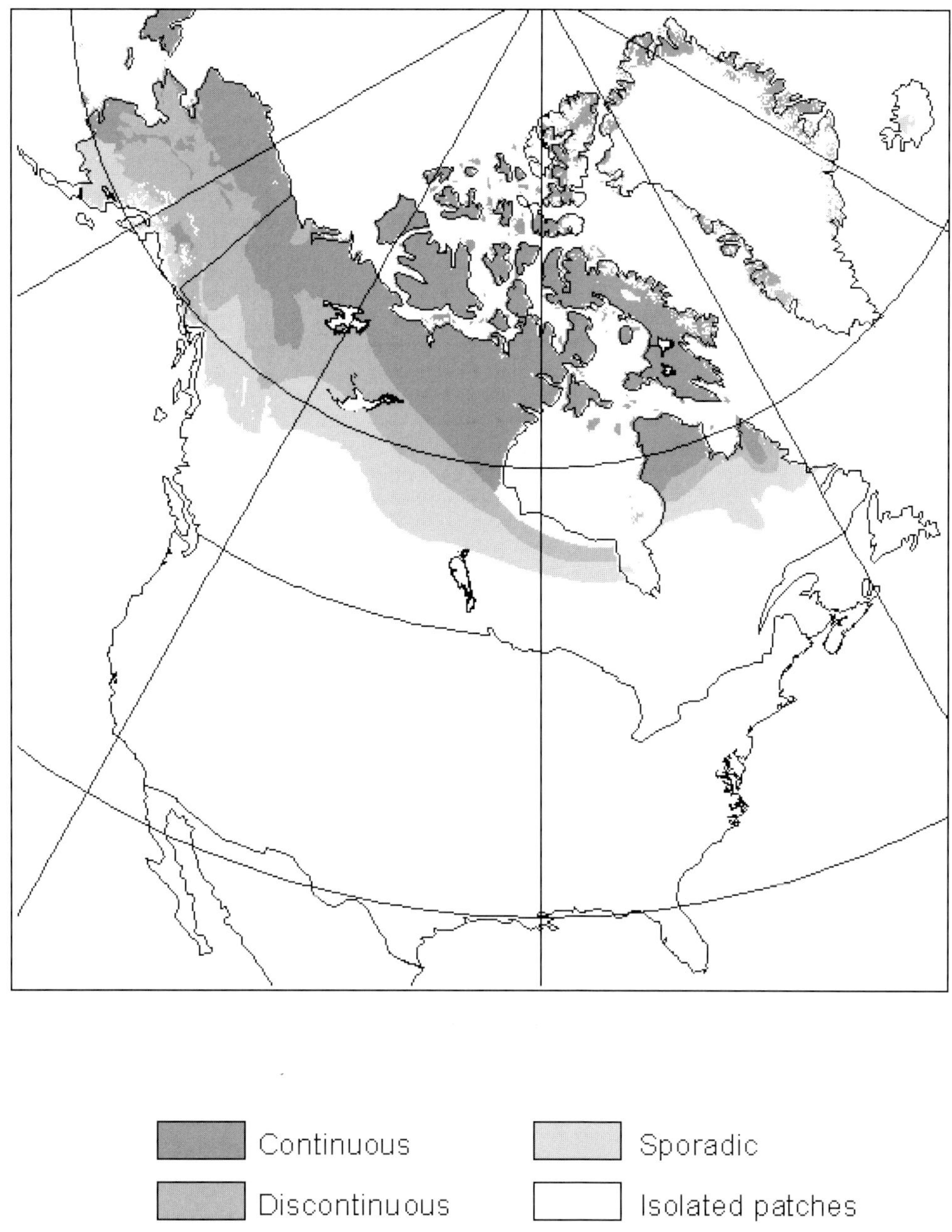

Figure 13.3 Map of permafrost zonation in North America. Zones of continuity are defined by percentage of area underlain by permafrost. Continuous: 90–100%. Discontinuous: 50–90%; Sporadic: 10–50%; Isolated patches: 0–10%. Adapted from Brown et al. (1997).

North America is the unusual thermal behavior of peat (section 13.2.3). Vast portions of the North American subarctic are occupied by peatlands (see chapter 8). The subarctic peatlands are an important component of the world's carbon pool and could play an important role in greenhouse warming (section 13.4). Peatlands are also an important influence on permafrost distribution. Because the thermal properties of peat favor the preservation of a frozen substrate, permafrost occurs over a wide area of taiga, extending from central Alaska through the northern parts of the Prairie Provinces, northern Ontario and Quebec, and even into the Maritime Provinces. In the "isolated patches" zone of figure 13.3, nonrelict permafrost at low elevation is confined almost exclusively to peatlands.

At the continental scale, one of the most striking features of permafrost in North America is the abrupt northward offset of zonal boundaries east of Hudson Bay. This pattern is a consequence of the juxtaposition of land and water, and it illustrates the climatic dependencies of permafrost. West of Hudson Bay, the climate is highly continental, with relatively low snowfall. Because Hudson Bay remains open through the early part of the winter, westerly winds bring moisture to northern Quebec early in the cold season and snowfall there is relatively high. This snow, which remains throughout most of the winter, acts as a layer of thermal insulation and decreases the severity of the soil climate, ultimately leading to the northward displacement of permafrost zones evident on the map (Brown, 1960, 1967). In northwestern North America, smooth zonal boundaries give way to a more complex pattern in the Cordilleran region, where an altitudinal zonation is superimposed on the latitudinal progression.

The configuration of permafrost zones is not static over time. Thie (1974) and Kwong and Gan (1994) detected northward movement of the southernmost permafrost limit in central Canada during this century. Zoltai's (1995) paleoecological analysis demonstrated southward movement of this limit at the close of the mid-Holocene warm interval. Simulation studies by Anisimov and Nelson (1996, 1997) indicate substantial potential for changes in the configuration of permafrost zones in response to global warming (section 13.3).

Permafrost mapping has also been carried out at regional scales in North America. Work has been concentrated in the discontinuous and sporadic zones, in which the presence or absence of permafrost is of great importance to engineering, resource extraction, and agricultural activities (section 13.4). Methods based on remote sensing include satellite imagery (Morrissey et al., 1986), synthetic aperture radar (Granberg,1994), ground penetrating radar (Judge et al., 1991), and sophisticated classification algorithms (Leverington and Duguay, 1997), as well as more traditional air-photo analysis (e.g., Thie, 1974; Halsey et al., 1995; Mollard, 1996).

Jorgenson and Kreig (1988) used a GIS approach, involving terrain information, soil thermal properties, a topoclimatic index, and climate data, to map permafrost distribution in a subarctic watershed in central Alaska. Riseborough and Smith (1998) outlined an extension to this work, involving calculations of the temperature at the top of permafrost, that may be useful in the context of regional mapping. Both methods rely on information about the ratio of temperatures in the air and at the surface (the *"n-factor"*), and their utility will be limited until more data on this parameter are available for broad categories of natural ground cover.

13. 2.2 Active Layer

Like permafrost, active-layer thickness is related closely to climatic parameters, especially the seasonal total of thawing degree days. It is not surprising, therefore, that the active layer decreases in thickness in a very general way over a latitudinal transect from south to north. Figure 13.4 is an idealized, general representation of how active-layer thickness would vary with geography in North America if the substrate comprised a uniform silt-textured soil with moderate soil moisture and without vegetation cover.

Because the depth of thaw is affected by many factors, however, the idealized representation of figure 13.4 is an extreme oversimplification. Vegetation, soil properties, and hydrological variation play very important roles in determining the depth of seasonal thaw and can easily obscure regional trends within local areas. Another extremely important factor is the well-known discrepancy between temperature at the surface and that measured at standard screen height in the air. Because surface temperatures are not measured routinely, air temperature is frequently substituted, which can introduce substantial error into analytic estimates of thaw depth.

As a result of these difficulties, detailed mapping of active-layer thickness at regional scales was not undertaken until relatively recently. Although remote sensing holds great potential for estimating the volume of thawed soil in a region, several recent experiments (Peddle and Franklin, 1993; Leverington and Duguay, 1996; McMichael et al., 1997) obtained only moderate correspondence between

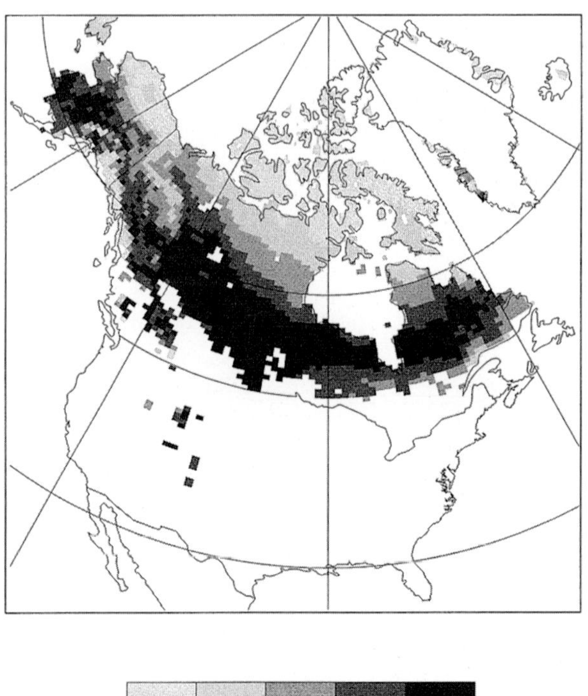

Figure 13.4 Active-layer thickness in North America, computed using contemporary climate data, with the assumption of homogeneous, moderately wet, silt-textured soil without vegetation cover. Values are in cm. Adapted from Anisimov et al. (1997).

measured active-layer thickness and those predicted using remotely sensed parameters.

A semiempirical approach to mapping, using GIS technology and an extensive program of field observations (Nelson et al., 1997), is based on determining thaw depth in control plots representative of vegetation and soil conditions in the target area. By combining climatic information with a digital elevation model of the study area, a digital representation of land cover characteristics, and observed active-layer thickness at several training sites, an estimate of the volume of thawed soil was achieved for a 26,278-km^2 area centered on the Kuparuk River basin in north-central Alaska (fig. 13.5). Volumetric estimates are important for deriving estimates of the potential for release of greenhouse gases to the atmosphere (section 13.3). Other recent work, also in the Kuparuk River basin, achieved comparable results using more purely computational methods. Hinzman et al. (1998) developed a complex, spatially distributed numerical model of heat and mass transfer. Shiklomanov and Nelson (1999) used a distributed analytic solution for thaw depth. Klene (1999) developed a solution based on n-factor data collected within representative naturally occurring ground cover categories. Similar fieldwork involving the n-factor is being carried out in the Mackenzie Valley in northwestern Canada (Nixon and Taylor, 1998).

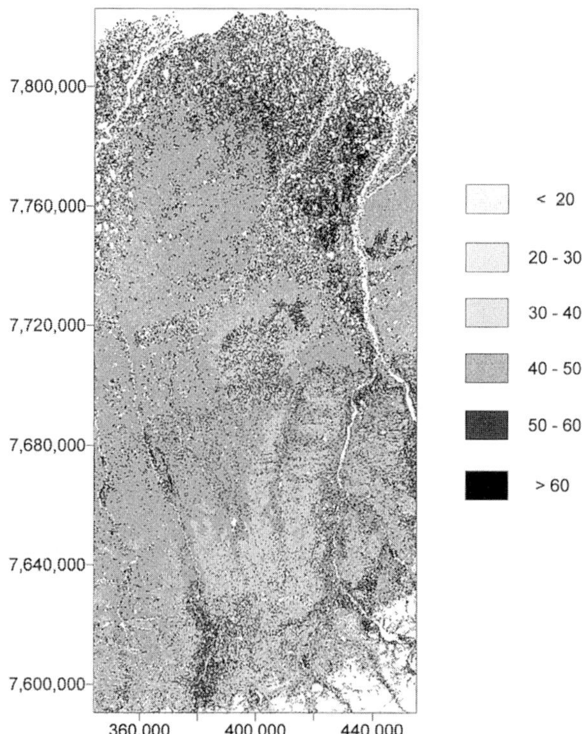

Figure 13.5 Map of active-layer thickness in Kuparuk River region, north-central Alaska, August 1995. Position is given by universal transverse Mercator coordinates, zone 6. Legend gives active-layer thickness in cm. Adapted from Nelson et al. (1997).

13.2.3 Ground Ice

Figure 13.6 provides a very generalized depiction of the relative abundance of ground ice in northern North America, following Brown et al. (1997). Terrain with high ice content (>20% by volume in the upper 10–20 m) occurs primarily in the western Arctic (portions of Alaska, Yukon, and westernmost Northwest Territories bordering the Beaufort Sea) and in the Canadian Arctic Archipelago. At this scale, ice content appears relatively low (<10%) throughout most of the Canadian Shield and in the complex terrain of the Western Cordillera.

Although at first glance it may seem paradoxical, in many parts of the North American Arctic there is an abundance of water at the surface, despite precipitation rates that would yield desert conditions under a warmer temperature regime. Because permafrost presents an impermeable layer at shallow depth, the unfrozen active layer in summer is often saturated, leading to ponded conditions at the surface. Low evaporation rates during the short, cool summers contribute to the situation. Viewed from the air, the most striking feature of the landscape in many parts of North America's western Arctic depicted in figure 13.6 with high ice content is the dominance of elliptically shaped *thaw lakes*, developed by thawing of ice-rich, near-surface permafrost. In the coastal parts of Alaska's North Slope, 50–70% of the land surface is covered either by lakes or by mires in the basins of former lakes (Hussey and Michelson, 1966). Most are shallow, and only those deeper than about 2 m freeze to the bottom in winter (Brewer, 1958). Regardless of size and depth, many of these lakes display a strikingly uniform orientation pattern, aligned nearly perpendicular to prevailing winds (fig. 13.7).

A qualitative model, known as the "thaw-lake cycle," was developed by Britton (1957) and refined by Billings and Peterson (1980). Coalescence of the small ponds contained in low-centered, ice-wedge polygons proceeds by thermal erosion of the wedges. Thaw subsidence occurs in supersaturated sediments and, when underlying ice-rich layers of permafrost thaw, the basin becomes a reservoir for meteoric water and for snowmelt runoff. Once water depth exceeds about 2 m, winter lake ice no longer reaches the lake sediments and a depression (a "thaw bulb") develops in the surface of the underlying permafrost. After reaching a sufficient size, the body of water becomes oriented by a two-cell circulation pattern induced by persistent, unidirectional sea breezes created by thermal contrasts along the coast (Kozo, 1979). Basins grow by thermoerosion processes acting primarily on the ends of the lakes, and they often develop an asymmetrical, elliptical outline (Carson and Hussey, 1962). Measured bank erosion rates at the

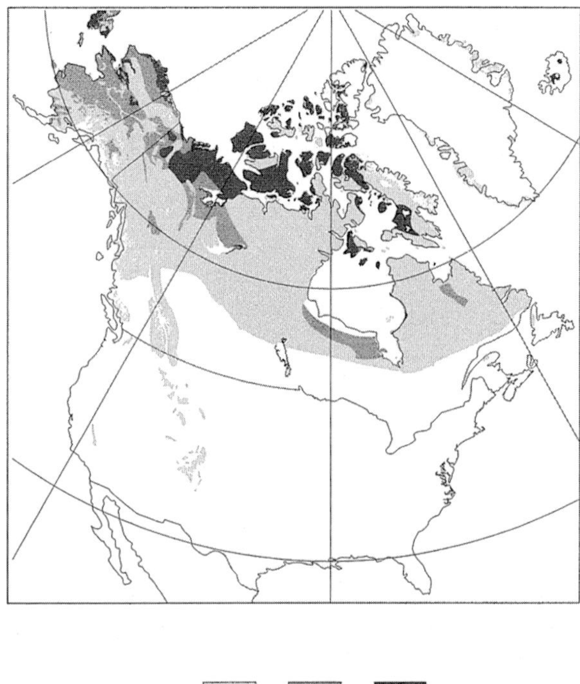

Figure 13.6 (*left*) Ground ice content in North America. Categories are defined by the percentage volume occupied by the ice in upper 10–20 m: High: >20%, Medium: 10–20%; Low: 0–10%. Adapted from Brown et al. (1997).

Figure 13.7 (*right*) Satellite image of active and drained thaw-lake basins in the vicinity of Barrow, Alaska.

degrading lake ends range from several cm to 25 cm yr^{-1} (Black, 1969; Tedrow, 1969). Undercutting of banks yields mats of fibrous organic material that break up and settle to the lake bottom, providing insulation for underlying systems of ice wedges. Exposed ice wedges near the margins of the lake are more vulnerable to thermal and mechanical erosion and breaching by headward erosion in small streams. Drainage of lakes through such mechanisms is common, as attested to by the many drained lake basins interspersed with existing lakes in figure 13.7. Drainage events have been well documented by Mackay (1988) and Brewer et al. (1993); the processes operating on a refreezing lake basin in the Canadian western Arctic were documented by Mackay (1997). Mackay (1988) estimated that one or two thaw lakes drain each year on the Tuktoyaktuk Peninsula of northwestern Canada. After drainage, thermokarst develops until recolonization by plants provides adequate insulation to establish thermal equilibrium between the new climate at the surface and the substrate. Ice-wedge growth yields low-centered polygons in the former lake basin, beginning a new chapter in the thaw-lake cycle.

Research in Canada (Mackay, 1992) and Alaska (Hopkins, 1949) indicates that the formation and drainage of thaw lakes is linked to changes in climate. Although still forming today, many thaw lakes came into existence during the warmer early Holocene, around 10,000 yr B.P. (radiocarbon years before AD 1950) (Ritchie et al., 1983; Hopkins and Kidd, 1988; Rampton, 1988). Their development is apparently associated with a regional thickening of the active layer (Burn, 1997) and development of thermokarst at regional scales. Lake drainage probably occurred preferentially in the colder late Holocene, beginning around 4500 yr B.P. In northwestern Canada, this period is associated with the accumulation of peat in drained lake basins and reactivation of ice-wedge cracking and growth (Mackay, 1992).

Although the distribution of thaw lakes has been mapped over regions of substantial size (Sellmann et al., 1975; Mellor, 1983), maps depicting the ages of these features have yet to be produced. Eisner and Peterson (1998) discussed the potential and some of the problems of dating drained lake basins near Barrow, Alaska. The structure of the thaw-lake cycle implies that a map of lake age covering an area of substantial size should contain both a regional signal attributable to climate change operating at low temporal frequencies, and a local component with substantial variability resulting from drainage of individual lakes.

Drained thaw-lake basins contain a considerable amount of organic carbon. Although estimates are not yet available, these features could make a significant contribution to global warming through release of greenhouse gases (section 13.3). Carbon accumulation rates were higher during the warm period encompassing the early to middle Holocene than during the cooler late Holocene, but whether this was due to changes in temperature and precipitation, plant succession, or local depositional factors is not known (Marion and Oechel, 1993).

Frost mounds form a widely discussed group of ice-cored, positive-relief features in the arctic lowlands of North America and other parts of the circumarctic region. Frost mounds have traditionally been classified by size and genesis, but recent research (e.g., Nelson et al., 1992; Gurney, 1998) indicates that a continuum exists between several categories of frost mound long thought to be distinct; terminological issues remain contentious, however. *Pingos* are ice-cored hills, often conical in shape, that attain heights of up to 100 m. As many as one-quarter of the world's pingos are found in the Tuktoyaktuk Peninsula of northwestern Canada (Mackay, 1998). These "closed system" pingos are formed in the basins of recently drained thaw lakes by expulsion of pore water as permafrost aggrades into the sediments that comprise the thaw bulb underlying the location of the former lake. Mackay (1979, 1998) provided comprehensive reviews of virtually all aspects of these remarkable features.

Figure 13.8 shows a *palsa*, a typical permafrost landform found throughout much of the subarctic. Palsas are medium-size permafrost mounds whose positive relief is attributable to the growth of subsurface ice. Although palsa-scale frost mounds can evolve by a variety of mechanisms, most in subarctic peatlands evolve through *ice segregation*, the formation of discreet ice lenses through migration of water to a freezing front. In environments with mean annual temperatures near 0°C, permafrost occurs preferentially in peat because the thermal properties of the material can vary radically by season. In winter, when peat is saturated and frozen, its thermal conductivity approaches that of ice and promotes heat flow from the ground to the atmosphere. In summer, however, when the peat is thawed and dry, it forms an effective insulating layer and tends to preserve underlying frozen layers. Because the ratio of peat's thermal conductivity in the frozen and thawed states is unusually high, the mean annual temperature at the base of the active layer can be substantially below that at the ground surface. This *thermal offset* allows permafrost to exist at locations with mean annual surface temperatures slightly above 0°C. As permafrost aggrades into the substrate at such locations, the formation of segregation ice initiates the growth of a palsa. The positive relief generated in this fashion keeps the incipient mound free of insulating snow cover, promoting further growth. The growth and deterioration of these and related forms in the peatlands of central Canada have provided valuable insights into changes in permafrost distribution through the use of sequential air photography (e.g., Thie, 1974; Zoltai, 1995).

Figure 13.8 Dome-shaped palsa, MacMillan Pass area, Yukon-Northwest Territories border, Canada. From Nelson et al. (1992).

13.2.4 Periglacial Slope Series

Quantitative investigations of slopes have not played as large a role in the development of periglacial geomorphology as they have in other branches of landform science. Some authorities (Davies, 1969; French, 1976) have characterized periglacial geomorphology as the study of "surface decorations." The neglect of larger landform elements is unfortunate because permafrost regions contain distinctive slope assemblages with the potential to provide substantial information about landscape evolution under cold, nonglacial conditions. A good review of the periglacial literature on slopes was provided by French (1996, chapter 10)

The upper parts of slope profiles in vast areas of unglaciated northwestern North America are composed of series of very large summit flats and terrace-like features (fig. 13.9). The lower portions of profiles are often made up of smoothly beveled surfaces. These slope profiles are usually attributed to a vaguely specified sequence of landscape leveling subsumed under the term *cryoplanation* (Bryan, 1946; Peltier, 1950). They are widespread in uplands of the "ice-free corridor" of western and central Alaska and Yukon Territory, where Pleistocene glaciation was confined to isolated cirques and valleys (Reger and Péwé, 1976, Lauriol and Godbout, 1988).

Cryoplanation is a collective term, attributed to Bryan (1946), that refers in its most general usage to leveling of the land surface under cold, nonglacial conditions. The distinctive stepped topography now associated with cryoplanation drew a great deal of commentary by earth scientists conducting exploratory work in Alaska and Yukon

Figure 13.9 Cryoplanation terraces in central Alaska: (a) Stepped terrace sequence with large-diameter sorted patterned ground on treads; (b) Sharp break in slope angle at junction of scarp and tread; (c) Block slope occupying terrace scarp; gently sloping tread in foreground; (d) Overview of cryoplanation terrace unit; road in foreground is approximately 3 m wide. From Nelson (1989a).

during and immediately following the Klondike gold rush (e.g., Moffit, 1905; Prindle, 1905). Because the landforms did not meet various tests proposed by proponents of the peneplain hypothesis, alternative interpretations involving frost action were devised before 1920 (e.g., Eakin, 1916). By midcentury, cryoplanation had achieved status as a climatically determined corollary to the normal cycle of erosion (see Peltier, 1950).

Cryoplanation is frequently ascribed to a suite of processes, known collectively as *nivation*, that is associated with the presence of snowbanks surviving well into the warm season. The structure and hydrology of late snowbanks were discussed by Lewkowicz and Harry (1991) and Lewkowicz and Young (1990). The presence of late-lying snow patches has been interpreted as intensifying erosion in several ways, including (1) increasing the number of freeze-thaw cycles, (2) suppressing vegetation in the core snowbank area, (3) providing moisture necessary for effective weathering, and (4) intensifying mass-movement downslope from the snow patch.

The primary morphological features ascribed to cryoplanation are *cryopediments* and *cryoplanation terraces*. Together, these forms can occupy entire slope sequences and are widespread in unglaciated sections of Alaska and Yukon. Cryopediments are relatively low-angle (1–10°) erosional forms, analogous to pediments in warm deserts. They function primarily as slopes of transportation in the lower part of valley-side profiles, with solifluction and rillwash moving weathered material to the fluvial network. French and Harry (1992) studied cryopediments in the Barn Mountains of northern Yukon, concluding that most are inactive at present.

Cryoplanation terraces are sequences of beveled, nearly flat areas ("treads") separated abruptly by much steeper

(10–30°) scarps. In central Alaska, the median area occupied by treads is 32,000 m^2, although examples ranging to nearly 1 km^2 are known (Reger, 1975; Nelson, 1989a). The terraces occupy the upper portions of slopes, particularly along ridge crests, where they form series often described as "giant stairsteps." Terrace series usually culminate in a summit flat. Treads are mantled with weathered debris ranging from silt to boulders; well-developed fields of sorted patterned ground frequently occupy much of a tread's area. Scarps are covered with coarse weathering debris to such an extent that they qualify as block slopes. Although lithological or structural controls over terrace morphology are discernable in some cases, numerous examples of cryoplanation terraces truncating geological structure have been documented (Reger, 1975). Evidence of solifluction is widespread on side slopes and on the cryopediments occupying the lower portions of slope profiles. Figure 13.9 illustrates some of the primary morphological attributes of cryoplanation landforms.

Cryoplanation terraces have usually been ascribed to nivation operating at the break in slope between the scarp and the tread, undercutting the slope and enlarging treads through parallel retreat of scarps. The entire topic of cryoplanation has been controversial throughout its history, however, and remains so today. The crux of the problem lies in the fact that very few process-oriented studies have been undertaken on cryoplanation forms. In the absence of process information, some authorities are reluctant to concede that the locally intensified erosion associated with late-lying snow patches can be associated with sculpting regional or subcontinental landscapes. Priesnitz (1988) and Thorn (1988) provide thoughtful discussions of problems related to the cryoplanation and nivation concepts.

Geographic evidence, however, supports the interpretation of cryoplanation features as climatically determined geomorphic features (Nelson, 1989a, 1998). Figure 13.10 depicts the median elevation of cryoplanation terraces over western and central Alaska. The gradual but steady increase in elevation away from moisture sources is consistent with that of glacial cirques in the same area (Péwé and Reger, 1972). This distinctive landform assemblage is ubiquitous throughout central Alaska and much of Yukon, contains a repeating sedimentological sequence, and occurs over an area extending nearly 2000 km from west to east. Its elevational trend is correlated strongly with the distribution of glacial cirques and with temperature and precipitation patterns in this subcontinental-scale area. Cryoplanation terraces and cryopediments have great potential for integrating periglacial studies more closely with the mainstream of geomorphology through treatment of slope geometry and Quaternary history. Until detailed, process-oriented information is obtained from studies of nivation, however, the origin of these fascinating landforms may remain controversial.

Figure 13.10. Correspondence between elevation of cryoplanation terraces and glacial cirques in Alaska. Note similarities in form and gradient of surfaces. (a) Cryoplanation terrace elevation, contoured from median values surrounding locations represented by triangles; data from Reger (1975). (b) Elevation of lowest north-facing glacial cirques in 1:63,360 USGS quadrangles; map redrawn from Bradley (1985), based on data from Péwé and Reger (1972).

13.3 Permafrost and Global Change

High-latitude regions are a central concern in the recent scientific focus on the possible effects of global warming. Most scenarios of climate change predict that warming will be pronounced in the polar regions, indicating that its impact on such cryospheric phenomena as glaciers, ice sheets, sea ice, and permafrost could be severe.

Permafrost plays three primary roles in the context of climatic change: (1) it provides a record of temperature change over the course of previous centuries; (2) by aggrading or degrading vertically, permafrost transmits changes of the climate signal to other components of the landscape, particularly when the volume of ground ice changes substantially; and (3) it can facilitate further changes by releasing below-ground carbon stocks to the atmosphere and hydrologic systems. An overview of the effects of climate change on permafrost was given by Fitzharris et al. (1996). A more comprehensive discussion, with specific reference to Canada, was provided by Woo et al. (1992; see also chapter 8).

Heat transfer in thick permafrost occurs primarily by conduction. This fact allows temperature changes to be inferred from careful analysis of temperature profiles in deep boreholes. In northern Alaska, curvature of the temperature profiles in the upper 100 m reveals a widespread, 2–4°C increase over the past century (Lachenbruch and Marshall, 1986; Clow et al., 1998). Information obtained from thermal measurements in boreholes has also been used to prepare maps of thaw susceptibility in the Mackenzie Valley (Aylsworth et al., 2001).

Cryostratigraphic studies have provided evidence for the development of widespread thermokarst during warm intervals (e.g., Burn, 1997, 1998; Murton et al., 1998). Because climate-change scenarios based on general circulation models indicate that warming will be most pronounced in the high-latitude regions, the potential for thermokarst and accompanying damage to the human infrastructure is an important concern of global-change science (Fitzharris et al., 1996). Substantial thickening of the active layer would be the primary immediate concern in the continuous permafrost zone of North America; thawing of thick, ice-rich permafrost would occur slowly because of the role of the latent heat of fusion. Farther south, in the discontinuous and sporadic zones, permafrost has already retreated substantially during the twentieth century. Thie (1974), Kwong and Gan (1994), and French and Egorov (1998) have documented the disappearance of permafrost from areas of central Canada, as well as a northward shift in the limit of its distribution. Osterkamp and Romanovsky (1999) have documented widespread warming of permafrost in central and southern Alaska since the late 1980s and identified the development of thermokarst terrain at several sites. Modeling indicated that these changes to Alaskan permafrost were a response to warming and changes in snow cover. Simulation experiments by Anisimov and Nelson (1996, 1997), based on general circulation model (GCM) predictions, indicate a long-term (century to millennial scale) potential for reductions of 12–28% in the total area of the Northern Hemisphere underlain by permafrost. The lower end of these areal estimates was developed using more recent transient GCM runs, therefore, it is thought to be more realistic than the higher values, which were based on 2 × CO_2 GCMs.

Carbon stocks in the permafrost regions are substantial (Billings, 1987). Gilmanov and Oechel (1995) estimated that the active layer of tundra ecosystems in North America and Greenland contains 91.3 Gt (gigatonnes) of soil organic matter. Recent work in northern Alaska (Michaelson et al., 1996; Bockheim et al., 1998) has revealed the presence of large amounts of carbon in the upper layers of permafrost, indicating that previous estimates of carbon stores in tundra regions will have to be revised upward. If a regional increase in thaw depth occurs in response to global warming, much of this carbon could reach the atmosphere in the form of greenhouse gases. This could produce a positive feedback in the climate system, although much remains to be resolved about the timing and magnitude of its effects (e.g., Waelbroeck et al., 1997). The volume edited by Kane and Reeburgh (1998) provides detailed information about field studies and modeling of the flux of trace gases from the tundra to the atmosphere.

13.4 Resource Extraction and Infrastructure

Much of the Far North is located on the Canadian Shield, an ancient craton containing tremendous mineral resources associated with igneous and metamorphic rock. Younger sedimentary strata, especially on the coastal plain bordering the Beaufort Sea, contain abundant petroleum reserves. Extraction of these riches in permafrost terrain and development of the associated infrastructure has led to a suite of problems requiring solutions unique to permafrost environments.

This section describes some of the engineering adjustments necessitated by the presence of frozen ground at depth. Often, technologies and strategies developed in temperate regions do not translate well into cold climates, and modifications are necessary. This section emphasizes man-land interactions, and illustrates the sensitivity of the landscape to human-induced change at the surface.

13.4.1 Roads, Runways, and Railways

Regions underlain by permafrost are generally wet and boggy since subsurface drainage of meltwater and rain is impeded by the frozen substrate. Transportation routes that traverse the muskeg and tundra are often developed on berms, so that roadbeds are above the soggy land. In the case of railroads, the berm is raised to establish a solid foundation and ensure a proper grade.

In the period between the World Wars, roads typically followed the path of least resistance and tended to stay on higher ground. Tracked vehicles used for exploration drilling on the North Slope of Alaska in the 1940s and 1950s destroyed the protective organic mat. Thermokarst of the underlying permafrost yielded linear scars, 3–5 m in width and up to 1 m in depth, that are still visible today on the tundra (Brown and Kreig, 1983).

In boggy areas, berms were often formed from local materials. On the Alaskan coastal plain, for example, roads were constructed by bulldozing the organic mat from either side of the road and stacking these layers to form a berm. Over the last half-century, thermokarst has developed where the protective organic was removed and ice-rich permafrost thawed (fig. 13.11). It is not uncommon to find streams or ponds adjacent to abandoned roads; where ice wedges have ablated, pools can be over 2 m deep.

Resource extraction led to the development of more permanent roads and airfields, with berms typically made from gravel. Because the berms are often 1–2 m in height, massive amounts of materials are imported. Stream beds can be mined for construction material, or rock can be blasted from nearby material sites to provide fill. The accumulation of rock aggregate atop the natural surface has several consequences.

The thermal conductivity of gravel embedded in a finer matrix is substantially greater than the natural organic mat. Given the same incident radiation load, annual thaw will penetrate deeper in the road berm compared to the natural surface (Lunardini, 1981). To prevent the thaw front from reaching the underlying permafrost table and causing subsidence, the road berm must be substantially thicker. Ice lensing and heave are maximized in fine-grained soils, and the effect increases with the depth of thaw. Thus, berms cannot be excessively thick because this promotes seasonal vertical displacement, although the use of gravel and sand reduces the incidence of ice-lens growth and consequent development of potholes during thaw. Despite precautions, railroad track misalignment and track elevation displacement require large annual maintenance costs.

Typically, the permafrost beneath roads, airstrips, and railways will aggrade upward into the berm. Bonding the unconsolidated materials helps strengthen the road foundation. Conversely, this prohibits the lateral water flow beneath the road. A road built along the base of a hill, for example, acts as a barrier for downhill surface and subsurface water flow. Road culverts can channel the discharge during summer, but in winter continued flow within coarse materials at depth is progressively constricted by the permafrost below and encroaching freezing front from above. As hydrostatic pressure increases in the narrowing cryoconduit on the upslope side of the road berm, the frozen overburden can deform and form mounds or an ice-cored plateau. If failure of the overburden occurs, supercooled water can rupture the surface through cracks and freeze to form a layer; repeated rupture yields a layered body of ice with younger material at the surface. Such features are known as *icings*, but are also frequently referred to by their Russian and German names (*naled* and *aufeis*, respectively). As the icing aggrades upward, it can overflow the road berm, causing structural damage and creating a traffic hazard. These icings can be removed with ablation by steam, at great expense.

Airport runway construction creates problems similar to those of road berms, but the effect can be aggravated by use of low-albedo paving materials. In this case, the heat flow to depth is increased in summer, and local subsidence can occur. Ice lens growth and differential heave in winter can become a severe problem, as was the case in the development of military airfields during World War II.

Both thermokarst terrain and icing formation are inadvertent results triggered by road and runway construction in regions underlain by permafrost. The secondary effects

Figure 13.11 Thermokarst terrain developed in winter road constructed in the winter of 1968–69 near Prudhoe Bay, Alaska. Photograph taken in August 1980. Detailed description of road is given in Nelson and Outcalt (1982).

of road dust on the surface albedo and vegetation near roads have also been studied (Moorhead et al., 1996).

13.4.2 *Buildings*

Building construction is problematic in permafrost environments (Ferrians et al., 1969). Heated dwellings built over ice-rich permafrost cause enhanced thaw and induce local differential subsidence and settling. Around Fairbanks, Alaska, tilted and cracked foundations were common in houses built after World War II, often resulting in abandonment (Péwé, 1983). One common response is to place the dwelling on wooden or concrete pilings embedded deep in the underlying permafrost. As a general rule, pilings must be driven to a depth of twice the thickness of the active layer to prevent heaving. This yields a layer of air that effectively decouples the elevated building from the underlying ground. This strategy requires that additional insulation be installed on the floors of buildings to inhibit heat loss.

Other strategies involve use of (1) gravel berms, (2) rigid board styrofoam, and (3) air vents. Styrofoam insulation 0.25 m thick is installed beneath heated buildings and along the building periphery. This material retards heat flow to depth at a cost of about half that of a gravel pad. Air vents, or openings under buildings, channel cold air to the subsurface and disperse heat.

13.4.3 *Municipal Services*

Strategies developed in temperate regions for providing municipal services (water, sewage, electric, gas, and phone lines) are not effective in the Far North (Johnston, 1981). Burial of utility lines in the frozen ground makes them very difficult to retrieve and service if necessary. In addition, water and sewage lines would freeze if buried, and would be subject to differential heave and subsidence if placed on the ground or in the active layer. The use of insulating materials is prohibitive because of the excessive thickness required to prevent freezing of the pipeline contents or thawing of the ground.

One method is to place utilities in enclosed, heated conduits above the ground (fig. 13.12). At Inuvik on the Mackenzie Delta, for example, water and sewage are carried in pipes suspended above the ground and surrounded by insulation encased in wood and steel. Known as utilidors, each building is linked to the central system. This system requires central planning and maintenance, so it is feasible only in larger permanent settlements.

At Barrow, Alaska, underground utilidors have also been constructed. Tunnels are excavated within permafrost, and they are sufficiently spacious to allow for all utility lines and to provide walkways for service access. Corridors are heated to prevent freezing and are insulated to prevent thawing of the surrounding ground. The municipal water supply comes from a local thaw lake that is sufficiently deep that it does not freeze to the bottom in winter. Sewage is expelled into another nearby thaw-lake basin and is eventually discharged into the nearby ocean.

13.4.4 *Oil and Gas Pipelines*

Large regions of the North have been made accessible in the last several decades, owing to the discovery of large petroleum deposits. Large-scale refinery operations are

Figure 13.12 Utilidor at Barrow, Alaska.

located in temperate regions farther south; for this reason, most of the crude oil pumped from the ground is transported via pipeline to ships for transport. In Alaska, the Trans-Alaska Pipeline System (TAPS) extends some 1300 km from the Prudhoe Bay oil fields to the shipping dock at Valdez. Pumping stations along the route provide both additional pressure and maintenance service.

Alaskan crude oil has a temperature of about 60°C at the well head and cools to 32–35°C at Valdez (Brown and Kreig, 1983). Clearly, burying a hot oil pipeline in ice-rich sediments would cause localized thawing and subsidence; liquefaction and loss of bearing strength would result in the eventual rupture of the pipeline. For this reason, about half the length of the TAPS is aboveground. On a per-unit basis, elevated pipeline costs about three times more than buried sections. Burial is possible only in southern Alaska, in stream valleys with unfrozen alluvium, at animal and road crossings, and in mountain passes where mass wasting would threaten an elevated structure.

The TAPS pipeline is 1.2 m in diameter and is surrounded by a 10-cm layer of fiberglass insulation impregnated with resin and encased within a galvanized steel jacket. It does not run in a straight line; viewed from above, it zigzags back and forth. The pipeline is composed of sections sitting atop an H-shaped structure made up of two vertical support members (VSM) connected by a horizontal brace. The VSMs, 45-cm-diameter steel pipes, are driven deep into the underlying permafrost. The pipe is clamped in a saddle assembly and nestles within the upper part of the "H" atop the horizontal brace. The contact surface, known as the shoe, is coated with a nonstick substance. Because the pipeline is subjected to a large temperature range, it expands and contracts. If it were straight, the accumulated movement could not be accommodated. Instead, portions of the flexible pipeline can slide along the surface of the horizontal brace. In this way, the zigzag arrangement converts expansion and contraction movement to lateral movement.

The TAPS was built during the period 1975–77, after 6 years of engineering and environmental planning. The primary concern during pipeline construction was the removal of unwanted heat so as to prevent enhanced ground thaw, subsidence, and rupture. The VSMs are filled with ammonia, and atop them are aluminum radiation fins, 1.2–1.8 m long and with a large surface area (fig. 13.13). During winter, the permafrost is warmer than the air. Ammonia within the pipe is vaporized and rises from depth to the top of the VSM. There it condenses and the heat of vaporization is released. This heat is conducted to the radiation fin, where it is dissipated to the atmosphere. The condensed ammonia settles to the base of the vertical pipe. Thus, the system is closed with respect to material but acts as a unidirectional "heat pipe" by channeling energy away from the support member. The system is so effective that temperatures in the nearby permafrost are reduced, which further stabilizes the VSM.

In Canada, a small-diameter buried oil pipeline was recently installed at Norman Wells, Northwest Territories. The oil is chilled to temperatures approaching the ground temperature, and the pipeline is buried in an embankment covered with wood chips. Experiments are currently under way to determine the optimal temperature of the oil so as to prevent ground thaw or ice lensing and heave.

Natural gas can be piped under pressure much like petroleum. However, it is far safer and more efficient to pump as liquid natural gas. In this case, it is possible to bury the gas pipeline, and the permafrost will act to keep the pipe insulated and chilled. The main concern of a

Figure 13.13 Portion of TAPS pipeline near Fairbanks showing vertical support members equipped with finned heat exchangers.

proposed buried chilled gas line is the formation of ice lenses, which cause differential heave and possible damage to the pipeline.

13.4.5 Mining Operations

Frozen ground is extremely difficult to mine. Dynamite is relatively ineffective as the ground tends to absorb much of the energy from the blast. Gold placer mines typically utilize hydraulic cannons (hydraulic giants) to erode and ablate ice-bonded sediments. Though effective, this method does tremendous damage to the water quality.

Open-pit mines can be excavated with explosives, dredges, and dozers. Problems arise, however, when mine tailings and ore are piled at the surface. In Schefferville, in central Quebec-Labrador, permafrost aggraded upward into the mined ore body over time and had to be remined for transport.

13.4.6 Agriculture

A recent study (Mills, 1994) indicates that global warming may provide enhanced climatic potential for agriculture in northern lands, particularly in .Alaska. This climatic potential is offset somewhat, however, by limitations imposed by the presence of ice-rich permafrost. Agriculture in the Far North typically involves forest clearing as the initial step. In ice-rich sediments, this can lead to thawing of the permafrost and subsidence. In Fairbanks, Alaska, for example, a field was cleared in 1908 at the University of Alaska Experimental Farm. Within 20 years, differential subsidence created mounds 3–8 ft (0.9–2.5 m) high and 20–50 ft (6–15 m) in diameter (Péwé, 1983).

Increases in the thickness of the active layer and development of thermokarst features occur after forest fires (Mackay, 1995) and in association with large mining or refining operations. In the case of the former, deforestation and destruction of the organic mat appear to be the primary causes. In the latter case, a reduction in the albedo can increase the flow of heat to depth. In all cases, thaw was increased for several years after the initial disturbance and may take several decades to stabilize.

13.5 Conclusion

This chapter began with an assertion that the intersection of geographic information science (Goodchild, 1992), data preservation and accessibility (Clark and Barry, 1998), and global-change research (Houghton et al., 1996) has become an important focus of research in permafrost regions. We conclude with a brief illustration of how these elements can be used to make general predictions about the geographical consequences of global change in the northern part of North America.

In concert with ever-increasing human activity in the Arctic, the prospect of global warming indicates the potential for environmental degradation and damage to human infrastructure in permafrost regions (sections 13.3 and 13.4). Although regional differences are apparent in results from different scenarios, general circulation models provide a basis for predicting areas in which climate change may be most pronounced. Scaling techniques (Root and Schneider, 1995) and regional climate models (e.g., Lynch et al., 1995) will produce increasingly detailed regional pictures of the potential for climate change. GIS technology provides a basis for using geocryological information with climate-change scenarios to conduct simulation experiments that examine the potential impact of climate change in cold regions.

Figure 13.14 provides a perspective on the degree of hazard potential associated with thickening of the active layer induced by climate change in the North American Arctic. Categories of risk were developed using a "settlement index" that takes into account ground-ice content and calculated increases in active-layer thickness, based on the ECHAM-1A climate-change scenario (Cubasch et al., 1992). Computations for changes in active-layer thickness were made using Kudryatsev's solution for the depth of thaw (Anisimov et al., 1997). The map indicates potential for severe settlement in many parts of the North American

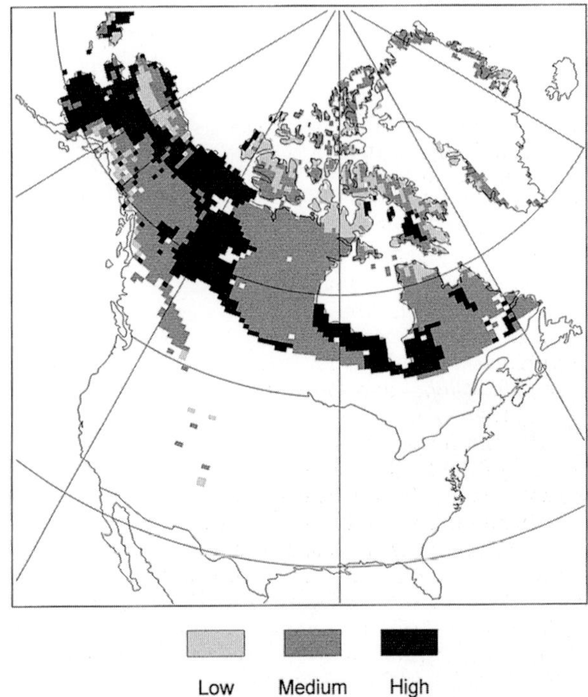

Figure 13.14 Hazard zonation map depicting potential for thaw consolidation and settlement, based on intersection of estimates about active layer increase (Anisimov et al., 1997) and ground ice content (Brown et al., 1997). Adapted from unpublished analysis by O.A. Anisimov and F.E. Nelson.

Arctic and could be used to assist development of engineering design and public policy. Although the presence of ground ice is a necessary prerequisite for initiation of thermokarst, areas containing large volumes of subsurface ice exhibit different degrees of hazard potential because the postulated climate warming is not homogeneous throughout the region.

Figure 13.14 provides a scale and a degree of regional detail that would have been impossible to achieve only a decade ago (Barry, 1988). The spatial perspective facilitated by the rise of geographic information theory and technology provides a basis for linking research with natural and human systems (ARCUS, 1998), and places geographers in an excellent position for addressing the most pressing research questions of the new millennium.

Acknowledgments We thank Nikolay Shiklomanov and Linda Parrish (University of Delaware) for preparing several of the figures. Oleg Anisimov (State Hydrological Institute, St. Petersburg, Russia) provided materials on which figure 13.14 is based. Jerry Brown (International Permafrost Association) gave permission to use the data and outlines employed to construct figures 13.3 and 13.6.

References

Andersland, O.B., and B. Ladanyi, 1994. *An Introduction to Frozen Ground Engineering*. Chapman & Hall, New York.

Anisimov, O.A., and F.E. Nelson, 1996. Permafrost distribution in the northern hemisphere under scenarios of climatic change. *Global and Planetary Change* 14: 59–72.

Anisimov, O.A., and F.E. Nelson, 1997. Permafrost zonation and climate change: Results from transient general circulation models. *Climatic Change* 35: 241–258.

Anisimov, O.A., N.I. Shiklomanov, and F.E. Nelson, 1997. Effects of global warming on permafrost and active-layer thickness: Results from transient general circulation models. *Global and Planetary Change* 61: 61–77.

ARCUS, 1998. *Toward and Arctic System Synthesis: Results and Recommendations*. Arctic Research Consortium of the United States, Fairbanks.

Associate Committee on Geotechnical Research—Permafrost Subcommittee, 1988. *Glossary of Permafrost and Related Ground-Ice Terms*. National Research Council of Canada, *Technical Memorandum* 142, Ottawa.

Aylsworth, J.M., M.M. Burgess, D.T. Desrochers, A. Duk-Rodkin, and J.A. Traynor, 2001. Surficial geology, subsurface materials, and thaw sensitivity of sediments. In: *The Physical Environment of the Mackenzie Valley: A Baseline for the Assessment of Environmental Change*. L. Dyke and G.R. Brooks (Eds.). *Geological Survey of Canada Bulletin*, Ottawa.

Barry, R.G., 1988. Permafrost data and information: Status and needs. In: *Proceedings of the Fifth International Conference on Permafrost*. K. Senneset (Ed.). Tapir Publishers, Trondheim, Norway, pp. 119–122.

Billings, W.D., 1987. Carbon balance of Alaskan tundra and taiga ecosystems: Past, present and future. *Quaternary Science Reviews* 6: 165–177.

Billings, W.D., and K.M. Peterson, 1980. Vegetational change and ice-wedge polygons through the thaw-lake cycle in arctic Alaska. *Arctic and Alpine Research* 12: 413–432.

Bird, J.B., 1967. *The Physiography of Arctic Canada*. The Johns Hopkins Press, Baltimore.

Black, R.F., 1969. Thaw depressions and thaw lakes—A review. *Biuletyn Peryglacjalny* 19: 131–150.

Bockheim, J.G., D.A. Walker, and L.R. Everett, 1998. Soil carbon distribution in nonacidic and acidic tundra of arctic Alaska. In: *Soil Processes and the Carbon Cycle*. R. Lal, J.M. Kimble, R.F. Follett, and B.A. Stewart (Eds.). CRC Press, Boca Raton, pp. 143–155.

Bostock, H.S., 1970. Physiographic subdivisions of Canada. In: *Geology and Economic Minerals of Canada*. R.J.W. Douglas (Ed.). Geological Survey of Canada, Economic Geology Report 1, Ottawa, pp. 10–30.

Bradley, R.S., 1985. *Quaternary Paleoclimatology*. Allen & Unwin, Boston.

Brewer, M.C., 1958. The thermal regime of an arctic lake. *Transactions of the American Geophysical Union* 39: 278–284.

Brewer, M.C., L.D. Carter, and R. Glenn, 1993. Sudden drainage of a thaw lake on the Alaskan Arctic coastal plain. In: *Proceedings of the Sixth International Conference on Permafrost*. South China University of Technology Press, Wushan Guangzhou, China, pp. 48–53.

Britton, M.E., 1957. Vegetation of the Arctic tundra. In: *Arctic Biology*. H.P. Hansen (Ed.). Oregon State University Press, Corvallis, pp. 26–61.

Brown, J., and R.A. Kreig, 1983. *Elliott and Dalton Highways, Fox to Prudhoe Bay, Alaska: Guidebook to Permafrost and Related Features*. Guidebook 4, Fourth International Conference on Permafrost, University of Alaska, Fairbanks.

Brown, J., O.J.J. Ferrians, J.A. Heginbottom, and E.S. Melnikov, 1997. *International Permafrost Association Circum-Arctic Map of Permafrost and Ground Ice Conditions*. Scale 1:10,000,000. U.S. Geological Survey, Circum-Pacific Map Series Map CP-45.

Brown, R.J.E., 1960. The distribution of permafrost and its relation to air temperature in Canada and the U.S.S.R. *Arctic* 13: 163–177.

Brown, R.J.E., 1967. *Permafrost in Canada*. Scale 1:7,603,200. Geological Survey of Canada, Map 1246–A.

Brown, R.J.E., 1970. *Permafrost in Canada: Its Influence on Northern Development*. University of Toronto Press, Toronto.

Bryan, K., 1946. Cryopedology—The study of frozen ground and intensive frost-action with suggestions on nomenclature. *American Journal of Science* 244: 622–642.

Burn, C.R., 1997. Cryostratigraphy, paleogeography, and climate change during the early Holocene warm interval, western Arctic coast, Canada. *Canadian Journal of Earth Sciences* 34: 912–925.

Burn, C.R., 1998. Field investigations of permafrost and climatic change in northwest North America. In: *Proceedings of the Seventh International Conference on Permafrost*. A.G. Lewkowicz and M. Allard (Eds.). Centre d'études Nordiques, Université Laval, Québec, pp. 107–120.

Carson, C.E., and K.M. Hussey, 1962. The oriented lakes of arctic Alaska. *Journal of Geology* 70: 417–439.

Carter, L.D., J.A. Heginbottom, and M.-K. Woo, 1987. Arctic lowlands. In: *Geomorphic Systems of North America*.

W.L. Graf (Ed.). Geological Society of America, Boulder, Colorado, pp. 583–627.

Clark, M.J., and R.G. Barry, 1998. Permafrost data and information: Advances since the Fifth International Conference on Permafrost. In: *Proceedings of the Seventh International Conference on Permafrost.* A.G. Lewkowicz and M. Allard (Eds.). Centre d'études Nordiques, Université Laval, Québec, pp. 181–188.

Clow, G.D., R.W. Saltus, A.H. Lachenbruch, and M.C. Brewer, 1998. Arctic Alaska Climate change estimated from borehole temperature: Past, present, future. *Eos, Transactions of the American Geophysical Union* 79: F883.

Cubasch, U., K. Hasselmann, H. Hock, E. Maier-Reimer, B.D. Santer, and R. Sausen, 1992. Time-dependent greenhouse warming computations with a coupled ocean-atmosphere model. *Climate Dynamics* 8: 55–69.

Davies, J.L., 1969. *Landforms of Cold Climates.* M.I.T. Press, Cambridge, Mass.

Eakin, H.M., 1916. The Yukon-Koyukuk region, Alaska. *USGS Bulletin* 631: 1–88.

Eisner, W.R. and K.M. Peterson, 1998. Pollen, fungi and algae as age indicators of drained lake basins near Barrow, Alaska. In: *Proceedings of the Seventh International Conference on Permafrost.* A.G. Lewkowicz and M. Allard (Eds.). Centre d'études Nordiques, Université Laval, Québec, pp. 245–250.

Ferrians, O., R. Kachadoorian, and G.W. Green, 1969. Permafrost and related Engineering Problems in Alaska. *USGS Professional Paper* 678: 1–37.

Fitzharris, B.B., I. Allison, R.J. Braithwaite, J. Brown, P.M.B. Foehn, W. Haeberli, K. Higuchi, V.M. Kotlyakov, T.D. Prowse, C.A. Rinaldi, P. Wadhams, M.-K. Woo, X. Youyu, O.A. Anisimov, A. Aristarain, R.A. Assel, R.G. Barry, R.D. Brown, F. Dramis, S. Hastenrath, A.G. Lewkowicz, E.C. Malagnino, S. Neale, F.E. Nelson, D.A. Robinson, P. Skvarca, A.E. Taylor, and A. Weidick, 1996. The cryosphere: Changes and their impacts. In: *Climate Change 1995: Impacts, Adaptations, and Mitigation of Climate Change—Scientific-Technical Analyses. Contribution of Working Group II to the Second Assessment Report of the Intergovernmental Panel on Climate Change.* R.T. Watson, M.C. Zinyowera, R.H. Moss, and D.J. Dokken (Eds.). Cambridge University Press, New York, pp. 241– 265.

French, H.M., 1976. *The Periglacial Environment.* Longman, New York.

French, H.M., 1996. *The Periglacial Environment.* Longman, Edinburgh.

French, H.M., 1998. An appraisal of cryostratigraphy in north-west arctic Canada. *Permafrost and Periglacial Processes* 9: 297–312.

French, H.M., and I.E. Egorov, 1998. 20th Century variations in the southern limit of permafrost near Thompson, northern Manitoba, Canada. In: *Proceedings of the Seventh International Conference on Permafrost.* A.G. Lewkowicz and M. Allard (Eds.). Centre d'études nordiques, Université Laval, Québec, pp. 297–304.

French, H.M., and D.G. Harry, 1992. Pediments and cold-climate conditions, Barn Mountains, unglaciated northern Yukon, Canada. *Geografiska Annaler* 74A: 145–157.

French, H.M., and O. Slaymaker, 1993. *Canada's Cold Environments.* McGill-Queen's University Press, Montreal.

Gilmanov, T.G., and W.C. Oechel, 1995. New estimates of organic matter reserves and net primary productivity of the North America tundra ecosystems. *Journal of Biogeography* 22: 723–741.

Goodchild, M., 1992. Geographical information science. *International Journal of Geographic Information Systems* 6: 31–45.

Goodrich, L.E., 1982. The influence of snow cover on the ground thermal regime. *Canadian Geotechnical Journal* 19: 421–432.

Granberg, H.B., 1994. Mapping heat loss zones for permafrost prediction at the northern alpine limit of the boreal forest using high-resolution C-band SAR. *Remote Sensing of Environment* 50: 280–286.

Gurney, S.D., 1998.Aspects of the genesis and geomorphology of pingos: Perennial permafrost mounds. *Progress in Physical Geography* 22A: 309–324.

Halsey, L.A., D.H. Vitt, and S.C. Zoltai, 1995. Disequilibrium response of permafrost in boreal continental western Canada to climate change. *Climatic Change* 30: 57–73.

Harlan, R.L., and J.F. Nixon, 1978. Ground thermal regime. In: *Geotechnical Engineering for Cold Regions.* O.B. Andersland and D.M. Anderson (Eds.). McGraw-Hill, New York, pp. 103–163.

Hinkel, K.M., S.I. Outcalt, and A.E. Taylor, 1997. Seasonal patterns of coupled flow in the active layer at three sites in northwest North America. *Canadian Journal of Earth Sciences* 34:667–678.

Hinzman, L.D., D.J. Goering, and D.L. Kane, 1998. A distributed thermal model for calculating soil temperature profiles and depth of thaw in permafrost. *Journal of Geophysical Research* 103: 28,975–28,991.

Hopkins, D.M., 1949. Thaw lakes and thaw sinks in the Imuruk Lake area, Seward Peninsula, Alaska. *Journal of Geology* 57: 119–131.

Hopkins, D.M., and J.G. Kidd, 1988. Thaw lake sediments and sedimentary environments. In: *Proceedings of the Fifth International Conference on Permafrost,* K. Senneset (Ed.). Tapir Publishers, Trondheim, Norway, pp. 790–795.

Houghton, J.T., L.G. Meira Filho, B.A. Callander, N. Harris, A. Kattenberg, and K. Maskell, 1996. *Climate Change 1995: The Science of Climate Change.* Cambridge University Press, Cambridge.

Hussey, K.M., and R.W. Michelson, 1966. Tundra relief features near Point Barrow, Alaska. *Arctic* 19: 162–184.

Johnston, G.H., 1981. *Permafrost: Engineering Design and Construction.* Wiley, New York.

Jorgenson, M.T., and R.A. Kreig, 1988. A model for mapping permafrost distribution based on landscape component maps and climatic variables. In: *Proceedings of the Fifth International Conference on Permafrost.* K. Senneset (Ed.). Tapir Publishers, Trondheim, Norway, pp. 176–182.

Judge, A., 1973. The prediction of permafrost thicknesses. *Canadian Geotechnical Journal* 10: 1–11.

Judge, A.S., C.M. Tucker, J.A. Pilon, and B.J. Moorman, 1991. Remote sensing of permafrost by ground-penetrating radar at two airports in Arctic Canada. *Arctic* 44: 40–48.

Kane, D.L., W.S. Reeburgh, and (Eds.), 1998. Land-Air-Ice Interactions Flux Study. *Journal of Geophysical Research* 103: 28,913–29,106.

Klene, A.E., 1999. *The n-Factor over Natural Surfaces: Relations Between Air and Surface Temperatures in the Kuparuk River Basin, Northern Alaska.* M.A. Thesis, State University of New York, Albany.

Kozo, T.L., 1979. Evidence for sea breeze on the Alaskan Beaufort Sea coast. *Geophysical Research Letters* 6: 849–852.

Kwong, Y.T.J., and T.Y. Gan, 1994. Northward migration of permafrost along the Mackenzie Highway and climatic warming. *Climatic Change* 26: 399–419.
Lachenbruch, A.H., 1962. Mechanics of Thermal Contraction Cracks and Ice-Wedge Polygons in Permafrost. *Geological Society of America Special Paper*, 70 pp.
Lachenbruch, A.H., and B.V. Marshall, 1969. Heat flow in the Arctic. *Arctic* 22: 300–311.
Lachenbruch, A.H., and B.V. Marshall, 1986. Changing climate: Geothermal evidence from permafrost in the Alaskan arctic. *Science* 234:689–696.
Lauriol, B., and L. Godbout, 1988. Les terrasses de cryoplanation dans le nord du Yukon: distribution, genese et age. *Géographie physique et Quaternaire* 42: 303–314.
Leffingwell, E. de K., 1915. Ground ice-wedges, the dominant forms of ground-ice on the north coast of Alaska. *Journal of Geology* 23:635–654.
Leffingwell, E. de K., 1919. The Canning River region, northern Alaska. *USGS Professional Paper* 109: 1–251.
Leverington, D.W., and C.R. Duguay, 1996. Evaluation of three supervised classifiers in mapping "depth to late-summer frozen ground," central Yukon Territory. *Canadian Journal of Remote Sensing* 22: 163–174.
Leverington, D.W., and C.R. Duguay, 1997. A neural network method to determine the presence or absence of permafrost near Mayo, Yukon Territory, Canada. *Permafrost and Periglacial Processes* 8: 205–215.
Lewkowicz, A.G., and D.G. Harry, 1991. Internal structure and environmental significance of a perennial snowbank, Melville Island, NWT. *Arctic* 44: 74–82.
Lewkowicz, A.G., and K.L. Young, 1990. Hydrology of a perennial snowbank in the continuous permafrost zone, Melville Island, Canada. *Geografiska Annaler* 72A: 13–21.
Lunardini, V.J., 1981. *Heat Transfer in Cold Climates.* Van Nostrand Reinhold, New York.
Lynch, A.H., W.L. Chapman, J.E. Walsh, and G. Weller, 1995. Development of a regional climate model of the western Arctic. *Journal of Climate* 8: 1555–1570.
Mackay, J.R., 1972. The world of underground ice. *Annals of the Association of American Geographers* 62: 1–22.
Mackay, J.R., 1979. Pingos of the Tuktoyaktuk Peninsula Area, Northwest Territories. *Géographie physique et Quaternaire* 33: 3–61.
Mackay, J.R., 1981. Active layer slope movement in a continuous permafrost environment, Garry Island, Northwest Territories, Canada. *Canadian Journal of Earth Sciences* 18: 1666–1680.
Mackay, J.R., 1988. Catastrophic lake drainage, Tuktoyaktuk Peninsula area, District of Mackenzie. In: *Current Research, Part D. Geological Survey of Canada Paper* 88–1D, Ottawa, pp. 83–90.
Mackay, J.R., 1992. Lake stability in an ice-rich permafrost environment: Examples from the western Arctic coast. In: *Aquatic Ecosystems in Semi-Arid Regions: Implications for Resource Management.* R.D. Robarts and M.L. Bothwell (Eds.). Environment Canada, Saskatoon, pp. 1–25.
Mackay, J.R., 1993. The sound and speed of ice-wedge cracking, Arctic Canada. *Canadian Journal of Earth Sciences* 30: 509–518.
Mackay, J.R., 1995. Active layer changes (1968 to 1993) following the forest-tundra fire near Inuvik, N.W.T., Canada. *Arctic and Alpine Research* 27: 323–336.
Mackay, J.R., 1997. A full-scale field experiment (1978–1995) on the growth of permafrost by means of lake drainage, western Arctic coast: A discussion of the method and some results. *Canadian Journal of Earth Sciences* 34: 17–33.
Mackay, J.R., 1998. Pingo growth and collapse, Tuktoyaktuk Peninsula area, western arctic coast, Canada: A long-term field study. *Géographie physique et Quaternaire* 52: 271–323.
Mackay, J.R., V.N. Konischev, and A.I. Popov, 1978. Geologic controls of the origin, characteristics, and distribution of ground ice. In: *Proceedings of the Third International Conference on Permafrost.* National Research Council of Canada, Ottawa, pp. 1–18.
Marion, G.M., and W.C. Oechel, 1993. Mid- to late-Holocene carbon balance in Arctic Alaska and its implications for future global warming. *Holocene* 3: 193–200.
McMichael, C.E., A.S. Hope, D.A. Stow, and J.B. Fleming, 1997. The relation between active layer depth and a spectral vegetation index in arctic tundra landscapes of the North Slope of Alaska. *International Journal of Remote Sensing* 18: 2371–2382.
Mellor, J.C., 1983. Use of seasonal windows for radar and other image acquisition and Arctic lake region management. In: *Proceedings of the Fourth International Conference on Permafrost.* National Academy Press, Washington, D.C., pp. 832–837.
Michaelson, G.J., C.L. Ping, and J.M. Kimble, 1996. Carbon storage and distribution in tundra soils of Arctic Alaska, U.S.A. *Arctic and Alpine Research* 28: 414–424.
Mills, P.F., 1994. The agricultural potential of northern Canada and Alaska and the impact of climatic change. *Arctic* 47: 115–123.
Moffit, F.H., 1905. The Fairhaven gold placers, Seward Peninsula, Alaska. *USGS Bulletin* 247: 1–85.
Mollard, J.D., 1996. *Landforms and Surface Materials of Canada: A Stereoscopic Airphoto Atlas and Glossary.* J.D. Mollard and Associates Limited, Regina, Saskatchewan.
Moorhead, D.L., A.E. Linkins, and K.R. Everett, 1996. Road dust alters extracellular enzyme activities in tussock tundra soils, Alaska, U.S.A. *Arctic and Alpine Research* 28: 346–351.
Moorman, B.J., F.A. Michel, and A. Wilson, 1996. C-14 dating of trapped gases in massive ground ice, western Canadian Arctic. *Permafrost and Periglacial Processes* 7: 257–266.
Morrissey, L.A., L.L. Strong, and D.H. Card, 1986. Mapping permafrost in the boreal forest with thematic mapper satellite data. *Photogrammetric Engineering and Remote Sensing* 52: 1513–1520.
Muller, S.W., 1947. *Permafrost or Permanently Frozen Ground and Related Engineering Problems.* J.W. Edwards, Ann Arbor, MI.
Murton, J.B., H.M. French, and M. Lamothe, 1998. The dating of thermokarst terrain, Pleistocene Mackenzie Delta, Canada. In: *Proceedings of the Seventh International Conference on Permafrost.* A. Lewkowicz and M. Allard (Eds.). Centre d'études Nordiques, Université Laval, Québec, pp. 777–782.
Nelson, F.E., 1986. Permafrost distribution in central Canada: Applications of a climate-based predictive model. *Annals of the Association of American Geographers* 76: 550–569.
Nelson, F.E., 1989a. Cryoplanation terraces: Periglacial cirque analogs. *Geografiska Annaler* 71A: 31–41.
Nelson, F.E., 1989b. Permafrost in eastern Canada: A review of published maps. *Physical Geography* 10: 233–248.

Nelson, F.E., 1998. Cryoplanation terrace orientation in Alaska. *Geografiska Annaler* 80A: 135–151.

Nelson, F.E., and S.I. Outcalt, 1982. Anthropogenic geomorphology in northern Alaska. *Physical Geography* 3: 17–48.

Nelson, F.E., K.M. Hinkel, and S.I. Outcalt, 1992. Palsa-scale frost mounds. In: *Periglacial Geomorphology.* J.C. Dixon and A.D. Abrahams (Eds.). Wiley, New York, pp. 305–325.

Nelson, F.E., S.I. Outcalt, J. Brown, N.I. Shiklomanov, and K.M. Hinkel, 1998. Spatial and temporal attributes of the active-layer thickness record, Barrow, Alaska, U.S.A. In: *Proceedings of the Seventh International Conference on Permafrost.* A. Lewkowicz and M. Allard (Eds.). Centre d'études Nordiques, Université Laval, Québec, pp. 797–802.

Nelson, F.E., N.I. Shiklomanov, G. Mueller, K.M. Hinkel, D.A. Walker, and J.G. Bockheim, 1997. Estimating active-layer thickness over a large region: Kuparuk River basin, Alaska, U.S.A. *Arctic and Alpine Research* 29: 367–378.

Nixon, F.M. and A.E. Taylor, 1998. Regional active layer monitoring across the sporadic, discontinuous and continuous permafrost zones, Mackenzie Valley, northwestern Canada. In: *Proceedings of the Seventh International Conference on Permafrost.* A.G. Lewkowicz and M. Allard (Eds.). Centre d'études Nordiques, Université Laval, Québec, pp. 815–820.

Nixon, J.F., 1998. Recent applications of geothermal analysis in northern engineering. In: *Proceedings of the Seventh International Conference on Permafrost.* A.G. Lewkowicz and M. Allard (Eds.). Centre d'études Nordiques, Université Laval, Québec, pp. 833–846.

Osterkamp, T.E., and M.W. Payne, 1981. Estimates of permafrost thickness from well logs in northern Alaska. *Cold Regions Science and Technology* 5: 13–27.

Osterkamp, T.E., and V.E. Romanovsky, 1999. Evidence for warming and thawing of discontinuous permafrost in Alaska. *Permafrost and Periglacial Processes* 10: 17–37.

Outcalt, S.I., F.E. Nelson, and K.M. Hinkel, 1990. The zero-curtain effect: Heat and mass transfer across an isothermal region in freezing soil. *Water Resources Research* 26: 1509–1516.

Peddle, D.R., and S.E. Franklin, 1993. Classification of permafrost active layer depth from remotely sensed and topographic evidence. *Remote Sensing of Environment* 44:67–80.

Peltier, L.C., 1950. The geographic cycle in periglacial regions as it is related to climatic geomorphology. *Annals of the Association of American Geographers* 40: 214–236.

Péwé, T.L., 1975. Quaternary geology of Alaska. *USGS Professional Paper* 835: 1–145.

Péwé T.L., 1983. Geologic Hazards of the Fairbanks Area, Alaska. *USGS Special Report* 15, 109 pp.

Péwé, T.L., and R.D. Reger, 1972. Modern and Wisconsinan snowlines in Alaska. In: *Proceedings of the 24th International Geological Congress*, pp. 187–197.

Plug, L.J., and B.T. Werner, 1998. A numerical model for the organization of ice-wedge networks. In: *Proceedings of the Seventh International Conference on Permafrost.* A.G. Lewkowicz and M. Allard (Eds.). Centre d'études Nordique, Université Laval, Québec, pp. 897–902.

Pollard, W.H., and R.O. van Everdingen, 1992. Formation of seasonal ice bodies. In: *Periglacial Geomorphology.* J.C. Dixon and A.D. Abrahams (Eds.). Wiley, New York, pp. 281–304.

Priesnitz, K., 1988. Cryoplanation. In: *Advances in Periglacial Geomorphology.* M.J. Clark (Ed.). Wiley, New York, pp. 49–67.

Prindle, L.M., 1905. The gold placers of the Fortymile, Birch Creek, and Fairbanks regions, Alaska. *USGS Bulletin* 251: 1–89.

Rampton, V.N., 1988. Quaternary Geology of the Tuktoyaktuk Coastlands, Northwest Territories. *Geological Survey of Canada Memoir* 423: 1–98.

Reed, J.C., 1969. The story of the Naval Arctic Research Laboratory. *Arctic* 22: 177–183.

Reger, R.D., 1975. *Cryoplanation Terraces of Interior and Western Alaska.* Ph.D. Thesis, Arizona State University, Tempe, 326 p.

Reger, R.D., and T.L. Péwé, 1976. Cryoplanation terraces: indicators of a permafrost environment. *Quaternary Research* 6: 99–109.

Richardson, J., 1839. Notice of a few observations which it is desirable to make on the frozen soil of British North America; drawn up for distribution among the officers of the Hudson's Bay Company. *Journal of the Royal Geographical Society* 9: 117–120.

Riseborough, D.W., 1990. Soil latent heat as a filter of the climate signal in permafrost. In: *Proceedings of the Fifth Canadian Permafrost Conference.* Centre d'études Nordiques, Université Laval, National Research Council of Canada, Québec, pp. 199–205.

Riseborough, D.W., and M.W. Smith, 1998. Exploring the limits of permafrost. In: *Proceedings of the Seventh International Conference on Permafrost.* A.G. Lewkowicz and M. Allard (Eds.). Centre d'études Nordiques, Université Laval, Québec, pp. 935–941.

Ritchie, J.C., L.C. Cwynar, and R.W. Spear, 1983. Evidence from north-west Canada for an early Holocene Milankovitch thermal maximum. *Nature* 305: 126–128.

Root, T.L., and S.H. Schneider, 1995. Ecology and climate: Research strategies and implications. *Science* 269: 334–341.

Sellmann, P.V., J. Brown, R.I. Lewellen, H.L. McKim, and C.J. Merry, 1975. The classification and geomorphic implications of thaw lakes on the Arctic Coastal Plain. *CRREL Research Report* 344, 20 pp.

Shiklomanov, N.I., and F.E. Nelson, 1999. Analytic representation of the active layer thickness field, Kuparuk River basin, Alaska. *Ecological Modelling* (in press).

Tedrow, J.C.F., 1969. Thaw lakes, thaw sinks and soils in northern Alaska. *Biuletyn Peryglacjalny* 20: 337–344.

Thie, J., 1974. Distribution and thawing of permafrost in the southern part of the discontinuous zone in Manitoba. *Arctic* 27: 189–200.

Thorn, C.E., 1988. Nivation: a geomorphic chimera. In: *Advances in Periglacial Geomorphology.* M.J. Clark (Ed.). Wiley, New York, pp. 3–31.

Trenhaile, A.S., 1998. *Geomorphology: A Canadian Perspective.* Oxford University Press, Toronto.

Tricart, J., and A. Cailleux, 1972. *Introduction to Climatic Geomorphology.* St. Martin's, New York.

Waelbroeck, C., 1993. Climate-soil processes in the presence of permafrost: A systems modelling approach. *Ecological Modelling* 69: 185–225.

Waelbroeck, C., P. Monfray, W.C. Oechel, S. Hastings, and G. Vourlitis, 1997. The impact of permafrost thawing on the carbon dynamics of tundra. *Geophysical Research Letters* 24: 229–232.

Wahrhaftig, C., 1965. Physiographic Divisions of Alaska. *USGS Professional Paper* 482: 1–52.

Washburn, A.L., 1980. *Geocryology: A Survey of Periglacial Processes and Environments.* Halsted Press, New York.

Williams, P.J., 1986. *Pipelines & Permafrost: Science in a Cold Climate.* The Carleton University Press, Don Mills, Ontario.

Williams, P.J., and M.W. Smith, 1989. *The Frozen Earth: Fundamentals of Geocryology.* Cambridge University Press, New York.

Woo, M.-k, A.G. Lewkowicz, and W.R. Rouse, 1992. Response of the Canadian permafrost environment to climatic change. *Physical Geography* 13: 287–317.

Wright, H.E.J., 1983. *Late-Quaternary Environments of the United States. Volume 2: The Holocene.* University of Minnesota Press, Minneapolis.

Wright, H.E.J., and S.C. Porter, 1983. *Late-Quaternary Environments of the United States. Volume 1: The Late Pleistocene.* University of Minnesota Press, Minneapolis.

Yershov, E.D., 1998. *General Geocryology.* Cambridge University Press, Cambridge.

Zoltai, S.C., 1971. Southern limit of permafrost features in peat landforms, Manitoba and Saskatchewan. *Geological Association of Canada Special Paper* 9: 305–310.

Zoltai, S.C., 1995. Permafrost distribution in peatlands of west-central Canada during the Holocene warm period 6000 years BP. *Géographie physique et Quaternaire* 49: 45–54.

14

The Boreal Forest

Glen M. MacDonald

The boreal forest of North America, extending unbroken from the Pacific coast of Alaska to the Atlantic coast of Newfoundland, encompasses some 356 million hectares (ha) (fig. 14.1). The forest reaches a latitude of almost 69°N in western Canada and has southern limits near 48°N in the Great Lakes region. The North American boreal forest is part of a larger circumpolar coniferous forest that extends across Eurasia. The global boreal biome covers over 1100 million ha and represents one-third of all forested land on the planet (Kauppi and Posch, 1985; Apps et al., 1993). The North American boreal forest is sparsely populated with an average of less than one person per square kilometer, and much of it remains in a natural state. Despite its massive geographic extent, the species diversity of the forest is very low. The tree flora is largely dominated by a dozen species, seven of these being conifers of the family Pinaceae. The modern distribution of the boreal biome can be related to a specific set of climatic conditions. As climate has changed in the past, so too has the location and nature of the boreal biome. Indeed, because much of the area now occupied by boreal forest was glaciated during the late Pleistocene and early Holocene, the present boreal forest is a relatively young feature. However, the forest itself exerts a strong influence on regional and global climate, and it likely was an important factor in past climatic changes. Clearly, changes in the boreal forest caused by the greenhouse effect will be an important aspect of future global warming.

This chapter will review the climate, soils, and vegetation of the boreal biome. The Quaternary history of the forest will be broadly outlined. Finally, the relationship between the forest and future climatic change will be considered.

14.1 Climate and Soils

14.1.1 Climate

Although to the casual observer the boreal biome appears monotonous from Alaska to Labrador, the boreal climate does display some diversity (fig. 14.1). Central regions, such as interior Alaska and western Canada, experience an increasingly severe continental climate toward the interior of the continent. The mean January and July temperatures can vary by as much as 50°C. The coldest temperature recorded at a permanent weather station in North America occurred at Snag, Yukon, where the extreme minimum daily temperature in February is –62.8°C (Canada, 1982b). In contrast, the average extreme maximum daily temperature during July at Snag is 31.7°C (Canada, 1982b). Despite the occurrence of warm summer days, the growing season is generally less than 6 months in the south and as short as

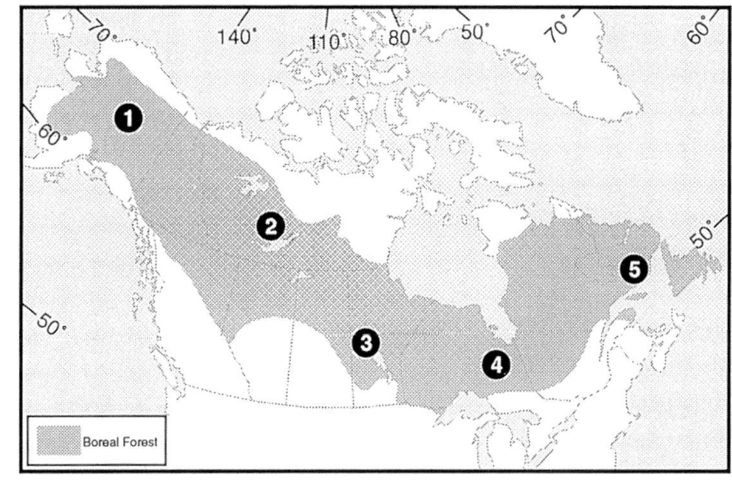

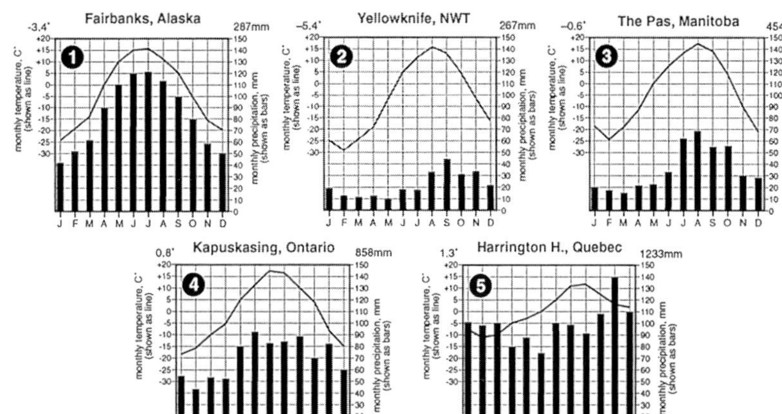

Figure 14.1 General location of the boreal biome in North America (based on synthesis of Rowe, 1972; National Atlas of Canada, 1972; Viereck and Little, 1982; Elliott-Fisk, 1988) and representative monthly temperature and precipitation from national weather service normals.

3 months in the north. Average annual temperatures tend to fall below 0°C. Precipitation is also low, averaging between 250 to 500 mm per year. The greatest precipitation occurs in the summer. In portions of the central boreal forest, potential evapotranspiration can exceed annual precipitation. Farther east in Ontario, there is an increase in precipitation and a decrease in seasonality. Increased Atlantic maritime influence leads to even lower seasonality and greater moisture in Quebec and Labrador. Annual precipitation in the boreal forest of southeastern Quebec and Newfoundland exceeds 1000 mm, and the annual distribution of precipitation is less seasonal.

Despite the geographic variations in temperature and precipitation within the boreal biome, some general climatic relations are apparent when looking at the Canadian boreal zone outside of the mountainous regions of Alaska. Boreal sites have short, frost-free periods of less than 120 days but usually more than 60 days (fig. 14.2a). Mean July temperatures generally do not fall below 10°C to 12.5°C in the north (fig. 14.2b). Sites with lower July temperatures are dominated by forest-tundra or tundra. Mean July temperatures above 17.5°C are generally not found at stations within the boreal forest. The annual growing degree days (> 5°C) in the forest range from 500 in the north to 1500 in the south (fig. 14.2c). Sites with 500 to 750 growing degree days are occupied by open woodlands and forest-tundra. Finally, the ratio of annual precipitation to potential evapotranspiration is highly variable, but generally remains above 0.50.

The thermal conditions that influence the location of the North American boreal biome are controlled by two main factors. The first is solar radiation. The northern extent of the boreal forest is limited to regions that receive more than 15 kly yr^{-1} (kilolangleys per year) net radiation (Hare and Ritchie, 1972; Ritchie, 1984). Extreme southern areas can receive up to 40 kly yr^{-1}. However, the northern and southern limits of the forest clearly do not simply parallel latitude, which would be the case if insolation angle and day length were the only factors affecting temperature. Rather, the northern and southern limits of the forest occur at lower latitudes in the central portion of the continent than in the west or the east (fig. 14.1). Bryson

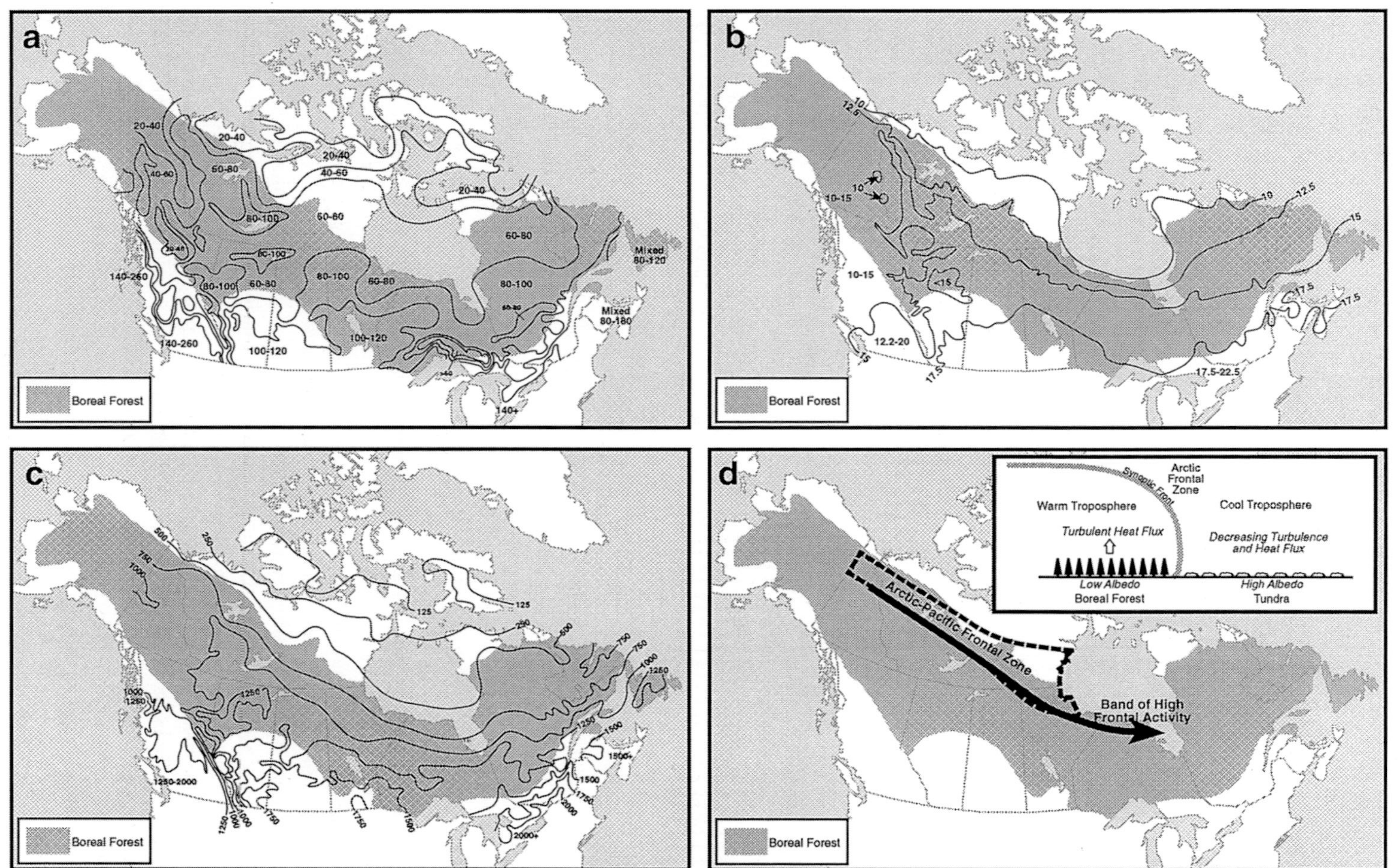

Figure 14.2 The boreal biome in relation to the following: (a) the annual number of frost-free days; (b) July daily mean temperature; (c) number of degree-days >5°C; and (d) mean July position of the Arctic Front. The inset shows the relationship between the boreal-tundra ecotone and the Arctic Front (sources include Bryson, 1966; National Atlas of Canada, 1986; Pielke and Vidale, 1995; MacDonald et al., 1998).

(1966) has shown that the northern and southern limits of the forest correspond with the mean or modal July and January positions of the Arctic Front (fig. 14.2d). Points north of the July position are dominated in summer by cold, dry polar air masses and cannot sustain forests. The front itself is often typified by a sharp change in surface temperatures (Scott, 1992). The influence of the front is manifest in sharp changes in annual growing degree days, growing season and precipitation that occur at the northern and southern edges of the boreal biome. The dipping configuration of the front is controlled by the high mountains of the Canadian Cordillera, which serve to anchor the summer frontal position in northwestern Canada, and by the cold temperatures of Hudson Bay, which depress the frontal position in central Canada (Bryson, 1966). The rapidity in the transition from winter to summer conditions and back again to winter conditions in the boreal forest has frequently been noted. Scott (1992) attributes such rapid seasonal transitions to the northward and southward propagation of the frontal position during fall and spring, respectively (see chapter 5).

14.1.2 Thermal Controls on Boreal Vegetation

Although there is a relationship between thermal conditions and the boreal forest, the exact nature of how these conditions control the distribution of the boreal vegetation remains poorly understood (Black and Bliss, 1980; Larsen, 1980; Elliott-Fisk, 1988). At its northern limits, the length of the growing season and the amount of available heat energy may be insufficient for trees to support foliar and bud growth and to produce viable pollen and seeds for reproduction. In addition, severe winter temperatures, coupled with blowing snow and ice, may promote destruction and desiccation of needles and buds (Scott et al., 1993). Along its southern limits, in the Canadian plains, the boreal forest gives way to parkland and prairie vegetation. In this area, the large-scale control on the forest limits is moisture stress caused by higher rates of potential evapotranspiration relative to precipitation. In eastern North America, the boreal forest is replaced by temperate deciduous forest. The deciduous species appear to be better competitors in the temperate regions. The short growing season of the boreal region may be insufficient for most temperate deciduous species to produce new leaves and derive enough carbohydrates from photosynthesis to grow and reproduce. Winter temperatures are likely also a factor in restricting the northern ranges of temperate deciduous species. After frost hardening, most boreal conifers can withstand temperatures below −70°C. In contrast, the buds of temperate deciduous species are damaged and destroyed at temperatures between −35° and 35°C (Sakai and Larcher, 1987). In the deciduous forest region of eastern North America, the southern limits of the boreal forest coincide with locations where degree growing days drop below 1750 and the mean daily minimum temperatures during cold January conditions drop to −20°C. Finally, the ectotrophic micorhizae associated with conifer roots may give the boreal trees an advantage in obtaining nutrients from the raw humus of the boreal soils (Walter, 1985). In the mountainous areas of western Canada and southern Alaska, the boreal forest is replaced by subalpine and coastal coniferous forests. The conifers in these forests are likely better adapted to the mild, moist coastal climate or the greater insolation intensities and diurnal variations of the mountains.

Positive feedbacks also exist between the presence of boreal forest vegetation and the location of the summer position of the Arctic Front. Hare and Ritchie (1972) pointed out that there is a strong gradient in thermal budget between the closed boreal forest and the tundra that is related to differences in albedo and surface conditions. Even in summer, the albedo of the boreal forest is low (0.10) compared to the tundra (0.15–0.20) (Monteith, 1975; Bonan et al., 1995). Lafleur et al. (1992) showed that the boreal forest has a greater sensible heat flux than the adjacent tundra even in areas where the local differences in the albedo of the forest and tundra were small. Pielke and Vidale (1995) calculated that the difference in heating rates between the boreal forest and adjacent tundra was on the order of 50 W m^{-2} (Watts per square meter). They concluded that this was sufficient to create a relatively deep boundary layer over the forest and a resulting steep thermal gradient in the atmosphere between the forest and tundra. This thermal gradient would place the Arctic Front in a preferred position. Analysis using climate models supports the view that the location of the boreal forest-tundra boundary can influence circulation and northern climate (Bonan et al., 1992; Foley et al., 1994).

14.1.3 Soils

As a result of low temperatures and an excess of precipitation relative to potential evapotranspiration, boreal soils are generally humus-rich podzols on mesic sites and organic soils on poorly drained areas (see chapter 9). Podzol nutrient levels are low in the A horizon due to leaching and subsequent illuviation in the B horizon. Calcium is particularly low at the surface. Low bioactivity in the cold soils results in conditions of low nitrogen availability and high acidity (Larsen, 1980; Elliott-Fisk, 1988). The acidic nature of the coniferous leaf litter may contribute to both the low bioactivity and acidity of the soils. These conditions contribute to slow decomposition rates and the development of thick organic soils in poorly drained sites. However, Pare and Bergeron (1996) have shown that nutrient levels are generally low under both coniferous and deciduous stands in the eastern boreal forest. Soils tend to thin northward, with tundra regions underlain by thin arctic brown soils and regosols. Beyond the southern limits of the boreal forest, chernozems are found in the temperate grasslands and brown soils are found in the deciduous forests.

Unlike central Alaska, most of the boreal region of Canada was glaciated (Dyke and Prest, 1987). The extensive areas of coarse to sandy glacial, glaciofluvial, and glaciolacustrine deposits provide a rapidly draining substrate that promotes podzol formation. However, glaciation has also produced drainage patterns that are highly deranged, creating numerous lakes and wetlands (see chapter 8). In addition, the southern limits of discontinuous permafrost are found in the forest, whereas the northern edge of continuous forest cover coincides with the northern limits of continuous permafrost (fig. 14.3a; see also chapter 13). The presence of permafrost contributes to the development of extensive peatlands and wetlands. The amount of land surface covered by wetlands in the boreal forest can exceed 76% of the total area in some regions (fig. 14.3b) and varies by substrate and topography. The depth of organic matter in boreal peatlands can reach several meters.

The low rates of biological activity and decomposition in boreal forests, coupled with the extent and depth of peat lands, is of profound importance to the carbon cycle. The low rates of decomposition promote high organic content in boreal soils. Apps et al. (1993) estimate that the majority of carbon in the boreal biome (216 Pg C total) of Alaska and Canada resides in the soils (75 Pg C) and peatlands (130 Pg C) (where Pg C = petagrams or 10^{15} grams of carbon).

The low nitrogen levels in the leaf litter from the spruce species *Picea glauca* and *Picea mariana* may play an important role in depressing the fertility of boreal soils. *Picea* leaf litter decays and mineralizes nitrogen very slowly. Low nitrogen levels appear to be a constraint on net primary productivity in the boreal forest (Van Cleve et al., 1983). Thus, the presence of spruce coupled with the low bioactivity of boreal soils may promote declines in spruce and predispose the nitrogen-stressed forest to die back from other factors (Pastor et al., 1987).

14.2 Modern Vegetation

14.2.1 Flora

The flora of the North American boreal forest is not very diverse in terms of species richness (table 14.1). This is apparent in the tree flora. The true boreal trees include the conifers *Picea glauca*, *Picea mariana*, *Pinus banksiana*, *Larix laricina*, and *Abies balsamea* of the family Pinaceae and the deciduous species *Betula papyrifera* of the family Betulaceae and *Populus balsamifera* and *Populus tremuloides* of the family Salicaceae. Two other conifers, *Pinus contorta* ssp. *latifolia* and *Abies lasiocarpa*, are present as quasi-boreal trees in the northern portions of Alberta and British Columbia and in the southern Yukon. Southern tree species such as *Thuja occidentalis* and *Fraxinus nigra* are found at the southeastern edge of the boreal zone.

The flora of large shrubs (≥ 1 m height) is also relatively small. A number of species belong to the Salicaceae, including some dozen or so willows (*Salix* spp.) and *Myrica gale*. From the Betulaceae, there are shrub birches (*Betula nana*, *Betula glandulosa*) and alders (*Alnus crispa*, *Alnus incana*). From the gooseberry family (Grossulariaceae), there are some half dozen species of *Ribes*. The Rosaceae are relatively well represented with several genera of woody plants, including *Spiraea*, *Sorbus*, *Amalanchier*, *Rubus*, *Potentilla*, and *Rosa*. However, species richness is not great within the genus. In Alaska, for example, there are 10 genera of woody plants from the Rosaceae, but a total of only 22 different shrubby species (Viereck and Little, 1972). Two widely distributed shrubs are in the Elaeagnaceae (*Shepherdia canadensis*, *Elaegnus commutata*). The family Ericaceae, represented by some 13 genera and about 30 species, is a particularly important component of the flora on peat lands and acidic sites. *Ledum* and *Vaccinium* are common shrubs.

The floras of small shrubs, herbs, and graminoids are similarly relatively small. Important components of the nonvascular flora include *Sphagnum* (peat moss), which often dominates the ground cover on bogs, and the feather mosses *Hylocmium splendens* and *Pleurozium schreberi*.

The degree of endemism in the North American boreal flora is relatively low. Some tree species such as *Picea glauca*, *Picea mariana*, *Abies balsamea*, and *Pinus banksiana* tend to be restricted to the boreal biome. Aspen, on the other hand, has a very wide distribution. All of the tree genera are represented in the flora of the Eurasian boreal forest. In addition, all of the genera have representatives in other North American conifer-dominated forests. Nearly all of the trees and large shrubs of Alaska can be found growing in the conterminous United States (Viereck and Little, 1972). Many of the shrub, herb, and other species in the North American boreal forest have circumarctic distributions.

14.2.2 Regional Vegetation and Plant Communities

The following discussion of the vegetation of the North American boreal region draws from a number of sources (La Roi, 1967; La Roi and Stringer, 1976; Kershaw, 1977; Rowe, 1977; Larsen, 1980, 1982; Elliott-Fisk, 1983, 1988; Payette, 1983; Van Cleve et al., 1986; Archibold, 1995) and from the observations of the author. The regional zonation and community structure of the boreal forest can be considered a four-tiered hierarchy. The first three tiers of this

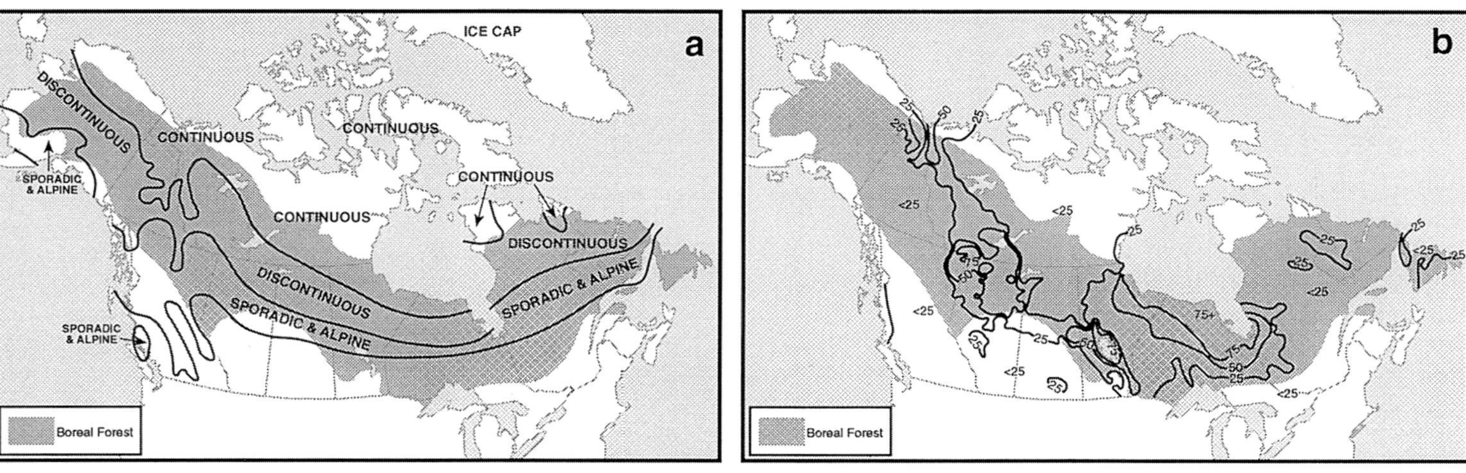

Figure 14.3 The boreal biome in relation to the following: (a) permafrost regions of North America; (b) percentage of wetland area of Canada (sources include Harris, 1986; National Atlas of Canada, 1986).

Table 14.1 Species and genera of vascular plants, and their common names, mentioned in the text, as well as plants commonly found in the boreal forest and boreal wetlands.

Abies balsamea (L.) Mill.	balsam fir
Abies lasiocarpa (Hook.) Nutt.	subalpine fir
Achillea spp.	yarrow
Agrostis spp.	bent grass
Alnus cripsa (Ait.) Pursh	mountain alder
Alnus incana (L.) Moench	western alder
Alnus rugosa (Du Roi) Spreng.	eastern alder
Andromeda spp.	bog rosemary
Arctostaphylos spp.	bearberry
Arnica spp.	
Artemisia spp.	sagebrush or wormwood
Aster spp.	aster
Astragalus spp.	milk vetch
Betula papyrifera Marsh.	tree birch
Betula nana L.	shrub birch
Betula glandulosa Michx.	glandular shrub birch
Bidens spp.	bur marigold
Brasenia schreberi Gmel.	water shield
Bromus spp.	brome grass
Calla palustris L.	wild calla
Carex spp.	sedge
Coptis groenlandica (Ceder) Fern.	goldthread
Cornus canadensis L.	dogwood
Cypripedium acaule Ait.	lady's slipper
Dodecatheon spp.	shooting star
Draba spp.	
Drosera spp.	sundew
Dryopteris spp.	shield fern
Elaeagnus commutata Bernh.	silverberry
Empetrum nigrum L.	crowberry
Epilobum spp.	fireweed
Equisetum spp.	horsetail
Erigeron spp.	fleabane
Eriophorum spp.	cotton grass
Fragaria spp.	strawberry
Fraxinus nigra Marsh.	black ash
Festuca spp.	fescue grass
Galium spp.	bedstraw
Gaultheria spp.	wintergreen-snowberry
Gentiana spp.	gentian
Juncus spp.	rush
Juniperus communis L.	mountain juniper
Juniperus horizontalis Moench	creeping juniper
Kalmia polifolia Wang.	bog laurel
Larix laricina (Du Roi) K. Koch	larch
Ledum palustre L.	labrador tea
Linnaea borealis L.	twinflower
Lonicera spp.	honeysuckle
Lycopus uniflorus Michx.	water horehound
Lysichiton americanum Hult. & St. John	skunk cabbage
Menyanthes trifoliata L.	buckbean
Mitella nuda L.	bishop's cap
Myrica gale L.	bog myrtle
Nuphar spp.	water lily
Nymphaea tetragona Georgi.	wwarf water lily
Pedicularis spp.	lousewort
Picea glauca (Moench) Voss	white spruce
Picea mariana (Mill.) BSP.	black spruce
Pinus banksiana Lamb.	jack pine
Pinus contorta Dougl. Ex Loud. var. *latifolia* Engelm.	lodgepole pine
Pinus strobus L.	white pine
Plantago spp.	plantain

Table 14.1 Continued

Poa spp.	blue grass
Polygonum spp.	knotweed
Populus balsamifera L.	balsam poplar
Populus tremuloides Michx.	aspen poplar
Potamogeton spp.	pondweed
Potentilla spp.	cinquefoil
Pyrola spp.	wintergreen
Ranunculus spp.	buttercup
Rhododendron spp.	rhododendron
Ribes spp.	currant
Rosa spp.	rose
Rubus spp.	cloudberry, raspberry, thimbleberry
Rumex spp.	sorrel
Salix spp.	willow
Sarracenia purpurea L.	pitcher-plant
Saxifraga spp.	saxifrage
Scirpus spp.	bulrush
Senecio spp.	ragwort
Shepherdia canadensis (L.) Nutt.	soapberry
Smilacina spp.	false Solomon's seal
Solidago canadensis L.	goldenrod
Sorbus spp.	mountain ash
Symphocarpus albus (L.) Blake	snowberry
Stellaria spp.	chickweed
Thalictrum spp.	meadow rue
Thuja occidentalis L.	white cedar
Trifolium spp.	clover
Typha latifolia L.	cat tail
Utricularia spp.	bladderwort
Vaccinium spp.	blueberry, cranberry
Veronica spp.	speedwell
Viburnum edule (Michx.) Raf.	high bush cranberry
Vicia spp.	vetch
Viola spp.	violet

proposed hierarchy can be translated into general geographic entities that serve to provide a regional zonation of the North American boreal forest (fig. 14.4a).

At the broadest tier, thermal conditions are the governing factors used to subdivide the boreal forest. At this level, the boreal region can be divided into three latitudinal components (fig. 14.4a). The first, in the south, is closed boreal forest, which includes both coniferous and deciduous tree species. Farther north, the canopy opens and the vegetation is typified by open conifer woodland with a lichen-dominated ground surface. At the northernmost extreme, the vegetation is forest-tundra, in which trees are restricted to small individuals and krummholz growing in small stands or as scattered single trees.

At the next tier, the regional divisions of the boreal forest mainly reflect differences in moisture and seasonality (fig. 14.4a). As discussed previously, moisture increases eastward from Ontario to Labrador, whereas seasonality decreases over the same region. Analysis of plant communities sampled along an east–west transect of the boreal forest demonstrates a longitudinal transition in understory vascular flora and bryophytes (La Roi, 1967; La Roi and Stringer, 1976). In addition, certain tree species such as *Thuja occidentalis* and *Fraxinus nigra* are restricted to eastern portions of the biome, whereas *Abies lasiocarpa* and *Pinus contorta* are found only in the west. It is not clear whether the restriction of certain species to the eastern or western portions of the biome reflects strict climatic control or the history of postglacial migration or perhaps some combination of these factors.

Large, interregional differences in edaphic conditions form the next tier (fig. 14.4b). The differences in relief and soil conditions produced by the topographic diversity of the Western Cordillera of Alaska and Canada, the gentle rolling terrain and sedimentary bedrock of the Canadian plains, the shallow and rocky soils of the Canadian Shield, and finally, the water-logged conditions of the flat, fine-grained marine deposits of the Hudson Bay lowlands all serve to produce distinct vegetation patterns. One of the most visible differences between these broad edaphic re-

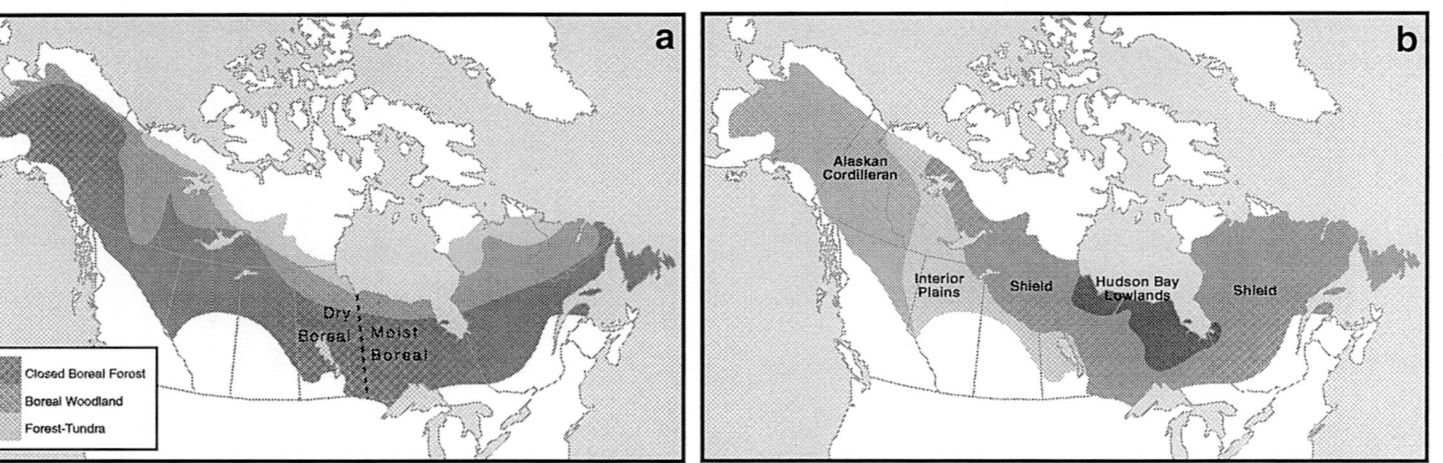

Figure 14.4 General divisions of the boreal biome by (a) closed forest, woodland, and forest-tundra latitudinal belts with dry western and moist eastern portions indicated, and (b) physiographic regions (based on data from National Atlas of Canada, 1986).

gions is in terms of drainage and the occurrence of wetlands (fig. 14.3b). In Alaska and the Cordilleran portions of Canada, the percentage of boreal forest dominated by wetlands ranges from 0 to 25%. The same situation is true for most of the area of the Canadian Shield. However, the area occupied by wetlands in the western Interior Plains and the Hudson Bay Lowlands can exceed 76%, with the largest area of wetland-dominated terrain being found along the southern shores of Hudson and James Bays (National Atlas of Canada, 1986). Although these divisions are very coarse compared to other boreal zonations, they serve to delineate the most important broad regional patterns.

Local plant communities within these broad climatic and edaphic regions reflect site-specific differences in topography, substrate, and drainage. Some of the most common of these local forest communities are described subsequently.

In southern portions of the forest, dry sites underlain by sandy soils are often occupied by *Pinus banksiana*. Vegetation, which can be extremely sparse under these stands, includes low shrubs, such as *Juniperus communis*, *Arctostaphylos uva-ursi*, and *Vaccinium* spp., as well as the fructose lichen *Polytrichium*. In western Alberta and the Yukon, *Pinus banksiana* is replaced by the closely related Rocky Mountain species *Pinus contorta* ssp. *latifolia*. Mesic sites are dominated by *Picea glauca* with *Abies balsamea*, *Betula papyrifera*, and *Populus tremuloides*. In portions of the Canadian prairie provinces, *Populus tremuloides* and *Betula papyrifera* can dominate very large areas of the southern boreal zone. East of central Alberta, *Abies balsamea* is replaced by *Abies lasiocarpa*, The understory on mesic sites consists of large shrubs such as *Cornus* spp., *Rosa* spp., *Alnus crispa*, *Salix* spp., *Shepherdia canadensis*, *Ribes* spp., and ericoids, including *Arctostaphylos*, *Vaccinium*, *Kalmia*, and *Ledum*. Smaller plants such as *Linnaea borealis*, *Clintonia* spp., *Listeria cordata*, *Lonicera dioica*, *Solidago multiradiata*, *Layhyrus ochroleucus*, and *Pyrola uniflora* are common. The ground cover is often dominated by mosses such as *Hylocomium splendens*, *Hypnum schreberi*, *Ptilium cristi-castrensis*, and *Pleurozium schreberi*. In eastern North America, shrub and herb species species such as *Acer spicatum*, *Vaccinium pennsylvanicum*, *Epigea repens*, *Gaultheria procumbens*, *Oxalis montana*, and *Trillium* spp. are found in addition to the other listed understory species. Moist sites are dominated by *Picea mariana*, with *Larix laricina* found on more nutrient-rich areas such as fens. On very moist sites, the typical ground cover is *Sphagnum* moss with *Betula glandulosa*, *Ledum groenlandicum*, *Rubus chamaemorus*, *Salix*, *Alnus* and ericoids. The understory on drier *Picea mariana* sites can be quite similar to that of sites dominated by *Picea glauca*. Among the mosses, *Pleurozium schreberi* tends to dominate over *Hylocmium splendens*. In large areas of Quebec and adjacent Labrador, lichens dominate the ground cover.

Three additional features of the southern forest should be noted. First, *Thuja occidentalis* occurs as a small tree or large shrub on bogs and on rocky sites in the southern boreal forest east of central Manitoba. Second, *Fraxinus nigra* can be found well into the boreal zone of Ontario along river valleys. Third, *Populus balsamifera* is often found as an element with *Picea glauca* on coarse substrates along the channels of streams and rivers.

To the north, both the structure of the vegetation and the tree floristics change. In the boreal woodland zone, the canopy becomes widely spaced and permanent openings occur on uplands. Lichens become increasingly important as ground cover. In the west, *Stereocaulon paschale* is dominant, whereas *Cladonia stellaris* characterizes the east. *Abies and Thuja occidentalis* are absent in the woodland zone. In the forest-tundra zone, tundra openings become dominant, and trees are generally restricted to valleys and sheltered sites along lakes. *Pinus banksiana*, *Betula papyrifera*, and *Populus* disappear from the vegetation in most areas. *Picea glauca* still tends to be more common on mesic to dry sites, whereas *Picea mariana* dominates on moist wooded sites. Near their northern limits, both species can occur as prostrate krummholz. Regeneration by seeds has been found to be extremely irregular, and both species of spruce commonly propagate by layering. The northern range limit of the biome is thus a relatively diffuse boundary (see Timoney et al., 1992). Exceptions to the occurrence of *Picea* as the northernmost tree species can be found in Alaska where *Populus balsamifera* occurs beyond the northern limits of spruce and in Labrador where *Populus tremuloides* can be found at a few tree-line sites.

A final feature of boreal vegetation that deserves mention is the extensive component of peatlands. Particularly impressive are the huge areas of the James Bay region dominated by string bogs and other forms of peatlands (see chapter 8). *Sphagnum* moss is the principal peat-forming species. Upon this acidic and nutrient-poor substrate, vegetation ranging from open woodlands to herb- and graminoid-dominated wetlands can be found. Woodlands are dominated by *Picea mariana* with some *Larix laricina* and an understory of shrub *Betula*, *Myrica gale*, *Rhododendron*, various ericoids, herbs, orchids, grasses, and sedges. Other sites may support only shrubs; others are dominated by herbs, orchids, grasses, and sedges.

14.2.3 Fire, Vegetation Organization, and Dynamics

The most important agent of vegetation disturbance in the North American boreal forest is fire (Kershaw, 1977; Larsen, 1980; Wein and MacLean, 1983; Dyrness et al., 1986; Johnson, 1992; Payette, 1992). Although wind, insects, and permafrost activity can be important causes of local vegetation disturbance, over large areas and long periods of time the impact of these forces is much less than that of fire. However, it should be noted that during a peak infestation of spruce budworm (*Choristoneura fumiferana*), in

1977–81 a total of 63.8 million m^3 yr^{-1} of Canadian timber were lost to this and other insects and diseases compared to 80 million m^3 yr^{-1} lost to fire (Honer and Bickerstaff, 1985). The spruce budworm, which has a cycle of about 29 years between major outbreaks, is perhaps the most destructive insect pest in the boreal forest (Blais, 1983). The impact of the budworm is most pronounced in the east where it can be particularly important in conjunction with fire in shaping the dynamics of sites where *Abies balsamea* is important (Bergeron and Dansereau, 1993; Bergeron and Charron, 1994).

Johnson (1992) estimated that from 0.5 to 2% of the upland boreal forest burns each year. The impact of fires on the boreal vegetation is particularly severe because these fires have high frontal intensities that may reach 30,000 kW m^{-1}, which results in frequent crown fires that destroy the forest cover. The nature of the boreal vegetation is intimately tied to fire (Payette, 1992).

Boreal fires occur frequently and can be extremely large. In 1950, a free-burning fire in northern British Columbia and adjacent Alberta lasted from 2 June to 31 October and extended over 1.4 million ha (Murphy and Tymstra, 1986). Even in areas where fire suppression has been widely practiced, fires in the boreal zone have been numerous and have burned very large areas. In Wood Buffalo National Park in Alberta, approximately one-quarter of the 44,000-km^2 park burned during the 1982 fire season (MacDonald et al., 1991; Larsen and MacDonald, 1995). The most frequent fires are often very small, but these limited fires account for only a small portion of burning in the boreal forest. Barney and Stocks (1983) found that in Alaska and the adjacent Canadian territories, 60 to 80% of the recorded fires were < 5 ha in size. In an analysis of Canadian boreal fires during the 1980–89 period, fires >200 ha accounted for only 3% of the total number of fires, but 97% of the total area burned (Johnson, 1992). Higgins and Ramsey (1992) calculated that fires of over 10,000 ha were responsible for 90% of the total area burned annually in Canada.

Two important factors influence the number and size of boreal fires: high rates of ignition events and high rates of spread. Fire ignition may be caused by human activity and lightning. Despite the frequency of human-ignited fires, lightning-caused fires actually account for 90% of the area burned in the boreal forest (Johnson, 1992). As the occurrence of summer frontal activity and lightning storms decreases toward treeline, so too does the number of ignition events (Johnson, 1992). Fire ignition tends to follow the northward and southward progression of the Arctic Front (Johnson and Rowe, 1975; Payette et al., 1989; Johnson, 1992). Fires ignite in southern portions of the forest as early as May and ignite near treeline in July. After July, the Artic Front moves southward, so fires that ignite in August are generally found in middle and southern sections of the forest.

Rates of spread are governed by fuel conditions and fire weather. The rate of spread of fast-advancing boreal fires is usually 5–10 m min^{-1} (Johnson, 1992), but can reach 100 m min^{-1} (Kiil and Grigel, 1969). The nature of boreal stands helps promote these rapid rates of fire spread. Conifer-dominated stands provide high levels of fuel in the form of dry litter, fine needles, twigs, bark flakes, and resinous products. In addition, the conical shape of boreal conifers promotes crown fires and rapid spread (Johnson, 1992). The ericoid shrubs and lichens are also important fuels because of their flammability and potential to carry fire between trees (Auclair, 1983). The ericoids have a high ether content and caloric values that promote combustion at low temperatures and burning at high temperatures (Auclair, 1983).

The largest and most severe boreal fires tend to occur when high-pressure ridges develop at the 50-kPa (kilo-Pascal) level and become stalled over the boreal forest (Johnson, 1992). These ridges have a usual duration of 5 to 10 days, but can persist much longer (Treidl et al., 1981). The ridge produces warm and dry conditions associated with subsiding air. During this phase, there is little chance of lightning igniting fires, nor are the winds strong enough to promote rapid spread. However, as the ridges break down into troughs, lightning and wind speed increase. These conditions combined with the dried fuel promote severe and extensive fires. There is a clear relationship between generally dry summers and large fire years (Flannigan and Harrington, 1988; Larsen and MacDonald, 1995). In most areas, these exceptional fire years account for the majority of areas burned in the boreal forest (Rowe et al., 1974; Viereck, 1983; Flannigan and Harrington, 1988; Larsen and MacDonald, 1995).

The decreased chance of lightning, decreased fuel loads, and decreased moisture stress toward the northern edge of the boreal forest lead to the lessened occurrence of fires (fig. 14.5). Payette et al. (1989) estimated that fires at sites in the forest-tundra zone of northern Quebec might occur as infrequently as less than once every 7800 years. At a more local scale, the average return time of fires in the boreal forest varies according to site type. *Picea glauca*–dominated stands in areas with natural fire breaks such as lakes and sharp topography may not burn for 100–400 years. However, on dry and open sites, *Picea glauca* stands may burn as often as every 60 years. *Picea mariana* stands seem to have fire-return intervals of around 60–100 years, but this can range to 200 years (Rowe et al., 1974; Yarie, 1981; Viereck, 1983). On well-drained substrates with little topographic relief, *Pinus banksiana* stands have average fire-return intervals as short as 28 years (Rowe and Scotter, 1973; Carroll and Bliss, 1982; Delisle and Dube, 1983). The frequency of fires may decline eastward in the moister areas of Quebec and Labrador (Wein and Moore, 1977; Viereck, 1983). Payette et al. (1989), however, found the

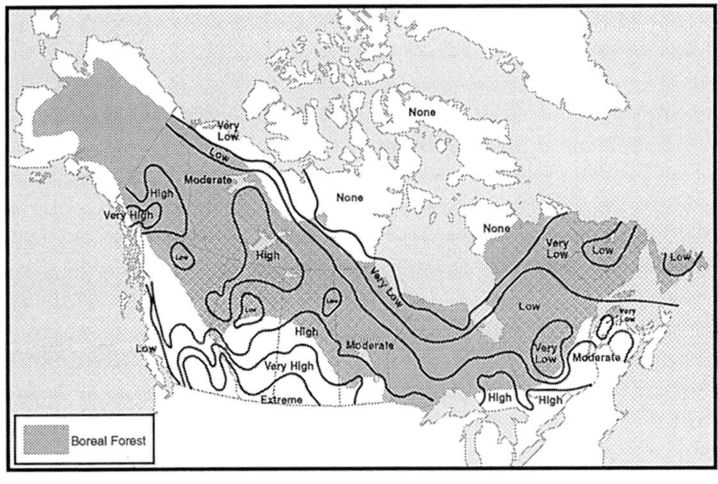

Figure 14.5 The boreal biome in relation to wildland fire occurrence in Canada (after Simmard, 1973).

fire incidence in southern Quebec to be comparable to sites in the western boreal zone. In southern Quebec, damage by spruce budworm may increase fire risks (Bergeron and Dansereau, 1993).

The high number of boreal fires coupled with their variability in size imparts a mosaic pattern to boreal vegetation in which stands of various ages, some large and some quite small, typify the landscape (Heinselman, 1970; Johnson, 1992). Johnson has calculated that about 10% of the boreal forest stands are less than 20 years old, whereas 75% are between 20 and 75 years old. This means the area of "old-growth" forest in the boreal region is very small. The frequency of fires and resulting youthfulness of the boreal stands has prompted most ecologists to regard the concept of a self-regenerating climax community as inappropriate for the boreal forest (e.g., Carlton and Maycock, 1978; Larsen, 1980; Cogbill, 1985; Bergeron and Dubuc, 1989; Bergeron and Charon, 1994). Most stands burn before significant regeneration by the dominant trees takes place. Kershaw (1977) has argued that if fire is excluded from conifer-lichen woodlands for periods of 200 years or more, a closed canopy forest of black spruce that regenerates by layering on feather moss substrate could develop. However, perhaps with the exception of sites in Labrador and adjacent portions of Quebec (Foster, 1983), it is unlikely that fire would be excluded from such closed-canopy sites for that length of time (Elliott-Fisk, 1988). Furthermore, it has been suggested that the removal of fire from *Picea mariana*–dominated sites may lead to a depletion of soil nutrients, an increase in permafrost, and the eventual replacement of forest by peatland rather than the establishment of a self-sustaining forest.

Studies of post-fire succession in the boreal forest suggest that most of the plants that will occupy the site during succession are established shortly after the fire (e.g., Methuen et al., 1978; Bergeron and Charron, 1994). Although soil seed banks are rare (Johnson, 1975), plants establish themselves after fire via sprouting (*Betula, Populus, Alnus, Salix*), release of seeds from serotinous cones (*Pinus banksiana, Pinus contorta*), release of seeds from non-serotinous cones that retain some seed crop (*Picea mariana*), and germination of lightweight, plumose, or winged seeds (*Picea, Pinus, Larix, Populus, Betula, Alnus, Salix*). Interestingly, Gauthier et al. (1996) have shown that fire history and age of a stand can influence the ratio of serotinous to nonserotinous cones in *Pinus banksiana* stands. A simple successional pattern is evident in boreal sites because different life-forms of plants attain dominance at different times following fires (Larsen and MacDonald, 1995). On mesic sites, the initial herb and graminoid vegetation is overgrown by shrubs, then by a mixture of deciduous and coniferous trees, and finally by coniferous trees. Among the bryophytes, mosses and liverworts such as *Polytrichum, Marachantia polymorpha, Certodon purpens*, and *Pohlia nutans* establish quickly, whereas the later-dominant feather mosses *Hylocomium splendens* and *Pleurozium shcreberi* come in as the tree canopy is established. Within the lichen-conifer forest, it is possible for lichens to dominate burned sites rapidly, initially precluding tree seedling establishment (Kershaw, 1977). Several factors determine which plants will dominate the site following fire: the soil conditions, the amount of duff that is burned to allow access to mineral soil, and the proximity of seed sources or the presence of sprouting taxa (Johnson, 1992). Of special importance is the release of nutrients that occurs after fires. Because of slow decomposition rates and the growth of organic substrates, many of the nutrients, particularly on *Picea mariana*–dominated sites, are tied up in the biomass. Burning produces significant increases in pH, nitrogen, calcium, magnesium, and phosphorus in the soil (Dyrness et al., 1986). Early successional plants appear to require high light intensities and high nutrient levels (Dyrness et al., 1986).

14.2.4 *Biomass and Productivity*

The biomass of the world's boreal forest has been roughly estimated to be around 200 t ha^{-1} with net primary productivity around 0.8 t ha^{-1} yr^{-1} (Whittaker and Likens, 1975), making the boreal forest the least productive of the world's major forest biomes. In comparison, the biomass and net primary productivity of the deciduous temperate forest average 300 t ha^{-1} and 1.2 t ha^{-1} yr^{-1}, respectively (Whittaker and Likens, 1975).

North American boreal forest biomass and productivity vary by latitude, stand type, and stand age. Aboveground biomass in the closed boreal forest in the south has been estimated at 82–163 t ha^{-1}, decreasing to 9–29 t ha^{-1} in the northern spruce-lichen woodlands (Larsen, 1980). Stand type and age play an important role in determining biomass and productivity (Black and Bliss, 1978; Zasada et al., 1978; Cannell, 1982; Viereck et al., 1986). In general, mature *Abies balsamea* and *Picea glauca* stands have the highest aboveground biomass and productivity, with *Picea mariana* and *Pinus banksiana* stands having the lowest. Aboveground biomass also decreases northward within stand types. For example, Black and Bliss (1978) found that 150-year-old treeline stands of *Picea mariana* had an aboveground volume of 3.9 m^3 ha^{-1} compared to 62.1 m^3 ha^{-1} for more southerly stands of the same age. With the exception of *Picea mariana*–dominated sites, the overstory is the largest component of the biomass. Within the overstory, trunks and branches appear to make up the bulk of the biomass. In general, there is an increase in biomass with stand age. However, Pare and Bergeron (1995) have shown that aboveground biomass can actually decrease in southern boreal stands after peaking at 75 years.

14.3 Late Quaternary Vegetation History of the Boreal Biome

In terms of its present geographic distribution, the boreal forest is relatively young. Aside from portions of Alaska, the Yukon, and mountains of the western Northwest Territories, the area occupied by the boreal biome was glaciated during the late Pleistocene (fig. 14.6). Portions of the Northwest Territories and Quebec that are occupied by boreal forest were not free of glacial ice until 7000–6000 B.P. (radiocarbon years before A.D. 1950). With the possible exception of *Populus* (Hopkins et al., 1981), there is no evidence that boreal tree genera such as *Picea* and *Pinus* were able to survive in the unglaciated portions of Alaska, the Yukon, and the Northwest Territories during the last glacial maximum. Fossil pollen evidence suggests that the glacial vegetation of these regions was a sparse tundra dominated by herbs and small shrubs with a climate that was too cold and dry to support the boreal conifers (Cwynar and Ritchie, 1980; Ager and Brubaker, 1985; MacDonald and Cwynar, 1985, 1991; Ritchie and MacDonald, 1986; MacDonald, 1987; Ritchie, 1987; Anderson and Brubaker, 1993; Szeicz et al., 1995). However, other boreal plants, including *Betula glandulosa*, a number of ericoids, and boreal herbs, likely survived the last glacial maximum in these regions. The refuge for boreal trees and many other components of the vegetation was south of the ice sheets, extending across the northern and central Mississippi valley to the southern Appalachians and the mid-Atlantic coast (see also chapter 3).

By 11,000 B.P., the western edges of the Northwest Territories and southern edges of the modern boreal zone in Manitoba were being deglaciated (fig. 14.6). In the Northwest Territories, a pioneer vegetation of herbs, shrubs, and *Populus* was established. The presence of *Populus* and nonarctic plant species such *Typha latifolia* suggests that the climate may have been dry, but not particularly cold, during this period (Ritchie et al., 1983; MacDonald, 1987; MacDonald and McLeod, 1996). The herb, small shrub, and *Populus* vegetation was replaced by a shrub birch–dominated vegetation by 12,000–13,000 B.P. in the unglaciated areas of Alaska and the Northwest Territories and by 10,000 B.P. in other portions of northwestern Canada (Cwynar and Ritchie, 1980; Ager and Brubaker, 1985; MacDonald, 1987; Ritchie, 1987; Anderson and Brubaker, 1993; Szeicz et al., 1995; MacDonald and McLeod, 1996).

The Holocene history of boreal forest development is mainly the history of northward tree migration into the area of the modern boreal biome. The histories of *Abies*, *Larix laricina*, and *Populus* migration are difficult to reconstruct from palynological records because of the low pollen production or preservation of these taxa. The migration history of *Betula papyrifera* is also difficult to reconstruct because of the difficulty in separating the pollen of tree and shrub birch species (Ives, 1977).

After *Populus*, *Picea* appears to have been the next tree taxon to expand into the present boreal region. Ritchie and MacDonald (1986) concluded that *Picea* rapidly migrated northwestward from the midcontinent after the ice retreat (fig. 14.6). Evidence indicates that *Picea* trees grew on the debris covering decaying glacial ice in the Great Lakes region (Florin and Wright, 1969; Szeicz and MacDonald, 1991). The genus moved rapidly northward along the western edge of the retreating ice sheet and reached the Arctic coast of the Northwest Territories by 9000 B.P. (fig. 14.6). Both *Picea glauca* and *Picea mariana* appear to have expanded northwestward at this time, and dense forest was established by 8000 B.P. (MacDonald, 1987, 1995; MacDonald and McLeod, 1996). Expansion of *Picea* into Alaska was somewhat slower, with spruce reaching its modern population densities and range limits in western Alaska at 4000 B.P. (Anderson and Brubaker, 1993). *Picea* forest was established in portions of northern Ontario by 9000 B.P. (Liu, 1990). Expansion of *Picea* northward in eastern Canada was somewhat slower, even along the Lab-

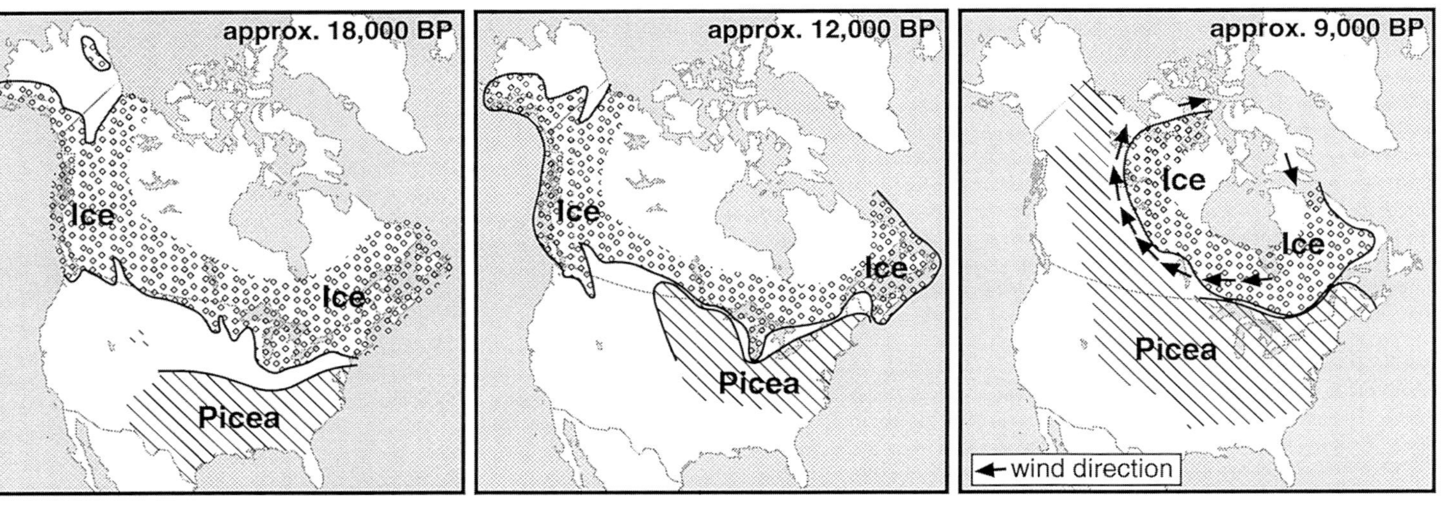

Figure 14.6 Early postglacial expansion of *Picea* into the area of the present boreal biome (after Ritchie and MacDonald, 1986).

rador coast, which was deglaciated relatively early (Ritchie and MacDonald, 1986). In northern Quebec, dense *Picea* forest was not established until 5000–4000 B.P. (Richard, 1981; Richard et al., 1982; Ritchie, 1987).

In the southern portions of the boreal biome from Ontario eastward, the expansion of *Pinus banksiana* closely followed that of *Picea*. In contrast, in western North America there is an increasing lag between the arrival of *Picea* and the arrival of *Pinus banksiana* and *Pinus contorta* at northern sites. Both pine species did not reach their modern range limits in the Northwest Territories and Yukon until the last millennium (MacDonald and Cwynar, 1985, 1991; MacDonald, 1987).

Peatlands are an important feature of the boreal landscape. Evidence for the establishment of peatlands in northwestern Canada can be gleaned from the depositional record of *Sphagnum* spores. These indicate that the early boreal forest lacked large areas of peatlands, but that these developed between 8000 and 6000 B.P. (MacDonald, 1987; MacDonald and McLeod, 1996). Radiocarbon dates from the base of peatlands in northwestern Canada (Zoltai and Tarnocai, 1975) are compatible with the interpretation from the fossil spore record. In addition, radiocarbon dates from peatlands in Alberta, Saskatchewan, and Manitoba indicate that peatlands at the southern edges of the forest did not form until after 6000 B.P., likely because of dry climatic conditions that prevailed prior to that time (Zoltai and Vitt, 1990).

Evidence of Holocene climatic changes and their impact on the boreal vegetation comes from the northern and southern edges of the biome. At the northern edge, there have been changes in treeline position or the density of trees in the forest-tundra zone in all areas except Alaska. The southern slopes of the east-to-west-trending Brooks Range coincide with the modern treeline in Alaska, and these mountains may have been a barrier to northward extensions of the boreal forest in the past. Fossil pollen records (fig. 14.7) and radiocarbon-dated stumps of *Picea* indicate that trees were growing 25 km north of their modern limits in the Mackenzie Valley of northwestern Canada between roughly 9000 and 4000 B.P. (Ritchie et al., 1983; Spear, 1993). Evidence from central Canada suggests that the population density of trees at the extreme northern limits of *Picea* was greater than at present between 5000 and 4000 B.P. (Moser and MacDonald, 1990; MacDonald et al., 1993). It is uncertain whether trees grew north of their modern range limits during this period. In northern Quebec and Labrador, pollen records suggest that the density of spruce in the forest tundra was greater than at present, and elevational treeline was higher between 4000 and 2000 B.P. (Richard, 1981; Lamb, 1985; Gajewski et al., 1993; Payette and Lavoie, 1994). It appears certain from pollen records, charcoal, and tree macrofossils that spruce did not actually extend its range limits significantly beyond its modern position at that time (Payette and Lavoie, 1994). The northward advance of the treeline and the increases in tree density in the forest tundra likely reflect warmer summer temperatures during the early and mid-Holocene. The asynchronous changes in treeline conditions from east to west (fig. 14.7) may reflect changes in the geometry of the Arctic Front over Canada (Moser and MacDonald, 1990; MacDonald et al., 1993; MacDonald and Gajewski, 1992).

Fossil pollen records provide evidence for a slight southward intrusion of aspen parkland vegetation into the bo-

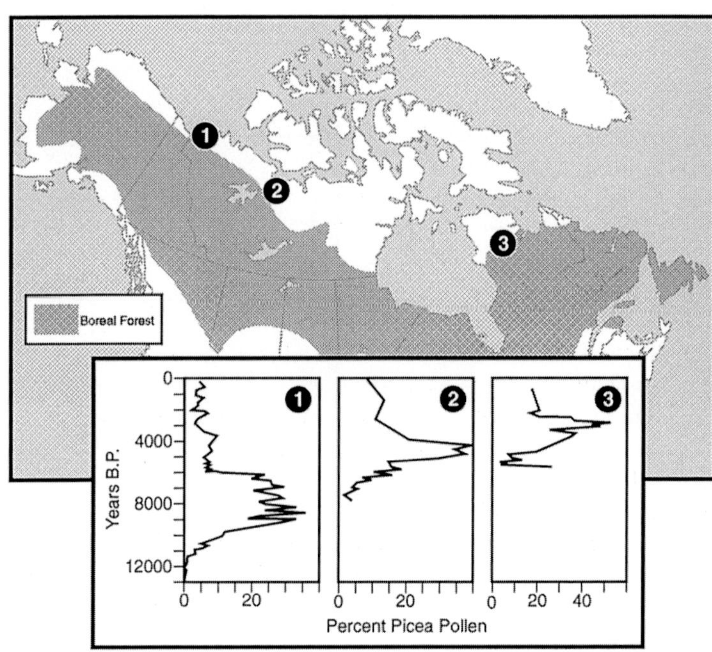

Figure 14.7 Fossil pollen records of the postglacial advance of the *Picea* treeline beyond its modern limits (after MacDonald and Gajewski, 1992).

real zone of western Canada between 9000 and 5000 B.P. (Lichti-Federovich, 1970; Ritchie, 1976; Mott, 1973; Hutton et al., 1994). The northward extent of vegetation change during this extension of parkland was only about 200 km (Hutton et al., 1994). The cause of this intrusion was probably drier conditions during the mid-Holocene (Lichti-Federovich, 1970; Ritchie, 1976; Mott, 1973; Hutton et al., 1994). In eastern North America, the range of *Pinus strobus* (white pine) extended several hundred kilometers north of its modern range limits in the southern boreal zone. Since about 5000 B.P., the range limits of *Pinus strobus* and its population density in the southern edges of the boreal forest have decreased (Anderson, 1985; Liu, 1990; Campbell and McAndrews, 1991). This retreat probably reflects climatic cooling during the late Holocene. During the past 2000 years, there is evidence for a southward expansion of boreal elements such a *Picea* and *Abies* in eastern North America, likely in response to a general cooling of climate (Webb et al., 1983, 1993).

14.4 Potential Impacts of Global Warming

The relationship between future global warming and the boreal forest is a key concern. The anticipated doubling of atmospheric CO_2 during the twenty-first century could lead to a 4°C increase in average annual temperature in the boreal region (Houghton et al., 1996). Changes in boreal vegetation could influence rates and magnitudes of global climate change. Increases in atmospheric CO_2 could also serve as a fertilizing agent, promoting greater primary productivity in the boreal forest (Mooney et al., 1991). However, CO_2 enrichment might not cause a significant increase in boreal productivity unless it were linked with increased enrichment of nutrients such as nitrogen, phosphorus, potassium, calcium, and magnesium (Kojima, 1994). Interestingly, increased temperatures associated with greenhouse warming could lead to increased availability of these soil nutrients. Van Cleve et al. (1990) conducted a soil-heating experiment in Alaska and found that soil-organic content increased by 20%, and the levels of available nitrogen, phosphorus, and potassium in the soils also increased.

Estimates of future temperatures based on computer climate models have been linked to both stand-scale and regional-scale models of boreal vegetation to gauge the potential impact of global warming. Stand-scale models of Alaskan boreal forest (Bonan et al., 1990) indicate that unless there is an unanticipated increase in precipitation or cloud cover, global warming could lead to a replacement of *Picea mariana* stands by mixed deciduous-spruce stands. Mesic *Picea glauca* forests would be transformed to *Populus tremuloides* stands, whereas present dry, forested sites would become steppe. A similar patch model (fig. 14.8) for northeastern British Columbia suggests that *Picea* stands could be replaced by *Pinus* (Burton and Cumming, 1995). A series of similar modeling exercises in eastern North America (Solomon, 1986; Pastor and Post, 1988) concluded that southern boreal forest could be replaced by temperate deciduous forest, whereas *Picea glauca* and *Betula papyrifera* could dominate the present boreal woodland zone.

Several efforts to predict the large-scale response of the boreal biome to global warming echo the findings from the stand-scale models in many ways (Emmanual et al., 1985; Rizzo and Wiken, 1992; Nielson, 1993). For example, the analysis by Rizzo and Wiken (1992) indicates a shift northward of the northern limits of the boreal biome to the Arctic coast in central Canada and perhaps onto Baffin Island (fig. 14.9). In central Canada, grassland and parkland could shift north to beyond 60° N latitude, thus fragmenting the closed boreal forest. The shift in the southern boreal limits appears less dramatic in Quebec and eastern Ontario.

Changes in the boreal forest may influence rates and magnitudes of global warming. A northward shift in forest cover would decrease high-latitude albedo and lead to increased global warming (Bonan et al., 1992). Boreal vegetation, soils, and peatlands are huge sinks and stores of carbon. Apps et al. (1993) estimate that the flux of carbon from the atmosphere into North American boreal regions is on the order of 104 Tg C yr^{-1} (where Tg C = teragrams or 10^{12} grams of carbon). Changes in boreal vegetation could either enhance or decrease the function of the boreal forest as a sink for atmospheric CO_2. Kojima (1994) calculated

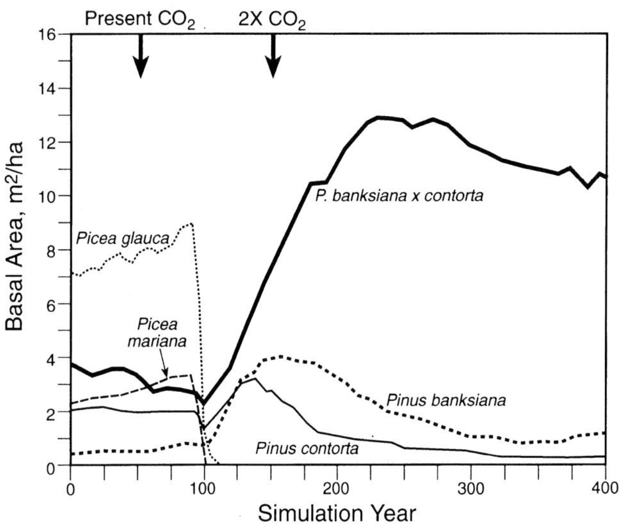

Figure 14.8 The potential change of *Picea*-dominated stands to *Pinus*-dominated stands after a doubling of atmospheric CO_2 based on a stand-simulation model for British Columbian boreal forest (after Burton and Cumming, 1995).

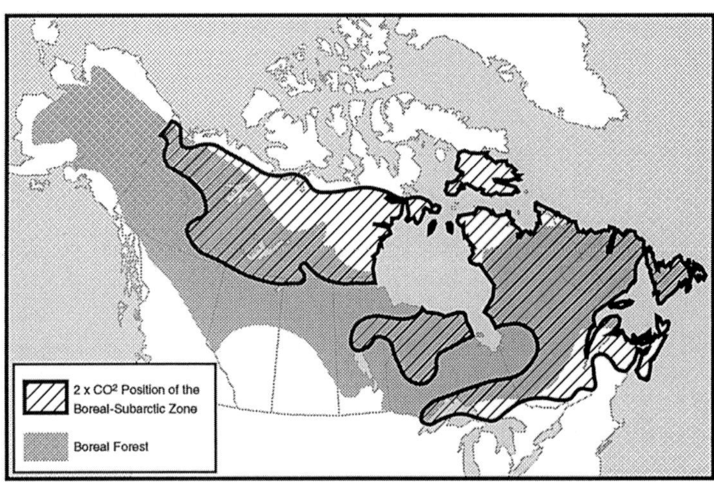

Figure 14.9 The potential northward shift and fragmentation of the Canadian boreal-subarctic ecoclimatic region after a doubling of atmospheric CO_2. The shift and fragmentation is based on the results from a linked climate and ecoclimatic regions model (after Rizzo and Wiken, 1992). The boreal ecoclimatic region includes the southern boreal forest, whereas the subarctic ecoclimatic region includes the northern boreal forest.

that a 5°C warming could increase the primary productivity of Alaskan and Canadian boreal forests by as much as 62%. Other estimates are much lower or suggest decreased net primary productivity (16%, Williams, 1985; 27% to −11%, Plochl and Cramer, 1995). Increased growth of plants would produce increased rates of carbon sequestration in boreal vegetation. However, such estimates depend on sufficient increases in moisture and soil nutrients to support the greater growth (Williams, 1985; Kojima, 1994).

In addition, heating of boreal soils could lead to increased decomposition rates and releases of carbon. Hom et al. (1990) found that boreal soils under *Picea mariana* stands would become a source of carbon if heated by 9°C or more. Boreal peatlands are a huge sink and store of carbon. The amount of carbon (C) in boreal peatlands of Alaska and Canada could be as high as 130 Pg C (Apps et al., 1993). The current flux of carbon from the atmosphere to North American boreal peatlands is around 28 Tg C yr^{-1} and could increase dramatically with climate warming. However, it is possible that rates of methane emissions might actually decline with increasing temperature and counter the impact of increased carbon (Gorham, 1988).

The impact of climate change on fire is important to our understanding of the effect of global warming on boreal vegetation. Much of the vegetation change seen in stand-scale models depends upon fire to initiate and maintain new vegetation conditions (e.g., Bonan et al., 1990). Flannigan and Van Wagner (1991) calculated that the annual area burned in the boreal forest could increase by 46% as a result of greenhouse warming. However, a recent analysis using the Canadian Climate Model suggests that fire activity in eastern Canada could actually decrease, whereas fires in boreal portions of western Canada would increase dramatically (Bergeron and Flannigan, 1995).

Estimating how the North American boreal forest will influence and respond to global warming remains a challenging problem (Apps et al., 1993). However, the confusing and often contradictory nature of estimates of future conditions should not be used as an excuse to avoid serious consideration of these problems. During the last 100 years, temperatures over the boreal forest in western Canada have risen by 1.7° C (Environment Canada, 1992). Tree-ring analysis and stand-age studies of treeline spruce from the eastern and western boreal forest have shown increased radial growth and recruitment over much of that period (e.g., Payette et al., 1985; Archambault and Bergeron, 1992; D'Arrigo and Jacoby, 1993; Lavoie and Payette, 1994; Szeicz and MacDonald, 1995a,b; MacDonald et al., 1998). In addition, during the last 20 years, increasing moisture stress appears to have limited the response of Alaskan *Picea glauca* trees to recent climate warming and increased levels of atmospheric CO_2 (Jacoby and D'Arrigo, 1995). It is not improbable that the boreal biome described in this chapter could be greatly changed over the course of the next 100 years.

14.5 Conclusion

The boreal forest of North America remains one of the most geographically extensive and pristine biomes in the world. Aside from its biological importance, the boreal forest plays a key role in global climate through its impact on albedo, the position of the Arctic Front, and global carbon balance. Despite its geographic enormity and relatively simple floristic composition, the boreal forest has been a mutable entity. Its modern form is essentially a product of postglacial warming and plant migration. The forest came to occupy the huge expanse of landscape stretching from Alaska to Newfoundland as glacial ice retreated between 12,000 and 6000 years ago. The forest did not occupy this landscape as a superorganismic whole, rather, different species of plants advanced at different rates and via different routes. The pioneering stands of *Populus* in northwestern Canada were replaced by the later-arriving *Picea*. In some portions of the Yukon and Northwest Territories, *Pinus* has become part of the vegetation only in the last few hundred years. The potential for continued climatic changes, particularly due to greenhouse warming, suggests that the boreal forest will also continue to change in terms of geographic distribution and plant community composition. Important questions remain about such changes in the boreal forest and how alteration of this massive biome may itself influence future climate.

References

Ager, T.A., and L.B. Brubaker. 1985. Quaternary palynology and vegetational history of Alaska. In *Pollen Records of Late Quaternary North American Sediments*, pp. 353–384, V.M. Bryant Jr. and R.G. Holloway, eds. Dallas: American Association of Stratigraphic Palynologists Foundation.

Anderson, M., and L.B. Brubaker, 1993. Holocene Vegetation and Climate Histories of Alaska. In *Global Climates: Since the Last Glacial Maximum*, pp. 386–400, H.E. Wright, J.E. Kutzbach, T. Webb III, W.F. Ruddiman, F.A. Street Perrott, and P.J. Bartlein, eds. Minneapolis: University of Minnesota Press.

Anderson, T.W., 1985. Late Quaternary pollen records from eastern Ontario, Quebec, and Atlantic Canada. In *Pollen Records of Late-Quaternary North American Sediments*, pp. 281–326, V.M. Bryant Jr. and R.G. Holloway, eds. Dallas: The American Association of Stratigraphic Palynologists.

Apps, M.J., W.A. Kurz, R.J. Luxmoore, L.O. Nilsson, R.A. Sedjo, R. Schmidt, L.G. Simpson, and T.S. Vinson, 1993. Boreal forests and tundra. *Water Air and Soil Pollution*, 70: 39–53.

Archambault, S., and Y. Bergeron, 1992. An 802 year tree-ring chronology from the Quebec Boreal Forest. *Canadian Journal of Forest Research* 22: 674–682.

Archibold, O.W., 1995. *Ecology of World Vegetation*. London: Chapman and Hall.

Auclair, A.N.D. 1983. The role of fire in lichen-dominated Tundra and Forest-Tundra. In *The Role of Fire in Northern Circumpolar Ecosystems*, pp. 235–256, R.W. Wein and D.A. MacLean, eds. Scope 18. Toronto: John Wiley & Sons, Canada.

Barney, R.J., and B.J. Stocks, 1983. Fire frequencies during the suppression period. In *The Role of Fire in Northern Circumpolar Ecosystems*, pp. 45–62, R.W. Wein and D.A. MacLean, eds. Scope 18. Toronto: John Wiley & Sons.

Bergeron, Y., and D. Charron, 1994. Postfire stand dynamics in a southern boreal forest (Quebec): A dendroecological approach. *Ecoscience* 1: 173–184.

Bergeron, Y., and P. Danserau, 1993. Predicting the composition of Canadian southern boreal forests in different fire cycles. *Journal of Vegetation Science* 4: 827–832.

Bergeron, Y., and M. Dubuc, 1989. Succession in the southern part of the Canadian boreal forest. *Vegetatio* 79: 51–63.

Bergeron, Y., and M.D. Flannigan, 1995. Predicting the effects of climate change on fire frequency in the southeastern Canadian boreal forest, *Water, Air and Soil Pollution* 82: 437–444.

Black, R.A., and L.C. Bliss, 1978. Recovery sequence of *Picea mariana/Vaccinium ultiginosum* forests after burning near Inuvik, Northwest Territories, Canada. *Canadian Journal of Botany* 56: 2020–2030.

Black, R.A., and L.C. Bliss, 1980. Reproductive ecology of *Picea mariana* (Mill.) BSP, at tree line near Inuvik, Northwest Territories, Canada. *Ecological Monographs* 50: 331–354.

Blais, J.R., 1983. Trends in the frequency, extent, and severity of spruce budworm outbreaks in eastern Canada. *Canadian Journal of Forest Research* 13: 539–547.

Bonan, G.B., F.S. Chapin III, and S.L. Thompson, 1995. Boreal forest and tundra ecosystems as components of the climate system. *Climate Change* 29: 145–167.

Bonan, G.B., D. Pollard, and S.L. Thompson, 1992. Effects of boreal forest vegetation on global climate. *Nature* 359: 716–718.

Bonan, G.B., H.H. Shugart, and D.L. Urban, 1990. The sensitivity of some high-latitude boreal forests to climatic parameters. *Climate Change* 16: 9–29.

Bryson, R.A, 1966. Air masses, streamlines and the boreal forest. *Geographical Bulletin* 8: 228–269.

Burton, P.J., and S.G. Cumming, 1995. Potential effects of climatic change on some western Canadian forests, based on phenological enhancements to a patch model of forest succession. *Water, Air and Soil Pollution* 82: 401–414.

Campbell, I.D., and McAndrews, J.H., 1991. Cluster analysis of late Holocene forest trends in Ontario. *Canadian Journal of Botany* 69: 1719–1730.

Canada, 1982a. *Climatic Normals: Precipitation*. Ottawa, Government of Canada.

Canada, 1982b. *Climatic Normals: Temperature*. Ottawa, Government of Canada.

Cannell, M.G.R., 1982. *World Forest Biomass and Primary Production Data*. London: Academic Press.

Carroll, S.B., and L.C. Bliss., 1982. Jack pine–lichen woodland on sandy soils in northern Saskatchewan and northeastern Alberta. *Canadian Journal of Botany* 60: 2270–2282.

Carlton, T.J., and P.F. Maycock, 1978. Dynamics of the boreal forest of James Bay. *Canadian Journal of Botany* 56: 1157–1173.

Carlton, T.J., and P.F. Maycock, 1980. Vegetation of the boreal forests south of James Bay: Non-centered component analysis of the vascular flora. *Ecology* 61: 1199–1212.

Cogbill, C.V., 1985. Dynamics of the boreal forests of the Laurentian Highlands, Canada. *Canadian Journal of Forest Research* 15: 252–261.

Cwynar, L.C., and J.C. Ritchie, 1980. Arctic steppe-tundra: A Yukon perspective. *Science* 208: 1375–1377.

D'Arrigo, R.D., and G.C. Jacoby Jr., 1993. Secular trends in high northern latitude temperature reconstructions based on tree rings. *Climatic Change* 25: 163–177.

Delisle, G.P., and D.E. Dube, 1983. *One and one-half centuries of fire in Wood Buffalo National Park*. Northern Forest Research Centre Forestry Report No. 28. Ottawa: Environment Canada, Canadian Forestry Service.

Dyke, A.S., and V.K. Prest, 1987. Late Wisconsinan and Holocene history of the Laurentide ice sheet. *Géographie Physique et Quaternaire* 41: 237–264.

Dyrness, C.T., L.A. Viereck, and K. Van Cleve, 1986. Fire in Taiga Communities of Interior Alaska. In *Forest Eco-*

systems in the Alaskan Taiga, pp. 74–86, K. Van Cleve, F.S. Chapin III, P.W. Flanagan, L.A. Viereck, and C.T. Dyrness, eds., New York: Springer-Verlag.

Elliott-Fisk, D.L., 1983. The stability of the northern Canadian tree limit. *Annals of the American Association of Geographers* 73: 560–576.

Elliott-Fisk, D.L., 1988. *The boreal forest. North American Terrestrial Vegetation.* Cambridge: Cambridge University Press.

Emmanual, W.R., H.H. Shugart, and M.P. Stevenson, 1985. Climate change and the broad-scale distribution of terrestrial ecosystem complexes. *Climatic Change* 7: 29–43.

Environment Canada, 1992. *The State of Canada's Climate: Temperature Change in Canada, 1895–1991.* Ottawa: Environment Canada.

Flannigan, M.D., and J.B. Harrington, 1988. A study of the relation of meteorological variables to monthly provincial area burned by wildfire in Canada. *Journal of Applied Meteorology* 27: 441–452.

Flannigan, M.D., and C.E. Van Wagner, 1991. Climate change and wildfire in Canada. *Canadian Journal of Forest Research* 21: 66–72.

Foley, J.A., J.E. Kutzbach, M.T. Coe, and S. Levis, 1994. Feedbacks between climate and boreal forests during the Holocene epoch. *Nature* 371: 52–54.

Foster, D.R., 1983. The history and pattern of fire in the boreal forest of southeastern Labrador. *Canadian Journal of Botany* 61: 2459–2471.

Florin, M.-B., and H.E. Wright Jr., 1969. Diatom evidence for the persistence of stagnant glacial ice in Minnesota. *Bulletin of the Geological Society of America* 80: 695–704.

Gajewski, K., S. Payette, and J.C. Ritchie, 1993. Holocene vegetation history at the boreal-forest-shrub-tundra transition in north-western Quebec. *Journal of Ecology* 81: 433–443.

Gauthier, S., Y. Bergeron, and J.-P. Simon. 1996. Effects of fire regime on the serotiny level of jack pine. *Journal of Ecology* 84: 539–548.

Gorham, E., 1988 Canada's peatlands: Their importance for the global carbon cycle and possible effects of "greenhouse" climatic warming. *Transactions of the Royal Society of Canada* Vol. III, Ser. V: 21–23.

Hare, F.K., and J.C. Ritchie, 1972. The boreal bioclimates. *Geographical Review* 62: 333–365.

Harris, S.A. 1986. *The Permafrost Environment.* London, Croon Helm.

Heinselman, M.L., 1970. The natural role of fire in northern conifer forests. *Naturalist* 21: 14–23.

Higgins, D.G., and G.S. Ramsey, 1992. *Canadian Forest Fire Statistics: 1988–1990.* Forestry Canada, Petawawa National Forestry Institute Information ReportPI-X-107E/F.

Hom, J.L., K. Van Cleve, and W.C. Oechel, 1990. *The effect of elevated soil temperature on the growth, nutrient content and photosynthetic response of black spruce* (Picea mariana *(Mill) B.S.P.) found on permafrost dominated soils in central Alaska.* Arkhangelsk, International Symposium on Effects of Climatic Change on Boreal Forests.

Honer, T.G., and A. Bickerstaff, 1985. *Canada's Forest Area and Wood Volume Balance 1977–81: An Appraisal of Change Under Present Levels of Management.* Canadian Forestry Service, Pacific Forestry Centre BC-X-272.

Hopkins, D.M., P.A. Smith, and J.V. Matthews Jr., 1981. Dated wood from Alaska and the Yukon: Implications for forest refugia in Beringia. *Quaternary Research* 15: 217–249.

Houghton, J.J., L.G. Meiro Filho, B.A. Callander, N. Harris, A. Kattenberg, and K. Maskell, 1996. *Climate Change 1995*, Cambridge, Cambridge University Press.

Hutton, M.J. , G.M. MacDonald, and R.J. Mott, 1994. Postglacial vegetation history of the Mariana Lake region, Alberta. *Canadian Journal of Earth Sciences* 31: 418–425.

Ives, J.W., 1977. Pollen separation of three North American birches. *Arctic and Alpine Research* 9: 73–80.

Jacoby, G.C., and R.D. D'Arrigo, 1995. Tree ring width and density evidence of climatic and potential forest change in Alaska. *Global Biogeochemical Cycles* 9: 227–234.

Johnson, E.A., 1975. Buried seed populations in the subarctic forest east of Great Slave Lake, Northwest Territories. *Canadian Journal of Botany* 53: 2933–2941.

Johnson, E.A., 1992. *Fire and Vegetation Dynamics: Studies From the North American Boreal Forest.* Cambridge, Cambridge University Press.

Johnson, E.A., and J.S. Rowe, 1975. Fire in the subarctic wintering ground of the Beverley caribou herd. *American Midland Naturalist* 94: 1–14.

Kauppi, K., and M. Posch, 1985. Sensitivity of boreal forests to possible climatic warming. *Climatic Change* 7: 45–54.

Kershaw, K.A., 1977. Studies on lichen-dominated systems. 20. An examination of some aspects of the northern boreal lichen woodlands in Canada. *Canadian Journal of Botany* 55: 393–410.

Kiil, A.D., and J.E. Grigel, 1969. *The May 1968 forest conflagrations in central Alberta—A review of fire weather, fuels and fire behavior.* Canadian Forestry Service, Northern Forest Research Centre Information Report A-X-24.

Kojima, S., 1994. Effects of Global Climatic Warming on the Boreal Forest. *Journal of Plant Research* 107: 21–97.

Lafleur, P.M., W.R. Rouse, and D.W. Carlson, 1992. Energy balance differences and hydrologic impacts across the northern treeline. *International Journal of Climatology* 12: 193–203.

Lamb, H.F., 1985. Palynological evidence for postglacial change in the position of tree limit in Labrador. *Ecological Monographs* 55: 241–258.

La Roi, G.H., 1967. Ecological studies in the boreal sprucefir forests of the North American taiga. *Ecological Monographs* 37: 220–253.

La Roi, G.H., and M.H.L. Stringer, 1976. Ecological studies in the boreal spruce-fir forests of the North American taiga. II. Analysis of the bryophyte flora. *Canadian Journal of Botany* 54: 619–643.

Larsen, C.P.S., and G.M. MacDonald, 1995. Relations between tree-ring widths, climate, and annual area burn in the boreal forest of Alberta. *Canadian Journal of Forestry Research* 25: 1746–1755.

Larsen, J.A., 1980. *The Boreal Ecosystem.* New York: Academic Press.

Larsen, J.A., 1982. *Ecology of the Northern Lowland Bogs and Conifer Forests.* New York: Academic Press.

Lavoie, C., and S. Payette, 1994. Recent fluctuations of the lichen-spruce forest limit in subarctic Quebec. *Journal of Ecology* 82: 725–734.

Lichti-Federovich, S., 1970. The pollen stratigraphy of a dated section of late-Pleistocene lake sediment from central Alberta. *Canadian Journal of Earth Sciences* 7: 938–945.

Liu, K.-B., 1990. Holocene Paleoecology of the boreal forest and Great Lakes-St. Lawrence forest in northern Ontario. *Ecological Monographs* 60: 179–212.

MacDonald, G.M., 1987. Postglacial development of the subalpine-boreal transition forest in western Canada. *Journal of Ecology* 75: 303–320.

MacDonald, G.M., 1995. Vegetation of the continental Northwest Territories at 6 ka *Géographie Physique et Quaternaire* 49: 37–43.

MacDonald, G.M., and L.C. Cwynar, 1985. A fossil pollen based reconstruction of the late Quaternary history of lodgepole pine (*Pinus contorta* ssp. *latifolia*) in the western interior of Canada. *Canadian Journal of Forest Research* 15: 1039–1044.

MacDonald, G.M., and L.C. Cwynar, 1991. Post-glacial population growth rates of *Pinus contorta* ssp. *latifolia* in western Canada. *Journal of Ecology* 79: 417–429.

MacDonald, G.M., and K. Gajewski, 1992. The northern treeline of Canada. *Geographical Snapshots of North America*. D.G. Janelle, ed. New York: The Guilford Press.

MacDonald, G.M. and McLeod, T.K., 1996. The Holocene closing of the "Ice-Free" Corridor. *Quaternary International*. 32: 87–95.

MacDonald, G.M., C.P.S. Larsen, J.M. Szeicz, and K.A. Moser, 1991. *Quaternary Science Reviews* 10: 53–71.

MacDonald, G.M., J.M. Szeicz, J. Claricoates, and K.A. Dale, 1998. Response of the central Canadian treeline to recent climatic changes. *Annals of the Association of American Geographers* 88: 183–208.

MacDonald, G.M., T.W.D. Edwards, K.A. Moser, R. Pienitz, and J.P. Smol, 1993. Rapid response to treeline vegetation and lakes to past climate warming. *Nature* 361: 243–246.

Methuen, I.R., C.F. van Wagner, and B.J. Stocks, 1978. The vegetation of four burned areas in northwestern Ontario. *Canadian Forestry Service Information Report*, PS-X-60.

Monteith, J.L., 1975. *Vegetation and the Atmosphere, vol. 2. Case Studies*, San Diego, Academic Press.

Mooney, H.A., B.G. Drake, R.L. Luxmoore, W.C. Oechel, and L.F. Petelka, 1991. Predicting ecosystem responses to elevated CO_2 concentrations. *BioScience* 41: 96–104.

Moser, K.A., and G.M. MacDonald, 1990. Holocene vegetation change at treeline north of Yellowknife, Northwest Territories. *Quaternary Research* 34: 227–239.

Mott, R.J., 1973. Palynological Studies in Central Saskatchewan: Pollen stratigraphy from lake sediment sequences, Ottawa. *Geological Survey of Canada Paper* 72-49.

Murphy, P.J., and C. Tymstra, 1986. *The 1950 Chinchaga River fire in the Peace River region of British Columbia, Alberta: Preliminary results of simulating forward spread distances*. Edmonton: Proceedings of the Third Western Region Fire Weather Committee Scientific and Technical Seminar, pp. 20–30.

National Atlas of Canada, 4th Edition, 1972. Ottawa.

National Atlas of Canada, 5th Edition, 1986. Ottawa.

Nielson, R.P., 1993. Vegetation redistribution: A possible biosphere source of CO_2 during climatic change. *Water, Air and Soil Pollution* 70: 659–673.

Pare, D., and Y. Bergeron, 1995. Above-ground biomass accumulation along a 230-year chronosequence in the southern portion of the Canadian boreal forest. *Journal of Ecology* 83: 1001–1007.

Pare, D., and Y. Bergeron, 1996. Effects of colonizing tree species on soil nutrient availability in a clay soil of the boreal mixedwood. *Canadian Journal of Forest Research* 26: 1022–1031.

Pastor, J., and W.M. Post, 1988. Response of northern forests to CO_2-induced climate change. *Nature* 334: 55–58.

Pastor, J., R.H. Gardner, V.H. Dale, and W.M. Post, 1987. Successional changes in nitrogen availability as a potential factor contributing to spruce declines in boreal North America. *Canadian Journal of Forest Research* 17: 1394–1400.

Payette, S., 1983. The forest tundra and present tree-lines of the northern Quebec-Labrador Peninsula. *Nordicana* 47: 3–23.

Payette, S., 1992. Fire as a controlling process in the North American boreal forest. In *A Systems Analysis of the Global Boreal Forest*, pp. 144–169, H.H. Shugart, R. Leemans, and G.B. Bonana, eds. Cambridge: Cambridge University Press.

Payette, S., and C. Lavoie, 1994. The arctic tree line as a record of past and recent climatic changes. *Environmental Reviews* 2: 78–90.

Payette, S., L. Filion, L. Gauthier, and Y. Boutin, 1985. Secular climate change in old-growth tree-line vegetation of northern Quebec. *Nature* 315: 135–138.

Payette, S., C. Morneau, L. Sirois, and M. Desponts, 1989. Recent fire history of the northern Quebec biomes. *Ecology* 70: 656–673.

Pielke, R.A., and P.L. Vidale, 1995. The boreal forest and the polar front. *Journal of Geophysical Research* 100: 25,755–25,758.

Plochl, M., and W. Cramer, 1995. Possible impacts of global warming on tundra and boreal forest ecosystems: Comparison of some biogeochemical models. *Journal of Biogeography* 22: 775–783.

Richard, P.J.H., 1981. Paleophytogéographie postglaciaire en Ungava par l'analyse pollinique. *Paleo-Québec 13*. Québec: Université du Québec.

Richard, P.J.H., A. Larouche, and M.A. Bouchard, 1982. Age de la déglaciation finale et histoire postglaciaire de la vegetation dans la partie centrale du Nouveau-Québec. *Géographie Physique et Quaternaire* 36: 63–90.

Ritchie, J.C., 1976. The late-Quaternary vegetational history of the western interior of Canada. *Canadian Journal of Botany* 54: 1793–1818.

Ritchie, J.C., 1984. *Past and Present Vegetation of the Far Northwest of Canada*. Toronto, University of Toronto Press.

Ritchie, J.C., 1987. *Postglacial Vegetation of Canada*. Cambridge, Cambridge University Press.

Ritchie, J.C., and G.M. MacDonald, 1986. The patterns of post-glacial spread of white spruce. *Journal of Biogeography* 13: 527–540.

Ritchie, J.C., L.C. Cywnar, and R.W. Spear, 1983. Evidence from northwest Canada for an early Holocene Milankovitch thermal maximum. *Nature* 305: 126–128

Rizzo, B., and E. Wiken, 1992. Assessing the sensitivity of Canada's ecosystems to climatic change. *Climatic Change* 21: 37–55.

Rowe, J.S., 1972. *Forest Regions of Canada*. Ottawa: Canadian Forestry Service.

Rowe, J.S., and G.W. Scotter, 1973. Fire in the boreal forest. *Quaternary Research* 3: 444–464.

Rowe, J.S., J.L. Bergstinsson, G.A. Padbury, and R. Hermesh, 1974. *Fire studies in the MacKenzie Valley*. Canadian

Department of Indian and Northern Affairs, ALUR 73-74-61.

Sakai, A., and W. Larcher, 1987. *Frost Survival of Plants.* Berlin: Springer-Verlag.

Scott, P.A., 1992. Annual development of climatic summer in northern North America: Accurate prediction of summer heat availability. *Climate Research* 2: 91–99.

Scott, P.A., R.I.C. Hansell, and W.R. Erickson, 1993. Influences of wind and snow on northern tree-line environments at Churchill, Manitoba, Canada. *Arctic* 46: 316–323.

Simmard, A.J., 1973. Wildland Fire Occurrence in Canada. Canadian Forest Service, Ottawa.

Solomon, A.M., 1986. Transient response of forests to CO_2-induced climatic change: Simulation modeling experiments in eastern North America. *Oecologia* 68: 567–579.

Spear, R.W., 1993. The palynological record of late-Quaternary arctic tree-line in northwest Canada. *Review of Palaeobotany and Palynology* 79: 99–111.

Szeicz, J.M., and G.M. MacDonald, 1991. Postglacial vegetation history of oak savanna in southern Ontario. *Canadian Journal of Botany* 69: 1507–1519.

Szeicz, J.M., and G.M. MacDonald, 1995a. Dendroclimatic reconstruction of summer temperatures in northwestern Canada since A.D. 1638 based on age-dependent modeling. *Quaternary Research* 44: 257–266.

Szeicz, J.M., and G.M. MacDonald, 1995b. Recent white spruce dynamics at the subarctic alpine treeline of north-western Canada. *Journal of Ecology* 83: 873–885.

Szeicz, J.M., G.M. MacDonald, and A. Duk-Rodkin, 1995. Late Quaternary vegetation history of the central MacKenzie mountains, Northwest Territories, Canada. *Palaeogeography, Palaeoclimatology, Palaeoecology* 113: 351–371.

Timoney, K.P., G.H. La Roi, S.C. Zoltai, and A.L. Robinson, 1992. The high subarctic forest-tundra of northwestern Canada: Position, width, and vegetation gradients in relation to climate. *Arctic* 45: 1–9.

Treidl, R.A., E.C. Birch, and P. Sajecki, 1981. Blocking action in the northern hemisphere: A climatological study. *Atmosphere-Ocean* 19: 1–23.

Van Cleve, K., W.C. Oechel, and J.L. Hom, 1990. Response of black spruce (*Picea mariana*) ecosystems to soil temperature modification in interior Alaska. *Canadian Journal of Forest Research* 20: 1530–1535.

Van Cleve K., F.S. Chapin III, P.W. Flanagan, L.A. Viereck, and C.T. Dyrness, 1986. *Forest Ecosystems in the Alaskan Taiga: A Synthesis of Structure and Function.* New York: Springer Verlag.

Van Cleve, K., L. Oliver, R. Schlenter, L.V. Viereck, and C.T. Dyrness, 1983. Productivity and nutrient cycling in the taiga forest ecosystems. *Canadian Journal of Forest Research* 13: 747–766.

Viereck, L.A., 1983. The effects of fire in the black spruce ecosystem of Alaska and northern Canada. In *The Role of Fire in Northern Circumpolar Ecosystems*, pp. 201–220, R. Wein and D.A. MacLean, eds. Chichester: John Wiley & Sons.

Viereck, L.A., and E.L. Little Jr., 1972. *Alaska Trees and Shrubs.* Washington, D.C.: Forest Service, U.S. Department of Agriculture.

Viereck, L.A., K. Van Cleve, C.T. Dyrness, 1986. Forest ecosystem distribution in the taiga environment. In *Forest Ecosystems in the Alaskan Taiga*, pp. 22–43, K. Van Cleve, F.S. Chapen III, P.W. Flanagan, L.V. Viereck, and C.T. Dyrness, eds. New York: Springer Verlag.

Walter, H., 1985. *Vegetation of the Earth.* 3rd ed. Berlin: Springer.

Webb, T., III, T.J.H. Richard, and R.J. Mott, 1983. A mapped history of Holocene vegetation in southern Quebec. *Syllogeus* 49: 273–336.

Webb, T., III, P.J. Bartlein, S.P. Harrison, and K.H. Anderson, 1993. Vegetation, Lake Levels, and Climate in eastern North America for the past 18,000 years. In *Global Climates since the Last Glacial Maximum*, pp. 415–467, H.E. Wright Jr., J.E. Kutzbach, T. Webb III, W.F. Ruddiman, F.A. Street-Perrott, and P.J. Bartlein, eds. Minneapolis: University of Minnesota Press.

Wein, R.W., and D.A. MacLean, eds., 1983. *The role of fire in northern circumpolar ecosystems.* Scope 18. New York: John Wiley & Sons.

Wein, R.W., and J.M. Moore, 1977. Fire history and rotations in the New Brunswick Acadian Forest. *Canadian Journal of Forest Research* 7: 285–94.

Whittaker, R.H., and G.E. Likens, 1975. The Biosphere and Man. In *Primary Productivity of the Biosphere*, pp. 305–328, H. Leith and R.H. Whittaker, eds. New York: Springer.

Williams, G.D.V., 1985. Estimated bioresource sensitivity to climatic change in Alberta, Canada. *Climatic Change* 7: 55–69.

Yarie, J., 1981. Forest fire cycles and life tables: A case study from interior Alaska. *Canadian Journal of Forest Research* 11: 554–562.

Zasada, J.C., K. Van Cleve, R.A. Werner, J.A. McQueen, and E. Nyland, 1978. Forest biology and management in high-latitude North American forests, pp. 137–195, In *North American forest lands at latitudes north of 60 degrees. Proceedings of a symposium, September 19–22, 1977.* Fairbanks: University of Alaska.

Zoltai, S.C., and C. Tarnocai, 1975. Perennially frozen peatlands in the Western Arctic and Subarctic of Canada. *Canadian Journal of Earth Sciences* 12: 28–43.

Zoltai, S.C., and D.H. Vitt, 1990. Holocene climatic change and the distribution of peatlands in western interior Canada. *Quaternary Research* 33: 231–240.

15

Appalachia and the Eastern Cordillera

David Shankman
L. Allan James

The uplands of eastern North America are commonly referred to as the Appalachian Highlands, based on the preeminence of the Appalachian Mountains within the region. However, the complex of Paleozoic fold belts found in the Appalachian Highlands also extends northeast, through Canada's Maritime Provinces to Newfoundland, and reappears in the Ouachita Mountains across the Mississippi River valley far to the southwest. For this larger region, the term Eastern Cordillera is appropriate in the present context. To the north and west, the region abuts the narrow St. Lawrence Lowland along the margin of the Canadian Shield, the Central Lowlands of the Great Lakes region, and the low plateaus of western Kentucky and Tennessee (fig. 15.1). To the south and east, the region extends to the Fall Line at the inner edge of the Coastal Plain. West of the Mississippi embayment are the Ouachita Mountains and Ozark Plateaus in Arkansas and southern Missouri. Whereas the latter areas, together with the Interior Low Plateaus, are not formally recognized as part of the Appalachian Highlands, they are included here because of similarities in structure, relief, and vegetation. The Eastern Cordillera thus defined, though dominated by the Appalachian Highlands, is not continuously mountainous but is sufficiently elevated to generate a distinctive physical and biotic regionalism quite different from that of the adjacent lowlands.

15.1 Physical Regions of the Appalachians

Based on Fenneman's (1938) widely accepted physical divisions of the United States, six provinces are identified within the Appalachian Highlands, namely, the Adirondack Mountains, Appalachian Plateaus, Ridge and Valley, Blue Ridge, Piedmont, and New England (fig. 15.1). Because these provinces differ in terms of tectonic evolution, they are distinguished from one another by structural and lithological differences that result in pronounced variations in topography (see chapter 1). Most provinces are, however, characterized by structures with a northeast-southwest strike attributable to Paleozoic orogenies that give the region some uniformity over more than 3000 km from Newfoundland to Alabama.

Though structurally an extension of the Canadian Shield and underlain by Precambrian igneous and metomorphic rock, the relatively small Adirondack Province in New York State is a mountainous region more appropriately

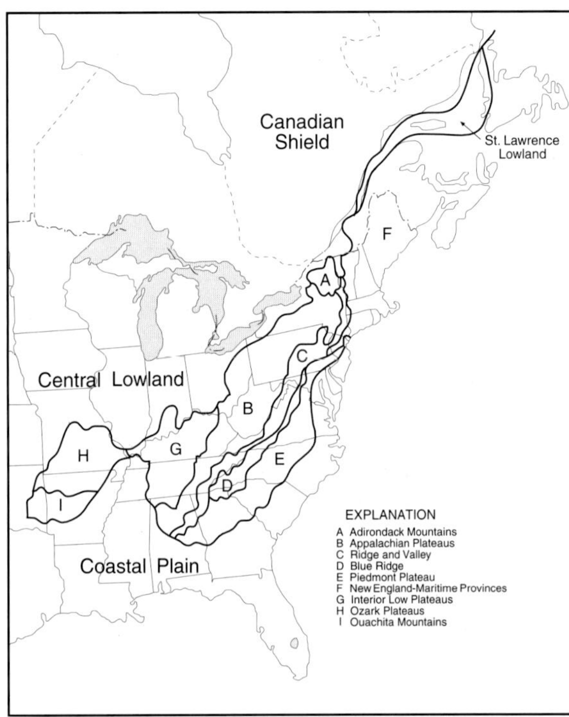

Figure 15.1 Physiographic regions of eastern North America.

allied in terms of its surface features with the mountains of the Appalachian fold belt. Its peaks range from 1000 to 1500 m in elevation, reaching 1629 m on Mt. Marcy. As elsewhere in the northern Appalachians, the Adirondacks were repeatedly glaciated during Pleistocene times, leaving prominent erosional features.

The Adirondacks are bounded on the south by the Appalachian Plateaus that form the westernmost province of the Appalachian Highlands. The plateaus occupy much of New York, Pennsylvania, and West Virginia, and extend southward in a narrow strip to northern Alabama for a total distance of 1600 km. This region is underlain by nearly horizontal to deformed Paleozoic clastic sedimentary strata whose conglomerates, sandstones, and shales are variably resistant to erosion. Some areas are deeply dissected, resulting in rugged terrain of fairly low elevation. The highest summits range from 1300 to 1370 m in West Virginia; mountain peaks in other parts of the region are much lower. The Allegheny Front, a high east-facing escarpment of erosion-resistant sandstones, separates the Appalachian Plateaus from the Ridge and Valley Province to the southeast.

The Ridge and Valley Province is a long, narrow zone extending southward more than 1900 km from New York to northern Alabama, and linking northward with the St. Lawrence Lowland in eastern Canada. Unlike the Appalachian Plateaus, the Ridge and Valley Province consists of a belt of tightly folded and faulted Paleozoic sedimentary rocks. Alternating synclines and anticlines in much of the region result in a northeast-southwest alternation of parallel ridges and valleys. Older limestones and shales tend to erode and form valleys, whereas younger Paleozoic sandstones and conglomerates form ridges. Folds and shallow thrust faults represent crustal warping and movement of rock toward the northwest. Along the eastern margin of the province is the Great Valley, a continuous lowland that includes Lebanon Valley in Pennsylvania, the Shenandoah Valley in Virginia, Hagarstown Valley in Maryland, and the Great Valley in eastern Tennessee.

To the east of the Ridge and Valley is the mountainous Blue Ridge Province. In contrast to the sedimentary rock underlying the Appalachian Plateaus and Ridge and Valley, the Blue Ridge consists largely of older Paleozoic and Precambrian metamorphic rock of a complex structure. The northern section of this 1000-km-long province is narrow, only 15–20 km in some places, but it broadens southward to include the highest and most rugged mountains in the Appalachian region. In the southern Blue Ridge along the North Carolina–Tennessee boundary are the Great Smoky Mountains, named for the blue haze occurring at high elevation. Mountain peaks range from 1500 m to greater than 1800 m in elevation. The highest peak in eastern North America is Mt. Mitchell (2037 m) in North Carolina. The Blue Ridge forms a natural barrier to transportation that long impeded the western expansion of European colonization.

To the east of these mountains, the Piedmont Province extends 1500 km from New York to Alabama. Its northwestern boundary is the foot of the Blue Ridge and from there it slopes gently toward the Atlantic Coastal Plain. Its boundary with the Coastal Plain is most readily identified by the Fall Line, a series of rapids or falls in stream channels that occur where rivers flow from the more resistant rock of the Piedmont onto less resistant sediments of the Coastal Plain. The Piedmont is a nonmountainous region incised by southeast-trending river valleys and characterized by gentle rolling topography. It is highest in the south where it attains an elevation of 550 m on the Dahlonega Plateau in Georgia. The Piedmont is underlain by Precambrian metamorphic and igneous rocks that dip beneath Cretaceous and Cenozoic marine sediments of the Coastal Plain.

The New England Province has structural similarities with the Piedmont, Blue Ridge, and Ridge and Valley provinces, and it was regarded by Fenneman (1938) as the northward continuation of these provinces. Indeed, the structural lineations and terranes of New England extend in turn northeast through the Maritime Provinces to Newfoundland and the adjacent continental shelf. The present geomorphic character of this province has been shaped largely by Pleistocene glaciation, distinguishing it from less glaciated and unglaciated terrain farther south. During the major glacial advances, ice sheets covered all but the highest peaks in this region. Deep soils and weathered rock are almost entirely absent, and much of the region is covered by a mantle of glacial and fluvioglacial deposits.

15.2 Landform Evolution

15.2.1 Origin of the Appalachians

The Appalachians have a complex geologic history, which, despite uncertainties, is best explained in the context of plate tectonics (see chapter 1). The evolutionary sequence began more than more than a billion years ago (>1 Ga) when the late Precambrian Grenvillian orogeny shaped the highly deformed rocks now found in the crystalline core of the Blue Ridge Mountains and left a series of shallow faults (fig. 15.2). Subsequent continental rifting, beginning about 700 Ma (million years before present), eventually created the Iapetus Ocean along the eastern margin of the Laurentian craton (Horton and Zullo, 1991). Throughout the Paleozoic, subduction and crustal accretion against this craton were associated with several orogenic events accompanied by granite plutonism, metamorphism, and island-arc volcanism (Hatcher, 1989; Rast, 1989). The Taconic orogeny (470–440 Ma) provided the framework for the Blue Ridge Mountains (Hatcher and Goldberg, 1991). The Acadian orogeny (400–350 Ma), generated by the crowding of several small plates against the Laurentian craton, is well represented in the folded terranes and postorogenic granitics of the northern Appalachians. The final major mountain-building episode, the Alleghenian orogeny (330–270 Ma) occurred as the West African plate moved against the Laurentian craton. Its effects are particularly well expressed in the central and southern Appalachians, mainly in the fold and thrust mountains of the Ridge and Valley Province (Hatcher and Goldberg, 1991; Horton and Zullo, 1991). Structures in the Appalachian Highlands are thought to involve a series of relatively shallow thrust faults that converge in a major horizontal fault zone on which the Blue Ridge and the Ridge and Valley crustal segments slid horizontally (Cook et al., 1979). During Triassic time (245–210 Ma), the enlarged North American plate began separating from Africa as Pangea broke up, with the Appalachian region forming the eastern mountainous margin of the new continent.

15.2.2 Cycles of Erosion and Denudation

The Appalachian Highlands have provided the setting for the development of a variety of influential geomorphic models (compare with Mills et al., 1987; Morisawa, 1989). William Morris Davis's concepts of landform evolution, which dominated geomorphology throughout the first half of the twentieth century, were largely derived from studies of the central Appalachians (Davis, 1899a). His writings, however, predate plate tectonic concepts and assume long periods of crustal stability. Although the theoretical framework of Davis' interpretations has fallen from favor, most modern concepts of denudation and drainage evolution draw heavily on observations he first applied to the region.

The concept of peneplain development, a fundamental assumption of Davis's cycle of erosion theory, was based largely on observations in the Ridge and Valley Province (Davis, 1899b,c). In Pennsylvania and New Jersey, the highest accordant ridge tops were interpreted as remnants of a broad former erosional surface named the Schooley Peneplain, and attributed to a Jurassic-Cretaceous erosional cycle (Davis, 1890, 1899c). Broad lowlands in the region were believed to be a later erosional surface referred to variously as the Kittatinny or Harrisburg Peneplain or the Tertiary base-level lowlands (Davis, 1890, 1899c; Morisawa, 1989). Davis interpreted narrow valleys cut into these Tertiary lowlands as the result of an ongoing Quaternary erosion cycle. Davis's cycle of erosion theory was immensely popular. Peneplains were soon thought to be recognized by Davis' disciples throughout the Appalachian region and beyond.

Davisian concepts of the progressive and spatially uniform lowering of large areas and the development of peneplain surfaces have long been criticized (Tarr, 1898; Denny, 1956; Hack, 1976, 1980). The advent of modern tectonic ideas negated the Davisian assumption of long periods of crustal stability required for the completion of a cycle of erosion, as did a growing appreciation of potentially rapid rates of geomorphic processes. Crustal activity has been substantial even on passive margins such as eastern North America and is characterized by high spatial and temporal variability. Stratigraphic evidence from marine sediments deposited after the opening of the Atlantic Ocean indicates differential uplift rates among the central Appalachians, the Adirondacks, and New England, events that do not support the classic Davisian model (Poag and Sevon, 1989).

15.2.3 Modern Models of Landform Development

The decline of Davisian concepts beginning in the 1940s paralleled a developing interest in very different perspec-

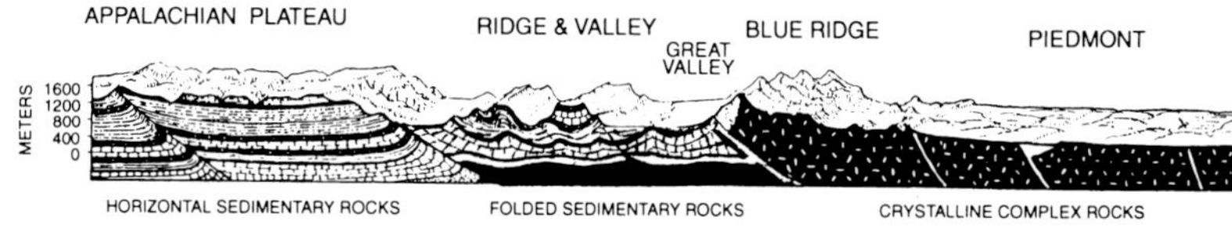

Figure 15.2 Cross section of the central Appalachians from the Piedmont westward to the Appalachian Plateaus.

tives in geomorphology. For a while, attention turned to local-scale surficial processes as opposed to landscape-scale processes (Mills et al., 1987; Soller and Mills, 1991). In recent decades, the accordant ridges and other topographic features thought to represent former peneplains have been reinterpreted in the context of improved landform models and erosion-resistant structures (Hack, 1960, 1976).

The Appalachian Piedmont was once considered one of the best examples of a peneplain because its underlying metamorphosed rocks had been worn down to a series of accordant drainage divides that were considered too flat-crested to have been caused by anything other than prolonged fluvial erosion (compare with Costa and Cleaves, 1984; Soller and Mills, 1991). Piedmont ridges are typically mantled by 10–20 m of saprolite overlain by a residual soil. Saprolite retains the underlying bedrock volume and structural features, although about one-third of the rock mass has been lost to weathering. Compaction and loss of about 70% of the total saprolite mass occurs during its conversion to soil near the surface. Recent pedogenic analysis indicates that surface lowering proceeds at the same average rate as the downward progression of the lower saprolite front. This suggests that the broad ridge tops resulted from pedogenesis and that the saprolite is Quaternary in age (Mills et al., 1987; Pavich, 1989; Mills and Delcourt, 1991). This interpretation contrasts with long-held assumptions that the thick residuum is much older, and it implies that dynamic equilibruim theories explain Piedmont landscape evolution better than peneplanation (Mills et al., 1987). The predominance of weathering processes indicates that Piedmont bedrock outcrops, such as Stone Mountain, Georgia, are not monadnocks in the Davisian sense of an erosional remnant of a former erosion cycle (Bates and Jackson, 1987), but survived because they were more resistant to weathering and erosion than the surrounding saprolite-covered surfaces (compare to Wahrhaftig, 1965).

15.2.4 Drainage Reversal and Divide Migration

Two classic problems of Appalachian geomorphology concern the long-term evolution of fluvial systems: (1) the question of drainage reversals and (2) explanations for transverse drainage. Some researchers have concluded that major streams originating in the Appalachian Highlands once flowed westward or northwestward across the region toward the interior of the continent (reviewed by Mills et al., 1987). Although Hack (1982) acknowledged that westward drainage-divide migration occurred in some local valleys, he did not consider the process to be dominant in the Appalachians. Others have argued, however, for pronounced westward retreat of the Blue Ridge escarpment in the southern Appalachians by headward erosion of Piedmont streams and piracy of Blue Ridge streams. Judson (1976) suggested that during the early Mesozoic, following the breakup of Pangea, many streams flowing northwest reversed and began to flow southeastward toward the Atlantic Ocean. According to Judson, during the breakup of Pangea, the continental crust lowered as it migrated away from the mid-Atlantic rift zone. This ultimately resulted in the reversal of many streams that began flowing east into the rift zone (fig. 15.2). These east-flowing streams had steeper gradients than channels flowing long distances westward, which resulted in headward erosion of east-flowing streams, capture of previously west-flowing streams and their tributaries, and the westward migration of drainage divides.

Another classic geomorphic question asks why there is an abundance of drainage systems that cut across transverse geologic structures. Incision of some of the major southeastern-flowing streams in the central Appalachians created spectacular *water gaps* with cliffs up to several hundred meters high. For example, the Delaware River flows through several water gaps along the Pennsylvania–New Jersey border in the Ridge and Valley Province before reaching Delaware Bay. To the south, the Susquehanna River flows through plunging anticlines and synclines in central Pennsylvania. The Potomac River, with headwaters in southern Pennsylvania, has cut water gaps in the Blue Ridge of West Virginia. In some cases, stream capture diverted streams; the abandoned water gaps, known as *wind gaps*, are common in the central part of the Ridge and Valley and the Blue Ridge Provinces. Although water and wind gaps are not common in the southern or glaciated Appalachians, their prominence in the central Appalachians had considerable influence on geomorphic theory, and they have posed a geomorphic enigma for generations.

Concepts of superpositioning and antecedence have become increasingly sophisticated. Oberlander (1985) recognized several modes by which drainage can evolve transverse to structures, that a variety of both superpositioning and antecedent processes can operate concurrently, and that permutations of both landform types may occur in the same region. He pointed out that thick, weak rock layers often lie conformably over erosion-resistant folded rocks in young orogenic zones such as the Zagros Mountains of Iran. If the weak rock layers are thick relative to fold amplitudes, there may be a complex evolution of drainage as anticlinal basins covered by weak rock elongate into valleys, enlarge, coalesce, and superpose channels onto competent rocks below (Oberlander, 1985). This process can result in the observed preferred location of water gaps in the high centers of broad anticlinoria.

A classic argument for transverse drainage in the Appalachians caused by superpositioning was based on stream incision of Coastal Plain sediments once extending much farther west than their present Fall Line boundary (Johnson, 1931). Channels incising down through the former Coastal Plain overburden eventually cut into and across

Piedmont structures (Staheli, 1976; compare with Hack, 1982; Battiau-Queney, 1989; Soller and Mills, 1991). Explanations of transverse drainge have also been based on transverse structural weakness (Epstein, 1966; compare with Mills et al., 1987; Clark, 1989). Clark (1989) presented a review of concepts about water gap and wind gap formation and their importance to geomorphic inquiry in the central Appalachians. He emphasized the importance of transverse structural weaknesses and described a process of *structural ensnarement* in which down-plunge lateral channel migration around anticlinal noses proceeds until weak transverse zones are encountered that allow more rapid incision rates.

15.3 Late Quaternary Landscape Evolution

Repeated glacial episodes during the Pleistocene had a notable effect on the northern Appalachian Highlands. Ice sheets from Canada advanced southward into the northeastern United States several times during the Pleistocene (see chapter 2). As a result of these ice sheets and local ice caps, glacial depositional and erosional landforms are common in the northern Appalachian Highlands. These features, however, generally represent only the most recent glacial events of late Quaternary age, on which considerable research has been conducted. Successive scouring of the surface has left few remnants of previous glaciation, and therefore early Quaternary glacial episodes in the Appalachians are not well documented (Braun, 1989).

15.3.1 Glaciation

Over the last 200,000 years, there were at least three major glacial advances of ice sheets in much of eastern North America. Although the dating of early events remains controversial, numeric ages given here are based on a commonly accepted method of combining marine oxygen-isotope data and Northern Hemisphere solar energy receipts based on Earth-sun orbital parameters (Mix, 1992). During the Illinoian glaciation in the late middle Pleistocene (Morrison, 1991; before 130 ka, thousand years before present), ice apparently covered areas of the Ohio River valley as far south as 36°40'N latitude, in New York and Pennsylvania well onto the Allegheny Plateau, and in New England out to Nantucket Island (Oldale et al., 1982; Oldale and Coleman, 1992). Continental ice presumably ablated completely during the Last Interglacial (130–115 ka).

Major glacial advances occurred at least twice during the Wisconsinan glaciation (80–10 ka). In early Wisconsinan time (80–65 ka), Laurentide ice reached the St. Lawrence Lowland, retreated, and then readvanced to its maximum extent in Pennsylvania and New York (Dreimanis and Goldthwaite, 1973). In mid-Wisconsinan time (65–35 ka), Laurentide ice retreated to a position well north of the St. Lawrence Lowland. The late Wisconsinan advance, beginning about 30 ka, has been studied in greatest detail because of the preservation of evidence.

There are two controversies regarding the source and the nature of the late Wisconsinan glaciation. First, the source of the Laurentide Ice Sheet that came into the region has been disputed (see review in Dyke et al., 1989). It had long been assumed there was one original ice sheet that extended over eastern North America. However, evidence has mounted for multiple origins of the Laurentide ice (Shilts et al., 1987; Vincent, 1989; Wright, 1989). The second controversy involves a debate about whether Laurentide ice crossed the northeastern Appalachians to the Atlantic Ocean. Many researchers concluded that ice extended beyond the present coastline and occupied much of the continental shelf exposed by the lower sea level (Hughes et al., 1985; Braun, 1989). Clearly, by 18 ka and possibly earlier, Laurentide ice extended across New England, northern Pennsylvania, New York, and New Jersey (fig. 15.3). However, it is not clear whether ice extended farther to the northeast, in Maine and the Maritimes. The evidence for glaciation in this region is primarily based on offshore sediments in addition to climate and ice-flow models. In contrast, arguments for only partial ice coverage in this region are supported by the lack of terrestrial evidence of glaciation, most notably the absence of glacial erratics (e.g., Boulton et al., 1985). The absence of glacial till has been explained by alternate means, however, such as dilution of till by local

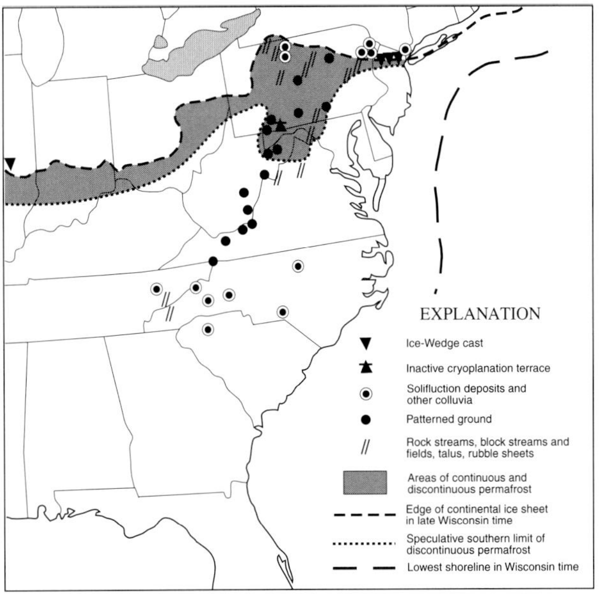

Figure 15.3 Extent of continental glaciation, areas of permafrost, and periglacial features in eastern North America during late Wisconsinan times. Source: Michelson et al., 1983.

lithologies and thermal changes in the glacial bed (Newman et al., 1985).

15.3.2 Deglaciation

Deglaciation was not a simple retreat of the ice margin to the northwest. Warming and thinning of the Laurentide Ice Sheet occurred between 18 and 16 ka, although an extensive readvance occurred about 15.5 ka (Mickelson et al., 1983; Hughes et al., 1985). The Laurentide Ice Sheet in the northeastern region was strongly influenced by marine-based margins. Calving embayments developed off the coast of Maine, and a calving margin advanced up the St. Lawrence Lowland between 14 and 12.8 ka (Borns, 1985; Hughes et al., 1985 Newman et al., 1985; Occhietti, 1989). Large ice streams accelerated into these calving margins, enabling rapid thinning and ablation of the adjacent ice sheet (Borns, 1985). Thus, glacial ice persisted in and emanated from local ice caps in Maine, Gaspé, New Brunswick, Nova Scotia, and Newfoundland after it had disappeared from the St. Lawrence Lowland (Vincent, 1989; Occhietti, 1989). Ice was largely gone from the region by 11 ka, except for small remnants in the highlands of Nova Scotia and Newfoundland (Hughes et al., 1985).

15.3.3 Periglacial Activity

Evidence of periglacial activity in the Appalachians indicates the development of deep frozen ground during Wisconsinan time that may have been continuous near the glacial boundary in New York and Pennsylvania (Péwé, 1983; Mills and Delcourt, 1991) (fig.15.3). Paleoperiglacial activity in the nonglaciated Appalachians was slow to be recognized relative to other regions, owing in part to thick forest cover and poor surface expression, but many periglacial features have now been identified in the central Appalachians and most are paleoclimatic relicts (Clark and Ciolkosz, 1988). In the central and southern Appalachians, small-scale features such as sorted polygons, circles, stripes, nets, and steps are common but tend to be active only where there is no ground cover. In contrast, unsorted patterned ground is relatively rare, with the exception of ice wedge casts seen in vertical exposures. Mesoscale periglacial features such as block fields and block streams are common throughout the central Appalachians (Clark and Ciolkosz, 1988). No periglacial activity has been noted in the outer Piedmont or on the Coastal Plain more than 160 km from the late Wisconsinan limit (Mills et al., 1987).

15.3.4 Fluvial and Colluvial Processes

As is common in mountainous regions, the dominant long-term geomorphic action of most stream systems has been incision. Studies of multiple stream terraces in the central Appalachians, particularly in the Ridge and Valley Province, have been carried out along the New and Tennessee Rivers. Along the New River in southwestern Virginia, upper terraces span a wide range of time from early Pleistocene or older to Sangamon, whereas a low group of terraces is of Wisconsinan age (Mills and Wagner, 1985; Mills and Delcourt, 1991). Young, low terraces along the lower Tennessee River suggest that maximum aggradation rates occurred during glacial-to-interglacial transitions (Mills and Delcourt, 1991).

A debate over the importance of rare episodic events to the evolution of fluvial landforms was largely framed around central Appalachian rivers. Opposing views were presented by Wolman and Miller (1960), who believed long-term processes dominated landform evolution, in contrast to Hack and Goodlett (1960), who believed episodic events were important in this region. Drastic effects of large floods on fluvial forms have been described by Clark (1987) and Miller (1995). Jacobson and others (1989) concluded that the magnitude and frequency of landslides in the central Appalachians vary with physiographic province. There has been little hillslope mass wasting in the Piedmont, but in the Blue Ridge and in the Ridge and Valley Provinces, it can be quite active. At the local scale, landslide activity varies with precipitation characteristics and geologic materials. Slopes with colluvium and shale residuum may experience many shallow slides during protracted low-intensity storms that saturate the fine-grained soils. In contrast, quartzite ridges tend to have the greatest slope failures during shorter, more intense storms as a result of more rapid drainage of the coarse materials (Jacobson et al., 1989). Clark (1987) documents several debris slides, debris slopes, and water blowouts in the central and southern Appalachians.

15.3.5 Climatic Conditions and Forested Landscapes

During the Wisconsinan glacial maximum, vegetation in eastern North America was displaced far to the south of its present latitudes (Delcourt et al., 1980; see also chapter 3). Tundra vegetation occurred along the edge of the ice sheet and extended several hundred kilometers farther south along the crests of the Appalachian Mountains (Watts, 1979). Boreal forest species, including black spruce (*Picea mariana*), red spruce (*P. rubens*), larch (*Larix laricina*), balsam fir (*Abies balsamea*), and jack pine (*Pinus banksiana*), now common in eastern Canada and northern New England, occupied large areas of the central eastern United States (Watts, 1970, 1979). Deciduous forest vegetation occurred primarily in the lower Mississippi River valley and northern Florida, and may have also grown with boreal species in some locations (Davis, 1981; Delcourt and Delcourt, 1977).

Between 15 and 11 ka, the warmer climate and retreat of ice initiated widespread vegetation changes in eastern

North America (Davis, 1981). Although northward migration of species may have begun as early as 15 ka, the most rapid changes occurred in the early Holocene between 11 and 7 ka (see also chapter 4). New surfaces previously under ice were initially colonized by tundra vegetation. Davis's (1981) reconstruction of tree migration patterns during the late Pleistocene and early Holocene shows that boreal species rapidly migrated northward into New England and Canada, reaching their current range within the past 7–5 ka in most cases. Temperate deciduous species of oak (*Quercus*), hickory (*Carya*), and elm (*Ulmus*), among others, also moved northward during the late Pleistocene and early Holocene. Northern migration rates were highly variable. Boreal species tended to migrate rapidly, as much as 300 m yr^{-1}. Chestnut (*Castanea dentata*) moved more slowly northward and may have arrived in New England only 2000 yr B.P. before becoming one of the dominant species throughout much of the Appalachian mountain region (Davis, 1981).

A few of the tree species that migrated northward during the Holocene have small populations that persist in isolated stands far to the south of their continuous ranges. For example, eastern hemlock (*Tsuga canadensis*) is common in New England and the high elevations of the Appalachians, but it also occurs in isolated stands on cool, moist north-facing river bluffs and ravines as far south as central Alabama and Georgia (Harper, 1943; Bormann and Platt, 1958). Balsam fir is common in the boreal forest in Labrador and northern Ontario, yet a few isolated stands also grow at high elevations in Virginia. Also, red spruce, which grows in Nova Scotia, occurs in extensive, noncontinuous stands at high elevations in the southern Appalachians (Oosting and Billings, 1951; Whittaker 1956).

15.4 Changing Landscapes in Historical Times

15.4.1 Old-growth Forests

Forests dominated the pre-European landscape of eastern North America. Many early explorers's descriptions of a pristine landscape helped develop the concept of the forest primeval with unbroken, undisturbed stands of ancient trees (e.g., Baird, 1832; Bartrum, 1958; Trautman, 1977). Whitney (1994) summarized the more prominent features of several early accounts of forests in eastern North America in terms of large, ancient trees with abundant woody debris and decayed organic matter on the forest floor, an abundance of epiphytes covering the trees and mosses in the understory, and dense stands of trees with luxuriant growth in the understory.

The many descriptions of these primeval forests indicate an abundance of old-growth stands. Yet, it is unlikely that massive, old-age trees dominated the entire landscape. Both natural and human-induced disturbances were common to this region, where they destroyed forest stands that covered areas of less than a hectare to many square kilometers. A more accurate description of many of these forests would invoke stands at different stages of recovery, depending on the period of time since they were disturbed. The New England coastal region, for example, is regularly hit by hurricanes that historically have leveled thousands of hectares of forest. Futhermore, tornadoes and other extreme climatic events have caused catastrophic windthrow across many parts of eastern North America. Also, periodic insect infestation caused widespread destruction of the forests. In the central Appalachian region, however, catastrophic disturbance is less common, and the primarily deciduous forests here naturally occur at later stages of successional development.

15.4.2 Forest Fire in Eastern North America

The forested areas in eastern North America have long been affected by fire, as shown by charcoal in lake sediments and alluvial deposits (Winkler, 1985; Patterson et al., 1987). Fire was probably most common in the extreme southern region of the Appalachians, where fire-dependent species commonly occur and lightning causes fires on xeric sites. Fires set by Native Americans, however, were likely more important compared to naturally occurring fires in the pre-European landscape (Harmon, 1982; Delcourt and Delcourt, 1997). In contrast to the southern Appalachians, lightning-caused fires were probably not as common in most of the colder, humid regions to the north, although fire remains an important ecological process even in the boreal forest beyond the Appalachian region (see chapter 14).

There are many historical accounts of Native Americans who regularly burned forests (Whitney, 1994). Early explorers in eastern North America noted that fire was used to drive game, enhance wildlife habitat, clear land for cultivation, and improve travel by reducing undergrowth. The extent and frequency of Native American burning, however, is not entirely clear and no doubt varied considerably in different regions. Native American and natural fires created openings in the forest, as did Native American agricultural practices. Accounts by early explorers indicate there were extensive areas of cleared land throughout much of the Appalachian region. During the sixteenth and seventeenth centuries, there was a massive decline in Indian populations as a result of diseases spread by European explorers. This resulted in rapid and widespread abandonment of large tracts of cleared land (Cowdrey, 1983; see also chapter 22). Some of these tracts in the coastal areas were reclaimed by early European colonists. Most abandoned farmland in the interior, however, began succeeding to forest as early as the sixteenth century.

15.4.3 Agricultural Settlement

Large-scale European colonization of eastern North America resulted in massive deforestation and conversion of cleared land to agriculture. The moist, typically fertile bottomland sites were usually the first areas to be cleared and settled. Because of steep slopes and poor soils, many areas within the Appalachians were not well suited for agriculture and were initially bypassed by settlers. As productive land became less available, settlers started farming on lower ridges and eventually on the flat rocky ridge tops. Industrial logging after the Civil War further contributed to the loss of forests, and by the early twentieth century, only small fragments of the region's old-growth virgin forests remained (Davis, 1996).

Poor farming practices in the Appalachians resulted in accelerated soil erosion and a decline in fertility that in many cases led to the abandonment of farmland. The most serious soil depletion typically occurred on upland sites that were too infertile or on slopes too steep for sustainable agriculture. Declining productivity led to abandonment of marginal agricultural sites and, in some areas of the Appalachians, to out-migration. Abandonment of farmland accelerated during the late nineteenth century, facilitated by increased mechanization and extensive development of more productive farmland toward the midcontinent. The alluvial valleys in the eastern United States continue to be intensively cultivated. These areas of fertile soils and low relief contain the most productive farmland in the region. These sites, however, represent only a small percentage of the land surface.

15.4.4 Invasion and Spread of Alien Species

Widespread deforestation and the abandonment of farmland favored expansion of aggressive, colonizing weedy species, including introduced species that in many cases are now common throughout much of eastern North America, such as Japanese honeysuckle (*Lonicera japonica*) and privet (*Ligustrum vulgare*). Both were brought in as ornamentals that later escaped cultivation and now grow in dense thickets from New England southward. Kudzu (*Pueraria lobata*) was introduced from southeast Asia for erosion control and now occupies large areas throughout the southern Appalachians, growing typically in dense layers that preclude regeneration of almost all native species. Exotic trees are also now common in eastern North America, including Lombardy poplar (*Populus nigra*), weeping willow (*Salix babylonica*), black locust (*Robinia pseudoacacia*), catalpa (*Catalpa speciosa*), Chinese mulberry (*Morus multicaulis*), Scots pine (*Pinus sylvestris*), and Norway spruce (*Picea abies*).

One of the most notable introductions to eastern North America was a fungal plant pathogen. Until the introduction of the chestnut blight (*Endothia parasitica*) at the beginning of the twentieth century, American chestnut (*Castanea dentata*) was one of the most common trees in many parts of eastern North America. American chestnut occurred throughout much of the central and southern Appalachian mountain region and in some mature forests may have accounted for 40% or more of the large trees (Keever, 1953). It was most abundant on moist, well-drained soils on mountain sides (Russell, 1987) and usually occurred in association with the most common oak species, such as white oak (*Quercus alba*), red oak (*Q. rubra*), and scarlet oak (*Q. coccinea*). The chestnut blight was first discovered in North America in 1904 in New York City, and it quickly spread. In the 1920s, infestations occurred as far south as Georgia and South Carolina. By the late 1940s, the chestnut blight had killed most of the mature chestnut trees throughout its range.

The chestnut blight had dramatic consequences for the forest landscape of the Appalachians. Many of the oaks formally found in association with chestnut increased in dominance. Hickory (*Carya*) species have increased in abundance, and in many areas the former chestnut-oak forests have been replaced by oak-hickory forests (Keever, 1953; Woods and Shanks, 1959; McCormick and Platt, 1980). Only a few mature individuals of the once dominant tree can still be found. The root systems of many individuals have survived, and root sprouting is common. But the large stems in almost all cases are killed by the blight. Chestnut now survives primarily as an understory tree (Paillet, 1984).

Other alien pathogens have also significantly affected forest community composition and structure. The American elm (*Ulmus americana*) is a common tree throughout the mesic and lowland forests of eastern North America, but many of the largest individuals were killed after the introduction of Dutch elm disease (*Ceratocystis ulmi*) from Europe about 1930 (Karnosky, 1979; Huenneke, 1983). Also, many of the large American beech (*Fagus grandifolia*) in some areas of the northeastern United States have been killed by beech bark disease caused by the interaction of insects and fungi (*Nectria*). The long-term effect of this disease on American beech throughout its broader range in eastern North America is not clear (Twey and Patterson, 1984).

15.5 Forest Regions

Today, the forest vegetation of the Appalachian region is highly complex. Forest communities include spruce-fir stands in the north and hardwood and mixed pine-hardwood communities farther south. Distinctive forest regions developed largely in response to climatic conditions. The severity of winter decreases and the length of the growing season increases significantly from north to south (fig. 15.4). The northern Appalachians, including New

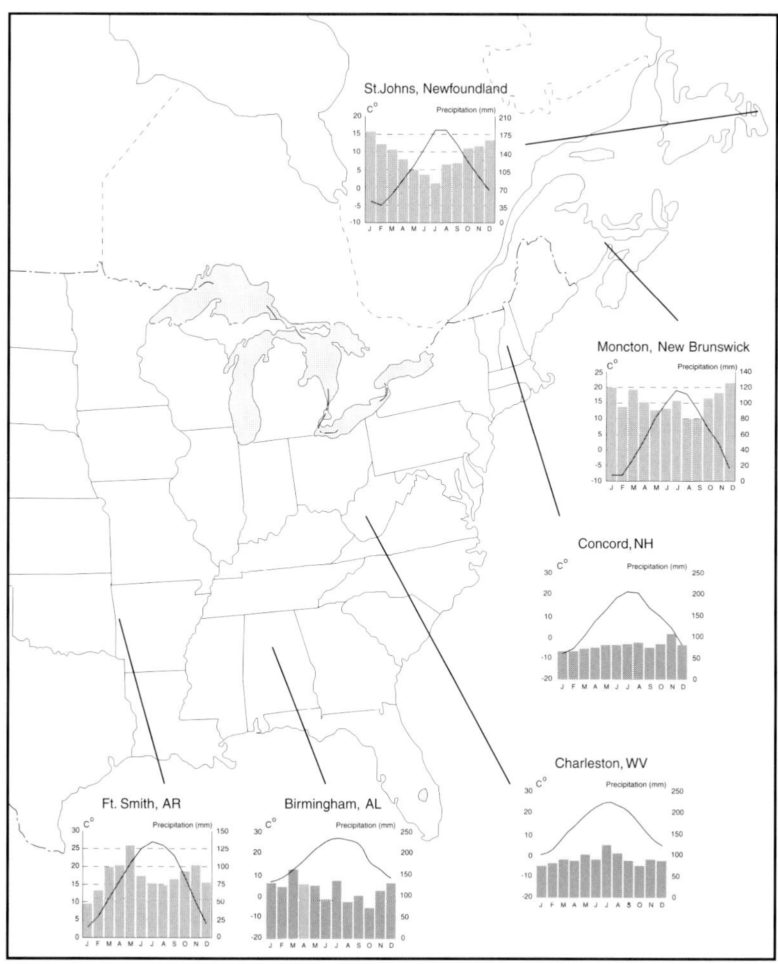

Figure 15.4 Climate diagrams showing monthly average temperature and precipitation for representative areas in the Appalachians.

Brunswick, Nova Scotia, and much of New England, have cold, continental climates. The short growing season of only 3–4 months and the cold winters favor coniferous forest species. In contrast, hardwood forest communities dominate the central Appalachians because of milder winters and longer growing seasons. The southern Appalachians of the western Carolinas, northern Georgia, and Alabama have a subtropical climate with a growing season of 7–8 months, which favors hardwood and mixed hardwood-pine forest communities.

The entire region is humid, and although some areas of the Appalachians have noticaby less summer and fall precipitation, there is not a distinct dry season. The summertime Bermuda High drives warm, humid air across much of eastern North America. Summer convection occurs throughout most the region, although it is much more common in the southern Appalachians. During the winter, the polar jet stream is displaced southward. Midlatitude cyclones and fronts, most common during the winter and spring, are the dominant precipitation mechanisms throughout the year for most of the region.

The classic works of Braun (1950) and Kuchler (1964) provide the basis for the forest classification throughout the Appalachian region. Within the Appalachian mountains are three distinctive forest regions (fig. 15.5). The northernmost is the Northern Transitional Forest, extending from southeast Canada and New England southward into central Pennsylvania. As its name states, this is transitional region between the conifer-dominated boreal forest to the north and deciduous forests to the south. The Central Hardwood Forest occupies much of the unglaciated Appalachian region of West Virginia, Kentucky, and Tennessee, in addition to the westernmost parts of Virginia and North Carolina. The term Central Hardwood Forest refers to three of Braun's centrally located forest regions, the oak-chestnut, mixed mesophytic, and western mesophytic forests (fig. 15.5). In this region, conifers are only occasionally dominant, except at the highest elevations of the Blue Ridge and the Ridge and Valley. The Southern Transitional Forest occupies most of the Piedmont and the southernmost mountain provinces, including the uplands west of the Mississippi River valley This is a transitional forest

Figure 15.5 Major forest associations in the Appalachian region. Abbreviations for the Central Hardwood Forest region follow: WM, western mesophytic; MM, mixed mesophytic; OC, oak-chestnut. Source: Braun, 1950.

region between the decidous forest region to the north and pine-dominated regions of the Coastal Plain farther south.

15.5.1 Northern Transitional Forests

The forest vegetation of the northern Appalachians is characterized by a large-scale mosaic of coniferous, hardwood, and mixed forest communities. The Maritime Provinces and Maine, the northernmost areas of the Appalachians, are occupied by spruce-hardwood stands that include white spruce (*Picea glauca*), red spruce (*P. rubens*), American beech (*Fagus grandifolia*), yellow birch (*Betula alleghaniensis*), and sugar maple (*Acer saccharum*). Hardwoods are more abundant on the favorable sites and generally decrease in dominance with distance north because of the harsher climate. Hardwood and mixed hardwood-hemlock (*Tsuga canadensis*) communities dominate the more southerly part of New England, in addition to lower elevations in the north.

There is a distinct altitudinal zonation of forest communities in the mountainous regions of the northern Appalachians. The Catskills and Green Mountains in New York and Vermont extend over 1200 m above sea level, the White Mountains in New Hampshire to more than 1800 m, and Mt. Katahdin in Maine to 1605 m. The general altitudinal sequence of these mountains, as well as mountains of lower elevation farther north, begins on the lower slopes with hardwoods sometimes mixed with spruce, followed by predominantly spruce forests, culminating in spruce-fir forests at the highest altitudes. These high-altitude forest communities are southern extensions of the spruce-fir boreal forest that extend unbroken from the northern margins of the Appalachians across to Alaska (Oosting and Billings, 1951; see also chapter 14). Lowland spruce-fir forest occurs below the hardwood forest in some areas because of cold-air drainage. In the high mountains of the northern Appalachians, alpine treeline is between 1100 and 1300 m. Growing season temperatures seem to be an important control on the altitude of treeline. In the northern Appalachians, treeline corresponds fairly well to the 13°C mean July isotherm (Cogbill and White, 1991).

The spruce-fir forests fairly common in eastern Canada and northern New England extend southward along the crests of the Appalachian Mountains, but at a progressively higher elevation with distance south because of the warmer climate. There is a fairly abrupt transition between the lower spruce-fir forests and the highest hardwood-dominated communities. The elevational zone where these forest communities meet is about 500–600 m in northern New England, although it is somewhat lower on cooler north-facing slopes and sheltered sites. The lower elevation of spruce-fir forests increases about 80 m for every 1° latitude to the south. The decrease in the spruce-fir–hardwood ecotone corresponds closely to the 17°C mean July isotherm (Cogbill and White, 1991).

As noted previously, forests of eastern North America are subject to a variety of natural and human disturbances. For example, New England forests have been much affected by hurricanes that strike the New England coast on average once every one or two decades (Simpson and Riehl, 1981). Although damage by hurricanes is greatest in the coastal regions, vegetation in the interior of New Hampshire, Vermont, Massachusetts, and Connecticut has also been significantly affected (Whitney, 1994). Hurricanes can level forests over thousands of hectares. Mature stands in which trees are less densely spaced and those on the more exposed sites are most vulnerable. Extensive blowdowns favor early colonizing species, including white birch (*Betula papyrifera*) and quaking aspen (*Populus tremuloides*), which are uncommon in mature stands. The existence of forest stands of differing age in response to windthrow has in some areas of New England created a landscape-scale mosaic of distinctive forest communities (Foster, 1988).

New England was the first part of North America heavily settled by European colonists. Almost all forests at low elevations were logged. The remaining old-growth forests consist mostly of subalpine spruce and spruce-fir stands at high elevations. Large areas of high-altitude, old-growth forests occur in the Catskills and Adirondack Mountains in New York, New Hampshire's White Mountains National Forest, and Baxter State Park in Maine, which includes Mt. Katahdin. The largest area of low-elevation, old-growth forest is the Big Reed Forest Preserve in Maine. However, many other smaller parcels of unlogged stands are scattered throughout the Northern Transitional Forest (Davis, 1996).

15.5.2 Central Hardwood Forests

A large portion of the central and southern Appalachian region is occupied by hardwood forest communities. These diverse forests contain many wide-ranging species. Oak (*Quercus*) is the largest genus in North America with 42 species, many of which occur throughout a large part of the Central Hardwood Forest region (White and White, 1996). Drier sites within this region are occupied by post oak (*Q. stellata*), chestnut oak (*Q. prinus*), and scarlet oak (*Q. coccinea*). In addition to other oaks, upland sites are also occupied by several wide-ranging hickories such as mockernut hickory (*Carya tomentosa*) and pignut hickory (*C. glabra*). Another dozen or so tree species occur on drier upland sites. Among the dominant tree species on mesic sites are white oak (*Quercus alba*), northern red oak (*Q. rubra*), and black oak (*Q. veluntina*), and on the most favorable sites tulip poplar (*Liriodendron tulipifera*), American beech (*Fagus grandifolia*), sugar maple (*Acer saccharum*), and hemlock, (*Tsuga canadensis*). Because of the range of environmental conditions related to slope aspect and soil conditions, fine-scale forest community composition and structure are highly variable.

Plant community patterns across most forested landscapes are largely a consequence of natural disturbance (White, 1979; Pickett and White, 1985). Major disturbances (including hurricanes, fire, insect infestation, avalanches) often destroy stands that cover areas ranging from less than a hectare to hundreds of square kilometers. Forest communities at early stages of recovery are dominated by early successional species. The Central Hardwood Forests, however, are affected less frequently by large-scale disturbance compared to the other forest regions in eastern North America (Runkle, 1990, 1996). Therefore, the hardwood stands in this region often reach a later stage of stand development during which older trees occupy the forest canopy.

Trees are typically closely spaced during early forest succession. At this stage of vegetation development, space created by the death of an individual is quickly filled by its neighbors. Progressively older stands have fewer and more widely spaced trees. The death of a dominant individual in mature stands creates a well-defined, and in some cases long-lived, canopy gap. These gaps are less likely to be filled by adjacent trees than in young stands (Oliver, 1981; Oliver and Larson, 1990). Instead, large canopy gaps facilitate the growth of individuals in the understory that can eventually occupy a canopy position. This process enables some species that are incapable of surviving a long time in dense shade in the central hardwood region to remain a component of the forest canopy. One of the best examples is the tulip poplar, which is a canopy dominant in many mesic sites in the central Appalachians.

Although much of the Central Hardwood Forest lacks major disturbances common to other regions, the forest vegetation here has been strongly modified by human impact, to some extent by Native Americans, but mostly since European settlement. Most forests have been logged, but there are surviving old-growth stands. Up to one-third of the Great Smoky Mountains National Park in North Carolina and Tennessee is covered with old-growth forest, including mixed mesic hardwoods, hemlock-dominated forests, and high-elevation spruce-fir stands. Smaller but substantial areas of old-growth forest are in the Joyce Kilmer Memorial Forest in the Nantahala National Forest in North Carolina with mesic hardwoods and hemlock-dominated stands, the Lilley-Cornett Woods in eastern Kentucky with oak-hickory and beech-hemlock stands, and the Chattahoochee National Forest in north Georgia with oak-dominated communities (White and White, 1996).

Some of the most interesting areas of the Central Hardwood Forest region are the cove forests in and around the Great Smoky Mountains and the loess hills of western

Tennessee and northern Mississippi. Cove forests occur on concave topography that has created cool, mesic sites. The number of species in coves is largely determined by soil type. Sites with rich calcareous soils support some the most diverse forest communities in eastern North America. Dominant species include sugar maple, buckeye (*Aesculus octandra*), basswood (*Tilia heterophylla*), and tulip poplar. In contrast, coves with more acidic soils support much less diverse forest communities often dominated by hemlock. The loess hills in western Tennessee and northwest Mississippi, adjacent to the lower Mississippi River alluvial valley, are also occupied by diverse forest communities. The loess is greater than 20 m thick in places at its western limits and rapidly thins with distance to the east. In contrast to the cove forests, the loess hills have less precipitation. This region, however, consists of deep calcareous soils that support luxuriant hardwood stands of American beech, tulip poplar, hickory, and white oak.

In contrast to the hardwood dominance throughout most of this region, forests of red spruce and Fraser fir (*Abies fraseri*) occur in the high mountains of Virginia, West Virginia, Tennessee, and North Carolina. These coniferous forests generally occur above 1500 m but are also present at lower elevations on the cooler, north-facing slopes. Red spruce is one of the most common species in the northern Appalachians, but farther south it occurs within a fairly narrow altitudinal range (1400–1800 m) and typically is most abundant on well-drained sites. In contrast, Fraser fir, which is endemic to the southern Appalachians, increases in dominance with higher elevation and forms almost pure stands on ridges and near mountain summits (Whittaker, 1956). During the 1980s and 1990s, almost all of the mature Fraser fir in the Great Smoky Mountains were killed by the balsam woolly adelgid (*Adelges piceae*), an insect introduced from Europe to the eastern United States early in the century (Busing et al., 1988).

15.5.3 Southern Transitional Forests

The Southern Transitional Forest extends from New Jersey to central Georgia and westward to the lower Mississippi River alluvial valley, where it is interrupted before reappearing in Arkansas, northern Louisiana, and east Texas (fig. 15.5). This includes much of the Piedmont, the southernmost sections of the Appalachian mountain provinces, and the Ozark Plateau and Ouachita Mountains. The southernmost portion of the Appalachians is a humid subtropical region with a long growing season and abundant precipitation throughout the year. This region supports hardwoods and mixed hardwood-pine forest communities. It contains many of the same hardwoods occurring in the Central Hardwood Forest region, but unlike that region, pines are common. Pine-dominated communities occur on dry upper slopes and ridge tops, sites generally less favorable for most hardwoods. Also, they rapidly colonize recently disturbed sites. This region does not reach high elevation, and the distinctive altitudinal vegetation zonation that occurs in the mountains to the north is generally lacking.

Most of the Southern Transitional Forest region was heavily settled during the ninteenth century, and large areas were cleared for cultivation. Much of the area has poor soils for agriculture, and early settlers gave little thought to soil conservation practices. Site productivity typically declined after several years of cultivation, and it was common practice for fields to be abandoned and new land cleared (Oosting, 1942). During the early 1900s, agriculture in the Piedmont declined, and large cultivated tracts were abandoned. This resulted in a patchwork of second-growth forest communities of various ages and composition (Braun, 1950). Few large areas of old-growth forest remain in this region.

Young pine stands dominate much of the region. Pines aggressively colonize abandoned fields and sites disturbed by fire. Many sites that can potentially support hardwood stands are now occupied by pine and in some cases mixed hardwood-pine forest communities (Billings, 1938; Coile, 1940; Oosting, 1942). Loblolly pine (*Pinus taeda*) and shortleaf pine (*P. echinata*) are common throughout most of the region. Shortleaf pine is more abundant on the drier upland sites, whereas loblolly pine is more common on lower slopes and valley bottoms because of its high tolerance for poor soil aeration. Both species, however, occur in a variety of topographic and soil conditions. Young pine forests are interrupted by narrow stream valleys that support bottomland forest communities. River birch (*Betula nigra*), cottonwood (*Populus deltoides*), and sycamore (*Platanus occidentalis*) occur on areas subject to frequent flooding. Other common bottomland species are willow oak (*Quercus phellos*), water oak (*Q. nigra*), and sugarberry (*Celtis laevigata*).

The Southern Transitional Forests extend beyond the Appalachian and Ozark-Ouchita region into eastern Oklahoma and Texas, but forest vegetation becomes less luxuriant in reponse to lower precipitation. With increasing aridity and more frequent droughts westward, trees are smaller with progressively fewer species. Forest cover decreases, and there is a gradual transition from forest vegetation to forest scrub and grassland, and eventually to continuous grassland (Bruner, 1931). Because of the drier conditions at the western extremes of the Southern Transitional Forest, continuous forest vegetation occurs mostly in river bottomlands, with the adjacent uplands occupied by grasslands. Farther west, bottomland trees also decrease in size and become more widely spaced.

15.6 Conclusion

Ranging through nearly 20° of latitude, the Appalachian region has diverse climates, vegetation, and physical landscapes. It is an amalgam of structural terranes involving Precambrian and Paleozoic rocks that were shaped by several orogenic episodes and separated collectively from Africa during the breakup of Pangea in early Mesozoic times. The region was subsequently subjected to prolonged weathering, erosion, and drainage reversal, and exposed to the rigors of frequent climate change during the Quaternary, including glaciation in the north and periglacial conditions throughout the latitudinal range at higher elevations. Eastern North America was strongly influenced by human activity during historic times, and it is not possible to understand the diverse physical landscapes of this region without considering human impacts. The vegetation was modified by Native Americans, who burned the forests and in some areas cleared land for cultivation. Europeans had a much greater effect on the vegetation through land conversion to agriculture, industrial logging, and introduction of alien species, including widely destructive insects and plant pathogens. Vegetation change has inevitably modified the variety of surface processes, most notably through the accelerated erosion of abandoned agricultural lands that were formerly forested (see chapter 23). Other direct impacts have resulted from dam construction and river flood-control projects, surface mining, extensive urbanization, and growth of the rural population. The physical geography of the Appalachians will continue to change, probably in more dramatic ways, in the near future as the population in eastern North America and the concomitant demand for resources increase.

References

Baird, F., 1832. View of the Valley of the Mississippi; or, the Emigrant and Traveller's Guide to the West. H.S. Tanner, Philadelphia.

Bartrum, W., 1958. The Travels of William Bartram, Naturalist's Edition. Edited with commentary and an annotated index by Francis Harper. Yale Unviversity Press, New Haven.

Bates, R.L., and J.A. Jackson, 1987. Glossary of Geology. American Geol. Inst., Alexandria, Va.

Battiau-Queney, Y., 1989. Constraints from deep crustal structure on long-term landform development of the British Isles and eastern United States. Geomorphology 2: 53–70.

Billings, D.W., 1938. The structure and development of old field shortleaf pine stands and certain associated properties of the soil. Ecological Monographs 8: 437–499.

Bormann, F.H., and R.B. Platt. 1958. A disjunct stand of hemlock in the Georgia piedmont. Ecology 39: 16–23.

Borns, H.W., Jr., 1985. Changing models of deglaciation in northern New England and adjacent Canada. In, pp. 135–138, Borns, H.W., Jr., P. LaSalle, and W.B. Thompson (eds.), Late Pleisocene History of Northeastern New England and Adjacent Quebec. Geological Society of America Special Paper 197.

Boulton, G.S., G.D. Smith, A.S. Jones, and J. Newsome, 1985. Glacial geology and glaciology of the last mid-latitude ice sheets. Geological Society of London Journal 142: 447–474.

Braun, D.D., 1989. Glacial and periglacial erosion of the Appalachians. Geomorphology 2: 233–256.

Braun, E.L., 1950. Deciduous Forests of Eastern North America. Blakiston, Philadelphia.

Bruner, W.E., 1931. The vegetation of Oklahoma. Ecological Monographs 1: 99–187.

Busing, R.T., E.E.C. Clebsch, C.C. Eagar, and E.R. Pauley, 1988. Two decades of change in a Great Smoky Mountains spruce-fir forest. Bulletin of the Torrey Botanical Club 115: 25–31.

Clark, G.M., 1987. Debris slide and debris flow historical events in the Appalachians south of the glacial border. In, pp. 125–138, Costa, J.E., and G.F. Wieczorek (eds.), Debris Flows/Avalanches: Process, Recognition, and Mitigation, Reviews in Engineering Geology Vol.VII. Geological Soc. America, Boulder, CO.

Clark, G.M., 1989. Central and southern Appalachian water and wind gap origins: Review and new data. Geomorphology 2: 209–232.

Clark, G.M., and E.J. Ciolkosz, 1988. Periglacial Geomorphology of the Appalachian Highlands and interior highlands south of the glacial border—A review. Geomorphology 1: 191–220.

Cogbill, C.V., and P.S. White, 1991. The latitude-elevation relationship for spruce-fir forest and treeline along the Appalachian mountain chain. Vegetatio 94: 153–175.

Coile, T.S., 1940. Soil changes associated with loblolly pine succession on abandoned land of the Piedmont plateau. Duke University School of Forestry Bulletin 5.

Cook, F.A., D.S. Albaugh, L.D. Brown, S. Kauffman, and J. Oliver, 1979. Thin-skinned tectonics in the crystalline southern Appalachians. Geology 7: 563–567.

Costa, J.E., and E.T. Cleaves, 1984. The Piedmont landscape of Maryland: A new look at an old problem. Earth Surface Processes and Landforms 9: 59–74.

Cowdrey, A.E., 1983. This Land, This South, an Environmental History. University Press of Kentucky, Lexington.

Davis, M.B., 1981. Quaternary History and the Stability of Forest Communities. In, pp. 132–151, West, D.C., H.H. Shugart, and D.B. Botkin (eds.), Forest Succession: Concepts and Applications, Springer-Verlag, New York.

Davis, M.B., 1996. Eastern Old-growth Forests: Prospects for Rediscovery and Recovery. Island Press, Washington, D.C.

Davis, W.M., 1890. The rivers of northern New Jersey. National Geographic Magazine 2: 81–110.

Davis, W.M., 1899a. The geographical cycle. Geographical Journal 14: 481–504.

Davis, W.M., 1899b. The peneplain. American Geologist 23: 207–239.

Davis, W.M., 1899c. The rivers and valleys of Pennsylvania. National Geographic Magazine 1: 183–253.

Delcourt, P.A., and H.R. Delcourt, 1977. The Tunica Hills, Louisiana-Mississippi: Late glacial locality for spruce and deciduous forest species. Quaternary Research 7: 218–237.

Delcourt, P.A., and H.R. Delcourt, 1997. Pre-Columbian native American use of fire on southern Appalachian landscapes. Conservation Biology 11: 1010–1014.

Delcourt, P.A., H.R. Delcourt, R.C. Bristor, and L.E. Lackey, 1980. Quaternary vegetation history of the Mississippi Embayment. Quaternary Research 13: 111–132.

Denny, C.S., 1956. Surficial geology and geomorphology of Potter County, Pennsylvania. U.S. Geological Survey Professional Paper 288.

Dreimanis, A., and R.P. Goldthwaite, 1973. Wisconsin glaciation in the Huron, Erie, and Ontario Lobes. In, pp. 71–106, Black, R.F., R.P. Goldthwaite, and H.B. Willman (eds.), The Wisconsin Stage, Geological Society of America Memoir 136, Boulder.

Dyke, A.S., J.-S. Vincent, J.T. Andrews, L.A. Dredge, and W.R. Cowan, 1989. The Laurentide Ice Sheet and an introduction to the Quaternary geology of the Canadian Shield. In, pp. 178–189, Fulton, R.J. (Coordinator), Quaternary Geology of the Canadian Shield, Chapter 3 in Fulton, R.J. (ed.), Quaternary Geology of Canada and Greenland, Geologic Survey of Canada, Geology of Canada, No. 1; also: Geological Soc. Amer., Geology of North America, Vol. K-1.

Epstein, J.B., 1966. Structural control of wind gaps and water gaps and of stream capture in the Stroudsburg area, Pennsylvania and New Jersey. U.S. Geological Survey Prof. Paper 550-B.

Fenneman, N.M., 1938. The Physiography of the Eastern United States. McGraw-Hill, New York.

Foster, D.R., 1988. Species and stand response to catastrophic wind in central New England, U.S.A. Journal of Ecology 76: 135–151.

Hack, J.T., 1960. Interpretation of erosional topography in humid temperate regions. American Journal of Science 258A: 80–97.

Hack, J.T., 1976. Dynamic equilibrium and landscape evolution. In, pp. 87–102, Melhorn, W.N., and R.C. Edgar (eds.), Theories of Landform Development. Publications in Geomorphology, Binghamton, N.Y.

Hack, J.T., 1980. Rock control and tectonism—Their importance in shaping the Appalachian Highlands. U.S. Geological Survey Prof. Paper 11265–B.

Hack, J.T., 1982. Physiographic divisions and differential uplift in the Piedmont and Blue Ridge. U.S. Geological Survey Prof. Paper 1265.

Hack, J.T., and J.C. Goodlett, 1960. Geomorphology and forest ecology of a mountain region in the Central Appalachians. U.S. Geological Survey Prof. Paper 347.

Harmon, M.E., 1982. Fire history of the westernmost portion of the Great Smoky Mountains National Park. Bulletin of the Torrey Botanical Club 109: 74–79.

Harper, R.M., 1943. Hemlock in the Tennessee Valley of Alabama. Castanea 8: 115–123.

Hatcher, R.D., Jr., 1989. Tectonic synthesis of the U.S. Appalachians. In, pp. 511–535, Hatcher, R.D., Jr., W.A. Thomas, and G.W. Viele (eds.), The Appalachian-Ouachita Orogen in the United States, The Geology of North America, Vol. F-2.

Hatcher, R.D., Jr., and S.A. Goldberg, 1991. The Blue Ridge Geologic Province, In, pp. 11–58, Horton, J.W., Jr., and V.A. Zullo (eds.), The Geology of the Carolinas, University of Tennessee Press, Knoxville.

Horton, J.W., Jr., and V.A. Zullo, 1991. An introduction to the geology of the Carolinas. In, pp. 1–10, Horton, J.W., Jr., and V.A. Zullo (eds.), The Geology of the Carolinas, University of Tennessee Press, Knoxville.

Huenneke, L.F., 1983. Understory response to gaps caused by the death of *Ulmus americana* in central New York. Bulletin of the Torrey Botanical Club 110: 170–175.

Hughes, T., H.W. Borns Jr., J.L. Fastook, J.S. Kite, M.R. Hyland, and T.V. Lowell, 1985. Models of glacial reconstruction and deglaciation applied to Maritime Canada and New England. In, pp. 139–150, Borns, H.W., Jr., P. LaSalle, and W.B. Thompson (eds.), Late Pleistocene History of Northeastern New England and Adjacent Quebec. Geological Society of America Special Paper 197.

Jacobson, R.B., A.J. Miller, and J.A. Smith, 1989. The role of catastrophic geomorphic events in central Appalachian landscape evolution. Geomorphology 2: 257–284.

Johnson, D.W., 1931. Stream sculpture on the Atlantic slope, a study in the evolution of Appalachian rivers. Columbia University Press, New York.

Judson, S. 1976. Evolution of Appalachian topography; In, pp. 29–42, Melhorn, W.N. and R.C. Flemal (eds.), Theories of Landform Development, Proceedings 6th Annual Geomorphology Symp., Binghamton, N.Y., 1975. Publications in Geomorphology, State University of New York, Binghamton.

Karnosky, D.F., 1979. Dutch elm disease: A review of the history, environmental implications, control, and research needs. Environmental Conservation 6: 311–322.

Keever, C., 1953. Present composition of some stands of the former oak-chestnut forest in the southern Blue Ridge Mountains. Ecology 34: 44–54.

Kuchler, A.W., 1964. Potential Natural Vegetation of the Conterminous United States. American Geographical Society Special Publication No. 36.

McCormick, J.F., and R.B. Platt, 1980. Recovery of an Appalachian forest following the chestnut blight or Catherine Keever—you were right. American Midland Naturalist 104: 264–273.

Mickelson, D.M., L. Clayton, D.S. Fullerton, and H.W. Borns Jr., 1983. The Late Wisconsin glacial record of the Laurentide Ice Sheet in the United States. In, pp. 3–37, Wright, H.E., Jr., and S.C. Porter (eds.), Late-Quaternary Environments of the United States. Univeristy of Minnesota Press.

Miller, A.J., 1995. Valley morphology and boundary conditions influencing spatial patterns of flood flow. In, pp. 57–81, Costa, J.E., A.J. Miller, K.W. Potter, and P.R. Wilcock (eds.), Natural and Anthropogenic Influences in Fluvial Geomorphology, The Wolman Volume, Geophysical Monograph 89, American Geophysical Union.

Mills, H.H., and P.A. Delcourt, 1991. Quaternary geology of the Appalachian Highlands and interior low plateaus. In, pp. 611–628, Morisawa, M. (ed.), Quaternary Non-Glacial Geology: Conterminous U.S. Geological Society of America, The Geology of North America, Vol. K-2, Boulder, Colo.

Mills, H.H., and J.R. Wagner, 1985. Long-term change in regime of New River indicated by vertical variation in extent and weathering intensity of alluvium. Journal of Geology 93: 131–142.

Mills, H.H., G.R. Brakenridge, R.B. Jacobson, W.L. Newell, M.J. Pavich, and J.S. Pomeroy, 1987. Appalachian mountains and plateaus; pp. 5–50, In Graf, W.L. (ed.), Geomorphic Systems of North America. Geological Soceity of America, Boulder.

Mix, A.C., 1992. The marine oxygen isotope record: Constraints on timing and extent of ice-growth events

(120–65 ka). In Clark, P.U., and P.D. Lea (eds.) The Last Interglacial-Glacial Transition in North America. Geol. Soc. Am. Spec. Paper 270.

Morisawa, M., 1989. Rivers and valleys of Pennsylvania, revisited. Geomorphology 2: 1–22.

Morrison, R.B., 1991. Introduction. In, pp. 1–12, Morrison, R.B. (ed.), Quaternary Nonglacial Geology: Conterminous U.S., Geological Soc. Amer., The Geology of North America, Vol. K-2; Boulder, Colo.

Newman, W.A., A.N. Genes, and T. Brewer, 1985. Pleistocene geology of northeastern Maine. In, pp. 59–70, Borns, H.W., Jr., P. LaSalle, and W.B. Thompson (eds.), Late Pleistocene History of Northeastern New England and Adjacent Quebec. GSA Spec. Paper 197.

Oberlander, T.M., 1985. Origin of drainage transverse to structures in orogens. In, pp. 155–182, Morisawa, M., and J.T. Hack (eds.), Tectonic Geomorphology, Proc. 15th Ann. Binghamton Geomorphology Symp., Sept. 1984. Allen & Unwin, Boston.

Occhietti, S., 1989. Quaternary geology of St. Lawrence Valley and adjacent Appalachian subregion. In, pp. 350–388, Karrow, P.F., and S. Occhietti (eds.), Quaternary Geology of the St. Lawrence Lowlands of Canada, Chapter 4, Fulton, R.J. (ed.), Quaternary Geology of Canada and Greenland, Geologic Survey of Canada, Geology of Canada, No. 1, also Geological Soc. Amer., Geology of North America, Vol. K-1.

Oldale, R.N., and S.M. Coleman, 1992. On the age of the penultimate full glaciation of New England. In, pp. 163–170, Clark, P.U., and P.D. Lea. (eds.) The Last Interglacial-Glacial Transition in North America. Geological Society of America Special Paper 270. Boulder, Colo.

Oldale, R.N., P.C Valentine, T.M. Cronin, E.C. Spiker, B.W. Blackwelder, D.F. Belknap, J.F. Wehmiller, and B.J. Szabo, 1982. Stratigraphy, structure, absolute age, and paleontology of the upper Pleistocene deposits at Sankaty Head, Nantucket Island, Massachusetts. Geology 10: 246–252.

Oliver, C.D., 1981. Forest development in North America following major disturbances. Forest Ecology and Management 3: 153–168.

Oliver, C.D., and B.C. Larson, 1990. Forest Stand Dynamics. McGraw-Hill, New York.

Oosting, H.J., 1942. An ecological analysis of the plant communities of Piedmont North Carolina. American Midland Naturalist 28: 1–126.

Oosting, H.J., and W.D. Billings, 1951. A comparison of virgin spruce-fir forests in the northern and southern Appalachian system. Ecology 32: 84–103.

Paillet, F.L., 1984. Growth-form and ecology of American chestnut sprout clones in northeastern Massachusetts. Bulletin of the Torrey Botanical Club 111: 316–328.

Patterson, W.A., III, K.J. Edwards, and D.J. Maguire, 1987. Microscopic charcoal as fossil indicators of fire. Quaternary Science Reviews 6: 3–23.

Pavich, M.J., 1989. Regolith residence time and the concept of surface age of the Piedmont "Peneplain." Geomorphology 2: 181–196.

Péwé, T.L., 1983. The paraglacial environment in North America during Wisconsin time. In: Porter, S.C. (ed.), The Late Pleistocene, Vol. 1, Late Quaternary Environments of the United States, Wright, H.E. Jr. (ed.), University of Minneapolis Press, Minneapolis.

Pickett, S.T.A., and P.S. White, 1985. The Ecology of Natural Disturbance and Patch Dynamics. Academic Press, San Diego.

Poag, C.W., and W.D. Sevon, 1989. A record of Appalachian denudation in postrift Mesozoic and Cenozoic sedimentary deposits of the U.S. Middle Atlantic continental margin. Geomorphology 2: 119–157.

Rast, N., 1989. The evolution of the Appalachian chain. In, pp. 323–348, Bally, A.W., and A.R. Palmer (eds.), The Geology of North America—An Overview, Geological Society of America, The Geology of North America, Vol. A, Boulder, Colo.

Runkle, J.R., 1990. Gap dynamics in an Ohio *Acer-Fagus* forest and speculations on the geography of disturbance. Canadian Journal of Forest Research 20: 632–641.

Runkle, J.R., 1996. Central Mesophytic Forests. In, pp. 161–177, Davis, M.D. (ed.), Eastern Old-Growth Forests: Prospects for Rediscovery and Recovery. Island Press, Washington, D.C.

Russell, E.W.B., 1987. Pre-blight distribution of *Castanea dentata* (March.) Borkh. Bulletin of the Torrey Botanical Club 114: 183–190.

Shilts, W.W., J.M. Aylsworth, C.A. Kaszycki, and R.A. Klassen, 1987. Canadian Shield. In, pp. 119–161, Chapter 5, Graf, W.L. (ed.), Geomorphic Systems of North America; Geological Soc. Amer., Centennial Spec. Vol. 2, Boulder, Colo.

Simpson, R.H., and H. Riehl, 1981. The Hurricane and Its Impact. Louisiana State University Press, Baton Rouge.

Soller, H.H., and H.G. Mills, 1991. Surficial Geology and Geomorphology. In, pp. 290–308, Horton, J.W., Jr., and V.A. Zullo (eds.), The Geology of the Carolinas. University of Tennessee Press, Knoxville.

Staheli, A.C., 1976. Topographic expression of superimposed drainage on the Georgia Piedmont. Geological Society of America Bulletin 87: 450–452.

Tarr, R.S., 1898. The Peneplain. American Geologist 21: 351–369.

Trautman, M.B., 1977. The Ohio country from 1750–1777 —A naturalist's view. Ohio Biological Survey, Biological Notes No. 10.

Twey, M.J., and W.A. Patterson III, 1984. Variations in beech bark disease and its effects on species composition and structure of northern hardwoods stands in central New England. Canadian Journal of Forest Research 14: 565–574.

Vincent, J.-S., 1989. Quaternary Geology of the southeastern Canadian Shield, In pp. 249–275, Fulton, R.J. (Coordinator), Quaternary Geology of the Canadian Shield, Chapter 3 in Fulton, R.J. (ed.), Quaternary Geology of Canada and Greenland, Geologic Survey of Canada, Geology of Canada, No. 1, also Geological Soc. Amer., Geology of North America, Vol. K-1.

Wahrhaftig, C.A., 1965. Stepped topography of the southern Sierra Nevada, California. Geol. Soc. Am. Bull. 76: 1165–1190.

Watts, W.A., 1970. The full-glacial vegetation of northwestern Georgia. Ecology 51: 19–33.

Watts, W.A., 1979. Late Quaternary vegetation of central Appalachia and the New Jersey coastal plain. Ecological Monographs 49: 427–469.

White, P.S., 1979. Pattern, process, and natural disturbance of vegetation. Botanical Review 45: 229–299.

White, P.S., and R.D. White, 1996. Old-growth oak and oak-hickory forests. In, pp. 178–198, Davis, M.B. (ed.), Eastern Old-growth Forests: Prospects for Rediscovery and Recovery. Island Press, Washington D.C.

Whitney, G.G., 1994. From Coastal Wilderness to Fruited Plain. Cambridge University Press, Cambridge.

Whittaker, R.H., 1956. Vegetation of the Great Smoky Mountains. Ecological Monographs 26: 1–80.

Winkler, M.G., 1985. A 12,000-year history of vegetation and climate of Cape Cod, Massachusetts. Quaternary Research 23: 301–312.

Wolman, M.G., and J.P. Miller, 1960. Magnitude and frequency of forces in geomorphic processes. Journal of Geology 68: 54–74.

Woods, F.W., and R.E. Shanks, 1959. Natural replacement of chestnut by other species in the Great Smoky Mountains National Park. Ecology 40: 349–361.

Wright, H.E., Jr., 1989. The Quaternary. In, pp. 513–536, Chapter 17, Bally, A.W., and A.R. Palmer (eds.), The Geology of North America—An Overview. Geological Soc. Amer, The Geology of North America, Vol. A, Boulder, Colo.

16

The Great Lakes Region

Randall J. Schaetzl
Scott A. Isard

The physical geography of the Great Lakes region is extremely diverse. The region lies on a continental interior plain, halfway between the equator and the North Pole, and as such is located within a latitudinal zone of steep climate and vegetation gradients. Yet five large lakes, for which the region is named, dramatically impact the landscapes in their vicinity. The strong influence that these lakes have on the physical geography, otherwise dominated by dramatic north–south environmental gradients, gives the Great Lakes region its unique and recognized character within North America (fig. 16.1).

The Great Lakes region spans the border of the United States and Canada and is centered on five, geologically young, freshwater lakes: Superior, Michigan, Huron, Erie, and Ontario. Together, they are referred to as the Laurentian Great Lakes and constitute the largest, fresh surface water system on Earth, with an area of 244,160 km^2 (Botts and Krushelnicki, 1988) and 20% of the world's fresh surface water (Quinn, 1988; table 16.1). Because large quantities of heat and moisture are transferred to overlying air masses, the Great Lakes strongly influence the climate, and consequently the vegetation, of the region. The region's physical character is rendered even more complex by a geologic transition from the crystalline rocks of the continental craton, overlain by thin glacial sediments, in the north to a series of sedimentary rock strata covered by deep unconsolidated deposits in the south.

The region stands alone as one united by hydrology, and it forms a major environmental ecotone between the intensively used lands to the south and west, with their large urban centers and vast agricultural tracts, and the seemingly endless forests and pristine waterways to the north. This chapter is designed to systematically review the major aspects of the physical geography of the Great Lakes region, with reference to both classical and modern scientific literature.

16.1 General Physiographic Setting

The Great Lakes region is, with rare exceptions, an area of low irregular plains or plains with hills (fig. 16.2). Lakes, rivers, wetlands, and fresh water abound under a climate that provides enough rainfall and snowfall to support hundreds of perennial streams. The southern part of the region, the Central Lowland province (Hunt, 1967), tends to have low relief and is more or less blanketed with glacial and aeolian sediments (Chapman and Putnam, 1984). In this province, the main relief features are occasional morainal ridges and isolated hills, also of glacial origin, as valley

Figure 16.1 Physiographic and cultural features of the Great Lakes region, including large urban centers, major hills, and ranges of durable bedrock.

incision by postglacial river systems is minimal. Expansive, monotonously flat plains, associated with ancestral glacial lakes, dominate parts of the southern Great Lakes region; these are now drained and provide fertile soils for agriculture. To the north, the Superior Upland province of the Canadian Shield (Hunt, 1967) includes most of the Superior and Georgian Bay basins, and is dominated by bedrock. In this area, bedrock structure has interacted with differential erosion by ice and water to leave a landscape of thin soils, bedrock-floored swamps, scour lakes, and shallow waterfalls. High hills, and even some low mountainous terrain near Lake Superior, comprise parts of the region. Although glacial sediments blanket the crystalline bedrock in the north, they tend to be thin and discontinuous (Fenneman, 1938). Across the entire region, the relative youth (< 18,000 years old) of the landscape has resulted in numerous swamps and a drainage network that is not well integrated (Folsom and Winters, 1970; fig. 16.3).

16.2 Geology

The geology of the Laurentian Great Lakes consists of a basement of Precambrian rocks, mostly igneous and metamorphic in nature, overlain in the southern half of the region by sedimentary sequences of Paleozoic sandstones, shales, and limestones (Hough, 1958) and, to a lesser extent, evaporites of gypsum, anhydrite, and halite (Bleimeister, 1961; fig. 16.4). The oldest rocks, to the north, comprise the Canadian Shield (Shilts et al., 1987, Card, 1990). The southern part of the shield, namely, the part within the Great Lakes region, contains rocks that vary widely in nature

Table 16.1 Hydrologic characteristics of the five Great Lakes and Lake St. Clair, and their drainage basins[a]

	Superior	Michigan	Huron	St. Clair	Erie	Ontario
Drainage basin area (land only) (km^2)	127,700	118,000	134,100	nd	78,000	64,030
Drainage basin population as of 1991	607,121	10,057,026	2,694,154	nd	11,682,169	8,150,895
Drainage basin land use (%)						
Agricultural	3	44	27	nd	68	39
Residential	1	9	2	nd	10	7
Forest	91	41	68	nd	21	49
Other	5	6	3	nd	1	5
Water area (km^2)	82,100	57,800	59,600	1,114	25,700	18,960
Base volume (km^3)	12,100	4,920	3,540	6	484	1,640
Average elevation (m)	183.1	176.3	176.3	174.7	173.9	74.6
Maximum depth (m)	407	281	229	nd	64	237
Bicarbonate + carbonate concentration[b] (ppm)	56	115	102	nd	117	119
Connecting channel	St. Marys River	Straits of Mackinac	St. Clair River	Detroit River	Niagara River and Welland Canal	St. Lawrence River
Connecting channel outflow (m^3 s^{-1})	2,218	2,520	5,289	5,408	5,812	6,962
Diversions-location	Long Lac-Ogoki[c]	Chicago[d]	none	none	Welland[e]	Welland[e]
Diversions (m^3 s^{-1})-discharge	154	−91	—	—	−260	260
Hydraulic residence time[f] (yr)	173	62	21	0.04	2.7	7.5

nd = no data

[a]After Quinn (1992b) and MSU CES (1985), unless otherwise indicated.

[b]After Pincus (1962).

[c]Water brought from the Hudson Bay watershed into Lake Superior, east of Thunder Bay (see Haywood, 1970, for a thorough discussion).

[d]Water taken by the City of Chicago and diverted to the Mississippi River basin.

[e]Water is removed from Lake Erie, used to generate hydroelectric power, and returned to Lake Ontario.

[f]Time required to "drain" the lake through its outlets, assuming all inflows are terminated and assuming no change in rate of outflow discharge.

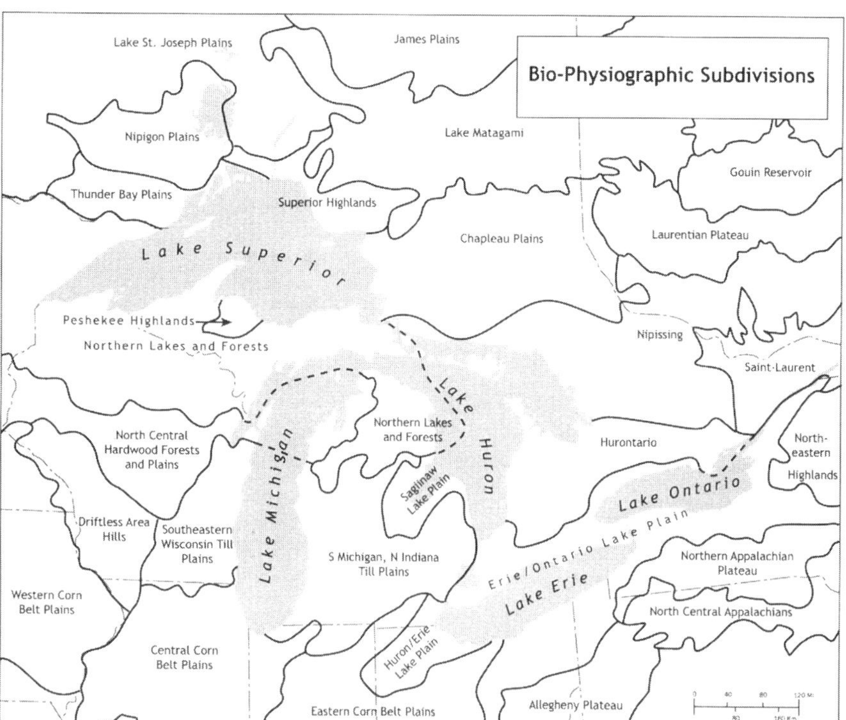

Figure 16.2 Bio-Physiographic subdivisions of the Great Lakes region (U.S. data modified from Hardwick and Holtgrieve, 1990, p. 162).

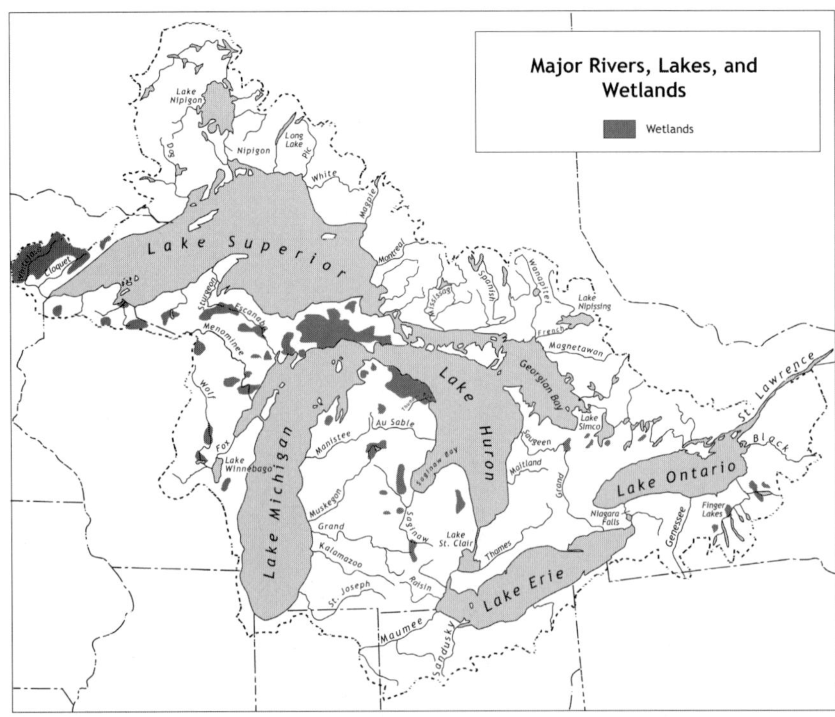

Figure 16.3 Major rivers, lakes, and wetlands of the Great lakes basin.

(Medaris, 1983; LaBerge, 1994). Metavolcanics and metasedimentary sequences, lava flows and associated greenstones, igneous intrusions, tuff and ash deposits, and various other sedimentary sequences, most of which have been slightly to highly metamorphosed or deformed, can all be found here (Cannon, 1973; Carter, 1979; Green, 1983). Granites, basalts, and slates are especially common. Many of these are intruded by sills and dikes of diabase, olivine, diorite, and granodiorite.

Many of the Precambrian rocks in the region have high contents of metallic ores of iron, copper, uranium, molybdenum, zinc, silver, gold, and platinum (Wharton, 1970; Barnes and Lalonde, 1973; Carter, 1979; Rickard, 1992). Many of the ores are found in association with igneous intrusions of, for example, magnetite (iron), quartz (gold, molybdenum, and copper), quartz-feldspar (gold and molybdenum), calcite (silver and gold), and syenite (uranium), or in fracture zones in the original crystalline rocks (copper and uranium) (Bright, 1979; Carter, 1979). The copper south of Lake Superior, which occurs in native metallic forms and sometimes in huge masses weighing many tons, is found in association with Precambrian lava flows (Bornhorst et al., 1988; Harringer, 1990). Nickel, gold, and copper are also abundant in the Precambrian rocks of the Canadian Shield.

Rare occurrences of diamonds in the glacial tills of Wisconsin and Michigan alerted early geologists to the possible presence of kimberlite ores in the Great Lakes region (Dawson, 1964; McGee and Hearn, 1984). These ores are found deep within the Precambrian rocks of northern Michigan, as well as in the Kapuskasing area of Ontario, and represent a possibly significant resource.

The break that exists between the Precambrian basement rocks of the Great Lakes region and the overlying Paleozoic rocks is a marked stratigraphic boundary (fig. 16.4). Not only does this boundary represent a period of nondeposition and surface erosion, it also marks a change from the crystalline rocks below to the sedimentary and unconsolidated strata above. The contact is gently undulating, as the Canadian Shield descends beneath the Paleozoic rocks to the south. The latter were, for the most part, laid down in shallow seas, beginning with thick sequences of sandstones of the Cambrian Period. Also during the Paleozoic, deepening of the Michigan basin and uplift of several areas to the south and west led to uneven amounts of deposition, such that unusual thicknesses of these Paleozoic sediments (≥ 4000 m) accumulated in the Michigan basin and parts of western Illinois.

Many of the Paleozoic rocks contain important oil and gas resources that are currently being tapped (McCaslin, 1987). Silurian reefs in the Niagara Formation are the most common sources of hydrocarbons, whereas Ordovician dolomites provide secondary targets (Mesolella, 1973; Bricker and Henderson, 1986). Devonian shales are currently a large supplier of shallow (<650 m) gas resources (Anonymous, 1994). The prospects of future natural gas discoveries in deep (>2500 m) Cambrian and Ordovician sedimentary rocks are also good (McCaslin, 1985). Penn-

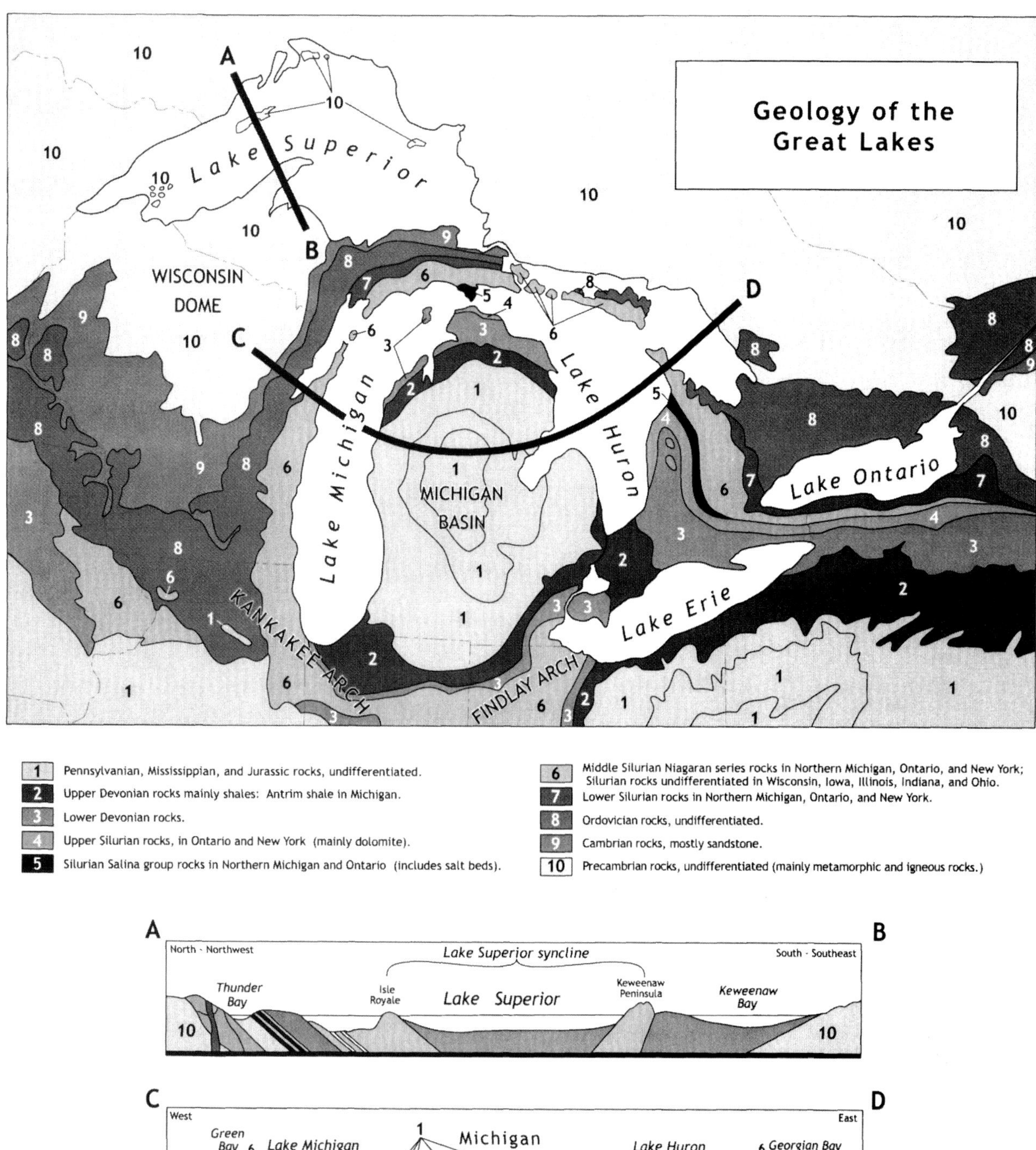
Geology of the Great Lakes
Lake Superior
Lake Michigan
Lake Huron
Lake Erie
Lake Ontario
WISCONSIN DOME
MICHIGAN BASIN
KANKAKEE ARCH
FINDLAY ARCH
A
B
C
D
1 Pennsylvanian, Mississippian, and Jurassic rocks, undifferentiated.
2 Upper Devonian rocks mainly shales: Antrim shale in Michigan.
3 Lower Devonian rocks.
4 Upper Silurian rocks, in Ontario and New York (mainly dolomite).
5 Silurian Salina group rocks in Northern Michigan and Ontario (includes salt beds).
6 Middle Silurian Niagaran series rocks in Northern Michigan, Ontario, and New York; Silurian rocks undifferentiated in Wisconsin, Iowa, Illinois, Indiana, and Ohio.
7 Lower Silurian rocks in Northern Michigan, Ontario, and New York.
8 Ordovician rocks, undifferentiated.
9 Cambrian rocks, mostly sandstone.
10 Precambrian rocks, undifferentiated (mainly metamorphic and igneous rocks.)
North - Northwest
South - Southeast
Lake Superior syncline
Thunder Bay
Isle Royale
Lake Superior
Keweenaw Peninsula
Keweenaw Bay
West
East
Green Bay
Lake Michigan
Michigan
Lake Huron
Georgian Bay

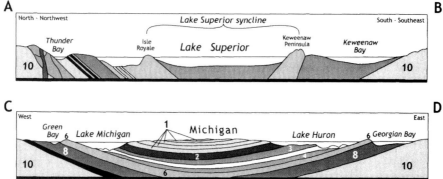

Figure 16.4 Geology of the Great Lakes region.

sylvanian coals, the youngest of the hard rock units in the region, are also widespread across parts of Illinois.

Silurian coral reefs are also highly resistant to erosion. Where these rocks are near the surface, such as at the Thornton Reef near Chicago (McGovney, 1989), they provide an important source of crushed rock and relatively pure $CaCO_3$, raw materials that helped fuel construction in the cities of the southern part of the region by providing the necessary materials for the manufacture of Portland cement (LaBerge, 1994).

Paleozoic salt beds, deposited as evaporites in the Michigan basin, occur in the eastern parts of the region, where they intrude into overlying Silurian rocks. These deposits, some as thick as 1500 m, provide a minable source of rock salt because they are relatively close to the surface as well as to major markets (Guillet, 1984). At one time, over one-third of the rock salt mined in the United States was located in the Great Lakes region (Bleimeister, 1961). The salt is used primarily for road deicing applications (Guillet, 1984). In other locations, caverns are created in the salt beds by brining; these caverns are then used temporarily to store refined petroleum products and natural gas (Wyrick, 1985).

Throughout the region, long periods of subaerial erosion, followed recently by glacial scour, have left only the most resistant rocks standing as positive features. Where quartzite is the bedrock, isolated hills or a series of hills remain as positive landform features. Examples include Rib Mountain in central Wisconsin, the Baraboo Range in southern Wisconsin, the Barron Hills in northwest Wisconsin, the Gogebic Iron Range, and some isolated knobs in Ontario near Lake Superior (Wharton, 1970; LaBerge et al., 1990; fig. 16.1). On the Paleozoic rocks, dolomite often forms ridges, with some of the harder sandstones forming low escarpments. Even where buried by thick deposits of glacial drift, the cuestas of the region form ridges (Rieck and Winters, 1982a). The Niagara dolomite, a particularly hard rock, forms a prominent cuesta throughout the region. Peninsulas such as the Door, Garden, and Bruce are underlain by Niagara dolomite, and all display prominent escarpments (Martin, 1965). This rock also forms the caprock over which pour the waters of Niagara Falls (Gilbert, 1907), and is home to small areas of karst topography where the overlying glacial drift is thin.

16.3 The Legacy of Glaciation

The Great Lakes region's distinctive physical geography is due, more than any other factor, to continental glaciation. Whereas Pleistocene ice sheets clearly formed most of the landscapes observable today, the oldest established "ice age" for this region occurred during Precambrian (Huronian) time (Crowell, 1980). The Gowganda Formation, part of the Huronian Supergroup of rocks, which is extensive across parts of Ontario (Roscoe, 1973), is probably a lithified Precambrian till (tillite; Schenk, 1965; Symons, 1975). In places, it is almost 1500 m thick, and almost everywhere yields rubidium-strontium (Rb-Sr) ages greater than 2.1×10^9 years (Keller, 1973; Symons, 1975). Young (1973) and LaBerge (1994) discuss the Gowganda and other Precambrian tillites that are observable in the Great Lakes region and elsewhere.

More recently, a sequence of Quaternary glaciations have sculpted the landscape, leaving behind landforms of erosion and deposition and drastically altering drainage systems (see chapter 2). Much of our understanding of these glacial landscapes is due to interpretations made by early geologists (e.g., Leverett, 1929; Leverett and Taylor, 1915; Bretz, 1951, 1955; White, 1973). Quaternary ice sheets undoubtedly advanced across the Great Lakes region before the last, or Wisconsinan, glaciation (Baker et al., 1983). Only deposits from Illinoian (oxygen-isotope stage 6) ice, however, have been documented in the region, and even they are at its southern and western fringes (Leighton and Brophy, 1961; Lineback, 1979; Baker et al., 1983; Johnson, 1986b). In most locations in the Great Lakes region, these deposits were later overridden by Wisconsinan ice or buried by Wisconsinan loesses (Rieck and Winters, 1982b; Winters et al., 1986).

The Wisconsinan ice (oxygen-isotope stages 4 through 2) advanced as distinct lobes, each centered around preexisting lowlands associated with major bedrock valley systems that existed before glaciation or with areas of relatively weak bedrock (Horberg and Anderson, 1956; Hough, 1958; McCartney and Mickelson, 1982; Attig et al., 1985; Rieck and Winters, 1991; fig. 16.5). The latter were usually strike valleys associated with shales or weak sandstones. Earlier ice sheets probably took similar paths, although evidence for this hypothesis is sketchy. Along its southern margins, the ice sheet was probably relatively thin and fast moving (Clark, 1992).

In any event, the basins of the Great Lakes owe their origin to both glacial scour and glacial deposition (Farrand, 1969). First, the repeated channeling of ice flow during the many Pleistocene glacial advances led to the development of deep lake basins (Horberg and Anderson, 1956). Parts of the Huron, Erie, and Michigan basins conform to the outcrop pattern of erodable Devonian shales and limestones, and these basins coincided with the movement of glacial ice lobes. Even the Superior basin, which lies almost wholly within the crystalline rocks of the Canadian Shield, is predominantly developed along a structural basin of relatively weak Cambrian and Upper Keweenawan (late Precambrian) sandstones. Second, processes of glacial deposition formed a series of high end-moraine loops on the margins of the lakes. In many instances, especially for Lakes Erie and Michigan, these moraines effectively dam the lakes and create higher lake levels than would otherwise be possible.

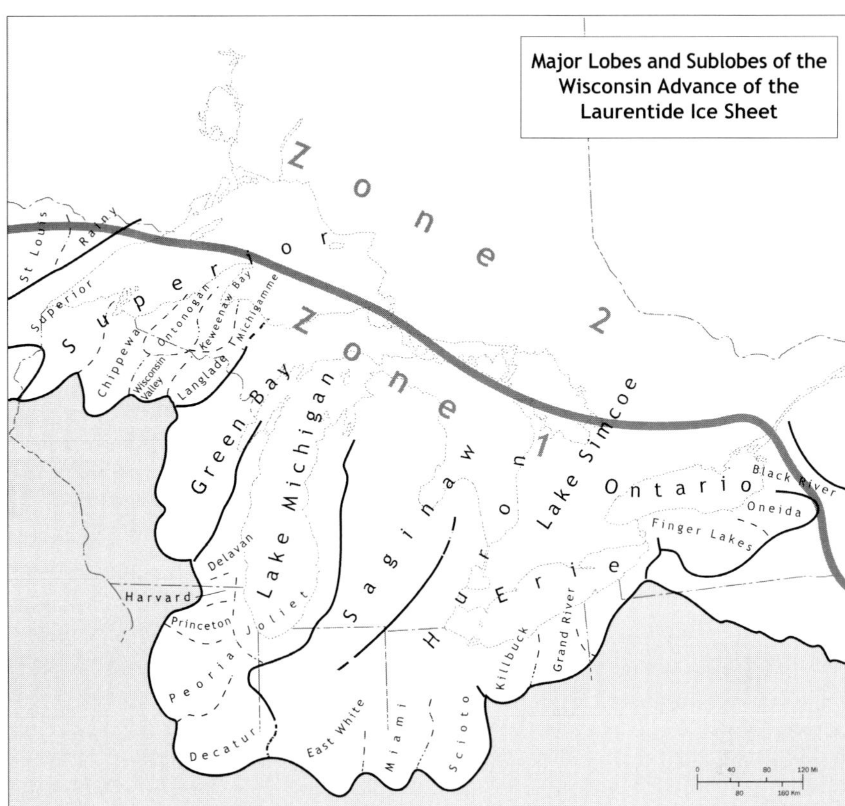

Figure 16.5 Major ice lobes of the Wisconsinan glaciation in the Great Lakes region. Zones 1 and 2 reflect two of Dyke and Prest's (1987) three major landscape zones of the Laurentide ice sheet. Zone 1 is a region of long eskers and ice flow lineaments. Zone 2 has extensive end moraines, both smooth and hummocky, and ice thrust masses, as well as some evidence of lineations (eskers and drumlins). Modified from various sources, including Mickelson et al. (1983), Attig et al. (1985), and Clayton et al. (1991).

The margin of the Wisconsinan ice sheet fluctuated widely as it advanced and retreated across the region (Clayton and Moran, 1982). Between 35,000 and 10,000 B.P. (radiocarbon years before 1950), the ice margin advanced as sublobes that eventually covered most of the region. The evidence for this period of ice-marginal advances and retreats comes from the stratigraphic record, as well as from the cross-cutting relationships of moraines (e.g., Leverett, 1929; Black, 1969; Farrand and Eschman, 1974; Black, 1978; Acomb et al., 1982; Johnson, 1986b; Johnson et al., 1997; Matsch and Schneider, 1986). Generally, though, the ice advanced during oxygen-isotope stage 4, the Altonian advance sometime after 75 ka (thousand years before present), retreated during the Farmdale interstade of stage 3, which terminated around 30–25 ka, and readvanced during later stadials (Dreimanis and Karrow, 1972; Dredge and Thorleifson, 1987; Vincent and Prest, 1987). By around 20,000 B.P. during stage 2, Woodfordian ice had reached its most southerly position in east-central Illinois, after coalescing with ice flowing westward from out of the Huron and Erie basins (Mickelson et al., 1983; Hansel and Johnson, 1992). Across much of Wisconsin and Michigan, the later stages of retreat and its associated readvances were quite rapid, with ice-front positions changing by hundreds of kilometers in only a few hundred years (Saarnisto, 1974; Dyke and Prest, 1987; Maher and Mickelson, 1996). By 9500 B.P., most of the Laurentide ice had withdrawn from the northernmost parts of the region.

Overall, constructional drift landforms are most common in the Great Lakes region. The only large-scale *erosional* landforms in drift are a few tunnel valleys (see subsequent discussion) and minor incisions of river valleys in postglacial times. The thicknesses of glacial drift deposited by the ice vary from zero to a few meters in the northern and northeastern parts of the region to greater than 50 m in the south, with preferred zones of deposition in interlobate areas of 200–400 m (Black, 1969; Rieck and Winters, 1993). Where the drift is thin, it commonly exists as till or glaciofluvial sands over eroded, scoured, and polished bedrock. Erosional features such as chattermarks, glacial grooves, roches mountonnées, and striations are therefore common in the Canadian part of the basin, where the drift is thin or absent (Kaszycki and Shilts, 1979). Here, drift is often preferentially deposited in the lee of resistant bedrock knobs, as crag-and-tail landforms or bedrock-cored drumlins (see chapter 2).

During glacial retreat, meltwaters drained down and annually filled several major river systems that exited the basin, most prominently the Mississippi, Wabash, and Illinois Rivers. Each winter, strong winds blew the silts that had accumulated in these valleys onto the uplands, whereas sands were left behind as dunes within the valleys proper. The end result was widespread deposits of windblown silt (loess) across uplands in the southern and western parts of the region (Leighton and Willman, 1950; Flint, 1971; Leigh, 1994; see also chapter 3). Loess deposition is associated most with periods of ice retreat (McKay, 1979). The loess deposits thin and became more finely textured farther from the valleys (Ruhe, 1984). In many Great Lakes landscapes, the uppermost surficial deposit is a late Wisconsinan loess, the Peoria loess, although in the northern and eastern parts of the region, loess is conspicuously thin or absent. In these areas, meltwaters ponded in front of the ice, allowing silts to settle out. Rivers that exited from these proglacial lakes, such as the Grand River in Michigan or the Ottawa River in Ontario, were therefore very poor loess sources. Soils formed in the calcareous and nutrient-rich loess are typically highly fertile, though easily erodible by both wind and water.

Looping end (recessional and terminal) moraines are perhaps the most distinctive and widespread glacial features in the region (Blewett, 1991; fig. 16.6). In the south, they are wide, low, and subdued features (Frye and Willman, 1960; Willman and Frye, 1970), perhaps because the ice sheet was melting (i.e., a wet bed glacier) as it advanced (Mickelson, 1987). To the north, end moraines are narrower, higher, and much more "hummocky," possibly due to frozen bed conditions during deposition or to the rapid sloughing of ice blocks into the drift. The interlobate moraines of the region are particularly irregular, with numerous short, steep slopes and many isolated, closed depressions or kettles (Black, 1969; Winters and Rieck, 1980). To be sure, some of these features, previously interpreted as active ice moraines, have been shown to be palimpsest (Totten, 1969) or ridges formed as stagnant ice melted, otherwise known as "heads of outwash" or morphosequences (Blewett and Rieck, 1987; Blewett and Winters, 1995). Land use on these moraines is less intensive than on other parts of the landscape, because of the potential for soil erosion on the steep slopes and the sandy or rocky character of the soils. Thus, forests often dominate morainic landscapes in the Great Lakes region.

The Great Lakes region also contains many vast and nearly featureless glaciolacustrine plains (Karrow and Calkin, 1985; fig. 16.6). As the ice receded, proglacial waters ponded between the ice margin and the higher ground beyond, forming an extensive series of proglacial lakes (Goldthwait, 1908; Bretz, 1951; Hough, 1963; Futyma, 1981). Often, these landscapes consist of wave-beveled till plains, in which case the soil texture can vary from clayey to sandy, or infilled plains with deep accumulations of clayey and silty sediments (Feenstra, 1979). Shorelines, most of which have been altered and tilted by isostatic rebound of the region after deglaciation (Tushingham, 1992), provide a valuable source of information on the spatial and temporal changes that occurred in these lakes (Farrand, 1969; Hansel et al., 1985; Larsen, 1985, 1994; Karrow, 1987). Additionally, landforms such as spits, deltas, and abandoned outlets, subsurface (lake bottom) stratigraphic information, and innumerable radiocarbon dates have been used to fill in the highly complex history of the Great Lakes over the last 16,000 years (Eschman and Karrow, 1985; Hansel et al., 1985; Colman et al., 1994).

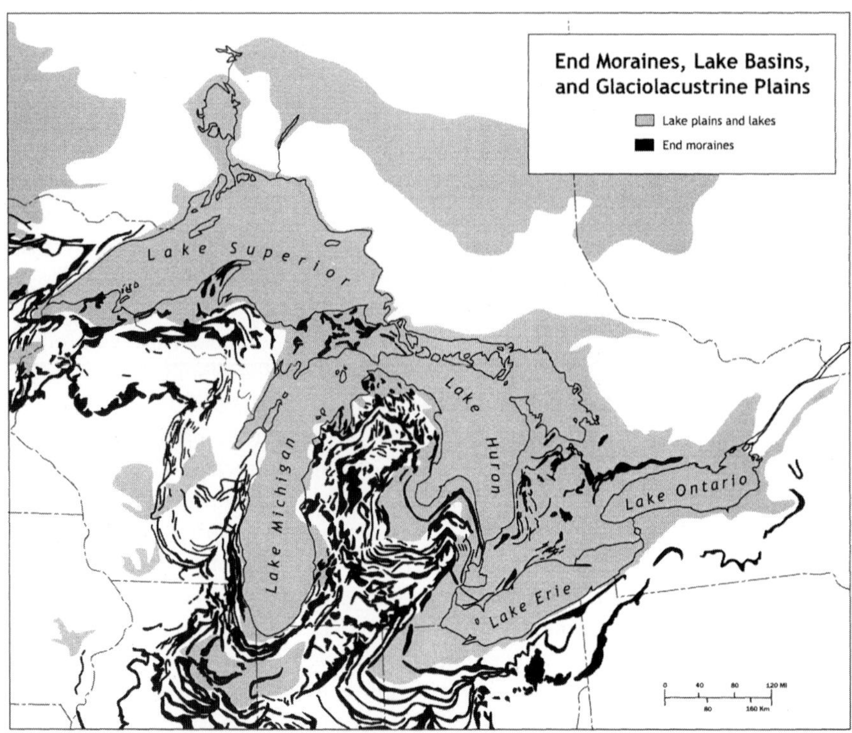

Figure 16.6 Major end moraines and glaciolacustrine plains of the Great Lakes region. In the lower St. Lawrence Valley, some of the area designated as lake plain is, in actuality, marine plain.

Located in central Wisconsin, glacial Lake Wisconsin is a good example of a large proglacial lake that was bounded by high topography on three sides and the ice front or its moraine on the other (Martin, 1965; Clayton and Attig, 1989). The sediments that comprise the former bed of this lake are today flat, wet, and sandy; they are used primarily for forestry and for cranberry production (Hole, 1976). Where the waters in proglacial lakes were deep, however, glaciolacustrine deposits tend to be clayey and can even be varved (Ernst and Hunter, 1987).

Landforms associated with glacial meltwater (glaciofluvial systems), such as outwash plains, eskers, and kames, are extremely commonplace in the Great Lakes region. Broad outwash plains, often containing abundant sand and gravel resources, dominate many Great Lakes' landscapes, especially in the northern parts of the region (Brubaker, 1975). In the south, they are frequently covered by a blanket of loess, rendering the character of the soils on them silty rather than sandy. These landscapes formed as meltwaters, flowing away from the ice, lost competence and deposited sands and gravels. Meltwaters issued forth by running off the ice or by exiting through subglacial tunnels; in the latter scenario, outwash fans may also have formed as the waters exited the ice margin (Attig et al., 1989). Many outwash plains are pitted with kettles and lakes, as a result of slump following the melting of partially buried ice blocks. Hence, the many lakes of southern Ontario, northern Wisconsin, Minnesota, and Michigan owe their origins to the rapid melting of the ice sheet, which left isolated fragments of ice out in front of the glacier (Scott, 1921; Attig et al., 1989). These ice blocks quickly became partially buried as meltwaters continually deposited sediments in the low areas between the blocks. In some regions, the rapid melting of the ice sheet led to widespread stagnation, such that rapid deposition of glaciofluvial sediments by debris-rich waters occurred between ice blocks (Johnson et al., 1995). This scenario produced classic kame and kettle, or hummocky, topography, with hills composed of varying combinations of flow till and stratified sands and gravels. Eskers are also common on Great Lakes landscapes where ice stagnation is inferred (Winters, 1961). Where stagnation combined with debris-poor waters, ice-walled lakes formed, ultimately producing plains when the lakes drained.

Drumlin fields are found throughout the Great Lakes region (Mickelson et al., 1983; fig. 16.7). The Green Bay lobe of the Woodfordian ice produced perhaps the largest drumlin assemblage, as it spread out of the lake basin and onto the uplands of southeastern Wisconsin (Borowiecka and Erickson, 1985). Large fields in southeastern Ontario (>6000 drumlins, Trenhaile, 1971), western New York, northwestern lower Michigan, and northeastern Minnesota are also noteworthy (Bergquist, 1941; Putnam and Chapman, 1943; Wright, 1957; Feenstra, 1979; Harry and Trenhaile, 1987; Francek, 1991). As was the case with the southeastern Wisconsin field, most of these fields exist downglacier from a deep lake basin, suggesting that a rapid, ascending ice advance favors the

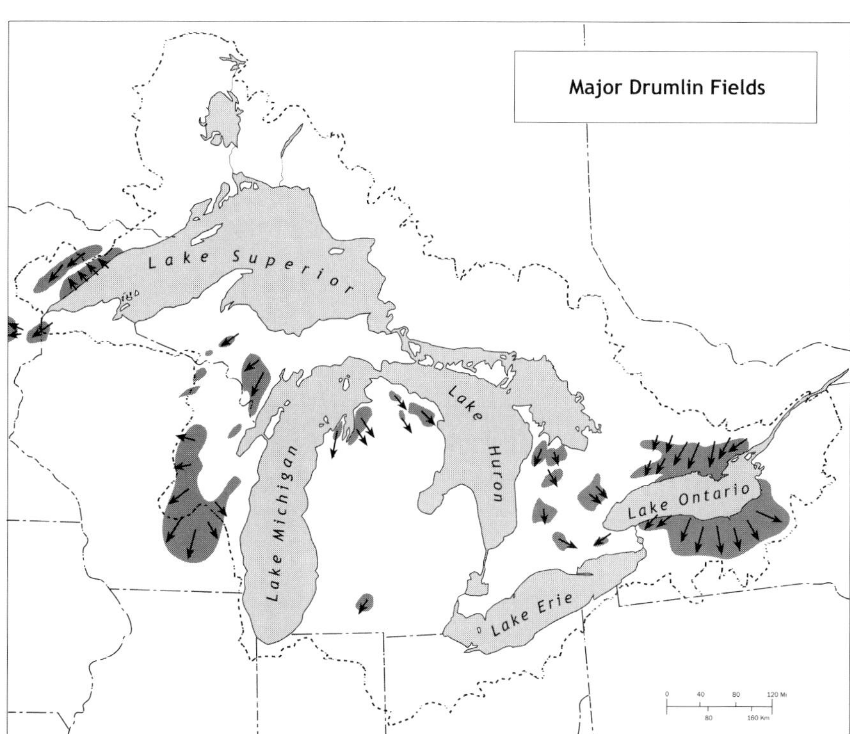

Figure 16.7 Major drumlin fields of the Great Lakes region. Modified from various sources, including Bergquist (1941), Borowiecka and Erickson (1985), Francek (1991), Putnam and Chapman (1943), Trenhaile (1971), and Wright (1957).

formation of these features or at least alters their form by affecting ice flow.

Permafrost conditions and periglacial activity were common on landscapes that were near the ice margin (Johnson, 1986a; Attig et al., 1989); some of the evidence for these processes comes from areas near, but outside of, the region (Smith, 1949). However, since permafrost conditions were short lived and confined spatially to a narrow zone near the ice front, they produced little surface expression and are difficult to find today.

16.4 Soils and Surficial Sediments

Soils of the Great Lakes region have formed primarily in transported sediments such as glacial drift, loess, dune sand, and alluvium (fig. 16.8). Only rarely can we find soils formed wholly in residuum (weathered bedrock), for in most cases, where bedrock is close to the surface, the weathered portions have been eroded either by ice or by running water. Soil parent materials in the southern half of the region are generally carbonate and nutrient rich, because most contain admixtures from glacially comminuted limestones. Most sediments also contain a significant component of detritus derived from the crystalline rocks of the Canadian Shield, and thus soils of the region are often high in micas and feldspars, and contain abundant amounts of other iron-rich minerals. In most locations south of outcropping dolomitic Paleozoic rocks, the soils contain high amounts of finely ground carbonates and are, therefore, strongly calcareous, for example, in Green Bay lobe deposits and eastward (Hole, 1976; Franzmeier et al.,

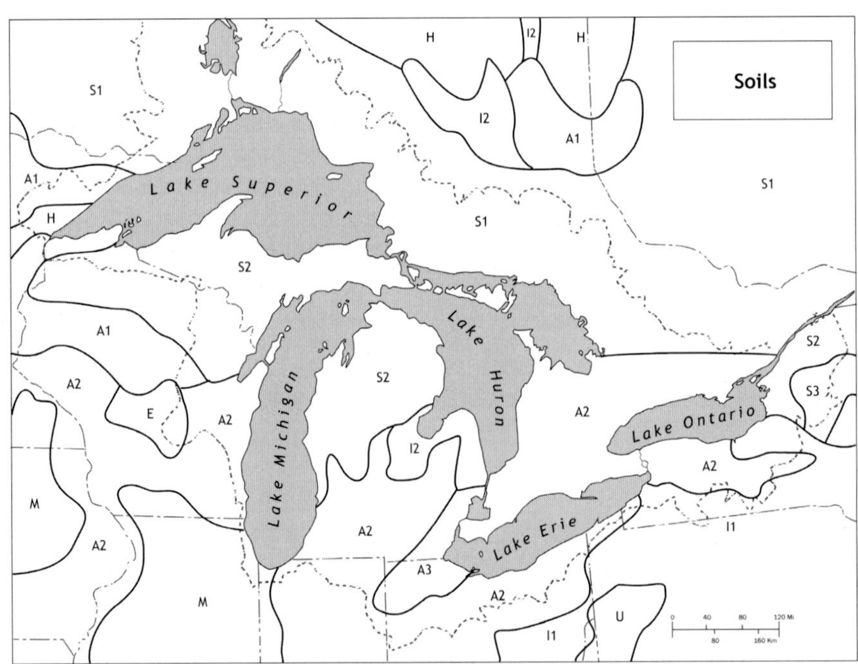

Figure 16.8 Major soils of the Great Lakes region, modified from U.S. Geological Survey (1987) and Agriculture Canada (1977):

- A1: Boralfs on loamy parent materials, with lesser amounts of Histosols and Psamments.
- A2: Alfisols on loamy parent materials, some wet.
- A3: Alfisols on loamy and clayey parent materials, mostly wet.
- E: Sandy Entisols on flat, wet landscapes.
- H: Organic soils (Histosols) overlying thin deposits of glacial drift, or bedrock.
- I1: Inceptisols on steeply sloping, bedrock-controlled landscapes.
- I2: Inceptisols on flat, wet landscapes.
- M: Mollisols, some wet.
- S1: Spodosols on sandy parent materials.
- S2: Spodosols on sandy parent materials, and Alfisols on loamy parent materials.
- S3: Spodosols, some wet.
- U: Ultisols and Alfisols on moderately sloping, bedrock-controlled landscapes.

1985; Schaetzl et al., 1996). Leaching of the carbonates and concomitant lowering of the pH is a prerequisite to many pedogenic processes in the region; hence, to the north and west of this "dolomite limit," where soil parent materials are inherently acidic, weathering and soil formation, especially those processes associated with fragipan formation or the translocation of clay, organic matter, iron, and aluminum, are generally accelerated (Olson and Hole, 1967; Habecker et al., 1990).

Soil patterns generally reflect soil-forming processes, which, depending on the scale of analysis, are conditioned by various soil-forming factors (see chapter 9). Regionally, soils reflect climate and vegetation patterns (Technical Committee on Soil Survey, 1960; Fehtenbacher et al., 1967; Schaetzl and Isard, 1996). From south to north across the region, as the climate gets colder and more snowy and as the vegetation changes from deciduous forest to mixed coniferous-deciduous forest, types and rates of soil processes change. In the south, weathering is faster, enhancing clay formation and leading to Alfisols with clay-rich Bt horizons in generally more finely textured parent materials (tills and loesses) (fig. 16.8). In the southwestern part of the region, the mid-Holocene invasion of base-cycling grasses, which replaced forested communities of the early Holocene (Webb et al., 1983), has maintained high pH levels and organic matter contents, leading to the formation of Mollisols with dark, thick A horizons (Smith et al., 1950; Wascher et al., 1960; fig. 16.8). Farther north, podzolization is more common on the increasingly more dominant, sandy parent materials (Gardner and Whiteside, 1952; Gaikawad and Hole, 1961; Schaetzl and Isard, 1991). Here, the coniferous character of the forest litter, combined with the tendency for the parent materials to be noncalcareous, creates acidic soil pH values that promote the translocation of iron, aluminum, and organic matter. Continuous infiltration of snowmelt has also been implicated as an important vector of soil development in snowbelt areas, where podzolization is especially prominent and Spodosols are common (Schaetzl and Isard, 1991, 1996).

Parent material affects soil patterns both regionally and locally. At the regional scale, the texture of parent materials changes from silty in the west and south, due to proximity to loess sources, to loamy in some south-central areas where loam tills are commonplace, to predominantly sandy and gravelly on the outwash deposits and dunes in the north (fig. 16.8). Sandy soils are ubiquitous on dunes around the shores of the Great Lakes (Olson, 1958) and on small inland dunes on sandy lake plains or outwash deposits (Clayton and Attig, 1989; Arbogast et al., 1997). Where soils are extremely calcareous (e.g., contain high amounts of limestone fragments), the high pH conditions promote calcium-humus bonds that tie up organic matter and lead to the formation of thick, dark A horizons, regardless of vegetation type (Gaikawad and Hole, 1965; Schaetzl, 1991). Entisols, which have thin profiles and are formed in a variety of sediments, are commonplace across the northernmost parts of the region, where glacial erosion has been dominant, drift is thin, and parent materials are the youngest. At wet sites, where high water tables have inhibited soil formation, wet Entisols are found. The very wettest sites exhibit organic matter accumulations of meters or more as slowly decomposing muck or peat deposits, now mapped as Histosols (Miller and Futyma, 1987). In the northern part of the region, lack of an integrated drainage net (i.e., many wet, poorly drained sites) coupled with cool temperatures that reduce decomposition has led to widespread areas of organic Histosols (Gates, 1942). In Michigan alone, over 1.6 billion tonnes of peat are estimated to exist in bogs and swamps (LeMasters and Jones, 1984), whereas Minnesota and Ontario contain still greater resources of this kind. Clayey parent materials occur on many of the glaciolacustrine plains associated with proglacial lakes. These landscapes are usually quite featureless and flat, and the soils are poorly drained. Many of these sites have since been drained by subsurface tiles and now comprise some of the best agricultural land in the region.

In general, soils in the region are considered young and are minimally weathered, with nonacid pH values and moderately high base saturation levels. Nutrient contents are high, especially in the soils developed under base-cycling vegetation such as grasses or some deciduous tree species. Favorable parent materials, such as loess or loam tills, are widespread across much of the southern half of the region. Although soil erosion is problematic on steeply sloping sites, many of these sites remain forested and are thus at little risk.

16.5 Climate

The location of the Great Lakes, within a lowland corridor midway between the equator and the North Pole, is central to its climatology. The corridor extends from the Gulf of Mexico to the Arctic Ocean and is bounded on the west by the Rocky Mountains and on the east by the Appalachian Mountains. The corridor broadens to the north, and the northeast portion of the wedge opens to the Atlantic Ocean. This topographic configuration provides a pathway for the frequent exchange of warm, humid subtropical and cold polar air masses.

The climate of the Great Lakes region is chiefly governed by latitude, continental location, large-scale circulation patterns, and the lakes themselves (Eichenlaub et al., 1990; fig. 16.9). Because virtually all atmospheric processes are driven by energy from the Sun, the region's latitude influences variations of incoming solar radiation in two important ways. First, its midlatitude position dictates that the annual cycle of solar radiation, and thus air temperature, has a large amplitude; the solar energy input is five to six times greater in early summer (≈24.1 MJ m^{-2} d^{-1}, or megajoules per square meter per day) than in early winter (≈4.3

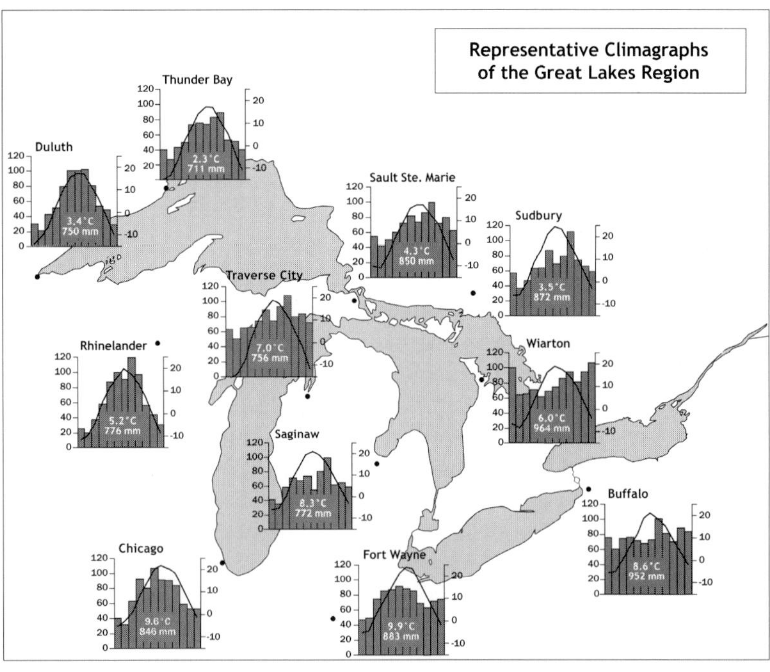

Figure 16.9 Representative climagraphs of the Great Lakes region. Months, from January through December, on the *x*-axis. Precipitation, in mm, on the (left) *y*-axis. Temperature, on the (right) *y*-axis, is in degrees Celsius.

MJ m^{-2} d^{-1}) (Bennett, 1975). Second, its location midway between the equator and the North Pole means that the region's 10° latitudinal spread results in a substantial difference in day length (≈1.5 hr on the December solstice) and solar inclination, and thus a 25% difference in solar energy input, between its southern and northern borders during winter. Land masses respond more rapidly to changes in solar energy inputs than do water bodies; consequently, the region's interior location on the continent enhances the seasonal air temperature variations due to latitude alone (Kupec, 1965). As a result, the mean monthly temperature for January is 17°C warmer at the region's southern borders than at its northern limits (Phillips and McCulloch, 1972). The difference in mean monthly air temperature across the region is much less during summer (e.g., 7°C in June) when south-to-north differences between day length and solar input are much less (Phillips and McCulloch, 1972; fig. 16.10a, b).

Large-scale atmospheric circulation also has an important control on the climate of the Great Lakes region and is primarily responsible for its interannual variations. During January, air masses that originate over the Pacific Ocean, Canadian Arctic, Atlantic Ocean, and Gulf of Mexico dominate the weather about 75%, 20%, 5%, and <1% of the time, respectively (Phillips and McColloch, 1972; see also chapter 5). However, by the time Pacific basin air masses reach the Great Lakes, their temperature and moisture characteristics often resemble those of continental origin (e.g., Schwartz, 1991). Air masses of Pacific origin also dominate in the region during summer; however, the frequency of air masses from the Gulf of Mexico ranges from 40% in the south to 10% in the north, and occasionally, cool arctic air masses and hot, dry air masses from the southwestern United States penetrate the region (Phillips and McColloch, 1972).

Low-pressure storm systems (extratropical cyclones), characterized by cloudy skies, windy conditions, and precipitation, often form along the frontal boundaries and track beneath the polar jet stream, across the Great Lakes. The position of the jet stream shifts with the seasons; it is typically north of the region during summer and south of it during winter (Barry and Chorley, 1987). Because cyclones traverse the Great Lakes region every 3 to 4 days on average (Angel and Isard, 1997a), they are responsible for a large proportion of the day-to-day variation in weather and have important impacts on shipping (Barcus, 1960), lake temperature profiles and associated vertical distributions of oxygen, pollutants, and biota (McCormick, 1990), ice cover (Assel, 1991), and wave and current motions and their associated shoreline damages (Changnon, 1987).

Extratropical cyclones are also the primary cause of precipitation in the Great Lakes region (Rodionov, 1994). Cyclones that originate over the Pacific Ocean and traverse the region typically precipitate most of their moisture as they cross the high mountains of western North America. However, these cyclones, as well as those that form to the lee of the Rockies, are usually accompanied by large-scale counterclockwise circulations that advect warm, moist subtropical air masses poleward through the corridor between the Rocky and Appalachian mountains. As this subtropical air crosses the Great Lakes region, it often mixes with colder polar air to produce ample precipitation. Occa-

a

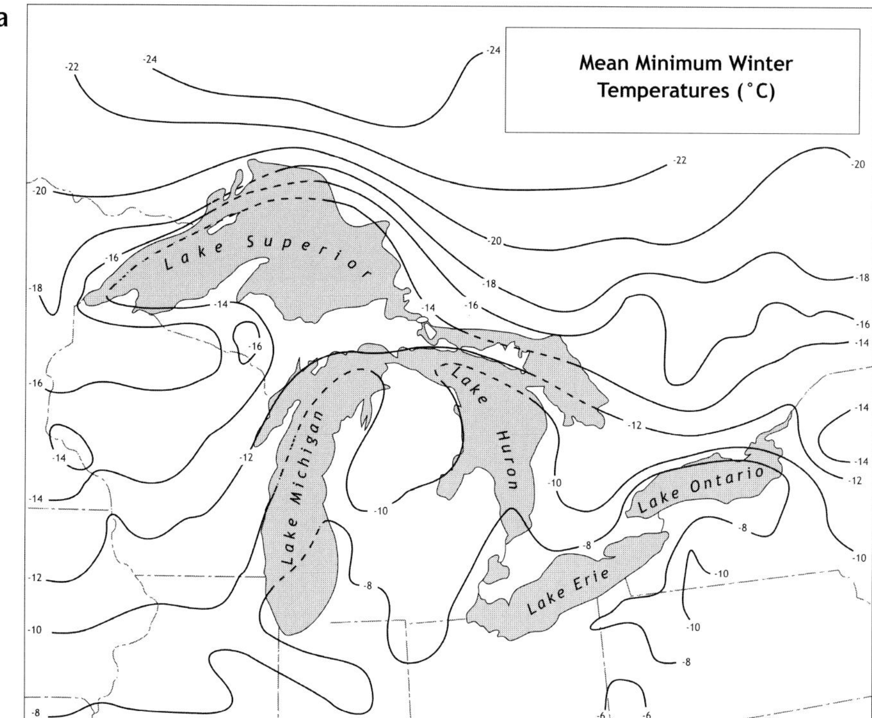

b

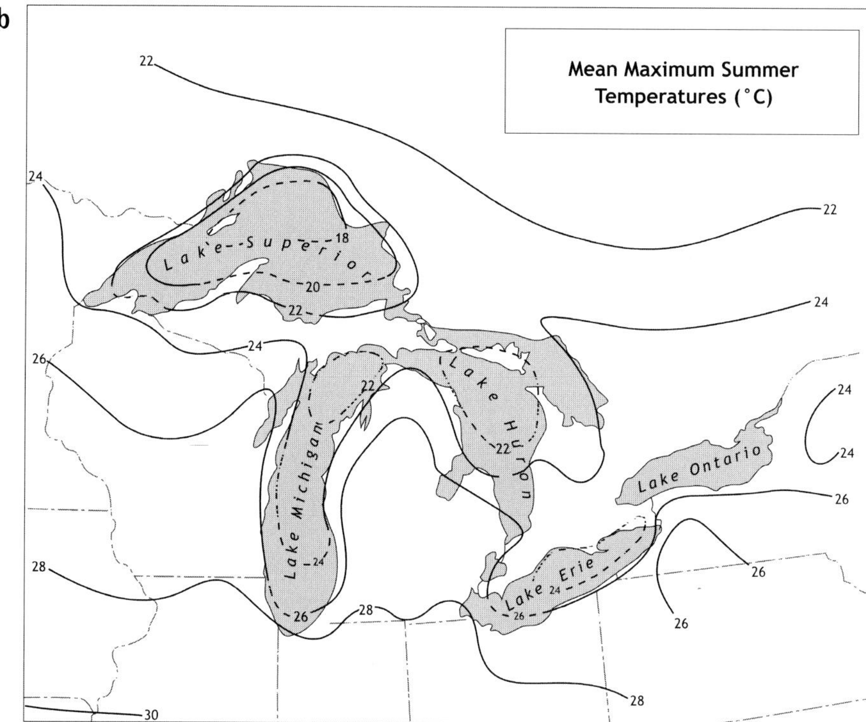

Figure 16.10 Climate maps of the Great Lakes region. a. mean minimum temperature computed as the mean of the daily minimum temperatures for December through February. Modified from Scott and Huff (1997); b. mean maximum temperature computed as the mean of the daily maximum temperatures for June through August. Modified from Scott and Huff (1997).

c

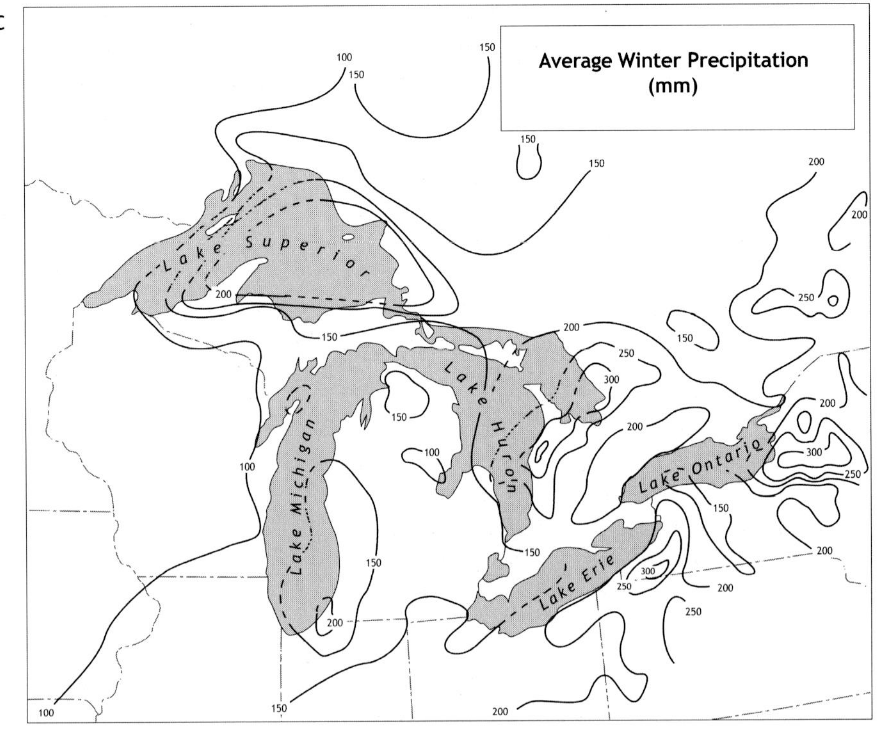

d

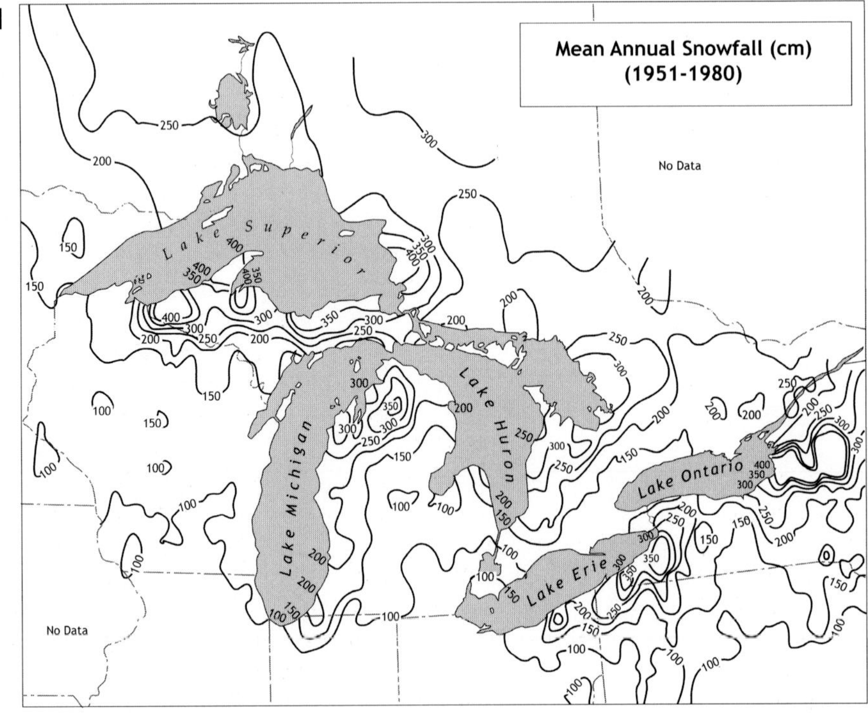

Figure 16.10 (*continued*) c. average winter precipitation. Modified from Scott and Huff (1997); and d. average snowfall, 1951–80. Modified from Norton and Bolsenga (1993).

sionally, cyclones that track northward along the east coast of North America also bring significant "wraparound" precipitation westward from the Atlantic Ocean to the eastern portion of the region (Kunkel et al., 1993).

During winter, when the polar jet stream is equatorward of the Great Lakes, frigid air masses from the Arctic can flow southward through the corridor, plunging air temperatures 30°C in 24 hours, to –40°C or lower, in northern portions of the region (U.S. Weather Bureau, 1959; Phillips and McCulloch, 1972). As the jet stream contracts around the North Pole in summer, the clockwise circulation of air associated with the semipermanent high-pressure system located over the subtropical Atlantic Ocean increasingly affects the weather. These air flows transport large amounts of heat and moisture poleward and are responsible for the high frequency of summer thunderstorms and warm humid days with air temperatures that occasionally reach 40°C in the southern portion of the region (U.S. Weather Bureau, 1959; Phillips and McCulloch, 1972).

Ice generally covers a portion of the Great Lakes between mid-December and late April (Saulesleja, 1986), but even in the coldest winters of record, the ice covers on Lakes Michigan, Huron, and Ontario have not been complete (Assel et al., 1983). A unique exception occurred on 17 February 1979, when all of the Great Lakes were nearly 100% ice covered (DeWitt et al., 1980). Normally, ice cover for the Great Lakes combined is approximately 17% during early January, reaching a high of 64% during late February and decreasing to 11% in early April (Angel and Isard, 1997b).

Over the Great Lakes basin as a whole, precipitation is maximal in June; February is the driest month (Pentland, 1968). The average number of days with thunderstorms decreases from more than 40 yr^{-1} south of Lake Michigan to about 25 yr^{-1} in the northern part of the region. Hail, a springtime phenomenon throughout much of the region, occurs on average twice a year, except on the lee shores of the lakes where most of the hail is graupel, a nondestructive soft hail, and the frequency of hail storms is 5 yr^{-1} with an autumn maximum (Changnon, 1978). Tornadoes are most frequent during spring and early summer and decrease in frequency from southwest to northeast across the region (e.g., Ojala and Ferrett, 1993) with a maximum of 5–6 yr^{-1} per 160-km^2 area in northern Illinois and Indiana to less than 0.5 yr^{-1} poleward of 44°N latitude in the region (Kelly et al., 1978). On average, the number of winter days with freezing rain exceeds 9 yr^{-1} throughout the region, reaching a maximum of more than 12 yr^{-1} in western Michigan and around Lake Superior (Eagleman, 1990).

The lakes themselves, because of their large surface area, have an important impact on the climate of the region (Changnon and Jones, 1972; Scott and Huff, 1997). These effects give the region its unique character. The modifications of larger-scale atmospheric conditions by the Great Lakes are collectively called *lake effects* and result from physical processes that occur within the atmosphere's surface boundary layer (Phillips, 1972; Wylie and Young, 1979). Lake effects stem from the water surface that the lakes expose to the atmosphere and from the contrasting physical properties of land and water that can result in pronounced surface temperature gradients at the shoreline. The most widely known climatic influence of the Great Lakes involves lake effect snowfall (fig. 16.10c, d); however, the lakes influence many other aspects of climate, including air temperature, summer precipitation, cloud cover, local winds, and cyclones (Thomas, 1964; Changnon, 1968; Norton and Bolsenga, 1993).

The mean annual air temperature and total precipitation in the Great Lakes region varies from about 2° to 10°C and from 700 to 1,000 mm, respectively (fig. 16.10). Typically, stations to the north and west exhibit a larger seasonal air temperature range than those to the south and east, whereas lakeshore stations have a smaller annual air temperature range than those farther from the lakes (Kunkel et al., 1993; fig. 16.9). As a result, the mean annual frost-free period ranges from approximately 180 days on the south shores of Lake Ontario to less than 80 days in the northern portion of the region (Fuller et al., 1995). Variations in precipitation amount among months are relatively small for stations in the southeast portion of the region and increase dramatically to the northwest. In the central and northern portions of the Great Lakes region, average monthly precipitation is greatest in late summer and early autumn (August and September), although there is also evidence of a bimodal profile of rainfall during the warm season (Trewartha, 1981; Keables, 1989). The gradients of air temperature and precipitation across the region are generally steeper in winter than summer.

Effects of the lakes on mean maximum summer and mean minimum winter air temperatures and total winter precipitation for the region are pronounced (fig. 16.10a, b). During both summer and winter, the lakes, and associated cloudiness, act to reduce mean maximum and increase mean minimum air temperatures. In contrast, effects on precipitation change with the season: the cool lake waters enhance atmospheric stability in summer, thus decreasing precipitation, and in winter, the warmer surface waters provide moisture and heat to the air passing over them, causing lake-effect snowstorms. The lake effects on temperature and precipitation extend in all directions during the summer, whereas in winter, the effects are most pronounced to the south (for air temperature) and to the east and south (for precipitation).

The lake effects on air temperature and precipitation during winter are generally greatest to the lee of those Great Lakes that present a long fetch to the wind (Reinking et al., 1993). As cold, arctic air masses sweep over the lakes, they warm from below and become saturated causing steam fog. The unstable air rises forming cumuliform clouds and producing heavy showers of snow. As this air moves over the lee shores of the lakes, it is lifted further by low hills and by convergence as air slows down over the rougher terrain (Strommen and Harman, 1978). Lake-effect snowfalls can account for 30% (Eichenlaub, 1970) to 60% (Braham and Dungey, 1984) of the total snowfall at these locations.

Cloud cover is also affected by the Great Lakes (Danard and McMillan, 1974) and in large part causes the contrast in air temperatures between inland locations on windward and leeward sides of the lakes. In general, the spatial and temporal variations of cloud cover in the region are similar to those for precipitation (Scott and Huff, 1997). The lake effect on cloud cover is most pronounced during winter, and as a result cloudiness is enhanced and extensive within the region. During summer, cloud cover is slightly reduced above and near the lakes. Clear skies occur directly over the lakes only about 20% of the time in winter and are more frequent over the eastern than western Great Lakes (Kristovich and Steve, 1995). Fog is more frequent in the Great Lakes region (≈30 days yr^{-1}) than at continental locations of similar latitudes in North America (Phillips and McCulloch, 1972). Nocturnal radiative cooling often produces fogs away from the lakes, whereas advection fogs during spring to early autumn and steam fogs during fall are common at lakeshore locations.

The Great Lakes have a pronounced influence on near-surface winds. Land-to-lake reduction in surface friction results in increased wind velocities and wind direction changes over the Great Lakes, whereas convergence (piling up) of air occurs on the downwind shore (Changnon and Jones, 1972). During the warm season when high pressure dominates the Great Lakes region, the contrast in temperature between lake and adjacent land surfaces can create horizontal air pressure differences, which in turn may produce local winds. The lake-land breeze circulation is the most prominent local wind in the Great Lakes region. This local circulation moderates warm summer temperatures over land near the shoreline (Estoque, 1981) and promotes the occurrence of high air pollution in cities adjacent to the lakes. Lake breezes that penetrate 80 km or more inland with lines of cumulus clouds and showers at the lake-breeze front are common in the region during calm, sunny days in summer (Harman and Hehr, 1972). At night, this pressure difference is sometimes reversed, causing a gentle land breeze on the shores of the Great Lakes (Simpson, 1994).

The Great Lakes can alter the speed and strength of passing cyclones by contributing latent and sensible heat, which intensify vertical motions within these storms (Sousounis and Fritsch, 1994). Many cyclones accelerate as they approach the region and deepen over the lakes (Angel and Isard, 1997b). The effect is most pronounced in late autumn when the lake surfaces are ice-free and warmer than the overlying air.

16.6 Vegetation

The Great Lakes region lies in a complex transition zone among the grasslands of the Great Plains, the mesic, broadleaf deciduous forests of the Ohio Valley, and the boreal, evergreen coniferous forests of the southern Canadian Shield (Braun, 1950; Shelford, 1963; fig. 16.11). The grassland-for-

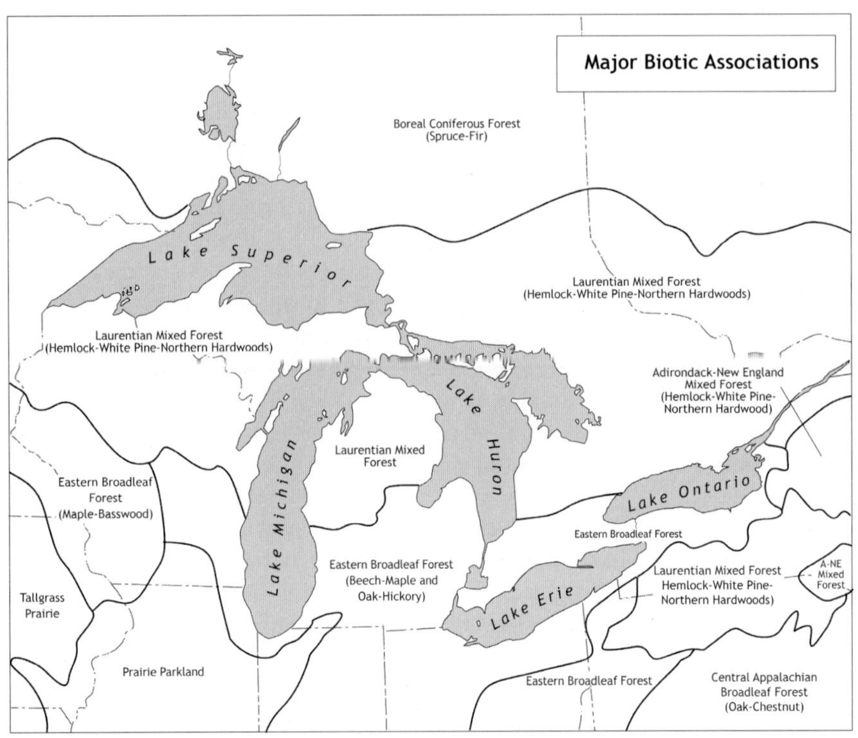

Figure 16.11 Natural vegetation assemblages of the Great Lakes region. Modified from Shelford (1963), and Bailey (1994).

est ecotone runs through the western and southwestern fringe of the region; grasslands in the west yield to oak savanna or aspen parkland, and finally to closed forest in the east (Barnes, 1989; Thomas and Anderson, 1990). The deciduous-coniferous forest ecotone is a broad band that runs from the northwest to the east-central parts of the region and is distinctive enough that it is known as the Lakes Forest (Whitney, 1987) or Laurentian Mixed Forest (McNab and Avers, 1994).

At the time of European settlement, the Great Lakes region was dominated by forest vegetation (Nichols, 1935; Curtis, 1959; Greller, 1988). Only on dry, flat landcapes in the western and southwestern parts of the region were there any sizeable expanses of tallgrass prairie; these areas represented the easternmost extension of the "prairie peninsula" (Curtis, 1959; Kilburn, 1959; Manogaran, 1983). On these landscapes, even though the climate and soils favored forest growth, frequent fires spreading eastward from the grasslands of the Great Plains inhibited the growth of woody plants, allowing grasses and forbs to flourish (Gleason, 1913; Clark, 1993). Fire frequency and intensity were also favored by the slightly drier climatic conditions and even occasional droughts of the western parts of the region (Heinselman, 1981; Manogaran, 1983; Grimm, 1984). In southwestern Michigan and parts of Wisconsin, Ohio, and Indiana, fires were favored on dry, flat, and sandy sites, where small outliers of grassland were found amidst an otherwise continuous canopy of forest (Gleason, 1917; Veatch, 1927; Curtis and Greene, 1949; Brewer et al., 1984). Today, these former prairie sites comprise some of the best agricultural lands in the region (Fenton, 1983).

The forests of the Great Lakes region vary in character based on north–south climatic and edaphic gradients. Broadleaf deciduous forests in the south yield to mixed coniferous-deciduous forests in the north, which in turn grade into boreal (coniferous) stands north of Lake Superior (Küchler, 1964; Merz, 1978; McNab and Avers, 1994). The boundaries between these three major forests are marked by distinct ecotones or floristic "tension zones" (Curtis, 1959). Both tension zones mark the southern limit of many tree species with more northern affinities and, conversely, the northernmost limit of many trees with more southern affinities (fig. 16.12). These floristic boundaries are driven primarily by climate, though in places soils also change abruptly at these locations (Curtis, 1959; Medley and Harman, 1987; Dodge, 1989; Schaetzl and Isard, 1991).

In the south, mesophytic deciduous forests are dominated by maple (*Acer* spp.), basswood (*Tilia americana*), yellow birch (*Betula alleghaniensis*), and beech (*Fagus grandifolia*). On drier sites, oak (*Quercus* spp.), hickory (*Carya* spp.), and chestnut (*Castanea dentata*) are or were more dominant. Farther north, conifers in the *Pinus*, *Abies*, and *Picea* genera become increasingly dominant, especially on dry, sandy and wet, organic soils (Pregitzer et al., 1983, Whitney, 1986; Host et al., 1987; Palik and Pregitzer, 1992; Barrett et al., 1995; Schaetzl and Brown, 1996). Within the Laurentian Mixed Forest vast, nearly pure stands of tall white pine (*Pinus strobus*) once existed. These stands fueled the economy of the Great Lakes region for much of the later nineteenth century, as the region became the top lumber producer in the United States (Larson, 1949; Eastman, 1986; Whitney, 1987). In Canada, north and east of Lake Superior, boreal forests of spruce (*Picea* spp.), fir (*Abies* spp.) and aspen (*Populus* spp.), with some larch (*Larix laricina*) and paper birch (*Betula papyrifera*), extend northward uninterrupted for hundreds of kilometers (Hare, 1950; Carleton and Maycock, 1978; Janke et al., 1978; Elliot-Fisk, 1988; see also chapter 14).

Locally, edaphic controls such as soil texture, which impacts water-holding capacity, nutrient status, pH, and depth to bedrock or to a water table dramatically impact forest composition (Wilde, 1933; Dodge and Harman, 1985; Medley and Harman, 1987; Barrett et al., 1995). Schaetzl and Brown (1996) provide data on the importance of soil wetness to forest composition in some presettlement stands; similarly, Whitney (1986) discusses the importance of texture. Generally, conifers are able to outcompete broadleaf trees on the poorest sites (e.g., nutrient-poor sands, wet sites, shallow soils, cold areas) in the northern parts of the region.

Superimposed on these regional and local patterns is a mosaic of disturbance, with fire and windthrow being two of the most common disturbance vectors (Heinselman, 1981; Leitner et al., 1991). On sandy sites, frequent fires historically have led to a type of "fire climax" forest of pines (*Pinus* spp.). Infertile, frequently burned sites in the northern parts of the region are dominated by jack pine (*Pinus banksiana*) or aspen (*Populus* spp.) (Abrams et al., 1985; Weber, 1987; Palik and Pregitzer, 1992), whereas more fertile but sandy sites once supported majestic white pine (*Pinus strobus*) forests before the extensive logging operations of the late nineteenth century (Maissurow, 1935; Jacobson, 1979; Whitney, 1986). Windthrow, the felling of trees by high winds, and snowdown, collapse under the weight of snow, lead to forest disturbances on a variety of scales, from single tree canopy gaps to large areas of blowdown (Runkle, 1982, 1985; Dunn et al., 1983; Schaetzl et al., 1989; Jonsson and Dynesius, 1993; Tyrrell and Crow, 1994). These disturbances are an important part of the ecology and the regeneration cycles for the forests of the region (Bormann and Likens, 1979; Pickett and White, 1985; Frelich and Lorimer, 1991).

16.7 Hydrology of the Great Lakes

The five Laurentian Great Lakes and the rivers and smaller lakes of the basin comprise a hydrologic system like no other on the planet. Together, these bodies of water essentially form a natural series of storage reservoirs linked by

a

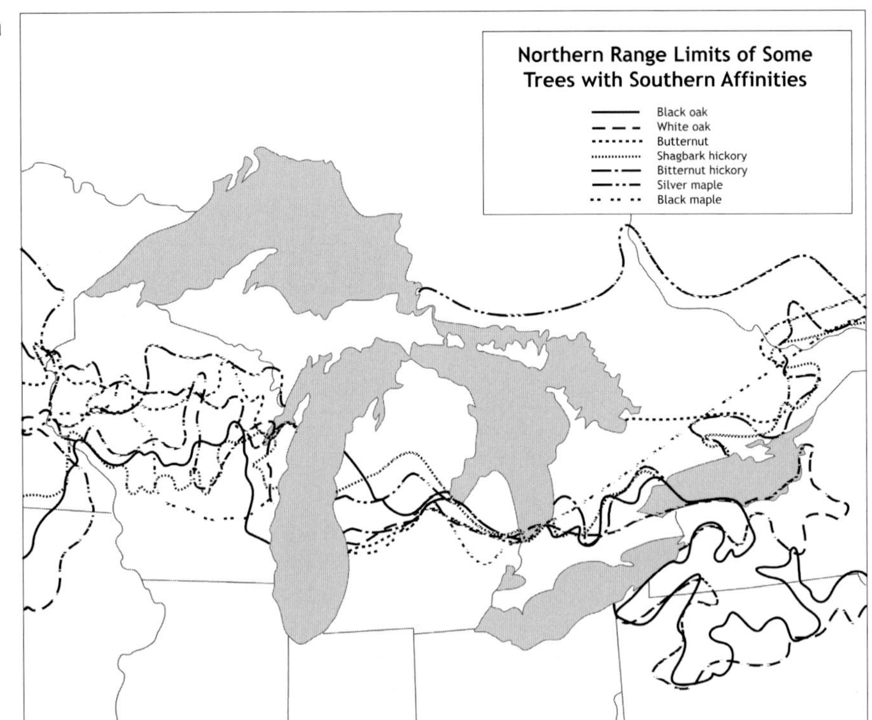

b

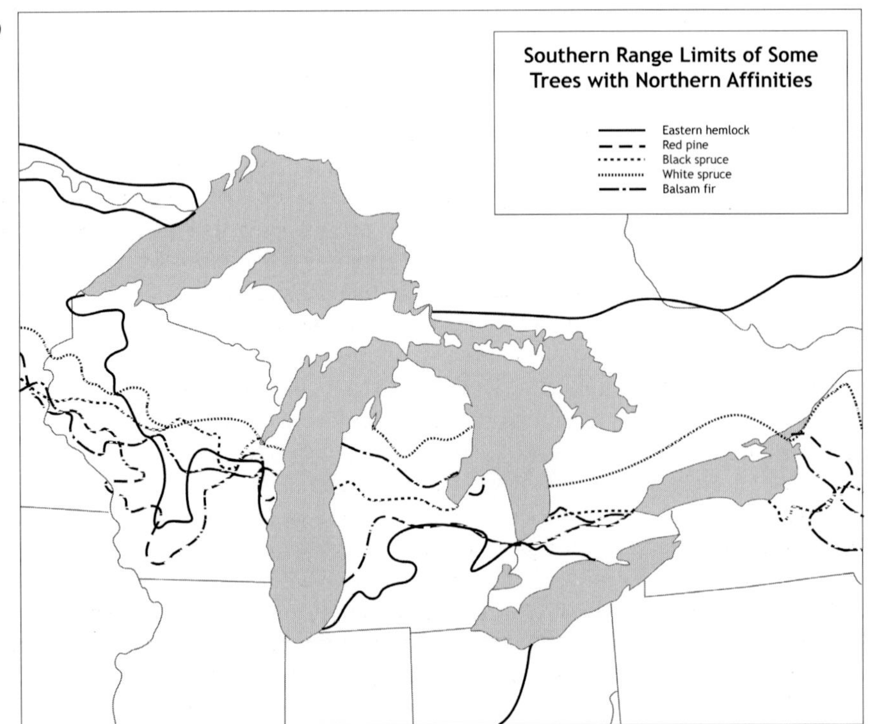

Figure 16.12 Northern (a) or southern (b) limits of the geographical ranges of selected trees in the Great Lakes region. To avoid excessive complexity, range outliers are not shown. Maps modified from Burns and Honkala (1990a, b).

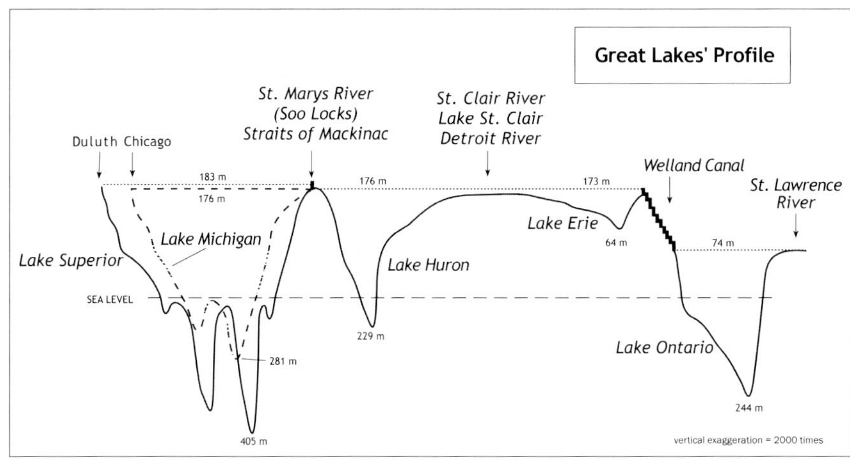

Figure 16.13 Profile of the hydrologic characteristics of the Great Lakes. Profiles are taken along the long axes of the lakes. Lake surface elevations are in meters above sea level. Depths shown are maximum depths in meters. Modified from Botts and Krushelnicki (1988).

connecting channels (fig. 16.3). Although after deglaciation they initially drained southward to the Mississippi and Ohio systems, all the lakes now drain through the St. Lawrence River. Two small, man-made diversions, however, do link the waters of the Great Lakes basin to adjoining basins: the Long Lac-Ogoki diversions bring waters into the Lake Superior watershed from the Hudson Bay region to the north, and the Chicago diversions allow some Great Lakes' water to flow into the Mississippi River basin (table 16.1). Each lake is unique and distinct: Lake Erie is the smallest, shallowest, and warmest (fig. 16.13), whereas Lake Superior, the largest freshwater lake on Earth, is the coldest, deepest, and most oligotrophic (Michigan Water Resources Commission, 1967; Botts and Krushelnicki, 1988; table 16.1).

Because of its humid climate, low relief, and recent glaciation, the Great Lakes region also contains a wealth of swamps, bogs, springs, and perennial streams (Scott, 1921; Fuller et al., 1995). With the exception of a few municipal water diversions, as at Chicago (Changnon and Changnon, 1996; table 16.1), most of the surface water in the basin exits through the St. Lawrence River to the Atlantic Ocean. To the north of the basin divide, waters flow into Hudson Bay and the Arctic Ocean, whereas rivers to the south of the basin are tributary, ultimately, to the Mississippi River. The Great Lakes themselves contain about 95% of the fresh surface water of the United States (Quinn, 1992a). These resources are utilized for, among others, hydropower, industrial applications, navigation, municipal water resources, recreation, and biotic habitats.

The Great Lakes are quite large in volume (23,000 km^3) when compared to the area of the basin that drains into them. This property, coupled with the relatively low amount of water that exits them (<1% per year), suggests a relatively slow turnover of, and long residence times for, water in the lakes proper (Botts and Krushelnicki, 1988; Quinn, 1992b). The "persistence" of the waters in the lakes (e.g., an estimated residence time of 173–191 years for Lake Superior; Botts and Krushelnicki, 1988) is also affected by the limited outlet capacity of the basin (Quinn, 1977; table 16.1). The long residence times of the lakes have implications for how quickly water quality might change as contaminant loadings vary, temporally and spatially (Steinhart et al., 1982; Quinn, 1992b).

The potential for water pollution is spatially variable across the basin. The Lake Superior basin has a low population density and therefore a more limited potential to accumulate water-borne pollutants. Conversely, tributaries to southern lakes, such as the Fox River, which enters Green Bay, or the Rouge, which drains into Lake Erie, have notoriously low water quality (Murray et al., 1997). Green Bay's watershed alone contains the world's largest concentration of pulp and paper mills (Botts and Krushelnicki, 1988) and poses potential long-term problems (Swackhamer and Armstrong, 1987). Saginaw Bay is also infamous for its concentrations of toxic materials and heavy metals (Dolan and Bierman, 1982). During the 1960s and 1970s, Lake Erie was badly polluted, in part because of its shallowness, but also because of the growth of cities and associated heavy industry within its basin (Allan et al., 1983; Sweeney, 1993). Today the lake is much cleaner, mostly as a result of tighter regulations, lowered amounts of phosphates in detergents, greater public awareness, and the introduction of zebra mussels (Hopson, 1975).

Most rivers in the region are geologically "young" and exhibit a deranged pattern with generally low gradients (fig. 16.3). For the most part, the rivers flow across a landscape that underwent the haphazard deposition of glacial drift. Today, these consequent rivers flow where the land surface is at a lower elevation than nearby areas. However, some of the larger river valleys that existed before the last glaciation contain rivers today, since these low spots on the landscape remained as such after a more-or-less even thickness of drift was deposited by the ice (Winters and Rieck,

1982). Waterfalls are found only in the north, where drift is thin and bedrock crops out along river courses (Penrose et al., 1988). Hydropower potential, therefore, is low, and only a few small dams are in operation across the region. Only at the Welland diversion, near Niagara Falls, is any appreciable hydroelectric power generated within the basin (table 16.1).

Runoff in the region peaks in April and May, as spring rains fall onto melting snowpacks in the north or frozen soil farther south (Browzin, 1964; Pentland, 1968). In the south, where snowpacks are thin or nonexistent in late winter, spring runoff is lower than in the snowy, northern regions. Both surface water and groundwater in the southern part of the basin tend to be "hard" (high in calcium and magnesium cations), because of the high amounts of carbonates in the glacial drift. Conversely, rivers flowing on the drift derived from the crystalline rocks of the Canadian Shield can contain very "soft" water; hence, Lake Superior waters have less than half the dissolved solids of any of the other four lakes (Michigan Water Resources Commission, 1967). Many surface waters are brown in color because of the addition of organic acids that have been translocated from soils in the basin. Tahquamenon Falls, in northern Michigan, have been nicknamed "root beer falls" because of the brown color of the water and the foam produced at the base of these two cataracts.

During historic times, water levels of the Great Lakes have fluctuated by about 2–2.5 m (Bruce, 1984; fig. 16.14). Over long timescales, these variations are primarily related to water balance factors (Quinn, 1992b). Natural inputs to the system (direct runoff, inflows from rivers and groundwater, precipitation, diversions) are generally balanced by natural discharges (evapotranspiration, losses at the lake outlets and to groundwater) and artificial withdrawals (by municipalities, diversions of water out of the basin) (Croley and Hunter, 1994; Lee and Southam, 1994). Artificial dredging of, and weeds and ice jams within, the connecting channels and the St. Lawrence River also affect lake levels quite dramatically (Brunk, 1968; Bishop, 1990; Quinn and Sellinger, 1990). Basin-related activities such as agriculture and urbanization have undoubtedly affected the runoff characteristics of the basin, although these impacts are difficult to quantify.

On decadal timescales, periods of abnormally high precipitation coupled with cooler than normal temperatures create high stands that may persist for months or years (Brinkmann, 1983; Fraser et al., 1990). Lake levels have been relatively high in recent years. Low stages occurred during the 1930s, as well as in the mid-1960s, with high levels in 1838, 1929, 1952, 1973, and 1986 (Quinn and Sellinger, 1990; Changnon, 1993). Because of its immense storage capacity and mostly forested watershed, Lake Superior currently exhibits the least amount of interannual variability in lake levels.

Annually, lake levels follow a sinusoidal cycle of about 0.5 m; they are usually highest in early to middle summer, following spring snowmelt and rains, and lowest in late winter, when much recent precipitation is held in snowpacks (Angel, 1995).

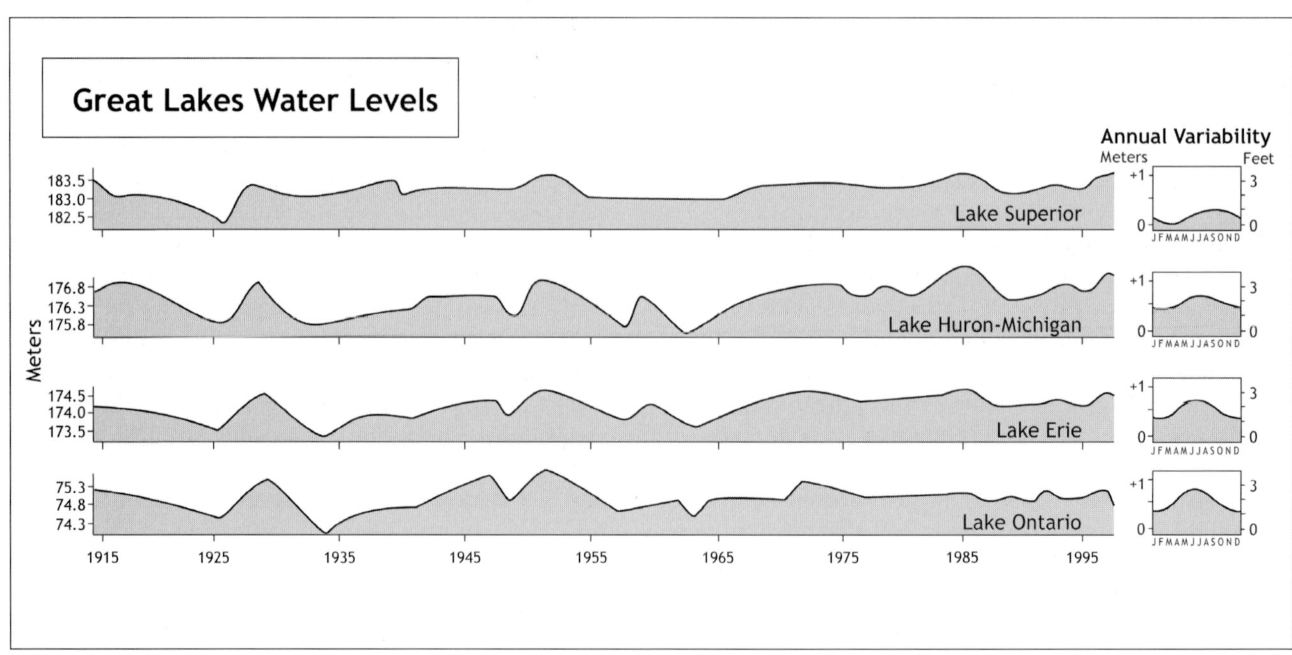

Figure 16.14 Levels of the Great Lakes from 1914 to 1986. Modified from Botts and Krushelnicki (1988).

On daily timescales, storms and strong winds may drive water to one side of the lake, causing a brief but potentially damaging surge (seiche) of waves if coupled with high water levels (Platzman, 1966; Hamblin, 1987). Seiches of over 2 m in height have been recorded on Lake Erie (Michigan Water Resources Commission, 1967); storm surges can reach 4 m or more in height (Mortimer, 1987). On shorter timescales, small variations in lake levels are driven by changes in barometric pressure, winds, waves, and minor tides (Dingman and Bedford, 1984). Tides on the lakes are seldom larger than 7 cm in magnitude (Michigan Water Resources Commission, 1967).

Coasts along the Great Lakes include the extensive marshes of Saginaw Bay and Green Bay, the erodable bluffs formed in glacial drift so common along the southern lakes, the wide sand beaches on the eastern coast of Lake Michigan, and the rocky and bedrock beaches along the western half of Lake Superior and parts of the Bruce Peninsula (Michigan Water Resources Commission, 1967; Hands, 1970). In some places, such as the Grand Sable dunes of northern Michigan or at Sleeping Bear Dunes National Lakeshore, sand dunes seemingly "rise" out of the lake and continue to great heights.

High lake levels coupled with storms often result in extensive flooding, destruction of structures near the shore, shoreline erosion and bluff damage, marina problems, and loss of wildlife habitat (Angel, 1995). During periods of below-average lake level, wide beaches develop; these beaches dissipate energy from storm waves and protect the shore from erosion.

Coastal erosion, an omnipresent process along some shore zones, has become a problem in recent years. Where development pressures impose on an eroding or erodable shore, problems quickly develop, as exemplified by some shore zones on Lakes Michigan, Erie, and Ontario (Buckler and Winters, 1975; Davidson-Arnott and Keizer, 1982). Damage is most severe when strong fall and winter storms, which drive waves and ice blocks against the shore, are coupled with periods of abnormally high lake levels (Barnes et al., 1994; Angel, 1995). Shore zones composed of unconsolidated materials such as dune sand or glacial drift are particularly at risk to erosion, as, for example, along the Scarborough Bluffs near Toronto. Bedrock coasts erode most slowly (Johnson and Johnston, 1995). Other factors that can contribute to faster-than-normal erosion rates include bluff height (lower bluffs erode faster), bluff lithology (layers of finer-textured material within bluffs lead to perched water and slumping and slope failure, and clayey bluffs are most resistant to erosion), bluff orientation (Buckler and Winters, 1983; Johnson and Johnston, 1995), and bluff geometry (headlands erode faster than sandy bays) (Jibson et al., 1994).

On many modern Great Lakes' beaches, shore protection structures such as riprap, tetrapods, seawalls or revetments, detached breakwaters, and groins are being used to offset the loss of beach, with mixed results (Buckler and Winters, 1975; Hadley, 1976). Whereas most of these features provide some degree of protection from rapid erosion, they require continual maintenance and seldom last for more than a few decades (Davidson-Arnott and Keizer, 1982). Jibson et al. (1994) argue that these features do not necessarily slow the regional rate of erosion on the shores of the Great Lakes, but only act to change its spatial distribution.

Not to be ignored are the consequences of dramatic falls in lake levels. During periods of low lake level, harbors infill with silt and otherwise become shallower, requiring increased dredging, tributary rivers incise their valleys, and lowered water supplies begin to affect hydroelectric utilities and municipalities, which rely on lake water (Colman et al., 1994). Some forecasting scenarios predict lower lake levels in the next century, attributable to a warmer (greenhouse-enhanced) climate and increased anthropogenic consumption of lake water.

16.8 Conclusion

Three factors have been paramount in shaping the physical character of the Great Lakes region. The first is its interior location on the North American continent, at 45°N latitude. Coincident with this location, the landscape has been repeatedly sculpted by continental ice sheets, and today, the region's weather is strongly influenced by repeated incursions of midlatitude cyclones. Second, the region is characterized by a pronounced north–south gradient in climate and vegetation. Third, the geologic transition from crystalline rocks in the north to sedimentary rocks in the south strongly impacts soils and the sediments that lie below.

However, it is the lakes themselves that unify the physical geography of the region, by dominating the landscape and modifying the climate so as to strongly influence vegetation, climate, and soil patterns in their vicinity. Even the human inhabitants of the region, who so often modify the environment to suit their own needs, recognize its dominant physical geography and thus focus their interests, either directly or indirectly, on the Great Lakes.

Acknowledgments We thank Jeff Andresen for a review of an early draft of our manuscript. The figures were drafted by Ellen White at the Center for Cartographic Research and Spatial Analysis and edited by personnel at the Center for Remote Sensing and GIS, both at Michigan State University.

References

Abrams, M.C., D.G. Sprugel, and D.I. Dickmann, 1985. Multiple successional pathways on recently disturbed

jack pine stands in Michigan. For. Ecol. Mgmt. 10: 31–48.
Acomb, L.J., D.M. Mickelson, and E.B. Evenson, 1982. Till stratigraphy and late glacial events in the Lake Michigan lobe of eastern Wisconsin. Geol. Soc. Am. Bull. 93: 289–296.
Agriculture Canada, 1977. Soils of Canada. 1:5,000,000 map accompanying *Soils of Canada*, Research Branch, Canadian Dept. of Agriculture.
Allan, R.J., A. Mudroch, and A. Sudar, 1983. An introduction to the Niagara River/Lake Ontario pollution problem. J. Great Lakes Res. 9: 111–117.
Angel, J.R., 1995. Large-scale storm damage on the U.S. shores of the Great Lakes. J. Great Lakes Res. 21: 287–293.
Angel, J.R., and S.A. Isard, 1997. An observational study of the influence of the Great Lakes on the speed and intensity of passing cyclones. Mon. Wea. Rev. 125: 2228–2237.
Angel, J.R., and S.A. Isard, 1998. The frequency and intensity of Great Lake cyclones. J. Clim. 11: 61–71.
Anonymous, 1994. Antrim gas play, production expanding in Michigan. Oil and Gas J. 92(22): 97–98.
Arbogast, A.F., P. Scull, R.J. Schaetzl, J. Harrison, T.P. Jameson, and S. Crozier, 1997. Concurrent stabilization of some interior dune fields in Michigan. Phys. Geogr. 18: 63–79.
Assel, R.A., 1991. Implication of CO_2 global warming on Great Lakes ice cover. Climatic Change 18: 377–395.
Assel, R.A., F.H. Quinn, G.A. Leshkevich, and S.J. Bolsenga, 1983. *Great Lakes Ice Atlas*. NOAA, Great Lakes Environ. Res. Lab., Ann Arbor, MI. 115 pp.
Attig, J.W., L. Clayton, and D.M. Mickelson, 1985. Correlation of late Wisconsin glacial phases on the western Great Lakes. Geol. Soc. Am. Bull. 96: 1585–1593.
Attig, J.W., D.M. Mickelson, and L. Clayton, 1989. Late Wisconsin landform distribution and glacier-bed conditions in Wisconsin. Sed. Geol. 62: 399–405.
Bailey, R.G., 1994. Ecoregions of the United States. 1:7,500,000 map. USDA-Forest Service, Washington, DC.
Baker, R.W., J.F. Diehl, T.W. Simpson, L.W. Zelazney, and S. Beske-Diehl, 1983. Pre-Wisconsin glacial stratigraphy, chronology and paleomagnetics of west-central Wisconsin. Geol. Soc. Am. Bull. 94: 1442–1449.
Barcus, R., 1960. *Freshwater Fury*. Wayne State Univ. Press, Detroit, MI. 166 pp.
Barnes, F.Q., and E.J. Lalonde, 1973. Lower Huronian stratigraphy Hyman and Drury Township, Sudbury District. Geol. Assoc. Canada Spec. Paper 12: 147–156.
Barnes, P.W., E.W. Kempema, E. Reimnitz, and M. McCormick, 1994. The influence of ice on southern Lake Michigan coastal erosion. J. Great Lakes Res. 20: 179–195.
Barnes, W.J., 1989. A case history of vegetation changes on the Meridean Islands of West-Central Wisconsin, USA. Biol. Cons. 49: 1–16.
Barrett, L.R., J. Liebens, D.G. Brown, R.J. Schaetzl, P. Zuwerink, T.W. Cate, and D.S. Nolan, 1995. Relationships between soils and presettlement vegetation in Baraga County, Michigan. Am. Midl. Nat. 134: 264–285.
Barry, R.G., and R.J. Chorley, 1987. *Atmosphere, Weather and Climate*. 5th ed., Methuen, NY. 460 pp.
Bennett, I., 1975. Variation of daily solar radiation in North America during the extreme months. Arch. Met. Geoph. Biokl. Ser. B, 23: 31–57.
Bergquist, S.G., 1941. The distribution of drumlins in Michigan. Papers Mich. Acad. Sci. Arts Letts. 27: 451–464.
Bishop, C.T., 1990. Historical variation of water levels in Lakes Erie and Michigan-Huron. J. Great Lakes Res. 16: 406–425.
Black, R.F., 1969. Glacial geology of northern Kettle Moraine State Forest, Wisconsin. Trans. Wisc. Acad. Sci. Arts Letters 57: 99–119.
Black, R.F., 1978. Quaternary geology of Wisconsin and contiguous upper Michigan. In: Mahaney, W.C. (ed.). *Quaternary Stratigraphy of North America*. Halsted Press, Wiley and Sons, New York. pp. 93–117.
Bleimeister, W.C., 1961. Rock salt mining operations in Michigan, Ohio, and Ontario. Mining Engineering 13: 467–471.
Blewett, W.L., 1991. Characteristics, correlations, and refinement of Leverett and Taylor's Port Huron Moraine in Michigan. East Lakes Geog. 26:52–60.
Blewett, W.L., and R.L. Rieck, 1987. Reinterpretation of a portion of the Munising moraine in northern Michigan. Geol. Soc. Am. Bull. 98: 169–175.
Blewett, W.L., and H.A. Winters, 1995. The importance of glaciofluvial features within Michigan's Port Huron moraine. Annals Assoc. Am. Geogs. 85: 306–319.
Bormann, F.H., and G.E. Likens, 1979. Catastrophic disturbance and the steady state in northern hardwood forests. Am. Sci. 67: 660–669.
Bornhorst, T.J., J.B. Paces, N.K. Grant, J.D. Obradovich, and N.K. Huber, 1988. Age of native copper mineralization, Keweenaw Peninsula, Michigan. Econ. Geol. 83: 619–625.
Borowiecka, B.Z., and R.H. Erickson, 1985. Wisconsin drumlin field and its origin. Zeit. Geomorph. 29: 417–438.
Botts, L., and B. Krushelnicki, 1988. *The Great Lakes: An Environmental Atlas and Resource Book*. U.S. Environ. Protection Agency and Environment Canada. 44 pp.
Braham, R.R.J., and M.J. Dungey, 1984. Quantitative estimates of the effect of Lake Michigan on snowfall. J. Clim. Appl. Met. 23:940–949.
Braun, E.L., 1950. *Deciduous Forests of Eastern North America*. The Free Press, New York, NY.
Bretz, J.H., 1951. The stages of Lake Chicago; their causes and correlations. Am. J. Sci. 259: 401–429.
Bretz, J.H., 1955. Geology of the Chicago region, Part II—The Pleistocene. Ill. St. Geol. Surv. Bull. 65. 132 pp.
Brewer, L.G., T.W. Hodler, and H.A. Raup, 1984. Presettlement vegetation of southwestern Michigan. Mich. Bot. 23: 153–156.
Bricker, D.M., and W.L. Henderson, 1986. Oil and gas developments in Michigan in 1985. Am. Assoc. Petroleum Geol. Bull. 70: 1280–1284.
Bright, E.G., 1979. No. 20 The Centre Lake Area, Haliburton and Hastings Counties. Ontario Geol. Surv. Misc. Paper 90: 86–88.
Brinkmann, W.A.R., 1983. Association between net basin supplies to Lake Superior and supplies to the lower Great Lakes. J. Great Lakes Res. 9: 32–39.
Browzin, B.S., 1964. Seasonal variations of flow and classification of rivers in the Great Lakes-St. Lawrence basin. Great Lakes Res. Division, Univ. of Michigan. Conf. Great Lakes Research, Publ. No. 11. pp. 179–204.
Brubaker, L.B., 1975. Postglacial forest patterns associated with till and outwash in north-central upper Michigan. Quat. Res. 5: 499–527.

Bruce, J.P., 1984. Great Lakes levels and flows: Past and future. J. Great Lakes Res. 10: 126–134.
Brunk, I.W., 1968. Evaluation of channel changes in St. Clair and Detroit Rivers. Water Resour. Res. 4: 1335–1346.
Buckler, W.R., and H.A. Winters, 1975. Rate of bluff recession at selected sites along the southeastern shore of Lake Michigan. Mich. Academ. 8:179–186.
Buckler, W.R., and H.A. Winters, 1983. Lake Michigan bluff recession. Ann. Assoc. Am. Geog. 73: 89–110.
Burns, R.M., and B.H. Honkala, eds., 1990a. *Silvics of North America.* Volume 1: Conifers. Agriculture Handbook 654. U.S. Forest Serv., U.S. Govt. Printing Office, Washington, DC. 675 p.
Burns, R.M., and B.H. Honkala, (eds.), 1990b. *Silvics of North America.* Volume 2: Hardwoods. Agriculture Handbook 654. U.S. Forest Serv., U.S. Govt. Printing Office, Washington, DC. 877 p.
Cannon, W.F., 1973. The Penokean Orogeny in northern Michigan. Geol. Assoc. Canada Spec. Paper 12: 251–271.
Card, K.D., 1990. A review of the Superior Province of the Canadian Shield, a product of Archean accretion. Precambrian Res. 48:99–156.
Carleton, T.J., and P.F. Maycock. 1978. Dynamics of the boreal forest south of James Bay. Can. J. Bot. 56: 1157–1173.
Carter, M.W., 1979. No. 11 Schreiber Area, District of Thunder Bay. Ontario Geol. Surv. Misc. Paper 90: 44–47.
Changnon, S.A., 1968. Precipitation from thunderstorms and snowfall around southern Lake Michigan. Proc. 11th Conf. Great Lakes Res., Intl. Assoc. Great Lakes Res. pp. 285–297.
Changnon, S.A., 1978. The climatology of hail in North America. Am. Meteor. Soc. Monogr. 38: 107–128.
Changnon, S.A., 1987. Climate fluctuations and record-high levels of Lake Michigan. Bull. Amer. Meteor. Soc. 68: 1394–1402.
Changnon, S.A., 1993. Changes in climate and levels of Lake Michigan: Shoreline impacts at Chicago. Climatic Change 23: 213–230.
Changnon, S.A., and J.M. Changnon, 1996. History of the Chicago diversion and future implications. J. Great Lakes Res. 22: 100–118.
Changnon, S.A., Jr., and D.M.A. Jones, 1972. Review of the influences of the Great Lakes on weather. Water Res. Res. 8: 360–371.
Chapman, L.J., and D.F. Putnam, 1984. *The Physiography of Southern Ontario.* Ontario Geol. Surv. Spec. Vol. 2. 270 pp.
Clark, J.S., 1993. Fire, climate change, and forest processes during the past 2000 years. In: Bradbury, J.P., and W.E. Dean (eds.). *Elk Lake, Minnesota: Evidence for Rapid Climatic Change in the North-Central United States.* Geol. Soc. Am. Spec. Paper 276: 295–308.
Clark, P.U., 1992. Surface form of the southern Laurentide Ice Sheet and its implications to ice-sheet dynamics. Geol. Soc. Am. Bull. 104:595–605.
Clayton, L., and J.W. Attig, 1989. *Glacial Lake Wisconsin.* Geol. Soc. Am. Memoir 173. 80 pp.
Clayton, L., and S.R. Moran, 1982. Chronology of Late Wisconsinan glaciation of middle North America. Quat. Sci. Revs. 1:55–82.
Clayton, L., J.W. Attig, D.M. Mickelson, and M.D. Johnson, 1991. Glaciation of Wisconsin. Wisc. Geol. Nat. Hist. Survey Educ. Series leaflet 36. 4 pp.
Colman, S.M., R.M. Forester, R.L. Reynolds, D.S. Sweetkind, J.W. King, P. Gangemi, G.A. Jones, L.D. Keigwin, and D.S. Foster, 1994. Lake-level history of Lake Michigan for the past 12,000 years: The record from deep lacustrine sediments. J. Great Lakes Res. 20:73–92.
Croley, T.E., II, and T.S. Hunter, 1994. Great Lakes monthly hydrologic data. NOAA Tech. Memo. ERL GLERL-83, Great Lakes Env. Res. Lab., Ann Arbor, MI. 83 pp.
Crowell, J.C., 1980. Continental glaciation through geological time. EOS 61: 251.
Curtis, J.T., 1959. *The Vegetation of Wisconsin.* Univ. of Wis. Press, Madison. 657 pp.
Curtis, J.T., and H.C. Greene, 1949. A study of relic Wisconsin prairies by the species-presence method. Ecology 30: 83–92.
Danard, M.B., and A.C. McMillan, 1974. Further numerical studies of the effects of the Great Lakes on winter cyclones. Mon. Wea. Rev. 102: 166–175.
Davidson-Arnott, R.G.D., and H.I. Keizer, 1982. Shore protection in the town of Stoney Creek, southwest Lake Ontario, 1934–1979: Historical changes and durability of structures. J. Great Lakes Res. 8: 635–647.
Dawson, J.B., 1964. An aid to prospecting for kimberlites. Econ. Geol. 59: 1385–1386.
DeWitt, B.H., D.F. Kahlbaum, D.G. Baker, J.H. Wartha, F.A. Keyes, D.E. Boyce, F.H. Quinn, R.A. Assel, A. Bakler-Blocker, and D.M. Kurdziel, 1980. Summary of Great Lakes weather and ice conditions, winter 1978–79. NOAA Tech. Memo. ERL GLERL-31, Great Lakes Env. Res. Lab., Ann Arbor, MI (PB80–179203). 123 pp.
Dingman, J.S., and K.W. Bedford, 1984. The Lake Erie response to the January 26, 1978 cyclone. J. Geophys. Res. 89: 6427–6445.
Dodge, S.L., 1989. Forest transitions and buried glacial outwash within the beech-maple region of Michigan, USA. Geografiska Annaler 71A: 137–144.
Dodge, S.L., and J.R. Harman, 1985. Soil, subsoil, and forest composition in south-central Michigan, USA. Phys. Geog. 6: 85–100.
Dolan, D.M., and V.J. Bierman Jr., 1982. Mass balance modeling of heavy metals in Saginaw Bay, Lake Huron. J. Great Lakes Res. 8: 676–694.
Dredge, L.A., and L.H. Thorleifson, 1987. The middle Wisconsinan history of the Laurentide ice sheet. Geog. Physique Quat. 41: 215–235.
Dreimanis, A., and P.F. Karrow, 1972. Glacial history of the Great Lakes-St. Lawrence region, the classification of the Wisconsin(an) stage, and its correlatives. 24th Intl. Geol. Cong. pp. 5–15.
Dunn, C.P., G.R. Guntenspergen, and J.R. Dorney, 1983. Catastrophic wind disturbance in an old-growth hemlock-hardwood forest, Wisconsin. Can. J. Bot. 61: 211–217.
Dyke, A.S., and V.K. Prest, 1987. Late Wisconsinan and Holocene history of the Laurentide ice sheet. Geog. Physique Quat. 41: 237–263.
Eagleman, J.R., 1990. *Severe and Unusual Weather.* Trimedia, Lenexa, KS. 394 pp.
Eastman, J., 1986. The ghost forest; nineteenth-century logging left Michigan's vast pine tracts stripped, stumped, and subject to fire. Nat. Hist. 95: 10–16.
Eichenlaub, V.L., 1970. Lake effect snowfall to the lee of the Great Lakes: Its role in Michigan. Bull. Amer. Meterol. Soc. 51: 403–412.
Eichenlaub, V.L., J.R. Harman, F.V. Nurnberger, and H.J. Stolle, 1990. *The Climatic Atlas of Michigan.* Univ. Notre Dame Press, South Bend, IN. 165 pp.

Elliot-Fisk, D.L., 1988. The boreal forest. In: Barbour, M.G., and W.D. Billings. (eds.). *North American Terrestrial Vegetation.* Cambridge Univ. Press, New York. pp. 33–62.

Ernst, J.E., and R.L. Hunter, 1987. Soil Survey of Sandusky County, Ohio. USDA Soil Conservation Service, U.S. Govt. Printing Office, Washington, DC.

Eschman, D.F., and P.F. Karrow, 1985. Huron Basin Glacial Lakes: A Review. In: Karrow, P.F., and P.E. Calkin (eds.). *Quaternary Evolution of the Great Lakes.* Geol. Soc. Canada Spec. Paper 30:79–93.

Estoque, M.A., 1981. Further studies of a lake breeze. Part I: Observational study. Mon. Wea. Rev. 109: 611–618.

Farrand, W.R., 1969. The Quaternary history of Lake Superior. Proc. 12th Conf. Great Lakes Res., Intl. Assoc. Great Lakes Res. pp. 181–197.

Farrand, W.R., and D.F. Eschman, 1974. Glaciation of the southern peninsula of Michigan: A review. Mich. Academ. 7: 31–56.

Feenstra, B.H., 1979. No. 23 Quaternary Geology of the Owen Sound (41A/10) Area, Grey County. Ontario Geol. Surv. Misc. Paper 90: 133–134.

Fehrenbacher, J.B., G.O. Walker, and H.L. Wascher, 1967. *Soils of Illinois.* Univ. Ill. Agric. Exper. Stn. Bull. 725. 47 pp.

Fenneman, N.M., 1938. *Physiography of Eastern United States.* McGraw-Hill, New York. 714 pp.

Fenton, T.E., 1983. Mollisols. In: Wilding, L.P., N.E. Smeck, and G.F. Hall (eds.). *Pedogenesis and Soil Taxonomy.* Elsevier, New York. pp. 125–163.

Flint, R.F., 1971. *Glacial and Pleistocene Geology.* Wiley and Sons, New York, 553 pp.

Folsom, M.M., and H.A. Winters, 1970. Drainage orders in Michigan. Mich. Academ. 2:79–91.

Francek, M.A., 1991. A spatial perspective on the New York drumlin field. Phys. Geog. 12: 1–18.

Franzmeier, D.P., R.B. Bryant, and G.C. Steinhardt, 1985. Characteristics of Wisconsinan glacial tills in Indiana and their influence on argillic horizon development. Soil Sci. Soc. Am. J. 49: 1481–1486.

Fraser, G.S., C.E. Larsen, and N.C. Hester, 1990. Climatic control of lake levels in the Lake Michigan and Lake Huron basins. In: *Late Quaternary History of the Lake Michigan Basin.* Geol. Soc. Am. Spec. Paper 251:75–90.

Frelich, L.E., and C.G. Lorimer, 1991. Natural disturbance regimes in hemlock-hardwood forests of the upper Great Lakes region. Ecol. Mon. 61: 145–164.

Frye, J.C., and H.B. Willman, 1960. Classification of the Wisconsin Stage in the Lake Michigan glacial lobe. Ill. Geol. Surv. Circ. 285. 16 pp.

Fuller, K., H. Shear, and J. Wittig (eds.), 1995. *The Great Lakes: An Environmental Atlas and Resource Book.* 3rd ed. U.S. EPA and Environment Canada. 46 pp.

Futyma, R.P., 1981. The northern limits of glacial Lake Algonquin in Upper Michigan. Quat. Res. 15: 291–310.

Gaikawad, S.T., and F.D. Hole, 1961. Characteristics and genesis of a Podzol soil in Florence County, Wisconsin. Trans. Wis. Acad. Sci. Arts Lett. 50: 183–190.

Gaikawad, S.T., and F.D. Hole, 1965. Characteristics and genesis of a gravelly Brunizemic regosol. Soil Sci. Soc. Am. Proc. 29:725–728.

Gardner, D.R., and E.P. Whiteside, 1952. Zonal soils in the transition region between the Podzol and Gray-Brown Podzolic regions in Michigan. Soil Sci. Soc. Am. Proc. 16: 137–141.

Gates, F.C., 1942. The bogs of northern lower Michigan. Ecol. Monogr. 12: 216–254.

Gilbert, G.K., 1907. Rate of recession of Niagara Falls. Bulletin U.S. Geol. Survey 306.

Gleason, H.A., 1913. The relation of forest distribution and prairie fires in the Middle West. Torreya 13: 173–181.

Gleason, H.A., 1917. A prairie near Ann Arbor, Michigan. Rhodora 19: 163–165.

Goldthwait, J.W., 1908. A reconstruction of water planes of the extinct glacial lakes in the Lake Michigan basin. J. Geol. 16: 459–476.

Green, J.C., 1983. Composition, origin and evolution of Keweenawan magmas; a review. In: Bornhorst, T.J., and J.F. Diehl (eds.). *Proceedings and Abstracts,* Twenty-ninth Annual Institute on Lake Superior Geology. p. 17.

Greller, A.M., 1988. Deciduous Forest. In: Barbour, M.G., and W.D. Billings (eds.). *North American Terrestrial Vegetation.* Cambridge Univ. Press, New York. pp. 287–316.

Grimm, E.C., 1984. Fire and other factors controlling the Big Woods vegetation of Minnesota in the mid-nineteenth century. Ecol. Mon. 54: 291–311.

Guillet, G.R., 1984. Salt in Ontario. In: *The Geology of Industrial Minerals in Canada.* Can. Inst. Mining Metallurgy Spec. Vol. 29: 143–147.

Habecker, M.A., K. McSweeney, and F.W. Madison, 1990. Identification and genesis of fragipans in Ochrepts of North Central Wisconsin. Soil Sci. Soc. Am. J. 54: 139–146.

Hadley, D.W., 1976. Shoreline erosion in southeastern Wisconsin. Wisc. Geol. Nat. Hist. Survey Spec. Rept. No. 5. 33 pp.

Hamblin, P.F., 1987. Meteorological forcing and water level fluctuations on Lake Erie. J. Great Lakes Res. 13: 436–453.

Hands, E.B., 1970. A geomorphic map of Lake Michigan shoreline. Proc. 13th Conf. Great Lakes Res., Intl. Assoc. Great Lakes Res. pp. 250–265.

Hansel, A.K., and W.H. Johnson, 1992. Fluctuations of the Lake Michigan lobe during the late Wisconsin subepisode. Sveriges Geologiska Unfersöking, Series Ca, 81: 133–144.

Hansel, A.K., D.M. Mickelson, A.F. Schneider, and C.E. Larsen, 1985. Late Wisconsinan and Early Holocene history of the Lake Michigan Basin. In: Karrow, P.F., and P. Calkins (eds.). *Quaternary Evolution of the Great Lakes.* Geol. Assoc. Canada Spec. Paper 30. pp. 39–53.

Hardwick, S.W., and D.G. Holtgrieve, 1990. *Patterns on our Planet. Concepts and Themes in Geography.* Merill Publ., New York. 414 pp.

Hare, F.K., 1950. Climate and zonal divisions of the boreal forest formation in eastern Canada. Geog. Rev. 40: 615–635.

Harman, J.R., and J.G. Hehr, 1972. Lake breezes and summer rainfall. Annals Assoc. Am. Geog. 62: 375–387.

Harringer, R.V., 1990. Copper Country. Lapidary J. 44: 65–74.

Harry, D.G., and A.S. Trenhaile, 1987. The morphology of the Arran drumlin field, southern Ontario, Canada. In: Menzies, J., and J. Rose (eds.). *Drumlin Symposium.* A.A. Balkema, Rotterdam. pp. 161–173.

Haywood, D., 1970. Lake Superior diversions and their effect. In: Humphrys, C.R. *Lake Superior—The only Great lake.* Student Water Publications, Mich. St. Univ. Vol. 1. pp. F1–F4.

Heinselman, M.A., 1981. Fire intensity and frequency as factors in the distribution and structure of northern ecosystems. In: Mooney, H.A., T.M. Bonnicksen, N.L. Christensen, J.E. Lotan, and W.A. Reiners (eds.). *Fire Regimes and Ecosystem Properties.* U.S. For. Serv. Gen. Tech. Rept. WO-26. pp. 7–57.

Hole, F.D., 1976. *Soils of Wisconsin.* Univ. of Wis. Press, Madison. 223 pp.

Hopson, N.E., 1975. Phosphorous removal by legislation. Water Resour. Bull. 11: 358.

Horberg, C.L., and R.C. Anderson, 1956. Bedrock topography and Pleistocene glacial lobes in central United States. J. Geol. 64: 101–116.

Host, G.E., K.S. Pregitzer, C.W. Ramm, J.B. Hart, and D.T. Cleland, 1987. Landform-mediated differences in successional pathways among upland forest ecosystems in northwestern lower Michigan. For. Sci. 33: 445–457.

Hough, J.L., 1958. *Geology of the Great Lakes.* Univ. Illinois Press, Urbana. 313 pp.

Hough, J.L., 1963. The prehistoric Great Lakes of North America. Am. Sci. 51: 84–109.

Hunt, C.B., 1967. *Physiography of the United States.* W.H. Freeman and Co., San Francisco. 480 pp.

Jacobson, G.L., Jr., 1979. The palaeoecology of white pine (*Pinus strobus*) in Minnesota. J. Ecol. 67: 697–726.

Janke, R.A., D. McKaig, and R. Raymond, 1978. Comparison of presettlement and modern upland boreal forests on Isle Royale National Park. For. Sci. 24: 115–121.

Jibson, R.W., J.K. Odum, and J.-M. Staude, 1994. Rates and processes of bluff recession along the Lake Michigan shoreline in Illinois. J. Great Lakes Res. 20: 135–152.

Johnson, B.L., and C.A. Johnston, 1995. Relationship of lithology and geomorphology to erosion of the western Lake Superior coast. J. Great Lakes Res. 21: 3–16.

Johnson, M.D., D.M. Mickelson, L. Clayton, and J.W. Attig, 1995. Composition and genesis of glacial hummocks, western Wisconsin, USA. Boreas 24:97–116.

Johnson, W.H., 1986a. Anderson site near Foosland. In: Follmer, L.R., D.P. McKenna, and J.E. King (eds.). *Quaternary records of central and northern Illinois.* Ill. St. Geol. Survey Guidebook 20, pp. 9–13.

Johnson, W.H., 1986b. Stratigraphy and correlation of the glacial deposits of the Lake Michigan Lobe prior to 14 ka BP. Quat. Sci. Revs. 5: 17–22.

Johnson, W.H., A.K. Hansel, E.A. Bettis III, P.F. Karrow, G.J. Larson, T.V. Lowell, and A.F. Schneider, 1997. Late Quaternary temporal and event classifications, Great Lakes Region, North America. Quat. Res. 47: 1–12.

Jonsson, B.G., and M. Dynesius, 1993. Uprooting in boreal spruce forests: Long-term variation in disturbance rate. Can. J. For. Res. 23: 2383–2388.

Karrow, P.F., 1987. Glacial and glaciolacustrine events in northwestern Lake Huron, Michigan and Ontario. Geol. Soc. Am. Bull. 98: 113–120.

Karrow, P.F., and P.E. Calkin (eds.), 1985. Quaternary Evolution of the Great Lakes, Geol. Assoc. Canada Spec. Paper 30.

Kaszycki, C.A., and W.W. Shilts, 1979. Average depth of glacial erosion, Canadian Shield. Geol. Surv. Canada Paper 79-1B: 395–396.

Keables, M.J., 1989. A synoptic climatology of the bimodal precipitation distribution in the upper Midwest. J. Clim. 2: 1289–1294.

Keller, B.M., 1973. Great glaciations in history of the Earth. Intl. Geology Rev. 15: 1067–1074.

Kelly, D.L., J.T. Schaefer, R.P. McNulty, C.A. Doswell III, and R.F. Abbey Jr., 1978. An augmented tornado climatology. Mon. Wea. Rev. 106: 1172–1183.

Kilburn, P.D., 1959. The forest-prairie ecotone in northeastern Illinois. Am. Midl. Nat. 62: 206–217.

Kristovich, D.A.R., and R.A. Steve III, 1995. A satillite study of cloud-band frequencies over the Great Lakes. J. Appl. Meteor. 34: 2083–2090.

Küchler, A.W., 1964. Potential natural vegetation of the conterminous United States. Spec. Publ. 36, Am. Geog. Soc., New York.

Kunkel, K.E., L.D. Mortsch, and P. Lewis, 1993. The climate of the Great Lakes-St. Lawrence River Basin. In: Lee, D.H., (ed.). *Climate, Climate Change, Water Level Forecasting and Frequency Analysis,* International Joint Commission Levels Reference Study, Phase II, Environment Canada. 197 pp.

Kupec, R.J., 1965. Continentality around the Great Lakes. Bull. Amer. Meteor. Soc. 46:54–57.

LaBerge, G.L., 1994. *Geology of the Lake Superior Region.* Geoscience Press, Tuscon, AZ. 313 pp.

LaBerge, G.L., J.S. Klasner, and P.E. Myers, 1990. New observations on the age and structure of Proterozoic quartzites in Wisconsin. U.S. Geol. Surv. Bull. 1904B.

Larsen, C.E. 1985. Lake level, uplift, and outlet incision, the Nipissing and Algoma Great Lakes. In: Karrow, P.F. and Calkin, P.E., (eds.). *Quaternary Evolution of the Great Lakes,* Geol. Assoc. Canada Spec. Paper 30: 63–77.

Larsen, C.E., 1994. Beach ridges as monitors of isostatic uplift in the upper Great Lakes. J. Great Lakes Res. 20: 108–134.

Larson, A.M., 1949. *History of the White Pine Industry in Minnesota.* Univ. Minnesota Press, Minneapolis. 432 pp.

Lee, D.H., and C. Southam, 1994. Effect and implications of differential isostatic rebound on Lake Superior's regulation limits. J. Great Lakes Res. 20: 407–415.

Leigh, D.L., 1994. Loess of the upper Mississippi Valley Driftless Area. Quat. Res. 42: 30–40.

Leighton, M.M., and J.A. Brophy, 1961. Illinoian glaciation in Illinois. J. Geol. 69: 1–31.

Leighton, M.M., and H.B. Willman, 1950. Loess formations of the Mississippi Valley. J. Geol. 58:599–623.

Leitner, L.A., C.P. Dunn, G.R. Guntenspergen, F. Stearns, and D.M. Sharpe, 1991. Effects of site, landscape features, and fire regime on vegetation patterns in presettlement southern Wisconsin. Landscape Ecol. 5: 203–217.

LeMasters, G.S., and E.A. Jones, 1984. State of Michigan Peat Resource Estimation. Mich. Tech. Univ. School of Forestry and Wood Products Tech. Bull. 84–1. Vols. I and II.

Leverett, F., 1929. Moraines and shorelines of the Lake Superior region. U.S. Geol. Survey Prof. Paper 154. 72 pp.

Leverett, F., and F.B. Taylor, 1915. The Pleistocene of Indiana and Michigan and the history of the Great Lakes. U.S. Geol. Surv. Mon. 53. 259 pp.

Lineback, J.A., 1979. The status of the Illinoian glacial stage. In: *Wisconsinan, Sangamonian, and Illinoian stratigraphy in central Illinois.* Midwest Friends of the Pleistocene Field Conference Guidebook. Illinois State Geol. Survey Guidebook 13. pp. 69–78.

Maher, L.J., and D.M. Mickelson, 1996. Palynological and radiocarbon evidence for deglaciation events in the Green Bay lobe, Wisconsin. Quat. Res. 46: 251–259.

Maissurow, D.K., 1935. Fire as a necessary factor in the perpetuation of white pine. J. For. 33: 373–387.

Manogaran, C., 1983. The prairie peninsula: A climatic perspective. Phys. Geog. 4: 153–166.
Martin, L.M., 1965. *The Physical Geography of Wisconsin*. Univ. Wisconsin Press, Madison. 608 pp.
Matsch, C.L., and A.F. Schneider, 1986. Stratigraphy and correlation of the glacial deposits of the glacial lobe complex in Minnesota and northwestern Wisconsin. Quat. Sci. Revs. 5:59–65.
McCartney, M.C., and D.M. Mickelson, 1982. Late Woodfordian and Greatlakean history of the Green Bay lobe, Wisconsin. Geol. Soc. Am. Bull. 93: 297–302.
McCaslin, J.C., 1985. West Michigan basin draws deep gas hunt. Oil and Gas J. 83(43): 139.
McCaslin, J.C., 1987. Michigan basin is a bright spot among east, southeast U.S. areas. Oil and Gas J. 85(23): 38–40.
McCormick, M.J., 1990. Potential change in the thermal strucutre and cycle of Lake Michigan due to global warming. Trans. Amer. Fish. Soc. 119: 183–194.
McGee, E.S., and B.C. Hearn Jr., 1984. The Lake Ellen Kimberlite, Michigan, U.S.A. In: Kornprobst, J. (ed.). *Kimberlites. I: Kimberlites and Related Rocks*. Proc. 3rd Intl. Kimberlite Conf., Developments in petrology series, Elsevier. pp. 143–154.
McGovney, J.E., 1989. Thornton Reef, Silurian, northeastern Illinois. In: Geldsetzer, H.H.J., James, N.P., and G.E. Tebbutt (eds.). *Reefs. Canada and Adjacent Area*. Can. Soc. Petr. Geol. Memoir 13. pp. 330–338.
McKay, E.D., 1979. Wisconsinan loess stratigraphy of Illinois. In: *Wisconsinan, Sangamonian, and Illinoian stratigraphy in central Illinois*. Midwest Friends of the Pleistocene Field Conference Guidebook. Illinois State Geol. Survey Guidebook 13. pp. 95–108.
McNab, W.H., and P.E. Avers (compilers), 1994. Ecological Subregions of the United States. U.S. For. Serv. Administrative Publ. WO-WSA-5. Section descriptions and maps. 267 pp.
Medaris, L.G., Jr., 1983. Early Proterozoic geology of the Great Lakes region. Geol. Soc. Am. Memoir 160.
Medley, K.E., and J.R. Harman, 1987. Relationships between the vegetation tension zone and soils distribution across central lower Michigan. Mich. Bot. 26:78–87.
Merz, R.W., 1978. *Forest Atlas of the Midwest*. NC For. Exp. St., St. Paul, MN; NE For. Exp. St., Upper Darby, PA; College of Forestry, Univ. of Minn., St. Paul. 48 pp.
Mesolella, K.J., 1973. Northern Michigan Silurian reef fairway potential giant. World Oil 176: 67–71.
MSU CES (Michigan State University Cooperative Extension Service), 1985. Bulletins E-1866-70, Sea Grant College Program. East Lansing, MI.
Michigan Water Resources Commission, 1967. *Michigan and the Great Lakes*. WRC Public Info. Series II. State of MI Water Resources Comm., Dept. of Conservation. 35 pp.
Mickelson, D.M., 1987. Central Lowlands. In: Graf, W.L. (ed.). *Geomorphic Systems of North America*. Geol. Soc. Am. Spec. Volume 2. pp. 111–118.
Mickelson, D.M., L. Clayton, D.S. Fullerton, and H.W. Borns, 1983. The Late Wisconsin glacial record of the Laurentide Ice Sheet in the United States. In: Wright, H.E., Jr., (ed.). *Late-Quaternary Environments of the United States. Vol. 1: The Late Pleistocene*. Univ. Minn. Press, Minneapolis. pp. 3–37.
Miller, N.G., and R.P. Futyma, 1987. Paleohydrological implications of Holocene peatland development in northern Michigan. Quat. Res. 27: 297–311.
Mortimer, C.H., 1987. Fifty years of physical investigations and related limnological studies on Lake Erie, 1928–1977. J. Great Lakes Res. 13: 407–435.
Murray, K.S., A. Farkas, M. Brennan, M. Czach, and M. Mayfield, 1997. Analysis of surface water quality in an urban watershed: Rouge River, southeastern Michigan. Mich. Academ. 29: 159–171.
Nichols, G.E., 1935. The hemlock-white pine-northern hardwood region of eastern North America. Ecology 16: 403–422.
Norton, D.C., and S.J. Bolsenga, 1993. Spatiotemporal trends in lake effect and continental snowfall in the Laurentian Great Lakes, 1951–1980. J. Climate 6: 1943–1956.
Ojala, C.F., and R.L. Ferrett, 1993. The decline of the tornado hazard in Michigan. Mich. Academ. 25: 397–410.
Olson, G.W., and F.D. Hole, 1967. The fragipan soils of north-eastern Wisconsin. Trans. Wis. Acad. Sci. Arts Lett. 56: 173–184.
Olson, J.S., 1958. Rates of succession and soil changes on southern Lake Michigan sand dunes. Bot. Gaz. 119: 125–170.
Palik, B.J., and K.S. Pregitzer, 1992. A comparison of presettlement and present-day forests on two bigtooth aspen-dominated landscapes in northern lower Michigan. Am. Midl. Nat. 127: 327–338.
Penrose, L., B.T. Penrose, and R. Penrose, 1988. *A Guide to 199 Michigan Waterfalls*. Friede Publications, Davison, MI. 184 pp.
Pentland, R.L., 1968. Runoff characteristics in the Great Lakes basin. Proc. 11th Conf. Great Lakes Res., Intl. Assoc. Great Lakes Res. pp. 326–359.
Phillips, D.W., 1972. Modification of surface air over Lake Ontario in winter. Mon. Wea. Rev. 100: 662–670.
Phillips, D.W., and J.A.W. McCulloch, 1972. The climate of the Great Lakes basin. Environment Canada, Climatological Studies No. 20. 40 pp.
Pickett, S.T.A., and P.S. White, (eds.), 1985. *The Ecology of Natural Disturbance and Patch Dynamics*. Academic Press, New York. 472 pp.
Pincus, H.J., (ed.), 1962. Great Lakes Basin. AAAS Publ. 71. Horn-Shafer Co., Baltimore, MD.
Platzman, G.W., 1966. The daily variation of water level on Lake Erie. J. Geophys. Res. 71: 2472–2483.
Pregitzer, K.S., B.V. Barnes, and G.D. Lemme, 1983. Relationship of topography to soils and vegetation in an Upper Michigan ecosystem. Soil Sci. Soc. Am. J. 47: 117–123.
Putnam, D.F., and L.J. Chapman, 1943. The drumlins of southern Ontario. Trans. Royal Soc. Canada (Series 3) 37:75–88.
Quinn, F.H., 1977. Annual and seasonal flow variations through the Straits of Mackinac. Water Resour. Res. 13: 137–144.
Quinn, F.H., 1988. Great Lakes water levels, past, present, and future. In: *The Great Lakes: Living with North America's Inland Waters*. Proc. Am. Water Resour. Assoc., Milwaukee. pp. 83–92.
Quinn, F.H., 1992a. Effects of climate change on the water resources of the Great Lakes. In: *Climate Change on the Great Lakes Basin*, Proc. Ann. Mtg. Am. Assoc. Advancement of Sci., Chicago, IL. pp. 10–16.
Quinn, F.H., 1992b. Hydraulic residence times for the Laurentian Great Lakes. J. Great Lakes Res. 18: 22–28.

Quinn, F.H., and C.E. Sellinger, 1990. Lake Michigan record levels of 1838, a present perspective. J. Great Lakes Res. 16: 133–138.

Reinking, R.F., R. Caiazza, R.A. Kropfli, B.W. Orr, B.E. Martner, T.A. Niziol, G.P. Byrd, R.S. Penc, R.J. Zamora, J.B. Snider, R.J. Ballentine, A.J. Stamm, C.D. Bedford, P. Joe, and A.J. Koscielny, 1993. The Lake Ontario winter storms (LOWS) project. Bull. Amer. Meteor. Soc. 74: 1828–1849.

Rickard, T.A., 1992. Mass copper. Mineralogical Record 23: 17–23.

Rieck, R.L., and H.A. Winters, 1982a. Characteristics of a glacially buried cuesta in southeast Michigan. Ann. Assoc. Am. Geog. 72: 482–494.

Rieck, R.L., and H.A. Winters, 1982b. Low altitude organic deposits in Michigan: Evidence for pre-Woodfordian Great Lakes and paleosurfaces. Geol. Soc. Am. Bull. 93:726–734.

Rieck, R.L., and H.A. Winters, 1991. Paleotopography and present terrain in southwestern Michigan. Mich. Academ. 23: 241–256.

Rieck, R.L., and H.A. Winters, 1993. Drift volume in the southern peninsula of Michigan—A prodigious Pleistocene endowment. Phys. Geog. 14: 478–493.

Rodionov, S.N., 1994. Association between winter precipitation and water level fluctuations in the Great Lakes and atmospehric circulation patterns. J. Clim. 7: 1693–1706.

Roscoe, S.M., 1973. The Huronian Supergroup, a Paleoaphebian succession showing evidence of atmospheric evolution. Geol. Assoc. Canada Spec. Paper 12: 31–47.

Ruhe, R.V., 1984. Loess derived soils, Mississippi valley region: I. Soil sedimentation system. Soil Sci. Soc. Am. J. 48: 859–867.

Runkle, J.R., 1982. Patterns of disturbance in some old-growth mesic forests of North America. Ecology 63: 1533–1556.

Runkle, J.R., 1985. Disturbance regimes in temperate forests. In: Pickett, S.T.A., and P.S. White. (eds.). *The Ecology of Natural Disturbance and Patch Dynamics.* Academic Press, New York. pp. 17–33.

Saarnisto, M., 1974. The deglaciation history of the Lake Superior region and its climatic implications. Quat. Res. 4: 316–339.

Saulesleja, A., 1986. *Great Lakes Climatological Atlas.* Environment Canada. 145 pp.

Schaetzl, R.J., 1991. Factors affecting the formation of dark, thick epipedons beneath forest vegetation, Michigan, USA. J. Soil Sci. 42:501–512.

Schaetzl, R.J., and D.G. Brown, 1996. Forest associations and soil drainage classes in presettlement Baraga County, Michigan. Great Lakes Geog. 3:57–74.

Schaetzl, R.J., and S.A. Isard, 1991. The distribution of Spodosol soils in southern Michigan: A climatic interpretation. Annals Assoc. Am. Geogs. 81: 425–442.

Schaetzl, R.J., and S.A. Isard, 1996. Regional-scale relationships between climate and strength of podzolization in the Great Lakes region, North America. Catena 28: 47–69.

Schaetzl, R.J., W.E. Frederick, and L. Tornes, 1996. Secondary carbonates in three fine and fine-loamy Alfisols in Michigan. Soil Sci. Soc. Am. J. 60: 1862–1870.

Schaetzl, R.J., S.F. Burns, D.L. Johnson, and T.W. Small, 1989. Tree uprooting: Review of impacts on forest ecology. Vegetatio 79: 165–176.

Schenk, P.E., 1965. Precambrian glaciated surface beneath the Gowganda Formation, Lake Timagami, Ontario. Science 149: 176–177.

Schwartz, M.D., 1991. An integrated approach to air mass classification in the North Central United States. Prof. Geog. 43:77–91.

Scott, I.D., 1921. *Inland Lakes of Michigan.* Mich. Geol. and Biol. Survey Publ. 30. Geological Series 25.

Scott, R.W., and F.A. Huff, 1997. Impacts of the Great Lakes on regional climate conditions. J. Great Lakes Rés. 22: 845–863.

Shelford, V.E., 1963. *The Ecology of North America.* Univ. Illinois Press, Urbana. 610 pp.

Shilts, W.W., J.M. Aylsworth, C.A. Kaszycki, and R.A. Klassen, 1987. Canadian Shield. In: Graf, W.L. (ed.). *Geomorphic Systems of North America.* Geol. Soc. Am. Centennial Special Vol. 2. pp. 119–161.

Simpson, J.E., 1994. *Sea Breeze and Local Wind.* Cambridge Univ. Press, Cambridge. 234 pp.

Smith, G.D., W.H. Allaway, and F.F. Riecken, 1950. Prairie soils of the upper Mississippi valley. Adv. Agron. 2: 157–205.

Smith, H.T.U., 1949. Periglacial features in the Driftless Area of southern Wisconsin. J. Geol. 57: 196–215.

Sousounis, P.J., and J.M. Fritsch, 1994. Lake-aggregate mesoscale disturbances. Part II: A case study of the effects on regional and synoptic-scale weather systems. Bull. Amer. Meteor. Soc. 75: 1793–1811.

Steinhart, C.E., L.J. Schierow, and W.C. Sonzogni, 1982. An environmental quality index for the Great Lakes. Water Resources Bull. 18: 1025–1031.

Strommen, N.D., and J.R. Harman, 1978. Seasonally changing patterns of lake-effect snowfall in western lower Michigan. Mon. Wea. Rev. 106: 504–509.

Swackhamer, D.L., and D.E. Armstrong, 1987. Distribution and characterization of PCBs in Lake Michigan water. J. Great Lakes Res. 13: 24–36.

Sweeney, R.A., 1993. "Dead" sea of North America?—Lake Erie in the 1960s and '70s. J. Great Lakes Res. 19: 198–199.

Symons, D.T.A., 1975. Huronian glaciation and polar wander from the Gowganda Formation, Ontario. Geology 3: 303–306.

Technical Committee on Soil Survey, 1960. *Soils of the North Central Region of the United States.* Univ. Wisconsin Agric. Exper. Station Bull. 544. 192 pp.

Thomas, M.K., 1964. A survey of Great Lakes snowfall. Great Lakes Res. Division, Univ. of Michigan. Conf. Great Lakes Research, Publ. No. 11. pp. 294–310.

Thomas, R., and R.C. Anderson, 1990. Presettlement vegetation of the Mackinaw River valley, central Illinois. Trans. Ill. State Acad. Sci. 83: 10–22.

Totten, S.M., 1969. Overridden recessional moraines in north-central Ohio. Geol. Soc. Am. Bull. 80: 1931–1946.

Trenhaile, A.S., 1971. Drumlins: Their distribution, orientation and morphology. Can. Geog. 15: 113–126.

Trewartha, G.T., 1981. The biomodal warm season precipitation profile of the upper middle west. Annals Assoc. Am. Geog. 71:566–571.

Tushingham, A.M., 1992. Postglacial uplift predictions and historical water levels of the Great Lakes. J. Great Lakes Res. 18: 440–455.

Tyrrell, L.E., and T.R. Crow, 1994. Structural characteristics of old-growth hemlock-hardwood forests in relation to age. Ecology 75: 370–386.

U.S. Geological Survey (USGS), 1987. Soils. In: *National Atlas of the United States of America*. U.S. Dept. of Interior, USGS, Reston, VA. Map: scale = 1:7,500,000.

U.S. Weather Bureau, 1959. Climatology and weather services of the St. Lawrence Seaway and the Great Lakes. U.S. Dept. of Commerce Tech. Paper No. 35. 75 pp.

Veatch, J.O., 1927. The dry prairies of Michigan. Mich. Acad. 8: 269–278.

Vincent, J.-S., and V.K. Prest, 1987. The early Wisconsinan history of the Laurentide ice sheet. Geog. Physique Quat. 41: 199–213.

Wascher, H.L., J.D. Alexander, B.W. Ray, A.H. Beavers, and R.T. Odell, 1960. Characteristics of soils associated with glacial tills in northeastern Illinois. Univ. of Illinois Agric. Exp. Sta. Bull. 665. 155 pp.

Webb, T., III, E.J. Cushing, and H.E. Wright, 1983. Holocene changes in the vegetation of the Midwest. In: Wright, H.E., Jr., (ed.). *Late-Quaternary Environments of the United States*, Vol. 2, The Holocene. Univ. of Minn. Press, Minneapolis. pp. 142–165.

Weber, M.G., 1987. Decomposition, litter fall, and forest floor nutrient dynamics in relation to fire in eastern Ontario jack pine ecosystems. Can. J. For. Res. 17: 1496–1506.

Wharton, R.J., 1970. Geology of the Gogebic Iron Range. In: Humphrys, C.R. *Lake Superior—The only Great lake*. Student Water Publications, Mich. St. Univ. Vol. 1. pp. A17–A22.

White, G.W., 1973. History of investigation and classification of Wisconsinan drift in north-central United States. In: Black, R.F., R.P. Goldthwait, and H.B. Willman, (eds). *The Wisconsinan Stage*. Geol. Soc. Am. Memoir 136. pp. 3–64.

Whitney, G.G., 1986. Relation of Michigan's presettlement pine forests to substrate and disturbance history. Ecology 67: 1548–1559.

Whitney, G.G., 1987. An ecological history of the Great Lakes forest of Michigan. J. Ecol. 75: 667–684.

Wilde, S.A., 1933. The relation of soils and forest vegetation of the lakes states region. Ecology 14: 94–105.

Willman, H.B., and J.C. Frye, 1970. Pleistocene stratigraphy of Illinois. Ill. St. Geol. Surv. Bull. 94. 204 pp.

Winters, H.A., 1961. Landforms associated with stagnant glacial ice. Prof. Geogr. l3: 19–23.

Winters, H.A., and R.L. Rieck, 1980. Significance of landforms in southeast Michigan. Ann. Assoc. Am. Geog. 70: 4l3–424.

Winters, H.A., and R.L. Rieck, 1982. Drainage reversals and transverse relationships of rivers to moraines in southern Michigan. Phys. Geog. 3: 70–82.

Winters, H.A., R.L. Rieck, and R.O. Kapp, 1986. Significance and ages of mid-Wisconsinan organic deposits in southern Michigan. Phys. Geog. 7: 292–305.

Wright, H.E., Jr., 1957. Stone orientation in Wadena drumlin field, Minnesota. Geografiska Annaler 39: 19–31.

Wylie, D.P., and J.A. Young. 1979. Boundary-layer observation of warm air modification over Lake Michigan using a tethered balloon. Bounday Layer Meteor. 17: 279–291.

Wyrick, J.E., 1985. Storage of gas in salt caverns. In: Schlitt, W. (ed.). *Proc. of the Symposium on Solution Mining of Salts and Brines*, Soc. Mining Engineers, New York. pp. 65–69.

Young, G.M., 1973. Tillites and aluminous quartzites as possible time markers for middle Precambrian (Aphebian) rocks of North America. Geol. Assoc. Canada Spec. Paper 12: 97–127.

17

The Central Lowlands and Great Plains

Vance T. Holliday
James C. Knox
Garry L. Running IV
Rolfe D. Mandel
C. Reid Ferring

The Central Lowlands and Great Plains represent the stable interior of North America south of the Boreal Forest (fig. 17.1). The two regions comprise the vast low-relief landscape between the Appalachian-Ouachita mountain system on the east and southeast, and the Rocky Mountains and eastern Basin and Range Province on the west. Most of the Central Lowlands are below 600 m above sea level and substantial segments are lower than 300 m. The Great Plains, drier and higher than the Lowlands, reach elevations between 600 and 1800 m. The Central Lowlands and Great Plains are stereotypically characterized as environmentally, topographically, and geologically homogeneous if not monotonous. As shown subsequently, however, these regions possess considerable physiographic diversity.

17.1 Geological and Environmental Setting

17.1.1 Bedrock Geology

The bedrock of the Central Lowlands and Great Plains consists mostly of terrestrial and shallow marine sedimentary material representing most of Phanerozoic time (Sloss, 1988; Stott and Aitken, 1993). In the Central Lowlands, the Paleozoic and Mesozoic rocks become generally younger to the south and west away from the Precambrian crystalline cratonic core of the continent exposed toward their northern margin. Some variation in the outcrop pattern is imposed by large-scale Paleozoic structures such as the Michigan and Illinois Basins and the Cincinnati and Wisconsin Arches in the northeastern Central Lowlands, the Anadarko Basin in the southern Central Lowlands, and the Williston Basin in the northern Great Plains (fig. 17.1) (Sloss, 1988; Bally, 1989). Lower Paleozoic rocks, more common in the northern Central Lowlands around the Great Lakes, are mostly marine sediment, usually limestone or shale. Upper Paleozoic rocks are more common in the southern half of the Central Lowlands and include both terrestrial deposits, typically sandstones and evaporites, and marine deposits, including limestones and shales.

Mesozoic deposits are most common along the western margin of the Central Lowlands and throughout the Great Plains. These deposits are mainly Cretaceous limestones and terrestrial clastics. Though often eroded or buried beneath the High Plains, these rocks are ubiquitous at or near

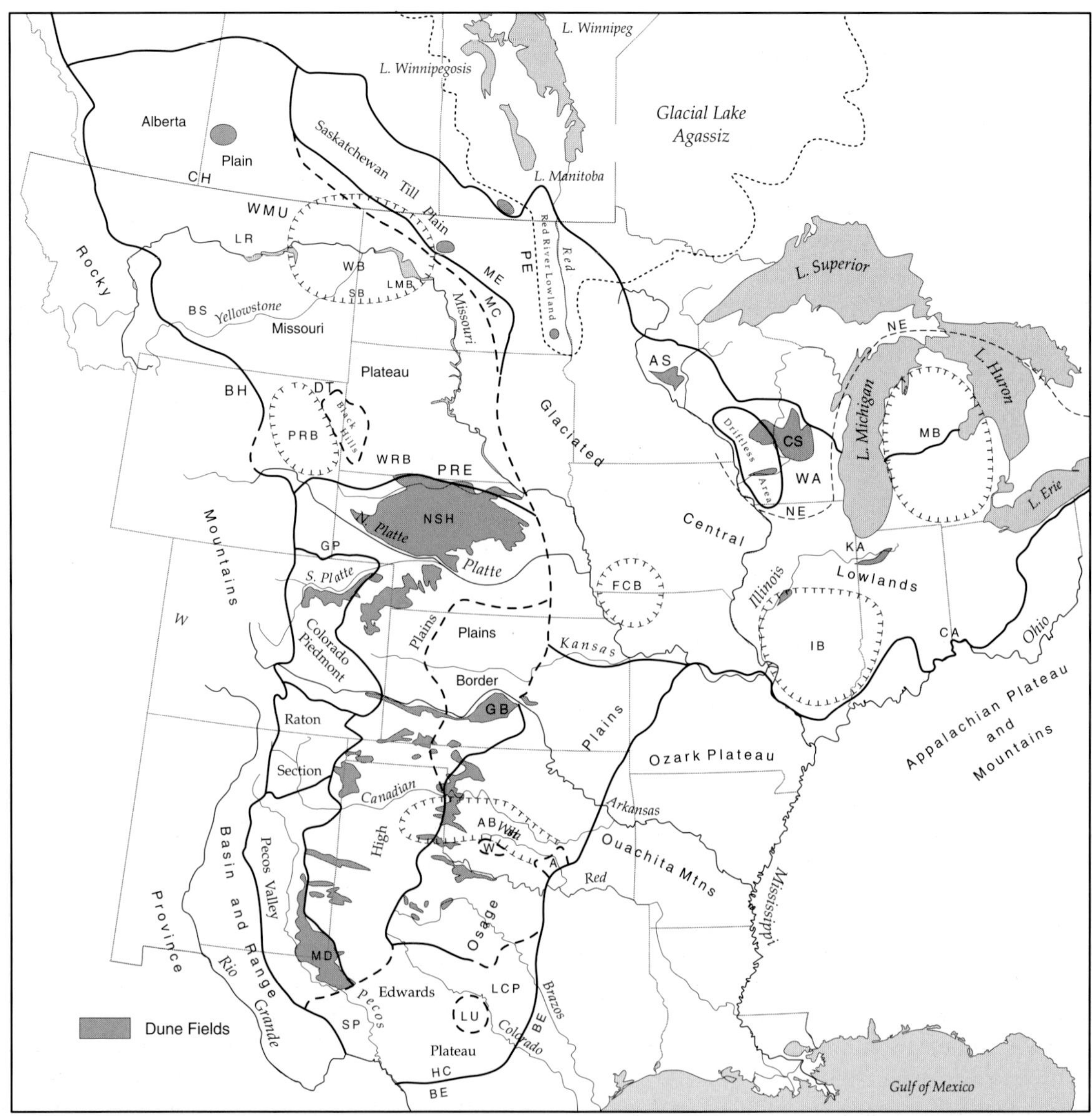

Figure 17.1 Central North America with the location of the Central Lowlands and Great Plains, and their physiographic subdivisions. The northern boundary is the approximate southern limit of the Boreal Forest. Also shown are the extent of Glacial Lake Agassiz (dashed line), selected structural basins and uplifts (AB = Anadarko Basin; A = Arbuckle Mountains; CA = Cincinnati Arch; FCB = Forest City Basin; IB = Illinois Basin; KA = Kankakee Arch; LU = Llano Uplift; MB = Michigan Basin; PRB = Powder River Basin; W = Wichita Mountains; WB = Williston Basin; WA = Wisconsin Arch), selected dune fields (AS = Anoka Sand Plain; CS = Central Sands of Wisconsin; GB = Great Bend Sand Prairie; MD = Mescalero Dunes; NSH = Nebraska Sand Hills), and other selected physiographic features (BE = Balcones Escarpment; BH = Big Horn Mountains; BS = Big Snowy Mountains; CH = Cypress Hills; DT = Devil's Tower; GP = Gangplank; HC = Hill Country; LCP = Lampasas Cut Plain; LMB = Little Missouri Badlands; LR = Little Rocky Mountains; MC = Missouri Coteau; ME = Missouri Escarpment; NE = Niagara Escarpment; PE = Pembina Escarpment; PRE = Pine Ridge Escarpment; SB = Sentinel Butte; SP = Stockton Plateau; WMU = Wood Mountain Upland; WRB = White River Badlands) (see text for relevant references).

the surface from the Edwards Plateau to the Arctic Ocean. Triassic and Jurassic rocks are less common, but extensive Triassic outcrops occur beneath the Osage Plains and Pecos River Valley. Mesozoic rocks are generally downwarped in a broad basin below the Great Plains, producing a gentle westward dip. Along the Rocky Mountain Front and around the Black Hills, both Mesozoic and Paleozoic rocks have been sharply upturned by orogenic uplift.

The Cenozoic is represented by lacustrine sediments and a variety of terrestrial clastic deposits (Sloss, 1988; Klassen, 1989; Gustavson et al., 1991; Madole et al., 1991; Wayne et al., 1991; Stott et al., 1993). Mostly Tertiary rocks underlie the Great Plains and are directly or indirectly related to the uplifts associated with the Laramide Orogeny and development of the Rio Grande Rift. Lower Tertiary strata are more common in the northern Great Plains; younger deposits are more common to the south. This southward progression reflects the timing of orogenic activity in the Rocky Mountains, with uplift beginning in the late Mesozoic in the north, progressing southward, and culminating in the Rio Grande Rift in middle to late Cenozoic time (Gabrielse and Yorath, 1991; Burchfiel et al., 1992). In the north, early to middle Tertiary deposits beneath the Alberta Plains and Missouri Plateau include alluvial and aeolian sediments derived from erosion of the Rocky Mountains, volcaniclastic rocks from volcanic centers in those mountains, and lacustrine sediments deposited in broad structural lows such as the Powder River Basin and Williston Basin. To the south, most Tertiary strata beneath the High Plains are Miocene-Pliocene alluvial and aeolian deposits, including the Ogallala Formation, derived from the Rocky Mountains and from the eastern flank of the Rio Grande Rift. Near the top of these sediments is an unusually well-developed pedogenic calcrete, the "caprock caliche" (fig. 17.2) (Reeves and Reeves, 1996). Quaternary sediments include a broad variety of alluvial, glacial, and aeolian deposits along with more localized lacustrine and volcanic deposits.

Erosion of the mountain systems and aggradation of the Great Plains has continued to be driven by epeirogenic uplift throughout late Cenozoic (post-Laramide) time (Gable and Hatton, 1983). Over the past 10 million years, the Rocky Mountain Front rose 1.5 to 2.0 km and the eastern Great Plains rose 100 to 500 m. Such uplift, in addition to building the Great Plains and enhancing the east-flowing drainage system, had a substantial impact on the environment by creating a rain shadow and drying the region. Localized structural and volcanic events also occurred.

Figure 17.2 The Caprock Escarpment separating the High Plains (left) from the Osage Plains (right). This view (looking northeast) is along the east-central southern High Plains in Texas. The rim-forming Ogallala "caprock caliche" is prominent just below the skyline at left.

17.1.2 Physiography and Modern Environment

The physiographic subdivisions discussed in this chapter follow accepted scientific usage (Fenneman, 1931, 1938; Slaymaker, 1989). The Central Lowlands are divided into the glaciated Central Lowlands, including the Driftless Area, and the unglaciated Osage Plains (fig. 17.1). The Great Plains are divided into the northern Great Plains (the Alberta Plains and Missouri Plateau) and southern Great Plains; the northern edge of the High Plains forms the boundary between the southern and northern Great Plains (fig. 17.1).

The physical configuration of the Central Lowlands and Great Plains is due largely to Cenozoic events and processes, particularly the Laramide Orogeny, late Cenozoic epeirogenesis, and Quaternary glaciation. Development of an east-flowing drainage across the continental interior (e.g., Witzke and Ludvigson, 1990) and aggradation of sediment created an extensive constructional surface. The High Plains are the best preserved example of this process. Development of the Mississippi-Missouri River system to the north and east isolated and dissected the High Plains, leaving them as a series of broad east-sloping plateaus. Steep escarpments are present along the margins of the plateau where it is held up by the Ogallala caprock (fig. 17.2). The caprock is weakly expressed or absent along the eastern Great Plains of Kansas, where a sloping surface cut on Ogallala sediments connects the High Plains surface with the Central Lowlands. This eroded "plains border" region is usually considered part of the High Plains, but is discussed here as part of the Central Lowlands. Cenozoic sediment did not cover the far southern end of the Great Plains. Instead, Cenozoic uplift elevated the Cretaceous limestone, forming the Edwards Plateau. During the Quaternary, the northern Great Plains were also influenced by glacial erosion and deposition along the southwest margin of the Laurentide Ice Sheet.

The Central Lowlands have undergone long-term erosion by the Mississippi-Missouri River system and substantial modification by the advance and retreat of glaciers. In the western and southern Central Lowlands, removal of Cenozoic sediment east of the High Plains and erosion of limestone on the northeastern Edwards Plateau demarks the boundary between the Plains and the Lowlands. Development of the upper Missouri River basin created the Missouri Plateau subsection of the Great Plains. In the Quaternary, glacial lobes on the southern margin of the Laurentide Ice Sheet repeatedly covered much of the northern Central Lowlands, but two areas were never glaciated: the Driftless Area in southwest Wisconsin and the Osage Plains on the southern Central Lowlands.

The size and location of the Central Lowlands and Great Plains result in a region with significant latitudinal and longitudinal environmental gradients and a classic continental climate (Borchert, 1950; Bryson and Hare, 1974; Court, 1974). With few topographic impediments, the climate of this vast region is subject to influences from several principal air masses: cold dry Arctic air, warm moist tropical air, and mild dry Pacific westerlies. Precipitation gradients decrease from the humid east (Dbf and Daf climate, 80–90 cm MAP) to the semiarid west (Bsk climate; 30–40 cm MAP), although temperature gradients decrease from the southern subtropics (Caf climate, 21°C MAT) northward, almost reaching the subarctic (Dbf and Dcf climates, 3°C MAT) (Trewartha, 1954; see also chapter 12). Average values for temperature and particularly precipitation are, however, relatively meaningless because of the strong continentality of the region.

Continentality is one of several climatic phenomena that characterize the region; others include severe thunderstorms, drought, and wind. The Great Plains and part of the Central Lowlands tend to be dominated by strong westerly flow, which serves to keep the Plains dry as a result of the rain-shadow effect and also operates as a wedge separating the Arctic and Gulf air masses. This zonal flow weakens in the spring and summer when meridional flows allow interaction of the warmer, wetter air with the colder, drier air. The result is frequent and often severe frontal thunderstorms. A strong jet aloft may result in tornadic storms. The eastern Great Plains and Central Lowlands comprise "tornado alley," where more tornadoes form than in any other region in the world (Eagleman, 1990).

Variability in the strength and moisture content of the air masses affecting the continental interior, as well as in the position and persistence of the jet stream, produce significant variability in annual precipitation. For example, at Lubbock, Texas, on the southwestern Great Plains, the mean annual precipitation is 48 cm, but annual totals range from 17 cm (1917) to 103 cm (1941). In Madison, Wisconsin, on the northern Central Lowlands, mean annual precipitation is 78 cm, but during a recent 5-year period varied from 62 cm (1988) to 110 cm (1993). A principal factor controlling humidity is the position and movement of tropical moisture from the Gulf of Mexico. These air masses move inland from the northern Gulf of Mexico and, in the spring and summer, dominate the central United States, including much of the Central Lowlands and occasionally the eastern Great Plains. Adequate precipitation throughout the region is dependent on frequent interaction of tropical air masses with either Arctic air or dry westerlies. Drought results from a persistence in either meridional or zonal flow, but no one type of circulation pattern accounts for all precipitation deficits in the region (Borchert, 1971; Barry, 1983).

Wind is one of the most conspicuous and persistent climatic phenomena on the Great Plains and western Central Lowlands (Johnson, 1965; Holliday, 1987). The treeless, low-relief landscape combined with the frequent passing of weather fronts through the region results in the almost constant movement of air. The Great Plains are the most persistently windy inland area of North America and have some of the highest average annual wind speeds of any nonmaritime region. Wind has had and continues to have a significant impact on the region, responsible in the past for the deposition of much surficial sediment and today for significant erosion.

The preagricultural vegetation pattern of the Central Lowlands and Great Plains largely reflected the climate pattern: the humid regions were forested and the drier areas were grasslands (Kuchler, 1964; Daubenmire, 1978; Harrington and Harman, 1991; Bragg, 1995). Continuous forest cover characterized the coolest, wettest portion of the Central Lowlands, whereas a complex pattern of oak savanna characterized the warmer and especially drier prairie-forest ecotone. Riparian forests penetrated the drier grassland regions to the west, following the floodplains of most major rivers. A wedge-shaped region in the western and northern Central Lowlands was covered with extensive areas of tallgrass prairie, the so-called Prairie Peninsula (Wright, 1968). To the south, the Osage Plains had mixed grasses, with oak in the east and mixed grasses with mesquite in the west. Most of the Great Plains was shortgrass prairie, although mixed grasses also were common on the eastern and northern Great Plains.

Several climatic characteristics of the region directly control the vegetation pattern. The average western margin of the Gulf of Mexico moisture defines the climatic transition between the humid, forested southeastern United States and the semiarid prairies of the plains (Borchert, 1950; Court, 1974). The prairie-forest ecotone and the tallgrass prairie, a region that climatically can support forest, probably were maintained by fire (Sauer, 1950; Curtis, 1956; Collins and Glenn, 1995). Many fires probably were started when dry grass was struck by the lightning that accompanies spring and summer storms. During Holocene times, human activity probably caused some grass fires as well, but the role played by prehistoric peoples in starting fires is difficult to determine. The grasslands of the plains also are characterized by periodic drought (Weaver, 1968), the most severe droughts occurring on the short-grass prairies of the southern and southwestern Great Plains (Bragg, 1995).

17.2 Central Lowlands: Glaciated and Driftless Area

17.2.1 Glaciated Central Lowlands

The Central Lowlands of North America were affected by a variety of geologic and geomorphic processes, but the Quaternary evolution of drainage systems and the repeated advance and retreat of glaciers created most of the present landforms in the region. The numerous glacial advances into the midcontinent (fig. 17.3) significantly altered drainage lines and the locations of drainage divides. Maps of the preglacial bedrock surface suggest a system of deeply incised valleys that differ significantly from the modern drainage configuration. One example is the Teays-Mahomet bedrock valley, a relict, deeply entrenched preglacial valley system in central Illinois, north-central Indiana, and southwestern Ohio (Melhorn and Kempton, 1991a,b).

The preglacial configuration of the upper Mississippi River is poorly understood, and segments of the modern upper Mississippi River appear to date from the early Pleistocene. The modern river is discordant with regional geologic structure, a condition that appears to be a result of Pleistocene glaciation (Willman and Frye, 1969; Anderson, 1988). The anticlinal, dome-like structure provided by the Wisconsin Arch (fig. 17.1), with nearby cratonic basins, results in a series of bedrock cuestas that form semicircular outcrop patterns that extend westward across the Mississippi River from southern Wisconsin and northern Illinois. Along the borders of southwestern Wisconsin and northwestern Illinois, the southward-flowing Mississippi River is entrenched through two of these prominent dolomite cuestas.

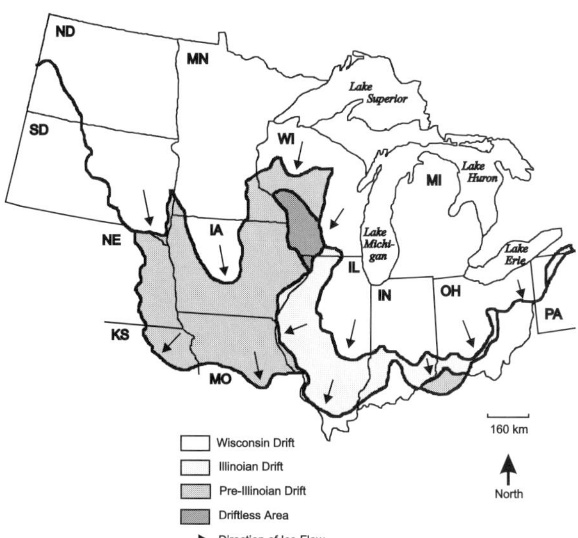

Figure 17.3 Extent of Pleistocene glaciations in the Central Lowlands of the United States (after Thornbury, 1965; Ruhe, 1969; Mickelson et al., 1983; Fullerton, 1986; Hallberg, 1986; Johnson, 1986; Dyke and Prest, 1987a,b; Mickelson, 1987; Knox and Attig, 1988).

The Central Lowlands probably represent one of the best known and most intensively studied glacial landscapes in North America. In the mid-nineteenth century, the region was among the first in the Western Hemisphere to be studied for the effects of continental glaciation, and, later in the same century, it was the site of pioneering research by T.C. Chamberlin and colleagues (Chamberlin, 1883; Chamberlin and Salisbury, 1885), who first recognized that North America was subjected to multiple Pleistocene glaciations. The resulting interpretation of glacial history—that there were four major stages of glaciation in the Central Lowlands (Nebraskan, Kansan, Illinoian, and Wisconsinan, oldest to youngest)—dominated thinking for well over half a century. Later evidence for at least six pre-Illinoian glaciations led to abandonment of the terms *Kansan* and *Nebraskan* as stratigraphic names (Boellstorff, 1978; Hallberg, 1986). All glacial deposits older than the Illinoian are currently referred to as pre-Illinoian (Richmond and Fullerton, 1986). Although convention still places the Pliocene-Pleistocene boundary at 1.65 Ma (million years before present) (Richmond and Fullerton, 1986), the oldest till in the Midwest predates this boundary at least by 0.5 Ma (Hallberg, 1986).

The Illinoian glaciation, about 300,000–130,000 years ago (Johnson, 1986; Richmond and Fullerton, 1986), was very extensive in the eastern Central Lowlands, and its southern limit almost reached the junction of the Ohio and Mississippi Rivers (fig. 17.3). Illinoian till is recognized also in a narrow band south of the late Wisconsinan drift margin near the eastern extremities of the glaciated Central Lowlands (Fullerton, 1986:28–29). The Illinoian ice sheet temporarily diverted the Mississippi River from its existing course, which before diversion was southeastward across northwestern Illinois to the present Illinois River and then southward along the present Illinois River to the present Mississippi River. The blockage of the Mississippi River along northwestern Illinois created a large lake on the upper Mississippi River that extended northward into the Driftless Area (Leigh and Knox, 1990). Spillover and upper Mississippi River drainage was then routed across the uplands of southeastern Iowa along the Illinoian ice front (Hallberg, 1980). After Illinoian glaciation, the Mississippi returned to its original course through central Illinois, following the course of the present Illinois River. About 20,000 years ago westward expansion of the late Wisconsinan glacier blocked the Mississippi River in north-central Illinois, resulting in a lake whose drainage incised the river in its present course along the western margin of Illinois (McKay, 1979:104).

Early Wisconsinan glaciation began about 79,000 years ago (Richmond and Fullerton, 1986), but no unequivocal early Wisconsinan till is clearly dated in the Central Lowlands (Curry and Pavich, 1996). Middle Wisconsinan glaciation in the Central Lowlands is inferred by glacial and aeolian deposits, but the till is poorly dated (Johnson, 1986; Johnson et al., 1997). Loess deposits derived from valley outwash, however, date the middle Wisconsinan between about 55,000 and 28,000 yr B.P. (Leigh and Knox, 1993, 1994; Leigh, 1994; McKay, 1979). After about 28,000 yr B.P., late Wisconsinan glaciers readvanced into the Central Lowlands and reached the headwaters of the Mississippi basin and into northern Iowa by about 25,000 yr B.P. (fig. 17.3) (McKay, 1979; Kemmis et al., 1981; Leigh and Knox, 1993). Maximum extent of late Wisconsinan glaciation occurred between about 21,000 and 18,000 yr B.P. (Fullerton, 1986; Johnson, 1986; Dyke and Prest, 1987a,b; Johnson et al., 1997).

Aeolian deposits of silt and sand are both well represented in the Central Lowlands (Thorpe and Smith, 1952). The most extensive aeolian deposit in the glaciated Central Lowlands is loess. Most of these loesses probably were derived from deflation of proglacial outwash, but deflation of periglacial and other cold dry environments also contributed to loess deposition in some areas of the region (Ruhe, 1983). For example, a significant part of the late Wisconsinan Peoria Loess in extreme southeastern Minnesota may be derived from very sparsely vegetated upland surfaces to the west (Mason et al., 1994). Nevertheless, mapping of regional loess thickness typically shows maximum values along the eastern margins of major glacial outwash valleys, such as the Mississippi and Missouri Rivers, and distinct fining downwind (eastward) (Ruhe, 1983). The thickness of late Wisconsinan loess varies from a maximum of about 20 m in western Iowa along the eastern margin of the Missouri River to about 4 to 5 m thick on much of the hilly pre-Illinoian drift region. Ruhe also showed that many areas of youngest late Wisconsinan glaciation, as in north-central Iowa, south-central, west-central, and northern Minnesota, northeastern Wisconsin, and most of Michigan, have insignificant loess cover.

Although loess of late Wisconsinan age dominates the Central Lowlands, Illinoian loess is present at many sites in the region. The pre-Illinoian loess record is fragmentary and missing at most localities. Only about six or seven distinctive loess or loess-derived units are reported (Leigh and Knox, 1994), none older than 790,000 yr B.P. (Hajic, 1986; Jacobs and Knox, 1994). The absence of early Pleistocene loess in the region probably is due to removal by erosion or to weathering, reworking, and mixing with bedrock residuum, which renders the sediment unrecognizable as loess in the modern record (Jacobs et al., 1997). Consequently, loess at most sites in the region is of late Wisconsinan age. Deposition of this loess ended between 14,000 and 12,000 yr B.P. (Ruhe, 1983).

Sand dunes and sheet sands occur along and especially adjacent to the eastern margins of most rivers that transported proglacial outwash (fig. 17.1) (e.g., Thorpe and Smith, 1952; Lineback, 1979; Gray, 1989). They also occur locally along the eastern sides of other rivers where poorly cemented sandstone formations crop out (fig. 17.1). Proglacial outwash in combination with local geologic factors in some cases produced extensive sand accumulations. Examples include the Anoka Sand Plain of east-central Minnesota (fig. 17.1), built as late Wisconsinan ice receded (Cooper, 1938; Keen and Shane, 1990), and the Central Sands of Wisconsin (fig. 17.1), formed during erosion and sedimentation from several causes, including wave action in a large proglacial lake, glacial ice, and glacial meltwater (Martin, 1932:316–344; Clayton and Attig, 1989).

On a continental scale, the till plains of the Central Lowlands may appear as a relatively homogeneous landscape of low relief and gentle slope. However, on a regional scale, great diversity emerges. The diversity can be grouped into two large categories. The first category recognizes differentiation on the basis of glaciological processes. Process differentiation is mainly restricted to areas of late Wisconsinan drift, because older drift surfaces have been too eroded by fluvial action to retain much of their landform genetic history. Consequently, the second category recognizes differentiation on the basis of the amount of time that subaerial processes have been modifying landscape surfaces since deglaciation. The Wisconsinan, Illinoian, and pre-Illinoian drift surfaces (fig. 17.3) have distinctly different landform, slope, and relief characteristics.

The type and intensity of glaciological process vary significantly across the glaciated Central Lowlands because of regional differences in factors such as preglacial topography, bedrock geology and regolith, latitudinal location, and proximity to ice source regions. Twelve types of glacial landform areas are recognized in the late Wisconsinan drift region of the Central Lowlands (Mickelson et al., 1983). Thermal conditions at the former glacier bed were a critical factor controlling this landscape diversity. A former frozen-bed environment probably explains much of the glacial landscape of eastern and northern Wisconsin, northeastern Minnesota, and much of west-central North Dakota, where drumlins, tunnel channels, ice-thrust features, and high-relief hummocky moraines are common (Mickelson, 1987). For example, in the drumlin landscape of eastern Wisconsin, ice may have advanced onto a permafrost surface that slowly decayed beneath the ice sheet, allowing large amounts of erosion and streamlining to take place. Farther south and southeast in Illinois, Indiana, and Ohio, and on the late Wisconsinan ice lobe of north-central Iowa, the former glacier bed was unfrozen, wet, and sliding (Mickelson, 1987). Consequently, the record left in the sliding zone by the retreating ice is one of flat or gently undulating till plain with classic end moraines. This landscape probably is composed of basal till; the transporting ice likely had little englacial or supraglacial sediment load.

Regional landscape diversity, which results from the age of the drift, reflects the tendency for drainage networks to increase their density with time on a Quaternary time scale. Among drift surfaces of pre-Illinoian, late Wisconsinan 20,000 yr B.P. (Tazewell) age, and late Wisconsinan 14,000 yr B.P. (Cary) age, drainage density on the pre-Illinoian surface is about 1.8 times greater than the drainage density on the Tazewell drift and 4.5 times greater than the drainage density on the Cary drift (Ruhe, 1969:108–111). Although the general tendency on a Quaternary time scale is for relief, slope steepness, and drainage density to increase with time after deglaciation, the evolutionary development can be episodic. Ruhe (1969:108–111) found that drainage evolution on late Wisconsinan drift in northwestern Iowa was relatively rapid between 20,000 and 14,000 yr B.P. but has been relatively slow since 14,000 yr B.P. The rapid expansion of the drainage network between 20,000 and 14,000 yr B.P. probably occurred under periglacial climatic conditions, whereas the very modest development since 14,000 yr B.P. reflects the stabilization of the landscape under postglacial forest and grassland vegetation.

Differences in watershed relief, slope steepness, and drainage density strongly influence watershed hydrology. For example, for a drainage area of 10 km^2 in small watersheds in Iowa, the mean annual flood on the hilly, dissected, and high drainage density pre-Illinoian drift is about 15 m^3s^{-1}, but the comparable mean annual flood on the gently sloping, little dissected, and low drainage density late Wisconsinan Cary drift is only about 1.2 m^3s^{-1} (fig. 17.4). Here, under very similar climatic conditions, the contribution of drainage evolution during Quaternary time accounts for an order of magnitude difference in size of the mean annual flood. The pre-Illinoian watersheds generally are represented by thick loess over glacial till, whereas the late Wisconsinan watersheds have little loess cover over till. If both watershed groups had thick loess cover, the differential in flood magnitudes would be even greater than that shown in figure 17.4. The differential would also be greater than it is at present if extensive artificial drainage of the late Wisconsinan drift surface had not occurred since the mid-nineteenth century, as discussed subsequently.

17.2.2 Human Impacts in the Glaciated Central Lowlands

Before agricultural settlement, the glaciated Central Lowlands were a mosaic of prairie and forest, with forest clearly dominant in most areas. Williams (1989:361), citing data from other sources, showed that over 12.3 million hectares (ha) of forest were cleared in the north-central United States from 1850 through 1909. Using Ohio as a microcosm of forest clearing, Williams (1989:361) indicated that about 96% of the state's approximately 9.8 million ha was covered with deciduous broadleaf forests before European settlement. By 1910, over 80% of the western one-half and between 50 and 80% of the eastern one-half of Ohio had been cleared and converted to agricultural land use (Williams, 1989:367–368). The conversion of prairie and forest lands to agricultural land normally accelerates surface runoff and soil erosion for three reasons: (1) rain impact on bare soil breaks down soil peds and impedes infiltration; (2) organic matter is depleted in the soil profile, further reducing infiltration; and (3) hydraulic roughness of the ground surface is reduced and rates of surface runoff are increased as bare or sparsely vegetated surfaces replace natural vegetation cover and associated organic litter. Consequently, much of the Central Lowlands, which represents the heart of the Corn Belt and which was also a major wheat-growing region during the late nineteenth century, experienced major increases in the magnitude and frequency of floods and soil erosion after the introduction of agriculture (see chapter 23). For example, average summer rain storms in the unglaciated Driftless Area of southwest Wisconsin, which is topographically and hydrologically similar to the pre-Illinoian landscape, have caused resulting runoff peak discharges to increase from three to five times more than those of the preagricultural period (Knox, 1977). Increased surface runoff of this type has, in turn, resulted in extensive soil erosion throughout the region (Happ et al., 1940; Knox, 1977, 1987; Trimble and Lund, 1982; Trimble, 1983; Magilligan, 1985; Meade et al., 1990; Beach, 1994; Argabright et al., 1996).

Human modification of the natural drainage networks has also greatly affected hydrologic and geomorphic processes in the Central Lowlands, especially where surficial sediment is late Wisconsinan drift younger than about 14,000 yr B.P. The poor natural drainage and extensive wetlands on much of the Late Wisconsinan landscape led to extensive channelization and artificial drainage. Accelerated drainage of wetlands in the Midwest during the mid-nineteenth century was the result of several factors, including increasing population pressure, legislation regarding the disposal of unsold public lands, increasing market accessibility in the interior of the country, the need for ever-increasing amounts of food to fuel the urban industrial revolution, land speculation, and the fear of malaria (McCorvie and Lant, 1993). Losses of wetland acreage due to human-initiated drainage are estimated at 90% or more for Ohio, Iowa, Indiana, and Illinois, and 32–71% for Wisconsin, Minnesota, and Michigan (McCorvie and Lant, 1993) . The lower percentage losses in the latter grouping apparently reflect lower agricultural potential because of their more northerly location and the fact that the morainic sediment and glacial landforms of these areas

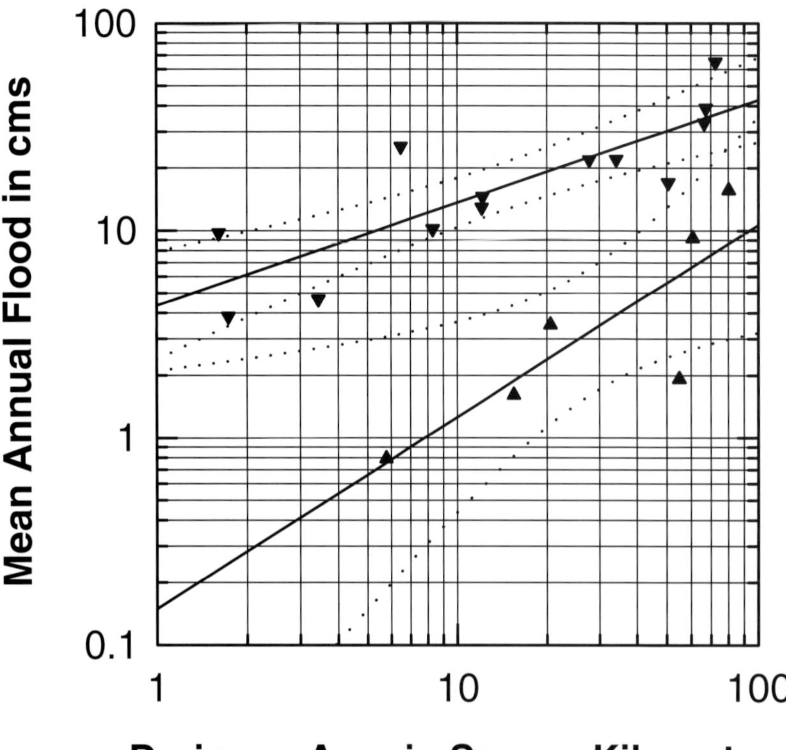

Figure 17.4 Physiographic influence on the magnitude of the mean annual flood in the glaciated Central Lowlands. Watersheds that developed in pre-Illinoian tills have steep hillslopes and high drainage densities compared to watersheds that developed in late Wisconsinan (Cary) till. Consequently, surface runoff as a fraction of total runoff is larger for the landscapes of pre-Illinoian age compared to the late Wisconsinan Cary-age landscapes (after U.S. Geological Survey, Iowa watersheds).

are less suitable for cultivation. The emplacement of artificial channels with large capacities and low hydraulic roughness has led to the replacement of many wetlands with agricultural fields, resulting in greatly enhanced surface runoff from these watersheds (Hirsch et al., 1990).

The acceleration of runoff has the greatest impact on the occurrence of small- and moderate-magnitude floods of relatively frequent occurrence (e.g., floods more frequent than the 50-year event). Extreme floods (e.g., less frequent than the 100-year event) typically involve runoff from rainfall or snowmelt that exceeds the thresholds of watershed surficial properties. During the Great Flood of 1993 on the upper Mississippi River, the media commonly speculated that drainage of wetlands and construction of flood control levées were major causal factors. However, careful analysis of rainfall-runoff relationships showed that the flooding was mainly the result of excessive rainfall on saturated soils (Pitlick, 1997; U.S. Army Corps of Engineers, 1996; see also chapter 7). Furthermore, the 1993 flood seems to be consistent with a shift in climatic conditions that have favored more frequent occurrences of large floods on the Mississippi River since about 1950 (Knox, 1984). Although agricultural land use, wetland drainage, and levée construction were not principally responsible for the Great Flood of 1993, these factors did contribute significantly to the transmission of agricultural chemicals (Goolsby et al., 1993).

17.2.3 The Driftless Area

The Driftless Area, as the name implies, is a region where ice-contact glacial deposits are absent. It is located mainly in southwestern Wisconsin, but also includes a small section of northwestern Illinois (fig. 17.3). Various lobes of continental glaciers abutted all sides of the Driftless Area, but the region was never completely encircled by glacial ice as shown on some early maps (Chamberlin, 1883, plate no. IX). Some early researchers thought the Driftless Area escaped glaciation because it was perceived to be higher than surrounding regions, but this hypothesis was subsequently refuted when topographic information showed the Driftless Area to be lower than many of the surrounding glaciated regions. Three principal factors contributed to the Driftless Area remaining unglaciated: (1) the southwestward diversion of ice flow into Minnesota by the deep trough associated with the structural syncline of Lake Superior; (2) the southward diversion of ice flow by the deep valley now occupied by Lake Michigan; and (3) ice flow resistance against the relatively high and erosionally resistant crystalline rocks of northern Wisconsin (Martin, 1932:113–118).

The "driftless" status of the Driftless Area has been challenged from time to time (Sardeson, 1897; Squire, 1897, 1898; Black, 1970), but no unequivocal evidence for glaciation has ever been presented (Knox, 1982; Mickelson et al., 1982). Given the enormous magnitude of erosion that has occurred in the Driftless Area during the Quaternary, it is possible that a pre-Illinoian glacier(s) covered the Driftless Area, but this seems unlikely based on known evidence. Similarly eroded topography in adjacent northeastern Iowa and southeastern Minnesota contains abundant evidence of pre-Illinoian glaciation. No evidence of this type is found in southwestern Wisconsin or northwestern Illinois except on and along the upland bluff tops at the western margin of the Driftless Area.

The hilly, steeply sloping, and highly dissected character of the Driftless Area has often been perceived to represent the character of the surrounding glaciated region before the continental glaciers invaded the area (e.g., Chamberlin and Salisbury, 1885:207). However, this perception is not strictly accurate because it is now apparent that much dissection in the region postdates the earliest glacial ice advance against its margins.

17.3 Unglaciated Central Lowlands: The Osage Plains

17.3.1 Physical Character

The Osage Plains occupy the southern part of the Central Lowlands (fig. 17.1). It is a region of physiographic transitions on both meridional and zonal axes, but its boundaries with adjacent divisions are generally defined by physical features. The western and northern perimeters of the Osage Plains are clearly delineated by the edge of the High Plains (fig. 17.2) and the southern limit of glacial deposits, respectively (fig. 17.1). On the east, the Ozark and Ouachita uplands are clear boundaries, but to the south, transitions to the Gulf Coastal Plain and the Edwards Plateau are poorly defined. The west-to-east increase in precipitation across the Osage Plains is expressed in the gradual changes from grassland in the west to mixed forests in the east (Blair and Hubbell, 1938; Blair, 1950).

Over most of the Osage Plains, local relief is quite low. Landforms primarily reflect differences in the relationship between bedrock lithology and erosion or dissolution. The most obvious geomorphic features are elongate low scarps, denoting changes in bedrock type. The principal subdivisions of the Osage Plains include the Gypsum Hills of Oklahoma and the Rolling Plains of Texas on the west, the Redbed Plains in the central region, and a mosaic of low hills and interspersed prairies along the eastern margins. The Wichita and Arbuckle Mountains in southern Oklahoma (fig. 17.1) are notable exceptions to the general low relief. These two hilly areas reflect Cambrian rifting, which included emplacement of granitic and mafic plutons and extrusives in the Wichita Mountains and rhyolites in the Arbuckle Mountains (McConnell and Gilbert, 1990). Neogene uplift of the latter is suggested by geomorphic data (see the following discussion); Madole (1988) and others show that large earthquakes accompanied faulting within the last two millennia in the northern Wichita Mountains, but only a few earthquakes were recorded historically over the broader region.

Rivers in the Osage Plains include those that head in the Rocky Mountains, such as the Arkansas and Canadian, and those that head either on the High Plains or within the Osage Plains. Most of the latter are younger drainages, based on terrace heights, terrace soil development, and lack of buried volcanic ashes in alluvium (fig. 17.5) (Madole et al., 1991). However, rivers such as the Brazos, Red, and North Canadian have gravel lithologies, indicating former connections to the Rocky Mountains (Gustavson, 1986a). The Brazos River exhibits deeply entrenched meanders west of the Balcones fault, indicating adjustment to epeirogenic uplift of the central Texas region beginning in the Miocene, which was coincident with initial construction of the High Plains (Ewing, 1991). The Red River also has entrenched meanders, but only south of the Arbuckle Mountains, where terraces along the entrenched meanders on the north side of the valley oppose bedrock scarps on the south side of the valley (Dalquest, 1965). The Washita River was superposed across the Arbuckle Mountains, forming a deep gorge. Fenneman (1938) argues that the Washita was superposed in the Cretaceous, but its entrenchment must have continued over much of the Cenozoic. However, the morphology of both the Red and Washita valleys attests to uplift of the Arbuckle Mountains since the Miocene. Valley fills in the Rolling Plains of Texas are dominated by late Pleistocene and Holocene sediments along the tributaries and headwaters of the Red (Caran and Baumgardner, 1990) and Brazos rivers (Blum et al., 1992). Downstream, most valleys have thick sediments below their broad floodplains, which were deposited mainly since the Last Glacial Maximum along low gradient streams (Ferring, 1991; Mandel, 1992a). These thick, young valley fills are characteristic in the Redbed Plains of Oklahoma, along major streams such as the Washita River, and as small tributaries. Terraces dating at least to the middle Pleistocene overlook broad late Pleistocene terraces and younger floodplains. Tributaries exhibit thick

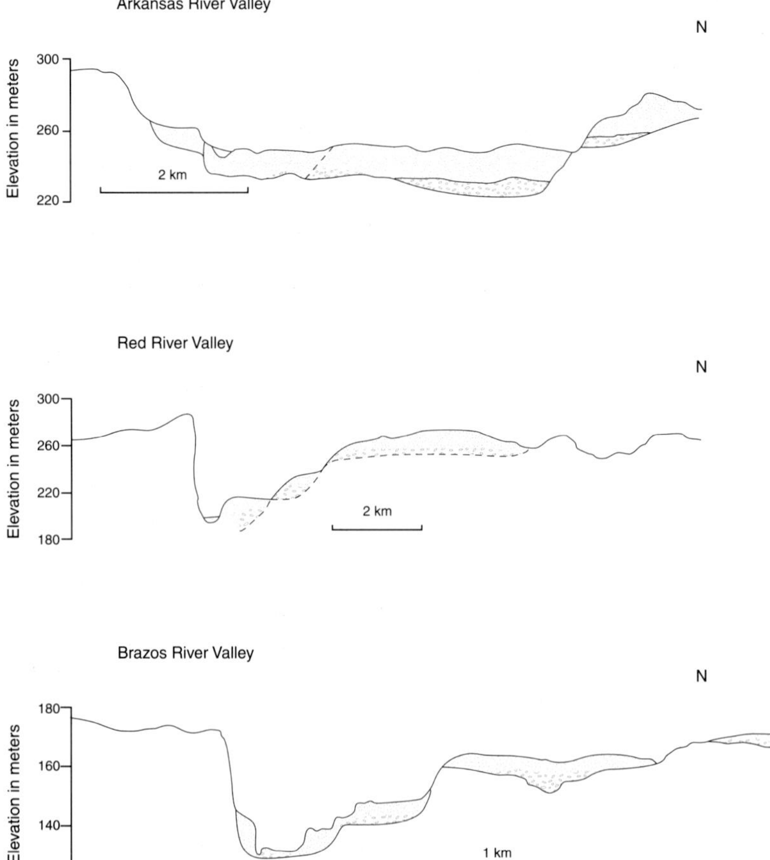

Figure 17.5 Geologic cross sections of three major streams in the Osage Plains (modified after Madole, 1991a).

late Quaternary fill with complexes of alluvium and buried soils that provide records of stream adjustment to climate change (Johnson and Martin, 1987; Mandel, 1992b), particularly during the dry middle Holocene.

Aeolian sediments in the Osage Plains are most common in the north, where sand dunes and loess deposits are widespread (fig. 17.1) (Thorpe and Smith, 1952; Madole, 1991a; Muhs and Holliday, 1995). Sand sheets interspersed with aeolian silts cover the Great Bend Sand Prairie on the inside of the Great Bend of the Arkansas. Sheet sand and scattered dunes also cover large areas south of the Arkansas River. Loess in the Kansas-Missouri portions of the Osage Plains includes the Loveland (middle Pleistocene), Peoria (Wisconsinan), and Bignell (Holocene), which had complex glacial and fluvial source areas (Welch and Hale, 1987; Johnson et al., 1990; Johnson and May, 1992; Maat and Johnson, 1996). The Bignell Loess and sand dunes of the same age probably reflect dry middle Holocene climates (Feng et al., 1994a,b).

Alluvium from the High Plains provided sand supplies that fostered the development of extensive sand sheets in western Oklahoma (Carter et al., 1990; Madole et al., 1991). Especially common in Oklahoma are dune belts, up to 20 km wide, that flank the northeast sides of major rivers (fig. 17.1). Based on topographic expression and soil morphology, the dunes range in age from late Pleistocene to late Holocene (Madole et al., 1991), although active dunes are also quite common. Along the Red River, sand in these riparian dunes grades to the east (downwind) to loesslike silt at the distal margins of broad terraces (Sellards, 1923). Riparian dunes and thin-sheet sand deposits are also common in the Rolling Plains of Texas (Gustavson, 1986b; Caran and Baumgardner, 1990).

Virtually all of the Osage Plains that overlie Permian bedrock have been influenced by karst processes and salt dissolution. Dissolution and collapse probably enhanced the widespread removal of Cretaceous bedrock from most of the central Redbed Plains and much of southern Kansas where Permian rocks crop out (Johnson, 1981). Subsidence from evaporite dissolution, seepage erosion, and sapping also probably promoted headward dissection of the Kansas and Smoky Hill Rivers (Osterkamp, 1987). Expansion of these drainages at the expense of the High Plains formed the east-sloping Plains Border region, where the Ogallala caprock is weakly expressed or absent. Dissolution of Permian salt also strongly influenced the geomorphic evolution of the Texas Rolling Plains (Gustavson, 1986a; Gustavson et al., 1991). The few, small playa lakes that occur on the Texas Rolling Plains are possibly the result of dissolution, as are the dolines and collapse features along the western Osage Plains in Oklahoma. Extensive salt plains occur along some streams below brine springs (Johnson, 1981). Archaeological sites (Ferring et al., 1976) and historic accounts, such as the 1839–40 exploration by Tixier (McDermott, 1940), illustrate the use of the Great Salt Flat in the central Cimarron Valley in Oklahoma by Native American populations. These flats still yield commercial salt.

17.3.2 Human Impacts in the Osage Plains

Human impacts to Osage Plains landscapes have primarily involved enhanced erosion and wind deflation as a result of plowing and overgrazing. Today, terraced hillslopes, planted windbreaks, and improved plowing techniques have reduced soil loss. The entire region is dotted with sediment control structures (small reservoirs) constructed by the U.S. Department of Agriculture to contain sediment before it blankets larger floodplains downstream. Also, many reservoirs have been constructed by the U.S. Army Corps of Engineers in attempts to control brines emanating from numerous seeps that contaminate surface waters on a grand scale. The government has been fighting these bitter waters since at least 1853, when Captain Marcy complained that even the "fresh" waters in the gypsum plains were "insipid, stagnant and muddy" (Foreman, 1937:103). In recent decades, population growth has been paralleled by reservoir construction. Large reservoirs supply all major cities with water, since potable groundwater resources are inadequate, and much surface water is too salty. Overuse of groundwater led to the formation of collapse features in Oklahoma (Madole et al., 1991) and to the loss of all of the once numerous artesian springs that formerly flowed in north-central Texas.

17.4 The Northern Great Plains and Adjacent Grasslands of the Central Lowlands

The northern Great Plains and adjacent areas of the Central Lowlands include portions of the north-central United States and south-central Canada (fig. 17.1). The northern Great Plains are bounded on the west by the Rocky Mountains, on the south by the Pine Ridge Escarpment, and on the east by the Missouri Escarpment. The northern boundary, for the purpose of this volume, is the relatively abrupt transition from grassland to boreal forest vegetation. At the northwestern end of the glaciated Central Lowlands, grasslands give way to aspen-birch parkland on the north and northeast (Kuchler, 1964; Trenhaile, 1990).

The gradual west-to-east slope of the region is interrupted by two prominent northeast- to east-facing, bedrock-controlled escarpments, which divide the region into three subregions (fig. 17.1). From west to east, the Alberta Plain–Missouri Plateau (the northern Great Plains) is separated from the Saskatchewan Plain–Glaciated Till Plain (Central Lowlands) by the Missouri Escarpment, and the Saskatchewan Till Plain is separated from the Manitoba Plain–Red River Valley Lowland by the Manitoba–Pembina Escarpment (Dawson, 1875; Fenneman, 1931, 1938; Trenhaile, 1990). Most of the region was glaciated during the Pleistocene, except for part of the Alberta Plain and parts of the Missouri Plateau south and west of the Missouri River trench. The Missouri Escarpment is traceable from east-central Alberta to south-central South Dakota. It is most prominent (up to 122 m high) from southeastern Saskatchewan through central North Dakota, but becomes less imposing northward where a separate, more westerly cuesta branches off to form the Wood Mountain Upland and Cypress Hills. In places, the escarpment is overlain by up to 150 m of Pleistocene glacial and proglacial sediments, forming relief features up to 45 m high such as the Missouri Coteau (Clayton and Freers, 1967; Klassen, 1989). The Manitoba–Pembina Escarpment extends from east-central Saskatchewan to southeastern North Dakota.

17.4.1 Structural Controls on the Landscape

In the unglaciated part of the region, broad alluvial valleys and highly dissected, bedrock plateaus are interspersed with conspicuous uplands, structural and erosional basins, and badlands (fig. 17.1). Isolated uplands in the western part of the region resulted from post-Laramide uplift (e.g., the Black Hills and the Big Horn Mountains) or igneous intrusion (Little Rocky Mountains and Devil's Tower) (Oldow et al., 1989; Wayne et al., 1991). Other uplands are erosional remnants of former regional constructional surfaces (e.g., Cypress Hills, Sentinel Butte, and Turtle Mountain) or thick accumulations of mid-Tertiary basaltic lava (Klassen, 1989; Oldow et al., 1989; Sauchyn, 1993). In the glaciated areas to the east, glacial and proglacial landforms dominate a younger landscape of generally lower relief (Fenneman, 1931, 1938; Klassen, 1989).

Paleozoic rocks, deeply buried across most of the region, crop out as hogbacks and cuestas around domal uplifts such as the Black Hills and Little Rocky Mountains (fig. 17.1) (Wayne et al., 1991). The Black Hills, the most extensive of these uplifts, rise 1200 m above the surrounding Missouri Plateau. Precambrian crystalline rocks and Tertiary intrusives are exposed in the center of the uplift. A nearly complete sequence of Paleozoic and Mesozoic formations dips away from the uplift (e.g., Rich, 1981; DeWitt et al., 1989). Physiography around the uplift is the result of differential erosion acting on the tilted rocks (Wayne et al., 1991). The crystalline core is ringed by a limestone plateau, a prominent valley formed in less resistant argillaceous shales and evaporites, and limestone- and sandstone-supported hogback ridges (Fenneman, 1938).

17.4.2 Glacial Landscapes

The Laurentide Ice-Sheet advanced across the glaciated part of the region repeatedly during the Pleistocene, possibly beginning as early as 2.6 Ma (e.g., Christiansen, 1992; Clayton and Moran, 1982; Dyke and Prest, 1987a,b). Landforms created during the most recent (Wisconsinan) episode of glacial growth and decay, largely unmodified by subsequent geomorphic processes, dominate the region north and east of the Missouri River (Klassen, 1989). Thin, discontinuous tills and scattered erratics indicate that a few small areas south and west of the Missouri River were also affected by Wisconsinan and older ice sheets. Extensive till plains are composed of relatively thick layers of mostly ice-contact deposits (e.g., Clayton and Moran, 1982; Klassen, 1989). Extensive hummocky moraines are also common and may have formed in contact with stagnant ice or perhaps subglacially; they are often associated with glaciotectonic landforms (e.g., Clayton and Freers, 1967; Aber, 1993; Shaw, 1983). Complex sets of end and recessional moraines and associated ice-contact stratified deposits, and glaciotectonic thrust moraines are common on the Missouri Coteau, Pembina Escarpment, and the Prairie Coteau (e.g., Clayton and Freers, 1967; Aber, 1993; Klassen, 1989).

Glaciolacustrine landforms such as strandlines, beaches, deltas, underflow fans, and level plains composed of fine-grained offshore sediments are widely recognized across the region (e.g., Fenton et al., 1983; Taylor et al., 1986; Christiansen, 1992; Teller, 1995). Ice dams along the Missouri River and its tributaries formed relatively small proglacial lakes (Perry, 1962; Taylor et al., 1986). Meltwater temporarily stored in most of these proglacial lakes ultimately drained to the Missouri River (Kehew and Teller, 1994; Teller, 1995).

Far more extensive proglacial lakes developed east of the Missouri Escarpment where meltwater was ponded among the retreating Laurentide Ice Sheet, drift dams, and bedrock ridges (e.g., Christiansen, 1992). Many proglacial lake basins from west-central Alberta to north-central North Dakota and adjacent Manitoba were linked by a system of ice-marginal meltwater channels and spillways (e.g., Brophy and Bluemle, 1983; Kehew and Clayton, 1983; Klassen, 1983, 1989; Kehew and Lord, 1986). These proglacial lakes were characteristically shallow and short-lived, though many inundated large areas. Their distribution and duration were controlled by periodic downcutting at their outlets and rapid drainage of proglacial lake basins upstream (Kehew and Lord, 1986). In addition, lower outlets periodically developed as the Laurentide Ice Sheet retreated. Most meltwater drained into Glacial Lake Agassiz, though some also fed into the Mississippi River system via the James River Spillway in North and South Dakota and into the Arctic Ocean via the Mackenzie River (e.g., Smith and Fischer, 1993; Lemmen et al., 1994; Teller, 1995).

Glacial Lake Agassiz (fig. 17.1) was the largest, deepest, and longest-lived (about 11,600–8,500 yr B.P.) of the terminal Wisconsinan proglacial lakes in the region, and also the most extensively studied (e.g., Upham, 1895; Mayer-Oakes, 1967; Teller and Clayton, 1983). Strandlines and beaches, deltas and underflow fans, and offshore sediments are found from southern and central Saskatchewan, and adjacent parts of Manitoba and eastern North Dakota, through north-central Ontario, northwestern Minnesota, and northeastern South Dakota. Drainage of the lake shifted among the Mississippi River, the ancestral Great Lakes, and the St. Lawrence River, and north into the Arctic Ocean as ice sheets advanced and retreated, outlets opened or closed, and lake levels rose and fell (e.g., Fenton et al., 1983; Smith and Fischer, 1993; Teller, 1995). Drainage into the North Atlantic via the St. Lawrence River (10,900–9,900 yr B.P.) may have profoundly affected climate by triggering the Younger Dryas episode, but the data are inconclusive (Broecker and Denton, 1989; Teller, 1990). Modern lakes Winnipeg, Manitoba, and Winnipegosis (fig. 17.1) are remnants of the Agassiz system (Last and Teller, 1983).

17.4.3 Regional Drainage

Uplift since the Laramide Orogeny and repeated rejuvenation of drainage systems by Pleistocene glacial advances resulted in profound fluvial incision and regional denudation (e.g., Perry, 1962; Wayne et al., 1991; Sauchyn, 1993). Tertiary and Pleistocene erosional surfaces, widely recognized within the unglaciated Alberta Plain–Missouri Plateau, suggest that regional incision and denudation occurred intermittently from mid-Tertiary through Pleistocene time.

Preglacial streams flowed northeastward across the region to Hudson Bay, but glacial advances diverted streams to their current locations. Today, the divide between the Missouri River drainage to the southwest and the Hudson Bay drainage to the northeast is represented by the Cypress Hills, Wood Mountain Upland, Missouri Escarpment, and low moraines to the east in southeastern North Dakota. Tributaries to the Missouri River in the unglaciated part of the region still occupy broad, northeast-trending, preglacial valleys (e.g., Perry, 1962; Wayne et al., 1991). Preglacial valleys in glaciated areas were buried by glacial and proglacial sediments (e.g., Christiansen, 1992; Last, 1984; Klassen, 1989; Evans and Campbell, 1995).

Regional drainage was diverted to the southeast along the ice margin with each glacial advance (Klassen, 1989). The southeasterly flow of the Missouri River across the region evolved with the formation of ice-marginal channels (Wayne et al., 1991). Pleistocene terraces preserved along its unglaciated tributaries indicate that southward diversion of the Missouri River, and rejuvenated fluvial incision, occurred repeatedly. This diversion also resulted in accelerated stream piracy, dissection, head cutting, and drainage network expansion in the unglaciated part of the region. Badlands are common landforms where drainage networks expanded into poorly consolidated bedrock (Wayne et al., 1991). The most extensive of these are the Little Missouri Badlands in North Dakota and the White and Cheyenne River Badlands in South Dakota (figs. 17.1, 17.6). Diversion of the Missouri River to its present course within the narrow, deep, steep-sided Missouri River Trench was not complete until the late-Pleistocene.

Glacial meltwater spillways are conspicuous features across the region. Many modern rivers flow as underfit streams in these deeply incised, wide, steep-sided valleys (Klassen, 1989; Beatty, 1990). The spillways underwent considerable postglacial geomorphic modification, including slope failure of the walls and filling by alluvium, debris flows, and alluvial fans (fig. 17.7) (e.g., Campbell and Evans, 1990; Sauchyn, 1993). Short, narrow, straight, comparatively steep, wind-aligned first- and second-order tributary valleys or coulées appear as head cuts in spillway walls as well (Beatty, 1975). Alluvial fans also occur where coulées and smaller tributaries adjoin spillways (Winder, 1965; Roed and Wasylyk, 1973; Artz, 1995; Running, 1995, 1996). Postglacial mass wasting also modified steep slopes formed in poorly consolidated bedrock in other parts of the region.

Stream valleys throughout the region underwent considerable modification through terminal Wisconsinan and Holocene time because of dramatic changes in deglacial hydrology and postglacial environment (Knox, 1983; Klassen, 1989). Evidence of these changes is preserved in terraces and erosion surfaces (e.g., O'Hara and Campbell, 1993;

Figure 17.6 The Little Missouri Badlands along the Missouri River in Theodore Roosevelt National Park, Medora, North Dakota, illustrating erosion of horizontally bedded, poorly consolidated Tertiary bedrock. Cliff-top loess (Oahe Formation) is exposed in the foreground.

Figure 17.7 The South Saskatchewan River valley (view upstream) in a glacial-meltwater spillway in southwestern Saskatchewan. Rotational slump blocks are common along the right valley wall.

Artz, 1995; Running, 1996). In some valleys, floodplains are blanketed by alluvial sediments resulting from European settlement (Artz, 1995; Running, 1996). These deposits are equivalent to the postsettlement alluvium observed throughout the Upper Midwest (e.g., Knox, 1972; Bettis, 1992; chapter 23).

Much of the Missouri Coteau in Saskatchewan and North Dakota is internally drained (Last, 1984). Along with a few larger lake basins, thousands of small playa lakes or "prairie potholes" are present in the area and to the east (Fenneman, 1938; Clayton and Freers, 1967; Last, 1984, 1992; Klassen, 1989; Evans and Campbell, 1995). The lakes occupy depressions in proglacial outwash, hummocky and stagnant ice moraines, glaciotectonic depressions, interlobate areas, and partially filled preglacial valleys. Many of these lakes and playas are saline or hypersaline. Their brine chemistry is in part controlled by the geochemistry of underlying Paleozoic and Mesozoic evaporites.

17.4.4 Aeolian Landscapes

Wind is an important geomorphic agent in the region (Clayton et al., 1976; Muhs and Maat, 1993). Persistent strong winds are a conspicuous component of the regional climate, and there are few natural barriers to slow their progress (Jensen, 1972; Muhs and Maat, 1993; Wolfe et al., 1994). Winds capable of entraining and transporting sand-size particles are common. Evidence of wind scour, blowouts, and erosion is widespread on unglaciated uplands (Clayton et al., 1980; Wayne et al., 1991). Wind erosion exacerbated by human mismanagement of regional soils is also a common problem (Gray, 1978; Wheaton and Chakravarti, 1990; Ashmore, 1993; chapter 23). Soil losses due to wind erosion on the Canadian Prairies are conservatively estimated at 160 million tonnes annually (Sparrow, 1984).

A variety of aeolian deposits and landforms are identified in the region (e.g., David, 1977; Catto, 1983). Relatively thin loess and sandy loess accumulations of Miocene and Pleistocene age are locally preserved in both glaciated and unglaciated settings (Catto, 1983; Vreeken, 1993). Windblown silty and loamy sediments of Holocene age are more widely observed (e.g., Clayton et al., 1976; Vreeken, 1993). In uplands adjacent to major river valleys in Saskatchewan and near the Missouri River Trench in North Dakota, Holocene loess derived from alluvium is up to 6 m thick and often thicker on the lee side of hillslopes (e.g., Clayton et al., 1976; Catto, 1983; Vreeken, 1993).

Dune fields and associated sand sheets are common landforms (e.g., David, 1977; Ahlbrandt et al., 1983). Sources of sand include glaciofluvial deposits, nearshore facies of glaciolacustrine sediments, and alluvial deposits. Sand sheets composed of finer, better-sorted sand are commonly found on the lee side of dune fields. The sand sheets evolve into a wide variety of parabolic dune forms. Some large parabolic dunes in southern Saskatchewan, up to 10 m high and 250 m across (fig. 17.8), formed over the past 200 years, indicating the rapid evolution of aeolian landscapes (Wolfe, 1996). Though less common, mound dunes, which have no slip face, are also well represented in dune fields (Halsey and Catto, 1994).

Aeolian activity and dune formation across the region peaked during the warmer, drier climatic conditions of the mid-Holocene (e.g., Clayton et al., 1976). However, episodes of increased aeolian activity during the late Holocene, including historic time, are also recognized (e.g., Muhs and Maat, 1993; Wolfe et al., 1994; Running, 1996). Today, most dune fields across the region are well vegetated and stable, but they could be easily remobilized by minor decreases in summer effective moisture or increased human disturbance.

Figure 17.8 A large crescentic parabolic dune in the Great Sandhills of Saskatchewan. The dune is only about 200 years old and has been moving at about 2m yr^{-1} since the 1960s (Wolfe, 1996). Note the small figure on the dune for scale.

17.5 The Southern Great Plains

The southern Great Plains were formed from a variety of both constructional and erosional geomorphic processes that resulted in a wide array of landscapes, contrary to the popular image of a vast, homogeneous plain. The region stretches from southern South Dakota to central Texas and is subdivided physiographically into the High Plains, Colorado Piedmont, Raton section, Pecos Valley, and Edwards Plateau (fig. 17.1). Erosion by the upper South Platte and upper Arkansas Rivers isolated the northern High Plains from the Rockies and formed the topographic depression known as the Colorado Piedmont. Development of the Pecos River system separated the southern High Plains from the southern Rockies and the mountains of the Rio Grande Rift. The Raton section is a volcanic pile between the Colorado Piedmont and Pecos River that also separates the High Plains from the mountains. The Edwards Plateau is a broad limestone upland whose eastern and southern margins are defined by the Balcones fault zone, which separates the Great Plains from the Gulf Coastal Plain.

17.5.1 Structural and Volcanic Landscapes

The effects of uplift and extrusive volcanic activity are commonly apparent at or near the margins of the southern Great Plains. The southern and southeastern edge of the region is defined by the Balcones Escarpment, which was formed by the Balcones fault zone (fig. 17.1). The fault zone, formed during uplift of the Edwards Plateau, consists of a series of en echelon short, normal northeast-southwest-trending faults, downthrown toward the southeast with local displacements up to 23 m (Grimshaw and Woodruff, 1986; Senger et al., 1990). Generally, the rocks exposed at the surface west of the fault zone are resistant limestones, dolomites, and marls; east of the zone, the rocks exposed are nonresistant chalks and calcareous clay units. This difference in resistance to erosion produced the prominent Balcones Escarpment.

Within the Edwards Plateau is the Llano Uplift or Central Mineral region (fig. 17.1). The area is composed mainly of Precambrian igneous and metamorphic rocks that were first pushed up through Paleozoic strata in the Mesozoic, then buried beneath Cretaceous sediment, and further raised and exposed by erosion during Cenozoic time (Petersen, 1988; Walker, 1992). Topographically, however, the Llano Uplift is a basin surrounded by high ridges of resistant Paleozoic and Cretaceous sedimentary rocks typical of the Edwards Plateau. The terrain is gently rolling, with barren granite knolls dotting the landscape. Standing

135 m above the floor of this basin is Enchanted Rock, the largest of several granitic domes in the region (fig. 17.9). Massive exfoliation sheets are common on the slopes of Enchanted Rock; these sheets, in turn, are often flanked by irregular castle tors. This terrain of domes, tors, and pedestal rocks was produced by differential weathering, much influenced by parallel joint sets in the granite. Valleys are remarkably straight because erosion has followed the linear fracture zones (McGehee, 1979; Petersen, 1988).

Volcanic landforms are found mostly in the Raton section, where basalts accumulated during episodic volcanic eruptions throughout the late Cenozoic (3.5–7.2 Ma, 2.5–1.8 Ma, and 18,000–4,500 yr B.P.), contemporaneous with and postdating accumulation of the Ogallala Formation (Baldwin and Muehlberger, 1959; Stormer, 1972). The volcanic field includes a variety of volcanic landforms, with relief in the region varying from 500 to 1200 m. Basalt forms prominent mesas such as Johnson Mesa and Clayton Mesa, either massively or as thinner caprock over Mesozoic and Cenozoic sedimentary rocks. Much of the mesa landscape formed as the upper Dry Cimarron drainage developed. Basaltic badlands or "malpais" also are common throughout the region. The more striking volcanic landforms include cinder cones such as Capulin Mountain, shield volcanoes such as Sierra Grande, and eroded stratovolcanoes, notably Robinson Mountain.

17.5.2 Aeolian Landscapes

Aeolian activity has been the dominant geomorphic process affecting the southern Great Plains during the Quaternary. Wind has been largely responsible for creating the surface of the High Plains. Throughout the Pleistocene, the entire region was repeatedly blanketed by aeolian sediments. On the southern High Plains, these aeolian deposits of the Blackwater Draw Formation probably were deflated from the Pecos River valley, aggrading incrementally throughout the Pleistocene (beginning >1.5 Ma and ending by 50,000 yr B.P.) and greatly modified by pedogenesis (Holliday, 1989). The central and northern High Plains are covered by loess. These layers of silt accumulated episodically throughout the Quaternary, but most of the deposits date to the last glaciation 21,000–13,000 yr B.P. (Martin, 1993; Madole, 1995; Pye et al., 1995), probably as a result of erosion of the older layers. The temporal and lithological relationships of the Blackwater Draw Formation and the loess have yet to be established.

Scattered across the surface of the High Plains are accumulations of aeolian sand resting on older aeolian and alluvial mantles. The Nebraska Sand Hills (50,000 km^2) cover the northern end of the High Plains (Swinehart, 1989). Smaller dune fields occur on the drier western side of the High Plains (Holliday, 1995a; Madole, 1995; Muhs and Holliday, 1995). The present size and shape of the dune fields are largely the result of later Holocene aeolian activity, which often reworked older deposits, including dunes and the Blackwater Draw Formation. Small lee dunes or lunettes formed downwind from desiccating lake basins (Holliday, 1997).

Aeolian deposits also are common in the Colorado Piedmont and Pecos Valley. In the Piedmont, sheets of loess, deposited intermittently from mid-Pleistocene to early Holocene times, are draped across uplands and lowlands alike, with late Pleistocene loess the most common (Madole, 1995). Sand sheets and dunes also are common along and near some of the rivers, including those that cut through the High Plains such as the Arkansas and Cimarron (fig. 17.1) (Madole, 1995; Muhs et al., 1996). The sand, derived from the floodplains, was deposited mostly during the Holocene. In the Pecos Valley, the Mescalero Dunes formed east of the Pecos River, where wind has stacked sand against the western escarpment of the southern High Plains.

Figure 17.9 Enchanted Rock in the Llano Uplift of the Edwards Plateau, Texas. Note the exfoliation sheets on this granitic dome.

17.5.3 Lake Basins

Lake basins are common on the High Plains landscape but are absent from the rest of the southern Great Plains. On the High Plains, thousands of playas contain seasonal, usually freshwater, lakes fed by runoff (fig. 17.10) (Sabin and Holliday, 1995; Holliday et al., 1996). Most of these basins are <5 km^2 and typically ≤1.5 km^2. On the southern High Plains, there are an estimated 25,000 playa basins. Playa occurrence may be locally higher elsewhere, for example, in western Kansas (Frye, 1950), but in general they are less numerous on the central and northern High Plains (Goudie and Wells, 1995), perhaps because the loess-covered landscape is younger and also less easily eroded by wind. The origin of playas has generated much controversy for decades, but there is probably no single process responsible for their formation (Reeves, 1966; Osterkamp

Figure 17.10 "Silver dollars": small playas of the southern High Plains, near Lubbock, Texas, after heavy rains. Photo courtesy High Plains Underground Water Conservation District No.1, Lubbock, Texas.

and Wood, 1987; Gustavson et al., 1995). Most playas probably result from erosive processes, including centripetal fluvial erosion and subsequent deflation of fluvial basin sediment. Other, more localized processes include subsidence as a result of solution of limestone or salt. Most playa basins began forming in the late Pleistocene, but older (early and mid-Pleistocene) filled basins are found below some modern basins.

On the southern High Plains, about 40 larger, less regularly shaped basins, called salinas, contain salt deposits, most commonly gypsum, and saline water (Reeves, 1991; Wood et al., 1992). Several of the basins are commercial sources of sodium sulfate. Salinas are larger and more irregularly shaped than the small playas. They are also groundwater discharge sites and may form as a result of a high water table combined with wind deflation. Salinas are generally older than the playa basins, with initial formation probably in the early and mid- Pleistocene.

Playas and salinas are unique wetland habitats, important to High Plains biodiversity (Bolen, 1982; Haukos and Smith, 1994). For example, they provide as much as 100,000 ha of wetlands for wildlife. Over a million waterfowl overwinter on playa wetlands, and as much as 90% of the midcontinental population of sandhill cranes utilizes this landscape. Moreover, the ephemeral nature of lakes in playa and salina basins enhances floristic diversity, which in turn leads to increased faunal diversity.

17.5.4 Drainage Development and Hydrology

The evolution of drainage systems across the Great Plains has defined most of the physiographic boundaries within the region. Throughout the Quaternary, fluvial processes have removed the far eastern edge of the late Tertiary alluvial and aeolian sediments derived from the Rockies. Farther west, development of the South Platte, Arkansas, and Pecos River systems and their tributaries separated the High Plains from the mountains of the Rocky Mountain Front Range and eastern Rio Grande Rift. Only one small area remains that preserves the original connection of the High Plains to the Rockies: the Gangplank around Cheyenne, Wyoming, on the northwestern High Plains (Mears, 1991).

Several types of drainage systems exist on the Great Plains: perennial rivers that originate in the Rocky Mountains; perennial rivers that originate on the Great Plains; and arroyos and dry valleys or draws that are largely contained within the Great Plains (Hadley and Toy, 1987; Holliday, 1995b). The large perennial systems that cut through the Great Plains (e.g., the Platte and Arkansas) and some of the rivers that head on the Osage Plains (e.g., the Brazos and Colorado) flow from a semiarid to humid environment, and they originated in the Tertiary as streams that helped to construct the Ogallala Formation (e.g, Gustavson and Finley, 1985; Reeves and Reeves, 1996). Arroyos and draws often developed across the High Plains as tributaries to larger drainage systems that border the region. Their present configuration is largely the result of late Pleistocene fluvial activity, but the location of many is related to the position of Tertiary paleodrainages. Most draws have been largely dry in the Holocene.

In the Colorado Piedmont, the early alluvial landscape history is represented by a series of late Pliocene to middle Pleistocene pediments, terraces, and isolated channel fills along the mountain front (Hunt, 1954; Scott, 1963; Madole, 1991b). The most prominent features are late Pleistocene to late Holocene fill terraces along the South Platte and Arkansas Rivers. The Pecos Valley contains late Tertiary pediments and a stepped sequence of Pleistocene and Holocene terraces (Leonard and Frye, 1962; Bachman, 1976). The alluvial landscapes of the Colorado Piedmont probably are related to climatic changes that directly influenced the drainage basin via episodic changes in precipitation runoff, stream discharge, erosion-sedimentation potential, groundwater recharge, and vegetation cover, as well as glaciation in the Rocky Mountain headwaters, which also affected discharge and erosion-sedimentation potential. In contrast, the alluvial landscapes of the middle and upper Pecos Valley were significantly influenced by the continuing subsidence-induced evolution of the river basin along with climatic and some glacial effects.

Drainage development has given the Edwards Plateau much of its unique character, notably in the Texas Hill Country along its southeast margin, where a deeply dissected landscape is characterized by rugged hills with steep slopes and narrow summits. High-gradient streams originating in the Hill Country flow to the Gulf of Mexico through narrow steep-sided canyons cut in resistant limestone. Between the Colorado and Brazos rivers is the Lampasas Cut Plain, characterized by broad valleys separated by wide, flat-topped hills and ridges capped by resistant limestone. The Pecos River cuts through the western Edwards Plateau, the isolated portion west of the river being the Stockton Plateau. The soft Cretaceous rocks that underlie this region have been deeply dissected, providing a terrain of rugged hills with steep slopes, canyons, and shallow soils.

Surface and subsurface hydrology sculpted the Edwards Plateau, but the landscape also affected hydrology and climate. The Balcones Escarpment is one of the most severely flooded regions of the United States (fig. 17.11) (Baker, 1975, 1977; Caran and Baker, 1986; Slade, 1986; Veenhuis, 1986). The climate and topography of the region, and its proximity to the Gulf of Mexico, produce frequent torrential rains when storms associated with easterly waves are orographically lifted along the Balcones Escarpment. Heavy rains also occur when moisture-laden air associated with an easterly wave collides with cold, dry air associated with a strong, polar surge into central Texas (Orton, 1966). Though rare, this scenario has resulted in severe flooding along the Balcones Escarpment (fig. 17.11). Storms over the steep bedrock slopes of the region produce high-percentage yield and rapid runoff (fig. 17.12). Urbanization also contributes to the drainage problems.

17.5.5 Subsidence and Karst Landscapes

Erosion by groundwater and subsurface removal of salts and limestone has been a significant factor in the geomorphic development of the southern Great Plains. The resistant Ogallala caprock caliche forms prominent escarpments around the margin of the High Plains in many areas, including the Caprock Escarpment and Mescalero Ridge of

Figure 17.11 A house going over the Austin Dam (now Tom Miller Dam) during the 15 June 1935 flood. View is toward west. Photograph courtesy of Austin History Center.

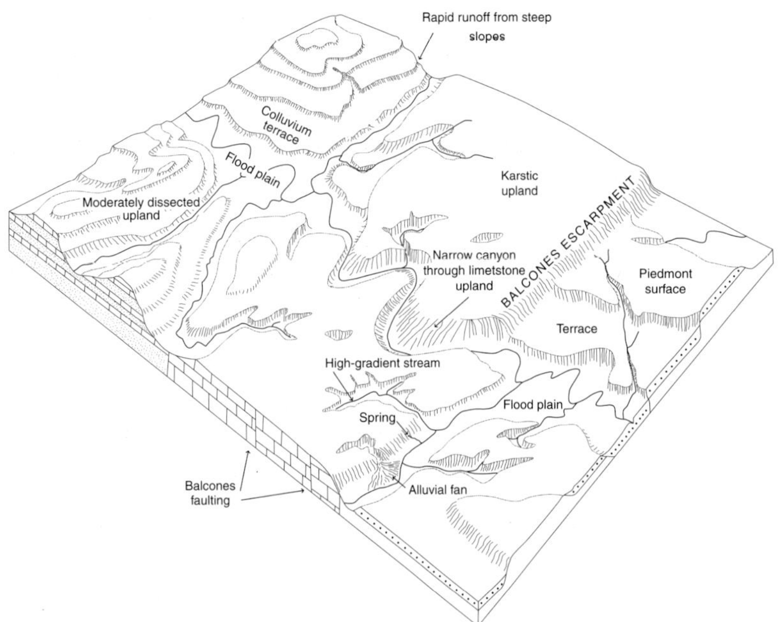

Figure 17.12 Block diagram representing geologic features that affect flood potential along the Balcones Escarpment, Texas (from Baker, 1975).

the eastern and western Llano Estacado, respectively, and the Hat Creek Breaks and Pine Ridge Escarpment at the northern margin of the High Plains (fig. 17.1). Scarp retreat is related generally to drainage development on soft sediment in the adjacent lowlands, but results specifically from groundwater processes such as seepage, sapping, and piping, as well as from cliff failure (Osterkamp, 1987; Gustavson and Simpkins, 1989). These processes are locally active throughout the region and likely produced the Goshen Hole, a large rectangular depression on the far northwestern High Plains (Mears, 1991).

Dissolution of salt in Paleozoic bedrock and subsequent subsidence was largely responsible for development of much of the Canadian and middle and upper Pecos river valleys, and possibly the upper Red River valley (Bachman, 1976; Gustavson, 1986a; Hawley, 1993). As the Pecos valley developed, it also pirated drainages, including the ancestral Brazos River, flowing from the mountains in the west across the High Plains to the east (Reeves, 1972; Gustavson and Finley, 1985). Solution and subsidence continues to affect the Pecos valley. Recent subsidence basins, deformed rocks, and karst landforms are locally common (Bachman, 1976; Sweeting, 1972; Hawley, 1993), and sinkholes occasionally open (Baumgardner et al., 1982). Carlsbad Caverns in New Mexico are a fine example of bedrock dissolution in the region.

The vast Cretaceous limestones of the Edwards Plateau provide an extensive area of continuous karst (Kastning, 1983, 1984; Veni, 1987; Smith and Veni, 1994). Sinkholes attributable to solution are rare in some areas because of the lithology of the plateau, low annual rainfall, and high fracture permeability of the bedrock. Subsidence sinkholes are very common throughout the region, however. Some of these are linear, but most are oval, circular, or irregular in shape. They range in size from less than 100 m in diameter to 15 km long by 7 km wide (Veni, 1994). Displacement of 12–15 m is common, but displacement of more than 70 m has been documented in some subsidence sinkholes (Kastning, 1987; Veni, 1994). Despite considerable subsidence, erosion has long since removed any topographic expression of the sinkholes. Collapse sinkholes, generally associated with cave entrances, vary considerably in size from <1 m to >25 m in diameter. Although many collapse sinkholes are marked by depressions at the surface, some of the largest have no topographic expression, having long since been filled by debris.

17.5.6 Groundwater and Subsurface Hydrology

Two of the better known sources of groundwater in North America—the Ogallala Aquifer and the Edwards Aquifer—lie beneath the southern Great Plains. Past and ongoing use and abuse of these important water bodies have played a critical role in the history of human activity throughout much of the region. The Ogallala or High Plains Aquifer underlies almost all of the High Plains and is the primary source of water for one of the principal agricultural regions of North America. The Ogallala Formation is the major water-bearing unit, but the aquifer comprises several hydraulically connected late Cenozoic deposits (Weeks and Gutentag, 1988). Groundwater is unconfined, and the flow is generally west to east. Recharge of the aquifer has been relatively minimal since the Ogallala Formation was isolated from the mountains to the west in the late Pliocene and early Pleistocene (Heath, 1988). Since then, recharge has been from precipitation, concentrated in areas with sandy surface deposits and in playa basins, and from seepage of streams (Weeks and Gutentag, 1988; Nativ, 1988). Recharge rates are quite variable (<1 mm yr^{-1} to 150 mm yr^{-1} from precipitation), but overall are lower than most other groundwater regions in the United States (Heath, 1988). Most recharge is on the northern High Plains, where evaporation demands are lower and sand dunes are common (Weeks and Gutentag, 1988). The aquifer discharges into rivers, notably the North Platte, Platte, and Canadian rivers, and through seeps and springs on the eastern escarpment (Gutentag et al., 1984).

The Cretaceous limestone of the Edwards Plateau contributed to the formation of the Edwards Aquifer, which consists of an unconfined (water-table) aquifer and a confined (artesian) aquifer (Abbott, 1975; Woodruff and Abbott, 1986; Buszka, 1987; Hovorka et al., 1996; Sharp and Banner, 1997). The unconfined aquifer receives about 80% of its recharge from influent streams. The remaining recharge is from direct precipitation on aquifer outcrops. Some of this groundwater discharges through seeps and springs that provide base flow for several headwater streams, which then cross the fractured cavernous limestones in the Balcones fault zone where water infiltrates into the confined aquifer. The confined aquifer provides much of the municipal and agricultural water for the Balcones fault zone. The most significant natural discharge points for the confined Edwards Aquifer are springs along the Balcones Escarpment (Guyton and Associates, 1979; Brune, 1981; Baker et al., 1986; Sharp and Banner, 1997). These springs were very important for both prehistoric and historic occupants of the region and continue to be major recreational areas for central Texas. The springs also provide habitat for several rare and endangered plant and animal species, including Edwards Plateau mock-orange (*Philadelphus ernestii*), sycamore-leaf snowbell (*Styrax plantifolia*), and the Texas blind salamander (*Typhlomolge rathbuni*) (Riskind and Diamond, 1986; Reddell, 1994).

17.5.7 Human Impacts

People have occupied the Great Plains for at least the past 11,000 radiocarbon years (Hofman, 1989; Frison, 1991), but obvious, deleterious effects of human occupation generally date to within the past century. Most problems, such as soil erosion, groundwater depletion, and water pollution are related to the growth of a large population and the development of an expansive, often irrigated agriculture industry in a semiarid, drought-prone setting. These problems were predicted long ago (Powell, 1878; Johnson, 1901).

Soil erosion is a recurring problem on both the High Plains and the Edwards Plateau, as well as in neighboring areas. On the High Plains, the combined effects of frequent drought, persistent wind, and easily eroded soils result in severe wind erosion and dust production (Holliday, 1987; Wheaton and Chakravarti, 1990; Lee and Tchakerian, 1996). During the dry seasons of 1954–1957, over 146,000 km^2 of land were damaged by wind erosion; in the 1975–76 dry season, over 32,000 km^2 were damaged (Kimberlin et al., 1977). At least 4,000 km^2 have been damaged each year on the Great Plains of the United States since the 1936–37 dry season. Wind erosion leads to aeolian sedimentation elsewhere—from local sand dunes to dust deposition far to the east (e.g., McCauley et al., 1981). Wind erosion and dust production are natural phenomena on the Great Plains, based on both historic and geologic data (Malin, 1946a,b,c; Holliday, 1991, 1995b), but human activity clearly has affected the intensity of wind erosion by loosening the soil and by leaving it bare during the windiest part of the year (spring) (Lee et al., 1994). The impact of wind erosion can be reduced, however, by following conservation practices (Ervin and Lee, 1994). Such measures may take on added importance because predicted greenhouse warming on the Great Plains may lead to increased temperature, reduced precipitation, and reactivation of sand dunes and sand sheets (Rosenzweig and Hillel, 1993; Muhs and Maat, 1993).

Soil erosion on the Edwards Plateau is due largely to grazing, the main agricultural use of the region because of its rugged terrain and thin soils. Grazing began with Spanish settlement in the 1700s and intensified in the 1800s with development of a sheep and goat industry (Wentworth, 1948; Bolton, 1970; Palmer, 1986). The cattle industry began expanding in the 1880s with development of deep wells and windmills, which allowed grazing throughout the year despite range conditions. By the 1930s, overgrazing and concomitant soil erosion had altered the landscape of the Edwards Plateau. Short grasses replaced midgrasses, juniper and mesquite invaded the uplands, and in some areas severe gullying and soil erosion transformed prairies to barren lands. Overgrazing is still a serious problem despite adoption of land management practices.

Groundwater levels beneath the High Plains and Edwards Plateau have declined significantly due to the introduction and growth of irrigation agriculture in the twentieth century and more recent growth of the cattle feedlot industry (Cronin, 1969; Weeks and Gutentag, 1988). The largest declines in the water table have been on the southern High Plains, where irrigation agriculture is most intense, evapotranspiration is high, and recharge is limited to playa basins. On the northern High Plains, evapotranspiration is lower and regional soils are sandy (Gutentag et al., 1984). Of the more than 2.03×10^{11} m^3 of water consumed by irrigation until 1980, 70% was used in Texas alone. In contrast, an estimated 65% of the water in storage in the aquifer is in Nebraska (Gutentag et al., 1984; Weeks and Gutentag, 1988).

On the Edwards Plateau, the demand for groundwater has caused the average pumpage rate to nearly equal the average yearly recharge rate of 8.33×10^8 m^3 (Sharp and Banner, 1997). Model-based predictions suggest that discharge from some springs could cease by the year 2020 due primarily to groundwater mining in the San Antonio area (Klemt et al., 1975; Guyton and Associates, 1979). Despite the varied and growing demand on the Edwards Aquifer, pumpage is unregulated because, according to Texas water law, groundwater is a property right vested with the landowner. Attempts to negotiate voluntary management plans to restrict pumpage have been largely unsuccessful, even though demand is projected to exceed average recharge by the year 2000.

The Ogallala and Edwards Aquifers are also threatened by pollution. The Edwards Aquifer is very susceptible to contamination from both surface and near-surface sources, including septic tanks, urban runoff, sewage discharges, and hazardous waste sites (Andrews et al., 1984; Veenhuis, 1986; Duffin and Musick, 1991; Sharp and Banner, 1997). The Edwards Aquifer is extremely permeable, which allows rapid movement of water and contaminants from the surface to the saturated portion of the aquifer (Duffin and Musick, 1991; Hovorka et al., 1996). Increasing density of development in the recharge zone along the Balcones Escarpment, especially between San Antonio and Austin, has placed a greater number and more varied types of contamination sources close to the Edwards Aquifer (Buszka, 1987; Duffin and Musick, 1991). On the High Plains, human activity near, or modification of, playa basins is also cause for concern. These sites collect runoff and therefore concentrate pollutants from farming and from petroleum and livestock production. These effects combined with cultivation on playa floors threaten or destroy their utility as wetlands (Bolen, 1982; Haukos and Smith, 1994). Moreover, because playa basins are recharge sites for the Ogallala Aquifer, contaminants may enter the regional water supply (Lehman, 1972).

17.6 Conclusion

The Central Lowlands and Great Plains are stereotyped as the geomorphically monotonous "heartland" of North America. The foregoing discussion clearly demonstrates, however, that though devoid of towering peaks, both of these vast areas contain a heterogeneous array of landscapes as well as other aspects of the environment. The region was constructed by tectonism, volcanism, alluviation, glaciation, dissolution and subsidence, and aeolian processes. The latitudinal gradient, especially for the Great Plains, also is striking, from below 30°N on the southern edge of the Edwards Plateau to above 50°N on the Canadian Prairies, providing a range of climates from near subarctic to subtropical. Beyond the internal variables, external processes, in particular the uplift of the Rocky Mountains, helped to sculpt the region. Uplift in western North America influenced development of the drainage systems that eroded the Central Lowlands and built the Great Plains, thus producing the rain shadow east of the Rockies. The result of all of these factors is a region that ranges in diversity from the till plains of the cool, humid Midwest bordering the craton and the Great Lakes to the alluvial plains and dune fields of the warm, dry Pecos River valley at the foot of the Basin and Range Province.

The legacy of geologic, geomorphic, and climatic variability is apparent in human terms. The Central Lowlands and Great Plains together are the site of the richest, most productive agricultural land in the world. The success of agriculture is in large measure due to the rolling to flat terrain of much of the area, the relatively mild climate and availability of water in most of the region, and the rich soils formed under grasslands and oak savannas in the vast glacial, alluvial, and aeolian deposits that blanket the landscape. This human legacy has also left its mark on the landscape as soil erosion, declining water tables, and pollution. Though spectacular, attention-getting environmental hazards such as volcanoes, earthquakes, and hurricanes rarely affect these two regions, the dangers of more subtle but significant processes such as erosion and pollution will continue to affect the region and likely accelerate as the population and demands for food continue to grow.

References

Abbott, P.L., 1975. On the hydrology of the Edwards Limestone, south-central Texas. *Journal of Hydrology* 24: 251–269.

Aber, J.S., 1993. Glaciotectonic landforms and structures. In: Quaternary and Late Tertiary Landscapes of Southwestern Saskatchewan and Adjacent Areas. D.J. Sauchyn (Ed.). Canadian Plains Research Center, University of Regina, Regina, Canada. pp. 20–26.

Ahlbrandt, T.S., J.B. Swinehart, and D.G. Maroney, 1983. The dynamic Holocene dune fields of the Great Plains and Rocky Mountain basins, USA. In: *Eolian Sediments and Processes*. M.E. Brookfield and T.S. Ahlbrandt (Eds.). Elsevier, Amsterdam, Netherlands, pp. 379–406.

Anderson, R.C., 1988. Reconstruction of preglacial drainage and its diversion by earliest glacial forebulge in the upper Mississippi Valley region. *Geology* 16: 254–257.

Andrews, F.L., T.L. Schertz, R.M. Slade Jr., and J. Rawson, 1984. *Effects of Storm-Water Runoff on Water Quality of the Edwards Aquifer near Austin, Texas*. U.S. Geological Survey Water-Resources Investigation Report 84-4124.

Argabright, M.S., R.G. Cronshey, J.D. Helms, G.A. Pavelis, and H.R. Sinclair Jr., 1996. *Historical changes in soil erosion, 1930–1992: The Northern Mississippi Valley loess hills*. U.S. Department of Agriculture, Natural Resources Conservation Service, Historical Notes 5.

Artz, J.A., 1995. Geological contexts of the Early and Middle Holocene archeological record in North Dakota and adjoining areas of the northern Great Plains. In: *Archeological Geology of the Archaic Period (8–3 ka) in North America*. E.A. Bettis III (Ed.). Geological Society of America Special Paper 297, pp. 67–86.

Ashmore, P., 1993. Contemporary erosion of the Canadian landscape. *Progress in Physical Geography* 17: 190–204.

Bachman, G.O., 1976. Cenozoic deposits of southeastern New Mexico and an outline of the history of evaporite dissolution. *Journal of Research of the U.S. Geological Survey* 4: 135–149.

Baker, E.T., R.M. Slade Jr., M.E. Dorsey, L.M. Ruiz, and G.L. Duffin, 1986. Geohydrology of the Edwards Aquifer in the Austin Area, Texas. Texas Water Development Board Report 293, Austin, Texas.

Baker, V.R., 1975. *Flood Hazards Along the Balcones Escarpment in Central Texas—Alternative Approaches to their Recognition, Mapping, and Management*. Bureau of Economic Geology Geological Circular 75-5, The University of Texas at Austin.

Baker, V.R., 1977. Stream-channel response to floods, with examples from Central Texas. *Geological Society of America Bulletin* 88: 1057–1071.

Baldwin, B., and W.R. Muehlberger, 1959. *Geologic Studies of Union County, New Mexico*. New Mexico Bureau of Mines and Mineral Resources Bulletin 63.

Bally, A.W., 1989. Phanerozoic basins of North America. In: *The Geology of North America—An Overview*. A. W. Bally and A.R. Palmer (Eds.). Geological Society of America Centennial Volume A, pp. 397–446.

Barry, R.G., 1983. Climatic environments of the Great Plains, past and present. In: *Man and the Changing Environments of the Great Plains*. W.W. Caldwell, C.B. Schultz, and T.M. Stout (Eds.). Transactions of the Nebraska Academy of Sciences 11, pp. 45–55.

Baumgardner, R.W., Jr., A.D. Hoadley, and A.G. Goldstein, 1982. *Formation of the Wink Sink, a Salt Dissolution and Collapse Feature, Winkler County, Texas*. Bureau of Economic Geology Report of Investigations 114, The University of Texas at Austin.

Beach, T., 1994. The fate of eroded soil: Sediment sinks and sediment budgets of agrarian landscapes in southern Minnesota, 1851–1988. *Annals of the Association of American Geographers* 84: 5–28.

Beatty, C.B., 1975. Coulee alignment and the wind in southern Alberta, Canada. *Geological Society of America Bulletin* 86: 119–128.

Beatty, C.B., 1990. Milk River in southern Alberta: A classic underfit stream. Canadian Landform Examples 16. *The Canadian Geographer* 34: 171–174.

Bettis E.A., III, 1992. Soil morphologic properties and weathering zone characteristics as age indicators in Holocene alluvium in the upper midwest. In: *Soils in Archaeology*. V.T. Holliday (Ed.). Smithsonian Institute Press, Washington D.C. pp. 119–144.

Black, R.F., 1970. Blue Mounds and the erosional history of southwestern Wisconsin. *Wisconsin Geological and Natural History Survey, Information Circular* 15, pp. H1–H11.

Blair, F.W., 1950. The biotic provinces of Texas. *The Texas Journal of Science* 2: 93–117.

Blair, F.W., and T.H. Hubbell, 1938. The biotic districts of Oklahoma. *The American Midland Naturalist* 20: 425–455.

Blum, M.D., J.T. Abbott, and S. Valastro Jr., 1992. Evolution of landscapes on the Double Mountain Fork of the Brazos River, West Texas: Implications for preservation and visibility of the archaeological record. *Geoarchaeology* 7: 339–370.

Boellstorff, J.D., 1978. North American Pleistocene stages reconsidered in light of probable Pliocene-Pleistocene continental glaciation. *Science* 202: 305–307.

Bolen, E.G., 1982. Playa wetlands of the U.S. Southern High Plains: Their wildlife values and challenges for management. In: *Wetlands: Ecology and Management*. B. Gopal, R.E. Turner, R.G. Wetzel, and D.F. Whigham, (Eds.). National Institute of Ecology, Jaipur, India, pp. 9–20.

Bolton, H.E., 1970. *Texas in the Middle Eighteenth Century: Studies in Spanish Colonial History and Administration*. University of Texas Press, Austin.

Borchert, J.R., 1950. The climate of the central North American grasslands. *Annals of the Association of American Geographers* 40: 1–39.

Borchert, J.R., 1971. The Dust Bowl in the 1970s. *Annals of the Association of American Geographers* 61: 1–22.

Bragg, T.B., 1995. The physical environment of Great Plains grasslands. In: *The Changing Prairie: North American Grasslands*, A. Joern and K.H. Keeler, (Eds.). Oxford University Press, New York, pp. 49–81.

Broecker, W.S., and G.H. Denton, 1989. The role of ocean-atmosphere reorganizations in glacial cycles. *Geochimica et Cosmochimica Acta* 53: 2465–2501.

Brophy, J.A., and J.P. Bluemle, 1983. The Sheyenne River: Its geological history and effects on Lake Agassiz. In: *Glacial Lake Agassiz*. J.T. Teller and L. Clayton (Eds.). Geological Association of Canada Special Paper 26, pp. 173–185.

Brune, G., 1981. *Springs of Texas*. Branch-Smith, Inc., Ft. Worth.

Bryson, R.A., and F.K. Hare, 1974. The climates of North America. In: *Climates of North America*. R.A. Bryson and F.K. Hare (Eds.), Elsevier Publishing, World Survey of Climatology 11, pp. 1–47.

Burchfiel, B.C., P.W. Lipman, and M.L. Zoback, (Eds.), 1992. *The Cordilleran Orogen: Conterminous U.S.* Geological Society of America Centennial Volume K-2.

Buszka, P.M., 1987. *Relation of Water Chemistry of the Edwards Aquifer to Hydrogeology and Land Use, San Antonio region, Texas*. U.S. Geological Survey Water-Resources Investigation Report 87-4116.

Campbell, I.A., and D.J.A. Evans, 1990. Glaciotectonism and landsliding in Little Sandhill Creek, Alberta. *Geomorphology* 4: 19–36.

Caran, S.C., and V.R. Baker, 1986. Flooding along the Balcones Escarpment, Central Texas. In: *The Balcones Escarpment*. P.L. Abbott and C.M. Woodruff Jr. (Eds.). Department of Geological Sciences, San Diego State University, San Diego, California, pp. 1–14.

Caran, S.C., and R.W. Baumgardner Jr., 1990. Quaternary stratigraphy and paleoenvironments of the Texas Rolling Plains. *Geological Society Of America Bulletin* 102: 768–785.

Carter, B.J., P.A. Ward III, and J.T. Shannon, 1990. Soil and geomorphic evolution within the Rolling Red Plains using Pleistocene volcanic ash deposits. *Geomorphology* 3: 471–488.

Catto, N.R., 1983. Loess in the Cypress Hills, Alberta, Canada. *Canadian Journal of Earth Sciences* 20: 1159–1167.

Chamberlin, T.C., 1883. *Geology of Wisconsin: Survey of 1873–1879*, Volume I. Commissioners of Public Printing, Wisconsin State Legislature, Madison.

Chamberlin, T.C., and R.D. Salisbury, 1885. Preliminary paper on the Driftless Area of the upper Mississippi Valley. *U.S. Geological Survey, 6th Annual Report*, pp. 201–322.

Christiansen, E.A. 1992. The Wisconsinan deglaciation of southern Saskatchewan and adjacent areas. *Canadian Journal of Earth Sciences* 29: 1767–1778.

Clayton, L., and J.W. Attig, 1989. *Glacial Lake Wisconsin.* Geological Society of America, Memoir 173.

Clayton, L., and T.F. Freers, (Eds.), 1967. *Glacial Geology of the Missouri Coteaux and Adjacent Areas.* North Dakota Geological Survey, Miscellaneous Series 30.

Clayton, L., and S.R. Moran, 1982. Chronology of Late Wisconsin glaciation in middle North America. *Quaternary Science Reviews* 1: 55–82.

Clayton, L., S.R. Moran, and B.W. Bickley Jr., 1976. *Stratigraphy, Origin, and Climatic Implications of Late-Pleistocene Upland Silts in North Dakota.* North Dakota Geological Survey, Miscellaneous Series 54.

Clayton, L., S.R. Moran, and J.P. Bluemle, 1980. *Explanatory Text to Accompany the Geologic Map of North Dakota.* North Dakota Geological Survey, Reports of Investigations 69.

Collins, S.L., and S.M. Glenn, 1995. Grassland ecosystem and landscape dynamics. In: *The Changing Prairie: North American Grasslands.* A. Joern and K.H. Keeler, (Eds.). Oxford University Press, New York, pp. 128–152.

Cooper, W.S., 1938. Ancient dunes of the Upper Mississippi Valley as possible climatic indicators. *American Meteorological Society Bulletin* 19: 193–204.

Court, A., 1974. The climate of the conterminous United States. In: *Climates of North America.* R.A. Bryson and F.K. Hare, (Eds.). Elsevier Publishing, World Survey of Climatology 11, pp. 193–343.

Cronin, J.G., 1969. *Ground Water in the Ogallala Formation in the Southern High Plains of Texas and New Mexico* (1: 500,000). U.S. Geological Survey Atlas HA-330.

Curry, B.B., and M.J. Pavich, 1996. Absence of glaciation in Illinois during marine isotope stages 3 through 5. *Quaternary Research* 46: 19–26.

Curtis, J.T., 1956. The modification of mid-latitude grasslands and forests by man. In: *Man's Role in Changing the Face of the Earth.* W.L. Thomas (ed.). University of Chicago Press, Chicago, pp. 721–736.

Dalquest, W.W., 1965. New Pleistocene formation and local fauna from Hardeman County, Texas. *Journal of Paleontology* 39: 63–79.

Daubenmire, R., 1978. *Plant Geography: With Special Reference to North America.* Academic Press, New York.

David, P.P., 1977. *Sand Dune Occurrences of Canada.* Indian and Northern Affairs, National Parks Branch, Ottawa.

Dawson, G.M., 1875. *Report on the Geology and Resources of the 49th Parallel.* Dawson Brothers, Montreal, Canada.

DeWitt, E., J.A. Redden, D. Buscher, and A.B. Wilson, 1989. *Geologic Map of the Black Hills Area, South Dakota and Wyoming* (1:250,000). U.S. Geological Survey Miscellaneous Investigation Series, Map I-1910.

Duffin, G., and S.P. Musick, 1991. *Evaluation of Water Resources in Bell, Burnet, Travis, Williamson and Parts of Adjacent Counties, Texas.* Texas Water Development Board Report 326, Austin.

Dyke, A.S., and V.K. Prest, 1987a. Late Wisconsinan and Holocene history of the Laurentide Ice Sheet. *Geographie Physique et Quaternaire* 41: 237–263.

Dyke, A.S. and V.K. Prest, 1987b. *Late Wisconsin and Holocene Retreat of the Laurentide Ice Sheet* (1:5,000,000). Geological Survey of Canada, Map 1702A, Ottawa.

Eagleman, J.R., 1990. *Severe and Unusual Weather* (2nd ed.). Trimedia Publishing Co., Lenexa, Kansas.

Ervin, R.T., and Lee, J.A., 1994. Impact of conservation practices on airborne dust in the Southern High Plains of Texas. *Journal of Soil and Water Conservation* 49: 430–437.

Evans, D.J.A., and I.A. Campbell, 1995. Quaternary stratigraphy of the buried valleys of the lower Red Deer River, Alberta, Canada. *Journal of Quaternary Science* 10: 125–148.

Ewing, T.E., 1991. *The Tectonic Framework of Texas* (text to accompany the tectonic map of Texas). Bureau of Economic Geology, The University of Texas at Austin.

Feng, Z.-D., W.C. Johnson, Y.-Lu, and P.A. Ward III, 1994a. Climatic signals from loess-soil sequences in the Central Great Plains, USA. *Palaeogeography, Palaeoclimatology, Palaeoecology* 110: 345–358.

Feng, Z.-D., W.C. Johnson, D.R. Sprowl, and Y.-L. Lu, 1994b. Loess accumulation and soil formation in Central Kansas, United States, during the past 400,000 years. *Earth Surface Processes and Landforms* 19: 55–67.

Fenneman, N.M., 1931. *Physiography of Western United States.* McGraw-Hill Book Company, New York.

Fenneman, N.M. 1938. *Physiography of Eastern United States.* McGraw-Hill Book Company, New York.

Fenton, M.M., S.R. Moran, J.T. Teller, and L. Clayton, 1983. Quaternary stratigraphy and history in the southern part of the Lake Agassiz basin. In: *Glacial Lake Agassiz.* J.T. Teller and L. Clayton (Eds.). Geological Association of Canada Special Paper 26, pp. 40–74.

Ferring, C.R., 1991. Upper Trinity River drainage basin, Texas. In: *Quaternary Non-Glacial Geology: Conterminous U.S.* R.B. Morrison (Ed.). Geological Society of America Centennial Volume K-2, pp. 526–531.

Ferring, C.R., D.J. Crouch, and T.D. Spivey, 1976. *An Archaeological Reconnaissance of the Salt Plains Areas of Northwestern Oklahoma.* Contributions of the Museum of the Great Plains 4, Museum of the Great Plains, Lawton, Oklahoma.

Foreman, G. (Ed.), 1937. *Adventure on the Red River, Report on the Exploration of the Red River by Captain Randolph B. Marcy and Captain G. B. McClellan.* University of Oklahoma Press, Norman.

Frison, G.C., 1991. *Prehistoric Hunters of the High Plains* (2nd ed.). Academic Press, San Diego.

Frye, J.C., 1950. Origin of Kansas Great Plains depressions. *State Geological Survey of Kansas Bulletin* 86: 1–20.

Fullerton, D.S., 1986. Stratigraphy and correlation of glacial deposits from Indiana to New York and New Jersey. In: *Quaternary Glaciations in the Northern Hemisphere.* V. Sibrava, D.Q. Bowen, and G.M. Richmond (Eds.). *Quaternary Science Reviews* 5: 23–36.

Gable, D.J., and T. Hatton, 1983. Maps of vertical crustal movements in the conterminous United States over the last 10 million years. *U.S. Geological Survey Miscellaneous Investigations, Map* I-1315.

Gabrielse, H., and C.J. Yorath, 1991. Tectonic synthesis. In: *Geology of the Cordilleran Orogen in Canada*. H. Gabrielse and C.J. Yorath, (Eds.). Geological Survey of Canada, Geology of Canada no. 4, pp. 677–705.

Goolsby, D.A., W.A. Battaglin, and E.M. Thurman, 1993. *Occurrence and Transport of Agricultural Chemicals in the Mississippi River Basin, July through August 1993*. U.S. Geological Survey Circular 1120-C.

Goudie, A.S., and G.L. Wells, 1995. The nature, distribution and formation of pans in arid zones. *Earth Science Reviews* 38: 1–69.

Gray, H.H., 1989. *Quaternary Geologic Map of Indiana* (1:500,000). Indiana Geological Survey Miscellaneous Map 49.

Gray, J.H., 1978. *Men Against the Desert*. Western Producer Prairie Books, Saskatoon, Saskatchewan, Canada.

Grimshaw, T.W., and C.M. Woodruff Jr., 1986. Structural style in an en echelon fault zone, central Texas: Geomorphologic and hydrologic implications. In: *The Balcones Escarpment*. P.L. Abbott and C.M. Woodruff Jr. (Eds.). Department of Geological Sciences, San Diego State University, San Diego, California, pp. 71–75.

Gustavson, T.C., 1986a. Geomorphic development of the Canadian River valley, Texas Panhandle: An example of regional salt dissolution and subsidence. *Geological Society of America Bulletin* 97: 459–472.

Gustavson, T.C. (Ed.), 1986b. *Geomorphology and Quaternary Stratigraphy of the Rolling Plains, Texas Panhandle*. Bureau of Economic Geology Guidebook 22, The University of Texas at Austin.

Gustavson, T.C., and R.C. Finley, 1985. *Late Cenozoic Geomorphic Evolution of the Texas Panhandle and Northeastern New Mexico*. Bureau of Economic Geology Report of Investigations 148, The University of Texas at Austin.

Gustavson, T.C., and W.W. Simpkins, 1989. *Geomorphic Processes and Rates of Retreat Affecting the Caprock Escarpment, Texas Panhandle*. Bureau of Economic Geology Report of Investigations 180, The University of Texas at Austin.

Gustavson, T.C., V.T. Holliday, and S.D. Hovorka. 1995. *Origin and Development of Playa Basins, Sources of Recharge to the Ogallala Aquifer, Southern High Plains, Texas and New Mexico*. Bureau of Economic Geology Report of Investigations 229, The University of Texas at Austin.

Gustavson, T.C., R.W. Baumgardner Jr., S.C. Caran, V.T. Holliday, H.H. Mehnert, J.M. O'Neill, C.C. Reeves Jr., 1991. Quaternary geology of the southern Great Plains and an adjacent segment of the Rolling Plains. In: *Quaternary Nonglacial Geology: Conterminous U.S.* R.B. Morrison (Ed.). Geological Society of America Centennial Volume K-2, pp. 477–501.

Gutentag, E.D., F.J. Heimes, N.C. Krothe, R.R. Luckey, and J.B. Weeks, 1984. *Geohydrology of the High Plains Aquifer in Parts of Colorado, Kansas, Nebraska, New Mexico, Oklahoma, South Dakota, Texas, and Wyoming*. U.S. Geological Survey Professional Paper 1400-B.

Guyton, W.F., and Associates, 1979. *Geohydrology of Comal, San Marcos, and Hueco Springs*. Texas Department of Water Resources Report 234, Austin, Texas.

Hadley, R.F., and T.J. Toy, 1987. Fluvial processes and river adjustments on the Great Plains. In: *Geomorphic Systems of North America*. W. Graf (Ed.). Geological Society of America Centennial Special Volume 2, pp. 182–188.

Hajic, E.R., 1986. Pre-Wisconsinan loesses and paleosols at Pancake Hollow, west-central Illinois. In: *Quaternary Records of Southwestern Illinois and Adjacent Missouri*. R.W. Graham, B.W. Styles, J.J. Saunders, M.D. Wiant, E.D. McKay, T.R. Styles, and E.R. Hajic (Eds.). Illinois State Geological Survey Guidebook 23, pp. 91–98.

Hallberg, G.R., 1980. *Pleistocene Stratigraphy in East-Central Iowa*. Iowa Geological Survey Technical Information Series 10.

Hallberg, G.R., 1986. Pre-Wisconsin glacial stratigraphy of the central plains region in Iowa Nebraska, Kansas, and Missouri. In: *Quaternary Glaciations in the Northern Hemisphere*. V. Sibrava, D.Q. Bowen, and G.M. Richmond (Eds.). *Quaternary Science Review* 5: 11–15.

Halsey, L.A., and N.R. Catto, 1994. Geomorphology, sedimentary structures, and genesis of dome dunes in western Canada. *Géographie Physique et Quaternaire* 48: 97–105.

Happ, S.C., G. Rittenhouse, and G. Dobson, 1940. *Some Principles of Accelerated Stream and Valley Sedimentation*. U.S. Department of Agriculture, Technical Bulletin 695.

Harrington, J.A., and J.R. Harman, 1991. Climate and vegetation in central North America: Natural patterns and human alterations. *Great Plains Quarterly* 11: 103–112.

Haukos, D.A., and L.M. Smith, 1994. The importance of playa wetlands to biodiversity of the Southern High Plains. *Landscape and Urban Planning* 28: 83–98.

Hawley, J.W., 1993. Overview of the geomorphic history of the Carlsbad area. *New Mexico Geological Society Guidebook, 44th Field Conference, Carlsbad Region, New Mexico and West Texas*, pp. 2–3.

Heath, R.C., 1988. Hydrogeologic setting of regions. In: *Hydrogeology*. W. Back, J.S. Rosenshein, and P.R. Seaber (Eds.). Geological Society of America Centennial Volume O-2, pp. 15–23.

Hirsch, R.M., J.F. Walker, J.C. Day, and R. Kallio, 1990. The influence of man on hydrologic systems. In: *Surface Water Hydrology*. M.G. Wolman and H.C. Riggs (Eds.). Geological Society of America Centennial Volume O-1, pp. 329–359.

Hofman, J.L., 1989. Prehistoric culture history—Hunters and gatherers in the Southern Great Plains. In: *From Clovis to Comanchero: Archeological Overview of the Southern Great Plains*. J.L. Hofman (Ed.). Arkansas Archeological Survey Research Series 35, pp. 26–60.

Holliday, V.T., 1987. Eolian processes and sediments of the Great Plains. In: *Geomorphic Systems of North America*. W. Graf (Ed.). Geological Society of America Centennial Special Volume 2, pp. 195–202.

Holliday, V.T., 1989. The Blackwater Draw Formation (Quaternary): A 1.4-plus m.y. record of eolian sedimentation and soil formation on the Southern High Plains. *Geological Society of America Bulletin* 101: 1598–1607.

Holliday, V.T., 1991. The geologic record of wind erosion, eolian deposition, and aridity on the Southern High Plains. *Great Plains Research* 1: 6–25.

Holliday, V.T., 1995a. Late Quaternary stratigraphy of the Southern High Plains. In: *Ancient Peoples and Landscapes*. E. Johnson (Ed.). Museum of Texas Tech University, Lubbock, pp. 289–313.

Holliday, V.T., 1995b. *Stratigraphy and Paleoenvironments of Late Quaternary Valley Fills on the Southern High Plains*. Geological Society of America Memoir 186.

Holliday, V.T., 1997. Origin and evolution of lunettes on the High Plains of Texas and New Mexico. *Quaternary Research* 47: 54–69.

Holliday, V.T., S.D. Hovorka, and T.C. Gustavson, 1996. Lithostratigraphy and geochronology of fills in small playa basins on the Southern High Plains. *Geological Society of America Bulletin* 108: 953–965.

Hovorka, S.D., A.R. Dutton, S.C. Ruppel, and J.S. Yeh, 1996. *Edwards Aquifer Ground-Water Resources: Geologic Controls on Porosity development in Platform carbonates, South Texas.* Bureau of Economic Geology Report of Investigations 238, The University of Texas at Austin.

Hunt, C.B., 1954. Pleistocene and Recent deposits in the Denver area, Colorado. *U.S. Geological Survey Bulletin* 996–C, pp. 91–140.

Jacobs, P.M., and J.C. Knox, 1994. Provenance and pedology of a long-term Pleistocene depositional sequence in Wisconsin's Driftless Area. *Catena* 22: 49–68.

Jacobs, P.M., J.C. Knox, and J.A. Mason, 1997. Preservation and recognition of middle and early Pleistocene loess deposits in the Driftless Area, Wisconsin. *Quaternary Research* 47, 147–154.

Jensen, J.E., 1972. *Climate of North Dakota.* National Weather Service, North Dakota State University, Fargo.

Johnson, K.S., 1981. Dissolution of salt on the east flanks of the Permian Basin in Southwestern U.S.A. *Journal of Hydrology* 54: 75–93.

Johnson, W.C., 1965. *Wind in the Southwestern Great Plains.* U.S. Department of Agriculture, Conservation Research Report 6.

Johnson, W.C., and C.W. Martin, 1987. Holocene alluvial-stratigraphic studies from Kansas and adjoining states of the East-Central Plains. In: *Quaternary Environments of Kansas.* W.C. Johnson (Ed.). Kansas Geological Survey, Guidebook Series 5. Lawrence, pp. 109–122.

Johnson, W.C., and D.W. May, 1992. The Brady Geosol as an indicator of the Pleistocene/Holocene boundary in the Central Great Plains. *American Quaternary Association, Program and Abstracts*, pp. 69.

Johnson, W.C., D.W. May, and V.L. Souders, 1990. Age and distribution of the Gilman Canyon Formation of Nebraska and Kansas. *Geological Society of America, Abstracts with Programs* 22:A87.

Johnson, W.D., 1901. The High Plains and their utilization. *U.S. Geological Survey 21st Annual Report for 1899–1900, part IV*, pp. 601–741.

Johnson, W.H., 1986. Stratigraphy and correlation of the glacial deposits of the Lake Michigan Lobe prior to 14 ka B.P. In: *Quaternary Glaciations in the Northern Hemisphere.* V. Sibrava, D.Q. Bowen, and G.M. Richmond (Eds.). *Quaternary Science Reviews* 5: 17–22.

Johnson, W.H., A.K. Hansel, E.A. Bettis, P.F. Karrow, G.J. Larson, T.V. Lowell, and A.F. Schneider, 1997. Late Quaternary temporal and event classifications, Great Lakes region, North America. *Quaternary Research* 47: 1–12.

Kastning, E.H., Jr., 1983. Relict caves as evidence of landscape and aquifer evolution in a deeply dissected carbonate terrain: southwest Edwards Plateau, Texas. *Journal of Hydrology* 61: 89–112.

Kastning, E.H., Jr., 1984. Hydrogeomorphic evolution of karsted plateaus in response to regional tectonism. In: *Groundwater as a Geomorphic Agent.* R.G. LaFleur (Ed.). George Allen and Unwin, London, pp. 351–382.

Kastning, E.H., Jr., 1987. Solution-subsidence-collapse in central Texas: Ordovician to Quaternary. In: *Karst Hydrogeology: Engineering and Environmental Applications.* B.F. Beck and W.L. Wilson (Eds.). A.A. Balkema, Boston, pp. 41–45.

Keen, K.L., and Shane, L.C.K., 1990. A continuous record of Holocene eolian activity and vegetation change at Lake Ann, east-central Minnesota. *Geological Society of America Bulletin* 102: 1646–1657.

Kehew, A.E., and L. Clayton, 1983. Late Wisconsin floods and the development of the Souris-Pembina spillway system in Saskatchewan, North Dakota, and Manitoba. In: *Glacial Lake Agassiz.* J.T. Teller and L. Clayton (Eds.). Geological Association of Canada Special Paper 26, pp. 188–209.

Kehew, A.E., and M.L. Lord, 1986. Origin and large-scale erosional features of glacial-lake spillways in the northern Great Plains. *Geological Society of America Bulletin* 97: 162–177.

Kehew, A.E., and J.T. Teller, 1994. History of late glacial runoff along the southwest margin of the Laurentide Ice Sheet. *Quaternary Science Reviews* 13: 859–877.

Kemmis, T.J., G.R. Hallberg, and A.J. Lutenegger, 1981. *Depositional environments of glacial sediments of landforms on the Des Moines Lobe, Iowa.* Iowa Geological Survey Guidebook 6.

Kimberlin, L.W., A.L. Hidlebaught, and A.R. Grunewald, 1977. The potential wind erosion problem in the United States. *Transactions of the American Society for Agricultural Engineering* 20: 873–879.

Klassen, R.W., 1983. Assiniboine delta and the Assiniboine-Qu'Appelle valley system—Implication concerning the history of Lake Agassiz in southwestern Manitoba. In: *Glacial Lake Agassiz.* J.T. Teller and L. Clayton (Eds.). Geological Association of Canada Special Paper 26, pp. 211–229.

Klassen, R.W., 1989. Quaternary geology of the southern Canadian Interior Plains. In: *Quaternary Geology of Canada and Greenland.* R J. Fulton, (Ed.). Geological Survey of Canada, Geology of Canada, no. 1, pp. 138–166.

Klemt, W.B., T.R. Knowles, G.R. Elder, and T.W. Sieh, 1975. *Ground Water Resources and Model Applications for the Edwards Balcones Fault Zone Aquifer in the San Antonio Region, Texas.* Texas Department of Water Resources Report 239, Austin, Texas.

Knox, J.C., 1972. Valley alluvium in southwestern Wisconsin. *Annals of the Association of American Geographers* 62: 401–410.

Knox, J.C., 1977. Human impacts on Wisconsin stream channels. *Annals Association of American Geographers* 67: 323–342.

Knox, J.C., 1982. Quaternary history of the Kickapoo and lower Wisconsin River valleys, Wisconsin. In: *Quaternary History of the Driftless Area.* J.C. Knox, L. Clayton, and D.M. Mickelson (Eds.). Wisconsin Geological and Natural History Survey Field Trip Guide Book 5, pp. 1–65.

Knox, J.C., 1983. Responses of river systems to Holocene climates. In: *Late-Quaternary Environments of the United States, vol. 2, The Holocene.* H.E. Wright (Ed.). University of Minnesota Press, Minneapolis, pp. 26–41.

Knox, J.C., 1984. Fluvial responses to small scale climate changes. In: *Developments and Applications of Geomorphology.* J.E. Costa and J.P. Fleisher (Eds.). Springer-Verlag, New York, pp. 318–342.

Knox, J.C., 1987. Historical valley floor sedimentation in the Upper Mississippi Valley. *Annals Association of American Geographers* 77: 224–244.

Knox, J.C., and J. Attig, 1988. Geology of the Pre-Illinoian sediments in the Bridgeport Terrace, lower Wisconsin River valley, Wisconsin. *Journal of Geology* 96: 505–514.

Kuchler, A.W., 1964. *Potential Natural Vegetation of the Conterminous United States.* American Geographical Society Special Publication 36 with separate map at 1:3,168,00.

Last, W.M., 1984. Sedimentology of playa lakes of the northern Great Plains. *Canadian Journal of Earth Sciences* 21: 107–125.

Last, W.M., 1992. Salt Lake Paleolimnology in the northern Great Plains: The facts, the fears, the future. In: *Aquatic Ecosystems in Semi-Arid Regions: Implications for Resource Management.* R.D. Roberts and M. Bothwell (Eds.). N.H.R.I. Symposium Series 7, Environment Canada, Saskatoon, pp. 51–62.

Last, W.M., and J.T. Teller, 1983. Holocene climate and hydrology of the Lake Manitoba basin. In: *Glacial Lake Agassiz.* J.T. Teller and L. Clayton (Eds.). Geological Association of Canada Special Paper 26, pp. 333–353.

Lee, J.A., and V.P. Tchakerian, 1996. Magnitude and frequency of blowing dust on the Southern High Plains of the United States, 1947–1989. *Annals of the Association of American Geographers* 85: 684–693.

Lee, J.A., B.L. Allen, R.E. Peterson, J.M. Gregory, and K.E. Moffett, 1994. Environmental controls on blowing dust direction at Lubbock, Texas, U.S.A. *Earth Surface Processes and Landforms* 19: 437–449.

Lehman, O.R., 1972. Playa water quality for groundwater recharge and use of playas for impoundment of feedyard runoff. In: *Playa Lake Symposium.* C.C. Reeves, Jr. (Ed.). International Center for Arid and Semi-Arid Land Studies, Special Publication 4, Texas Tech University, Lubbock, pp. 25–30.

Leigh, D.S., 1994. Roxana Silt of the Upper Mississippi Valley: Lithology, source, and paleoenvironment. *Geological Society of America Bulletin* 106: 430–442.

Leigh, D.S., and J.C. Knox, 1990. The Bridgeport Terrace and Hegery site 1. In: *Current Perspectives on Illinois Basin and Mississippi Arch Geology.* W. Hammer and D.F. Hess (Eds.). Geological Society of America, North Central Section Geology Field Guidebook, Macomb, Western Illinois University, p. F35–F42.

Leigh, D.S., and J.C. Knox, 1993. AMS radiocarbon age of the Upper Mississippi Valley Roxana Silt. *Quaternary Research* 39: 282–289.

Leigh, D.S., and J.C. Knox, 1994. Loess of the Upper Mississippi Valley Driftless Area. *Quaternary Research* 42: 30–40.

Lemmen, D.S., A. Duk-Rodkin, and J.M. Bednarski, 1994. Late glacial drainage systems along the northwestern margin of the Laurentide Ice Sheet. *Quaternary Science Reviews* 13: 805–828.

Leonard, A.B., and J.C. Frye, 1962. *Pleistocene molluscan faunas and physiographic history of Pecos Valley in Texas.* Bureau of Economic Geology, Report of Investigations 45, The University of Texas at Austin.

Lineback, J.A., 1979. *Quaternary Deposits of Illinois* (1:500,000). Illinois Geological Survey Map.

Maat, P.B., and W.C. Johnson, 1996. Thermoluminescence and new ^{14}C age estimates for late Quaternary loesses in southwestern Nebraska. *Geomorphology* 17: 115–128.

Madole, R.F., 1988. Stratigraphic evidence of Holocene faulting in the Midcontinent: The Meers fault, southwestern Oklahoma. *Geological Society of America Bulletin* 100: 392–401.

Madole, R.F., 1991a. Osage Plains. In: *Quaternary Nonglacial Geology: Conterminous U.S.* R.B. Morrison (Ed.). Geological Society of America Centennial Volume K-2, pp. 503–515.

Madole, R.F., 1991b. Colorado Piedmont section. In: *Quaternary Nonglacial Geology: Conterminous U.S.* R.B. Morrison (Ed.). Geological Society of America Centennial Volume K-2, pp. 456–462.

Madole, R.F., 1995. Spatial and temporal patterns of late Quaternary eolian deposition, eastern Colorado, U.S.A. *Quaternary Science Reviews* 14: 155–177.

Madole, R.F., C.R. Ferring, M.J. Guccione, S.A. Hall, W.C. Johnson, and C.J. Sorenson, 1991. Quaternary geology of the Osage Plains and Interior Highlands. In: *Quaternary Nonglacial Geology: Conterminous U.S.* R.B. Morrison (Ed.). Geological Society of America Centennial Volume K-2, pp. 503–546.

Magilligan, F.J., 1985. Historical floodplain sedimentation in the Galena River basin, Wisconsin and Illinois. *Annals of the Association of American Geographers* 75: 583–594.

Malin, J.C., 1941a. Dust storms, part one, 1850–1860. *The Kansas Historical Quarterly* 14: 129–145.

Malin, J.C., 1941b. Dust storms, part two, 1861–1880. *The Kansas Historical Quarterly* 14: 265–297.

Malin, J.C., 1941c. Dust storms, part three, 1881–1900. *The Kansas Historical Quarterly* 14: 391–413.

Mandel, R., 1992a. Geomorphology. In: *An Archaeological Survey of the Proposed South Bend Reservoir Area: Young, Stephens and Throckmorton Counties, Texas.* J.W. Saunders, C.S. Mueller-Wille, and D.L. Carlson (Eds.). Archaeology Research Laboratory, Texas A&M University, College Station, pp. 53–83.

Mandel, R., 1992b. Soils and landscape evolution in central and southern Kansas: Implications for archaeological research. In: *Soils and Archaeology.* V.T. Holliday (Ed.). Smithsonian Institution Press, Washington, pp. 41–100.

Martin, C.W., 1993. Radiocarbon ages on late Pleistocene loess stratigraphy of Nebraska and Kansas, Central Great Plains, U.S.A. *Quaternary Science Reviews* 12: 179–188.

Martin, L., 1932. *The Physical Geography of Wisconsin.* Wisconsin Geological and Natural History Survey Bulletin 36, 2nd ed., Madison.

Mason, J.A., E.A. Nater, and H.C. Hobbs, 1994. Transport direction of Wisconsinan loess in southeastern Minnesota. *Quaternary Research* 41: 44–51.

Mayer-Oakes, W.J. (Ed.), 1967. *Life, Land, and Water.* Occasional Papers of the Anthropology Department 1, University of Manitoba Press, Winnipeg, Canada.

McCauley, J.F., C.S. Breed, M.J. Grolier, and D.J. Mackinnon, 1981. The U.S. dust storm of February, 1977. In: *Desert Dust: Origin, Characteristics, and Effect on Man.* T.L. Pewe (Ed.). Geological Society of America Special Paper 186, pp. 123–147.

McConnell, D.A., and M.C. Gilbert, 1990. Cambrian extensional tectonics and magmatism within the Southern Oklahoma aulacogen. *Tectonophysics* 174: 147–157.

McCorvie, M.R., and C.L. Lant, 1993. Drainage district formation and the loss of Midwestern wetlands, 1850–1930. *Agricultural History* 67: 13–39.

McDermott, J.F. (Ed.), 1940. *Tixier's Travels on the Osage Prairies*. University of Oklahoma Press, Norman, OK.

McGehee, R.V., 1979. *Precambrian Rocks of the Southeastern Llano Region, Texas*. Bureau of Economic Geology Geological Circular 79-3, The University of Texas at Austin.

McKay, E.D. 1979. Wisconsinan loess stratigraphy in Illinois. In: *Wisconsinan, Sangamonian, and Illinoisan Stratigraphy in Central Illinois*. L. Follmer, E. McKay, J.A. Lineback, and D.A. Gross (Eds.). Illinois State Geological Survey Guidebook Series 14, pp. 37–67.

Meade, R.H., T.R. Yuzyk, and T.J. Day, 1990. The movement and storage of sediment in rivers of the United States and Canada. In: *Surface Water Hydrology*. M.G. Wolman and H.C. Riggs (Eds.). Geological Society of America, The Geology of North America, v. O-1, pp. 255–280.

Mears, B., Jr., 1991. The High Plains in Wyoming. In: *Quaternary Nonglacial Geology: Conterminous U.S.* R.B. Morrison (Ed.). Geological Society of America Centennial Volume K-2, pp. 450–452.

Melhorn, W.N., and J.P. Kempton, 1991a. The Teays System; A summary. In: *Geology and Hydrogeology of the Teays-Mahomet Bedrock Valley System*. W.N. Melhorn and J.P. Kempton (Eds.). Geological Society of America Special Paper 258, pp. 125–128.

Melhorn, W.N., and J.P. Kempton (Eds.), 1991b. *Geology and Hydrogeology of the Teays-Mahomet Bedrock Valley System*. Geological Society of America Special Paper 258.

Mickelson, D.M., 1987. Central Lowlands. In: *Geomorphic Systems of North America*. W.L. Graf (Ed.). Geological Society of America Centennial Special Volume 2, pp. 111–118.

Mickelson, D.M., J.C. Knox, and L. Clayton, 1982. Glaciation of the Driftless Area: An evaluation of the evidence. In: *Quaternary History of the Driftless Area*. J.C. Knox, L. Clayton, and D.M. Mickelson (Eds.). Wisconsin Geological and Natural History Survey Field Trip Guide Book 5, pp. 155–169.

Mickelson, D.M., L. Clayton, D.S. Fullerton, and H.W. Borns Jr., 1983. The Late Wisconsin glacial record of the Laurentide Ice Sheet in the United States. In: *Late Quaternary Environments of the United States, v. 1, The Late Pleistocene*. H.E. Wright Jr. and S.C. Porter (Eds.). University of Minnesota Press, Minneapolis, pp. 3–37.

Muhs, D.R., and V.T. Holliday, 1995. Active dune sand on the Great Plains in the 19th century: Evidence from accounts of early explorers. *Quaternary Research* 43: 198–208.

Muhs, D.R., and P.B. Maat, 1993. The potential response of eolian sands to greenhouse warming and precipitation reduction on the Great Plains of the U.S.A. *Journal of Arid Environments* 25: 351–361.

Muhs, D.R., T.W. Stafford, S.D. Cowherd, S.A. Mahan, R. Kihl, P.B. Maat, C.A. Bush, and J. Nehring, 1996. Origin of the late Quaternary dune fields of northeastern Colorado. *Geomorphology* 17: 129–149.

Nativ, R., 1988. *Hydrogeology and Hydrochemistry of the Ogallala Aquifer, Southern High Plains, Texas Panhandle and Eastern New Mexico*. Bureau of Economic Geology Report of Investigations 177, The University of Texas at Austin.

O'Hara, S.L., and I.A. Campbell. 1993. Holocene geomorphology and stratigraphy of the lower Falcon valley, Dinosaur Provincial Park, Alberta, Saskatchewan. *Canadian Journal of Earth Sciences* 30: 1846–1852.

Oldow, J.S., A.W. Bally, H.G. Ave Lallemant, and W.P. Leeman, 1989. Phanerozoic evolution of the North American Cordillera; United States and Canada. In: *The Geology of North America—An Overview*. A.W. Bally and A.R. Palmer (Eds.). Geological Society of America Centennial Volume A, pp. 139–232.

Orton, R., 1966. Characteristic meteorology of some large flood-producing storms in Texas—Easterly waves. In: *Symposium on Consideration of Some Aspects of Storms and Floods in Water Planning*. Texas Water Development Board Report 33, pp. 1–18, Austin, Texas.

Osterkamp, W.R., 1987. Groundwater—An agent of geomorphic change. In: *Geomorphic Systems of North America*. W. Graf (Ed.). Geological Society of America Centennial Special Volume 2, pp. 188–195.

Osterkamp, W.R., and W.W. Wood, 1987, Playa-lake basins on the Southern High Plains of Texas and New Mexico: Part I. Hydrologic, geomorphic, and geologic evidence for their development. *Geological Society of America Bulletin* 99: 215–223.

Palmer, E.C., 1986. Land use and cultural change along the balcones Escarpment: 1718–1986. In: *The Balcones Escarpment*. P.L. Abbott and C.M. Woodruff Jr. (Eds.). Department of Geological Sciences, San Diego State University, San Diego, California, pp. 153–161.

Perry, E.S., 1962. *Montana in the Geologic Past*. State of Montana Bureau of Mines and Geology Bulletin 26.

Petersen, J.F., 1988. *Enchanted Rock State Natural Area: A Guide to the Landforms* (2nd Ed.), Terra Cognita Press, San Marcos, Texas.

Pitlick, J., 1997. A regional perspective of the hydrology of the 1993 Mississippi River basin floods. *Annals of the Association of American Geographers* 87: 135–151.

Powell, J.W., 1878. *Report on the Lands of the Arid Region of the United States*. U.S. Geographical and Geological Survey of the Rocky Mountain Region. U.S. Government Printing Office, Washington, D.C.

Pye, K., Winspear, N.R., and Zhou, L.-P., 1995. Thermoluminescence ages of loess and associated sediments in central Nebraska, USA. *Palaeogeography, Palaeoclimatology, Palaeoecology*, 118: 73–87.

Reddell, J.R., 1994. The cave fauna of Texas with special reference to the western Edwards Plateau. In: *The Caves and Karst of Texas*, W.R. Elliot and G. Veni (Eds.). National Speleological Society, Huntsville, Alabama, pp. 31–49.

Reeves, C.C., Jr., 1966. Pluvial lake basins of west Texas. *Journal of Geology* 74: 269–291.

Reeves, C.C., Jr., 1972. Tertiary-Quaternary stratigraphy and geomorphology of west Texas and southeastern New Mexico. In: *Guidebook for East-central New Mexico*. V.C. Kelley and F.D. Trauger (Eds.). New Mexico Geological Society Guidebook 24, pp. 108–117.

Reeves, C.C., Jr., 1991. Origin and stratigraphy of alkaline lake basins, Southern High Plains. In: *Quaternary Nonglacial Geology: Conterminous U.S.* R.B. Morrison (Ed.). Geological Society of America Centennial Volume K-2, pp. 484–486.

Reeves, C.C., Jr., and J.A. Reeves, 1996. *The Ogallala Aquifer (of the Southern High Plains), vol. 1—Geology*. Estacado Books, Lubbock, TX.

Rich, F.J. (Ed.), 1981. *Geology of the Black Hills, South Dakota and Wyoming*. American Geological Institute, Falls Church, VA.

Richmond, G.M., and D.S. Fullerton, 1986. Introduction to Quaternary glaciations in the United States of America. In: *Quaternary Glaciations in the Northern Hemisphere.* V. Sibrava, D.Q. Bowen, and G.M. Richmond (Eds.). *Quaternary Science Reviews* 5: 3–10.

Riskind, D.H., and D.D. Diamond, 1986. Plant communities of the Edwards Plateau of Texas. In: *The Balcones Escarpment.* P.L. Abbott and C.M. Woodruff Jr. (Eds.). Department of Geological Sciences, San Diego State University, San Diego, California, pp. 21–32.

Roed, M.A., and D.J., Wasylyk, 1973. Age of inactive alluvial fans—Bow River, Alberta. *Canadian Journal of Earth Science* 10: 1834–1940.

Rosenzweig, C., and D. Hillel, 1993. The Dust Bowl of the 1930s: Analog of greenhouse effect in the Great Plains? *Journal of Environmental Quality* 22: 9–22.

Ruhe, R.V., 1969. *Quaternary Landscapes in Iowa.* Iowa State University Press, Ames.

Ruhe, R.V., 1983. Depositional environment of Late Wisconsin loess in the midcontinental United States. In: *Late Quaternary Environments of the United States, v. 1, The Late Pleistocene.* H.E. Wright Jr. and S.C. Porter (Eds.). University of Minnesota Press, Minneapolis, pp. 130–137.

Running, G.L., IV, 1995. Archaeological geology of the Rustad Quarry Site (32RI775): An Early Archaic site in southeastern North Dakota. *Geoarchaeology* 10: 183–204.

Running, G.L., IV, 1996. The Sheyenne Delta from the Cass Phase to the present—Landscape evolution and paleoenvironment. In: *Quaternary Geology of the Southern Lake Agassiz Basin.* K.L. Harris, M.R. Luther, and J.R. Reid (Eds.). North Dakota Geological Survey, Miscellaneous Series 82, pp. 136–151.

Sabin, T.J., and V.T. Holliday, 1995. Morphometric and spatial relationships of playas and lunettes on the Southern High Plains. *Annals of the Association of American Geographers* 85: 286–305.

Sardeson, F.W., 1897. On glacial deposits in the Driftless Area. *American Geologist* 20: 392–403.

Sauchyn, D.J., 1993. Quaternary and Late Tertiary landscape evolution in the western Cypress Hills. In: *Quaternary and Late Tertiary Landscapes of Southwestern Saskatchewan and Adjacent Areas.* D.J. Sauchyn (Ed.). Canadian Plains Research Center, University of Regina, pp. 27–45.

Sauer, C.O., 1950. Grassland climax, fire and man. *Journal of Range Management* 3: 16–20.

Scott, G.R., 1963. *Quaternary Geology and Geomorphic History of the Kassler Quadrangle, Colorado.* U.S. Geological Survey Professional Paper 421-A.

Sellards, E.H., 1923. Geologic and soil studies on the alluvial lands of the Red River Valley. *University of Texas Bulletin* 2327: 27–87.

Senger, R.K., E.W. Collins, and C.W. Kreitler, 1990. *Hydrogeology of the Northern Segment of the Edwards Aquifer, Austin Area.* Bureau of Economic Geology Report of Investigations 192, The University of Texas at Austin.

Sharp, J.M., Jr., and J.L. Banner, 1997. The Edwards Aquifer: A Resource in Conflict. *GSA Today* 7(8): 1–9.

Shaw, J., 1983. Drumlin formation related to inverted melt-water erosional marks. *Journal of Glaciology* 29: 461–479.

Slade, R.M., Jr., 1986. Large rainstorms along the Balcones Escarpment in Central Texas. In: *The Balcones Escarpment.* P.L. Abbott and C.M. Woodruff Jr. (Eds.). Department of Geological Sciences, San Diego State University, San Diego, California, pp. 15–19.

Slaymaker, H.O., 1989. Physiography of Canada and its effects on geomorphic processes. In: *Quaternary Geology of Canada and Greenland.* R.J. Fulton, (Ed.). Geological Survey of Canada, Geology of Canada, no. 1, pp. 581–583.

Sloss, L.L. (Ed.), 1988. *Sedimentary Cover—North American Craton; U.S.* Geological Society of America Centennial Volume D-2.

Smith, A.R., and G. Veni, 1994. Karst regions of Texas. In: *The Caves and Karst of Texas.* W.R. Elliott and G. Veni (Eds.). National Speleology Society, Huntsville, Alabama, pp. 7–12.

Smith, D.G., and T.G. Fisher, 1993. Glacial Lake Agassiz: The northwestern outlet and paleoflood. *Geology* 21: 9–12.

Sparrow, H.O., 1984. *Soils at Risk, Canada's Eroding Future.* A Report on Soil Conservation by the Standing Committee on Agriculture, Fisheries and Forestry to the Senate of Canada, Ottawa.

Squire, G.H., 1897. Studies in the Driftless region of Wisconsin. *Journal of Geology* 5: 825–836.

Squire, G.H., 1898. Studies in the Driftless region of Wisconsin. *Journal of Geology* 6: 182–192.

Stormer, J.C., 1972. Ages and nature of volcanic activity on the Southern High Plains, New Mexico and Colorado. *Geological Society of America Bulletin* 83: 2443–2448.

Stott, D.F., and J.D. Aitken (Eds.), 1993. *Sedimentary Cover of the Craton in Canada.* Geological Survey of Canada, Geology of Canada no. 5.

Stott, D.F., J. Dixon, J.R. Dietrich, D.H. McNeil, L.S. Russell, and A.R. Sweet, 1993. Tertiary. In: *Sedimentary Cover of the Craton in Canada.* D.F. Scott and J.D. Aitken, (Eds.). Geological Survey of Canada, Geology of Canada no. 5, pp. 439–465.

Sweeting, M.M., 1972. Karst and solution phenomena in the Santa Rosa area, New Mexico. In: *Guidebook for East-Central New Mexico.* V.C. Kelley and F.D. Trauger (Eds.). New Mexico Geological Society Guidebook 24, pp. 168–170.

Swinehart, J.B., 1989. Wind-blown deposits. In: *An Atlas of the Sand Hills.* A. Bleed and C. Flowerday (Eds.). University of Nebraska-Lincoln, Conservation and Survey Division Resource Atlas 5, pp. 43–56.

Taylor, R.L., J.M. Ashley, R.A. Chadwick, S.G. Custer, D.R. Lageson, W.W. Locke III, D.W. Mogk, J.G. Schmitt, and J.B. Erickson, 1986. *Geological Map of Montana* (1:1,900,800). Department of Earth Sciences, Montana State University, Bozeman.

Teller, J.T., 1990. Meltwater and precipitation runoff to the North Atlantic, Arctic, and Gulf of Mexico from the Laurentide Ice Sheet and adjacent regions during the Younger Dryas. *Paleooceanography* 5: 897–905.

Teller, J.T., 1995. History and drainage of large ice-dammed lakes along the Laurentide Ice Sheet. *Quaternary International* 28: 83–92.

Teller, J.T., and L. Clayton (Ed.), 1983. *Glacial Lake Agassiz.* Geological Association of Canada Special Paper 26.

Thornbury, W.D., 1965. *Regional Geomorphology of the United States.* John Wiley, New York.

Thorpe, J., and Smith, H.T.U., 1952. *Pleistocene Eolian Deposits of the United States, Alaska, and Parts of Canada* (1: 2,500,000). Geological Society of America Map.

Trenhaile, A.S., 1990. *The Geomorphology of Canada: An Introduction.* Oxford University Press, Toronto, Canada.

Trewartha, G.T., 1954. *An Introduction to Climate.* McGraw-Hill Book Co., New York.

Trimble, S.W., 1983. A sediment budget for Coon Creek basin in the Driftless Area, Wisconsin, 1853–1977. *American Journal of Science* 283: 454–474.

Trimble, S.W., and S.W. Lund, 1982. *Soil Conservation and the Reduction of Erosion and Sedimentation in the Coon Creek Basin, Wisconsin.* U.S. Geological Survey Professional Paper 1234, pp. 1–35.

U.S. Army Corps of Engineers, 1996. *The Great Flood of 1993 Post-flood Report.* U.S. Army Corps of Engineers, North Central Division, Chicago.

Upham, W., 1895. *Glacial Lake Agassiz.* U.S. Geological Survey Monograph 25.

Veenhuis, J., 1986. *The Effects of Urbanization on Floods in the Austin Metropolitan Area, Texas.* U.S. Geological Survey Water-Resources Investigation 86–4069.

Veni, G., 1987. Fracture permeability: Implications on cave and sinkhole development and their environmental assessments. In: *Karst Hydrogeology: Engineering and Environmental Applications.* B.F. Beck and W.L. Wilson (Eds.). A.A. Balkema, Boston, pp. 101–105.

Veni, G., 1994. Hydrogeology and evolution of caves and karst in the southwestern Edwards Plateau, Texas. In: *The Caves and Karst of Texas.* W.R. Elliott and G. Veni (Eds.). National Speleology Society, Huntsville, Alabama, pp. 13–30.

Vreeken, W.J., 1993. Loess and associated paleosols in southwestern Saskatchewan and southern Alberta. In: *Quaternary and Late Tertiary Landscapes of southwestern Saskatchewan and Adjacent Areas.* D.J. Sauchyn (Ed.). Canadian Plains Research Center, University of Regina, Regina, Canada. pp. 27–45.

Walker, N., 1992. Middle Proterozoic geologic evolution of the Llano uplift, Texas: Evidence from U-Pb zircon geochronometry. *Geological Society of America Bulletin* 103: 494–504.

Wayne, W.J., J.S. Aber, S.S. Agard, R.N. Bergantino, J.P. Bluemle, D.A. Coates, M.E. Cooley, R.F. Madole, J.E. Martin, B. Mears, Jr., R.B. Morrison, and W.M. Sutherland, 1991. Quaternary geology of the northern Great Plains. In: *Quaternary Nonglacial Geology: Conterminous U.S.* R.B. Morrison (Ed.). Geological Society of America Centennial Volume K-2, pp. 441–476.

Weaver, J.E., 1968. *Prairie Plants and their Environments: A Fifty-Year Study in the Midwest.* University of Nebraska Press, Lincoln.

Weeks, J.B., and E.D. Gutentag, 1988. Region 17, High Plains. In: *Hydrogeology.* W. Back, J.S. Rosenshein, and P.R. Seaber (Eds.). Geological Society of America Centennial Volume O-2, pp. 157–164.

Welch, J.E., and J.M. Hale, 1987. Pleistocene Loess in Kansas—Status, Present Problems and Future Considerations. In: *Quaternary Environments of Kansas.* W.C. Johnson (Ed.). Guidebook Series 5, Kansas Geological Survey, Lawrence, pp. 67–84.

Wentworth, E.H., 1948. *America's Sheep Trails.* Iowa State College Press, Ames.

Wheaton, E.E., and A.K. Chakravarti, 1990. Dust storms in the Canadian Prairies. *International Journal of Climatology* 10: 829–837.

Williams, M., 1989. *Americans and their Forests: A Historical Geography.* Cambridge University Press, Cambridge.

Willman, H.B., and J.C. Frye, 1969. *High-level Glacial Outwash in the Driftless Area of Northwestern Illinois.* Illinois Geological Survey Circular 440.

Winder, C.G., 1965. Alluvial cone construction by alpine mudflow in a humid environment. *Canadian Journal of Earth Science* 2: 270–277.

Witzke, B.J., and G.A. Ludvigson, 1990. Petrographic and stratigraphic comparisons of sub-till and inter-till alluvial units in western Iowa: Implications for development of the Missouri River drainage. In: *Holocene Alluvial Stratigraphy and Selected Aspects of the Quaternary History of Western Iowa.* E.A. Bettis (Ed.). Midwest Friends of the Pleistocene, 37th Field Conference Guidebook, pp. 119–143.

Wolfe, S.A., 1996. Stops 10, 24, 25, 27. In: *Landscapes of the Palliser Triangle.* D.S. Lemmen (Ed.). Guidebook for the Canadian Geomorphology Research Group Field Trip, University of Saskatchewan, Saskatoon, pp. 38–39, 66, 67, 72.

Wolfe, S.A., D.J., Huntley, and J. Ollerhead, 1994. Recent and late Holocene sand dune activity in southwestern Saskatchewan. *Current Research 1995-B, Geological Survey Canada*, pp. 131–140.

Wood, W.E., W.E. Sanford, and C.C. Reeves, Jr., 1992. Large lake basins of the southern High Plains: Ground-water control of their origin? *Geology* 20: 535–538.

Woodruff, C.M., Jr., and P.L. Abbott, 1986. Stream piracy and evolution of the Edwards Aquifer along the Balcones Escarpment, central Texas. In: *The Balcones Escarpment.* P.L. Abbott and C.M. Woodruff, Jr. (Eds.). Department of Geological Sciences, San Diego State University, San Diego, California, pp. 77–89.

Wright, H.E., Jr., 1968. History of the Prairie Peninsula. In: *The Quaternary of Illinois.* R.E. Bergstrom (Ed.). University of Illinois, College of Agriculture Special Publication 14, pp. 78–88.

18

The Western Cordillera

George P. Malanson
David R. Butler

A significant portion of the research on the physical geography of North America has been conducted within the Western Cordillera (see the classics in Ives and Barry, 1974; Ives, 1980). The reasons for this richness are varied, but we can identify three. First, many physical geographers, and colleagues in other disciplines contributing to physical geography, have been drawn to the mountains and to the science simultaneously or have found that the science suited their predilection to spend time in the mountains. Second, physical geography is often revealed on environmental gradients, conceptually distinct from spatial transects, but gradients are relatively evident in mountain areas because they are compressed on and correlated with spatial transects. Also, in the mountains, processes can be accelerated because of the steep slopes. Third, the Western Cordillera is a rich source of data for the study of physical geography (Price, 1978). The area contains many distinct slopes, valleys, ridges, and streams; spatially and temporally distinct evidence of glaciation, vegetation, and climatic changes exists; human impacts have been limited or at least relatively discrete in time and space; and agencies with funding (U.S. Forest Service, National Park Service, U.S. Geological Survey, Environment-Canada) have been interested in research in the Western Cordillera.

Given the abundance of scientific information about the Western Cordillera, the following discussion highlights selected areas of information and research through the lens of hierarchy theory as conceptualized in ecology (Allen and Hoekstra, 1992). Briefly, hierarchy theory holds that processes and patterns at different scales have typical relations: those at a given scale are created by the combinations of processes at the next lower scale while constrained by the patterns of the next higher scale. The physical geography of the Western Cordillera is thus reviewed by examining process and pattern at four scales: *continental*, focusing on the whole system as the spatial context for processes operating at smaller scales; *across ranges*, but actually spanning the continent; *within ranges*, with a focus on valley heterogeneity; and *within valleys*, with a focus on both elevational and transverse heterogeneity (fig. 18.1). In this chapter, we emphasize the last scale because this is the scale at which most research activity occurs; we also recognize the constraints that coarser scale phenomena place on finer scale phenomena.

18.1 Continental Scale

The existence of the Western Cordillera can be ascribed to a single process: plate tectonics. The Western Cordillera is the result of two major tectonic mechanisms that have led to mountain building, namely, collision and subduc-

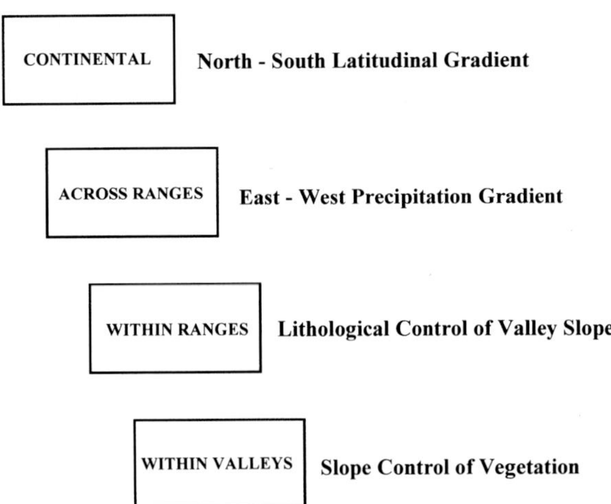

Figure 18.1 Schematic hierarchy of the Western Cordillera; coarser scale features constrain the processes at the finer scales that create them.

tion (see chapter 1). A succession of plates, plate fragments, and island arcs have collided with the North American Plate, resulting in the accretion of terranes, such as Wrangellia (fig. 18.2), onto the continent. Subduction has been responsible for the uplift and augmentation of much of the cordilleran mass. As successive oceanic plates have moved under the western margin of the North American plate, the increased mass has led to isostatic response. The relationship between isostasy and denudation makes mountains—not folding, faulting, or volcanism alone (Pinter and Brandon, 1997). Folding, faulting, and volcanic activity within the Western Cordillera are all expressions of repeated orogenic events that have characterized the region throughout Mesozoic and Cenozoic time. These processes may be viewed as responses of the crust to forces generated by compression, extension, and isostatic disequilibrium.

Fold structures are evident throughout the Cordillera and are well expressed in the Rocky Mountains, in the western ranges of British Columbia where the Porcupine Creek anticlinorium is a dominant feature (Lickorish, 1993), and in the Coast Ranges of Oregon and California. Fault structures include simple normal faults generated by crustal extension, complex overthrust faults produced by massive compression, and, particularly along the western margins of the region, strike-slip faults as a response to lateral shear. Normal faulting is primarily responsible for the fault-block terrain of the Great Basin where similarities in the climatology, geomorphology, and biogeography of discrete mountain ranges have been induced by a common structural geology. Northern extensions of this relationship occur in Idaho and Montana, where the Lost River, Lemhi, and Beaverhead mountains are primarily fault-tilted blocks that have been created by upward pressure and crustal extension induced by subduction of oceanic crust beneath the North American plate. As a result of the upward pressure, the crust has fractured, and alternating horsts and grabens have formed. Overthrust faults are best exemplified by the Lewis Overthrust in the northern Rockies, where Precambrian mudstone and limestone were thrust over weaker Cretaceous rock (fig. 18.3). Farther north, in the Front Ranges of the Rockies in Alberta, the McConnell Thrust represents a complex of movements reflected in duplex structures (Maurel, 1991).

Cenozoic volcanism, related to repeated plate subduction, is expressed at intervals throughout the Western Cordillera but nowhere better than in the Cascade Range, where a string of individual volcanoes, dominated by Mt. Rainier in size but by Mount St. Helens in recent activity, exhibit steep, conical composite forms (Lippman and Mullineaux, 1982). These volcanoes acquired their present form during the Quaternary, and nearly all have been active during Holocene time, but their foundations lie in Tertiary volcanism. Tertiary outpourings of basalt also formed extensive

Figure 18.2 (*top*) Unnamed peaks form part of Wrangellia in south-central Alaska.

Figure 18.3 (*bottom*) The Lewis Overthrust fault at Yellow Mountain, Glacier National Park, Montana. Rock slides occur as younger, weaker rocks beneath retreat against older, harder rocks above.

lava plains on the Columbia Plateau and continued intermittently throughout the Quaternary on the Snake River Plain of Idaho. In the latter region, volcanism has been associated with the migration of the Yellowstone hot spot relative to the overriding North American plate. Volcanism continues to be expressed today in geothermal activity on the Yellowstone Plateau and neighboring areas.

The spatial juxtaposition of the Pacific and the North American plates gives the Western Cordillera as a whole its predominant spatial structure, that is, its north–south axis. This axis is a fundamental constraint on the physical geography of the cordillera at finer scales because it is aligned on the latitudinal gradient of insolation, transverse to the zonal gradient of atmospheric circulation. This orientation provides the template for, and the constraints on, the gradients across ranges.

18.2 Gradients Across Ranges

The primary gradients across the major ranges are climatic. The constraint of the continental-scale latitudinal extent of the Western Cordillera creates the north–south gradient expressed in the regional pattern of insolation, temperature (fig. 18.4), and length of growing season. However, the alignment of the ranges perpendicular to the primary zonal flow of the atmosphere creates the other climatic gradient: a combination of orographic precipitation and rain shadow (fig. 18.5). These climatic gradients of energy and moisture are actually expressed at different scales. The energy gradient is generally a single latitudinal gradient for the entire cordillera, whereas the moisture gradient is spatially complex because of repetition in the alternation of orographic uplift and rain shadow in the context of variability in the atmospheric flow. These climatic gradients create the two other gradients in the landscape at this scale, glaciation and vegetation.

18.2.1 Glaciation and Snowfall

Now as in the past, patterns of glaciation reflect the combined effects of solar energy and atmospheric precipitation. Glaciation is greatest in regions that receive the most precipitation, ideally but not necessarily as snow, while being cold enough to maintain glaciers. Thus, during Pleistocene cold stages, the Cordilleran Ice Sheet was most vigorously developed over southern Alaska, southwest Yukon, and British Columbia. Farther north, central Alaska was ice-free because of low precipitation, although the Brooks Range carried a small ice cap. To the south, higher areas such as the Sierra Nevada and the Yellowstone Plateau had ice caps, but higher temperatures and modest precipitation limited glaciation. Comparing Pleistocene and Holocene glaciation, we see that the general pattern has changed

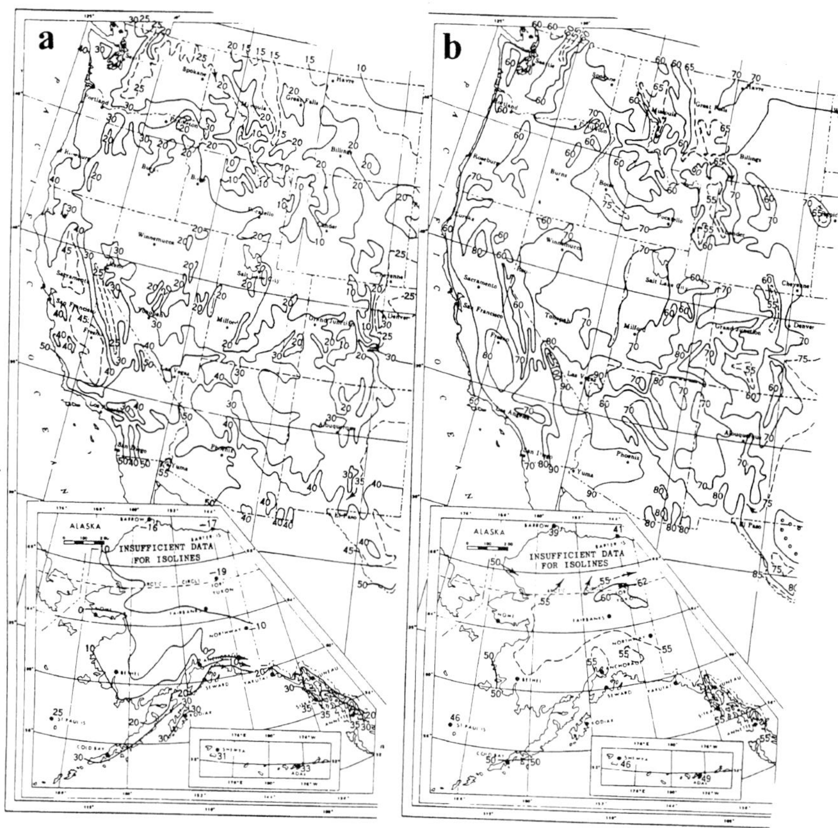

Figure 18.4 Isotherms (mean daily temperature, °F) reveal the major temperature gradients in the Western Cordillera: (a) January; (b) July.

Figure 18.5 Isohyets (mean annual precipitation, inches) reveal the major precipitation gradients in the Western Cordillera.

little, although the extent and intensity of Holocene glaciation have been much reduced. Summary chronologies for the Pleistocene glaciation of much of the Western Cordillera may be found in Volume 5 of *Quaternary Science Reviews*, 1986.

Moran (1974) illustrated the concept of hierarchical constraint when he suggested that air-mass climatology throughout the Pleistocene was influenced by the mountains to the extent that this effect was greater than the forces of climatic change. Leonard (1984) also found similar patterns in the San Juan Mountains of Colorado,where equilibrium line altitudes (ELAs) of Pleistocene glaciers correlate with present-day snow accumulation. Similarly, Locke (1990) concluded that late Pleistocene regional controls of ELA in western Montana were parallel to those of the present. In contrast, Zielinski and McCoy (1987) found different patterns between past and present for ELAs in the Great Basin, and Ellis and Calkin (1979) found differences in orientations between Pleistocene and modern glaciers in the Brooks Range of Alaska. Seltzer (1994) concluded that, over long time scales, the snow line responds only to relatively large changes in annual accumulation, and temperature is probably the most important single variable.

The distribution and effects of Holocene glaciation in the Western Cordillera have also been extensively studied (Burke and Birkeland, 1983; Luckman et al., 1993). Controversy exists as to the extent of early Holocene glacier advances (Benedict, 1981). In the Canadian Rockies and northern Montana, the Crowfoot Advance is generally assumed to be early Holocene because its deposits are overlain by tephra related to the eruption of Mt. Mazama, now Crater Lake, around 6,800 yr B.P. (Luckman et al., 1993). Later Holocene glacial episodes in the American Rockies typically reveal three advances: between 5000 and 3000 yr B.P., around 2000 yr B.P., and finally around 500 yr B.P., the so-called Little Ice Age (Burke and Birkeland, 1983). A late Holocene advance in British Columbia, designated the Peyto Advance, and correlated with the Tiedemann Advance farther west, occurred between about 3100 and 2500 yr B.P. The Little Ice Age in the Canadian Rockies and northern Montana saw glaciers reach their greatest downvalley extent since the Pleistocene (Luckman and Kearney, 1986). Dates of the onset of the Little Ice Age vary, with earliest dates (800–600 yr B.P.) from the Peyto and Robson glaciers in British Columbia (Luckman et al., 1993). The Little Ice Age ended around 1850 in most locations (Smith et al., 1995).

Currently active glaciers are most common in coastal Alaska and the coastal ranges of British Columbia. Farther north, in the Brooks Range, modern glaciers are quite small because of low precipitation. Farther south, higher temperatures now restrict glaciers to small sheltered hollows, mostly on north- and east-facing slopes. Many of these current glaciers are now strongly in retreat (McCarthy and Smith, 1994) (fig. 18.6). We have observed, however, that small glaciers and summer snow patches near fast-retreating glaciers have remained unchanged over the past 70 years in Glacier National Park, perhaps because these smaller features responded quickly to climate warming in the nineteenth century, whereas the larger glaciers had a significant time lag.

The regional patterns of modern snowfields across ranges have also been studied with emphasis on ther response to synoptic weather patterns (Yarnal, 1984). Although a connection may be made to El Niño–Southern Oscillation events, a more direct and extensive link to synoptic systems was made by Chagnon et al. (1990). One of the more

Figure 18.6 The retreat of this glacier on Mt. Blackburn, south-central Alaska, is marked by a higher kame terrace and a large volume of rock debris on the ice.

intriguing contrasts relates higher snowfall over the upper Colorado River basin to northeasterly flow (Klein et al., 1965) and southwesterly flow (Barry 1992). McGinnis (1997) has elucidated current patterns of snowfall in the upper Colorado River drainage. The intersection of synoptic patterns, controlled by the interaction of the mountains and Rossby waves, and regional topography accounts for snow distribution at a regional scale. In a similar vein, heavy snowfalls and snow avalanche activity are readily related to synoptic pressure surfaces, illustrating regional constraints (Mock, 1995).

18.2.2 Vegetation

The pattern of vegetation across the ranges is a topic of broad-scale biogeography. Both the overall north–south gradient and how this gradient has changed in the Holocene are areas of current research. Peet (1988) contrasted the differences in the results of gradient analyses of vegetation on elevational and topographic gradients over this broader scale gradient. On the gradient analysis diagrams that he presented, we can see definite shifts in both the elevation of given species and their position on the moisture gradient (fig. 18.7). For example, in the Santa Catalina Mountains, *Abies lasiocarpa* (subalpine fir) is limited to the wettest sites at the highest elevations (>2600 m). In the Bitterroot Mountains, this species extends across a wide range of elevations and the entire moisture gradient, to higher elevations under drier conditions. In the northern Rockies of Jasper National Park, Alberta, subalpine fir is confined to a small elevational range on the drier portion of the moisture gradient. A general expansion of fen vegetation indicates a gradient of increased limits to growth attributable to saturated soils.

Baker (1983) examined the composition of alpine vegetation in a mountain range in relation to biogeographic patterns across ranges while also considering temporal change. He found that species widespread in the western United States are found in protected snowy sites, which occur across all ranges, whereas those restricted to the Rockies are most common in open, windy low-radiation sites, which may be indicative of the drier conditions of the Rockies relative to other ranges in the cordillera. Allen et al. (1991), examining the relations between latitudinal variation in species composition and their structure along local environmental gradients in the Rockies, found similar local patterns within an overall gradient, especially in the composition of low-elevation forests.

Harper et al. (1978) described plant species distributions across the mountains of the Great Basin. They considered the extent to which the floras of ranges were related to their isolation from one another and from the Sierra Nevada and the Rockies. Surprisingly, they found that the floras are more similar to those of the Rockies, despite being downwind of the Sierra Nevada, probably because mountains in the Great Basin were previously less isolated from the Rockies by intervening desert habitat than is the case today. Billings (1978) analyzed alpine plant distributions in the Great Basin and found that more northerly ranges shared taxa with arctic-alpine zones to the north and east. Hadley (1987) extended these analyses, noting that the overall latitudinal gradient was affected by barriers to migration, alpine "island" size, and isolation. Interregional comparisons include those of Parker (1987), who compared forest structures of Yosemite and Glacier National Parks in California and Montana, respectively.

Although many of the individual studies are within ranges, relating vegetation to climatic gradients, and especially climatic change, is best considered at the scale across ranges. Treeline dynamics have been linked to climatic change (LaMarche, 1973; Scuderi et al., 1993), although whether treeline movement can be used as an indicator of climatic change is debatable (Kupfer and Cairns, 1996). The common trend is a decrease in the elevation of treeline with latitude, until the alpine treeline coincides with the Arctic treeline (Arno and Hammerly, 1984) (fig. 18.8). At this scale few comparisons have been made. Using topographic maps, Becwar and Burke (1982) estimated that 80% of the transition from forest to tundra in Glacier National Park, Montana, occurs over a 550-m range. In contrast, in Rocky Mountain National Park, Colorado, 80% of the transition occurs over 200 m, but more extensive comparisons would indicate where to look for processes of control (discussed subsequently). Brown (1994) compared the relationships between environmental gradients, four tree-line vegetation types, and the normalized difference vegetation index in Glacier National Park with observations in Rocky Mountain National Park. Differences in the observed relationships were attributable primarily to latitude, as well as to rock type and precipitation regimes. Spatial patterns of the four vegetation types were very similar in both study areas.

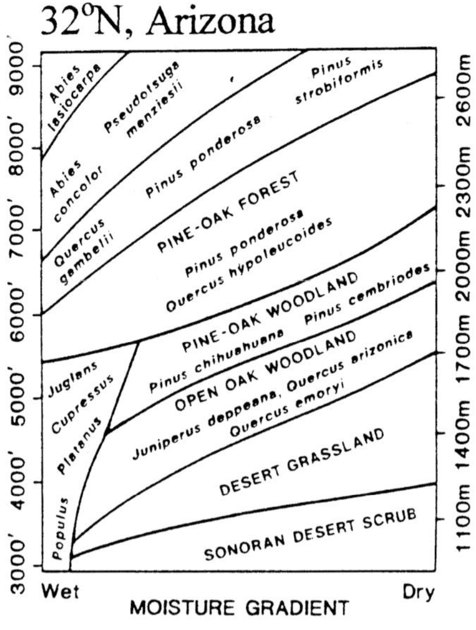

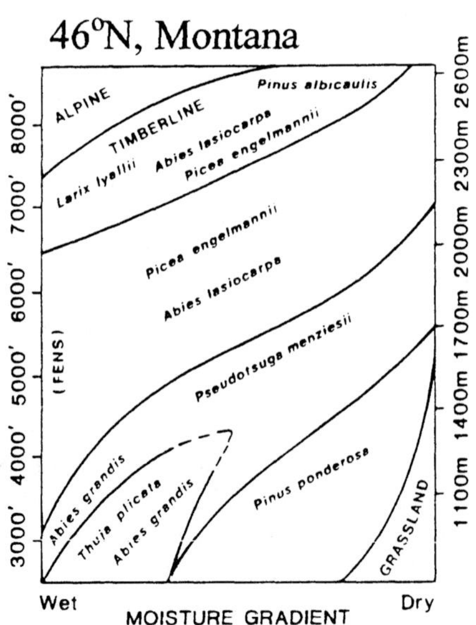

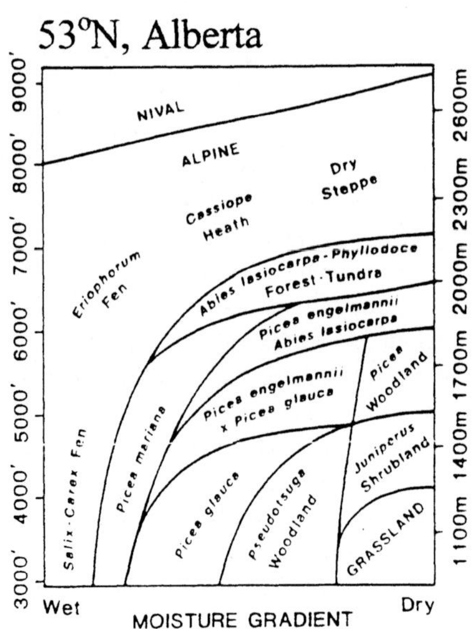

Figure 18.7 Gradient diagrams of the distribution of vegetation on elevation and moisture gradients at three latitudes; from Peet (1988).

18.3 Within Ranges

At this scale, we will examine the pattern created by the heterogeneity of ridge and valley systems within a range or portions of the larger ranges. The overall constraining patterns of climate, glaciation, and vegetation are smooth gradients that are made complex and given heterogeneity at this scale by differences among valleys and ridges. The central elements that provide this heterogeneity among valleys are differences in geology, topography, and temporal uniqueness such as fire history. Additionally, biogeographic patterns are studied at this scale using geographic information systems (GIS), remote sensing, and regional simulation models.

18.3.1 Geology, Topography, and Vegetation

The geologic structure within a range determines the spatial arrangement of major ridges and valleys. There is considerable variation in how this constraint is imposed. For example, in Glacier National Park, Montana, folding and faulting created the Lewis and Livingstone ranges, which have north–south axes, resulting in major east–west glacial troughs. In the Great Basin, the fault block mountains and graben valleys also have north–south axes. The spatial arrangements of valleys determines the heterogeneity of climates within a range through their effect on insolation and relative exposure to regional and local winds. The differential receipt of solar radiation between north- and south-facing slopes is obvious in many mountain ranges, resulting in differences in available moisture for plants, which then display great contrasts over a short distance across a ridge.

Lithologic constraints on geomorphic processes are common, for example, in landslide occurrence along the Lewis Overthrust in Montana (Oelfke and Butler, 1985). Geologic differences also affect vegetation patterns within ranges. Parker (1991) found that geologic substrate was a secondary control of vegetation composition in the mountains of southern Arizona, with local controls, as discussed subsequently, being more important. Her ordination diagrams indicate some separation by substrate that may be considered to be a coarser scale constraint. Butler and Walsh (1990) found that lithology and structure were important in determining the location of avalanche paths (fig. 18.9), the vegetation within which is discussed next.

The environmental consequences of the 1980 eruptions of Mount St. Helens in the Cascades, during which swift-moving lahars were generated in tephra and other volca-

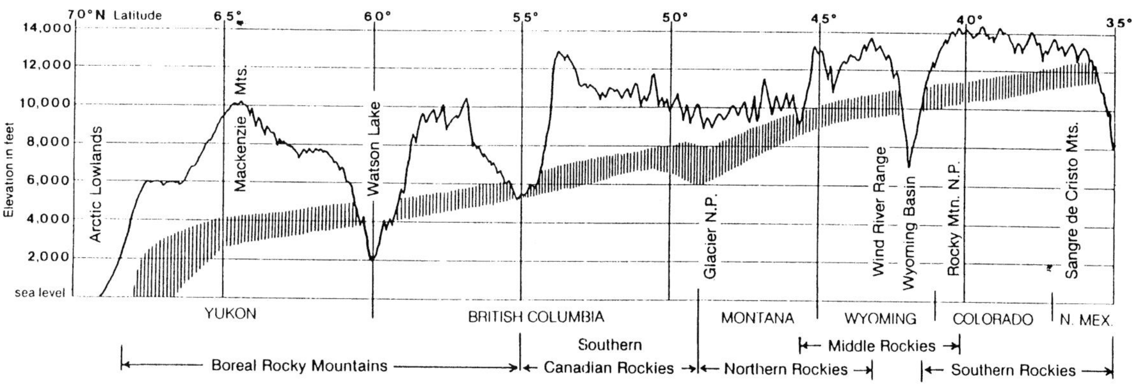

Figure 18.8 (*top*) The elevation of treeline declines with latitude and longitude; from Arno and Hammerly (1984).

Figure 18.9 (*at left*) The location of avalanche paths can be controlled by geological features such as a sill.

nic debris, illustrate that large-scale mountain building produces specific lithologies, which lead to local processes such as distinctive soil development (Clayton, 1995). Basic concepts about vegetation change in particular have been informed by these studies. For example, del Moral (1993) has advanced concepts of' primary succession through the focus on the barren post-eruption surface there. The eruption also had specific effects on glaciers (Brugman and Meier, 1981), geomorphic features (Voight et al., 1981), and local hydrology, from stream geomorphology (Lehre et al., 1983) to lake characteristics (Dion, 1981).

The relationship of emergent landscape pattern types to topography within ranges is readily observable at treeline in the Rocky Mountains. In Glacier National Park, Montana, for example, five general groups of pattern variables characterize emergent pattern types, namely, diversity, texture, patch density, edge complexity, and abruptness (Allen and Walsh, 1996). The three most common pattern types are (1) an irregular spatial pattern with high patch edge density and juxtaposition of alpine and subalpine types, (2) a zonal pattern with moderately sorted patches, and (3) a highly contrasting pattern with numerous krummholz and tundra islands. Three other patterns consist of dominance by a single vegetation type, an abrupt treeline transition, and high patch density. This and similar studies establish a link between tree-line pattern and process by showing how site conditions vary among different treeline pattern types.

Regional variation in vegetation has also been assessed through combinations of remote sensing, GIS, and simulation models. Frank and Thorn (1985) used a digital elevation model to stratify alpine tundra because of pronounced differences in geomorphic processes that varied with topography and vegetation. The most advanced models at the regional scale are those developed by Running (Running and Nemani, 1991). More specific characterizations of relations within ranges have also addressed albedo (Duguay and LeDrew, 1991), lake turbidity levels and their relation to basin morphometry (Brown and Walsh, 1992), mapping and modeling the alpine treeline ecotone (Brown, 1994; Baker et al., 1995); and examination of scale dependencies of plant pattern and topography (Allen and Walsh, 1996). Methodological work in remote sensing has addressed the effect of topographic orientation on the reflectance at a site using multiple methods for altering measured radiance on the basis of slope angle and slope aspect information calculated from digital elevation models (Meyer et al., 1993).

18.3.2 Fire History and Behavior

Although fire history may seem an unusual association with geology and topography, it can be used to differentiate landscapes at the same scale. Johnson and Wowchuk (1993) have argued that fire history in the southern Canadian Rockies is related to synoptic-scale climatology. At this scale, there are patterns (frequency) among, rather than within, ranges (Fowler and Asleson, 1984), but most work has focused on patterns at a finer scale (Goldblum and Veblen, 1992). For example, McCune and Allen (1985) have elucidated the effects of fire on vegetation in the Bitterroot Mountains on the Idaho–Montana border. The valleys are geologically similar but have very different vegetation communities, which are a result of their responses to the unique fire histories of each valley. In general, these differences in fire history may result from chance and the vagaries of ignition or they may also reflect the patterns of topography, local climate, and vegetation within the valleys. In general, there is a consistent pattern of decreasing fire frequency with increasing elevation and more mesic aspect, and an indirect relation between fire frequency, size, and intensity (Taylor, 1993).

Fire behavior, a process of the biophysical environment, has particular geographical characteristics. The actual patterns of fire spread and intensity for a given fire depend essentially on the local weather, topography, and fuels. Fire behavior modeling is still imprecise, but Malanson and Butler (1985a) used a model to assess the role of avalanche paths in limiting fire spread, which they may do, especially if firefighting efforts are focused on these paths (Suffling, 1993). Keane et al. (1996) developed a forest-dynamics model that includes a fire-spread submodel for a valley in Glacier National Park, but this type of modeling is probably best used to examine variation across ranges because of the high degree of stochasticity.

18.4 Within Valleys

The geological, climatological, and historical conditions of the valley within its range constrain the heterogeneity of its hydrology, geomorphology, biogeography, and soils, which are created by the finer scaled processes of water movement, weathering and erosion, plant and animal dynamics, and soil development. General descriptions of the climate of a range or within a range are often detailed documentations of gradients (Greenland, 1989), histories (Naftz et al., 1996), or processes (Saunders and Bailey, 1994).

18.4.1 Hydrology

Good records of precipitation patterns within mountain ranges are rare because long-term records are from meteorological stations in valley bottoms. Efforts have been made to extrapolate synoptic observations from valley stations into mountains, with limited but valued success (Band et al., 1993).

Basin shape has been shown to affect annual runoff but not in simple terms (McArthur and Hope, 1993). Conditions of runoff in basins have been described (DeGraff, 1981; Bajewsky and Gardner, 1989; Grimm et al., 1995), extending to glacial hydrology and related runoff (Marston, 1983; Weirich, 1986). Using remote sensing and GIS-derived landscape-scale data, Brown and Walsh (1992) were able to relate lake water quality to basin morphometry—an application of the concept of hierarchical constraint. Other work on water quality tends to be much more focused on temporal patterns (Watras et al., 1995).

Water infiltration in mountain areas is extremely variable because of the abundance of both bare rock and coarse soils. Once water has infiltrated, the hydraulic conductivity is usually high and downslope movement relatively rapid. An overall model incorporating hillslope hydrology has been examined for mountains in Montana. Band et al. (1993) presented a combination of TOPMODEL and FOREST-BGC that combined evapotranspiration with hillslope hydrology. The model is sensitive to soil conductivity and indicates that saturated overland flow in portions of hillslopes can differentiate slope hydrology.

The distribution of snow and snowpack is an ongoing area of investigation (Woolridge et al., 1996), but most of this research in recent years has been oriented toward water resources. The spatial distribution of snow in the mountains of North America has been modeled for basins. Hamilton and Lahey (1982) statistically related snow depth, water equivalent, and areal extent to topography. Bales and Harrington (1995) reviewed progress in snow hydrology with an emphasis on runoff modeling. Recent progress has been made in understanding the spatial distributions of snow in alpine basins (e.g., Cline, 1992) and its translation to runoff (e.g., Kite, 1991). Remote sensing and GIS have been applied to assessing snow cover and characteristics. Microwave remote sensing and radar are proving as useful in alpine basins as they are over more extensive areas (e.g., Shi et al., 1994).

Avalanche activity is a related aspect of snow hydrology. Some studies address the climatological conditions under which avalanches occur, whereas others deal with the distribution of avalanche paths in a region, the role of avalanches in moving snow, and the prediction of avalanches (Ives and Bovis, 1978; Butler and Malanson, 1985). DeScally (1996) found that late-spring basin discharge is greater from basins with avalanche activity because of the later melting of avalanche snow. In terms of the overall morphology of avalanche paths, the source area may be a critical determinant of runout distance in some cases, but track width is more important in others (Butler and Malanson, 1992). Track width creates patterns of vegetation within the path (Malanson and Butler, 1985b; see next section), which in turn affect geomorphic processes. The geomorphic work of avalanches has been variously reported (Matthes, 1938). Gardner (1970) considered avalanches to be significant shapers of the landscape in the Lake Louise area, but farther south, Butler (1985) found little ongoing geomorphic work being accomplished, and Butler and Malanson (1990) concluded that the forms of avalanche paths probably reflected a more active past. One aspect of avalanche-path geomorphology is impact pools that result from episodic, high-magnitude avalanches (Smith et al., 1994).

18.4.2 Geomorphology

Overall mountain geomorphology has been reviewed (Ives, 1987), relevant specifics of regions have been reviewed (Graf, 1987), and general models of mountain geomorphology have been developed (Caine, 1971). Barsch and Caine (1984) presented a conceptual figure of research themes in mountain geomorphology, a sketch of the basic elements of relief in the Rocky Mountains, and a diagram of sediment fluxes within morphological units and sediment systems (fig. 18.10).

Specific studies of slope processes and resulting mass-movement and fluvial deposits are numerous. Examples include the study of periglacial processes ranging from solifluction to rock glaciers, rapid mass movements, and fluvial processes and resulting alluvial fan and deltaic deposits. Studies of mass movement, for example, link climatic and geologic constraints with the details of turbulent flow (Butler et al., 1998). At a single scale, process

Sediment System	Morphologic Units	Sediment Fluxes
Glacial & Coarse Debris Systems	Glacierized Valleys Steep Bedrock Slopes & Talus	Glacial Transport Rockfall, Snow Avalanches, Mudflow, Rock Glaciers Talus Creep
Fine Sediment System	Waste Mantled Slopes	Solifluction & Soil Creep Slope Wash
Fluvial & Geochemical Systems	Stream Channels, Valley Floors, Fans & Lakes	Fluvial Transport & Deposition Solute Transport Lake Sedimentation

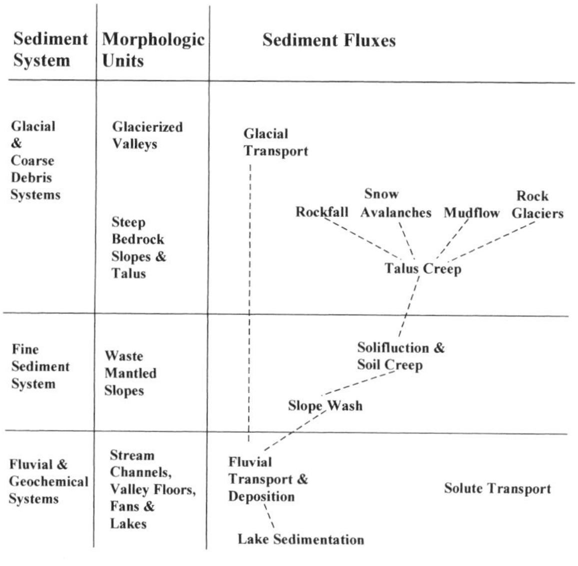

Figure 18.10 Derived from Barsch and Caine (1984), this figure represents the major geomorphic systems of mountain environments.

and pattern can be linked as in tying debris production and flow processes to alluvial fan form (Orme, 1989).

Although these studies are on slopes, they might best be considered, along with vegetation response to climatic change, as part of a picture best interpreted across ranges; in some cases, these studies have been used to infer such change (Ploufle and Jette, 1997). In particular, the presence of periglacial landforms in environments that do not currently support widespread periglacial processes has been used to infer periods of more severe climate during earlier parts of the Holocene and, more obviously, during Pleistocene cold stages (Dixon, 1991).

18.4.3 Soils

Studies of soils in the Western Cordillera have involved the development of general models of soil formation, the impact of sediment influx and outflow in alpine environments, and the use of paleosols to interpret paleoenvironmental change. Burns and Tonkin (1982) examined the spatial and temporal relationships of soils and landscapes with respect to topography, climate, vegetation, parent material, and time in the Colorado Front Range. They divided this alpine environment, defined as the ecological zone above treeline, into three geomorphic areas (ridge top, valley side, and valley bottom) and discussed whether time or space was a more significant pedogenic factor in each area. From these discussions, Burns and Tonkin (1982) developed a synthetic alpine slope model that identified seven microenvironments, from extremely windblown sites to wet meadows and semipermanent snowpacks, and described the typical soils characterizing each. Topography was considered to be the most important pedogenic factor because it controls both snow cover and loess distribution, which in turn influence plant community development and the rate of soil development, including its tendency toward a "climax state."

The role of wind in sediment delivery to alpine soils and subsequent chemical weathering has been revealed by plot-scale studies (Thorn et al., 1989; Caine, 1992). Fine-grained particles are moved from higher to lower topographic positions on hillslopes via lateral movement, in association with enhanced weathering of biotites. Vertical movement of solutes is also clearly illustrated. Pedogenesis along environmental gradients has been described for sites near treeline at Boreas Pass, Colorado (Olgeirson, 1974), in the Mosquito Range of Colorado (Stanton et al., 1994), and in the Chugach Mountains of Alaska (Sveinbjornsson et al., 1995).

Studies of paleosols in the Western Cordillera also provide insight into the timing of past glacial and interglacial stages (e.g., Hall and Shroba, 1995), as well as into paleoenvironmental conditions prevalent during such intervals (e.g., Caine, 1969; Butler, 1984). Many paleosol studies examine soils developed on, and therefore younger than, glacial moraines and rock glaciers (Benedict, 1973). Pohl's (1995) study in the Ajo Mountains of Arizona was distinguished by her examination of paleosols developed on piedmont alluvium, rather than glacial or periglacial landforms.

18.4.4 Biogeography

Studies in the Western Cordillera have contributed greatly to the development of ecological biogeography. At this scale, vegetation patterns are examined with an emphasis on several topics of current research, namely fire, riparian and meadow vegetation, avalanche paths, and multiple environmental gradients.

After Merriam's pioneering research (Merriam, 1898), much recent work has focused on plant communities related to environmental gradients (Peet, 1988). The primary gradient is elevation (Habeck, 1987). Numerous studies document the distribution of species with elevation, but most of these address multiple gradients because, although simplifying gradients has been worthwhile, plants are simultaneously affected by a variety of forces. The examination of the distribution of plant species along multiple gradients brings together several elements and illustrates the importance of a comprehensive systems approach in biogeography.

Gradient analyses of vegetation do not examine the changes in abundance along spatial transects. Instead, abstract gradients of factors that might closely affect plant reproduction, growth, and mortality are constructed. In mountain environments, these abstract gradients may be correlated with spatial transects, as in the case in which temperature gradients and moisture gradients may be correlated with location on a slope, for example, in the Santa Catalina Mountains of Arizona (Whittaker and Niering, 1965). These studies found that elevation, a surrogate for moisture availability, structured the arrangement of species and the composition of communities and elucidated the individualistic distribution of species on this gradient.

The primary elevation gradient is complicated by geomorphology and substrate and by grazing at low elevations and fire at high elevations (Barton, 1993). Other research has also found elevation to be an important surrogate gradient. Elevational differences in moisture availability and associated soil chemistry and texture have led to structured forests in Colorado and California (Allen and Peet, 1990; Taylor, 1990a). Aspect influences evapotranspiration through radiation and wind, and structures meadow, fell-field, and snow-patch vegetation in Colorado (Isard, 1986). Variation in the structure of forest in the southern Cascades relates to slope position, but position also relates to multiple physical, biological, and disturbance processes (Parker, 1993). Other work has linked vegetation to specific geomorphic surfaces such as talus (Kershaw and Gardner, 1986).

Parker (1991) illustrated how differences in processes at different scales could be examined using these methods. She found that vegetation was ordered primarily by soil texture, related to geomorphic position on slopes at a local scale, but that, at a coarser scale, vegetation was secondarily structured by differences in geologic substrate. This approach can also be applied at very fine scales. Statistical differences in the distribution of vegetation adjacent to snow patches is associated with distance to the center of the patch, a gradient in part related to differences in soils (Helm, 1982). Johnson (1996) found that subtleties of topography, leading to differences in water table depth, shading, and groundwater temperature and conductivity, accounted for much of the variance in species distributions in a treed fen in Colorado.

One subset of studies of multiple gradients regards disturbance as a gradient. Although the primary disturbance is fire (Baker, 1992), and thus the hierarchical constraint of fire size and severity is determined by regional patterns discussed previously, other disturbances include insect pests (Romme et al., 1986), blowdown (Savage et al., 1992), simple gap dynamics (Taylor and Halpern, 1991), and human activity (Savage, 1991). Multiple disturbances are common and important. Grazing, combined with climatic change, has led to change in forest structure in mountains along the Arizon–New Mexico border Savage (1991). Fire and spruce beetle activity are linked (Baker and Veblen, 1990), and fire is also linked to avalanches (Veblen et al., 1994). Avalanches influence the spread of fires, and resulting stand devastation precludes outbreaks of spruce beetle populations. Again, coarser scale phenomena, including atmospheric circulation events such as the recurrence of El Niño–Southern Oscillation phenomena, constrain the conditions for these disturbances and the ecological response.

One field that has been extensively studied is the pattern of riparian vegetation, which has ties to fluvial geomorphology (e.g., Marston, 1982) and thus to coarser scale landscape constraints (Malanson, 1993). Although Baker (1989) and Bendix (1994) have elucidated the multiscale nature of patterns of montane riparian vegetation, particular patterns within valleys are most evident. Key elements that differentiate riparian vegetation are establishment events. These are responses to the frequent high-energy disturbances that occur in mountain river valleys, such as the Animas River in Colorado (Baker, 1990), which can lead to multiple states (Baker and Walford, 1995). Malanson and Butler (1990) examined the topographic, geomorphic, sediment, and woody debris factors that might interact and determine the distribution of species on gravel bars in the northern Rockies (fig. 18.11). Using a systems model, they found that, although significant relations between vegetation and the hypothesized influences did exist in general (stepwise regression of vegetation ordination scores on the independent variables), when considered as a system, path analysis did not reveal significant relations.

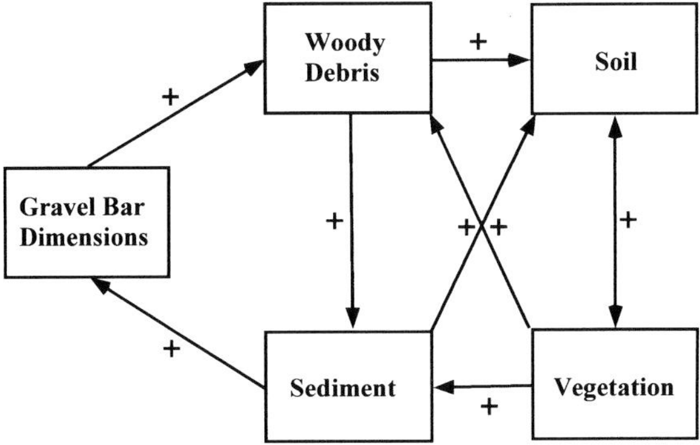

Figure 18.11 Systems of interactions on montane floodplain gravel bars create multiple feedback loops; adapted from Malanson and Butler (1990).

These events are, however, constrained by the paraglacial conditions of mountain valleys—the disequilibrium that occurs between the time of most weathering and the time of transport of glacial sediments (Church and Ryder, 1972).

Many riparian environments in the Western Cordillera supported extensive beaver (*Castor canadensis*) populations before European contact (Butler and Malanson, 1994). Extensive fur-trapping and land-use changes nearly drove the beaver to extinction, but isolated pockets remained in mountain locations such as in northwestern Montana and Alberta. Even in such locations, beaver populations expanded impressively during the twentieth century. The beaver's reoccupation of its native range is reintroducing beaver dams and ponds along riparian corridors (Meentemeyer and Butler, 1995). The absence of those ponds during the latter half of the nineteenth and early twentieth centuries produced a disequilibrium situation that has not yet been redressed by recent beaver expansions. Beaver ponds act as sediment sinks along riparian corridors, with significant amounts of sediment retention behind beaver dams. The ponds elevate local water tables and decrease net flow velocity, reducing stream erosion and providing valuable habitat for wetland-dependent plants and animals.

The biogeography of avalanche paths is an extension of riparian studies. Avalanche studies have focused on the patterns within paths. Malanson and Butler (1985b) described a transverse pattern consisting of three zones constrained by the size and shape of a given path: an inner zone, responding to the conditions along an incised stream, a flanking zone, and a drier outer zone. Not all three zones occur on every path. The longitudinal pattern of vegetation reflects the gradient in elevation but also in disturbance frequency (Butler, 1979). Avalanche paths in turn may affect the dynamics of vegetation in subalpine forests by being a source area for early successional species (Malanson and Cairns, 1995).

Studies of the development or invasion of mountain meadows also provide insights into cordilleran biogeography. Most of this research involves dendrochronology and tries to determine the influences of climate versus human impact, such as cattle introductions, on meadow change. Dunwiddie (1977) found that grazing favored tree establishment in the Wind River Mountains, but Butler (1986) and Jakubos and Romme (1993) attributed similar invasions in Idaho and in Yellowstone to climatic change. In the Olympic Mountains of northwest Washington, Woodward et al. (1995) found that subalpine fir (*Abies lasiocarpa*) invaded dry meadows during wetter climatic episodes, whereas mountain hemlock (*Tsuga mertensiana*) was established in wetter meadows during drier episodes; no pattern was evident in intermediate meadows. In many cases, the effects of fire, grazing, and climatic change are difficult to separate because they have occurred simultaneously and the primary evidence is the date of establishment of the invading trees (Taylor, 1990b). Other than that, because temporal change in meadows is constrained by coarser scale events such as climatic change and possibly their position relative to ecotones (Hessl and Baker, 1997), their place in a hierarchy is less clear than other ecological features.

18.4.5 Integration Across Scales: Treeline as Exemplar

No study yet links many of the different components of physical geography across scales. Studies of treeline, already examined across and within ranges, could be a focus for such an integration (fig. 18.12). At a local scale, several studies have shown a relationship among summer temperatures, treeline elevation, and the density and width of tree rings. Hansen-Bristow (1986) found that soil temperature at a 50-cm depth was a better determinant than air temperature of the inception of summer growth. The bulk of treeline research indicates a strong relationship between summer temperature and the elevation and latitude of treeline. Although treeline has been addressed primarily as an ecological pattern responding to the major climatic gradient of temperature, the limitations on tree growth at treeline are probably more complex (Habeck, 1969). Alpine treeline may be determined by multiple interacting factors, such as fire, wind, moisture, wind and snow, or substrate, including soil depth, fertility, and stability, and by feedbacks in the biological system through microclimate. Although the overall gradient of decreasing treeline elevation with latitude is clear (fig. 18.8), the variation around this trend may be informative. For example, although the variability of treeline elevation increases with latitude in the few cases where it has been quantified, lower variability in the Yellowstone area contrasts with the more rugged Tetons at about the same latitude (Becwar and Burke, 1982; fig. 18.12). Differences in topography, related to more intensive glaciation, may provide an explanation. The variabil-

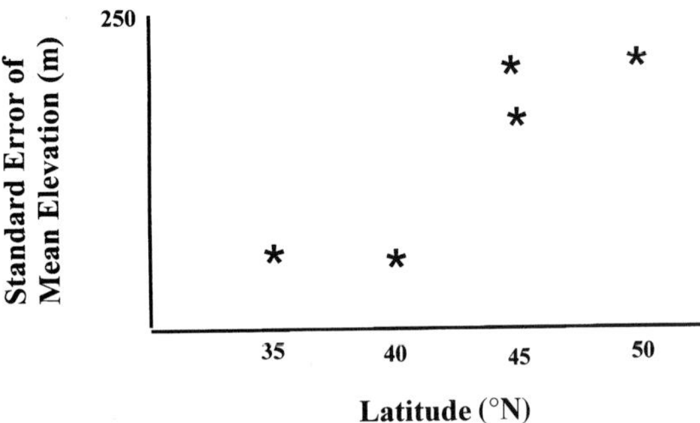

Figure 18.12 The standard error of mean tree-line elevation increases with latitude, but topography linked to intense glaciation may be causal; data from Becwar and Burke (1982).

ity in the elevation of the alpine treeline ecotone in the more rugged terrain may be due to a combination of variability in macroclimate, microclimate, and snow and debris avalanches, and to competition with tundra species (Walsh et al., 1992; Malanson and Butler, 1994). The latter suggests that more fertile tundra soils may allow alpine tundra plants to outcompete tree seedlings along the ecotone, extending tundra vegetation to lower-than-expected elevations and precluding the upward advance of treeline. In a way similar to the nutrient averaging hypothesis (Stevens and Fox, 1991), krummholz and trees may also suffer more from periglacial disturbance of the substrate because of their more extensive roots. Slopes of lower angle may have had active solifluction processes in the ecotone during the Holocene, which would probably prevent tree establishment (Hansen-Bristow and Ives, 1985), the critical phase of tree life history in this environment (Weisberg and Baker, 1995a,b).

Stevens and Fox (1991) concluded that treeline was largely a phenomenon expressing the carbon balance of trees, so that trees ended where their carbon balance was zero. Explanations of abrupt treelines fall on a gradient from abiotic control, namely, abrupt differences in resources and climate, to biotic interactions such as competition. Although a well-regarded mechanistic model of forest carbon balance could not accurately predict the location of treeline without edge-specific factors, inclusion of winter injury and additional exposure factors enabled the prediction of treeline in actual areas in Glacier National Park within the limits of available data on soil depth (Cairns and Malanson, 1997). The treeline is sensitive to several factors influencing carbon balance, especially topographically determined edaphic factors.

Abiotic control of carbon balance alone may not, however, explain an abrupt treeline on a gradual abiotic gradient; perhaps trees modify their environment, making the establishment of additional trees more likely (Malanson, 1997). Wilson and Agnew (1992) detailed the possibility that positive feedback in the biological processes within the plant communities could account for an abrupt ecotone. In these cases, the plants modify the environment that they experience, so that what was a continuous abiotic gradient becomes discontinuous in terms of the potential carbon balance. Malanson (1997) used a simulation to represent an ecotone at a mountain treeline. Seed rain and seedling survival modified the dominant patterns determined by the strength of feedback and the steepness of the abiotic gradient. The feedbacks were spatially autocorrelated, so they created waves of mortality and regeneration that have been observed on mountain slopes. These dynamics mean that the pattern at the ecotone at any point in time is ephemeral and may respond differently to environmental change.

Brown et al. (1994) suggested that a combination of empirical and physical models is needed to describe vegetation patterns at the treeline ecotone because of the multiscale influences on the patterns. They combined the findings of studies on topographically based modeling and plant physiological modeling to understand vegetation patterns at treeline. The combination of remotely sensed observations of patterns and mechanistic explanatory models provided a more complete picture of the environmental influences on treeline pattern. This approach allows the adaptation of mechanistic plant growth models (Cairns and Malanson, 1997) within the context of constraints imposed by the hydrology and geomorphology of the slope, the topography and geology of the valley, and the climate of the range.

18.5 Conclusion

Although considerable effort in physical geography has been directed toward increasingly reductionist research on processes, with good results, we believe in the value of working across scales—even when that means moving to coarser scales. This type of work not only enriches our knowledge, but provides a basis for a fundamentally geographical contribution to understanding the physical environment. The Western Cordillera provides an excellent setting for such research because the scales are relatively discrete with a basis in the fundamental environmental processes of interest.

Acknowledgments We wish to thank the personnel of Glacier National Park, particularly Debbie Vale and Debbie Fenner, whose help over the years has made our mountain research easier. Research incorporated in this chapter was supported by the National Science Foundation (grants SES-91 11853, SES-9109837, and SBR-9709810), the Association of American Geographers, the National Geographic Society, and the Burlington Northern Foundation.

References

Allen, R.B., and R.K. Peet, 1990. Gradient analysis of forests of the Sangre de Cristo Range, Colorado. Canadian Journal of Botany, 68: 193–201.

Allen, R.B., R.K. Peet, and W.L. Baker, 1991. Gradient analysis of latitudinal variation in southern Rocky Mountain forests. Journal of Biogeography, 18: 123–139.

Allen, T.H.F., and T. Hoekstra, 1992. Toward a Unified Ecology. Columbia University Press, New York.

Allen, T.R., and S.J. Walsh, 1996. Spatial composition and pattern of alpine treeline, Glacier National Park, Montana. Photogrammetric Engineering and Remote Sensing, 62: 1261–1268.

Amo, S.F., and R.P. Hammerly, 1984. Timberline, Mountain and Arctic Forest Frontiers. The Mountaineers, Seattle.

Bajewsky, I., and J.S. Gardner, 1989. Discharge and sediment-load characteristics of the Hilda rock-glacier stream, Canadian Rocky Mountains, Alberta. Physical Geography, 10: 295–306.

Baker, W.L., 1983. Alpine vegetation of Wheeler Peak, New Mexico, USA: Gradient analysis, classification, and biogeography. Arctic and Alpine Research, 15: 223–240.

Baker, W.L., 1989. Macro- and micro-scale influences on riparian vegetation in western Colorado. Annals of the Association of American Geographers, 79: 65–78.

Baker, W.L., 1990. Climatic and hydrologic effects on the regeneration *of Populus angustifolia* James along the Animas River, Colorado. Journal of Biogeography, 17: 59–73.

Baker, W.L., 1992. Structure, disturbance, and change in the bristlecone pine forests of Colorado, USA. Arctic and Alpine Research, 24: 17–26.

Baker, W.L., and T.T. Veblen, 1990. Spruce beetles and fires in the nineteenth-century subalpine forests of western Colorado, USA. Arctic and Alpine Research, 22: 65–80.

Baker, W.L., and G.M. Walford, 1995. Multiple stable states and models of riparian vegetation succession on the Animas River, Colorado. Annals of the Association of American Geographers, 85: 320–338.

Baker, W.L., J.J. Honaker, and P.J. Weisberg, 1995. Using aerial photography and GIS to map the forest-tundra ecotone in Rocky Mountain National Park, Colorado, for global change research. Photogrammetric Engineering and Remote Sensing, 61: 313–320.

Bales, R.C., and R.F. Harrington, 1995. Recent progress in snow hydrology. Review of Geophysics: 1011–1020.

Band, L.E., P. Patterson, R. Nemani, and S.W. Running, 1993. Forest ecosystem processes at the watershed scale: Incorporating hillslope hydrology. Agricultural and Forest Meteorology, 63: 93–126.

Barry, R.G., 1992. Mountain Weather and Climate. Routledge, New York

Barsch, D., and N. Caine, 1984. The nature of mountain geomorphology. Mountain Research and Development, 4: 287–298.

Barton, A.M., 1993. Factors controlling plant distributions: Drought, competition, and fire in montane pines in Arizona. Ecological Monographs, 63: 367–397.

Becwar, M.R., and M.J. Burke, 1982. Winter hardiness limitations and physiography of woody timberline flora. In: P.H. Li and A. Sakai (Editors), Plant Cold Hardiness and Freezing Stress: Mechanisms and Crop Implications. Vol 2. Academic Press, New York, pp. 307–323.

Bendix, J., 1994. Scale, direction, and pattern in riparian vegetation-environment relationships. Annals of the Association of American Geographers, 84: 652–665.

Benedict, J.B., 1973. Chronology of cirque glaciation, Colorado Front Range. Quaternary Research, 3: 584–599.

Benedict, J.B., 1981. The Fourth of July Valley: Glacial Geology and Archaeology of the Timberline Ecotone. Research Report, 2. Center for Mountain Archaeology, Ward, CO.

Billings, W.D., 1978. Alpine phytogeography across the Great Basin. Great Basin Naturalist Memoirs, 2: 105–117.

Brown, D.G., 1994. Comparison of vegetation-topography relationships at the alpine treeline ecotone. Physical Geography, 15: 125–145.

Brown, D.G., and S.J. Walsh, 1992. Relationships between the morphometry of alpine and subalpine basins and remotely-sensed estimates of lake turbidity, Glacier National Park, Montana. Physical Geography, 13: 250–272.

Brown, D.G., D.M. Cairns, G.P. Malanson, S.J. Walsh, and D.R. Butler, 1994. Remote sensing and GIS techniques for spatial and biophysical analyses of alpine treeline through process and empirical models. In: W.K. Michener, S. Stafford, and J. Brunt (Editors), Environmental Information Management and Analysis: Ecosystem to Global Scales. Taylor and Francis, Philadelphia.

Brugman, M.M., and M.F. Meier, 1981. Response of glaciers to the eruptions of Mount St. Helens. In: P.W. Lippman and D.R. Mullineaux (Editors), The 1980 Eruption of Mount St. Helens, Washington. U.S. Geological Survey Professional Paper 1250, pp. 743–756.

Burke, R.M., and P.W. Birkeland, 1983. Holocene glaciation in the mountain ranges of the western United States. In: H.E. Wright Jr. (Editor), Late Quaternary Environments of the United States, Vol. 2. University of Minnesota Press, Minneapolis, pp. 3–11.

Burns, S.F., and P.J. Tonkin, 1982. Soil-geomorphic models and the spatial distribution and development of alpine soils. In: C.E. Thorn (Editor), Space and Time in Geomorphology. George Allen and Unwin, London, pp. 25–43.

Butler, D.R., 1979. Snow avalanche path terrain and vegetation, Glacier National Park, Montana. Arctic and Alpine Research, 11: 17–32.

Butler, D.R., 1984. An early Holocene cold climatic episode in eastern Idaho. Physical Geography, 4: 86–98.

Butler, D.R., 1985. Vegetational and geomorphic change on snow avalanche paths, Glacier National Park, Montana, USA. Great Basin Naturalist, 45: 313–317.

Butler, D.R., 1986. Conifer invasion of subalpine meadows, central Lemhi Mountains, Idaho. Northwest Science, 60: 166–173.

Butler, D.R., 1995. *Zoogeomorphology: Animals as Geomorphic Agents.* Cambridge University Press, Cambridge, 232 pp.

Butler, D.R., and G.P. Malanson, 1985. A history of high-magnitude snow avalanches, southern Glacier National Park, Montana, USA. Mountain Research & Development, 5: 175–182.

Butler, D.R., and G.P. Malanson, 1990. Non-equilibrium geomorphic processes and patterns on avalanche paths in the northern Rocky Mountains, USA. Zeitschrift fur Geomorphologie, 34: 257–270.

Butler, D.R., and G.P. Malanson, 1992. Effects of terrain on excessive travel distance by snow avalanches. Northwest Science, 66: 77–85.

Butler, D.R., and G.P. Malanson, 1994. Canadian landform examples—Beaver landforms. Canadian Geographer, 38: 76–79.

Butler, D.R., and S.J. Walsh, 1990. Lithologic, structural, and topographic influences on snow-avalanche path location, eastern Glacier National Park, Montana. Annals of the Association of American Geographers, 80: 362–378.

Butler, D.R., G.P. Malanson, F.D. Wilkerson, and G.L. Schmid, 1998. Late Holocene sturzstroms in Glacier National Park, Montana, USA. In: J. Kalvoda (Editor), Geomorphological Hazards in High Mountain Areas. Kluwer, Dordrecht, pp. 149–166.

Caine, N., 1969. A model for alpine talus slope development by slush avalanche. Journal of Geology, 77: 92–101.

Caine, N., 1971. A conceptual model for alpine slope process study. Arctic and Alpine Research, 3: 319–329.

Caine, N., 1992. Spatial patterns of geochemical denudation in a Colorado alpine environment. In: J.C. Dixon and A.D. Abrahams (Editors), Periglacial Geomorphology. Wiley, Chichester, pp. 63–88.

Cairns, D.M., and G.P. Malanson, 1997. Examination of the carbon balance hypothesis of alpine treeline location, Glacier National Park, Montana. Physical Geography, 18: 125–145.

Chagnon, D., T.B. McKee, and N.J. Doesken, 1990. Hydroclimatic variability in the Rocky Mountain region. Climatology Report 90-3, Atmospheric Science Paper No. 475. Department of Atmospheric Sciences, Colorado State University, Ft. Collins.

Church, M., and J.M. Ryder, 1972. Paraglacial sedimentation: A consideration of fluvial processes conditioned by glaciation. Geological Society of America Bulletin, 83: 3059–3072.

Clayton, J.L., 1995. Incipient weathering and soil development in ash deposits from the Mount Saint Helens eruption of 1980, U.S. Forest Service General Technical Report RM, pp. 84–86.

Cline, D.W., 1992. Modeling the redistribution of snow in alpine areas using geographic information processing techniques. In: M. Ferrick (Editor), Proceedings of the 1992 Eastern Snow Conference, pp. 13–24.

DeGraff, J.V., 1981. A stream-gradient index applied to subhumid montane streams. Physical Geography, 2: 174–183.

del Moral, R., 1993. Early primary succession on a barren volcanic plain at Mount saint Helens, Washington. American Journal of Botany, 80: 981–991.

deScally, F.A., 1996. Avalanche snow melting and summer streamflow differences between high-elevation basins, Cascade Mountains, British Columbia, Canada. Arctic and Alpine Research, 28: 25–34.

Dion, N.P., 1981. Effects of Mount Saint Helens eruptions on selected lakes in Washington: Hydrologic effects of the eruptions of Mount Saint Helens, Washington, 1980. U.S. Geological Survey Circular 850–G.

Dixon, J.C., 1991. Alpine and subalpine soil properties as paleoenvironmental indicators. Physical Geography, 12: 370–384.

Duguay, C.R., and E.F. LeDrew, 1991. Mapping surface albedo in the east slope of the Colorado Front Range, USA, with Landsat Thematic Mapper. Arctic and Alpine Research, 23: 213–223.

Dunwiddie, P.W., 1977. Recent tree invasion of subalpine meadows in the Wind River Mountains, Wyoming. Arctic and Alpine Research, 9: 393–399.

Ellis, J.M., and P.E. Calkin, 1979. Nature and distribution of glaciers, neoglacial moraines, and rock glaciers, east-central Brooks Range, Alaska. Arctic and Alpine Research, 11: 403–420.

Fowler, P.M., and D.O. Asleson, 1984. The location of lightning-caused wildland fires, northern Idaho. Physical Geography, 5: 240–252.

Frank, T.D., and C.E. Thorn, 1985. Stratifying alpine tundra for geomorphic studies using digitized aerial imagery. Arctic and Alpine Research, 17: 179–188.

Gardner, J., 1970. Geomorphic significance of avalanches in the Lake Louise area, Alberta, Canada. Arctic and Alpine Research, 2: 135–144.

Goldblum, D., and T.T. Veblen, 1992. Fire history of a Ponderosa pine/Douglas fir forest in the Colorado Front Range. Physical Geography, 13: 133–148.

Graf, W.L. (Editor), 1987. Geomorphic Systems of North America. Geological Society of America, Boulder, CO.

Greenland, D., 1989. The climate of Niwot Ridge, Front Range, Colorado, USA. Arctic and Alpine Research, 21: 380–391.

Grimm, M.M., E.W. Wohl, and R.D. Jarrett, 1995. Coarse-sediment distribution as evidence of an elevation limit for flash flooding, Bear Creek, Colorado. Geomorphology, 14: 199–210.

Habeck, J.R., 1969. A gradient analysis of a timberline zone at Logan Pass, Glacier Park, Montana. Northwest Science, 43: 65–73.

Habeck, J.R., 1987. Present-day vegetation in the northern Rocky Mountains. Annals of the Missouri Botanical Garden, 74: 804–840.

Hadley, K.S., 1987. Vascular alpine plant distributions within the central and southern Rocky Mountains, USA. Arctic and Alpine Research, 19: 242–251.

Hall, R.D., and R.R. Shroba, 1995. Soil evidence for a glaciation intermediate between the Bull Lake and Pinedale glaciations at Fremont lake, Wind River Range, Wyoming, USA. Arctic and Alpine Research, 27: 89–98.

Hamilton, W.L., and J.F. Lahey, 1982. Mountain snowcover model for Crater Lake National Park Oregon and vicinity. Physical Geography, 3: 83–95.

Hansen-Bristow, K.J., 1986. Influence of increasing elevation on growth characteristics at timberline. Canadian Journal of Botany, 64: 2517–2523.

Hansen-Bristow, K.J., and J.D. Ives, 1985. Composition, form and distribution of the forest-alpine tundra ecotone, Indian Peaks, Colorado, USA. Erdkunde, 39: 286–295.

Harper, K.T., D.C. Freeman, W.K. Ostler, and L.G. Klikoff, 1978. The flora of Great Basin mountain ranges: Diversity, sources, and dispersal ecology. Great Basin Naturalist Memoirs, 2: 81–103.

Helm, D., 1982. Multivariate analysis of alpine snow-patch vegetation cover near Milner Pass, Rocky Mountain National Park, Colorado, USA. Arctic and Alpine Research, 14: 87–95.

Hessl, A.E., and W.L. Baker, 1997. Spruce and fir regeneration and climate in the forest-tundra ecotone of Rocky Mountain National Park, Colorado, USA. Arctic and Alpine Research, 29: 173–183.

Isard, S.A., 1986. Factors influencing soil moisture and

plant community distribution on Niwot Ridge, Fiynt Range, Colorado, USA. Arctic and Alpine Research, 18: 83–96.

Ives, J.D. (Editor), 1980. Geoecology of the Colorado Front Range. Westview Press, Boulder.

Ives, J.D., 1987. The mountain lands. In: M.J. Clark, K.J. Gregory, and A.M. Gurnell (Editors), Horizons in Physical Geography. Macmillan, London, pp. 232–249.

Ives, J.D., and R.G. Barry, 1974. Arctic and Alpine Environments. Methuen, London.

Ives, J.D., and M.J. Bovis, 1978. Natural hazards maps for land-use planning, San Juan Mountains, Colorado. Arctic and Alpine Research, 10: 185–212.

Jakubos, B. and W.H. Romme, 1993. Invasion of subalpine meadows by lodgepole pine in Yellowstone National Park, Wyoming, USA. Arctic and Alpine Research, 25: 382–390.

Johnson, E.A., and D.R. Wowchuk, 1993. Wildfires in the southern Canadian Rocky Mountains and their relationship to mid-tropospheric anomalies. Canadian Journal of Forest Research, 23: 1213–1222.

Johnson, J.B., 1996. Phytosociology and gradient analysis of a subalpine treed fen in Rocky Mountain National Park, Colorado. Canadian Journal of Botany, 74: 1203–1218.

Keane, R.E., K.C. Ryan, and S.W. Running, 1996. Simulating effects of fire on northern Rocky Mountain landscapes with the ecological process model FIRE-BGC. Tree Physiology, 16: 319–331.

Kershaw, L.J., and J.S. Gardner, 1986. Vascular plants of mountain talus slopes, Mt. Rae area, Alberta, Canada. Physical Geography, 7: 281–230.

Kite, G.W., 1991. Watershed modeling using satellite data applied to a mountain basin in Canada. Journal of Hydrology, 128: 157–169.

Klein, W.H., C.W. Crockett, and J.F. Andrews, 1965. Objective predictions of daily precipitation and cloudiness. Journal of Geophysical Research 70: 801–813.

Kupfer, J.A., and P.M. Cairns, 1996. The suitability of montane ecotones as indicators of global climatic change. Progress in Physical Geography, 19: 18–34.

LaMarche, V.C., Jr., 1973. Holocene climatic variations inferred from treeline fluctuations in the White Mountains, California. Quaternary Research, 3: 632–660.

Lehre, A.K., B. Collins, and T. Dunne, 1983. Post-eruption sediment budget for the North Fork Toutle River drainage, June 1980–June 1981. Zeitschrift fur Geomorphologie, Supplementband, 46: 143–163.

Leonard, E.M., 1984. Late Pleistocene equilibrium-line altitudes and modern snow accumulation patterns, San Juan Mountains, Colorado, USA. Arctic and Alpine Research, 16: 65–76.

Lickorish, W.H., 1993. Structural evolution of the Porcupine Creek anticlinorium, western Main Ranges, Rocky Mountains, British Columbia. Journal of Structural Geology, 15: 477–490.

Lippman, P.W., and D.R. Mullineaux (Eds.), 1982. The 1980 eruption of Mount St. Helens, Washington. U.S. Geological Survey Professional Paper 1250.

Locke, W.W., 1990. Late Pleistocene glaciers and the climate of western Montana, USA. Arctic and Alpine Research, 22: 1–13.

Luckman, B.H., G. Holdsworth, and G.D. Osborn, 1993. Neoglacial glacier fluctuations in the Canadian Rockies. Quaternary Research, 39: 144–153.

Luckman, B.K., and M.S. Kearney, 1986. Reconstruction of Holocene changes in alpine vegetation and climate in the Maligne Range, Jasper National Park, Alberta. Quaternary Research, 26: 244–261.

Malanson, G.P., 1993. Riparian Landscapes. Cambridge University Press, Cambridge.

Malanson, G.P., 1997. Effects of feedbacks and seed rain on ecotone patterns. Landscape Ecology, 12: 27–38.

Malanson, G.P., and D.R. Butler, 1985a. Avalanche paths as fuel breaks: Implications for fire management. Journal of Environmental Management, 19: 229–238.

Malanson, G.P., and D.R. Butler, 1985b. Transverse pattern of vegetation on avalanche paths in the northern Rocky Mountains, Montana. Great Basin Naturalist, 44: 453–458.

Malanson, G.P., and D.R. Butler, 1990. Woody debris, sediment, and riparian vegetation of a subalpine river, Montana, USA. Arctic and Alpine Research, 22: 183–194.

Malanson, G.P., and D.R. Butler, 1994. Tree–tundra competitive hierarchies, soil fertility gradients, and the elevation of treeline in Glacier National Park, Montana. Physical Geography, 15: 166-180.

Malanson, G.P., and D.M. Cairns, 1995. Effects of increased cloud-cover on a montane forest landscape. Ecoscience, 2: 75–82.

Marston, R.A., 1982. The geomorphic significance of log steps in forest streams. Annals of the Association of American Geographers, 72: 99–108.

Marston, R.A., 1983. Supraglacial stream dynamics on the Juneau icefield. Annals of the Association of American Geographers, 73: 597–608.

Matthes, F.E., 1938. Avalanche sculpture in the Sierra Nevada of California. Bulletin of the International Association of Scientific Hydrology, 23: 631–637.

Maurel, L.E., 1991. Out-of-sequence thrusting at the leading-edge of the McConnell Thrust, Front Ranges of the Rocky Mountains, Alberta, Canada (abstract). Tectonophysics, 191: 428.

McArthur, D.S., and A.S. Hope, 1993. The role of perimeter shape in estimating annual runoff from small Sierra Nevada basins. Physical Geography, 14: 394–403.

McCarthy, D.P., and D.J. Smith, 1994. Historical glacier activity in the vicinity of Peter Lougheed Provincial Park, Canadian Rocky Mountains. Western Geography, 4: 94–109.

McCune, B., and T.F.H. Allen, 1985. Will similar forests develop on similar sites? Canadian Journal of Botany, 63: 367–376.

McGinnis, D.L., 1997. Estimating climate-change impacts on Colorado snowpack using downscaling methods. Professional Geographer, 49: 117–125.

Meentemeyer, R.K., and D.R. Butler, 1995. Temporal and spatial changes in beaver pond locations, eastern Glacier National Park, Montana, USA. Geographical Bulletin, 37: 97–104.

Merriam, C.H., 1898. Life zones and crop zones in the United States. U.S. Biological Survey Bulletin, 10: 1–79.

Meyer, P., K.I. Itten, T. Kellenberger, S. Sandmeier, and R. Sandmeier, 1993. Radiometric corrections of topographically induced effects on Landsat TM data in an alpine environment. ISPRS Journal of Photogrammetry and Remote Sensing, 48: 17–28.

Mock, C.J., 1995. Avalanche climatology of the continental zone in the southern Rocky Mountains. Physical Geography, 16: 165–187.

Moran, J.M., 1974. Possible coincidence of a modern and a glacial-age climatic boundary in the montane west, United States. Arctic and Alpine Research, 6: 319–321.

Naftz, D.L, R.W. Klusman, and E.A. McConnaughey, 1996. Little Ice Age evidence from a south-central North American ice core, USA. Arctic and Alpine Research, 28: 35–41.

Oelfke, J.G., and D.R. Butler, 1985. Landslides along the Lewis Overthrust Fault, Glacier National Park, Montana. Geographical Bulletin, 27: 7–15.

Olgeirson, E., 1974. Parallel conditions and trends in vegetation and soil on a bald near treeline, Boreas Pass, Colorado. Arctic and Alpine Research, 6: 185–203.

Orme, A.R., 1989. The nature and rate of alluvial fan aggradation in a humid temperate environment, northwest Washington. Physical Geography, 10: 131–146.

Parker, A.J., 1987. Morphological divergence between conifer forests of Yosemite and Glacier National Parks, USA. Arctic and Alpine Research, 19: 252–260.

Parker, A.J., 1993. Structural variation and dynamics of lodgepole pine forests in Lassen Volcanic National Park, California. Annals of the Association of American Geographers, 84: 613–629.

Parker, K.C., 1991. Topography, substrate, and vegetation patterns in the northern Sonoran Desert. Journal of Biogeography, 17: 151–163.

Peet, R.K., 1988. Forests of the Rocky Mountains. In: M.G. Barbour and W.D. Billings (Editors), North American Terrestrial Vegetation. Cambridge University Press, Cambridge, pp. 63–101.

Pinter, N., and M.T. Brandon, 1997. How erosion builds mountains. Scientific American: 74–79.

Ploufle, A, and H. Jette, 1997. Middle Wisconsinan sediments and paleoecology of central British Columbia: Sites at Necoslie and Nautley Rivers. Canadian Journal of Earth Sciences, 34: 200–208.

Pohl, M.M., 1995. Radiocarbon ages on organics from piedmont alluvium, Ajo Mountains, Arizona. Physical Geography, 16: 339–353.

Price, L.W., 1978. Mountains of the Pacific Northwest, USA: A study in contrasts. Arctic and Alpine Research, 10: 465–478.

Romme, W.H., D.H. Knight, and J.B. Yavitt, 1986. Mountain pine beetle outbreaks in the Rocky Mountains: Regulators of primary productivity? American Naturalist, 127: 484–494.

Running, S.W., and R.R. Nemani, 1991. Regional hydrologic carbon balance responses of forests resulting from potential climatic change. Climatic Change, 19: 349–368.

Saunders, I.R., and W.G. Bailey, 1994. Radiation and energy budgets of alpine tundra environments of North America. Progress in Physical Geography, 18: 517–538.

Savage, M., 1991. Structural dynamics of a southwestern pine forest under chronic human influence. Annals of the Association of American Geographers, 81(2): 271–289.

Savage, M., M. Reid, and T.T. Veblen, 1992. Diversity and disturbance in a Colorado subalpine forest. Physical Geography, 13: 240–249.

Scuderi, L.A., C. Barker Schaaf, K.U. Orth, and L.E. Band, 1993. Alpine treeline growth variability: Simulation using an ecosystem process model. Arctic and Alpine Research, 25: 175–182.

Seltzer, G.O., 1994. Climatic interpretation of alpine snowline variations on millennial time scales. Quaternary Research, 41: 154–159.

Shi, J., J. Dozier, and H. Rott, 1994. Snow mapping in alpine regions with synthetic aperture radar. IEEE Transactions on Geoscience and Remote Sensing, 32: 152–158.

Smith, C.J., D.P. McCarthy, and M.E. Colenutt, 1995. Little Ice Age glacial activity in Peter Lougheed Provincial Park and Elk Lakes Provincial Park, Canadian Rocky Mountains. Canadian Journal of Earth Sciences, 32: 579–589.

Smith, D.J., D.P. McCarthy, and B.H. Luckman, 1994. Snow-avalanche impact pools in the Canadian Rocky Mountains. Arctic and Alpine Research, 26: 116–127.

Stanton, M.L., M. Rejmanek, and C. Galen, 1994. Changes in vegetation and soil fertility along a predictable snowmelt gradient in the Mosquito Range, Colorado, USA. Arctic and Alpine Research, 26: 364–374.

Stevens, G.C., and J.F. Fox, 1991. The causes of treeline. Annual Review of Ecology and Systematics, 22: 177–191.

Suffling, R., 1993. Induction of vertical zones in sub-alpine valley forests by avalanche-formed fuel breaks. Landscape Ecology, 8: 127–138.

Sveinbjornsson, J., J. Davis, W. Abadie, and A. Butler, 1995. Soil carbon and nitrogen mineralization at different elevations in the Chugach Mountains of south-central Alaska, USA. Arctic and Alpine Research, 27: 29–37.

Taylor, A.H., 1990a. Habitat segregation and regeneration patterns of red fir and mountain hemlock in ecotonal forests, Lassen Volcanic National Park, California. Physical Geography, 11: 36–48.

Taylor, A.H., 1990b. Tree invasion in meadows of Lassen Volcanic National Park, California. Professional Geographer, 42: 457–470.

Taylor, A.H., 1993. Fire history and structure of red fir forests, Swan Mountain Experimental Forest, Cascade Range, northeastern California. Canadian Journal of Forest Research, 23: 1672–1678.

Taylor, A.H., and C.B. Halpern, 1991. The structure and dynamics of *Abies magnifica* forests in the southern Cascade Range, USA. Journal of Vegetation Science, 2: 189–200.

Thorn, C.E., J.C. Dixon, R.G. Darmody, and J.M. Rissing, 1989. Weathering trends in fine debris beneath a snow patch, Niwot Ridge, Front Range, Colorado. Physical Geography, 10: 307–321.

Veblen, T.T., K.S. Hadley, and E.M. Nel, 1994. Disturbance regime and disturbance interactions in a Rocky Mountain subalpine forest. Journal of Ecology, 82: 125–135.

Voight, B., H. Glicken, R.J. Janda, and P.M. Douglass, 1981. Catastrophic rockslide avalanche of May 18. In: P.W. Lipman and D.R. Mullineaux (Editors), The 1980 Eruptions of Mount St. Helens, Washington. U.S. Geological Survey Professional Paper 1250, pp. 347–377.

Walsh, S.J., G.P. Malanson, and D.R. Butler, 1992. Alpine treeline, Glacier National Park, Montana, USA. In: D.G. Janelle (Editor), Geographical Snapshots of North America, Guilford Press, New York, pp. 167–171.

Watras, C.J., K.A. Morrison, and N.S. Bloom, 1995. Mercury in remote Rocky Mountain lakes of Glacier National Park, Montana, in comparison with other temperate North American regions. Canadian Journal of Fisheries and Aquatic Science, 52: 1220–1228.

Weirich, F. H., 1986. A study of the nature and incidence of density currents in a shallow glacial lake. Annals of the Association of American Geographers, 76: 396–413.

Weisberg, P.J., and W.L. Baker, 1995a. Spatial variation in tree regeneration in the forest-tundra ecotone, Rocky Mountain National Park, Colorado. Canadian Journal of Forest Research, 25: 1326–1339.

Weisberg, P.J., and W.L. Baker, 1995b. Spatial variation in tree seedling and krummholz growth in the forest-tundra ecotone of Rocky Mountain National Park, Colorado, USA. Arctic and Alpine Research, 27: 116–129.

Whittaker, R.H., and W.A. Niering, 1965. Vegetation of the Santa Catalina Mountains, Arizona. II. A gradient analysis of the south slope. Ecology, 46: 429–452.

Wilson, J.B., and A.D.Q. Agnew, 1992. Positive-feedback switches in plant communities. Advances in Ecological Research, 23: 263–336.

Woodward, A., E.G. Schreiner, and D.G. Silsbee, 1995. Climate, geography, and tree establishment in subalpine meadows of the Olympic Mountains, Washington, USA. Arctic and Alpine Research, 27: 217–225.

Woolridge, G.L., R.C. Musselman, R.A. Sommerfeld, D.G. Fox, and B.H. Connell, 1996. Mean wind patterns and snow depths in an alpine–subalpine ecosystem as measured by damage to coniferous trees. Journal of Applied Ecology, 33: 100–108.

Yarnal, B., 1984. Synoptic-scale atmospheric circulation over British Columbia in relation to the mass balance of Sentinel Glacier. Annals of the Association of American Geographers, 74: 375–392.

Zielinski, G.A., and W.D. McCoy, 1987. Paleoclimatic implications of the relationship between modern snowpack and Late Pleistocene equilibrium-line altitudes in the mountains of the Great Basin, western USA. Arctic and Alpine Research, 19: 127–134.

19

Desert Environments

Julie E. Laity

The core deserts of southwestern North America stretch from southeastern California to western Texas and from Nevada and Utah to the Mexican states of Sonora, Chihuahua, and Coahuila, and much of the peninsula of Baja California (fig. 19.1). Beyond these areas, semidesert conditions extend north to eastern Washington, south onto the central Mexican plateau, and east to link with the steppes of the High Plains (see chapter 12). In area, about 55% of the North American deserts are considered semiarid, 40% arid, and only 5% hyperarid. This chapter, however, focuses on the core deserts of the southwestern United States and northwest Mexico, namely, the Chihuahuan, Sonoran, Mojave, and Great Basin deserts.

North American deserts owe their aridity to a rain-shadow effect caused by mountains blocking moisture of Pacific origin in the winter and from the Gulf of Mexico in the summer, the occurrence of a subtropical high-pressure cell, and cold currents along the western coast. Rain-shadow effects are greatest in the Great Basin and Mojave deserts, whereas the effect of high pressure is important in the Chihuahuan and Sonoran deserts. Cold coastal currents play a role in the western deserts, particularly along the outer coast of Baja California.

Demarcation of the deserts has been based on multiple criteria, including climate, geology, vegetation physiognomy, and floristic composition, with the result that the defined boundaries of the desert may vary according to the criteria selected. From a climatic perspective, the deserts can be classified according to their temperature and the seasonality of their precipitation (fig. 19.2). The Great Basin, characterized by its northern position, high altitude, the receipt of 60% of its winter moisture as snow, mean monthly temperatures below 0°C from December through February, and mean annual temperature of 9°C, is considered a "cold" desert. The distinction between "cold" and "warm" deserts is based principally on winter temperatures, as all deserts are characterized by hot summer temperatures. The mean annual temperature of the more southern warm deserts is 20°C, and essentially all of their precipitation occurs as rain. The number of frost-free days averages 80 to 150 in the Great Basin and 210 to 365 in the Mojave, Sonoran, and Chihuahuan deserts (McMahon and Wagner, 1985). The Great Basin, Mojave, and western Sonoran deserts receive winter and spring precipitation derived from frontal systems that move in from the Pacific. Winter precipitation, falling from November to February, makes up 44% of the annual total in the Mojave Desert, 38% in the Great Basin Desert, 28% in the Sonoran Desert, and 18% in the Chihuahuan Desert. Moving from west to east from the Mojave to Chihuahuan Desert, the ratio of winter to summer rainfall decreases. The eastern Sonoran Desert mainly has a bimodal rainfall regime, with a strong primary maximum in July and August and a secondary maximum in February. The Chihuahuan Desert receives summer rain-

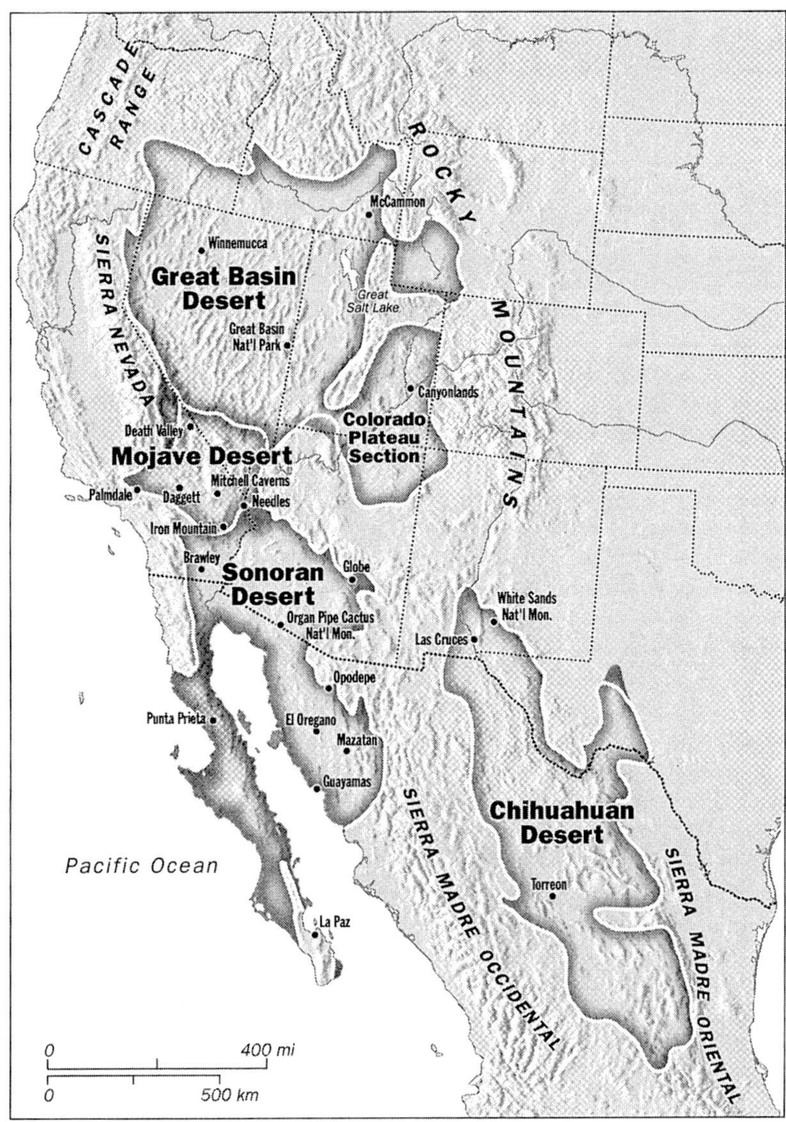

Figure 19.1 The core deserts of southwestern North America.

fall. Summer precipitation, which falls from June to September, comprises 59% of the annual total in the Chihuahuan Desert, 53% in the Sonoran Desert, 27% in the Mojave Desert, and 25% in the Great Basin Desert.

The North American monsoon, known also as the Southwest United States, Mexican, or Arizona monsoon, is centered over northern Mexico and acts to increase summer rainfall over the southern deserts, but the effects tend to be both spatially and temporally variable (fig. 19.3). In addition to regional differences in precipitation, there are also significant elevational gradients, with lower altitudes receiving rainfall that is lower in amount and more episodic than at moister, higher elevations. Precipitation approaches 500 mm yr^{-1} in the higher mountains but is less than 100 mm yr^{-1} at low elevations near the Gulf of California and Death Valley.

Freezing weather and the seasonal and altitudinal differences in rainfall between the deserts are reflected in vegetation structure and floristic composition. The cold desert areas are primarily semiarid with steppe vegetation, dominated by the evergreen *Artemisia* spp. (sagebrush) complex. The lower and midelevation bajadas of the Mojave Desert are dominated by a creosote bush–bursage (*Larrea tridentata–Ambrosia dumosa*) community. Extensive areas of stabilized and active dunes support an insular flora that is approximately 95% indigenous. The Sonoran Desert, characterized by infrequent freezing temperatures and bimodal annual rainfall, has a unique character and contains subtropical arborescent flora, including leguminous trees and large columnar cacti (fig. 19.4). The species diversity of some areas of the Sonoran Desert is higher than in the eastern deciduous forests of North America. The Chihua-

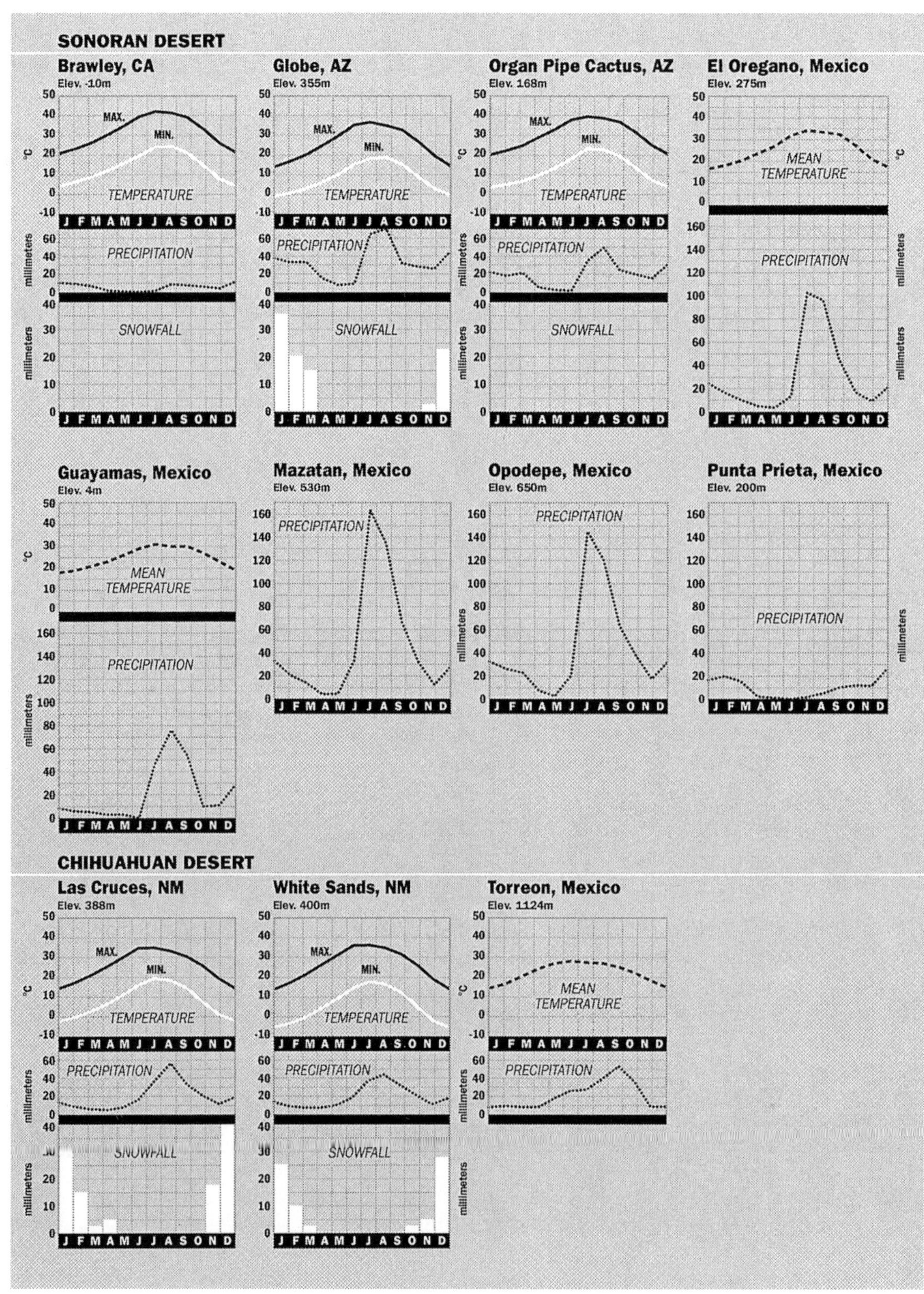

Figure 19.2 Climate diagrams for the Great Basin, Mojave, Sonoran, and Chihuahuan deserts, with locations of associated meteorological stations. The deserts differ according to their temperature, which is affected by latitude and altitude, and the seasonality and nature of their precipitation.

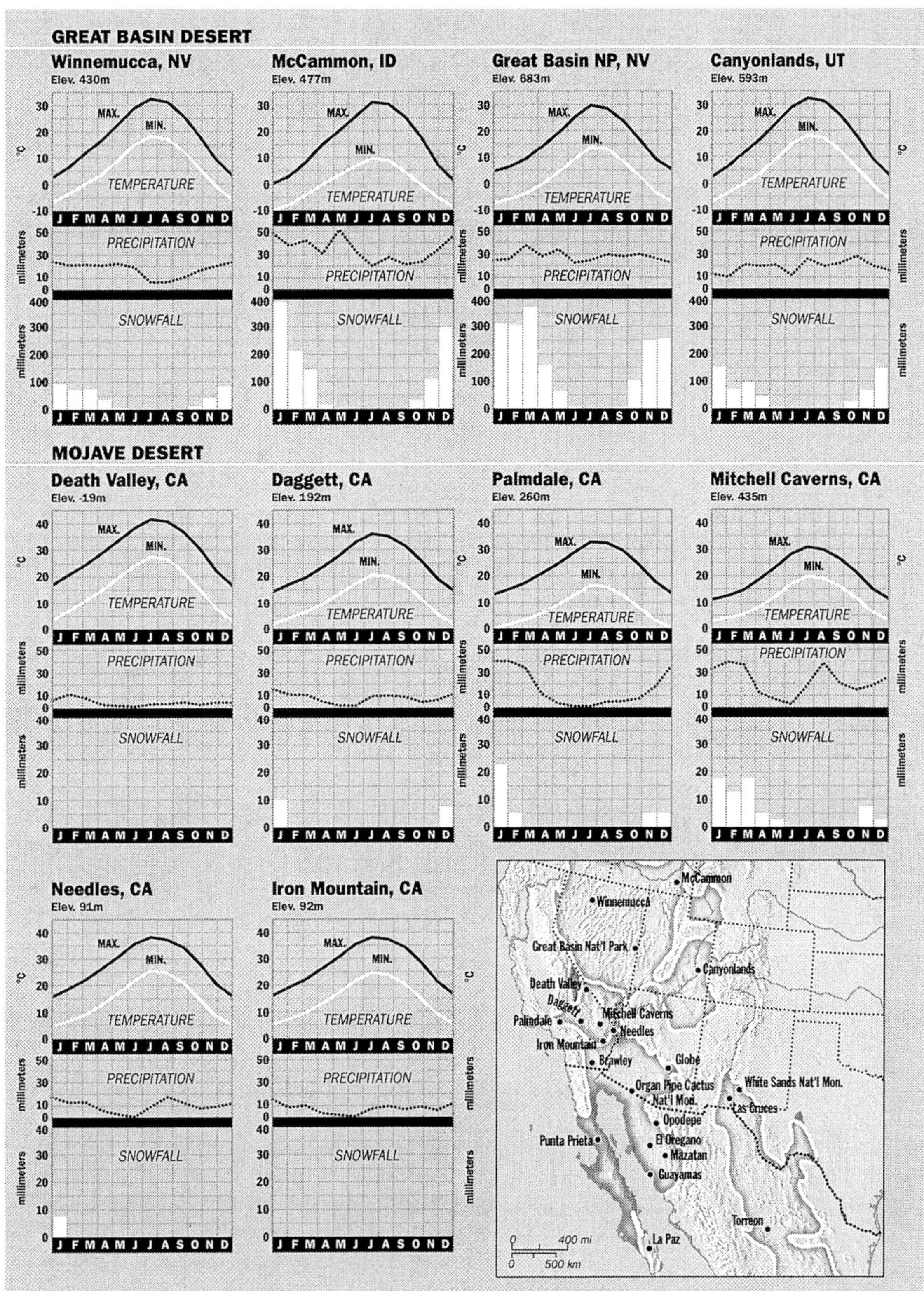
GREAT BASIN DESERT
Winnemucca, NV
Elev. 430m
McCammon, ID
Elev. 477m
Great Basin NP, NV
Elev. 683m
Canyonlands, UT
Elev. 593m
MAX.
MIN.
TEMPERATURE
PRECIPITATION
SNOWFALL
J F M A M J J A S O N D
°C
millimeters
MOJAVE DESERT
Death Valley, CA
Elev. -19m
Daggett, CA
Elev. 192m
Palmdale, CA
Elev. 260m
Mitchell Caverns, CA
Elev. 435m
Needles, CA
Elev. 91m
Iron Mountain, CA
Elev. 92m
McCammon
Winnemucca
Great Basin Nat'l Park
Canyonlands
Death Valley
Daggett
Palmdale
Mitchell Caverns
Needles
Iron Mountain
Brawley
Globe
Organ Pipe Cactus Nat'l Mon.
White Sands Nat'l Mon.
Las Cruces
Opodepe
Punta Prieta
El Oregano
Mazatan
Guaymas
Torreon
La Paz
0 400 mi
0 500 km

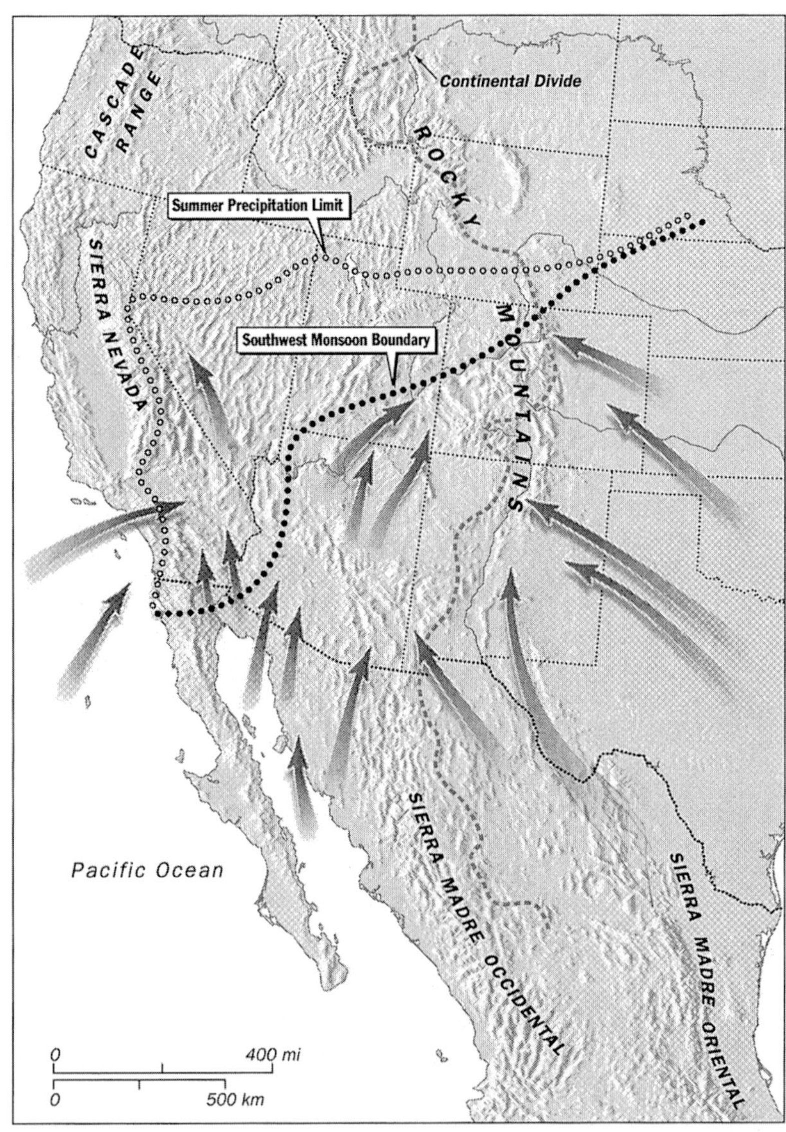

Figure 19.3 (*top*) The North American monsoon is centered over northern Mexico and acts to increase summer rainfall over the deserts. East and south of the southwestern monsoon boundary, more than half of the annual precipitation occurs during the summer months. North of the boundary, the amount decreases until the summer precipitation limit is reached.

Figure 19.4 (*at right*) Organ Pipe Cactus National Monument contains a variety of tall shrubs, short trees, and large columnar cacti characteristic of the Sonoran Desert.

huan Desert, owing to its higher elevations, cooler temperatures, and greater precipitation, is dominated more by perennial grasslands than the other deserts. The desert ecology of North America has been dramatically and irrevocably altered by the introduction of exotic, invasive plants such as *Bromus tectorum* (cheatgrass), *Salsola* spp. (Russian thistle, tumbleweed), and the woody phreatophytes *Tamarix* spp. (saltcedar). These invasions were hastened by habitat disturbance, including overgrazed rangelands and riparian zones affected by river regulation and damming. During the climatic oscillations of Pleistocene and earlier Holocene times, each desert experienced considerable changes in temperature and available moisture and, therefore, in ecological response. Repeated wetting and drying cycles caused the cyclic expansion and contraction of desert vegetation in the region.

The core deserts of North America are contained mainly within the Basin and Range Province, a region characterized by an alternating pattern of long, narrow subparallel mountain ranges separated by alluvial basins, trending north or northwest, the collective result of late Cenozoic crustal extension and faulting that began 12–15 million years ago (see chapter 1). Contrasts in range trend, area, altitude, and relief allow the Basin and Range Province to be subdivided into five sections (fig. 19.5): the Great Basin, the Mojave-Sonoran Desert, the Salton Trough, the Mexican Highlands, and the Sacramento Mountains (Fenneman, 1931). The Great Basin, the largest of these sections, is an enormous crustal bulge with a central area of elevated basins flanked by significantly lower terrain. Drainage within the Great Basin is largely internal, with more than 100 closed basins, many of which supported perennial lakes during wetter intervals of Pleistocene time. The Mojave-Sonoran section displays relatively subdued topography: drainage within the Mojave Desert is internal and its closed basins are largely undissected, whereas Sonoran Desert drainage is integrated into the Gila, Salt, or Bill Williams river systems and basin dissection is more continuous and widespread (Dohrenwend, 1987). The southwestern margin of the Mojave-Sonoran section is bounded by the Salton Trough, a structural depression containing as much as 6000 m of late Cenozoic sedimentary deposits. The Mexican Highlands are transitional between the Sonoran Desert and the Colorado Plateau and dominate northwestern Mexico. The ranges trend north to northwest, and drainage is better integrated than in neigh-

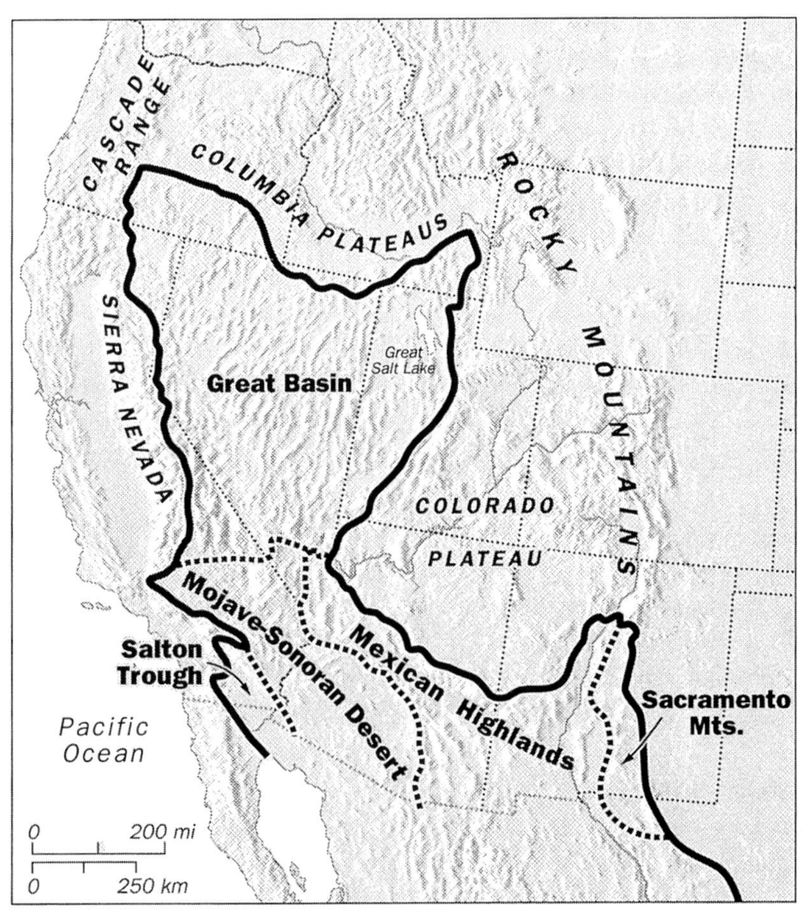

Figure 19.5 Subdivisions of the Basin and Range Geomorphic Province.

boring areas. The Sacramento Mountains lie to the east of the Mexican Highlands and are transitional between the Rio Grande rift zone and the Great Plains farther east. To the east of the Basin and Range Province lies the Colorado Plateau, a semiarid region of high plateaus and isolated mountains, 90% of which are drained by the Colorado River.

Within this century, technological developments have resulted in rapid exploitation of North American deserts for agriculture, grazing, minerals, urban development, recreation, and military purposes. Human impact has resulted in increased levels of wind and water erosion, ground subsidence due to groundwater overdraft, the spread of salinity as a result of irrigation, ecological effects consequent to the diversion and damming of rivers, the desiccation of inland lakes because of interbasin water transfers, and rising levels of air pollution.

This chapter examines some of the significant hydrologic, biologic, and geomorphic aspects of the deserts of southwestern North America, commencing with an overview of selected aspects of their physical geography and concluding with a discussion of the four principal deserts defined previously. Following a summary of their climatic and tectonic settings, the discussion focuses specifically on the geomorphology of these deserts and on studies that have contributed significantly to their understanding.

19.1 Selected Aspects of North American Deserts

19.1.1 Surface Flow and Groundwater Geomorphology

Despite their implicit dryness, most arid areas of North America have fluvial forms clearly imprinted on the landscape for three principal reasons: (1) the action of flash floods, (2) the presence of allogenic streams such as the Colorado River with sources in the snowmelt of high mountains, and (3) past pluvial conditions. In contrast with the perennial flows of most humid-region rivers, semiarid and arid climatic patterns dictate a discontinuity in fluvial dynamics (Graf, 1988). Because they are relatively bare of vegetation and commonly have low infiltration capacities, arid and semiarid soils are susceptible to surface runoff and, given sufficiently intense rainfall, to flooding. In arid-region rivers, losses to seepage and evapotranspiration result in downstream decreases in discharge. Total stream power, therefore, decreases downstream, reducing the energy available for sediment transport and channel change.

Large streams may have floods that are an order of magnitude or more larger than the mean annual flow. For example, the 50-year flood on the Gila River in Arizona (fig. 19.6) is 280 times the mean annual discharge (Graf, 1988). As a result, drastic changes in river morphology may occur during large flood events. In settled areas, such as California and Arizona, surface runoff may be modified further by human activity in stream channels. The rapid rise of streamflow that is characteristic of arid and semiarid regions may seriously disrupt and damage urbanized areas and farmland (Graf, 1988; Kresan, 1988).

Widespread dams have changed both the water and the sediment load of rivers. The Colorado River once carried 18.5 km^3 of annual flow, moving 125–160 million tonnes of suspended sediment annually to the Gulf of California. At present, extraction of the water for human use has resulted in little or no water or sediment reaching the sea (Schwarz et al., 1991; see also chapter 22).

Groundwater plays an important role in hydrologic, geomorphic, and biologic processes in deserts. Sustained seepage supports minor streamflow in some of the most arid regions, such as the Darwin Falls area of the Argus Range, west of Panamint Valley, California. Seepage also maintains wetlands on playa margins and marshy areas along channels. Small artesian springs or *pozos* provide potable water at various points in the salt flats at Adair Bay in the southern part of the Gran Desierto, one of the driest areas of the Sonoran Desert. The pozos provide essential freshwater for birds and mammals, and support a flora markedly different from the rest of the Sonoran Desert (Ezcurra et al., 1988). On the Colorado Plateau, groundwater sapping has maintained sandstone cliffs (Bryan, 1928; Ahnert, 1960; Schumm and Chorley, 1964; Howard and Selby, 1994) and plays an important role in the headward retreat of deeply entrenched, theater-headed valleys (Laity and Malin, 1985; Howard and Kochel, 1988; Baker et al., 1990). Overdraft of groundwater reserves is an environmental problem in many regions. Continuous declines in groundwater levels in the Las Vegas Valley, Nevada, by as much as 90 m in some areas, have resulted in land subsidence and fissuring (Bell et al., 1992; Morris et al., 1997).

19.1.2 Desert Surface Stability

Over vast areas of North American deserts, wind and water erosion are reduced by agents that impart surface soil stability. These include stony pavements and lag materials, soil impregnation by calcium carbonate and calcrete development, and cryptobiotic surface crusts. On older dunes, vegetation is an effective stabilizer.

Stone pavements, which are widespread, developed in a variety of environmental settings, including abandoned alluvial surfaces, stabilized sand sheets and sand ramps, aeolian mantles on lava flows, pluvial lake beach ridges, and residual mantles on low-angle hillslopes. An increasing body of evidence suggests that stone pavements are not formed simply by deflation, but rather by a diverse variety of geomorphic processes that result in stone sorting, surface creep, clast disintegration, and syndepositional lifting of surface clasts by the accumulation of salt- and carbonate-rich aeolian fines (Dohrenwend, 1987).

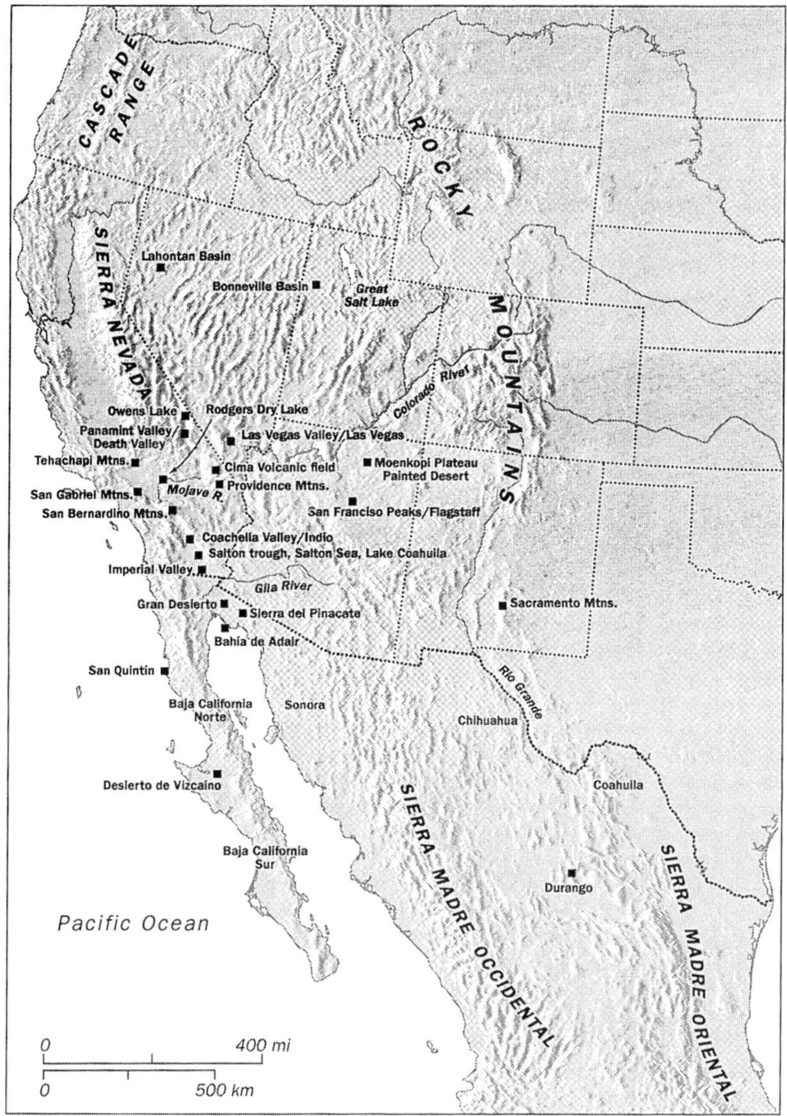

Figure 19.6 Locations cited in the text.

Aeolian dust accelerates mechanical fragmentation of rock and, by accumulating in the pore spaces beneath clasts, promotes soil development and a vertically accreting mantle. The suggestion that pavement clasts are "born at the surface" has been supported by cosmogenic surface-exposure dating in the Cima volcanic field (Wells et al., 1994). Pavement surfaces in the Panamint Valley, California, appear dynamic even in maturity (Haff and Werner, 1996). Plats were cleared of stones and the recovery of the stony area documented by repeat photography. Displacement of stones by animal activity was a major component of the resurfacing process.

Calcretes are widely distributed throughout the warm desert environments of North America. Rates of carbonate accumulation vary widely and are affected by climatic, lithologic, and topographic factors. The carbonate is principally derived from external sources, chiefly solid carbonate in aeolian dust and Ca^{2+} dissolved in rainwater (Gile et al., 1979). Pedogenic accumulations are precipitated from carbonates originally present in rainwater or leached by infiltrating rainwater from aeolian fines present in upper soil horizons. Carbonate accumulations proceed from coatings on pebbles, to segregated filaments and nodules, to massive accumulations that gradually plug the calcic horizon. Eventually, dense pedogenic calcretes may develop. Nonpedogenic accumulations include gully-bed cementation, laminar layers on interfluves and hillslopes, and case-hardened surfaces. These are usually associated with lateral surface or subsurface water flow (Lattman, 1973). Isotopic studies of pedogenic carbonate provide both chronological and paleoenvironmental information on alluvial deposits. The carbon and oxygen isotopic composition of soil carbonates in alluvial piedmont deposits along the western flank of the Providence Mountains, California, suggests that the climate has become warmer and drier during the Holocene (Wang et al., 1996). The accumulation of carbonate material affects both aeolian and fluvial erosion. For example, on rock-mantled falling dunes in southernmost Death Valley, carbonate accumulation in dune sands reduces aeolian erosion, but promotes gullying of the lower dune surfaces. Drainage basins with significant calcrete development have a high runoff potential, and stream hydrographs tend to be flashy.

Cryptobiotic (microphytic, cryptogamic, biogenic, or microfloral) crusts are found on exposed sandy or silty soils in open shrub and grassland communities throughout the North American deserts and beyond. The crusts are composed of nonvascular plants, including algae, fungi, lichens, and bryophytes, and affect soil surface stability, water infiltration, and plant succession (Eldridge, 1993). The distribution, ecology, and role of nonvascular plants in arid and semiarid environments is not well known. On the Colorado Plateau, cryptobiotic soil crusts provide up to 70% of the living cover (Belnap and Gardner, 1993). These crusts significantly reduce aeolian erosion (Williams et al., 1995b) and influence infiltration rates, the depth of water penetration, runoff, and sediment production. Crusted soils show increased depth of water penetration and less soil movement than noncrusted soils. Their ability to absorb and hold more water is due in part to their microtopography, with well-developed crusts showing two to three centimeters of relief across rough surfaces. Ponding takes place in small depressions, increasing the time for infiltration and decreasing runoff (Williams et al., 1995a). Trampling by cattle and off-road vehicles can severely damage cryptobiotic crusts. Grazing near Navajo National Monument, Arizona, has a more pronounced effect on cryptobiotic cover and diversity than on vascular plants (Brotherson et al., 1983). A single disturbance of the crust by hooves may not be significant, but multiple disturbances eliminate the crust.

19.1.3 Aeolian Processes

Active sand dunes cover small areas (<1%) of arid North America. Much broader regions are mantled by older windblown sediments, including both sand and dust, that have been repeatedly reworked during Quaternary times. Many of these windblown sediments can be seen from Earth-orbiting satellites, and their study has illuminated sediment sources, transportation pathways, and wind regimes (Blount et al., 1990; Zimbelman et al., 1995; Blumberg and Greeley, 1996).

Aeolian processes appear particularly sensitive to climate changes. For example, vertical sequences of buried soils within sand ramps and climbing and falling dunes, buried soils on dated lava flows of the Cima volcanic field (McFadden et al., 1984), and microchemical laminations in rock varnish (Dorn, 1986) indicate periods of aeolian activity and inactivity during Quaternary time.

North American dunes are among the best studied in the world as a result of the relatively close proximity of dunes to research institutions, the availability of high-altitude and satellite imagery, NASA studies of aeolian activity on Earth as analogues for Venus and Mars (Greeley, 1995), oil and gas exploration in sandstones leading to detailed studies of aeolian sediments and processes, interest in paleoclimatology, and vexing environmental problems, such as dune destabilization and dust generation (fig. 19.7).

Wind erosion results from both deflation (removal of loosened material and its transport as fine grains in atmospheric suspension) and abrasion (mechanical wear of coherent material). Landforms include ventifacts, wind-eroded rocks of varying size, form, and material composition, and yardangs, streamlined landforms that commonly occur as subparallel arrays with the major axes parallel to the wind. Yardang formation is inconsequential in North American deserts, occurring only in small areas at Rogers Lake, California (Ward and Greeley, 1984), and in Arizona, but ventifacts are relatively widespread. Ventifact formation is influenced by factors similar to those that affect dunes: wind frequency, magnitude, and persistence, as well as sediment supply, basin geomorphology, and vegetation cover. Ventifacts occur where strong winds are combined with abundant moving sand, such as near lake shorelines, downwind of alluvial rivers, adjacent to dune sands, and in corridors of former sand transit. In the Mojave Desert, ventifacts occur in proximity to the Dumont Dunes, Death Valley Dunes, Algodones Dunes, Panamint Dunes, the Devils Playground, and the Kelso Dunes (Laity, 1995). Most ventifacts in North America are fossil in nature, characterized by weathered, dulled, fretted, or partly exfoliated faces (H.T.U. Smith, 1967; R.S.U. Smith, 1984; Laity, 1992) and by rock varnish and rock coatings such as silica glaze and oxalate coatings (Dorn, 1986, 1995). Aspects of ventifact form (facets, grooves, flutes, and pits) enable a determination of palaeowind direction, and their presence provides clues to the nature of earlier environments (Laity, 1992). Published works on ventifacts in the southwestern United States date to the mid-nineteenth century (Blake, 1855; Gilbert, 1875). Blackwelder (1929) examined fossil ventifacts along the eastern base of the Sierra Nevada, and Maxson (1940) worked in northern Death Valley. This early research provided valuable observations on the role of sand as an abrading agent, the nature of surface microfeatures, and the role of topography in enhancing wind action. Sharp (1980), investigating ventifacts in the Coachella Valley, California, demonstrated that abrasion rates are in part determined by time-dependent particle flux.

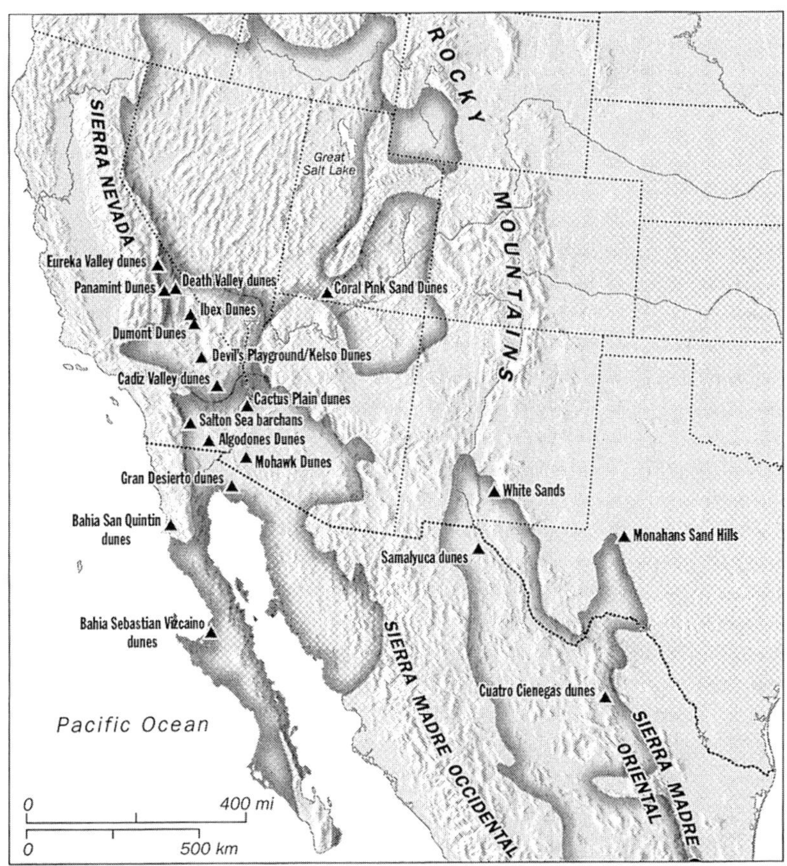

Figure 19.7 Dune fields of the core North American deserts.

Although aeolian dust is less obvious than dune sand, it is nonetheless ubiquitous in distribution, and its significance is manifested in the weathering of rocks (Villa et al., 1995), the nature of rock coatings, soil formation, and the gradation of surfaces though stone-pavement formation. Reheis et al. (1995) studied changes in soil-accumulation rates on alluvial fans to provide insights into paleoclimate, dust supply, and soil-forming processes in eastern California and southern Nevada. A consistently large increase in soil-accumulation rates occurred around the Pleistocene-Holocene boundary. Near late Pleistocene pluvial lakes, high early Holocene accumulation rates of silt, clay, and $CaCO_3$ were associated with incorporation of dust deflated from desiccating playas.

Dust levels in arid regions of Arizona, Texas, and California deserts have increased as a result of human activity, including off-road vehicle use, lake drainage, and agriculture (Bach et al., 1996). The most infamous dust storms are derived from Owens Lake at the southern end of Owens valley. Dust emission poses numerous problems, including low visibility that triggers car accidents, loss of topsoil and soil nutrients, sediment deposition on highways, railways, and irrigation and drainage ditches, and health hazards associated with inhaling fine particles (Gregory et al., 1995).

19.2 North American Desert Subregions

19.2.1 Chihuahuan Desert

The Chihuahuan is the largest desert in southwestern North America (518,000 km^2), with three-fourths in northern Mexico and the remainder in western Texas and southern New Mexico. It lies from 600 to 1500 m above sea level astride the Rio Grande valley in the north and from 1100 to 1500 m across the high plateaus of northern Mexico farther south. The Chihuahuan Desert receives an average of 235 mm of rain, with a range of 150 to 400 mm (Schmidt, 1979). Rain falls mostly in the summertime as local, drenching thunderstorms associated with monsoonal conditions originating with the moist air from the Gulf of Mexico. Annual temperatures average 18.6°C. Summer temperatures range from 25° to 30°C, with temperature extremes higher than 50°C being rare. Winter temperatures are low (e.g., January <10°C), accounting for the cool average annual temperature (Schmidt, 1989). Freezing weather occasionally lasts for several days. Throughout much of the desert, runoff drains not into rivers, but into interior basins ringed by hills and mountains. Valley basins are predominately grassland and the upper bajadas are dominated by desert scrub or arborescent woodlands. Characteristic plants are tarbush (*Flourensia cernua*), creosote bush (*Larrea tridentata*), whitethorn (*Acacia vernicosa*), marriola (*Parthenium incanum*), candelilla (*Euphorbia antisyphilitica*), guayule (*Parthenium argentatum*), and lechuguilla (*Agave lecheguilla*).

Dunes of the Chihuahuan Desert are largely active today. Many are formed of gypsum sand or include a significant component of gypsum. White Sands National Monument in south-central New Mexico is the most famous of the gypsum dune fields. Dunes in the Cuatro Cienegas Basin in Coahuila, Mexico, are also composed largely of gypsum.

Lake Lucero, a former pluvial lake that had an area of about 466 km^2 and a depth of nearly 20 m (Morrison, 1991), provided sediment for the 700 km^2 gypsum dune field of White Sands National Monument. The dunes move east-northeast, driven by winds that blow mostly from the west to southwest (McKee, 1966). From downwind to upwind, the dunes change as follows: (1) low, broad dome-shaped dunes; (2) transverse dunes up to 12 m in height; (3) barchan dunes up to 8 m in height; and (4) U-shaped dunes at the northeastern and eastern margins of the field.

Southeast of Albuquerque, New Mexico, on the eastern margin of Laguna del Parro in Estancia Valley, large U-shaped dunes up to 2.5 km wide and 24 to 39 m in height are formed of gypsum, clay, and calcium carbonate. The most recent dunes are thought to postdate the last pluvial stage of Lake Meinzer at 4000–5000 B.P. (radiocarbon years before AD 1950).

Other Chihuahuan Desert dune fields are the Samalyuca Dunes of northern Mexico and the Monahans Sandhills of western Texas. The 250-km^2 Samalyuca field, consisting of peaked dunes up to 165 m in height, is fed by sand from extensive Pleistocene lake beds to the west (Webb, 1969). The Monahans Sandhills, largely active today, consist of barchanoid, transverse, parabolic, and coppice dunes.

19.2.2 Sonoran Desert

The Sonoran Desert covers about 275,000 km^2 around the northern two-thirds of the Gulf of California, mostly in the Mexican state of Sonora but extending across the narrow peninsula of Baja California to the open Pacific coast. Elevations range from below sea level in the Imperial Valley, California, to about 1500 m in mountain foothills. Like the Great Basin Desert, the Sonoran Desert is broken by mountains that stand like islands in a sea of desert. More than half of the Sonoran Desert is within 80 km of a coast. Although two-thirds of it lies in northwestern Mexico, about half of Arizona and the southeastern corner of California are also in the Sonoran Desert. It is separated from the higher Mojave Desert along a line between Needles and Indio, California, and from the Chihuahuan Desert by the Sierra Madre Occidental.

The Sonoran Desert flora is unique for its diversity and the abundance of trees and columnar cacti. Characteristic plants of the Sonoran Desert include saguaro (*Carnegiea gigantea*), blue palo verde (*Cercidium floridum*), little-leaf palo verde (*Cercidium microphyllum*), triangle-leaf bursage (*Ambrosia deltoidea*), ironwood (*Olneya tesota*), and creosote bush (*Larrea tridentata*). Analysis of fossil packrat middens indicates that desert scrub vegetation has been present for most of the last 13,370 years (Cole, 1986). During periods of low sea level in Pleistocene times, an even larger area of core desert for xerophytic species existed. During these times, lower areas of the Sonoran Desert probably acted as a refugium for Mojave Desert plant species (Cole, 1986). The establishment of the modern Mojave–Sonoran Desert boundary dates from 8000–9000 yr B.P., when junipers and oaks disappeared from desert lowlands (Van Devender, 1990; Rinehart and McFarlane, 1995).

The region surrounding the lower Colorado River Valley, termed the Colorado Desert (Yeager, 1957), is the hottest and driest part of the Sonoran Desert (Ezcurra and Rodrigues, 1986). Total vegetative cover of shrubs is less than 6%, and the flora is characterized by a low abundance of trees and cacti, and a relatively high abundance of microphyllous shrubs and ephemerals (Shreve, 1964; Cole, 1986). The somewhat higher terrain farther east, in southwestern Arizona, is wetter and cooler, and is floristically more complex.

The international border between the United States and Mexico in the Sonoran Desert is marked by a sharp discontinuity in albedo and grass cover, the result of long-term overgrazing in Mexico. Compared to adjacent land in southern Arizona, the Mexican landscape has an albedo approximately 5% higher, with 29% more bare soil and 28% less grass cover (Balling, 1988). Maximum surface and air temperatures in Mexico are about 4°C warmer, principally because of differential evapotranspiration rates (Balling, 1989).

The Sonoran Desert differs from the Chihuahuan, Great Basin, and Mojave deserts in that it is subtropical, rarely experiencing freezing temperatures for more than 24 consecutive hours. Summer temperatures are among the highest in North America: daily highs of 49°C or more are not uncommon in the western half of the desert. Rainfall in the Sonoran Desert is more evenly distributed throughout the year than in the Chihuahuan Desert, resulting in more verdant vegetation. Winter rainfall is derived from Pacific frontal systems, and, for most locations, this season is the time of maximum rainfall. Median winter precipitation ranges from 30 mm in southeastern California to 93 mm in southeastern Arizona (Woodhouse, 1997). Winter climate is also affected by the reoccurrence of El Niño–Southern Oscillation events, causing quasi-periodic increases in precipitation associated with above-normal sea-surface temperatures over the eastern Pacific Ocean (Douglas and Englehart, 1981; Cayan and Peterson, 1989). Little rain falls in late winter, spring, and early summer, when the Sonoran Desert is affected by the eastern edge of the Pacific high pressure cell. During high summer, monsoonal circulation, associated with the import of moist air from the Pacific Ocean across the Gulf of California and northwestern Mexico, results in summer thunderstorms, but precipitation is less than that of winter (Ezcurra and Rodrigues, 1986; Schmidt, 1989). However, there are significant regional and elevational differences in precipitation patterns. Precipitation in the western Sonoran Desert is chiefly from winter rain, whereas in the eastern Sonoran Desert of Arizona, summer monsoonal precipitation is more significant. Precipitation decreases with elevation, averaging 50–100 mm yr^{-1} at low elevations near the Gulf of California and the Salton Sea.

The Gulf of California was initiated in Neogene times by complex tectonic motions involving the appearance of the East Pacific Rise spreading center and its associated transforms beneath the gulf and the consequent transfer of the Peninsular Ranges miniplate to the Pacific plate around 5 Ma (million years before present) (see chapter 1). The Salton Trough, the landward extension of this rift, is a rapidly subsiding basin bounded by two major strike-slip fault zones: the San Andreas fault and the San Jacinto fault (Lonsdale, 1989; Dohrenwend and Smith, 1991; Dohrenwend et al., 1991). The trough, whose surface descends to −70 m below sea level, is filled with late Cenozoic sediments derived when the Colorado River episodically shifted its course away from the Gulf of California to the Salton Trough. A series of late Quaternary lakes, collectively referred to as Lake Coahuilla, occupied the basin and are recorded by abandoned shorelines, lake deposits, and archaeological sites. The lake reached a maximum depth of 95 m and covered 5700 km^2 (Waters, 1983). Distinct high stands occurred in the late Holocene at ≈2300, ≈1300, ≈1200, ≈900–600, and ≈500–400 yr B.P. (Waters, 1983). Lake development and height were controlled by geomorphic rather than climatic events, specifically by channel switching around sediment introduced by the Colorado River. Based on the accounts of early Spanish explorers, the Colorado River has been flowing into the Gulf of California at least since 1540, leading to desiccation of Lake Coahuilla (Sykes, 1937; Waters, 1983).

The present day Salton Sea was accidentally created in 1905 when Colorado River floodwaters washed out canal headgates south of Yuma, Arizona, causing the river to change course and flow into the Imperial Valley, California. The breach was repaired, and by 1907 the river had returned to its normal channel, but the Salton Sea remained. It is a saline lake with a surface area of 930 km^2 and a maximum depth of 16 m. Its salinity exceeds that of the ocean as a result of the dissolution of preexisting basin salts and high evaporation rates. The lake has no outlet and is sustained principally by agricultural drainage from the intensively farmed Imperial and Coachella valleys. The Salton Sea is a key stop on the Pacific flyway for migratory birds and provides an important habitat for five endangered species. However, it is beset by a number of environmental problems, including elevated salinity, rising lake elevations, high nutrient loading and elevated levels of selenium from irrigation drainwater, the discharge of industrial pollutants and poorly treated sewage from Mexico via the New River, and massive fish and bird die-offs, including the deaths of over 140,000 eared grebes (*Podiceps nigricollis*) in 1992.

During late Cenozoic time, strike-slip displacement along the San Andreas and related fault zones and ocean-floor spreading caused subcontinental magma to well up into the rift. Most of the rocks underlying the Gran Desierto, Mexico, are of igneous origin. The Sierra Pinacate of Mexico is a large volcanic field composed of basaltic lava flows, cinder cones, and maars mostly of Quaternary age. Low Pleistocene sea levels significantly expanded the area of the Colorado Desert and the Colorado delta, affected islands within the Gulf of California, and provided expanded source areas for the accumulation of sand in the Gran Desierto.

North America's largest field of active sand dunes, the Gran Desierto sand sea of northwest Sonora, lies immediately northeast of the head of the Gulf of California. North of the sand sea and southeast of the Salton Sea are the Algodones and East Mesa dunes, which parallel the San Andreas fault in southernmost California and northernmost Baja California Norte. Other sizable Sonoran Desert dune fields include the Mohawk Dunes of Arizona, and dunes along the desert coasts of Sonora and Baja California in Mexico.

The source of the material comprising the Algodones, East Mesa, and Gran Desierto dunes has been the subject of numerous studies. Merriam (1969) analyzed sand from seven sites in the Algodones dunes and East Mesa and four sites in the Gran Desierto and found little difference between them in mineral composition, carbonate content, and size of material. He concluded that the dunes had a common source and mode of origin, with the bulk of the dune sand derived from Colorado River delta sediment (Merriam, 1969). Subsequent studies have confirmed this source. For example, Muhs et al. (1995) used mineralogical compositions and trace element concentrations to infer that the Algodones dunes were derived from sandy shoreline sediments of Lake Coahuilla, which were derived, in turn, from the Colorado River during its periodic incursions into the Salton Trough. As the present-day extent of the Lake Coahuilla shoreline is limited, Muhs et al. (1995) suggested that the Algodones dunes probably formed during multiple episodes of accumulation.

The Algodones dune field is one of the most well-studied aeolian deposits in North America, and it has served as an Earth analogy for dunes on Mars (fig. 19.8). Aerial imagery indicates that the dominant forms are transverse and barchanoid ridges (draa). A transect from west to east across the northwestern end of the dune field reveals sand sheet and zibars, linear dunes and crescentic dunes, compound-complex crescentic dunes or draa (farther to the southeast),

Figure 19.8 Algodones Dunes, California (photo from the Spence Collection, UCLA).

and an eastern sand sheet (Kocurek and Nielson, 1986; Nielson and Kocurek, 1986). Coppice dunes occur where sand sheets are dominant, notably on the eastern and western margins of the dune field (Muhs et al., 1995). Dune orientations of the transverse and barchanoid ridge draas are coincident with a dominant northwest wind direction. Superimposed features on the transverse ridge draas are the result of secondary air flow (Havholm and Kocurek, 1988).

The dunes and sand sheets of East Mesa are principally stabilized by creosote bush (*Larrea tridentata*). Pits dug in aeolian deposits revealed no evidence of soil formation, suggesting the dunes were stabilized within the past few hundred years (Muhs et al., 1995). When Lancaster's (1988) dune mobility index is applied to the East Mesa area, it predicts that the dunes should be fully active. Muhs et al. (1995) surmise that the recent stability of the East Mesa dunes has resulted from a rise in the water table between 1939 and 1960 coincident with leakage from irrigation canals.

The Gran Desierto sand sea covers an area of 5700 km^2 whose western parts are underlain by deposits of former courses of the Colorado River. The sand sea comprises many different dune morphologies, including chains and clusters of star dunes, a range of active and relict crescentic dune forms, reversing dunes, linear and parabolic dunes, and sand sheets. The star dunes are a prominent feature, reaching 80–100 m in height and occurring in chains up to 20 km long, with a spacing of 2–3.5 km. The regional-scale pattern of dune morphology results from changes in wind regimes across the sand sea (Lancaster, 1995). To the west and north, northerly and westerly winds influence the southeastward migration of crescentic dunes. To the east and south, southerly winds cause crescentic dunes to migrate north and northwestward. The star dunes occur at the intersection of these two wind regimes, reflecting the association of this dune type with complex and seasonally reversing winds.

The Gran Desierto sand sea has accumulated as a result of fluctuations in sediment supply. In addition to climate change, fluvial and tectonic reorganizations of the Colorado River delta region and eustatic changes in sea level have influenced sediment input. Up to five generations of dunes and three periods of sand-sheet formation are recognized, deposited adjacent to or superposed on one another. For example, star dunes are superimposed on two older generations of crescentic dunes (Lancaster, 1995). Distinct periods of dune formation are separated by erosional unconformities, deflation lag surfaces, or weakly developed soils. The sand sea is characterized by relatively abrupt changes in dune morphology, with dunes of different colors, composition, and grain size and sorting adjacent to each other.

Blount and Lancaster (1990) used grain size and sorting, mineralogy, and spectral signature on Landsat TM imagery to conclude that sand was derived principally from the ancestral Colorado River delta, the present lower Colorado River valley, and the coast of the Bahia del Adair. The different dune generations most likely resulted from changes in sediment source areas. For example, during the Last Glacial Maximum, when sea level was ≈120 m lower than at present, the coastline of the Bahia del Adair was located 150 km south of its present position, exposing a 50- to 100-km-wide zone of sandy sediment. Similarly, the Colorado River delta front was located 100 km southeast of its modern position (van Andel, 1964), allowing for southerly sand input to the Gran Desierto. Periods of climatic change also influenced the development of the sand sea. Increased summer rainfall in the mid-Holocene (Van Devender, 1990) probably acted to reduce sediment transport rates and stabilize parts of the sand sea.

Immediately west of the Salton Sea in southern California is a small, 15-km^2 dune field consisting of about 70 dunes, dominated by barchans and distorted barchans but including parabolic and vegetated dunes (Haff and Presti, 1995). Data available from this dune field include long-term rates and directions of movement of individual barchans and changes in dune height, width, and length. For example, Long and Sharp (1964) analyzed dune motion in the region between 1941 and 1963, observing 34 dunes moving eastward at an average of 15.3 m y^{-1} between 1941 and 1956 and 25.0 m y^{-1} between 1956 and 1963. Haff and Presti (1995) estimate an average rate of 16.2 m y^{-1} for the period 1963–81. Their work suggests that the dune field is declining in area and population, that new barchans are no longer being formed in source areas, and that some larger dunes are starved for sand and declining in size.

In southwestern Arizona, parallel dune ridges 100 m apart cover much of the Cactus Plain, east of Parker. The eastern margin of the field includes climbing dunes. A single ridge of dunes, 5 km in width, extends about 25 km along the floor of lower Mohawk Valley (Bryan, 1925).

Summer monsoons bring "haboobs" or dust storms to Arizona in July and August, carrying clay- and silt-size material to heights of 3000 m or more. The source of the fine material includes abandoned or fallow farm fields and construction sites. Winter precipitation is effective in stimulating vegetation growth and reducing spring and summer dust loads in southeastern Arizona (MacKinnon et al., 1990).

Several dune fields also occur on the arid Pacific coast of central Baja California (Orme, 1973; Orme and Tchakerian, 1986). Some of these are strictly coastal dunes emplaced on the desert margin, for example, at El Socorro near San Quintín where several stabilized dune phases are mantled by active dunes moving up to 8 km inland along a 2-km front. Three hundred kilometers to the southeast, beyond the coastal dunes around Bahía Sebastian Vizcaíno, true desert dunes occupy much of the Desierto de Vizcaíno in Baja California Sur with linear forms 100 m apart and dunes extending up to 100 km southeastward parallel with the prevailing wind (Inman et al., 1966).

19.2.3 The Mojave Desert

The Mojave Desert is the smallest North American desert, covering 140,000 km^2 in southeastern California and southernmost Nevada. It is roughly triangular in shape and bounded on the north by the Tehachapi Mountains, on the southwest by the San Gabriel and San Bernardino Mountains, and on the east by the valley of the lower Colorado River. Differences in tectonic activity give rise to distinct terrains within the desert, but, in general, the Mojave landscape is more subdued than the southwest Great Basin to the north or the Salton Trough to the south. Elevations, which commonly lie above 1000 m and reach 2000 m in some isolated mountain ranges, distinguish the Mojave or "high desert" from the neighboring Sonora or "low desert" to the southeast. The impacts of late Quaternary climate change are conspicuous in the surficial stratigraphy of the area. Alluvial fan morphology indicates a relatively wet environment in the late Pleistocene, followed by an arid Holocene, when sparse soils and vegetation allowed more rapid hillslope runoff and higher stream power. Holocene fans commonly have a distinctive bar and swale topography, which contrasts with the smooth pavement-covered surfaces of late Pleistocene fans (Wells et al., 1987).

Rainfall averages about 76–102 mm annually across the desert floor, increasing with rising elevation to about 279 mm. Most precipitation is derived from frontal systems during the winter months, although strong, localized thunderstorms enter the easternmost part of the Mojave Desert in July through September. During the summer, temperatures often exceed 40°C for 100 consecutive days or more. Winters are colder than in the Sonoran Desert: in valley bottoms where cold air settles at night, temperatures may drop below −18°C in winter and may be close to freezing even in summer.

Lying between the Great Basin Desert to the north and the Sonoran Desert to the southeast, the Mojave incorporates vegetational elements from each of these regions, in addition to some endemic species (Rowlands et al., 1982). The occasional hard frosts in winter exclude many of the subtropical, arborescent forms that are common in the Sonoran Desert. Characteristic Mojave Desert plants include white bursage (*Ambrosia dumosa*), creosote bush (*Larrea tridentata*), Mojave yucca (*Yucca schidigera*), Joshua tree (*Yucca brevifolia*), Mojave sage (*Salvia mohavensis*), and Fremont dalea (*Psorothamnus fremontii*).

Drainage within the Mojave Desert is mostly internal, with most of the closed basins relatively undissected. During wetter intervals of the Pleistocene, the Mojave Desert was characterized by a more integrated drainage system than today. Most notably, the Mojave River flowed north and then east from the San Bernardino Mountains, forming Lakes Manix and Mojave, before spilling northward toward Death Valley, where it was joined by waters from the Owens and Amargosa drainages to form Lake Manly (see chapter 3). The nature of the regional topography and the distribution of vertebrate fish faunas in widely separated, isolated basins in the Mojave Desert have led to suggestions that the Mojave River, joined perhaps by waters from Death Valley, may have spilled eastward through what are now Bristol, Cadiz, and Danby dry lakes to the lower Colorado River (Blackwelder, 1933, 1954; Hubbs and Miller, 1948; Miller, 1981; Hale, 1985). The highest point along the proposed route reaches 594 m near Ludlow, California. However, surface and subsurface indicators do not support a fluvial-lacustrine connection between these drainage systems and the Colorado River via the Soda-Bristol route during the past 4 Ma (Brown and Rosen, 1995). The highest shoreline features in Death Valley are nearly 500 m lower than the height necessary for overflow at Ludlow. Furthermore, neither Lake Mojave nor the basins east of Ludlow have seen large deep-water perennial lakes during the Pleistocene. It seems that high mountain catchments provided sufficient water to sustain large lakes in the Mojave Desert region during pluvial intervals, but not enough to promote overflow to the Colorado River.

Evidence from both dunes and ventifacts indicates that aeolian activity in the Mojave Desert has been both prolonged and episodic, but more extensive and more intense in the past than it is today (H.T.U. Smith, 1967; R.S.U. Smith, 1984; Tchakerian, 1991; Laity, 1992; Lancaster, 1994; Zimbelman et al., 1995). The principal active dune fields are in the eastern Mojave, including the Kelso Dunes of the Devils Playground area, the Ibex and Dumont Dunes of the northern Silurian Valley (fig. 19.9), and the dunes of the Cadiz and Palen valleys. Field studies in the Kelso Dunes (Sharp, 1966), the Ibex Dunes (Garrett, 1966), and the Dumont Dunes (Nielson and Kocurek, 1987) document little net movement of dunes ridges, despite short-term changes induced by seasonal wind reversals. Inactive dunes are extensive and include sand ramps and climbing and falling dunes, many of which are now mantled by coarse talus and fluvial debris and are fluvially dissected. Sand ramps are as much as 6 km long and often bury the lower flanks of ranges. Rendell et al. (1994) used luminescence dating techniques on sand-ramp sediments to suggest two main periods of late Pleistocene aeolian deposition, between >35 and 25 ka and between 15–10 ka.

Human use of the Mojave Desert has influenced both sand and dust movement. Along the Mojave River east of Barstow, California, groundwater withdrawal for agriculture, domestic use, industry, and recreational lakes, combined with removal of trees and the use of off-road vehicles, have destabilized dunes and enhanced dust storm activity. Annual precipitation in the lower Mojave Valley averages only 112 mm, evapotranspiration rates are 2120 mm yr^{-1}, and strong westerly winds exceed 5 m sec^{-1} between 45 to 65% of the time from March through June. The Mojave River is the main source of groundwater recharge for the basin, with about 80% of the recharge derived from flood flows along a 160-km reach between the San Bernardino Mountains and Afton Canyon. Additionally, some of the groundwater was recharged when the climate was cooler and wetter than at present (Izbicki et al., 1995).

A study of phreatophytic vegetation in 1929 recorded over 3000 ha of tules, willows, cottonwoods, and other plants growing along the Mojave River from the Forks to Cady Mountain (Blaney and Ewing, 1935). Groundwater reached the surface at the Calico fault, marked by a line of sand dunes, and near hard rock barriers in the vicinity of Camp Cady. Thompson (1929) reported year-round stream

Figure 19.9 The Ibex Dunes of southeastern California are a small group of active star dunes in southern Death Valley National Park. Note the figures on the dune crest for scale.

flow and ponds at these sites. The vegetation that flourished in response to this groundwater anchored large coppice dunes. Groundwater, which was at or near the surface in the 1930s, is presently 16–30 m below the surface, well beyond the reach of vegetation. In addition to causing the death of mature phreatophytes, groundwater decline also prevents the establishment of mesquite seedlings that survive the summer on underground water. Dieback of the vegetation, in combination with the strong regional winds, has caused destabilization of floodplain dunes, which are now actively migrating and encroaching on homes and agricultural properties. Additionally, blowing dust in the lower Mojave Valley commonly obscures downwind areas. Under natural conditions, short-term recovery of the environment is unlikely as most groundwater was recharged more than 20,000 years ago, when climatic conditions were different than today (Izbicki et al., 1995). Presently, the Mojave River Pipeline is being constructed to bring water from the California Aqueduct to replenish groundwater within the basin.

19.2.4 The Great Basin Deserts

The Great Basin was named by John Fremont in 1844, who recognized that the region has no outlet to the sea. The second largest desert in size, it covers about 400,000 km^2 of the northern Basin and Range Province. Much of Nevada and Utah are in the Great Basin Desert, as are parts of Oregon, Washington, Idaho, California, and Wyoming. The region is bounded by the Columbia Plateau to the north, the Sierra Nevada to the west, the Mojave Desert to the south, and the Wasatch Range and Colorado Plateau to the east. The central area of the Great Basin consists of elevated basins and ranges flanked on three sides by lower terrain: the 140,000-km^2 Bonneville basin on the east, the 115,000-km^2 Lahontan basin on the west, and a poorly defined lowland to the south (Dohrenwend, 1987). The desert floor is broken by numerous north–south-trending mountain ranges, the highest (White Mountains) reaching 4340 m above sea level. The lowest basin elevation (85 m below sea level) is the floor of Death Valley, California. Most of the drainage within the Great Basin is internal and many basins contain playa remnants of former Pleistocene lakes (Morrison, 1991; Mifflin and Wheat, 1979; see also chapter 3). Perennial rivers include the Truckee, Carson, Walker, and Owens rivers, flowing from the east flank of the Sierra Nevada, and the Humboldt River of Nevada.

Lying in the rain shadow of the Sierra Nevada–Cascade chain, the Great Basin receives from 100 to 300 mm of precipitation annually, a little less than half of which falls in the summer. The Great Basin Desert is a temperate desert, with extremes appropriate to its midlatitude, interior location. Winters bring freezing temperatures and snow; the average temperature in January is −2° C. Nonetheless, summer temperatures in the southern Great Basin can be very hot. The highest temperature ever recorded in the United States (57°C) was in Death Valley.

Great Basin ecosystems are a function of temperature, latitude, elevation, rainfall, and geology. Characteristic plants include big sagebrush (*Artemisia tridentata*), shadscale (*Artiplex confertifolia*), rubber rabbitbrush (*Chrysothamnus nauseosus*), blackbrush (*Coleogyne ramosissima*), and greasewood (*Sarcobatus vermiculatus*). Grazing has caused various changes in several plant communities, notably a reduction in grasses.

The Great Basin Desert of the northern Basin and Range Province merges southeastward with the semidesert terrain of the Colorado Plateau, a region of quite different tectonic style and landscape which, at higher elevations, passes in turn into open coniferous woodland. The Colorado Plateau is a roughly circular area of about 384,000 km^2 that consists of plateaus and isolated mountains encompassing parts of Utah, Colorado, New Mexico, and Arizona. Although it lies mostly above 1500 m and is a definable tectonic unit, the plateau shows considerable internal variation (Hunt, 1974). Its landscape illustrates the effects of differential erosion on nearly horizontal or only moderately deformed strata of varying strength (fig. 19.10; see also chapter 1). The landscape is dominated by canyons, cuesta scarps, and plains stripped to bedrock. Volcanic fields form a discontinuous belt along the southern and western margins of the Colorado Plateau, with lava flows ranging from 9.9 Ma to late Holocene in age (Patton et al., 1991).

Investigations of the Colorado Plateau have given rise to many contributions to geomorphology from the classic work of Gilbert (1877) and Bryan (1928), to slope development and mass wasting (Ahnert, 1960; Schumm and Chorley, 1964, 1966; Oberlander, 1977), fluvial erosion (Graf, 1978, 1983; Hereford, 1984; Patton and Boison, 1986), volcanism (Hamblin, 1970), tectonic geomorphology (McGill and Stromquist, 1979), and to a lesser degree, aeolian processes (Hack, 1941). Furthermore, studies of the origin of segmented cliffs in sandstones (Oberlander, 1977), of graben formation, joint inheritance, and astrobleme formation, and of sapping in the development of theater-headed valley networks (Laity and Malin, 1985) illustrate the role of the plateau as a geomorphic laboratory for terrestrial and planetary studies.

Large-scale mass wasting has modified many of the escarpments, canyons, and mountain flanks of the Colorado Plateau. The largest landslides occur where plastic, commonly montmorillonitic, shale units underlie more competent rock. Many are composite in morphology, origin, and age, and may be very large. The Coleman slide on the Aquarius Plateau is estimated to be as much as 60-m thick, with a 0.4-km-long head scarp, and a 2-km-downslope extent (Patton et al., 1991). In addition to large landslides, rockfalls and slab failures are common where resistant cliff-forming strata crop out (Bradley, 1963). Landslide activity on the Colorado Plateau has been related to climate change

Figure 19.10 The Colorado Plateau section of the Great Basin Desert is characterized by differential erosion on nearly horizontal strata. Monument Valley appears in the distance.

(Schumm and Chorley, 1964), and episodes of enhanced landslide activity at higher altitudes of the plateau may be coeval with glacial episodes (Shroder and Sewell, 1985).

Theater-headed valleys formed by a combination of groundwater, fluvial, and mass-wasting process are among the most prominent geomorphic features of the Colorado Plateau. Many such valleys occur in the Jurassic Navajo Sandstone, a highly transmissive aquifer underlain by essentially impermeable rocks. Groundwater, moving laterally down the hydraulic gradient, emerges in concentrated zones of seepage at valley headwalls. Small-scale erosional processes slowly reduce the support of steep cliffs, and the valleys grow headward by successive slab failures (Laity and Malin, 1985; Laity, 1988). Rates of valley development were probably greater during wetter intervals of the Pleistocene when recharge rates were two to three times higher than today and water tables as much as 60 m higher (Zhu et al., 1998).

Much of the Great Basin Desert is mantled by a thin veneer of sand, which in turn is well anchored by vegetation. Deeper, locally extensive dune fields occur in almost every basin of the southwestern Great Basin. For example, star dunes are developed at the southern end of Eureka Valley, where the highest dune in California is found (208 m), at the northern end of Panamint Valley, and in central and southern Death Valley (R.S.U. Smith, 1982). The location of these dune fields suggests substantial topographic control of near-surface winds throughout the region.

Dune fields on the Colorado Plateau include the Little Sahara Sand Dunes in central Utah and the Coral Pink Dunes in southern Utah, which lie on the boundary between desert and woodland. Dunes of volcanic ash occur east of the San Francisco Peaks, a volcanic field of Quaternary age, near Flagstaff, Arizona.

Dunes also occur over about 65,000 km^2 of northeastern Arizona, in a region of high plateaus and mesas. The dunes were mapped by Hack (1941) as mostly northeast-trending U-shaped and longitudinal dunes, with isolated barchans, transverse dunes, falling dunes, and climbing dunes. The wind regime is very active, in part because topographic effects of the San Francisco volcanic field cause funneling and severe turbulence. Unidirectional peak winds of 15 to 25 m sec^{-1} are common. Sand grains, which originate in the flood plain of the Little Colorado River, travel northeast across the Painted Desert toward the Moenkopi Plateau, forming fields of barchan dunes. Some reform as climbing dunes that funnel up abandoned watercourses, climbing 60 m to the top of the Red Rock Cliffs, where the sand recombines to form dome dunes and linear sand dunes (Breed and Breed, 1979). The region has experienced multiple periods of dune formation, subject to episodic destruction by periods of heightened fluvial activity. During the Holocene, erosional retreat of the plateau cliffs, diminished winds, and a lessened sand supply reduced dune formation. At present, linear dunes of the Moenkopi Plateau are being worn away by deflation and fluvial erosion.

Aeolian deposits cover a large expanse of the southeastern Colorado Plateau in the Chaco River drainage basin of northwestern New Mexico (Hack, 1941; Wells et al., 1990). Deposits consist principally of sand sheets, parabolic and modified parabolic dunes, barchan-barchanoid dune complexes, linear dunes, and ridge dunes. Sand sheets are the most areally extensive, comprising approximately 88% of aeolian deposits and ranging up to 2 m in thickness. Three major late Quaternary phases of aeolian activity are recognized, based on soil development, radiocarbon and cultural dates, and degree of preservation of sedimentary structures (Wells et al., 1990). The most significant deposition occurred during the late Pleistocene to Holocene transition, when widespread sand sheets and modified parabolic dunes were developed in an environment more mesic than today.

Middle Holocene deposition was relatively insignificant in the basin, but the youngest period of aeolian activity (after 1.5 ka, thousand years before present) resulted in a diverse suite of eolian landforms (Wells et al., 1990).

Wind erosion features of the Colorado Plateau range from such large-scale forms as deflation hollows, yardangs, and wind-fluted cliffs, to smaller features, including blowouts and scour grooves. Yardangs are developed in Mesozoic claystones and siltstones in the Painted Desert, with the largest individuals about 50 m long.

Owens Lake, at the southern end of Owens valley, is the site of dust storms that in most years generate the highest 24-hour fine-dust levels in the United States, often posing severe health hazards to humans. The lake was desiccated in the mid-1920s after the Los Angeles Department of Water and Power began diverting the tributaries of the Owens River into the Los Angeles River. Evaporation left a 280-km^2 dry alkaline lake bed, underlain by a water table that is a few centimeters to several meters below the surface. Frequent surface flooding occurs in the winter and spring after storms. Salt crusts result from hydration and dehydration of salts brought to the lake-bed surface by groundwater discharge or by precipitation and evaporation recycling of surficial evaporite deposits (Cochran et al., 1988). Efflorescent crusts on the eastern and southern sides of the lake are vulnerable to wind-blown saltating particles (Barone et al., 1981; St. Amand et al., 1986; Cochran et al., 1988). Owens dry lake is one of the largest sources of very fine particulate matter (PM_{10}) in the western hemisphere (Cahill et al., 1996). The largest and most frequent storms occur in the spring and autumn. Wind flow in the Owens Valley is highly turbulent. Steep valley walls and variations in valley width help to create surface eddies ≈5–9 km in diameter (Reid et al., 1994). Dust plumes have been carried to 2000 m or higher over the lake bed within 5 km of the dust source. Dust storms produced in the southern regions follow the Sierra Nevada into Indian Wells Valley, south of the Coso Range. Northern plumes follow the local topography to Darwin, where they either move south along the Argus Range or empty into Panamint Valley.

19.3 Conclusion

The deserts of North America exhibit considerable diversity in their climate, vegetation, geomorphology, and tectonic history. Because deserts are close to major research institutions and as a result of advances in remote sensing, air photo interpretation, field instrumentation, and dating techniques, desert research has experienced an upsurge in recent decades. Deserts are also valuable for paleoclimatic study and as planetary analogues. In addition, growing environmental problems require a better understanding of the nature and resources of arid regions. Arid zone research in North America has made important contributions to our understanding of dust generation, soil-pavement formation, rock coatings and varnishes, climatic history, dune development, planetary processes, groundwater geomorphology, aeolian abrasion, slope formation, fluvial processes, including flooding and fan formation, and soil formation.

Despite advances in our understanding of desert processes and the role and nature of climate change, however, the evolution of North American deserts is not well understood. Whereas the late Pleistocene and Holocene events are reasonably decipherable, the relative importance of tectonic and climatic forcing of desert environments over the longer term is much less well known. Likewise, research suggests that deposition in the larger dune fields probably occurred as multiple phases of aeolian activity, but the timing of such events remains poorly constrained. Furthermore, the increasing human impact on the desert in recent decades, through military and recreational activity, groundwater depletion, surface water diversion, irrigation and other activities, indicates that studies of the natural system can rarely ignore the human factor. Thus, the deserts of southwestern North America will continue to provide fascinating challenges for talented researchers in decades to come.

Acknowledgments I thank David Fuller, California State University Northridge, for his careful preparation of the maps and diagrams that accompany this chapter, and Antony Orme for his many useful comments and suggestions.

References

Ahnert, F., 1960. The influence of Pleistocene climates upon the morphology of cuesta scarps on the Colorado Plateau. *Annals of the Association of American Geographers* 50: 139–156.

Bach, A.J., A.J. Brazel, and N. Lancaster, 1996. Temporal and spatial aspects of blowing dust in the Mojave and Colorado Deserts of southern California. *Physical Geography* 17 (4): 329–353.

Baker, V.R., 1990. Spring sapping and valley network development, with case studies by R.C. Kochel, V.R. Baker, J.E. Laity, and A.D. Howard. In: *Groundwater Geomorphology; The role of subsurface water in Earth-surface processes and landforms.* C.G. Higgins and D.R. Coates (Eds.). Geological Society of America Special Paper 252, Boulder, Colorado, pp. 235–265.

Balling, R.C., Jr., 1988. The climatic impact of a Sonoran vegetation discontinuity. *Climatic Change* 13: 99–109.

Balling, R.C., Jr., 1989. The impact of summer rainfall on the temperature gradient along the United States–Mexico border. *Journal of Applied Meteorology* 28: 304–308.

Barone, J.B., L.L. Ashbaugh, B.H. Kusko, and T.A. Cahill, 1981. The effect of Owens Dry Lake on air quality in the Owens Valley with implications to the Mono Lake Area. In: *Atmospheric Aerosols: Source Air Quality*

Relationships (ACS Symposium Series, No. 167). P. Radke (Ed.). No. 18: 327–346.

Bell, J.W., J.G. Price, and M.D. Mifflin, 1992. Subsidence-induced fissuring along preexisting faults in Las Vegas Valley, Nevada. *Proceedings, Association of Engineering Geologists*, 35th Annual Meeting, Los Angeles: 66–75.

Belnap, J., and J.S. Gardner, 1993. Soil microstructure in soils of the Colorado Plateau: The role of Cyanobacterium *Microcoleus Vaginatus*. *Great Basin Naturalist* 53: 40–47.

Blackwelder, E., 1929. Sandblast action in relation to the glaciers of the Sierra Nevada. *Journal of Geology* 37: 256–260.

Blackwelder, E., 1933. Lake Manly, an extinct lake in Death Valley. *Geographical Review* 23: 464–471.

Blackwelder, E., 1954. Pleistocene lakes and drainages in the Mojave region, southern California. In: *The Geology of Southern California*. R.H. Jahns (Ed.). Division of Mines and Geology Bulletin 170: 35–40.

Blake, W.P., 1855. On the grooving and polishing of hard rocks and minerals by dry sand. *American Journal of Science* 20: 178–181.

Blaney, H.F., and P.A. Ewing, 1935. *Utilization of the waters of Mojave River, California*. United States Department of Agriculture, Washington, D.C.

Blount, G., and N. Lancaster, 1990. Development of the Gran Desierto sand sea. *Geology* 18: 724–728.

Blount, H.G., M.O. Smith, J.B. Adams, R. Greeley, and P.R. Christensen, 1990. Regional aeolian dynamics and sand mixing in the Gran Desierto: Evidence from Landsat Thematic Mapper images. *Journal of Geophysical Research* 95: 15,463–15,482.

Blumberg, D.G., and R. Greeley, 1996. A comparison of general circulation model predictions to sand drift and dune orientations. *Journal of Climate* 9: 3248–3259.

Bradley, W.C., 1963. Large-scale exfoliation in massive sandstones of the Colorado Plateau. *Geological Society of America Bulletin* 74 (5): 519–527.

Breed, C.S., and W.J. Breed, 1979. Windforms of central Australia and a comparison with some linear dunes on the Moenkopi Plateau, Arizona, In: *Scientific results of the Apollo-Soyuz Missions*. F. El-Baz (Ed). NASA Special Paper 412, pp. 319–358.

Brotherson, J.D., S.R. Rushforth, and J.R. Johansen, 1983. Effects of long-term grazing on cryptogam crust cover in Navajo National Monument. *Journal of Range Management* 36: 579–581.

Brown, W.J., and M.R. Rosen, 1995. Was there a Pliocene-Pleistocene fluvial-lacustrine connection between Death Valley and the Colorado River? *Quaternary Research* 43: 286–296.

Bryan, K., 1925. The Papago country, Arizona. *U.S. Geological Survey Water Supply Paper* 499.

Bryan, K., 1928. Niches and other cavities in sandstone at Chaco Canyon, New Mexico: *Zeitschrift Für Geomorphologie* 3: 128–140.

Cahill, T.A., T.E. Gill, J.S. Reid, E.A. Gearhart, and D.A. Gillette, 1996. Saltating particles, playa crusts, and dust aerosols at Owens "dry" lake, California. *Earth Surface Processes and Landforms* 21, 621–639.

Cayan, D.R., and D.H. Peterson, 1989. The influence of North Pacific atmospheric circulation on streamflow in the West: *Geophysical Monograph* 55: 375–395.

Cochran, G.F., T.M. Mihevic, S.W. Tyler, and T.J. Lopes, 1988. Study of salt crust formation mechanisms on Owens (Dry) Lake, California. *Desert Research Institute Publication* No. 41108, Reno, Nevada.

Cole, K.L., 1986. The lower Colorado River Valley: A Pleistocene desert. *Quaternary Research* 25: 392–400.

Dohrenwend, J.C., 1987. Basin and Range. In: *Geomorphic Systems of North America*. W.L. Graf (Ed.). Geological Society of America, Centennial Special Volume 2, Boulder, Colorado, pp. 303–341.

Dohrenwend, J.C., and R.S.U. Smith, 1991. Quaternary geology and tectonics of the Salton Trough. In: *Quaternary Nonglacial Geology; Conterminous U.S.* R.B. Morrison (Ed.). The Geology of North America, v. K-2. Geological Society of America, Boulder, Colorado, pp. 334–337.

Dohrenwend, J.C., W.B. Bull., L.D. McFadden, G.I. Smith, R.S.U. Smith, and S.G. Wells, 1991. Quaternary geology of the Basin and Range Province in California. In: *Quaternary Nonglacial Geology; Conterminous U.S.* R.B. Morrison (Ed.). The Geology of North America, v. K-2. Geological Society of America, Boulder, Colorado, pp. 321–352.

Dorn, R.I. 1986. Rock varnish as an indicator of aeolian environmental change. In: *Aeolian Geomorphology*. W.G. Nickling (Ed.). Allen and Unwin, London, pp. 291–307.

Dorn, R.I., 1995. Alterations of ventifact surfaces at the glacier/desert interface. In: *Desert Aeolian Processes*. V.P. Tchakerian (Ed.). Chapman & Hall, London, pp. 199–217.

Douglas, A.V., and P.J. Englehart, 1981. On a statistical relationship between rainfall in the central equatorial Pacific and subsequent winter precipitation in Florida. *Monthly Weather Review* 114: 1716–1738.

Eldridge, D.J., 1993. Cryptogams, vascular plants, and soil hydrological relations: some preliminary results from the semiarid woodlands of Eastern Australia. *Great Basin Naturalist* 43: 48–58.

Ezcurra, E., and V. Rodrigues, 1986. Rainfall patterns in the Gran Desierto, Sonora, Mexico. *Journal of Arid Environments* 10: 13–28.

Ezcurra, E., R.S. Felger, A.D. Russell, and M. Equihua, 1988. Freshwater Islands in a Desert Sand Sea: The hydrology, flora, and phytogeography of the Gran Desierto Oases of northwestern Mexico. *Desert Plants*: 9 (2): 36–63.

Fenneman, N.M., 1931. *Physiography of the western United States*. McGraw-Hill, New York, 543 p.

Garrett, D.M., 1966. Geology of the Saratoga Springs sand dunes, Death Valley National Monument, California. M.S. thesis, University of Southern California, Los Angeles.

Gilbert, G.K., 1875. Report on the Geology of Portions of Nevada, Utah, California, and Arizona examined in the years 1871 and 1872. *Report of the U.S. Geographical and Geological Surveys West of the 100th Meridian*, 3, Geology, 17–187, Washington, D.C.

Gilbert, G.K., 1877. *Report on the geology of the Henry Mountains*. Geographical and Geological Survey of the Rocky Mountain Region, Washington, D.C., 160 p.

Gile, L.H., Peterson, F.F., and Grossman, R.B., 1979. *The Desert Soil Project Monograph*; Washington, D.C., U.S. Soil Conservation Service, 984 p.

Graf, W.L., 1978. Fluvial adjustments to the spread of tamarisk in the Colorado Plateau region: *Geological Society of America Bulletin* 86: 1491–1501.

Graf, W.L., 1983. The arroyo problem; Paleohydrology and paleohydraulics in the short term. In: *Background to*

Paleohydrology. K.G. Gregory (Ed.). John Wiley and Sons, London, pp. 279–302.

Graf, W.L., 1988. Definition of flood plains along arid-region rivers. In: *Flood Geomorphology*. V.R. Baker, R.C. Kochel, and P.C. Patton (Eds.). Wiley, pp. 231–242.

Greeley, R., 1995. Geology of terrestrial planets with dynamic atmospheres. *Earth, Moon, and Planets* 67: 13–29.

Gregory, J.M., J.A. Lee, G.R. Wilson, and U.B. Singh, 1995. Modeling seasonal patterns of blowing dust on the southern High Plains. In: *Desert Aeolian Processes*. V.P. Tchakerian (Ed.). Chapman & Hall, London, pp. 233–250.

Hack, J.T., 1941. Dunes of the western Navajo Country. *Geographical Review* 31: 240–263.

Haff, P.K., and D.E. Presti, 1995. Barchan dunes of the Salton Sea region, California. In: *Desert Aeolian Processes*. V.P. Tchakerian (Ed.). Chapman & Hall, London, pp. 153–177.

Haff, P.K., and B.T. Werner, 1996. Dynamical processes on desert pavements and the healing of surficial disturbances. *Quaternary Research* 45: 38–46.

Hale, G.R., 1985. Mid-Pleistocene overflow of Death Valley toward the Colorado River. In: *Quaternary Lakes of the eastern Mojave Desert, California*. G.R. Hale (Ed.). Field Trip Guidebook, Friends of the Pleistocene, Pacific Cell, pp. 36–81.

Hamblin, W.K., 1970. Late Cenozoic basalt flows of the western Grand Canyon. *Utah Geological and Mineral Survey Guidebook to the Geology of Utah*, no. 23, pp. 21–38.

Havholm, K.G., and G. Kocurek, 1988. A preliminary study of the dynamics of a modern draa, Algodones, southeastern California, USA. *Sedimentology* 35: 649–669.

Hereford, R., 1984. Climate and ephemeral-stream processes; Twentieth-century geomorphology and alluvial stratigraphy of the Little Colorado River, Arizona. *Geological Society of America Bulletin* 95: 654–668.

Howard, A.D., and R.C. Kochel, 1988. Introduction to cuesta landforms and sapping processes on the Colorado Plateau. In: *Sapping features on the Colorado Plateau*. A.D. Howard, R.C. Kochel, and H.E. Holt (Eds.). Proceedings and Field Guide for the NASA Groundwater Sapping Conference: National Aeronautic and Space Administration Special Publication SP-491, pp. 6–56.

Howard, A.D., and M.J. Selby, 1994. Rock Slopes. In: *Geomorphology of Desert Environments*. A.D. Abrahams and A.J. Parsons (Eds.). Chapman & Hall, London, pp. 123–172.

Hubbs, C.L., and R.R. Miller, 1948. The Great Basin with emphasis on glacial and post-glacial times. The zoological evidence: Correlation between fish distribution and hydrographic history in the desert basins of the western United States. *Utah University Bulletin* 38: 18–166.

Hunt, C.B., 1974. *Natural regions of the United States and Canada*: San Francisco, W.H. Freeman, 726 p.

Inman, D.L., G.C. Ewing, and J.B. Corliss, 1966. Coastal sand dunes of Guerrero Negro, Baja California, Mexico. *Geological Society of America Bulletin* 77: 787–802.

Izbicki, J.A., P. Martin, and R.L. Michel, 1995. Source, movement and age of groundwater in the upper part of the Mojave River basin, California, USA. *Application of Tracers in Arid Zone Hydrology* (Proceedings of the Vienna Symposium, August 1994). IAHS Publ. no. 232, pp. 43–56.

Kocurek, G., and J. Nielson, 1986. Conditions favorable for the formation of warm-climate aeolian sand sheets. *Sedimentology* 33: 795–816.

Kresan, P.L., 1988. The Tucson, Arizona, flood of October 1983: Implications for land management along alluvial river channels. In: *Flood Geomorphology*. V.R. Baker, R.C. Kochel, and P.C. Patton (Eds.). Wiley, New York, pp. 465–89.

Laity, J.E., 1988. The role of groundwater sapping in valley evolution on the Colorado Plateau. In A.D. Howard, R.C. Kochel, and H.E. Holt (Eds.). *Sapping features of the Colorado plateau, a comparative planetary geology field guide*: Washington, D.C., National Aeronautics and Space Administration Special Publication 491, pp. 63–80.

Laity, J.E., 1992. Ventifact evidence for Holocene wind patterns in the east-central Mojave Desert. *Zeitschrift für Geomorphologie, Supplement Band* 84: 1–16.

Laity, J.E., 1995. Wind abrasion and ventifact formation in California. In: *Desert Aeolian Processes*. V.P. Tchakerian (Ed.). Chapman & Hall, London, pp. 295–321.

Laity, J.E., and M.C. Malin, 1985. Sapping processes and the development of theater-headed valley networks on the Colorado Plateau. *Geological Society of America Bulletin* 96: 203–217.

Lancaster, N., 1988. Development of linear dunes in the southwestern Kalahari, southern Africa. *Journal of Arid Environments* 14: 233–244.

Lancaster, N., 1994. Controls on aeolian activity: New perspectives from the Kelso Dunes, Mojave Desert, California. *Journal of Arid Environments* 27: 113–124.

Lancaster, N., 1995. Origin of the Gran Desierto sand sea, Sonora, Mexico: Evidence from dune morphology and sedimentology. In: *Desert Aeolian Processes*. V.P. Tchakerian (Ed.). Chapman & Hall, London, pp. 11–35.

Lattman, L.H., 1973. Calcium carbonate cementation of alluvial fans in southern Nevada. *Geological Society of America Bulletin* 84: 3013–3028.

Long, J.T., and R.P. Sharp, 1964. Barchan-dune movement in Imperial Valley, California. *Geological Society of America Bulletin* 75: 149–156.

Lonsdale, P., 1989. Geology and tectonic history of the Gulf of California. In: *The Eastern Pacific Ocean and Hawaii*. E.L. Winterer, D.M. Hussong, and R.W. Decker (Eds.). Decade of North American Geology, v. N, Geological Society of America, Boulder, Colorado, pp. 499–521.

MacKinnon, D.J., D.F. Elder, P.J. Helm, M.F. Tuesink, and C.A. Nist, 1990. A method of evaluating effects of antecedent precipitation on duststorms and its application to Yuma, Arizona, 1981–1988. *Climatic Change* 17: 331–360.

Maxson, J.H., 1940. Fluting and faceting of rock fragments. *Journal of Geology* 48: 717–751.

McFadden, L.D., S.G. Wells, J.C. Dohrenwend, B.D. Turrin, 1984. Cumulic soils formed in eolian parent materials on flows of the Cima volcanic field, Mojave Desert, California: In: *Surficial geology of the eastern Mojave Desert, California*. J.C. Dohrenwend (Ed.). Geological Society of America 1984 Annual Meeting Guidebook, pp. 134–149.

McGill, G.E., and A.W. Stromquist, 1979. The grabens of Canyonlands National Park, Utah: Geometry, mechanics, and kinematics. *Journal of Geophysical Research*, 84: 4547–4563.

McKee, E.D., 1966. Structures of dunes at White Sands National Monument, New Mexico (and a comparison with structures of dunes from other selected areas). *Sedimentology* 7: 1–69.

McMahon, J.A., and F.H. Wagner, 1985. The Mojave, Sonoran, and Chihuahuan Deserts of North America. In: *Hot deserts and arid shrublands*. D.D. Evans and J.L. Thames (Eds.). Elsevier, Amsterdam, pp. 105–202.

Merriam, R., 1969, Source of sand dunes of southeastern California and northwestern Sonora, Mexico. *Geological Society of America Bulletin* 80: 531–534.

Mifflin, M.D., and M.M. Wheat, 1979. Pluvial lakes and estimated pluvial climates of Nevada: *Nevada Bureau of Mines and Geology, Bulletin* 94, 57 p.

Miller, R.R., 1981. Co-evolution of deserts and pupfishes (genus *Cyprinodon*) in the American Southwest. In: *Fishes in North American Deserts*. R.J. Naiman and D.L. Soltz (Eds.). Wiley, New York, pp. 39–94.

Morris, R.L, D.A. Devitt, A.M. Crites, G. Borden, and L.N. Allen, 1997. Urbanization and water conservation in Las Vegas Valley, Nevada. *Journal of Water Resources Planning and Management*, pp. 189–195.

Morrison, R.B., 1991. Quaternary geology of the southern Basin and Range province. In: *Quaternary Nonglacial Geology; Conterminous U.S.* R.B. Morrison (Ed.). The Geology of North America, v. K-2. Geological Society of America, Boulder, Colorado, pp. 353–371.

Muhs, D.R., C.A. Bush, S.D. Cowherd, and S. Mahan, 1995. Geomorphic and geochemical evidence for the source of sand in the Algodones dunes, Colorado Desert, southeastern California. In: *Desert Aeolian Processes*. V.P. Tchakerian (Ed.). Chapman & Hall, London, pp. 36–74.

Nielson, J., and G. Kocurek, 1986. Climbing zibars of the Algodones. *Sedimentary Geology* 48: 1–15.

Nielson, J., and G. Kocurek, 1987. Surface processes, deposits, and development of star dunes; Dumont dune field, California. *Geological Society of America Bulletin* 99: 177–186.

Oberlander, T.M., 1977. Origin of segmented cliffs in massive sandstones of southeastern Utah. In: *Geomorphology in Arid Regions*. D.O. Doehring (Ed.). Allen and Unwin, Boston, pp. 79–114.

Orme, A.R., 1973. *Coastal dune systems of northwest Baja California, Mexico*. U.S. Navy, Office of Naval Research, Washington, D.C. 43 p.

Orme, A.R., and V.P. Tchakerian, 1986. Quaternary dunes of the Pacific coast of the Californias. In: *Aeolian Geomorphology*, W.G. Nickling (Ed.), Allen and Unwin, London, pp. 149–175.

Patton, P.C., and P.J. Boison, 1986. Process and rates of Holocene terrace formation in Harris Wash, Escalante River Basin, south-central Utah. *Geological Society of America Bulletin* 97: 369–378.

Patton, P.C., N. Biggar, C.D. Condit, M.L. Gillam, D.W. Love, M.N. Machette, L. Mayer, R.B. Morrison, and J.N. Rosholt, 1991. Quaternary geology of the Colorado Plateau. In: *Quaternary Nonglacial Geology; Conterminous U.S.* R.B. Morrison (Ed.). The Geology of North America, v. K-2. Geological Society of America, Boulder, Colorado, pp. 373–406.

Reheis, M.C., J.C. Goodmacher, J.W. Harden, L.D. McFadden, T.K. Rockwell, R.R. Shroba, J.M. Sowers, and E.M. Taylor, 1995. Quaternary soils and dust deposition in southern Nevada and California. *Geological Society of America Bulletin* 107(9): 1003–1022.

Reid, J.S., R.G. Flocchini, T.A. Cahill, and R.S. Rut, 1994. Local meteorological, transport, and source aerosol characteristics of late autumn Owens Lake (dry) dust storms. *Atmospheric Environment* 28 (9): 1699–1706.

Rendell, H.M., N. Lancaster, and V.P. Tchakerian, 1994. Luminescence dating of Late Quaternary aeolian deposits at Dale Lake and Cronese Mountains, Mojave Desert, California. *Quaternary Geochronology* (*Quaternary Science Reviews*) 13: 417–422.

Rinehart, R.B., and D.A. McFarlane, 1995. Early Holocene vegetation record from the Salton Basin, California. *Quaternary Research* 43: 259–262.

Rowlands, P., H. Johnson, E. Ritter, and A. Endo, 1982. The Mojave Desert. In: *Reference Handbook on the Deserts of North America*. G.L. Bender (Ed.). Greenwood Press, pp. 103–145.

Schmidt, R.H., Jr., 1979. A climatic delineation of the "real" Chihuahuan Desert. *Journal of Arid Environments* 2: 243–250.

Schmidt, R.H., Jr., 1989. The arid zones of Mexico: Climatic extremes and conceptualization of the Sonoran Desert. *Journal of Arid Environments* 16: 241–256.

Schumm, S.A., and R.J. Chorley, 1964. The fall of Threatening Rock. *American Journal of Science* 262: 1041–1054.

Schumm, S.A., and R.J. Chorley, 1966. Talus weathering and scarp recession in the Colorado Plateaus. *Zeischrift für Geomorphologie* 10: 11–36.

Schwarz, H.E., J. Emel, W.J. Dickens, P. Rogers, and J. Thompson, 1991: Water quality and flows. In: *The Earth as Transformed by Human Action*. B.L. Turner (Ed.). Cambridge University Press, Cambridge, pp. 253–270.

Sharp, R.P., 1966. Kelso Dunes, Mojave Desert, California. *Geological Society of America Bulletin* 77: 1045–1074.

Sharp, R.P., 1980. Wind-driven sand in Coachella Valley, California: Further data. *Bulletin of the Geological Society of America* 91: 724–730.

Shreve, F., 1964. Vegetation of the Sonoran Desert. In: *Vegetation and Flora of the Sonoran Desert*, v. 1. F. Shreve and I.L. Wiggins (Eds.). Stanford University Press, Stanford, pp. 1–186.

Shroder, J.F., and R.E. Sewell, 1985. Mass movement in the La Sal Mountains, Utah. In: *Contributions to Quaternary geology of the Colorado Plateau*. G.E. Christenson, C.G. Oviatt, J.F. Shroder, and R.E. Sewell (Eds.). Utah Geological and Mineralogical Survey Special Studies 64, pp. 49–85.

Smith, H.T.U., 1967. *Past versus present wind action in the Mojave Desert region, California*. Air Force Cambridge Research Laboratories Publication AFCRL-67-0683, Bedford, Massachusetts, 34 p.

Smith, R.S.U., 1982. Sand dunes in North American deserts. In: *Reference Handbook on the Deserts of North America*. G.L. Bender (Ed.). Greenwood Press, Westport, Connecticut, pp. 481–526.

Smith, R.S.U., 1984. Eolian geomorphology of the Devils Playground, Kelso Dunes, and Silurian Valley, California. In: *Surficial geology of the eastern Mojave Desert, California*. J.C. Dohrenwend (Ed.). Geological Society of America 97th Annual Meeting Guidebook, Field Trip 14, pp. 162–174.

St. Amand, P., L. Mathews, C. Gains, and R. Reinking, 1986. Dust storms from Owens and Mono Lakes. *Naval Weapons Center Technical Publication* 6731.

Sykes, G., 1937. *The Colorado Delta*. American Geographical Society Special Publication No. 19.

Tchakerian, V.P., 1991. Late Quaternary aeolian geomorphology of the Dale Lake sand sheet, southern Mojave

Desert, California. *Physical Geography* 12: 347–369.

Thompson, D.G., 1929. The Mohave Desert Region, California. A Geographic, Geologic, and Hydrologic Reconnaissance. *United States Geological Survey Water Supply Paper* 578, 759 p.

van Andel, T.H., 1964. Recent marine sediments of the Gulf of California. In: *Marine geology of the Gulf of California.* T.H. van Andel and G.C. Shor (Eds.) American Association of Petroleum Geologists, Memoir 3, pp. 216–310.

Van Devender, T.R., 1990. Late Quaternary vegetation and climate of the Sonoran Desert, United States and Mexico. In: *Packrat Middens: The Last 40,000 Years of Biotic Change.* J.L. Betancourt, T.R. Van Devender, and P.S. Martin (Eds.). University of Arizona Press, Tucson, pp. 134–166.

Villa, N., R.I. Dorn, and J. Clark, 1995. Fine material in rock fractures: Aeolian dust or weathering? In: *Desert Aeolian Processes.* V.P. Tchakerian (Ed.). Chapman & Hall, London, pp. 219–231.

Wang, Y., E. McDonald, R. Amundson, L. McFadden, and O. Chadwick, 1996. An isotopic study of soils in chronological sequences of alluvial deposits, Providence Mountains, California. *Geological Society of America Bulletin* 108 (4): 379–391.

Ward, A.W., and R. Greeley, 1984. Evolution of the yardangs at Rogers Lake, California. *Bulletin of the Geological Society of America* 95: 829–837.

Waters, M.R., 1983. Late Holocene lacustrine chronology and archaeology of ancient Lake Cahuilla, California. *Quaternary Research* 19: 373–387.

Webb, E.L., 1969. Geology of Sierra de Samalyuca, Chihuahua, Mexico. In: *Guidebook of the border region: Albuquerque.* D.A. Cordoba, S.A. Wengard, and J. Shomaker (Eds.). New Mexico Geological Society, 20th field conference, pp. 176–181.

Wells, S.G., L.D. McFadden, and J.C. Dohrenwend, 1987. Influence of late Quaternary climatic changes on geomorphic and pedogenic processes on a desert piedmont, eastern Mojave Desert, California. *Quaternary Research* 27: 130–146.

Wells, S.G., L.D. McFadden, and J.D. Schultz, 1990. Eolian landscape evolution and soil formation in the Chaco dune field, southern Colorado Plateau, New Mexico. *Geomorphology* 3: 517–546.

Wells, S.G., L.D. McFadden, C.T. Olinger, and J. Poths, 1994. Use of cosmogenic ^{3}HE to understand desert pavement formation. In: *Geological Investigations of an Active Margin.* S.F. McGill and T.M. Ross (Eds.). Geological Society of America Cordilleran Section Guidebook, pp. 201–205.

Williams, J.D., J.P. Dobrowolski, and N.E. West, 1995a. Microphytic crust influence on interrill erosion and infiltration capacity. *Transactions of the ASAE* 38: 139–146.

Williams, J.D., J.P. Dobrowolski, N.E. West, and D.A. Gillette, 1995b. Microphytic crust influence on wind erosion. *Transactions of the ASAE* 38: 131–137.

Woodhouse, C.A., 1997. Winter climate and atmospheric circulation patterns in the Sonoran Desert region, USA. *International Journal of Climatology* 17: 859–873.

Yeager, E.C., 1957. *The North American Deserts.* Stanford University Press, Stanford.

Zhu, C., R.K. Waddel, Jr., I. Star, and M. Ostrander, 1998. Responses of ground water in the Black Mesa basin, northeastern Arizona, to paleoclimatic changes during the late Pleistocene and Holocene. *Geology* 26 (2): 127–130.

Zimbelman, J.R., S.H. Williams, and V.P. Tchakerian, 1995. Sand transport paths in the Mojave Desert, southwestern United States. In: *Desert Aeolian Processes.* V.P. Tchakerian (Ed.). Chapman & Hall, London, pp. 101–129.

20

The Mediterranean Environment of Greater California

Amalie Jo Orme

North America's Mediterranean-type environment occupies a long narrow portion of the continent, bordered on the west by the Pacific Ocean and extending eastward into mountainous terrain that reaches over 4000 m above sea level in the Sierra Nevada. Covering more than 350,000 km^2 in area and 1600 km in latitude from southern Oregon (43°N) to Baja California Norte (30°N), the region, herein termed Greater California, never exceeds 300 km in width and in the south is little more than 100 km wide. Because of the rain shadow imposed by its high eastern mountains, the region abuts directly eastward against desert terrain of the Basin and Range Province, namely, the temperate Nevada Desert in the north and the subtropical Mojave and Sonoran Deserts farther south (fig. 20.1). To the north, the region merges into the temperate, mostly coniferous, forests of the Pacific coast; to the south, into the coastal desert of central Baja California.

Unlike the classic Mediterranean region of southern Europe and northern Africa, Greater California has no extensive east-west seaway, no large islands, and no internal peninsulas or enclosed seas. Instead, the dominant tectonic grain has imposed a series of northwest to southeast-trending mountain ranges and valleys that parallel the Pacific coast, modifying the inland penetration of oceanic influences. In this respect, it is more comparable to the Mediterranean region of central Chile, which also lies along an active mountainous plate margin. In other respects, however, Greater California is typically Mediterranean. It experiences warm, dry summers and cool, wet winters, with variability within the region reflecting latitude, elevation, aspect, and proximity to the ocean. In general, precipitation increases from south to north and from the coast to the mountains, whereas coastal fog is generated by advection across the cool California current. The adaptation and diversity of plant and animal life within this environment are also noteworthy, although, since European settlement in the late eighteenth century, many plant species have become rare or endangered and some are now extinct. Animal life has experienced a similar fate, with shrinkage of specific niches and expansion of others.

This chapter examines those features that give Greater California its distinctive physical geography, namely, the strongly seasonal climate and hydrologic regime, and the related ecological responses expressed in the vegetation cover and related ecosystem diversity. Greater California is the only region of North America characterized by winter rains and summer drought, where most streams flow vigorously in winter but atrophy in summer, and the only region where the vegetal response is adapted to such seasonality. Other features of this landscape are less unique and not discussed in detail. For example, the region's tectonic framework and rugged terrain, though spectacular, are more or less shared with neighboring parts of the continent's Pacific Rim. Certainly, the transfer of a narrow

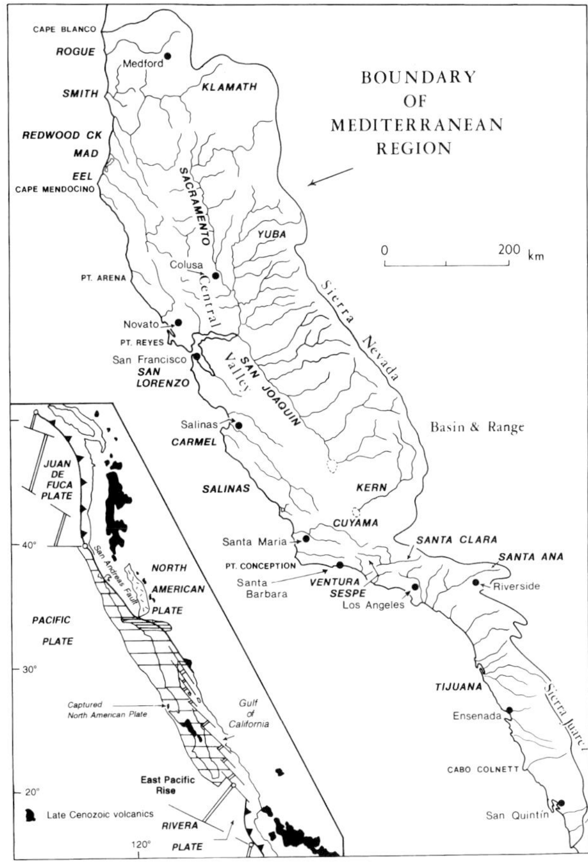

Figure 20.1 The Mediterranean region of Greater California with (inset) tectonic setting.

sliver of the North American plate to the Pacific plate and related seismic activity are dramatic, but these do not give the region its specific Mediterranean character. However, the landscape's response to strongly seasonal climatic and hydrologic forcing, reflected in drought, fire, floods, erosion, and mass movement, is distinctly Mediterranean. These natural hazards and certain distinctive human impacts on this environment are visited in conclusion.

20.1 Evolution of the Environment

20.1.1 Tectonic Framework

The climatic and biotic character of the Mediterranean region is linked closely to the Cenozoic tectonic evolution of western North America (see chapter 1). Located along the leading western edge of a shifting continent, the region has experienced appreciable uplift and deformation during Cenozoic time. Significantly, over the past 30 Ma (million years before present), the North American plate has progressively overridden the subducting Farralon plate and, as the East Pacific Rise has in turn passed beneath the continent, so a sliver of the Californias has been captured by the Pacific plate (fig. 20.1, inset). With the Rise now located in the Gulf of California and the captured plate bounded to the east by the San Andreas fault zone, Baja California and much of coastal California are rafting northwest at about 6 cm yr^{-1} relative to the North American plate. The Peninsular Ranges, western Transverse Ranges, southern Coast Ranges, and Point Reyes Peninsula are part of this rafting sliver, rising and deforming in response to stresses that are most severely expressed in the western Transverse Ranges (Nicholson et al., 1994; Orme, 1998).

North of Cape Mendocino, however, the East Pacific Rise still lies offshore such that the Juan de Fuca plate, a remnant of the Farallon plate, continues to subduct beneath the North American plate, giving rise to active volcanism in the Cascade volcanic arc, notably expressed in the volcanic piles of Mt. Shasta and Mt. Lassen. Farther south, the northern Coast Ranges, the Central Valley, and the Sierra Nevada remain part of the North American plate, but their proximity to the plate boundary during later Cenozoic time has given them much structural complexity. Neogene uplift raised the Sierra Nevada from around 1000 m above sea level at 25 Ma to 3500 m by 2 Ma, and added a further 500 m during the Quaternary (Unruh, 1991), forming a massive precipitation barrier along the region's eastern margin. Meanwhile, the shallow Pliocene sea in the Central Valley was confined by uplift of the Coast Ranges to the west and transformed by debris from the rising Sierra Nevada into an alluvial plain whose remnant lakes have been reclaimed only in historic time.

20.1.2 Climate History

Tectonic evolution has modified and often enhanced the Mediterranean nature of the region's climate. Just as today's pattern of cool, moist winters and warm, dry summers is influenced by latitude and relief, there is evidence for similar effects during the Neogene. Fossil records of Mediterranean-type vegetation can be traced back at least into the Miocene as shown by warm chaparral species in the San Francisco Bay area that have been dated to ≈6 Ma (Axelrod, 1989). Similarly, flora dated at 4.5 Ma from the Central Valley contain taxa that formed an oak-laurel woodland on floodplains and nearby Sierra Nevada foothills (Axelrod, 1980).

However, the nature and persistence of Mediterranean conditions have been profoundly affected by the dramatic climate shifts of the past two million years. With multiple warm-cold oscillations during this time, the region's higher and more northerly areas, namely, the Sierra Nevada, Cascade Range, Trinity Alps, and perhaps a small part of the Transverse Ranges, repeatedly experienced glacier ice, frost climates, and related hydrologic changes during cold stages. The ecology of these areas postdates the glacier retreat and at higher elevations has had to adapt to an edaphic legacy of glacial and periglacial deposits (see chapters 3 and 4).

Although most of the Mediterranean region was not directly affected by glacier ice or frost climates, it did not escape the attendant climatic fluctuations of the Quaternary Period. During cold stages, mean annual temperatures were lower, annual precipitation higher, and seasonality less pronounced than today. Such conditions favored hygrophilous vegetation at the expense of drought-tolerant forms. With more assured water supplies, there may also have been an increase in faunal diversity and abundance, as suggested by the rich Rancholabrean fauna of southern California (see chapter 3). During warm stages, conditions were likely similar to those of today, perhaps slightly warmer during the Last Interglacial (≈130–115 ka), and the sclerophyllous vegetation so typical of the modern chaparral developed.

20.2 Climatic Character

20.2.1 General Atmospheric Circulation

The atmospheric circulation over Mediterranean North America is strongly influenced by the proximity of the Hawaiian high-pressure cell. Summers are warm and dry, especially in the interior, as high pressure dominates the region. By contrast, when the cell's influence diminishes in winter, cyclonic activity over the north Pacific Ocean moves south and east with the polar jet stream, bringing cool, moist conditions to the region. Winter storms may be quite intense, yielding moderate to heavy rainfall to coastal and foothill locations, and snow to higher elevations. Those winter storms that stall offshore draw in further moisture from Pacific tropical waters, enhancing the magnitude and intensity of precipitation.

In addition to normal winter moisture sources, the region also experiences water years related to El Niño–Southern Oscillation (ENSO) events during which precipitation may double average annual values. When a plume of warm water migrates eastward from the western Pacific Ocean near Indonesia to the west coast of South America on an erratic 4- to 7-year cycle, sea-surface temperatures in the eastern Pacific may rise 3–7°C above normal. Winter storms gain additional moisture from this influx of tropical warm water into cyclonic systems that are migrating south and east from the north Pacific. This combination of warm sea-surface temperatures and traveling winter cyclones enhances the region's precipitation, as seen during the 1982–83 and 1997–98 water years. Although numerous El Niño events have been documented or inferred over the past century, the net spatial effect of such events varies considerably. Although flooding and mass movement commonly accompany major El Niño events in this part of North America, the erratic nature of the associated storms yields highly variable erosional and depositional responses. Further, even though cumulative rainfall associated with El Niño may be quite high, some of the largest floods of the past century, notably in 1968–69, have in fact been associated with relatively weak or even non–El Niño conditions (fig. 20.2).

Further complicating the temporal pattern of precipitation are years linked to La Niña, an ocean-atmosphere phenomenon associated with the reversal of El Niño conditions, which yields below-normal moisture in California, for example, during the 1998–99 water year. Such circumstances often tip the balance toward drought, with diminishing water availability affecting agriculture and domestic usage. Longer-term records from tree-ring chronologies

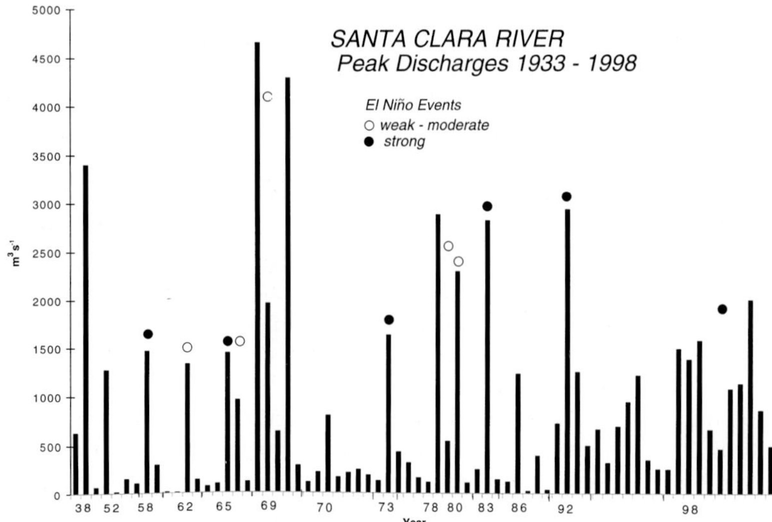

Figure 20.2 Santa Clara River peak discharges, 1933–98: discrete discharge events exceed the average peak discharge for the period by 1.5 standard deviations; thus there may be several peaks in a "wet" water year such as 1997–98, whereas "dry" years do not appear (data from U.S. Geological Survey, 1999a).

(Haston et al., 1988) and offshore sediment cores (Schimmelmann et al., 1998) suggest that alternations of excess precipitation and protracted drought may have affected the region throughout the late Holocene.

20.2.2 Climatic Diversity

The Mediterranean climate of Greater California has long been classified as Cs in the Köppen system; this climate has dry warm to hot summers, mild modestly wet winters, and abundant sunshine especially in summer (Trewartha, 1968). This climatic province has in turn been divided locally into a coastal (Csb) and an interior (Csa) region. The Csb region has cool summers, mild winters, reduced annual temperature range, and precipitation augmented by coastal advection fog. Average temperatures for the warmest summer month are <22°C but at least 4 months of the year have mean temperatures >10°C. The Csa region has warmer summers, cooler wetter winters and greater annual temperature range than the Csb region. Although little advection fog penetrates the interior, the Central Valley is prone to dense radiation fogs. Average temperatures for the warmest summer month are >22°C, and may exceed 30°C, but average winter days are still >0°C.

Although the previous description is useful, closer examination of climate records reveals much variability in average, maximum, and minimum temperatures on both an annual and seasonal basis. Moreover, the magnitude and intensity of moisture imposed on the region are also more variable. Though winter precipitation derived from Pacific cyclonic systems still dominates, it is often concentrated in powerful but discrete storm cells and strongly enhanced by relief. These facts favor a more sophisticated approach to climatic regionalization.

For present purposes, seven distinct climatic regions are defined for Greater California, each with its own expression of a Mediterranean-type climate (fig. 20.3). Records for two weather stations from each region for the period 1961–90 illustrate average monthly temperature and average monthly precipitation, with temperature maxima, minima, and range discussed subsequently. These stations cover a range of latitude, elevation, and distance from the sea. Mean monthly solar radiation for four stations from southern Oregon to southern California is shown in figure 20.4.

The Klamath-Cascade region extends from southern Oregon near 43°N to inland northern California at about 40°N. Removed from direct marine influence, this rugged region has average maximum temperatures >18.0°C and average minimum temperatures >–1.2°C. Annual temperature ranges from 12.4°C at Roseburg, the lowest and most northerly station, to 19.4°C at Burney, the highest and most southerly station. Rainfall is winter-dominated and ranges from 488 mm yr^{-1} at Ashland to 699 mm yr^{-1} at Burney.

The North Coast region, extending from 42°N to below 39°N, is dominated by marine influences and frequent fogs. On average, it experiences maximum temperatures >14.6°C and minimum temperatures >4.0°C. Annual temperature ranges are between 6.4°C at Eureka on the coast and 18.3°C at Willits, 35 km inland. Annual precipitation is often heavy, reaching a coastal maximum of 1656 mm yr^{-1} at Crescent City, with more at higher elevations.

The Central Coast region, in contrast, has a distinct marine component and a notably drier inland component. Extending from 39°N near San Francisco to below 35°N at Point Arguello, the Central Coast region has average annual maximum temperatures of >18°C, whereas average minimum temperatures are >13°C. Annual temperature ranges at the coast are about 9°C, whereas those inland may reach 20°C. Precipitation decreases southward along the coast, from 500 mm yr^{-1} at San Francisco to 400 mm yr^{-1} at Morro Bay (35°N), and to 316 mm yr^{-1} at Point Arguello. Annual rainfall varies inland with relief, as shown by the difference between King City, at 97 m above sea level in the Salinas Valley in the lee of the Santa Lucia Range, which averages 290 mm yr^{-1}, and Priest Valley, only 40 km to the east, at 701 m in the west-facing foothills of the Diablo Range, which averages 510 mm yr^{-1}.

The Transverse Ranges, unique within California as an east- to west-trending assemblage of mountains and valleys, extend east from Pt. Arguello (121°W) to the northern margin of the Salton Trough (116°W). In the San Bernardino Mountains, which rise to 3505 m, and in the San Emigdio Mountains, rising to 2692 m, both elevation and interior versus coastal contrasts are expressed in the climate. The average annual temperature range at Santa Barbara on the coast is <12°C, whereas that of interior Ojai at 228 m in elevation approaches 17.5°C. Average annual maximum temperatures in the region are between 15°C and 19°C. Clear orographic effects and expected increases in aridity with more interior locations are somewhat muted. Because winter storms enter the region mainly from the west and southwest, seaward-facing slopes experience orographic enhancement of rainfall but, because of the broken nature of many ranges, amounts are highly variable. For example, Redlands, lying at 401 m on south-facing foothills of the San Bernardino Mountains 75 km inland from the Pacific Ocean, receives an average annual precipitation of 325 mm. Ojai, nestled at 228 m within a small valley 20 km north of the Ventura coast, receives 525 mm of rainfall annually. Rainfall variability is explained in part by storm tracks and locations where storms apparently stall, gather more moisture, and experience rapid temperature changes. Although snow contributes to water budgets at higher elevations, snowfall in the Transverse Ranges is limited because of relatively warm winter temperatures.

The Mediterranean part of the Peninsular Ranges extends from 34°N southward to ≈30°N in Baja California. Coastal locations such as San Diego (32°N, 4 m elevation) reflect oceanic influences with an annual temperature

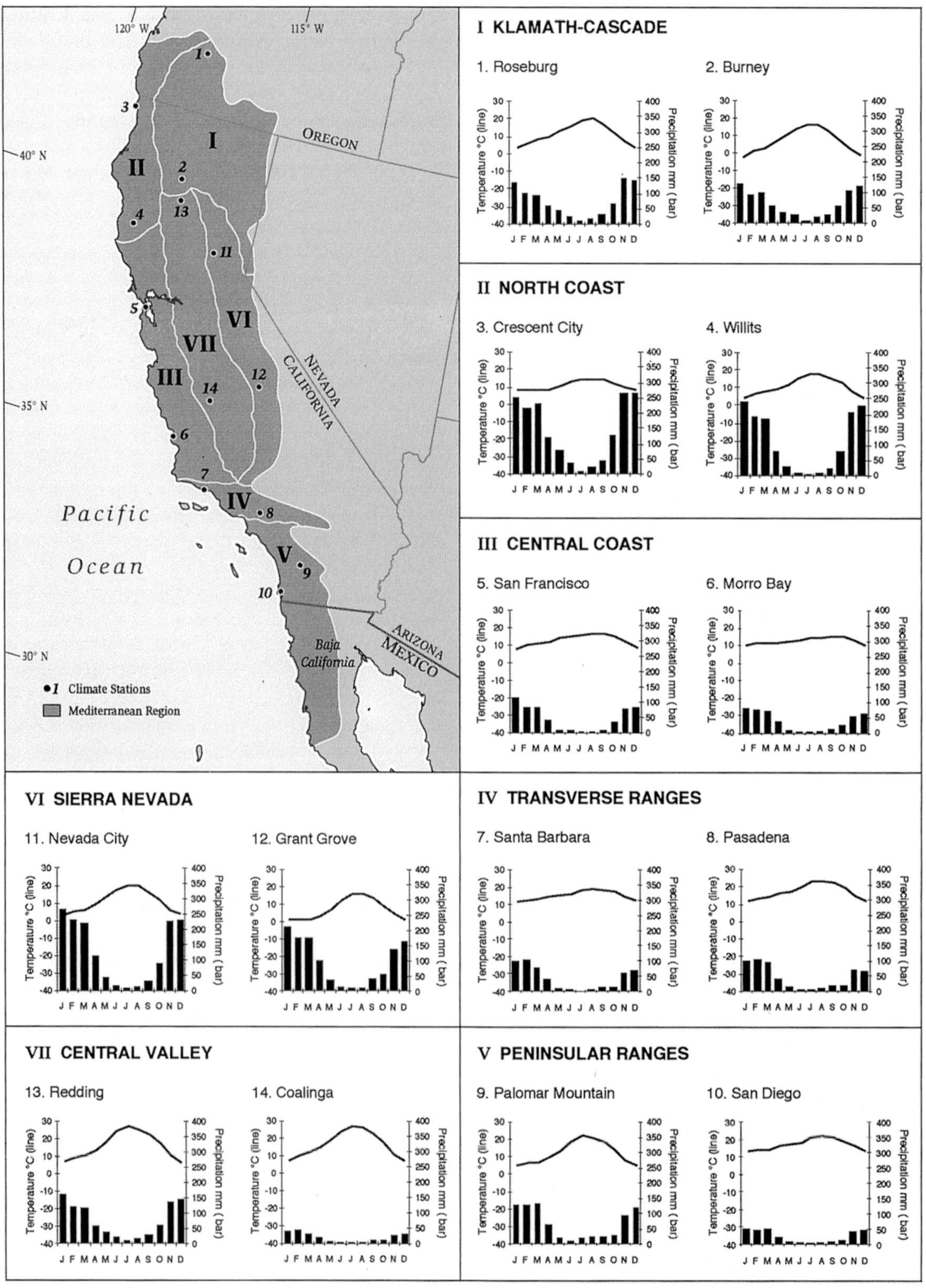

Figure 20.3 Climate regions and characteristic temperature and precipitation regimes for Greater California for the period 1961–90 (data from Worldclimate, 1999).

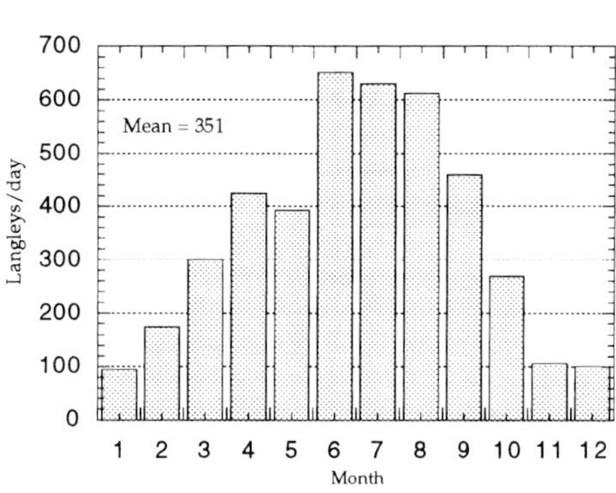

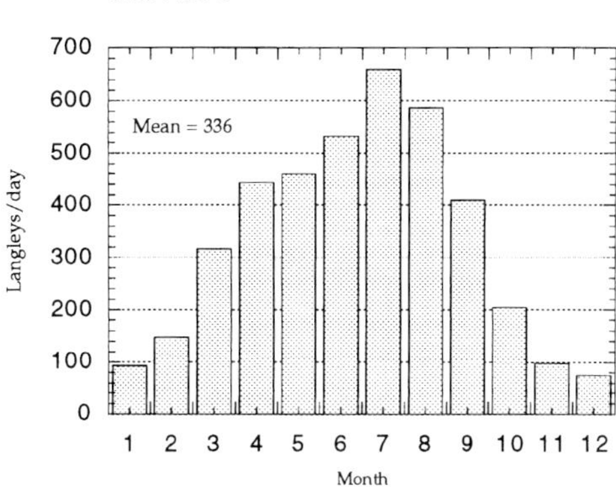

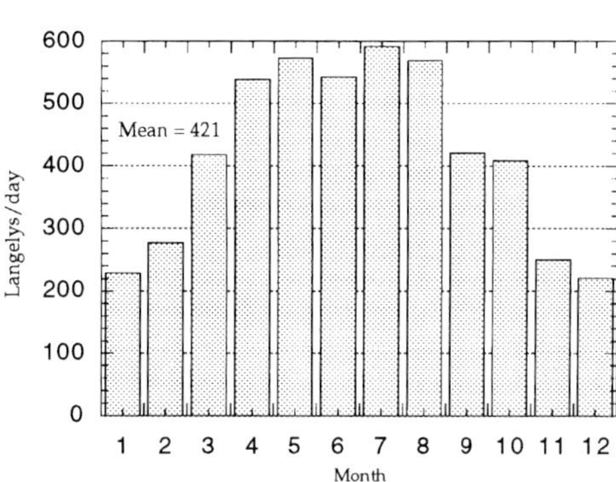

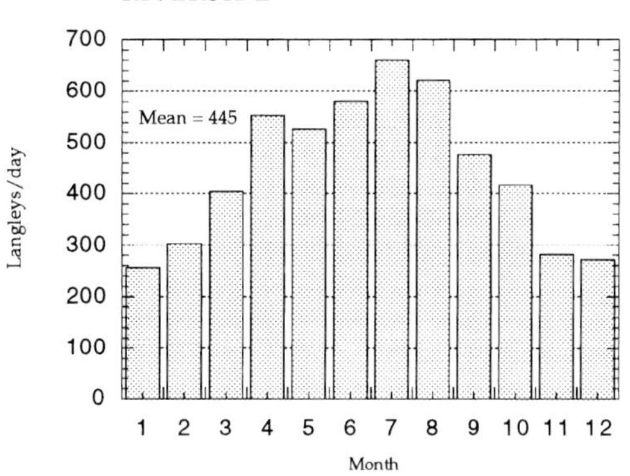

Figure 20.4 Monthly solar radiation for four stations in Greater California. Note change of scale (data from CIMIS, 1999).

range of 7.3°C, from a maximum of 21.5°C to a minimum of 14.2°C, and an annual precipitation of 251 mm. In contrast, Palomar Mountain (49 km inland, 1691 m), with an annual temperature range of 13°C and 713 mm of precipitation, is typical of more interior locations, reflecting distance from the sea and orographic enhancement of Pacific storms. The lee side of the Peninsular Ranges is much drier, and a rapid transition toward desert is reflected in thinning vegetation cover.

Extending from 35°N to 40°N, the Sierra Nevada offers dramatic contrasts in temperature and precipitation. Grant Grove, located near 37°N at 2011 m, has on average a maximum temperature of 13.3°C, a minimum temperature of 2.3°C, and annual precipitation of 1053 mm. Farther north, within the western foothills, Nevada City at 29°N and 847 m, experiences on average an annual maximum temperature of 19.3°C, an annual minimum temperature of 4.4°C, and precipitation of 1421 mm yr^{-1}. Yet such figures do not adequately characterize the effects of elevation within the Sierra Nevada. Annual snowpack above 3000 m increases toward the central and northern parts of the range, thus contributing additional water to the region. Moreover, summer orographic rainfall and snow may add substantial moisture within a single storm event. By contrast, the eastern foothills of the Sierra Nevada, in the temperate desert just beyond the Mediterranean region, display a profound rainshadow effect, with annual precipitation less than 400 mm.

Throughout its length, the 750-km-long Central Valley lies within the rain shadow of the Coast Ranges. Thus, precipitation totals are never as high as in the ranges to the west. Further, as the summer drought season lengthens southward, precipitation decreases from 845 mm y^{-1} at Redding (40°N, 153 m elevation) to only 198 mm yr^{-1} at Coalinga (36°N, 204 m elevation). Most moisture, when it comes, is in the form of rain rather than snow. Annual differences in minimum and maximum temperatures from the northern to the southern ends of the valley are limited, from 14°C at Redding to 16°C at Coalinga. However, almost uniform annual maximum temperatures are found throughout the region: Coalinga (25.7°C), Fresno (36°N, 102 m, 24.7°C), Modesto (37°N, 27 m, 23.7°C), and Redding (23.7°C). Annual minimum temperatures average 9–10°C throughout.

20.3 Water Resources

20.3.1 Hydroclimate

With such diverse precipitation and evaporation patterns, surface and subsurface water resources vary across the region. As may be expected, there is more surface water in the north and west, whereas water yields decrease and become more erratic to the south. Superimposed on this pattern, however, are deviations from the average in the form of major El Niño flood events, at least ten of which have occurred during the past century (NCDC, 1999) along with episodes of drought. Such deviations rarely occur simultaneously across the region, while some unusually rainy years appear to be independent of ENSO forcing. Moreover, over the longer term before instrumentation, tree-ring data indicate several periods of extended drought. Water resources are further modified by human activity. Water transfers and the presence of hundreds of dams and thousands of debris basins and other control structures have altered average and peak discharges, sediment loads, and groundwater flows in many watersheds (see chapter 22).

The availability of water for river discharge depends on a number of factors, including rainfall, evapotranspiration, drainage basin shape, seepage, and diversion for human use. Clearly, in an environment with distinct wet and dry seasons, the occurrence of drought-related conditions figures prominently in surface water supplies. Lynch (1931) recognized the region's propensity for drought in a compendium of data gleaned from late eighteenth and early nineteenth century mission records and later Euroamerican explorers' diaries. He estimated that 30% of all seasons between 1769 and 1930 had rainfall less than 80% of average. Although a modified Palmer Drought Severity Index is not without its flaws (the lag between precipitation and runoff is not factored, and runoff is underestimated), the 1895–1998 record is highly informative. For example, the dry periods 1897–99, 1959–62, and 1984–90 appear throughout the region. Similarly, the very wet 1905–09, 1941–42, and 1982–83 periods are found regionwide. Moreover, the 1997–98 El Niño event, which saw a combined land and ocean anomaly of 0.62°C (Livezey et al., 1997), affected the entire region by raising rainfall and consequent streamflow significantly above average. However, the variability in moisture and surface runoff that characterizes a region spanning 13° of latitude and 4300 m of elevation is also expressed in this record. For example, the above-average rainfall conditions of 1937–38 and 1968–69, which brought record floods to southern California, are seen as near normal moisture conditions in the more northerly and interior portions of the region.

All factors considered, records of streamflow throughout the region still closely reflect the distinct seasonal patterns in rainfall and drought. Figure 20.5 illustrates annual surface discharges for four rivers from northernmost California to northern Mexico. Several factors are apparent, including the difference between the perennial flow regimes of the more northerly rivers and the intermittent flows of the more southerly streams, the dramatic range of discharges experienced by each river during the period of record, and the periodic though disjunct recurrence of

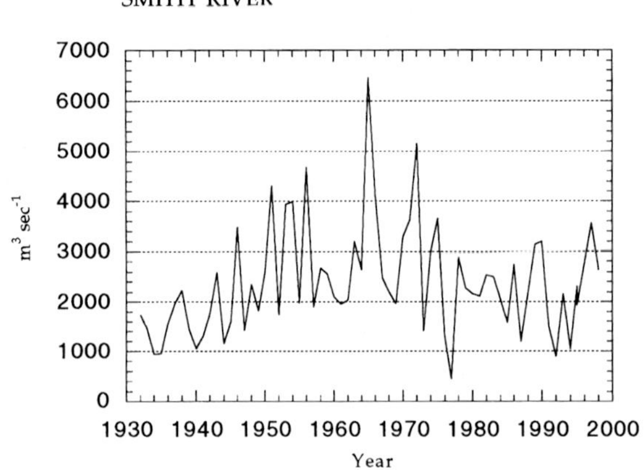

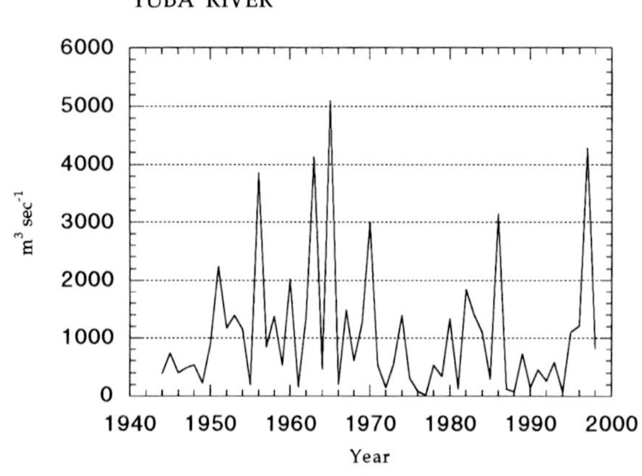

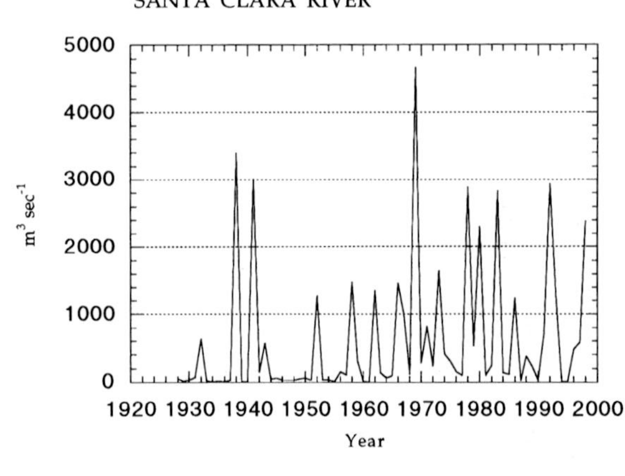

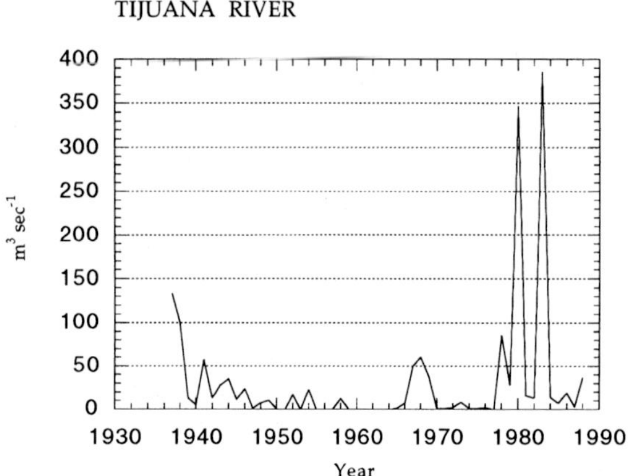

Figure 20.5 Annual discharge for four rivers in Greater California, 1920–98. Note change of scale (data from U.S. Geological Survey, 1999a).

major flow events. Figure 20.6 illustrates the profound difference between above-average and below-average discharge years for the Santa Clara River. These differences clearly are more pronounced in the drier parts of the region, where, for example, during the brief drought of the mid-1980s, the discharges of the Santa Clara and Tijuana rivers shrank to mere trickles during the wet season. The Santa Clara River (fig. 20.7), is typical of rivers in the subhumid south where neglible summer flow shrinks within the relict, anastomosing channels of the winter floodplain, beyond which citrus orchards of the higher floodplain are also prone to occasional winter inundation.

Suspended sediment loads from rivers in the region in turn reflect the dependence of discharge on seasonal inputs of water (table 20.1). As expected, large volumes of material are carried during exceptionally wet years, producing sediment plumes at the coast in which silt and clay move far offshore while sand and gravel are reworked onshore toward the beaches. The load transported is highly variable, especially for those rivers with the distinct "flashy" component found in the south where the Ventura River's total sediment yield for a dry year, 960 tonnes in 1977, may be compared with that for a wet year, 3.6 million tonnes in 1978 (Hill and McConaughy, 1988). The bulk of the load in both years was in the form of suspended sediment, and in 1978 over 96% of the sediment load occurred in response to just two storm events in February and March. Similarly, the nearby Santa Clara River discharged 47.6 million tonnes of sediment during the winter floods of 1969, much of it from hillslope failures and temporary storage within the floodplain (fig. 20.7), compared with >2 million tonnes during relatively dry years.

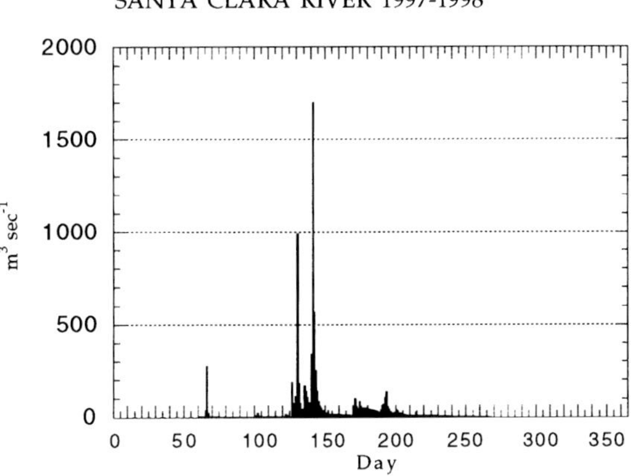

Figure 20.6 Discharge for Santa Clara River, California, for a "wet" year (1997–98) and a "dry" year (1986–87). Note change of scale (data from U.S. Geological Survey, 1999a).

20.3.2 *Water Availability*

The provision of water in a region with a distinctive dry season, a growing population, and a major commitment to agriculture, clearly represents an important issue. Water availability, especially for areas under cultivation or with large urban populations, is severely impacted during the dry season and has long required a special commitment to water transfers (see chapter 22). Winter rainfall in the Central Valley, for example, is not enough to sustain the many fruits and vegetables that are grown year round. Developments such as the federally sponsored Central Valley Project (with 4 dams, 24 reservoirs, and nearly 1000 km of canals) and the state-sponsored State Water Project (with 25 dams and reservoirs, 1100 km of aqueducts, and 8 hydroelectric power plants) represent major modifications to surface water supplies designed to maintain both agriculture and urban growth (State of California, 1990). The total irrigated area in the region approaches 800,000 hectares (ha) (U.S. Department of Interior, 1992).

20.4 Ecosystem Diversity and Regionalism

20.4.1 *Diversity and Adaptation Strategies*

As a result of the climate's seasonal nature, ecosystem diversity tends to be high, with plant adaptation reflecting both existing conditions and longer term climate changes. Mediterranean North America contains over 5800 individual plant species, of which at least 58% are endemic to the region (Barbour et al., 1993). A more conservative estimate recognizes 5100 species with ≈50% endemic (Cowling et al., 1996). Regardless of the precise figures, this is a remarkably rich list of species and their encompassing communities in a relatively small area. Greater California has a species density of about 12,000 per million km² compared with a density of 2500 found in the

Figure 20.7 Santa Clara River, California: a summer view looking upstream from the confluence with Sespe Creek. The river drains a 4219-km² basin in the Transverse Ranges (photo: Spence Collection, UCLA).

classical Mediterranean region. However, at least 15% of those species are now rare or endangered, compared with 11% in Mediterranean Europe, and 34 species are now extinct. The distribution of these many species reflects climate, topography, substrate, and human occupance, but clearly the evolution of the Mediterranean ecosystem has also been a story of adaptation to drought, wildfire, and geographic uniqueness associated with the region's geologic history.

Adaptive strategies of plants to a climate with summer drought include the evolution of sclerophyllous leaves (thick leaves with waxlike cuticles that minimize moisture loss), heat-evading foliage (leaves oriented vertically to decrease sunlight receipt, or leaves whose light colors increase albedo from the plant surface), subsurface food and water storage (the presence of a bulb), the evolution of tap roots (to obtain moisture during initial plant growth), shallow subsurface rootlets (which increase the ability to obtain moisture from minimal precipitation and fog), and germination response to wildfires (the development of sprouting systems from root crowns and seeders that depend on the heat or chemicals released by fire).

These adaptive strategies are particularly important in the more drought-prone areas of the region. For example, leaf orientation is critical for avoiding direct solar radiation in certain species of *Arctostaphylos* and *Ceanothus* in southern California. Other plants such as *Salvia leucophylla* have adapted to drought through seasonal dimorphism, that is by developing two leaf types (winter versus summer) that maximize water use and minimize water loss (Westman, 1983). The evolution of shallow, lateral root systems adapted to gathering moisture from ephemeral sources is an important survival strategy used by *Adenostoma fasciculatum* and *Arctostaphylos glauca* (Kummerow, 1981).

Table 20.1 Suspended sediment loads for selected rivers (tonnes × 10^6)

Drainage	Area (km²)	1972–73	1973–74	1976–77	1977–78
Klamath River	21,950	0.25	10.50	0.12	1.00
Redwood Creek	717	0.70	2.00	0.05	0.88
San Lorenzo River	275	0.40	0.08	0.04	0.30
Salinas River	10,764	1.40	0.50	0.04	14.00
Sespe Creek	650	2.60	0.20	0.25	8.00
Santa Clara River	4128	4.00	0.80	0.25	26.00

Source: U.S. Geological Survey (1999b).

Although wildfires may occur anywhere in the region, their recurrence in the drier areas has forced some plants to adapt for survival using either seed sprouting or root-crown sprouting after fire (fig. 20.8). Sweeney (1956) recognized the relationship between fire and the germination of many herb species within the chaparral community. The viability of seeds dormant for several decades and essentially unaffected by moisture was documented by Keeley and Keeley (1986) for many plants, though it is clear that some herbs germinate by contact with burned shrubs, some by contact with chemicals released by plants during and following a fire, and others by contact with ash. Further, with the removal of overstory shrubs and the release of shrub minerals during a fire, the germination of herbaceous species increases significantly (Keeley et al., 1981). The postfire regeneration of shrubs is complicated, with much variation of sprouting versus nonsprouting behavior found in species of *Arctostaphylos*, *Ceanothus*, and *Rhus*. Although heat-activated seed germination has long been known to exist (Hadley, 1961), this behavior varies between species and may depend on other factors such as soil temperature, antecedent moisture, and carbohydrate storage in root systems. Root-crown sprouting represents another regeneration strategy with sprouting at the base, of such plants as *Adenostoma fasciculatum*, *Heteromeles arbutifolia*, *Pinus muricata*, *Quercus*, and *Rhamnus*.

20.4.2 Biotic Communities

The organization and distribution of plants in Mediterranean North America is complex, with considerable variability based on temperature, moisture constraints, substrate, and human disturbance. Dallman (1998) recognizes four major plant forms within the world's Mediterranean regions (shrubland, coastal scrub, woodland, and forest), each expressed in unique communities. In Greater California, these communities are represented by chaparral, coastal sage scrub, oak woodland, and redwood forest (Barbour and Major, 1988). However, the vegetation of the region is not confined to these four communities. For example, though now restricted, native grasslands are still an important component of the vegetation (see chapter 25). Early classification schemes, such as that of Clements (1916), focused on the link between vegetation and climate zones, whereas Cooper (1922), working exclusively in California's sclerophyll groups, proposed the concept of associations based on the dominance of a particular species. Still other systems were based on species units (Wieslander, 1935; Jensen, 1947), climatically inferred communities (Munz and Keck, 1949), habitat types (Thorne, 1976), and plant series (Sawyer and Keeler-Wolf, 1995). More recent schemes identify rare natural communities (the California Natural Diversity Data Base, 2000) and the relationship of vegetation to ecosystem types (Bailey, 1997; see chapter 12).

The richness of the animal communities in the region is also apparent, with some 214 mammals, 540 birds, 77 reptiles, 47 amphibians, and 83 freshwater fish present (Schoenherr, 1992). However, 134 species are listed as endangered (EPA, 1999). Clearly, the impact of humans on animal communities is wide ranging and includes changes in habitat range and quality, alterations in water courses and water chemistry, introduction of nonnative food sources, and predation. Although many habitats and wildlife corridors in or near developed areas have disappeared, some animals such as coyote have adapted to the human presence. Recognizing the influence of climate, distance from the sea, topography, and substrate, discussion of these biotic communities is now linked to the climatic regions outlined previously.

Figure 20.8 Chaparral fire in the Santa Monica Mountains, October 1978 (photo: L.L. Loeher).

20.4.3 Klamath-Cascade (I)

The Klamath-Cascade region contains three distinct subregions based on proximity to the coast, elevation, and dissection by the Klamath, Salmon, and Trinity Rivers: the Klamath Mountains, Cascade Range, and Modoc Plateau. The Klamath subregion includes the Siskiyou Mountains along the California-Oregon border, the Klamath, Salmon and Marble mountains in the center, and the Trinity Alps farther south. Elevations are typically 1500–2100 m and reach 2725 m on Thompson Peak. The Cascade subregion is dominantly volcanic and boasts major cones such as Mt. Shasta (4317 m), which erupted in 1786, and Mt. Lassen (3187 m), which was active from 1914 through 1921. The Cascades receive more precipitation on their western slopes but are increasingly arid east of the main ridgelines. The Modoc Plateau is a volcanic tableland, 1200–1500 m in elevation.

Lying along the western edge of the Klamath region, below 300–600 m on north-facing slopes and on soils derived from sandstone and schist, is the redwood forest. Often in distinct groves, coast redwood (*Sequoia sempervirens*) averages 60 m high and may reach 120 m. Supported by ample rainfall, fog drip (up to 250 mm yr^{-1}), and sheltered conditions, these long-lived trees (500–2000+ years) produce a closed canopy, which limits the undergrowth. Associated with this forest are Douglas fir (*Pseudotsuga menziesii*), western hemlock (*Tsuga heterophylla*), grand fir (*Abies grandis*), and Sitka spruce (*Picea sitchensis*).

Farther inland, mixed evergreen forest interfingers with the redwood forest and rises to 1400 m. Dominated by Douglas fir, with ponderosa pine (*Pinus ponderosa*) and white fir (*Abies concolor*) in moister habitats, this mixed forest contains elevation-related gradients, substrate-related communities, and blends into sagebrush scrub, chaparral, and oak woodlands in the drier east. Typically found on granitic soils are tanoak (*Lithocarpus densiflora*), bush chinquapin (*Chrysolepis sempervirens*), giant chinquapin (*C. chrysophylla*), and canyon live oak (*Quercus chrysolepis*). More xeric areas support fewer trees and favor shrubs such as huckleberry oak (*Quercus vaccinifolia*), California coffeeberry (*Rhamnus californica*), hoary-leaf manzanita (*Arctostaphylos canescens*), wedgeleaf ceanothus (*Ceanothus cuneatus*), and tobacco brush (*Ceanothus velutinus*).

Grading eastward into the Cascade Range and Modoc Plateau, between 300 and 1200 m, sagebrush scrub includes big sagebrush (*Artemisia tridentata*), low sagebrush (*A. arbuscula*), rabbitbrush (*Chrysothamnus nauseosus*), and open stands of ponderosa pine, Jeffrey pine (*Pinus jeffreyi*), and western juniper (*Juniperus occidentalis*). A moist, yellow pine forest dominated by white fir (*Abies concolor*) with sugar pine (*Pinus lambertiana*), ponderosa pine, Jeffrey pine, incense cedar (*Calocedrus decurrens*), Baker cypress (*Cupressus bakeri*), Port Orford cedar (*Chamaecyparis lawsoniana*), and Pacific yew (*Taxus brevifolia*) occurs between 1200–1500 m.

Eastward, the area grades into drier conditions, supporting ponderosa pine, sugar pine, California black oak (*Quercus kelloggii*), and several chaparral species. Within the Cascade Range, Washoe pine (*Pinus washoensis*) appears. At higher elevations (1400–1950 m) in the Klamath region and up to 3100 m in the Cascade Range, lodgepole pine (*Pinus contorta murrayana*), red fir (*Abies magnifica*), noble fir (*Abies procera*), white pine (*Pinus monticola*), mountain hemlock (*Tsuga mertensiana*), and curlleaf mountain mahogany (*Cercocarpus ledifolius*) occur. Above 2100 m in the Klamath Mountains and to 3600 m in the Cascade Range, subalpine vegetation consists of mountain hemlock, foxtail pine (*Pinus balfouriana*), and greenleaf manzanita (*Arctostaphylos patula*). Cascade alpine associations above 3500 m include bilberry (*Vaccinium caespitosum*) and mountain heather (*Phyllodoce* spp).

In a region whose habitats range from moist year round to semiarid, animal life is diverse. Mammal populations include wapiti (*Cervus elaphus roosevelti*), now confined to the coastal redwood forest, black-tailed deer (*Odocoileus hemionus columbianus*), mountain beaver (*Aplodontia rufa*), white-footed vole (*Arborimus albipes*), red tree vole (*Arborimus longicadus*), northern flying squirrel (*Glaucomys sabrinus*), western gray squirrel (*Sciurus griseus*), several chipmunks (*Tamias* spp.) porcupine (*Erethizon dorsatum*), ermine (*Mustela erminea*), wolverine (*Gulo gulo*), badger (*Taxidea taxus*), and river otter (*Lutra canadensis*). The bird population of this forested region includes Oregon ruffed grouse (*Bonasa umbellus*), varied thrush (*Ixoreus naevius*), Vaux's swift (*Chaetura vauxi*), bald eagle (*Haliaeetus leucocephalus*), chestnut-backed chickadee (*Parus rufescens*), and various nuthatches and warblers. Amphibians are dominated by various salamanders, including *Aneides flavipunctutatus*, *Plethodon stormi*, *Hydromantes shastae*, and *Ambystoma macrodactylum*.

20.4.4 North Coast (II)

North Coast ridges and valleys are the wettest part of the Mediterranean region, with over 900 mm of annual rainfall, frequent fog, and perpetually wet soils derived from Franciscan serpentinites, ophiolites, and other rocks that contain abundant magnesium, iron, nickel, and cobalt, and are inherently unstable. This unusual environment supports major stands of coast redwood at lower elevations, joined by Douglas-fir, western hemlock, and Pacific madrone (*Arbutus menziesii*) up to 1200 m. Eastward lies an oak woodland with specific microclimatic preferences. Coast live oak (*Quercus agrifolia*) prefers moister habitats, whereas valley oak (*Q. lobata*) and interior live oak (*Q. wislizenii*) favor drier, more southerly locations. This community, which also includes blue oak (*Q. douglasii*) and Garry oak (*Q. garryana*), passes eastward into foothill woodlands bordering the northern Central Valley.

The "staircase" of plant communities found on marine terraces in Mendocino County illustrates another preference associated with soil age and elevation (Schoenherr, 1992). Grasses and annuals such as owl clover (*Orthocarpus erianthus*) and California poppy (*Eschscholzia californica*) populate the lowest terrace exposed to salt spray and wind. Just beyond the spray zone, a coastal scrub of sticky monkey flower (*Mimulus aurantiacus*), coyote brush (*Baccharis pilularis*), California blackberry (*Rubus vitifolius*), and several lupins is found. Higher still, redwood forest occurs on sandy substrates, pine forest on serpentinite soils, with *Sphagnum* bogs and carnivorous plants including the California pitcher plant (*Darlingtonia californica*) and sundew (*Drosera rotundifolia*) in wetter sites. The highest terraces (120–160 m) are populated by closed-cone Bishop pine (*Pinus muricata*), a probable Pliocene relict on wet nutrient-poor soils, and the slow-growing pygmy cypress (*Cupressus pygmaea*) and Bolander pine (*Pinus contorta bolanderi*).

Mammals are similar to those of the Klamath-Cascade

region and include several species of squirrel, chipmunk, and vole. Additionally, marten (*Martes americana*), ermine (*Mustela erminea*), skunk (*Mephitis mephitis*), and fisher (*Martes pennanti*) appear. The largest mammals also include wapiti, black-tailed deer, and another subspecies of mule deer (*Odocoileus hemionus*). Amphibians include salamanders from the Klamath-Cascade region, plus the California slender salamander (*Batrachoseps attenuatus*), the rough-skinned newt (*Taricha granulosa*), and the red-bellied newt (*T. rivularis*). The most notable reptile in the region is the western pond turtle (*Clemmys marmorata*). Among the gastropods, the large, yellow banana slug (*Ariolimax* spp.) is an important forest floor resident. Birds resemble those of the Klamath-Cascade region, and the coastal forests also harbor spotted owl (*Strix occidentalis*), northern spotted owl (*S. occidentalis caurina*), and marbled murrelet (*Brachyrampus marmoratus*).

20.4.5 Central Coast (III)

The Central Coast from north of San Francisco to Pt. Arguello comprises several ranges and valleys developed in granitic, metamorphic, and sedimentary rocks, defined by en echelon fault systems, and flanked by coastal embayments and marine terraces (fig. 20.9). The coastal strip experiences about 135 fog days annually, half of them during the summer months, but summer drought becomes more pronounced inland. Vegetation thus reflects distance from the coast. Coastal sage scrub, partly dependent on summer fog, comprises black sage (*Salvia mellifera*), buckwheat (*Eriogonum fasciculatum*), sagebrush (*Artemisia californica*), and encelia (*Encelia californica*). With increasing elevation, this scrub grades subtly into a coastal chaparral dominated by manzanita (*Arctostaphylos montereyensis*, *A. pajaroensis*, *A. hookeri*) and ceanothus (*Ceanothus rigidis*, *C. dentatus*). Woodland pockets reveal distinct edaphic, slope, and aspect preferences. Coast live oak prefers the moister coastal slopes, whereas blue oak, interior live oak, and valley oak prefer the more xeric inland sites. Other unique stands include Monterey cypress (*Cupressus macrocarpa*), Gowen Cypress (*C. goveniana*), and several closed-cone pines, namely, Monterey pine (*Pinus radiata*) found on well-drained soils around Monterey and farther south near Cambria, knobcone pine (*P. attenuata*), a fire-dependent tree that prefers fine-grained ultramafic substrates in the Santa Cruz Mountains, and patchy stands of Bishop pine in the Monterey Peninsula and elsewhere. With increasing distance from the sea, more drought-tolerant plant communities appear, especially in interior valleys, including annual grasses such as needlegrass (*Stipa lepida*, *S. pulchra*), California brome (*Bromus carinatus*), and blue wildrye (*Elymus glaucus*), as well as California juniper (*Juniperus californica*), narrowleaf goldenbush (*Ericameria linearifolia*), and California joint fir (*Ephedra californica*).

Figure 20.9 The Big Sur coast of Central California showing descent of rugged Santa Lucia Range onto fragmentary marine terraces and pocket beaches (photo: A.R. Orme).

The larger mammals of this region include bobcat (*Lynx rufus*), mountain lion (*Felis concolor*), gray fox (*Urocyon cinereoargenteus*), wild pig (*Sus scrofa*), coyote (*Canis latrans*), and black-tailed deer. Smaller mammals include squirrel, shrew, bat, mouse, and woodrat (Henson and Usner, 1996). At least 427 bird species live in this region, including nearshore species such as grebe, cormorant, heron, egret, turnstone, plover, tern, and gull, and inland species such as rufous-sided towhee, California thrasher, western meadowlark, western kingbird, purple martin, northern harrier, prairie falcon, golden eagle, and black-shouldered kite (Roberson, 1985). Reptiles and amphibians include several species of salamander (tiger, arboreal, California, Pacific slender), frog (red-legged, foothill yellow-legged, Pacific tree), lizard (coast horned, sagebrush, side-blotched), and snake (racer, sharp-tailed, ringneck, long-nosed, western rattlesnake).

20.4.6 Transverse Ranges (IV)

The Transverse Ranges include several ranges, such as the Santa Ynez, San Gabriel, San Bernardino, and Santa Monica mountains, defined by active strike-slip and thrust fault systems such as the San Gabriel, Santa Susana–Oak Ridge, and Malibu Coast faults. The San Andreas fault slices obliquely through the region (fig. 20.1 inset, fig. 1.9). Rocks within this tectonic complex range from the 1.2-Ga (billion years before present) San Gabriel batholith to emergent marine Neogene and Quaternary sediment whose rapid uplift and dissection have yielded some of the least developed soils in the region. The coastal ranges are dominated by scrub or chaparral vegetation, which at higher elevations grades into woodland and forest. Though diverse, these plant communities have one feature in common—adaptation to seasonal drought.

Coastal sage scrub dominates exposed south-facing locations and some inland valleys. Drought tolerant and generally unable to withstand frost, its plants are deciduous, many with small leaves that are shed by high summer, thus compounding water stress. The community is populated by California sagebrush, black sage, purple sage (*Salvia leucophylla*), white sage (*S. apiana*), California buckwheat, bush penstemon (*Keckiella cordifolia*), deerweed (*Lotus scoparious*), chamise (*Adenostoma fasciculatum*), monkey flower (*Mimulus* spp.), and poison oak (*Toxicodendron diversilobum*). Succulents also are present, largely as prickly pear (*Opuntia* spp.) and live-forever (*Dudleya pulverulenta, D. cymosa, D. lanceolata,*). North-facing species tend to be evergreen with large root systems and include laurel sumac (*Malosma laurina*), lemonadeberry (*Rhus integrifolia*), and Christmas berry (*Heteromeles arbutifolia*).

At higher elevations more distant from the coast, chaparral communities dominate the landscape (fig. 20.10). Schoenherr (1992) and Sawyer and Keeler-Wolf (1995) describe a lower (warm) and upper (cold) chaparral. Unlike coastal sage associations, lower chaparral is somewhat frost tolerant and is characterized by plants with small, evergreen sclerophyllous leaves. On south-facing slopes, it includes chamise, hoary leaf ceanothus (*Ceanothus crassifolia*), and whitethorn (*C. leucodermis*). North-facing slopes include scrub oak (*Quercus berberidifolia*), holly-leaved redberry (*Rhamnus ilicifolia*), hairyleaf ceanothus (*Ceanothus oliganthus*), bigpod ceanothus (*C. megacarpus*), wild sweet pea (*Lathyrus vestitus*), man-root (*Marah macrocarpus*), and honeysuckle (*Lonicera* spp). Many members of the lower chaparral community are "fire-dependent" and, after a fire, initial successional species may be short lived. These include bleeding heart (*Dicentra ochroleuca*), golden yarrow (*Eriophyllum confertiflorum*), star lily (*Zygadenus fremontii*), Catalina mariposa lily (*Calochortus catalinae*), whispering bells (*Emmenanthe penduliflora*), yerba santa (*Eriodictyon* spp.), and phacelia (*Phacelia* spp.). Riparian habitats are also important in coastal sage and chaparral communities, providing habitats for several large trees such as western sycamore (*Platanus racemosa*), big leaf maple (*Acer macrophyllum*), white alder (*Alnus rhombifolia*), and Fremont cottonwood (*Populus fremontii*). The upper chaparral community tolerates frost and drought and is dominated on south-facing slopes by eastwood manzanita (*Arctostaphylos glandulosa*), bigberry manzanita (*A. glauca*), greenleaf manzanita (*A. patula*), silk-tassel bush (*Garrya* spp.), and mountain mahogany (*Cercocarpus betuloides*). North-facing slopes are characterized by coniferous trees, including big-cone Douglas-fir (*Pseudotsuga macrocarpa*) and Coulter pine (*Pinus coulteri*).

As a region with distinct moisture and elevation contrasts, the animal population is quite diverse. Home to mountain lion, bobcat, coyote, mule deer, bighorn sheep (*Ovis canadensis californiana*), black bear (*Ursus americanus*), woodrat (*Neotoma* spp.), and several mice and voles, this region also hosts the endangered California condor (*Gymnogyps californianus*), a vulture with a 3-m wingspan, which in historic time also inhabited the southern Coast Ranges and Sierra Nevada. A victim of lead poisoning found in the remains of hunted deer, thinning of egg shells because of DDT absorption, and ingestion of other poisons used in animal extermination, the condor population diminished from about 150 birds in the 1950s to 27 (6 in the wild, 21 in captivity) in 1986. Subject to a complex and sometimes controversial program of breeding in captivity in southern California, the condor was carefully reintroduced into the 26,500-ha Sespe Condor Sanctuary in the early 1990s, offering hope that the species may survive. The region provides habitat to hundreds of bird species, including the acorn woodpecker (*Melanerpes formicivorus*), which

Figure 20.10 Chaparral and coastal sage environment, with fire breaks, in the Santa Monica Mountains in 1936 (photo: Spence Collection, UCLA).

is partial to valley oak habitat, and numerous chaparral dwellers, including scrub jay (*Aphelocoma coerulescens*), California quail (*Callipepla californica*), rufous-sided towhee (*Pipilo erythophthalmus*), northern flicker (*Colaptes auratus*), red-tailed hawk (*Buteo jamaicensis*), and the great horned owl (*Bubo virginianus*). The chaparral community in the region is home to several reptiles, including the western fence lizard (*Sceloporous occidentalis*), coast horned lizard (*Phrynosoma coronatum*), western whiptail lizard (*Cnemidophorus tigris*), common kingsnake (*Lampropeltis getulus*), gopher snake (*Pituophis melanoleucus*), and western rattlesnake (*Crotalus viridis*). During long summer droughts, amphibians are scarce, but some canyons offer habitat for the Monterey salamander (*Ensatina eschscholtzii*).

20.4.7 Peninsular Ranges (V)

The Peninsular Ranges are characterized by northwest-trending ranges, which include the San Jacinto and Laguna mountains and, in Baja California Norte, the Sierra Juarez and Sierra San Pedro Martir. Cretaceous batholiths and metamorphic rocks are widely exposed, flanked by Cretaceous and later marine sediments, all of which have been pervasively fractured by faults such as the San Jacinto, Elsinore, and Agua Blanca faults. Broad marine terraces flank the Pacific coast (fig. 20.11). Like the Transverse Ranges, the region displays an elevational sequence of coastal sage, chaparral, and woodland. However, more xeric communities begin to the dominate the region southward, while, like the Central Coast, some mesic trees survive as relicts of past climate or with specific edaphic associations.

The coastal sage scrub, which includes black and purple sage, California buckwheat, prickly pear, and scrub oak, grades in interior valleys into a southern oak woodland dominated by coast live oak, Engelmann oak (*Q. engelmannii*), and California walnut (*Juglans californica*). Lower chaparral communities include chamise, redshank (*Adenostoma sparsifolium*), wedgeleaf ceanothus (*Ceanothus cuneatus*), deerbrush (*C. integerrimus*), whitethorn (*C. leucodermis*), basket bush (*Rhus trilobata*), fragrant sage (*Salvia clevelandii*), and bluecurls (*Tricostema parishii*). The upper chaparral, which ranges from about 1400 m to 1700 m in the Santa Ana Mountains to 2000 m in the San Jacinto Mountains, is dominated by bigberry and greenleaf manzanita, rare Otay manzanita (*Arctostaphylos otayensis*), Fremontia (*Fremontodendron californicum*), and mountain mahogany (*Cercocarpus betuloides*). Farther east and in the Sierra Juárez, pinyon-juniper woodland appears with pinyons (*Pinus monophylla, P. quadrifolia*) and western juniper. Interior uplands between 1700 and 2400 m support forest with ponderosa and sugar pine, white fir, and bigcone Douglas-fir on moister western slopes, and Jeffrey and Coulter pine, black oak, and canyon live oak on drier leeward slopes. The peaks of the San Jacinto Mountains support lodgepole and limber pine, with sagebrush (*Artemisia rothrockii, A. arbuscula*) in the alpine zone above 3000 m.

Figure 20.11 Pacific coast of Baja California Norte at the Mediterranean-Desert transition south of San Quintín. The broad marine terrace is backed by a degraded sea cliff and supports a paleodune hummock at its seaward edge (photo: A.R. Orme).

Several unexpected trees occur within the Peninsular Ranges, including disjunct distributions of knobcone pine, Torrey pine (*Pinus torreyana*), Tecate cypress (*Cupressus guadalupensis forbesii*), and Cuyamaca cypress (*C. arizonica stephensonii*). Specific habitats support these plants, some of which are limited relict stands whereas others have phenologic connections elsewhere in western North America. The knobcone pine, which prefers acidic nutrient-deficient soils, periodic fire, and fog drip, is found along the south face of the San Bernardino Mountains, near peaks in the Santa Ana Mountains, and 500 km farther south near Ensenada. The Torrey pine, a relict population now numbering less than 9000 but which was more widespread during moister Pleistocene time, is restricted to coastal cliffs north of San Diego and to Santa Rosa Island. The Tecate cypress occurs on deep gabbroic soils and north-facing slopes along the Mexico–United States border, on Tecate Peak and Otay Mountain in Baja California, and in the Santa Ana Mountains. These rare trees, some 200 years old, are now threatened by the increased frequency of high-intensity fires (Skinner and Pavlik, 1994). Pine cones in late Quaternary sediment suggest that Tecate cypress once extended from the Los Angeles Basin to the Colorado Desert (Zedler, 1981). The Cuyamaca cypress also occurs on deep gabbroic soils, often with Coulter pine. Found in two stands near Cuyamaca Peak in the Laguna Mountains, these trees have been almost eliminated by recent fires (Dunn, 1988).

Animals of the region include an array of mammals ranging in size from the larger mountain lion, coyote, bobcat, mule deer, and ringtail (*Bassariscus astutus*) to the smaller woodrat (*Neotoma* spp.), deer mouse (*Peromyscus maniculatus*), pinyon mouse (*P. truei*), desert shrew (*Notiosorex crawfordi*), and kangaroo rat (*Dipodomys agilis*). As aridity increases southward, amphibians become fewer but include large-blotched salamander (*Ensatina eschscholtzii klauberi*) and Monterey salamander (*Ensatina eschscholtzii eschscholtzii*). Reptiles include the coast horned lizard (*Phrynosoma coronatum blainvillii*), which is now nearing endangerment, orange-throated whiptail (*Cnemidophorus hyperythrus*), granite spiny lizard (*Sceloporus orcutti*), sagebrush lizard (*S. graciosus*), rosy boa (*Lichanura trivirgata*), red-diamond rattlesnake (*Crotalus ruber*), speckled rattlesnake (*C. mitchelli*), common kingsnake (*Lampropeltis getulus*), and California mountain kingsnake (*L. zonata*). Adaptation to drought is also evident among bird species that tolerate heat and conserve moisture. Scrub jay (*Aphelocoma coerulescens*), California quail (*Callipepla californica*), thrasher (*Toxostoma redivivum*), cactus wren (*Campylorhynchus brunneicapillus*), greater roadrunner (*Geococcyx californianus*), California gnatcatcher (*Polioptila californica*), and phainopepla (*Phainopepla nitens*) occur. Larger birds include red-shouldered hawk (*Buteo lineatus*), long-eared owl (*Asio otus*), lesser nighthawk (*Chordeiles acutipennis*), and Harris' hawk (*Parabuteo unicinctus*) in the far south.

20.4.8 Sierra Nevada (VI)

The 80-km-wide Sierra Nevada stretches nearly 700 km from north to south, a massive asymmetric block of granitic, metasedimentary, and volcanic rocks with relatively gentle west-facing slopes and a much steeper eastern face, both much scored by repeated Pleistocene glaciation. Latitude, elevation, substrate, rain shadow, and slope affect the diverse biotic communities that have developed most recently since the melting of the last ice cap between 18 ka (thousand years before present) and 12 ka and the ensuing return of warmer conditions during the Holocene. Major contrasts exist between the Sierra's western (windward) slopes and its eastern (rain-shadow) slopes (fig. 11.2).

The lower western foothills of the Sierra Nevada support chaparral, generally on south-facing slopes, and woodland communities. Adapted both to drought and fire, the chaparral is diverse, displaying distinct areas dominated by chamise, various ceanothus (including deerbrush, *Ceanothus integerrimus*), manzanita (including Ione, *Arctostaphylos myrtifolia*), and oak (interior live, scrub, and Brewer, *Quercus garryana brewerii*), and huckleberry (*Q. vaccinifolia*). The foothill woodland is dominated by blue oak, interior live oak, and valley oak, which interfinger with birchleaf mahogany, coffeeberry (*Rhamnus* spp.), and manzanita in the chaparral and with pine associations at higher elevations.

Yellow pine forest, found at 1000–2100 m along the west side and 2109–2800 m on the east side of the Sierra Nevada, includes ponderosa, Jeffrey, Washoe, and gray (*Pinus sabiniana*) pine, winter-deciduous black oak, incense cedar, manzanita, wild currant, and kit-kit-dizze (*Chamaebatia foliolosa*). These trees favor specific microclimates, with ponderosa pine occurring in areas with more than 620 mm precipitation (snow-dominated) annually, Jeffrey pine on more xeric sites with colder winters, and gray pine on granite domes and open south-facing slopes. On the west side and toward the south, the giant sequoia (*Sequoiadendron giganteum*) occurs with sugar and lodgepole pine, and white and red fir. The giant sequoia, found on well-drained granitic soils between 1400 and 2600 m, is fast growing, long lived, and big. The largest living tree is 83 m in height and over 11 m in diameter at its base. Its present distribution is disjunct, with 75 groves today ranging from 1 to 1620 ha in size, mostly south of the Kings River (Sawyer and Keeler-Wolf, 1995). The distribution of the giant sequoia was quite different in the past, with fossil plants found in the Great Basin, Colorado Plateau, and Santa Monica Mountains.

The lodgepole pine–red fir forest at higher elevations (2100–2700 m) is dominated by lodgepole, sugar, and western white pine (*Pinus monticola*), red and white fir, and mountain hemlock. Although these trees may live with one another, each represents a member of a changing habitat. Lodgepole pine generally serves as a pioneer in high-altitude meadows, eventually being replaced by red fir as soils become better drained and more developed. Again, at higher elevations, western white pine invades areas of sugar pine as soils become better drained.

The subalpine (2700–3600 m) area of the Sierra Nevada supports several five-needle pine associations. Cold, severe wind, blowing ice, and snowpack cause these trees to suffer cellular dehydration and slow growth, though longevity is common. Whitebark pine (*Pinus albicaulis*) favors open ground, often in groups around meadows. Limber pine (*P. flexilis*) is long lived (250–2500 years) and favors granitic soils and dry restricted sites, generally on the eastern face of the range. Foxtail pine is also long lived (up to 3300 years), but its distribution is disjunct, its two primary locations in the Klamath Mountains and Sierra Nevada being 500 km apart. Adapted to drought and intense solar radiation, the deeply rooted foxtail pine closely resembles the longer lived (>4000 years) bristlecone pine (*P. longaeva*), found east of the Sierra Nevada in the White and Inyo mountains and the Last Chance Range, and the distant Rocky Mountain bristlecone pine (*P. aristata*).

The alpine zone (3400–4400 m) experiences intense insolation ($\approx$1.8 ly min^{-1}), large diurnal and seasonal temperature ranges, strong winds, lightning strikes, and edaphic aridity (most moisture is snowfall). Plants here are broadly adapted to either wet or dry meadows, open rocky areas, fell-fields, and heath. Wet meadows are home to sedges (*Carex* spp.) and grasses, whereas dry meadows are populated with grasses (*Calaogrostis*, *Festuca*) and annuals such as phlox (*Phlox* spp.), lupine (*Lupinus* spp.), paintbrush (*Castilleja* spp.), and scarlet gilia (*Ipomopsis aggregata*). Fell-fields support cushion plants such as lupines and buckwheat, which create their own microhabitats. Heathlands, which occur on felsic, well-drained soils alongside small lakes and meadows and in riparian habitats, include white heather (*Cassiope mertensiana*), red mountain heather (*Phyllodoce breweri*), and alpine laurel (*Kalmia polifolia*).

Mediterranean conditions end quite abruptly over the crest of the Sierra Nevada as rain-shadow aridity on its eastern slopes introduces temperate desert conditions more typical of the Great Basin. Pine forests descend around 2800 m into pinyon-juniper woodland characterized by single-leaf pinyon, California juniper, and Sierra juniper (*Juniperus occidentalis* var. *australis*), tolerant of both snow and drought, and thus adapted to free-draining rocky slopes and alluvial fans. Plant communities at lower elevations are quite different from their counterparts on the western slopes. From 1200 m downslope, sagebrush scrub dominates, with rubber rabbitbrush (*Chrysothamnus nauseosus*), Great Basin sagebrush (*Artemisia tridentata*), blackbrush (*Coleogyne ramosissima*), shadscale (*Atriplex confertifolia*), and various grasses.

Indicator animals of the Sierra Nevada are strikingly linked with specific plant communities and elevation. Lower western foothill communities harbor western gray squirrel, California ground squirrel (*Spermophilus beecheyi*), badger, California quail, several woodpeckers, burrowing owl (*Athene cunicularia*), yellow-billed magpie (*Pica nutalli*), western bluebird (*Sialia mexicana*), black phoebe (*Sayornis nigricans*), and nighthawk (*Chordeiles* spp.).

In the forest communities, food availability with elevation is increasingly affected by snow and winter cold. Mammals of the yellow pine forest include mule deer and black bear, for whom past management policies have been disastrous. The deer population was severely impacted by unregulated hunting and the introduction of sheep in the nineteenth century. In response to declining numbers, a series of protective laws were enacted between 1883 and 1907. With limits on deer hunting and bounties on mountain lion, the deer's natural predator, herd populations increased with such vigor that new foraging grounds were needed. As herds migrated over larger ranges, native plant seedlings were affected (Schoenherr, 1992). Ultimately, many deer herds were moved to new locations to address the problems of food supply and environmental guardianship. Black bear have been similarly impacted by people, notably where these inherently private creatures are lured to campsites by the prospect of readily available food. As more and more people use the region's resources for recreation, so bears must be translocated to areas far from campsites. This forest also supports a limited array of amphibians and reptiles, including salamander (*Ensatina eschscholtzii*), western skink (*Eumeces skiltonianus*), northern alligator lizard (*Elgaria coerulea*), California mountain kingsnake (*Lampropeltis zonata*), and rubber boa (*Charina bottae*). Among the many birds are northern flicker, black-headed grosbeak (*Pheucticus melanocephalus*), western tanager (*Piranga ludoviciana*), dark-eyed junco (*Junco hyemalis*), and pileated woodpecker (*Dryocopus pileatus*).

At higher elevations in the lodgepole pine–red fir forest, the mammals include chipmunk, marten, golden-mantled ground squirrel (*Spermophilus lateralis*), and chickaree (*Tamiasciurus douglasii*). Birds include the Sierra grouse (*Dendragapus obscurus*), great gray owl (*Strix nebulosa*), hermit thrush (*Catharus guttatus*), and northern goshawk (*Accipiter gentilis*). In the subalpine and alpine zones are California bighorn sheep (*Ovis canadensis californiana*), yellow-bellied marmot (*Marmota flaviventris*), Belding's ground squirrel (*Spermophilus beldingi*), long-tailed weasel (*Mustela frenata*), pika (*Ochotona princeps*), snowshoe hare (*Lepus americanus*), and white-tailed jackrabbit (*Lepus townsendii*). Birds include Clark's nutcracker (*Nucifraga columbiana*), mountain bluebird (*Sialia currucoides*), and gray-crowned rosy finch (*Leucosticte arctoa tephrocotis*).

20.4.9 Central Valley (VII)

The Central Valley is an elongate area 700 km long and 120 km wide, the locus of marine sedimentation from Jurassic to late Cenozoic time and, more recently, of alluvium derived from the adjacent mountains. The region's climate is subhumid to semiarid because mountains to the west siphon off much of the incoming precipitation, only partly offset by streamflow and snowmelt from the Sierra Nevada to the east. Divided into two sections—the Sacramento Valley in the north and the San Joaquin Valley in the south—the region supports foothill woodland, valley grassland, scrubland, freshwater marshes, and riparian forest. Because of the region's dramatic agricultural transformation over the past 150 years, many indigenous grasses, marsh plants, and riparian woodlands have succumbed to tillage, grazing, water diversions for irrigation, and aggressive nonnative plants.

The foothill woodland community (at 1200–1550 m on the west and 900–1500 m on the east edge of the Valley) is populated by blue oak, interior live oak, foothill pine, and California buckeye (*Aesculus californica*). Valley grasslands are still extensive, though many native species have been superseded by introduced grasses. Although areas of true untouched grasslands are now limited, some bunchgrasses such as needlegrass, triple-awned grass, and bluegrass survive in association with annuals such as mariposa lily (*Calochortus* spp.), blue dick (*Dichelostemma pulchella*), purple owl clover (*Orthocarpus purpurascens*), lupine, and California poppy. Riparian environments support valley oak, California sycamore, red and white alder, Fremont cottonwood, and arroyo willow (*Salix lasiolepis*). Freshwater wetlands, often associated with desiccated Holocene lakes, have been decimated by human impacts, but an impressive variety of plants, including spikerush, bulrush, cattail, duckweed, buttonwillow, and mulefat, survives.

The animals of the Central Valley have experienced substantial changes in number and distribution since the influx of Europeans. With the conversion of most native grassland to agriculture and diversion of water resources, herds of larger grazing species such as pronghorn antelope (*Antilocapra americana*), tule elk, and mule deer, as well as predatory mountain lion, grizzly bear, and gray wolf (*Canis lupus*), have disappeared. Today, mammals include coyote, the San Joaquin kit fox (*Vulpes macrotis mutica*), and several species of rat, squirrel, mice, and rabbit. Birds are diverse, though many wetland species have all but disappeared as a result of habitat loss or hunting. Predators include golden eagle (*Aquila chrysaetos*), white-tailed kite (*Elanus caeruleus*), kestrel (*Falco sparverius*), and several species of hawk. Other birds present are turkey vulture (*Cathartes aura*), burrowing owl (*Athene cunicularia*), crow (*Corvus brachyrhynchos*), black-billed magpie (*Pica pica*), and yellow-billed magpie. Among the remaining wetland species are wood duck (*Aix sponsa*), red-breasted merganser (*Mergus serrator*), common merganser (*Mergus merganser*), and western yellow-billed cuckoo (*Coccyzus americanus*).

20.5 Natural Hazards and Human Impacts

20.5.1 Extreme Natural Hazards

The impacts of natural events such as seismic activity, lightning-triggered fire, heavy winter rains and related flooding, erosion, and mass movement are etched on the region. Such events are extreme because of their potential to cause enormous damage in a short period of time. In an environment that for much of the year is bathed in dry warmth and tranquility, these events develop quite quickly—instantaneously in the case of earthquakes, within several minutes to a few hours in terms of wildfires driven by Santa Ana winds, and within hours to a few days for floods and debris flows, depending in part on antecedent moisture conditions. Many of these hazards, notably fire and flood, are naturally part of Mediterranean-type systems, common to similar environments elsewhere with their propensity for potentially devastating, low-frequency, high-magnitude events. And, whereas the seismic hazard is neither unique to nor shared by all Mediterranean environments, the impact of fire and flood are certainly enhanced by steep tectonic relief.

Fire potential is highest after prolonged drought, specifically in late summer and autumn when convective thunderstorms generate frequent lightning strikes and when high pressure over the western United States establishes strong outflows: the desiccating Santa Ana winds of southern California, which reduce humidity and drive desert dust and searing fires toward the coast (fig. 20.12). Human responses to fire are varied. In areas with no fire suppression policy, as in Sierra San Pedro Mártir of Baja California, 865 distinct fires occurred between 1925 and 1995, but the total area burned was one-tenth that of neighboring southern California, which has a long tradition of fire suppression (Minnich and Franco-Vizcaino, 1997b). When they do occur, larger fires usually rage in older stands of chaparral and conifers, more common in southern California. Eventually, both areas acquire a vegetation mosaic of old growth adjacent to burned areas and recovering cover of various ages.

High-magnitude and high-intensity winter precipitation is well documented in the region's hydrologic records. For example, February 1998 rainfall totals related to El Niño conditions broke all-time records at many locations (table 20.2). Related flooding was widespread, with Clear Lake in northern California reaching its highest level since 1909. Earlier, during and after the record rainfall of 1969, soil slips, debris flows, and deeper-seated landslides were important throughout southern California. Scott and Williams (1978) measured thousands of slope failures in 72 watersheds within the Transverse Ranges, with some debris yields as high as 63,000 m^3 km^{-2}. In the Santa Monica Mountains, Campbell (1975) found that a total seasonal antecedent rainfall of 254 mm coupled with a rainfall rate of 5 mm hr^{-1} was sufficient to initiate shallow soil slips. The response to these events in urban areas has been to erect a series of flood-control structures that in theory trap sediment and convey excess water rapidly seaward through systems of concrete channels (see chapter 22). In reality, rarely does a winter of heavy rainfall pass without damage from floodwaters and debris flows somewhere in the region (fig. 20.13).

Seismic hazards are not confined to Mediterranean regions, but, because such regions are often favored for human settlement, they may have serious consequences. For reasons discussed in chapter 1, Greater California is earthquake country where major events of Magnitude 7 (M7) or larger may be expected somewhere in the region every few years and even smaller events may cause serious problems in urban areas. Many such events are associated with the San Andreas fault zone (Wallace, 1990). Historically, the most noteworthy of these have been the M8.3 Fort Tejon earthquake of 1857, which offset a central stretch of the fault 10 m but fortunately occurred in mostly open country (Sieh, 1978), the M8.3 San Francisco earthquake of 1906, which ruptured 430 km of the fault farther north and, with the fires it spawned in the city, led to much death and destruction (Lawson, 1908), and the M7.1 Loma Prieta earthquake of 1989, which revisited death and destruction on the San Francisco Bay area (Ellsworth, 1990). Seismic activity is not confined, however, to the San Andreas fault zone. Many earthquakes are associated with ruptures along other strands of the broad system of dextral shear that characterizes the plate margin, notably the

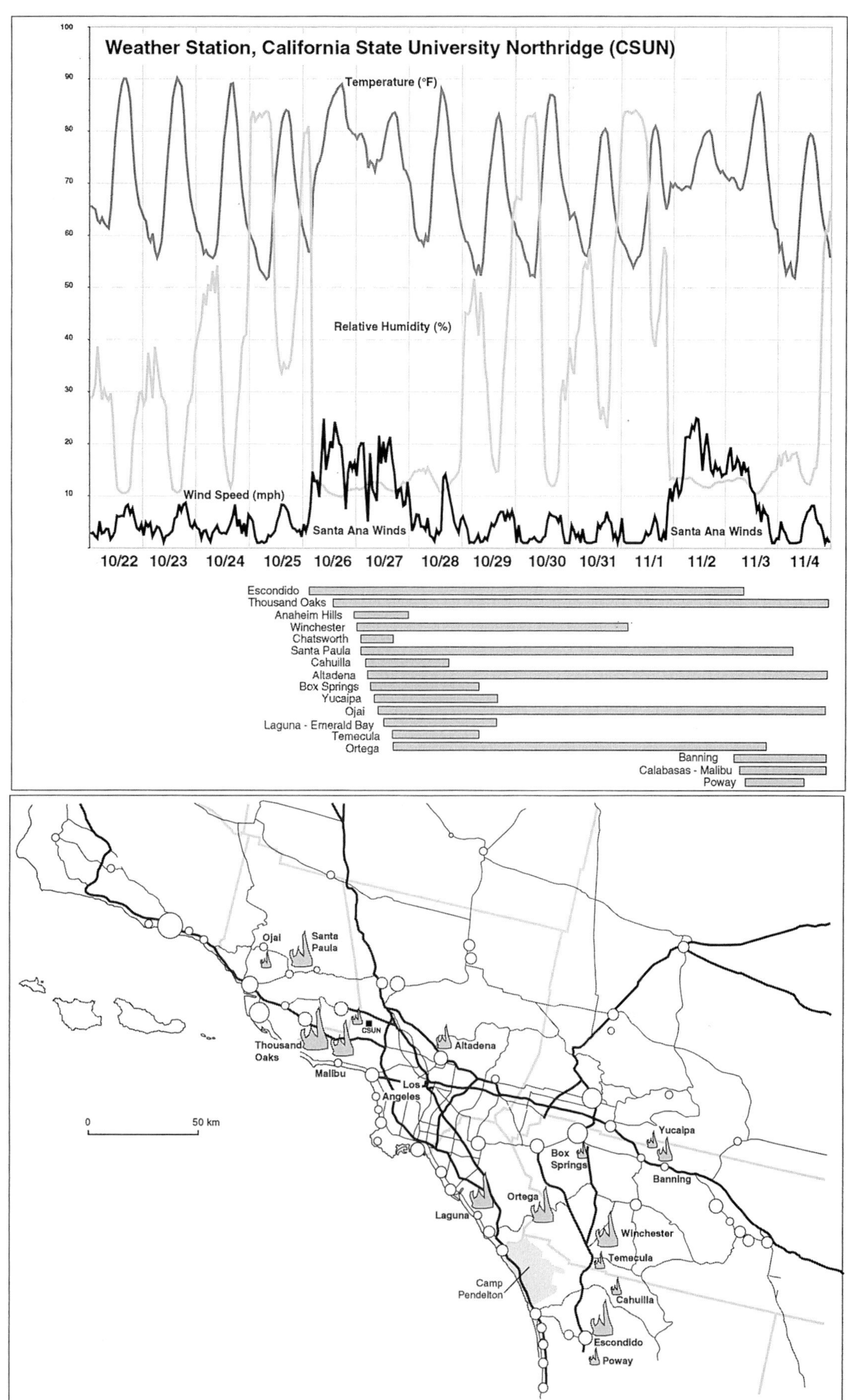

Figure 20.12 Relationship of fire occurrence to temperature, relative humidity, and wind speed during autumn 1993 (courtesy of William Bowen).

Table 20.2 California precipitation (mm): average annual and average February values compared with El Niño values, 1997–98

Location	Annual Average	El Niño 1997–98	Percent Average	February Average	February 1998	Percent Average
Colusa	400	835	209	72	316	439
Santa Rosa	749	1410	188	122	488	400
Salinas	338	851	252	58	278	479
Santa Maria	314	828	264	64	294	459
Santa Barbara	431	1194	277	95	552	581
Los Angeles	380	778	205	76	347	457
Riverside	265	544	205	53	241	455
Oceanside	255	508	199	45	293	651

Sources: CIMIS (1999), Worlclimate (1999).

Newport-Inglewood fault zone in the urbanized Los Angeles Basin and the Hayward-Calaveras fault zone east of San Francisco Bay. Ruptures on thrust faults within the Transverse Ranges are also troubling because of their proximity to urban areas (Woods and Seiple, 1995). Movements on the Santa Susana thrust, which caused the M6.4 San Fernando (Sylmar) earthquake of 1971, and on a buried thrust nearby, which caused the M6.7 Northridge earthquake of 1994, were responsible for many deaths and widespread dislocation in the Los Angeles area (fig. 20.14). Although major earthquakes may cause minor and temporary out-migration, the region's population continues to grow, such that seismic safety is a significant issue being addressed through building codes and, less effectively, through land-use restrictions.

20.5.2 Human Impacts on the Natural System

Human impacts on historic and modern landscapes are expressed, for example, through ecological adjustments to fire and fire suppression, the spread of exotic plants and animals, the conversion of native grasslands to cropland and pasture, changes in water supply and quality, and engineered responses to natural events, such as flooding, mass movement, fire, and seismicity.

Figure 20.13 Destruction of bridges across Big Tujunga Wash at the base of the western San Gabriel Mountains during floods of winter 1978 (photo: U.S. Geological Survey).

Figure 20.14 Freeway collapse during the M6.4 San Fernando earthquake of February 1971 (photo: Los Angeles Department of Building and Safety).

The role of people in triggering fire, by accident or arson, is a recurrent problem in modern Mediterranean environments, as illustrated by the devastating Calabasas-Malibu arson fire of November 1993 (fig. 20.12). Whether caused by people or nature, however, once fires develop they tend to respond to comparable dynamics, becoming especially serious where abundant natural fuel, low humidity, and strong desiccating winds drive the flames. Thus the controversial issues of fire suppression, landscape preservation, and encouragement of native vegetation become important over the long term. There are significant differences in fire suppression policy between those areas under the separate jurisdictions of the United States and Mexico. Fire suppression, an active policy within California for nearly a century, may produce over-mature stands of trees and shrubs that gather fuel over time so that when fires occur the resulting conflagration is intense (soil temperatures may reach 840°C; De Bano, 1977) and may thoroughly destroy large areas of vegetation. Fire also favors the development of hydrophobic (water-repellant) soils as organo-silicates and other volatized materials move down through the soil, impeding infiltration and increasing surface erosion (see chapter 22). Fire suppression coupled with the draining of meadows in Yosemite Valley since the mid-nineteenth century has led to the growth of dense stands of ponderosa pine, incense-cedar, white fir, and Douglas-fir at the expense of black oak, sedges, and grasses. During the severe drought of the 1930s, many conifers were invaded by western pine beetle (*Dendroctonus brevicomis*), subsequently died, and were then felled, peeled, or burned to prevent further beetle spread. This in turn led to infestation of fungal root disease by *Heterobasidion annosum* and the development of extensive vegetation gaps, of which 68 exist today and which range in extent from a few square meters to over 2500 m^2 (Slaughter and Rizzo, 1999). Ironically, such gaps provide open areas for grasses and herbs, which increase the wildlife food base and restore park vistas to their premanagement state.

The intentional and accidental introduction of exotic plants, especially aggressive weedy species, represents perhaps the largest threat to native vegetation. At least 1045 exotic species or 17.7% of the total population now grow wild in California (Randall et al., 1998). Frenkel (1970) estimated that at least 16 exotics were introduced during the Spanish period (1769–1821), a further 63 during the Mexican period (1821–48), with the remainder following Euroamerican occupation. Even though all parts of the region are affected by invasive species, specific areas show particularly higher densities of exotic vascular plants. Exotic plants clearly are more numerous within coastal areas and at elevations below 1800 m. One area with the lowest percentage of exotics is Punta Banda in Baja California, a peninsula remote from ports of entry, whose 50 exotics (19.4% of the population) constitute a low number by coastal standards (Randall et al., 1998). Moreover, because of fire and grazing policies in Baja California Norte, plant species found today may be very similar to those recorded over 200 years ago (Minnich and Franco-Vizcaino, 1997a). In addition to the replacement of native species, exotic plants generate other implications. For example, exotic grasses often germinate more efficiently in response to fire, outcompeting natives and promoting more frequent fires. European wild oats, South African veld grass, and Pampas grass often destabilize hillslopes, and soil chemistry is altered by such exotics as ice plant, which transmits salts into the soil, and European brooms, which accelerate nitrogen enrichment (D'Antonio and Haubensak, 1998). Beyond this, invasion of riparian and estuarine communities by the giant reed (*Arundo donax*) and perennial peppergrass (*Lepidium latifolium*) decreases native diversity. As sites become increasingly monocultural, wildlife habitat is altered so that critical elements such as insect diversity, vital to migratory bird groups, is also affected (Dudley, 1998).

The conversion of native vegetation to cropland, pasture, and urban areas has had direct impacts on animal populations in the region. There are, however, interesting scenarios that point to critical ecological links, which, with careful planning, have partially restored vital breeding niches. The greater sandhill crane (*Grus canadensis tabida*), a listed sensitive species, uses a migratory path through the Central Valley and Klamath-Cascade regions on its way to breeding farther north. With many of the staging areas under cultivation, the importance of managing the landscape with "wild hay" adjacent to wetlands has been recognized as critical to the foraging habits of the crane and other birds. Studies initiated in 1986 in the Ash Creek Wildlife Area of northern California have shown that populations of crane, and of other species such as the song sparrow (*Melospiza melodia*), have increased significantly with the practice of "late haying." This practice, which delays hay cutting to permit fledgling success, increases nesting, brood-rearing and foraging time while reducing alert time (Epperson et al., 1999).

Human impacts on water supply and quality are a continuing issue in the region, especially in urban and agricultural areas (see chapter 22). Common problems in urban areas include contamination of water in natural and concrete channels and nearshore waters with pathogenic viruses, bacteria, and fungi. Although constituent mass emissions and concentrations of effluents from municipal wastewater facilities offer a partial profile of water chemistry, errant constituents from uncontrolled sources complicate the picture of water quality. Studies of effluents discharged through controlled facilities into the Southern California Bight from 1971 to 1996 showed reductions in cadmium, chromium, lead, mercury, zinc, nickel, selenium, and arsenic (Raco-Rands, 1998). Recently, however, there have been increases in cyanide and nonchlorinated phenols, and more significantly in fecal coliform and streptococci from uncontrolled sources.

In agricultural areas, the introduction of pesticides and growth-enhancement chemicals into soil and water poses problems. Further, with irrigation, naturally occurring elements may concentrate in runoff and groundwater. In the western Central Valley, the water table in irrigated lands rises and concentrates salts and other elements such as selenium and boron in the crop root zone (see chapter 22). To address this problem, more irrigation water was initially introduced, but this led to further concentration of minerals in groundwater (Quinn, 1998). Subsequent mitigation of the problem has included a monitoring program with target reduction goals using diversion drainage channels. Results for 1995–97 showed a 33–44% reduction in selenium, suggesting that with diligence and funding, such problems may be brought under control.

Human impacts on wetlands in the region are far reaching, with the disappearance of over 80% of these areas since European occupation (see chapter 22). For example, the 3000-km^2 Sacramento–San Joaquin Delta contains the largest wetland in the western United States, with a contributing watershed of 148,000 km^2. However, the delta has experienced dramatic physical and biological changes over the past 150 years, initially from hydraulic gold-mining debris from the Sierra Nevada, which raised stream beds and increased flooding, and later from levees and 1800 separate water diversions for agriculture and urban use (Shelton and Fridirici, 1997). Although freshwater releases from upstream reservoirs now prevent the extensive saltwater intrusion seen before 1940, low inflows still inhibit wetland health by limiting the dilution of pollutants from agricultural, industrial, and urban uses, this at a time when natural seasonal inflows appear to be changing in response to climate changes (Vaux, 1991; Shelton and Fridirici, 1997).

20.6 Conclusion

The Mediterranean-type environment of Greater California is unique in North America for its summer drought and winter rains, and the distinctive ecological response to this seasonality imposed over 13° of latitude and a 4000-m elevational range. Despite the hazards of fire, flood, and drought associated with this climate regime, the region's prolonged sunshine and warmth, together with its mineral, timber, and land resources, have long attracted human immigrants. For example, California's population, most of whom live within the Mediterranean region, has risen rapidly from 93,000 in 1850, 1.5 million in 1900, and 10.5 million in 1950, to perhaps 35 million in 2000 (an uncertain figure because of the many undocumented immigrants). During the same period, the proportion of the urban population has grown from 8% in 1850 to nearly 95% in 2000. The greater Los Angeles area now contains 16 million people, the San Francisco Bay area 7 million, and the San Diego-Tijuana area astride the United States–Mexico border a further 4 million.

The significance of this growth to the physical environment is degradation of the region's land, air, water, plant, and animal resources (Orme and Orme, 1998). Such impacts are greatest in and near urban and industrial areas, where much of the natural landscape has been obliterated or severely modified, where air and water pollution is widespread, and where the built landscape has introduced artificial hydrologies and habitats. To feed the population of this region and beyond, the expansion of agriculture over the past 150 years has similarly transformed most natural lowlands into farmland. Some 10% of the Mediterranean region is now irrigated. And support for this population, especially in the subhumid south, has necessitated vast transfers of water from the Sierra Nevada, northern California, and the Colorado River. Further, the recreational quest of urban people impacts coasts and mountains far beyond the cities. Concern for these areas and their resources has seen the creation of many federal and state parks, notably the Yosemite (1890), Sequoia (1890), and Redwood (1968) National Parks in California, and the Sierra Juarez and Sierra San Pedro Mártir National Parks in Baja California Norte. Similarly, ocean resources have been given some protection by marine sanctuaries around Monterey Bay and the Channel Islands. Nonetheless, this region's persistent attraction to people over the past 150 years has come to threaten the very qualities that distinguish its Mediterranean environment.

References

Axelrod, D.I., 1980. Contributions to the Neogene paleobotany of central California. *University of California Publications in Geological Sciences*, 121, 1–212.

Axelrod, D.I., 1989. Age and origin of chaparral. In: S.C. Keeley (Editor), *The California chaparral: Paradigms reexamined.* Natural History Museum of Los Angeles County Science Series 34, 7–19.

Bailey, R.G., 1997. *Map: Ecoregions of North America.* USDA Forest Service, The Nature Conservancy, and U.S. Geological Survey, 1: 15,000,000.

Barbour, M., and J. Major, 1988. *Terrestrial vegetation of California.* California Native Plant Society, Sacramento.

Barbour, M., B. Pavlik, F. Drysdale, and S. Lindstrom, 1993. *California's changing landscapes: Diversity and conservation of California vegetation.* California Native Plant Society, Sacramento.

California Natural Diversity Database, 2000. California Department of Fish and Game. http://www.dfg.ca.gov/whdab/html.cnddb.html

Campbell, R.H., 1975. Soil slips, debris flows, and rainstorms in the Santa Monica Mountains and vicinity, southern California. *U.S. Geological Survey Professional Paper*, 851.

CIMIS (California Irrigation Management Information System), 1999. Monthly weather data. http://www.dpla.water.ca.gov.

Clements, F.E., 1916. Plant succession: An analysis of the development of vegetation. *Carnegie Institute of Washington Publication* 242, Washington, D.C.

Cooper, W.S., 1922. The broad-sclerophyll vegetation of California. *Carnegie Institute of Washington Publication* 319, Washington, D.C.

Cowling, R.M., P.W. Rundel, B.B. Lamont, M.K. Arroyo, and M. Arianoutsou, 1996. Plant diversity in mediterranean-climate regions. *Trends in Ecology and Evolution*, 11, 362–366.

Dallman, P.R., 1998. *Plant life in the world's Mediterranean climates: California, Chile, South Africa, Australia, and the Mediterranean Basin.* California Native Plant Society and University of California Press, Berkeley.

D'Antonio, C.M., and K. Haubensak, 1998. Community and ecosystem impacts of introduced species. *Fremontia*, 26, 13–18.

DeBano, L.F., P.H. Dunn, and C.E. Conrad, 1977. Fire's effect on physical and chemical properties of chaparral soils. In: H.A. Mooney and C.E. Conrad (Editors), Proceedings of environmental consequences of fire and fuel management in Mediterranean ecosystems Symposium. *U.S.D.A. Forest Service General Technical Report*, WO-3, 65–74.

Dudley, T., 1998. Exotic plant invasions in California riparian areas and wetlands. *Fremontia*, 26, 24–29.

Dunn, A.T., 1988. The biogeography of the California floristic province. *Fremontia*, 15, 3–9.

Ellsworth, W.L., 1990. Earthquake history, 1769–1989. In: R.E. Wallace (Editor), The San Andreas Fault system, California. *U.S. Geological Survey Professional Paper* 1515, 153–181.

EPA (Environmental Protection Agency), 1999. Endangered species project. http://www.epa.gov/oppfead1/endanger/statesx.htm.

Epperson, W.L., J.M. Eadie, D.B. Marcum, E.L. Fitzhugh, R.E. Delmas, 1999. Late season harvest provides habitat for marshland birds. *California Agriculture*, 53, 12–17.

Frenkel, R.E., 1970. Ruderal vegetation along some California roadsides. *University of California Publications in Geography*, 20, 1–163.

Hadley, E.B., 1961. Influence of temperature and other factors on *Ceanothus Megacarpus* seed germination. *Madroño*, 16, 132–138.

Haston, L., F.W. Davis, J. Michaelsen, 1988. Climate response functions for bigcone spruce: A mediterranean-climate conifer. *Physical Geography*, 9, 81–97.

Henson, P., and D.J. Usner, 1996. *The natural history of Big Sur.* University of California Press, Berkeley, 416.

Hill, B.R., and C.E. McConaughy, 1988. Sediment loads in the Ventura River Basin, Ventura County, California 1969–1981. *U.S. Geological Survey Water Resources Investigations Report* 88-4149, 23.

Jensen, H.A., 1947. A system for classifying vegetation in California. *California Fish and Game* 33, 199–266.

Keeley, J.E., and S.C. Keeley, 1986. Chaparral and wildfires. *Fremontia*, 14, 18–21.

Keeley, S.C., J.E. Keeley, S.M. Hutchinson, and A.W. Johnson, 1981. Postfire succession of the herbaceous flora in southern California chaparral. *Ecology*, 62, 1608–1621.

Kummerow, J., 1981. Structure of roots and root systems. In: F. DiCastri, D.W. Goodall, and R.L. Specht (Editors), Ecosystems of the World II: Mediterranean-type shrublands, Elsevier, Amsterdam, 269–288.

Lawson, A.C., 1908. The California earthquake of April 19, 1906: Report of the State Earthquake Investigation Commission. *Carnegie Institute of Washington Publication* 87, Washington, D.C.

Livezey, R.E., M. Masutani, A. Leetmaa, H. Rui, M. Ji, and A. Kumar, 1997. Teleconnective response of the Pacific-North American region atmosphere to large central equatorial Pacific SST anomalies. *Journal of Climate*, 10, 1787–1819.

Lynch, H.B., 1931. *Rainfall and stream run-off in southern California since 1769.* The Metropolitan Water District of Southern California, Los Angeles.

Minnich, R.A., and E. Franco-Vizcaino, 1997a. Mediterranean vegetation of northern Baja California, *Fremontia*, 25, 3–12.

Minnich, R.A., and E. Franco-Vizcaino, 1997b. Protecting vegetation and fire regimes in the Sierra San Pedro Mártir of Baja California, *Fremontia*, 25, 13–21.

Munz, P.A., and D.D. Keck, 1949. California plant communities. *Aliso* 2, 87–105.

NCDC (National Climate Data Center), 1999. The top 10 El Niño events of the 20th century. http://www.ncdc.noaa.gov/ol/climate/research/1998/enso/10elnino.

Nicholson, C., C.C. Sorlein, T. Atwater, J.C. Orwell, and B.P. Luyendyk, 1994. Microplate capture, rotation of the western Transverse Ranges, and initiation of the San Andreas transform as a low-angle fault system, *Geology*, 22, 491–495.

Orme, A.R., 1998. Late Quaternary tectonism along the Pacific Coast of the Californias: A contrast in style. In: I.S. Stewart and C. Vita-Finzi (Editors), *Coastal Tectonics.* Geological Society Special Publications 146, London, 179–197.

Orme, A.R., and A.J. Orme, 1998. Greater California. In: A.J. Conacher and M. Sala (Editors), *Land degradation in Mediterranean environments of the world: Nature and extent, causes and solutions.* 109–122. John Wiley and Sons, Chichester.

Quinn, N.W.T., J.C. McGahan, and M.L. Delamore, 1998. Innovative strategies reduce selenium in grasslands drainage. *California Agriculture*, 52, 12–18.

Raco-Rands, V., 1998. Characteristics of effluents from large municipal wastewater facilities in 1996. In: M. J. Allen, C. Francisco, and D. Hallock (Editors), *Southern California Coastal Water Research Project Annual Report 1996*, 10–20.

Randall, J.M., M. Rejmánek, and J.C. Hunter, 1998. Characteristics of the exotic flora of California. *Fremontia*, 26, 3–12.

Roberson, D., 1985. *Monterey birds: Status and distribution of birds in Monterey county.* Monterey Audubon Society, Carmel.

Sawyer, J.O., and T. Keeler-Wolf, 1995. *A manual of California vegetation.* California Native Plant Society, Sacramento.

Schimmelmann, A., M. Zhao, C.C. Harvey, and C.B. Lange, 1998. A large California flood and correlative global climatic events 400 years ago. *Quaternary Research*, 49, 51–61.

Schoenherr, A.A., 1992. *A natural history of California.* University of California Press, Berkeley, 772.

Scott, K.M., and R.P. Williams, 1978. Erosion and sediment yields in the Transverse Ranges, Southern California. *U.S. Geological Survey Professional Paper* 1030.

Shelton, M.L., and R.M. Fridirici, 1997. Decadal changes of inflow to the Sacramento-San Joaquin Delta, California. *Physical Geography*, 18, 215–231.

Sieh, K.E., 1978. Slip along the San Andreas fault associated with the great 1857 Earthquake. *Seismological Society of America Bulletin*, 68, 1421–1448.

Skinner, M.W., and B.M. Pavlik, 1994. *Inventory of rare and endangered vascular Plants of California*, 5th edition. California Native Plant Society, Sacramento.

Slaughter, G.W., and D.M. Rizzo, 1999. Past forest management promoted root disease in Yosemite Valley. *California Agriculture*, 53, 17–24.

State of California, 1990. Management of the California State Water Project. *Bulletin 132-90*, Department of Water Resources.

Sweeney, J.R., 1956. Responses of vegetation to fire: A study of the herbaceous vegetation following fires. *University of California Publications in Botany*, 28, 143–250.

Thorne, R.F., 1976. The vascular plant communities of California. In: J. Latting (Editor), *Plant communities of southern California.* California Native Plant Society, Sacramento, 1–31.

Trewartha, G.T., 1968. *An introduction to climate.* McGraw-Hill Book Co., New York.

Unruh, J.R., 1991. Uplift of the Sierra Nevada and implications for late Cenozoic epeirogeny in the western Cordillera. *Geological Society of America Bulletin*, 103, 1395–1401.

U.S. Department of Interior, 1992. *Summary statistics: Water, land and related data*, Bureau of Reclamation, 321.

U.S. Geological Survey, 1999a. Historical stream flow daily mean values. http://waterdata.usgs.gov/nwis-w/.

U.S. Geological Survey, 1999b. Suspended sediment database. http://webserver.cr.usgs.gov/sediment/.

Vaux, H.J., 1991. Global climate change and California's water resources. In: J.B. Knox and A.F. Scheuring (Editors), *Global climate change and California: Potential impacts and responses.* University of California, Berkeley, 69–96.

Wallace, R.E., 1990. The San Andreas Fault System, California. *U.S. Geological Professional Paper* 1515.

Westman, W.E., 1983. Plant community structure—Spatial partitioning of resources. In: F.J. Kruger, D.T. Mitchell, and J.U.M. Jarvis (Editors), *Mediterranean-type ecosystems: The role of nutrients*, Springer-Verlag, Berlin, 417–445.

Wieslander, A.E., 1935. A vegetation map of California. *Madrono* 3, 140–144.

Woods, M.C., and W.R. Seiple (Editors), 1995. The Northridge, California, earthquake of 17 January 1994. *Special Publication* 116, California Department of Conservation Division of Mines and Geology, Sacramento.

Worldclimate, 1999. Monthly weather data. http: //www.worldclimate.com.

Zedler, P.H., 1981. Vegetation change in chaparral and desert communities. In: D.C. West, H.H. Shugart, and D.B. Botkin (Editors), *Forest succession: Concepts and application*, Springer-Verlag, Berlin, 406–430.

21

Ocean Coasts and Continental Margins

Antony R. Orme

As befits a continent that extends from the tropics to the High Arctic, the coasts of North America experience a wide range of physical and biological processes, which are reflected in a variety of environments. Furthermore, except along the Pacific rim, the mainland shore is often fronted by broad continental shelves and backed by coastal plains, notably around the Gulf of Mexico, the unglaciated Atlantic coast, Hudson Bay, and portions of the Arctic coast. Because these shelves and plains have been influenced by eustatic and isostatic sea-level changes during late Cenozoic time, they are integral components of the coastal zone. Along such coasts, quite subtle fluctuations in sea level may have major impacts on coastal processes, such as sedimentation, biotic migration, and wetland evolution. Such is not the case along the mountainous Pacific coast where the effects of sea-level fluctuations are limited to a narrow rim, glaciated in the north, that descends steeply onto narrow shelves or fault-controlled borderlands. Nor is it true of Greenland, eastern Baffin Island, and Labrador where recent glaciation and crustal adjustments have left a legacy of rugged coastal terrain. Such contrasts between steep rugged coasts and gently shelving coastal zones reflect the continent's tectonic and geomorphic evolution, but details within this framework are due mostly to processes at work during and since the large major eustatic rise of sea level that culminated less than 5000 years ago.

Unlike Africa or Australia where long stretches of coast are unbroken by sizable inlets, the North American coast is augmented by many gulfs, estuaries, and islands, sufficiently numerous to render coastline measurements rather meaningless. However, a useful perspective is gained by noting that the continental mainland is bordered by about 4000 km of temperate Atlantic coast from Newfoundland to the Florida Strait, by 4500 km of subtropical coast around the Gulf of Mexico, by 7000 km of temperate Pacific coast from Baja California to the Alaska Peninsula, and by long cold coasts on the Bering Sea, Arctic Ocean, Hudson Bay, and Labrador Sea.

Beyond the mainland shore, continental shelves add 6.74×10^6 km^2, or 28%, to the continent's land area of 24.06×10^6 km^2 (table 1.1), a large addition with major physical and biological implications during eustatic sea-level changes. Off the northern coast, a vast shelf supports the Arctic Archipelago, including such large islands as Baffin (476×10^3 km^2), Ellesmere (213×10^3 km^2), and Victoria (212×10^3 km^2). Beyond these lies Greenland (2176×10^3, km^2), the world's largest island below continental size, tenuously attached to the continent's tectonic framework. Farther south, off Newfoundland (111×10^3 km^2), the continental shelf covers 345×10^3 km^2. The shelves underlying 600×10^3 km^2 or 38% of the Gulf of Mexico, widest in the carbonate platforms off Florida and Yucatan, and 1120×10^3

km^2 or 50% of the Bering Sea were shallow enough to be widely exposed by late Cenozoic marine lowstands.

This chapter offers an integrated perspective on North America's coastal zone, examining first the various tectonic, eustatic, and isostatic controls over coastal evolution, then the relevant climatic, marine, hydrologic, and ecologic factors, and finally the variety of coasts within the context of six dominant regional categories.

21.1 Tectonic Origins and Coastal Evolution

The evolution of the North American coast can be evaluated at three temporal and spatial scales. The coarsest scale, measurable in millions of years and regional contrasts, relates to the tectonic forces that have shaped the continent's outline after the breakup of Pangea over the past 180 million years. Eustatic and isostatic fluctuations of sea level across the continental margin are observed at intermediate scales, which cover thousands of years and the variable space between the outer continental shelf and the inner coastal plain. The finer scales, discussed later, are the changes related to coastal processes functioning within specific tectonic and sea-level contexts, measurable on timescales varying from seconds to decades or more, and in space ranging from sand grains to seacliffs, beaches, and wetlands.

21.1.1 Tectonic Origins

Early in the plate-tectonic revolution, Inman and Nordstrom (1971) sought to relate the world's coasts to tectonic forces that had shaped the gross outlines of the continents. They distinguished between collision coasts near the leading edge of shifting plates, trailing-edge coasts in the wake of these plates, and marginal coasts behind island arcs. Although problematic and subject to change with increased knowledge, their scheme focused fresh attention on the structural origins of coasts. In a sense, their classification was a lineal descendant, within a renewed paradigm, of the distinction between Pacific and Atlantic coastal types recognized much earlier by Suess (1892). Pacific-type coasts were characterized by coast-parallel structures and mountainous terrain, what Supan (1930) later called concordant coasts. Atlantic-type coasts were typified by structures more or less perpendicular to the shoreline, Supan's discordant coasts. Advances in plate tectonics allow these schemes to be redefined in terms of active or convergent coastal margins on the one hand, mostly along the Pacific, and passive or divergent coastal margins elsewhere (see chapter 1).

The Pacific coast lies near the continent's active margin where the North American plate has long been converging with, or shearing alongside, oceanic plates farther west, leading to repeated terrane accretion and the shaping of coast-parallel mountains along the continent's leading edge. Gross variations along this margin reflect varying plate interactions. From Cape Mendocino to Vancouver Island, where the Cascadia subduction zone lies close offshore, mountain ranges typically parallel the coast, with active volcanoes farther inland. A similar tectonic setting occurs in peninsular Alaska, but here volcanic activity linked to subduction in the Aleutian Trench is still building an island-arc system. The coast from the Gulf of Alaska to Vancouver Island is structurally more complex because of the presence of the Queen Charlotte transform system. Similarly, the open coast of the Californias between Cape Mendocino and Cabo San Lucas, though reflecting late Cenozoic plate convergence in its coast-parallel mountains, is now responding on a captured plate sliver to stresses generated in the wake of the East Pacific Rise (fig. 21.1). In contrast, the rifting of the Gulf of California along this rise over the past 6 Ma (million years before present) has produced fresh passive-margin coasts, similar to those bounding the Red Sea.

The Atlantic and Gulf coasts reflect their origins along the continent's passive or divergent margins. In the south, the lengthy period of plate separation and crustal flexuring,

Figure 21.1 Active-margin tectonism along the Pacific coast. The San Andreas fault, which defines much of the northern California coast, extends here from Bolinas Lagoon (bottom left), landward of the Point Reyes Peninsula, to beyond Bodega Bay. The sediment plume of the Russian River is also prominent (photo: NASA).

initiated when the Atlantic Ocean began opening around 180 Ma, can be observed in the continuum from broad continental shelf to coastal plain and in the massive carbonate platforms beneath the Bahamas, Florida, and Yucatan, a reflection of the convoluted opening of the Gulf of Mexico to tropical seas. Farther north, the progressive "unzipping" of the North Atlantic off the Grand Banks (≈150–120 Ma), Labrador (≈90–70 Ma), and Greenland (≈60–40 Ma) is reflected in a closer relationship between the plate margin and the present coast. However, only in Newfoundland and the Gulf of St. Lawrence is there any true similarity between discordant structures and the classic Atlantic-type coast beloved of early European scientists. The cold northern coast is less easily categorized but, lacking active orogens, mostly reflects a passive margin of long duration. Across all these passive margins, continued post-Pangean crustal flexuring is reflected at the coast, for example, in rugged Labrador and deformed continental shelves. These passive-margin coasts are also far from the active orogens of the west, thus needing long rivers, such as the Mackenzie and Mississippi, to transport their large sediment loads seaward across low gradients.

21.1.2 Relative Sea-Level Change

The behavior of continental and oceanic plates combines with the fluctuating volumes of ocean water to yield a complex equation that determines relative sea level and thus the location of coastal processes. Within predictable limits, a high continental freeboard or low ocean volume exposes much of the continental shelf, expanding the subaerial extent of continents with major implications for physical and biological activity. Conversely, a low freeboard or high ocean volume not only drowns the shelf but penetrates inland across the coast, restricting terrestrial activity at the expense of marine processes. Changes in continental freeboard and ocean volume have long been a part of Earth history, but coastal studies usually focus on recent and impending sea-level changes relevant to the contemporary landscape.

Marine deposits onlapping the continent's margins indicate that, in the absence of crustal deformation, maximum eustatic submergence occurred during Cretaceous time and ocean waters have since fluctuated within 200 m of present sea level. Eustatic highstands during Quaternary interglacials rarely rose more than 10 m above the present, whereas glacial lowstands probably fell to no more than –150 m. It is likely that the zone between 0 and –80 m was frequently reworked during these oscillations, allowing the legacies of higher sea levels to survive onshore while permitting thicker, more continuous marine sedimentation on the outer continental shelf. More certainly, the Last Interglacial highstand (≈125 ka, thousand years before present) rose to about +6 m and the Last Glacial Maximum lowstand (≈20 ka) fell to around –130 m.

Ideally, on stable shores, evidence for repeated Quaternary highstands is expressed in seacliffs and relict beaches near present sea level, whereas lowstands are reflected in features submerged on the shelf. But the situation is not ideal. Because of continuing tectonism and crustal loading by water, ice, and sediment, coastal zones are rarely stable. Quaternary shorelines along the Pacific coast have been massively deformed by orogenesis, whereas similar features elsewhere have been displaced by persistent epeirogenesis and isostatic adjustments (fig. 21.2). In this complex

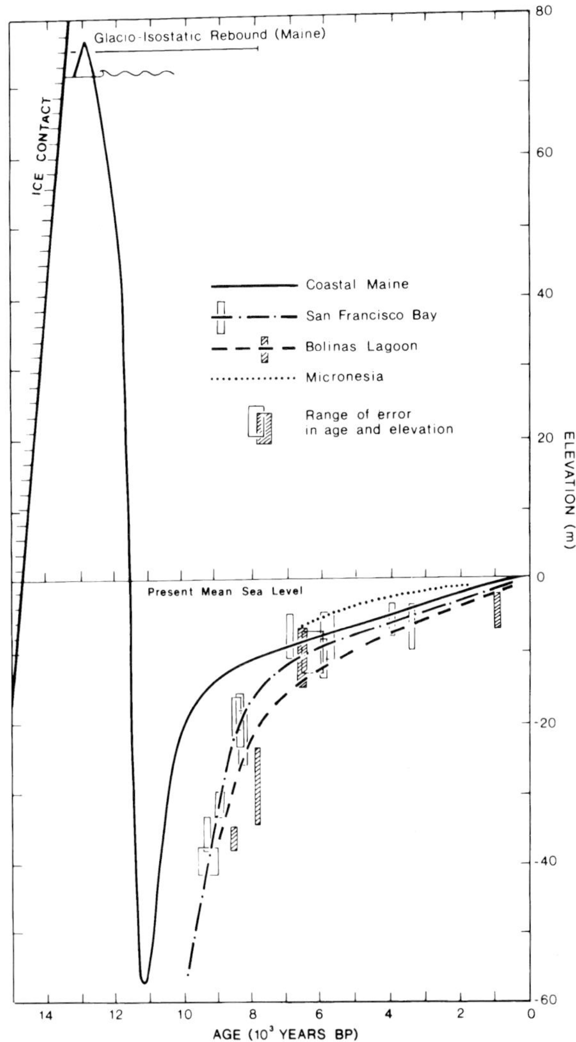

Figure 21.2 Contrasting late Quaternary sea-level changes within and beyond the Pleistocene ice front. Glacioisostatic crustal rebound carried late Pleistocene shorelines above present sea level in formerly glaciated Maine, whereas in California glacioeustatic sea level was rising from its lowstand around –130 m. The subsequent Holocene transgression was comparable in both areas, but shorelines in San Francisco Bay and Bolinas Lagoon have been prone to tectonic subsidence within the San Andreas fault zone.

scenario, the six coastal regions defined subsequently each contain a typical set of sea-level signatures attributable to global eustatic forcing, modified within each region by local tectonic and isostatic forcing. There is now little support for the past practice of correlating marine stillstands solely on the basis of elevation, but where stillstands can be dated, their elevation offers a measure of the nature and rate of tectonic and isostatic deformation and thus valuable insight into crustal geophysics (e.g., Orme, 1998; Peltier, 1998). Ultimately, the most important sea-level change for the present coast has been the Flandrian transgression because its culmination in mid-Holocene time set the scene for subsequent isostatic adjustments and coastal processes.

Tide-gauge data from many locations indicate that sea level has continued to rise modestly over the past century, at rates of around 1 to 2 mm yr^{-1}, or about one-tenth the rate of the Flandrian transgression. This recent rise is attributable in part to hydroisostatic loading of continental shelves by high late Holocene seas, and in part to global warming, which causes polar ice to melt and ocean water to expand, and for which both natural change and human impacts are invoked. However, where glacioisostatic rebound or tectonic uplift persist, notably around Hudson Bay and northernmost California, respectively, sea level continues to fall relative to the emergent land.

21.2 Modern Coastal Processes

North America's latitudinal spread provides its coasts with distinctive suites of climatic, marine, hydrologic, and ecologic conditions. Arctic coasts are dominated by cold-region processes and frequent sea ice. Atlantic and Pacific coasts are affected by the pulsating atmospheric high-pressure systems and related oceanic gyres of their adjacent oceans, and by the eastward progression of storms embedded within the net westerly circulation of the midlatitudes. The Gulf of Mexico and Gulf of California are impacted by warmth and moisture exported from the nearby tropics, emphatically so during the hurricane season. To these must now be added the impacts of human activity, negligible in the Arctic and greatest near urban centers, and of global warming and rising sea level related less certainly to human activity.

21.2.1 Climatic Factors

Atmospheric systems affect the coast directly by influencing temperature, precipitation, and wind, and indirectly by generating waves and currents. The Atlantic and Pacific coasts lie near the semipermanent Bermuda and Hawaiian high-pressure systems, respectively (see chapter 5). Expansion of the Bermuda High in summer allows warm, moist air from the tropical Atlantic to move into the Atlantic and Gulf coasts from the south and east, while midlatitude cyclonic systems are displaced northward. Conversely, contraction in winter allows these cyclones more latitude along the Atlantic coast, extending the effect of northeasters far to the south. Similarly, summer expansion of the Hawaiian High bathes the Californias in warm dry air and northwesterly winds, reflected in coastal dune orientation, while weaker cyclonic systems move eastward across the Gulf of Alaska. Again, contraction of this High in winter enables rain-bearing cyclones to impact most of the coast, but more often in the north, bringing drenching rains that are transformed by mountainous relief into raging sediment-laden torrents.

The northern coast from Labrador to Alaska, so distant from sources of warmth and moisture, is mostly dry and cold, exceedingly so under winter anticyclonic conditions, less so in the short summer when modest warmth and moisture may penetrate the region. Mean January temperatures in this area typically fall below –20°C, whereas mean July temperatures rarely exceed 10°C. Perennial coastal ice typifies the High Arctic, but disappears for a few months in most summers from the shores of the Labrador Sea, Baffin Bay, Hudson Bay, and the mainland farther west (fig. 21.3).

The coasts of the Gulf of Mexico and Gulf of California are subtropical, as shown by the northward extent of mangroves, which flourish in frost-free conditions where the mean temperature of the coldest month exceeds 20°C. Whereas the former gulf is exposed to moderating oceanic influences, the blistering heat of the Gulf of California is reflected in mean July temperatures above 30°C and in xeric vegetation. The climates of both areas may be tempered by the strong winds and heavy rains of tropical cyclones (hurricanes) during late summer and autumn (fig. 21.4).

21.2.2 Marine Factors

Marine factors function across a range of conditions, from the influence of large-scale ocean currents on the temperature and salinity of coastal waters, through the effect of wind-generated storm waves and swells on erosion and nearshore circulations, to the role of tides in setting water levels and directing tidal currents.

Except in the Arctic, the North American coast is strongly influenced by the great subtropical current gyres of the North Atlantic and North Pacific oceans (fig. 21.3). Allied to the overlying anticyclonic wind systems, these gyres are massive clockwise flows of upper ocean water that are linked with far-reaching meridional circulations and compensating deep-water flows. Collectively, these currents form the thermohaline conveyor belts that distribute the excess solar-induced warmth of the tropics poleward and denser, cold northern waters equatorward and beyond.

In simple terms, within the constraints of fluid motion on a rotating sphere and ocean-basin geometry, the North Atlantic gyre is triggered by persistent northeast trade

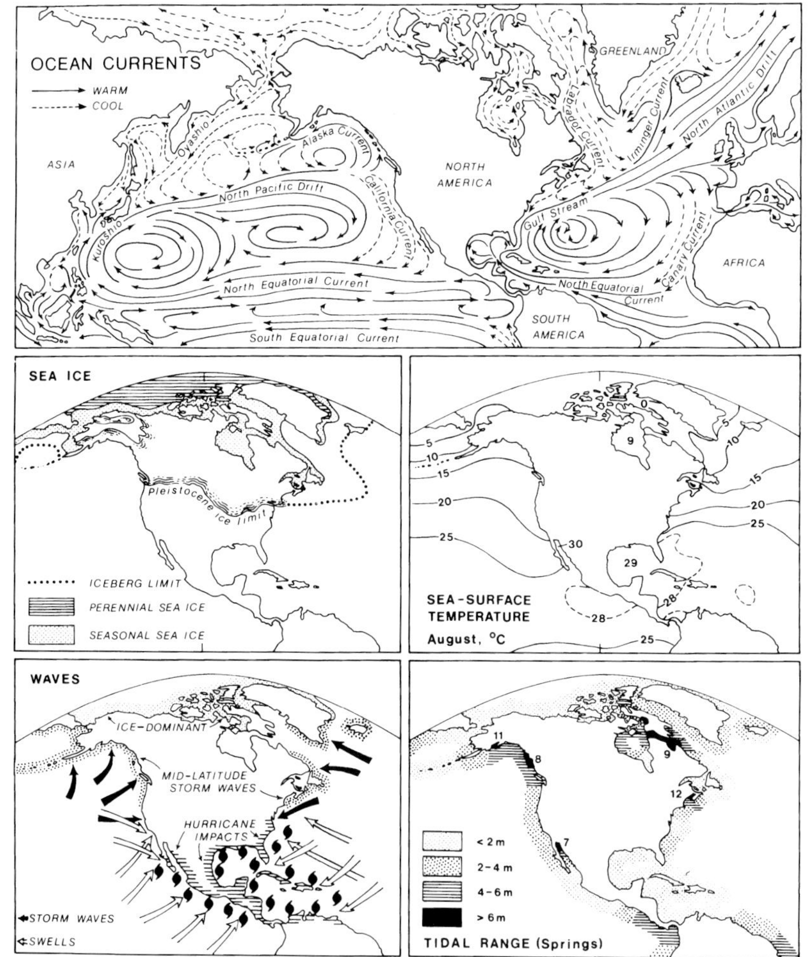

Figure 21.3 (*at left*) Selected environmental variables that affect North American coasts. Temporal changes in these variables reflect ocean-atmosphere forcing.

Figure 21.4 (*bottom*) Atlantic and Pacific hurricanes, August 7, 1980. Hurricane Allen is entering the Gulf of Mexico en route to the Texas coast, while Hurricane Howard is developing off Baja California (photo: GOES image, NOAA, 1980).

winds that blow toward the Intertropical Convergence Zone. Frictional coupling between the atmosphere and the ocean surface, aided by turbulent transmission within the oceans, sets up the North Equatorial Current, which, joined by part of the South Equatorial Current, flows westward off northern South America and then either into the Caribbean and Gulf of Mexico and out through the Florida Strait or along the outer coast of the Antilles toward Florida. Off eastern Florida, these currents link to form the Gulf Stream. Initially, the Gulf Stream is a narrow intense jet of warm (≈20–22°C), saline (>36‰) water, ≈100 km wide, with surface velocities up to 2–3 m s^{-1}, and a depth limited to about 800 m by the underlying Blake Plateau. As the Gulf Stream moves toward Cape Hatteras, its flow is augmented by lateral influxes from the North Atlantic gyre to ≈85 × 10^6 m^3 s^{-1}. Moving farther offshore beyond the cape, it reaches a maximum flow of ≈150 × 10^6 m^3 s^{-1}, forming meanders and eddies en route to the Grand Banks. It then becomes the more diffuse North Atlantic Drift, flowing mostly northeastward but also northward to Iceland as the warm Irminger Current and linking with the West Greenland Current in the Labrador Sea.

Such a massive export of warm tropical water is compensated by return flows at the surface by the Canary Current off northwestern Africa and the East Greenland and Labrador currents, which move cool water southward, and at depth by the export of North Atlantic Deep Water. These cool currents have a chilling effect on adjacent coasts and produce dense fog when warm air advects from the south over their surface waters. The warm West Greenland Current and the cool Labrador Current involve the exchange of about 6 × 10^6 m^3 s^{-1} of water across the Labrador Sea. More extensive exchange with the Arctic Ocean, where a weak clockwise gyre occurs beneath perennial sea ice, is inhibited by narrow straits and submarine ridges.

The North Pacific gyre is similar but, within a broader ocean, weaker and less penetrating. Again, a North Equatorial Current triggers a well-defined stream, the Kuroshio (≈20°C, ≈35‰ salinity, ≈50 × 10^6 m^3 s^{-1} flow), which in turn diffuses off Japan into the North Pacific Drift (fig. 21.3). Moving east toward North America, this drift splits into a warm Alaska Current and a cool California Current. The Alaska Current forms part of an anticlockwise subpolar gyre in the Gulf of Alaska, part of which enters the Bering Sea and feeds the cool Oyashio off Kamchatka. The Bering Strait, only 45 m deep and 50 km wide, restricts exchange of water with the Arctic Ocean to less than 1 × 10^6 m^3 s^{-1}. The slow diffuse California Current, 1000 km wide but with a velocity of <0.25 m s^{-1}, brings cool water and advection fog to the Californias, augmented by coastal upwelling of colder waters, all of which generate ecologically important responses. Episodic perturbations in the ocean-atmosphere system of the equatorial Pacific also generate El Niño conditions wherein warm ocean waters from the western Pacific gravitate eastward toward the Americas (see chapter 4). El Niño's anomalously high sea-surface temperatures and increased moist-air convection lead in the Californias to warm storms and floods, as in the 1982–83 and 1997–98 winters, but in Florida to drier than normal conditions.

Anticyclonic winds emanating from the Bermuda and Hawaiian high pressure cells generate southeasterly swells off the east coast and northwesterly swells off California (fig. 21.3). To the north, midlatitude cyclones generate storm-wave environments in response to Atlantic northeasters and Pacific southwesters. Storm-wave frequency and magnitude thus increase northward, wave heights often exceeding 5 m from Cape Cod to Newfoundland, and in the Gulf of Alaska. In contrast, except during hurricanes, wave energy is much less in the sheltered Gulf of Mexico and is further reduced by friction across the continental shelf such that mean wave heights rarely exceed 1 m. Wave action in the Arctic depends on the duration of ice-free conditions. How wave energy translates into wave-induced current systems and littoral drift depends largely on coastal orientation and nearshore bathymetry. Thus northwesterly swells along the California coast usually promote a net southward littoral drift, but reversals occur when southerly swells are generated by distant storms. Along the east coast, seasonal reversals in the relative importance of storm waves and swells may also reverse littoral drift, southward under the influence of northeasters, northward with southeasterly swells, but these patterns are also affected by feedbacks related to shifting sand bodies shaped by these currents.

Underpinning these effects is the tidal range over which other marine processes operate (fig. 21.3). For example, storm waves and surges are at their most damaging in terms of coastal erosion when superimposed on high tides. Tides along the open Atlantic coast are typically semidiurnal (two high and two low tides of similar magnitude daily), with the mean range of spring tides along the open coast increasing erratically northward from <1 m in south Florida to >3 m in south Greenland. Tides along the open Pacific coast are mixed (two high and two low tides of unequal magnitude) and again, propagating northward, the mean range of spring tides increases from <2 m in Baja California to >4 m in the Gulf of Alaska. Tides in the Gulf of Mexico are mostly diurnal (one high and one low tide daily) with a mean range of spring tides of <1 m. Arctic tides are mostly semidiurnal and microtidal at springs (<2 m). However, spring ranges rise dramatically where tidal forces are constrained by narrowing embayments, notably in the Bay of Fundy (≈12 m), the Gulf of California (≈7 m), Cook Inlet (≈11 m), and various Arctic straits and inlets (≈6–11 m). Tidal currents are relatively weak along open coasts but may reach velocities of ≈7 m s^{-1} in straits between water bodies whose tidal cycles are wholly or partly out of phase. Such hydraulic currents range from those of the Bering Strait to those between islands off British Columbia and Alaska. Swift reversing tidal currents also occur where the flood is forced through narrow straits into bays and the ebb is augmented by stream discharge, notably in San Francisco's Golden Gate. Reversing currents also shape estuarine bedforms and flood-tidal and ebb-tidal deltas between barrier islands.

21.2.3 Hydrologic Factors

Terrestrial drainage affects coastal environments in many ways—by modifying the thermal and chemical properties of coastal waters, by forming density currents in and beyond river mouths, and by yielding terrigenous sediment to coastal systems. Such impacts reflect the hydroclimates and erodibility of contributing basins. Sediment inputs are important where debris-laden streams and glacier meltwaters descend quickly from the erodible Pacific mountain rim, less so where streams must run out across the Gulf and Atlantic coastal plains and sediment-starved rivers from the Canadian Shield reach the sea, and least important where terrestrial drainage seeps seaward through the karst terrains of Florida and Yucatan. Suspended sediment inputs to southwestern estuaries range from 10 to 50 000 mg 1^{-1}, but in the northeast range from only 1 to 25 mg 1^{-1} (Orme, 1990b). Major drainage systems function at a different scale, however, as shown by the massive contributions that the Mississippi and Mackenzie rivers have made to delta growth and shelf sedimentation.

The contrast between inflowing streams and ambient ocean waters is reflected in thermal and chemical transition zones, such as the brackish waters (salinity, S, <30‰) that push the Gulf Stream off the Atlantic continental shelf. Conversely, where there is little freshwater influx and high evaporation, notably in the Gulf of California and southern California during the summer, hypersaline waters (S >40‰) and salt pans characterize coastal wetlands. The mixing of freshwater inflows and ambient seawater is everywhere important to the physical and ecological behavior of estuaries and wetlands (Orme, 1990b).

21.2.4 Other Ecologic Factors

Climatic, marine, and hydrologic factors combine with biotic processes such as evolution, migration and competition to produce distinctive coastal ecologies. At regional scales, upland biomes reaching the coast are modified by reduced thermal extremes and frequent exposure to high winds, salt spray, and fog drip. At local scales, littoral communities develop in symbiotic relationships to the specialized edaphic and hydrologic conditions of beaches, dunes, and wetlands. With the exception of tropical mangroves and trailing beach vines in the far south, North America's littoral vegetation is temperate to subpolar. Grasses (Poaceae) such as *Elymus*, *Ammophila*, and *Sporobolus* are typical of the beach-dune continuum, stabilizing shifting sand and favoring further deposition. Salt-tolerant grasses such as *Spartina*, *Puccinellia*, and *Distichlis*, chenopods (Chenopodiaceae) like *Salicornia*, *Suaeda*, and *Atriplex*, sedges (Cyperaceae) like *Carex* and *Scirpus*, and rushes (Juncaceae) perform similar roles in salt and brackish marshes, trapping sediment, forming peats, and thereby extending wetland environments seaward. Most of these are native plants, though usually related to genera of global extent, but some exotics such as the sand-fixing *Ammophila arenaria* have displaced local plants.

Where siliciclastic sedimentation is minimal, notably on the Florida and Yucatan carbonate platforms, corals appear and biogenic sediments formed from skeletal and pelletal carbonates occur. On Andros Island in the Bahamas, intertidal lime muds are formed largely from the fecal pellets of the gastropod *Batillaria*. Even with siliciclastic inputs, marine animals may form extensive colonies, such as those formed by the oyster *Crassostrea virginica* in the marshes of Georgia (Edwards and Frey, 1977) and by the exotic mud-boring clam *Geukensia demissa*, which ripraps creek banks in San Francisco Bay, where the native oyster *Ostrea lurida* has been decimated by dredging and pollution (Pestrong, 1972).

21.3 Coastal Regions

Many criteria have been invoked in attempts to explain the world's coastal regions. The scheme used here combines location, tectonism, process, form, and ecology to identify six coastal regions for North America. Like the continental landscape as a whole, a major distinction exists between those coasts that have been glaciated during late Cenozoic time and those that escaped the direct impacts of glaciation. Glaciation affected about half the North American coastline, exposing bedrock, reshaping coastal valleys for later fjords, transporting rock waste toward glacier margins and continental shelves where it was later reworked by rising postglacial seas, and causing repeated glacioisostatic adjustments. Like the rivers of glaciated North America, these coasts are postglacial in form. In contrast, nonglaciated coasts, though affected by glacioeustatic changes, reveal a much longer record of erosion and deposition, often extending far back in the Cenozoic. Glaciated and nonglaciated coasts may be further divided by location, with all it implies for tectonism, geomorphic process, and ecology. Pacific coasts near the continent's active margin are mostly mountainous, subject to rapid sediment inputs from short rivers and exposed northward to increasing storminess. Atlantic and Gulf coasts near passive margins are more subdued, subject to widespread shelf sedimentation and barrier-lagoon formation beyond the mainland shore and to storminess derived from midlatitude cyclones in the north and tropical cyclones farther south. Where nearshore clastic sedimentation is minimal, carbonate platforms and coral reefs have formed in warm seas around 5% of the continent's shoreline.

21.3.1 Arctic and Subarctic Coasts

The cold coasts from Labrador to western Alaska are framed by six lithotectonic styles (fig. 1.4). Precambrian basement rocks reach the outer mainland shore from Labrador to the Coppermine River, and the coasts of Baffin, east Devon, and Ellesmere islands, and southeastern and western Greenland. Paleozoic platform covers, mostly Ordovician carbonates, underlie much of the southern Arctic Archipelago from western Baffin to the Beaufort Sea and the south shores of Hudson Bay. Deformed rocks of the Innuitan orogen form much of the northern Arctic Archipelago (Queen Elizabeth Islands) and northern Greenland, whereas the Caledonian fold belt outlines northeast Greenland. The Cordilleran orogen reaches the Beaufort Sea and forms peninsulas that jut into the Bering Sea farther west. Finally, late Cenozoic sediments form a coastal plain across the outer archipelago to the North Slope of Alaska, and thence locally southward along the Bering Sea.

Whereas these rocks form mostly subdued coastal relief, post-Pangean crustal uplift is reflected in coastal mountains linked to the partial opening of the Labrador Sea and Baffin Bay between 90 and 35 Ma and early Cenozoic rifting of Greenland from Scandinavia. Secondary fault structures are reflected in major inlets along the the archipelago's eastern margins in Hudson Strait, Frobisher Bay, Cumberland Sound, Lancaster Sound, and its westward extension as the Parry Channel (Bird, 1985). Cenozoic fluvial erosion and later glacial sculpture of these uplifted margins shaped corridors, which were in turn flooded by postglacial seas, forming spectacular fjords such as those around Scoresby Sound (325 km long, 1450 m deep) in northeastern Greenland, Nansen Sound on Ellesmere Island, and in Labrador (fig. 21.5). Most channels through the Arctic Archipelago were also deepened by glacial erosion. Although Holocene warming has seen the retreat of glacial ice from most of the coast, a 500-km stretch of Melville Bay and 100 km of Kane Basin in northwestern Greenland are backed by ice cliffs up to 40 m high, and valley glaciers elsewhere in Greenland and Ellesmere Island often reach tidewater (Nielson, 1985).

The crust was also repeatedly depressed by and relieved of the weight of Quaternary ice sheets, leaving a legacy of emergent glacioisostatic shorelines and glaciomarine sediment. The south-central part of the archipelago has experienced continuous (but not constant) emergence during postglacial time, even outpacing the Flandrian transgression (Bird, 1985). Beyond this core area, notably on east Baffin and Labrador, shoreline records are more complex because initial isostatic emergence was confounded by later eustatic submergence until, as the transgression waned, isostatic uplift reasserted itself. Beyond the margins of the Wisconsinan ice sheet along the Beaufort Sea, tectonic subsidence, sediment loading, forebulge collapse, and the Flandrian transgression have combined to favor persistent postglacial submergence. However, the 80-km-wide Bering Strait and its shelves are shallow enough to limit the exchange of all but near-surface waters between the Pacific Ocean and the Arctic Ocean, thereby reinforcing the latter's coldness. Farther south, beneath the center of the former Laurentide Ice Sheet, Hudson Bay continues to drain in response to glacioisostatic uplift, leaving behind a bathtub-ring legacy of Holocene shorelines (fig. 2.14). To the east, though Greenland is still depressed beneath its ice cap, partial rebound from its larger Pleistocene ice mass has raised the southern tip some 60 m and the central east and west coasts locally more than 200 m (Nielson, 1985).

These are cold coasts, where periglacial features reach the shore, where seasonal sea ice occurs for 7 to 9 months of the year and at higher latitudes perennial sea ice lies

Figure 21.5 Nachvak Fjord, Labrador (photo: J. Gallagher, 1988).

offshore, and where cold surface waters (0° to −2°C) descend 100 m or more before sinking to generate outflowing deep water. Cold winters, cool summers (but long summer days), low precipitation (usually <250 mm yr^{-1} in the High Arctic), and low evaporation cause these coasts to have an unusually harsh climate. Wave action is muted by coastal ice but may be significant in late summer and early autumn when this ice moves offshore, notably along the west side of Baffin Bay and Davis Strait. Even then, ice returns with onshore winds, reshaping shelf sediment and forming pressure ridges on beaches. Microtidal conditions also limit ice impacts, but where mean tidal ranges increase, notably to 9.3 m in Ungava Bay, boulders are moved by ice floes. Permafrost and ground ice result in coastal features now unique to this region, such as thermoerosional niches and thaw lakes, which form arcuate shores when breached (see chapter 13). Beyond the High Arctic, the Labrador and Bering Sea coasts are less harsh and precipitation rises southward 1000 mm yr^{-1}, but frequent fogs and cold drizzle occur as warmer southerly air advects over cold surface waters.

Despite deflation potential, coastal dunes are not well developed in the High Arctic because suitable material is limited and normal littoral processes can function only for short intervals when the shore is free of ice and snow (Taylor and McCann, 1983; Ruz and Allard, 1994). Along the subarctic shores of Hudson Bay, however, niveo-aeolian deposition is more common and active parabolic dunes up to 4 m high become stabilized by *Elymus mollis* (wild lyme grass), *Honkenya peploides* (sandwort), and *Mertensia* spp. (Saint-Laurent and Filion, 1992).

The lowland coasts of the Beaufort, Chukchi, and Bering seas are different for two reasons: (1) low coastal gradients favor barrier beaches and cuspate forelands, and (2) sediment-laden rivers from the Western Cordillera lead to delta growth under low-energy microtidal conditions. Major deltas on the Beaufort Sea include those of the Mackenzie (8506 km^2), Sagavanirktok (1178 km^2), and Colville (1687 km^2) rivers (Walker, 1985), whereas the vast Yukon River has at one time or another during the Quaternary reached the Bering Sea along a 600-km front between Norton Sound (its present delta) and Kuskokwim Bay, creating a legacy of wetlands. Coastal dunes form where sand is available. Seacliffs occur where Cordilleran structures reach the coast, notably on the Seward Peninsula, and also where the Brooks Range reaches the Chukchi Sea in 300-m-high Cape Lisburne and the Kuskokwim Mountains meet the Bering Sea in Cape Newenham (Walker, 1985).

Persistent cold, lengthy frosts, coastal ice, and low precipitation create an ecological polar desert for these coasts, dominated by a treeless tundra of mosses, lichens, grasses, sedges, herbs, and shrubs (Godfrey et al., 1982). The sparse vegetation cover of the High Arctic includes *Dryas integrifolia*, *Cassiope tetragona*, *Saxifraga oppositifolia*, and *Carex*. Farther south, along the subarctic Labrador coast, these are joined by *Betula nana*, *Empetrum nigrum*, *Arctostaphylos alpina*, *Vaccinium uliginosum*, *Ledum groenlandicum*, and numerous mosses and lichens, with salt meadows of *Carex*, *Glyceria*, *Juncus*, *Ranunculus*, *Stellaria*, *Potentilla anserina*, and *Plantago maritima* in sheltered locations.

Sandy beaches throughout the region are dominated by *Elymus mollis* and *Arenaria peploides* (sea chickweed), often joined by *Armeria maritima*, *Juncus arcticus*, and *Plantago eripoda*. Once these pioneers have begun to trap shifting sand, *Lathyrus japonicus* and *Senecio pseudoarnica* appear, and these pave the way for grassland and heath tundra. Moving coastal ice limits salt marshes in the High Arctic, but such marshes as occur include *Puccinellia phryganodes*, *Stellaria humifusa*, *Cochlearia officinalis*, and, at higher levels, *Primula borealis* and *Carex ursina*. However, brackish and freshwater wetlands commonly reach toward the shore (see chapter 8). Outliers of the boreal forest locally reach the shores of southern Labrador, the Mackenzie delta, and the Bering Sea.

21.3.2 North Atlantic Glaciated Coast

The fog-shrouded ruggedness of subarctic Labrador passes southward into cool temperate conditions from Newfoundland to the Hudson River (fig. 21.6). The coast now becomes lower and less rugged, and glaciated bedrock is increasingly mantled by glaciogenic deposits that have been reworked into varied coastal depositional forms. The latter reflect both increasing sediment availability and gentler nearshore gradients associated with the extensive continental shelf, 400 km wide on the Grand Banks southeast of Newfoundland and 200 km wide farther south.

This is a passive-margin coast where Grenvillian structures parallel the north shore of the Gulf of St. Lawrence and Appalachian structures reach the open coast. Precambrian and Paleozoic igneous and metamorphic rocks give the northern parts of this coast much of its character in the form of seacliffs and rocky bays, but as these are masked southward by Quaternary deposits coastal barriers backed by lagoons and salt marshes become more important. The Wisconsinan terminal moraine and proglacial outwash deposits, though widely breached, define the outer coast from Nantucket Island, through Martha's Vineyard and Block Island, to central Long Island and Staten Island in the Hudson estuary. Similarly, a major recessional moraine outlines the coast from Cape Cod through western Rhode Island to northern Long Island where it fuses with the earlier morainic ridge.

Present coastal location reflects complex late Quaternary changes in relative sea level whose magnitude and timing vary along the coast because of the asynchronous nature of deglaciation and isostatic rebound (Kelley et al., 1992; Knebel et al., 1992). In essence, the Wisconsinan ice sheet began retreating from southern New England around

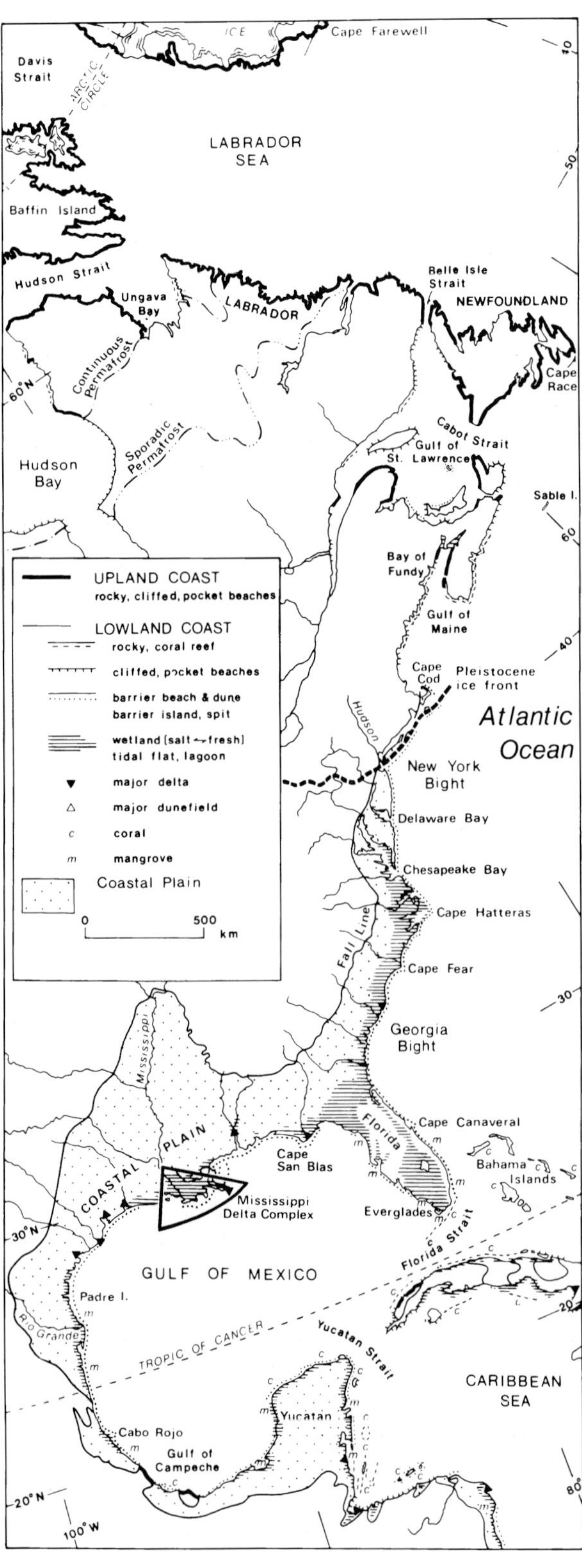

Figure 21.6 Dominant morphology of the Atlantic and Gulf of Mexico coasts.

16 ka and from the Maritime Provinces around 14 ka, and was accompanied by a marine transgression with ice and sea in contact, forming the Champlain Sea in the St. Lawrence Valley (Hillaire-Marcel and Ochietti, 1980). Outwash sands and gravels were deposited on underlying bedrock and earlier tills, and were in turn overlain by glaciomarine muds, both inland and seaward of the present shore. As a result of glacioisostatic depression, relative sea level around 13–12 ka reached 18 m above present in Boston Harbor and nearly 80 m in coastal Maine (fig. 21.2). Later, as isostatic rebound led to coastal emergence, local sea level fell at 11 ka to −22 m in Boston Harbor and −60 m in Maine, allowing streams to dissect exposed glaciomarine sediment and form deltas beyond the present coast. The Flandrian eustatic transgression began crossing this shore around 9 ka, eroding earlier sediment, depositing marine sands and gravels across a time-transgressive unconformity, and infilling stream channels with Holocene estuarine deposits. Despite this transgression, emergent glaciomarine sediment of late Pleistocene age still occurs onshore, notably in the Presumpscot Formation (14–11.5 ka) in Maine (Bloom, 1963; Kelley et al., 1992), as well as offshore beneath the Gulf of Maine.

Wave action along this coast is associated in winter with midlatitude cyclones moving offshore to generate northeasters, which in turn promote littoral drift toward the south. Summer waves are less powerful and normally associated with cyclonic systems farther north, which generate southwesterly winds and northward littoral drift. Swells generated by hurricanes farther south commonly affect exposed south-facing shores during late summer and autumn. More rarely, hurricanes may cross the coast attended by storm surges and much damage. Long Island Sound and the Gulf of St. Lawrence are sheltered from most storm-wave activity. Mean tidal ranges increase dramatically northward, from 1 m off Long Island and 3 m off Boston, to the spectacular tides of the Bay of Fundy, from 6 m at the mouth to almost 12 m in the Minas Basin, where a maximum fair-weather range of 16.3 m has been recorded (McCann, 1985). Tidal ranges diminish farther north but increase in the funnel-shaped St. Lawrence estuary to more than 4 m near Québec city. Sea ice may persist for 6–7 months in the Strait of Belle Isle, 3–4 months in the Gulf of St. Lawrence, and for lesser periods elsewhere, but, except in shallow bays, the open coast southward from southeast Newfoundland is essentially ice free. Permafrost, though reaching northern Labrador shores, does not affect this coast.

Where sediment availability combines with gentle nearshore gradients, barrier islands, baymouth barriers, spits, and cuspate forelands have developed. Barrier islands occur along the southern margins of the Gulf of St. Lawrence in New Brunswick, Prince Edward Island, and the Magdalen Islands (McCann, 1979), and along the morainic and outwash terrain from Cape Cod to Long Island (fig. 21.7). Ebb

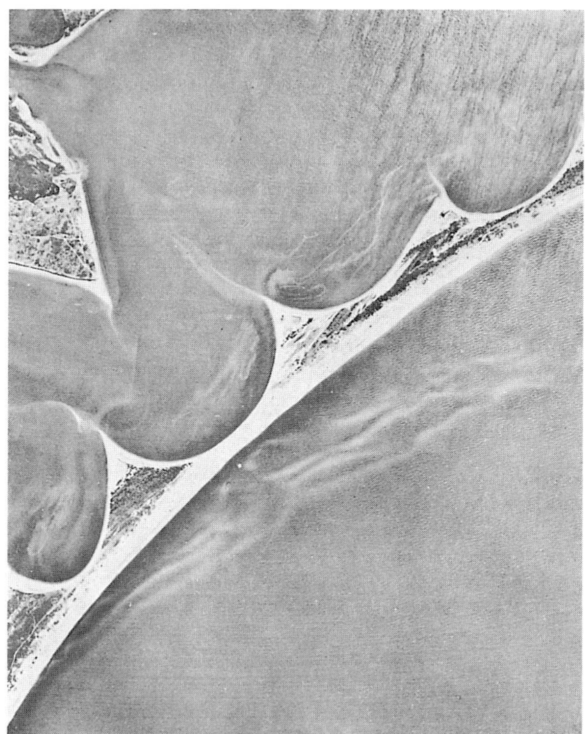

Figure 21.7 Barrier-lagoon system and subtidal bars off Nantucket Island, Massachusetts. The barrier's smoothly curved outer coast contrasts with the cusped inner coast (photo: U.S. Geological Survey, 1938).

and flood tidal deltas often develop between barriers, whereas partly submerged drumlins in Boston Harbor have been reworked by spit and tombolo growth. Tidal flats and back-barrier lagoons favor salt-marsh accretion, offering opportunities for reclamation, which in the Bay of Fundy began as early as the seventeenth century. Seasonal ice rafting introduces erratic boulders and tears up salt marshes in the St. Lawrence estuary (Dionne, 1972). Near the shelf edge 200 km off Nova Scotia lies Sable Island, a small crescentic structure, 40 km long and <1.5 km wide, composed of highly mobile beach sand capped by active and vegetated parabolic dunes shaped by strong westerly winds (McCann and Byrne, 1994). Elsewhere, bedrock control is evident in cliffed headlands separated by rocky embayments and pocket beaches.

Winters remain cold, but, compared with subarctic coasts, summers are warmer and longer, precipitation is better distributed throughout the year, fogs are less persistent, and warm Gulf Stream eddies dissipate cold waters from the north. Significantly, the annual average frost-free period increases from 100 days in northern Newfoundland to 220 days in southern New England.

The boreal forest that reaches the north shores of the Gulf of St. Lawrence and Newfoundland is dominated by black spruce (*Picea mariana*), white spruce (*P. glauca*), balsam fir (*Abies balsamea*), tamarack (*Larix laricina*), paper birch (*Betula papyrifera*), and quaking aspen (*Populus tremuloides*). Across the Gulf, however, the mixed conifer–hardwood Acadian forest reflects a transition to cool temperate, damp, and often foggy conditions. Conifers are augmented by red spruce (*P. rubens*), northern white-cedar (*Thuja occidentalis*), jack pine (*Pinus banksiana*), eastern white pine (*P. strobus*), and eastern hemlock (*Tsuga canadensis*); hardwoods include northern red oak (*Quercus rubra*), beech (*Fagus grandifolia*), and sugar maple (*Acer saccharum*). Many of these trees spread onto stable dunes. Spruce and fir disappear from the coastal forest in southern Maine, pitch pine (*P. rigida*) and eastern redcedar (*Juniperus virginiana*) become more important, and many deciduous trees common farther south make their first appearance in southern New England, notably stave oak (*Quercus alba*), black oak (*Q. velutina*), and various cherries (*Prunus* spp.) and serviceberries (*Amelanchier* spp.) of the rose family.

Southward warming is also reflected in littoral ecology. The dominant beach plants of the High Arctic and Labrador, *Elymus mollis* and *Arenaria peploides*, become patchy in Newfoundland and are gradually replaced southward by the beach grass *Ammophila breviligulata* and by *Cakile edentulata* and *Salsoli kali* (Godfrey et al., 1982). *Ammophila* thrives in loose sand and soon traps foredunes, where it is joined by *Solidago sempervirens*, *Lathyrus japonicus*, *Artemisia stelleriana*, *Festuca rubra*, and others. As dunes stabilize, a tundra-like heathland may develop with *Hudsonia tomentosa* and *Arctostaphylos uva-ursi* as ground covers, with *Vaccinium* spp. in damp hollows. Elsewhere, stable dunes and coastal hills carry shrublands dominated in the north by *Myrica pensylvanica*, *Rosa virginiana*, *Spiraea latifolia*, *Rhus toxicodendron*, and *Rubus* spp., interspersed with many of the trees noted previously, and farther south by *Prunus maritima*, *Andropogon scoparius*, and *Deschampsia flexuosa*. Stable dunes in the Province Lands of Cape Cod support a remnant of primeval beech forest.

Likewise, coastal salt marshes now begin to assume a temperate appearance that changes little from New England to the Gulf of Mexico. Arctic and subarctic marsh plants are replaced by *Spartina* cordgrasses. *Spartina alterniflora* dominates the intertidal marsh; *S. patens*, *Distichlis spicata*, *Suaeda maritima*, and *Limonium* spp. dominate the high marsh; pickleweed (*Salicornia* spp.) occurs around salt pans; and *Spartina pectinata*, *Myrica gale*, *Typha angustifolia*, and *Juncus* spp. grow in northern brackish and freshwater marshes. *Plantago maritima* and *Triglochin maritima* may also dominate northern high marshes but diminish southward. Many other genera characterize the high marsh, with southward warming reflected in species changes.

21.3.3 North Atlantic Coast Beyond the Ice Front

From the Hudson River to Florida, the coast meanders across the low-gradient, 200–300-km-wide continuum of the Atlantic Coastal Plain and continental shelf, impinging on Appalachian uplands in the north, approaching the shelf edge at Cape Hatteras and southeast Florida, and elsewhere ranging from 100 to 200 km seaward of the Fall Line (fig. 21.6). Gone is the glacial veneer of the coastal strip between Cape Cod and Long Island, replaced by features shaped by prolonged fluvial and marine processes. Nevertheless, the evolution of this coast still owes much to the indirect impacts of glaciation through eustatic and isostatic changes superimposed onto longer term flexuring of the continent's Atlantic freeboard.

This is an archetypal passive-margin coast, where quiescent Appalachian orogens emerge far inland behind the Fall Line or lie at depth beneath Mesozoic and Cenozoic sediments, and where the rifted margins of North America's continental plate lie 100–500 km offshore beyond the continental slope. The emergence of the Atlantic Coastal Plain here, compared to its submergence farther north, probably reflects post-Pangean flexural warping related to a hinge zone oblique to the coast (Steckler et al., 1988), but added subtleties are provided by recent glacioisostatic responses. Thus, whereas the glaciated coast to the north has been subjected to glacioisostatic emergence during the Holocene, the northern part of the middle Atlantic coast has seen forebulge collapse, which helps to explain the submergence of Delaware Bay and Chesapeake Bay. Whereas the coastal zone mostly reflects the fluvial delivery of siliciclastic sediment from the Appalachians during post-Pangean time, the warmer waters of the Florida platform, separated from this debris by a structural trough between the Atlantic and the Gulf of Mexico until at least Miocene time, saw prolonged carbonate sedimentation. Although Florida has been attached to North America throughout the Quaternary and siliciclastic sediment has been moving south, carbonate deposition continues in the far south and in the Bahamas. The Florida Keys are remnants of a Last Interglacial reef tract and ooid sand shoals, the Key Largo Limestone and Miami Oolite, respectively, that are built on earlier biogenic carbonates and now rise above active coral reefs, dominated by *Acropora* coral, along their southeast margin (Davis et al., 1992).

With gentle seaward gradients, this coast contains a rich legacy of late Cenozoic sea-level changes caused by tectonic flexuring, sediment loading, and eustatic and isostatic forcing (Cronin et al., 1981). Miocene and early Pliocene shores lie mostly landward of the present coast, raised by tectonic flexuring on structures such as the Cape Fear Arch in the Carolinas and the Peninsular Arch in Florida. Thus, from the Delmarva Peninsula to North Carolina, the Yorktown (4.5–3 Ma) and Chowan River (2.8 Ma) formations reflect dominantly marine deposition under warm temperate conditions with low fluvial sediment inputs (Krantz, 1991; Ramsey, 1992). In contrast, the late Pliocene saw increased sediment inputs from the Appalachians under cooler, wetter climates, reflected in the mostly terrigenous Bacons Castle and upper Beaverdam (2.3–2.0 Ma) formations. An early Pleistocene marine phase then produced the coast-parallel Surry scarp and Windsor Formation (≈1.5 Ma) (Ramsey, 1992). Farther south, the Georgia Bight is lined by a stacked sequence of late Pliocene and Pleistocene shelf, barrier, and lagoonal facies, and deltaic deposits related to the Pee Dee, Santee, and Savannah rivers. The Florida peninsula has a low spine of emergent Pliocene and early Pleistocene marine deposits.

During Quaternary glaciations, low sea levels allowed land drainage and glacial meltwaters to extend across the exposed continental shelf, with the Hudson, Delaware, and Susquehanna rivers carving deep channels and flushing terrigenous sediment far offshore. During high interglacial sea levels, these drainages were partly flooded, terrigenous and marine sediment was reworked into offshore shoals, transgressive barriers, and estuarine fills, and cliffs were cut into the coastal plain. During these highstands, coastal deposition shifted the mouth of Chesapeake Bay progressively southward, as it is doing today, leaving the Susquehanna's ancestral paleochannels beneath the prograding tip of the Delmarva Peninsula (Colman et al., 1990). On the peninsula's seaward side, shoals and barriers related to marine oxygen isotope stage (OIS) 5 (125–75 ka) indicate eustatic changes ranging from a high of 6 m above sea level at 125 ka (OIS 5e) to a low of −23 m at 115 ka (OIS 5d) (Toscano, 1992). Bluffs cut during the OIS 5e highstand also occur on New Jersey's Cape May and in the Suffolk scarp of Virginia. Farther south, 18 Quaternary highstands are preserved in coastal deposits 60 m thick beneath Albemarle and Pamlico Sounds, North Carolina (Riggs et al., 1992). Late Pleistocene barriers also form the nucleus of the Sea Islands along the Georgia Bight, and beach ridges of similar age line Atlantic Florida.

The intricacies of the present coast reflect erosion and deposition during and since the main Flandrian transgression. Chesapeake Bay below Annapolis has filled with 52×10^9 m^3 of Holocene deposits, leaving a water volume of 48×10^9 m^3, which, at Holocene rates of sedimentation (5.2×10^6 m^3 yr) and in the absence of sea-level change, would be filled in 9200 years (Colman et al., 1992). At its maximum around 6–4 ka, the Flandrian transgression rose 1–2 m above present sea level, as shown by beach ridges in Florida. This was perhaps a response to mid-Holocene warmth and its effect was similar to that predicted for future greenhouse-warming scenarios. At present, for complex reasons, sea level is again rising against this coast, in North Carolina at a rate of 10–25 cm a century (Riggs et al., 1992). Because of low coastal gradients, small sea-level fluctuations may cause large lateral shoreline changes.

Because prevailing westerly winds blow offshore, this coast is dominated by southeasterly swells but these are often augmented by storm waves generated by northeasters and hurricanes. The former, more common in the north and in winter, develop as midlatitude cyclones pass eastward across the coast. The latter, more common in the south and in late summer and autumn, are triggered by tropical cyclones that strengthen to hurricane force (fig. 21.8). Very low atmospheric pressure and steep pressure gradients generate high winds and storm surges at the coast, with major impacts on natural and human systems alike. Category 5 hurricanes (on the 1-to-5 Saffir-Simpson Scale, see chapter 6) generate winds >250 km hr^{-1} and storm surges >5.5 m above normal sea level, causing massive erosion, overwash, barrier breaching, and widespread flooding. On average, direct hurricane impacts may be expected somewhere along the Florida coast more or less annually, along the North Carolina coast once every 3 years, and along the Delaware coast once every 15 years. Mean wave height decreases southward from 1.5 m at Cape Hatteras to 0.8 m at Cape Canaveral, but remains at 0.7 m in southeast Florida, where the partial shelter offered by the Bahamas Islands is offset by a narrower, more steeply sloping continental shelf. Because this mix of southeasterly swells and northeasterly storm waves is affected by varying coastal orientation and bathymetry, nearshore current circulations are complex, and littoral drift, though seasonally high, is subject to frequent reversal. This helps to explain the cuspate forelands, notably at Capes Hatteras, Lookout, Fear, and Canaveral. As wave heights diminish southward, tidal ranges increase from around 1 m off North Carolina to 2.8 m for spring tides at St. Helena Sound, South Carolina, before declining to 2 m off the St. Johns River, Florida, and less than 1 m farther south. The central Georgia bight thus tends to be tide-dominated (Hayes, 1985), compared with wave-dominated North Carolina where wave-induced onshore migration of longshore bars may reach 24 m d^{-1} (Sallenger et al., 1985).

This is a classic barrier-island coast, a dynamic system of shifting offshore bars and shoals, migrating tidal inlets with ebb and flood deltas, back-barrier lagoons, accreting tidal flats and wetlands, and distant mainland estuaries (fig. 21.9). Barrier islands are typically very long (5–100 km) relative to width (<1 km). Distal ends near tidal inlets are unstable, and ocean shores are frequently reworked by storm waves, but core areas may support high dunes. Most back-barrier lagoons are similarly long and narrow, and have often become choked with sediment and reclaimed, or otherwise re-engineered as links in the Intracoastal Waterway system. At the Outer Banks of North Carolina, however, barrier islands leading toward Cape Hatteras are separated from a swampy mainland by up to 50 km of water across Pamlico Sound.

The origin of barrier islands has attracted many hypotheses, harking back to the ideas of de Beaumont (1845)

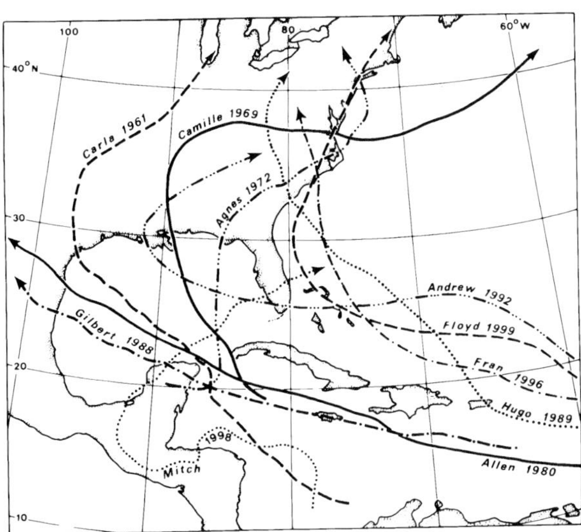

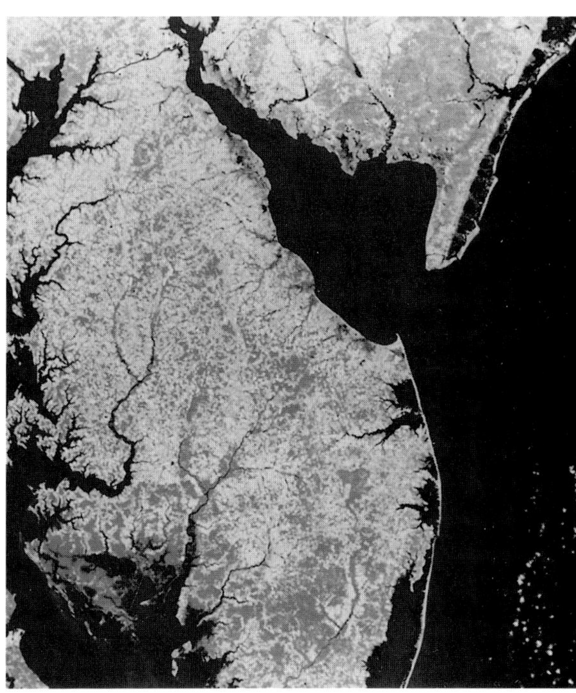

Figure 21.8 Tracks of selected Atlantic hurricanes. Most such storms begin as tropical disturbances over warm ocean water, intensify to hurricane force on approach to the coast, and weaken quickly over land, but some are revitalized on again reaching warm water (source: NOAA).

Figure 21.9 Barrier island-lagoon coast north and south of Delaware Bay (photo: NASA).

on offshore bar accretion, Gilbert (1885) on spit migration, and McGee (1890) on ridge submergence. More recently, barrier-lagoon systems of the Atlantic and Gulf coasts have been the focus of much research (e.g., Hoyt, 1967; Kraft et al., 1979; Leatherman, 1979; Oertel, 1985: Dolan and Lins, 1987; Oertel et al., 1992). In essence, such barriers form ideally on low-gradient nearshore bottoms with abundant sediment and strong constructive wave action. When sea level rises, barriers are normally reworked upslope, transgressing their own back-barrier deposits, overrunning landscapes shaped during previous regressions, and absorbing relict coastal features of earlier transgressions. As a transgression ends, barriers become more stable, though still subject to wave overwash and reworking of their distal ends, back-barrier lagoons become sediment sinks, and muds in turn provide substrate for marsh development. When sea level falls, barriers are stranded on emergent strandplains, back-barrier lagoons and marshes are converted to dry land, and flanking tidal inlets are scoured by outflowing terrestrial streams. This general scenario explains many, but not all, of the features found along the Atlantic and Gulf coasts. Modern barriers are essentially forms initiated during and after the last major transgression, reworked under present conditions even as their back-barrier lagoons are converted to wetlands. Subtle differences between barriers reflect such variables as local rates of sea-level change, the bathymetry of nearshore ramps, the relative roles of wave and tidal forces and river discharge, and the availability of sediment. For example, coasts with higher tidal ranges favor barriers flanked by tidal inlets, ebb and flood-tidal deltas, and back-barrier tidal channels, whereas wave-dominated coasts favor overwash and landward migration of barriers even with minimal sea-level rise. Further, although abundant sediment existed for barrier growth during the main Flandrian transgression, the more stable sea levels of the late Holocene have reduced sediment availability and, with increased storminess, have prompted beachface erosion and barrier retreat along many fronts.

Ecologically, this coast is temperate in the north and subtropical in the south. Under the warming influence of the Gulf Stream, the frost-free period increases from 230 days in the north to 310 days in coastal Georgia. Frost is almost unknown along the coast south of Cape Canaveral, but cold snaps still limit many tropical species (Godfrey et al., 1982). Precipitation is distributed throughout the year, with midlatitude cyclonic rain and snow diminishing southward as rain from tropical storms and convective thunderstorms increases. Convective thunderstorms develop over land and move eastward, explaining why annual rainfall averages 1567 mm at West Palm Beach but only 1016 mm at Key West.

Much of the Atlantic Coastal Plain is dominated by pine forest, with pitch pine giving way southward to loblolly pine (*P. taeda*), dominant on coastal plain and barrier island alike, longleaf pine (*P. palustris*), and, farther south, slash pine (*P. elliottii*). Eastern redcedar and coastal heath reach their southern coastal limits in Virginia. The region also contains many oaks: deciduous stave oak, black oak, and southern red oak (*Quercus falcata*); water oak (*Q. nigra*), laurel oak (*Q. laureliana*), and willow oak (*Q. phellos*) that become increasingly evergreen to the south; and the dominant evergreen live oak (*Q. virginiana*). Other hardwoods include sweetgum (*Liquidambar styraciflua*) on drier sites, and sweetbay (*Magnolia virginiana*), black tupelo (*Nyssa sylvatica*), and green ash (*Fraxinus pennsylvanica*) on wetter sites and riverine swamp forests. The forest's evergreen nature is enhanced southward by many smaller trees, whereas the approaching humid tropics are heralded by vines, mosses, and lichens. Palms also appear, with cabbage palm (*Sabal palmetto*) growing in sandy habitats from Florida to North Carolina. In shrublands, northern plants give way to southern species such as *Aralia spinosa* and *Baccharis halimifolia*.

Beach vegetation is typified by *Cakile edentula*, *Salsola kali*, and *Euphorbia polygonifolia*, joined southward by *Croton punctatus*, *Sesuvium portulacastrum*, and in the far south by subtropical *Euphorbia* and *Ipomoea*. Dunes in the north are still dominated by *Ammophila* and *Solidago*, but *Lathyrus japonicus* and *Artemisia* yield to more southern plants such as *Strophostyles helvola* and *Panicum* spp. However, *Uniola paniculata* appears at the south end of Assateague Island off the Delmarva Peninsula and gradually becomes the dominant dune grass as *Ammophila* disappears south of Cape Fear. Dunes along the Georgia Bight support many grasses, including *Panicum amarum*, *Cenchrus tribuloides*, *Eragrostis pilosa*, and *Sporobolus virginicus*, various herbs such as *Diodea virginica*, *Solidago virginicus*, and *Oenothera humifusa*; the woody vines *Ampelopsis* and *Parthenocissus*, and the dune-building shrub *Iva imbricata*. Once stabilized, the dunes are invaded by shrubs and trees from the coastal forest. Interdune hollows may support freshwater wetlands, with *Typha* and *Juncus* often extensive.

Grasslands occur on back-barrier overwash surfaces and old inlets between the dunes and the high salt marsh, notably on the Outer Banks of North Carolina. *Spartina patens* var. *monogyna* dominates, along with *Andropogon*, *Solidago*, *Eragrostis*, *Muhlenbergia*, and *Hydrocotyle*, and in wetter areas *Typha*, *Juncus*, and *Scirpus* spp. *Spartina patens* grows decumbent in high salt marshes north of Cape Cod but, as the variety *monogyna*, becomes more upright and aggressive southward, invading barrier flats and low dunes along the central Atlantic coast.

Salt marshes are widely developed along this coast behind barrier islands and against mainland shores and estuaries, grading inland into brackish and freshwater marshes. The low marsh is dominated by *Spartina alterniflora*, above which *Salicornia virginica*, *S. bigelovii*, and, farther south, *Batis maritima* occur. The high marsh is dominated by *Spartina patens*, with *Distichlis spicata* and *Scirpus robustus* in the north, and *Juncus roemerianus, Fimbrystylis castanea, Limonium carolinum*, and others farther south. Red mangrove (*Rhizophora mangle*) and black mangrove (*Avicennia germinans*) begin their tropical ranges on Florida's east coast.

21.3.4 The Gulf of Mexico Coast

The Gulf coast stretches in a broad arc from the Florida Keys to the Yucatan Peninsula, a shoreline distance of 4500 km although the end points are little more than 600 km apart (fig. 21.6). In many respects, this coast is a continuation of the Atlantic coast but with less energy, more mud, and warmer habitats. The coastal plain and continental shelf around the northern Gulf are even wider and flatter, the former extending 800 km up the Mississippi valley, the latter 200 km wide off west Florida. Farther south, where the Sierra de los Tuxtlas approaches the Vera Cruz coast, the plain-shelf continuum is less than 50 km wide. Beyond that, the karst landscape and broad shelf of Yucatan resemble the Florida carbonate platform.

The Gulf coast is also a passive-margin coast, initiated by the Jurassic separation of North America from Gondwana, but complicated by the suturing of Florida basement, the growth of oceanic crust beneath the Gulf, and the rotation of Yucatan, as well as coast-normal structural arches and basins, coast-parallel normal faults, and salt-dome tectonics. Post-Pangean sea-level changes are reflected in seaward-dipping marine and terrestrial strata beneath the coastal plain. These usually become younger toward the coast, although erosion and reworking of older deposits during transgressions have created gaps in the stratigraphic record (Walker and Coleman, 1987). Thus Paleogene limestones lie at shallow depth beneath Florida's Big Bend, Neogene deposits are exposed in coastal bluffs in Mobile Bay, and Quaternary sediments, though extensive, are often thin.

As elsewhere, the Gulf coast reflects the changing sea levels, sediment budgets, and geomorphic processes of the Quaternary, but under low-energy conditions the responses are subtle. Events during and since the Last Interglacial are particularly relevant to the shaping of the present coast. Thus the marine transgression that introduced the Last Interglacial is reflected in the widespread Biloxi Formation, a suite of shallow-water and estuarine deposits plastered across the shelf-plain continuum. The OIS-5e highstand is represented by a barrier complex, the Gulfport Formation, which rises 3–12 m above the mainland shore of the northeast Gulf, and the similar Ingleside Formation of Texas and Tamaulipas (Anderson et al., 1992; Otvos, 1992). The respective floodplain facies inland are represented by the Prairie and Beaumont formations, which spread across the emerging continental shelf during regressions, augmented by fluvioglacial meltwaters draining down the Mississippi valley. Wisconsinan lowstands saw fluvial dissection of the widely exposed continental shelf, forming valleys that were later backfilled during the Flandrian transgression and became bays on the modern coast. Thus the submerged extensions of the Trinity and Sabine river valleys in Texas became Galveston Bay and Sabine Lake, respectively (Anderson et al., 1992).

This is a low-energy coast sheltered from Atlantic swells, and even storm waves generated in the Gulf expend much energy crossing the shallow continental shelf. Indeed, the Big Bend of northwest Florida has been termed a "zero-energy" coast because the shelf slopes at only $0°1'$ and coastal concavity causes wave rays to diverge and dissipate (Tanner, 1985). Wave heights average less than 0.1 m here and rarely exceed 1 m across the Gulf as a whole. Low wave energy in turn limits longshore current velocities and littoral drift. The latter moves weakly from east to west along the northern Gulf coast in response to circulations generated by prevailing southeast winds, but frequent reversals and localized cells are common. South of Padre Island, littoral drift is northward in response to changing coastal orientation. The range of mixed and diurnal tides is less than 2 m, and usually less than 1 m—just sufficient to impose tidal dominance on coasts with little wave action and for tidal currents to scour "passes" between barrier islands.

Exceptions to these low-energy conditions occur when hurricanes hit the coast with strong winds and accompanying storm surges (fig. 21.8). Many passes between barrier islands close during hurricanes, whereas others open, for example, Hurricane Pass and Redfish Pass which sliced through west-central Florida's barriers in 1921 (fig. 21.10). In 1900, Galveston Island, Texas, was inundated by a 6-m hurricane surge, with the loss of 6000 lives. More recently, in 1969, Hurricane Camille, at category 5 the most severe historic event along the Gulf coast, cut a broad swath through Ship Island, Mississippi, and caused widespread damage and loss of life.

With negligible coastal gradients, low barrier islands often alternate with or front extensive muddy wetlands, the former reflecting wave-dominance, and the latter tide-dominance. High seacliffs are lacking, but erosion of old coral, worm, and oyster reefs may produce low rocky shores. Barriers and beach-ridge plains form most readily where clastic sediment is available from nearby rivers and bluffs, or where shelf sands were worked onshore during the Flandrian transgression. Such barriers occur in west-central Florida, farther west where Appalachian rivers reach the Gulf, in the oak-crested (chenier) beach ridges west of the Mississippi delta, along the western Gulf coast where Padre Island is nearly 200 km long, and around the Gulf of Campeche. Low dunes atop these barriers rarely exceed 10 m in height. Variable waves and currents locally combine to build cuspate forelands, notably at Cape San Blas west of the Apalachicola delta and at Cabo Rojo in northern Vera Cruz. The shallow lagoons and wetlands landward of these barriers include Laguna Madre behind

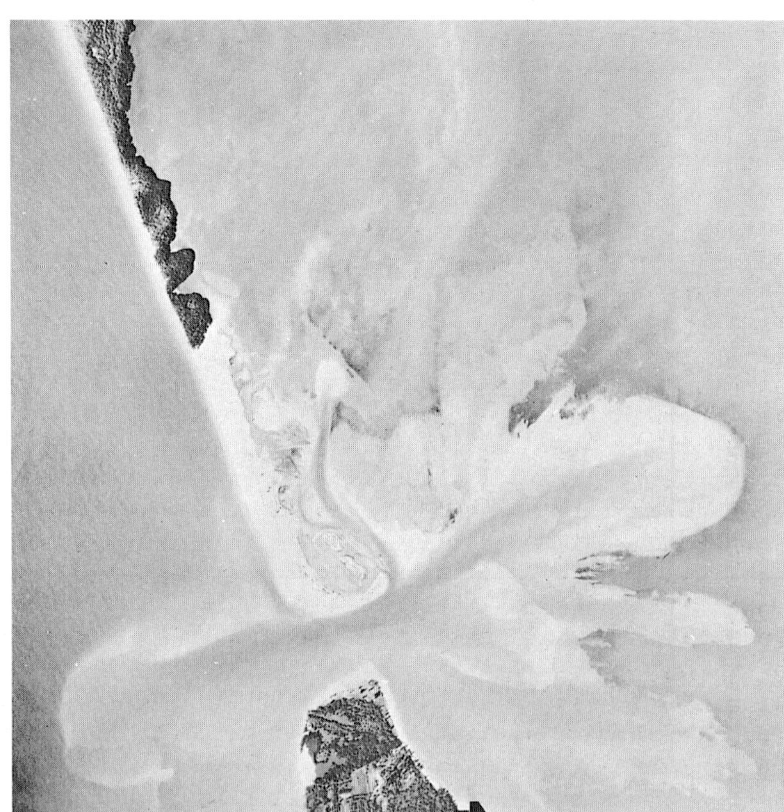

Figure 21.10 Redfish Pass cut by hurricane activity through western Florida's barrier-island system in 1921. A flood-tidal delta diffuses into Charlotte Harbor to the right; a narrower ebb-tidal jet flows into the Gulf of Mexico to the left (photo: U.S. Department of Agriculture, 1944).

Padre Island and its namesake in Tamaulipas. In the absence of barriers, wetlands reach the open coast, notably along Florida's frayed shores off the Big Bend and the Everglades, and west of the Mississippi delta. At Florida's southern tip, where siliciclastic debris is lacking, beaches form mostly from reef and shell detritus.

Under such low-energy conditions, deltas often form at the mouths of major rivers. Some, such as the Apalachicola, Mobile, Sabine, and Rio Grande deltas, form in estuaries behind barriers; others like the Mississippi complex and the lesser Brazos delta reach the open coast. The 29,000- km^2 Mississippi delta was the focus of much early research (e.g., Russell, 1936; Fisk, 1944), and study of its evolution, ecology, hydrocarbon potential, and engineering problems continues (e.g., Coleman, 1982, 1988; Penland et al., 1988). As the first such feature to be studied by modern techniques, the Mississippi delta was once regarded as the typical delta but is now viewed as a strongly river-dominated form in a spectrum that encompasses varying degrees of river, wave, and tide dominance.

In essence, the modern birdfoot and Atchafalaya deltas are but the most recent in a sequence of lobes that have formed off the Mississippi river mouth during later Holocene time, successors to earlier systems that reflect the river's major roles in waste removal and shelf progradation. Each lobe has evolved in a similar way (Nummedal et al., 1985). An initial constructional phase sees rapid shoreline progradation with a thick wedge of terrigenous clastics. After 1000–1500 years of active delta growth, a major channel switch occurs in response to hydraulic gradient advantages and a new delta lobe develops. The abandoned lobe is then slowly destroyed by erosion and dispersal of fine sediment, while coarser sediment is reworked into barrier islands. Concurrent subsidence allows the sea to flood back-barrier marshes, widening lagoons and reducing barriers to offshore shoals until, at some future time, the area may see a new cycle of delta growth. Present subsidence rates in the Mississippi delta range from 0.3 to 2 m yr^{-1} (Walker and Coleman, 1987).

The Holocene delta complex comprises six major lobes and associated barrier-lagoon systems dating back over 7000 years, each reflecting rapid deposition of the river's dominantly muddy load, subsequent channel switching, abandonment, and subsidence. These are the Maringouin (≈7–6 ka), Teche (≈6–4 ka), St. Bernard (≈4–1 ka), Lafourche (≈3.5–0 ka), Plaquemines-Balize or modern (≈1–0 ka), and, since the 1973 floods, the Atchafalaya delta lobes. The barrier-island phase can be observed in the rapidly eroding Isles Dernieres and Timbalier Islands over the subsiding Lafourche lobe, whereas the Chandeleur Islands barely survive over the older St. Bernard lobe, and former barriers off the Maringouin delta now form the Ship and Trintity shoals offshore. The modern birdfoot delta, with its fingering distributaries, natural levees, crevasse splays, and marginal wetlands, is the essence of river dominance. Farther offshore, the loading of prodelta clays by fluvial silt and sand causes submarine landslides, widespread soft-sediment deformation, diapiric intrusion, and thrusting of mud lump islands to the surface (Prior and Colman, 1978). Were it not for flood-control structures upstream, the Mississippi River would by now have switched its entire discharge to the Atchafalaya lobe, abandoning its course past New Orleans!

Ecologically, the Gulf coast is warm temperate in the north and humid tropical in the south. Despite outbreaks of cold polar air, the frost-free period exceeds 300 days along the north coast. Frost is unknown farther south. Rainfall occurs throughout the year, mostly from summer convective thunderstorms and tropical cyclones. It exceeds 2000 mm annually in the Gulf of Campeche, but diminishes to 600–1000 mm along the western Gulf coast where winters are relatively dry.

The southern pine and oak forests and riverine swamp forests of the Atlantic Coastal Plain continue westward from Florida into east Texas. Loblolly, longleaf, and slash pines are widespread, evergreen live oak and other Atlantic oaks remain important, whereas myrtle oak (*Quercus myrtifolia*) in Florida, and Nuttall oak (*Q. nuttallii*) and post oak (*Q. stellata*) in the lower Mississippi valley are locally dominant. However, as the climate becomes drier to the west, forest gives way to prairie grassland, which, typified by short perennial grasses (*Andropogon*, *Cenchrus*, *Chloris*, *Paspalum*), extends from central Texas southward into Tamaulipas. With more rainfall in southern Tamaulipas and northern Yucatan, this grassland gives way to thorn forest (*Acacia*, *Agave*, *Cordia*, *Prosopis*, *Fiscus*, *Randia*, *Yucca*) and cactus (*Opuntia*, *Pachycereus*) (Psuty and Mizobe, 1982). Farther south, deciduous woodland passes into the evergreen forests and savannas of the humid tropics.

Beach and dune vegetation of the northern Gulf coast is dominated by many of the grasses found along the southern Atlantic coast, notably *Uniola paniculata* and *Sesuvium portulacastrum*, with *Andropogon scoparius* on secondary dunes, and *Fimbrystylis castanea*, *Spartina patens*, *Scirpus americanus*, and others in interdune hollows and flats where water nears the surface. Except for a few groves of live oaks and planted exotics, trees are rare on Texas barriers, although mesquite (*Prosopis glandulosa*) grows well farther south (Godfrey et al., 1982). Southward along the Mexican Gulf coast, these grasses are joined on the backshore by creeping vines such as *Ipomoea pes-caprae* and *Canavalia maritima*.

With low-energy mudflats widely available, salt marshes like those of the Atlantic seaboard are well developed along the coast from the Florida panhandle to east Texas. Farther west, however, increasing aridity and fewer tidal inlets promote hypersaline conditions behind barriers, notably in Laguna Madre behind Padre Island. Salt-marsh vegetation is sparse, with halophytes such as *Salicornia virginica* and *Suaeda linearis* patchily present, and vast *Spartina alterniflora* marshes are lacking (Godfrey et al., 1982). Mangroves extend northward along Florida's west coast and beyond, but they succumb to winter frost. Black mangrove reappears toward the south end of Laguna Madre in Texas, and swamps of red, black, and white (*Laguncularia racemosa*) mangroves dominate Mexico's wet coastal ecosystems.

The Everglades of southern Florida are an unusual watery ecosystem, nowhere more than 3 m above sea level, where subtle changes of water level, salinity, and substrate have created a wide range of habitats (fig. 21.11). These include emergent pinelands dominated by the local slash pine (*P. elliottii* var. *densa*), sawgrass prairies, and hammocks where tropical hardwoods such as mahogany (*Swietenia mahogani*), gumbo limbo (*Bursera simaruba*), and cocoplum (*Chrysobalanus icaco*) grow alongside live oak, red maple (*Acer rubrum*), and sugarberry (*Celtis laevigata*). Cypress swamps are dominated by *Taxodium*, and toward the coast are swamps and islands formed from red, black, and white mangroves. These habitats support a variety of fish, birds, mammals, amphibians, and reptiles, including endangered wood stork (*Mycteria americana*), Florida panther (*Felis concolor coryi*), manatee (*Trichechus manatus*), American crocodile (*Crocodylus acutus*), and several turtles. Although declared a national park in 1947, the Everglades are still threatened by hydrologic impacts arising from past diversions and canalization, and by modern problems of water quality, exotic species, and human activity.

Figure 21.11 Shrub-covered hammocks rising from a "sea of grass" in the Florida Everglades (photo: A.R. Orme, 1995).

21.3.5 *North Pacific Coast Beyond the Ice Front*

The unglaciated temperate coast of North America extends from the Gulf of California to near Juan de Fuca Strait (fig. 21.12). Consistent with its active-margin location, this is a relatively straight, often mountainous coast with small coastal lowlands, narrow shelves or borderlands, frequent submarine canyons, few large islands, and short, steep sediment-laden rivers augmented by the Columbia and Colorado drainages from the western interior. Rocky headlands enclose pocket beaches, barrier beaches block many small estuaries, dune fields are locally prominent, but wetlands are restricted and barrier islands are absent. Although direct impacts of glaciation linger in the north and indirect effects of past wetter climates are widely expressed in relict landslides and coastal deposits, related eustatic sea-level changes are confused by widespread tectonic deformation.

This coast's tectonic setting divides broadly into three units. First, from Juan de Fuca Strait to Cape Mendocino, the coast is one of classic convergence wherein remnants of the Farallon plate (Juan de Fuca and Gorda plates) continue to subduct offshore beneath the overriding North American plate, producing coast-parallel mountains, high seismicity, and active volcanism. Second, from Cape Mendocino to Cabo San Lucas, the North American plate sliver captured by the Pacific plate around 5 Ma is now shearing northwest alongside its former host, deforming as it goes. Third, the Gulf of California, a peculiarity to this active-margin setting, was initiated when the North American plate began overriding the East Pacific Rise in late Miocene time and then pulled apart around 6 Ma, introducing the extensional Basin and Range Province in Sonora and Sinaloa to the transtensional tectonics of this spreading center (Stock and Hodges, 1989). In tectonic terms, its coasts line a fledgling passive-margin system that is separating Baja California from mainland Mexico at a mean rate of 6 cm yr^{-1}, but active faults, frequent earthquakes, and volcanic activity around the Gulf belie any passivity.

The pattern of post-Pangean shoreline locations and sea-level changes, reflected in this active margin's accretionary terranes and subduction scenarios, is difficult to read. Suffice it to say that several generations of oceanic crust, igneous activity, marine sediment, and terrigenous waste have been deformed, exposed or consumed along the continent's margin and are now expressed in bold headlands and subsiding basins alike. Miocene marine deposits and submarine volcanics have been raised into rugged coastal ranges, such as the Santa Monica Mountains. As late as Pliocene time, the sea still occupied much of California's Central Valley, the Ventura and Los Angeles basins, and the lower Colorado valley. Fragments of relative calm survive, however, along Baja California's outer coast where successive Cretaceous and Cenozoic shorelines run closely parallel (Orme, 1972, 1980).

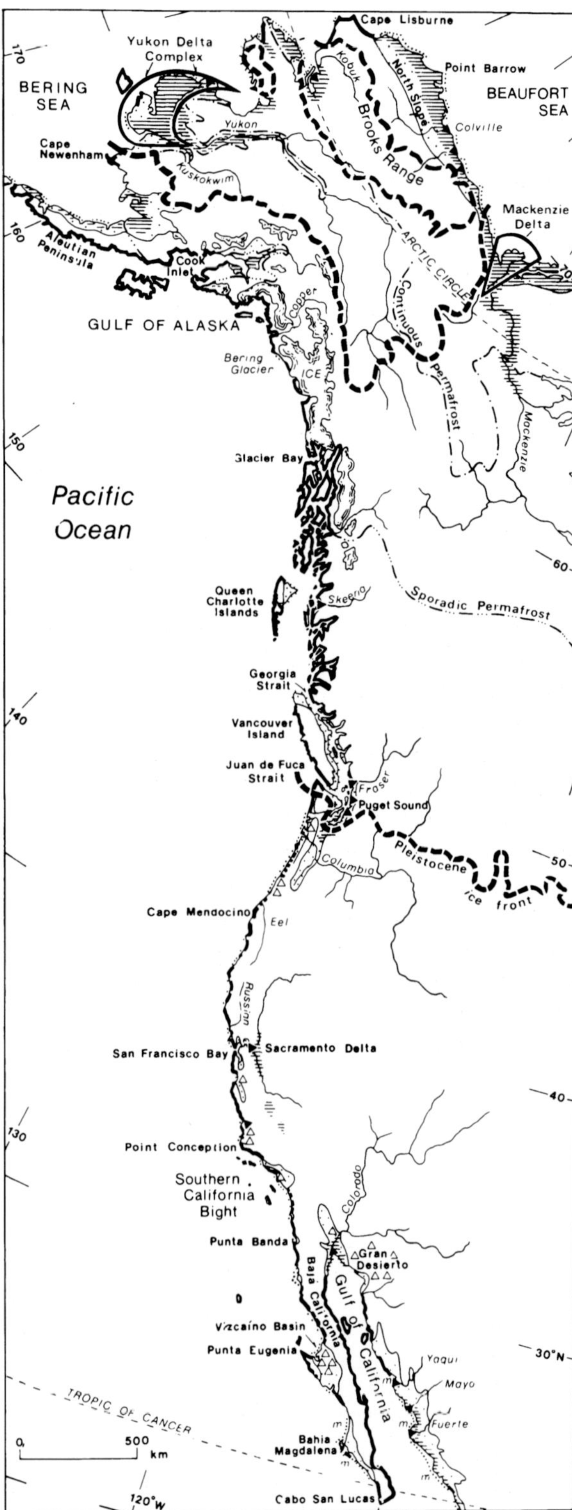

Figure 21.12 Dominant morphology of the Pacific coast (see figure 21.6 for key).

The Quaternary sea-level record is more evident but often deformed and not easily interpreted. Dated shoreline sequences do, however, offer excellent measures of the nature and rate of tectonic deformation (fig. 21.13; Orme, 1998). Early to middle Pleistocene marine terraces have been massively uplifted in several localities. In Baja California these terraces rise to 150 m above sea level on Punta Eugenia (Ortleib, 1987), 357 m above Valle del Rosario, and 345 m on Punta Banda (Orme, 1972, 1980; Rockwell et al., 1989). In southern California, they occur to 580 m on San Nicolas Island (Lawson, 1893; Muhs, 1985), to 411 m on the Palos Verdes peninsula (Woodring et al., 1946; Muhs et al., 1992), and to 600–700 m near Ventura (Yeats and Rockwell, 1991). Farther north, terraces have been uplifted to 247 m in the San Luis Range (Hanson et al., 1994). Tectonic subsidence also occurs, with one of the world's thickest known sequences of Pleistocene sediment carried down nearly 5000 m below sea level in the Ventura Basin (Yeats, 1977).

Rates of coastal deformation since the Last Interglacial (OIS 5e) highstand around 125 ka are no less impressive. Relatively stable coasts far from plate margins reveal that the OIS 5e shoreline rose to 6 m above present sea level, but along the unglaciated North Pacific coast it reaches 43 m on Punta Banda (Orme, 1980; Rockwell et al., 1989), 46 m on the Palos Verdes peninsula (Muhs et al., 1992), 33 m on the San Luis Range, and 90 m against the Santa Lucia Mountains (fig. 21.14; Hanson et al., 1994). Even the OIS 5c and 5a highstands, whose eustatic maxima reached –5 m at 105 ka and 85 ka, respectively, have been raised above present sea level, to as much as 175 m near Ventura, where mid-Holocene shorelines reach 37 m (Lajoie et al., 1982). The uplift rates indicated by these deformed shorelines, from 0.1 to as much as 10 m ka^{-1}, are truly remarkable, and field data and tide-gauge records show that uplift and subsidence continue, although at lower rates than present eustatic changes (Orme, 1998). Uplift rates up to 0.3 m ka^{-1} have been measured from Santa Cruz northward (Bradley and Griggs, 1976). Although gradual but cumulative tectonic forcing may have caused much of this deformation, instantaneous vertical movements may occur during earthquakes. Such coseismic sea-level changes probably caused the submergence of Holocene salt marshes fronting the Cascadia subduction zone (Nelson, 1992; Atwater, 1996).

The Flandrian transgression has been defined at several localities along this coast. The sea entered the Golden Gate about 10 ka and, rising at a rate of 2 cm yr^{-1}, flooded laterally across San Francisco Bay as rapidly as 30 m yr^{-1} until 8 ka (fig. 21.2; Atwater et al., 1977). The rise then slowed and for the past 6 ka has averaged 1 to 2 mm yr^{-1}. During this time however, Holocene salt marsh deposits have undergone 5 m of tectonic and hydroisostatic subsidence, a rate of 0.8 mm yr^{-1} superimposed onto the transgression's eustatic effects. Holocene sea-level changes measured at Bolinas Lagoon, Morro Bay, and Malibu, though similar, have been augmented by coseismic subsidence associated with the San Andreas, Los Osos, and Malibu Coast faults, respectively (Berquist, 1978; Gallagher, 1996; Orme, 1999). Tide-gauge data show that sea level has continued to rise along the southern California coast at rates of 1 to 2 mm yr^{-1} over the past century (Orme, 2000).

Winds and waves along this coast are greatly influenced by the clockwise outflow of air from the Hawaiian high-pressure system. In summer, when the system expands, northwesterly swells dominate much of the coast and southwesterly swells retreat to Washington and farther north. In

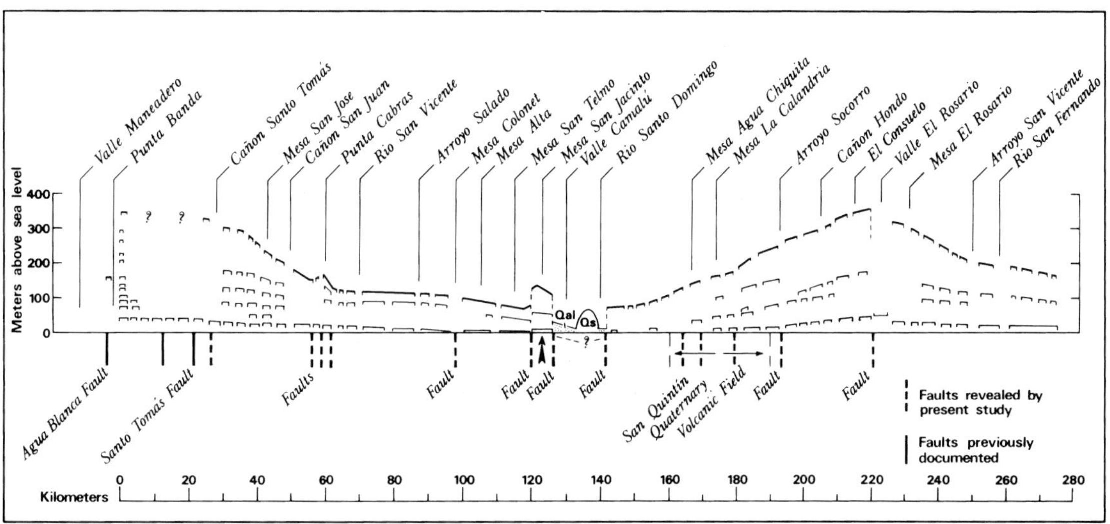

Figure 21.13 Tectonic deformation of Plio-Pleistocene shorelines, northwestern Baja California, Mexico (Orme, 1980).

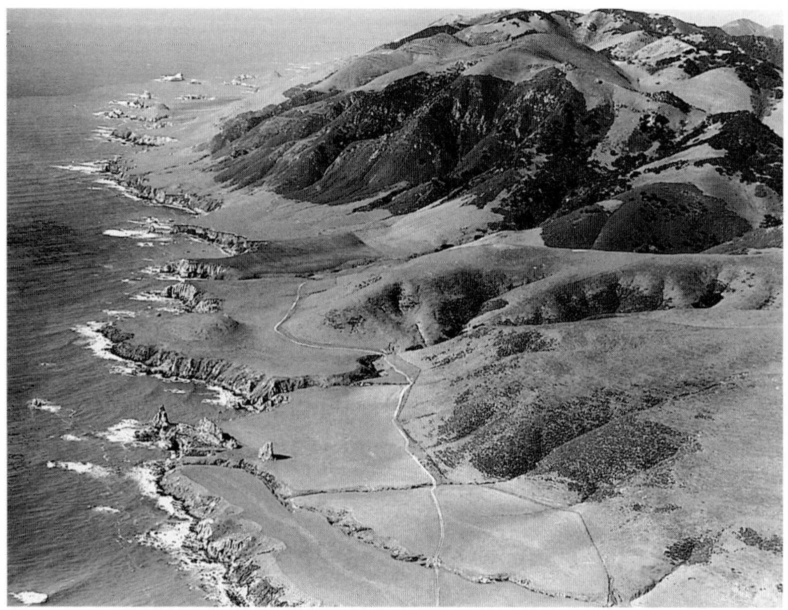

Figure 21.14 Marine terraces flanking the San Luis Range, south-central California coast. The low terrace with its prominent sea stacks relates to the 125-ka (OIS 5e) and 80-ka (OIS 5a) seas with shoreline angles at 23–33 m and 7–11 m, respectively. Older terraces rise to 247 m, and the skyline reaches 431 m (photo: Spence Collection, UCLA, 1947; see Orme, 1998).

winter, when the system contracts, southwesterly swells extend southward along the Oregon coast and storm waves may affect the entire coast, but especially the north, in response to the eastward progression of midlatitude cyclones. From Cape Mendocino to Point Conception, 50% of swells approach from the northwest, the remainder from west-northwest or west. In the Southern California Bight, changing coastal orientation, refraction, and offshore islands cause 70% of swells to pass up the Santa Barbara Channel from due west, whereas 80% of swells approach Los Angeles from west-southwest (Orme, 1985). In addition, southerly swells set up by late summer tropical cyclones off western Mexico, by Southern Hemisphere winter storms, and by local winter cyclones passing along more southerly tracks may erode south-facing shores (fig. 21.3). Mean wave heights increase northward from 1 to 3 m, but 5- to 7-m breakers may occur anywhere during storm events, and 14-m breakers have been recorded off Crescent City (Orme, 1982). The persistent swells generate sympathetic longshore currents of 1–2 m s^{-1}, northward in Washington, reversing seasonally in Oregon, and southward in California, but storm events and coastal geometry may combine locally to change these patterns. Tsunamis may be generated by seismic activity in the Cascadia subduction zone and other offshore faults, and may reach the coast from distant events, such as the 7-m waves that damaged Crescent City after the 1964 M9.2 Anchorage earthquake. This coast has a mixed mesotidal regime with spring ranges increasing northward from <2 m to >4 m. Reversing tidal currents through the Golden Gate reach 1.7 m s^{-1} on the flood and, augmented by stream discharge through San Francisco Bay, 2.3 m s^{-1} on the ebb. Northward propagation of tides in the narrowing Gulf of California raises mean tidal range from 1 m in the south to 7 m off the Colorado delta, where 10-m spring tides and swift tidal currents occur.

Stream-sediment discharge, so important for beach and dune nourishment, reflects the availability of rock waste, precipitation and runoff patterns, and the extent to which fluvial transport is blocked by dams. Thus northern California's erodible rocks, steep slopes, heavy precipitation, and storm runoff combine to produce frequent landslides and high erosion rates. The Klamath, Eel, and Russian rivers account for 77% of suspended sediment discharged to the California coast north of Point Conception (fig. 21.1; Orme, 1985). The Eel River has a remarkable sediment yield of 1750 tonnes km^2 yr^{-1}, and its annual load of 14×10^6 tonnes was similar to the pre-dam Columbia River, despite the Eel having little more than 1% of the Columbia's basin area (table 1.3; Milliman and Sivitski, 1992). However, the Columbia River was a major source of sediment to the continental shelf during late Pleistocene superfloods (see chapter 3), and much of this was reworked onshore into dune fields (Cooper, 1958). Indeed, extensive dunes commonly occur near major river mouths such as the Salinas, Santa Maria, and Santa Ynez estuaries, and the former outlet of the Los Angeles River to Santa Monica Bay.

Dam construction has had a negative impact on sediment reaching the coast, more so in the south of the region where water supplies are more tenuous and rivers more likely to be dammed (see chapter 22). Because of upstream dams, little sediment now reaches the Colorado River estuary, whereas the sediment load of the Columbia River has been reduced by one-third (see table 1.3). Among smaller rivers, California's 4219-km^2 Santa Clara basin, 37% of which is dammed, discharges $<2 \times 10^6$ tonnes of sediment to the sea in most years, although such is the nature of the hydrologic regime that it discharged 48×10^6 tonnes during the 1969 floods.

Despite the linearity of this rocky coast, bold headlands and nearshore submarine canyons limit littoral drift and favor pocket beaches whose sediment supply, much curtailed during the present high sea level, depends more on nearby seacliffs. Seacliff retreat is a complex response of incompetent rocks to episodic mass movement, including reactivation of Pleistocene landslides and basal marine erosion. Recent retreat rates near San Diego, Santa Barbara, and Santa Cruz range from negligible in harder Cretaceous sandstone to 0.6 m yr^{-1} in softer Quaternary deposits (Griggs and Johnson, 1979; Orme, 1991). Mass movement is a management problem where relict Pleistocene landslides and coastal benches underlain by weak seaward-dipping rocks have been invaded by suburbia, notably along the Palos Verdes and Malibu coasts near Los Angeles (fig. 21.15; Orme, 1991), and the Newport area of Oregon. Recurrent mass movement also typifies Washington's 20–100 m sea cliffs. Submarine canyons often occur off major river mouths, notably Monterey Canyon off the Salinas estuary, but others, unrelated to present streams, have been shaped by turbidity currents directed offshore by headlands and structural lineaments.

Most river mouths along this coast are more or less blocked by barrier beaches. These are more durable toward the south as stream discharge and its ability to remove these barriers diminish in response to increasing summer drought. Even major rivers, such as California's Santa Maria and Santa Clara, may be closed by shifting sand during dry summer months, whereas along the arid coast of central Baja California, several generations of late Quaternary barrier beaches survive in the absence of modern streamflow (Orme, 1980). Farther north, however, increasing perennial flows maintain open estuaries, partly blocked by spits that are subject to much seasonal reworking. Along the Oregon coast, for example, the Nestucca and Siletz spits were subject to massive erosion and overwash by storm seas during the 1970s (Komar and Rea, 1976; Komar, 1978).

Steep coastal relief usually limits the size of estuaries and lagoons behind these coastal barriers. Of the larger estuaries, Grays Harbor and Willapa Bay north of the Columbia River may reflect isostatic forebulge collapse. With low wave energy and gentle relief along the Gulf of California's eastern shore, however, high sediment inputs from the Sierra Madre Occidental have led to barrier-lagoon growth, notably off the Yaqui, Mayo, and Fuerte deltas. Similar features have formed around Bahia Magdalena on Baja California's outer coast, but along the peninsula's arid Gulf coast the paucity of terrigenous sediment allows calcareous skeletal debris to line rocky shores. At the hyperarid head of the Gulf, the Colorado River delta, now starved of sediment, presents broad expanses of tidal flats, salt pans, and underfit distributary channels (Thompson, 1968).

San Francisco Bay, this coast's most distinctive inlet, occupies a late Cenozoic structural trough bounded by active strike-slip faults, a sump for the 80,000-km^2 Sacramento–San Joaquin drainage system, deepened by fluvial erosion and repeatedly flooded by Quaternary seas. The 90-km-long bay is linked to California's Central Valley through the Carquinez Strait, whose bedrock channel lies 60 m below sea level, and to the Pacific Ocean through the

Figure 21.15 Via de las Olas landslide, Pacific Palisades, California. On 31 March 1958, this abandoned but unstable seacliff collapsed after heavy rains. Rather than remove the landslide toe, the coast highway was later relocated seaward (photo: Spence Collection, UCLA, 1958; see Orme, 1991).

Golden Gate, whose channel is scoured to over –100 m and off which a lunate mass of terrigenous sediment mantles the shelf. Bay waters, which 200 years ago occupied 1800 km^2, now cover 1100 km^2, 70% of which is less than 4 m deep. This size reduction is due in part to massive fluvial inputs of hydraulic-gold-mining debris from the Sierra Nevada foothills during the later nineteenth entury, in part to natural sedimentation, and in part to reclamation. Gilbert (1917) estimated that 1816 × 10^6 m^3 of debris were flushed from the foothills between 1850 and 1914, of which 50% was redeposited downstream, 48% reached the bay, but only 2% entered the ocean.

Persistent onshore winds and high fluvial sediment inputs have long favored sand dunes along this coast, notably where subsiding lowland basins provide space for dune growth. However, the region's most extensive dune fields, in the Vizcaíno Basin of central Baja California and the Gran Desierto east of the Colorado delta, are more correctly desert dunes that approach the coast. Cooper (1958, 1967) early distinguished between coastal dunes of differing but uncertain ages. More recently, aided by various dating methods, these dunes have been placed within a timeframe for the late Quaternary (Orme and Tchakerian, 1986; Orme, 1990a, 1992). In essence, transverse dunes form during periods of sediment abundance, ideally when sea level falls and large quantities of sand are exposed on emergent continental shelves, for example, during the stage 3–2 transition around 40–25 ka. Such features, since modified by weathering and erosion, form extensive paleodunes in the Salinas, Santa Maria, Los Angeles, and San Quintín basins. Conversely, parabolic dunes develop during periods of sediment deficiency, for example, as sea level rose across the stage 2–1 transition, 15–5 ka. Dunes may also be reactivated by vegetation changes induced by fire and human land use such as the grazing activities that accompanied early European colonization of this coast after 1769.

This region's 25° latitudinal spread is reflected in its ecology. Mean temperatures decline northward, in summer from >30°C in the Gulf of California to 15°C at Juan de Fuca Strait, and in winter from 20°C to 5°C. Precipitation increases northward from <250 mm yr^{-1} in southern Baja California to >3000 mm yr^{-1} in the Olympic Peninsula. However, the cooling effect of the California Current and related fogs helps to explain the survival of conifers far to the south, including closed-cone pine (*Pinus muricata*) on Cedros Island and Tecate cypress (*Cupressus guadalupensis*) on Guadeloupe Island off central Baja California, relicts of Pleistocene migration. In other respects, there is a latitudinal transition from subtropical scrub in Sinaloa and southern Baja California, through the deserts of central Baja California and the northern Gulf (see chapter 19) and the chaparral communities from northern Baja California to beyond San Francisco Bay (see chapter 20), to coniferous forests in the north. The fog belt from central California to southern Oregon favors such distinctive local species as Monterey pine (*Pinus radiata*), Monterey cypress (*Cupressus macrocarpa*), coast redwood (*Sequoia sempervirens*), and Port Orford cedar (*Chamaecyparis lawsoniana*). Here also, Douglas-fir (*Pseudotsuga menziesii*), western hemlock (*Tsuga heterophylla*), western redcedar (*Thuja plicata*), lodgepole pine (*Pinus contorta*), Sitka spruce (*Picea sitchensis*), and various firs (*Abies* spp.) begin their range that, except for Douglas-fir, extends northward to Alaska.

Beach and dune vegetation partly reflect this latitudinal gradient, with a shift from forb to grass communities as warmth and summer aridity decrease north of Point Conception, but local conditions and introduced species also influence community structure (Barbour et al., 1976). Tropical species such as *Ipomoea brasiliensis*, *I. stolonifera*, and *Scaevola plumeri*, and warmth-loving *Sesuvium portulacastrum* and *Sporobolus virginicus* occur in Sinaloa and southern Baja California. *Abronia maritima* and the the introduced *Carpobrotus chilense* and *Cakile maritima* extend from northern Baja California throughout southern California (fig. 21.16). From central California to the Juan de Fuca Strait, *Elymus mollis*, *Ambrosia chamissonis*, and the introduced *Ammophila arenaria* are dominant. *Abronia maritima* is the main foredune pioneer in the south, whereas *Ammophila arenaria* has assumed this role in the north since 1930, displacing *Abronia latifolia*, and is widely used for dune stabilization. Dune stabilization is favored by xeric shrubs in the south and by grasses (e.g., *Festuca rubra*) and composites (e.g., *Solidago spathulata*) farther north, paving the way for shrubs like *Vaccinium ovatum* and *Gaultheria shallon*, and eventually for conifers.

Wetlands are restricted by the smallness of most estuaries and lagoons, but are locally important, notably in San Francisco Bay, where, despite reclamation, some 200 km^2 of marshes and mudflats are still exposed at mean tide level. In the south, mangroves extend northward to southern Sonora and Magdalena Bay. Farther north, *Spartina foliosa* dominates many low salt marshes from San Quintín to Humboldt Bay, whereas higher marshes are dominated by *Salicorina virginica*, *S. bigelovii*, *Jaumea carnosa*, *Limonium californicum*, *Batis maritima*, and *Distichlis spicata* (fig. 21.16; Purer, 1942; Orme, 1973b; Ferren et al., 1996). Oregon and Washington marshes are transitional to the colder marshes farther north. Many southern plants become less important as species like *Carex lyngbyei*, *Deschampsia caespitosa*, and *Scirpus maritimus* invade the marshes (Godfrey et al., 1982). The northern limit of *Carex lyngbyei* marks the transition to the subarctic marshes dominated by *Carex ursina*. However, tolerant plants like *Salicornia virginica*, *Jaumea carnosa*, *Scirpus maritimus*, and *Distichlis spicata* range from Baja California north to Alaska.

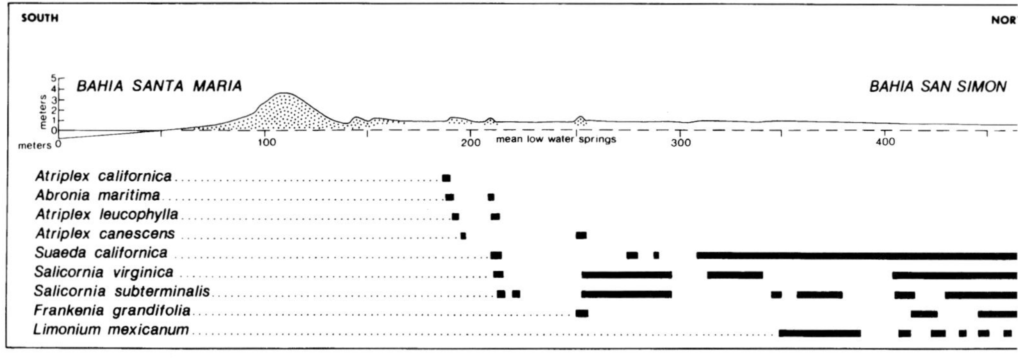

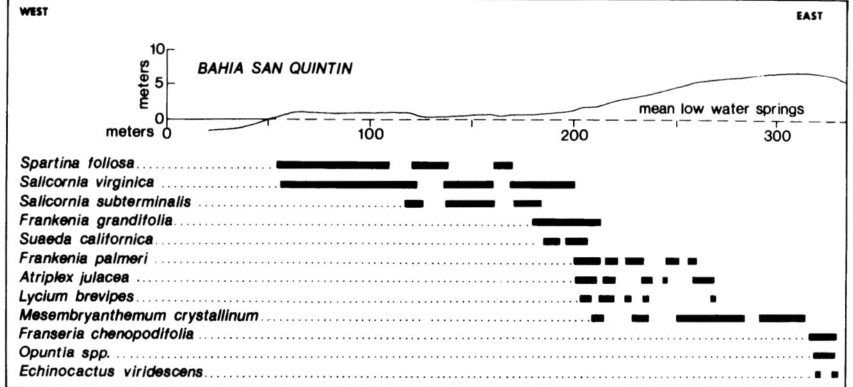

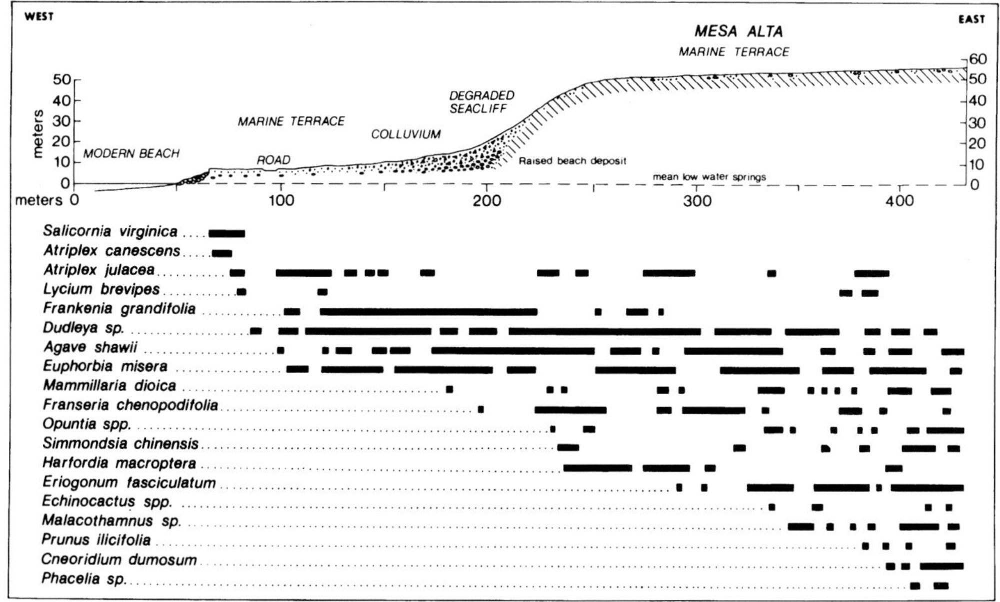

Figure 21.16 Vegetation transects across the outer coast of Baja California. Upper: Open coast with active unvegetated foredunes protecting back-barrier wetland; Middle: Back-barrier wetland passing into semiarid scrub; Lower: Marine-terrace assemblage showing progressive colonization by shrubs at the chaparral-desert transition (see Orme, 1973a, 1973b).

21.3.6 North Pacific Glaciated Coast

This glaciated coast extends from the Alaska Peninsula southward to Juan de Fuca Strait and Puget Sound (fig. 21.12). It is a deeply dissected, mountainous coast of fjords, archipelagoes, and tidewater glaciers. Continuing tectonism, high seismicity, and frequent earthquakes along this active margin are linked to the 3000-km-long Aleutian Trench, up to 7500 m deep, and farther south to the Queen Charlotte transform system and Cascadia subduction zone. The Lituya Bay earthquake of 1958, with a 7-m lateral and 11-m vertical displacement, triggered a massive landslide that dumped over 30×10^6 m^3 of rock into the bay and released 750 m of ice from Turner Glacier into Yakutat Bay, initiating a tsunami. The M9.2 Anchorage earthquake of 1964 displaced much of the coast around the Gulf of Alaska and generated a 30-m-high tsunami in Valdez Harbor (Plafker, 1965). Some 60 volcanic centers, two-thirds of them historically active, are linked to subduction in the Aleutian Trench between Anchorage and the west end of the 2200-km-long chain of Aleutian Islands. Such volcanism is reflected in coasts modified by explosive collapse, lava flows, and widespread tephra. Uplift, shown by glaciomarine deposits 230 m above sea level near Juneau, reflects both tectonic forcing and glacioisostatic rebound.

Consistent with active-margin tectonism, the coast is dominated by coast-parallel structures, formed generally into a series of high coastal mountains and rugged offshore islands separated by sheltered tidal passages, except between Prince William Sound and Cape Spencer where the Chugach and St. Elias mountains (Mt. Logan, 5971 m; Mt. St. Elias, 5489 m) reach directly to the outer coast. These features were originally dissected by fluvial erosion and repeatedly modified by glacial erosion. Fjords, deepened by valley glaciers moving more or less perpendicular to the coast, flow into massive structural troughs such as Cook Inlet, Hecate Strait, Queen Charlotte Sound, and Georgia Strait, which were also modified by glaciation. Lowland coasts are rare but occur where sediment-laden rivers form deltas, such as those of the Skeena, Fraser, Nooksack, and Skagit rivers, and where emergent glacial, glaciomarine, and fluvioglacial deposits occur, for example, at the head of Cook Inlet, on northeast Graham Island and Vancouver Island, and around Puget Sound.

Much interest along this coast centers on the continuing relationship between glaciation and coastal form. During Pleistocene maxima, many Cordilleran glaciers ran onto the continental shelf and calved in deep water near its edge (fig. 21.17). Wisconsinan glaciers began retreating between 15 ka and 12 ka, and earlier Holocene time saw continuing net retraction until, after the postglacial thermal maximum, glaciers readvanced in response to the colder, snowier coastal climates of late Holocene time, the Neoglacial interlude (Molnia, 1986). Field evidence, historic maps, ship logs, and remote sensing reveal much about the behavior of these coastal glaciers over the past 200 years. For example, the Bering Glacier, at 191 km the largest and longest glacier in mainland North America, was in a retracted position from 8000 to 1500 B.P., allowing the sea to flood 15 km inland of the present shore (Molnia and Post, 1995). A Neoglacial advance began after 1500 B.P. (radiocarbon years before 1950), culminating with the glacier reaching its Holocene maximum 1000–500 years ago. Retraction from this Neoglacial maximum has been interrupted by six glacier surges over the past 100 years alone. Bering Glacier's piedmont lobe and ice-marginal Vitus Lake are now separated from the ocean by a 3-km-wide beach-outwash-moraine complex.

Farther south, the glaciers of Glacier Bay have also retreated from a Neoglacial maximum in Icy Strait, dramatically so over the past 200 years. From 1794, when observed by explorer George Vancouver, to 1860, the single ice front in Glacier Bay retreated 40 km (Hall et al., 1995). Since then, splitting into two, the Muir Glacier has retreated another 48 km and the Grand Pacific Glacier in Tarr Inlet another 58 km before readvancing after 1935. Whereas these historic recessions are due largely to climatic warming since the culmination of the Little Ice Age, glacier behavior also depends on tidewater relations. Tidewater glaciers may retreat rapidly when they calve icebergs into deep water

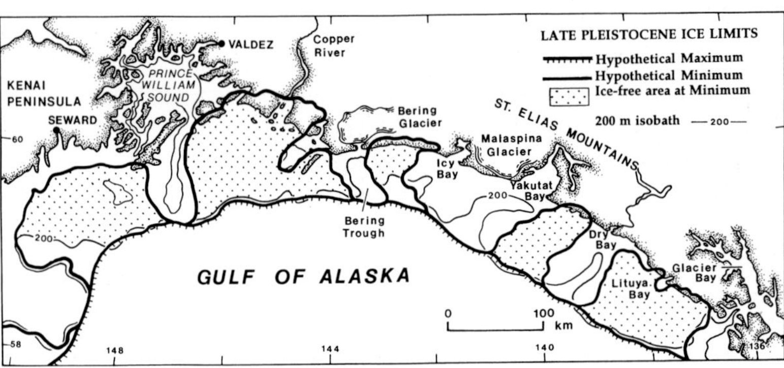

Figure 21.17 Gulf of Alaska showing hypothetical maximum and minimum late Pleistocene glacier positions (from Molnia and Post, 1995).

(fig. 21.18), whereas grounded glaciers retreat at much slower rates. This may explain the anomalous behavior of the Taku Glacier, near Juneau, which retreated 8 km from 1750 to 1890 because it calved into a 100-m-deep fjord, but which has since advanced 7.3 km over the proglacial debris now filling the fjord—this at a time when nearby glaciers have been retreating (Post and Motyka, 1995).

Wave energy along this coast reflects the frequent cyclonic activity over the north Pacific Ocean. Storm waves are common, although their impact is muted in the shelter of islands and inland passages. This is also a macrotidal coast, where tidal ranges increase northward and up inlets to maxima of 8 m at Prince Rupert and 11 m at the head of Cook Inlet (fig. 21.19). The many islands also promote out-of-phase tidal cycles, which generate swift tidal currents through intervening channels, reaching 7 m s^{-1} in Discovery Passage and the Seymour Narrows in Georgia Strait, and in Deception Pass north of Whidbey Island, Washington. Despite high wave and tidal energy, however, major constructional forms are limited by a paucity of beach-forming materials and by steep nearshore gradients, although such features as Dungeness spit on the south shore of Juan de Fuca Strait are noteworthy.

This coast experiences cool summers (12–14°C in July), moderately cold winters (–5 to 2°C), frequent storms, extended cloud cover, heavy preciptation (1400–4000 mm yr^{-1}), and low evaporation. With mountains often reaching the shore, coniferous forest approaches sea level and distinctive littoral vegetation zones are restricted. A pollen record for the past 13,000 years from Langara Island, Queen Charlotte Islands, indicates that the grass-sedge tundra of late Pleistocene times was invaded by an open community of Sitka alder (*Alnus crispa sinuata*) and lodgepole pine around 12 ka, although the latter declined during the thermal maximum between 9 ka and 6 ka (fig. 21.20; Heusser, 1995). The onset of cooler, wetter conditions during late Holocene time has led to the growth of a more closed forest of Sitka spruce, western hemlock, western redcedar, and various firs. Northward, lodgepole pine and mountain hemlock (*T. mertensiana*) become important, but forest continuity decreases as a muskeg of sedge and heath expands. This temporal sequence is reflected in modern spatial distributions of vegetation near retreating glaciers. In Icy Bay and the Bering Glacier foreland, for example, moss, sedge, fireweed (*Epilobium latifolium*), and horsetail (*Equisetum variegatum* var. *alaskanum*) occur closest to glaciers, then willow and alder, with black cottonwood (*Populus balsamifera*) and Sitka spruce, occur farthest away (Heusser, 1995). Beach and dune vegetation has limited opportunity in this harsh environment. Salt marshes are normally dominated by *Puccinellia phryganodes* in Alaska and by *Carex ursina* farther south. Other *Puccinellia* and *Carex* species occur, together with those species noted previously that range from Baja California to Alaska.

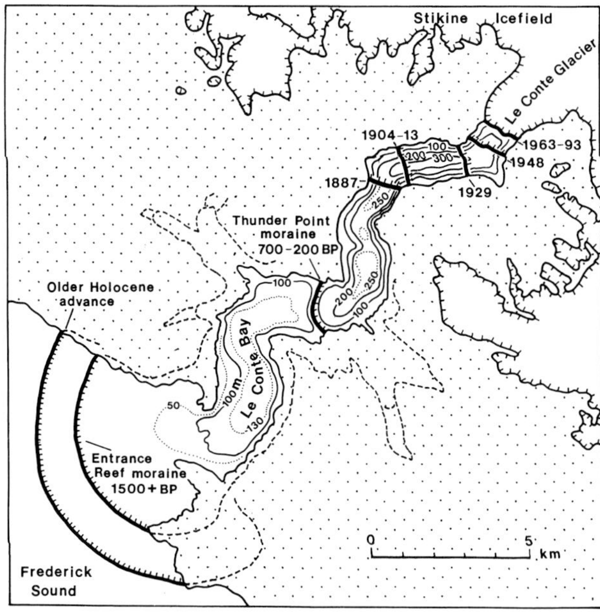

Figure 21.18 Late Holocene retreat of Le Conte tidewater glacier, southeast Alaska (from Post and Motyka, 1995).

21.4 Conclusion

The North American coast retains much of its natural grandeur, but a major challenge for modern science and management alike is provided by the human use and misuse of coastal resources, especially over the past century. The problems are often acute along the coasts of the conterminous United States, where half the population now lives within 75 km of tidewater or the Great Lakes. The situation is less acute in Canada and Alaska, essentially because there are fewer people and much of the coast is harsh and remote, but even here problems posed by overfishing, mineral development, and pollution need careful attention.

Human attitudes and thus management responses to coastal resources divide into three main periods: before 1960, from 1960 to 1980, and since 1980. In the lengthy historic period before about 1960, once mere survival ceased to be an issue, those agencies and individuals concerned with the coast sought mostly to encourage and facilitate resource development. Hazard mitigation was an early priority, leading in the United States to the creation of a national lighthouse system in 1789, the Coast Survey in 1807, shore-protection works in New England and Lake Erie in 1824, and a weather service in 1870 designed to provide storm warnings to mariners. Meanwhile, coastal lands were progressively reclaimed for harbor development. During the economic depression of the 1930s, civilian labor was widely employed in coastal construction and, in attempts to combat erosion, stimulated by the formation

Figure 21.19 The perils of a large tidal range. The Canadian Pacific steamship *Princess May* ran onto the rocks off Sentinel Island, Alaska, at high tide on 5 August 1910. Despite its precarious position at low tide, the ship was later refloated without major damage (photo: W.H. Case, 1910, from Maritime National Historical Park collection, San Francisco).

of the Beach Erosion Board of the U.S. Army Corps of Engineers in 1930. After World War II, increases in population, paid vacations, car ownership, modern highways, and international trade saw long stretches of coast manipulated and reclaimed. Increasingly, wetlands were reclaimed for industrial and harbor facilities, dunes were leveled for airports, power plants gravitated to the coast, barrier islands were developed for recreation and second homes, sand and gravel were mined for use in construction, and pollution and waste disposal became wider problems. Between 1888 and 1973, some 20×10^9 m^3 of solid waste were dumped into the New York Bight, filling part of the Hudson River's offshore channel and forming hills 10–15 m high on the shelf (Williams, 1975). Between 1945 and 1965, California lost two-thirds of its remaining coastal wetlands to reclamation. In short, by around 1960, much of the coast had succumbed to industrial and commercial development or had become a haven for the more privileged, managed, if at all, by an ineffective collage of federal, state, and local controls.

Between 1960 and 1980, sooner in some areas, attitudes toward coastal resources began to change, led more from below than above by younger people concerned about coastal mismanagement. In 1960, citizens founded a Save-the-Bay Association to oppose further infilling of San Francisco Bay. In 1963, the Jones Act began to protect Massachusetts' remaining wetlands. At the United States federal level, a plethora of environmental legislation culminated in the National Environment Policy Act of 1970, with its call for environmental impact analyses, and the Coastal Zone Management Act of 1972. However, such is the political structure that this legislation was designed to assist rather than replace the role of state and local governments; it still fell to the latter to enact the policies and procedures for dealing with the mounting problems of coastal management. Thus the California Coastal Act of 1976 and similar legislation in other states came to provide the framework for addressing such issues as development, marine resources, and recreation.

Since around 1980, with responsible legislation in place, the pace of coastal degradation has slowed but not ceased. Development proposals are now subject to lengthy agency hearings and often to court proceedings. Sometimes proposals wither and die; more often they are implemented later in modified form, after some perception of public in-

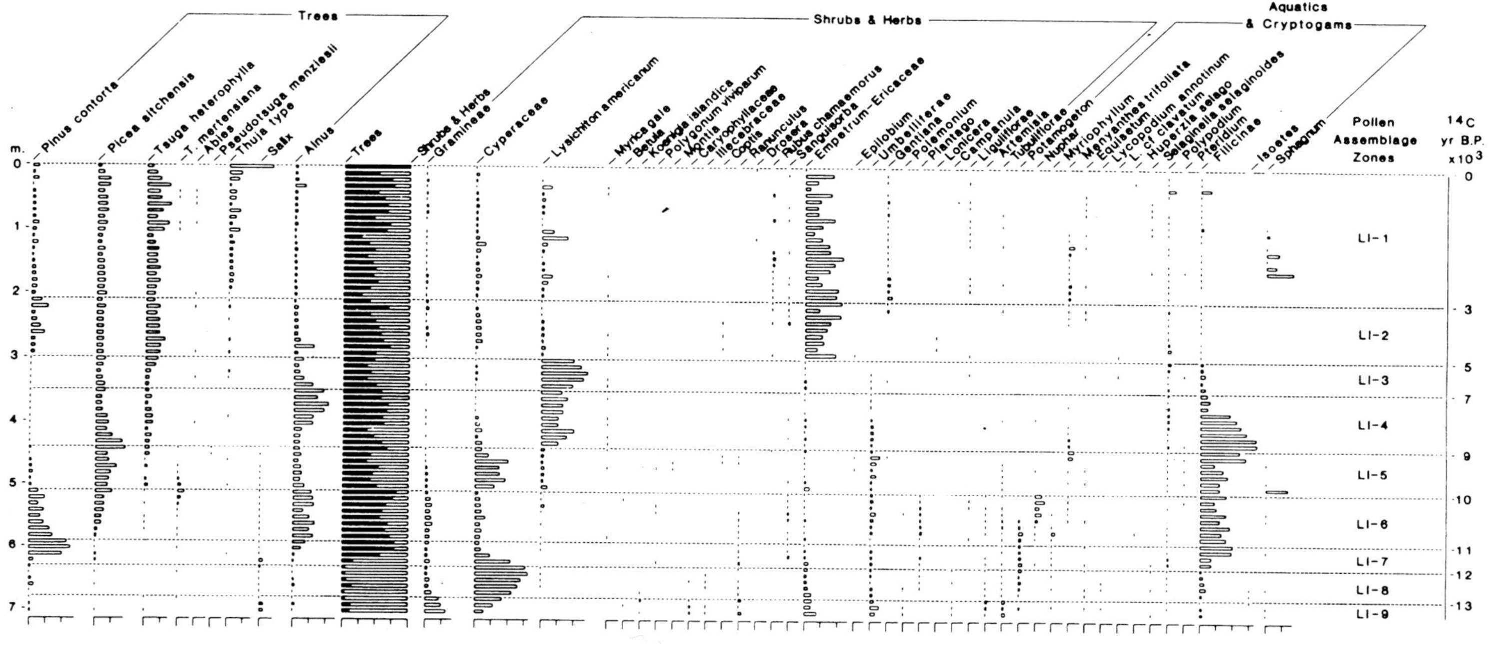

Figure 21.20 Late Quaternary vegetation sequence, Langara Island, Queen Charlotte Islands, British Columbia (from Heusser, 1995).

volvement. However, such legislation does little or nothing to redress problems created by earlier activities, such as the construction of homes on shifting barrier islands and unstable coastal bluffs whose unsuitable locations may only be revealed to an unthinking public during hurricane surges and landslides. Nor does it truly address the rights of property owners, so vigorously defended in the United States, to do whatever they wish with their land. And only rarely can reclaimed wetlands be restored to their original state. Nevertheless, some useful attempts have been made by government and individuals alike to conserve and protect some of the best coastal resources, for example, in the Everglades National Park, various protected seashores, and marine sanctuaries such as those around Stellwagen Bank, the Florida Keys, California's Channel Islands, and the Olympic Peninsula. Also, helped by such agencies as the Coastal Engineering Research Center, which replaced the Beach Erosion Board in 1963, erosion problems are now met less often by "hard" engineering solutions such as seawalls and breakwaters and more often by "soft" solutions involving artificial beach nourishment (fig. 21.21).

Coastal management remains a continuing challenge in the face of human impacts superimposed on a naturally changing system. Although this chapter has outlined the character of the North American coast, the necessary science must be applied diligently to the mutual benefit of all who may use the coast in the future.

References

Anderson, J.B., M.A. Thomas, F.P. Siringan, and W.C. Smyth, 1992. Quaternary evolution of the east Texas coast and continental shelf. In: C.H. Fletcher and J.F. Wehmiller (Editors), *Quaternary coasts of the United States: Marine and lacustrine systems.* Society for Sedimentary Geology, SEPM Special Publication 48, 253–263.

Atwater, B.F., 1996. Coastal evidence for great earthquakes in western Washington. *U.S. Geological Survey Professional Paper* 1560, 77–90.

Atwater, B.F., C.W. Hedel, and E.J. Helley, 1977. Late Quaternary depositional history, Holocene sea-level changes, and vertical crustal movement, southern San Francisco Bay, California. *U.S. Geological Survey Professional Paper* 1014.

Barbour, M.G., T.M. de Jong, and A.F. Johnson, 1976. Synecology of beach vegetation along the Pacific coast of the United States of America: A first approximation. *Journal of Biogeography*, 3, 55–69.

Berquist, R., 1978. Depositional history and fault-related studies, Bolinas Lagoon, California. *U.S. Geological Survey Open-File Report* 78-802.

Bird, J.B., 1985. Arctic Canada. In: E.C. Bird and M.L. Schwartz (Editors), *The world's coastline.* Van Nostrand Reinhold, New York, 241–251.

Bloom, A.L., 1963. Late Pleistocene fluctuations of sea level and postglacial crustal rebound in coastal Maine. *American Journal of Science*, 261, 862–879.

Bradley, W.C., and G.B. Griggs, 1976. Form, genesis, and deformation of central California wave-cut platforms. *Geological Society of America Bulletin*, 87, 433–449.

Coleman, J.M., 1982. *Deltas: Processes of deposition and models for exploration.* 2nd Edition, International Human Resources Development Corporation, Boston.

Coleman, J.M., 1988. Dynamic changes and processes in the Mississippi delta. *Geological Society of America Bulletin*, 100, 999–1015.

Colman, S.A., J.P. Halka, and C.H. Hobbs, 1992. Patterns and rates of sediment accumulation in the Chesapeake Bay during the Holocene rise in sea level. In: C.H. Fletcher and J.F. Wehmiller (Editors), *Quaternary coasts of the United States: Marine and lacustrine systems.* Society for Sedimentary Geology, SEPM Special Publication 48, 101–111.

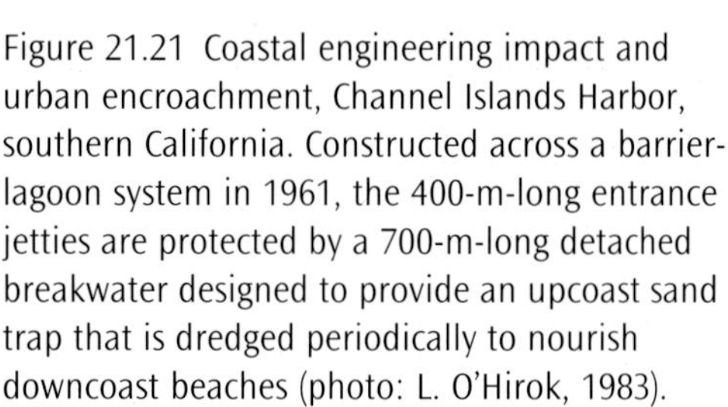

Figure 21.21 Coastal engineering impact and urban encroachment, Channel Islands Harbor, southern California. Constructed across a barrier-lagoon system in 1961, the 400-m-long entrance jetties are protected by a 700-m-long detached breakwater designed to provide an upcoast sand trap that is dredged periodically to nourish downcoast beaches (photo: L. O'Hirok, 1983).

Colman, S.A., J.P. Halka, C.H. Hobbs, R.B. Mixon, and D.S. Foster, 1990. Ancient channels of the Susquehanna River beneath Chesapeake Bay and the Delmarva Peninsula. *Geological Society of America Bulletin*, 102, 1268–1279.

Cooper, W.S., 1958. *Coastal sand dunes of Oregon and Washington*. Geological Society of America Memoir 72.

Cooper, W.S., 1967. *Coastal dunes of California*. Geological Society of America Memoir 104.

Cronin, T.M., B.J. Szabo, T.A. Ager, J.E. Hazel, and J.P. Owens, 1981. Quaternary climates and sea levels of the U.S. Atlantic Coastal Plain. *Science* 211, 233–240.

Davis, R.A., A.C. Hine, and E.A. Shinn, 1992. Holocene coastal development on the Florida Peninsula. In: C.H. Fletcher and J.F. Wehmiller (Editors), *Quaternary coasts of the United States: Marine and lacustrine systems*. Society for Sedimentary Geology, SEPM Special Publication 48, 193–212.

de Beaumont, E., 1845. *Leçons de géologie pratique*. 7me Leçon-levées de sables et galets, Paris.

Dionne, J.C., 1972. Caractéristiques des schorres des regions froides, en particulier de l'estuaire du Saint-Laurent. *Zeitschrift für Geomorphologie*, 15, 137–180.

Dolan R., and H. Lins, 1987. Beaches and barrier islands. *Scientific American*, 255, 67–77.

Edwards, J.M., and R.W. Frey, 1977. Substrate characteristics within a Holocene salt marsh, Sapelo Island, Georgia. *Senckenbergiana Maritima*, 9, 215–259.

Ferren, W.R., P.L. Fiedler, R.A. Leidy, K.D. Lafferty, and L.A.K. Mertes, 1996. Wetlands of California, *Madroño*, 43, 105–233.

Fisk, H.N., 1944. *Geological investigation of the alluvial valley of the Lower Mississippi River*. Mississippi River Commission, U.S. Army Corps of Engineers, Vicksburg.

Gallagher, J., 1996. *Late Holocene evolution of the Chorro delta, Morro Bay, California*. Ph.D. dissertation, University of California, Los Angeles.

Gilbert, G.K., 1885. The topographic features of lake shores. *Fifth Annual Report*, U.S. Geological Survey, 69–123.

Gilbert, G.K., 1917. Hydraulic mining debris in the Sierra Nevada. *U.S. Geological Survey Professional Paper* 105.

Godfrey, P.J., M.M. Godfrey, and D. Disraeli, 1982. North America, coastal ecology. In: M.L. Schwartz (Editor), *The encyclopeadia of beaches and coastal environments*, Hutchinson Ross, New York, 580–593.

Griggs, D.B., and R.E. Johnson, 1979. Coastline erosion, Santa Cruz County. *California Geology*, 32, 67–76.

Hall, D.K., C.S. Benson, and W.O. Field, 1995. Changes of glaciers in Glacier Bay, Alaska, using ground and satellite measurements. *Physical Geography*, 16, 27–41.

Hanson, K.L., J.R. Wesling, W.R. Lettis, K.I. Kelson, and L. Mezger, 1994. Correlation, ages, and uplift rates of Quaternary marine terraces: South-central coastal California. In: I.B. Alterman, R.B. McMullen, L.S. Cluff, and D.B. Slemmons (Editors), *Seismotectonics of the central California Coast Ranges*. Geological Society of America Special Paper, 292, 45–71.

Hayes, M.O., 1985. Atlantic USA—South. In: E.C.F. Bird and M.L. Schwartz (Editors), *The world's coastline*. Van Nostrand Reinhold, New York, 207–211.

Heusser, C.J., 1995. Late Quaternary vegetation response to climatic-glacial forcing in North Pacific America. *Physical Geography*, 16, 118–149.

Hillaire-Marcel, C. and S. Ochietti, 1980. Chronology, paleogeography and paleoclimatic significance of the late and post-glacial events in eastern Canada. *Zeitschrift für Geomorphologie*, 24, 373–392.

Hoyt, J.H., 1967. Barrier island formation. *Geological Society of America Bulletin*, 78, 1125–1136.

Inman, D.L., and C.E. Nordstrom, 1971. On the tectonic and morphologic classification of coasts. *Journal of Geology*, 79, 1–21.

Kelley, J.T., S.M. Dickson, D.F. Belknap, and R. Stuckenrath, 1992. Sea-level change and late Quaternary sediment accumulation on the southern Maine inner continental shelf. In: C.H. Fletcher and J.H. Wehmiller (Editors), *Quaternary coasts of the United States: Marine and lacustrine systems*, Society for Sedimentary Geology, SEPM Special Publication 48, 24–34.

Knebel, H.J., R.R. Rendigs, R.N. Oldale, and M.H. Bothner, 1992. Sedimentary framework of Boston Harbor, Massachusetts. In: C.H. Fletcher and J.F. Wehmiller (Editors), *Quaternary coasts of the United States: Marine and lacustrine systems*, Society for Sedimentary Geology, SEPM Special Publication 48, 35–43.

Komar, P.D., 1978. Wave conditions on the Oregon coast during the winter of 1977–78 and the resulting erosion of Nestucca spit. *Shore and Beach*, 46, 3–8.

Komar, P.D., and C.C. Rea, 1976. Erosion of Siletz Spit, Oregon. *Shore and Beach*, 44, 9–15.

Kraft, J.C., E.A. Allen, D.F. Belknap, C.J. John, and E.M. Maurmeyer, 1979. Processes and morphologic evolution of an estuarine and coastal barrier system. In: S.P. Leatherman (Editor), *Barrier islands from the Gulf of St. Lawrence to the Gulf of Mexico*. Academic Press, New York, 149–184.

Krantz, D.E., 1991. A chronology of Pliocene sea-level fluctuations: The U.S. Middle Atlantic Coastal Plain record. *Quaternary Science Reviews*, 10, 163–174.

Lajoie, K.R., A.M. Sarna-Wojcicki, and R.F. Yerkes, 1982. Quaternary chronology and rates of crustal deformation in the Ventura area, California. In: J.D. Cooper (Editor), *Neotectonics in southern California*. Geological Society of America Cordilleran Section Annual Meeting Guidebook, 43–51.

Lawson, A.C., 1893. The post-Pliocene diastrophism of the coast of southern California. *University of California Department of Geological Sciences Bulletin*, 1, 115–160.

Leatherman, S.P., 1979. Migration of Assateague Island, Maryland, by inlet and overwash processes. *Geology*, 7, 104–107.

McCann, S.B., 1979. Barrier islands in the southern Gulf of St. Lawrence. In: S.P. Leatherman (Editor), *Barrier islands from the Gulf of St. Lawrence to the Gulf of Mexico*. Academic Press, New York, 29–63.

McCann, S.B., 1985. Atlantic Canada. In: E. C. Bird and M.L. Schwartz (Editors), *The World's Coastline*. Van Nostrand Reinhold, New York, 235–240.

McCann, S.B., and M.L. Byrne, 1994. Dune morphology and the evolution of Sable Island, Nova Scotia, in historic times. *Physical Geography*, 15, 342–357.

McGee, W.J., 1890. Encroachment of the sea. *The Forum*, 9, 437–449.

Milliman, J.D., and J.P.M. Sivitski, 1992. Geomorphic/tectonic control of sediment discharge to the ocean: The importance of small mountainous rivers. *Journal of Geology*, 100, 525–544.

Molnia, B.F., 1986. Late Wisconsin glacier history of the Alaskan continental margin. In: T.D. Hamilton, K.M. Reed, and R.M. Thorson (Editors), *Glaciation in Alaska*

—*The geological record*. Alaska Geological Society, 219–236.

Molnia, B.F., and A. Post, 1995. Holocene history of Bering Glacier, Alaska: A prelude to the 1993–94 surge. *Physical Geography*, 16, 87–117.

Muhs, D.R., 1985. Amino acid age estimates of marine terraces and sea levels on San Nicolas Island, California. *Geology*, 13, 58–61.

Muhs, D.R., G.H. Miller, J.F. Whelan, and G.L. Kennedy, 1992. Aminostratigraphy and oxygen isotope stratigraphy of marine-terrace deposits, Palos Verdes Hills and San Pedro areas, Los Angeles County, California. In: C.H. Fletcher and J.F. Wehmiller (Editors), *Quaternary coasts of the United States: Marine and lacustrine systems*. Society for Sedimentary Geology, SEPM Special Publication, 48, 363–376.

Nelson, A.R., 1992. Holocene tidal-marsh stratigraphy in south-central Oregon—Evidence for localized sudden submergence in the Cascadia subduction zone. In: C.H. Fletcher and J.F. Wehmiller (Editors), *Quaternary coasts of the United States: Marine and lacustrine systems*. Society for Sedimentary Geology, SEPM Special Publication, 48, 287–301.

Nielsen, N., 1985. Greenland. In: E.C. Bird and M.L. Schwartz (Editors), *The world's coastline*. Van Nostrand Reinhold, New York, 261–265.

Nummedal, D., J.M. Coleman, R. Boyd, and S. Penland, 1985. Louisiana. In: E.C.F. Bird and M.L. Schwartz (Editors), *The world's coastline*, Van Nostrand Reinhold, New York, 147–153.

Oertel, G.F., 1985. The barrier island system. *Marine Geology*, 63, 1–18.

Oertel, G.F., J.C. Kraft, M.S. Kearney, and H. J. Woo, 1992. A rational theory for barrier-lagoon development. In: C.H. Fletcher and J.F. Wehmiller (Editors), *Quaternary coasts of the United States: Marine and lacustrine systems*. Society for Sedimentary Geology, SEPM Special Publication, 48, 77–87.

Orme, A.R., 1972. Quaternary deformation of western Baja California, Mexico, as indicated by marine terraces and associated deposits. *Tectonics*, 24th International Geological Congress, Montreal, 3, 627–634.

Orme, A.R., 1973a. *Coastal dune systems of northwest Baja California, Mexico*. Technical Report 0-73-1, U.S. Navy, Office of Naval Research, Washington, D.C.

Orme, A.R., 1973b. *Coastal salt marshes of northwest Baja California, Mexico*. Technical Report 0-73-2, U.S. Navy, Office of Naval Research, Washington, D.C.

Orme, A.R., 1980. Marine terraces and Quaternary tectonism, northwest Baja California, Mexico. *Physical Geography*, 1, 138–161.

Orme, A.R., 1982. Temporal variability of a summer shorezone. In: C. E. Thorn (Editor), *Space and time in geomorphology*, Allen and Unwin, London, 285–313.

Orme, A.R., 1985. California. In: E.C.F. Bird and M.L. Schwartz (Editors), *The world's coastline*, Van Nostrand Reinhold, New York, 27–36.

Orme, A.R., 1990a. The instability of Holocene coastal dunes: The case of the Morro dunes, California. In: K.F. Nordstrom, N.P. Psuty, and R.W.G. Carter (Editors), *Coastal dunes: Form and Process*, Wiley, New York, 315–336.

Orme, A.R., 1990b. Wetland morphology, hydrodynamics and sedimentation. In: M. Williams (Editor), *Wetlands: A threatened landscape*, Institute of British Geographers Special Publication Series 25, Blackwell, Oxford, 42–94.

Orme, A.R., 1991. Mass movement and seacliff retreat along the southern California coast. *Southern California Academy of Sciences Bulletin*, 90, 58–79.

Orme, A.R., 1992. Late Quaternary deposits near Point Sal, south-central California: A time frame for coastal-dune emplacement. In: C.H. Fletcher and J.F. Wehmiller (Editors), *Quaternary coasts of the United States: Marine and lacustrine systems*. Society for Sedimentary Geology, SEPM Special Publication, 48, 309–315.

Orme, A.R., 1998. Late Quaternary tectonism along the Pacific coast of the Californias: A contrast in style. In: I.S. Stewart and C. Vita-Finzi (Editors), *Coastal Tectonics*. The Geological Society, London, Special Publications, 146, 179–197.

Orme, A.R., 2000. Evolution and historical development. In: R.A. Ambrose and A.R. Orme (Editors), *Lower Malibu Creek and barrier-lagoon system, resource enhancement and management*. California State Coastal Conservancy, 1–37.

Orme, A.R., and V.P. Tchakerian, 1986. Quaternary dunes of the Pacific coast of the Californias. In: W.G. Nickling (Editor), *Aeolian geomorphology*, Allen and Unwin, London, 149–175.

Ortlieb, L., 1987. *Neotectonique et variations du niveau marin au Quaternaire dans la région du Golfe de Californie, Mexique*. Institut Française de Recherche Scientifique pour la Développement en Coopération. Collection Etudes et Thèses, Paris.

Otvos, E.G., 1992. Quaternary evolution of the Apalachicola coast, northeastern Gulf of Mexico. In: C.H. Fletcher and J.F. Wehmiller (Editors), *Quaternary coast of the United States: Marine and lacustrine systems*. Society for Sedimentary Geology, SEPM Special Publication, 48, 221–232.

Peltier, W.R., 1998. Global glacial isostatic adjustment and coastal tectonics. In: I.S. Stewart and C. Vita-Finzi (Editors), *Coastal Tectonics*, The Geological Society, London, Special Publications, 146, 1–29.

Penland, S., R. Boyd, and J.R. Suter, 1988. Transgressive depositional systems of the Mississippi delta plain: A model for barrier shoreline and shelf sand development. *Journal of Sedimentary Petrology*, 58, 932–949.

Pestrong, R., 1972. San Francisco Bay tidelands. *California Geology*, 25, 27–40.

Plafker, G., 1965. Tectonic deformation associated with the 1964 Alaskan earthquake. *Science*, 148, 1675–1687.

Post, A., and R.J. Motyka, 1995. Taku and Le Conte glaciers, Alaska: Calving-speed control of late Holocene asynchronous advances and retreats. *Physical Geography*, 16, 59–82.

Prior, D.B., and J.M. Coleman, 1978. Disintegrating retrogressive landslides on very-low-angle subaqueous slopes, Mississippi delta. *Marine Geotechnology*, 3, 37–60.

Psuty, N.P., and C. Mizobe, 1982. Central and South America, coastal ecology. In: M.L. Schwartz (Editor), *The Encyclopedia of Beaches and Coastal Environments*, Hutchinson Ross, New York, 191–201.

Purer, E.A., 1942. Plant ecology of the coastal salt marshlands of San Diego County, California. *Ecological Monographs*, 12, 81–111.

Ramsey, K. W., 1992. Coastal response to late Pliocene climate change: Middle Atlantic Coastal Plain, Virginia and Delaware. In: C.H. Fletcher and J.F. Wehmiller

(Editors), *Quaternary coasts of the United States: Marine and lacustrine systems.* Society for Sedimentary Geology, SEPM Special Publication, 48, 121–127.

Riggs, S.R., L.L. York, J.F. Wehmiller, and S.W. Snyder, 1992. Depositional patterns resulting from high-frequency Quaternary sea-level fluctuations in northeastern North Carolina. In C.H. Fletcher and D.F. Wehmiller (Editors), *Quaternary coasts of the United States: Marine and lacustrine systems.* Society for Sedimentary Geology, SEPM Special Publication, 48, 141–153.

Rockwell, T.K., D.R. Muhs, G.L. Kennedy, M.E. Hatch, S.H. Wilson, and R.E. Klinger, 1989. Uranium-series ages, faunal correlations and tectonic deformation of marine terraces within the Agua Blanca fault zone at Punta Banda, northern Baja California, Mexico. In: P.L. Abbott (Editor), *Geological Studies in Baja California*, Pacific Section, Society of Economic Paleontologists and Mineralogists, 63, 1–16.

Russell, R.J., 1936. Physiography of lower Mississippi River delta, Louisiana. *U.S. Geological Survey Bulletin*, 8, 3–193.

Ruz, M.-H., and M. Allard, 1994. Coastal dune development in cold-climate environments. *Physical Geography*, 15, 372–380.

Saint-Laurent, D., and L. Filion, 1992. Interprétation paléoécologique des dunes à la limite des arbres, secteur nord-est de la mer d'Hudson, Québec. *Géographie Physique et Quaternaire*, 46, 209–220.

Sallenger, A.H., R.A. Holman, and W.A. Birkemeier, 1985. Storm-induced response of a nearshore bar system. *Marine Geology*, 64, 237–257.

Steckler, M., A.B. Watts, and J.A. Thorne, 1988. Subsidence and basin modeling at the U.S. passive margin. In: R.E. Sheridan and J.A. Grow (Editors), *The Atlantic continental margin*, Geological Society of America, Boulder, 399–416.

Stock, J.M., and K.V. Hodges, 1989. Pre-Pliocene extension around the Gulf of California, and the transfer of Baja California to the Pacific plate. *Tectonics*, 8, 99–115.

Suess, E., 1892. *Das Antlitz der Erde*, 2. Vienna. Translated by H.B. Solas, 1906. *The face of the Earth*, Oxford University Press, Oxford.

Supan, A., 1930. *Grundzüge der Physichen Erdkunde*, 2 (1): *Das Land (Allgemeine Geomophologie)*, De Gruyter, Berlin.

Tanner, W.F., 1985. Florida. In: E.C.F. Bird and M.L. Schwartz (Editors), *The world's coastline*, Van Nostrand Reinhold, New York, 163–167.

Taylor, R.B., and S.B. McCann, 1983. Coastal depositional landforms in northern Canada. In: D.E. Smith and A.G. Dawson (Editors), *Shorelines and isostasy.* Institute of British Geographers Special Publication, 16, Academic Press, London, 53–75.

Thompson, R.W., 1968. *Tidal-flat sedimentation on the Colorado River delta, northwestern Gulf of California.* Geological Society of America Memoir 107.

Toscano, M.A., 1992. Record of oxygen isotope stage 5 on the Maryland inner shelf and Atlantic Coastal Plain—A post-transgressive-highstand regime. In: C.H. Fletcher and J.F. Wehmiller (Editors), *Quaternary coasts of the United States: Marine and lacustrine systems. Society for Sedimentary Geology*, SEPM Special Publication, 48, 89–99.

Walker, H.J., 1985. Alaska. In: E.C. Bird and M.L. Schwartz (Editors), *The world's coastline.* Van Nostrand Reinhold, New York, 1–10.

Walker, H.J., and J. M. Coleman, 1987. Atlantic and Gulf Coastal Province. In: W.L. Graf (Editor), *Geomorphic systems of North America*, Geological Society of America, Centennial Special Volume 2, Boulder, 51–110.

Williams, S.J., 1975. Anthropogenic filling of the Hudson River (shelf) channel. *Geology*, 3, 597–600.

Woodring, W.P., M.N. Bramlette, and W.S.W. Kew, 1946. Geology and paleontology of the Palos Verdes Hills, California. *U.S. Geological Survey Professional Paper* 207.

Yeats, R.S., 1977. High rates of vertical crustal movement near Ventura, California. *Science*, 196, 295–298.

Yeats, R.S., and T.K. Rockwell, 1991. Quaternary geology of the Ventura and Los Angeles basins, California. In: R.B. Morrison (Editor), *Quaternary nonglacial geology: Conterminous U.S.* Geological Society of America, Boulder. The Geology of North America, K-2, 185–189.

III

NATURE IN THE HUMAN CONTEXT

22

Human Imprints on the Primeval Landscape

Antony R. Orme

No physical geography is complete without some evaluation of the changes wrought on the natural landscape by human activity. Such impacts reflect the practical needs, cultural ethos, and technical abilities of people, all of which change over time. For North America, as elsewhere, a useful distinction may be made between the role of prehistoric peoples with simple needs and limited skills, and historic peoples who, armed with more advanced technology, have made increasing demands on the landscape and its resources. Unlike in Asia and Europe, however, the dividing time line between prehistoric and historic cultures came quite late in North America, where only in the last few centuries, more recently in some areas, have human imprints become truly significant.

The theme of this chapter is vast, so the approach here is necessarily selective. After reviewing the peopling of the continent, the discussion moves from an evaluation of impacts on plants and animals, through the effects of mining, to impacts on water resources. This order has a certain logic because plants and animals are the first resources that, for reasons of survival, people normally use. Later, as technology improves, these resources, with minerals and water, are exploited more thoroughly until in many areas the primeval landscape all but disappears, replaced by the agricultural, industrial, and urban landscapes. However, the sequence is not rigid. For example, twentieth-century oil drillers in Alaska invaded the wilderness directly, whereas water resources were impacted from the earliest human encounters with plants and animals, but not to the extent that they would be later.

Although eschewing environmental determinism, it is stressed that human impacts on the North American landscape have been staged against a changing natural backcloth that has variably influenced human activity. Thus fluctuating ice sheets and sea levels clearly influenced late Pleistocene human immigration to North America. Later, the early Holocene retreat of tundra before advancing forest and the prairie expansion into former forest areas challenged human survival skills, just as later Holocene changes in forest and grassland distributions called for further adjustments. A warm interlude between A.D. 800 and 1200, linked to the climatic anomaly so well documented in medieval Europe, favored Norse settlements in Greenland that later succumbed to advancing permafrost and glaciers. Climatic variability also increased along the Pacific coast around this time, reflected in higher sea-surface temperatures and alternations of rainy periods with prolonged drought, which affected prehistoric subsistence patterns and social organization in California. Similarly, increasing drought in the Great Plains and the Southwest after 1200, and cold stormy conditions along the eastern seaboard during the Little Ice Age (≈1500–1850), challenged the survival of settled farming communities, regardless of other human impacts. Recent human impacts on climate, through the emission of greenhouse gases, have been superimposed in turn on a natural warming trend.

Human impacts on the land have long been the focus of scholarly enquiry. From a North American perspective, one notable milestone was the publication in 1864 of *Man and Nature* by George Perkins Marsh, a theme since revisited many times (e.g., Thomas, 1956; Turner, 1990). In addition, many historical geographies have examined evolving interactions between people and landscape in a broader context (e.g., Harris and Warkentin, 1974; Meinig, 1986, 1993; Mitchell and Groves, 1987).

22.1 The Peopling of North America—An Overview

The distinction between the human imprints of prehistoric and historic peoples is based on the gradual colonization of North America by Europeans and others following Columbus' accidental landfall on the Bahamas in 1492. From a landscape perspective, the Columbian encounter of 1492 was a watershed, separating a long, preceding period of, at best, modest human impacts from a short, subsequent period of increasingly severe impacts that led to the landscapes of today. But it was not a sharp watershed because, even though indigenous impacts withered from conflict and disease, early European colonists were little better equipped for survival and resource exploitation than their predecessors. Gradually, however, as immigrant numbers increased and technology improved, people with a variety of resource penchants came from Europe and elsewhere to reshape much of the landscape.

22.1.1 Pre-Columbian Peoples

Much has been written about the pristine landscapes allegedly encountered by early European immigrants to North America and about how indigenous peoples lived in harmony with nature (e.g., Sale, 1990). Both beliefs are misleading. Many landscapes observed by early Europeans had been modified by centuries of human-induced fire, by hunting and fishing, by farming, often with irrigation, and by the construction of villages and temples. Further, although respecting the spirituality of their links with nature, most native peoples were able to live "in harmony" with these landscapes because their numbers were small and they exerted little pressure on available resources. Denevan (1992a) has estimated that around 3.8 million people inhabited North America in 1492. Even allowing for some margin of error, this is a small number compared with the 50.1 million then living elsewhere in the Americas, mainly in Mexico (17.2 million) and the Andes (15.7 million). It is also small compared with the 312 million people now living in the United States (281 million) and Canada (31 million). But 3.8 million is no mean number and this, together with field evidence, led Denevan (1992b) to term the perception of North America in 1492 as a sparsely populated wilderness the "pristine myth." Nonetheless, because of the small population, much of the prehistoric landscape must have been but lightly etched by human impacts.

Most archeological evidence shows that the earliest human immigrants to North America traveled by land across Beringia from Asia during late Pleistocene time, when seas were below present levels, summers were warming, and parkland was replacing open tundra. They then moved southward through the widening corridor between the Laurentide and Cordilleran ice sheets, spread across the vast interior lowlands, and kept coming throughout the Holocene. These early people were nomadic hunters and foragers whose main impacts were on the animals they pursued. Abundant evidence reveals the presence of campsites across Beringia during the closing millennia of the Pleistocene and of an unusually rapid diffusion of hunting groups southward across North America during the early Holocene. Others probably moved into North America along Pacific coastal routes, relying more on marine life (Fladmark, 1975), but rising postglacial seas, coastal erosion, earthquakes and volcanic activity have erased or buried much of this record. The later Holocene movement of Inuit (Eskimo) and Aleut hunters in skin and wood boats across the Bering Strait shows that even the partial submergence of Beringia was no barrier to migration. Some early peoples may also have entered the continent from other directions, accidental tourists driven onshore by favorable winds and currents. Still others may have passed through Beringia much earlier, leaving records of their passage that have yet to be found, perhaps buried beneath glacial deposits, loess, and tephra (Chlachula, 1996). However, it is still likely that most immigrants entered by way of Beringia and its coastal waters during the terminal Pleistocene. Indeed, for our purposes, it matters less how and when they came than that they arrived, armed with certain skills, pushed by the need for survival, pulled by the opportunities perceived ahead, or simply victims of chance.

The earliest immigrants to North America worked flint, chert, and obsidian into functional tools that define the Lithic period. This onlapping sequence of nomadic hunters included the Clovis culture, of the terminal Pleistocene, whose distinctively fluted spear points are widely associated with mammoth kill sites (Sellards, 1938; Holliday et al., 1994), and the early Holocene Folsom and later Plano cultures, whose spear points and spear-throwing atlatls occur in bison kill sites across the Great Plains (Holliday, 1997). Although hunting long persisted, more versatile hunter-foragers emerged during the early Holocene (10,000–7000 B.P., radiocarbon years before 1950) to define the onset of the Archaic period (Bettis, 1995). They hunted, trapped, fished, gathered edible plants, shaped specialized weapons and tools, wove baskets and cloth from plants, and constructed boats. Although they were still largely migratory, their campsites and refuse heaps suggest some seasonal stability. As such, Archaic peoples became locally distinctive, from the Red Paint culture of Atlantic Canada (≈5000–2500 B.P.), through the Old Copper culture of the Great Lakes (≈6000–3500 B.P.), who fashioned copper tools and ornaments, to the Cochise culture of the Southwest (≈9000–3000 B.P.), who, perhaps through contact with more advanced Mesoamerican peoples, began cultivating maize around 5500 B.P. (Waldman, 1985).

Further cultural transitions around 3000 B.P. initiated the Formative period, which passed into the Classic lineage later encountered by Europeans, notably in Mesoamerica. In general, the Formative period involved the spread of agriculture, permanent settlements, advanced weaving and pottery, and ceremonial activities; during this period, the bow and arrow led to improved hunting skills. These peoples included the burial-mound builders of the eastern woodlands, notably the Adena of the Ohio Valley (3000–1800 B.P.) and the more widespread Hopewell people (2300–1300 B.P.). In the southern Appalachians, where late Archaic peoples had been collecting hickory nuts and growing squash (*Cucurbita pepo*) and gourds (*Lagenaria siceraria*) around 4000 B.P., these woodland peoples began more widespread forest clearance after 2800 B.P. and were cultivating maize (*Zea mays*) by 1500 B.P. (Delcourt et al., 1986). People of the later Mississippian culture (1300–500 B.P.), builders of palisaded villages and temple mounds, spread farming practices based on maize and beans (*Phaseolus* spp.) across much of the Southeast. Related clearances probably decreased the proportion of landscapes dominated by hardwoods in favor of pines, which thrive in more open conditions (Whitehead and Sheehan, 1985; Shankman and Wills, 1995). The larger settlements, including Cahokia on the Illinois River, which contained 85 burial and temple mounds and was the focus for perhaps 75,000 people, had been abandoned before European contact around A.D. 1600.

In the Southwest, a succession of settled societies reached a cultural climax around A.D. 1200–1400 but shortly thereafter abandoned their lands and settlements, either to disperse into smaller groups or to be absorbed by neighbors and newcomers. These peoples included early Mogollon forager-farmers, later Hohokam irrigation farmers, and subsequently dominant Anasazi peoples. Meanwhile, quite advanced societies existed along the north Pacific coast without agriculture, while the peoples of the far north, including Inuit and Aleut immigrants after 5000 B.P., continued to hunt and fish in the Archaic tradition into modern time.

At the time of European contact, North America contained about a dozen culture areas which can be further grouped in terms of environmental impact (fig. 22.1). The far north, boreal forest, interior plateaus, and Pacific coast were occupied mainly by hunters, foragers, and fisherfolk, more numerous and more settled amid the rich marine resources of coastal areas. The eastern woodlands, from the St. Lawrence and Mississippi valleys to the Gulf of Mexico, were peopled by farmers who augmented their diet from trapping, fishing, and foraging. The Great Plains were more thinly occupied by farming communities clustered along the rivers that crossed the prairies from the western mountains (although the horse, reintroduced by Spaniards, was soon to recreate a seminomadic culture based on bison hunting). In the Southwest, settled peoples with farming traditions lingered on but, as drought intensified, dependence of foraging and hunting increased.

Pre-Columbian peoples have left a collective legacy in the many indigenous names that were adopted by later immigrants, including toponyms ranging from the Mississippi (Algonquian "big river") to Yucaipa (Shoshonean "wetland"). Other names have been widely accepted for scientific usage, including Inuit words like *nunatak*, *pingo*, and *talik*, and Algonquian words like *muskeg* and *pocosin*.

22.1.2 Post-Columbian Impacts and Their Significance

The past 500 years have seen a massive transformation of the primeval landscape of North America by successive waves of immigrants, initially from Europe but later from Africa and elsewhere. The impact on existing inhabitants has been devastating. As a result of conflict and introduced diseases against which there was no natural immunity, the native population fell from 3.8 million in 1500 to around 1 million by 1800, a 74% drop (Denevan, 1992a). During the eighteenth century, new immigrants came to outnumber native inhabitants who today, despite some resurgence, still number less than 3 million. The effect of these changes, though complex, may be divided in terms of landscape impacts into two time periods: from 1500 to 1800, and from 1800 to the present.

During the earlier period, successful European immigration, as distinct from exploratory and military ventures, was strongly regional in terms of its ethnic roots and resource impacts (fig. 22.2). From bases in Mexico and the Caribbean, Spaniards penetrated the south from the sixteenth century onward, establishing settlements in Florida (San Agustín, 1565), the Southwest (Santa Fé, 1609), and Alta California (San Diego, 1769), bringing a penchant for livestock rearing and mining. Toponyms, such as *alamo* (cottonwood), *ciénaga* (marsh), and *encina* (evergreen oak), suggest the landscapes they found, many now lost. In the early seventeenth century, French subsistence farmers colonized Acadia (Port Royal, 1605) and the St. Lawrence and Ottawa valleys (Québec, 1608). Fur traders (*voyageurs*, *coureurs de bois*) and missionaries then spread thinly through the Great Lakes (Sault Ste-Marie, 1668) toward the prairies (Fort La Reine, 1738; La Corne, 1753) and down the Mississippi to the delta (Fort de la Boulage, 1700; New Orleans, 1718). Meanwhile, English, Welsh, Scots, and Irish

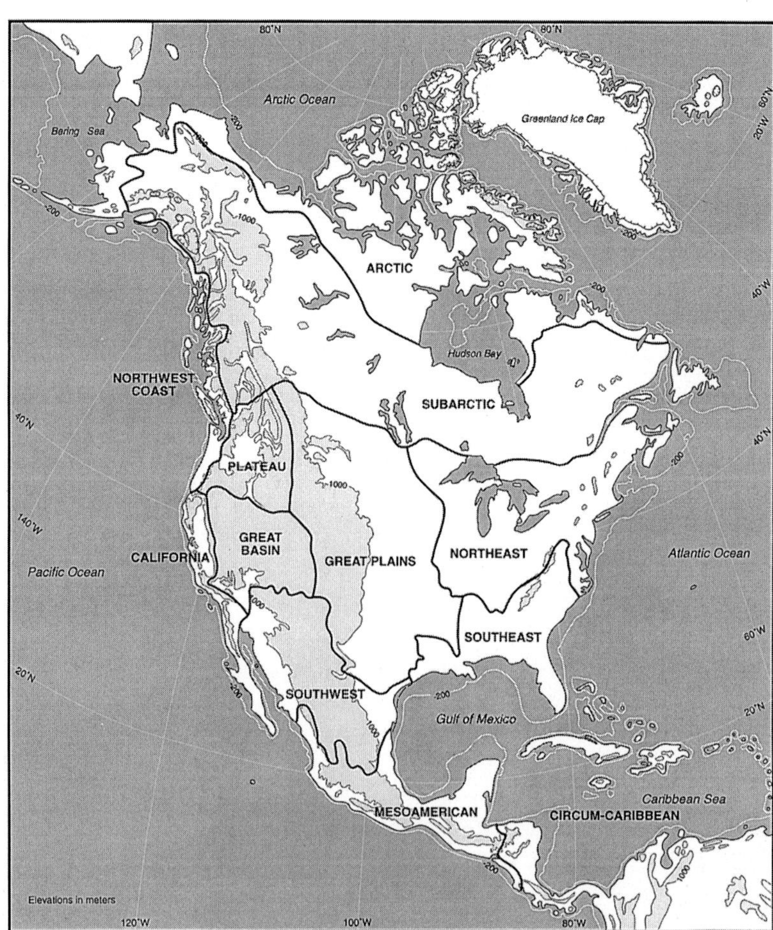

Figure 22.1 Areas of indigenous culture in North America at the time of European contact around 1500.

were colonizing the Atlantic seaboard, establishing fish camps in Newfoundland (1610), subsistence farms in New England (Plymouth, 1620), and "plantations" in Virginia (Jamestown, 1607) and the Carolinas (Charles Town, 1670). They were joined on the Grand Banks fisheries by Bretons and Portuguese, onshore by Dutch (Fort Nassau, 1614) and Swedish colonists (Fort Christina, 1638), and later by Germans and other Europeans. Helped by African slaves, commercial agriculture came to the Atlantic Coastal Plain and, after 1700, settlers spilled onto the Piedmont and later across the Appalachians. Kentucky had 70,000 settlers by 1790 (Goetzmann and Williams, 1992). In the continuing pursuit of furs, individuals and partnerships such as the Hudson's Bay Company (1670) and the North West Company (1783) founded outlying trading posts (e.g., Rupert House, 1668; Fort York, 1684; Churchill, 1717). They sponsored expeditions to the northwest, reaching the Arctic Ocean down the Coppermine River in 1772 and the Mackenzie River in 1789, and across the Rockies in 1793 to the Pacific coast whose marine resources were already being taken by Russian fur traders (*promyshlenniki*) with bases from Kodiak Island (Three Saints, 1784) to California (Fort Ross, 1812).

Much knowledge of the contemporary landscape may be gleaned from the diaries and crude maps of early traders and missionaries, and from naturalists like Catesby, Cuming, Kalm, Michaux, and Steller, surveyors like de Léry and Byrd, cartographers like Cárdenas, Evans, and de L'Isle, and numerous artists (Goetzmann and Williams, 1992). A growing awareness of the continent's geography was in turn disseminated through fledgling colleges such as Harvard and Yale, and in textbooks being written at the time by local, as distinct from European, authors, notably Jedidiah Morse's *American Geography* of 1789 (Martin, 1998).

By 1800, nature's resources were coming under increasing pressure, and, as the native population was decimated by disease and conflict, the earlier landscape mosaic of the eastern woodlands reverted to scrub and forest, more nearly primeval in 1750 than it had been in 1492 (Williams, 1989; Denevan, 1992b).

Changing political arrangements throughout much of North America between 1763 and 1821 coincided with the onset of the industrial revolution in Europe, which in turn increased demand for raw materials and stimulated technical advances. In response, from 1800 onward, the North

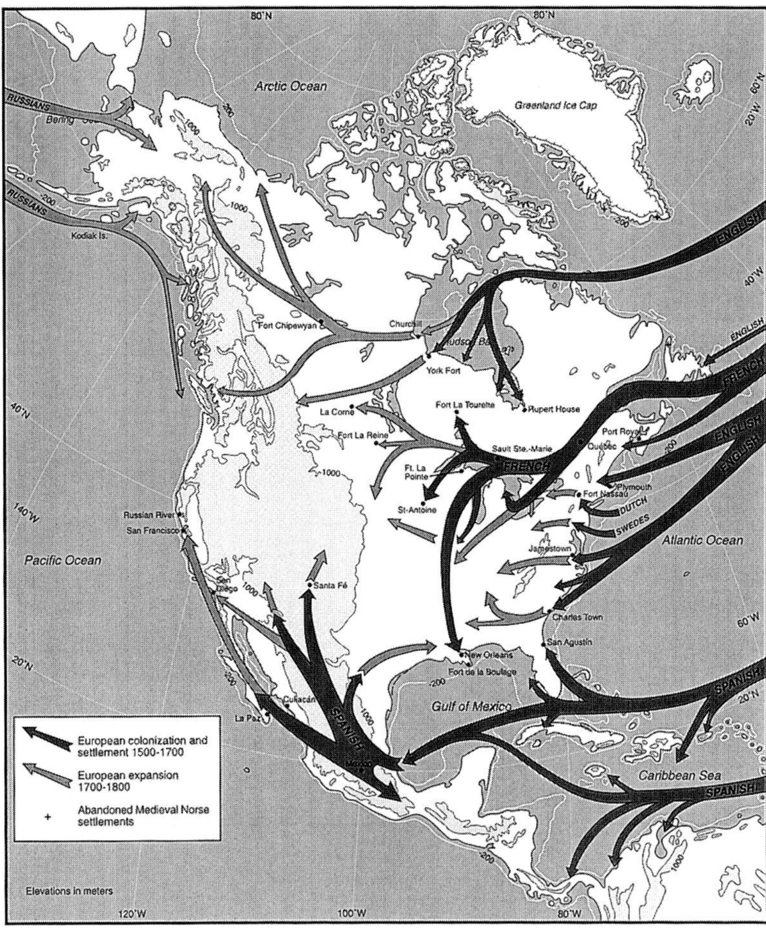

Figure 22.2 Principal routes of European immigration to North America, 1500–1800.

American landscape was to change more rapidly as trees were felled, minerals exploited, commercial agriculture extended, local industries developed, and canals and later railways built to facilitate transport. The frontiers of European settlement pushed westward across the Mississippi and the Great Lakes onto the prairies and beyond. The more remote areas of the far north and interior west, hitherto still the realm of native peoples and fur trappers, were now invaded by mineral prospectors, railway engineers, and government agents, variously driven by prospects ranging from individual gain through commercial opportunity to national expansion.

As the nineteenth century progressed, the North American landscape became better mapped and its resources better known from trading, military, and naval expeditions across the continent and along its more remote coasts, from early geological surveys and from naturalists like Audubon, Nuttall, and Engelmann. Expeditions such as those of Lewis and Clark (1805–06), Pike (1806–07), Frémont (1843–48), and Palliser (1857–59), combined with the work of topographical engineers and railway surveyors in the 1840s and 1850s, paved the way for increased resource exploitation and settlement, particularly after present political affiliations were confirmed between 1845 and 1867. However, not all such surveys were optimistic. Long's foray of 1819–20, for example, concluded that the Great Plains were "almost wholly unfit for cultivation, and of course, uninhabitable by a people depending on agriculture for their subsistence. . . . The scarcity of wood and water, almost uniformly prevalent, will prove an insuperable obstacle in the way of settling the country . . ." This view perpetuated the myth of "The Great American Desert," which long deterred potential immigration to the region, and was to some extent echoed by Palliser's views on the Canadian prairies (see chapter 6).

During the later nineteenth century, scientific knowledge of the American West was improved by several surveys, notably those led by King (1867–73), Wheeler (1867–72), Powell (1869–78), and Hayden (1873–76), competition between which led to the formation of the U.S. Geological Survey in 1879. These surveys were conducted just ahead of the frontiers of settlement approaching from the east, now aided by railways after completion of the first transcontinental route in 1869. Nevertheless, shortage of water became a major challenge for agriculture and urban growth alike as time elapsed, subject to many solutions and much controversy. Farther north, the Geological Survey of Canada, founded in 1842 to encourage mineral exploration, became, like its American counterpart, a much broader resource-oriented organization (Vodden, 1992). Its role, and human impacts generally, grew with political confederation in 1867, purchase of the vast lands of Hudson's Bay Company in 1870, completion of the Canadian Pacific Railway in 1885, the Yukon gold rush of 1896–99, and discovery of nickel, copper, and iron in the Canadian Shield.

By 1890, according to the director of the U.S. census and Frederick Jackson Turner (1920), the American frontier of settlement, of reality and myth, had ceased to exist. The Canadian prairies were being settled and the far north was being probed by naval surveys and scientific expeditions. People of African heritage, long the backbone of southern agriculture, were moving north and west. As the nineteenth century ended, new waves of immigrants were reaching the east coast from south and east Europe, while immigrants from eastern Asia were coming to the west coast. As the twentieth century progressed and travel became easier, people came from everywhere, including those of mixed indigenous and Spanish heritage who again moved north from Mesoamerica. Each immigrant wave brought new impetus to the processes of landscape change. Nevertheless, the cold north and the arid west continue to challenge human determination, engineering ingenuity, and scientific enquiry, while agriculture and urbanization nearly everywhere generate specific problems for the environment and its management.

Figure 22.3 Contrast between old-growth coniferous forest (dark tones) and second-growth mixed forest (light tones reflect abundant alder in full growth) after logging in Whatcom County, Washington, 1984 (photo: A.R. Orme).

Figure 22.4 Urbanization replaces native oak chaparral and riparian woodland, Santa Monica Mountains, California, 1955 (photo: Fairchild Collection, UCLA).

22.2 Impacts of Vegetation Change

People impact native vegetation directly by setting fires, cutting, girdling, peeling, trampling, and machining, and indirectly by introducing nonnative plants and animals, exotic diseases, and exotic pests, and by influencing water resources and air quality. Such impacts trigger a range of environmental responses, from modest changes in habitat diversity to wholesale replacement of natural systems by agricultural, urban, and industrial landscapes (figs. 22.3, 22.4). The consequences of such responses range far beyond the vegetation to influence hydrologic regimes, geomorphic processes, and climate change. This discussion focuses on the role of vegetation in physical systems and on hydrogeomorphic responses to vegetation change. Ecological impacts of vegetation change are discussed in chapter 12 and elsewhere.

22.2.1 The Physical Role of Vegetation

Vegetation influences hydrologic regimes through its influence on interception, infiltration, evapotranspiration, snowmelt, and runoff, each of which affects the availability of running water for erosion and sediment transport. The amount of precipitation intercepted by plants and returned to the atmosphere as evaporation varies from negligible in open thin-foliaged shrubs to around 50% in dense hemlock forests. Coniferous trees intercept more rainfall than deciduous trees in full leaf (Kittredge, 1948). Deciduous trees intercept two or three times more rainfall in summer than in winter, whereas during optimal growth, dense grasses and herbs may intercept as much as leafy deciduous trees. For precipitation penetrating the vegetation cover, directly or as leaf drip and stemflow, the rate at which it infiltrates the ground is influenced by the sponge-like effect of plant litter, by root systems, and by organic matter incorporated into the soil. Temperate forests with thick protective litter and organic-rich soils may have a water-absorbing capacity 50 times greater than arable land. Because plants shelter the soil and reduce wind velocity, direct evaporation from the ground is reduced beneath a vegetation canopy, especially in the growing season, but this is offset by increased transpiration from leaves. With low relative humidity, warm temperatures, or high winds, evapotranspiration may deprive drainage basins of much or all the water that would otherwise be available for runoff. Forest canopies also retard snowmelt by reducing solar radiation intensity and maximum surface temperatures. Thus a reduction in vegetation cover will likely decrease interception, transpiration, and infiltration, and increase snowmelt and the amount of water directly available for evaporation and runoff.

Vegetation also directly influences runoff and thus surface erosion. Plant litter discourages concentration of flow into rills, whereas surface vegetation increases surface roughness and thereby reduces overland flow velocities. Because forests produce abundant litter and flow velocities vary as the square root of slope, the effect of forest clearance on steep slopes may be dramatic. Vegetation regulates flow most effectively for small streams during normal rainfall, and, at a local scale, forests almost eliminate discharge peaks from small storms. At larger scales during higher intensity rainstorms, vegetation is less effective, and, as the flood history of the Ohio River shows, forest cover has limited the influence of great floods within major river systems. Vegetation also affects hillslope stability by transpiring subsurface water and providing root strength that binds topsoil to subsoil and subsoil to parent material. Forests and dense chaparral are most effective in this context; shallow-rooted grasses, notably introduced annuals such as European wild oats (*Avena* spp.), less so. Conversely, the spread of the European giant reed, *Arundo donax*, has promoted sedimentation along riparian corridors. Vegetation also inhibits wind deflation and traps wind-borne particles.

22.2.2 The Role of Fire

Lightning strikes and volcanic eruptions have occurred throughout Earth's history, but related fires, augmented by spontaneous combustion, became a recurrent landscape process with the development of terrestrial vegetation. Many Pleistocene deposits, for example, contain charcoal before human arrival (e.g., MacDonald et al., 1991; Orme, 1992). Fires are also linked to climatic cycles, being more frequent when very wet years, which promote growth, are followed by prolonged drought, when fuel is readily ignited. Such a correlation between fire and El Niño–La Niña cycles has been suggested for the montane coniferous forests of the Southwest (Swetnam and Betancourt, 1990). However, from archaeological evidence of early immigrants making campfires and using fire to drive game, it is often assumed that fire became an even more potent force in landscape change during the prehistoric peopling of North America. Much later, at the time of first European contact, the Powatan, Algonquin, and Iroquois peoples of the eastern forests were burning vegetation to create more open ground for hunting and farming, and to improve the berry harvest, whereas the prairie grasslands were often burned to drive game (Day, 1953; Stewart, 1956). In the mixed-conifer forests of the Sacramento Mountains, New Mexico, nomadic Mescalero Apache and their pursuers seem to have increased fire frequency over the century preceding their eventual pacification in the 1870s (Kaye and Swetnam, 1999). Farther west, Juan Cabrillo's log recorded "great smokes" as he sailed off California in autumn 1542—presumably fires driven by Santa Ana winds (Moriarty, 1968). Juan Crespi, diarist to the first overland Spanish expedition along the California coast in 1769, noted "good land covered with grass and well supplied with water but without trees" (Bolton, 1927) in a region that should have been clothed in oak chaparral woodland.

Two fundamental questions thus arise. To what extent have natural fire frequency and magnitude been altered by the peopling of the continent? What effect did these changes have on the landscape? The first question is not easily answered. On one hand, historic fire patterns in the Californias and the Southwest suggest that before 1900 the natural regime was one of frequent small fires, similar to those in modern Baja California Norte, where, lacking a fire-suppression policy, a fine-grained vegetation mosaic inhibits larger fires (Minnich, 1983; Kaye and Swetnam, 1999). When fire suppression was introduced to southern California in the twentieth century, fire frequency declined, and, with fuel accumulating, when fires did occur, they were larger, hotter, and more damaging. In the mixed conifer forests north of the Grand Canyon, fire intervals were longer under native American influence before 1870 than during the Euroamerican period, 1870–1919, but shorter than during the suppression era from 1920 on (Wolf and Mast, 1998). Conversely, charcoal from a 560-year record (1425–1985) in varved sediment in the Santa Barbara Channel shows no distinct change in fire frequency across three land-use periods: the native Chumash period of frequent burning, the Spanish and early Euroamerican period with little fire control, and the twentieth century with active fire suppression (Mensing et al., 1999). This suggests that fire frequency is influenced by fuel and weather conditions, especially transitions from wet phases to drought and desiccating Santa Ana winds, rather than by human practices. Nevertheless, though still fuel dependent, accidental ignitions and arson fires near urban centers and in recreational areas have increased in recent decades.

The effects of fire on the physical landscape are dramatic, complementing those on the biological landscape evaluated in chapters 12 and 14. As with other forms of vegetation change, the removal of plants and plant litter by fire exposes the surface to dry ravel and to increased raindrop impact and runoff. But fire is unusual for the suite of changes that it initiates before actual erosion. For example, fire modifies soil structure and texture, and thereby infiltration and runoff, by generating very high surface temperatures (≈600–1000°C) that destroy organic matter, consume nutrients, and fuse soil particles (DeBano et al., 1979). Fire also modifies water repellency, present to some degree in most soils, by vaporizing the waxy organic substances responsible for hydrophobicity at the surface, and then translocating and condensing this matter farther down the soil profile (fig. 22.5). In forest and chaparral soils in particular, this can lead to the accumulation of dense water-repellent layers beneath the surface that inhibit infiltration and lead to increased rilling and shallow debris flows (DeBano, 1981; Henderson and Golding, 1983; Wells, 1987). Thus, when rains arrive, fire-ravaged landscapes typically suffer accelerated rates of surface erosion and soil slippage, at least until the water-repellent layers are modified or eroded and infiltration improves.

22.2.3 Hydrogeomorphic Impacts of Vegetation Change

The net effect of vegetation change is a disruption of the water cycle and of biological barriers to erosion and mass movement. The magnitude of these disruptions depends

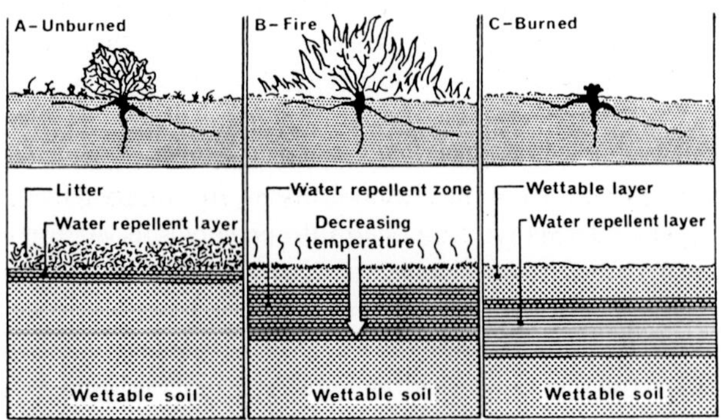

Figure 22.5 Water repellency in soil before, during, and after fire: A, Before fire, hydrophobic substances accumulate in surface litter and subjacent mineral soil; B, Fire burns litter, causing hydrophobic substances to translocate downward along temperate gradients; C, After fire, a dense water-repellent layer occurs at greater depth, inhibiting infiltration and promoting surface runoff (from DeBano, 1981).

on the severity and persistence of the changes, which are in turn related to time and technology. Minor changes in water budgets and erosion regimes may be inferred from prehistoric resource usage, notably in the eastern woodlands, but their magnitude is masked by the more dramatic legacies of historic land-use practices, imposed by peoples who were technically more advanced, more demanding, but not necessarily wiser.

To the early colonists along the Atlantic seaboard, the eastern woodlands must have seemed inexhaustible, extending beyond the farthest horizon they knew. There was wood for lumber, shingles, barrel staves, and, from southern pines in particular, naval stores; and there was wood to burn (Stilgoe, 1982; Cowdrey, 1983). As a consequence of burning, cutting, and girdling, agriculture and settlement spread westward across the continent at the expense, first, of the woodlands and then, later, of the interior grasslands (Whitney, 1994). Some 90% of the eastern deciduous woodlands were cleared, and in some areas up to 95% of the native tallgrass prairie was cultivated (Brown, 1993; see chapter 23). Invasion by nonnative species, including fire-prone Eurasian grasses, changed both floristic composition and the ecosystem's hydrogeomorphic response. In the American West, forest resources came under additional stress from federal legislation in 1850 granting settlers title to 130 hectares (ha) after living on and farming the land for four years, and in 1878 allowing intending citizens to buy 65 ha of timberland at $1 ha^{-1}. Commercial logging eventually invaded nearly every forest region, but, from the 1870s onward as resources were depleted and habitats disrupted, calls for conservation and replacement of cut-and-run logging with sustained-yield timber harvesting and even forest preservation began to ripple westward across the continent.

The hydrogeomorphic effects of forest clearance have long attracted scientific interest, generating controlled experiments in drainage basins before and after logging or, by substituting space for time, in adjacent logged and unlogged basins with comparable environments. One celebrated approach was initiated in the 1960s in the Hubbard Brook Experimental Forest, in the White Mountains of New Hampshire (Bormann and Likens, 1970). After water and nutrient budgets had been assessed for six contiguous basins, all trees within one 16-ha basin were cut and dropped in place, and regrowth was inhibited by herbicide. Annual runoff from the basin increased by 40%, whereas dissolved concentrations of nitrate, calcium, sodium, magnesium, potassium, and aluminum rose sharply; this loss of soil fertility led to downstream eutrophication. Particulate matter eroded from the deforested basin increased ninefold compared with nearby undisturbed ecosystems. Reduced losses of organic matter were massively offset by increased mineral sediment because of the removal of biological barriers to soil and stream-bank erosion.

The effects of commercial timber harvesting have generated much controversy. Such concerns pit environmentalists and many indigenous peoples against timber companies and private property owners, with government agencies caught in the middle. The loss of forest habitat for wildlife such as bear, spotted owl, and fish is particularly contentious. From a physical perspective, timber harvesting generates major changes in slope stability, water yields, nutrient budgets, and sediment yields. Much depends on the silvicultural practices used—clearcutting, patch cutting, and the like—and on the rotation sought to sustain a yield of forest products over the years. Much also depends on how felled trees are removed or "yarded." In the early days, this was accomplished by oxen and horses with disturbance limited to smallish areas. Later, from the 1890s to the 1930s, steam power and railways were used, to the limits imposed by gradient. From the 1920s, however, impacts increased as tractor yarding formed skid trails and logging roads were built for truck haulage. Timber demands during and after World War II combined with better technology, such as the power saw, to exacerbate logging impacts. Eventually, unrestrained logging became subject to stricter controls, notably with the passage of stricter forest-practice acts, such as those in California in 1973 and Washington in 1974. These acts required logging companies to submit timber harvesting plans for multiagency approval prior to logging, addressing such issues as silvicultural practices, yarding methods, road construction, and post-harvest management. Meanwhile, more sophisticated yarding methods, such as high-lead cable and skyline units, balloons and helicopters, began reducing the physical effects of yarding and road construction, though the biological effects mostly remained.

Figure 22.6 Slope failure following heavy rain and logging-road failure in coniferous forest, Skagit County, Washington, 1985 (photo: A.R. Orme).

Logging of the old-growth forests of the Pacific Rim, from northern California to Alaska, has long posed major problems. Here, Douglas-fir (*Pseudotsuga menziesii*), western hemlock (*Tsuga heterophylla*), coast redwood (*Sequoia sempervirens*), western redcedar (*Thuja plicata*), Ponderosa pine (*Pinus ponderosa*), and Sitka spruce (*Picea sitchensis*) are commercially important. However, the combination of steep slopes, pervasively fractured bedrock, cohensioness colluvium, high rainfall, and snowmelt causes many forested hillslopes to approach failure thresholds. As long as pioneer logging during the nineteenth century was confined to accessible lowland forests, this was not a problem, but, as these forests gave way to farmland, logging penetrated rugged foothills and roads were built with scant regard for drainage and maintenance. The response was increased mass movement on steeper slopes, mostly as debris flows and swifter debris avalanches, which fed mineral and organic waste into streams augmented by reduced transpiration and deprived of bank-stabilizing vegetation (fig. 22.6). Debris torrents in these channels in turn flushed this slurry downstream onto farmlands, whereas the introduction of fine debris impacted the preferred gravel habitats of anadromous and resident fish alike. Deep-seated landslides and earthflows occurred less commonly. Such responses were not always immediate, dependent as they were on the slow decay of anchoring root systems and the onset of a triggering climatic event. However, a common factor in such movements was the failure of abandoned or ill-maintained logging roads and landings, especially where fill was poorly constructed or road drains and choked culverts were unable to cope with overland flow (e.g., Swanson and Dyrness, 1975; Jones and Grant, 1996; Madej and Ozaki, 1996).

Despite the cumulative effects of timber harvesting (e.g., Geppert et al., 1984), it is important to maintain a sense of perspective. First, hydrogeomorphic impacts of logging tend to diminish over time as new forests arise from seed and stump sprouting, and from reforestation now mandated by forest practice acts. Second, many forest regions have long been prone to mass movement and debris torrents resulting from heavy rains and rapid snowmelt, especially after natural fires and windthrows. Appalachian forests have a long record of rainfall-induced debris avalanches that predate or are independent of human activity (e.g., Hack and Goodlett, 1960; Eschner and Patric, 1982). Forested mountains in British Columbia reveal a litany of massive earthflows for the Holocene (Bovis and Jones, 1992), whereas the Cascade Volcanic Arc is prone to devastating lahars (e.g., Dragovich et al., 1994). Furthermore, sedimentary sequences from lakes, alluvial fans, and terrace deposits, dated by ^{14}C methods and dendrochronology, show that climatic events triggering widespread debris avalanches have occurred at 30–50 year intervals in the Cascade foothills throughout the Holocene (Orme, 1989, 1990). In 1983, for example, massive rain-on-snow events generated debris avalanches within old-growth and second-growth forests alike. Nevertheless, if past errors are to be avoided, prudent management of forest resources requires that careful attention be given to the choice of logging sites, silvicultural practices, road location and construction, waste treatment, and reforestation. Conservation of remaining old-growth forests in parks is also an option—contentious and costly where private land ownership is involved but nevertheless important.

Conversion of chaparral to grass, a common feature in California, also impacts the physical system, as shown by a study in the San Gabriel Mountains. One basin, Monroe Canyon, was cleared of riparian woodland and partly converted from chaparral to grass after a lightning wildfire in 1960. Its formerly intermittent stream became perennial, and exceptional rainstorms in 1969 led to a sevenfold increase in mass movement and an eightfold increase in channel scour compared with nearby basins that had been allowed to revert naturally to chaparral (Orme and Bailey, 1970, 1971). The main channel of Monroe Canyon widened from 1–2 m in 1958 to as much as 55 m in 1969 (fig. 22.7).

These studies show that vegetation change may have a dramatic effect on mass movement and erosion rates, nutrient budgets, and downstream sediment delivery, especially as stabilizing root systems decay in the years after change. Over longer periods, these effects appear to diminish if the natural vegetation is allowed to revert, but a permanent change in land cover will generate prolonged adjustments, subject only to nutrient and sediment availability.

A

Figure 22.7 Monroe Canyon, San Gabriel Mountains, California: A, In 1959 after riparian woodland was cleared to increase water yield (photo: Forest Service, USDA); B (*opposite*), In 1969 after a lightning-triggered wildfire in 1960 and partial conversion of watershed cover from chaparral to grass, followed by heavy rains and debris flows in January and February 1969 (photo: A.R. Orme).

22.3 Impacts on Animals

Human activities have affected North America's native fauna in many ways: by hunting, trapping, domestication, competition, and habitat conversion. The effect has been a dramatic reduction in the number and range of most native species, leading many to endangerment and some to extinction. Significant though these impacts have been, they were superimposed on an animal population that was also responding to evolutionary and ecological changes. The Cenozoic Era has seen many animals evolve and migrate, only to disappear, and the recent past has been little different. Thus, the venerable order of crocodilians retreated from the Arctic to the Great Plains and then to the Atlantic and Gulf coastal plain during the Cenozoic. It is wise to gauge human impacts against this changing environmental backcloth.

22.3.1 Pre-Columbian Impacts

Explanations for the megafaunal extinctions across the Pleistocene-Holocene transition have long pitted hypotheses of environmental change (e.g., Guilday, 1967) against those involving human predation (e.g., Martin, 1967; see chapter 3). Proponents of environmental change point to the survival of "Pleistocene" taxa in sites whose ^{14}C-age is Holocene. Those favoring human predation dismiss many of these ages, notably those based on bone, and emphasize the clustering of butchering sites, especially of mammoth, during the terminal Pleistocene when Clovis hunters were entering North America. Although the final answer may be elusive, human predation certainly occurred but was probably superimposed on a megafauna already in demographic decline, weakened by genetic senescence and ecological changes. For example, Clovis hunters may have found mammoth and bison easier prey when concentrated at water holes during the drought interludes that seemingly occurred in the Southwest during the terminal Pleistocene (Haynes, 1991). Other large animals did, however, survive but became restricted in range and variety. Thus two late Pleistocene bison species (*Bison antiquus, B. latifrons*) succumbed to these changes but the genus survives, just, as the American bison (*B. bison*).

Range contractions among smaller mammals during Holocene prehistory also pose problems. Certainly the dramatic environmental changes of the terminal Pleistocene were long past, but fluctuating Holocene temperatures and water resources undoubtedly caused range adjustments among many animals, even as human predation intensified. Thus, although their remains occur in prehistoric

B

human sites, the retreat of the fisher (*Martes pennanti*) and porcupine (*Erythizon dorsatum*) into the boreal forest and western mountains, and the eastward retreat of the groundhog (*Marmota monax*) from the increasing aridity of the Great Plains may be explained by Holocene climatic and ecological changes (Semken, 1983). Conversely, villages and new habitats may have favored other animals such as the rice rat (*Oryzomys palustris*). Prehistoric peoples also began domesticating certain animals, notably the dog (*Canis familiaris*), in the terminal Pleistocene. Molecular data, including comparative DNA, suggest that its immediate ancestor was the gray wolf (*Canis lupus*), and, just as gray wolves still hunt large animals in packs, so packs of domesticated dogs may have helped human hunters (Wayne, 1993).

22.3.2 Post-Columbian Impacts

The impact of European and later immigrants on North America's wildlife has been devastating—directly so from hunting and trapping, indirectly through a vast array of habitat changes. Among the larger animals, the bison (buffalo) has been severely affected. Early explorers found bison across most of the continent, although only the interior had enough grass for large herds. Coronado's soldiers of 1540–42 were probably the first Europeans to see the vast bison herds of the Great Plains and the peoples who relied on them for much of their material culture. At that time, there were perhaps 60 million bison in North America. However, the acquisition and trade in horses and firearms reinvigorated the indigenous "buffalo culture" of the plains. A precipitous decline in the number and range of bison ensued until by 1884 fewer than a thousand remained, mostly in small groups in Wyoming, Montana, and Alberta (fig. 22.8; Hornaday, 1887). These formed the nucleus for subsequent conservation efforts (fig. 22.9).

Most other large mammals have also suffered reductions in range and number during historic time. The grizzly bear (*Ursus arctos horribilis*), once widespread across western North America, is now restricted to wilderness in Alaska and western Canada with perhaps 800 surviving in a few localities farther south (IGBC, 1987). The elk (wapiti, *Cervus canadensis*) formerly ranged across the continent, but, because it was widely hunted, is now mostly restricted to western mountains and Canadian prairies. Bighorn sheep (*Ovis canadensis*) are likewise now confined to western mountains, and pronghorn (*Antilocapra americana*) to the High Plains and sagebrush country. The gray wolf became extinct in the United States but has been recently reintroduced to Yellowstone from Canada, to the concern of local ranchers. In the far north, caribou (*Rangifer tarandus*), muskox (*Ovibos moschatus*), polar bear (*Ursus maritimus*), and Arctic fox (*Alopex lagopus*) have long provided for Inuit people, but the use of high-power rifles now threatens this resource base.

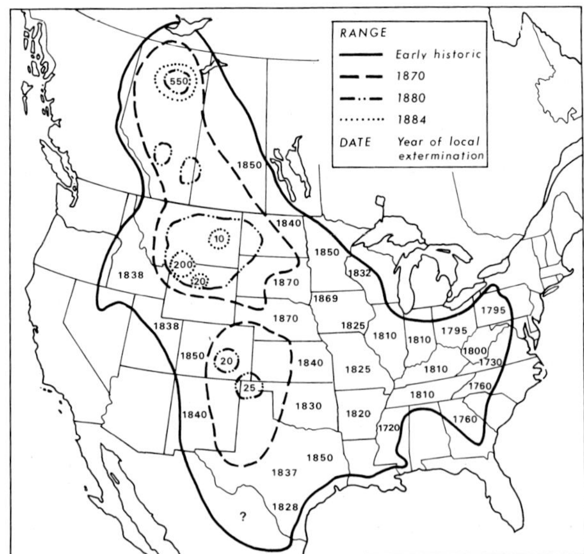

Figure 22.8 (*top*) Range contraction and extermination of bison in North America, including number surviving in 1884 (modified from Hornaday, 1887).

Figure 22.9 (*bottom*) Protected bison grazing in the Madison River floodplain, Yellowstone National Park, Wyoming, 1996. The coniferous forest backcloth had been ravaged by fires that burned over 3000 km^2 or 36% of the park in summer 1988 (photo: A.R. Orme).

Of the smaller mammals, beaver (*Castor canadensis*) and to a lesser extent fisher, marten (*Martes americana*), river otter (*Lutra canadensis*), and muskrat (*Ondatra zibethica*) were widely trapped for their fur. Over three centuries, from the onset of the fur trade around 1600, the beaver was pursued almost to extinction, mostly by native peoples trading with *voyageurs*, *coureurs de bois*, and various companies. Then, as fashions changed, the beaver recovered and now occurs across the continent, except in the High Arctic, the arid Southwest, and Florida. The decline and later recovery of dam-building beaver and burrowing otter and muskrat probably caused adjustments in stream hydrology and riparian ecology. Beaver dams favor upstream aggradation and wetlands, whereas beaver removal and the subsequent decay or collapse of such dams lead to fluvial incision, lower water tables, and wetland loss.

Many links in the food chain have been disrupted by habitat changes, the most notorious of which was the widespread use of the pesticide DDT, which, ingested through prey, led the bald eagle (*Haliaeetus leucocephalus*) and California condor (*Gymnogyps californianus*) to near extinction. Reductions in wetland habitat have affected both migratory and resident birds and fish, for example, in California where, since 1850, 96% of the Central Valley's wetlands have been reclaimed for agriculture and 80% of coastal wetlands have been lost to development, to the detriment of the pelican (*Pelicanus erythrorhyncus, P. occidentalis*) and other species (Orme and Orme, 1998). Road construction has disrupted wildlife migration and foraging patterns —the armadillo (*Dysypus novemcinctus*) is often road kill in Texas. The native fauna must also now compete for food with domestic and other exotic animals introduced during historic times. Some of the latter have become feral, notably swine (*Sus scrofa*) and nutria (coypu, *Myocastor coypus*), whereas others such as rats (*Rattus rattus, R. norvegicus*) and the house mouse (*Mus musculus*) have seized niches formerly used by natives.

During the twentieth century, concern for human impacts on animals grew, leading eventually to such actions as the banning of DDT, water quality controls, and the protection of endangered species. For example, when the Continental Congress adopted the bald eagle for the Great Seal of the United States in 1782, there were perhaps half a million such eagles in the contiguous states. By 1963, only 417 breeding pairs remained, but, declared endangered in 1967, the eagle later recovered to 5800 pairs in 1999. In the 1980s, the surviving 27 California condors were removed from the wild and nurtured in zoos, until several breeding pairs could be returned to nature and an uncertain future in the 1990s (Orme and Orme, 1998). Along the coast, the elephant seal (*Mirounga angustirostris*), reduced by hunting to a few score in 1892 but protected by Mexico in 1911, now numbers 50,000 along the coast south of San Francisco (Kreismann, 1991). Other animals that have recovered under protective legislation include the gray whale (*Eschrichtius gibbosus*) and sea otter (*Enhydra lutris*) of the Pacific coast and the alligator (*Alligator mississippiensis*) of the Gulf and Atlantic coastal plain. Despite protection as early as 1893, the manatee (sea cow, *Trichechus manatus*) of the Atlantic and Gulf coasts remains endangered from habitat loss, water pollution, and collisions with watercraft and locks (EPA, 1999). Even less fortunate, North America's only other sirenian of recent times, Steller's sea cow (*Hydrodamatis gigas*) was hunted to extinction within 27 years of its discovery during Bering's second voyage to Alaska in 1741.

22.4 Impacts of Mining

Mining impacts the landscape directly through the surface working of accessible minerals, and indirectly through the downstream effects of these and related processing operations and by subsidence and waste disposal associated with underground mining. Mining also has extended impacts through its need for timber, water, and power, and mining communities may stimulate local farming. In North America, the impacts of mining during the past 150 years result from the surge in demand and technical advances that accompanied the industrial revolution. Earlier, prehistoric peoples had worked stone and developed annealing techniques for copper and lead, notably around the Great Lakes, but their mining impacts were negligible. During early historic times, while Spaniards were extending their largely vain search for gold and silver northward from Mexico, other Europeans were modestly mining and quarrying their way westward from the eastern seaboard, leading to conflicts with indigenous peoples, for example, with the Winnebago in the Galena lead-mining area of the upper Mississippi valley in the 1820s, that were a foretaste of later friction between native land rights and mining company leases. Around 1850, however, the scale and impact of mining operations began to expand, in part because of gold rushes but more as a reflection of the growing demand for coal, base metals, and construction materials in eastern cities and overseas, and of the canals and railways now available for their transport.

22.4.1 Direct Impacts of Surface Mining

The effect of surface mining depends in part on available technology and in part on the mineral's mode of occurrence. For most of historic time, technology limited these effects to small shallow surface workings, whereas deeper minerals were mined underground. As the twentieth century progressed, however, surface mining expanded rapidly as larger earth-moving equipment and better processing techniques became available (Toy, 1984). In addition to actual excavation, surface mining also usually involves vegetation clearance, site preparation, railway and road construction, overburden removal, and waste disposal—in short, a massive reshaping of local landscapes.

In terms of surface impacts, a distinction may be made between open pit mining, which accounts for one-third of all disturbance, and strip mining and dredging, which explain most other disruption. Open pit mining is usually conducted for clay, sand, gravel, and stone, but pits exploiting more localized metals, such as iron and copper, can be spectacular, notably in the iron-rich Mesabi Range of Minnesota and the copper-rich Bingham Canyon, Utah (fig. 22.10). Although different from the dragline excavators now used in most open pits, hydraulic mining that involves powerful jets of water was once widely used to recover gold

Figure 22.10 Bingham Canyon copper mine, a vast open pit in the Oquirrh Mountains, Utah (photo: Kennecott Copper Corporation).

and other precious metals from alluvial deposits. "Hydraulicking" shaped many gold "diggings" in Cenozoic fluvial deposits in the Sierra Nevada foothills, fed by a system of reservoirs and about 10,000 km of canals, ditches, and pipelines. The Stewart mine in Placer County, California, a pit 5 km long, 2 km wide, and 120 m deep, yielded \$6 million worth of gold between 1865 and 1878 (Clark, 1979).

In relatively flat terrain, shallow sedimentary minerals such as little deformed coal and phosphate are reclaimed by area strip mining, which involves a succession of parallel trench cuts with overburden removed to earlier trenches, creating a washboard effect (fig. 22.11). In steeper terrain, notably in the coal-mining country of Appalachia, contour strip mining is used whereby bench cuts follow desirable seams along the contour using mechanical excavators or augers (fig. 22.12). Dredging suctions or scoops mineral-bearing sediment from rivers, lakes, and coastal areas into barges; this method is widely used for gold-bearing placer and sand and gravel deposits.

Perhaps 20,000 km^2 of the North American landscape have been directly impacted by surface mining, mostly linked to just 7 of the 50 or more minerals worked. Coal accounts for ≈40% of this area, sand and gravel ≈30%, and iron, phosphate, gold, clay, and stone 25%. Appalachia alone contains 60% of the area disturbed by coal mining, whereas the Canadian Shield contains large areas disturbed by mining for iron, copper, nickel, and other metals. The impact of sand and gravel workings is more or less proportional to population, with most pits found near growing cities because of their insatiable demand for construction aggregate.

Figure 22.11 (*top*) Area strip mining for coal consists of removing overburden to expose coal in pit floor, Raleigh County, West Virginia, 1964. Later reclamation involved replacing spoil and landscaping (photo: Natural Resources Conservation Service, USDA).

Figure 22.12 (*bottom*) Contour strip mining for coal in Kanawha County, West Virginia, 1967 (photo: Natural Resources Conservation Service, USDA).

Waste disposal and exhausted sites pose major problems for mining and related processing. Gold mining creates vast quantities of waste for the amount of gold recovered, and coal seams are often thin relative to the volume of overburden and other waste generated. While mining remained small in scope, these problems were of little concern, but the growth of activities during the later nineteenth century led to recognition of the need for controls and reclamation. In the 1920s, some coal companies in Appalachia attempted reforestation on mining waste. Ensuing decades saw legislation adopted by most state and provincial authorities requiring regrading and landscaping of mining waste, with varying results. Many former surface mines, especially those with artificial lakes, have been reclaimed for recreation or used for solid waste disposal and then for residential and commercial development. Land-use planning is now an integral part of mining operations across North America, with mining itself being treated as an interim land use subject to strict regulation. But the legacies of past mining practices linger.

22.4.2 Downstream Impacts of Surface Mining

Surface mining often generates downstream impacts from accelerated sediment yields that cause flooding and habitat disruption. Acid drainage and other effluents from mines and on-site processing plants are of particular concern for their impact on surface and groundwater alike, and thus on aquatic habitats and human health, notably where sulfur-bearing minerals like coal react with water to produce sulfuric acid and other toxic substances. Acid mine drainage dissolves iron and other minerals which, as the polluted water is diluted by cleaner water, precipitate as an orange biotoxic slime in stream channels. This is a frequent problem in the coal mining country of Appalachia. Also damaging are tailing-dam failures such as the release of 500,000 m^3 of water and coal waste that killed 118 persons at Buffalo Creek, West Virginia, in 1972, and of 370,000 m^3 of radioactive water and 1000 tonnes (t) of sediment at Church Rock, New Mexico, in 1979 that contaminated Rio Puerco for 110 km downstream (U.S. Committee on Large Dams, 1994). Airborne dust, mine fires, and smelter fumes may pollute soil and inhibit vegetation recovery for decades, requiring remedial action by government and industry alike (fig. 22.13).

Figure 22.13 The "copper desert" near Copperhill, Tennessee, 1955. The vegetation was destroyed by fumes from copper smelting earlier in the twentieth century, resulting in widespread surface erosion. Here, in 1955, a two-year growth of kudzu (*Pueraria thunbergiana*), a perennial vine, is being used for erosion control (photo: Natural Resources Conservation Service, USDA).

After the California Gold Rush, hydraulic gold mining in the Sierra Nevada foothills flushed fossil river gravels downstream, causing extensive sedimentation and flooding in lowland areas. As the Yuba River bed rose at a rate of 0.3 m yr^{-1}, the city of Marysville built protective levées but was still flooded. By 1880, 20,000 ha of the Sacramento Valley had been inundated by gold tailings; by 1884, 990 $\times 10^6$ m^3 of debris had been washed from the foothills in just 35 years (Orme and Orme, 1998). Litigation between upstream miners and downstream farmers was resolved in favor of the latter in 1884, leading to regulation of hydraulic mining in 1893. Thereafter, hydraulic mining declined, bucket dredging increased, notably along the American, Feather, and Yuba rivers, and the displaced gravels became a source of construction material. But the damage had been done, notably in San Francisco Bay, which had filled rapidly with sediment from the gold fields (see chapter 21). In his classic work on hydraulic mining in the Sierra Nevada, Gilbert (1917) thought it would take until 1950 for the debris to be flushed from the system. As Gilbert's biographer stated, "hydraulicking was strip mining at its most rapacious" (Pyne, 1980, p. 208).

Mining in rivers also affects channel morphology and pattern (e.g., Graf, 1979; James, 1989). Hydraulic mining and dredging for placer gold between 1859 and 1957 substantially changed the Middle Fork of the South Platte River in Colorado (Hilmes and Wohl, 1995). Leached of fines, riparian vegetation could not reestablish itself on coarse tailings; this led to increased channel mobility and decreased sinuosity. Unlike placer gold mining where waste is returned to the channel, sand and gravel mining creates voids in river channels and floodplains, which, if not replenished by sediment inputs, may cause streams to degrade both upstream and downstream (Collins and Dunne, 1990).

22.4.3 *Subsidence Caused by Underground Mining*

Underground mining for coal, salt, and metals may lead to surface subsidence, notably when the supporting pillars in the room-and-pillar method of recovery collapse, often decades after mining has ceased. Abandoned coal mines in Appalachia and Wyoming have caused major subsidence problems, now countered to some extent by backfilling of subsurface voids. An unusual type of surface subsidence was also created in the Nevada desert by the underground detonation of nuclear devices during Project Plowshare in the 1960s (fig. 22.14; Gerber et al., 1966).

More extensive in impact, mining for underground fluids has led to subsidence in many areas, notably in California and the arid Southwest. Thus the expansion of irrigation agriculture based initially on groundwater pumping led to widespread subsidence in California's San Joaquin Valley after 1900. By 1964, 12.3×10^9 m^3 of water—or one-quarter of all groundwater pumped for irrigation in the United States—were being pumped annually from 40,000 wells in the valley (Lofgren, 1975). Between 1943 and 1968, when surface water from the California Aqueduct began replacing groundwater, subsidence of up to 9 m occurred in a 120-km-long trough along the valley's west side, and more than 11,000 km^2 of farmland had subsided over 0.3 m (fig. 22.15). Such subsidence is caused in part by the loss of groundwater support for underground aquifers and in part by compaction of clay-rich aquicludes (Ireland et al., 1984). Although groundwater levels recovered and subsidence rates declined after 1968, droughts in 1976–77 and 1987–92 reactivated the problem (California Department of Water Resources, 1998). However, natural limits to groundwater availability and aquitard compaction suggest that future subsidence values will be only 20–40% of earlier peak values (Basagaoglu et al., 1999).

Although the problem is scarcely evident to the casual observer, damage to wells and drainage systems has had a major impact on farming practices in the valley (Orme and Orme, 1998). Farther north, some 650 km^2 of the Santa Clara Valley subsided up to 4 m from groundwater pumping between 1912 and 1967, requiring groundwater recharge and levée construction along stream channels and around southern San Francisco Bay to inhibit flooding and salt-water intrusion (Poland, 1981). Similar problems have occurred from depletion of the Beaumont aquifer beneath Galveston Bay, Texas, and from the thirst of growing communities around Phoenix and Las Vegas (Bell and Price, 1991). Groundwater mining also changes underground stress fields and may lead to increased seismicity and minor earthquakes. In Canada, with its generally humid climate and lower demand, groundwater subsidence is not a major problem.

Petroleum recovery from California's marine sedimentary basins has caused similar subsidence problems. In one San Joaquin Valley oil field where petroleum production began in 1894 and peaked in 1914, subsidence of up to 0.6 m was measured between 1935 and 1965—a small part of the total amount during the past century. Within the Los Angeles basin, petroleum recovery from the Long Beach-Wilmington oil field caused subsidence of up to 9 m over

Figure 22.14 Surface effect of small underground thermonuclear explosions, Yucca Flats, Nevada, during the early 1960s. The craters overlie chimneys of fractured rock. During this period, Project Plowshare was designed to test nuclear explosions for what was termed "geographical engineering," for example, the excavation of a sea-level canal across the Central American isthmus (photo: U.S. Atomic Energy Commission).

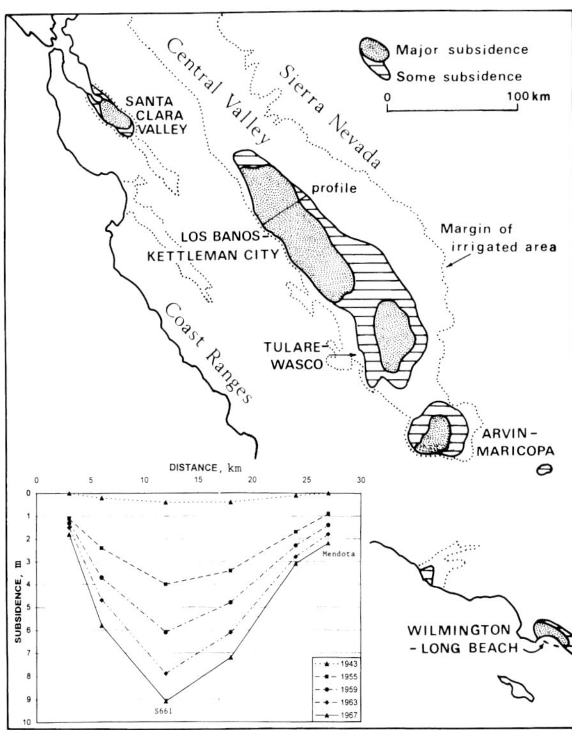

Figure Figure 22.15 Areas of major subsidence caused by underground fluid withdrawal in California (inset graph from Basagaoglu et al., 1999).

50 km² between 1940 and 1958. Port facilities subsided to below sea level, requiring remedial water injections and costly engineering to alleviate the problem (Orme and Orme, 1998).

22.5 Impacts on Water Resources

Water is an essential but much abused resource, which, in North America as elsewhere, has been impacted by generations of use for human consumption, agriculture, industry, power generation, flood control, navigation, and recreation. Direct impacts have involved controlling, diverting, damming, and polluting surface waters, pumping and polluting groundwater, and draining wetlands, as well as their upstream and downstream effects. Indirect impacts usually involve disruption of the water cycle and related biogeochemical cycles through such processes as vegetation change and urbanization (see chapter 24), which in turn affect flood magnitude and timing, erosion and sedimentation, and aquatic habitats.

22.5.1 Early Impacts

Prehistoric peoples across North America saw the value of river floodplains for agriculture, and it is near such sites that most early farming villages were located. Then as now, however, the peoples of the arid Southwest were among the first to seek more active control over scarce water supplies. Around A.D. 600, the Hohokam developed sophisticated irrigation systems involving diversion dams, siltation basins, and canals designed to transfer water from the Gila and Salt rivers of Arizona to fields of corn, beans, squash, and cotton (Nobel, 1991). When Hohokam culture later collapsed, perhaps because of drought, salinization, or human impacts, these irrigation systems fell into disuse but were sufficiently well conceived to influence later Euroamerican irrigation farming.

Early European colonization of the Atlantic seaboard was inevitably linked with the construction of small dams and leats designed to assure water supplies for domestic use and water-powered mills. Such mills were often sited where streams flowed swiftly from uplands onto nearby lowlands, notably along the Fall Line above the Atlantic Coastal Plain and flanking the St. Lawrence Valley. The first water-powered sawmill west of the Allegheny Mountains was built in 1776, and by 1830 there were 40 such mills in what is now West Virginia (Cowdrey, 1983).

As the nineteenth century progressed, narrow-boat canals were built to facilitate commerce between the Atlantic seaboard and the interior. These included the 584-km-long Erie Canal (1825) between Lake Erie and the Hudson River, 2000 km of feeder canals linking Lake Erie with the Ohio River, and by 1840 a further 3000 km of canals in the United States. To the north, the Lachine Canal (1825) bypassed rapids on the St. Lawrence River at Montreal, the Rideau Canal (1832) linked Ottawa with Kingston, and the Champlain (1819) and Chambly (1843) canals helped to move Québec forest products to coastal markets. Although many, such as the Blackstone Canal between Worcester and Providence (1828–48), fell into disuse with the growth of railways, canals often disrupted natural drainage more or less permanently. Others were enlarged, paving the way for later ship canals, notably those on the 3800-km-long St. Lawrence Seaway, opened throughout from Montreal to Duluth in 1959. Other rivers were canalized or improved to facilitate navigation or to reduce flood hazards. In this way, the Mississippi and its tributaries came to abandon many primeval meanders and shallow reaches in favor of straightened levéed sections—with serious consequences for future flood regimes. Indeed, navigation works from 1837 onward have reduced the Mississippi's channel capacity by about one-third, such that the 1973 flood stage reached the 200-year recurrence level with a flow that had but a 30-year recurrence interval (Belt, 1975).

Farther west, water, or rather the lack of it, was long seen as a limiting factor to the development of the Great Plains, with their recurrent droughts, and the severely arid Southwest. Many nineteenth century surveys thus focused on water resources, wetland reclamation, and the prospects for irrigated agriculture. In 1878, John Wesley Powell presciently advocated communal ranching and irrigation farming on a limited scale supported by systems of dams throughout the West (Powell, 1878). By 1900, water resources in California's Sacramento Valley had been severely impacted by levée construction designed to counter flood hazards posed by river beds raised by mining debris and to reclaim wetlands for agriculture. By 1918, 3000 km^2 of wetlands were enclosed behind 560 km of levées. Today, some 1800 km of levées characterize the valley and its downstream delta.

22.5.2 Impacts over the Past Century

Earlier interference with water resources pales before the dramatic impacts of the past century. Of these, the construction of dams along most major rivers has had far-reaching effects, not only nearby but in landscape impacts related to distant urban growth and economic activity. Thus, the Tennessee Valley Authority, established in 1933 with socioeconomic goals for a depressed region, came to embrace more than 20 dams in a comprehensive management scheme for flood control, hydroelectric power, navigation, soil conservation, recreation, and public health. Among the Columbia River's many dams (fig. 22.16), Mica, British Columbia (243 m high, 1972), became the continent's highest, whereas Grand Coulée, Washington (168 m, 1942), provided the most hydroelectric capacity (6494 MW). Daniel Johnson Dam (214 m, 1968) on the Manicouagan River, Québec, and Bennett WAC (183 m, 1967) on the Peace River, British Columbia, impounded the largest reservoirs, 142 km^3 and 70 km^3, respectively (World Register of Dams, 1998). The thirst of the Great Plains and the Southwest was partly met by series of large dams and accompanying reservoirs on the Missouri and other Mississippi tributaries, and in the Colorado River basin. Whereas the 75,000 dams in the continental United States are capable of storing a volume of water almost equaling one year's mean runoff, dams in the Great Plains and Western Cordillera may store more than three times the mean annual runoff (Graf, 1999).

Large dams have had major physical and ecological impacts (fig. 22.17). Upstream, reservoir sedimentation and valley fills have lessened reservoir life expectancy, raised water tables, reduced slope stability, caused flooding, and changed riparian ecology. Concentrations of pesticides, herbicides, and heavy metals in reservoir sediment also pose public health hazards. For example, since completion of the Hoover Dam (221 m, 1936), the Colorado River has built a massive delta into its reservoir, Lake Mead, whereas alluvial fills along the Virgin River have been choked by phreatophytes. Water loading by Lake Mead, which at capacity holds 40 km^3 of water over an area of 660 km^2, has also generated isostatic crustal responses (18 cm of downwarping had occurred by 1950) and increased seismicity. Upstream alluviation caused by the small Gillespie Dam (1921) on the Gila River led to invasion by tamarisk (*Tamarix chinensis*), frequent channel switching, and flooding to the detriment of nearby farmland (Graf, 1985). Conversely, the new habitats provided by reservoirs and alluvial wedges greatly expand the opportunities for aquatic and riparian biota.

Downstream impacts of dams are even more dramatic. Below dams, flood peaks are usually reduced and low flows are increased by the release of impounded water. Deprived of sediment trapped by dams, sediment loads decrease, and surplus stream energy entrains finer bed and bank materials, causing channels to become armored with residual coarse sediment. Such armoring extends 100 km downstream below Hoover Dam (Williams and Wolman, 1984). Farther downstream, significant changes in channel shape

Figure 22.16 Bonneville Dam, the lowest dam on the Columbia River where it cuts through the Cascade Range 200 km inland from the Pacific Ocean. This view, taken in 1938 shortly after dam completion, shows the fish hatchery, fish ladder, and navigation lock designed to minimize the dam's disruptive potential (photo: Spence Collection, UCLA).

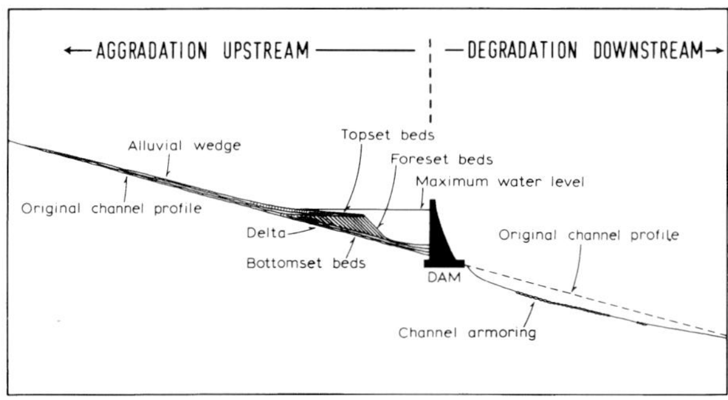

Figure 22.17 Effect of dam construction on river-channel processes.

may occur as bedrock and alluvial bedforms adjust to reductions in flood peaks and sediment delivery, just as riparian habitats respond to less extreme fluctuations of water level. Thus the Colorado River has changed from a natural system with large spring floods, low summer flows, and variable sediment loads, to a controlled system of modest spring floods, higher summer flows, and much reduced sediment loads (Graf, 1985). Before Glen Canyon Dam (216 m, 1964) affected the river flowing through the Grand Canyon, median discharge was 210 $m^3 s^{-1}$, mean annual floods were 2440 $m^3 s^{-1}$ (with peaks from 500 to 8500 $m^3 s^{-1}$), and mean sediment concentration was 1500 parts per million (ppm) (Howard and Dolan, 1981; Lucchitta and Leopold, 1999). With flood peaks of 3700 $m^3 s^{-1}$, around 4.5×10^6 t of suspended sediment passed downstream daily. After the dam was built, the respective figures changed to 350 $m^3 s^{-1}$, 760 $m^3 s^{-1}$ (700–990 $m^3 s^{-1}$ range), and 7 ppm. Consequently, there has been much downstream erosion and little replenishment of bars and beaches. With a reduction in flood scour, the formerly sparse riparian vegetation of the canyon floor changed to dense communities of tamarisk, camelthorn (*Alhagi pseudalhagi*), sandbar willow (*Salix sessilifolia*), desert broom (*Baccharis sarothroides*), and cattail (*Typha* spp.), with Russian olive (*Elaeagnus angustifolia*) in former high flood zones (Graf, 1985).

Similarly, the formerly flood-prone Roanoke River in North Carolina is now controlled by seven major dams built in the upper basin between 1951 and 1966. These dams have reduced discharge variability downstream, but, in replacing formerly transient high-magnitude floods with prolonged low-magnitude events from controlled water releases, mesic plants such as American beech (*Fagus americana*) have colonized the floodplain, whereas regeneration of flood-tolerant species such as baldcypress (*Taxodium distichum*) has been hindered (Konrad, 1998). Similar responses of baldcypress to changing flood regimes are inferred elsewhere across the Atlantic and Gulf Coastal Plain (Shankman and Kortright, 1994).

Sediment loads for many rivers show significant reductions as a result of dam construction (table 22.1). Indeed, the Colorado River's formerly perennial lower reaches in Mexico became intermittent as a result of upstream impoundments, causing a problem for international relations. Because of early unrestrained farming practices, the four Atlantic coastal rivers probably had even higher pre-dam loads before the period of measurement (table 22.1). Reduced sediment loads at river mouths also limit the natural replenishment of coastal beaches, leading to erosion, for example, in southern California where dams now restrict sediment delivery from 37% of the Santa Clara River basin and 42% of the Ventura River basin.

The useful life of dams is limited by reservoir sedimentation, structural integrity, and economic constraints, which, together with ecological considerations, have made the removal of older dams increasingly attractive; several small dams in New England and Washington have been dismantled. Proposals for dam removal in turn raise the question of whether to preserve often valuable artificial habitat, such as upstream wetlands, induced by the original dam construction. Dams and irrigation works may also fail. In 1905, floods diverted the Colorado River into the Alamo section of the old Imperial Canal (1901), which had been designed to irrigate California's Imperial Valley, and thence into the Salton Sink, 85 m below sea level, creating the Salton Sea in part of former Pleistocene Lake Coahuilla. In 1889, failure of a neglected earthen dam in the Appalachian headwaters of the Ohio River near Johnstown, Pennsylvania, led to a catastrophic flood in which 2209 people died. In 1928, the collapse of the 60-m-high St. Francis Dam in San Francisquito Canyon, southern California, released 47×10^6 m^3 of water in a wall 20 m high down the Santa Clara Valley to the sea, drowning 450 people (fig. 22.18; Outland, 1977). In 1976, failure of the 94-m-high Teton Dam flooded 1625 km^2 of southeast Idaho.

Although water transfers alone raise contentious issues, they have also spawned remarkable urban growth, which,

Table 22.1 Sediment loads for selected North American rivers before and after dam construction

River	Basin Area (10^6 km^2)	Sediment Load (10^6 t yr^{-1}) Pre-Dam	Post-Dam
Mississippi	3.27	400	210
Columbia	0.67	15	10
Rio Grande	0.67	20	0.8
Colorado	0.63	120	0.1
Roanoke, VA-NC	0.025	2	<1
Santee, NC-SC	0.027	1	trace
Savannah, SC-GA	0.025	2.8	<1
Altamaha, GA	0.035	2.5	<1

Source: Selected from Milliman and Sivitski, 1992.

Figure 22.18 St. Francis Dam, originally 60 m high, the day after its failure and subsequent devastating flood, San Francisquito Canyon, Los Angeles County, California, March 13, 1928 (photo: Spence Collection, UCLA).

in the relatively dry Southwest, has often spread indiscriminately across floodplains and hillslopes that are still prone to seasonal fire, floods, and mass movement. This has created an anomalous situation whereby cities favored by imported water must now also address problems of flood control. Urban growth in the Los Angeles basin, for example, was stimulated by the import of water from the Owens Valley after 1913, leading to desiccation of Owens Lake, and from the Colorado River after 1941 and the California Aqueduct after 1973. To protect this urban sprawl from winter storm hazards, however, an intricate network of 300 check dams, 106 debris basins, 20 large storage reservoirs, 1000 km of concrete channels, and over 2000 km of storm drains was built in Los Angeles County alone between 1920 and 1970. During this same period, 300×10^6 m^3 of debris were produced, or 120 times the debris-basin capacity of the system. Many structures are now choked with debris, which is denied to local beaches, whereas others function only through periodic debris removal and costly maintenance. Rapid runoff from impermeable surfaces and down flood-control channels also restricts groundwater recharge, leading to subsidence and seawater intrusion, and requiring the provision of spreading grounds and further engineering works. Thus, the myopia of local government, the cynicism of developers, and the ignorance of the general public have collectively increased flood and mass movement hazards, requiring much engineering ingenuity and, often, federal assistance when disaster strikes. Similar scenarios are played out in many other areas, including desert communities such as Las Vegas, Nevada, where monsoonal rains during summer 1999 shed a year's rainfall in a few hours.

Finally, soil-water and groundwater resources have often been impacted by irrigated agriculture, notably by salinization and other forms of chemical contamination caused by the downward leaching of irrigation waters under seasonally arid conditions. This poses serious problems from Texas to California, where one-quarter of all farms were irrigated as early as 1890 and where 35,000 km^2 are now irrigated. High salt content in soil water encourages zoospores to attach to plant roots; these in turn use the sugars and amino acids leaked by root-cell membranes, decreasing the root's ability to resist disease and, with root-rot present, discouraging new roots (Orme and Orme, 1998). Hyperconcentrations of selenium may also result from the accumulation of irrigation water. In its liquid phase, it occurs as selenate (SeO_4^{-2}) or selenite (SeO_3^{-2}) and is often associated with perched water tables, absorbed by plants such as mustard and milkvetch, and transmitted up the food chain. In California's Central Valley, selenium concentrations often exceed 4000 parts per billion (ppb) in contrast to 0.2 ppb in pristine freshwater. Selenium is toxic in the 3000–5000-ppb range and recommended limits for domestic water are 10 ppb. High concentrations may cause embryonic malformation in animals: 20% of waterfowl nests in one wetland within the valley were found to contain deformed embryos (Tanji et al., 1986).

22.6 Conclusion

Human imprints on the primeval North American landscape excite many emotions, yet they are the price paid for progress as human needs and ingenuity have combined over the years to spur technical advances that are increasingly invasive of the continent's resources. Lest the previous examples paint a picture of unrelieved gloom, note that recent decades have witnessed a growing awareness of environmental impacts and a desire to address them (see chapter 25). Whereas early attempts at environmental management were mostly concerned with resource use and hazard mitigation (e.g., wetland reclamation, improved navigation), there is now a legacy of attempts at more rational management extending back over a century, augmented more recently by a flurry of sound environmental legislation. In both Canada and the United States, local and federal laws now address such issues as forest practices, endangered wildlife, hazardous waste, and water quality, whereas industry and society at large are now sensitized, if not always responsive, to the need for reducing the deleterious impacts of their activities.

The landscapes inherited from nature have been much changed by human impacts, but large areas of the continent, especially in the north and west, still retain much of their primeval grandeur. It is the responsibility of present and future generations, even as they pursue life and livelihood, to minimize their imprints and to ensure protection for the best that nature has bequeathed us.

References

Basagaoglu, H., M.A. Mariño, and T.M. Botzan, 1999. Land subsidence in the Los Banos-Kettleman City area, California: Past and future occurrence. *Physical Geography*, 20, 67–82.

Bell, J.W., and J.G. Price (Editors), 1991. *Subsidence in Las Vegas Valley, 1990–1991*. Nevada Bureau of Mines and Geology, Reno.

Belt, C.B., 1975. The 1973 flood and man's constriction of the Mississippi River. *Science*, 189, 681–684.

Bettis, E.A., (Editor), 1995. Archaeological geology of the Archaic Period in North America. *Geological Society of America Special Paper* 297.

Bolton, H.E., 1927. *Fray Juan Crespi, missionary explorer of the Pacific coast, 1769–1774*. University of California Press, Berkeley.

Bormann, F.H., and G.E. Likens, 1970. The nutrient cycles of an ecosystem. *Scientific American*, 223, 92–101.

Bovis, M.J., and P. Jones, 1992. Holocene history of earthflow mass movements in south-central British Columbia: The influence of hydroclimatic changes. *Canadian Journal of Earth Sciences*, 29, 1746–1755.

Brown, D.A., 1993. Early nineteenth-century grasslands of the mid-continent Plains. *Annals of the Association of American Geographers*, 83, 589–612.

California Department of Water Resources, 1998. *Compaction recorded by extensometer wells since 1984 in the west San Joaquin Valley, California*. State of California, Sacramento.

Chlachula, J., 1996. Geology and Quaternary environment of the first preglacial paleolithic sites found in Alberta, Canada. *Quaternary Science Reviews*, 15, 285–313.

Clark, W.B., 1979. Fossil river beds of the Sierra Nevada. *California Geology*, 32, 143–149.

Collins, B., and T. Dunne, 1990. *Fluvial geomorphology and river-gravel mining*. California Department of Conservation, Division of Mines and Geology, Sacramento.

Cowdrey, A.E., 1983. *This Land, this South*. University of Kentucky Press, Lexington.

Day, G.M., 1953. The Indian as an ecological factor in the northeastern forest. *Ecology*, 34, 329–346.

DeBano, L.F., 1981. Water repellent soils: A state-of-the-art. *U.S. Forest Service, Pacific Southwest Forest and Range Experiment Station*, General Technical Report PSW-46.

DeBano, L.F., R.M. Rice, and C.E. Conrad, 1979. Soil heating in chaparral fires: Effects on soil properties, plant nutrients, erosion and runoff. *U.S. Forest Service, Pacific Southwest Forest and Range Experiment Station*, Research Paper PSW-145.

Delcourt, P.A., H.R. Delcourt, P.A. Cridlebaugh, and J. Chapman, 1986. Holocene ethnobotanical and paleoecological record of human impact on vegetation in the Little Tennessee River valley, Tennessee. *Quaternary Research*, 25, 330–349.

Denevan, W.M. (Editor), 1992a. *The native population of the Americas in 1492*. 2nd edition, University of Wisconsin Press, Madison.

Denevan, W.M., 1992b. The pristine myth: The landscape of the Americas in 1492. *Annals of the Association of American Geographers*, 82, 369–385.

Dragovich, J.D., P.T. Pringle, and T.J. Walsh, 1994. Extent and geometry of the mid-Holocene Osceola mudflow in the Puget Lowland—Implications for Holocene sedimentation and paleogeography. *Washington Geology*, 22, 3–26.

EPA (Environmental Protection Agency), 1999. Sea-Stats No. 13—Manatees, Environmental Protection Agency website, www.epa.gov/gumpo/seast13.html.

Eschner, A.R., and J.H. Patric, 1982. Debris avalanches in eastern upland forests. *Journal of Forestry*, 80, 343–347.

Fladmark, K.R., 1975. A paleoecological model for Northwest Coast prehistory. National Museum of Man, *Mercury Series, Archaeological Survey of Canada*, Paper 43, Ottawa.

Geppert, R.R., C.W. Lorenz, and A.G. Larson, 1984. *Cumulative effects of forest practices on the environment: A state of the knowledge*. Washington Forest Practices Board, Department of Natural Resources, Olympia.

Gerber, C., R. Hamburger, and E.W. Seabrook Hull, 1966. *Plowshare*. U.S. Atomic Energy Commission, Division of Technical Information.

Gilbert, G.K., 1917. Hydraulic mining debris in the Sierra Nevada. *U.S. Geological Survey Professional Paper* 105.

Goetzmann, W.H., and G. Williams, 1992. *The atlas of North American exploration from the Norse voyages to the race to the Pole*. Prentice Hall, New York.

Graf, W.L., 1979. Mining and channel response. *Annals of the Association of American Geographers*, 69, 262–275.

Graf, W.L., 1985. *The Colorado River: Instability and Basin Management*. Association of American Geographers, Washington, D.C.

Graf, W.L., 1999. Dam nation: A geographic census of American dams and their large-scale hydrologic impacts. *Water Resources Research*, 35, 1305–1311.

Guilday, J.E., 1967. Differential extinction during late Pleistocene and recent times. In P.S. Martin and H.E. Wright (Editors), *Pleistocene extinctions: The search for a cause*. Yale University Press, New Haven, 121–154.

Hack, J.T., and J.C. Goodlett, 1960. Geomorphology and forest ecology of a mountainous region in the central Appalachians. *U.S. Geological Survey Professional Paper* 437.

Harris, R.C., and J. Warkentin, 1974. *Canada before Confederation: A study in historical geography*. Oxford University Press, New York.

Haynes, C.V., 1991. Geoarchaeological and paleohydrological evidence for Clovis-age drought in North America and its bearing on extinction. *Quaternary Research*, 35, 438–450.

Henderson, G.S., and D.L. Golding, 1983. The effect of slash burning on the water repellency of forest soils at Vancouver, British Columbia. *Canadian Journal of Forest Research*, 13, 353–355.

Hilmes, M.M., and E.E. Wohl, 1995. Changes in channel morphology associated with placer mining. *Physical Geography*, 16, 223–242.

Holliday, V.T., 1997. *Paleoindian geoarchaeology of the southern High Plains*. University of Texas Press, Austin.

Holliday, V.T., C.V. Haynes, J.L. Hofman, and D.J. Meltzer, 1994. Geoarchaeology and geochronology of the Miami (Clovis) site, southern High Plains, Texas. *Quaternary Research*, 41, 234–244.

Hornaday, W.T., 1887. The extermination of the American bison with a sketch of its discovery and life history. In *Report of the U.S. National Museum for 1887*, Washington, D.C., 367–548.

Howard, A., and R. Dolan, 1981. Geomorphology of the Colorado River in the Grand Canyon. *Journal of Geology*, 89, 269–298.

IGBC (Interagency Grizzly Bear Committee), 1987. *Grizzly bear compendium*. National Wildlife Federation, Washington, D.C.

Ireland, R.L., J.F. Poland, and F.S. Riley, 1984. Land subsidence in the San Joaquin Valley, California. *U.S. Geological Survey Professional Paper* 437-I.

James, L.A., 1989. Sustained storage and transport of hydraulic gold mining sediment in the Bear River, California. *Annals of the Association of American Geographers*, 79, 570–592.

Jones, J.A., and G.E. Grant, 1996. Peak flow responses to clearcutting and roads in small and large basins, western Cascades, Oregon. *Water Resources Research*, 32, 959–974.

Kaye, M.W., and T.W. Swetnam, 1999. An assessment of fire, climate, and Apache history in the Sacramento Mountains, New Mexico. *Physical Geography*, 20, 270–286.

Kittredge, J., 1948. *Forest influences*. McGraw-Hill, New York.

Konrad, C.E., 1998. A flood climatology of the Lower Roanoke River Basin in North Carolina. *Physical Geography*, 19, 15–34.

Kreismann, B., 1991. *California: An environmental atlas and guide*. Bear Klaw Press, Davis.

Lofgren, B.E., 1975. Land subsidence due to groundwater withdrawal, Arvin-Maricopa area, California. *U.S. Geological Survey Professional Paper* 437D.

Lucchitta, I., and L.B. Leopold, 1999. Floods and sandbars in the Grand Canyon. *GSA Today*, 9 (4), 1–7. Geological Society of America, Boulder.

MacDonald, G.M., C.P.S. Larsen, J.M. Szeicz, and K.A. Moser, 1991. The reconstruction of boreal forest history from lake sediments: A comparison of charcoal, pollen, sedimentological and geochemical indices. *Quaternary Science Reviews*, 10, 53–71.

Madej, M.A., and Ozaki, V., 1996. Channel response to sediment wave propagation and movement, Redwood Creek, California, USA. *Earth Surface Processes and Landforms*, 21, 911–927.

Marsh, G.P., 1864. *Man and Nature; or, Physical Geography as Modified by Human Action*. Charles Scribner and Company, New York; Sampson Low, London.

Martin, G.J., 1998. The emergence and development of geographic thought in New England. *Economic Geography*, xx, 1–13.

Martin, P.S., 1967. Pleistocene overkill. In P.S. Martin and H.E. Wright (Editors), *Pleistocene extinctions: The search for a cause*. Yale University Press, New Haven, 75–120.

Meinig, D.W., 1986. *The shaping of America: A geographical perspective on 500 years of history*. Volume 1, Atlantic America, 1492–1800. Yale University Press, New Haven.

Meinig, D.W., 1993. *The shaping of America: A geographical perspective on 500 years of history*. Volume 2, Continental America, 1800–1867. Yale University Press, New Haven.

Mensing, S.A., J. Michaelson, and R. Byrne, 1999. A 560–year record of Santa Ana fires reconstructed from charcoal deposited in the Santa Barbara basin, California. *Quaternary Research*, 51, 295–305.

Milliman, J.D., and J.P.M. Syvitski, 1992. Geomorphic/tectonic control of sediment discharge to the ocean: The importance of small mountainous rivers. *Journal of Geology*, 100, 525–544.

Minnich, R.A., 1983. Fire mosaics in southern California and northern Baja California. *Science*, 219, 1287–1294.

Mitchell, R.D., and P.A. Grove (Editors), 1987. *North America: The historical geography of a changing continent*. Rowan and Littlefield, Totowa.

Moriarty, J.R., 1968. *Cabrillo's log, 1542–1543, a voyage of discovery, a summary by Juan Paez*. The Western Explorer, 5.

Nobel, D.G. (Editor), 1991. *The Hohokam: Ancient people of the desert*. School of American Research Press, Santa Fe.

Orme, A.R., 1989. The nature and rate of alluvial fan aggradation in a humid temperate environment, northwest Washington. *Physical Geography*, 10, 131–146.

Orme, A.R., 1990. Recurrence of debris production under coniferous forest, Cascade foothills, northwest United States. In J.B. Thornes (Editor), *Vegetation and erosion*. Wiley, Chichester, 67–84.

Orme, A.R., 1992. Late Quaternary deposits near Point Sal, south-central California: A timeframe for coastal dune emplacement. In: C.H. Fletcher and J.M. Wehmiller (Editors), *Quaternary coasts of the United States: Marine and lacustrine systems*, Society for Sedimentary Geology, Special Publication 48, 309–315.

Orme, A.R., and R.G. Bailey, 1970. The effect of vegetation conversion and flood discharge on stream channel geometry: The case of southern California watersheds. *Proceedings of the Association of American Geographers*, 2, 101–106.

Orme, A.R., and R.G. Bailey, 1971. Vegetation conversion and channel geometry in Monroe Canyon, southern California. *Yearbook of the Association of Pacific Coast Geographers*, 33, 65–82.

Orme, A.R., and A.J. Orme, 1998. Greater California. In A.J. Conacher and M. Sala (Editors), *Land degradation in Mediterranean environments of the world: Nature and extent, causes and solutions*. Wiley, Chichester, 109–122, and contributions to other chapters.

Outland, C.F., 1977. *Man-made disaster: The story of St. Francis Dam*. Arthur H. Clark Company, Glendale.

Poland, J.F., 1981. *The occurrence and control of land subsidence due to groundwater withdrawal with special reference to the San Joaquin and Santa Clara valleys, California*. Ph.D. dissertation, Stanford University, Palo Alto.

Powell, J.W., 1878. *Report on the lands of the arid region of the United States*. Department of the Interior, Washington, D.C. [reprinted 1962 (W. Stegner, Editor), Belknap Press, Harvard University Press, Cambridge].

Pyne, S.J., 1980. *Grove Karl Gilbert: A great engine of research*. University of Texas Press, Austin.

Sale, K., 1990. *The conquest of paradise: Christopher Columbus and the Columbian legacy*. Knopf, New York.

Sellards, E.H., 1938. Artifacts associated with fossil elephant. *Geological Society of America Bulletin*, 49, 999–1010.

Semken, H.A., 1983. Holocene mammalian biogeography and climatic change in the eastern and central United States. In S.C. Porter (Editor), *Late-Quaternary environments of the United States, Volume 2, The Holocene*, University of Minnesota Press, Minneapolis, 182–207.

Shankman, D., and R.M. Kortright, 1994. Hydrogeomorphic conditions limiting the distribution of baldcypress in the southeastern United States. *Physical Geography*, 15, 282–295.

Shankman, D., and K.M. Wills, 1995. Pre-European settlement forest communities of the Talladega Mountains, Alabama. *Southeastern Geographer*, 35, 118–131.

Stewart, O.C., 1956. Fire as the first great force employed by man. In W.L. Thomas (Editor), *Man's role in changing the face of the Earth*. University of Chicago Press, Chicago, 115–133.

Stilgoe, J.R., 1982. *Common landscapes of America, 1580–1845*. Yale University Press, New Haven.

Swanson, F.J., and C.T. Dyrness, 1975. Impact of clear-cutting and road construction on soil erosion by landslides in the western Cascade Range, Oregon. *Geology*, 3, 393–396.

Swetnam, T.W., and J.L. Betancourt, 1990. Fire–Southern Oscillation relations in the southwestern United States. *Science*, 249, 1017–1020.

Tanji, K., A. Lauchli, and J. Meyer, 1986. Selenium in the San Joaquin Valley: A challenge to western irrigation. *Environment*, 28, 8–16.

Thomas, W.L. (Editor), 1956. *Man's role in changing the face of the Earth*. University of Chicago Press, Chicago.

Toy, T.J., 1984. Geomorphology of surface-mined lands in the western United States. In J.E. Costa and P.J. Fleisher (Editors), *Developments and applications of geomorphology*. Springer-Verlag, Berlin, 133–170.

Turner, B.L., (Editor), 1990. *The Earth as transformed by human action*. Cambridge University Press, New York.

Turner, F.J., 1920. The significance of the frontier in American history. Address to the American Historical Association, 1893; reprinted in *The Frontier in American history*.

U.S. Committee on Large Dams, 1994. *Tailings dam incidents*. Denver.

Vodden, C., 1992. *No stone unturned: The first 150 years of the Geological Survey of Canada*. Energy, Mines and Resources Canada.

Waldman, C., 1985. *Atlas of the North American Indian*. Facts on File, New York.

Wayne, R.K., 1993. Molecular evolution of the dog family. *Trends in Genetics*, 9, 218–224.

Wells, W.G., 1987. The effects of fire on the generation of debris flows in southern California. In J.E. Costa and G.F. Wieczorek (Editors), Debris flows/avalanches: Process, recognition, and mitigation. Geological Society of America, *Reviews in Engineering Geology*, 7, 105–114.

Whitehead, D.R., and M.C. Sheehan, 1985. Holocene vegetational changes in the Tombigbee River valley, eastern Mississippi. *American Midland Naturalist*, 113, 122–137.

Whitney, G.G., 1994. *From coastal wilderness to fruited plain: A history of environmental change in temperate North America from 1500 to the present*. Cambridge University Press, New York.

Williams, G.P., and M.G. Wolman, 1984. Downstream effects of dams on alluvial rivers. *U.S. Geological Survey Professional Paper*, 1286.

Williams, M., 1989. *Americans and their forests: A historical geography*. Cambridge University Press, Cambridge.

Wolf, J.J., and J.N. Mast, 1998. Fire history of mixed-conifer forests on the North Rim, Grand Canyon National Park, Arizona. *Physical Geography*, 19, 1–14.

World Register of Dams, 1998. International Commission on Large Dams.

23

Agriculture, Erosion, and Sediment Yields

James C. Knox

Agricultural land use normally accelerates rates and magnitudes of natural erosion and sedimentation processes because the stability of the land surface is disrupted when the protective benefits of natural vegetation are depleted through overgrazing or during crop cultivation. Accelerated erosion and sedimentation mostly result from processes related to the actions of running water and/or wind. Although erosion and sedimentation from running water are accelerated nearly everywhere when the vegetation cover is depleted, wind erosion and sedimentation are most common in arid and semiarid environments and in areas of sandy soils. The general objective of this chapter is to review examples that illustrate the influence of agricultural activities on erosion and sedimentation and on the degradation of soil resources for selected regions of North America.

It is evident from the stratigraphic record that Native American agriculture had at least local and sometimes regional impact on accelerating erosion and sedimentation. However, the arrival of European agriculture in North America accelerated erosion and sedimentation rates to magnitudes that are unparalleled in the preceding postglacial record. For Canada and the United States, the great impact of agriculture generally begins in the 1600s in the eastern sector but dates only from the 1800s in the central and much of the western sector (Happ et al., 1940; Daniels, 1966; Wolman, 1967; Knox, 1972, 1977, 1987; Butzer, 1974; Trimble, 1974, 1975, 1983; Costa, 1975; Cooke and Reeves, 1976; Davis, 1976; Meade, 1982; Sparrow, 1984; Trimble and Lund, 1982; Magilligan, 1985; Jacobson and Coleman, 1986; Barnhardt, 1988; Williams, 1989; Balling and Wells, 1990; Ashmore, 1993; Miller et al., 1993, Phillips, 1993, 1997; Orme and Orme, 1998). However, in Mexico and the southwestern United States, European agricultural influences, including cropping and livestock raising, occurred in the 1500s (Butzer, 1992a,b).

23.1 Agriculture and Accelerated Erosion and Sedimentation

23.1.1 Effects of Runoff

Most agricultural land uses increase the surface runoff fraction of total runoff from a watershed, resulting in accelerated flooding and soil erosion. Increases in the surface runoff fraction result from direct and indirect responses of reducing vegetation cover, and from destruction of the natural soil structure associated with cultivation and compaction by heavy farm equipment. Changes in the character of vegetation can be expected to have particularly important hydrologic influences because vegetation cover prevents direct raindrop impact on soil particles, greatly increases the hydraulic roughness at the soil surface, and determines the level of organic matter in the soil solum, all factors that reduce surface runoff and promote higher rates of infiltration. Figure 23.1 schematically summarizes the common hydrologic response after conversion of natural cover to agricultural land use. Note that a shift from natural cover to agriculture causes surface runoff to become a much larger fraction of the total runoff, and peak runoff to become much larger and to occur more quickly. Consequently, with less infiltration, recharge of the water table is decreased, and the magnitude of base flow is lowered (fig. 23.1).

Vegetation intercepts and reduces the erosive energy of falling rain. Raindrops strike the ground surface at velocities of about 9 m s^{-1} and exert sufficient force to break up soil aggregates and spatter soil particles as much as 60 cm into the air (Storey et al., 1964). The dislodged and broken soil particles plug infiltration routes and become susceptible to erosion by overland flow. Without vegetation cover, overland flow encounters few obstacles to slow its velocity. Increases in the velocity of surface runoff produce geometric increases in erosive energy. Sartz and Tolsted (1974) compared runoff behavior between two small watersheds in southwestern Wisconsin and showed that even grazing of humid region landscapes can have very important hydrologic consequences. They found that runoff was similar when both watersheds were grazed, but after 3 years of nongrazing in one of the watersheds, the ungrazed-to-grazed ratio for mean total flow had dropped from 1.17 to 0.10 and the mean peak flow from 0.82 to 0.03.

Enlow and Musgrave (1938, p. 622) quote Henry A. Wallace, former United States Secretary of Agriculture, as stating that ". . .the aim in protecting our fields against erosion is to make the water 'walk off' and not 'run off'." Enlow and Musgrave (1938) found that dense vegetation, such as grass, causes runoff to move in thin sheets across the land surface at lower velocities than runoff across cultivated cropland. They observed that concentrated runoff moving downslope between rows of cultivated crops typically attains relatively high velocities and becomes very erosive. They indicated that runoff from a 3% slope of standard plot length of about 22 m for bluegrass cover was only 10% of that for the same slope and plot length for corn. The shifts in recent years to closer spacing of corn rows and higher densities of corn plants in the rows, coupled with minimum tillage and other conservation measures, have resulted in greatly lowered soil losses from corn fields. Figure 23.2, which shows extensive rill erosion between widely spaced corn plants in a nineteenth-century northwestern Illinois corn field, illustrates the high susceptibility to erosion that was common before the middle to late 1950s when herbicides began to make cross-row cultivation unnecessary.

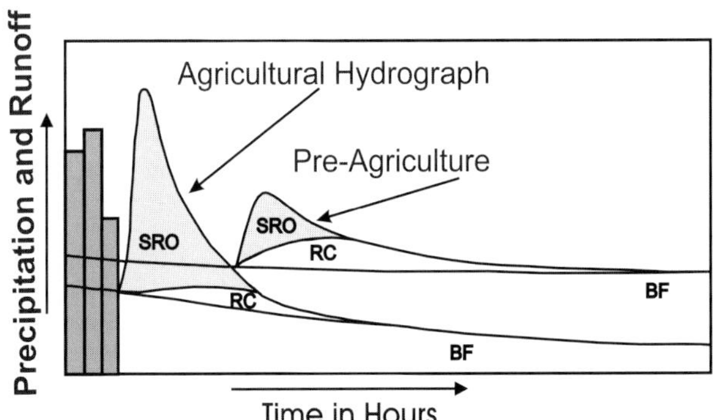

Figure 23.1 Schematic hydrographs showing effects of agricultural land use on surface runoff and base-flow runoff after a rainstorm (bar graph). The conversion of natural land cover to cropland and pasture tends to result in more surface runoff, higher magnitude peak flows, smaller recharge, and consequently a lower magnitude low-stage flow (groundwater-fed base flow). Conversion also shortens the length of time to reach peak stage.

Improved conservation practices, temporary diversions of cropland to federal land-idling programs, and conversion of erodible cropland to other land uses were judged to be the principal reasons for average annual sheet and rill erosion on U.S. cropland, declining from about 9.64 tonnes per hectare per year (t ha^{-1} y^{-1}) in 1982 to about 8.52 t ha^{-1} y^{-1} in 1987 (George and Choate, 1989; Lee, 1990). The United States (U.S.) Department of Agriculture National Resources Inventory data for 1992 indicate average annual cropland soil loss by sheet and rill erosion continued to decline below the 1987 rates to about 6.95 t ha^{-1} y^{-1} (U.S. Department of Agriculture, 1994). The 1992 rate also included the Caribbean region, where rates have been increasing. Although these recent continued declines are impressive, the soil losses pale by comparison to rates of soil loss that were occurring before conservation methods were practiced. For example, Argabright et al. (1996), who compared historical soil erosion rates for five counties in the loess-covered hilly region along the Mississippi River in northeast Iowa, southwest Wisconsin, and southeast Minnesota, found dramatic decreases in rates since the 1930s. They found that the average annual rate of soil loss in 1930 on land in row crops, small grains, and rotation meadow was about 33.4 t ha^{-1} y^{-1}, but because of better land-conservation practices the losses decreased to 17.5 t ha^{-1} y^{-1} by 1982 and decreased further to 14.1 t ha^{-1} y^{-1} by 1992. Despite great improvements in agricultural soil losses, during times of tillage and before plant cover is sufficient to intercept the falling rain, most row crops yield high rates of surface runoff and soil erosion.

Good vegetation cover also promotes the inclusion of organic matter and related humus in the soil, which in turn increases porosity and permeability. Consequently, organic matter is an important influence on soil infiltration capacity and water storage. Storey et al. (1964) cite earlier work by Auten that showed that the upper 23 cm of soil under

Figure 23.2 Soil erosion in a late nineteenth century northwestern Illinois corn field. The stumps of widely spaced former trees still remain in the field, indicating that the original landscape probably was an open savanna grassland. Before the introduction of herbicides in the 1950s, corn was typically planted on a grid with rows spaced slightly greater than 1 m. The grid arrangement allowed cultivation lengthwise and crosswise in the field for weed control, but any rainstorm occurring when cultivation was up and down the hillslope resulted in extreme soil erosion. (Photo: Wisconsin State Historical Society Collection)

several old-growth stands in oak-hickory and other hardwood forests in the Ohio Valley were 13% lighter at oven dryness than equal volumes of soil from adjacent cultivated fields and heavily grazed pastures. The difference was attributed to decreasing porosity as organic matter became depleted in the cultivated fields and to increasing soil compression in the heavily grazed and trampled pastures. Musgrave and Holtan (1964) found that pore sizes and pore-size distribution are greatly affected by the content of soil organic matter, and that, although effects are more pronounced in silt- and clay-dominated soils, the effects are nevertheless widespread and often of large magnitude. Intensive cultivation and oxidation significantly reduce the organic content of soils. In central Missouri during the early 1930s, Hans Jenny compared organic matter in an undisturbed virgin prairie soil with organic matter in an adjoining field cropped to corn, wheat, and oats for 60 years without the addition of manure or fertilizer (Albrecht, 1938). Jenny found the field had lost 38% of the organic matter represented in the prairie soil. A similar southwestern Wisconsin study in 1995, but comparing soil organic carbon in native prairie with organic carbon in an adjoining field, also showed major degradation of soil organic content with cultivation. In this case, rotations of corn, wheat, oat, and hay crops on a very low-gradient upland field since about the mid–nineteenth century accounted for approximately a 60% loss of organic carbon from the top 10–15 cm of the soil profile (Knox and Hudson, 1995). When averaged over the top 20 cm, organic carbon in the prairie ranged from 3% to 5%, but was only 1.5–2.5% in the adjoining field (Knox and Hudson, 1995). The depletion of organic content by oxidation and erosion reduced the porosity about 10% below its original value for the prairie soil.

23.1.2 Effects of Wind

Agriculture has accelerated wind erosion and deposition of surficial sediment by influencing three principal controls: vegetation cover, surface roughness, and degree of soil cloddiness (Chepil and Woodruff, 1954; Péwé, 1981). Lyles (1977, p. 820) noted, "Erosion may be expected wherever the surface soil is finely divided, loose, and dry; the surface is smooth and bare; and the field is unsheltered, wide, and improperly oriented with respect to prevailing wind direction." Bennett and Lowdermilk (1938) indicated that soil cultivation causes a soil to become less cohesive because it depletes the binding effect of grass roots and also depletes the spongy organic matter that normally accumulates under a cover of grass. They found that cultivation therefore causes soil to become a loose, dry, powdery material that can be easily eroded by wind action during periods of drought. The most serious wind erosion tends to be associated with periods of low precipitation, high temperatures, and strong wind velocities (Chepil et al., 1963). For these reasons, agriculturally induced wind erosion and sedimentation are most serious in semiarid and arid regions, but may occasionally occur even in humid regions during times of severe drought. Although many perceive that wind erosion as a serious problem ended with the "dust bowl" days of the drought-stricken Great Plains during the 1930s (Hovde, 1934), wind erosion remains a serious problem in most semiarid and arid regions (Kimberlin et al., 1977) (fig. 23.3). An investigation by McCauley et al., (1981) of erosion and deposition due to strong winds during February 1977 in the middle of a drought on the southern High Plains, showed that plowed fields were locally eroded to depths of greater than 1 meter. They also found that fine sand winnowed from vulnerable soils was deposited in lobate sheets from several centimeters to more than a meter deep extending several kilometers downwind from plowed fields and blowouts. Wheaton's (1984a, 1984b) reviews of wind-related soil erosion since the 1930s indicates that wind erosion continues to be a serious problem and will likely remain so in the future. Indeed, Changnon (1983) reported a general upward trend in the frequency of dust storms in intensely farmed Illinois. And although U.S. Na-

Figure 23.3 Cultivation and overgrazing causes soils to become loose, dry, powdery material that is highly susceptible to deflation during droughts. The Great Plains experienced a severe drought in the early 1930s and dust storms occurred frequently during that period. This particular dust storm swept across Springfield, southeastern Colorado, in May 1937. (Photo: United States Soil Conservation Service)

tional Resource Inventory surveys showed that sheet and rill erosion decreased slightly on cropland in the United States between surveys of 1982 and 1987, loss of soil by wind erosion increased slightly from 6.95 t ha^{-1} y^{-1} to 7.40 t ha^{-1} y^{-1} (George and Choate, 1989). The increase was attributed in part to the high frequency of droughty conditions during the 1980s in the Great Plains. National Resources Inventory data for 1992, when climate conditions were more moist, indicate average annual U.S. cropland soil loss by wind erosion had decreased about 22% from the 1987 level (U.S. Department of Agriculture, 1994).

23.2 The Eastern Woodlands

Doolittle (1992) reviewed the evidence for pre-European agriculture in North America and he used descriptive accounts of early European explorers to provide evidence that the native population in the eastern woodlands caused locally extensive agricultural disturbance. However, in acknowledging the diversity of natural environments in the eastern woodlands, as well as the diversity of population density, Doolittle found it difficult to generalize about the agricultural landscape of the region. The evidence compiled by Doolittle supports the idea that pre-European agriculturally accelerated erosion and sedimentation could have been significant in restricted areas of villages and locally where fields were present. However, early native influences on erosion and sedimentation are insignificant compared to what followed after European settlement in the region.

Hugh Hammond Bennett (1939) in his classic book *Soil Conservation* frequently used the Piedmont and Coastal Plain region of the southeastern United States to illustrate the destructive influence of agriculture on the nation's soil resources. Although sheet, rill, and gully erosion have occurred extensively is this region, deep gullies are perhaps the most obvious evidence of natural landscape degradation by soil erosion. An extreme example is gullying at Providence Canyon in southwestern Georgia (fig. 23.4). Magilligan and Stamp (1997) noted that cultivation began during the 1830s in the Providence Canyon area and probably was responsible for the deep gully system that was well developed by 1859.

Early colonial settlements on the Coastal Plain and Piedmont of what later became the eastern United States involved extensive forest clearance to facilitate tobacco cultivation. Costa (1975) noted that by the late 1690s cultivation extended from the Coastal Plain onto the Appalachian Piedmont as soils became exhausted and a need for new lands developed. He described the typical sequence of cultivation as growing only three to four crops of tobacco in a 1–2 year period after forest clearing, then either abandoning the land to second-growth forest or shifting to corn or wheat cultivation. Costa's estimates of agriculturally induced erosion of Piedmont upland soils of Virginia and Maryland, along with his review of other investigators' estimates for historical erosion of other Piedmont soils, showed historical soil losses ranging from about 7.6 cm to 30.5 cm after initial forest clearance. Happ (1945) concluded that cultivation of the Carolina Piedmont resulted in about 15.2 cm of erosional stripping over a time span of about 150 years. Trimble (1974, 1975) estimated that the region of the southern Piedmont extending from eastern Alabama to southern Virginia experienced an average soil loss of about 18 cm after forest clearance. Using U.S. De-

Figure 23.4 Providence Canyon gully system, Stewart County, southwestern Georgia. The gully system formed after land clearance for cotton production began in the 1830s. By the late 1850s, the gully was deeply incised into sandy clay and clayey sand deposits of Tertiary and upper Cretaceous age (Magilligan and Stamp, 1997). (Photo: J.C. Knox)

partment of Agriculture estimates of historical truncation of soil profiles, Trimble (1974) estimated the average depth of soil loss related to agriculture in the southern Piedmont was about 18 cm for Alabama, 19 cm for Georgia, 24 cm for South Carolina, and 14 cm for North Carolina and Virginia. He reported that European settlement of the Piedmont started about 1700 in Virginia and was completed in Alabama by the 1830s.

Phillips (1993), using soil-profile truncations as a basis for evaluating historical soil erosion in the lower Neuse Basin of North Carolina, estimated upland soil losses of greater than 9.5 t ha^{-1} y^{-1} for the entire agricultural period compared to negligible soil loss during pre-Colonial times. Phillips et al. (1993) and Phillips (1997) found that on the low relief and permeable soils of the North Carolina Coastal Plain, interfluve soil erosion of 15–25 cm and a mean regional loss of 14.5 cm of soil erosion have occurred in response to agricultural activity. This loss was translated to an average rate of about 9.3 t ha^{-1} y^{-1} since the beginning of agricultural settlement in the early 1700s (Phillips et al., 1993). The high erosion rates for the North Carolina Coastal Plain are not unique because Phillips (1993) cited work by Lowrance et al. that indicated a Georgia Coastal Plain watershed experienced long-term historical agricultural upland erosion rates of 15 t ha^{-1} y^{-1}. Phillips (1993) also cited research by Dendy and by Beasley that indicated erosion rates in the Mississippi Coastal Plain ranging between 5.0 and 12.0 t ha^{-1} y^{-1} in a cotton field and between 12.5 and 14.2 t ha^{-1} y^{-1} at sites where forestry planting preparation was underway.

Compared with the southern sector of the eastern woodlands, relatively little is known about the history of agriculturally accelerated erosion and sedimentation in New England and eastern Canada. Ashmore (1993) reviewed the works of others and indicates that sediment yields from river systems draining the Laurentian highlands of the Canadian Shield tend to be very low because of the combined effects of stable geologic conditions and very limited agricultural influence. However, Ashmore reports that in areas of glacial and marine deposits in the St. Lawrence valley and coastal provinces where agriculture is common, considerable slope wash from fields and bank erosion from accelerated flooding are characteristic.

Throughout North America, much of the upland soil loss from agricultural land use was transported relatively short distances and stored elsewhere in the drainage system (Meade et al., 1990). Massive storage of sediment produced by agricultural erosion was especially prominent in Piedmont and Atlantic drainage watersheds of the eastern woodlands. Meade (1982) reported that more than 90% of the soil eroded from uplands of the southern Piedmont in the last 200 years is still stored on hillslopes and in the valleys of the Piedmont. Costa (1975) estimated that about 66% percent of the soil eroded from a 155-km^2 watershed in the Maryland Piedmont since about 1700 remains stored on hillslopes and in the valleys. Phillips (1991) found that sediment budgets for the (> 1000 km^2) upper Tar, upper Neuse, Haw, and Deep River basins in the North Carolina Piedmont showed that about 90% of the gross upland erosion was being stored in the watersheds, mainly as colluvium. Meade (1982) pointed out that although upland soil erosion has decreased dramatically in the Piedmont and Coastal Plain regions of the United States, there has not been a corresponding decrease in the sediment loads of the major rivers that drain the region. Although extensive urban development in the northeastern United States might have contributed to accelerated sediment loads of rivers there, Meade (1982) acknowledged that it clearly does not account for the persistence of large river sediment loads elsewhere, especially in the southern states where pasture and woodland have replaced cropland. Meade (1982) concluded that the persistence of high-magnitude sediment loads is explained by erosion of agriculture sediment that had been stored in the drainage system. Nevertheless, statistical comparisons of the 1982 and 1987 U.S. National Resources Inventory show that the Appalachian and southeastern states of the eastern woodland have experienced average reductions in cropland sheet and rill erosion amounting to 9.0% and 11.5%, respectively (Lee, 1990). The reduction appears to reflect in large part a conversion of 1982 cropland to other uses and to better land-use management. National Resources Inventory data by states for 1992 show a continued reduction of sheet and rill erosion from 1987 levels for the eastern woodland region (U.S. Department of Agriculture, 1994).

23.3 Midcontinental North America: Woodlands and Grasslands

23.3.1 Fluvial Erosion and Sedimentation

Williams (1989) indicated that westward expansion beyond the Appalachian Mountains was slow from the 1600s until about 1810 when it rapidly accelerated. Conversion of prairie and forest to agricultural cropland in the midcontinent of North America followed the conversion of the eastern woodlands by about a century. Williams (1989) reported that during the decade 1860–69, approximately 8 million hectares (ha) of U.S. forestlands were cleared for agricultural expansion, nearly half of this being in the midwestern states of Minnesota, Wisconsin, Michigan, Illinois, Indiana, and Ohio. By the 1880s, agricultural clearing began to focus on the open prairies and plains rather than on the forested regions. Fueling the agricultural expansion into the midcontinent was the availability of rich farmlands to replace the badly eroded and nutrient-depleted soils of the eastern states. Early agriculture in the Midwest was destined to repeat the poor land-use practices of the eastern woodlands. Johnson (1991, p. 3) quoted a statement made around 1880 by W. W. Daniells, the first professor of agriculture at the University of Wisconsin: "The early agriculture of Wisconsin was mere land-skimming. Good cultivation was never thought of. The same land was planted successively to one crop, as long as it yielded enough to pay for cultivation."

The consequences of nearly two centuries of accelerated soil erosion caused by agricultural land use are expressed today in the region's truncated soil profiles. Trimble (1983) estimated that 9–15 cm of soil truncation has occurred in the Coon Creek watershed located along the Mississippi River on the western edge of Wisconsin. Farther south in Wisconsin, Benedetti (1993) found that an average of about 16 cm of soil loss is representative of a small 30-ha tributary watershed in the Platte River system. Elsewhere, Beach (1992, 1994) estimated that 11–15 cm of soil truncation occurred in response to agriculture in the hilly and mostly loess-covered landscapes of southeastern Minnesota. As in the river systems of the eastern woodlands, enormous quantities of the eroded sediment remain stored as colluvium and alluvium in the drainages. Although the available data on sediment delivery are biased toward small watersheds of the upper Mississippi valley where delivery is normally expected to be higher due to the steep terrain, they consistently show more than 75% of the historically eroded sediment remains within their respective watersheds (Trimble, 1983; Benedetti, 1993; Beach, 1994). Trimble (1983) found that less than 7% of the agriculture-related soil erosion has left the basin of Coon Creek, but Beach (1994) showed that such low values for small watersheds here reflect the base level and back water effects of the Mississippi River at the mouth of Coon Creek. Other small basins farther removed from the Mississippi River have higher delivery ratios (Benedetti, 1993; Beach, 1994).

Although the focus of subsequent examples emphasizes the upper Mississippi valley, the historical development of cropping and grazing produced major increases in soil erosion and sedimentation throughout midcontinental North America. Sediment loads of rivers draining the northeastern Great Plains in Saskatchewan and Manitoba are strongly dominated by erosion and reworking of deltaic deposits associated with the higher levels of relict Pleistocene lakes. Nevertheless, a distinct shift in sedimentation characteristics occurs in Lake Manitoba deposition after agricultural settlement of the region (Last, 1984). Ashmore (1993), in reviewing many studies of contemporary erosion in the Canadian landscape, observed that, despite dense agricultural land use on the Canadian prairies, the rivers have relatively modest agriculturally accelerated sediment loads. Ashmore attributed this phenomenon to late Wisconsinan glaciation, which is responsible for a poorly integrated drainage network and a landscape characterized by numerous closed basins that prevent sediment delivery to the larger streams and rivers.

Elsewhere in the Great Plains of North Dakota, Hamilton (1967) and Artz (1995) have reported relatively thick accumulations of agriculturally derived sediment on floodplains. Baker et al. (1993) found that European agricultural settlement and attendant forest clearance in northeastern Iowa caused changes in the landscape, vegetation, insect fauna, and water quality unequaled in rate and magnitude since the melting of late Wisconsinan glaciers. Floodplain aggradation rates were estimated to be one to two orders of magnitude over those of the pre-European settlement period. Ruhe (1969) described thick alluvial deposits in southwestern Iowa river valleys, concluding that these deposits have resulted from the previous 115 years of cultivation. Brice (1966) showed that agriculturally accelerated erosion and sedimentation also characterized the thick loess region of Medicine Creek watershed in southwestern

Nebraska, where extensive gullying had contributed massive amounts of sediment to downstream channels and floodplains. Reporting that nearly 2 m of sediment had accumulated on the valley flat of the Dry Creek tributary between 1920 and 1953, Brice (1966) concluded that an average rate of accumulation of about 2.5 cm y^{-1} probably was representative of the valley flats of Medicine Creek drainage basin for this period. Entrenched streams and gully systems become efficient conveyors of sediment through the watershed (Happ et al., 1940). Entrenched channels and common gullies apparently explain why Dry Creek is delivering about 63% of the estimated upland eroded sediment (Spomer et al., 1986).

Elsewhere in the central midcontinental region, Jacobson and Primm (1994) examined the history of land use in the Missouri Ozarks. They found that cropping, grazing, and logging caused the region's rivers to experience aggradation by substantial quantities of gravel, accelerated channel migration and avulsion, and growth of gravel point bars. Farther south, in the lower Mississippi valley in the state of Mississippi and adjacent areas, agriculturally accelerated soil erosion and sedimentation are legendary. The classic monograph by Happ et al. (1940) on accelerated stream and valley sedimentation emphasized two watersheds in northern Mississippi where gullying, sheet erosion, and valley trenching resulted from conversion of woodlands to agricultural cropland of cotton and corn. The 1987 U.S. National Resources Inventory showed that Tennessee, Kentucky, and Mississippi were experiencing the highest sheet and rill erosion on United States cropland of 19.7, 19.1, and 16.4 t ha^{-1} y^{-1}, respectively (George and Choate, 1989).

Lee's (1990) comparison of cropland sheet and rill erosion estimates for the 1982 and 1987 U.S. National Resources Inventory showed that north-central states of the midcontinent experienced significant reductions in soil loss. The region with the largest reduction, 19.1%, was the corn belt states of Illinois, Indiana, Iowa, Missouri, and Ohio. Lee (1990) attributed this large reduction to improvements in land cover and land-use management factors. The National Resources Inventory data by states for 1992 show this region experienced a continued reduction of sheet and rill erosion below 1987 levels (U.S. Department of Agriculture, 1994). These examples suggest that wherever natural land cover was converted to agricultural land use, some degree of accelerated soil erosion followed. However, the relative response magnitude depended on the physiographic setting and to some degree on the soil-husbandry traditions of cultural groups (Trimble, 1985).

The severe erosion and sedimentation in the forest and prairie regions of midcontinental North America began with the initial land clearance for agriculture, but the most destructive period occurred during the latter half of the nineteenth and the first half of the twentieth centuries. Flood peaks from typical summer storm runoff during this period of poor land-conservation methods exceeded the presettlement flood peaks of storm runoff by 5–6 times in the hilly region of southwestern Wisconsin (Knox, 1977). Hayes et al. (1949, p. 48) summarized results of erosion control methods developed at the Upper Mississippi Valley Conservation Experiment Station as follows: "No control measure has been developed for this problem area that will prevent the loss of soil from a cultivated sloping field." The accelerated surface runoff from agricultural land use led to very severe sheet and rill erosion as shown in figure 23.2. However, in some areas of thick loess and sandy loam valley terraces, gully erosion also became a severe problem. Gully erosion has been particularly severe in the thick loess regions of western Iowa, southern and eastern Nebraska, and Missouri (Brice, 1966; Daniels and Jordan, 1966; Piest et al., 1976) and in the loess-covered uplands of the lower Mississippi valley (Happ et al., 1940).

Although gully cutting and filling episodes are a natural process known to have also characterized pre-European settlement time in the thick loess regions (Ruhe, 1969), agriculturally accelerated runoff and stream channelization projects have been key contributing causes of gullying during the nineteenth and twentieth centuries (Conservation and Natural Resources Subcommittee, 1971). Two examples illustrate the magnitude of gully erosion that has occurred in the midcontinent. Figure 23.5, from Jo Daviess County, northwestern Illinois, shows a large gully system about 1910 that was described as typical of many valley headwaters in the county. These gullies developed mainly from increased surface runoff from cropland on the upland ridges. Runoff was concentrated in formerly grassy, unchanneled headwater swales that soon experienced channel trenching from large magnitude runoff caused by cropland replacement of prairie ridge-top grasslands. Figure 23.6 shows an example of gullies that formed on a Pleistocene terrace in west-central Wisconsin after acceleration of upland agricultural runoff across the terrace surface. This severe gullying began in the late nineteenth century and was prevalent through the 1930s and 1940s in several counties of western Wisconsin. The gully shown in figure 23.6 was reported to have advanced about 300 m during runoff from a single rainstorm in 1922 (Zeasman, 1963).

Studies of Midwest erosion and sedimentation responses to agricultural land use have provided a number of detailed estimates of temporal variations in sedimentation rates. Historical rates of floodplain sedimentation in this region tend to be about an order of magnitude or more higher than the long-term natural Holocene rate when averaged over the entire agricultural period (Happ et al., 1940; Knox, 1972, 1977, 1987; Davis, 1976; Trimble and Lund, 1982; Trimble, 1983, 1993; Magilligan, 1985; Miller et al., 1993; Beach, 1994; Knox and Hudson, 1995). Studies by the aforementioned authors indicate the long-term historical rates of floodplain sedimentation have averaged from about 0.5–0.6 cm y^{-1} in small watersheds of less than 100 km^2 to as

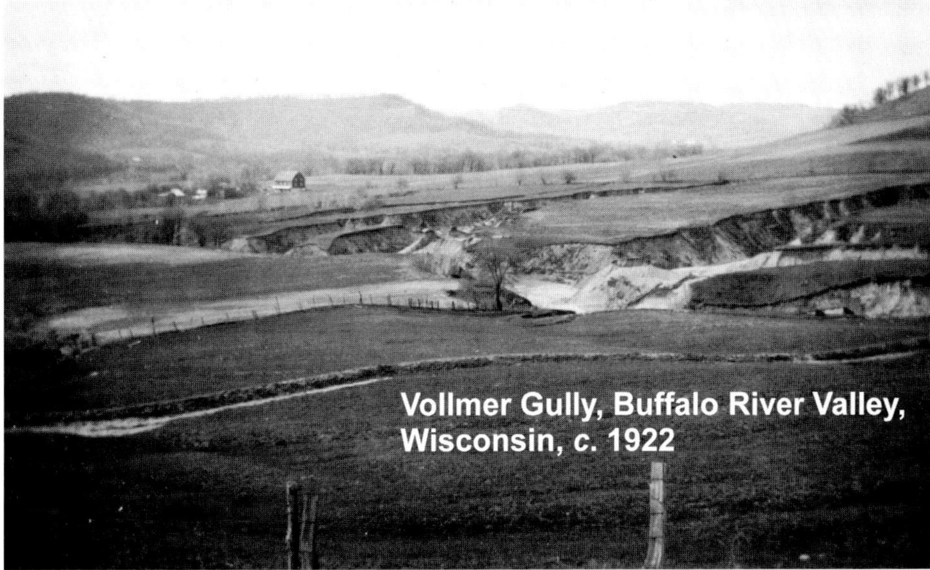

Figure 23.5 (*top*) Gully in Jo Daviess County, Illinois, circa 1910. Gullies of the type shown in this photo were described as typical of a large number of valley heads in Jo Daviess County about 1910. At the time of settlement in the early nineteenth century, the location of the gully was a grass-covered unchanneled swale, but the conversion of the prairie upland to cropland accelerated surface runoff and resulted in destabilization of the alluvial surface. Gully formation led to a positive feedback process in which the energy of runoff that once was dissipated across the valley floor then became concentrated in the deep gully. The concentrated energy enlarged the gully, which in turn allowed it to contain even more runoff with still higher energy and so on. (Photo: Trowbridge and Shaw, 1916)

Figure 23.6 (*bottom*) The Vollmer Gully in central-western Wisconsin represents the common, severe land degradation that began in this region during the late nineteenth century and continued as a major problem through the first three to four decades of the twentieth century. (Photo: Archives, University of Wisconsin, Madison; O.R. Zeasman, Photographer).

much as 2.0–3.0 cm y^{-1} in larger watersheds. By comparison, for southwestern Wisconsin tributaries of similar size, the natural Holocene rate of floodplain vertical accretion estimated from radiocarbon ages of stream bank alluvium was about 0.02 cm y^{-1} (Knox, 1985). Sediment delivery tends to occur episodically, which contributes to a tendency for calculated average sedimentation rates to decline as the length of the averaging period increases.

Nevertheless, it is clear that sedimentation rates during the agricultural period greatly exceed natural background rates. Webb and Webb (1988) found that the average rate of postsettlement sedimentation during the past 330 years for about 300 small lakes in eastern North America has been about four to five times faster than natural rates before settlement. Davis (1976) found that forest removal followed by agriculture in a southern Michigan watershed caused sedimentation rates in a small (6.7-ha) lake to increase by a factor of 10 to 30 in the early years of settlement and subsequently level off at about 10 times the presettlement rate. Davis concluded that, during brief intervals, the historical rate at the southern Michigan lake may have been 70 times the natural presettlement rate. Knox and Hudson (1995) reported even larger historical increases in sedimentation rates for the Platte River system of southwestern Wisconsin (fig. 23.7). There, on a historical short-term decadal scale, average floodplain sedimentation rates of 1–3 cm y^{-1} exceed the natural Holocene rate by 50 to 150 times or more. Knox and Hudson reported rates as high as 20 cm y^{-1} for 2 years during the early 1950s, when extreme summer floods occurred, but Trimble and Lund (1982) reported a rate of 15 cm y^{-1} for an 8-year period in the 1930s in the Coon Creek watershed in southwestern Wisconsin. The high-magnitude, short-term variability in sedimentation rates of the type shown in figure 23.7 does not appear to be present in the vertical accretion of the natural Holocene floodplain of the region (Knox, 1985; Knox and Hudson, 1995).

The magnitude of agricultural impact on sedimentation rates in the large Mississippi River floodplain is similar to that documented for the tributaries. McHenry et al. (1984) found that rates of sedimentation in backwater lakes and pools of the upper Mississippi River averaged 3.4 cm y^{-1} for the period 1954–64, but decreased to an average of 1.8 cm y^{-1} for the period 1965–75. A longer estimate of historical sedimentation in a backwater lake of the upper Mississippi River floodplain between Wisconsin and Minnesota has been determined by Eckblad et al. (1977), and their information indicated that the average sedimentation rate was about 1 cm y^{-1} from 1896 to 1973. Although existing data are inadequate to generalize natural sedimentation rates for backwater sloughs and floodplain lakes, limited sedimentation data for stream banks and island levées on the Mississippi River floodplain between northeastern Iowa and southwestern Wisconsin suggest that the historical rates are probably also an order of magnitude greater than natural Holocene rates. For example, in western Wisconsin on an upper Mississippi River floodplain levée with radiocarbon age control, levée-crest sedimentation occurred at a modest rate of about 0.07 cm y^{-1} between about 700 B.C. and the beginning of agricultural disturbance about 150 years ago (Knox, unpublished data).

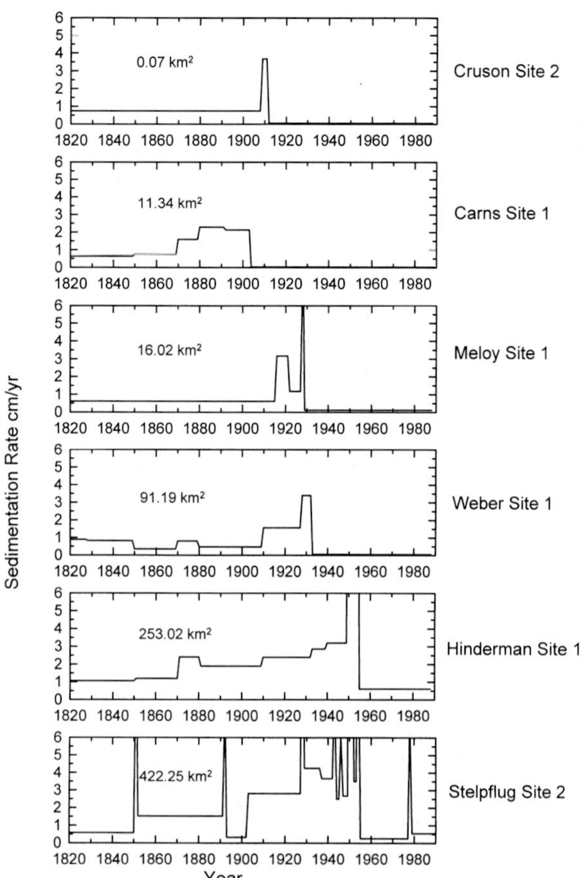

Figure 23.7 Downstream variation in historical overbank sedimentation rates for six drainage areas in the Platte River system, Grant County, Wisconsin. Lateral expansion of large-capacity meander belts during the historical period resulted in minimal valley-floor flooding and sedimentation for small tributaries from about 1900 and intermediate size tributaries from about the 1930s. The reduction in sedimentation rates since the mid-1950s for the two larger drainage areas farther downstream is due mainly to introduction of land-conservation practices. The preagriculture postglacial rate of overbank sedimentation averaged 0.02 cm y^{-1} in this watershed (after Knox and Hudson, 1995).

23.3.2 Aeolian Erosion and Sedimentation

Agriculturally accelerated wind erosion also has been a major problem within the midcontinental region, mainly in the Great Plains (Hagen, 1991). Larney et al. (1995, p. 91) reported "... wind erosion is one of the most serious soil degradation processes affecting agricultural sustainability on the semi-arid Canadian prairies." They cited summer fallow cropland as a principal contributor to the pervasive wind erosion problem. Holliday et al. (chapter 17, this volume; and in Ostercamp et al., 1987) and Livingstone and Warren (1996) have reviewed the evidence for recent wind erosion in the Great Plains. Citing data from others, Holliday et al. indicated that a dry episode from 1954 to 1957 resulted in wind erosion damages to 14.6 million hectares of land and that wind erosion during the anomalously dry 1975–76 period damaged 3.2 million hectares of land. Wind erosion during 1954–57 exceeded that of the better known "dust bowl" period of the 1930s, although Livingstone and Warren (1996) used information from Gillette and Hanson (1989) to suggest that dust-storm frequencies have never reached even one-quarter of the frequencies observed in the 1930s.

Gill (1996) provided an extensive review of evidence showing that water-table drawdown by irrigation of agricultural lands has exposed many playas to severe wind erosion. Because playas represent deposits of fine sediment and saline particulates, the dessication and erosion of playa sediment affect soil properties and human health downwind of the erosion area (Gill, 1996). Playas are especially characteristic of the southern Great Plains. Gill (1996) indicated that an extensive region of saline lakes of late Pleistocene origin occurs in the northern Great Plains of Saskatchewan, where drought and water usage have combined to accelerate dust plumes composed of sodium sulfate salts and clastic silt and clay particles. Lee et al. (1993, 1994) and Ervin and Lee (1994) show that agriculture tillage and irrigation practices have greatly increased the region's susceptibility to wind erosion; they found that nondrought years have been characterized by magnitudes of wind erosion comparable to that of drought years. The widespread introduction of center-pivot irrigation in the western Great Plains has been associated with large fields with few windbreaks, and these conditions favor wind erosion (Breed and McCauley, 1986). Cropland soil losses in 1977 for the southern High Plains in Colorado, New Mexico, and Texas averaged 26.4 t ha^{-1} y^{-1} in 1977, whereas sheet and rill erosion from water on the same cropland averaged only 6.0 t ha^{-1} y^{-1} in 1977 (Batie, 1983: 30–31). Ervin and Lee (1994) cited evidence that suggests about 1.9 million hectares have been damaged annually by wind erosion in the Great Plains since 1955.

The climatic setting of the Great Plains is close to the natural threshold separating stable versus unstable dune conditions, and this proximity accounts for the region's sensitivity to both natural and human influences on wind erosion (Madole, 1994; Muhs and Holliday, 1995; Lee and Tchakerian, 1996). Muhs and Maat (1993) found that temperature increases and precipitation reductions predicted from global circulation models have the potential for reactivating significant areas of now stable or mostly stable sand dunes and sand sheets on the Great Plains. Wheaton (1984a) reached a similar conclusion for Saskatchewan's northern Great Plains. The delicate sensitivity of this region was also demonstrated by Wolfe et al. (1995), who examined late Holocene and recent dune activity in southwestern Saskatchewan. They found that over the last 50 years dune activity has varied in accordance with temperature and precipitation trends, with even moderate droughts causing dune reactivation. Livingstone and Warren (1996) concluded that relationships among wind speed, drought, and land use are hard to disentangle because the relative contributions vary in different ways, in different areas, at different times, and at different scales.

23.4 The Southwest and Mexico

23.4.1 Land Cover Modification

The Southwest and Mexico have a long, albeit controversial, history of erosion and sedimentation accelerated by humans (Rich, 1911; Butzer, 1993). Accelerated runoff and soil erosion in this region have been influenced by cropland cultivation, grazing, and irrigation. Overgrazing has been particularly singled out as a contributing cause of erosion (Denevan, 1967). Some investigators have argued that pre-Spanish native populations had important localized impacts on accelerating erosion and sedimentation (Butzer, 1992a,b). Denevan (1992) suggested that by 1519 pressures of food production for the native Aztec population of central Mexico led to deforestation and land degradation, including severe, widespread soil erosion. Whitmore and Turner (1992) described denudation of the tropical forests of the Maya lowlands that occurred before A.D. 1000 and the complete transformation of the Basin of Mexico during Aztec times as examples of major impacts on natural landscapes by native populations. O'Hara et al. (1993), using lacustrine sediment cores from Lake Pátzcuaro, Michoacán, central Mexico, recognized three periods of agriculturally related soil erosion since 4000 yr B.P. (radiocarbon years before 1950). O'Hara et al. (1993) and Street-Perrott et al. (1989) suggested that erosion rates between about 2500–1200 yr B.P. and 850–350 yr B.P. were comparable to those that followed Spanish settlement when plow agriculture was introduced. Pre-Spanish accelerated soil erosion also was found by Baldwin et al. (1954), who identified agriculturally derived sediments that buried soil profiles in the intermountain basins of southern and central Mexico.

Butzer (1992a) reported that the Spanish influence on Indian depopulation of Mexico began around A.D. 1520 and that land grants were awarded in numbers during the early 1560s. By the 1640s in the west and north of Mexico, Spaniards controlled about 25% of the land on which they ran as many as 6–8 million sheep and 1.5–2 million cattle. Denevan (1992) cited the work of Melville, who attributed the abrupt beginning of active soil erosion about A.D. 1570 in the region north of the Valley of Mexico to overgrazing by Spanish livestock that began then. However, Denevan (1992) concluded that relative contributions of Indian and Spanish land degradation in Mexico are quite varied in time and place, making generalizations difficult.

Crops of maize, squash, gourd, and beans were widely introduced in the Southwest between about 3000 and 2000 years ago, and these spread onto the Colorado Plateau by about 200 B.C. (Butzer, 1990). The lowlands of the Gila and Salt Rivers in Arizona were at least locally intensely cultivated and irrigated, apparently achieving maximum development around A.D. 1400 (Butzer, 1990). Cooke and Reeves (1976) observed that an episode of arroyo cutting in the Southwest between the tenth and fourteenth centuries A.D. coincided with development and spread of Pueblo irrigation channels, and they speculated about a possible relationship. Graf (1988) showed that irrigation channels greatly alter natural drainage lines and that irrigation channels do not represent equilibrium configurations for large flood flows. Graf (1994) later summarized the history of irrigation on the northern Rio Grande. He indicated that Pueblo irrigation systems were slightly modified and expanded by Spanish and Mexican immigrants, who arrived in the sixteenth century, and by Anglo-American settlers, who arrived in increasing numbers after 1848. Graf noted that diversion structures became more numerous and elaborate in the late nineteenth century, but that by 1880 problems of upland erosion and arroyo formation in tributaries coupled with downstream river aggradation caused poor drainage conditions that led to curtailment of further irrigation. The Middle Rio Grande Conservancy District, established in 1925, ultimately led to the construction of dams and levees and channelization to address problems of controlling the river, developing river diversions, and improving field drainage (Graf, 1994). Maintenance and improvements to these engineering works continue to the present.

Irrigation may also cause subsidence of the land surface and contribute to other problems such as wind erosion. Orme and Orme (1998) explained that modest subsidence may result from hydrocompaction when loose moisture-deficient alluvial deposits are wetted for the first time, and that major subsidence can occur from compaction of the aquifer system in response to excessive groundwater extraction. Lofgren (1975) has documented major land subsidence as a result of ground-water withdrawal in the Arvin-Maricopa area of California. He indicated that a maximum subsidence of 9 m occurred along a 120-km distance on the western margin of the California's San Joaquin Valley (see chapter 22). Gill (1996) reviewed the results of many investigators who have studied the combined effects of drought and overdraft of water in the western United States. These results showed that withdrawal of water, in many cases for agricultural use, has led to massive drawdown of lakes and exposure of desiccated surfaces to severe wind erosion. The work of Smith et al. (1989) on the geology of the Great Basin attributed the desiccation of Nevada's Winnemucca Lake and the 30-m lowering of the level of Walker Lake to agricultural practices. Historical data compiled by Milne (1987) and Lebo et al. (1994) indicated that diversion of 42% of the Truckee River inflow to agricultural purposes led to a 19-m drop in the level of Pyramid Lake, Nevada, between 1904 and 1992. Dust plumes now rise from the exposed former lake bed.

Unlike in the eastern woodlands, the midcontinent, and Great Plains regions, few quantitative data documenting the magnitudes and rates of agriculturally induced soil erosion and sedimentation are available for the Southwest and Mexico. Even in California, where there are several large research universities and federal and state agencies concerned with sediment issues, there are relatively few data-based studies of preagricultural erosion rates against which modern rates may be compared. Orme and Orme (1998), however, cited a U.S. Department of Agriculture (USDA) study of Sand Canyon watershed in Ventura County, southern California, which indicated that the shift from conditions of the Native American period to present-day agriculture resulted in approximately an eightfold increase in annual erosion and approximately a tenfold increase in annual sediment yield.

Orme and Orme (1998) also presented USDA estimates of historical changes in flood peak discharges for rural Grimes Canyon watershed in Ventura County. These data indicate that the 2-year probability flood of the Native American period was doubled during the period of Spanish-Mexican settlement and then under present-day conditions further increased by about seven times the Native American magnitude. Comparable statistics for the 100-year-probability peak discharge are 1.5 times during the period of Spanish-Mexican settlement and 2.1 times for present-day conditions. Therefore, it is not surprising that the survey by Baldwin et al. (1954) of soil erosion in Mexico showed a very high percentage of the Mexican landscape, including most of the Baja California peninsula, as having 10–25% of the land classed as severely eroded. Little of Mexico was mapped as showing slight or no erosion from agriculture. Severe erosion implies that more than 75% of the topsoil has been lost, and that numerous or deep gullies might also be present.

Aeolian erosion also has been a major problem in the Southwest and Mexico, and continues to be a major concern on cultivated lands. The U.S. Department of Agriculture (1994) National Resources Inventory showed that soil loss by wind erosion on cultivated land in Arizona more than doubled between the 1982 and 1992 surveys. Elsewhere in Nevada and New Mexico, magnitudes of wind erosion on cultivated land remained high. On a more positive note, the 1992 National Resources Inventory showed that wind erosion on noncultivated nonfederal land tended to be relatively low and has been decreasing in magnitude since 1982.

23.4.2 Sensitivity to Climatic Change

The influence of climate and land use on erosion and sedimentation has been particularly difficult to determine in the Southwest because of the high sensitivity of the region's landscape, which responds to even modest shifts in climate. Much of the debate has concerned arroyos (fig. 23.8). Tuan (1966) observed that many early investigators assumed that arroyos resulted from accelerated flooding caused by overgrazing, but the later realization that arroyos also were characteristic of periods that predated agricultural grazing forced a more multivariate explanation. Cooke and Reeves (1976, p. 6) made the sobering observation that ". . . there is a certain correlation between the professional interests of investigators and the conclusions they reach on the causes of arroyo cutting." Graf (1983), after reviewing the voluminous arroyo literature, declared the "paradigm of origin" for arroyos to be in crisis because available knowledge was then incapable of providing an answer in general terms.

The sensitivity of the region to even modest climatic shifts was difficult for many early researchers to appreciate. For example, Thornthwaite et al. (1942) acknowledged numerous studies by Kirk Bryan and his students that suggested that the Southwest arroyo-cutting episode initiated in the late 1800s was dominantly a response to climate change. Nevertheless, they also commented that it was hard to reconcile Bryan's views ". . . with the known fact that a land surface which has been irritated by overgrazing and livestock and wagon trails will be gullied by a less intense rainstorm than one which has suffered no such irritation" (Thornthwaite et al., 1942, p. 46). Defining the contributions of agricultural land use versus climatic factors has been difficult because of limited quantitative data that show how surface runoff, soil erosion, and flooding respond to changes in ground cover in the Southwest.

The views of two early investigators, Ellsworth Huntington and Kirk Bryan, illustrate the uncertainty (Cooke and Reeves, 1976; Knox, 1983). Huntington (1914) hypothesized that valley alluviation occurred during dry episodes when vegetation was minimal and sediment yields were high. Bryan (1928), however, hypothesized that minimal vegetation during dry episodes would be associated with greater runoff and large floods that, in turn, would promote channel entrenchment and valley floor erosion. Both Huntington and Bryan considered overgrazing and other human effects of land use to be catalysts that initiated responses in fluvial systems already highly sensitive to climatic change.

Recent research results of Hereford (1984, 1993), Hereford et al. (1996), and Graf et al. (1987) showed clearly that both climate change and land use have played important roles in erosion and sedimentation episodes in the Southwest, but that climate change has been a key control in the long-term perspective. Hereford (1993), for example, showed that changes of both climate and land use were contributing causes of large floods that were key factors leading to the entrenchment of the San Pedro River in southern Arizona between about 1890 and 1908. Hereford et al. (1996) recognized two well-defined late Holocene episodes of arroyo cutting in most southern Colorado Plateau stream systems, the first between about A.D. 1200 and 1400 or slightly earlier in some watersheds, and the second beginning in the 1880s and ending by about A.D 1940.

Figure 23.8 The Kanab Arroyo near Kanab, Utah, was initiated by a series of large floods in the 1880s in a landscape that was stressed by agricultural activities. Webb et al. (1992) suggested cattle and sheep grazing that began in 1863, and other agricultural developments that involved construction of dams and ditches for drainage control, and the occurrence of livestock trails along alluvial channels probably all contributed to make the Kanab channel vulnerable to active entrenchment by large floods. (Photo: J.C. Knox)

The widespread synchronous response of these channel erosion episodes was shown to be associated with times of greater regional stream runoff.

In the Virgin River system of southwestern Utah, Hereford et al. (1996) found that both recent historical and late Holocene episodes of channel erosion corresponded with periods of unusually high streamflow, whereas periods of deposition and valley aggradation occurred during periods of relatively low streamflow. Their results support Huntington's hypothesis of valley aggradation during dry phases of climate, but a detailed quantitative study linking climate and hillslope processes in semiarid New Mexico led Leopold et al. (1966, p. 240) to comment that "considerable evidence points to a coincidence of increasing aridity with degradation and increasing humidity with aggradation, but . . . a firm conclusion will have to await continuation of just the kinds of observations being here reported."

23.4.3 Dams

Dams of all sizes, from small stock-watering ponds to large multipurpose structures, have been built on river systems throughout North America (see also chapter 22). Since many of these structures were built to provide irrigation to agricultural fields, a few comments are in order. Small dams on very small upland tributaries probably have a net positive effect on total watershed erosion because they attenuate flood runoff and in some cases may stop headward gully erosion. However, the impact of dams on erosion and sedimentation in arid and semiarid environments is particularly significant because natural rivers of these regions carry high sediment loads, much of which is trapped in reservoirs behind the dams (Meade et al., 1990). The combined effects of greatly reduced downstream flood magnitudes and the downstream release of water of low sediment concentration have resulted in a major metamorphosis of the region's river channels (Burkham, 1972; Schumm, 1977; Williams and Wolman, 1984; Hirsch et al., 1990; Collier et al., 1996). Graf (1994) concluded that nearly all major rivers of the Southwest had broad, sandy channels with braided configurations and meandering low-flow channels in the early nineteenth century. Graf indicated that most of these rivers today have changed to single-thread or compound channels that flood less often. Graf (1994, p. 97–100) attributes the channel metamorphosis to "removal of beavers, the development of gullies and arroyos, land-management schemes, changes in climate, and the construction of dams."

Despite the obvious direct physical influences of dams on the erosion, sedimentation, and morphology of the region's rivers, hydrologic responses to climate change may also be a contributing factor as in the case of the arroyo episodes. Graf (1994) observed that the upper Rio Grande, which is not controlled by a dam, and many other Southwest streams without major dams also experienced historical shifts from wide-shallow braided patterns to single-thread patterns. Schumm and Lichty (1963) have shown that the recurrence frequency of large floods can be critical in determining whether channels are braided or meandering in sensitive environments such as the Southwest. This relationship suggests that it would be useful to determine whether climatic conditions favored a decreased frequency of large floods during the first half of the twentieth century when much of this metamorphosis was under way.

23.5 Northwest

The Northwest region is defined here as the northern part of the Western Cordillera, including the Cascade Range and northern Rocky Mountains, with their intermontane basins and valley systems. Agricultural cropland is abundant at lower elevations, whereas logging, once prevalent in the lowlands and foothills, now occurs mostly at higher elevations and in the more northerly regions. Grazing may occur at all elevations in much of central and southern sectors of this region. In the northern and high-elevation landscapes, the effect of agriculture on erosion and sedimentation is minor or absent. On the other hand, dryland agriculture is prevalent on the Columbia Plateau in the northwestern United States, where soils are very fine grained and have experienced serious erosion by deflation processes (Stetler and Saxton, 1996). For example, the Palouse region of eastern Washington and northern Idaho is a highly productive region of dryland farming where wheat and fallow crop rotations are practiced. Busacca et al. (1993) characterized the Palouse region as underlain by thick loess deposits often with steep slopes that average between 5° and 17° (also see chapter 3). As a consequence, water erosion rates of 200–400 t ha^{-1} have been documented for a single winter season, and the region is characterized by an average rate of soil erosion that is one the highest in the United States (Busacca et al., 1993).

Soil losses from agricultural land are especially high because most precipitation is delivered during the winter season, November to May, when there is little plant cover to protect the ground surface. Farther south, in the now dry pluvial basin of Fort Rock Lake, central Oregon, clearance of natural vegetation from the pluvial basin floor by homesteaders between 1890 and 1910 resulted in up to 1 m of erosion by deflation (Allison, 1966; Gill, 1996). Aeolian erosion has also been accelerated in response to water-table lowering as a result of irrigation demands. For example, McDowell et al. (1991) showed that lunettes in the northern part of the Great Basin have been reactivated partly in response to withdrawal of water for agricultural irrigation.

Erosion and sedimentation problems in the Northwest have also been attributed to stream bank destabilization related to cattle grazing (Kauffman et al., 1983). Through trampling, rubbing, and browsing of vegetation on streambanks, cattle contribute to bank caving, adjustments in channel shape, and changes in sediment load (Platts et al., 1983). Clifton (1987) surveyed channel cross sections in grazed and ungrazed reaches of Wickiup Creek in the Blue Mountains of northeastern Oregon to determine channel response to long-term exclosure from cattle grazing. She found that during the 53-year absence of livestock and progressive revegetation of a minimally forested reach of the creek, there was a 94% reduction in the capacity of channel cross sections. Comparisons with channel reaches immediately upstream and downstream of the exclosure showed that the exclosed channel was about 40–60% narrower and 40–45% deeper. Thick channel-bank vegetation in the ungrazed reach had stabilized the banks, efficiently trapping sediment. However, the results showed that where forest cover was more dense, with logs and other organic debris common in the channel, the effects of grazing were not strongly apparent because the channel was mainly responding to the effects of the organic debris.

The northern Northwest sector, in British Columbia, the Yukon, and Alaska, comprises largely natural landscapes where agricultural land use has had little impact. Slaymaker (1987) found that, in British Columbia's coastal mountains, agricultural land use had no measurable impact on large- and intermediate-scale river systems, but on small systems, where natural background erosion and sedimentation rates were low, agricultural land use produced significant impacts. Church and Slaymaker (1989), Church et al. (1989), and Church (1990) have shown that most valley bottoms and valley sides adjacent to streams contain large volumes of relatively unstable stored sediment derived from the most recent glaciation of the region. This stored sediment in the riparian zone is easily remobilized and represents the dominant source of sediment for many rivers (Church et al., 1989). Elsewhere, in the Peace River valley, a large system flowing north and east from the Rocky Mountains across northern Alberta is bordered by extensive agricultural lands and consequently experiences a significantly accelerated sediment load (Slaymaker, 1972). Agricultural contributions to accelerated erosion and sedimentation become relatively insignificant farther north. For example, in describing the processes that control the physical properties of rivers and lakes in the delta of the Mackenzie River, whose drainage includes runoff from all of the eastern slope of the Rocky Mountains between latitude 54°N in central Alberta and 69°N at the Arctic Ocean, Lapointe (1990) made no mention of human forces.

Sorting out human influences from natural causes of erosion and sedimentation in the dynamic mountainous environments that typify much of the Northwest is a difficult exercise. Lisle (1990) noted that, excluding watersheds affected by glaciers or volcanic activity, the Eel River in northwestern California's Coast Ranges has the highest recorded average suspended sediment yield per drainage area in the conterminous United States. The exceptionally high sediment load reflects widespread tectonic deformation, recent rapid uplift, weak pervasively fractured bedrock, high seasonal rainfall, and widespread disturbance of the landscape by logging and grazing during the last 100 years. Other researchers have concluded that loss of tree-root strength in noncohesive soils, as forest is replaced by grassland, has probably helped to destabilize hillslopes and contributed to gully formation. Certainly, sediment yields have increased severalfold since intensive timber harvesting and associated road building began in the Eel basin in the 1950s, but the relative contributions of logging and grazing are difficult to separate from the effects of a series of large floods that have also occurred during the past 50 years. The effects of timber harvesting and associated forest practices are discussed further in chapter 22.

The effects of overgrazing on significantly accelerating erosion and sedimentation seem to have declined in the Northwest during the last 50 years, much in the same way that was previously noted for the midcontinent and Southwest regions. For example, in the upper Salmon River area of Idaho, Emmett (1975) found that cattle and sheep numbers have been reduced until the balance of stock and forage are about in equilibrium. He estimated that the number of sheep in the early 1970s was only about 10% of the number that existed in the peak years of 1905–10. Overgrazing in these early years created serious erosion problems in the watershed, but today such erosion is evident only in isolated areas.

23.6 Conclusion

Most agricultural land uses accelerate rates of erosion and sedimentation well above natural rates because the balance between climate and vegetation is disrupted. Removal of natural vegetation and replacement with cropland and pasture typically expose soil to raindrop impact, reduce surface hydraulic roughness, reduce soil organic content, and contribute to the destruction of natural soil structure. The consequences of these changes result in greatly accelerated soil erosion and high-magnitude sediment yields from watersheds. Much of the agriculturally eroded sediment remains stored as colluvium and alluvium elsewhere in watersheds, and this sediment is often subject to remobilization. The sensitivity of the stored sediment to remobilization indicates that watershed sediment yields will continue to experience accelerated loads for a long period into the future, despite the greatly improved land-use practices of the last several decades. Watershed erosion by fluvial action is a direct response to reduced soil infiltration capacities and increased overland flow runoff. Accelerated landscape erosion by aeolian processes also typically results from agricultural land use when the protection of vegetation is removed. Cultivation commonly causes many soils to become loose, dry, and powdery, and therefore susceptible to deflation by wind.

Agricultural disturbance on a large scale is a relatively recent phenomenon for most of North America. Elsewhere in the world, Starkel (1987) has indicated that agricultural influence on erosion and sedimentation occurred as early as 9500 B.P. in steppe regions and that some forested regions were cleared for cultivation as early as 6000 B.P. But in North America, major disturbance apparently began sometime after about 4000 years ago in Mexico and between about 3000 and 2000 years ago in the southwestern United States. Although there is evidence that agriculture associated with native populations may have caused localized erosion and sedimentation, the arrival of European agriculture accelerated erosion and sedimentation rates to magnitudes and geographic coverage that are unparalleled in the preceding postglacial record. The European agricultural influence began in the 1500s in Mexico, in the 1600s and 1700s in the eastern woodlands, and in the 1800s for most of the midcontinent and the western and northwestern regions of North America.

Soil erosion surveys in Mexico show that a very high percentage of the landscape is severely eroded with more than 75% of the topsoil lost. Much of the topsoil has also been lost from agricultural soils in the United States. Many soils of eastern and southeastern states have lost about 15–18 cm of topsoil, and losses of 10–15 cm of topsoil have been documented for watersheds in the northern Midwest. These topsoil losses were largely the result of increased surface runoff. However, in the Great Plains from Texas to the Canadian prairies and in the region of dryland farming where a crop rotation of wheat and fallow is practiced, wind erosion has been equally destructive to the topsoil.

Present-day agricultural land use is increasingly oriented to the idea of "sustainability" (Lee, 1996). This concept, as applied to soil erosion, assumes that a certain loss is tolerable if that loss does not exceed the rate at which weathering and other natural processes lead to new soil formation. Friend (1992) stated that the idea of soils as a renewable resource first appeared in the English literature in the 1970s, and since then a prevailing acceptance of that idea has existed. He pointed out that in many regions soils seem to be forming at rates of about 2.5 cm per 100–200 years, but that soil erosion in many of these regions has historically exceeded the formation rate. In some localities, soil erosion still exceeds the formation rate despite greatly reduced erosion rates since the 1950s. It is also noteworthy that many modern soils are forming in parent material such as loess or other silt-rich sediment that was deposited in response to late Pleistocene glacial climatic conditions. Older, residual parent material underlying these late Pleistocene deposits often is low in nutrients and highly weathered. Therefore, the soil losses viewed as tolerable for present parent material may be inapplicable to deposits that underlie the thin covering of late Pleistocene sediment. All of this leads us to wonder whether tolerable soil losses for sustainability are properly calibrated to reflect the complexity of parent material resources and true rates of soil formation. Ecologist Aldo Leopold stated, "We have no way to restore the soil to lands that have washed away. Soil is the fundamental resource, and its loss the most serious of all losses" (Flader and Callicott, 1991, p. 109).

Acknowledgments The National Science Foundation supported my research on historical erosion and sedimentation in the upper Mississippi valley and contributed to the collection and analysis of information that made this manuscript possible. I thank David Leigh for assistance in the field and laboratory in conjunction with my upper Mississippi valley research. I thank Robert H. Meade, Antony Orme, and Karl Zimmerer for their reviews of this manuscript.

References

Albrecht, W.A., 1938, Loss of soil organic matter and its restoration, in *Yearbook of Agriculture 1938*, 75th Congress, 2nd Session, House Document No. 398: 347–360.

Allison, I.S., 1966, *Fossil Lake, Oregon: Its Geology and Fossil Faunas*. Corvallis, Oregon State University Press, 48 p.

Argabright, M.S., R.G. Cronshey, J.D. Helms, G.A. Pavelis, and H.R. Sinclair Jr., 1996, *Historical Changes in Soil Erosion, 1930–1992: The Northern Mississippi Valley Loess Hills*, Washington, U.S. Department of Agriculture, Natural Resources Conservation Service, Historical Notes Number 5, 92 p.

Artz, J.A., 1995, Geological contexts of the early and middle Holocene archeological record in North Dakota and adjoining areas of the northern Great Plains, in E.A. Bettis III, (Ed.), *Archeological Geology of the Archaic Period in North America*, Geological Society of America Special Paper 297: 67–86.

Ashmore, P., 1993, Contemporary erosion of the Canadian landscape. *Progress in Physical Geography* 17: 190–204.

Baker, R.G., D.P. Schwert, E.A. Bettis III, and C.A. Chumbley, 1993, Impact of Euro-American settlement on a riparian landscape in northeast Iowa, midwestern USA: An integrated approach based on historical evidence, floodplain sediments, fossil pollen, plant macrofossils and insects. *The Holocene* 3: 314–323.

Baldwin, M., and Field Staff, Foreign Agricultural Service, U.S. Department of Agriculture, 1954, Soil erosion survey of Latin America. *Journal of Soil and Water Conservation* 9: 158–168.

Balling, R.C., and S.G. Wells, 1990, Historical rainfall patterns and arroyo activity within the Zuni River drainage basin, New Mexico. *Annals of the Association of American Geographers* 80: 603–617.

Barnhardt, M.L., 1988, Historical sedimentation in west Tennessee gullies. *Southeastern Geographer* 28(1): 1–18.

Batie, S.S., 1983, *Soil Erosion: Crisis in America's Croplands?* Washington, D.C., The Conservation Foundation, 136 p.

Beach, T., 1992, Estimating erosion in medium-size drainage basins. *Physical Geography* 13: 206–224.
Beach, T., 1994, The fate of eroded soil: Sediment sinks and sediment budgets of agrarian landscapes in southern Minnesota, 1851–1988. *Annals of the Association of American Geographers* 84: 5–28.
Benedetti, M.M., 1993, Sediment budget response to land use changes: Big Jack Watershed, Grant County, Wisconsin. M.S. Thesis, Department of Geography, University of Wisconsin, Madison, 71 p.
Bennett, H.H., 1939, *Soil Conservation*. New York, McGraw-Hill Publishing Co., 993 p.
Bennett, H.H., and W.C. Lowdermilk, 1938, General aspects of the soil-erosion problem, in *Yearbook of Agriculture 1938*, 75th Congress, 2nd Session, House Document No. 398: 581–608.
Breed, C.S., and J.F. McCauley, 1986, Use of dust storm observations on satellite images to identify areas vulnerable to severe wind erosion. *Climatic Change* 9: 243–251.
Brice, J.C., 1966, *Erosion and Deposition in the Loess-Mantled Great Plains: Medicine Creek Drainage Basin, Nebraska*. U.S. Geological Survey Professional Paper 352-H: 255–339.
Bryan, K., 1928, Historic evidence on changes in the channel of the Rio Puerco, a tributary of the Rio Grande in New Mexico. *Journal of Geology* 36: 265–282.
Burkham, D.E., 1972, *Channel Changes of the Gila River in Safford Valley, Arizona 1846–1970*. U.S. Geological Survey Professional Paper 655-G, 24 p.
Busacca, A.J., C.A. Cook, and D.J. Mulla, 1993, Comparing landscape-scale estimation of soil erosion in the Palouse using Cs-137 and RUSLE. *Journal of Soil and Water Conservation* 48: 361–367.
Butzer, K.W., 1974, Accelerated soil erosion: A problem of human-land relationships, in I.R. Manners, and M. Mikesell, (Eds.), *Perspective on Environment*, Association of American Geographers Commission on College Geography 13: 57–78.
Butzer, K.W., 1990, The Indian legacy in the American landscape, in M.P. Conzen, (Ed.), *The Making of the American Landscape*, London: Harper Collins Academic, 27–50.
Butzer, K.W., 1992a, The Americas before and after 1492: Current geographical research. *Annals of the Association of American Geographers* 82: 343–565.
Butzer, K.W. (Ed.), 1992b, The Americas before and after 1492: An introduction to current geographical research. *Annals of the Association of American Geographers* 82: 345–368.
Butzer, K.W., 1993, No Eden in the New World. *Nature* 362: 15–17.
Changnon, S.A., 1983, Record dust storms in Illinois: Causes and implications. *Journal of Soil and Water Conservation* 38: 58–63.
Chepil, W.S., and N.P. Woodruff, 1954, Estimations of wind erodibility of field surfaces. *Journal of Soil and Water Conservation* 9: 257–265, 285.
Chepil, W.S., F.H. Siddoway, and D.V. Armbrust, 1963, Climatic index of wind erosion conditions in the Great Plains. *Soil Science Society of America Proceedings* 27: 449–452.
Church, M., 1990, Fraser River in central British Columbia, in M.G. Wolman, and H.C. Riggs, (Eds.), *Surface Water Hydrology*, Boulder, Geological Society of America, The Geology of North America O-1: 282–287.
Church, M., and H.O. Slaymaker, 1989, Disequilibrium of Holocene sediment yield in glaciated British Columbia. *Nature* 337: 452–454.
Church, M., R. Kellerhals, and T.J. Day, 1989, Regional clastic sediment yield in British Columbia. *Canadian Journal of Earth Sciences* 26: 31–45.
Clifton, C.F., 1987, Effects of vegetation and land use on the channel morphology of Wickiup Creek, Blue Mountains, Oregon. M.S. Thesis, Department of Geography, University of Wisconsin, Madison, 107 p.
Collier, M., R.H. Webb, and J.C. Schmidt, 1996, *Dams and Rivers: A Primer on the Downstream Effects of Dams*. U.S. Geological Survey Circular 1126, 94 p.
Conservation and Natural Resources Subcommittee, 1971, *Stream Channelization: Parts 1–4*. Washington D.C., Ninety-Second Congress, Committee on Government Operations, Report of Hearings.
Cooke, R.U., and R.W. Reeves, 1976, *Arroyos and Environmental Change in the American South-West*. Oxford, Oxford University Press, 213 p.
Costa, J.E., 1975, Effects of agriculture on erosion and sedimentation in the piedmont province, Maryland. *Geological Society of America Bulletin* 86: 1281–1286.
Daniels, R.B., and R.H. Jordan, 1966, *Physiographic history and the soils, entrenched stream systems, and gullies, Harrison County, Iowa*. U.S. Department of Agriculture Technical Bulletin 1348: 1–116.
Davis, M.B., 1976, Erosion rates and land-use history in southern Michigan. *Environmental Conservation* 3(2): 139–148.
Denevan, W.M., 1967, Livestock numbers in nineteenth-century New Mexico, and the problem of gullying in the Southwest. *Annals of the Association of American Geographers* 57: 691–703.
Denevan, W.M., 1992, The pristine myth: The landscape of the Americas in 1492. *Annals of the Association of American Geographers* 82: 369–385.
Doolittle, W.E., 1992, Agriculture in North America on the eve of contact: A reassessment. *Annals of the Association of American Geographers* 82: 386–401.
Eckblad, J.W., N.L. Peterson, and K. Ostlie, 1977, The morphometry, benthos and sedimentation rates of a floodplain lake in Pool 9 of the Upper Mississippi River. *American Midland Naturalist* 97: 433–443.
Emmett, W.W., 1975, *The Channels and Waters of the Upper Salmon River Area, Idaho*. U.S. Geological Survey Professional Paper 870-A: 1–116.
Enlow, C.R., and G.W. Musgrave, 1938, Grass and other thick-growing vegetation in erosion control, in *Yearbook of Agriculture 1938*, 75th Congress, 2nd Session, House Document No. 398: 615–633.
Ervin, R.T., and J.A. Lee, 1994, Impact of conservation practices on airborne dust in the Southern High Plains of Texas. *Journal of Soil and Water Conservation* 49: 430–437.
Flader, S.L., and J.B. Callicott, 1991, *The River of the Mother of God and Other Essays by Aldo Leopold*. Madison, The University of Wisconsin Press, 384 p.
Friend, J.A., 1992, Achieving soil sustainability. *Journal of Soil and Water Conservation* 47: 156–157.
George, T.A., and J. Choate, 1989, A first look at the 1987 National Resources Inventory. *Journal of Soil and Water Conservation* 44: 555–556.
Gill, T.E., 1996, Eolian sediments generated by anthropogenic disturbance of playas: Human impacts on the

geomorphic system and geomorphic impacts on the human system. *Geomorphology* 17: 207–228.

Gillette, D.A., and K.J. Hanson, 1989, Spatial and temporal variability of dust production caused by wind erosion in the United States. *Journal of Geophysical Research* 94: 2197–2206.

Graf, W.L., 1983, The arroyo problem: Paleohydrology and paleohydraulics in the short term, in K.J. Gregory, (Ed.), *Background to Paleohydrology*, London, John Wiley and Sons, 279–302.

Graf, W.L., 1988, *Fluvial Processes in Dryland Rivers.* New York, Springer-Verlag, 346 p.

Graf, W.L., 1994, *Plutonium and the Rio Grande: Environmental Change and Contamination in in the Nuclear Age.* New York, Oxford University Press, 329 p.

Graf, W.L., R. Hereford, J. Laity, and R.A. Young, 1987, Colorado Plateau, in W.L. Graf, (Ed.), *Geomorphic Systems of North America.* Boulder, Geological Society of America, Centennial Special Volume 2: 259–302.

Hagen, L.J., 1991, A wind erosion prediction system to meet user needs. *Journal of Soil and Water Conservation* 46: 106–111.

Hamilton, T.M., 1967, Late-recent alluvium in western North Dakota, in L. Clayton, and T.F. Freers, (Eds.), *Glacial Geology of the Missouri Coteaux and Adjacent Areas*, North Dakota Geological Survey Miscellaneous Series 30: 151–158.

Happ, S.C., 1945, Sedimentation in South Carolina Piedmont valleys. *American Journal of Science* 243: 113–126.

Happ, S.C., G. Rittenhouse, and G.C. Dobson, 1940, *Some Principles of Accelerated Stream and Valley Sedimentation*, Washington, D.C., United States Department of Agriculture, Technical Bulletin No. 695, 134 p.

Hayes, O.E., A.G. McCall, and F.G. Bell, 1949. *Investigations in Erosion Control and the Reclamation of Eroded Land at the Upper Mississippi Valley Experiment Station near LaCrosse, Wisconsin, 1933–1943.* Washington, D.C., United States Department of Agriculture, Technical Bulletin 973, 87 p.

Hereford, R., 1984, Climate and ephemeral-stream processes: Twentieth-century geomorphology and alluvial stratigraphy of the Little Colorado River, Arizona. *Geological Society of America Bulletin* 95: 654–668.

Hereford, R., 1993, *Entrenchment and Widening of the Upper San Pedro River, Arizona.* Boulder, Geological Society of America Special Paper 282, 46 p.

Hereford, R., G.C. Jacoby, and V.A.S. McCord, 1996, *Late Holocene Alluvial Geomorphology of the Virgin River in the Zion National Park Area, Southwest Utah.* Boulder, Geological Society of America Special Paper 310, 41 p.

Hirsch, R.M., J.F. Walker, J.C. Day, and R. Kallio, 1990, The influence of man on hydrologic systems, in M.G. Wolman, and H.C. Riggs, (Eds.), *Surface Water Hydrology*, Boulder, Geological Society of America, The Geology of North America, Volume O-1: 329–359.

Hovde, M.R., 1934, The great dust storm of November 12, 1933. *Monthly Weather Review* 62: 12–13.

Huntington, E., 1914, *The Climatic Factor as Illustrated in Arid America.* Carnegie Institution Publication 192.

Jacobson, R.B., and D.J. Coleman, 1986, Stratigraphy and recent evolution of Maryland Piedmont flood plains. *American Journal of Science* 286: 617–637.

Jacobson, R.B., and A.T. Primm, 1994, *Historical Land-Use Changes and Potential Effects on Stream Disturbance in the Ozark Plateaus, Missouri*, Washington, U.S. Geological Survey, Open-File Report 94-333, 95 p.

Johnson, L.C., 1991, *Soil Conservation in Wisconsin: Birth to Rebirth.* Madison: University of Wisconsin Department of Soil Science, 332 p.

Kauffman, J.B., W.C. Kreuger, and M. Vavra, 1983, Impacts of cattle on streambanks in northeastern Oregon. *Journal of Range Management* 36: 683–685.

Kimberlin, L.W., A.L. Hidlebaught, and A.R. Grunewald, 1977, The potential wind erosion problem in the United States. *Transactions of the American Society for Agricultural Engineering* 20:873–879.

Knox, J.C., 1972, Valley alluviation in southwestern Wisconsin. *Annals of the Association of American Geographers* 62: 401–410.

Knox, J.C., 1977, Human impacts on Wisconsin stream channels. *Annals of the Association of American Geographers* 67: 323–342.

Knox, J.C., 1983, Responses of rivers to Holocene climates, H.E. Wright, Jr., (Ed.), *Late-Quaternary Environments of the United States: Volume 2, The Holocene.* Minneapolis, University of Minnesota Press, 26–41.

Knox, J.C., 1985, Responses of floods to Holocene climatic change in the Upper Mississippi Valley. *Quaternary Research* 23: 287–300.

Knox, J.C., 1987, Historical valley floor sedimentation in the upper Mississippi Valley. *Annals of the Association of American Geographers* 77: 224–244.

Knox, J.C., and J.C. Hudson, 1995, Physical and cultural change in the Driftless Area of southwest Wisconsin, in M.P. Conzen, *Geographical Excursions in the Chicago Region*, Association of American Geographers, Ninety-First Annual Meeting, Chicago, 107–131.

Lapointe, M., 1990, The Mackenzie Delta, Northwest Territories, in M.G. Wolman, and H.C. Riggs, (Eds.), *Surface Water Hydrology*, Boulder, Geological Society of America, The Geology of North America O-1: 292–295.

Larney, F.J., M.S. Bullock, S.M. McGinn, and D.W. Fryrear, 1995. Quantifying wind erosion on summer fallow in southern Alberta. *Journal of Soil and Water Conservation* 50: 91–95.

Last, W.M., 1984, Modern sedimentology and hydrology of Lake Manitoba, Canada. *Environmental Geology* 5: 177–190.

Lebo, M.E., J.E. Reuter, and P.A. Meyers, 1994, Historical changes in sediments of Pyramid Lake, Nevada, USA: Consequences of changes in the water balance of a terminal desert lake. *Journal of Paleolimnology* 12: 87–101.

Lee, J.A., and V.P. Tchakerian, 1996, Magnitude and frequency of blowing dust on the Southern High Plains of the United States, 1947–1989. *Annals of the Association of American Geographers* 85: 684–693.

Lee, J.A., K.A. Wigner, and J.M. Gregory, 1993, Drought, wind and blowing dust on the Southern High Plains of the United States. *Physical Geography* 14: 56–67.

Lee, J.A., B.L. Allen, R.E. Peterson, J.M. Gregory, and K.E. Moffett, 1994, Environmental controls on blowing dust direction at Lubbock, Texas, United States. *Earth Surface Processes and Landforms* 19: 437–449.

Lee, L.K., 1990, The dynamics of declining soil erosion rates. *Journal of Soil and Water Conservation* 45: 622–624.

Lee, L.K., 1996, Sustainability and land-use dynamics. *Journal of Soil and Water Conservation* 51: 295.

Leopold, L.B., W.W. Emmett, and R.M. Myrick, 1966, *Channel and Hillslope Processes in a Semiarid Area, New*

Mexico. U.S. Geological Survey Professional Paper 352-G: 193–253.

Lisle, T.E., 1990, The Eel River, northwestern California: High sediment yields from a dynamic landscape, in M.G. Wolman, and H.C. Riggs, (Eds.), *Surface Water Hydrology*, Boulder, Geological Society of America, The Geology of North America O-1: 311–314.

Livingstone, I., and A. Warren, 1996, *Aeolian Geomorphology*. Essex, U.K., Addison Wesley Longman Limited, 211 p.

Lofgren, B.E., 1975, *Land Subsidence Due to Ground-Water Withdrawal, Arvin-Maricopa Area, California*. U.S. Geological Survey Professional Paper 437-D.

Madole, R.F., 1994, Stratigraphic evidence of desertification in the west-central Great Plains within the past 1000 yr. *Geology* 22: 483–486.

Magilligan, F.J., 1985, Historical floodplain sedimentation in the Galena River basin, Wisconsin and Illinois. *Annals of the Association of American Geographers* 75: 583–594.

Magilligan, F.J., and M.L. Stamp, 1997, Historical land-cover changes and hydrogeomorphic adjustment in a small Georgia watershed. *Annals of the Association of American Geographers* 87: 614–635.

McCauley, J.F., C.S. Breed, M.J. Grolier, and D.J. Mackinnon, 1981, The U.S. dust storm of February 1977, in T.L. Péwé, (Ed.), *Desert Dust: Origin, Characteristics, and Effect on Man*. Geological Society of America Special Paper 186: 123–147.

McDowell, P.F., P.J. Bartlein, and S.P. Harrison, 1991, Environmental controls of playa status and processes, western U.S. *Geological Society of America Abstracts* 23(5): A283.

McHenry, J.R., J.C. Ritchie, C.M. Cooper, and J. Verdon, 1984, Recent rates of sedimentation in the Mississippi River, in J.G. Wiener, R.V. Anderson, D.R. McConville, (Eds.), *Contaminants in the Upper Mississippi River: Proceedings of the 15th Annual Meeting of the Mississippi River Research Consortium*. Butterworth Publishers, Boston, 99–117.

Meade, R.H., 1982, Sources, sinks, and storage of river sediment in the Atlantic drainage of the United States. *Journal of Geology*, 90: 235–352.

Meade, R.H., T.R. Yuzyk, and T.J. Day, 1990, Movement and storage of sediment in rivers of the United States and Canada, in M.G. Wolman, and H.C. Riggs, (Eds.), *Surface Water Hydrology*, Boulder, Geological Society of America, The Geology of North America O-1: 255–280.

Miller, S.O., D.F. Ritter, R.C. Kochel, and J.R. Miller, 1993, Fluvial responses to land-use changes and climatic variations within the Drury Creek watershed, southern Illinois. *Geomorphology* 6: 309–329.

Milne, W.S., 1987, A comparison of reconstructed lake-level records since the mid-1800's of some Great Basin lakes. M.S. Thesis, Golden, Colorado School of Mines, Department of Geology.

Muhs, D.R., and V.T. Holliday, 1995, Active dune sand on the Great Plains in the 19th century: Evidence from accounts of early explorers. *Quaternary Research* 43: 198–208.

Muhs, D.R., and P.B. Maat, 1993, The potential response of eolian sands to greenhouse warming and precipitation reduction on the Great Plains of the U.S.A. *Journal of Arid Environments* 25: 351–361.

Musgrave, G.W., and H.N. Holtan, 1964. Infiltration, in V.T. Chow (Ed.) *Handbook of Applied Hydrology*. New York, McGraw-Hill Book Company, Chapter 12, 30p.

O'Hara, S.L., F.A. Street-Perrott, and T.P. Burt, 1993, Accelerated soil erosion around a Mexican highland lake caused by prehispanic agriculture. *Nature* 362: 48–51.

Orme, A.R., and A.J. Orme, 1998, Greater California, in A.J. Conacher, and M. Sala, (Eds.) *Land Degradation in Mediterranean Environments of the World*. Chichester, John Wiley and Sons, Ltd.: 109–122, 254–255, 299–300, 331–333, 369–372, 389–390, 405–406, 428–433.

Osterkamp, W.R., M.M. Fenton, T.C. Gustavson, R.F. Hadley, V.T. Holliday, R.B. Morrison, and T.J. Toy, 1987, Great Plains, in W.L. Graf, (Ed.) *Geomorphic Systems of North America*. Boulder, Geological Society of America, Centennial Volume 2: 163–210.

Péwé, T.L., 1981. Desert dust: An overview, in T.L. Péwé, (Ed.), *Desert Dust: Origin, Characteristics, and Effect on Man,* Geological Society of America Special Paper, 186: 1–10.

Phillips, J.D., 1991, Fluvial sediment budgets in the North Carolina Piedmont. *Geomorphology* 4: 231–241.

Phillips, J.D., 1993, Pre- and post-colonial sediment sources and storage in the lower Neuse basin, North Carolina. *Physical Geography* 14: 272–284.

Phillips, J.D., 1997, A short history of a flat place: Three centuries of geomorphic change in the Croatan National Forest. *Annals of the Association of American Geographers* 87: 197–216.

Phillips, J.D., M. Wyrick, G. Robbins, and M. Flynn, 1993, Accelerated erosion on the North Carolina coastal plain. *Physical Geography* 14: 114–130.

Piest, R.F., C.E. Beer, and J. Spomer, 1976, Entrenchment of drainage systems in western Iowa and northwestern Missouri. Proceedings of the Third Federal Inter-Agency Sedimentation Conference, Denver, United States Water Resources Council, Section 5: 48–60.

Platts, W.S., W.F. Megahan, and G.W. Minshall, 1983, *Methods for Evaluating Stream, Riparian, and Biotic Conditions*. U.S. Department of Agriculture Forest Service General Technical Report INT-138, 70 p.

Rich, J.L., 1911, Recent stream trenching in the semi-arid portion of southwestern New Mexico, a result of removal of vegetation cover. *American Journal of Science* 32: 237–245.

Ruhe, R.V., 1969, *Quaternary Landscapes in Iowa*. Ames, Iowa State University Press, 255 p.

Sartz, R.S., and D.N. Tolsted, 1974, Effect of grazing on runoff from two small watersheds in southwestern Wisconsin. *Water Resources Research* 10: 354–356.

Schumm, S.A., 1977, *The Fluvial System*. New York, John Wiley and Sons, 338 p.

Schumm, S.A., and R.W. Lichty, 1963, Channel Widening and Floodplain Construction Along Cimarron River in Southwestern Kansas. U.S. Geological Survey Professional Paper 352-D: 71–88.

Slaymaker, H.O., 1972, Sediment yield and sediment control in the Canadian Cordillera, in H.O. Slaymaker, and H.J. McPherson, (Eds.), *Mountain Geomorphology*, British Columbia Geographical Series 14: 235–245.

Slaymaker, H.O., 1987, Sediment and solute yields in British Columbia and Yukon: Their geomorphic significance reexamined, in V. Gardiner, (Ed.), *International Geomorphology 1986 Part I*, Chichester, John Wiley and Sons Ltd., 925–945.

Smith, G.I., L. Benson, and D.R. Currey, 1989, *Quaternary Geology of the Great Basin*. Washington, American

Geophysical Union, 28th International Geological Congress Field Trip Guidebook T117, 78 p.

Sparrow, H.O., 1984, *Soils at Risk: Canada's Eroding Future.* A Report on Soil Conservation by the Standing Committee on Agriculture, Fisheries, and Forestry to the Senate of Canada, Ottawa.

Spomer, R.G., R.L. Mahurin, and R.F. Piest, 1986. Erosion, deposition, and sediment yield from Dry Creek basin, Nebraska. *Transactions of the American Society of Agricultural Engineers*, 29, 489–493.

Starkel, L., 1987, Man as a cause of sedimentological changes in the Holocene: Anthropogenic sedimentological changes during the Holocene. *Striae* 26: 5–12.

Stetler, L.D., and K.E. Saxton, 1996, Wind erosion and PM(10)emissions from agricultural fields on the Columbia Plateau. *Earth Surface Processes and Landforms* 21: 673–685.

Storey, H.C., R.L. Hobba, and J.M. Rosa, 1964, Hydrology of forest lands and rangelands, in V.T. Chow, (ed.), *Handbook of Applied Hydrology*, New York, McGraw-Hill Book Company, 22-1–22-52.

Street-Perrott, F.A., R.A. Perrott, and D.D. Harkness, 1989, Anthropogenic soil erosion around Lake Pátzcuaro, Michoacán, Mexico, during the Preclassic and Late Postclassic-Hispanic Periods. *American Antiquity* 54: 759–765.

Thornthwaite, C.W., C.F.S. Sharpe, and E.F. Dosch, 1942, *Climate and Accelerated Erosion in the Arid and Semi-Arid Southwest, With Special Reference to the Polacca Wash Drainage Basin, Arizona.* Washington, U.S. Department of Agriculture, Technical Bulletin 808, 134 p.

Trimble, S.W., 1974, *Man-Induced Soil Erosion on the Southern Piedmont 1700–1970.* Soil Conservation Society of America, 180 p.

Trimble, S.W., 1975, A volumetric estimate of man-induced soil erosion on the southern Piedmont. U.S. Department of Agriculture, Agricultural Research Service Publication S-40:142–154.

Trimble, S.W., 1983, A sediment budget for Coon Creek basin in the Driftless Area, Wisconsin, 1853–1977. *American Journal of Science* 283: 454–474.

Trimble, S.W., 1985, Perspectives on the history of soil erosion control in the eastern United States. *Agricultural History* 59: 162–180.

Trimble, S.W., 1993, The distributed sediment budget model and watershed management in the Paleozoic Plateau of the upper Midwestern United States. *Physical Geography* 14: 285–303.

Trimble, S.W., and S.W. Lund, 1982, Soil conservation and the reduction of erosion and sedimentation in the Coon Creek basin, Wisconsin. U.S. Geological Survey Professional Paper 1234: 1–35.

Trowbridge, A.C., and E.W. Shaw, 1916, *Geology and Geography of the Galena and Elizabeth Quadrangles.* Illinois State Geological Survey Bulletin No. 26, 233 p.

Tuan, Yi-Fu, 1966, New Mexican gullies: A critical review and some recent observations. *Annals of the Association of American Geographers* 56: 573–597.

U.S. Department of Agriculture, 1994, Web Site: http://www.nhq.nres.usda.gov/NRI/tables/1992/table.9.html and http://www.nhq.nres.usda.gov/NRI/tables/1992/table.10.html.

Webb, R.H., S.S. Smith, and V.A.S. McCord, 1992, Arroyo cutting at Kanab Creek, Kanab, Utah, in J. Martinez-Goytre, and W.M. Phillips, (Eds.), *Paleoflood Hydrology of the Southern Colorado Plateau*, University of Arizona, Tucson, ALPHA Contribution Number 12: 17–38.

Webb, R.S., and T. Webb III, 1988, Rates of sediment accumulation in pollen cores from small lakes and mires of eastern North America. *Quaternary Research* 30: 284–297.

Wheaton, E.E., 1984a, *Climatic Change Impacts on Wind Erosion in Saskatchewan, Canada.* Saskatchewan Research Council, Publication E-906-16-B-84, Technical Report 153, 27 p.

Wheaton, E.E., 1984b, *Wind erosion—Impacts, causes, models and controls: A literature review and bibliography.* Saskatchewan Research Council, Publication E-906-48-E-84, 19 p.

Whitmore, T.M., and B.L. Turner, II, 1992, Landscapes of cultivation in Mesoamerica on the eve of the conquest. *Annals of the Association of American Geographers* 82: 402–425.

Williams, M., 1989, *Americans and Their Forests: A Historical Geography.* Cambridge, Cambridge University Press, 599 p.

Williams, G.P., and M.G. Wolman, 1984, *Downstream Effects of Dams on Alluvial Rivers.* U. S. Geological Survey Professional Paper 1286.

Wolfe, S.A., D.J. Huntley, and J. Ollerhead, 1995. Recent and late Holocene sand dune activity in southwestern Saskatchewan. *Current Research 1995-B*, Geological Survey of Canada, 131–140.

Wolman, M.G., 1967, A cycle of sedimentation and erosion in urban river channels. *Geografiska Annaler* 49a: 385–395.

Zeasman, O.R., 1963, First large structure—Vollmer Dam. Wisconsin State Soil and Water Conservation Committee, *A Brief History of Soil Erosion Control in Wisconsin*, 12–13.

24

Urbanization and Environment

Jon Harbor
Budhendra Bhaduri
Matt Grove
Martha Herzog
Shankar Jaganayapathy
Marie Minner
John Teufert

The US model of metropolitan growth is driven by a loosely regulated real estate market and conspicuous consumption of land, building materials, and privately owned vehicles. It is premised on the seemingly unlimited availability of capital, energy resources and land.

(Angotti, 1995, p. 635).

North America is the only continent where the majority of the population lives in metropolitan areas. In the United States, 87% of the population lives in metropolitan areas and their hinterlands (Golden, 1981; Angotti, 1995; fig. 24.1). Infilling between cities has resulted in the development of major megalopolises, such as the Philadelphia, Boston, Washington, D.C., New York urban corridor. This pattern of urbanization has had major environmental impacts, both within urban areas themselves and throughout much larger zones of influence. Urban development has altered local and regional climates, and has degraded surface and groundwater quantity and quality, and the ecosystems. Although there have been severe impacts at a range of scales, management initiatives in recent decades have reduced or curbed current impacts through regulation and voluntary measures. There has also been growing emphasis on reducing future impacts through environmentally sensitive urban planning.

24.1 Urban Growth in North America

The extent and range of environmental impacts due to urbanization in North America are closely related to the continent's unique history of urban growth. Modern North American urban centers began to develop after European settlement and the emergence of colonial economic systems. Early nineteenth-century North American cities operated mainly as commercial centers that supplied primary products to markets outside the continent, and received finished goods, capital, and labor in return. However, by the mid-1800s, national urban systems emerged, domi-

A

B

C

Figure 24.1 North American cities Montreal (A) and New York City (B) include high rise cores surrounded by extensive and expanding suburban areas, and industrial complexes (C).

nated by major cities such as New York, Vancouver, and Mexico City, replacing the previous structure of loosely connected regional economies (Richardson, 1982). Cities and towns became the centers of government, society, and economy, and by 1920 over half of the population of the United States was living in designated urban areas (Johnston, 1982).

Almost every major North American city had been founded by 1900. Since then, almost all new urban developments have been satellites of existing cities (new areas growing on the outer edges of the city or beyond the city limits.) During the 1900s, cities grew in population and in area; their occupational and economic structures became increasingly diversified, and land use became more specialized (Richardson, 1982). The basic pattern of twentieth-century urban growth in North America was characterized by decentralization, suburbanization, and sprawl(Johnston, 1982). This was exemplified in the United States where the population increasingly became urban, but its urban areas rapidly decentralized (Angotti, 1995). Most metropolitan areas in North America are still growing in population, although there are some exceptions in the northern United States. However, even metropolitan areas that are stagnating or declining in population are still increasing their total built area through low-density suburban growth (sprawl) (Johnston, 1982).

By 1920, suburbanization was well established in most North American cities because developments in technology had greatly improved the mobility of people and goods, and improved communications had decreased the need for people to be clustered in high-density central areas (Chinitz, 1991). Suburban life offered more spacious living, lower crime rates, and better schools, which outweighed the inconvenience and cost of commuting to the workplace (Richardson, 1982). Through the course of the twentieth century, money, talent, and jobs moved from the city to the suburbs, and suburbs became increasingly independent. Suburbs became less dependent on the central city, and are now typically able to finance their own services (Johnston, 1982).

Suburbanization of employment had become so prevalent by the 1980s that the volume of intrasuburb commuting greatly exceeded city-to-suburb commuting, and by 1986, 57% of office space in the United States was located outside urban downtown areas (Chinitz, 1991). A critical factor in suburbanization for much of North America was the automobile (fig. 24.2). Current sprawling land-use patterns could not have developed without the personal mobility provided by low-cost automobiles. In turn, now that these land-use patterns have developed, they make development of effective mass transit systems difficult and render the automobile a necessity of metropolitan life in much

A

B

Figure 24.2 Heavy reliance on car transportation in many North American cities is a major contributor to air pollution, to frustrating traffic jams such as this in Mexico City (A), and to significant increases in impervious area coverage (B).

of North America. In the 1980s, the automobile population in the United States increased three times faster than the human population, and the commuting population grew twice as fast as the average population growth (Chinitz, 1991). In addition to its role in enabling land-use change, the automobile is also an important element of a number of urban environmental impacts, including air pollution and surface water quality.

In summary, North American urban areas have grown, have become more diversified, and have decentralized through suburbanization and exurban growth. Compared to central cities, the density of population within suburban areas has been too low to reduce the dependency on automobiles, despite growing suburban employment (Chinitz, 1991). The urban population in North America increased from 154 million in 1965 to 223 million in 1995, and is predicted to rise to 313 million by 2025 (World Resources, 1996).

24.2 Urbanization and the Atmosphere

Urban centers have a major impact on climate and air quality over a range of spatial scales (fig. 24.3). For example, in addition to direct impacts within the city, airborne sulfur produced by Chicago's industries has damaged forests in northern Canada, 2000 km away, while less than 40 km away, Gary, Indiana's residents are exposed to low air quality from Chicago's steel mills (USDA Forest Service, 1992). Greenhouse gases released by activities concentrated in urban areas contribute to global warming and the growth of the ozone hole in the upper atmosphere, whereas the surplus of ozone in the lower atmosphere harms the respiratory systems of urban residents. Even though several successful approaches have been used during the last few decades to reduce air pollution, the issue remains a major concern in urban areas.

24.2.1 Air Quality

Health problems related to air pollution have been a major concern in North American cities since the early nineteenth century, when the major source of atmospheric pollution was the coal that powered the industrial revolution (Petulla, 1988). Despite regulatory programs, such as those established in the United States under the Clean Air Act (1963, with major amendments in 1970), air quality continues to be a major and costly problem throughout North America. For example, the majority of Canadians are exposed regularly each summer to ozone levels above regulatory "acceptable" levels, and in Mexico City ozone levels in 1991 exceeded World Health Organization guidelines on 192 days (Elsom, 1996). Major pollutants of concern include ozone (O_3), particulates, sulfur dioxide (SO_2), nitrogen oxide (NO_2), and carbon monoxide and dioxide (CO, CO_2).

Ozone In the lower atmosphere, ozone is produced by a photochemical reaction between nitrogen dioxides and volatile organic compounds (VOCs) in the presence of ultraviolet radiation. Motor vehicle emissions are a major source of VOCs in all urban areas, although in Mexico City the primary source of ozone pollution is liquefied petroleum gas used for heating (Pick and Butler, 1997). Health impacts related to high ozone levels include headaches, chest pains, nausea, shortness of breath, and significant increase in deaths from asthmatic attacks. Ozone levels in major North American cities continue to be a major concern, with Mexico City, Los Angeles, and New York City leading the world in ambient ozone concentrations (World Resources, 1996). Ozone levels in the Los Angeles basin

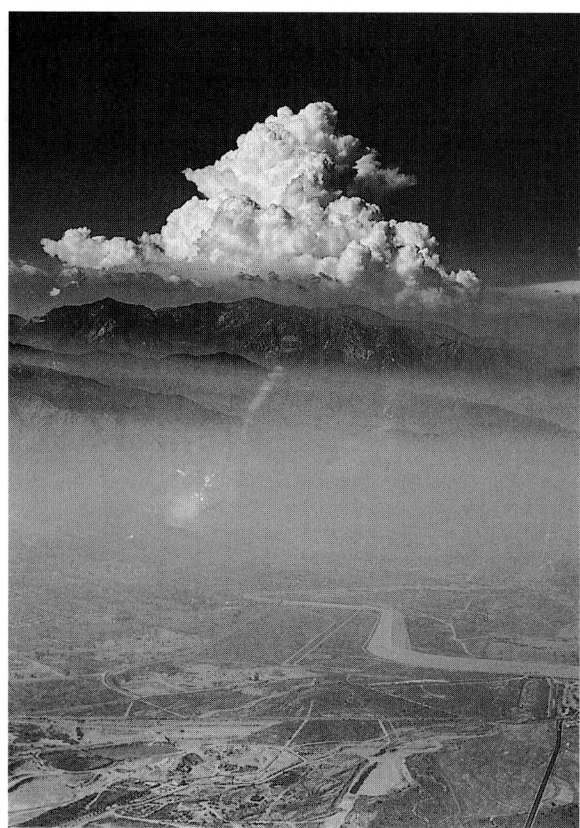

Figure 24.3 Air pollution in the Los Angeles basin, southern California. With onshore winds, smog is driven inland and trapped against the San Gabriel Mountains (photo: Spence Collection, UCLA).

exceed regulatory acceptable levels at least 100 days per year, leading to health advisories suggesting that asthmatic and elderly residents stay indoors (Elsom, 1996). The primary management approach to combat high ozone levels is to try to reduce automobile emissions by encouraging car pooling and the use of other forms of transportation. However, as few North American cities have well-developed mass transit systems and development patterns reflect easy access to individual car transport, reducing car use is difficult in practice. Efforts to introduce alternate fuels and electric cars are under way, with regulations and economic incentives in place in an attempt to offset the higher costs of these alternatives to consumers.

Particulates Atmospheric particulates are generated by soil erosion by wind, automobile emissions, and industrial processes, including fossil fuel combustion. Particulates can be transported over considerable distances and may include heavy metals and other chemical pollutants. Particulates can act as condensation seeds, contributing to hazardous smog conditions. When particulates reach the ground as wet or dry deposition, they become a source for potential surface and groundwater pollution. As health hazards, particulates smaller than 0.5 μm reach and settle in the alveoli of the lungs, causing health problems such as asthma and cancer (Stroker and Seger, 1976). Efforts to reduce particulates have included requiring more stringent erosion control in parts of North America, increasing use of daily cover on landfill sites, and more strictly regulating emissions from industrial facilities and power plants using fossil fuels. As a result, atmospheric particulate loads in the United States decreased by 18% in the 1980s (World Resources, 1996).

Sulfur dioxide Sulfur produced by emissions from fuel combustion, particularly coal, combines with oxygen to produce sulfur dioxide (SO_2), which has direct effects on human health and contributes to acid rain. Health impacts linked to SO_2 include bronchial constriction, and increased respiratory and heart problems. When it reacts with water in the atmosphere, SO_2 contributes to acid rain, which can be deposited more than 2000 km away from its source, causing environmental damage such as increased lake, stream, and soil acidity, and increased tree stress and mortality. Increasingly strict emissions regulations encourage the use of low-sulfur coal, fuel oil and diesel, and the development of advance "scrubber" systems in exhaust stacks. As a result, atmospheric SO_2 loads in the United States and Canada decreased by 14% in the 1980s (World Resources, 1996).

Greenhouse gases Many greenhouse gases, such as CO_2, O_3, NO_X, and methane, are released from urban areas. The dominant greenhouse gas, CO_2, is produced by the combustion of fossil fuels that contribute approximately 5 billion metric tonnes of CO_2 to the atmosphere annually and is linked to global warming (Elsom, 1996; Idso et al., 1998). Chicago alone releases approximately 1.6 million metric tonnes of CO_2 annually (USDA Forest Service, 1992). Although reduction of CO_2 emissions by more efficient use of fossil fuels and conversion to alternative energy sources are the goals of national regulatory programs and international agreements, tree planting programs are also used to combat high CO_2 levels in urban areas. Trees act as carbon sinks, but also reduce demand for fossil-fuel-based energy through their role in local climate regulation by reducing heating and cooling needs in urban areas (McPherson et al., 1994).

Management Many air quality indicators have shown steady improvement during recent decades, despite urban growth, partly because of regulatory measures, including the U.S. Clean Air Act of 1963 (and amendments) and the Canadian Clean Air Act of 1969. For example, in Indiana over the period 1988–97, CO levels fell by 29%, soot and dust were reduced by 35%, SO_2 emissions were reduced by 60%, and the number of days with poor ozone levels fell from 52 (1988) to 10 (1997) (IDEM, 1998). However, for cities throughout North America, O_3, particulates, and SO_2 reduce air quality, causing significant health impacts. During the 1990s, the Environmental Protection Agency (EPA) recommended progressively stricter controls on particulates and O_3 levels in urban areas, particularly on automobile use. Strategies and technologies that decrease automobile emissions and efforts to reduce dependency on automobiles are major elements of the strategy to improve urban air quality.

24.2.2 Urban Heat Islands

Air temperatures in urban areas are typically warmer than surrounding rural areas, an effect known as the urban heat island (fig. 24.4). The average annual temperature in Chicago is 1.85°C higher than its surrounding rural areas, as a result of increased heat storage of buildings and pavement in the city (Cutler, 1976), with daily differences as high as 15°C. At sunset, urban areas cool more rapidly than surrounding rural areas because large, dry, artificial surfaces release more energy than rural ones. In Chicago, these differences in cooling rates produce maximum temperature differences between urban and rural areas approximately 3 to 5 hours after sunset. After sunrise, however, urban areas warm up more slowly than surrounding rural areas. If weather conditions are relatively calm, a rural-urban breeze system develops, producing a typical urban heat island (Oke, 1997). The typical heat island is separated from surrounding areas by a temperature cliff, which follows the city's perimeter, and includes a plateau of higher temperatures and a peak of maximum temperatures located above the city center (fig. 24.4). Local cool islands can occur within the heat island, for example, where there are parks and other open spaces.

Overheating associated with urban heat islands causes an increase in energy consumption because of the use of air-conditioning, costing $1 billion annually in the United States (USDA Forest Service, 1992). In addition to thermal impacts, elevated urban temperatures are also linked to air quality problems through the creation of a heat inversion dome that traps atmospheric pollutants and increases the frequency of smoggy days.

24.3 Urbanization and Surface Water

Water reaches the land surface in the form of rain, sleet, hail, and snow, then is returned to the atmosphere, runs off the surface, or infiltrates the soil (fig. 24.5). Approximately 50% of precipitation on forested land is returned back to the atmosphere through evaporation from the surface and transpiration by plants (Hough, 1995). The remaining water either goes into groundwater storage after infiltrating the soil or flows over the surface as runoff, which ends up in water bodies such as ponds, lakes, streams, rivers, and oceans. The amount of runoff that occurs during or immediately after a storm event (stormwater runoff) is primarily determined by surface infiltration properties, slope, and vegetation. As other landscapes are transformed into urban areas, a new hydrologic environment is created that changes both the quantity and quality of runoff and other components of the hydrologic cycle.

Impervious surfaces common in urban areas (fig. 24.2) decrease infiltration capacity, leading to flooding and loss of groundwater recharge. Moreover, urban activities gen-

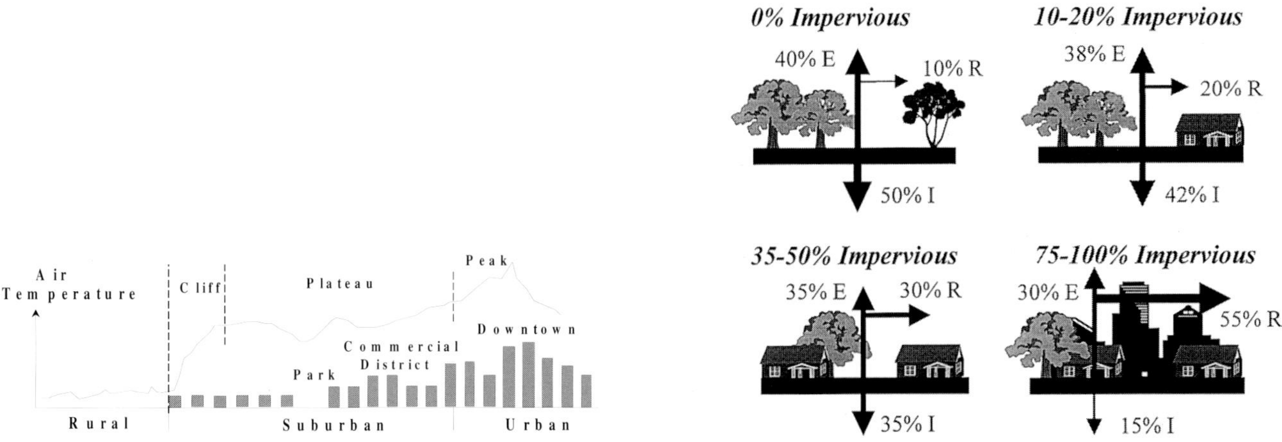

Figure 24.4 (*left*) A generalized cross section of an urban heat island (modified after Oke, 1987).

Figure 24.5 (*right*) Modification of the hydrologic balance by urbanization (based on Hough, 1995). E: Evapotranspiration, R: Runoff, I: Infiltration.

erate numerous pollutants that are transported through surface runoff to lakes, streams, and rivers, causing degraded water quality. In addition to concerns about water supply, water management in urban areas has focused on controlling peak discharges that cause flooding and on controlling the impacts of point sources of pollution on water quality. This has involved local flood control ordinances coupled with national regulations aimed at reducing discharges of sanitary and industrial waste. However, continued flooding and water quality problems have led to increasingly stringent flood control regulations (the requirement that peak flows from new developments must be less than or equal to predevelopment flows) and an emphasis on reducing nonpoint source pollution inputs from urban areas. In the planning field, "percent impervious area" has recently been identified as a simple yet practical indicator of environmental impact associated with urbanization (Schueler, 1994).

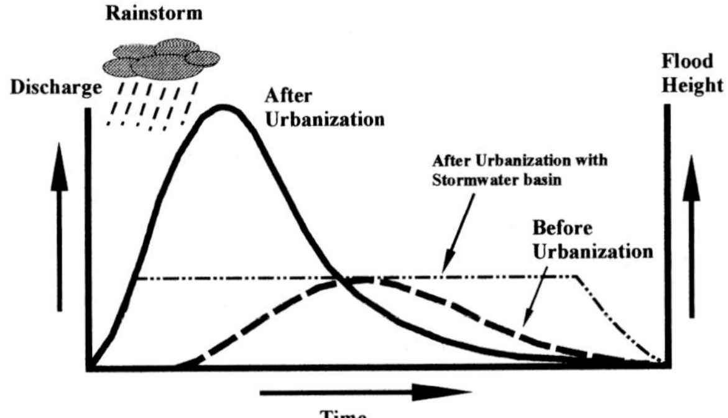

Figure 24.6 Modification of a flood hydrograph as a result of urbanization. The postdevelopment hydrograph rises to a peak faster, has a higher peak discharge, and has a greater total runoff volume than a predevelopment hydrograph. Stormwater management basins act to reduce the peak flow to predevelopment levels, but do not typically reduce the total volume of runoff (modified from Harbor, 1994).

24.3.1 Imperviousness and Flood Hazard

As vegetation is cleared and the land is covered with impervious surface materials, less water can infiltrate the ground (loss of groundwater recharge) or return to the atmosphere through evapotranspiration. This leaves more water available for runoff. Up to 85% of precipitation can become stormwater runoff in completely paved and roofed urban areas (Hough, 1995; fig. 24.5). Drainage systems and modified channels (ditched, straightened, and piped), designed to expedite disposal of excess runoff in urban areas, typically cause flooding by delivering more runoff at a faster rate to the receiving lakes, ponds, wetlands, and streams (fig. 24.6). Thus urbanization leads to increases both in the volume of surface or stormwater runoff and in peak discharges, leading to local and downstream flooding. For example, in the early 1980s, rainfall monitoring indicated increases in flood magnitude and frequency around Chicago over the last 50 years, during which there was significant urbanization in the area (Changnon, 1984, cited in Walesh, 1989). However, small-scale urbanization does not always cause increases in peak discharges. For instance, in a study of suburbanization in Ohio, McClintock et al. (1995) showed that transformation of some forms of agricultural land to low-density residential areas actually reduced peak discharge because of an increase in infiltration rates.

Flood hazards can also be magnified by a number of other factors that are directly influenced by urbanization. Precipitation volumes from thunderstorms have increased with large-scale urbanization, which in turn increases runoff volumes (Marsh and Grossa, 1996). The addition of particulate matter to the atmosphere by urban activities provides increased formation of raindrop nuclei, and the heating and mechanical mixing of air due to urban structures increases rising air and moisture, causing cooling and condensation. All these factors contribute to increased rainfall and increased frequency and magnitude of floods. For example, rainfall records for Chicago over the last 50 years indicate a 76% increase in the frequency of 2.54-centimeters (1-inch) storm events in the city compared to surrounding nonurban areas (Walesh, 1989).

24.3.2 Long-Term Hydrologic Impact of Imperviousness

The traditional focus in urban surface-water management has been the control of peak discharges from individual, high magnitude storm events that cause flooding. However, scientists now realize that there is a long-term hydrologic impact associated with urbanization, which is dominated by runoff generated from smaller storm events. A study of the relationship between imperviousness and runoff from over 40 runoff monitoring sites in the United States showed that the total runoff volume for a 0.4-hectare (ha) parking lot was about 16 times more than that of an undeveloped meadow (Schueler, 1994). Similarly, conversion of a watershed in Hudson, Ohio, from a predevelopment agricultural to a high-density residential condition resulted in a 26% increase in runoff volume for individual high-magnitude, low-frequency storms (McClintock et al., 1995). However, a hydrologic analysis based on the long-term climatic record predicts a 195% increase in average annual runoff volume, which translates to a loss of groundwater recharge and potential severe impacts on residential and municipal water supplies if groundwater is not adequately replenished (McClintock et al., 1995). The health of wetlands, which maintain their wet condition from ground and soil water during dry periods of the year, can potentially be threatened from a loss of groundwater recharge.

24.3.3 Management of Hydrologic Impacts

Efforts to manage surface water quantity problems are implemented through a number of structural, vegetative, or managerial approaches known as best management practices (BMPs). Flood control structures such as stormwater detention and retention basins are common BMPs designed to hold temporary excess runoff generated as a result of urbanization and to release this water at a discharge rate typical of the pre-urbanized condition (fig. 24.6; Harbor, 1994). These structures are also sometimes modified to act as groundwater recharge basins while also temporarily holding runoff to reduce flood hazard. In most urban areas in the United States, flood control structures are required to meet strict functional requirements. For example, the city of West Lafayette, Indiana, requires stormwater management basins to be built in new developed areas, with design criteria that limit postdevelopment discharge from a 100-year recurrence interval storm to the predevelopment discharge rate from a 10-year recurrence interval storm.

24.3.4 Erosion, Sedimentation, and Urbanization

Increased runoff volumes and peak discharges can have a significant impact on stream and river channels, primarily through increased erosion. In addition to changes in runoff, as land-use practices modify "natural" conditions, soil erosion is often accelerated (fig. 24.7). This modification of the landscape includes removal of vegetation and topsoil (for logging and construction), altering drainage conditions (by diverting rivers and creating reservoirs), and construction (of buildings, roads, and pipelines) (Cooke and Dornkamp, 1990). Topography is also often altered with urbanization as hills are leveled or reduced to provide level building sites, and valleys, gullies, and swales are filled (Craul, 1992; fig. 22.4). Soil erosion rates, which are related to slope steepness and length, amount of vegetation, and erosivity and erodibility, are enhanced (Cooke and Dornkamp, 1990). Topographic modification can also result in slope failure, and landslides and slumps are a frequent sight along poorly designed road cuts.

The construction phase of land modification produces high rates of soil erosion because a large amount of land is exposed and disturbed by excavation and vehicular movement (fig. 24.7). In Virginia, for example, erosion rates during construction are 10 times the rates from agricultural land, 200 times the rates from grassland, and 2000 times the rates from forestland (Goudie, 1990). Most of the displaced sediment eventually reaches waterways, where sedimentation can alter the physical and biological characteristics of channels and reservoirs by increasing sediment concentration and deposition (Wolman and Schick, 1967; fig. 24.7). For example, the Newhalem Creek in Washington has become a poor habitat for salmon spawning because of change in the size of its substrate (Morris, 1992). Excess sedimentation can result in flooding (Dunne and Leopold, 1978) and reduces the capacity and life of reservoirs (Goudie, 1990, see chapter 22).

24.3.5 Urbanization and Surface Water Quality

Percent impervious area is an approximate measure of the extent of various human activities in urban areas that generate, expose, or make available at or near the land surface various types of potential pollutants (Walesh, 1989). Impervious area as a measure of urbanization has also been linked to ecological impacts, for example, in the Maryland Piedmont, where a strong correlation was found between percent watershed imperviousness and diversity of fish species (Schueler, 1994). Larger percent impervious areas are crudely correlated with both increased point sources of water pollution (sewage treatment facilities, industrial facilities) and increased nonpoint source pollution (road and parking lot runoff, construction sites, and herbicides, pesticides and fertilizers). Increasing imperviousness in urban areas not only removes natural filtration systems, such as soil and vegetative cover, but also ensures a better efficiency of flushing surface pollutants. Increased velocity and flow rate of runoff resulting from urbanization facilitates the mobilization and transport of pollutants to receiving waters. The increase in urban stormwater pollution also mirrors the growth of automobile ownership in North America. Automobiles release a wide range of pollutants to paved roads, parking lots, driveways, and garages, and these pollutants directly contribute to runoff.

Important urban pollutants include the following:

- Oxygen-demanding wastes that use up oxygen essential for aquatic life
- Nutrients (nitrogen and phosphorus) that cause harmful accelerated growth of aquatic plants, including algae blooms (eutrophication)
- Inorganic compounds, including road salt from deicing, heavy metals (e.g., copper, lead, zinc, cadmium, chromium, and mercury), and organic compounds that can have toxic or even lethal effects on organisms
- Sediment eroded from construction and demolition activities at significantly higher rates than the normal erosion rate increase the turbidity of surface water, quickly choke lakes and streams, thus impairing their normal functions that support aquatic life (Marsh and Grossa, 1996).
- Pathologic pollutants such as bacteria, viruses, and other disease-causing microorganisms associated with human and animal wastes, which severely hinder recreational usage of lakes and streams. Pathogens such as *Giardia* and *Cryptosporidium* have become particularly problematic.

A

B

C

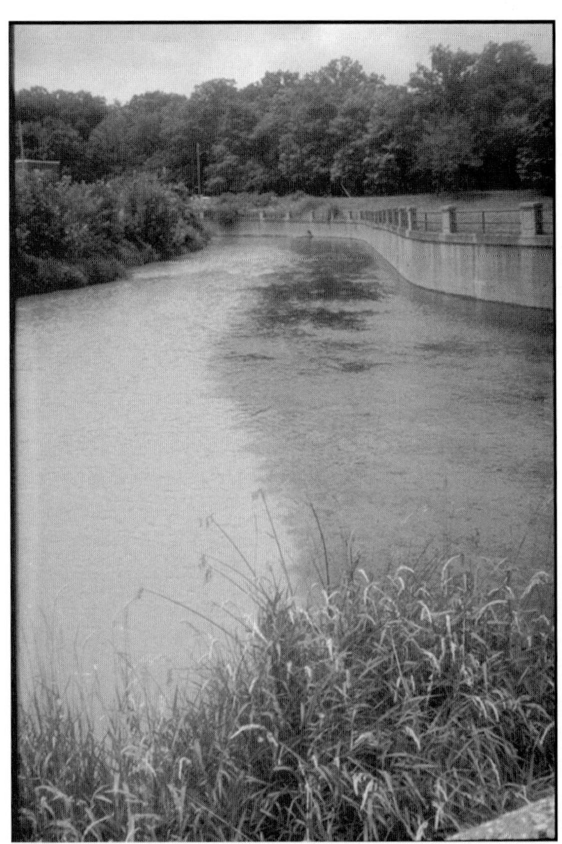

Figure 24.7 Soil erosion on construction sites in urbanizing areas (A and B) is a leading cause of nonpoint source pollution. Downstream impacts can be seen in C at the confluence of a main stem river (right) with a tributary that has a high sediment load from nonpoint source pollution, including construction activity.

24.3.6 Sources of Water Pollution in Urban Areas

Water pollution in urban areas occurs from both point and nonpoint sources. Point sources of pollution occur from identifiable point outlets, such as pipes, drains, drums, and tanks, enter the waterways at discrete, identifiable locations, and can usually be measured. Sewered municipal and industrial wastewater and effluents from solid waste disposal sites are major point sources of pollution. Nonpoint sources (NPS) of pollution, on the other hand, are more difficult to identify and measure because nonpoint source pollution originates from diffuse sources spread over a larger areal extent, including road-salt runoff from roads, sediment-laden runoff from construction sites, and wet and dry atmospheric deposition.

Point sources of pollution Point sources, in most cases, have continuous, uninterrupted discharges and enter the receiving water bodies at identifiable points. The flow and amount of pollutants from these sources do not vary with rainfall, except in the case of combined sewer systems. About 66% of point sources of water pollution in the United States come from urban areas (Marsh and Grossa, 1996). In the United States, under the Clean Water Act (1972), point sources are regulated, their control is mandated, and a permit is required for the waste discharges. The Clean Water Act helped to establish the National Pollutant Discharge Elimination System (NPDES), which enforces the implementation of pollution abatements of point sources such as industrial and municipal wastewater, and discharges from storm sewers. Generally, point-source pollution has decreased significantly in response to regulation. For example, sediment records from two depositional subbasins within Lake Ontario show that heavy metals (mercury, lead, chromium, and copper) and organic pollutant accumulation rates were at their peak between 1968 and 1970 and that accumulation rates have been decreasing steadily since (Long, 1996).

Many North American urban areas are served by combined sewer systems that transport both stormwater and municipal wastewater sewage. During wet weather conditions or after heavy rainfall events, the large volume of combined stormwater and wastewater cannot be handled by wastewater treatment plants and has to be released to waterways, causing a significant amount of water pollution. For example, approximately 800 million liters of untreated sewage were released to the streams and rivers in

North Carolina after hurricane Fran hit the area in September 1996 (fig. 21.8). In presently urbanizing areas, separate sewer systems are used for stormwater and sanitary sewage to remove the problem of combined sewer overflows (CSO). But pollution from combined sewers is a major concern in older cities, where the existing combined sewer systems have become inadequate to support the increasing growth of the city and its increased wastewater and stormwater volumes. Cities such as Cleveland, Ohio, and St. Paul, Minnesota, are switching from combined to separate sewer systems to control water pollution from CSOs.

Another advanced approach to reducing pollution associated with urbanization is the increasing use of constructed wetland systems that effectively combine physical, chemical, and biological processes to remove both particulate and dissolved pollutants from runoff. In urban areas, constructed wetlands and biofiltration ponds (with resource-efficient plants that reduce nutrient and pesticide loads from runoff) are used to simulate the function and the effectiveness of natural wetlands in reducing water pollution.

In Mexico City, all of the wastewater treatment plants use at least secondary treatment (Pick and Butler, 1997), but most wastewater is not treated at all. The primary approach to disposing of wastewater in Mexico City is through an open sewage canal, which transports untreated residential, commercial, and industrial wastewater to a neighboring agricultural area where it is used to fertilize and irrigate over 65,000 ha of farmland.

Nonpoint sources of pollution Pollution from nonpoint sources occurs mainly as a response to storm events and thus is usually intermittent in nature. Consequently, the flow and the amount of nonpoint source pollution vary significantly. Soon after the implementation of the NPDES system in the United States, it was realized that point source pollution control could not eliminate all water quality problems and that nonpoint sources were a major contributor to degraded water quality (Marsh and Grossa, 1996). The Nationwide Urban Runoff Program (EPA, 1984) indicated that in urban areas stormwater runoff is the most important nonpoint source of pollution (Novotny and Olem, 1994). In response to this, the Clean Water Act (1972) was amended to expand NPDES to include stormwater runoff from urban and urbanizing areas. This expansion has involved treatment of stormwater runoff to reduce sediment and associated chemical pollutants, as well as toxicity caused by dissolved chemical pollutants (Scholze et al., 1993); it has also increased the emphasis on source reduction, including better control of erosion on construction sites (fig. 24.8). Another popular approach is to integrate water quantity and quality control approaches by modifying infiltration and flood control structures such as detention and retention basins to enhance sedimentation and removal of sediment-associated chemical pollutants. The bulk of the pollutant loads is typically carried in the first few centimeters of rainfall and commonly referred to as the "first flush." Effective treatment of first flush can significantly improve runoff quality.

A

B

C

Figure 24.8 New regulations aimed at decreasing erosion impacts during construction in urban areas has led to increased efforts at erosion control. Examples include use of concrete and gabion linings to prevent channel erosion (A), use of sediment control dams (B), and widespread use of silt fence (C) (failing at one end in this example).

24.4 Urbanization and Groundwater

Nearly 75% of the major cities in the United States depend heavily on groundwater for their water supply, about 50% of drinking water in the United States comes from groundwater supplies, and consumptive uses of groundwater have been increasing at a rate of 3–4% annually (Hayes, 1984). North America experienced a threefold increase in the volume of water withdrawn for urban needs during the nineteenth century. By the second half of the twentieth century, increased demand tied to accelerated rates of population growth in urban areas accompanied by technological advances in extracting and supplying groundwater, promoted a rapid rise in the amount of water used and distributed. On a global scale, during the past 300 years, the urban use of groundwater has increased almost 40-fold (L'Vovich and White, 1990).

24.4.1 Links Between Groundwater and Surface Water

With urbanization, impermeable roofs, paved roads, and sidewalks replace permeable soil areas, resulting in increased runoff and reduced groundwater recharge (fig. 24.6). To compensate for this loss of recharge, recharge wells that route runoff directly down to aquifers and recharge basins that trap runoff and provide time for infiltration into groundwater have been constructed. Although recharge basins and wells are beneficial in restoring potentially lost recharge to the underlying aquifers, they can be sources of contamination. Runoff diverted into recharge basins in Nassau and Suffolk counties, close to the City of New York, contained contaminants such as fertilizers, pesticides, deicing salts, organic debris, grease, road oil, rubber, asphaltic materials, hydrocarbons, animal feces, and food wastes (Koppelman, 1987). Many of these contaminants are not biodegradable and may continue to persist in groundwater at levels dangerous to health for long periods of time. Similarly, private wells, although not unique to urbanization, can be problematic in urban areas where high population densities tap into shallow aquifers. If not installed properly, these wells may serve as conduits for flow and seepage of pollutants from surface water into groundwater.

24.4.2 Urbanization and Groundwater Quantity

Effects of excessive withdrawal include declining groundwater levels and in some cases declining surface water levels and dessication of wetlands. In coastal areas, overpumping may induce landward movement of the fresh–saltwater interface, causing saltwater intrusion of coastal freshwater wells. Saltwater intrusion has occurred in each of the 21 coastal states in the United States (Atkinson et al., 1986). This problem is primarily associated with high water demand from large population centers rather than agricultural activities, and it is clear that saltwater intrusion is a potential threat when areas proximal to the sea are developed (Atkinson et al., 1986).

Excessive withdrawals of groundwater can also cause ground subsidence, which has resulted in damage to building structures, highways, pipelines, and tunnels (Fetter, 1988). There are numerous cases of groundwater mining and associated land subsidence, notably in Houston, Baton Rouge, and Phoenix (Keller, 1996). Many parts of Mexico City have experienced extensive subsidence (as much as 7 m between 1880 and 1970) as groundwater has been extracted from sand and gravel aquifers in the ancient lake basin beneath the city (Pick and Butler, 1997). Subsidence of up to 1.1 m has been reported from Tucson, and 0.2 m from Las Vegas (Costa and Backer, 1981). Subsidence of 3 to 3.5 m in the greater Houston area had occurred by the year 2000, resulting in nearly 80 km^2 of land lost to permanent saltwater inundation.

24.4.3 Urbanization and Groundwater Quality

A popular misconception is that groundwater is secure from pollution and degradation because of its subsurface location and the filtering effect of overlying soil and geological materials. This assumption is valid under certain circumstances, but natural protective mechanisms are not always effective against the combined assaults that have accompanied urbanization and industrial expansion. In North America, improper disposal of industrial and municipal wastes, such as pesticides and road salt, leaking underground storage tanks (petroleum and chemical), and fertilizers, have polluted and degraded groundwater. Industries in the United States make use of more than 60,000 chemicals and produce more than 500 new chemicals each year. In addition, there has been a significant shift toward production of more organic chemicals, and more than 90% of the 200 contaminants in groundwater listed by the U.S. Office of Technology Assessment (USOTA) in 1984 as potentially hazardous were organic chemicals (Watson and Burnett, 1993). Pollutants may seep into groundwater when spills and incidental discharges occur. Some of the principal causes of spills that have been documented in Long Island, New York, include mishandling of delivered or stored chemicals, accidents involving trucks containing industrial chemicals, and miscellaneous discharges into groundwater (Koppelman, 1987).

Landfills are also a growing problem in urban areas, because if not properly constructed and maintained they can leak and pollute groundwater. For example, the Fresh Kills landfill receives more than 17,000 tonnes of refuse a day from New York City and, until recent remediation work was completed, was generating more than 3.8 million liters (1 million gallons) of leachate a day (Watson and Burnett, 1993).

24.4.4 *Managing the Impacts of Urbanization on Groundwater*

In North America, strong regulations have been implemented to deal with threats to groundwater quality and supply. In the United States, for example, a wide range of federal and state laws and programs includes elements designed to protect groundwater from contamination. These programs include the National Environmental Policy Act of 1969, the Federal Water Pollution Control Act of 1972, the Clean Water Act Amendments of 1977, the Safe Drinking Water Act of 1974, the Resource Conservation and Recovery Act of 1976 (RCRA), the Comprehensive Environmental Response, Compensation and Liability Act of 1980 (CERCLA), the Superfund Amendments and Reauthorization Act of 1986 (SARA), and the Toxic Substances Control Act (TOSA). The Safe Drinking Water Act was passed in 1974 to protect main source aquifers from contamination by underground injection of industrial and other waste. This act requires the EPA to set drinking-water standards to protect public health. Maximum contaminant levels were required to be established for toxic and carcinogenic substances and are now listed by the EPA. RCRA is a management system designed to deal primarily with hazardous solid waste. Under this act, waste generators, transporters, managers, treatment plants, and disposal facilities are all regulated. For example, facilities cannot place their operations on recharge areas of main source aquifers, and land disposal of liquid hazardous waste is prohibited. TOSCA was enacted to protect human health and the environment. This act requires testing and restricting chemicals that may present an "unreasonable" risk to human health and environment. This act also regulates disposal and storage of toxic substances where groundwater contamination has occurred or is likely to occur.

The regulatory approach is aimed primarily at protecting groundwater from present-day and future contamination, but there still exists the problem of cleaning up past contamination. Detecting, monitoring, and managing groundwater contamination is inherently difficult because of its physical location and the complexities of aquifer geology. Critical issues that plague effective management and decision making are incomplete geological and environmental information, technological limitations, and high remediation costs. Another issue is who should pay for cleanup. Typically, governments bear the major cost of cleanup because most companies cannot afford the investment cleanup requires.

Urbanization and Groundwater: The Case of Long Island Nassau and Suffolk counties, on Long Island close to New York City, have been two of the fastest growing counties in the United States since the 1940s (Koppelman, 1987). Tourism, agriculture, seasonal homes, and residential communities thrive in areas of seemingly healthy and aesthetically attractive natural settings. However, it became apparent in the 1980s that rapid urbanization was having severe impacts on the environment and groundwater.

There were many sources and causes of groundwater contamination on Long Island. Domestic on-site waste disposal systems, including cesspools, septic tanks, and leach fields, carried effluent into the groundwater system. These effluents had high concentrations of nitrate, chemicals, detergents, metals, bacteria, and viruses. Other compounds found in groundwater in this area include chemicals such as chloroform, carbon tetrachloride, trichloroethylene, and others that are common in industry as degreasers and solvents. Approximately 450 million liters per day of raw sewage flows through thousands of kilometers of sanitary sewers in the area. Sewers frequently leaked into the groundwater. About 34 million liters of sewage treatment plant effluents and industrial waste discharge totaling approximately 80 million liters of permitted industrial wastes were discharged into the ground every day (Koppelman, 1987).

Health implications of exposure to these contaminants are unknown in Nassau and Suffolk counties. In general, the greatest concern from long-term exposure to organic contaminants in groundwater is cancer. Epidemiological studies in other areas show links between contaminated water supplies and elevated cancer mortality rates. Because there is usually a delay of 20–30 years between exposure to a carcinogen found in contaminated groundwater and the incidence of cancer (Page, 1987), it is possible that the exposure to contaminants may in the future produce increased incidences of cancer.

After proper planning, extensive public support, and the involvement of government organizations, a groundwater management plan was developed and implemented on Long Island. This plan requires, for example, designated state and local agencies to plan for disposal of pollutants on land and subsurface locations to protect groundwater quality. Suffolk County established the Suffolk County Pine Barrens Commission, which reviews all actions within the major deep recharge area on the island. Thus every zoning change, subdivision, and industrial location are now scrutinized for their compatibility with the groundwater management plan. This plan has been institutionalized at all levels as the guide and standard to be followed (Koppelman, 1987). Although very significant steps have been taken to deal with groundwater concerns on Long Island, we will not know until the next few decades whether this level of management can be sustained and whether the plan has had a significant long-term impact on groundwater in the area.

24.5 Municipal Solid Waste

Households and commercial establishments in North America produce solid waste at a rate equivalent to 1 to 2 kilograms per person per day (EPA, 1996; Pick and Butler, 1997). Traditional landfills have had significant impacts on air and water quality, although throughout North America there has been a progressive shift from open to sanitary landfills. For example, by 1994 in Mexico City, 85% of the landfill area was sanitary landfill (Pick and Butler, 1997). However, throughout North America, it has become increasingly difficult to find locations for new landfills in urban areas, which has created a crisis in waste disposal management.

In the United States, the EPA has placed restrictions on landfill siting and design that are intended to prevent new landfills from becoming a nuisance to local areas. The siting criteria restrict proximity to airports, floodplains, wetlands, fault zones, seismic impact zones, and unstable areas. The design criteria require use of clay layers, composite liners, and leachate collection systems to prevent the leaking of liquid from the landfill into surrounding areas and groundwater. Also required is continuous monitoring of the landfill, even after it closes, to ensure the safety of the local community. In addition to siting requirements, resistance by local residents against placing landfills within their community and land shortages in urban areas have made it increasingly difficult to obtain approval for new landfills, and landfill capacities are decreasing. More than 80% of the garbage generated each year in the United States is put into landfills, and the growing concern about landfill capacity has created a push for alternative methods for disposal of waste. The main solution to the disposal problem has been to decrease the amount of landfilled solid waste by increasing the use of alternative means of disposal, such as recycling and combustion.

Compared to mid-1990s levels, a reduction in the waste stream by 50% through integrated waste management could be achieved by decreasing the amount of packaging materials through better designs, by implementating recycling programs, and by composting. In 1988, Seattle, Washington, was faced with the closure of its only landfill. A recycling and waste-reduction program was implemented to reduce the amount of landfilled waste by 60% by 1998. In only one year, Seattle was able to reduce its solid waste stream by 37% (Keller, 1996).

Materials such as glass, aluminum, batteries, cardboard, metals, newspaper, paper, tin cans, and tires can be recycled. Although many people are under the impression that recycling is a free or inexpensive route to waste disposal, in the short term it costs more to recycle items than to dispose of them immediately in a landfill. If long-term costs were identified and charged to those who dispose, the incentive to recycle would improve.

Combustion is another alternative to reduce the volume and weight of solid waste. Some incinerators or resource recovery facilities are able to generate electricity and steam for their own needs and to sell the remaining electricity and steam to local utilities. However, siting combustion facilities is very difficult due to local resistance even though the EPA has set very stringent rules and guidelines restricting the emittance of pollutants. These guidelines require, for example, the continuous monitoring of emissions of sulfur dioxide, NO_x, opacity, CO, lead, and temperature to verify that they do not exceed the limits set by the EPA (Holmes et al., 1993).

24.6 Urbanization and Noise Pollution

Noise pollution is defined as any unwanted sound that has a negative impact on the surrounding environment (Holmes et al., 1993). Noise pollution does not produce any visible effects on the environment and is commonly overlooked as a potentially serious problem in urban areas. However, in the United States, it is estimated that 25 million people are subjected to dangerous noise levels at work or during their daily activities, often without realizing it. Excessive levels of noise can have serious effects on humans, ranging from irritation and increased stress to permanent hearing loss (Holmes et al., 1993). Similar impacts on urban wildlife seem likely, although these effects are less well understood.

Increased levels of noise are a natural consequence of urbanization. Major contributors to noise pollution include automobile and airplane traffic and industrial machinery and processes. For residents of urban areas, these sources generate a continuous level of background noise that over long periods of time can cause significant problems. The hum of an air conditioning unit can make conversation difficult, and extended exposure to the sound produced by highway traffic can cause damage to hearing. Loud noises also produce short-term stress reactions that raise the heart rate, increase blood pressure and cholesterol levels, and interfere with sleep patterns (Holmes et al., 1993).

The effects of noise pollution are not limited to the human inhabitants of urban areas. If exposure to noise produces adverse effects on people, then it is reasonable to assume that urban wildlife will be affected in similar ways. It is much more difficult to measure the impacts of noise on animals because, compared to humans, animal communication and auditory responses are poorly understood. In addition, the many different animal species that exist in urban areas make it nearly impossible to identify and measure the effects of noise in terms that can be applied to all animals in general.

Depending on whether a noise pollution problem already exists or whether a new urban development is being planned, a number of noise control methods are available. Existing noise pollution problems can be eliminated or minimized by reducing noise sources, controlling the sound path, and/or protecting the receiver with earplugs or muffs (Davis, 1991). For new urban developments, careful planning can reduce noise problems by minimizing the conflicts between noise sources, such as industry or automobile traffic, and areas that require low noise levels.

Reducing noise sources and protecting the receiver can be effective in the workplace or when operating machinery, but these methods are often impossible or impractical for controlling many urban noise problems. Instead, sound barriers, in the form of topographic highs, walls, or vegetation, are commonly used to attenuate outdoor noise generated by automobiles, airplanes, and construction activity. Sound barriers have varying degrees of effectiveness, depending on their size, location, and frequency of noise (Davis, 1991). Barriers must be high enough and long enough to completely block the line of sight between the source and the receiver to significantly reduce sound levels. In addition, barriers are more effective against higher frequencies, reducing the effectiveness of barriers in situations where the sound source produces lower frequencies (Davis, 1991).

In new urban developments, proper land-use planning is the most effective method for minimizing the effects of noise. Land-use planning can reduce the impacts of noise by isolating residences, schools, and hospitals from major highways, airports, and industrial centers. Several regulatory agencies in the United States, including the Environmental Protection Agency, the Department of Housing and Urban Development, the Federal Aviation Administration, and the Federal Highway Administration, have set criteria for acceptable noise levels for various land uses and construction activities (Davis, 1991).

24.7 Urbanization and Ecology

24.7.1 Biodiversity

Biodiversity refers to the level of variety in animals, plants, and microorganisms in a given area. Urbanization causes habitat fragmentation, which reduces biodiversity and may result in extinction of native species. The introduction of alien species into newly developed urban areas can also result in changes in biodiversity. Biodiversity is also used as one measure of the degree to which humans have had an impact on the balance of ecosystems.

Land modification on the fringes of cities has resulted in habitat loss and alteration due to the drainage of wetlands and the conversion of both natural and agricultural areas to urban land-use types. Habitat fragmentation increases as natural habitats are broken into more isolated patches that support less diversity of species of plants and animals (Orians, 1995). Urban growth has also created a large demand for more agricultural production and for more resources such as wood for buildings, furniture, and paper, thus impacting natural resources far beyond the city (see chapter 22).

The introduction of alien species of plants and animals to urban areas, for example, for landscaping and as pets, also impacts native species. As urbanization increases, these new species compete for food and space with the native populations and may cause the latter to dwindle and disappear. This changes the biological balance of an area and allows dominance by new animals and plants, altering the integration and balance of the pre-urban ecosystem. For example, in the south Florida Everglades, the *Malaleuca* tree was introduced to provide a low-cost approach to solving high water table problems. *Malaleuca* seeds originally scattered in one development spread rapidly, because the species was well suited to the Everglades environment. The *Malaleuca* tree covered more than 200 thousand hectares (500 thousand acres) in 1995 and is spreading rapidly (Outcalt, 1995).

Biodiversity can be used as an environmental indicator to assist in evaluating the impact urbanization has had on the environment. Measures of biodiversity, including numbers of fish species and indices of biological integrity based on macroinvertebrate diversity and numbers, have been related to measures of urbanization such as percent watershed imperviousness (Schueler, 1994). This linking of biological indicators to physical characteristics of land-use change provides new tools for use in environmental planning (see section 24.8). In addition, communities can set goals based on biodiversity. For example, in the Puget Lowland area in Washington, a Water Quality Management Plan was implemented in 1987 and expanded in 1994, with the goal of protecting the biological diversity and health of the area. The plan involved classification of environmental and performance indicators relevant to protection of the waters of Puget Sound. Such plans provide a basis for management activities, as well as an assessment of how well goals are being met by continuous monitoring and characterization of the environment. Monitoring also provides a baseline for future reference in assessing the impacts of urbanization (Minsch et al., 1996).

24.7.2 Management

Habitat fragmentation reduces biodiversity as areas are isolated and linkages to other similar areas are reduced. Urban planning can be approached constructively from a biological perspective if goals are established that include maintenance of sustainable natural ecosystems appropriate to urban conditions and maintenance of linkages between fragmented ecosystems.

In Irvine, California, for example, after an urban monoculture of elm trees was destroyed by Dutch elm disease, city planners sought to guard against future failures by planting a variety of trees (Gangloff, 1995). They were able to create a sustainable culture that would tolerate urban stresses and form a colorful urban forest. Similarly, biological preservation and enhancement were key elements in the design of Heritage Park in Edmonton, Canada. Park construction included diversion of potentially polluted urban runoff around a natural wetland (instead of into the wetland, as would be typical of urban stormwater design), saving the habitat of 30 bird species as well as native vegetation (Mandel, 1998). On the margins of the Los Angeles urban area, freeway construction has included a provision for underpasses designed to allow native animals such as mountain lion to move readily between extensive, fragmented habitats.

However, management also has to include recognition of concerns about occasional negative impacts of plants and animals on human health in urban areas. For example, increasing concerns about the aesthetic and health impacts of droppings from large populations of Canada Geese in Mississuaga, Canada, as well as the degradation of park and lawn grass caused by geese scavenging, led to largely unsuccessful attempts to relocate thousands of geese from Lake Ontario to New Brunswick (Shilts, 1998). Concerns about animal droppings as a cause for elevated levels of the bacterium *Escherichia coli* (*E. coli*) in surface waters are likely to increase as more stringent surface water quality standards are applied to urban areas.

24.8 Environmental Urban Planning

From modest beginnings, North American cities and their impact on the environment have grown considerably over the past 150 years, first in response to railroad and tramway construction, then, during the later twentieth century in particular, in response to vastly increased car ownership and extensive highway development. This increased accessibility has led in turn to an explosion of commercial and residential development around and between older urban cores. Such rapid growth has significantly increased the potential for environmental change by increasing impervious surface area, introducing sources of pollutants, and altering natural ecosystems. The complexity and interconnectedness of modern urban areas make environmental problems some of the most difficult to solve. To manage this complex process, comprehensive plans have become an essential tool for minimizing the environmental impacts of urbanization (Gangloff, 1995).

One goal of urban planning is to minimize potential hazards and develop an efficient and sustainable urban environment. The key components of an environmentally sound urban development plan include sensible land-use patterns, efficient automobile and pedestrian traffic, and energy-saving and environmentally sound design practices (Gangloff, 1995; Arendt, 1996). By controlling the location, type, and intensity of development, the negative impacts of urbanization on the environment can be minimized. For instance, good land-use planning should presumably discourage intense industrial development along the edges of valuable wetland habitat and should seek to retain elements of the natural landscape as recreational space for the population. Controlling the locations of commercial centers can improve the flow of automobile traffic, thereby decreasing fuel consumption. Urban planning can also reduce the environmental effects of urbanization by incorporating open spaces to increase infiltration and decrease stormwater runoff.

New and innovative concepts in urban development are also being developed to improve energy efficiency and sustainability. The primary goals of ecological sustainability are to maximize energy efficiency, minimize waste and pollution, and incorporate natural processes while providing for social and economic growth (Gangloff, 1995). A number of alternative designs for urban development that take advantage of the natural environment have been proposed. An example of this type of design is the Pedestrian Pocket, created by Peter Calthorpe. The Pedestrian Pocket design directs urban growth into a collection of small, relatively self-contained areas that include both residential and commercial space. Each area is small enough for all the components to be within a short walking distance, and individual pockets are connected by efficient mass transit systems to reduce automobile traffic and energy consumption. Implementation of these types of designs could produce sustainable environments that maximize continued growth while minimizing the environmental impacts of urbanization.

24.9 Conclusion

Urban growth and industrial development have had wide-ranging environmental impacts, particularly in North America because of the intensity of urban development. Beyond high-density cores and satellite centers, low-density urban growth evolved in response to relatively cheap land and easy access to low-cost individual transportation. Urbanization has impacts across all areas of physical geography, from climate and biogeography to hydrology and geomorphology. Increasing realization of the impacts that urban areas have on human health and "natural" ecosystems, has led, especially during the later twentieth century, to regulatory approaches that have been more or less effective in controlling the increase of impacts with continued urban growth, and in some cases they have reduced impacts. Currently, impact reduction and avoidance are high priorities, and there is increasing emphasis on exploring alternative patterns of urban design to avoid environmental problems in the first place, as opposed to simply trying to reduce problems after development has taken place.

References

Angotti, T., 1995. "The metropolis revisited." *Futures*. 27: 627–639.

Arendt, R., 1996. *Conservation design for subdivisions, a practical guide to creating open space network*. Washington D.C: Island Press.

Atkinson, F., D. Miller, S. Curry, and S. Lee, 1986. "Areas

of Potential Problems." In S. Atkinson (Ed.), *Saltwater Intrusion, Status and Potential in the Contiguous United States*, Ann Arbor: Lewis Publishers.

Chinitz, B., 1991. "A framework for speculating about future urban growth patterns in the US." *Urban Studies*. 28: 939–959.

Cooke, R., and J. Dornkamp, 1990. *Geomorphology in Environmental Management*. Oxford: Clarendon Press.

Costa, E., and R. Backer, 1981. *Surficial Geology: Building with the Earth*. New York: Wiley.

Craul, P., 1992. *Urban Soil in Landscape Design*. New York: Wiley.

Cutler, I., 1976. *Chicago: Metropolis of the mid-continent*. Dubuque, IA: Kendall/Hunt.

Dunne, T., and L. Leopold, 1978. *Water in Environmental Planning*. New York: Freeman.

Elsom, D., 1996. *Smog Alert: Managing Urban Air Quality*. London: Earthscan.

EPA (Environmental Protection Agency), 1984. Nonpoint source pollution in the United States. Washington, D.C.

EPA (Environmental Protection Agency), 1996. *Municipal Solid Waste Factbook*. Version 3.0. Washington, D.C.

Fetter, C., 1988. *Applied Hydrogeology*, 2nd Edition. New York: MacMillan.

Gangloff, D., 1995. "The sustainable city." American Forests. May/June.

Golden, H., 1981. *Urbanization and Cities*. Lexington, KY: Heath.

Goudie, A., 1990. *The Human Impact on the Natural Environment*. Cambridge: MIT Press.

Harbor, J., 1994. "A practical method for estimating the impact of land use change on surface runoff, groundwater recharge and wetland hydrology." *Journal of the American Planning Association*, 60: 91–104.

Hayes, G., 1984. "Introduction." In H. Kahn (Ed.), *Virginia's groundwater*, Environmental Defense Fund, Proceedings, November 9–10, 1983. Blacksburg: Virginia Water Resources Research Center.

Holmes, G., B. Singh, L. Theodore, 1993. *Handbook of Environmental Management & Technology*. New York: Wiley.

Hough, M., 1995. *Cities and Natural Process*. New York: Routledge.

IDEM (Indiana Department of Environmental Management), 1998. State of Environment Report, Indianapolis, Indiana.

Idso, C.D., S.B. Idso, and R.C. Balling, 1998. "The urban CO_2 dome of Phoenix, Arizona." *Physical Geography*. 19: 95–108.

Johnston, R., 1982. *The American Urban System: A Geographical Perspective*. New York: St. Martin's Press.

Keller, E., 1996. *Environmental Geology*. New Jersey: Prentice-Hall.

Koppelman, E., 1987. "Long Island Case Study." In G. Page (Ed.), *Planning for Groundwater Protection*, Orlando: Academic Press, 157–203.

Long, D., 1996. "Anatomy of pollution: Loading chronology of selected contaminants in the Great Lakes." Abstracts with program, 1996 Annual meeting of the Geological Society of America, Denver, CO, A-201.

L'Vovich, I., and F. White, 1990. "Use and Transformation of Terrestrial Water Systems." In B. Turner II (Ed.), *The Earth as Transformed by Human Action: Global and Regional Changes in the Biosphere over the Past 300 Years*. New York: Cambridge University Press, 236–252.

Mandel, C., 1998. "Soothing the urban soul." *Canadian Geographic*, May/June 1998: 31–40.

Marsh, W., and J., Grossa, 1996. *Environmental Geography. Science, Land Use, and Earth Systems*. New York: Wiley.

McClintock, K., J. Harbor, and T. Wilson, 1995. "Assessing the hydrologic impact of land use change in wetland watersheds, a case study from northern Ohio, USA." In D. McGregor and D. Thompson (Eds.), *Geomorphology and Land Management in a Changing Environment*, London: Wiley, 107–119.

McPherson, E., D. Nowak, and R. Rowntree, 1994. *Chicago's Urban Forest ecosystem: Results of the Chicago urban forest climate project*. United States Department of Agriculture, General Technical Report NE-186.

Minsch, K., K. Anderson, J. Dohrmann, M. Tyler, 1996. Environmental Indicators in the Puget Sound. In: *Assessing the Cumulative Impacts of Watershed Development on Aquatic Ecosystems and Water Quality*, United States Environmental Protection Agency, Washington D.C.

Morris, S. 1992. "Geomorphic Assessment of the Effects of Flow Diversion on Anadromous fish spawning habitat: Newhalem Creek, Washington." *Professional Geographer*. 44: 444–452.

Novotny, V., and H. Olem, 1994. *Water Quality*. New York: Van Nostrand Reinhold.

Oke, T.R., 1997. *Boundary Layer Climates*. New York: Methuen.

Orians, G., 1995. "Thought for the morrow: Cumulative threats to the environment." *Environment*. Sept. 1995: 6–14, 33–36.

Outcalt, A., 1995. "Noxious plants." *Land and Water*. 6: 8.

Page, G.W., 1987. "Drinking water and health." In G.W. Page (Ed.), *Planning for Groundwater Protection*, Orlando: Academic Press, 73–74.

Petulla, J., 1988. *American Environmental History*, 2nd Edition. Columbus: Merrill.

Pick, J., and E. Butler, 1997. *Mexico Megacity*. Boulder, CO: Westview Press.

Richardson, J., 1982. "The evolving dynamics of American urban development," In G. Gappert and R. Knight, (Eds.), *Cities in the 21st Century*. Urban Affairs Annual Reviews, 23.

Scholze, R., V. Novotny, and R. Schonter, 1993, "Efficiency of best management practices for controlling priority pollutants in runoff." *Water Science Technology*, 28: 215–224.

Schueler, T., 1994. "The importance of imperviousness." *Watershed Protection Techniques*, 1: 100–111.

Shilts, E., 1998. "What's good for the goose." *Canadian Geographic*, May/June 1998: 70–76.

Stroker, H., and S. Seger, 1976, *Environmental Chemistry: Air and Water Pollution*, 2nd. edition, Glenview, IL: Scott, Foresman and Co.

USDA Forest Service, 1992, *Chicago's Evolving Urban Forest*. Chicago, IL: United States Department of Agriculture.

Walesh, S., 1989. *Urban surface water management*. New York: Wiley.

Watson, I., and D. Burnett, 1993. *HYDROLOGY An Environmental Approach*. Ft. Lauderdale: Buchanan Books.

Wolman, M., and A. Schick, 1967. "Effects of construction on fluvial sediment, urban and suburban areas of Maryland." *Water Resources Research*. 3: 451–464.

World Resources, 1996. *World Resources 1996–97, database diskette*. World Resources Institute, Washington, D.C.

25

Environmental Management and Conservation

Mark A. Blumler

We have met the enemy, and he is us
—Pogo

Environmental management and conservation is an enormous topic, encompassing a huge and ever-expanding list of issues, controversies, and debates over both science and policy. Some topics, such as soil erosion, are covered elsewhere in this volume, but an inordinately long list remains. Consequently, this chapter cannot be comprehensive. Instead, it focuses on biological resources and conservation, and emphasizes those aspects that I find illuminating or exemplary—notably the profound influence of changing paradigms over the past century or so, and the interplay among science, environmentalism, and policy as it has played out over North American landscapes. The general historical outline of environmental management and conservation is well known, though its interpretation is a matter of contentious debate; a brief review follows. There are some excellent treatments of the history of environmental thought and policy (e.g., Glacken, 1967; Botkin, 1990), and others that, though less judicious, contain much relevant information. In addition, excellent regional or otherwise more narrowly focused studies exist (e.g. Cronon, 1983; Botkin, 1995). But given the emotionally charged nature of environmental debates, the highly idealistic, utopian attitudes of many scholars toward nature, and the generally dualistic quality of discourse in the Western world, it should not be surprising—in fact, it is entirely understandable—if there has been some distortion of the reality of North Americans' changing relationship with the environment. In this chapter, I offer my own interpretation. In particular, I emphasize that both the environmental movement and the science of ecology originated within the Western world and consequently have been strongly constrained by Western modes of envisioning reality. One outcome of this is a tendency in the environmental literature to view conservation as a contest between despoilers and conservers, that is, to blame others for environmental problems. In my view, we who call ourselves environmentalists are as culpable as anyone.

Geographically, most of my discussion and examples relate to the United States, although Canada and northern Mexico also are included within this volume's purview (extratropical North America). Like the United States, Canada and Mexico confront important environmental problems and have a long history of engagement with resource management and conservation. In addition, certain specific issues are of lesser concern to the United States than to one or the other of these countries, because of differences in natural resource availability, economic conditions, or environment. Nonetheless, it is difficult to overstate the influence of the United States on environmental thought and management in North America, or, for that matter, the entire globe. Environmentalism began in the United States, as did a majority of the important schools of environmental thought (the rest originated in Europe). Moreover, Americans dominate the list of ecologists who have significantly improved scientific understanding of ecosystems. Canadian environmental thought was largely derived from its neighbor to the south. The history of resource and conservation management followed similar trajectories in the two countries, with developments almost always occurring first in the United States (Brown, 1968; Nash, 1968; Dubasak, 1990; Foster, 1998). French and Spanish beliefs about the environment had some influence on Mexico, but the latter is distinguishable from its northern neighbors primarily in the weakness of its environmentalism and environmental laws (Simonian, 1995).

25.1 Historical Overview: The United States

Initially, Europeans settling in North America were profligate in their use of resources, much more so than was characteristic of the lands from which they had migrated. Despite the presence of native inhabitants, Europeans saw an empty land, bursting with resources in seemingly inexhaustible supply (see chapter 22). And there was always the frontier beckoning when settlement in a given region began to press against environmental limits. Most Americans seem to have shared Jefferson's view that the conversion of natural ecosystems to human-dominated, mostly agrarian landscapes was desirable.

> Precisely planted rows of corn and wheat, neat white farm houses, and a one-room country school represented civilization, democracy, and progress. (Koppes, 1988: 230)

Such environmental management as occurred during these early years was mostly concerned with hazard control and resource inventory. As the frontier receded and resources became depleted, however, an increasing emphasis on conservation and environmental protection resulted, which continues to this day (see table 25.1 for a time line of important events). The official closing of the frontier in the late nineteenth century is often correlated with the birth of the environmental movement—which certainly became prominent shortly thereafter, associated with the political success of the Progressives—but there were additional, related factors at work, such as the transition from an agrarian to an urban and industrial society. Nonetheless, it is probably fair to say that the deep emotional and symbolic significance of the frontier to Americans was a major factor in the early and continuing prominence of that branch of the environmental movement concerned with the protection of "wilderness," in other words, with the prevention of further development of wild lands (Nash, 1971; Dubasak, 1990).

25.1.1 *Conservation and Preservation*

From the start, the American environmental movement comprised individuals who emphasized the conservation and wise use of natural resources and others who championed the preservation of wild nature (Koppes, 1988; Meine, 1995). Gifford Pinchot, who had been trained in Europe in the nascent tradition of scientific forestry and was the first chief of the U.S. Forest Service, exemplified the conservationist wing, whereas John Muir led the preservationist contingent. In practice, the two groups shared many beliefs and goals, and they got along rather well until the battle over Hetch Hetchy Dam in Yosemite National Park exposed their philosophical differences. Despite the powerful, positive feelings that wilderness evoked in Americans, the utilitarian philosophy of the conservationists was more acceptable politically, and they wielded much greater power than the preservationists. Conservationists were able to formulate policy through their control of government land-use agencies such as the U.S. Department of Agriculture Forest Service and, later, the Bureau of Land Management (BLM), especially during the presidencies of the two Roosevelts. Preservationists were more likely to rely on nongovernmental organizations (NGOs) to achieve their occasional successes in setting aside lands as national parks, and later, wilderness areas. Initially, the emphasis was on scenic beauty, but Aldo Leopold and others began to shift the preservationist focus toward biotic communities in the 1930s. In turn, this created an increasing opposition to roads and amenities in wilderness areas (Koppes, 1988). In recent decades, as the rural population has continued to decline relative to that in urban agglomerations, Americans have become increasingly preservationist, shifting the power balance within the environmental movement.

25.1.2 *Pollution*

Awareness of and increasing concern about anthropogenic toxic substances in the environment have their own history, largely separate until recently from that of resource conservation and biodiversity preservation. Many of the pioneers in the field of toxics research were also Progressives (e.g., Alice Hamilton), but their concern was with public health, not with wild nature. Widespread awareness that pollutants could pose significant health hazards developed relatively late, whereas the extension of concern to the potential negative effects of pollutants on other species and natural ecosystems did not occur until even later. The story is complex (Sellers, 1994), and I will note only a few salient aspects.

Table 25.1 Some landmarks in the history of environmental management and conservation in North America

1861	Mexico's first national forestry law enacted
1862	Homestead Act opens up federal lands in the western United States to settlement, but fails to recognize that large acreages are necessary for viable agriculture in much of the region
	Morrill Act creates the land-grant college system, with a view to enhancing agricultural productivity, scientifically
1864	George Perkins Marsh publishes *Man and Nature, or, Physical Geography as Modified by Human Action*, Charles Scribner and Company, New York
1871	Bureau of Fisheries created by U.S. Congress
1872	Yellowstone, the world's first national park, created
1876	Audubon Society founded
1885	Adirondack Forest Preserve created
	Core of what would become Banff National Park set aside by Canadian government
1890	U.S. Census Bureau declares the frontier closed
	U.S. Congress protects what would become Yosemite and Sequoia National Parks, the first parks set aside for preservationist reasons
1891	First federal forest reserve (later to be called national forests) created at Yellowstone
1892	Sierra Club founded under the leadership of John Muir
1893	Ontario's Algonquin Park created
1903	First federal wildlife refuge created, on Pelican Island, Florida
1905	U.S. Forest Service created with Gifford Pinchot as its director
1911	Weeks Act allows the federal government to acquire lands in the eastern United States for the creation of new national forests
	Mexico protects the elephant seal
1913	Hetch Hetchy Dam controversy splits utilitarian and preservationist wings of the conservation movement
1914	U.S. Congress provides funds for predator control on federal land
1916	Migratory Bird Treaty grants United States and Canada greater power to protect waterfowl and other migratory birds
	National Park Service created by U.S. Congress
	Clements' highly influential views on succession published (F.E. Clements, Plant succession: an analysis of the development of vegetation, *Carnegie Institute of Washington Publications*, 242, 1–512)
1917	Mexico's first national park, Desierto de los Leones, created
1922	Mexico's first wildlife refuge, Isla Guadelupe, created
1924	First wilderness area set aside at Gila, New Mexico
1926	Gleason's Darwinian critique of Clementsian succession published (H.A. Gleason, The individualistic concept of the plant association, *Bulletin of the Torrey Botanical Club*, 53, 7–26)
	Hamilton's report on lead poisoning in effect puts research into chemical (pollutant) causes of sickness on a par with that into bacterial diseases (Aub, J.C., L.T. Fairhall, A.S. Minot, P. Reznikoff, and A. Hamilton, *Lead poisoning*, Williams and Wilkins Co., Baltimore)
1930	Canadian National Parks Act, modeled on U.S. act of 1916 though less preservationist, enacted
	Transfer of Resources Agreement gives Canadian Provinces control over lands within their borders; consequently, subsequent national parks would be created mostly within the Territories
1934	Taylor Grazing Act ends open range in the western United States and creates the Grazing Service (later to become the Bureau of Land Management)
1935	The Dust Bowl causes U.S. Congress to create the Soil Conservation Service
	Mexican President Cardenas creates Department of Forestry, Fish, and Game, leading to the creation of forest preserves
1937	United States and Mexico sign Treaty for the Protection of Migratory Birds and Game Mammals
1938	Northern Mexico's first national parks, El Sabinal and Nuevo León, created
1940	U.S. Fish and Wildlife Service created by merging two agencies
	Mexican President Cardenas abolishes Department of Forestry, Fish, and Game
1942	Lindeman's article gives "ecosystem" its modern meaning (R.L. Lindeman, The trophic-dynamic aspect of ecology, *Ecology*, 23, 399–418)
1945	Smokey the Bear campaign initiated; soon convinces most Americans that wildland fire is bad
1949	Publication of Aldo Leopold's most influential work (A. Leopold, *A Sand County Almanac and Sketches Here and There*, Oxford University Press, New York)
1962	*Silent Spring* gives rise to the modern environmental movement (R. Carson, *Silent Spring*, Houghton Mifflin, Boston)
	Mexican conservationist Enrique Beltran proposes zoning as a solution to preservation-exploitation conflicts in national parks, an approach subsequently adopted throughout much of the Third World
1963	Leopold Report recommends that American national parks should be managed to resemble conditions at the time of first European contact
	National and Provincial Parks Association of Canada, the first Canadian NGO with a preservationist orientation, created, modeled on the National Parks Association of the United States
1964	Wilderness Act passed by U.S. Congress
	Canadian federal government Policy Statement on National Parks makes preservation a management goal
1968	Publication of Hardin's influential paper (G. Hardin, The tragedy of the commons, *Science*, 162, 1243–1248)
1969	National Environmental Policy Act (NEPA) passed by U.S. Congress

(*continued*)

Table 25.1 (*continued*)

1970	Earth Day initiated U.S. Environmental Protection Agency created Clean Air Act passed by U.S. Congress Canada enacts Arctic Waters Pollution Prevention Act
1971	Mexican Law for the Prevention and Control of Pollution is enacted, but poorly enforced Canadian branch of Audubon Society becomes the Canadian Nature Federation, and increases rapidly in membership Canadian Clean Air Act enacted
1972	*Sierra Club v. Morton* gives NGOs the right to bring suit to ensure that U.S. government agencies enforce environmental statutes Clean Water Act passed by U.S. Congress Preservationists defeat Village Lake Louise development proposal for Banff National Park
1973	Endangered Species Act passed by U.S. Congress Publication of Schumacher's book generates interest in "appropriate technology" (E.F. Schumacher, *Small Is Beautiful*, Harper & Row, New York)
1977	Publication of seminal review on succession signals the end of acceptance of Clementsian theory by most ecologists (Connell, J.H., and R.O. Slatyer, Mechanisms of succession in natural communities and their role in community stability and organization, *The American Naturalist*, 111, 1119–1144)
1978	Publication of review on causes of species diversity marks the acceptance of nonequilibrium ecological theory by many top ecologists (Connell, J.H., Diversity in tropical rain forests and coral reefs, *Science* 199, 1302–1210) Mexico's first biosphere reserve, Montes Azules, created
1979	Environment Canada, a department of the federal government, established
1982	Mexico enacts an antipollution law that is stronger than the 1971 statute, though still poorly enforced Mexican government establishes a cabinet-level environmental ministry
1987	Montreal Protocol controlling ozone-damaging chlorofluorocarbons
1989	Publication of article that applies nonequilibrium ecological theory to range management signals acceptance of the nonequilibrium paradigm by many range management professionals (Westoby, M., B.H. Walker, and I. Noy-Meir, Opportunistic management for rangelands not at equilibrium, *Journal of Range Management*, 42, 266–273)
1992	U.S. Forest Service officially adopts "ecosystem management" as its means of achieving multiple use
1993	North American Free Trade Agreement (NAFTA) exacerbates pollution problems along the Mexico–United States border
1997	Kyoto Agreement attempts to regulate the production of greenhouse gases

By the late nineteenth century, the Industrial Revolution had transformed the workplaces and cities of the United States, and concern arose over water and, secondarily, air quality. Medical scientists researching the former were strongly influenced by the discoveries of Pasteur and Koch, who had convinced many that the old miasmatic theory of disease should be replaced with a new paradigm, the germ theory, connecting disease to bacteria. Consequently, researchers investigating the effects of industrial and other effluents on water quality focused almost exclusively on bacteria, largely ignoring possible direct chemical impacts on health. Meanwhile, factory workers were suffering from a number of dramatic ailments, which they themselves often blamed on chemicals in the factory air. Research scientists such as Hamilton who began to investigate these ailments systematically in the early decades of the twentieth century were less bound to the bacterial paradigm. In addition, the biomedical researchers collaborated well with engineers, who could model the movement of chemicals within the factory and locate possible sources of toxins. Consequently, the new field of "industrial hygiene" made rapid progress, and by the 1930s some practitioners had moved out of the factory to document the negative impacts of pollutants in the atmosphere. On the other hand, scientists responsible for investigating and managing water quality took little or no note of the discoveries in industrial hygiene, and they did not recognize a significant role for purely chemical causes of disease until the 1950s. After all, Koch's postulates could not apply to such diseases! It is not surprising, then, that Rachel Carson's *Silent Spring* did not appear until 1962. By then, the ground had been prepared for public acceptance of her thesis by the increasing publicity over nuclear fallout, smog episodes in Los Angeles and London, and gradually increasing awareness that chemicals could cause disease.

This story illustrates some of the complexities in the relationship between science and environmental management. The germ theory of disease unquestionably represented a major advance. It enabled medical scientists and sanitary engineers to attack serious diseases, such as typhoid, that fit the model. But precisely because it was such an undeniable breakthrough, a bandwagon effect caused scientists to ignore other possible causes of disease for decades. Moreover, as Sellers (1994: 75) pointed out, scientific specialization, which often today is regarded as hindering the study of the environment, in this case both "enhanced and impeded our understanding of environmental problems." On one hand, it allowed great strides to be made in identifying and controlling specific waterborne bacterial diseases as well as airborne toxins; but on the other hand, lack of communication between the disciplines prevented additional progress from being made on waterborne toxins until much later. Sellers noted also that a contributing factor was a "professional etiquette of specialization," which caused the industrial hygienists to hold their collective tongue rather than criticize the unwillingness of water quality professionals to investigate possible toxins. Unfortunately, such politeness continues to bedevil transdisciplinary research today (Blumler, 1992a, 1993).

25.1.3 The Modern Environmental Movement

Silent Spring was largely responsible for creating a national consciousness about, and concern over, toxic substances in the environment. This intersected with the burgeoning interest in nature and its preservation, and the modern environmental movement took off. Garrett Hardin's (1968) essay on "the tragedy of the commons" was also highly influential, in setting the tone of environmentalists' attitudes about the relationship among population growth, resource use, and the environment. On the other hand, several scholars severely and cogently criticized Hardin's thesis, with the result that a more nuanced understanding of commons resources is now beginning to develop in some sectors of the environmental movement (Ostrom et al., 1999). During the 1970s, the environmental movement gained considerable strength throughout much of the developed world, and even began to gain some small degree of influence in the Third World.

In the United States, environmental NGOs proliferated during the 1960s and 1970s (Cutler, 1995). They gained ever more power to influence American policy, as a series of laws and judicial decisions (e.g. the National Environmental Policy Act) forced the Forest Service and BLM to open up their policy-making process, in effect redefining "multiple use." In theory, the federal agencies had always been committed to multiple use, but in practice, the form that this took was very different from the current mandate. Until recently, the Forest Service regarded itself as employing scientifically trained experts who were best qualified to determine policy, whereas now it has no choice but to allow various interest groups a significant role in policy making (Pyne 1982, 1997; Hess, 1992; Brunson and Kennedy, 1995; Parker, 1995).

Today, land-use planning on federal lands frequently involves several highly emotional interest groups with disparate agendas, and government agency officials are caught in the crossfire. Environmentalists tend to view the Forest Service and BLM as captives of logging and ranching interests; ranchers and loggers, on the other hand, have always regarded the government agencies as unsympathetic to their concerns, and more interested in bureaucratic control than proper ecological management (Hess, 1992; Brunson and Kennedy, 1995). As a result, morale is declining in all the government land agencies, including even the National Park Service, which has been comparatively well insulated from criticism (Kessler and Salwasser, 1995; Pyne, 1997). On the other hand, the agencies seem to be responding to criticism by incorporating ecological concerns into management to a greater degree (Brunson and Kennedy, 1995). For instance, the Forest Service recently adopted "ecosystem management" as its official policy (Kessler and Salwasser, 1995); in doing so, the Forest Service shifted itself in the direction of the environmental management paradigm, termed "adaptive management," that currently prevails among scientists (Holling, 1978).

Recently, environmentalists' concerns have expanded yet again to encompass the possible threats posed by atmospheric and other global change, such as acid rain, greenhouse warming, and stratospheric ozone depletion. Today, most environmental NGOs, and probably most individual environmentalists, are especially concerned about some specific subset of the total number of environmental issues (Cutler, 1995). It remains useful, if undeniably oversimplified, to distinguish among conservationists, with an interest in sustainable development (paralleling Pinchot's earlier emphasis on sustained yield and equal opportunity for all); preservationists, interested especially in biodiversity (in contrast to the early preservationists' focus on aesthetically pleasing wild landscapes); and environmentalists, primarily concerned about pollution, and frequently also anti-technology or supportive only of "appropriate technology" along the lines of Schumacher's (1973) influential book, *Small Is Beautiful.*

25.2 Historical Overview: Canada and Mexico

> In the United States, the [environmental] movement developed as a creature of its culture, without any strong, outside influences. In contrast, there was substantial American influence on Canadian conservation.
>
> Dubasak, 1990: 202

25.2.1 Canada

In Canada, the general trajectory of resource use or abuse, followed by attempts at conservation, was similar to that of the United States. However, events generally lagged in Canada because of its much lower population density. This meant that whenever an environmental problem began to become significant in Canada, Canadians could (and did) look to the United States for an example of how to attack the problem. On the other hand, Canadian feelings about wilderness historically were far more negative than in the

United States, and consequently, the environmental movement in Canada lacked a major preservationist element until recently. Muir, Pinchot, and Leopold all had some influence in Canada, but at first only Pinchot's wise-use philosophy was adopted widely. Significant support for preservation did not develop until the 1970s (Nash, 1968; Foster, 1998). In part, this was because Canada had so much remaining wilderness that it was difficult to perceive the need to preserve it. But there was more to it:

> Canada did not share the romantic, almost mythic, identification Americans had with the land. Americans used land to redefine and distinguish themselves from Europeans; Canadians were emphatically not separated from British and European heritage. (Dubasak, 1990: 42)

Wilderness is an essential part of America. In Canada, on the other hand, wilderness was often much more threatening than in the United States, so much so that government assistance would be necessary to open it up. Even today, considerable ambivalence remains regarding wild nature, and Canada still has no equivalent of the Wilderness Act. Moreover, the greater legal rights afforded Canadian provinces and aborigines, compared to their American counterparts, has strongly constrained the federal government's ability to preserve wild land. Nonetheless, since the 1960s, Canadians have increasingly aligned themselves with the American view on wilderness (Henderson, 1992).

With the globalization of the environmental movement around 1970, Canadians became very concerned about pollution, acid rain, and similar issues. From this time forward, Canadians have been influenced not only by American environmentalists and resource managers, but also by European and United Nations pronouncements. For instance, the Brundtland Commission report (WCED, 1987) had more impact in Canada than in the United States (Mitchell, 1995). On the other hand, economic problems and ethnic fractures have to a significant extent countered the influence of environmentalism among Canadians in the past two decades (Foster, 1998).

25.2.2 Mexico

Mexico is a representative Third World nation, in that environmental management and protection have generally taken, and continue to take, a back seat to economic development. This is noticeable, for instance, in industrial areas along the border with the United States, where pollution is severe—a situation that was further exacerbated by the North American Free Trade Agreement. Given that environmentalism is on the whole a middle-class phenomenon, it is not surprising that the movement has never had many adherents in Mexico. Initially, influences from the western Mediterranean affected environmental thought and management, notably the belief that forests prevent desertification (Simonian, 1995). Miguel Quevedo, who was responsible for many early environmental laws and programs, was educated in France. Consequently, he believed that deforestation causes climatic desiccation, and he also shared many of the views that Pinchot acquired during his European training. Since 1970, the environmental movement has become stronger, as it has everywhere; moreover, pollution problems in Mexico City are so severe that the government has had no choice but to attempt mitigation. But overall, the history of environmental and resource management in Mexico seems chaotic rather than progressive, with periods during which environmental laws were enacted followed by periods during which many were abrogated. Most environmental laws were without teeth and poorly, if at all, enforced. For instance, on paper Mexico has many parks and preserves, but in practice economic exploitation and resource extraction in such areas often continue unabated (Simonian, 1995). As is true of many Third World nations, data on resource stocks and environmental degradation are so sketchy that it is difficult to evaluate the current condition of the Mexican environment with confidence.

25.3 Paradigms and Environmental Management

> Plus ça change, plus c'est la même chose—a French proverb with a surprisingly Eastern feel

25.3.1 Environmentalism in the Context of Western Thought

The environmental movement originated in the Western world and still reflects that context. In particular, beliefs and ideals from the Enlightenment strongly influenced the conservationist wing, whereas, as is widely recognized, the antecedents of preservationism lay in Romanticism. Pinchot's conservationists believed that science and rational management would improve nature, to the benefit of all Americans. The preservationists were particularly attracted to spectacular scenery, such as at Yosemite, Yellowstone, and Grand Canyon—that is, their appreciation of nature was more aesthetic than ecological. Aestheticism remained pronounced into the 1970s, as evidenced by the Sierra Club publications of the nature photography of Ansel Adams, in which rocks, water, and other nonliving things are at least as prominent as biota. Environmentalist leaders such as David Brower (e.g., in McPhee, 1971) were still basing their arguments for preservation on aesthetic grounds in the 1970s.

Although superficially polar opposites, the Enlightenment and Romantic movements had a great deal in common. Both were characterized by idealism, a conceptual separation of humans and nature, and a specific, dualistic way of analyzing reality. Much of this can be traced back to Descartes. Descartes treated mind and body, or reason and emotion, as dualistic opposites, separate and distinct, and antagonistic. I accept Toulmin's (1990) thesis that his motivation for doing so was idealistic: he was horrified by the extreme emotionalism of the Thirty Years War, during which Catholics and Protestants massacred each other across much of Europe. In response, he turned to rationalism as a means to overcome prejudice, eventually comprehend everything, and consequently, be able to determine whether the Catholics or the Protestants were correct. This *modus operandi* gave rise to the Enlightenment. Subsequently, Comte's positivism outlined a formal methodology for conducting supposedly emotion-free investigations, which dominated the practice of science in the nineteenth and early twentieth centuries. A whole series of related dualisms developed out of the mind–body, reason–emotion split, such as humans–nature. The fundamental flaw in this approach was that reason and emotion (or humans and nature) cannot really be separated. Given the Enlightenment's emphasis on reason and suppression of emotion, it was inevitable that there would be a reaction, Romanticism, that glorified emotion and downgraded reason. And whereas the Enlightenment had elevated humans over nature, the Romantics did the reverse. But they still regarded the two as antithetical (McIntosh, 1985). Thus, although the Romantics appreciated Nature, they did not understand it any better than Enlightenment philosophers. Some of the most widely read historians of environmental thought have come from the movement's currently dominant preservationist wing (e.g., Worster, 1977; Oelschlaeger, 1991); hence they have tended to miss the fundamental similarities in Enlightenment and Romantic beliefs about nature.

Dualistic thinking does not have to incorporate the notions of separate and antagonistic entities. For example, consider the yin-yang symbol: dualistic opposites are depicted as associated nonhierarchically and nonlinearly, each incorporating the other, embedded in each other, and flowing into each other, that is, in a *relationship*. It seems to me that the picture of humans and nature currently emerging among ecologists is consonant with this symbol, rather than with the antithetical picture characteristic of Western thought (compare with Botkin 1990, 1995).

25.3.2 Ecology in the Context of Western Thought

> Ecological and conservation thought at the turn of the century was nearly all in what might be called closed systems of one kind or another. In all of them some kind of balance or near balance was to be achieved. The geologists had their peneplain; the ecologists visualized a self-perpetuating climax; the soil scientists proposed a thoroughly mature soil profile, which eventually would lose all trace of its geological origin and become a sort of balanced organism in itself. It seems to me that social Darwinism, and the entirely competitive models that were constructed for society by the economists of the nineteenth century, were all based on a slow development towards some kind of equilibrium. I believe there is evidence in all of these fields that the systems are open, not closed, and that probably there is no consistent trend towards balance. Rather, . . . we should think in terms of massive uncertainty, flexibility and adjustability.
>
> —H.M. Raup, 1964: 19

The science of ecology also developed in the West and reflects that background. Many highly influential ideas came into ecology and related fields such as evolution from the surrounding society, that is, they are not truly scientific concepts (table 25.2). For a long time, they remained essentially untested, though they are increasingly receiving critical attention and are often found wanting (Blumler, 1996). For instance, the "balance of nature" is a Christian belief that goes back at least to the time of Aquinas (McIntosh, 1985). Incorporated into ecology, it gave rise in turn to a series of "equilibrium" assumptions, such as that ecosystems are naturally stable, are predictable, and have a number of ecological niches each of which is occupied by one and only one species. The ecosystem itself is an equilibrium concept in the models of many prominent ecologists, in that it borrows presumptions of homeostasis and negative feedback from systems theory. Over the decades, a great mass of ecological theory has built upon equilibrium foundations, for instance, the classical models of population growth in relation to carrying capacity, and the theory of island biogeography (MacArthur and Wilson, 1967). Note that "equilibrium" as used in ecology is not identical to its meaning in geomorphology. As the previous

Table 25.2 Some nonscientific ecological concepts

- Balance of nature
  - stability
  - equilibrium concept
    - predator-prey curves (Lotka-Volterra)
    - theory of island biogeography
  - carrying capacity (K)
  - ecological niche
  - ecosystem properties (Odum)
- Nature undisturbed (stable)/humans disturb nature
  - humans outside of nature
  - humans active, nature passive
- Succession
  - climax
  - ruderals
  - annual plants = disturbance
- Community concept
- Typological species concept

quote illustrates, to ecologists the Davisian cycle of erosion is an example of equilibrium theory; in geomorphology, equilibrium is considered to be the paradigm that replaced Davis and is bound up with an emphasis on process-oriented studies.

Formerly, environmentalists and ecologists both held to an equilibrium view of nature, reflecting their common intellectual heritage and ideals; most environmentalists still do. But in the past few decades, the perspective of many ecologists has altered drastically, largely because their empirical observations could not be aligned with the equilibrium view. This change in ecological thought is usually described as a paradigm shift, from a belief in a nature that is normally in equilibrium, governed by gradual, developmental processes, and in which disturbance is rare to one that is governed by nonequilibrium processes in which natural disturbance is pervasive, communities are kaleidoscopic, and human disturbance has complex but not necessarily degradational impacts (Connell, 1978; Blumler, 1984, 1992b, 1993, 1996, 1998; Westoby et al., 1989; Botkin, 1990; Worster, 1990; Sprugel, 1991; Glenn-Lewin et al., 1992; Zimmerer, 1994; Pickett and Ostfeld, 1995). "Paradigm shift" is grossly overused today, but in this case the term is apt. Moreover, the change within ecology is emblematic of a more fundamental shift going on throughout the sciences, from viewing nature as static to seeing it as dynamic. Two well-known examples, the acceptance of plate tectonics and, recently, of catastrophic dinosaur extinctions by a majority of scientists, both reflect (and reinforce) this change in how nature is perceived. As a result of the shift to nonequilibrium thinking, considerable tension exists between ecologists and the environmental NGOs today. For instance,

> The perception of balance in nature is a damnable heresy that persists in most fields of applied ecology and resource management to the detriment of establishing realistic goals and guides.—H.B. Johnson (quoted in Brown, 1993)

Because of the Romantic heritage of preservationism, many environmentalists hold on to a view of nature that is increasingly in conflict with ecologists' empirical findings. At the same time, because most ecologists also are preservationists at heart, their shift away from traditional views has been very slow in coming. In fact, I have argued elsewhere that nonequilibrium ecologists do not yet realize the full implications of recent research because even they are still tied to some extent to the old paradigm (Blumler, 1993, 1996). This is seen most clearly in beliefs about succession, generally regarded as the central concept in plant ecology. Drury and Nisbet (1971) pointed out that Clementsian succession theory conformed to a general tendency at the beginning of the twentieth century to create developmental models of nature, in a wide range of disciplines, in accord with Herbert Spencer's highly influential beliefs (see also McIntosh, 1985; Blumler, 1996). In geomorphology, for example, we have the aforementioned Davisian cycle of erosion. However, geomorphology and most other fields later abandoned such models. Ecologists took a much longer time to become critical of traditional succession theory and, even today, tend to be less critical than they might be. Ecosystem ecologists such as Odum (1969) have been particularly uncritical in their incorporation of traditional concepts, and consequently they have become the darlings of the environmental movement while losing status within the ecological science community. This has occurred despite the fact that some undeniably significant contributions have resulted from the ecosystem approach (compare with McIntosh, 1985; Hagen, 1992).

Clements' 1916 model of vegetation succession is in essence both simple and questionable, although textbooks usually depict succession in a beautiful, idealized manner, fostering its uncritical acceptance (Blumler, 1996). According to Clements, after disturbance removes the vegetation, there is a gradual, progressive return of ever taller and woodier plants until eventually the stable "climax" is reached. The linear, progressive nature of the model echoes Enlightenment ideals, whereas the climax is really a restatement of the Edenic myth, or, as the Romantic poet Longfellow would have it, "the forest primeval." Thus, climax is a bio-utopian concept. Consequently, even nonequilibrium ecologists have trouble moving away from the concept completely. They also have been loathe to give up on the notion that succession is unidirectional, despite the fact that the ecological literature is now replete with striking, well-documented counterexamples. Succession can even run backward, from perennials to annuals (Blumler, 1993)! I believe that such "exceptions" outnumber the rule in semiarid and arid regions, and that succession theory is a major impediment to understanding how ecosystems in such climates work.

25.3.3. *Management Implications of the New Paradigm*

> Worster (1977) claimed that 'where the climax is ignored or distorted as an ideal, the only criterion left is the marketplace'. This view is not only short-sighted . . . but also wrong.
>
> —Sprugel, 1991: 14

The nonequilibrium view of nature, if correct, considerably complicates environmental management. Under the old paradigm, we could take a "nature knows best" approach and manage for specific, predictable climax communities. Now, it is believed that "[climax] is an imaginary condition . . . never attained in nature" (Botkin and Keller, 1994: 169) and that no single community of species is inherently natural to a given site (Christensen, 1988; Sprugel,

1991). Rather, communities are constantly changing in composition, in both space and time (Whittaker, 1975; Davis, 1981; Sprugel, 1991), which is another way of saying that communities do not really exist. This further implies that vegetation and biome mapping is extremely problematic and inherently misleading, since it unavoidably incorporates notions of stability and repeatability (Brown, 1993). Yet it is difficult to see how one can manage without such maps. Perhaps the most widely recognized failure of the traditional paradigm concerns fisheries, which until recently were managed as if carrying capacity is a fixed equilibrium value that, therefore, is calculable. Now, increasingly, we recognize that this is not so (McEvoy, 1988; Botkin, 1995; Harris, 1995; Mitchell, 1997).

One example related to succession theory is the revegetation of barren lands, such as after fire, volcanic eruptions, or mining. Often, we wish to cover the surface with plants to prevent catastrophic soil erosion (fig. 22.13). This is particularly desirable if toxic chemicals may be present in the regolith, such as in mine spoils. Traditionally, revegetation has been carried out with introduced herbs, which are fast growing and responsive to fertilizer and irrigation. Under Clementsian theory, it was predicted that these species would facilitate the establishment of later successional native plants. In practice, however, the herbs prove to be highly competitive, and their successful establishment often greatly slows or prevents altogether the return of native tree species (Brown, 1996).

The saguaro cactus (*Cereus giganteus*) of the southwestern United States and northern Mexico illustrates the contrasting management implications of the two paradigms. Niering et al. (1963), writing when the equilibrium view was still dominant, analyzed the saguaro within the context of a presumably stable ecosystem, that is, in terms of interactions among the species currently present within its environment, and attempted to determine the specific conditions that favor successful establishment of seedlings. Parker (1993), on the other hand, looked at establishment over the long term and related it to changing climatic and other conditions. The equilibrium view tended to ignore climatic variation and to assume that regeneration occurs at a constant rate in the absence of human disturbance. The new nonequilibrium view recognizes that climate is enormously variable and accepts that seedling establishment may be highly episodic and dependent on unusual circumstances. Indeed, Parker found that saguaro regeneration was episodic, occurring under specific climatic conditions. In addition, and what I find most interesting, she reported that, although regeneration was poor under livestock grazing, it was excellent immediately after the cessation of grazing but not thereafter. These results have a parallel in a study by Bartolome and Gemmill (1981) of the effects of livestock grazing on a perennial bunchgrass, *Nasella pulchra*, in California. When the animals were removed from the study pasture, *N. pulchra* initially established new individuals. But after 2 years, this establishment ceased completely. Bartolome and Gemmill argued that it was the spread of more competitive annual grasses, which had been kept down by the cattle, that prevented further establishment of bunchgrasses. The equilibrium interpretation would be simply that livestock are bad for cacti and bunchgrasses, because there is little or no seedling establishment under grazing; but in the nonequilibrium view, it appears possible that grazing facilitated subsequent regeneration by eliminating competing plants. To put it another way, for many species, regeneration may often occur not under some particular set of conditions, but in the *transitions* between conditions—in these two cases, between the grazed and ungrazed states.

If so, fluctuating management would be preferable to the normal approach of attempting to determine a specific (equilibrium) condition that favors the species of concern and then imposing that condition. There is great concern today about many long-lived plant species that do not appear to be regenerating in sufficient numbers to sustain themselves. Given the nonequilibrium reality that seedling establishment is likely to be highly variable, however, this concern often may be misplaced. On the other hand, poor regeneration, if continued for a sufficiently long period of time, could entirely eliminate a species, so it does not seem appropriate to ignore evidence of poor regeneration either.

25.4 The Example of Fire Management

25.4.1 History

The history of fire management in the United States illustrates many of the complexities, ironies, and trends that have occurred in environmental management and conservation (Pyne, 1982, 1997). Some Europeans brought traditional fire practices with them to North America, whereas many others adapted Native American practices, especially in the South and the West, so that a majority of Americans came to regard burning as appropriate and ecologically sound if done properly. Meanwhile in Europe, intensification of farming practices and the rise of silviculture as a scientific discipline, which began in Germany in the nineteenth century in line with the Enlightenment program of bringing order to nature, thus improving it, and spread rapidly to neighboring countries, created an anti-fire bias that persists to the present day. In colony after colony around the globe, European powers attempted to impose a policy of fire suppression on peoples who used fire for many purposes. Even after independence, scientists and government land managers typically continued to recommend fire suppression because of their European training. And in country after country, the policy was eventually abandoned because it did not work. The same was true in the United States. Despite the fact that the United States had achieved independence long before scientific forestry arose in Europe, American academics still looked to Europe for intellectual leadership during the nineteenth century. Pinchot's training in Germany and France in scientific forestry accorded him considerable authority in the United States. Thus, the conservationist wing of the environmental movement became staunchly anti-fire, and the Forest Service and other government land agencies became committed to a policy of fire suppression. Pinchot went so far as to equate allowing fire to allowing slavery, that is, he saw fire suppression in idealistic terms.

Opposition to this policy arose especially in two regions, California and the South. In California, a "let-burn-and-light-burn" policy was proposed: let fires burn when they are far from settled areas, and set light fires intentionally in forests that are located near human habitation. This is similar to the policy that the Forest Service and other agencies follow today. The proponents of this policy were timber owners, ranchers, settlers, the Southern Pacific Railroad, a few civil engineers, and two prominent writers, Joaquin Miller and Stewart Edward White (who owned some timber). They justified light burning on ecological grounds and also because it was the tried and true Native American way. Pinchot's two successors as Forest Service Chief, in contrast, dismissed light burning as "Paiute forestry," the implication being that aboriginal practices were not scientifically justified. Another opponent of fire suppression was Los Angeles water czar William Mulholland, whose arguments were surprisingly modern and ecologically sound (Minnich, 1987). In short, the people who supported light burning were those whom environmentalists today tend to view as the "bad guys," whereas not only Pinchot's contingent but also preservationists supported fire control. For instance, John Muir bitterly opposed the setting of fires in Sierran forests (Muir, 1897; Pyne, 1982, 1997).

By the 1920s, the Forest Service had "won" this debate by publishing its research purporting to demonstrate the advantages of complete fire suppression. The focus of opposition then shifted to the South, where the Forest Service had been acquiring abandoned lands since the passage of the Weeks Act in 1911. There, opposition to the Forest Service's attempts to impose fire suppression was far more widespread, though concentrated among livestock owners and other individuals who had burned the piney woods for generations. Imposition of fire control had pronounced negative effects on the ability of these people to obtain their traditional livelihoods. Possibly many Southerners considered Forest Service officials to be carpetbaggers. Unlike in California, the Forest Service did not have a monopoly on southern fire research, and consequently, it was never able to impose its viewpoint on the academic community. Important articles by Yale forester H.H. Chapman (1926), Biological Survey wildlife biologist H.L. Stoddard (1931), and S.W. Greene (1931) of the Bureau of Animal Husbandry were sufficiently persuasive that the Forest Service in 1932 modified its policy to allow intentional, or prescribed, fire under certain tightly circumscribed circumstances. Thereafter, the Forest Service's opposition to prescribed fire in the South gradually loosened, as more and more information accumulated that fire was essential to the management of longleaf and other commercial pines. At the same time, however, the Forest Service continued to promote fire suppression as national policy. It initiated its extremely persuasive public relations campaign using Smokey the Bear in 1945. Smokey was soon adopted in Canada, and in Mexico became Simon le Oso.

Nonetheless, with the South as an alternative to the prevailing paradigm, managers and researchers in other regions began to consider the possibility that their ecosystems also might benefit from fire. For instance, in 1947, H.H. Biswell brought his experience with fire in Georgia to the University of California, Berkeley, and became such an advocate of prescribed fire (e.g., Biswell, 1958) that successive cohorts of forestry majors knew him as Harry the Torch. Although he was apparently somewhat irascible, which caused some individuals to dig in their heels in opposition to his proposals, Biswell's status at Berkeley gave his ideas the weight of authority, and ultimately he was highly influential. For instance, he probably had a major influence on A. Starker Leopold, also a Berkeley professor and the primary architect of the 1963 Leopold Report, which recommended that the National Park Service employ fire systematically to recreate the landscapes that had existed at the time of the first European contact. Until then, the Park Service had largely followed the lead of the Forest Service, which had always had a reputation for greater scientific expertise than the other land management agencies. But by 1968, the Park Service had a policy of allowing natural fires to burn, if possible, and sometimes setting intentional fires; the Forest Service would not follow suit for another 10 years, though many individual managers had been modifying strict fire suppression for some time. In general, by the 1970s, fire suppression was in full retreat, as ecologists and land agency managers recognized that the policy had, in many ecosystems, caused a buildup of fuels to levels that eventually created conflagrations that overwhelmed firefighting capacity. Scientists, land managers, and environmentalists recognized that fire is often good, not bad, for ecosystems.

Then came the great Yellowstone fires of 1988, which the National Park Service initially allowed to burn (fig. 22.9). This created a huge public relations problem because the Forest Service had decades earlier been so successful, through its Smokey the Bear campaign (the graphic fire scenes in *Bambi* also contributed), in convincing Americans that fires should be stamped out as soon as they are discovered. In contrast, environmentalists assumed that the fires got out of hand because of the failings of the previous policy, that is, because of a buildup of fuel under fire suppression. But as fire ecologists pointed out, Yellowstone's lodgepole pine ecosystem appears to be one that burns naturally every 200 years or so, building up huge amounts of fuel in the interim. In other words, conflagrations are natural in the Yellowstone environment, and the 1988 fires did no harm to park ecosystems (e.g., Romme and de Spain, 1989; Sprugel, 1991). Of course, the problem is that the fires spread out of the park onto adjoining private land, where they damaged human livelihoods. On the other hand, the Forest Service, which did not have Yellowstone Park's extreme let-burn policy, was able to suppress all but one of the over 200 fires that started that summer on national forest lands adjacent to the park.

25.4.2 Some Implications

The history of American fire management is exemplary in several respects. First, it illustrates that environmentalist policies are not always environmentally sound. The Smokey the Bear policy was an environmentalist policy; so also was the let-burn policy that replaced it, and it also was flawed (though less so), as the Yellowstone fires demonstrated. Second, the Yellowstone fires illustrate that a "nature knows best" or "leave it alone" policy often produces untoward effects (compare with Botkin, 1990). Third, it is difficult to fit American environmental history into the Procrustean bed of a contest between despoilers and conservers. It was the timber barons, stockmen, railroad companies, and Los Angeles water czar who urged a fire policy that today seems ecologically sound, whereas it was the environmentalists who pushed for complete fire suppression. A fourth implication of American fire management history is that changes in paradigms often, perhaps always, must come from without, and sometimes from surprising sources. Who could have predicted that Southerners would play an important role in overthrowing what seemed to be a scientifically established policy? Or that a contributing factor would be lingering southern resentment of the North, and hence of the federal government and its officials?

This leads to another point, the importance of respecting local- or regional-scale folk knowledge. To cite another example, Newfoundland fishermen warned Canadian government officials that their once remarkable fisheries were in danger of being depleted, but the government officials and scientists did not believe them until it was too late (Cameron, 1990; Neis, 1992; Mitchell, 1997). (It also is important to recognize that folk knowledge is neither perfect nor unbiased; see, for example, Aageson, 1998). Additional examples of environmental degradation due in part to the failure of Western scientists and government officials to pay attention to folk knowledge are legion in the Third World (Richards, 1985).

Finally, there are some interesting parallels between beliefs about fire and those about grazing on western rangelands. In Clementsian terms, both fire and grazing are disturbances that interfere with establishment of the presumed climax. Consequently, it is not surprising that a certainty exists today regarding the negative impacts of livestock "overgrazing" that is similar to the certainty that formerly existed regarding fire. Even the presumed negative effects of fire—vegetation destruction, threats to wildlife, soil erosion and its impacts downstream (Pyne, 1982)—were almost identical to the (presumed) negative effects of grazing that concern environmentalists today. The major substantive difference is that it was eventually accepted that fire is natural, whereas domestic livestock may never be regarded as such, given that they are introduced from the Old World. Nonetheless, I believe that eventually it will be accepted that grazing is no more of a problem in the western United States than in Africa or the Middle East, where careful empirical studies have demonstrated that for the most part overgrazing is a myth (Behnke and Scoones, 1992; Warren, 1995; Blumler, 1998). Ironically, it has been easier to disprove the overgrazing myth in Africa than in North America, for the following reasons: (1) there are few antigrazing environmentalists in Third World countries; (2) crucial scientific research was conducted by outsiders, who were not entangled in government bureaucracies; and (3) Americans readily accept that in the past they may have been patronizing toward African pastoralists, therefore overestimating the environmental degradation such peoples have produced, whereas it is more difficult for Americans to see that they might be equally patronizing today toward the mostly white ranchers of their own western range. In an article that greatly influenced range ecologists, Westoby et al. (1989) illustrated how a nonequilibrium perspective alters what is seen as proper grazing management, compared to the traditional viewpoint. However, the environmental NGOs concerned with the western range remain firmly within the equilibrium paradigm.

25.4.3 The Current Situation

In a sense, the Yellowstone fires were an example of what Pyne terms an "intermix" fire, one that spreads from wildlands into places where people live. The Oakland hills fire of 1991 and the frequent chaparral fires in southern California are the best known examples, but less dramatic fires of this sort have occurred in recent years in several other parts of the country. Obviously, a let-burn policy is hazardous in such places. On the other hand, light burning to reduce the likelihood of conflagration is also problematic, because the government agency setting the fire would be liable if it were to escape and destroy dwellings. This problem was highlighted during the summer of 2000 by the controlled burn that escaped from the forest around Los Alamos, New Mexico. Minnich (1983) contrasted fires in southern California and the similar environment of northern Baja California, Mexico. The former has been under a regime of active fire suppression throughout most of this century, whereas in the latter, fires are usually allowed to burn. Average fire size in northern Baja California has been small, in southern California very large; the change in size occurs right along the border. Thus, it is unlikely to be due to environmental differences between the two regions, but is probably due instead to human management. The implication is that southern California would have fewer conflagrations if it adopted a let-burn policy; unfortunately, that is unlikely to happen given the liability issue.

In the future, fire managers probably will find themselves increasingly constrained by competing environmental concerns. For instance, air pollution laws prevent prescribed fire in many places where it might otherwise be useful. Moreover, fire has become a global climate change issue, since it emits greenhouse gases to the atmosphere (Levin, 1991; Crutzen and Goldammer, 1993). Global change modelers have become interested in determining the extent of anthropogenic fire before modern times, a question formerly of concern primarily to biogeographers, to serve as a baseline for comparison with current rates of biomass burning (Clark and Royall, 1995; Robock and Graf, 1994; Woodcock and Wells, 1994). I believe Pyne's (1997) assessment that the global change modelers have a serious anti-fire bias is overstated but not entirely invalid. Thus it is possible that fire suppression will make a comeback as a component of greenhouse warming regulations, depending on how current debates over premodern burning are resolved (e.g., by the Iroquois in Ontario; Campbell and McAndrews, 1995; Clark, 1995; Clark and Royall, 1995; Byrne and Finlayson, 1998).

Fire also interacts with pathogens. For instance, I believe that it might be possible to use prescribed fire to reduce the incidence of Lyme disease. Fire is thought to reduce the density of ticks, which spread the disease (caused by a spirochete). It also would eliminate much of the preferred vegetation ticks use to disperse onto their large mammalian hosts. In addition, fire almost certainly would reduce the density of deer mice, an important component of the spirochete's life cycle in the eastern United States. Unfortunately, given the complex interfingering of tick-harboring woods and suburban tracts in many centers of Lyme disease today, prescribed fire often is not an option. Even so, it might be possible, by creating firebreaks in rural areas adjoining Lyme disease foci, to employ fire to prevent the dispersal of the disease to new areas from its currently restricted range. Similar arguments may apply to many pests that attack native vegetation.

Despite the numerous mistakes in fire policy, our forests are in reasonably good shape. North American forests are growing back more rapidly than they are being cut and may be serving as a major carbon sink, mitigating greenhouse warming (MacCleery, 1993; Moffat, 1998). In part this is due to resource substitution—trees no longer have to be chopped down to supply wood to heat homes through long, northern winters—and in part due to agricultural intensification, which has rendered much subsistence farmland in the eastern United States and Canada inviable and caused its abandonment. North American rangelands also are in better shape than generally recognized, again despite some noticeable errors in management (Hess, 1992). In the old equilibrium view of nature, human impacts are almost inevitably destructive; but under the newer, nonequilibrium paradigm, the picture becomes far more complicated.

25.5 Another Example: California Grasslands

Domestic livestock arrived in California in 1769, along with the Spanish. Thus, livestock grazing has a longer history in California than in most of the West. At times, livestock built up to high numbers, usually followed by a collapse triggered by drought, flood, or changes in market conditions. Nonetheless, range productivity appears to be at least as high today as it was initially, despite the fact that the fertile bottomlands, the most productive part of the original rangeland, are now largely converted to other uses. Burcham (1956) attempted to make the case that grassland productivity had declined due to overgrazing, but his data belie his conclusion. Productivity today (Janes, 1969; Blumler, 1992a) is considerably higher than the figures he calculated for the nineteenth century.

Initially, grazing was concentrated in what is now known as "valley grassland," as well as the interfingering oak park forests, which together covered much of the lower elevation portions of the state (Griffin, 1977; Heady, 1977). These ecosystems sport an extremely high number of endemic plant species, most of them flowering annuals (Raven and Axelrod, 1978; Blumler, 1992b). But today the herbaceous cover is dominated by introduced plants from the Mediterranean region, also mostly annuals, which have been spreading in California since before Spanish settlement (Blumler, 1995; Mensing and Byrne, 1997). The invaders—wild oats (*Avena fatua* and *A. barbata*), filaree (*Erodium cicutarium*, *E. botrys*, and *E. obtusiplicatum*), brome grasses (numerous *Bromus* spp.), bur clover (*Medicago polymorpha*), and many, many others—have been so succcessful that they typically comprise 80–90% of the vegetative cover on both disturbed (grazed) and undisturbed sites (Bartolome, 1979). Thus, they pose a major threat to native plant species. However, some current attempts to conserve native species seem likely to prove counterproductive, because of misapplications of ecological theory and the overarching influence of landscape ideals (Blumler, 1992b).

The invasion and replacement of the native vegetation was so rapid that it may be impossible to determine with any high degree of confidence the makeup of the pre-European ecosystem. Paleobotanical (phytolith) investigation of a single site suggests that perennial bunchgrasses such as the needlegrasses (*Nasella* spp.) may have been more common than they are today (Bartolome et al., 1986). In contrast, geographers examining the earliest accounts left by Spanish and other explorers have concluded that native annuals dominated many areas (Wester, 1981; Blumler, 1992b, 1995; Mattoni and Longcore, 1997; Minnich and Dezzani, 1998; Minnich and Franco Vizcaino, 1998). Current habitat preferences of native species also suggest that native annuals have been displaced to a greater extent than native perennials. Moreover, although spectacular displays of native annual wildflowers still occur, they have been declining for at least the past century (Roof, 1971). In contrast, there is little evidence for significant decline in bunchgrass abundance during the same time period. Nonetheless, environmental NGOs, such as the Nature Conservancy that are concerned with biodiversity conservation, and government agencies overseeing parks and preserves have almost unanimously assumed that the pre-European ecosystem was bunchgrass-dominated. Management tends to focus on expunging alien species and "restoring" a bunchgrass landscape. This can be achieved through the careful use of fire, grazing, mowing, herbicides, or other disturbance, though it is unlikely that aliens ever can be eliminated entirely (e.g., Menke, 1992).

Clements (1920, 1934) first claimed that perennial bunchgrasses were the natural dominants of California grasslands, based on his model of succession. Evidence he presented to support his opinion was weak (Blumler, 1995), whereas under today's paradigm about succession, it is plausible that annual plants could have dominated (Blumler, 1984). Before Clements came to California, a consensus had long existed that the grassland was naturally dominated by annuals. Yankee settlers in the mid–nineteenth century quickly accepted this as fact and managed their livestock accordingly. They understood many facets of California range ecology that scientists had to relearn in the twentieth century, in part because the scientists were misled by Clementsian theory. Scientists spent considerable effort attempting to establish the presumed climax bunchgrasses on the range, with little success (Kay et al., 1981). The current efforts to establish bunchgrass landscapes seem more successful, probably because they do not need to be economically viable, as a livestock operation must.

The bunchgrasses used in current conservation efforts (notably purple needlegrass, *Nasella pulchra*) are not rare species. The emphasis is on restoring a hypothetical, idealized landscape rather than on biodiversity conservation. Similar projects entail planting and protecting oaks (*Quercus* spp.), which are generally believed to be less abundant today than formerly, despite some evidence to the contrary (Mensing, 1992). There seems to be an unconscious assumption underlying these projects that all native plants are coevolved, and so the elimination of aliens and the establishment of one or two key native species will favor the rest (Blumler, 1992b). This assumption makes more sense under an equilibrium view of nature than under a nonequilibrium one. Under the latter perspective, it is likely that at least a few native species will be hurt by the manipulations that also harm aliens (Blumler, 1992b). Moreover, if bunchgrasses or oaks shade the ground too densely, it is possible that additional native species will suffer. Since some of these natives may be rare and endangered, this is a cause for concern.

Most environmentalists concerned with California grasslands share a common goal of preserving as many native plant species as possible, yet they can become embroiled in contentious arguments because their visions of what the pristine landscape looked like differ and because management sometimes focuses on landscapes rather than on biodiversity. The following is an example of the emotionalism that characterizes these debates:

> the once-great [Bay Area] wildflower fields have . . . been utterly ruined by ill-advised 'forestation' . . . carried out by people who have no remote concept of what they are doing . . . The end result is the rendering of once open, sunny, wildflower-bearing land into foreboding and pestilential jungles, of no earthly use to man, bird or beast.
>
> the perversion of the land continues as an annual ritual, featuring mental defectives with armloads of . . . trees on the search for 'barren places' to 'reforest' and convert to useless woodlands in need of costly fire protection.
>
> If one tries, [wildflower enthusiasts] should be authorized to strangle him with one of his own cheap transplants—a truly appropriate demise.—Roof, 1971: 6, 13

In this sort of atmosphere, I have found it difficult to convey the simple point that our emphasis should be on biodiversity, not on landscape and on native species number, not native purity. Prior to "restoration," the area should be surveyed and the native species already present identified. Native species should be carefully monitored to ensure that none are seriously harmed by the manipulations used to favor bunchgrasses or oaks. Analogous attempts to impose idealized visions upon the land have long bedeviled conservation and resource management throughout the West (Hess, 1992).

25.6 Current Debates

> We must first relinquish our hopes for utopia if we really wish to save the earth.—M. Lewis, 1992: 250

25.6.1 Sustainability

Currently, there is enormous interest in achieving "sustainable development," especially in the Third World, but also in North America. There also is considerable debate about whether this is even feasible, or whether the term is an oxymoron. In addition, some feel that sustainable development may be achievable, but only at the expense of biodiversity. Sustainable development is often considered a new concept, but Pinchot and the Progressives also attempted to manage for sustained yields (e.g., Bunting, 1994).

One difference between the Pinchot conservationists and today's advocates of sustainable development is that many of the latter accept the Marxist critique of capitalism (Lewis, 1992). Today, theorizing about sustainable development often incorporates an emphasis on the importance of ensuring equity. It is often argued that sustainable development (and conservation) cannot succeed without ensuring that all "stakeholders" have a voice in making policy, and that all ethnicities, genders, classes, and so on are equally empowered and have equal opportunities for advancement (e.g., WCED, 1987). However, there seems to be precious little empirical evidence that equity actually is relevant (Kiss, 1998). Whatever one's opinion of Marxism, for instance, there is no disputing that pollution levels in the Soviet Union and its eastern European satellites were extraordinarily high (Lewis, 1992), despite a distribution of wealth that was more equitable than in capitalist countries such as the United States and Canada. Nonetheless, one of the points I tried to make in the section on fire management is that top-down approaches tend to be flawed; it is important to consider the viewpoints of those who have a history of involvement with the resource in question, no matter how unscientific they may appear.

25.6.2 Species Extinctions

Diamond (1989) listed four causes of the global extinction crisis: habitat destruction, invasions, overexploitation, and global climate change. But, given a temperature increase of only about 0.5°C in the twentieth century, well within the magnitude of natural climatic change in some past centuries, global climate change may not have caused any extinctions as yet. At the same time, it must be considered a potential future cause of extinctions on an unprecedented scale. Similarly, it seems likely that pollution, and the alteration of the biogeochemical cycling of nutrients such as nitrogen, may be additional factors causing a small but increasing number of extinctions (Vitousek et al., 1996; Minnich and Dezzani, 1998). Overexploitation primarily affects species of economic importance; there are well-known examples from North America, such as the now-extinct passenger pigeon, as well as cases of species that were nearly driven extinct but then salvaged, such as bison, tule elk, and several marine mammals. Although exploitation can easily drive population numbers down close to the point of extinction, under normal circumstances it becomes increasingly uneconomic to continue the exploitation as the species becomes increasingly rare; presumably, such negative feedback has helped to save many species. Exceptions occur when the species being exploited commands a high market value. This is currently so, for instance, of American ginseng (*Panax quinquefolium*), which apparently is being depleted throughout its range. It may also pose a threat to black bear, because bear gall bladder commands a very high price in east Asian markets (Gauthier, 1995).

Habitat destruction is unquestionably the major cause of species extinctions in much of North America, though there are a few localized regions where invading aliens may be a more important factor. The effects of habitat destruction on species loss are generally modeled according to the tenets of island biogeography theory (MacArthur and Wilson, 1967), specifically the well-known relationship between land area and number of species. This relationship predicts that approximately 50% of the extant species will be eliminated if 90% of the original habitat is destroyed, and it is generally assumed that this will occur instantaneously (e.g., Wilson, 1992). However, this assumption ignores the likelihood that habitat islands will remain "supersaturated," that is, that they will hold more species than one would predict from the theoretical equation, after destruction of surrounding areas. Such places should eventually undergo "relaxation" down to the equilibrium number of species, but this may take thousands of years (Brown and Gibson, 1983). An excellent example is the tallgrass prairie, which has suffered greater habitat destruction than any other American ecosystem. Approximately 98% of the original tallgrass prairie has been destroyed. One would predict, then, an eventual loss of almost three-fourths of the native species, but thus far there has been very little extinction. Local extinctions are frequent, but since different species go extinct in different prairie patches, overall species loss is mitigated (Leach and Givnish, 1996). On one hand, then, ecologists have a window of time to save species after habitat destruction; on the other hand, if habitat is not restored, in the long run extinction may be major.

Although habitat destruction is the major focus of biodiversity conservation concerns, invading aliens sometimes pose a more serious threat (Drake and Mooney, 1986). Invading species are a far more serious problem in the United States than in Mexico, reflecting the fact that developed nations intentionally introduce organisms as garden ornamentals or pets, for pasture improvement, and so on, at a far higher rate than Third World nations do. For instance, a number of plant species are currently spreading into northern Baja California from southern California, where they were first introduced (Minnich and Franco Vizcaino, 1998). Buffelgrass (*Cenchrus ciliaris*), a major invader in Sonora, was intentionally introduced into northern Mexico but only after it had been intentionally imported into Arizona. Canada also is less affected by introduced species than the United States, presumably because of its less hospitable climate and much smaller population.

Invading aliens currently pose the most serious threat

to aquatic ecosystems, probably because of the ease with which humans inadvertently transport such organisms (Cohen and Carlton, 1998; Sawicki, 1998). In terrestrial habitats, alien species are especially problematic in California and Florida, two states where development is proceeding apace and where there are large numbers of vulnerable endemics. In general, habitat destruction poses the more severe threat to many local endemics, whereas invading species may be a more serious danger, in the long run, to natives with extensive geographical ranges. For instance, many annual plants and some perennials native to California grasslands that are not currently considered endangered have been declining in abundance since the mid–nineteenth century (Blumler, 1992b, 1995). They are declining in abundance even in parks, preserves, and pastures, primarily as a result of their progressive replacement by aliens. If this trend continues, it is only a matter of time before many of these species are classsified as endangered. Precisely because they are not local endemics, their rarity may not be recognized until it is too late.

Myths abound about invading species, for instance, that they require disturbance and that they lower ecosystem productivity and biomass. Although it is certainly true that aliens often benefit from disturbance, it is not always the case (Sauer, 1988; Blumler, 1995). And even though they probably sometimes decrease productivity, invaders also may increase it. The classic example is the myrtle (*Myrica faya*), introduced into Hawaii, where it colonizes fresh lava flows and increases soil fertility through the ability of its symbiont to fix nitrogen. This in turn changes the entire course of succession, generally favoring other introduced species (Vitousek et al., 1987). Attempts to favor natives over aliens in California grasslands foundered for a long time on the assumption that the latter species must require disturbance (Blumler, 1992a). Fortunately, however, California preservationists began to abandon this assumption several years ago.

Invading species are a diverse group; but science is concerned with discovering general relationships or "laws." To state that invaders require disturbance is probably fine as a generalization. But when taken as a hard and fast rule for management purposes, the inevitable exceptions rear their heads and pose problems. Similarly, it was proposed recently that invading grasses succeed by shortening natural fire intervals, thereby replacing native plants that are adapted to less frequent fire (D'Antonio and Vitousek, 1992). Although this may well be valid for many alien grasses, such as cheatgrass (*Bromus tectorum*) in the intermountain West, it certainly is not true of others, such as the characteristic dominant alien grasses in California: wild oats, soft chess (*B. mollis*), and ripgut brome (*B. diandrus*). In fact, frequent fire is successfully employed today to favor native grasses, legumes, and forbs over these intruders. Fortunately, California preserve managers had already adopted fire as a management strategy before the D'Antonio and Vitousek article was published, else they might have wasted time attempting to favor natives by suppressing fire (as was attempted earlier in this century in accord with Clementsian expectations). In short, management must be empirical and flexible, rather than overly reliant on sweeping theories. It is important to recognize that not all alien species are alike and that some are more dangerous than others. Some managers attempt to eliminate all aliens indiscriminately, but it is preferable to attempt to determine which species pose the most serious threats and to concentrate on them. The purported detrimental impacts of a few celebrated cases, such as purple loosestrife (*Lythrum salicaria*), apparently have never been documented empirically, but rather are assumed (Sawicki, 1998).

25.6.3 Responding to Global Change

Concern over global climate change, as well as related changes such as in biogeochemical cycling, is increasing rapidly. It is clear that the magnitude of human-induced changes to the atmosphere and interacting systems will continue to increase for the forseeable future. It is also clear that the potential impacts could be major. However, the complexity of Earth's ocean–atmosphere–biosphere system and the diversity of chemicals that we are producing make it difficult to predict exactly what these impacts will be, and therefore, exactly what sort of policy should be enacted. The basic procedure for attempting to predict global change involves three steps: progressive modeling, with the aid of powerful computers; monitoring what is actually happening in the real world, with the aid of satellite remote sensing; and mapping the results in a Geographical Information System. Results are compared with computer model predictions, and if there are discrepancies, the model is modified. This procedure has allowed climatologists to make great strides in comprehending our climate system, but it is unreasonable to expect that we will ever be able to predict future climate with absolute confidence. It will be necessary to make predictions, and formulate policy, in the face of considerable scientific uncertainty (Hare, 1995).

Currently, policies focus on reducing the magnitude of future global change, by reducing greenhouse gases and other atmospheric contaminants. Although this is probably appropriate, it illustrates that our society is vulnerable to environmental change. Since natural climate change can be extremely rapid, and since we almost certainly will induce significant changes in the climate system even if we do reduce greenhouse gas emissions (Warrick, 1993), perhaps we should also attempt to transform our society so that it can better tolerate change.

At a local and regional scale, planning and policy response to global climate change for the most part is based on accepting computer model predictions as though they are completely reliable, which they are not. It would be preferable if planners were given information concerning the reliability of the various forecasts, but climate modelers thus far have been so concerned about persuading people that global warming is real that they often have glossed over the uncertainty in their forecasts in their communications with government officials and social scientists. In fact, some portions of the outputs from global circulation models are more reliable than others. For instance, most if not all climate modelers accept that the predictions for temperature are more reliable than those for precipitation. However, this sort of information has not yet filtered down to those involved in planning. A shift to presenting climate predictions in the form of confidence intervals rather than as single values is necessary. Then, policy makers could plan for those eventualities that are most certain to occur, while the recognition that there is considerable uncertainty about some facets of global climate change might encourage policies that allow flexibility in the face of the inevitable future surprises.

25.7 Conclusion

Some individuals view environmental problems in apocalyptic terms, and it certainly is not out of the question that we might unwittingly produce a major global environmental disaster in the near future. Nonetheless, as we enter the twenty-first century, North American biological resources are in reasonably good shape. Forests and rangelands have proven remarkably resilient to our numerous errors in management—or perhaps, in a nonequilibrium perspective, such resiliency is not so remarkable after all. The major concern should be with fisheries, many of which are endangered. In addition, there is legitimate concern over likely future species extinctions, even though the number of species eliminated since European settlement has not been great. Extrapolating from island biogeography theory, it is likely that many species indeed will be threatened with extinction as we continue to destroy natural habitat; on the other hand, since the process is unlikely to be instantaneous, we probably have a considerable window of time to devise effective mitigation strategies.

A complicating factor, likely to become an increasingly problematic aspect of environmental management, is conflict between goals. For instance, the legitimate desire to prevent soil erosion after mining conflicts with the goal of reestablishing the natural vegetation (Brown, 1996); the desire to let fire have its natural place in ecosystems can conflict with air pollution laws, and so on. Our society tends to deal with such conflicts through the creation of complex laws and regulations, with bureaucracies to enforce those regulations. This can produce an ossified sort of environmental management, which sometimes becomes highly counterproductive (Hess, 1992; Lovich and de Gouvenain, 1997).

In addition, since environmental problems are real-world problems and since the human imprint on the land today is so spatially complex as well as pervasive, the incorporation of geography into environmental planning will become ever more critical (Marsh, 1998). Unfortunately, geographical considerations often are given short shrift in planning today, because of the lack of training in the relevant subject areas. On the other hand, many scientists in other disciplines are devising methods of spatial analysis to aid them in their study of the environment. Perhaps the best known of these approaches is landscape ecology, which in my opinion has thus far produced something of a mixed bag; but at least it incorporates the realization that spatial pattern matters.

This chapter has sought to illustrate that there is a real danger in thinking we know best or are more pure in our motives than others. This certainly was true of Pinchot and his colleagues, and it created the problems in fire management outlined previously. However, it seems equally true today of the preservationist wing of the environmental movement, which has managed to gain so much political clout. It also can be true of scientists, when they dismiss experientially obtained but nonscientific knowledge such as folk tradition, and it is true of those postmodernists and critical social theorists who dismiss science as "positivism." As we enter the next millennium, it would be helpful to accept how little we actually know and that we never will understand the environment with certainty. I have also tried to suggest that the Cartesian tendency to view issues in idealized, highly polarized terms is on the whole inaccurate. In another context, Toulmin (1990) has called for a return to the skeptical humanism of Montaigne and Shakespeare, put forth during the time period just before Descartes. This should encourage the flexibility of thought and action that I believe will be at a premium in the coming years of massive environmental change.

Of course, science has a very important role to play in achieving appropriate environmental and resource management goals. The general public seems to think that science is about certainty, but this is no longer true. Ever since the quantum mechanics revolution, the scientific world view has become increasingly probabilistic. Contra Einstein, God *does* play dice with the universe! Probabilism is particularly acute in ecology and environmental science, because nature is so complex. What scientific study of these fields is about, or should be about, is the estimation of the odds (and the confidence interval around that estimate) that a thing will happen in a certain way, and the constant revision of those odds as new data become available. It seems to me that probabilism can be taken as the formal, twentieth and twenty-first century scientific equivalent of Montaigne's skeptical humanism. Consonant with these thoughts is the increasingly popular approach to environmental problems known as "adaptive environmental management" (Holling, 1978; Walters, 1986; Kessler and

Salwasser, 1995; Pickett and Ostfeld, 1995; Mitchell, 1997). The emphasis in this approach is on management as a series of experiments, with the presumption that there will be repeated surprises and failures to which managers must respond by modifying their actions. Thus, the approach is experiential, flexible, and pragmatic, rather than positivist. It is this sort of approach, I believe, that will characterize the twenty-first century.

Acknowledgments I wish to express my gratitude to the educators in the College of Natural Resources, University of California, Berkeley, especially R. Wakimoto, A. Schultz, J. Bartolome, P. Zinke, C. Turner, J. McBride, A. Sylvester, and E. Stone, who gave me a grounding in historical, applied, and policy aspects of ecological issues to which ecology majors are seldom exposed. They also introduced me to alternative paradigms, enabling me to be more critical of prevailing views within the science of ecology than I would have been otherwise. The perspectives of these scholars varied, which in itself was good. But they straddled the divide between the traditional, Clementsian, equilibrium ecology, and the new nonequilibrium perspective that was beginning to gain adherents, and, generally speaking, they respected both.

References

Aageson, D.L., 1998. Indigenous resource rights and conservation of the monkey-puzzle tree (*Araucaria araucana*, Araucariaceae): A case study from southern Chile. *Economic Botany*, 52, 146–160.

Bartolome, J.W., 1979. Germination and seedling establishment in California annual grassland. *Journal of Ecology*, 67, 273–281.

Bartolome, J.W., and B. Gemmill. 1981. The ecological status of *Stipa pulchra* (Poaceae) in California. *Madroño*, 28, 172–184.

Bartolome, J.W., S.E. Klukkert, and W.J. Barry, 1986. Opal phytoliths as evidence for displacement of native Californian grassland. *Madroño*, 33, 217–222.

Behnke, R.H., and I. Scoones, 1992. Rethinking range ecology: Implications for rangeland management in Africa. *Overseas Development Institute Dryland Networks Programme Issues Paper*, International Institute for Environment and Development, London.

Biswell, H.H., 1958. Prescribed burning in Georgia and California compared. *Journal of Range Management*, 11, 293–297.

Blumler, M.A., 1984. *Climate and the annual habit.* M.A. thesis, University of California, Berkeley.

Blumler, M.A., 1992a. *Seed weight and environment in mediterranean-type grasslands in California and Israel.* Ph.D. thesis, University of California, Berkeley.

Blumler, M.A., 1992b. Some myths about California grasslands and grazers. *Fremontia*, 20(3), 22–27.

Blumler, M.A., 1993. Successional pattern and landscape sensitivity in the Mediterranean and Near East. *In* Thomas, D.S.G., and R.J. Allison (eds.) *Landscape Sensitivity.* John Wiley and Sons, Chichester, 287–305.

Blumler, M.A., 1995. Invasion and transformation of California's valley grassland, a Mediterranean analogue ecosystem. *In* Butlin, R., and N. Roberts (eds.) *Human Impact and Adaptation: Ecological Relations in Historical Times.* Blackwell, Oxford, 308–332.

Blumler, M.A., 1996. Ecology, evolutionary theory, and agricultural origins. *In* Harris, D.R., (ed.) *The Origins and Spread of Agriculture and Pastoralism in Eurasia.* UCL Press, London, 25–50.

Blumler, M.A., 1998. Biogeography of land use impacts in the Near East. *In* Zimmerer, K.S., and K.R. Young, eds., *Nature's Geography: New Lessons for Conservation in Developing Countries.* University of Wisconsin Press, Madison, 215–236.

Botkin, D.B., 1990. *Discordant Harmonies: A New Ecology for the Twenty-First Century.* Oxford University Press, New York.

Botkin, D.B., 1995. *Our Natural History: The Lessons of Lewis and Clark.* Grosset/Putnam, New York.

Botkin, D.B., and E. Keller, 1994. *Environmental Science: Earth as a Living Planet.* John Wiley and Sons, New York.

Brown, D.A., 1993. Early nineteenth-century grasslands of the midcontinent plains. *Annals of the Association of American Geographers*, 83, 589–612.

Brown, J.H., and A.C. Gibson, 1983. *Biogeography.* The C.V. Mosby Co., St. Louis.

Brown, R., 1996. *Theory versus reality in mine reclamation.* M.A. thesis, SUNY-Binghamton, Binghamton, NY.

Brown, R.C., 1968. The doctrine of usefulness: Natural resource and National Park policy in Canada, 1887–1914. In Nelson, J.G., and R.C. Scace (eds.) *The Canadian National Parks: Today and Tomorrow.* Vol. 1. University of Calgary, 94–110.

Brunson, M.W., and J.J. Kennedy, 1995. Redefining "multiple use": Agency responses to changing social values. In Knight, R.L., and S.F. Bates (eds.) *A New Century for Natural Resources Management.* Island Press, Washington, DC, 143–158.

Bunting, R., 1994. Abundance and the forests of the douglas-fir bioregion, 1840–1920. *Environmental History Review*, 18(4), 41–62.

Burcham, L.T., 1956. Historical geography of the range livestock industry of California. Ph.D. dissertation, University of California, Berkeley.

Byrne, R., and W.D. Finlayson, 1998. Iroquoian agriculture and forest clearance at Crawford Lake, Ontario. In Finlayson, W.D. *Iroquoian Peoples of the Land of Rocks and Water, A.D. 1000–1650: A Study in Settlement Archaeology.* Vol. 1. London Museum of Archaeology, London, Ontario, 94–107.

Cameron, S.D., 1990. Net losses: The sorry state of our Atlantic fishery. *Canadian Geographic*, 110(2), 28–37.

Campbell, I.D., and J.H. McAndrews, 1995. Charcoal evidence for Indian-set fires: A comment on Clark and Royall. *The Holocene*, 5, 369–370.

Carson, R., 1962. *Silent spring.* Houghton Mifflin, Boston.

Chapman, H.H., 1926. Factors determining natural reproduction of longleaf pine on cut-over lands in LaSalle Parish, Louisiana. *Yale School of Forestry Bulletin*, 16.

Christensen, N.L., 1988. Succession and natural disturbance: Paradigms, problems, and preservation of natural ecosystems. In Agee, J.K., and D.R. Johnson (eds.) *Ecosystem Management for Parks and Wilderness.* University of Washington Press, Seattle, 62–86.

Clark, J.S., 1995. Climate and Indian effects on southern Ontario forests: A reply to Campbell and McAndrews. *The Holocene*, 5, 371–379.

Clark, J.S., and P.D. Royall, 1995. Transformation of a northern hardwood forest by aboriginal (Iroquois) fire: Charcoal evidence from Crawford Lake, Ontario, Canada. *The Holocene*, 5, 1–9.

Clements, F.E., 1916. Plant succession: An analysis of the development of vegetation. *Carnegie Institute of Washington Publications*, 242, 1–512.

Clements, F.E., 1920. Plant indicators. *Carnegie Institute of Washington Publications*, 290, 1–388.

Clements, F.E., 1934. The relict method in dynamic ecology. *Journal of Ecology*, 22, 39–68.

Cohen, A.N., and J.T. Carlton, 1998. Accelerating invasion rate in a highly invaded estuary. *Science*, 279, 555–558.

Connell, J.H., 1978. Diversity in tropical rain forests and coral reefs. *Science*, 199, 1302–1310.

Cronon, W., 1983. *Changes in the Land*. Farrar, Straus, and Giroux, New York.

Crutzen, P.J., and J.G. Goldammer (eds.) 1993. *Fire in the Environment*. John Wiley and Sons, Chichester.

Cutler, M.R., 1995. Old players with new power: The nongovernmental organizations. In Knight, R.L., and S.F. Bates (eds.) *A New Century for Natural Resources Management*. Island Press, Washington, DC, 189–208.

D'Antonio, C.M., and P.M. Vitousek, 1992. Biological invasions by exotic grasses, the grass/fire cycle, and global change. *Annual Review of Ecology and Systematics*, 23, 63–87.

Davis, M.B., 1981. Quaternary history and the stability of forest communities. In West, D.C., H.H. Shugart, and D.B. Botkin (eds.) *Forest Succession*. Springer-Verlag, New York, 132–153.

Diamond, J., 1989. Overview of recent extinctions. In Western, D., and M.C. Pearl. *Conservation for the Twenty-First Century*. Oxford University Press, New York.

Drake, J.A., and H.A. Mooney (eds.) 1986. *Ecology of Biological Invasions of North America and Hawaii*. Springer-Verlag, New York.

Drury, W.H. and I.C.T. Nisbet, 1971. Inter-relations between developmental models in geomorphology, plant ecology, and animal ecology. *General Systems*, 16, 57–68.

Dubasak, M., 1990. *Wilderness Preservation: A Cross-Cultural Comparison of Canada and the United States*. Garland Publishing, New York.

Foster, J., 1998. *Working for Wildlife: The Beginning of Preservation in Canada*. 2nd ed. University of Toronto Press, Toronto.

Gauthier, D.A., 1995. The sustainability of wildlife. In Mitchell, B. (ed.) *Resource and Environmental Management in Canada*. 2nd ed. Oxford University Press, Toronto, 207–235.

Glacken, C., 1967. *Traces on the Rhodian Shore*. University of California Press, Berkeley.

Glenn-Lewin, D.C., R.K. Peet, and T.T. Veblen, 1992. Prologue. In Glenn-Lewin, D.C., R.K. Peet, and T.T. Veblen (eds.) *Plant Succession*. Chapman and Hall, London, 1–10.

Greene, S.W., 1931. The forest that fire made. *American Forests*, 37, 583–584, 618.

Griffin, J.R., 1977. Oak woodland. In Barbour, M.G., and J. Major (eds.) *Terrestrial Vegetation of California*. John Wiley and Sons, San Francisco, 383–415.

Hagen, J.B., 1992. *An Entangled Bank: The Origins of Ecosystem Ecology*. Rutgers University Press, New Brunswick, NJ.

Hardin, G., 1968. The tragedy of the commons. *Science*, 162, 1243–1248.

Hare, F.K., 1995. Contemporary climatic change: The problem of uncertainty. In Mitchell, B., (ed.) *Resource and Environmental Management in Canada*. 2nd ed. Oxford University Press, Toronto, 10–28.

Harris, L., 1995. The east coast fisheries. In Mitchell, B., (ed.) *Resource and Environmental Management in Canada*. 2nd ed. Oxford University Press, Toronto, 130–150.

Heady, H.F., 1977. Valley grassland. In Barbour, M.G., and J. Major (eds.) *Terrestrial Vegetation of California*. John Wiley and Sons, San Francisco, 491–514.

Henderson, N., 1992. Wilderness and the nature conservation ideal: Britain, Canada, and the United States contrasted. *Ambio*, 21, 394–399.

Hess, K., Jr., 1992. *Visions upon the Land: Man and Nature on the Western Range*. Island Press, Washington, DC.

Holling, C.S. (ed.) 1978. *Adaptive Environmental Assessment and Management*. John Wiley and Sons, Chichester.

Janes, E.B., 1969. Botanical composition and productivity in the California annual grassland in relation to rainfall. M.S. thesis, University of California, Berkeley.

Kay, B.L., R.M. Love, and R.D. Slayback, 1981. Revegetation with native grasses. I. A disappointing history. *Fremontia*, 9(1), 11–15.

Kessler, W.B., and H. Salwasser. 1995. Natural resource agencies: transforming from within. In Knight, R.L., and S.F. Bates (eds.) *A New Century for Natural Resources Management*. Island Press, Washington, DC, 171–187.

Kiss, A., 1998. [Letter to Editor]. *Science*, 281, 347–348.

Koppes, C.R., 1988. Efficiency, equity, esthetics: Shifting themes in American conservation. In Worster, D., (ed.) *The Ends of the Earth*. Cambridge University Press, Cambridge, 230–251.

Leach, M.K., and T.J. Givnish. 1996. Ecological determinants of species loss in remnant prairies. *Science*, 273, 1555–1558.

Levin, J., (ed.) 1991. *Global Biomass Burning: Atmospheric, Climatic, and Biospheric Implications*. MIT Press, Cambridge, MA.

Lewis, M.W., 1992. *Green Delusions: An Environmentalist Critique of Radical Environmentalism*. Duke University Press, Durham, NC.

Lovich, J.E., and R.C. de Gouvenain, 1997. Salt cedar invasion in desert wetlands of the southwestern United States: Ecological and political implications. In Majumdar, S.K., E.W. Miller, and F.J. Brenner (eds.) *Ecology of Wetlands and Associated Systems*. The Pennsylvania Academy of Science, Philadelphia, 447–467.

MacArthur, R., and E.O. Wilson, 1967. *The Theory of Island Biogeography*. Princeton University Press, Princeton, NJ.

MacCleery, D.W., 1993. *American Forests: A History of Resiliency and Recovery*. Forest History Society, Durham, NC.

Marsh, W.M., 1998. *Landscape Planning: Environmental Applications*. 3rd ed. John Wiley and Sons, New York.

Mattoni, R., and T.R. Longcore, 1997. The Los Angeles coastal prairie: A vanished community. *Crossosoma*, 23(2), 71–102.

McEvoy, A.F., 1988. Toward an interactive theory of nature and culture: Ecology, production, and cognition in the California fishing industry. In Worster, D., (ed.) *The Ends of the Earth.* Cambridge University Press, Cambridge, 211–229.

McIntosh, R.P., 1985. *The Background of Ecology.* Cambridge University Press, Cambridge.

McPhee, J., 1971. *Encounters with the Archdruid.* Farrar, Straus and Giroux, New York.

Meine, C.D., 1995. The oldest task in human history. In Knight, R.L., and S.F. Bates (eds.) *A New Century for Natural Resources Management.* Island Press, Washington, DC, 7–35.

Menke, J., 1992. Grazing and fire management for native perennial grass restoration on California grasslands. *Fremontia*, 20(2), 22–25.

Mensing, S.A., 1992. The impact of European settlement on blue oak (*Quercus douglasii*) regeneration and recruitment in the Tehachapi Mountains, California. *Madroño*, 39, 36–46.

Mensing, S.A., and R. Byrne, 1997. Pre-mission invasion of *Erodium cicutarium* in California. *Journal of Biogeography*, 25, 757–762.

Minnich, R.A., 1983. Fire mosaics in southern California and northern Baja California. *Science*, 219, 1287–1294.

Minnich, R.A., 1987. Fire behavior in southern California chaparral before fire control: The Mount Wilson burns at the turn of the century. *Annals of the Association of American Geographers*, 77, 599–618.

Minnich, R.A., and R.J. Dezzani, 1998. Historical decline of coastal sage scrub in the Riverside-Perris plain, California. *Western Birds*, 29, 366–391.

Minnich, R.A., and E. Franco Vizcaino, 1998. Land of chamise and pines: Historical accounts and current status of northern Baja California's vegetation. *University of California Publications in Botany*, 80, 1–166.

Mitchell, B., (ed.) 1995. *Resource and Environmental Management in Canada.* 2nd ed. Oxford University Press, Toronto.

Mitchell, B., 1997. *Resource and environmental management.* Addison Wesley Longman, Harlow, UK.

Moffat, A.S., 1998. Temperate forests gain ground. *Science*, 282, 1253.

Muir, J., 1897. The American forests. *Atlantic Monthly*, 80, 145–157.

Nash, R., 1968. Wilderness and man in North America. In Nelson, J.G., and R.C. Scace (eds.) *The Canadian National Parks: Today and Tomorrow.* Vol. 1. University of Calgary, 66–93.

Nash, R., 1971. *Wilderness and the American Mind.* Yale University Press, New Haven, CT.

Neis, B., 1992. Fishers' ecological knowledge and stock assessment in Newfoundland. *Newfoundland Studies*, 8(2), 155–178.

Niering, W.A., R.H. Whittaker, and C.H. Lowe, 1963. The saguaro: A population in relation to environment. *Science*, 142, 15–23.

Odum, E. P. 1969. The strategy of ecosystem development, *Science*, 164, 262–270.

Oelschlaeger, M., 1991. *The Idea of Wilderness: From Prehistory to the Age of Ecology.* Yale University Press, New Haven.

Ostrom, E., J. Burger, C.B. Field, R.B. Norgaard, and D. Policansky, 1999. Revisiting the commons: Local lessons, global challenges. *Science*, 284, 278–282.

Parker, K.C., 1993. Climatic effects on regeneration trends for two columnar cacti. *Annals of the Association of American Geographers*, 83, 452–474.

Parker, V., 1995. Natural resources management by litigation. In Knight, R.L., and S.F. Bates (eds.) *A New Century for Natural Resources Management.* Island Press, Washington, DC, 209–220.

Pickett, S.T.A., and R.S. Ostfeld, 1995. The shifting paradigm in ecology. In Knight, R.L., and S.F. Bates (eds.) *A New Century for Natural Resources Management.* Island Press, Washington, DC, 261–278.

Pyne, S.J. 1982. *Fire in America: A Cultural History of Wildland and Rural Fire.* Princeton, University Press, Princeton, NJ.

Pyne, S.J., 1997. *World Fire: The Culture of Fire on Earth.* University of Washington Press, Seattle.

Raup, H.M., 1964. Some problems in ecological theory and their relation to conservation. *Journal of Ecology*, 52(Suppl.), 19–28.

Raven, P.H., and D.I. Axelrod, 1978. Origin and relationships of the California flora. *University of California Publications in Botany*, 72, 1–134.

Richards, P., 1985. *Indigenous Agricultural Revolution: Ecology and Food Production in West Africa.* Westview Press, Boulder, CO.

Robock, A., and H.-F. Graf, 1994. Effects of pre-Industrial human activities on climate. *Chemosphere*, 29, 1087–1098.

Romme, W., and D. De Spain, 1989. The Yellowstone fire. *Scientific American*, 261(5), 36–46.

Roof, J., 1971. Growing California's field wildflowers. *The Four Seasons*, 3(4), 1–24.

Sauer, J.D., 1988. *Plant Migration: The Dynamics of Geographic Patterning in Seed Plant Species.* University of California Press, Berkeley.

Sawicki, V., 1998. Investigation of the distribution of *Lythrum salicaria* and its effect on biodiversity of North American flora: Broome County, New York. M.A. thesis, SUNY-Binghamton, Binghamton, NY.

Schumacher, E.F., 1973. *Small Is Beautiful.* Harper and Row, New York.

Sellers, C., 1994. Factory as environment: Industrial hygiene, professional collaboration and the modern sciences of pollution. *Environmental History Review*, 18(1), 55–83.

Simonian, L., 1995. *Defending the Land of the Jaguar: A History of Conservation in Mexico.* University of Texas Press, Austin.

Sprugel, D.G., 1991. Disturbance, equilibrium, and environmental variability—what is natural vegetation in a changing environment? *Biological Conservation*, 58, 1–18.

Stoddard, H.L., 1931. *The Bobwhite Quail: Its Habits, Preservation and Increase.* Charles Scribner's Sons, New York.

Toulmin, S., 1990. *Cosmopolis: The Hidden Agenda of Modernity.* University of Chicago Press, Chicago.

Vitousek, P.M., C.M. D'Antonio, L.L. Loope, and R. Westbrooks, 1996. Biological invasions as global environmental change. *American Scientist*, 84, 468–478.

Vitousek, P.M., L.R. Walker, L.D. Whiteaker, D. Mueller-Dombois, and P.A. Matson, 1987. Biological invasion by *Myrica faya* alters ecosystem development in Hawaii. *Science*, 238, 802–804.

Walters, C.J., 1986. *Adaptive Management of Renewable Resources.* McGraw-Hill, New York.

Warren, A., 1995. Changing understandings of African pastoralism and the nature of environmental paradigms. *Transactions of the Institute of British Geographers*, 20, 193–203.

Warrick, R.A., 1993. Slowing global warming and sea level rise: The rough road from Rio. *Transactions of the Institute of British Geographers*, 18, 140–148.

WCED (World Commission on Environment and Development), 1987. *Our Common Future*. Oxford University Press, Oxford.

Wester, L.L., 1981. Composition of native grasslands in the San Joaquin Valley, California. *Madroño*, 28, 231–241.

Westoby, M., B.H. Walker, and I. Noy-Meir, 1989. Opportunistic management for rangelands not at equilibrium. *Journal of Range Management*, 42, 266–273.

Whittaker, R.H., 1975. *Communities and Ecosystems*. 2nd ed. Macmillan, New York.

Wilson, E.O., 1992. *The Diversity of Life*. Belknap Press of the Harvard University Press, Cambridge, MA.

Woodcock, D.W., and P.V. Wells, 1994. The burning of the New World: The extent and significance of broadcast burning by early humans. *Chemosphere*, 29, 935–949.

Worster, D., 1977. *Nature's Economy: The Roots of Ecology*. Sierra Club, San Francisco.

Worster, D., 1990. The ecology of order and chaos. *Environmental History Review*, 14(1–2), 1–18.

Zimmerer, K.S., 1994. Human ecology and the "new ecology": The prospect and promise of integration. *Annals of the Association of American Geographers*, 84, 108–125.

Index